Periodic Table of th...

(handwritten:) distilled water 7   neutral

(handwritten:) sodium hydroxide 14

(handwritten:) Base

(handwritten:) PH

1								
1 **H** 1.008								

2

Metals		...als

3 **Li** 6.941	4 **Be** 9.012

◄————————— Transition Metals ———————————

11 **Na** 22.99	12 **Mg** 24.30

		3	4	5	6	7	8	9
19 **K** 39.10	20 **Ca** 40.08	21 **Sc** 44.96	22 **Ti** 47.87	23 **V** 50.94	24 **Cr** 52.00	25 **Mn** 54.94	26 **Fe** 55.85	27 **Co** 58.93
37 **Rb** 85.47	38 **Sr** 87.62	39 **Y** 88.91	40 **Zr** 91.22	41 **Nb** 92.91	42 **Mo** 95.94	43 **Tc** (98)	44 **Ru** 101.1	45 **Rh** 102.9
55 **Cs** 132.9	56 **Ba** 137.3		72 **Hf** 178.5	73 **Ta** 180.9	74 **W** 183.8	75 **Re** 186.2	76 **Os** 190.2	77 **Ir** 192.2
87 **Fr** (223.0)	88 **Ra** (226.0)		104 **Rf** (265)	105 **Db** (268)	106 **Sg** (271)	107 **Bh** (270)	108 **Hs** (277)	109 **Mt** (276)

Lanthanides

57 **La** 138.9	58 **Ce** 140.1	59 **Pr** 140.9	60 **Nd** 144.2	61 **Pm** (145)	62 **Sm** 150.4

Actinides

89 **Ac** (227.0)	90 **Th** 232.0	91 **Pa** 231.0	92 **U** 238.0	93 **Np** (237.0)	94 **Pu** (244)

Note: Atomic masses shown here are March 2013 IUPAC values, rounded to 4 digits.

						17	**18**
						1 **H** 1.008	2 **He** 4.003

			13	**14**	**15**	**16**		
			5 **B** 10.81	6 **C** 12.01	7 **N** 14.01	8 **O** 16.00	9 **F** 19.00	10 **Ne** 20.18
			13 **Al** 26.98	14 **Si** 28.09	15 **P** 30.97	16 **S** 32.07	17 **Cl** 35.45	18 **Ar** 39.95

10	**11**	**12**						
28 **Ni** 58.69	29 **Cu** 63.55	30 **Zn** 65.39	31 **Ga** 69.72	32 **Ge** 72.61	33 **As** 74.92	34 **Se** 78.96	35 **Br** 79.90	36 **Kr** 83.80
46 **Pd** 106.4	47 **Ag** 107.9	48 **Cd** 112.4	49 **In** 114.8	50 **Sn** 118.7	51 **Sb** 121.8	52 **Te** 127.6	53 **I** 126.9	54 **Xe** 131.3
78 **Pt** 195.1	79 **Au** 197.0	80 **Hg** 200.6	81 **Tl** 204.4	82 **Pb** 207.2	83 **Bi** 209.0	84 **Po** (209.0)	85 **At** (210.0)	86 **Rn** (222.0)
110 **Ds** (281)	111 **Rg** (280)	112 **Cn** (285)	113 ***** (284)	114 **Fl** (289)	115 ***** (288)	116 **Lv** (293)	117 ***** (294)	118 ***** (294)

63 **Eu** 152.0	64 **Gd** 157.2	65 **Tb** 158.9	66 **Dy** 162.5	67 **Ho** 164.9	68 **Er** 167.3	69 **Tm** 168.9	70 **Yb** 173.0	71 **Lu** 175.0

95 **Am** (243)	96 **Cm** (247)	97 **Bk** (247)	98 **Cf** (251)	99 **Es** (252)	100 **Fm** (257)	101 **Md** (258)	102 **No** (259)	103 **Lr** (262)

***** These elements have been discovered but not authenticated by the IUPAC.

( ) Indicates the mass number of the longest-lived isotope.

Chemistry

Principles and Reactions

William L. Masterton
University of Connecticut

Cecile N. Hurley
University of Connecticut

 CENG

anada • Mexico • Singap   om • United States

Chemistry: Principles and Reactions,
Eighth Edition
William L. Masterton, Cecile N. Hurley

Product Director: Mary Finch

Product Manager: Lisa Lockwood

Content Developer: Ed Dodd

Associate Content Developer: Elizabeth Woods

Product Assistant: Karolina Kiwak

Media Developer: Brendan Killion, Lisa Weber

Marketing Manager: Janet del Mundo

Content Project Manager: Jennifer Risden

Art Director: Maria Epes

Manufacturing Planner: Judy Inouye

Production Service: MPS Limited

Photo Researcher: Lumina Datamatics

Text Researcher: Lumina Datamatics

Copy Editor: Chris Sabooni

Text Designer: Delgado and Company

Cover Designer: Irene Morris

Cover Image: © Dmitriev Lidiya/
Shutterstock.com

Interior Design Image: © wongwean/
Shutterstock.com

Compositor: MPS Limited

For product information and technology assistance, contact us at
**Cengage Customer & Sales Support, 1-800-354-9706
or support.cengage.com.**

For permission to use material from this text or product, submit all requests online at **www.cengage.com/permissions**.

Library of Congress Control Number: 2014943694

ISBN: 978-0-357-67075-0

Cengage
200 Pier 4 Boulevard
Boston, MA 02210
USA

Cengage is a leading provider of customized learning solutions with employees residing in nearly 40 different countries and sales in more than 125 countries around the world. Find your local representative at: **www.cengage.com**.

To learn more about Cengage platforms and services, register or access your on-line learning solution, or purchase materials for your course, visit **www.cengage.com**.

Printed in Mexico
Print Number: 01   Print Year: 2020

To Jim, Joe, and Regina

They also serve who only stand and wait.
—John Milton
On His Blindness

Brief Contents

Contents

v

Preface

It is always difficult for an author to praise the virtues of one's own book. I could tell the instructors that the book is so inspiring that students will be turned on to chemistry with little or no effort on the instructor's part. I doubt you would believe that. I could also tell you that the text is so clearly written, so attuned to the students in the twenty-first century that your students will learn chemistry with little or no effort on their part. You certainly would not believe that. I can tell you that the two goals in writing this edition have been to make it as clear and as interesting as possible. I hope you believe that, because it is true.

Today's freshmen are quite different from those of a few years ago. Text messaging and Twitter™ have strongly influenced sentence length and structure. In current writing and conversation, short sentences or sentence fragments convey straight-to-the-point information. Multimedia presentations are a way of life. This edition, like the seventh, is written to be fully in tune with today's technology and speech.

Why Write a Short Book?

Rising tuition costs, depleted forests, and students' aching backs have kept me steadfast in my belief that it should be possible to cover a text completely (or at least *almost* completely) in a two-semester course. The students (and their parents) justifiably do not want to pay for 1000-page books with material that is never discussed in the courses taught with those texts.

The common perception is that a short book is a low-level book. I believe, however, that treating general concepts in a concise way can be done without sacrificing depth, rigor, or clarity. The criterion for including material continues to be its importance and relevance to the student, not its difficulty. To achieve this, the following guidelines are used.

1. Eliminate repetition and duplication wherever possible. Like its earlier editions, this text uses

   ■ Only one method for balancing redox reactions, the half-equation method introduced in Chapter 17.
   ■ Only one way of working gas-law problems, using the ideal gas law in all cases (Chapter 5).
   ■ Only one way of calculating ΔH (Chapter 8), using enthalpies of formation.
   ■ Only one equilibrium constant for gas-phase reactions (Chapter 12), the thermodynamic constant K, often referred to as K_p. This simplifies not only the treatment of gaseous equilibrium but also the discussion of reaction spontaneity (Chapter 16) and electrochemistry (Chapter 17).

2. Relegate to the Appendices or Beyond the Classroom essays topics ordinarily covered in longer texts. Items in this category include

   ■ MO (molecular orbital) theory (Appendix 4). Experience has shown (and continues to show) that although this approach is important to chemical bonding, most general chemistry students do not understand it but only memorize the principles discussed in the classroom.
   ■ Nomenclature of organic compounds. This material is of little value in a beginning course and is better left to a course in organic chemistry.
   ■ Qualitative analysis. This is summarized in a few pages in an essay in Chapter 15 in the Beyond the Classroom section. An extended discussion of the qualitative scheme and the chemistry behind it belongs in a laboratory manual, not a textbook.
   ■ Biochemistry. This material is traditionally covered in the last chapter of general chemistry texts. Although there are several biochemical topics included in the text (among them a discussion of heme in Chapter 19 and

carotenoids in Chapter 6), an entire chapter is not devoted to biochemistry. Interesting as this material is, it requires a background in organic chemistry that first-year students lack.

3. Avoid superfluous asides, applications to the real world, or stories about scientists in the exposition of principles. There are many applications incorporated in the context of problems and some of the exposition of general principles. In general, however, a bare-bones approach is used. Students can easily be distracted by interesting but peripheral tidbits while they are striving hard to understand the core concepts. Favorite real-world applications and personal stories about scientists are in separate sections, Beyond the Classroom and Chemistry: The Human Side. Students say that they read these two sections first and that these are the parts of the book that "we really enjoy the most." (Talk about faint praise!) They do admit to enjoying the marginal notes too.

What Changes Have Been Made?

The eighth edition has not been as radically changed as the seventh. I talked to students, instructors, and TAs and listened to suggestions and complaints.

While all the changes made to the seventh edition were enthusiastically received, there were areas where making small changes would make them better. For the eighth edition, the following changes were made:

- The Example format has been revised, so that the strategy, analysis, and solution follow each part of the example. The most common comment was: "Show me first how to do part (a) before asking me about part (b)."
- More flowcharts have been added. There was unanimous support and requests for more of them. We revised some of the existing ones and added a few more.
- The discussion of balancing redox equations has been moved from Chapter 4 to Chapter 17. Instructors comment that they have had to reintroduce redox equations in Chapter 17 and treat it like new material. Students and TAs both agree that Chapter 4 is a dense and heavy chapter. Thus, redox reactions are treated in Chapter 4 only as far as stoichiometric calculations are involved. Balanced equations are provided for these reactions.

Detailed List of Changes by Chapter

Global Changes
- Changes in about 25% of the topical end-of-chapter problems
- Almost all of the summary problems have either been revised or are completely new to this edition
- Revised artwork with enhanced labeling and several new photos
- Almost all of the chapter opening art is new

Chapter 1
- Redrawn flowchart for matter classification
- New Examples 1.3 and 1.6

Chapter 2
- Redrawn Figure 2.5 (Rutherford Experiment)
- New Examples 2.1 and 2.5
- New Beyond the Classroom essay on the origin of some elements' names
- Redrawn flowchart for naming molecular compounds

Chapter 3
- New Examples 3.1 (b), 3.5, 3.7, and 3.11

Chapter 4
- Redrawn Figure 4.1
- New Example 4.5
- Revised discussion of redox reactions excluding balancing of half and complete reactions
- New flowchart for determining oxidation number
- New figure summarizing differences between oxidation and reduction
- New Beyond the Classroom essay on antacids

Chapter 5
- New Figure 5.6
- New Examples 5.2 and 5.4 (a) and (b)
- Discussion of the relation of time and molar mass to the rate of effusion

continued

Chapter 6
- New Example 6.1
- New discussion on electron capacities for principal levels and subshells
- New table summarizing electron capacities in principal levels and subshells

Chapter 7
- New Figure 7.5

Chapter 8
- Replaced Figure 8.10 with a new table

Chapter 9
- Figure 9.3 redrawn

Chapter 10
- New Examples 10.9 and 10.10
- Figure 10.14 redrawn

Chapter 11
- New Figure 11.4
- Example 11.3 redone

Chapter 12
- Examples 12.1 and 12.6 rewritten

Chapter 13
- New Example 13.3
- Examples 13.2 and 13.9 rewritten

- Revised Figure 13.14

Chapter 14
- Example 14.3 rewritten

Chapter 15
- Example 15.3 rewritten
- Example 15.1 reclassified as nongraded
- Examples 15.8 and 15.9 reclassified as graded

Chapter 16
- Example 16.2 rewritten

Chapter 17
- New Section 17.1 on balancing redox half and complete reactions
- Review of oxidation, reduction, oxidizing agents, and reducing agents
- New schematic on balancing half-reactions
- New Examples 17.1 and 17.2

Chapter 18
- New Table 18.1—summary of different modes of radioactive decay
- New paragraph about SPECT (single proton emission computer tomography) scans

Chapters 19–23
No changes

Alternate Editions

Chemistry: Principles and Reactions, Eighth Edition Hybrid Version with Access (24 months) to OWLv2 with MindTap Reader

ISBN: 978-1-305-08215-1

This briefer, paperbound version of *Chemistry: Principles and Reactions*, Eighth Edition does not contain the end-of-chapter problems, which can be assigned in OWLv2, the online homework and learning system for this book. Access to OWLv2 and the MindTap Reader eBook is included with the Hybrid version. The MindTap Reader is the full version of the text, with all end-of-chapter questions and problem sets.

Supporting Materials

Please visit **http://www.cengage.com/chemistry/masterton/CPAR8e** for information about student and instructor resources for this text, including custom versions and laboratory manuals.

Acknowledgments

Many people who have used this book—instructors, teaching assistants, students, and former students now teaching general chemistry—have e-mailed, written, and called with suggestions on how to improve the exposition. I am grateful to them all.

Reviewers who have helped in the preparation of this edition include the following:

Mamoun Bader (Penn State University)
Nancy Bryson (Berry College)
Andrea Gorczyca (Brookhaven College)
Arlin Gyberg (Augsburg College)
James Harris (Monadnock Regional High School)
Isaac Hon (Albertus Magnus College)
James Mack (University of Cincinnati)
Lawrence Mavis (St. Clair County Community College)
Alexander Nazarenko (SUNY Buffalo State)
Lorna Pehl (Eastern Wyoming College)
Richard Roberts (Des Moines Area Community College)
Joseph Sinski (Bellarmine University)
Jessica Thomas (Purdue University North Central)
John Wilterding (Olivet College)

Special thanks to Professor Fatma Selampinar (University of Connecticut) for her accuracy reviews. Her thoroughness and absolute attention to detail are incredible. She not only solved every new problem but was a sounding board and uncomplaining listener to a harried author.

This edition would not have been possible without the superb guidance of my content developer, Ed Dodd. He was a real gift. He smoothed rough patches and demanded perfection from everyone on the team. It was a real pleasure working with him.

Many people worked on the editorial and production team for this text. They took pages of manuscript, rough ideas, crude sketches, and long wish lists and put them together to create this edition. They prodded, cajoled, and set impossible deadlines. They are:

Mary Finch, Product Director
Maureen Rosener, Product Manager
Lisa Lockwood, Product Manager
Peter McGahey, Managing Developer
Elizabeth Woods, Associate Content Developer
Karolina Kiwak, Product Assistant
Lisa Weber, Media Developer
Brendan Killion, Media Developer
Nicole Hamm, Marketing Director
Janet del Mundo, Marketing Manager
Jennifer Risden, Senior Content Project Manager
Maria Epes, Art Director
Judy Inouye, Manufacturing Planner
John Sarantakis, Project Manager, Intellectual Property Acquisition
Jill Traut, Project Manager at MPS Limited
Dhanalakshmi Singaravelu, and Padmapriya Soundararajan, Image Researchers at Lumina Datamatics
Pinky Subi, Text Researcher at Lumina Datamatics

One person who does not belong to any team deserves special recognition. Jim Hurley picked up the slack when time was short, deadlines were imminent, and the list of tasks was long. He listened to endless complaints and commiserated. Thank you once again for continuing on this journey with me.

Cecile N. Hurley
University of Connecticut
Storrs, CT
November 2014

To the Student

You've probably already heard a lot about your general chemistry course. Many think it is more difficult than other courses. There may be some justification for that opinion. Besides having its very own specialized vocabulary, chemistry is a quantitative science—which means that you need mathematics as a tool to help you understand the concepts. As a result, you will probably receive a lot of advice from your instructor, teaching assistant, and fellow students about how to study chemistry. We would, however, like to acquaint you with some of the learning tools in this text. They are described in the pages that follow.

Learning Tools in *Chemistry: Principles and Reactions*, Eighth Edition

Examples

In a typical chapter, you will find ten or more examples each designed to illustrate a particular principle. These examples are either general (green bars), graded (purple bars), or conceptual (blue bars). These have answers, screened in color. They are presented in a two-column format. Most of them contain three parts:

- **Analysis**, which lists
  1. The information given.
  2. The information implied—information not directly stated in the problem but data that you can find elsewhere.
  3. What is asked for.
- **Strategy**
  This part gives you a plan to follow in solving the problem. It may lead you through a schematic pathway or remind you of conversion factors you have to consider or suggest equations that are useful.
- **Solution**
  This portion shows in a stepwise manner how the strategy given is implemented.
- Many of the examples end with a section called **End Points.** These are either checks on the reasonableness of your answer or relevant information obtained from the problem.

You should find it helpful to get into the habit of working all problems this way.

EXAMPLE	
Calculate the wavelength in nanometers of the line in the Balmer series that results from the transition $n = 4$ to $n = 2$.	
ANALYSIS	
Information given:	$n = 2$; $n = 4$
Information implied:	speed of light (2.998×10^8 m/s) Rydberg constant (2.180×10^{-18} J) Planck constant (6.626×10^{-34} J · s)
Asked for:	wavelength in nm

continued

STRATEGY

1. Substitute into Equation 6.4 to find the frequency due to the transition.

$$\nu = \frac{R_H}{h}\left(\frac{1}{n_{lo}^2} - \frac{1}{n_{hi}^2}\right)$$

Use the lower value for n as n_{lo} and the higher value for n_{hi}.

2. Use Equation 6.1 to find the wavelength in meters and then convert to nanometers.

SOLUTION

1. Frequency

$$\nu = \frac{2.180 \times 10^{-18}\ J}{6.626 \times 10^{-34}\ J \cdot s}\left(\frac{1}{(2)^2} - \frac{1}{(4)^2}\right) = 6.169 \times 10^{14}\ s^{-1}$$

2. Wavelength

$$\lambda = \frac{2.998 \times 10^8\ m/s}{6.169 \times 10^{14}\ s^{-1}} \times \frac{1\ nm}{1 \times 10^{-9}\ m} = 486.0\ nm$$

END POINT

Compare this value with that listed in Table 6.2 for the second line of the Balmer series.

Graded Examples

Throughout the text, you will encounter special *graded* examples. Note that they are the problems with the purple bars. A typical graded example looks like the following:

EXAMPLE   GRADED

For the reaction

$$A + 2B \rightarrow: C$$

determine

(a) the number of moles of A required to react with 5.0 mol of B.

(b) the number of grams of A required to react with 5.0 g of B.

(c) the volume of a 0.50 M solution of A required to react with 5.0 g of B.

(d) the volume of a 0.50 M solution of A required to react with 25 mL of a solution that has a density of 1.2 g/mL and contains 32% by mass of B.

There are two advantages to working a graded example:

1. By working parts (a) through (d) in succession, you can see how many different ways there are to ask a question about mass relations in a reaction. That should cushion the shock should you see only part (d) in an exam.
2. The parts of the graded example do not just progress from an easy mass relations question to a more difficult one. The value of the graded example is that the last question *assumes the ability to answer the earlier ones.* You may be able to answer parts (a) and (b) with a limited understanding of the material, but to answer part (d) you need to have mastered the material.

Use the graded example as you review for exams. Try to skip the earlier parts [in this case (a), (b) and (c)] and go directly to the last part (d). If you can solve (d), you do not need to try (a), (b), and (c); you know how to do them. If you can't, then try (c) to see where you may have a problem. If you can't do (c), then try (b). As a last resort, start at (a) and work your way back through (d).

Marginal Notes

Sprinkled throughout the text are a number of short notes in the margin. Many of these are of the "now, hear this" variety, others are mnemonics, and still others make points that we forgot to put in the text. (These were contributed by your fellow students.) Some—probably fewer than we think—are supposed to be humorous.

Chemistry: The Human Side

Throughout the text, short biographies of some of the pioneers of chemistry appear in sections with this heading. They emphasize not only the accomplishments of these individuals but also their personalities.

Chemistry: Beyond the Classroom

Each chapter contains a Beyond the Classroom feature. It is a self-contained essay that illustrates a current example either of chemistry in use in the world or an area of chemical research. It does not intrude into the explanation of the concepts, so it won't distract you. But we promise that those essays—if you read them—will make you more scientifically literate.

Chapter Highlights

At the end of each chapter, you will find a brief review of its concepts. A review is always helpful not only to refresh yourself about past material but also to organize your time and notes when preparing for an examination. The chapter highlights include

- The *Key Terms* in the chapter. If a particular term is unfamiliar, refer to the index at the back of the book. You will find the term in the glossary that is incorporated in the index and also the pages in the text where it appears (if you need more explanation).

- The *Key Concepts* and *Key Equations* introduced in the chapter. These are indexed to the corresponding examples and end-of-chapter problems. End-of-chapter problems available on OWLv2 are also cross-referenced. If you have trouble working a particular problem here, it may help to go back and reread the example that covers the same concept.

Summary Problem

Each chapter is summarized by a multistep problem that covers all or nearly all of the key concepts in the chapter. You can test your understanding of the chapter by working this problem. A major advantage of the summary problems is that they tie together many different ideas, showing how they correlate with one another. An experienced general chemistry professor always tells his class, "If you can answer the summary problem without help, you are ready for a test on its chapter."

Questions and Answers

At the end of each chapter is a set of questions and problems that your instructor may assign for homework. They are also helpful in testing the depth of your knowledge about the chapter. These sets include

- Conceptual problems that test your understanding of principles. A calculator is not (or should not be) necessary to answer these questions.
- Questions that test your knowledge of the specialized vocabulary that chemists use (e.g., write the names of formulas, write the chemical equation for a reaction that is described).
- Quantitative problems that require a calculator and some algebraic manipulations.

Classified problems start the set and are grouped by type under a particular heading that indicates the section and/or topic from the chapter that they address. The classified problems occur in matched pairs, so the second member illustrates the same principle as the first. This allows you more than one opportunity to test yourself. The second problem (whose number is even) is answered in Appendix 5. If your instructor assigns the odd problems without answers for homework, wait until the problem solution is discussed and solve the even problem to satisfy yourself that you understand how to solve the problem of that type.

Each chapter also contains a smaller number of **Unclassified** problems, which may involve more than one concept, including, perhaps, topics from a preceding chapter.

The section of **Challenge** problems presents problems that may require extra skill and/or insight and effort. They are all answered in Appendix 5.

Even-numbered questions and Challenge Problems answered in Appendix 5 have fully worked solutions available in the *Student Solutions Manual*. Please visit **http://www.cengage.com/chemistry/masterton/CPAR8e** for information about the *Student Solutions Manual*.

Appendices

The appendices at the end of the book provide not only the answers to the even-numbered problems but also additional materials you may find useful. Among them are

- Appendix 1, which includes a review of SI base units as well as tables of thermodynamic data and equilibrium constants.
- Appendix 3, which contains a mathematical review touching on just about all the mathematics you need for general chemistry. Exponential notation and logarithms (natural and base 10) are emphasized.

Other Resources to Help You Pass Your General Chemistry Course

Besides the textbook, several other resources are available to help you study and master general chemistry concepts. Please visit **http://www.cengage.com/chemistry/masterton/CPAR8e** for information about student resources for this text, including custom versions and laboratory manuals.

Matter and Measurements

The painting shows measuring instruments used in the Middle Ages. We still use many of them today.

Museum of the History of Science in Oxford, England

There is measure in everything.

—HORACE

Almost certainly, this is your first college course in chemistry; perhaps it is your first exposure to chemistry at any level. Unless you are a chemistry major, you may wonder why you are taking this course and what you can expect to gain from it. To address that question, it is helpful to look at some of the ways in which chemistry contributes to other disciplines.

If you're planning to be an engineer, you can be sure that many of the materials you will work with have been synthesized by chemists. Some of these materials are organic (carbon-containing). They could be familiar plastics like polyethylene (Chapter 23) or the more esoteric plastics used in unbreakable windows and nonflammable clothing. Other materials, including metals (Chapter 20) and semiconductors, are inorganic in nature.

Perhaps you are a health science major, looking forward to a career in medicine or pharmacy. If so, you will want to become familiar with the properties of aqueous solutions (Chapters 4, 10, 14, and 16), which include blood and other body fluids. Chemists today are involved in the synthesis of a variety of life-saving products. These range from drugs used in chemotherapy (Chapter 19) to new antibiotics used against resistant microorganisms.

Beyond career preparation, an objective of a college education is to make you a better-informed citizen. In this text, we'll look at some of the chemistry-related topics that make the news:

- depletion of the ozone layer (Chapter 11).
- alternative sources of fuel (Chapter 17).
- the pros and cons of nuclear power (Chapter 18).

Another goal of this text is to pique your intellectual curiosity by trying to explain the chemical principles behind such recent advances as

- "self-cleaning" windows (Chapter 1).
- "the ice that burns" (Chapter 3).
- "maintenance-free" storage batteries (Chapter 17).
- "chiral" drugs (Chapter 22).

We hope that when you complete this course you too will be convinced of the importance of chemistry in today's world. We should, however, caution you on one point.

Chapter Outline

1-1 Matter and Its Classifications

1-2 Measurements

1-3 Properties of Substances

Chemistry deals with the properties and reactions of substances.

Although we will talk about many of the applications of chemistry, *our main concern will be with the principles that govern chemical reactions.* Only by mastering those principles will you understand the basis of the applications mentioned above.

This chapter begins the study of chemistry by

- considering the different types of matter: pure substances versus mixtures, elements versus compounds (Section 1-1).
- looking at the kinds of measurements fundamental to chemistry, the uncertainties associated with those measurements, and a method to convert measured quantities from one unit to another (Section 1-2).
- focusing on certain physical properties, including density and water solubility, which can be used to identify substances (Section 1-3).

1-1 Matter and Its Classifications

Matter is anything that has mass and occupies space. It can be classified either with respect to its physical phases or with respect to its composition (Figure 1.1).

The three phases of matter are solid, liquid, and gas. A **solid** has a fixed shape and volume. A **liquid** has a fixed volume but is not rigid in shape; it takes the shape of its container. A **gas** has neither a fixed volume nor a shape. It takes on both the shape and the volume of its container.

Matter can also be classified with respect to its composition:

- pure substances, each of which has a fixed composition and a unique set of properties.
- mixtures, composed of two or more substances.

Pure substances are either elements or compounds (Figure 1.1), whereas mixtures can be either homogeneous or heterogeneous.

Elements

An **element** is a type of matter that cannot be broken down into two or more pure substances. There are 118 known elements, of which 91 occur naturally.

Many elements are familiar to all of us. The charcoal used in outdoor grills is nearly pure carbon. Electrical wiring, jewelry, and water pipes are often made from copper, a metallic element. Another such element, aluminum, is used in many household utensils.

Some elements come in and out of fashion, so to speak. Sixty years ago, elemental silicon was a chemical curiosity. Today, ultrapure silicon has become the basis for the multibillion-dollar semiconductor industry. Lead, on the other hand, is an element moving in the other direction. A generation ago it was widely used to make paint pigments, plumbing connections, and gasoline additives. Today, because of the toxicity of lead compounds, all of these applications have been banned in the United States.

In chemistry, an element is identified by its **symbol**. This consists of one or two letters, usually derived from the name of the element. Thus the symbol for carbon is C; that for aluminum is Al. Sometimes the symbol comes from the Latin name of the element or one of its compounds. The two elements copper and mercury, which were known in ancient times, have the symbols Cu *(cuprum)* and Hg *(hydrargyrum)*.

Table 1.1 (p. 4) lists the names and symbols of several elements that are probably familiar to you. In either free or combined form, they are commonly found in the laboratory or in commercial products. The abundances listed measure the relative amount of each element in the earth's crust, the atmosphere, and the oceans.

Curiously, several of the most familiar elements are really quite rare. An example is mercury, which has been known since at least 500 B.C., even though its abundance is only 0.00005%. It can easily be prepared by heating the red mineral cinnabar (Figure 1.2, p. 4).

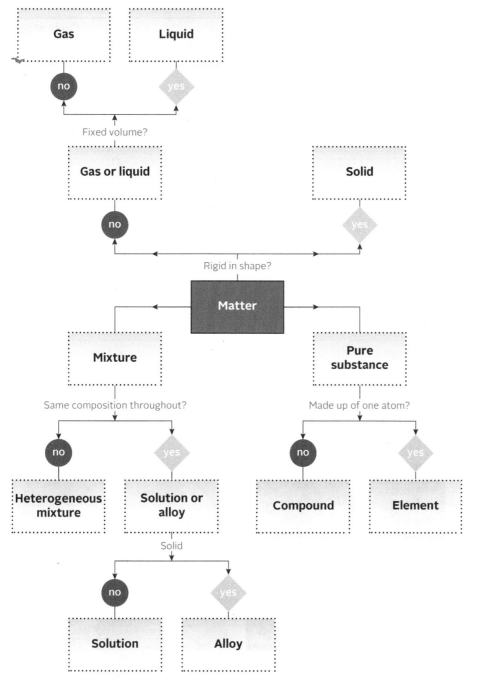

Figure 1.1 **Classification of matter into solid, liquid, and gas.**

Mercury is the only metal that is a liquid at room temperature. It is also one of the densest elements. Because of its high density, mercury was the liquid extensively used in thermometers and barometers. In the 1990s all instruments using mercury were banned because of environmental concerns. ◄ 𝒊 Another useful quality of mercury is its ability to dissolve many metals, forming solutions (amalgams). A silver-mercury-tin amalgam is still used to fill tooth cavities, but many dentists now use tooth-colored composites because they adhere better and are aesthetically more pleasing.

In contrast, aluminum (abundance = 7.5%), despite its usefulness, was little more than a chemical curiosity until about a century ago. It occurs in combined form in clays and rocks, from which it cannot be extracted. In 1886 two young chemists, Charles Hall in the United States and Paul Héroult in France, independently worked out a process for extracting aluminum from a relatively rare ore,

𝒊 _____
Mercury thermometers, both for laboratory and clinical use, have been replaced by digital ones.

Table 1.1 Some Familiar Elements with Their Percentage Abundances

Element	Symbol	Percentage Abundance	Element	Symbol	Percentage Abundance
Aluminum	Al	7.5	Manganese	Mn	0.09
Bromine	Br	0.00025	Mercury	Hg	0.00005
Calcium	Ca	3.4	Nickel	Ni	0.010
Carbon	C	0.08	Nitrogen	N	0.03
Chlorine	Cl	0.2	Oxygen	O	49.4
Chromium	Cr	0.018	Phosphorus	P	0.12
Copper	Cu	0.007	Potassium	K	2.4
Gold	Au	0.0000005	Silicon	Si	25.8
Hydrogen	H	0.9	Silver	Ag	0.00001
Iodine	I	0.00003	Sodium	Na	2.6
Iron	Fe	4.7	Sulfur	S	0.06
Lead	Pb	0.0016	Titanium	Ti	0.56
Magnesium	Mg	1.9	Zinc	Zn	0.008

The mineral cinnabar, from which mercury is obtained

Mercury, an element

© Cengage Learning/Charles D. Winters

(a)   (b)

Figure 1.2 **Cinnabar and mercury.**

bauxite. That process is still used today to produce the element. By an odd coincidence, Hall and Héroult were born in the same year (1863) and died in the same year (1914).

Compounds

A **compound** is a pure substance that contains more than one element. Water is a compound of hydrogen and oxygen. The compounds methane, acetylene, and naphthalene all contain the elements carbon and hydrogen, in different proportions.

Compounds have fixed compositions. That is, a given compound always contains the same elements in the same percentages by mass. A sample of pure water contains precisely 11.19% hydrogen and 88.81% oxygen. In contrast, mixtures can vary in composition. For example, a mixture of hydrogen and oxygen might contain 5, 10, 25, or 60% hydrogen, along with 95, 90, 75, or 40% oxygen.

Sodium, a metallic element that is soft enough to be cut with a knife.

Chlorine, a nonmetallic element that is a gas.

Sodium chloride, the crystalline chemical compound formed when sodium combines with chlorine.

© Cengage Learning/Charles D. Winters (Photos a–c)

a   b   c

Figure 1.3 **Sodium, chlorine, and sodium chloride.**

The properties of compounds are usually very different from those of the elements they contain. Ordinary table salt, sodium chloride, is a white, unreactive solid. As you can guess from its name, it contains the two elements sodium and chlorine. Sodium (Na) is a shiny, extremely reactive metal. Chlorine (Cl) is a poisonous, greenish-yellow gas. Clearly, when these two elements combine to form sodium chloride, a profound change takes place (Figure 1.3).

Many different methods can be used to resolve compounds into their elements. Sometimes, but not often, heat alone is sufficient. Mercury(II) oxide, a compound of mercury and oxygen, decomposes to its elements when heated to 600°C. Joseph Priestley, an English chemist, discovered oxygen more than 200 years ago when he carried out this reaction by exposing a sample of mercury(II) oxide to an intense beam of sunlight focused through a powerful lens. The mercury vapor formed is a deadly poison. Sir Isaac Newton, who distilled large quantities of mercury in his laboratory, suffered the effects in his later years.

Another method of resolving compounds into elements is *electrolysis*, which involves passing an electric current through a compound, usually in the liquid state. By electrolysis it is possible to separate water into the gaseous elements hydrogen and oxygen. Several decades ago it was proposed to use the hydrogen produced by electrolysis to raise the *Titanic* from its watery grave off the coast of Newfoundland. It didn't work.

© Cengage Learning/Charles D. Winters

Figure 1.4 **A heterogeneous mixture of copper sulfate crystals (blue) and sand.**

Mixtures

A **mixture** ◀ *i* contains two or more substances combined in such a way that each substance retains its chemical identity. When you shake copper sulfate with sand (Figure 1.4), the two substances do not react with one another. In contrast, when sodium is exposed to chlorine gas, a new compound, sodium chloride, is formed.

There are two types of mixtures:

1. Homogeneous or uniform mixtures are ones in which the composition is the same throughout. Another name for a homogeneous mixture is a **solution,** which is made up of a solvent, usually taken to be the substance present in largest amount, and one or more solutes. Most commonly, the **solvent** is a liquid, whereas

i Most materials you encounter are mixtures.

A homogeneous mixture of copper and zinc

A piece of granite, a heterogeneous mixture that contains discrete regions of different minerals (feldspar, mica, and quartz)

© Cengage Learning/Charles D. Winters

(a)   (b)

Figure 1.5 Two mixtures.

All gaseous mixtures, including air, are solutions.

GLC is a favorite technique in the forensics labs of many TV shows.

a **solute** may be a solid, liquid, or gas. Soda water is a solution of carbon dioxide (solute) in water (solvent). Seawater is a more complex solution in which there are several solid solutes, including sodium chloride; the solvent is water. It is also possible to have solutions in the solid state. Brass (Figure 1.5a) is a solid solution containing the two metals copper (67%–90%) and zinc (10%–33%).◄ⓘ

2. Heterogeneous or nonuniform mixtures are those in which the composition varies throughout. Most rocks fall into this category. In a piece of granite (Figure 1.5b), several components can be distinguished, differing from one another in color.

Many different methods can be used to separate the components of a mixture from one another. A couple of methods that you may have carried out in the laboratory are

■ **filtration**, used to separate a heterogeneous solid-liquid mixture. The mixture is passed through a barrier with fine pores, such as filter paper. Copper sulfate, which is water-soluble, can be separated from sand by shaking with water. On filtration the sand remains on the paper and the copper sulfate solution passes through it.

■ *distillation*, used to resolve a homogeneous solid-liquid mixture. The liquid vaporizes, leaving a residue of the solid in the distilling flask. The liquid is obtained by condensing the vapor. Distillation can be used to separate the components of a water solution of copper sulfate (Figure 1.6).

A more complex but more versatile separation method is *chromatography*, a technique widely used in teaching, research, and industrial laboratories to separate all kinds of mixtures. This method takes advantage of differences in solubility and/or extent of adsorption on a solid surface. In *gas-liquid chromatography*, a mixture of volatile liquids and gases is introduced into one end of a heated glass tube. As little as one microliter (10^{-6} L) of sample may be used. The tube is packed with an inert solid whose surface is coated with a viscous liquid. An unreactive "carrier gas," often helium, is passed through the tube. The components of the sample gradually separate as they vaporize into the helium or condense into the viscous liquid. Usually the more volatile fractions move faster and emerge first; successive fractions activate a detector and recorder.

Gas-liquid chromatography (GLC)◄ⓘ (Figure 1.7) finds many applications outside the chemistry laboratory. If you've ever had an emissions test on the exhaust system of your car, GLC was almost certainly the analytical method used. Pollutants such as carbon monoxide and unburned hydrocarbons appear as peaks on a graph. A computer determines the areas under these peaks, which are proportional to the concentrations of pollutants, and prints out a series of numbers that tells the inspector whether your car passed or failed the test. Many of the techniques used to test people for drugs (marijuana, cocaine, and others) or alcohol also make use of gas-liquid chromatography.

Ultra-high-speed gas chromatography (GC) fitted with an odor sensor is a powerful tool for analyzing the chemical vapors produced by explosives or other chemical or biological weapons.

In this section we will look at four familiar properties that you will almost certainly measure in the laboratory: length, volume, mass, and temperature. Other physical and chemical properties will be introduced in later chapters as they are needed.

Figure 1.6 **Apparatus for a simple distillation.**

Digital thermometer

1 Distilling flask with solution of copper sulfate where the two components, water and copper sulfate, are being separated.

2 Water vaporizes and is cooled in the condenser.

Cooling water out

3 Distilled water collects here.

4 The copper sulfate, a blue solid, does not vaporize and remains in the distillation flask.

Cooling water in

Figure 1.7 **Mass spectrometer and gas chromatograph.** Airport security uses these instruments to separate mixtures (chromatograph) and to detect the presence of nitrogen containing explosives (mass spectrograph). Larger versions of these instruments are used to scan checked baggage.

1-2 Measurements

Chemistry is a quantitative science. The experiments that you carry out in the laboratory and the calculations that you perform almost always involve measured quantities with specified numerical values. Consider, for example, the following set of directions for the preparation of aspirin (measured quantities are shown in italics).

Add *2.0 g* of salicylic acid, *5.0 mL* of acetic anhydride, and *5 drops* of *85%* H_3PO_4 to a *50-mL* Erlenmeyer flask. Heat in a water bath at *75°C* for *15 minutes*. Add cautiously *20 mL* of water and transfer to an ice bath at *0°C*. Scratch the inside of the flask with a stirring rod to initiate crystallization. Separate aspirin from the solid-liquid mixture by filtering through a Buchner funnel *10 cm* in diameter.

Scientific measurements are expressed in the **metric system**. As you know, this is a decimal-based system in which all of the units of a particular quantity are related to one another by factors of 10. The more common prefixes used to express these factors are listed in Table 1.2.

Table 1.2 Metric Prefixes

Factor	Prefix	Abbreviation	Factor	Prefix	Abbreviation
10^6	mega	M	10^{-3}	**milli**	m
10^3	**kilo**	k	10^{-6}	micro	µ
10^{-1}	deci	d	10^{-9}	**nano**	n
10^{-2}	**centi**	c	10^{-12}	pico	p

Figure 1.8 Measuring volume. A buret (*left*) delivers an accurately measured variable volume of liquid. A pipet (*right*) delivers a fixed volume (e.g., 25.00 mL) of liquid.

Figure 1.9 Weighing a solid. The solid sample plus the paper on which it rests weighs 144.998 g. The pictured balance is a single-pan analytical balance.

ⓘ Writing "m" in upper case or lower case makes a big difference.

ⓘ Many countries still use degrees centigrade.

Figure 1.10 Relationship between Fahrenheit and Celsius scales. This figure shows the relationship between the Fahrenheit and Celsius temperature scales. Note that there are 180 degrees F for 100 degrees C (1.8 F/C) and 0°C = 32°F.

Instruments and Units

The standard unit of *length* in the metric system is the meter, which is a little larger than a yard. The meter was originally intended to be 1/40,000,000 of the earth's meridian that passes through Paris. It is now defined as the distance light travels in a vacuum in 1/299,792,458 of a second.

Other units of length are expressed in terms of the meter, using the prefixes listed in Table 1.2. You are familiar with the centimeter, the millimeter, and the kilometer:

$$1 \text{ cm} = 10^{-2} \text{ m} \qquad 1 \text{ mm} = 10^{-3} \text{ m} \qquad 1 \text{ km} = 10^{3} \text{ m}$$

The dimensions of very tiny particles are often expressed in nanometers:

$$1 \text{ nm} = 10^{-9} \text{ m}$$

Volume is most commonly expressed in one of three units

- cubic centimeters $\qquad 1 \text{ cm}^3 = (10^{-2} \text{ m})^3 = 10^{-6} \text{ m}^3$
- liters (L) $\qquad\qquad 1 \text{ L} = 10^{-3} \text{ m}^3 = 10^{3} \text{ cm}^3$
- milliliters (mL) $\qquad 1 \text{ mL} = 10^{-3} \text{ L} = 10^{-6} \text{ m}^3$

Notice that a milliliter is equal to one cubic centimeter:

$$1 \text{ mL} = 1 \text{ cm}^3$$

The device most commonly used to measure volume in general chemistry is the graduated cylinder. A pipet or buret (Figure 1.8) is used when greater accuracy is required. A pipet is calibrated to deliver a fixed volume of liquid—for example, 25.00 mL—when filled to the mark and allowed to drain. Different volumes can be delivered accurately by a buret, perhaps to ±0.01 mL.

In the metric system, *mass* is most commonly expressed in grams, kilograms, or milligrams:

$$1 \text{ g} = 10^{-3} \text{ kg} \qquad 1 \text{ mg} = 10^{-3} \text{ g}$$

This book weighs about 1.5 kg. The megagram, more frequently called the *metric ton*, is

$$1 \text{ Mg} ◂ⓘ = 10^{6} \text{ g} = 10^{3} \text{ kg}$$

Properly speaking, there is a distinction between mass and weight. *Mass* is a measure of the amount of matter in an object; *weight* is a measure of the gravitational force acting on the object. Chemists often use these terms interchangeably; we determine the mass of an object by "weighing" it on a balance (Figure 1.9).

Temperature is the factor that determines the direction of heat flow. When two objects at different temperatures are placed in contact with one another, heat flows from the one at the higher temperature to the one at the lower temperature.

Thermometers used in chemistry are marked in degrees **Celsius** (referred to as degrees centigrade until 1948). ◂ⓘ On this scale, named after the Swedish astronomer Anders Celsius (1701–1744), the freezing point of water is taken to be 0°C. The normal boiling point of water is 100°C. Household thermometers in the United States are commonly marked in *Fahrenheit* degrees. Daniel Fahrenheit (1686–1736) was a German instrument maker who was the first to use the mercury-in-glass thermometer. On this scale, the normal freezing and boiling points of water are taken to be 32° and 212°, respectively (Figure 1.10). It follows that (212°F − 32°F) = 180°F covers the same temperature interval as (100°C − 0°C) = 100°C. This leads to the general relation between the two scales:

$$t_{°F} = 1.8 \, t_{°C} + 32° \tag{1.1}$$

The two scales coincide at −40°; as you can readily see from equation 1.1:

At −40°C: $t_{°F} = 1.8(-40°) + 32° = -72° + 32° = -40°$

For many purposes in chemistry, the m~ ~ient unit o~ the **kelvin** (K); note the absence of ~ ~he kelvin i~ 1/273.16 of the difference be~ ~ tempera~ the triple point of water~ ~ween tem~ and in °C is

This scal~ ~tist wh~ 1848, at t~ lower ~

[handwritten notes overlaid: "Cannot +/− protons"; "get rid of electron Cation"; "Protons dont change"; "Isotopes can have diff n°"; "combo of isotopes makes up molar mass"; "get electron Anion (octet rule)"]

EXAMPL~

Mercury ther~ mercury in gla~ point in~

Information given.

Asked for: ~ °F and K

STRATEGY

1. S~ ~ for (a).
2. S $\dfrac{m}{V} = \rho$

(a)

(b

[handwritten: $726\ kg/m^3$; $0.726\ g/mL$]

~OLUTION = 1.~

~ = 5~

~ber of d~ system. T~ a self-con~ ~em of Units~ have discuss~

kilog~
~ature: kelvir~

[handwritten table:

subatomic particle	relative charge	AmU
p^+	$1+$	1.0073
e^-	$1-$	0.0005486
$n°$	0	1.0087

]

[handwritten: $0.726\ g/mL = \dfrac{15.08}{V}$]

[handwritten: $= 20.66\ mL \rightarrow 20.7\ mL$]

Uncertainties in Measureme~ ~ificant Figure

Every measurement carries with it a degree of uncertainty. Its magnitude depends on the nature of the measuring device and the skill of its operator. Suppose, for example, you measure out 8 mL of liquid using the 100-mL graduated cylinder shown in Figure 1.11. Here the volume is uncertain to perhaps ±1 mL. With such a crude measuring device, you would be lucky to obtain a volume between 7 and 9 mL. To obtain greater precision, you could use a narrow 10-mL cylinder, which has divisions in small increments. You might now measure a volume within 0.1 mL of the desired value, in the range of 7.9 to 8.1 mL. By using a buret, you could reduce the uncertainty to ±0.01 mL.

Anyone making a measurement has a responsibility to indicate the uncertainty associated with it. Such information is vital to someone who wants to repeat the

---

*The triple point of water (Chapter 9) is the one unique temperature and pressure pair at which ice, liquid water, and water vapor can coexist in contact with one another.

Figure 1.11 Uncertainty in measuring volume. The uncertainty depends on the nature of the measuring device. Eight mL of liquid can be measured with less uncertainty in the 10-mL graduated cylinder than in the 100-mL graduated cylinder.

experiment or judge its precision. The three volume measurements referred to earlier could be reported as

8 ± 1 mL	(large graduated cylinder)
8.0 ± 0.1 mL	(small graduated cylinder)
8.00 ± 0.01 mL	(buret)

In this text, we will drop the ± notation and simply write

8 mL   8.0 mL   8.00 mL ◀ⓘ

There's a big difference between 8 mL and 8.00 mL, perhaps as much as half a milliliter.

When we do this, it is understood that there is an *uncertainty of at least one unit in the last digit*—that is, 1 mL, 0.1 mL, 0.01 mL, respectively. This method of citing the degree of confidence in a measurement is often described in terms of **significant figures**, the meaningful digits obtained in a measurement. In 8.00 mL there are three significant figures; each of the three digits has experimental meaning. Similarly, there are two significant figures in 8.0 mL and one significant figure in 8 mL.

Frequently we need to know the number of significant figures in a measurement reported by someone else (Example 1.2).

EXAMPLE 1.2

Using different balances, three different students weigh the same object. They report the following masses:

(a) 1.611 g   (b) 1.60 g   (c) 0.001611 kg

How many significant figures does each value have?

STRATEGY

Assume each student reported the mass in such a way that the last number indicates the uncertainty associated with the measurement.

SOLUTION

(a) 1.611 g	4	
(b) 1.60 g	3	The zero after the decimal point is significant. It indicates that the object was weighed to the nearest 0.01 g.
(c) 0.001611 kg	4	The zeros at the left are not significant. They are only there because the mass was expressed in kilograms rather than grams. Note that 1.611 g and 0.001611 kg represent the same mass.

END POINT

If you express these masses in exponential notation as 1.611×10^0 g, 1.60×10^0 g, and 1.611×10^{-3} kg, the number of significant figures becomes obvious.

Unfortunately, the uncertainty here is uncertain.

Sometimes the number of significant figures in a reported measurement is ambiguous. Suppose that a piece of metal is reported to weigh 500 g. ◀ⓘ You cannot be sure how many of these digits are meaningful. Perhaps the metal was weighed to the nearest gram (500 ± 1 g). If so, the 5 and the two zeros are significant; there are three significant figures. Then again, the metal might have been weighed only to the nearest 10 g (500 ± 10 g). In this case, only the 5 and one zero are known accurately; there are two significant figures. About all you can do in such cases is to wish the person who carried out the weighing had used exponential notation. The mass should have been reported as

$$5.00 \times 10^2 \text{ g} \qquad \text{(3 significant figures)}$$

or

$$5.0 \times 10^2 \text{ g} \qquad \text{(2 significant figures)}$$

or

$$5 \times 10^2 \text{ g} \qquad \text{(1 significant figure)}$$

In general, *any ambiguity concerning the number of significant figures in a measurement can be resolved by using exponential notation* ◄ (often referred to as "scientific notation"), discussed in Appendix 3.

Most measured quantities are not end results in themselves. Instead, they are used to calculate other quantities, often by multiplication or division. The precision of any such derived result is limited by that of the measurements on which it is based. **When measured quantities are multiplied or divided, the number of significant figures in the result is the same as that in the quantity with the smallest number of significant figures.** ◄ *ⓘ*

The number of significant figures is the number of digits shown when a quantity is expressed in exponential notation.

The rule is approximate, but sufficient for our purposes.

EXAMPLE 1.3

A patient is given an antibiotic intravenously. The rate of infusion is set so that the patient receives 1.15 mg of antibiotic per minute. How many milligrams of antibiotic are received after 35 minutes of infusion?

ANALYSIS

Information given:	rate of infusion (1.15 mg/min) time elapsed (35 minutes)
Asked for:	amount of antibiotic infused

STRATEGY

1. Substitute into the formula.

$$\text{rate} = \frac{\text{amount infused}}{\text{time}}$$

2. Recall the rules for significant figures.

SOLUTION

amount infused	amount infused = rate × time = (1.15 mg/min)(35 min) = 40.25 mg
significant figures	rate: 3; time: 2 The answer should have 2 significant figures.
amount infused	4.0×10^1 mg

The rules for "rounding off" a measurement, which were applied in Example 1.3, are as follows:

1. *If the digits to be discarded are less than − − 500 . . ., leave the last digit unchanged.* Masses of 23.315 g and 23.487 g both round off to 23 g if only two significant digits are required.

2. *If the digits to be discarded are greater than – – 500 . . ., add one to the last digit.* Masses of 23.692 g and 23.514 g round off to 24 g.

3. *If, perchance, the digits to be discarded are – – 500 . . . (or simply – – 5 by itself), round off so that the last digit is an even number.* Masses of 23.500 g and 24.5 g both round off to 24 g (two significant figures). ◄ ⓘ

When measured quantities are added or subtracted, the uncertainty in the result is found in a quite different way than when they are multiplied and divided. It is determined by counting the number of decimal places, that is, the number of digits to the right of the decimal point for each measured quantity. ***When measured quantities are added or subtracted, the number of decimal places in the result is the same as that in the quantity with the greatest uncertainty and hence the smallest number of decimal places.*** ◄ ⓘ

To illustrate this rule, suppose you want to find the total volume of a vanilla latté made up of 2 shots of espresso (1 shot = 46.1 mL), 301 mL of milk, and 2 tablespoons of vanilla syrup (1 tablespoon = 14.787 mL).

	Volume	Uncertainty	
Espresso coffee	92.2 mL	±0.1 mL	1 decimal place
Milk	301 mL	±1 mL	0 decimal place
Vanilla syrup	29.574 mL	±0.001 mL	3 decimal places
Total volume	423 mL		

Because there are no digits after the decimal point in the volume of milk, there are none in the total volume. Looking at it another way, we can say that the total volume, 423 mL, has an uncertainty of ±1 mL, as does the volume of milk, the quantity with the greatest uncertainty.

In applying the rules governing the use of significant figures, you should keep in mind that certain numbers involved in calculations are exact rather than approximate. To illustrate this situation, consider the equation relating Fahrenheit and Celsius temperatures:

$$t_{°F} = 1.8 t_{°C} + 32°$$

The numbers 1.8 and 32 are exact. Hence they do not limit the number of significant figures in a temperature conversion; that limit is determined only by the precision of the thermometer used to measure temperature.

A different type of exact number arises in certain calculations. Suppose you are asked to determine the amount of heat evolved when *one* ◄ ⓘ *kilogram* of coal burns. The implication is that because "one" is spelled out, *exactly* one kilogram of coal burns. The uncertainty in the answer should be independent of the amount of coal.

Conversion of Units

It is often necessary to convert a measurement expressed in one unit to another unit in the same system or to convert a unit in the English system to one in the metric system. To do this we follow what is known as a **conversion factor** approach or dimensional analysis. For example, to convert a volume of 536 cm³ to liters, the relation

$$1 \text{ L} = 1000 \text{ cm}^3 \ ◄ ⓘ$$

is used. Dividing both sides of this equation by 1000 cm³ gives a quotient equal to 1:

$$\frac{1 \text{ L}}{1000 \text{ cm}^3} = \frac{1000 \text{ cm}^3}{1000 \text{ cm}^3} = 1$$

ⓘ This way, you round up as often as you round down.

ⓘ In a multi-step calculation, round off only in the final step.

ⓘ A number that is spelled out (one, two, . . .) does not affect the number of significant figures.

ⓘ There are exactly 1000 cm³ in exactly 1 L.

The quotient 1 L/1000 cm^3, which is called a conversion factor, is multiplied by 536 cm^3. Because the conversion factor equals 1, this does not change the actual volume. However, it does accomplish the desired conversion of units. The cm^3 in the numerator and denominator cancel to give the desired unit: liters.

$$536 \text{ cm}^3 \times \frac{1 \text{ L}}{1000 \text{ cm}^3} = 0.536 \text{ L}$$

To convert a volume in liters, say 1.28 L to cm^3, you must use a different form of the conversion factor. Use the units as a guide.

$$1.28 \text{ L} \times \frac{1000 \text{ cm}^3}{1 \text{ L}} = 1280 \text{ cm}^3 = 1.28 \times 10^3 \text{ cm}^3$$

Notice that a single relation (1 L = 1000 cm^3) gives two conversion factors:

$$\frac{1 \text{ L}}{1000 \text{ cm}^3} \quad \text{and} \quad \frac{1000 \text{ cm}^3}{1 \text{ L}}$$

Always check the units of your final answer. If you accidentally use the wrong form of the conversion factor, you will not get the desired unit. For example, if in your conversion of 1.28 L to cm^3 you used the conversion factor 1 L/1000 cm^3, you would get

$$1.28 \text{ L} \times \frac{1 \text{ L}}{1000 \text{ cm}^3} = 1.28 \times 10^{-3} \frac{\text{L}^2}{\text{cm}^3}$$

In general, when you make a conversion *choose the factor that cancels out the initial unit:*

$$\text{initial unit} \times \frac{\text{wanted unit}}{\text{initial unit}} = \text{wanted unit}$$

Conversions between English and metric units can be made using Table 1.3. We will call these "bridge conversions." They allow you to move from one system to another.

Table 1.3 Relations Between Length, Volume, and Mass Units

Metric		English		Metric-English	
Length					
1 km	= 10^3 m	1 ft	= 12 in	1 in	= 2.54 cm*
1 cm	= 10^{-2} m	1 yd	= 3 ft	1 m	= 39.37 in
1 mm	= 10^{-3} m	1 mi	= 5280 ft	1 mi	= 1.609 km
1 nm	= 10^{-9} m = 10 Å				
Volume					
1 m^3	= 10^6 cm^3 = 10^3 L	1 gal	= 4 qt = 8 pt	1 ft^3	= 28.32 L
1 cm^3	= 1 mL = 10^{-3} L	1 qt (U.S. liq)	= 57.75 in^3	1 L	= 1.057 qt (U.S. liq)
Mass					
1 kg	= 10^3 g	1 lb	= 16 oz	1 lb	= 453.6 g
1 mg	= 10^{-3} g	1 short ton	= 2000 lb	1 g	= 0.03527 oz
1 metric ton	= 10^3 kg			1 metric ton	= 1.102 short ton

*This conversion factor is exact; the inch is defined to be exactly 2.54 cm. The other factors listed in this column are approximate, quoted to four significant figures. Additional digits are available if needed for very accurate calculations. For example, the pound is defined to be 453.59237 g.

EXAMPLE 1.4

A red blood cell has a diameter of 7.5 μm (micrometers). What is the diameter of the cell in inches?
(1 inch = 2.54 cm)

ANALYSIS	
Information given:	cell diameter (7.5 μm) bridge conversion (1 in = 2.54 cm)
Information implied:	relation between micrometers and centimeters
Asked for:	7.5 μm in inches
STRATEGY	
Follow the plan:	μm $\longrightarrow$ cm $\longrightarrow$ inch
SOLUTION	
7.5 μm in inches	$7.5 \ \mu m \times \dfrac{1 \times 10^{-6} \ m}{1 \ \mu m} \times \dfrac{100 \ cm}{1 \ m} \times \dfrac{1 \ in}{2.54 \ cm} = 3.0 \times 10^{-4} \ in$

Sometimes the required conversion has units raised to a power. To obtain the desired unit, you must remember to raise *both* the unit and the number to the desired power. Example 1.5 illustrates this point.

EXAMPLE 1.5

The beds in your dorm room have extra-long mattresses. These mattresses are 80 inches (2 significant figures) long and 39 inches wide. (Regular twin beds are 72 inches long.)

What is the area of the mattress top in m^2? (1 inch = 2.54 cm)

ANALYSIS	
Information given:	mattress length (80 in) and width (39 in) bridge conversion (1 inch = 2.54 cm)
Information implied:	centimeter to meter conversion
Asked for:	area in m^2
STRATEGY	
1. Recall equation for finding the area of a rectangle:	area = length $\times$ width
2. Follow the plan:	in^2 $\longrightarrow$ cm^2 $\longrightarrow$ m^2
SOLUTION	
area in in^2	80 in $\times$ 39 in = 3.12×10^3 in^2 (We will round off to correct significant figures at the end.)
area in m^2	$3.12 \times 10^3 \ in^2 \times \dfrac{(2.54)^2 \ cm^2}{(1)^2 \ in^2} \times \dfrac{(1)^2 \ m^2}{(100)^2 \ cm^2} = 2.0 \ m^2$
END POINT	

There are 36 inches in one yard, so the dimensions of the mattresses are approximately 1 yd wide and 2 yd long or 2 yd^2. A meter is almost equivalent to a yard (see Table 1.3) so the calculated answer is in the same ball park.

CHEMISTRY THE **HUMAN** SIDE

The discussion in Section 1-2 emphasizes the importance of making precise numerical measurements. Chemistry was not always so quantitative. The following recipe for finding the philosopher's stone was recorded more than 300 years ago.

Take all the mineral salts there are, also all salts of animal and vegetable origin. Add all the metals and minerals, omitting none. Take two parts of the salts and grate in one part of the metals and minerals. Melt this in a crucible, forming a mass that reflects the essence of the world in all its colors. Pulverize this and pour vinegar over it. Pour off the red liquid into English wine bottles, filling them half-full. Seal them with the bladder of an ox (*not* that of a pig). Punch a hole in the top with a coarse needle. Put the bottles in hot sand for three months. Vapor will escape through the hole in the top, leaving a red powder. . . .

One man more than any other transformed chemistry from an art to a science. Antoine Lavoisier was born in Paris; he died on the guillotine during the French Revolution. (Lavoisier was executed because he was a tax collector; chemistry had nothing to do with it.) Above all else, Lavoisier understood the importance of carefully controlled, quantitative experiments. These were described in his book *Elements of Chemistry*. Published in 1789, it is illustrated with diagrams by his wife. (E. I. DuPont (1772–1834) was a student of Lavoisier.)

Antoine Lavoisier
(1743–1794)

The results of one of Lavoisier's quantitative experiments are shown in Table A; the data are taken directly from Lavoisier. If you add up the masses of reactants and products (expressed in arbitrary units), you find them to be the same, 510. As Lavoisier put it, "In all of the operations of men and nature, nothing is created. An equal quantity of matter exists before and after the experiment."

This was the first clear statement of the law of conservation of mass (Chapter 2), which was the cornerstone for the growth of chemistry in the nineteenth century. Again, to quote Lavoisier, "it is on this principle that the whole art of making experiments is founded."

Table A Quantitative Experiment on the Fermentation of Wine (Lavoisier)

Reactants	Mass (Relative)	Products	Mass (Relative)
Water	400	Carbon dioxide	35
Sugar	100	Alcohol	58
Yeast	10	Acetic acid	3
		Water	409
		Sugar (unreacted)	4
		Yeast (unreacted)	1

1-3 Properties of Substances

Every pure substance has its own unique set of properties that serve to distinguish it from all other substances. A chemist most often identifies an unknown substance by measuring its properties and comparing them with the properties recorded in the chemical literature for known substances.

The properties used to identify a substance must be **intensive properties;** that is, they must be independent of amount. The fact that a sample weighs 4.02 g or has a volume of 229 mL tells us nothing about its identity; mass and volume are **extensive properties;** that is, they depend on amount. Beyond that, substances may be identified on the basis of their

■ **chemical properties,** observed when the substance takes part in a chemical reaction, a change that converts it to a new substance. For example, the fact that mercury(II) oxide decomposes to mercury and oxygen on heating to 600°C can be used to identify it. Again, the chemical inertness of helium helps to distinguish it from other, more reactive gases, such as hydrogen and oxygen.

- **physical properties,** observed without changing the chemical identity of a substance. ◄ ⓘ Two such properties particularly useful for identifying a substance are
  - **melting point,** the temperature at which a substance changes from the solid to the liquid state.
  - **boiling point,** the temperature at which bubbles filled with vapor form within a liquid. If a substance melts at 0°C and boils at 100°C, we are inclined to suspect that it might just be water.

Figure 1.12 shows gold and describes its different properties. In the remainder of this section we will consider a few other physical properties that can be measured without changing the identity of a substance.

Density

The **density** of a substance is the ratio of mass to volume:

$$\text{density} = \frac{\text{mass}}{\text{volume}} \qquad d = \frac{\text{mass}}{V} \qquad (1.3)$$

Note that even though mass and volume are extensive properties, the ratio of mass to volume is intensive. Samples of copper weighing 1.00 g, 10.5 g, 264 g, . . . all have the same density, 8.94 g/mL at 25°C.

For liquids and gases, density can be found in a straightforward way by measuring independently the mass (using a scale) and the volume (using a pipet or graduated cylinder) of a sample. (Example 1.6 illustrates the process.)

Taste is a physical property, but it is never measured in the lab.

© Cengage Learning/Charles D. Winters

Figure 1.12 Properties of gold. The color of gold is an *intensive* property. The quantity of gold in a sample is an *extensive* property. The fact that gold can be stored in the air without undergoing any chemical reaction with oxygen in the air is a *chemical* property. The temperature at which gold melts (1063°C) is a *physical* property.

EXAMPLE 1.6

Glycerol is a viscous liquid used by both the pharmaceutical and food industries as a sweetener, thickener, and stabilizer. To determine its density, a student delivers a 15.0 mL sample by pipet into a flask with a mass of 28.45 g. The mass of the flask and glycerol sample is 47.37 g. What is the density of glycerol?

ANALYSIS	
Information given:	mass of empty flask (28.45 g) mass of flask + sample (47.37 g) volume of sample (15.0 mL)
Asked for:	density of the sample

STRATEGY	

1. Find the mass of the sample by difference.

   mass of sample = (mass of flask + sample) − (mass of flask)

2. Recall the formula for density.

   $$\text{density} = \frac{\text{mass}}{\text{volume}}$$

SOLUTION	
1. mass of sample	mass of sample = (mass of flask + sample) − (mass of flask) = 47.37 g − 28.45 g = 18.92 g
2. density	$d = \dfrac{\text{mass}}{V} = \dfrac{18.92\ \text{g}}{15.0\ \text{mL}} = 1.26\ \text{g/mL}$

END POINT	

The density is expressed in three significant figures because the volume is measured only to 3 significant figures.

For solids, the mass of the sample can be obtained directly by weighing it. The volume of a regular solid (a cube, for example) can be calculated using the given dimensions of the sample. The volume of an irregular solid, like a rock, is obtained by displacement (Example 1.7).

EXAMPLE 1.7

Consider two samples of palladium (Pd), an element used in automobile catalytic converters. Sample A is a cylindrical bar with a mass of 97.36 g. The bar is 10.7 cm high and has a radius of 4.91 mm. Sample B is an irregular solid with a mass of 49.20 g. A graduated cylinder has 10.00 mL of water. When sample B is added to the graduated cylinder, the volume of the water and the solid is 14.09 mL. Calculate the density of each sample.

SAMPLE A:

ANALYSIS

Information given:	mass (97.36 g), radius, r (4.91 mm), height, h (10.7 cm)
Asked for:	density of Pd

STRATEGY

1. Recall the formula to obtain the volume of a cylinder.

$V = \pi r^2 h$

2. Substitute into the definition of density.

$d = \dfrac{\text{mass}}{V}$

SOLUTION

V	$V = \pi r^2 h = \pi \left(4.91 \text{ mm} \times \dfrac{1 \text{ cm}}{10 \text{ mm}} \right)^2 \times 10.7 \text{ cm} = 8.10 \text{ cm}^3$
d	$d = \dfrac{\text{mass}}{V} = \dfrac{97.36 \text{ g}}{8.10 \text{ cm}^3} = 12.0 \text{ g/cm}^3$

SAMPLE B:

ANALYSIS

Information given:	mass: (49.20 g) volume of water before Pd addition: (10.00 mL) volume of water and Pd: (14.09 mL)
Asked for:	density of Pd

STRATEGY

1. The volume of the Pd is the difference between the volume of the Pd and water and the volume of the water alone.

2. Substitute into the definition of density.

$d = \dfrac{\text{mass}}{V}$

SOLUTION

V	$V = V_{H_2O + Pd} - V_{H_2O} = 14.09 \text{ mL} - 10.00 \text{ mL} = 4.09 \text{ mL}$
d	$d = \dfrac{\text{mass}}{V} = \dfrac{49.02 \text{ g}}{4.09 \text{ mL}} = 12.0 \text{ g/mL}$

END POINT

The units for density are g/cm^3 and g/mL. Since $1 \text{ cm}^3 = 1$ mL, these can be used interchangeably.

Figure 1.13 Density. The wood block has a lower density than water and floats. The ring has a higher density than water and sinks.

Figure 1.13 shows the relation between substances with a density greater than that of water (1.0 g/mL) and those with a density less than that of water.

In a practical sense, density can be treated as a conversion factor to relate mass and volume. Knowing that mercury has a density of 13.6 g/mL, we can calculate the mass of 2.6 mL of mercury:

$$2.6 \text{ mL} \times 13.6 \frac{\text{g}}{\text{mL}} = 35 \text{ g}$$

or the volume occupied by one kilogram of mercury:

$$1.000 \text{ kg} \times \frac{10^3 \text{ g}}{1 \text{ kg}} \times \frac{1 \text{ mL}}{13.6 \text{ g}} = 73.5 \text{ mL}$$

Solubility

Solubility is referred to as the process by which a solute dissolves in a solvent, and is ordinarily a physical rather than a chemical change. The extent to which it dissolves can be expressed in various ways. A common method is to state the number of grams of the substance that dissolves in 100 g of solvent at a given temperature.

EXAMPLE 1.8 GRADED

Sucrose is the chemical name for the sugar we consume. Its solubility at 20°C is 204 g/100 g water, and at 100°C is 487 g/100 g water. A solution is prepared by mixing 139 g of sugar in 33.0 g of water at 100°C.

a What is the minimum amount of water required to dissolve the sugar at 100°C?

ANALYSIS

Information given:	sucrose solubility at 100°C (487 g/100 g water) composition of solution: sucrose (139 g), water (33.0 g)
Asked for:	minimum amount of H_2O to dissolve 139 g sucrose at 100°C

STRATEGY

Relate the mass H_2O required to the mass sucrose to be dissolved at 100°C by using the solubility at 100°C as a conversion factor.

SOLUTION

mass H_2O required	$139 \text{ g sucrose} \times \dfrac{100 \text{ g } H_2O}{487 \text{ g sucrose}} = 28.5 \text{ g } H_2O$

b What is the maximum amount of sugar that can be dissolved in the water at 100°C?

ANALYSIS

Information given:	sucrose solubility at 100°C (487 g/100 g water) composition of solution: sucrose (139 g), water (33.0 g)
Asked for:	maximum amount of sucrose that can be dissolved in 33.0 g H_2O at 100°C

STRATEGY

Relate the mass of H_2O required to the mass of sucrose to be dissolved at 100°C by using the solubility at 100°C as a conversion factor.

SOLUTION

mass sucrose	$33.0 \text{ g } H_2O \times \dfrac{487 \text{ g sucrose}}{100 \text{ g } H_2O} = 161 \text{ g sucrose}$

continued

c	The solution is cooled to 20°C. How much sugar (if any) will crystallize out?

ANALYSIS	
Information given:	sucrose solubility at 20°C (204 g/100 g water) composition of solution: sucrose (139 g), water (33.0 g)
Asked for:	mass of sucrose in the solution that will not dissolve at 20°C

STRATEGY
1. The question really is: How much sucrose will dissolve in 33.0 g water at 20°C? Relate the mass of sucrose that can be dissolved by 33.0 g H$_2$O at 20°C by using the solubility at 20°C as a conversion factor.
2. Take the difference between the calculated amount that can be dissolved and the amount of sucrose that was in solution at 100°C.

SOLUTION	
1. mass sucrose	$33.0 \text{ g H}_2\text{O} \times \dfrac{204 \text{ g sucrose}}{100 \text{ g H}_2\text{O}} = 67.3$ g sucrose will dissolve at 20°C.
2. undissolved sucrose	139 g in solution − 67.3 g can be dissolved = **72 g undissolved**

Figure 1.14 shows the solubility of sugar in water as a function of temperature. Alternatively, we can say that it gives the concentration of sugar in a **saturated solution** at various temperatures. For example, at 20°C, we could say that "the solubility of sugar is 204 g/100 g water" 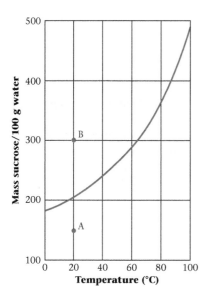 or that "a saturated solution of sugar contains 204 g/100 g water."

At any point in the area below the curve in Figure 1.14, we are dealing with an **unsaturated solution**. Consider, for example, point A (150 g sugar per 100 g water at 20°C). This solution is unsaturated; if we add more sugar, another 54 g will dissolve to give a saturated solution (204 g sugar per 100 g water at 20°C).

At any point in the area above the curve, the sugar solution is **supersaturated solution**. This is the case at point B (300 g sugar per 100 g water at 20°C). Such a solution could be formed by carefully cooling a saturated solution at 60°C to 20°C, where a saturated solution contains 204 g sugar per 100 g water. The excess sugar stays in solution until a small seed crystal of sugar is added, whereupon crystallization quickly takes place. At that point the excess sugar

$$300 \text{ g} - 204 \text{ g} = 96 \text{ g}$$

comes out of solution.

The "100 g water" in the solubility expression is an exact quantity.

When the temperature changes, the amount of solute in solution changes, but the mass of water stays the same.

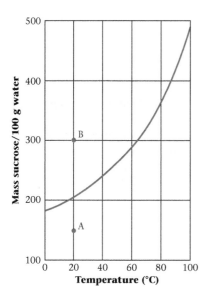

Figure 1.14 Solubility of table sugar (sucrose). The solubility of sugar, C$_{12}$H$_{22}$O$_{11}$, in water increases exponentially with temperature.

CHEMISTRY **BEYOND** THE CLASSROOM

Arsenic

An element everyone has heard about but almost no one has ever seen is arsenic, symbol As. It is a gray solid with some metallic properties, melts at 816°C, and has a density of 5.78 g/mL. Among the elements, arsenic ranks 51st in abundance. It is about as common as tin or beryllium. Two brightly colored sulfides of arsenic, realgar and orpiment (Figure A), were known to the ancients. The element is believed to have been isolated for the first time by Albertus Magnus in the thirteenth century. He heated orpiment with soap. The alchemists gave what they thought to be arsenic (it really was an oxide of arsenic) its own symbol (Figure B) and suggested that women rub it on their faces to whiten their complexion.

The principal use of elemental arsenic is in its alloys with lead. The "lead" storage battery contains a trace of arsenic along with 3% antimony. Lead shot, which are formed by allowing drops of molten lead to fall through the air, contains from 0.5 to 2.0% arsenic. The presence of arsenic raises the surface tension of the liquid and hence makes the shot more spherical.

In the early years of the twentieth century, several thousand organic compounds were synthesized and tested for medicinal use, mainly in the treatment of syphilis. One of these compounds, salvarsan, was found to be very effective. Arsenic compounds fell out of use in the mid-twentieth century because of the unacceptable side effects that occurred at the dosages that were thought to be necessary. In the 1970s, Chinese medicine tried a highly purified oxide of arsenic in a low-dose regimen. It was shown to be effective in the treatment of some leukemias. Western medicine has confirmed these results. Molecular studies and clinical trials are ongoing and suggest that arsenic oxides show great promise in the treatment of malignant disease.

Industry and farming have used arsenic compounds. However, because of its great toxicity and ability to leach into wells and streams, its is no longer produced in the United States, but it is still imported from other countries. Until the 1940s, arsenic compounds were used as agricultural pesticides. Today, most uses of arsenic in farming are banned in the United States, and its use as a preservative in pressure-treated wood has been greatly reduced.

The "arsenic poison" referred to in crime dramas is actually an oxide of arsenic rather than the element itself. Less than 0.1 g of this white, slightly soluble powder can be fatal. The classic symptoms of arsenic poisoning involve various unpleasant gastrointestinal disturbances, severe abdominal pain, and burning of the mouth and throat.

In the modern forensic laboratory, arsenic is detected by analysis of hair samples. A single strand of hair is sufficient to establish the presence or absence of the element. The technique most commonly used is neutron activation analysis, described in Chapter 18. If the concentration is found to be greater than about 0.0003%, poisoning is indicated.

This technique was applied in the early 1960s to a lock of hair taken from Napoleon Bonaparte (1769–1821) on St. Helena. Arsenic levels of up to 50 times normal suggested he may have been a victim of poisoning, perhaps on orders from the French royal family. More recently (1991), U.S. President Zachary Taylor (1785–1850) was exhumed on the unlikely hypothesis that he had been poisoned by Southern sympathizers concerned about his opposition to the extension of slavery. The results indicated normal arsenic levels. Apparently, he died of cholera, brought on by an overindulgence in overripe and unwashed fruit.

© Cengage Learning/Charles D. Winters

Figure A Realgar and orpiment.

Figure B Alchemist symbol for arsenic.

The crystallization of excess solute is a common problem in the preparation of candies and in the storage of jam and honey. From these supersaturated solutions, sugar separates either as tiny crystals, causing the "graininess" in fudge, or as large crystals, which often appear in honey kept for a long time (Figure 1.15).

Marna G. Clarke

Figure 1.15 Rock candy. The candy is formed by crystallization of sugar from a saturated solution that is cooled slowly.

Key Concepts

1. Convert between °F, °C, and K.
   (**Example 1.1; Problems 13–16**)
2. Determine the number of significant figures in a measured quantity.
   (**Example 1.2; Problems 17, 18, 25, 26**)
3. Determine the number of significant figures in a calculated quantity.
   (**Example 1.3; Problems 27–30**)
4. Use conversion factors to change the units of a measured quantity.
   (**Example 1.4; Problems 31–44, 60**)
5. Relate density to mass and volume.
   (**Example 1.5; Problems 45–52, 61, 63**)
6. Given its solubility, relate mass of solute to that of solvent.
   (**Example 1.6; Problems 53–56, 57**)

Key Equations

Fahrenheit temperature $t_{°F} = 1.8\, t_{°C} + 32°$

Kelvin temperature $T_K = t_{°C} + 273.15$

Density $\dfrac{mass}{V}$

Key Terms

centi	element	mixture	significant figures
compound	extensive properties	—heterogeneous	solution
conversion factor	intensive properties	—homogeneous	—saturated solution
chemical properties	kilo	nano	—supersaturated solution
density	milli	physical properties	—unsaturated solution

Summary Problem

Potassium permanganate, a purple crystal, can be used as a mild antiseptic to prevent infection from cuts and scrapes. It is made up of three atoms: potassium, oxygen, and manganese. Consider the following information about potassium permanganate:

- density at 20°C: 2.703 g/cm³
- melting point: 2.40×10^2°C
- solubility at 20°C: 6.38 g/100 g water
- solubility at 60°C: 25 g/100 g water

(a) What are the symbols of the three elements in potassium permanganate?

(b) List the physical properties of potassium permanganate described above.

(c) What is the mass of a sample of potassium permanganate with a volume of 48.7 cm³?

(d) Express the density in pounds per cubic foot.

(e) Express the melting point of potassium permanganate in °F and K.

(f) How many grams of potassium permanganate can be dissolved in 38.5 g of water at 20°C?

(g) An aqueous solution of potassium permanganate is prepared by dissolving 15.0 g of potassium permanganate in 65.0 g of water at 60°C. The solution is carefully cooled to 20°C. A homogeneous solution is obtained. Is the solution at 60°C saturated, unsaturated, or supersaturated? What about the solution at 20°C?

(h) A solution is prepared by dissolving 10.0 g of potassium permanganate in 55.0 g of water at 60°C. Will all the solute dissolve? How many grams of solute will crystallize out of solution if it is rapidly cooled to 20°C?

Express your answers to the correct number of significant figures; use the conversion factor method throughout.

Answers

(a) K, Mn, O
(b) density, melting point, solubility, color
(c) 132 g
(d) 168.7 lb/ft³
(e) 464°F; 513 K
(f) 2.46 g
(g) at 60°C: unsaturated; at 20°C: supersaturated
(h) yes; 6.5 g

Questions and Problems

Even-numbered questions and Challenge Problems have answers in Appendix 5 and fully worked solutions in the *Student Solutions Manual.*

The questions and problems listed here are typical of those at the end of each chapter. Some are conceptual. Most require calculations, writing equations, or other quantitative work. The headings identify the primary section and topic of each set of questions or problems, such as "Symbols and Formulas" or "Significant Figures." Those in the "Unclassified" category may involve more than one concept, including, perhaps, topics from a preceding chapter. "Challenge Problems," listed at the end of the set, require extra skill and/or effort. The "Classified" questions and problems (Problems 1–56 in this set) occur in matched pairs, one below the other, and illustrate the same concept. For example, Questions 1 and 2 are nearly identical in nature; the same is true of Questions 3 and 4, and so on.

1-1 Matter and Its Classifications

Types of Matter

1. Classify each of the following as element, compound, or mixture.
    (a) air   (b) iron
    (c) soy sauce   (d) table salt
2. Classify each of the following as element, compound, or mixture.
    (a) gold
    (b) milk
    (c) sugar
    (d) vinaigrette dressing with herbs
3. Classify the following as solution or heterogeneous mixture.
    (a) normal urine
    (b) gasoline
    (c) batter for chocolate chip cookies
4. Classify the following as solution or heterogeneous mixture.
    (a) iron ore
    (b) chicken noodle soup
    (c) tears
5. How would you separate into its components
    (a) a homogeneous solution of table salt and water?
    (b) sand from gasoline?
    (c) carbon dioxide gas from butane gas?
6. How would you separate into its different components
    (a) a mixture of the volatile gases propane, butane, and isopropane?
    (b) a solution of rubbing alcohol made up of isopropyl alcohol and water?
7. Write the symbol for the following elements.
    (a) titanium   (b) phosphorus
    (c) potassium   (d) magnesium
8. Write the symbol for the following elements.
    (a) copper   (b) carbon
    (c) bromine   (d) aluminum

9. Write the name of the element represented by the following symbols.
    (a) Hg   (b) Si   (c) Na   (d) I
10. Write the name of the element represented by the following symbols:
    (a) Mn   (b) Au   (c) Pb   (d) Ag

1-2 Measurements

Measurements

11. What instrument would you use to determine
    (a) the mass of a head of lettuce?
    (b) whether your refrigerator is cooling water to 10°C?
    (c) the volume of a glass of orange juice?
12. What instrument would you use to
    (a) measure the length of your bed?
    (b) measure the amount of acid delivered by a beaker?
    (c) determine whether you have a fever?
13. A glass of lukewarm milk is suggested for people who cannot sleep. Milk at 52°C can be characterized as lukewarm. What is the temperature of lukewarm milk in °F? In K?
14. A recipe for apple pie calls for a preheated 350°F (three significant figures) oven. Express this temperature setting in °C and in K.
15. Gallium is one of the few metals that can melt at room temperature. Its melting point is 29.76°C. If you leave solid gallium in your car on an early summer morning when the temperature is 75.0°F, what physical state is the gallium in when you return to your car and the interior car temperature is 85.0°F?
16. Computers are not supposed to be in very warm rooms. The highest termperature tolerated for maximum performance is 308 K. Express this temperature in °C and °F.

Significant Figures

17. How many significant figures are in each of the following?
    (a) 0.890 cm   (b) 210°C
    (c) 2.189×10^6 nm   (d) 2.54 cm = 1 in
    (e) 140.00 g
18. How many significant figures are there in each of the following?
    (a) 0.136 m   (b) 0.0001050 g
    (c) 2.700×10^3 nm   (d) 6×10^{-4} L
    (e) 56003 cm^3
19. Round off the following quantities to the indicated number of significant figures.
    (a) 7.4855 g (three significant figures)
    (b) 298.693 cm (five significant figures)
    (c) 11.698 lb (one significant figure)
    (d) 12.05 oz (three significant figures)
20. Round off the following quantities to the indicated number of significant figures.
    (a) 17.2509 cm (4 significant figures)
    (b) 168.51 lb (3 significant figures)
    (c) 500.22°C (3 significant figures)
    (d) 198.500 oz (3 significant figures)

21. Express the following in scientific notation:
    (a) 132.5 cm
    (b) 0.000883 km
    (c) 6,432,000,000 nm

22. Express the following measurements in scientific notation.
    (a) 4020.6 mL   (b) 1.006 g   (c) 100.1°C

23. Which of the following statements use only exact numbers?
    (a) The little boy weighs 74.3 lbs.
    (b) The boiling point of ethanol is 78°C.
    (c) The mailman delivered 3 letters, 2 postcards and one magazine.

24. Which of the following statements use only exact numbers?
    (a) The temperature in our dorm room is kept at 72°F.
    (b) I bought 6 eggs, 2 cookies, and 5 tomatoes at the farmers' market.
    (c) There are 1×10^9 nanometers in 1 meter.

25. A basketball game at the University of Connecticut's Gampel Pavilion attracted 10,000 people. The building's interior floor space has an area of 1.71×10^5 ft². Tickets to the game sold for $22.00. Senior citizens were given a 20% discount. How many significant figures are there in each quantity? (Your answer may include the words *ambiguous* and *exact*.)

26. A listing of a house for sale states that there are 5 bedrooms, 4000 ft² of living area, and a living room with dimensions 17×18.5 ft. How many significant figures are there in each quantity? (Your answer may include the words *ambiguous* and *exact*.)

27. Calculate the following to the correct number of significant figures. Assume that all these numbers are measurements.
    (a) $x = 17.2 + 65.18 - 2.4$
    (b) $x = \dfrac{13.0217}{17.10}$
    (c) $x = (0.0061020)(2.0092)(1200.00)$
    (d) $x = 0.0034 + \dfrac{\sqrt{(0.0034)^2 + 4(1.000)(6.3 \times 10^{-4})}}{(2)(1.000)}$
    (e) $x = \dfrac{(2.998 \times 10^8)(3.1 \times 10^{-7})}{6.022 \times 10^{23}}$

28. Perform the indicated calculations. Write your answers with the correct number of significant figures.
    (a) $\dfrac{13.01 \text{ g}}{1.23 \text{ L}}$
    (b) $17.2 \text{ lb} + 65.18 \text{ lb} - 2.3 \text{ lb}$
    (c) $(0.003240 \text{ cm})(1.25 \text{ cm})(1.0 \times 10^2 \text{ cm})$
    (d) $\dfrac{0.024 + \sqrt{(0.024)^2 + 4(6.3 \times 10^{-3})}}{2}$
       Where 4 and 2 are exact numbers, and all other quantities are measurements.
    (e) $\dfrac{(4.63 \times 10^4)(1.8 \times 10^{-6})}{(6.022 \times 10^{23})}$

Assume all quantities are measurements.

29. The volume of a sphere is $4\pi r^3/3$, where r is the radius. One student measured the radius to be 4.30 cm. Another measured the radius to be 4.33 cm. What is the difference in volume between the two measurements?

30. The volume of a square pyramid is $(1/3)Bh$ where B is the area of the square base and h is the height. One student's measurements of a pyramid are B = 1.33 m², h = 2.79 m. Another student's measurements of the same pyramid are B = 1.31 m², h = 2.81 m. What is the difference in the volumes of the pyramid calculated by the two students?

Conversion Factors

31. Write the appropriate symbol in the blank ($>$, $<$, or $=$).
    (a) 303 m _____ 303×10^3 km
    (b) 500 g _____ 0.500 kg
    (c) 1.50 cm³ _____ 1.50×10^3 nm³

32. Write the appropriate symbol in the blank ($>$, $<$, or $=$).
    (a) 37.12 g _____ 0.3712 kg
    (b) 28 m³ _____ 28×10^2 cm³
    (c) 525 mm _____ 525×10^6 nm

33. Convert 22.3 mL to
    (a) liters   (b) in³   (c) quarts

34. Convert 0.2156 L to
    (a) mL
    (b) in³
    (c) quarts

35. The height of a horse is usually measured in hands. One hand is exactly 1/3 ft.
    (a) How tall (in feet) is a horse of 19.2 hands?
    (b) How tall (in meters) is a horse of 17.8 hands?
    (c) A horse of 20.5 hands is to be transported in a trailer. The roof of the trailer needs to provide 3.0 ft of vertical clearance. What is the minimum height of the trailer in feet?

36. At sea, distances are measured in nautical miles and speeds are expressed in knots.

    1 nautical mile = 6076.12 ft
    1 knot = 1 nautical mi/h (exactly)

    (a) How many miles are in one nautical mile?
    (b) How many meters are in one nautical mile?
    (c) A ship is traveling at a rate of 22 knots. Express the ship's speed in miles per hour.

37. The unit of land measure in the English system is the acre, while that in the metric system is the hectare. An acre is 4.356×10^4 ft². A hectare is ten thousand square meters. A town requires a minimum area of 2.0 acres of land for a single-family dwelling. How many hectares are required?

38. A gasoline station in Manila, Philippines, charges 38.46 pesos per liter of unleaded gasoline at a time when one U.S. dollar (USD) buys 47.15 pesos (PHP). The car you are driving has a gas tank with a capacity of 14 U.S. gallons and gets 24 miles per gallon.
    (a) What is the cost of unleaded gasoline in Manila in USD per gallon?
    (b) How much would a tankful of unleaded gasoline for your car cost in USD?
    (c) Suppose that you have only PHP 1255 (a day's wage for an elementary school teacher) and the car's tank is

almost empty. How many miles can you expect to drive if you spend all your money on gasoline?

39. A lap in most tracks in the United States is 0.25 mi (English lap). In most countries that use the metric system, a metric lap is 0.40 km. A champion marathon runner covers a mile in about 5.0 min. How many minutes will the runner take to run an English lap? A metric lap?

40. Cholesterol in blood is measured in milligrams of cholesterol per deciliter of blood. If the unit of measurement were changed to grams of cholesterol per milliliter of blood, what would a cholesterol reading of 185 mg/dL translate to?

41. Some states have reduced the legal limit for alcohol sobriety from 0.10% to 0.080% alcohol by volume in blood plasma.

(a) How many milliliters of alcohol are in 3.0 qt of blood plasma at the lower legal limit?

(b) How many milliliters of alcohol are in 3.0 qt of blood plasma at the higher legal limit?

(c) How much less alcohol is in 3.0 qt of blood plasma with the reduced sobriety level?

42. The area of the 48 contiguous states is 3.02×10^6 mi². Assume that these states are completely flat (no mountains and no valleys). What volume, in liters, would cover these states with a rainfall of two inches?

43. In Europe, nutritional information is given in kilojoules (kJ) instead of nutritional calories (1 nutritional calorie = 1 kcal). A packet of soup has the following nutritional information:

$$250 \text{ mL of soup} = 235 \text{ kJ}$$

How would that same packet be labeled in the United States if the information has to be given in nutritional calories per cup? (There are 4.18 joules in one calorie and 2 cups to a pint.)

44. In the old pharmaceutical system of measurements, masses were expressed in grains. There are 5.760×10^3 grains in 1 lb. An old bottle of aspirin lists 5.000 grains of active ingredient per tablet. How many milligrams of active ingredient are there in the same tablet?

1-3 Properties of Substances

Physical and Chemical Properties

45. The cup is a measure of volume widely used in cookbooks. One cup is equivalent to 225 mL. What is the density of clover honey (in grams per milliliter) if three quarters of a cup has a mass of 252 g?

46. The egg whites from four "large" eggs occupy a volume of 112 mL. The mass of the egg whites from these four eggs is 1.20×10^2 g. What is the average density of an egg white from one "large" egg?

47. A metal slug weighing 25.17 g is added to a flask with a volume of 59.7 mL. It is found that 43.7 g of methanol ($d = 0.791$ g/mL) must be added to the metal to fill the flask. What is the density of the metal?

48. A solid with an irregular shape and a mass of 11.33 g is added to a graduated cylinder filled with water ($d = 1.00$ g/mL) to the 35.0-mL mark. After the solid sinks to the bottom, the water level is read to be at the 42.3-mL mark. What is the density of the solid?

49. A waterbed filled with water has the dimensions 8.0 ft × 7.0 ft × 0.75 ft. Taking the density of water to be 1.00 g/cm³, how many kilograms of water are required to fill the waterbed?

50. Wire is often sold in pound spools according to the wire gauge number. That number refers to the diameter of the wire. How many meters are in a 10-lb spool of 12-gauge aluminum wire? A 12-gauge wire has a diameter of 0.0808 in. Aluminum has a density of 2.70 g/cm³. ($V = \pi r^2 \ell$)

51. Air is 21% oxygen by volume. Oxygen has a density of 1.31 g/L. What is the volume, in liters, of a room that holds enough air to contain 55 kg of oxygen?

52. The unit for density found in many density tables is kg/m³. At a certain temperature, the gasoline you pump into your car's gas tank has a density of 732.22 kg/m³. If your tank has a capacity of 14.0 gallons, how many grams of gasoline are in your tank when it is full? How many pounds?

53. Magnesium sulfate ($MgSO_4$) has a solubility of 38.9 g/100 g H_2O at 30°C. A solution is prepared by adding 9.50 g of $MgSO_4$ to 25.0 g of water at 40°C. A homogeneous mixture is obtained. Is the solution saturated, unsaturated, or supersaturated? One gram of magnesium sulfate is added to the solution cooled to 30°C. Would you expect some of the $MgSO_4$ to precipitate? If so, how much? If not, how much more $MgSO_4$ can be added before precipitation takes place?

54. Potassium sulfate has a solubility of 15 g/100 g water at 40°C. A solution is prepared by adding 39.0 g of potassium sulfate to 225 g of water, carefully heating the solution, and cooling it to 40°C. A homogeneous solution is obtained. Is this solution saturated, unsaturated, or supersaturated? The beaker is shaken, and precipitation occurs. How many grams of potassium sulfate would you expect to crystallize out?

55. Sodium bicarbonate (baking soda) is commonly used to absorb odor. Its solubility is 9.6 g/100 g H_2O at 30°C and 16 g/100 g H_2O at 60°C. At 60°C, 9.2 g of baking soda are added to 46 g of water.

(a) Is the resulting mixture homogeneous at 60°C? If not, how many grams of baking soda are undissolved?

(b) The mixture is cooled to 30°C. How many more grams of water are needed to make a saturated solution?

56. Magnesium chloride is an important coagulant used in the preparation of tofu from soy milk. Its solubility in water at 20°C is 54.6 g/100 g. At 80°C, its solubility is 66.1 g/100 g. A mixture is made up of 16.2 g of magnesium chloride and 38.2 g of water at 20°C.

(a) Is the mixture homogeneous? If it is, how many more grams of magnesium chloride are required to make a saturated solution? If the mixture is not homogeneous, how many grams of magnesium chloride are undissolved?

(b) How many more grams of magnesium chloride are needed to make a saturated solution at 80°C?

Unclassified

57. The solubility of lead nitrate at 100°C is 140.0 g/100 g water. A solution at 100°C consists of 57.0 g of lead nitrate in 64.0 g of water. When the solution is cooled to 10°C, 25.0 g of lead nitrate crystallize out. What is the solubility of lead nitrate in g/100 g water at 10°C?

58. Radiation exposure to human beings is usually given in rems (radiation equivalent for man). In SI units, the exposure is measured in sieverts (Sv). One rem equals 0.0100 Sv. At one time, the exposure due to the nuclear reactors in Japan was measured to be 8217 mSv/h. How many rems would a person exposed to the radiation for 35 min have absorbed? If one mammogram gives off 0.30 rems, how many mammograms would that exposure be equivalent to?

59. The following data refer to the element phosphorus. Classify each as a physical or a chemical property.
    (a) It exists in several forms, for example, white, black, and red phosphorus.
    (b) It is a solid at 25°C and 1 atm.
    (c) It is insoluble in water.
    (d) It burns in chlorine to form phosphorus trichloride.

60. A supersaturated sugar solution (650.0 g sugar in 150.0 g water) is prepared to make "rock candy." The solution is kept at 25°C, where the solubility of sugar is 220.0 g/100 g water. A string is dipped into the solution, and in time, large crystals of sugar (rock candy) form. How many grams of solid are obtained when the solution just reaches saturation?

61. The density of wind-packed snow is estimated to be 0.35 g/cm³. A flat roof that is 35 by 43 feet has 28 inches of snow on it. How many pounds of snow are on the roof?

62. The dimensions of aluminum foil in a box for sale in supermarkets are 66⅔ yards by 12 inches. The mass of the foil is 0.83 kg. If its density is 2.70 g/cm³, then what is the thickness of the foil in inches?

63. The Kohinoor Diamond ($d = 3.51$ g/cm³) is 108 carats. If one carat has a mass of 2.00×10^2 mg, what is the mass of the Kohinoor Diamond in pounds? What is the volume of the diamond in cubic inches?

64. A pycnometer is a device used to measure density. It weighs 20.455 g empty and 31.486 g when filled with water ($d = 1.00$ g/cm³). Pieces of an alloy are put into the empty, dry pycnometer. The mass of the alloy and pycnometer is 28.695 g. Water is added to the alloy to exactly fill the pycnometer. The mass of the pycnometer, water, and alloy is 38.689 g. What is the density of the alloy?

65. Titanium is used in airplane bodies because it is strong and light. It has a density of 4.55 g/cm³. If a cylinder of titanium is 7.75 cm long and has a mass of 153.2 g, calculate the diameter of the cylinder. ($V = \pi r^2 h$, where V is the volume of the cylinder, r is its radius, and h is the height.)

Conceptual Questions

66. Label each of the properties of iodine as intensive (I) or extensive (E).
    (a) Its density is 4.93 g/mL.
    (b) It is purple.
    (c) It melts at 114°C.
    (d) One hundred grams of melted iodine has a volume of 122 mL.
    (e) Some iodine crystals are large.

67. How do you distinguish
    (a) chemical properties from physical properties?
    (b) distillation from filtration?
    (c) a solute from a solution?

68. What is the difference between
    (a) mass and density?
    (b) an extensive and an intensive property?
    (c) a solvent and a solution?

69. Mercury, ethyl alcohol, and lead are poured into a cylinder. Three distinct layers are formed. The densities of the three substances are

$$\text{mercury} = 13.55 \text{ g/cm}^3$$

$$\text{ethyl alcohol} = 0.78 \text{ g/cm}^3$$

$$\text{lead} = 11.4 \text{ g/cm}^3$$

Sketch the cylinder with the three layers. Identify the substance in each layer.

70. How many significant figures are there in the length of this line?

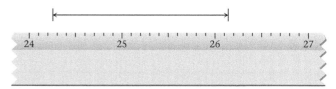

71. Consider the following solubility graph.

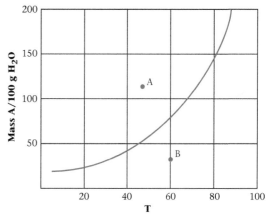

    (a) At point *A*, how many grams of the compound are dissolved in 100 g of water? Is the solution saturated, unsaturated, or supersaturated?
    (b) At point *B*, how many grams of the compound are dissolved in 100 g of water? Is the solution saturated, unsaturated, or supersaturated?
    (c) How would you prepare a saturated solution at 30°C?

72. Given the following solubility curves, answer the following questions:
    (a) In which of the two compounds can more solute be dissolved in the same amount of water when the temperature is decreased?
    (b) At what temperature is the solubility of both compounds the same?

(c) Will an increase in temperature always increase solubility of a compound? Explain.

Challenge Problems

73. A different civilization on a distant planet has developed a new temperature scale based on ethyl alcohol. The freezing point of ethyl alcohol (−117°C) is designated as 0°J, and its boiling point (78°C) is designated as 100°J. Assuming that the relationship between °C and °J is linear,

(a) draw a graph of the line using the data above.
(b) what is the slope of the line?
(c) what is the y-intercept of the line?
(d) Write an equation to convert °J to °C.

74. At what point is the temperature in °F exactly twice that in °C?

75. Oil spreads on water to form a film about 100 nm thick (two significant figures). How many square kilometers of ocean will be covered by the slick formed when one barrel of oil is spilled (1 barrel = 31.5 U.S. gal)?

76. A laboratory experiment requires 12.0 g of aluminum wire ($d = 2.70$ g/cm^3). The diameter of the wire is 0.200 in. Determine the length of the wire, in centimeters, to be used for this experiment. The volume of a cylinder is $\pi r^2 \ell$, where r = radius and ℓ = length.

77. An average adult breathes about 8.50×10^3 L of air per day. The concentration of lead in highly polluted urban air is 7.0×10^{-6} g of lead per one m^3 of air. Assume that 75% of the lead is present as particles less than 1.0×10^{-6} m in diameter, and that 50% of the particles below that size are retained in the lungs. Calculate the mass of lead absorbed in this manner in 1 year by an average adult living in this environment.

78. A student determines the density of a metal by taking two cylinders of equal mass and volume. Cylinder A is filled with mercury and the metal. The mercury and metal together have a mass of 92.60 g. Cylinder B is filled only with mercury. The mass of the mercury in cylinder B is 52.6 g more than the mass of mercury and metal in cylinder A. What is the density of the metal? The density of mercury is 13.6 g/cm^3.

2

Atoms, Molecules, and Ions

An artist's view of an atom.

Atom from atom yawns as far
As moon from earth, or star
from star.

—RALPH WALDO EMERSON
Atoms

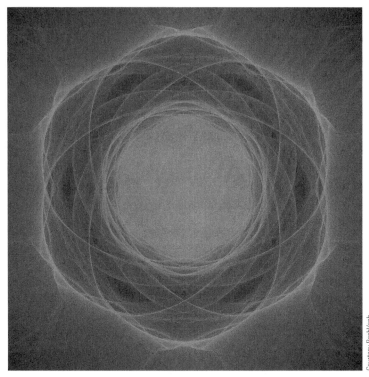

Courtesy RochVanh

To learn chemistry, you must become familiar with the building blocks that chemists use to describe the structure of matter. These include

- *atoms* (Section 2-1), composed of electrons, protons, and neutrons (Section 2-2) and their quantitative properties—atomic mass and atomic number (Section 2-3).
- *molecules*, the building blocks of several elements and a great many compounds. Molecular substances can be identified by their formulas (Section 2-5) or their names (Section 2-7)
- *ions*, species of opposite charge found in all ionic compounds. Using relatively simple principles, it is possible to derive the formulas (Section 2-6) and names (Section 2-7) of ionic compounds.

Early in this chapter (Section 2-4), we will introduce a classification system for elements known as the *periodic table*. It will prove useful in this chapter and throughout the remainder of the text.

2-1 Atoms and the Atomic Theory

In 1808, an English scientist and schoolteacher, John Dalton, developed the atomic model of matter that underlies modern chemistry. Three of the main postulates of modern **atomic theory**, all of which Dalton suggested in a somewhat different form, are stated below and illustrated in Figure 2.1.

1. *An element is composed of tiny particles called atoms.* All atoms of a given element have the same chemical properties. Atoms of different elements show different properties.
2. *In an ordinary chemical reaction, atoms move from one substance to another, but no atom of any element disappears or is changed into an atom of another element.*
3. *Compounds are formed when atoms of two or more elements combine.* In a given compound, the relative numbers of atoms of each kind are definite and constant. In general, these relative numbers can be expressed as integers or simple fractions.

On the basis of Dalton's theory, the **atom** can be defined as the smallest ◄ ⓘ particle of an element that can enter into a chemical reaction.

Atoms of element 1

Atoms of element 2

—Compound 1
—Compound 2

Different combinations produce different compounds.

2-2 Components of the Atom

Like any useful scientific theory, the atomic theory raised more questions than it answered. Scientists wondered whether atoms, tiny as they are, could be broken down into still smaller particles. Nearly 100 years passed before the existence of subatomic particles was confirmed by experiment. Two future Nobel laureates did pioneer work in this area. J. J. Thomson was an English physicist working at the Cavendish Laboratory at Cambridge. Ernest Rutherford, at one time a student of Thomson's (Figure 2.2), was a native of New Zealand. Rutherford carried out his research at McGill University in Montreal and at Manchester and Cambridge in England. He was clearly the greatest experimental physicist of his time, and one of the greatest of all time.

Atoms of different elements have different masses.

No atom disappears or is changed in a chemical reaction.

Electrons

The first evidence for the existence of subatomic particles came from studies of the conduction of electricity through gases at low pressures. When the glass tube shown in Figure 2.3 is partially evacuated and connected to a spark coil,

Figure 2.1 Some features of Dalton's atomic theory.

ⓘ
Atoms are indeed tiny (they have diameters of about 10^{-10} m).

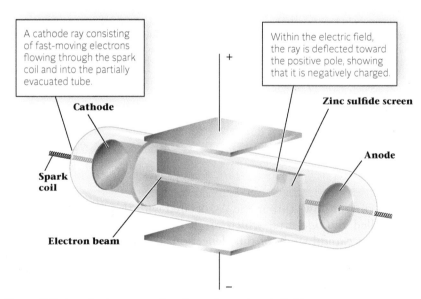

A cathode ray consisting of fast-moving electrons flowing through the spark coil and into the partially evacuated tube.

Within the electric field, the ray is deflected toward the positive pole, showing that it is negatively charged.

Cathode

Zinc sulfide screen

Anode

Spark coil

Electron beam

Figure 2.3 A cathode ray passing through an electric field.

Figure 2.2 J. J. Thomson and Ernest Rutherford (*right*). They are talking, perhaps about nuclear physics, but more likely about yesterday's cricket match.

CHEMISTRY THE **HUMAN** SIDE

John Dalton was a quiet, unassuming man and a devout Quaker. When presented to King William IV of England, Dalton refused to wear the colorful court robes because of his religion. His friends persuaded him to wear the scarlet robes of Oxford University, from which he had a doctor's degree. Dalton was color-blind, so he saw himself clothed in gray.

Dalton was a prolific scientist who made contributions to biology and physics as well as chemistry. At a college in Manchester, England, he did research and spent as many as 20 hours a week lecturing in mathematics and the physical sciences. Dalton never married; he said once, "My head is too full of triangles, chemical properties, and electrical experiments to think much of marriage."

Dalton's atomic theory explained three of the basic laws of chemistry:

The *law of conservation of mass:* This states that *there is no detectable change in mass in an ordinary chemical reaction.* If atoms are conserved in a reaction (postulate 2 of the atomic theory), mass will also be conserved.

The *law of constant composition:* This tells us that *a compound always contains the same elements in the same proportions by mass.* If the atom ratio of the elements in a compound is fixed (postulate 3), their proportions by mass must also be fixed.

The *law of multiple proportions:* This law, formulated by Dalton himself, was crucial to establishing atomic theory. It applies to situations in which two elements form more than one compound. The law states that in these compounds, *the masses of one element that combine with a fixed mass of the second element are in a ratio of small whole numbers.*

The validity of this law depends on the fact that atoms combine in simple, whole-number ratios (postulate 3). Its relation to atomic theory is further illustrated in Figure A.

John Dalton
(1766–1844)

Marna G. Clarke

Figure A Chromium-oxygen compounds and the law of multiple proportions. Chromium forms two different compounds with oxygen, as shown by their different colors. In the green compound there are two chromium atoms for every three oxygen atoms ($2\,Cr:3\,O$) and 2.167 g of chromium per gram of oxygen. In the red compound there is one chromium atom for every three oxygen atoms ($1\,Cr:3\,O$) and 1.083 g of chromium per gram of oxygen. The ratio of the chromium masses, 2.167 : 1.083, is that of two small whole numbers, 2.167 : 1.083 = 2 : 1, an illustration of the law of multiple proportions.

an electric current flows through it. Associated with this flow are colored rays of light called *cathode rays,* which are bent by both electric and magnetic fields. From a careful study of this deflection, J. J. Thomson showed in 1897 that the rays consist of a stream of negatively charged particles, which he called **electrons.** We now know that electrons are common to all atoms, carry a unit negative charge (-1), and have a very small mass, roughly 1/2000 that of the lightest atom.

Every atom contains a definite number of electrons. This number, which runs from 1 to more than 100, is characteristic of a neutral atom of a particular element. All atoms of hydrogen contain one electron; all atoms of the element uranium contain 92 electrons. We will have more to say in Chapter 6 about how these electrons are arranged relative to one another. Right now, you need only know that they are found in the outer regions of the atom, where they form what amounts to a cloud of negative charge.

Protons and Neutrons; the Atomic Nucleus

After J. J. Thomson discovered the negatively charged particles (electrons) in the atom, he proposed the structure of the atom to consist of a positively charged sphere with the negatively charged electrons embedded in that sphere. It was known at the time to be the "plum pudding model." Today, we would probably call it the "raisin bread model," where the raisins are the electrons distributed in the positively charged nucleus, the bread (Figure 2.4).

A series of experiments carried out under the direction of Ernest Rutherford in 1911 shaped our ideas about the nature of the atom. He and his students bombarded a piece of thin gold foil (Figure 2.5, p. 26) with α particles (helium atoms minus their electrons). With a fluorescent screen, they observed the extent to which the α particles were scattered. Most of the particles went through the foil unchanged in direction; a few, however, were reflected back at acute angles. This was a totally unexpected result, inconsistent with the model of the atom in vogue at that time. In Rutherford's words, "It was as though you had fired a 15-inch shell at a piece of tissue paper and it had bounced back and hit you." By a mathematical analysis of the forces involved, Rutherford showed that the scattering was caused by a small, positively charged **nucleus** at the center of the gold atom. Most of the atom is empty space, ◄🛈 which explains why most of the bombarding particles passed through the gold foil undeflected.

Since Rutherford's time scientists have learned a great deal about the properties of atomic nuclei. For our purposes in chemistry, the nucleus of an atom can be considered to consist of two different types of particles (Table 2.1):

1. The **proton,** which has a mass nearly equal to that of an ordinary hydrogen atom. The proton carries a unit positive charge ($+1$), equal in magnitude to that of the electron (-1).
2. The **neutron,** an uncharged particle with a mass slightly greater than that of a proton.

Because protons and neutrons are much heavier than electrons, most of the mass of an atom ($>99.9\%$) is concentrated in the nucleus, even though the volume of the nucleus is much smaller than that of the atom. ◄🛈

Figure 2.4 **Representation of Thomson's model.** The raisins are representative of the electrons distributed according to Thomson's model.

🛈 If the nucleus were the size of your head, the electron would be 5 miles away.

🛈 The diameter of an atom is 10,000 times that of its nucleus.

Table 2.1 Properties of Subatomic Particles

Particle	Location	Relative Charge	Relative Mass*
Proton	Nucleus	+1	1.00728
Neutron	Nucleus	0	1.00867
Electron	Outside nucleus	−1	0.00055

*These are expressed in atomic mass units.

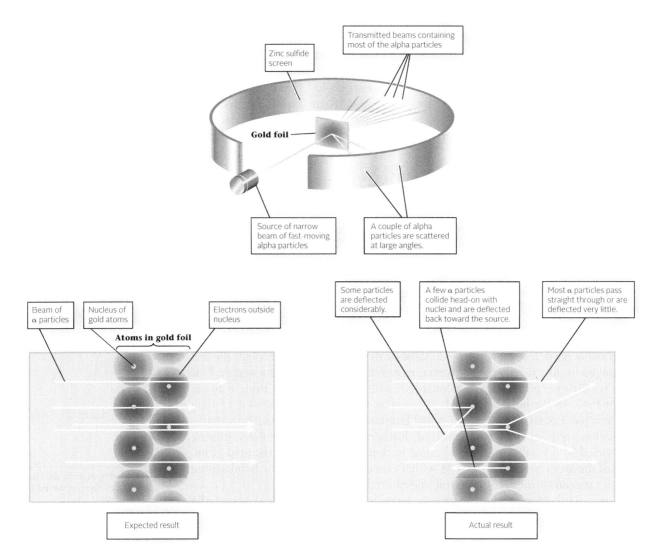

Figure 2.5 **Rutherford's α particle scattering experiment.**

| 2-3 | Quantitative Properties of the Atom |

Atomic Number

All the atoms of a particular element have the same number of protons in the nucleus. This number is a basic property of an element, called its **atomic number** and given the symbol Z:

$$Z = \text{number of protons}$$

In a neutral atom, the number of protons in the nucleus is exactly equal to the number of electrons outside the nucleus. Consider, for example, the elements hydrogen ($Z = 1$) and uranium ($Z = 92$). All hydrogen atoms have one proton in the nucleus; all uranium atoms have 92. In a neutral hydrogen atom there is one electron outside the nucleus; in a uranium atom there are 92.

H atom: 1 proton, 1 electron $Z = 1$

U atom: 92 protons, 92 electrons $Z = 92$

Mass Numbers; Isotopes

The **mass number** of an atom, given the symbol A, is found by adding up the number of protons and neutrons in the nucleus:

$$A = \text{number of protons} + \text{number of neutrons}$$

All atoms of a given element have the same number of protons, hence the same atomic number. They may, however, differ from one another in mass and therefore in mass number. This can happen because, although the number of protons in an atom of an element is fixed, the number of neutrons is not. It may vary and often does. Consider the element hydrogen ($Z = 1$). There are three different kinds of hydrogen atoms. They all have one proton in the nucleus. A light hydrogen atom (the most common type) has no neutrons in the nucleus ($A = 1$). Another type of hydrogen atom (deuterium) has one neutron ($A = 2$). Still a third type (tritium) has two neutrons ($A = 3$).

Number of neutrons = $A - Z$

An atom that contains the same number of protons but a different number of neutrons is called an **isotope**. The three kinds of hydrogen atoms just described are isotopes of that element. They have masses that are very nearly in the ratio $1:2:3$. Among the isotopes of the element uranium are the following:

Isotope	Z	A	Number of Protons	Number of Neutrons
Uranium-235	92	235	92	143
Uranium-238	92	238	92	146

The composition of a nucleus is shown by its **nuclear symbol**. Here, the atomic number appears as a subscript at the lower left of the symbol of the element. The mass number is written as a superscript at the upper left.

$$\text{Mass number} \longrightarrow {}^{A}_{Z}X \longleftarrow \text{element symbol}$$
$$\text{Atomic number} \longrightarrow$$

The nuclear symbols for the isotopes of hydrogen and uranium referred to above are

$$^{1}_{1}H, \, ^{2}_{1}H, \, ^{3}_{1}H \qquad ^{235}_{92}U, \, ^{238}_{92}U$$

Quite often, isotopes of an element are distinguished from one another by writing the mass number after the symbol of the element. The isotopes of uranium are often referred to as U-235 and U-238.

$^{2}_{1}H_2O$ ice and $^{1}_{1}H_2O$ ice. Solid deuterium oxide (*bottom*) is more dense than liquid water and sinks, whereas the ordinary water ice (*top*) is less dense than liquid water and floats.

© Cengage Learning/Charles D. Winters

EXAMPLE 2.1

a   An isotope of antimony (Sb, $Z = 51$) is used as a tracer to detect leaks in oil pipelines. This isotope has 73 neutrons in its nucleus. What is its nuclear symbol?

ANALYSIS

Information given:	Z (51); number of neutrons, n(73)
Asked for:	nuclear symbol

STRATEGY

1. Note that Z stands for the atomic number or the number of protons p^+.

2. Recall that a nuclear symbol is written $^{A}_{Z}X$ where A stands for the number of neutrons (n) and protons (p^+).

3. Use this strategy for parts (b) and (c) also.

continued

	SOLUTION
nuclear symbol	$Z = p^+ = 51$; $A = p^+ + n = 51 + 73 = 124$; $^A_Z Sb = ^{124}_{51}Sb$

b One of the most harmful components of nuclear waste is a radioactive isotope of strontium, $^{90}_{38}Sr$; it can be deposited in your bones, where it replaces calcium. How many protons are there in the nucleus of Sr-90? How many neutrons?

	ANALYSIS
Information given:	nuclear symbol: $^{90}_{38}Sr$
Asked for:	p^+; n

	SOLUTION
protons	$Z = p^+ = 38$
neutrons	$A = p^+ + n = 90$; $90 = 38 + n$; $n = 90 - 38 = 52$

c Write the symbol for the element used as a medical tool for scanning lungs. It has 54 protons and 79 neutrons.

	ANALYSIS
Information given:	$p^+ = 54$; $n = 79$
Information implied:	identity of the element
Asked for:	nuclear symbol

	SOLUTION
nuclear symbol	$Z = p^+ = 54$ (placed on bottom left of the element)
	$A = p^+ + n = 54 + 79 = 133$
	The element (X) is xenon identified by its atomic number Z.
	nuclear symbol: $^A_Z X = ^{133}_{54}Xe$

Atomic Masses

Individual atoms are far too small to be weighed on a balance. However, as you will soon see, it is possible to determine quite accurately the relative masses of different atoms and molecules. Indeed, it is possible to go a step further and calculate the actual masses of these tiny building blocks of matter.

Relative masses of atoms of different elements are expressed in terms of their *atomic masses* (often referred to as *atomic weights*). The **atomic mass** of an element indicates how heavy, on the average, one atom of that element is compared with an atom of another element.

To set up a scale of atomic masses, it is necessary to establish a standard value for one particular species. The modern atomic mass scale (sometimes called the **Carbon-12 scale**) is based on the most common isotope of carbon, $^{12}_6C$. This isotope is assigned a mass of exactly 12 **atomic mass units (amu):** ◀ **ⓘ**

1 amu = 1/12 mass C atom ≈ mass H atom.

$$\text{mass of C-12 atom} = 12 \text{ amu (exactly)}$$

It follows that an atom half as heavy as a C-12 atom would weigh 6 amu, an atom twice as heavy as C-12 would have a mass of 24 amu, and so on.

Isotopic Abundances

Relative masses of individual atoms can be determined using a mass spectrometer (Figure 2.6). Gaseous atoms or molecules at very low pressures are ionized by removing one or more electrons. The cations formed are accelerated by a potential

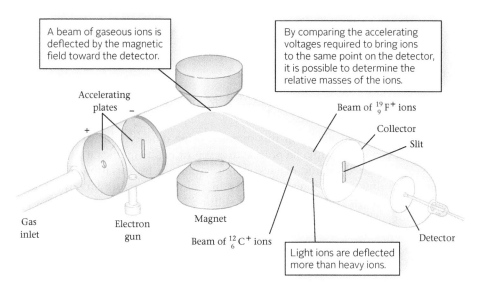

A beam of gaseous ions is deflected by the magnetic field toward the detector.

By comparing the accelerating voltages required to bring ions to the same point on the detector, it is possible to determine the relative masses of the ions.

Accelerating plates

Beam of $^{19}_{9}F^+$ ions

Collector

Slit

Gas inlet

Electron gun

Magnet

Beam of $^{12}_{6}C^+$ ions

Detector

Light ions are deflected more than heavy ions.

Figure 2.6 **The mass spectrometer.**

of 500 to 2000 V toward a magnetic field, which deflects the ions from their straight-line path. The extent of deflection is inversely related to the mass of the ion. By measuring the voltages required to bring two ions of different mass to the same point on the detector, it is possible to determine their relative masses. For example, using a mass spectrometer, it is found that a $^{19}_{9}F$ atom is 1.583 times as heavy as a $^{12}_{6}C$ atom and so has a mass of

$$1.583 \times 12.00 \text{ amu} = 19.00 \text{ amu}$$

As it happens, naturally occurring fluorine consists of a single isotope, $^{19}_{9}F$. It follows that the atomic mass of the element fluorine must be the same as that of F-19, 19.00 amu. The situation with most elements is more complex because they occur in nature as a mixture of two or more isotopes. To determine the atomic mass of such an element, it is necessary to know not only the masses of the individual isotopes but also their atom percents (*isotopic abundances*) in nature.

Fortunately, isotopic abundances as well as isotopic masses can be determined by mass spectrometry. The situation with chlorine, which has two stable isotopes, Cl-35 and Cl-37, is shown in Figure 2.7. The atomic masses of the two isotopes are determined in the usual way. The relative abundances of these isotopes are proportional to the heights of the recorder peaks or, more accurately, to the areas under these peaks. For chlorine, the data obtained from the mass spectrometer are

	Atomic Mass	Abundance
Cl-35	34.97 amu	75.53%
Cl-37	36.97 amu	24.47%

We interpret this to mean that, in elemental chlorine, 75.53% of the atoms have a mass of 34.97 amu, and the remaining atoms, 24.47% of the total, have a mass of 36.97 amu. With this information we can readily calculate the atomic mass of chlorine using the general equation

Figure 2.7 **Mass spectrum of chlorine.** Elemental chlorine (Cl_2) contains only two isotopes: 34.97 amu (75.53%) and 36.97 amu (24.47%).

atomic mass Y =

$$(\text{atomic mass Y}_1) \times \frac{\% \text{ Y}_1}{100\%} + (\text{atomic mass Y}_2) \times \frac{\% \text{ Y}_2}{100\%} + \cdots \qquad (2.1)$$

where $Y_1, Y_2, \ldots$ are isotopes of element Y.

$$\text{atomic mass Cl} = 34.97 \text{ amu} \times \frac{75.53}{100.0} + 36.97 \text{ amu} \times \frac{24.47}{100.0} = 35.46 \text{ amu}$$

Atomic masses calculated in this manner, using data obtained with a mass spectrometer, can in principle be precise to seven or eight significant figures. The accuracy of tabulated atomic masses is limited mostly by variations in natural abundances. Sulfur is an interesting case in point. It consists largely of two isotopes, $^{32}_{16}\text{S}$ and $^{34}_{16}\text{S}$. The abundance of sulfur-34 varies from about 4.18% in sulfur deposits in Texas and Louisiana to 4.34% in volcanic sulfur from Italy. This leads to an uncertainty of 0.006 amu in the atomic mass of sulfur.

If the atomic mass of an element is known *and* if it has only two stable isotopes, their abundances can be calculated from the general equation cited above.

EXAMPLE 2.2

Bromine is a red-orange liquid with an average atomic mass of 79.90 amu. Its name is derived from the Greek word *bromos* (βρομος), which means stench. It has two naturally occurring isotopes: Br-79 (78.92 amu) and Br-81 (80.92 amu). What is the abundance of the heavier isotope?

ANALYSIS

Information given:	Br-81 mass (80.92 amu); Br-79 mass (78.92 amu) average atomic mass (79.90)
Asked for:	abundance of Br-81

STRATEGY

1. All abundances must add up to 100%

2. Recall the formula relating abundance and atomic mass (Equation 2.1)

$$\text{atomic mass Y} = \left(\text{atomic mass Y}_1 \times \frac{\% \text{ Y}_1}{100\%}\right) + \left(\text{atomic mass Y}_2 \times \frac{\% \text{ Y}_2}{100\%}\right) + \cdots$$

SOLUTION

1. % abundances	Br-81: x; Br-79: $100 - x$
2. Substitute into Equation 2.1.	$79.90 \text{ amu} = 78.92 \text{ amu} \left(\dfrac{100 - x}{100}\right) + 80.92 \text{ amu} \left(\dfrac{x}{100}\right)$
3. Solve for x.	$79.90 = 0.7892(100 - x) + 0.8092\,x$ $79.90 = 78.92 - 0.7892\,x + 0.8092\,x$ $x = 49\%$

END POINT

The atomic mass of Br, 79.90, is just about halfway between the masses of the two isotopes, 78.92 and 80.92. So, it is reasonable that it should contain nearly equal amounts of the two isotopes.

The inside cover or opening pages of your text show the symbols of all the elements arranged in a particular way. This is called a **periodic table.** We will have more to say about this table in the next section (Section 2-4). For now, you can use it to find the element's average atomic mass ◄ⓘ (rounded to four digits and written below the element's symbol). Its atomic number, Z, is above the symbol.

The average atomic mass shown in the periodic table is not equal to the mass number.

Masses of Individual Atoms; Avogadro's Number

For most purposes in chemistry, it is sufficient to know the relative masses of different atoms. Sometimes, however, it is necessary to go one step further and calculate the mass in grams of individual atoms. Let us consider how this can be done.

To start with, consider the elements helium and hydrogen. A helium atom is about four times as heavy as a hydrogen atom (He = 4.003 amu, H = 1.008 amu). It follows that a sample containing 100 helium atoms weighs about four times as much as a sample containing 100 hydrogen atoms. Again, comparing samples of the two elements containing a million atoms each, the masses will be in a 4 (helium) to 1 (hydrogen) ratio. Turning this argument around, it follows that a sample of helium weighing four grams must contain very nearly the same number of atoms as a sample of hydrogen weighing one gram. ◀ ⓘ More precisely

ⓘ If a nickel weighs twice as much as a dime, there are equal numbers of coins in 1000 g of nickels and 500 g of dimes.

no. of He atoms in 4.003 g helium = no. of H atoms in 1.008 g hydrogen

This reasoning is readily extended to other elements. A sample of any element with a mass in grams equal to its atomic mass contains the same number of atoms, N_A, regardless of the identity of the element.

The question now arises as to the numerical value of N_A; that is, how many atoms are in 4.003 g of helium, 1.008 g of hydrogen, 32.07 g of sulfur, and so on? As it happens, this problem is one that has been studied for at least a century. Several ingenious experiments have been designed to determine this number, known as **Avogadro's number** and given the symbol N_A. As you can imagine, it is huge. (Remember that atoms are tiny. There must be a lot of them in 4.003 g of He, 1.008 g of H, and so on.) To four significant figures,

$$N_A = 6.022 \times 10^{23}$$

To get some idea of how large this number is, suppose the entire population of the world were assigned to counting the atoms in 4.003 g of helium. If each person counted one atom per second and worked a 48-hour week, the task would take more than 10 million years. ◀ ⓘ

ⓘ Most people have better things to do.

The importance of Avogadro's number in chemistry should be clear. *It represents the number of atoms of an element in a sample whose mass in grams is numerically equal to the atomic mass of the element.* Thus there are

6.022×10^{23} H atoms in 1.008 g H   atomic mass H = 1.008 amu

6.022×10^{23} He atoms in 4.003 g He   atomic mass He = 4.003 amu

6.022×10^{23} S atoms in 32.07 g S   atomic mass S = 32.07 amu

Knowing Avogadro's number and the atomic mass of an element, it is possible to calculate the mass of an individual atom (Example 2.3a). You can also determine the number of atoms in a weighed sample of any element (Example 2.3b).

Avogadro's number has been given a special name—mole. We will have more to say about the mole in Section 3-1.

EXAMPLE 2.3

Consider arsenic (As), a favorite poison used in crime stories. This element is discussed at the end of Chapter 1. Avogadro's number is 6.022×10^{23}.

a	What is the mass of an arsenic atom?

ANALYSIS	
Information given:	Avogadro's number (6.022×10^{23})
Information implied:	atomic mass
Asked for:	mass of an arsenic atom

continued

Change atoms to grams (atoms → g) by using the conversion factor

$$\frac{6.022 \times 10^{23} \text{ atoms}}{\text{atomic mass}}$$

SOLUTION

mass of an As atom	$1 \text{ atom As} \times \dfrac{74.92 \text{ g As}}{6.022 \times 10^{23} \text{ atoms As}} = 1.244 \times 10^{-22} \text{ g}$

b How many atoms are there in ten grams of arsenic?

ANALYSIS

Information given:	mass of sample (10.00 g) from (a) mass of one As atom (1.244×10^{-22} g/atom)
Asked for:	number of atoms in a ten-gram sample

STRATEGY

Change grams to atoms (g → atom) by using the conversion factor

$$\frac{1 \text{ atom}}{1.244 \times 10^{-22} \text{ g}}$$

SOLUTION

atoms of As	$10.00 \text{ g As} \times \dfrac{1 \text{ atom As}}{1.244 \times 10^{-22} \text{ g As}} = 8.038 \times 10^{22} \text{ atoms As}$

c How many protons are there in 0.1500 lb of arsenic?

ANALYSIS

Information given:	mass of sample (0.1500 lbs) from (a) mass of one As atom (1.244×10^{-22} g/atom)
Information implied:	atomic number pounds to grams conversion factor
Asked for:	number of protons in 0.1500 lb As

STRATEGY

Change pounds to grams, grams to atoms, and atoms to protons by using the conversion factors

$$\frac{453.6 \text{ g}}{1 \text{ lb}} \qquad \frac{\text{no. of protons (Z)}}{1 \text{ atom}} \qquad \frac{1 \text{ atom}}{1.244 \times 10^{-22} \text{ g}}$$

and follow the plan:

lb → g → atom → proton

SOLUTION

number of protons	$0.1500 \text{ lb} \times \dfrac{453.6 \text{ g}}{1 \text{ lb}} \times \dfrac{1 \text{ atom}}{1.244 \times 10^{-22} \text{ g}} \times \dfrac{33 \text{ protons}}{1 \text{ atom As}} = 1.805 \times 10^{25} \text{ protons}$

END POINT

Because atoms are so tiny, we expect their mass to be very small: 1.244×10^{-22} g sounds reasonable. Conversely, it takes a lot of atoms, in this case, 8.038×10^{22} atoms, to weigh ten grams.

2-4 Introduction to the Periodic Table

From a microscopic point of view, an element is a substance all of whose atoms have the same number of protons, that is, the same atomic number. The chemical properties of elements depend upon their atomic numbers, which can be read from the periodic table. A complete periodic table that lists symbols, atomic numbers, and atomic masses is given on the inside front cover or opening pages of this text. For our purposes in this chapter, the abbreviated table in Figure 2.8 will suffice.

Periods and Groups

The horizontal rows in the table are referred to as **periods**. The first period consists of the two elements hydrogen (H) and helium (He). The second period starts with lithium (Li) and ends with neon (Ne).

The vertical columns are known as **groups**. Historically, many different systems have been used to designate the different groups. Both Arabic and Roman numerals have been used in combination with the letters A and B. The system used in this text is the one recommended by the International Union of Pure and Applied Chemistry (IUPAC) in 1985. The groups are numbered from 1 to **18**, starting at the left.

Elements falling in Groups 1, 2, 13, **14, 15, 16, 17,** and **18*** are referred to as **main-group elements**. The ten elements in the center of each of periods 4 through 6 are called **transition metals**; they fall in Groups 3 through 12. The first transition series (period 4) starts with Sc (Group 3) and ends with Zn (Group 12).

The metals in Groups 13, 14, and 15, which lie to the right of the transition metals (Ga, In, Tl, Sn, Pb, Bi), are often referred to as **post-transition metals**.

Certain main groups are given special names. The elements in Group 1, at the far left of the periodic table, are called **alkali metals**; those in Group 2 are

Other periodic tables label the groups differently, but the elements have the same position.

You'll have to wait until Chapter 7 to learn why the second digit of some group numbers are in bold type.

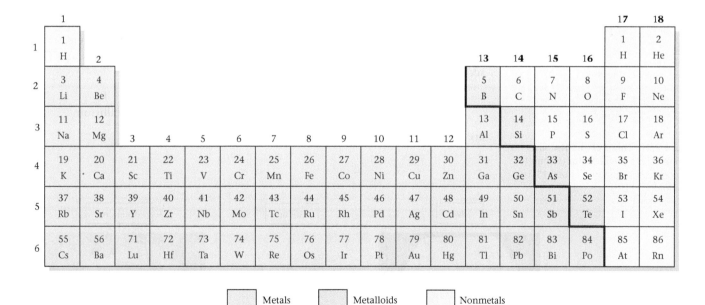

Metals   Metalloids   Nonmetals

Figure 2.8 Periodic table. The group numbers stand above the columns. The numbers at the left of the rows are the period numbers. The black line separates the metals from the nonmetals. (*Note:* A complete periodic table is given inside the cover or opening pages of your text.)

*Prior to 1985, Groups **13** to **18** were commonly numbered 3 to 8 or 3A to 8A in the United States.

Figure 2.9 Inert gases. Neon is used in advertising signs. The gas in tubes adds both color and light. The test kit is used to detect radon in home basements.

referred to as **alkaline earth metals**. As we move to the right, the elements in Group 17 are called **halogens**; at the far right, the **noble gases** (also called "unreactive gases") constitute Group 18.

Elements in the same main group show very similar chemical properties. For example,

- lithium (Li), sodium (Na), and potassium (K) in Group 1 all react vigorously with water to produce hydrogen gas.
- helium (He), neon (Ne) (Figure 2.9), and argon (Ar) in Group 18 do not react with any other substances.

On the basis of observations such as these, we can say that *the periodic table is an arrangement of elements, in order of increasing atomic number, in horizontal rows of such a length that elements with similar chemical properties fall directly beneath one another in vertical groups.*

The person whose name is most closely associated with the periodic table is Dmitri Mendeleev (1834–1907), a Russian chemist. In writing a textbook of general chemistry, Mendeleev devoted separate chapters to families of elements with similar properties, including the alkali metals, the alkaline earth metals, and the halogens. Reflecting on the properties of these and other elements, he proposed in 1869 a primitive version of today's periodic table. Mendeleev shrewdly left empty spaces in his table for new elements yet to be discovered. Indeed, he predicted detailed properties for three such elements (scandium, gallium, and germanium). By 1886 all of these elements had been discovered and found to have properties very similar to those he had predicted.

Metals and Nonmetals

The diagonal line or stairway that starts to the left of boron in the periodic table (Figure 2.8) separates metals from nonmetals. The more than 80 elements to the left and below that line, shaded in blue in Figure 2.8, have the properties of **metals** (Figure 2.10a); in particular, they have high electrical conductivities. Elements above and to the right of the stairway shaded in yellow in Figure 2.8 are **nonmetals** (Figure 2.10b); about 18 elements fit in that category.

Along the stairway (zig-zag line) in the periodic table are several elements that are difficult to classify exclusively as metals or nonmetals. They have properties between those of elements in the two classes. In particular, their electrical conductivities are intermediate between those of metals and nonmetals. ◀ *i* The six elements shaded in green in Figure 2.8

These elements, particularly Si, are used in semiconductors.

B	Si	Ge	As	Sb	Te
boron	silicon	germanium	arsenic	antimony	tellurium

are often called **metalloids** (Figure 2.10c).

Figure 2.10 Three elements. Silver (Group 11) is a metal. Sulfur (Group 16) is a nonmetal; antimony (Group 15) is a metalloid.

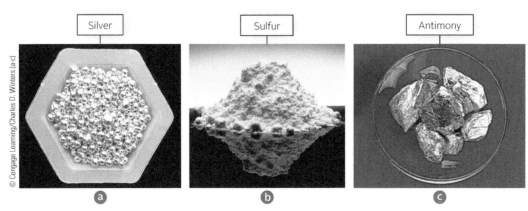

Silver (a)   Sulfur (b)   Antimony (c)

1																17	18
H (10)	2												13	14	15	16	H (10)	
	Be													C (18)	N (3)	O (65)	F	
Na (0.1)	Mg (0.05)	3	4	5	6	7	8	9	10	11	12			P (1.2)	S (0.3)	Cl (0.1)	
K (0.2)	Ca (1.5)			V	Cr	Mn	Fe	Co		Cu	Zn			As	Se		
					Mo						Cd				Te	I	
											Hg	Tl	Pb				

Major elements (percent in body)   Trace elements   Highly toxic elements

Figure 2.11 Biologically important elements and highly toxic elements. For the major elements in the body, their percent abundance in the body is given below the symbols.

Figure 2.11 shows the biologically important elements. The "good guys," essential to life, include the major elements (yellow), which account for 99.9% of total body mass, and the trace elements (green) required in very small quantities. In general, the abundances of elements in the body parallel those in the world around us, but there are some important exceptions. Aluminum and silicon, although widespread in nature, are missing in the human body. The reverse is true of carbon, which makes up only 0.08% of the earth's crust but 18% of the body, where it occurs in a variety of organic compounds including proteins, carbohydrates, and fats.

The "bad guys," shown in pink in Figure 2.11, are toxic, often lethal, even in relatively small quantities. Several of the essential trace elements *become* toxic if their concentrations in the body increase. Selenium is a case in point. You need about 0.00005 g/day to maintain good health, but 0.001 g/day can be deadly. That's a good thing to keep in mind if you're taking selenium supplements.

2-5 Molecules and Ions

Isolated atoms rarely occur in nature; only the noble gases (He, Ne, Ar, . . .) consist of individual, nonreactive atoms. Atoms tend to combine with one another in various ways to form more complex structural units. Two such units, which serve as building blocks for a great many elements and compounds, are molecules and ions.

Molecules

Two or more atoms may combine with one another to form an uncharged **molecule.** The atoms involved are usually those of nonmetallic elements. Within the molecule, atoms are held to one another by strong forces called *covalent bonds,* ◄ ⓘ which consist of shared pairs of electrons (Chapter 7). Forces between neighboring molecules, in contrast, are quite weak.

ⓘ Super Glue™ is weak compared with covalent bonds.

A **molecular substance** is most often represented by a **molecular formula,** in which the number of atoms of each element is indicated by a subscript written after the symbol of the element. Thus we interpret the molecular formulas for water (H_2O), ammonia (NH_3), and methane (CH_4) to mean that in

- the water molecule, there is one oxygen atom and two hydrogen atoms.
- the ammonia molecule, there is one nitrogen atom and three hydrogen atoms.
- the methane molecule, there is one carbon atom and four hydrogen atoms.

Structural formulas help us predict chemical properties.

The structure of a molecule is sometimes represented by a **structural formula,** which shows the bonding pattern within the molecule. The structural formulas of water, ammonia, and methane are

$$H-O-H \qquad H-\underset{\underset{H}{|}}{N}-H \qquad H-\underset{\underset{H}{|}}{\overset{\overset{H}{|}}{C}}-H$$

The dashes represent covalent bonds. The geometries of these molecules are shown in Figure 2.12.

Sometimes we represent a molecular substance with a formula intermediate between a structural formula and a molecular formula. A **condensed structural formula** suggests the bonding pattern in the molecule and highlights the presence of a reactive group of atoms within the molecule. Consider, for example, the organic compounds commonly known as methyl alcohol and methylamine. Their structural formulas are

methyl alcohol   methylamine

The condensed structural formulas of these compounds are written as

$$CH_3OH \qquad CH_3NH_2$$

Figure 2.12 Space filling (top) and ball-and-stick (bottom) models of water (H_2O), ammonia (NH_3), and methane (CH_4). The "sticks" represent covalent bonds between H atoms and O, N, or C atoms. The models illustrate the geometry of the molecules (discussed in Chapter 7).

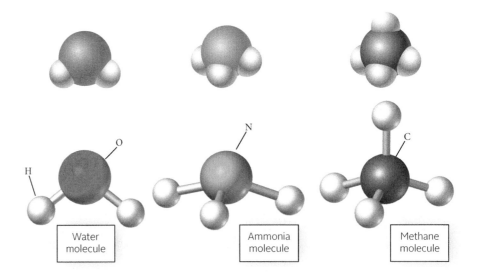

These formulas take up considerably less space and emphasize the presence in the molecule of

■ the OH group found in all *alcohols,* including ethyl alcohol, CH_3CH_2OH, the alcohol found in intoxicating beverages such as beer and wine.
■ the NH_2 group found in certain *amines,* including ethylamine, $CH_3CH_2NH_2$.

EXAMPLE 2.4	
Give the molecular formulas of ethyl alcohol, CH_3CH_2OH, and ethylamine, $CH_3CH_2NH_2$.	
ANALYSIS	
Information given:	structural formula
Asked for:	molecular formula
STRATEGY	
Add up the atoms of each element and use the sums as the subscripts for the element.	
SOLUTION	
ethyl alcohol	C: 1 + 1 = 2; H: 3 + 2 + 1 = 6; O: 1 molecular formula: C_2H_6O
ethylamine	C: 1 + 1 = 2; H: 3 + 2 + 2 = 7; N: 1 molecular formula: C_2H_7N
END POINT	
Note that although molecular formulas give the composition of the molecule, they reveal nothing about the way the atoms fit together. In that sense they are less useful than the structural formulas.	

Elements as well as compounds can exist as discrete molecules. In hydrogen gas, the basic building block is a molecule consisting of two hydrogen atoms joined by a covalent bond:

$$H—H$$

Other molecular elements are shown in Figure 2.13.

Ions

When an atom loses or gains electrons, charged particles called **ions** are formed. Metal atoms typically tend to lose electrons to form positively charged ions called **cations** (pronounced cát-ahy-uhn). Examples include the Na^+ and Ca^{2+} ions, formed from atoms of the metals sodium and calcium:

$$Na\ atom \longrightarrow Na^+\ ion + e^-$$
$$(11p^+, 11e^-) \qquad (11p^+, 10e^-)$$
$$Ca\ atom \longrightarrow Ca^{2+}\ ion + 2e^-$$
$$(20p^+, 20e^-) \qquad (20p^+, 18e^-)$$

(The arrows separate *reactants,* Na and Ca atoms, from *products,* cations and electrons.)

Nonmetal atoms form negative ions called **anions** (pronounced án-ahy-uhn) by gaining electrons. Consider, for example, what happens when atoms of the nonmetals chlorine and oxygen acquire electrons:

$$Cl\ atom + e^- \longrightarrow Cl^-\ ion$$
$$(17p^+, 17e^-) \qquad (17p^+, 18e^-)$$

Group 15	Group 16	Group 17
		H_2 (*g*)
N_2 (*g*)	O_2 (*g*)	F_2 (*g*)
P_4 (*s*)	S_8 (*s*)	Cl_2 (*g*)
		Br_2 (*l*)
		I_2 (*s*)

Figure 2.13 Molecular elements and their physical states at room temperature: gaseous (*g*), liquid (*l*), or solid (*s*).

$$\text{O atom} + 2e^- \longrightarrow \text{O}^{2-} \text{ ion}$$
$$(8p^+, 8e^-) \qquad (8p^+, 10e^-)$$

C, P, and the metalloids do not form monatomic ions.

Notice that *when an ion is formed, the number of protons in the nucleus is unchanged*. It is the number of electrons that increases or decreases.

EXAMPLE 2.5

a Scandium, a transition metal, is added to aluminum to create an alloy used in sports equipment, like aluminum bats and bicycle frames. How many protons, neutrons, and electrons are there in this scandium ion: $^{45}_{21}\text{Sc}^{3+}$?

ANALYSIS

Information given:	nuclear symbol and charge
Information implied:	A, Z
Asked for:	p^+, n, e^-

STRATEGY (for all parts)

1. Recall the placement of Z and A in the nuclear symbol.
2. $Z = p^+$; $A = p^+ + n$; $e^- = p^+ -$ charge

SOLUTION

p^+, e^-	$^{45}_{21}\text{Sc}^{3+}$: $Z = p^+ = 21$; $e^- = p^+ - (\text{charge}) = 21 - (+3) = 18$;
n	$A = n + p^+$; $45 = n + 21$; $n = 24$

b Sulfur is present in an ore called chalcocite. The ion in the ore has 16 neutrons and 18 electrons. Write the nuclear symbol for the ion.

ANALYSIS

Information given:	e^-, n
Information implied:	Z
Asked for:	nuclear symbol for S

SOLUTION

Z,	From the periodic table, sulfur has atomic number 16 so $Z = 16 = p^+$.
A and charge	$A = p^+ + n = 16 + 16 = 32$; Charge $= p^+ - e^- = 16 - 18 = -2$
Nuclear symbol	$^{32}_{16}\text{S}^{2-}$

c A rare earth metal has varied uses from super computers to cancer therapy. It has an ion with a +2 charge. It has 37 electrons and 51 neutrons. Write its nuclear symbol.

ANALYSIS

Information given:	e^-, n, charge
Asked for:	nuclear symbol

SOLUTION

Z, element's identity	$Z = p^+ = e^- + \text{charge} = 37 + 2 = 39$; The element is Y(yttrium).
A	$A = p^+ + n = 39 + 51 = 90$
	nuclear symbol: $^{A}_{Z}\text{X} = ^{90}_{39}\text{Y}^{2+}$

END POINT

A cation always contains more protons than electrons; the reverse is true of an anion. Figure 2.14 shows the species described in this example.

Figure 2.14 **The ions of the elements shown are (a) Sc³⁺, (b) S²⁻, (c) Y²⁺.**

The ions dealt with to this point (e.g., Na^+, Cl^-) are *monatomic;* that is, they are derived from a single atom by the loss or gain of electrons. ◄ ⓘ Many of the most important ions in chemistry are *polyatomic,* containing more than one atom. Examples include the hydroxide ion (OH^-) and the ammonium ion (NH_4^+). In these and other polyatomic ions, the atoms are held together by covalent bonds, for example,

$$(O-H)^- \qquad \left(\begin{array}{c} H \\ | \\ H-N-H \\ | \\ H \end{array} \right)^+$$

ⓘ We write +2 when describing the charge but 2+ when using it as a superscript in the formula of an ion.

In a very real sense, you can think of a polyatomic ion as a "charged molecule."

Because a bulk sample of matter is electrically neutral, any **ionic compound** always contains both cations (positively charged particles) and anions (negatively charged particles). Ordinary table salt, sodium chloride, is made up of an equal number of Na^+ and Cl^- ions. The structure of sodium chloride is shown in Figure 2.15. Notice that

■ there are two kinds of structural units in NaCl, the Na^+ and Cl^- ions.
■ there are no discrete molecules; Na^+ and Cl^- ions are bonded together in a continuous network.

Ionic compounds are held together by strong electrical forces between oppositely charged ions (e.g., Na^+, Cl^-). These forces are referred to as **ionic bonds**.

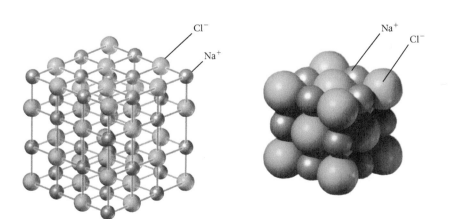

Figure 2.15 **Sodium chloride structure.** In these two ways of showing the structure, the small spheres represent Na^+ ions and the large spheres Cl^- ions. Note that in any sample of sodium chloride there are equal numbers of Na^+ and Cl^- ions, but no NaCl molecules.

Typically, ionic compounds are solids at room temperature and have relatively high melting points (mp NaCl = 801°C, CaCl$_2$ = 772°C). To melt an ionic compound requires that oppositely charged ions be separated from one another, thereby breaking ionic bonds.

When an ionic solid such as NaCl dissolves in water, the solution formed contains Na$^+$ and Cl$^-$ ions. Since ions are charged particles, the solution conducts an electric current (Figure 2.16) and we say that NaCl is a strong *electrolyte*. In contrast, a water solution of sugar, which is a molecular solid, does not conduct electricity. Sugar and other molecular solutes are *nonelectrolytes*.

Then there are weak electrolytes, which dissolve mostly as molecules with a few ions.

The solution of pure water does not contain ions and thus does not light the bulb.

The solution of sucrose (table sugar) and pure water also lacks ions, and fails to light the bulb.

The solution of sodium chloride (NaCl) and pure water does contain ions, and thus lights the bulb.

© Cengage Learning/James Maynard (a-c)

Figure 2.16 Electrical conductivity test. For electrical current to flow and light the bulb, the solution in which the electrodes are immersed must contain ions, which carry electrical charge.

EXAMPLE 2.6 CONCEPTUAL

The structure of a water solution of KNO$_3$, containing equal numbers of K$^+$ and NO$_3^-$ ions, may be represented as

(H$_2$O molecules are not shown.) Construct a similar beaker to show the structure of a water solution of potassium sulfate, K$_2$SO$_4$. Use ⚪ to represent a sulfur atom.

STRATEGY

Once you deduce the formula of the ionic compound, it's all downhill. Remember, though, that you have to show the relative numbers of cations and anions.

SOLUTION

2-6 Formulas of Ionic Compounds

When a metal such as sodium (Na) or calcium (Ca) reacts with a nonmetal such as chlorine (Cl_2), the product is ordinarily an ionic compound. The formula of that compound (e.g., NaCl, $CaCl_2$) shows the simplest ratio between cation and anion (one Na^+ ion for one Cl^- ion; one Ca^{2+} ion for two Cl^- ions). In that sense, the formulas of ionic compounds are simplest formulas. Notice that the symbol of the metal (Na, Ca) always appears first in the formula, followed by that of the nonmetal.

To predict the formula of an ionic compound, you need to know the charges of the two ions involved. Then you can apply the principle of **electrical neutrality**, which requires that *the total positive charge of the cations in the formula must equal the total negative charge of the anions.* Consider, for example, the ionic compound calcium chloride. The ions present are Ca^{2+} and Cl^-. For the compound to be electrically neutral, there must be two Cl^- ions for every Ca^{2+} ion. The formula of calcium chloride must be $CaCl_2$, indicating that the simplest ratio of Cl^- to Ca^{2+} ions is $2:1$.

Cations and Anions with Noble-Gas Structures

The charges of ions formed by atoms of the main-group elements can be predicted by applying a simple principle:

> *Atoms that are close to a noble gas* (Group 18) *in the periodic table form ions that contain the same number of electrons as the neighboring noble-gas atom.*

This is reasonable; noble-gas atoms must have an extremely stable electronic structure, because they are so unreactive. Other atoms might be expected to acquire noble-gas electronic structures by losing or gaining electrons.

Applying this principle, you can deduce the charges of ions formed by main-group atoms:

Group	No. of Electrons in Atom	Charge of Ion Formed	Examples
1	1 more than noble-gas atom	+1	Na^+, K^+
2	2 more than noble-gas atom	+2	Mg^{2+}, Ca^{2+}
16	2 less than noble-gas atom	−2	O^{2-}, S^{2-}
17	1 less than noble-gas atom	−1	F^-, Cl^-

Two other ions that have noble-gas structures are

Al^{3+} (Al has three more e^- than the preceding noble gas, Ne)

N^{3-} (N has three fewer e^- than the following noble gas, Ne)

Table 2.2 Some Common Polyatomic Ions

+1	−1	−2	−3
NH_4^+ (ammonium)	OH^- (hydroxide)	CO_3^{2-} (carbonate)	PO_4^{3-} (phosphate)
Hg_2^{2+} (mercury I)	NO_3^- (nitrate)	SO_4^{2-} (sulfate)	
	ClO_3^- (chlorate)	CrO_4^{2-} (chromate)	
	ClO_4^- (perchlorate)	$Cr_2O_7^{2-}$ (dichromate)	
	CN^- (cyanide)	HPO_4^{2-} (hydrogen phosphate)	
	$C_2H_3O_2^-$ (acetate)		
	MnO_4^- (permanganate)		
	HCO_3^- (hydrogen carbonate)		
	$H_2PO_4^-$ (dihydrogen phosphate)		

Potassium dichromate

Potassium permanganate

Potassium nitrate

© Cengage Learning/Charles D. Winters

Figure 2.17 Ionic compounds containing polyatomic ions. Potassium dichromate ($K_2Cr_2O_7$), potassium permanganate ($KMnO_4$), and potassium nitrate (KNO_3).

Cations of the Transition and Post-Transition Metals

Several metals that are farther removed from the noble gases in the periodic table form positive ions. These include the transition metals in Groups 3 to 12 and the post-transition metals in Groups 13 to 15. The cations formed by these metals typically have charges of $+1$, $+2$, or $+3$ and ordinarily do not have noble-gas structures. We will postpone to Chapter 4 a general discussion of the specific charges of cations formed by these metals.

Many of the transition and post-transition metals form more than one cation. Consider, for example, iron in Group 8. This metal forms two different series of compounds with nonmetals. In one series, iron is present as a $+2$ cation

$$Fe^{2+}: \quad FeCl_2, FeBr_2, \ldots$$

In the other series, iron exists as a $+3$ cation

$$Fe^{3+}: \quad FeCl_3, FeBr_3, \ldots$$

Polyatomic Ions

Table 2.2 lists some of the polyatomic ions that you will need to know, along with their names and charges. Figure 2.17 shows some of the compounds that have polyatomic ions. Notice that

- there are only two common polyatomic cations, NH_4^+ and Hg_2^{2+}. *All other cations considered in this text are derived from individual metal atoms* (e.g., Na^+ from Na, Ca^{2+} from Ca, . . .).
- most of the polyatomic anions contain one or more oxygen atoms; collectively these species are called **oxoanions**.

EXAMPLE 2.7

Predict the formula of the ionic compound

(a) formed by barium with iodine.

(b) containing a transition metal with a $+1$ charge in period 4 and Group 11 and oxide ions.

(c) containing an alkaline earth in period 5 and nitrogen.

(d) containing ammonium and phosphate ions.

STRATEGY

1. Recall charge of metals: group 1 ($+1$); group 2 ($+2$); Al ($+3$)

2. Recall charge of nonmetals: group 16: -2; group 17: -1; N: -3

3. The formula has to be electrically neutral.

4. Use Table 2.2 for polyatomic ions.

SOLUTION

(a)	Charges	Ba is in group 2; thus its charge is $+2$. I is in group 17, thus its charge is -1.
	electrical neutrality	$Ba^{2+}I^{1-}$; $2I^-$ are needed. The formula is BaI_2.
(b)	Charges	Period 4, group 4 is Cu with a given charge of $+1$. O is in group 16, thus its charge is -2.
	electrical neutrality	$Cu^{1+}O^{2-}$; 2 Cu^{1+} are needed. The formula is Cu_2O.
(c)	Charges	The alkaline earth (group 2) in period 5 is Sr; thus its charge is $+2$. N in an ionic compound is always -3.
	electrical neutrality	$Sr^{2+}N^{3-}$; 3 Sr^{2+} and 2 N^{3-} are needed. The formula is Sr_3N_2.
(d)	Charges	Ammonium is a polyatomic ion: NH_4^+ (Table 2.2). Phosphate is a polyatomic ion: PO_4^{3-} (Table 2.2).
	electrical neutrality	$NH_4^+ PO_4^{3-}$; 3 NH_4^+ are needed. The formula is $(NH_4)_3PO_4$.

continued

2-7   Names of Compounds

A compound can be identified either by its formula (e.g., NaCl) or by its name (sodium chloride). In this section, you will learn the rules used to name ionic and simple molecular compounds. ◄ⓘ To start with, it will be helpful to show how individual ions within ionic compounds are named.

ⓘ To name a compound, you have to decide whether it is molecular or ionic; the rules are different.

Ions

Monatomic cations take the name of the metal from which they are derived. Examples include

$$Na^+ \text{ sodium} \qquad K^+ \text{ potassium}$$

There is one complication. As mentioned earlier, certain metals in the transition and post-transition series form more than one cation, for example, Fe^{2+} and Fe^{3+}. To distinguish between these cations, the charge must be indicated in the name. This is done by putting the charge as a Roman numeral in parentheses after the name of the metal:

$$Fe^{2+} \text{ iron(II)} \qquad Fe^{3+} \text{ iron(III)}$$

(An older system used the suffixes *-ic* for the ion of higher charge and *-ous* for the ion of lower charge. These were added to the stem of the Latin name of the metal, so that the Fe^{3+} ion was referred to as ferric and the Fe^{2+} ion as ferrous.)

Monatomic anions are named by adding the suffix *-ide* to the stem of the name of the nonmetal from which they are derived.

				H^-	hydride
N^{3-}	nitride	O^{2-}	oxide	F^-	fluoride
		S^{2-}	sulfide	Cl^-	chloride
		Se^{2-}	selenide	Br^-	bromide
		Te^{2-}	telluride	I^-	iodide

Polyatomic ions, as you have seen (Table 2.2), are given special names. Certain nonmetals in Groups 15 to 17 of the periodic table form more than one polyatomic ion containing oxygen (oxoanions). The names of several such oxoanions are shown in Table 2.3. From the entries in the table, you should be able to deduce the following rules:

1. When a nonmetal forms two oxoanions, the suffix *-ate* is used for the anion with the larger number of oxygen atoms. The suffix *-ite* is used for the anion containing fewer oxygen atoms.
2. When a nonmetal forms more than two oxoanions, the prefixes *per-* (largest number of oxygen atoms) and *hypo-* (fewest oxygen atoms) are used as well.

Ionic Compounds

The name of an ionic compound consists of two words. The first word names the cation and the second names the anion. This is, of course, the same order in which the ions appear in the formula.

Cl, Br, and I form more than two oxoanions.

The oxoanions of bromine and iodine are named like those of chlorine.

Table 2.3 Oxoanions of Nitrogen, Sulfur, and Chlorine

Nitrogen	Sulfur	Chlorine
		ClO_4^- *perchlorate*
NO_3^- nitr*ate*	SO_4^{2-} sulf*ate*	ClO_3^- chlor*ate*
NO_2^- nitr*ite*	SO_3^{2-} sulf*ite*	ClO_2^- chlor*ite*
		ClO^- *hypo*chlor*ite*

In naming the compounds of transition or post-transition metals, we ordinarily indicate the charge of the metal cation by a Roman numeral:

$Cr(NO_3)_3$ chromium(III) nitrate $SnCl_2$ tin(II) chloride

In contrast, we never use Roman numerals with compounds of the Group 1 or Group 2 metals; they always form cations with charges of +1 or +2, respectively.

EXAMPLE 2.8

Name the following ionic compounds:

(a) CaS   (b) $Al(NO_3)_3$ (c) $FeCl_2$

STRATEGY

Recall symbols for elements, symbols for polyatomic ions (Table 2.2), and suffixes for nonmetals.

SOLUTION

(a) CaS	Ca = calcium; S = sulfur → sulfide; calcium sulfide
(b) $Al(NO_3)_3$	Al^{3+} = aluminum; NO_3^- = nitrate; aluminum nitrate
(c) $FeCl_2$	Fe^{2+} = iron, which is a transition metal so (II) should be written after the name of the metal; Cl^- = chlorine → chloride; iron(II) chloride

© Cengage Learning/Charles D. Winters

Nitrogen dioxide (NO_2), a binary molecular compound. It is a reddish-brown gas at 25°C and 1 atm.

Binary Molecular Compounds

When two nonmetals combine with each other, the product is most often a binary molecular compound. There is no simple way to deduce the formulas of such compounds. There is, however, a systematic way of naming molecular compounds that differs considerably from that used with ionic compounds.

The systematic name of a binary molecular compound, which contains two different nonmetals, consists of two words.

1. The first word gives the name of the element that appears first in the formula; a Greek prefix (Table 2.4) is used to show the number of atoms of that element in the formula.

2. The second word consists of

   ■ the appropriate Greek prefix designating the number of atoms of the second element. (See Table 2.4.)
   ■ the stem of the name of the second element.
   ■ the suffix *-ide*.

Table 2.4 Greek Prefixes Used in Nomenclature

Number*	Prefix	Number	Prefix	Number	Prefix
2	di	5	penta	8	octa
3	tri	6	hexa	9	nona
4	tetra	7	hepta	10	deca

*The prefix mono (1) is seldom used.

To illustrate these rules, consider the names of the several oxides of nitrogen:

N_2O_5 dinitrogen *penta*oxide N_2O_3 dinitrogen *trioxide*
N_2O_4 dinitrogen *tetra*oxide   NO   nitrogen ox*ide*
NO_2 nitrogen *dioxide* N_2O *di*nitrogen ox*ide*

When the prefixes tetr*a*, pent*a*, hex*a*, . . . are followed by the letter "o," the *a* is often dropped. For example, N_2O_5 is often referred to as *dinitrogen pentoxide*.

Many of the best-known binary compounds of the nonmetals have acquired common names. These are widely—and in some cases exclusively—used. Examples include

H_2O water PH_3 phosphine
H_2O_2 hydrogen peroxide AsH_3 arsine
NH_3 ammonia   NO   nitric oxide
N_2H_4 hydrazine N_2O nitrous oxide
C_2H_2 acetylene CH_4 methane

EXAMPLE 2.9

Give the names of the following molecules:
(a) SF_4 (b) PCl_3 (c) N_2O_3 (d) Cl_2O_7

STRATEGY

1st element: subscript → prefix (Table 2.4) + element name

2nd element: subscript → prefix + element name ending in *ide*

SOLUTION

(a) SF_4	S: subscript = 1; no prefix = sulfur F: subscript = 4 = tetra; F = fluorine → fluoride; sulfur tetrafluoride
(b) PCl_3	P: subscript = 1; no prefix = phosphorus Cl: subscript = 3 = tri; Cl = chlorine → chloride; phosphorus trichloride
(c) N_2O_3	N: subscript = 2 = di; dinitrogen O: subscript = 3 = tri; O = oxygen → oxide; dinitrogen trioxide
(d) Cl_2O_7	Cl: subscript = 2 = di; dichlorine O: subscript = 7 = hepta; O = oxygen → oxide; dichlorine heptaoxide

Acids

A few binary molecular compounds containing H atoms ionize in water to form H^+ ions. These are called *acids*. One such compound is hydrogen chloride, HCl; in water solution it exists as aqueous H^+ and Cl^- ions. The water solution of hydrogen chloride is given a special name: It is referred to as *hydrochloric acid*. A similar situation applies with HBr and HI:

Pure Substance		Water Solution	
HCl(g)	Hydrogen chloride	H$^+$(aq), Cl$^-$(aq)	Hydrochloric acid
HBr(g)	Hydrogen bromide	H$^+$(aq), Br$^-$(aq)	Hydrobromic acid
HI(g)	Hydrogen iodide	H$^+$(aq), I$^-$(aq)	Hydriodic acid

Most acids contain oxygen in addition to hydrogen atoms. Such species are referred to as **oxoacids.** Two oxoacids that you are likely to encounter in the general chemistry laboratory are

$$HNO_3 \quad \text{nitric acid} \qquad H_2SO_4 \quad \text{sulfuric acid}$$

The names of oxoacids are simply related to those of the corresponding oxoanions. The *-ate* suffix of the anion is replaced by *-ic* in the acid. In a similar way, the suffix *-ite* is replaced by the suffix *-ous*. The prefixes *per-* and *hypo-* found in the name of the anion are retained in the name of the acid.

ClO$_4$$^-$	*perchlorate* ion	HClO$_4$	*perchloric* acid
ClO$_3$$^-$	*chlorate* ion	HClO$_3$	*chloric* acid
ClO$_2$$^-$	*chlorite* ion	HClO$_2$	*chlorous* acid
ClO$^-$	*hypochlorite* ion	HClO	*hypochlorous* acid

Figures 2.18 and 2.19 summarize the rules for naming ionic compounds, molecules, and acids.

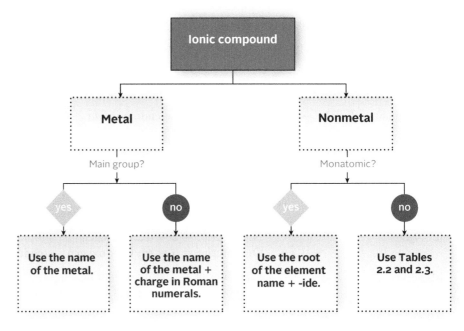

Figure 2.18 Flowchart for naming binary ionic compounds.

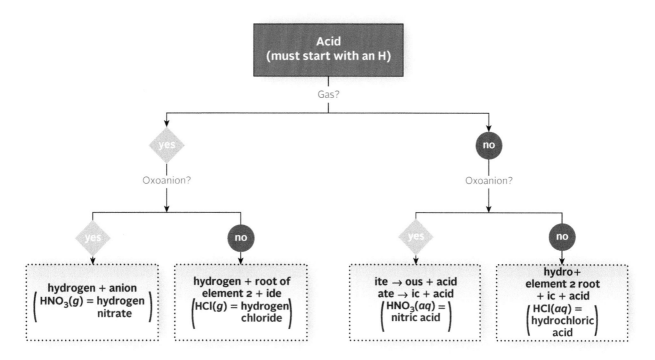

Figure 2.19 Flow chart for determining an acid.

EXAMPLE 2.10	
Give the names of	

a   HBr(g)

STRATEGY	
Gases follow the rules for naming binary molecules.	

SOLUTION	
HBr(g)	No prefixes for both elements (subscripts are both 1).
	H = hydrogen; Br = bromine → bromide; hydrogen bromide

b   ClO₂(aq)

STRATEGY	
Name of oxoanion (Table 2.3) (change *ite* to *ous*) + the word "acid."	

SOLUTION	
ClO₂(aq)	ClO₂⁻ = chlorite → chlorous + acid; chlorous acid

c   H₂SO₄(aq)

STRATEGY	
Name of oxoanion (Table 2.3) (change *ate* to *ic*) + the word "*acid*."	

SOLUTION	
H₂SO₄(aq)	SO₄²⁻ = sulfate → sulfuric + acid; sulfuric acid

continued

d	HIO(*aq*)

STRATEGY

Name of oxoanion (Table 2.3) (change *ite* to *ous*) + the word "acid." The oxoanion's name is analogous to the naming of the chlorine oxoanions.

SOLUTION

HIO(*aq*)	IO⁻ = hypoiodite → hypoiodous + acid; hypoiodous acid
	Hypoiodous is analogous to hypochlorous, the name for ClO⁻.

END POINT

You need to learn the names of the oxoanions listed in Table 2.3.

CHEMISTRY **BEYOND** THE CLASSROOM

Mastering the Peri'god'ic Table

by Rob Kyff

In ancient Rome, Mercury (the messenger of the gods) was constantly zipping around from deity to deity. So when the Romans needed a word for the poisonous metallic element that flowed quickly at room temperature, they named it "Mercury" for their speedy courier (Figure A).

In fact, when it comes to naming elements, Greek and Roman gods pretty much ran the (periodic) table.

When a new planet swam into the ken of British astronomer William Herschel in 1781, for instance, he named it "Uranus," after the Roman god of the sky. Seven years later, when the German chemist Martin Klaproth discovered a radioactive new element, he decided it would be trendy and hip to name it after the new planet—voila!—"uranium" (Figure B).

And when English clergyman William Gregor detected another new metallic element in 1791, Klaproth was again on the case. He wanted to name the newcomer for a child of Uranus, but somehow "Uranus, Jr." just didn't work.

So he named it for all of Uranus' children: a race of giants known as "the Titans." And so "titanium" was born.

Seven years later, obviously on a roll, Klaproth dubbed another new element "tellurium," for Tellus, a.k.a. "Terra," the Roman goddess of earth.

In 1801, Sicilian astronomer Giuseppi Piazzi spotted what seemed to be a strange, small planet orbiting the sun between Mars and Jupiter. Because Ceres, the Roman goddess of agriculture, had been viewed as the protector of Piazzi's native Sicily, he named the new object, which we now know to be an asteroid, "Ceres."

So just two years after Piazzi's discovery, the existence of another new element was verified. You can guess what hap-

Source: By permission of Rob Kyff and Creators Syndicate, Inc

Figure A (*Left*) Statue of the Roman god Mercury by the German sculptor Ludwig von Hofer, on the top of the Old Chancellery Building in Stuttgart, Germany. (*Right*) The element mercury.

Figure B (*Left*) A false-color image of the planet Uranus, as seen from the Hubble Space Telescope. (*Right*) Uranium ore conglomerate (also called pitchblend). The dark areas between the large quartz pebbles are uranium.

pened next. Klaproth, always eager to mimic the god-given nomenclature of astronomers, called the new element "Cerium."

Then in 1868, with Klaproth gone but his naming methods not forgotten, British astronomer Norman Lockyer noticed that one element detected in the sun's spectrum could not be correlated with any element found on Earth. So he named this new, solar-generated element "helium" for "helios," the Greek god of the sun.

When Tantalus, a mortal son of Zeus, shared his father's secrets with ordinary humans, his father punished him by placing him in fresh water up to his chin and dangling luscious fruits above his head. Doesn't sound so bad, right? But every time Tantalus stooped to quaff his thirst, the water drained away. And when he reached for the fruit, it moved upward beyond his grasp. This excruciating torture inspired the English word "tantalize."

In 1814, the Swedish chemist Anders Eckeberg discovered a new metal that could be placed in strong acid without "drinking" it, that is, reacting with it. So his fellow Swedish chemist Jöns Berzelius, remembering the perpetually parched Tantalus, dubbed the new element "tantalum."

At about the same time, the English chemist Charles Hatchett discovered a new metal, in Connecticut. He dubbed it "columbium," after "Columbia," a term for the United States. But other scientists claimed that columbium was identical to tantalum and wasn't new at all.

This dispute was resolved in 1846 when German chemist Heinrich Rose determined that columbium was indeed a new element. But because it resembled tantalum so closely, Rose dubbed it "niobium" after Tantalus' weeping daughter, Niobe. American scientists stuck with "columbium" for a while, but "niobium" eventually prevailed.

Another mythological daughter also gave her name to an element. Selene, the daughter of Hyperion, was the Greek goddess of the moon.

In 1818 the aforementioned Berzelius discovered a new element very similar to tellurium. Because "tellurium" had been named for Tellus, the Roman goddess of the earth, Berzelius strove for balance by calling its sister element "selenium" for the goddess of the moon.

Finally, in 1940 chemists at the University of California produced two new radioactive elements derived from uranium (which had been named in 1788 for Uranus, the Roman god of the sky).

So, following the order of the planets, they named the newcomers "neptunium" (for Neptune, Roman god of the sea) and "plutonium" (for Pluto, Greek god of the underworld).

Though Pluto is no longer considered a planet, its namesake element abides.

Key Concepts

1. Relate a nuclear symbol to the number of protons and neutrons.
   (**Example 2.1; Problems 7–16**)
2. Relate atomic mass, isotopic abundance, and average mass of an element.
   (**Example 2.2; Problems 17–30**)
3. Relate atomic mass to Avogadro's number.
   (**Example 2.3; Problems 31–38**)
4. Relate elements and the periodic table.
   (**Problems 39–46**)
5. Relate structural, condensed, and molecular formulas.
   (**Example 2.4; Problems 47, 48**)
6. Relate the ionic charge to the number of electrons.
   (**Example 2.5; Problems 49–52**)
7. Predict formulas of ionic compounds from charge of ions.
   (**Example 2.7; Problems 59, 60**)
8. Relate names to formulas
   ■ Ionic compounds
   (**Example 2.8; Problems 61–64**)
   ■ Binary molecular compounds
   (**Example 2.9; Problems 55–57**)
   ■ Oxoacids and oxoanions
   (**Example 2.10; Problems 65, 66**)

Key Equation

$$\text{Atomic Mass Y} = (\text{atomic mass Y}_1) \times \frac{\% \text{ Y}_1}{100\%} + (\text{atomic mass Y}_2) \times \frac{\% \text{ Y}_2}{100\%} + \cdots$$

Key Terms

atom	—molecular formula	metalloids	—groups
atomic mass	—structural formula	molecule	—periods
atomic mass units (amu)	ions	neutron	proton
atomic number	—anions	nonmetals	transition metals
Avogadro's number	—cations	nuclear symbol	
electrons	isotope	nucleus	
formula	main-group elements	oxoacids	
—condensed structural	mass number	oxoanions	
formula	metals	periodic table	

Summary Problem

Sulfur is a bright yellow solid at room temperature. It is one of the components in gunpowder, matches, and fireworks. It is added to rubber to make it tougher (vulcanization) and to topical ointments for the treatment of skin disorders.

(a) What is its molecular formula?

(b) How many protons and electrons are there in an atom of sulfur?

(c) If aluminum ions combine with sulfide ions, is the resulting compound molecular? What is the compound's name and formula?

(d) Is the compound described in (c) an electrolyte?

(e) Two atoms of sulfur combine with two atoms of chlorine. Is this compound a molecule? What is its name and formula?

(f) Write the nuclear symbol for a sulfur atom with 16 protons and 18 neutrons.

(g) To what group and period in the periodic table does sulfur belong?

(h) How many neutrons are there in the isotope S-36?

(i) The naturally occurring isotopes of sulfur, their atomic masses, and abundances are:
S-32: 31.97207 amu - 94.93%
S-33: 32.97146 amu - 0.76%
S-34: 33.96787 amu - 4.29%
S-36: 35.96708 amu - 0.02%
What is the average atomic mass for sulfur?

(j) How many atoms are there in 12.55 g of elemental sulfur (S) ?

(k) What is the mass of a billion (1×10^9) atoms of sulfur?

(l) Sulfur can combine with oxygen in many different ways. Write the names of the following compounds and ions: SO_3, H_2SO_3 (aq), $SO_4{}^{2-}$, Na_2SO_3.

Answers

(a) S_8
(b) 16 protons, 16 electrons
(c) no; Al_2S_3 - aluminum sulfide
(d) yes
(e) yes; S_2Cl_2 - disulfur dichloride
(f) $^{34}_{16}S$
(g) group 16, period 3
(h) 20 neutrons
(i) 32.07 amu
(j) 2.357×10^{23} atoms
(k) 5.325×10^{-14} g
(l) SO_3: sulfur trioxide
    H_2SO_3: sulfurous acid
    $SO_4{}^{2-}$: sufate ion
    Na_2SO_3: sodium sulfite

Questions and Problems

Even-numbered questions and Challenge Problems have answers in Appendix 5 and fully worked solutions in the *Student Solutions Manual*.

2-1 Atoms and the Atomic Theory

Atomic Theory and Laws

1. State in your own words the law of conservation of mass. State the law in its modern form.
2. State in your own words the law of constant composition.

3. Two basic laws of chemistry are the law of conservation of mass and the law of constant composition. Which of these laws (if any) do the following statements illustrate?
(a) Lavoisier found that when mercury(II) oxide, HgO, decomposes, the total mass of mercury (Hg) and oxygen formed equals the mass of mercury(II) oxide decomposed.
(b) Analysis of the calcium carbonate found in the marble mined in Carrara, Italy, and in the stalactites of the Carlsbad Caverns in New Mexico gives the same value for the percentage of calcium in calcium carbonate.

(c) Hydrogen occurs as a mixture of two isotopes, one of which is twice as heavy as the other.

4. Two basic laws of chemistry are the law of conservation of mass and the law of constant composition. Which of these laws (if any) do the following statements illustrate?

(a) The mass of phosphorus, P, combined with one gram of hydrogen, H, in the highly toxic gas phosphene, PH_3, is a little more than twice the mass of nitrogen, N, combined with one gram of hydrogen in ammonia gas, NH_3.

(b) A cold pack has the same mass before and after the seal between two reactants is broken to allow reaction to occur.

(c) It is highly improbable that carbon monoxide gas found in Los Angeles is $C_{1.2}O_{1.1}$.

2-2 Components of the Atom

5. Who discovered the electron? Describe the experiment that led to the deduction that electrons are negatively charged particles.

6. Who discovered the nucleus? Describe the experiment that led to this discovery.

7. Selenium is widely sold as a dietary supplement. It is advertised to "protect" women from breast cancer. Write the nuclear symbol for naturally occurring selenium. It has 34 protons and 46 neutrons.

8. Radon is a radioactive gas that can cause lung cancer. It has been detected in the basement of some homes. In an atom of Rn-220,

(a) how many protons are there?

(b) how many neutrons are there?

2-3 Quantitative Properties of the Atom

Nuclear Symbols and Isotopes

9. How do the isotopes of argon, Ar-36, Ar-38, and Ar-40, differ from each other? Write the nuclear symbol for all three isotopes.

10. Consider two isotopes Fe-54 and Fe-56.

(a) Write the nuclear symbol for both isotopes.

(b) How do they differ from each other?

11. Uranium-235 is the isotope of uranium commonly used in nuclear power plants. How many

(a) protons are in its nucleus?

(b) neutrons are in its nucleus?

(c) electrons are in a uranium atom?

12. An isotope of americium (Am) with 146 neutrons is used in many smoke alarms.

(a) How many electrons does an atom of americium have?

(b) What is the isotope's mass number A?

(c) Write its nuclear symbol.

13. Consider the following nuclear symbols. How many protons, neutrons, and electrons does each element have? What elements do R, T, and X represent?

(a) $^{30}_{14}R$ (b) $^{89}_{39}T$ (c) $^{133}_{55}X$

14. Consider the following nuclear symbols. How many protons, neutrons, and electrons does each element have? What elements do A, L, and Z represent?

(a) $^{75}_{33}A$ (b) $^{51}_{23}L$ (c) $^{131}_{54}Z$

15. Nuclei with the same mass number but different atomic numbers are called *isobars*. Consider Ca-40, Ca-41, K-41 and Ar-41.

(a) Which of these are isobars? Which are isotopes?

(b) What do Ca-40 and Ca-41 have in common?

(c) Correct the statement (if it is incorrect): Atoms of Ca-41, K-41, and Ar-41 have the same number of neutrons.

16. See the definition for isobars in Question 15. Consider Cr-54, Fe-54, Fe-58, and Ni-58.

(a) Which of these are isobars? Which are isotopes?

(b) What do Fe-54 and Fe-58 have in common?

(c) Which atoms have the same number of neutrons?

Atomic Masses and Isotopic Abundances

17. Calculate the mass ratio of a bromine atom to an atom of

(a) neon   (b) calcium   (c) helium

18. Arrange the following in the order of increasing mass.

(a) a potassium ion, K^+

(b) a phosphorus molecule, P_4

(c) a potassium atom

(d) a platinum atom

19. Cerium is the most abundant rare earth metal. Pure cerium ignites when scratched by even a soft object. It has four known isotopes: ^{136}Ce (atomic mass = 135.907 amu), ^{138}Ce (atomic mass = 137.905 amu), ^{140}Ce (atomic mass = 139.905 amu), and ^{142}Ce (atomic mass = 141.909 amu). Ce-140 and Ce-142 are fairly abundant. Which is the more abundant isotope?

20. Consider the three stable isotopes of oxygen with their respective atomic masses: O-16 (15.9949 amu), O-17 (16.9993 amu), O-18 (17.9992 amu). Which is the most abundant?

21. Bromine has two occuring isotopes: ^{79}Br with atomic mass 78.9183 and ^{81}Br with atomic mass 80.9163. Without using a calculator, what would you estimate the % abundance of Br-79 to be?

22. Rubidium has two naturally occurring isotopes: ^{85}Rb (atomic mass = 84.9118 amu) and ^{87}Rb (atomic mass = 86.9092 amu). The percent abundance of ^{87}Rb can be estimated to be which of the following?

(a) 0%   (b) 25%   (c) 50%   (d) 75%

23. Strontium has four isotopes with the following masses: 83.9134 amu (0.56%), 85.9094 amu (9.86%), 86.9089 amu (7.00%), and 87.9056 amu (82.58%). Calculate the average atomic mass of strontium.

24. Neon is an inert gas with three stable isotopes. It is used in gas lasers and in advertising signs. Its isotopes and their abundances are:

Ne-20   19.9924 amu   90.51%

Ne-21   20.9938 amu   0.27%

Ne-22   21.9914 amu   9.22%

What is the average atomic mass of neon?

25. Naturally occurring silver (Ag) consists of two isotopes. One of the isotopes has a mass of 106.90509 amu and 51.84% abundance. What is the atomic mass of the other isotope?

26. Copper has two naturally occurring isotopes. Cu-63 has an atomic mass of 62.9296 amu and an abundance of 69.17%. What is the atomic mass of the second isotope? What is its nuclear symbol?

27. Silicon (average atomic mass = 28.0855 amu) has three isotopes. Their masses are 27.9769 amu, 28.9765 amu, and 29.9738 amu. The abundance of the heaviest isotope is 2.96%. Estimate the abundances of the first two isotopes.

28. Magnesium (average atomic mass = 24.305 amu) consists of three isotopes with masses 23.9850 amu, 24.9858 amu, and 25.9826 amu. The abundance of the middle isotope is 10.00%. Estimate the abundances of the other isotopes.

29. Zinc has four stable isotopes: Zn-64, Zn-66, Zn-67, and Zn-68. Their natural abundances are 48.6%, 27.9%, 4.1%, and 18.8%, respectively. Sketch the mass spectrum for zinc.

30. Chlorine has two isotopes, Cl-35 and Cl-37. Their abundances are 75.53% and 24.47%, respectively. Assume that the only hydrogen isotope present is H-1.
(a) How many different HCl molecules are possible?
(b) What is the sum of the mass numbers of the two atoms in each molecule?
(c) Sketch the mass spectrum for HCl if all the positive ions are obtained by removing a single electron from an HCl molecule.

31. Lead is a heavy metal that remains in the bloodstream, causing mental retardation in children. It is believed that 3×10^{-7} g of Pb in 1.00 mL of blood is a health hazard. For this amount of lead how many atoms of lead are there in one mL of a child's blood?

32. Silversmiths are warned to limit their exposure to silver in the air to 1×10^{-8} g Ag/L of air in a 40-hour week. What is the allowed exposure in terms of atoms of Ag/L/week?

33. Determine
(a) the number of atoms in 0.185 g of palladium (Pd).
(b) the mass of 127 protons of palladium.

34. For bismuth (Bi), determine
(a) the number of atoms in 0.243 g.
(b) the mass of 139 protons.

35. The isotope Si-28 has a mass of 27.977 amu. For ten grams of Si-28, calculate
(a) the number of atoms.
(b) the total number of protons, neutrons, and electrons.

36. Myocardial perfusion imaging (MPI) is the latest tool used in the detection of coronary artery disease. It uses thallium-201 chloride. How many neutrons are in
(a) seventy atoms of thallium-201?
(b) one picogram (1×10^{-12} g) of thallium-201?

37. A cube of sodium has length 1.25 in. How many atoms are in that cube? (Note: d_{Na} = 0.968 g/cm³.)

38. A cylindrical piece of pure copper (d = 8.92 g/cm³) has diameter 1.15 cm and height 4.00 inches. How many atoms are in that cylinder? (Note: the volume of a right circular cylinder of radius r and height h is $V = \pi r^2 h$.)

2-4 Introduction to the Periodic Table

Elements and the Periodic Table

39. Give the symbols for
(a) potassium   (b) cadmium   (c) aluminum
(d) antimony   (e) phosphorus

40. Name the elements represented by
(a) S   (b) Sc   (c) Se   (d) Si   (e) Sr

41. Classify the elements in Question 39 as metals (main group, transition, or post-transition), nonmetals, or metalloids.

42. Classify the elements in Question 40 as metals (main group, transition, or post-transition), nonmetals, or metalloids.

43. How many metals are in the following groups?
(a) Group 1   (b) Group 13   (c) Group 17

44. How many nonmetals are in the following periods?
(a) period 2   (b) period 4   (c) period 6

45. Which group in the periodic table
(a) has one metalloid and no nonmetals?
(b) has no nonmetals or transition metals?
(c) has no metals or metalloids?

46. Which period of the periodic table
(a) has no metals?
(b) has no nonmetals?
(c) has one post-transition metal and two metalloids?

2-5 Molecules and Ions

47. Given the following condensed formulas, write the molecular formulas for the following molecules.
(a) dimethylamine $(CH_3)_2NH$
(b) propyl alcohol $CH_3(CH_2)_2OH$

48. Write the condensed structural formulas and molecular formulas for the following molecules. The reactive groups are shown in red.

(a)
$$
\begin{array}{c}
\quad\; H \\
\quad\; | \\
H - C - C = O \quad\quad \text{(acetic acid)}\\
\quad\; | \quad\;\; | \\
\quad\; H \;\; O - H \\
\quad\; H
\end{array}
$$

(b)
$$
\begin{array}{c}
\quad\; H \\
\quad\; | \\
H - C - Cl \quad\quad \text{(methyl chloride)} \\
\quad\; | \\
\quad\; H
\end{array}
$$

49. Give the number of protons and electrons in
(a) an N_2 molecule (identified in 1772).
(b) an N_3^- unit (synthesized in 1890).
(c) an N_5^+ unit (synthesized in 1999).
(d) an N_5N_5 salt (a U.S. Air Force research team's synthesis project).

50. Give the number of protons and electrons in the following:
(a) S_8 molecule.   (b) SO_4^{2-} ion.
(c) H_2S molecule.   (d) S^{2-} ion.

51. Complete the table given below. If necessary, use the periodic table.

Nuclear Symbol	Charge	Number of Protons	Number of Neutrons	Number of Electrons
_____	0	9	10	_____
^{31}P	0	_____	16	_____
_____	+3	27	30	_____
_____	_____	16	16	18

52. Complete the table given below.

Nuclear Symbol	Metal, Nonmetal, Metalloid	Group	Period	Number of Neutrons
_____	_____	13	3	14
_____	metalloid	16	_____	76
Xe-134	_____	_____	_____	_____
_____	nonmetal	14	_____	8

53. Classify the following compounds as electrolytes or nonelectrolytes.
  (a) potassium chloride, KCl
  (b) hydrogen peroxide, H_2O_2
  (c) methane, CH_4
  (d) barium nitrate, $Ba(NO_3)_2$

54. Which (if any) of the following compounds are nonelectrolytes?
  (a) citric acid ($C_6H_8O_7$)
  (b) calcium nitrate, $Ca(NO_3)_2$
  (c) ammonium carbonate, $(NH_4)_2CO_3$
  (d) iodine tribromide (IBr_3)

2-6 Formulas of Ionic Compounds
2-7 Names of Compounds

Names and Formulas of Ionic and Molecular Compounds

55. Write the formulas for the following molecules.
  (a) methane   (b) carbon tetraiodide
  (c) hydrogen peroxide (d) nitrogen oxide
  (e) silicon dioxide

56. Write the formulas of the following molecules
  (a) silicon disulfide
  (b) phosphine
  (c) diphosphorus pentoxide
  (d) radon tetrafluoride
  (e) nitrogen trichloride

57. Write the names of the following molecules.
  (a) ICl_3 (b) N_2O_5 (c) PH_3 (d) CBr_4 (e) SO_3

58. Write the names of the following molecules.
  (a) $TeCl_2$ (b) I_4O_9 (c) XeO_3 (d) S_4N_4

59. Give the formulas of all the compounds containing no ions other than K^+, Ca^{2+}, Cl^-, and S^{2-}.

60. Give the formulas of compounds in which
  (a) the cation is Ba^{2+}, the anion is I^- or N^{3-}.
  (b) the anion is O^{2-}, the cation is Fe^{2+} or Fe^{3+}.

61. Write the formulas of the following ionic compounds.
  (a) iron(III) acetate
  (b) calcium nitrate
  (c) potassium oxide
  (d) gold(III) chloride
  (e) barium nitride

62. Write formulas for the following ionic compounds:
  (a) potassium hydrogen phosphate
  (b) magnesium nitride
  (c) lead(IV) bromide
  (d) scandium(III) chloride
  (e) barium acetate

63. Write the names of the following ionic compounds.
  (a) $K_2Cr_2O_7$ (b) $Cu_3(PO_4)_2$ (c) $Ba(C_2H_3O_2)_2$
  (d) AlN (e) $Co(NO_3)_2$

64. Write the names of the following ionic compounds.
  (a) $Pb(C_2H_3O_2)_2$
  (b) $Al(OH)_3$
  (c) $Sr(HCO_3)_2$
  (d) $Cu_3(PO_4)_2$
  (e) Rb_2O

65. Write the names of the following ionic compounds.
  (a) HCl(aq)   (b) $HClO_3$(aq)   (c) $Fe_2(SO_3)_3$
  (d) $Ba(NO_2)_2$ (e) $NaClO$

66. Write formulas for the following ionic compounds.
  (a) nitric acid   (b) potassium sulfate
  (c) iron(III) perchlorate (d) aluminum iodate
  (e) sulfurous acid

67. Complete the following table.

Name	Formula
nitrous acid	_____
_____	$Ni(IO_3)_2$
gold(III) sulfide	_____
_____	H_2SO_3(aq)
nitrogen trifluoride	_____

68. Complete the following table.

Name	Formula
bromous acid	_____
_____	$Co(IO_4)_3$
platinum(IV) chloride	_____
_____	HNO_2(aq)
dinitrogen trioxide	_____

Unclassified

69. Write the formulas and names of the following:
  (a) An ionic compound whose cation is a transition metal with 25 protons and 22 electrons and whose anion is an oxoanion of nitrogen with two oxygen atoms.
  (b) A molecule made up of a metalloid in Group 13 and three atoms of a halogen in period 2.
  (c) An ionic compound made up of an alkaline earth with 20 protons, and an anion with one hydrogen atom, a carbon atom, and 3 oxygen atoms.

70. Identify the following elements:
  (a) A member of the same period as selenium but with two fewer protons than selenium.
  (b) A transition metal in group 6, period 6.
  (c) An alkaline earth with 38 protons.
  (d) A post-transition metal in group 15.

71. Identify the following elements.
  (a) A member of the same period as tellurium but with three fewer protons than tellurium.
  (b) A post-transition metal in Group 14
  (c) An alkali with 19 protons
  (d) A metalloid in Group 15 with more than 40 protons

72. A molecule of ethylamine is made up of two carbon atoms, seven hydrogen atoms, and one nitrogen atom.

(a) Write its molecular formula.

(b) The reactive group in ethylamine is NH_2. Write its condensed structural formula.

73. Criticize each of the following statements:

(a) The Rutherford experiment confirmed the presence of electrons in an atom.

(b) A period is that segment of the periodic table in which elements are arranged by similar chemical properties.

(c) Isotopes have the same mass number.

(d) The molecule Be_3N_2 is triberylium dinitride.

74. Which of the following statements is/are always true? Never true? Usually true?

(a) Compounds containing chlorine can be either molecular or ionic.

(b) An ionic compound always has at least one metal.

(c) When an element in a molecule has a "di" prefix, it means that the element has a +2 charge.

75. Some brands of salami contain 0.090% sodium benzoate ($NaC_7H_5O_2$) as a preservative. If you eat 6.00 oz of this salami, how many atoms of sodium will you consume, assuming salami contains no other source of that element?

76. Carbon tetrachloride, CCl_4, was a popular dry-cleaning agent until it was shown to be carcinogenic. It has a density of 1.589 g/cm^3. What volume of carbon tetrachloride will contain a total of 6.00×10^{25} molecules of CCl_4?

Conceptual Problems

77. Which of the following statements are true?

(a) Neutrons have neither mass nor charge.

(b) Isotopes of an element have an identical number of protons.

(c) C-14 and N-14 have identical neutron/proton (n/p^+) ratios.

(d) The vertical columns in a periodic table are referred to as "groups."

(e) When an atom loses an electron, it becomes positively charged.

78. A student saw the following nuclear symbol for an unknown element: $^{23}_{11}X$. Which of the following statements about X and $^{23}_{11}X$ are true?

(a) X is sodium.

(b) X is vanadium.

(c) X has 23 neutrons in its nucleus.

(d) X^{2+} has 13 electrons.

(e) $^{23}_{11}X$ has a proton/neutron ratio of about 1.1.

79. Using the laws of constant composition and the conservation of mass, complete the molecular picture of hydrogen molecules ($\bigcirc\!-\!\bigcirc$) reacting with chlorine molecules ($\square\!-\!\square$) to give hydrogen chloride ($\square\!-\!\bigcirc$) molecules.

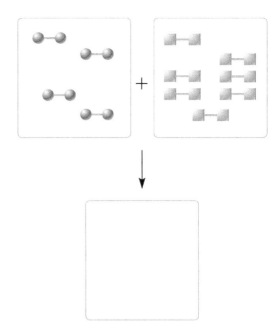

80. Use the law of conservation of mass to determine which numbered box(es) represent(s) the product mixture after the substances in the unnumbered box undergo a reaction.

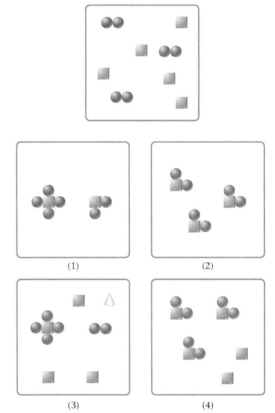

81. If squares represent carbon and spheres represent chlorine, make a representation of liquid CCl_4.

82. If a square represents a sulfur atom and a circle represents a sodium atom, sketch a representation of a sodium sulfide crystal.

83. Scientists are trying to synthesize elements with more than 114 protons. State the expected atomic number of
(a) the newest inert gas.
(b) the new element with properties similar to those of the alkaline earth metals.
(c) the new element that will behave like the halogens.
(d) the new (nontransition) metal whose ion will have a +2 charge.
(e) the new element that will start period 8.

84. Write the nuclear symbol for the element whose mass number is 234 and has 60% more neutrons than protons.

85. Mercury(II) oxide, a red powder, can be decomposed by heating to produce liquid mercury and oxygen gas. When a sample of this compound is decomposed, 3.87 g of oxygen and 48.43 g of mercury are produced. In a second experiment, 15.68 g of mercury is allowed to react with an excess of oxygen and 16.93 g of red mercury(II) oxide is produced. Show that these results are consistent with the law of constant composition.

86. Write the atomic symbol for the element whose ion has a −2 charge, has 20 more neutrons than electrons, and has a mass number of 126.

87. Consider the elements oxygen, fluorine, argon, sulfur, potassium, and strontium. From this group of elements, which ones fit the descriptions below?
(a) Two elements that are metals
(b) Four elements that are nonmetals
(c) Three elements that are solid at room temperature
(d) An element that is found in nature as X_8
(e) One pair of elements that may form a molecular compound
(f) One pair of elements that may form an ionic compound with formula AX
(g) One pair of elements that may form an ionic compound with formula AX_2
(h) One pair of elements that may form an ionic compound with formula A_2X
(i) An element that can form no compounds
(j) Three elements that are gases at room temperature

Challenge Problems

88. Three compounds containing only carbon and hydrogen are analyzed. The results for the analysis of the first two compounds are given below:

Compound	Mass of Carbon (g)	Mass of Hydrogen (g)
A	28.5	2.39
B	34.7	11.6
C	16.2	—

Which, if any, of the following results for the mass of hydrogen in compound C follows the law of multiple proportions?
(a) 5.84 g   (b) 3.47 g   (c) 2.72 g

89. Ethane and ethylene are two gases containing only hydrogen and carbon atoms. In a certain sample of ethane, 4.53 g of hydrogen is combined with 18.0 g of carbon. In a sample of ethylene, 7.25 g of hydrogen is combined with 43.20 g of carbon.
(a) Show how the data illustrate the law of multiple proportions.
(b) Suggest reasonable formulas for the two compounds.

90. Calculate the average density of a single Al-27 atom by assuming that it is a sphere with a radius of 0.143 nm. The masses of a proton, electron, and neutron are 1.6726×10^{-24} g, 9.1094×10^{-28} g, and 1.6749×10^{-24} g, respectively. The volume of a sphere is $4\pi r^3/3$, where r is its radius. Express the answer in grams per cubic centimeter. The density of aluminum is found experimentally to be 2.70 g/cm³. What does that suggest about the packing of aluminum atoms in the metal?

91. The mass of an atom of N-14 is 2.3440×10^{-23} g. Using that fact and other information in this chapter, find the mass of an N-14^{3-} ion.

92. Each time you inhale, you take in about 500 mL (two significant figures) of air, each milliliter of which contains 2.5×10^{19} molecules. In delivering the Gettysburg Address, Abraham Lincoln is estimated to have inhaled about 200 times.
(a) How many molecules did Lincoln take in?
(b) In the entire atmosphere, there are about 1.1×10^{44} molecules. What fraction of the molecules in the earth's atmosphere was inhaled by Lincoln at Gettysburg?
(c) In the next breath that you take, how many molecules were inhaled by Lincoln at Gettysburg?

93. Hydrogen gas is prepared in a lab experiment. In this experiment, 18.00 g of aluminum metal are mixed with 25 mL of HCl ($d = 1.025$ g/cm³). After the experiment, there are 12.00 g of aluminum and 30.95 g of a solution made up of water, aluminum ions, and chloride ions. Assuming no loss of products, how many liters of H_2 gas are obtained? The density of hydrogen gas at the temperature and pressure of the experiment is 0.0824 g/L.

Mass Relations in Chemistry; Stoichiometry

Balanced chemical equations make it possible to relate masses of reactants and products. Scales are devices commonly used to measure mass.

Ex umbris et imaginibus in veritatem! (From shadows and symbols into the truth!)

—JOHN HENRY CARDINAL NEWMAN

To this point, our study of chemistry has been largely qualitative, involving few calculations. However, chemistry is a quantitative science. From Chapter 2 we see that atoms differ from one another not only in composition (number of protons, electrons, neutrons) but also in mass. Chemical formulas, which we learned to write and name, tell us not only the atom ratios in which elements are present but also mass ratios.

The general topic of this chapter is **stoichiometry** (stoy-key-OM-e-tree), the study of mass relations in chemistry. Whether dealing with molar mass (Section 3-1), chemical formulas (Section 3-2), or chemical reactions (Section 3-3), you will be answering questions that ask "how much?" or "how many?". For example,

- how many grams of acid are needed to prepare a solution required for an experiment? (Section 3-1)
- how much iron can be obtained from a ton of iron ore? (Section 3-2)
- how much nitrogen gas is required to form a kilogram of ammonia? (Section 3-3)

3-1 The Mole

People in different professions often use special counting units. You and I eat eggs one at a time, but farmers sell them by the dozen. We spend dollar bills one at a time, but Congress distributes them by the billion. Chemists have their own counting unit, Avogadro's number (Section 2-3).

Figure 3.1 One-mole quantities of sugar ($C_{12}H_{22}O_{11}$), baking soda ($NaHCO_3$), and copper nails (Cu).

The quantity represented by Avogadro's number is so important that it is given a special name, the **mole**. It is mentioned briefly on page 31. A mole represents 6.022×10^{23} items, whatever they may be (Figure 3.1).

$$1 \text{ mol H atoms} = 6.022 \times 10^{23} \text{ H atoms}$$
$$1 \text{ mol O atoms} = 6.022 \times 10^{23} \text{ O atoms}$$
$$1 \text{ mol } H_2 \text{ molecules} = 6.022 \times 10^{23} \text{ } H_2 \text{ molecules}$$
$$1 \text{ mol } H_2O \text{ molecules} = 6.022 \times 10^{23} \text{ } H_2O \text{ molecules}$$
$$1 \text{ mol electrons} = 6.022 \times 10^{23} \text{ electrons}$$
$$1 \text{ mol pennies} = 6.022 \times 10^{23} \text{ pennies}$$

(One mole of pennies is a lot of money. It's enough to pay all the expenses of the United States for the next billion years or so without accounting for inflation.)

A mole represents not only a specific number of particles but also a definite mass of a substance as represented by its formula (O, O_2, H_2O, NaCl, . . .). *The **molar mass**, MM, in grams per mole, is numerically equal to the sum of the masses (in amu) of the atoms in the formula.*

Formula	Sum of Atomic Masses		Molar Mass, MM
O		16.00 amu	16.00 g/mol
O_2	2(16.00 amu) =	32.00 amu	32.00 g/mol
H_2O	2(1.008 amu) + 16.00 amu =	18.02 amu	18.02 g/mol
NaCl	22.99 amu + 35.45 amu =	58.44 amu	58.44 g/mol

Notice that the formula of a substance must be known to find its molar mass. It would be ambiguous, to say the least, to refer to the "molar mass of hydrogen." One mole of hydrogen atoms, represented by the symbol H, weighs 1.008 g; the molar mass of H is 1.008 g/mol. One mole of hydrogen molecules, represented by the formula H_2, weighs 2.016 g; the molar mass of H_2 is 2.016 g/mol.

Figure 3.2 Aspirin (acetylsalicylic acid), $C_9H_8O_4$.

Mole-Gram Conversions

As you will see later in this chapter, it is often necessary to convert from moles of a substance to mass in grams or vice versa. Such conversions are readily made by using the general relation

$$mass = \text{MM} \times n \tag{3.1}$$

The mass is in grams, MM is the molar mass (g/mol), and n is the amount in moles. In effect, the molar mass, MM, is a conversion factor that allows you to calculate *moles* from *mass* or vice versa.

EXAMPLE	3.1

Acetylsalicylic acid (ASA), $C_9H_8O_4$, is the active ingredient of aspirin (Figure 3.2).

a	What is the mass in grams of 0.509 moles of acetylsalicylic acid (ASA)?

ANALYSIS	
Information given:	moles of acetylsalicylic acid (0.509) formula for acetylsalicylic acid ($C_9H_8O_4$)
Information implied:	molar mass (MM) of acetylsalicylic acid
Asked for:	mass of acetylsalicylic acid (ASA)

continued

STRATEGY

Substitute into Equation 3.1
$$mass = MM \times n$$

SOLUTION

MM of $C_9H_8O_4$	$9(12.01) + 8(1.008) + 4(16.00) = 180.15$ g/mol
mass	$mass = MM \times n = 0.509 \text{ mol} \times \dfrac{180.15 \text{ g}}{1 \text{ mol}} = 91.7 \text{ g}$

b   How many moles are present in 28.00 g of $C_9H_8O_4$?

ANALYSIS

Information given:	mass of ASA (28.00 g) formula for acetylsalicylic acid ($C_9H_8O_4$)
Information implied:	molar mass (MM) of acetylsalicylic acid
Asked for:	moles of ASA in the sample

STRATEGY

Substitute into Equation 3.1.
$$mass = MM \times n$$

SOLUTION

moles ASA	$n = \dfrac{mass}{MM} = \dfrac{28.00 \text{ g}}{180.15 \text{ g/mole}} = 0.1554 \text{ mol ASA}$

c   How many molecules of $C_9H_8O_4$ are there in 12.00 g of acetylsalicylic acid? How many carbon atoms?

ANALYSIS

Information given:	mass of ASA (12.00 g) formula for acetylsalicylic acid ($C_9H_8O_4$)
Information implied:	molar mass (MM) of acetylsalicylic acid Avogadro's number (N_A)
Asked for:	number of molecules of ASA number of carbon atoms

STRATEGY

Follow the plan outlined in Figure 3.3 (p. 54).
$$mass \xrightarrow{MM} moles \xrightarrow{N_A} molecules \xrightarrow{9 \text{ C atoms}} \text{C atoms}$$

SOLUTION

number of molecules	$12.00 \text{ g ASA} \times \dfrac{6.022 \times 10^{23} \text{ molecules ASA}}{180.15 \text{ g ASA}} = 4.011 \times 10^{22}$
number of C atoms	$4.011 \times 10^{22} \text{ molecules ASA} \times \dfrac{9 \text{ C atoms}}{1 \text{ molecule}} = 3.610 \times 10^{23}$

END POINT

The mass of the sample in grams is always larger than the number of moles because every known substance has a molar mass greater than 1 g/mol.

We have now considered several types of conversions involving numbers of particles, moles, and grams. They are summarized in Figure 3.3. Conversions of the type we have just carried out come up over and over again in chemistry. They will be required in nearly every chapter of this text. Clearly, you must know what is meant by a mole. Remember, a mole always represents a certain number

Almost every calculation that you will make in this chapter requires that you know precisely what is meant by a mole.

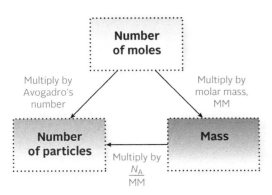

Figure 3.3 Flowchart for the interconversion of particles, mass, and moles.

[NH$_3$] means "concentration of ammonia in moles per liter." ℹ

of items, 6.022×10^{23}. Its mass, however, differs with the substance involved: A mole of H$_2$O, 18.02 g, weighs considerably more than a mole of H$_2$, 2.016 g, even though they both contain the same number of molecules. In the same way, a dozen bowling balls weigh a lot more than a dozen eggs, even though the number of items is the same for both.

Moles in Solution; Molarity

To obtain a given amount of a pure solid in the laboratory, you would weigh it out on a balance. Suppose, however, the solid is present as a solute dissolved in a solvent such as water. In this case, you ordinarily measure out a given volume of the water solution, perhaps using a graduated cylinder. The amount of solute you obtain in this way depends not only on the volume of solution but also on the **concentration** of solute, i.e., the amount of solute in a given amount of solution. The concentration of a solute in solution can be expressed in terms of its **molarity:**

$$\text{molarity } (M) = \frac{\text{moles of solute}}{\text{liters of solution}} \qquad (3.2)$$

The symbol [ ] is commonly used to represent the molarity of a species in solution. ◄ℹ For a solution containing 1.20 mol of substance A in 2.50 L of solution,

$$[\text{A}] = \frac{1.20 \text{ mol}}{2.50 \text{ L}} = 0.480 \text{ mol/L} = 0.480 \, M$$

One liter of such a solution would contain 0.480 mol of A; 100 mL of solution would contain 0.0480 mol of A, and so on.

The molarity of a soluble solute can vary over a wide range. With sodium hydroxide, for example, we can prepare a 6 M solution, a 1 M solution, a 0.1 M solution, and so on. The words "concentrated" and "dilute" are often used in a qualitative way to describe these solutions. We would describe a 6 M solution of NaOH as concentrated; it contains a relatively large amount of solute per liter. A 0.1 M NaOH solution is dilute, at least in comparison to 1 M or 6 M.

To prepare a solution to a desired molarity, you first calculate the amount of solute required. This is then dissolved in enough solvent to form the required volume of solution. Suppose, for example, you want to make one liter of 0.100 M K$_2$CrO$_4$ solution (Figure 3.4). You first weigh out 19.4 g (0.100 mol) of K$_2$CrO$_4$ (MM = 194.20 g/mol). Then stir with enough water to form one liter (1000 mL) of solution.

The molarity of a solution can be used to calculate

■ the number of moles of solute in a given volume of solution.
■ the volume of solution containing a given number of moles of solute.

Here, as in so many other cases, a conversion factor approach is used (Example 3.2).

EXAMPLE	3.2

Nitric acid, HNO$_3$, is extensively used in the manufacture of fertilizer. A bottle containing 75.0 mL of nitric acid solution is labeled 6.0 M HNO$_3$.

a	How many moles of HNO$_3$ are in the bottle?

ANALYSIS	
Information given:	V (75.0 mL) and M (6.0 M) of HNO$_3$ in the bottle
Information implied:	molar mass (MM) of HNO$_3$
Asked for:	moles of HNO$_3$ in the bottle

continued

1. Do not forget to change the volume unit given (mL) to L.

2. Use the molarity of HNO_3 as a conversion factor:

$$\frac{6.0 \text{ mol } HNO_3}{1 \text{ L solution}}$$

SOLUTION

moles HNO_3	$75.0 \text{ mL solution} \times \dfrac{1 \text{ L}}{1000 \text{ mL}} \times \dfrac{6.0 \text{ mol } HNO_3}{1 \text{ L solution}} = \boxed{0.45 \text{ mol}}$

b A reaction needs 5.00 g of HNO_3. How many milliliters of solution are required?

ANALYSIS

Information given:	V (75.0 mL) and M (6.0 M) of HNO_3 in the bottle mass of HNO_3 required (5.00 g)
Information implied:	molar mass (MM) of HNO_3
Asked for:	volume of HNO_3 required

STRATEGY

1. Use the molarity and the molar mass of HNO_3 as conversion factors.

$$\frac{6.0 \text{ mol } HNO_3}{1 \text{ L solution}} \qquad \frac{1 \text{ mol } HNO_3}{63.02 \text{ g } HNO_3}$$

2. Follow the plan:

$$\text{mass} \xrightarrow{MM} \text{moles} \xrightarrow{M} \text{volume}$$

SOLUTION

volume HNO_3	$5.00 \text{ g } HNO_3 \times \dfrac{1 \text{ mol } HNO_3}{63.02 \text{ g } HNO_3} \times \dfrac{1 \text{ L solution}}{6.0 \text{ mol } HNO_3} = 0.013 \text{ L} = \boxed{13 \text{ mL}}$

c Ten mL of water are added to the solution. What is the molarity of the resulting solution? (Assume volumes are additive.)

ANALYSIS

Information given:	V (75.0 mL) and M (6.0 M) of HNO_3 in the bottle volume of water added (10.00 mL) Assume volumes are additive.
Information implied:	molar mass (MM) of HNO_3
Asked for:	molarity (M) of diluted solution

STRATEGY

1. Adding water does not change the number of moles of solute. It does change the volume of solution.

2. Substitute into Equation 3.2. Use (75.0 mL + 10.0 mL) as the total volume.

SOLUTION

M	$\dfrac{0.45 \text{ mol}}{0.0750 \text{ L} + 0.0100 \text{ L}} = 5.3 \text{ mol/L} = \boxed{5.3 \text{ M}}$

END POINT

The molarity of a solution decreases when water is added to the solution, but the moles of solute in solution remain the same.

Figure 3.4 **Preparing one liter of 0.100 *M* potassium chromate.** A 0.100 *M* potassium chromate solution is made by adding enough water to 19.4 g of K_2CrO_4 to make one liter of solution. The weighed K_2CrO_4 (19.4 g) is transferred to a 1000-mL volumetric flask. Enough water is added to fully dissolve all of the solid by swirling. The final step is to shake the flask repeatedly until a homogeneous solution is formed.

Knowing the molarity of a solution, you can readily obtain a specified amount of solute. All you have to do is to calculate the required volume, as in Example 3.2b. Upon measuring out that volume, you should obtain the desired number of moles or grams of solute. Concentrations of reagents in the general chemistry laboratory are most often expressed in molarities. We will have more to say about molarity and other concentration units in Chapter 10.

As pointed out in Chapter 2, when an ionic solid dissolves in water, the cations and anions separate from each other. This process can be represented by a chemical equation in which the reactant is the solid and the products are the positive and negative ions in water (aqueous) solution. For the dissolving of $MgCl_2$, the equation is

$$MgCl_2(s) \longrightarrow Mg^{2+}(aq) + 2Cl^-(aq)$$

This equation tells us that one mole of $MgCl_2$ yields *one* mole of Mg^{2+} ions and *two* moles of Cl^- ions in solution. It follows that in any solution of magnesium chloride:

molarity Mg^{2+} = molarity $MgCl_2$ molarity Cl^- = 2 × molarity $MgCl_2$

Similar relationships hold for ionic solids containing polyatomic ions (Table 2.2) or transition metal cations (Figure 3.5). ◄ ⓘ

[ion] = ion subscript × [parent compound] ⓘ

Figure 3.5 **Charges of transition metal cations commonly found in aqueous solution.**

3	4	5	6	7	8	9	10	11	12
			Cr^{3+}	Mn^{2+}	Fe^{2+} Fe^{3+}	Co^{2+} Co^{3+}	Ni^{2+}	Cu^{2+}	Zn^{2+}
								Ag^+	Cd^{2+}
									Hg^{2+}

$$(NH_4)_3PO_4(s) \longrightarrow 3NH_4^+(aq) + PO_4^{3-}(aq)$$

molarity NH_4^+ = 3 × molarity $(NH_4)_3PO_4$
molarity PO_4^{3-} = molarity $(NH_4)_3PO_4$

$$Cr_2(SO_4)_3(s) \longrightarrow 2Cr^{3+}(aq) + 3SO_4^{2-}(aq)$$

molarity Cr^{3+} = 2 × molarity $Cr_2(SO_4)_3$
molarity SO_4^{2-} = 3 × molarity $Cr_2(SO_4)_3$ ◀ *i*

i ——————————
For aluminum sulfate:
$[Al^{3+}] = 2[Al_2(SO_4)_3]$,
$[SO_4^{2-}] = 3[Al_2(SO_4)_3]$.

EXAMPLE 3.3

Potassium dichromate, $K_2Cr_2O_7$, is used in the tanning of leather. A flask containing 125 mL of solution is labeled 0.145 M $K_2Cr_2O_7$.

a | What is the molarity of each ion in solution?

ANALYSIS

Information given:	volume, V (125 mL), and molarity, M (0.145 M), of solution
Information implied:	number of each ion in the parent compound
Asked for:	molarity of each ion in solution

STRATEGY

1. Distinguish between the parent compound and the ions.

2. Count the number of ions in the parent compound (recall that it is the subscript of the ion), and use it as a conversion factor.

$$\frac{2 \text{ mol } K^+}{1 \text{ mol } K_2Cr_2O_7} \qquad \frac{1 \text{ mol } Cr_2O_7^{2-}}{1 \text{ mol } K_2Cr_2O_7}$$

SOLUTION

$[K^+]$	$\dfrac{0.145 \text{ mol } K_2Cr_2O_7}{1 \text{ L}} \times \dfrac{2 \text{ mol } K^+}{1 \text{ mol } K_2Cr_2O_7} = 0.290 \text{ mol/L} = \boxed{0.290 \ M}$
$[Cr_2O_7^{2-}]$	$\dfrac{0.145 \text{ mol } K_2Cr_2O_7}{1 \text{ L}} \times \dfrac{1 \text{ mol } Cr_2O_7^{2-}}{1 \text{ mol } K_2Cr_2O_7} = 0.145 \text{ mol/L} = \boxed{0.145 \ M}$

b | A sample containing 0.200 moles of K^+ is added to the $K_2Cr_2O_7$ solution. Assuming no volume change, what is the molarity of the new solution?

ANALYSIS

Information given:	volume, V (125 mL), and molarity, M (0.145 M), of solution moles of K^+ added (0.200)
Information implied:	number of each ion in the parent compound
Asked for:	molarity of $K_2Cr_2O_7$ after the addition of K^+ ions

STRATEGY

1. Use the molarity of K^+ found in (a) and substitute into Equation 3.2 to find the moles of K^+ initially.

2. Find the moles of K^+ after the addition.

3. Find $[K^+]$ again by substituting into Equation 3.2.

4. Find $[K_2Cr_2O_7]$ by relating $[K^+]$ to $K_2Cr_2O_7$.

5. The overall plan is:

$$[K^+_{initial}] \xrightarrow{V} (mol \ K^+)_{initial} \xrightarrow{+ \ 0.200 \ mol \ K^+} (mol \ K^+)_{final} \xrightarrow{V} [K^+]_{final} \xrightarrow{2 \ K^+/K_2Cr_2O_7} [K_2Cr_2O_7]_{final}$$

continued

SOLUTION	
1. $(\text{mol K}^+)_{\text{initial}}$	$0.125 \text{ L} \times \dfrac{0.290 \text{ mol K}^+}{1 \text{ L}} = 0.03625$
2. $(\text{mol K}^+)_{\text{final}}$	$0.03625 + 0.200 = 0.236$
3. $[\text{K}^+]_{\text{final}}$	$\dfrac{0.236 \text{ mol K}^+}{0.125 \text{ L}} = 1.89 \text{ mol/L} = 1.89 \text{ } M$
4. $[\text{K}_2\text{Cr}_2\text{O}_7]_{\text{final}}$	$\dfrac{1.89 \text{ mol K}^+}{1 \text{ L}} \times \dfrac{1 \text{ mol K}_2\text{Cr}_2\text{O}_7}{2 \text{ mol K}^+} = 0.945 \text{ mol/L} = 0.945 \text{ } M$

END POINT
The concentration of K^+ should be twice that of $Cr_2O_7^{2-}$ in either solution. It is!

3-2 Mass Relations in Chemical Formulas

As you will see shortly, the formula of a compound can be used to determine the mass percents of the elements present. Conversely, if the percentages of the elements are known, the simplest formula can be determined. Knowing the molar mass of a molecular compound, it is possible to go one step further and find the molecular formula. In this section we will consider how these three types of calculations are carried out.

Percent Composition from Formula

The **percent composition** of a compound is specified by citing the mass percents of the elements present. For example, in a 100-g sample of water there are 11.19 g of hydrogen and 88.81 g of oxygen. Hence the percentages of the two elements are

$$\frac{11.19 \text{ g H}}{100.00 \text{ g}} \times 100\% = 11.19\% \text{ H} \qquad \frac{88.81 \text{ g O}}{100.00 \text{ g}} \times 100\% = 88.81\% \text{ O}$$

We would say that the percent composition of water is 11.19% H, 88.81% O.

Knowing the formula of a compound, Fe_2O_3, you can readily calculate the mass percents of its constituent elements. It is convenient to start with one mole of compound (Example 3.4a). The formula of a compound can also be used in a straightforward way to find the mass of an element in a known mass of the compound (Example 3.4b).

Figure 3.6 **Hematite ore.**

Jo Edkins, 2007. Used with permission.

EXAMPLE 3.4 GRADED
Metallic iron is most often extracted from hematite ore (Figure 3.6), which consists of iron(III) oxide mixed with impurities such as silicon dioxide, SiO_2.

a	What are the mass percents of iron and oxygen in iron(III) oxide?

ANALYSIS	
Information given:	formula of the iron oxide (Fe_2O_3)
Information implied:	molar mass (MM) of Fe_2O_3
Asked for:	mass % of Fe and O in Fe_2O_3

continued

STRATEGY

1. To find the mass percent of Fe follow the plan outlined below. Start with one mole of Fe_2O_3.

$$n_{Fe_2O_3} \xrightarrow{\text{subscript}} n_{Fe} \xrightarrow{\text{MM Fe}} \text{mass Fe} \xrightarrow{\text{MM } Fe_2O_3} \% \text{ element}$$

2. Find the mass percent of oxygen by difference. Note that the compound is made up of only two elements, Fe and O.

SOLUTION

mass % Fe	$1 \text{ mol } Fe_2O_3 \times \dfrac{2 \text{ mol Fe}}{1 \text{ mol } Fe_2O_3} \times \dfrac{55.85 \text{ g}}{1 \text{ mol Fe}} = 111.7 \text{ g Fe}$
	$\dfrac{111.7 \text{ g Fe}}{159.7 \text{ g } Fe_2O_3} = 69.94\%$
mass % O	mass % O = 100% − mass % Fe = 100.00% − 69.94% = 30.06%

b How many grams of iron can be extracted from 1.000 kg of Fe_2O_3?

ANALYSIS

Information given:	mass of Fe_2O_3 (1.000 kg = 1.000×10^3 g)
Information implied:	from (a): mass % of Fe in Fe_2O_3 (69.94% = 69.94 g Fe/100.00 g Fe_2O_3)
Asked for:	mass of Fe obtained from 1.000 kg of Fe_2O_3

STRATEGY

mass of Fe = (mass Fe_2O_3)(massFe/100 g Fe_2O_3)

SOLUTION

mass Fe	$1.000 \times 10^3 \text{ g } Fe_2O_3 \times \dfrac{69.94 \text{ g Fe}}{100 \text{ g } Fe_2O_3} = 699.4 \text{ g Fe}$

c How many metric tons of hematite ore, 66.4% Fe_2O_3, must be processed to produce 1.000 kg of iron?

ANALYSIS

Information given:	mass % of Fe_2O_3 in the ore (66.4% = 66.4 g Fe_2O_3/100.0 g ore) mass of Fe needed (1.000 kg)
Information implied:	from (a): mass % of Fe in Fe_2O_3 (69.94% = 69.94 g Fe/100.00 g Fe_2O_3) factor for converting metric tons to grams.
Asked for:	mass of hematite needed to produce 1.000 kg of Fe

STRATEGY

1. Distingush between the mass % of Fe_2O_3 in the ore (66.4%) and the mass % of Fe in Fe_2O_3 (69.94%).

2. Start with 1000 g of Fe and use the mass percents as conversion factors to go from mass of Fe to mass of the ore.

$$\frac{69.94 \text{ g Fe}}{100.00 \text{ g } Fe_2O_3} \quad \frac{66.4 \text{ g } Fe_2O_3}{100.00 \text{ g ore}}$$

3. Convert grams to metric tons.
   1 metric ton = 1×10^6 g

SOLUTION

mass of hematite needed	$1000 \text{ g Fe} \times \dfrac{100 \text{ g } Fe_2O_3}{69.94 \text{ g Fe}} \times \dfrac{100 \text{ g ore}}{66.4 \text{ g } Fe_2O_3} \times \dfrac{1 \text{ metric ton}}{1 \times 10^6 \text{ g}} = 2.15 \times 10^{-3}$ metric tons

The calculations in Example 3.4 illustrate an important characteristic of formulas. In one mole of Fe_2O_3 there are *two* moles of Fe (111.7 g) and *three* moles of O (48.00 g). This is the same as the atom ratio in Fe_2O_3, 2 atoms Fe : 3 atoms O. In general, ***the subscripts in a formula represent not only the atom ratio in which the different elements are combined but also the mole ratio.*** ◄ ⓘ

ⓘ Can you explain why the mole ratio must equal the atom ratio?

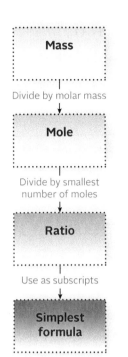

Figure 3.7 Flowchart for determining the simplest formula of a compound.

Simplest Formula from Chemical Analysis

A major task of chemical analysis is to determine the formulas of compounds. The formula found by the approach described here is the **simplest formula**, which gives the simplest whole-number ratio of the atoms present. For an ionic compound, the simplest formula is ordinarily the only one that can be written (e.g., $CaCl_2$, Cr_2O_3). For a molecular compound, the molecular formula is a whole-number multiple of the simplest formula, where that number may be 1, 2, ...

Compound	Simplest Formula	Molecular Formula	Multiple
Water	H_2O	H_2O	1
Hydrogen peroxide	HO	H_2O_2	2
Propylene	CH_2	C_3H_6	3

The analytical data leading to the simplest formula may be expressed in various ways. You may know

- the masses of the elements in a weighed sample of the compound.
- the mass percents of the elements in the compound.
- the masses of products obtained by reaction of a weighed sample of the compound.

The strategy used to calculate the simplest formula depends to some extent on which of these types of information is given. The basic objective in each case is to find the number of moles of each element, then the simplest mole ratio, and finally the simplest formula.

Figure 3.7 shows a schematic pathway for this process and Example 3.5 illustrates how a simple formula is obtained when the masses of the elements in the compound are given. The masses of the elements can also be calculated when the results of a combustion experiment (discussed on page 61) are given.

EXAMPLE 3.5

A 30.00-g sample of a white crystal used in water purification contains 4.731 g of aluminium, 8.436 g of sulfur, and 16.83 g of oxygen. Find the simplest formula of the compound.

ANALYSIS

Information given:	mass of sample (30.00 g) mass of each element in the compound: Al (4.731 g); S (8.436 g); O (16.83 g)
Information implied:	atomic masses of the elements
Asked for:	simplest formula for the compound

STRATEGY

Follow the flowchart for determining the simplest formula (Figure 3.7).

SOLUTION

moles of each element	Al: $\dfrac{4.731 \text{ g Al}}{26.98 \text{ g/mol}} = 0.1753 \text{ mol}$ S: $\dfrac{8.436 \text{ g S}}{32.07 \text{ g/mol}} = 0.2630 \text{ mol}$ O: $\dfrac{16.83 \text{ g O}}{16.00 \text{ g/mol}} = 1.052 \text{ mol}$

continued

ratios	$Al: \dfrac{0.1753 \text{ mol}}{0.1753 \text{ mol}} = 1; \quad S: \dfrac{0.2630 \text{ mol}}{0.1753 \text{ mol}} = 1.5 \quad O: \dfrac{1.052 \text{ mol}}{0.1753 \text{ mol}} = 6.0$
	Since we need smallest *whole number* ratios, we multiply all ratios by 2. 2 Al : 3 S : 12 O
simplest formula	$Al_2S_3O_{12}$

END POINTS

1. A mole ratio of 1.00 A : 1.00 B : 3.33 C would imply a formula $A_3B_3C_{10}$. If the mole ratio were 1.00 A : 2.50 B : 5.50 C, the formula would be $A_2B_5C_{11}$. In general, multiply through by the smallest whole number that will give integers for all the subscripts.

2. Note that in this case, the mass of the sample was not used to find the simplest formula.

Sometimes you will be given the mass percents of the elements in a compound. If that is the case, one extra step is involved. **Assume a 100-g sample and calculate the mass of each element in that sample.**

Suppose, for example, you are told that the percentages of K, Cr, and O in a compound are 26.6%, 35.4%, and 38.0%, respectively. It follows that in a 100.0-g sample there are

26.6 g K, 35.4 g Cr, and 38.0 g O

Working with these masses, you can go through the same procedure followed in Example 3.5 to arrive at the same answer. (Try it!)

The most complex problem of this type requires you to determine the simplest formula of a compound given only the raw data obtained from its analysis. Here, an additional step is involved; you have to determine the masses of the elements from the masses of the new compounds that are obtained from the experiment.

Simple organic compounds such as hexane (containing C and H only) or ethyl alcohol (containing C, H, and O) can be analyzed using the apparatus shown in Figure 3.8. A weighed sample of the compound is burned in oxygen in a process called *combustion*. In combustion, the C in the compound is converted to CO_2 while H is converted to H_2O. The masses of CO_2 and H_2O can be determined by measuring the increase in mass of the absorbers. If oxygen is originally present, its mass is determined by difference.

mass O = mass of sample − (mass of C + mass of H)

To determine the mass of C (in CO_2) and H (in H_2O), recall that there is one atom of C (12.01 g/mol) in a molecule of CO_2 (44.01 g/mol) and two atoms of H (2 × 1.008 g/mol) in a molecule of H_2O (18.02 g/mol). Thus the following conversion factors apply:

$$\text{For C: } \frac{12.01 \text{ g C}}{44.01 \text{ g CO}_2} \qquad \text{For H: } \frac{2(1.008) \text{ g H}}{18.02 \text{ g H}_2O}$$

Figure 3.8 Combustion train used for carbon-hydrogen analysis. The absorbent for water is $Mg(ClO_4)_2$. Carbon dioxide is absorbed by finely divided NaOH. Alternatively, H_2O and CO_2 can be determined by gas phase chromatography.

O_2

Furnace

C_xH_y

H₂O absorber

H_2O

CO₂ absorber

CO_2

H₂O is absorbed by magnesium perchlorate, CO₂ passes through

CO₂ is absorbed by finely divided NaOH supported on asbestos

Sample containing hydrogen and carbon

EXAMPLE   3.6

The compound that gives vinegar its sour taste is acetic acid, which contains the elements carbon, hydrogen, and oxygen. When 5.00 g of acetic acid are burned in air, 7.33 g of CO_2 and 3.00 g of water are obtained. What is the simplest formula of acetic acid?

ANALYSIS

Information given:	elements in acetic acid (C, H, and O) mass of acetic acid (5.00 g) result of combustion analysis (7.33 g CO_2, 3.00 g H_2O)
Information implied:	molar masses (MM) of CO_2 and H_2O
Asked for:	simplest formula of acetic acid

STRATEGY

1. Find the mass of C and H by using the conversion factors:

$$\frac{12.01 \text{ g C}}{44.01 \text{ g } CO_2} \qquad \frac{2(1.008) \text{ g H}}{18.02 \text{ g } H_2O}$$

2. Find the mass of O by difference:

mass of sample = mass of C + mass of H + mass of O

3. Follow the schematic pathway shown in Figure 3.7.

SOLUTION

1. mass of C	$7.33 \text{ g } CO_2 \times \dfrac{12.01 \text{ g C}}{44.01 \text{ g } CO_2} = 2.00 \text{ g C}$
mass of H	$3.00 \text{ g } H_2O \times \dfrac{2(1.008) \text{ g H}}{18.02 \text{ g } H_2O} = 0.336 \text{ g H}$
2. mass of O	mass of O = mass of sample − (mass of C + mass of H) $= 5.00 \text{ g} - (2.00 \text{ g} + 0.336 \text{ g}) = 2.66 \text{ g}$
3. mol of each element	C: $\dfrac{2.00 \text{ g C}}{12.01 \text{ g/mol}} = 0.167$; H: $\dfrac{0.336 \text{ g H}}{1.008 \text{ g/mol}} = 0.333$; O: $\dfrac{2.66 \text{ g O}}{16.00 \text{ g/mol}} = 0.166$
ratios	C: $\dfrac{0.167 \text{ mol}}{0.166 \text{ mol}} = 1$; H: $\dfrac{0.333 \text{ mol}}{0.167 \text{ mol}} = 2$; O: $\dfrac{0.166 \text{ mol}}{0.166 \text{ mol}} = 1$
simplest formula	Using the ratios as subscripts, the simplest formula is CH_2O.

Molecular Formula from Simplest Formula

Chemical analysis always leads to the simplest formula of a compound because it gives only the simplest atom ratio of the elements. As pointed out earlier, the molecular formula is a whole-number multiple of the simplest formula. That multiple may be 1 as in H_2O, 2 as in H_2O_2, 3 as in C_3H_6, or some other integer. ◄ ⓘ To find the multiple, one more piece of data is needed: the molar mass.

ⓘ
Sometimes it's not so easy to convert the atom ratio to simplest formula (see Problem 42 at the end of the chapter).

EXAMPLE   3.7

Vitamin C is found in many fruits and vegetables, particularly in citrus fruit. Its simplest formula is $C_3H_4O_3$ and its molar mass (determined by a mass spectrometer) is 176 g/mol. What is the molecular formula of citric acid?

continued

ANALYSIS	
Information given:	simplest formula ($C_3H_4O_3$)
	MM_a—actual molar mass (176 g/mol)
Asked for:	molecular formula

STRATEGY	

1. Determine the molar mass of the simplest formula, MM_s.

2. Find the ratio

$$\frac{\text{actual molar mass}}{\text{simplest formula molar mass}} = \frac{MM_a}{MM_s}$$

3. To get the molecular formula, multiply all subscripts in the simplest formula by the ratio.

SOLUTION	
1. MM_s	$3(12.01)$ g C $+ 4(1.008)$ g H $+ 3(16.00)$ g O $= 88.06$ g $C_3H_4O_3$ g/mol
2. ratio	$\dfrac{MM_a}{MM_s} = \dfrac{176}{88.06} = 2$
3. molecular formula	$C_{3\times2}H_{4\times2}O_{3\times2}$ The molecular formula of Vitamin C is $C_6H_8O_6$.

3-3   Mass Relations in Reactions

A chemist who carries out a reaction in the laboratory needs to know how much **product** can be obtained from a given amount of starting material (**reactant**). To do this, he or she starts by writing a balanced **chemical equation**.

Writing and Balancing Chemical Equations

Chemical reactions are represented by chemical equations, which identify reactants and products. Formulas of reactants appear on the left side of the equation; those of products are written on the right. In a balanced chemical equation, there are the same number of atoms of a given element on both sides. The same situation holds for a chemical reaction that you carry out in the laboratory; atoms are conserved. For that reason, *any calculation involving a reaction must be based on the balanced equation for that reaction.*

Beginning students are sometimes led to believe that writing a chemical equation is a simple, mechanical process. Nothing could be further from the truth. One point that seems obvious is often overlooked. *You cannot write an equation unless you know what happens in the reaction that it represents.* All the reactants and all the products must be identified. Moreover, you must know their formulas and physical states.

To illustrate how a relatively simple equation can be written, consider liquid hydrazine (N_2H_4). It was used to fuel the rocket engines for the Mars rover Curiosity, which landed on Mars in August 2012. It is still there (Figure 3.9). Hydrazine can react with liquid dinitrogen tetroxide to produce gaseous nitrogen and water vapor. To write a balanced equation, proceed as follows:

1. *Write a "skeleton" equation in which the formulas of the reactants appear on the left and those of the products on the right.* In this case,

$$N_2H_4 + N_2O_4 \longrightarrow N_2 + H_2O$$

2. *Indicate the physical state of each reactant and product, after the formula, by writing*

(g)   for a gaseous substance
(l)   for a pure liquid
(s)   for a solid
(aq)   for an ion or molecule in water (aqueous) solution

Figure 3.9 Mars rover Curiosity.

Using the letters (s), (l), (g), (aq) lends a sense of physical reality to the equation.

In this case ◄ *i*

$$N_2H_4(l) + N_2O_4(l) \longrightarrow N_2(g) + H_2O(g)$$

3. Balance the equation. To accomplish this, start by writing a coefficient of 4 for H_2O, thus obtaining 4 oxygen atoms on both sides:

$$N_2H_4(l) + N_2O_4(l) \longrightarrow N_2(g) + 4H_2O(g)$$

Now consider the hydrogen atoms. There are $4 \times 2 = 8$ H atoms on the right. To obtain 8 H atoms on the left, write a coefficient of 2 for N_2H_4:

$$2N_2H_4(l) + N_2O_4(l) \longrightarrow N_2(g) + 4H_2O(g)$$

Finally, consider nitrogen. There are a total of $(2 \times 2) + 2 = 6$ nitrogen atoms on the left. To balance nitrogen, write a coefficient of 3 for N_2:

$$2N_2H_4(l) + N_2O_4(l) \longrightarrow 3N_2(g) + 4H_2O(g)$$

This is the final balanced equation for the reaction of hydrazine with dinitrogen tetraoxide.

Three points concerning the balancing process are worth noting.

1. Equations are balanced by adjusting coefficients in front of formulas, never by changing subscripts within formulas. On paper, the equation discussed above could have been balanced by writing N_6 on the right, but that would have been absurd. Elemental nitrogen exists as diatomic molecules, N_2; there is no such thing as an N_6 molecule. ◄ *i*

6N is no good either.

2. In balancing an equation, it is best to start with an element that appears in only one species on each side of the equation. In this case, either oxygen or hydrogen is a good starting point. Nitrogen would have been a poor choice, however, because there are nitrogen atoms in both reactant molecules, N_2H_4 and N_2O_4.

3. In principle, an infinite number of balanced equations can be written for any reaction. The equations

$$4N_2H_4(l) + 2N_2O_4(l) \longrightarrow 6N_2(g) + 8H_2O(g)$$
$$N_2H_4(l) + \tfrac{1}{2}N_2O_4(l) \longrightarrow \tfrac{3}{2}N_2(g) + 2H_2O(g)$$

are balanced in that there are the same number of atoms of each element on both sides. Ordinarily, the equation with the simplest whole-number coefficients

$$2N_2H_4(l) + N_2O_4(l) \longrightarrow 3N_2(g) + 4H_2O(g)$$

is preferred.

Frequently, when you are asked to balance an equation, the formulas of products and reactants are given. ◄ *i* Sometimes, though, you will have to derive the formulas, given only the names (Example 3.8).

It's time to review Table 2.2.

EXAMPLE 3.8

Crystals of sodium hydroxide (lye) (Figure 3.10) react with carbon dioxide from air to form a colorless liquid, water, and a white powder, sodium carbonate, which is commonly added to detergents as a softening agent. Write a balanced equation for this chemical reaction.

STRATEGY

1. Translate names to formulas and write a "skeleton" equation. Recall the discussion in Sections 2-6 and 2-7.

2. Write the physical states.

3. Balance by starting with an element that appears only once on each side of the equation.

4. Check that you have the same number of atoms of each element on both sides of the equation.

continued

	SOLUTION
1. Skeleton equation	$NaOH + CO_2 \rightarrow Na_2CO_3 + H_2O$
2. Physical states	$NaOH(s) + CO_2(g) \rightarrow Na_2CO_3(s) + H_2O(l)$
3. Balance	$2NaOH(s) + CO_2(g) \rightarrow Na_2CO_3(s) + H_2O(l)$
4. Check H	2 on left 2 on right
Check C	1 on left 1 on right
Check Na	2 on left 2 on right
Check O	$2 + 2 = 4$ on left $3 + 1 = 4$ on right
	The balanced equation is
	$2NaOH(s) + CO_2(g) \rightarrow Na_2CO_3(s) + H_2O(l)$

Mass Relations from Equations

The principal reason for writing balanced equations is to make it possible to relate the masses of reactants and products. Calculations of this sort are based on a very important principle:

The coefficients of a balanced equation represent numbers of moles of reactants and products.

To show that this statement is valid, recall the equation

$$2N_2H_4(l) + N_2O_4(l) \longrightarrow 3N_2(g) + 4H_2O(g)$$

The coefficients in this equation represent numbers of molecules, that is,

2 molecules N_2H_4 + 1 molecule $N_2O_4 \longrightarrow$ 3 molecules N_2 + 4 molecules H_2O

A balanced equation remains valid if each coefficient is multiplied by the same number, including Avogadro's number, N_A:

$2N_A$ molecules N_2H_4 + $1N_A$ molecule $N_2O_4 \longrightarrow 3N_A$ molecules N_2 + $4N_A$ molecules H_2O

Recall from Section 3-1 that a mole represents Avogadro's number of items, N_A. Thus the equation above can be written

$$2 \text{ mol } N_2H_4 + 1 \text{ mol } N_2O_4 \longrightarrow 3 \text{ mol } N_2 + 4 \text{ mol } H_2O$$

The quantities 2 mol of N_2H_4, 1 mol of N_2O_4, 3 mol of N_2, and 4 mol of H_2O are chemically equivalent to each other in this reaction. Hence they can be used in conversion factors such as

$$\frac{2 \text{ mol } N_2H_4}{3 \text{ mol } N_2}, \frac{3 \text{ mol } N_2}{1 \text{ mol } N_2O_4}, \text{ or a variety of other combinations}$$

These are also called *stoichiometric ratios*. Use the coefficients of the balanced equations for the ratios.

If you wanted to know how many moles of hydrazine were required to form 1.80 mol of elemental nitrogen, the conversion would be

$$n_{N_2H_4} = 1.80 \text{ mol } N_2 \times \frac{2 \text{ mol } N_2H_4}{3 \text{ mol } N_2} = 1.20 \text{ mol } N_2H_4$$

To find the number of moles of nitrogen produced from 2.60 mol of N_2O_4,

$$n_{N_2} = 2.60 \text{ mol } N_2O_4 \times \frac{3 \text{ mol } N_2}{1 \text{ mol } N_2O_4} = 7.80 \text{ mol } N_2$$

Figure 3.10 **Sodium hydroxide pellets.**

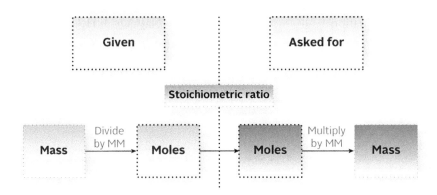

Figure 3.11 **Flowchart for mole-mass conversions.**

Simple mole relationships of this type are readily extended to relate moles of one substance to grams of another or grams of one substance to moles or molecules of another (Example 3.9). A schematic pathway that you can follow is shown in Figure 3.11.

EXAMPLE 3.9

Ammonia is used to make fertilizers (Figure 3.12) for lawns and gardens by reacting nitrogen gas with hydrogen gas. The balanced equation for the reaction is

$$N_2(g) + 3H_2(g) \longrightarrow 2NH_3(g)$$

STRATEGY

For all parts of this example, follow the schematic pathway shown in Figure 3.11.

a How many moles of ammonia are formed when 1.34 mol of nitrogen react?

ANALYSIS

Information given:	balanced equation: $[N_2(g) + 3H_2(g) \rightarrow 2NH_3(g)]$ moles N_2 (1.34)
Asked for:	mol NH_3 formed

SOLUTION

mol NH_3	mol $H_2 \rightarrow$ mol NH_3 $1.34 \text{ mol } H_2 \times \dfrac{2 \text{ mol } NH_3}{1 \text{ mol } N_2} = 2.68 \text{ mol } NH_3$

b How many grams of hydrogen are required to produce 2.75×10^3 g of ammonia?

ANALYSIS

Information given:	mass of ammonia (2.75×10^3 g) balanced equation: $[N_2(g) + 3H_2(g) \rightarrow 2NH_3(g)]$
Information implied:	molar masses of NH_3 and H_2
Asked for:	mass of H_2 needed

SOLUTION

mass H_2	mass $NH_3 \rightarrow$ mol $NH_3 \rightarrow$ mol $H_2 \rightarrow$ mass H_2 $2.75 \times 10^3 \text{ g } NH_3 \times \dfrac{1 \text{ mol } NH_3}{17.03 \text{ g } NH_3} \times \dfrac{3 \text{ mol } H_2}{2 \text{ mol } NH_3} \times \dfrac{2.016 \text{ g } H_2}{1 \text{ mol } H_2} = 488 \text{ g } H_2$

continued

c	How many molecules of ammonia are formed when 2.92 g of hydrogen react?	
	ANALYSIS	
Information given:	mass of H_2 (2.92 g) balanced equation: $[N_2(g) + 3H_2(g) \rightarrow 2NH_3(g)]$	
Information implied:	molar mass of H_2 Avogadro's number (N_A)	
Asked for:	molecules of NH_3 produced	
	SOLUTION	
molecules NH_3	mass $H_2 \rightarrow$ mol $H_2 \rightarrow$ mol $NH_3 \xrightarrow{N_A}$ molecules NH_3 $2.92 \text{ g } H_2 \times \dfrac{1 \text{ mol } H_2}{2.016 \text{ g } H_2} \times \dfrac{2 \text{ mol } NH_3}{3 \text{ mol } H_2} \times \dfrac{6.022 \times 10^{23} \text{ molecules}}{1 \text{ mol } NH_3} = 5.81 \times 10^{23}$	

d	How many grams of ammonia are produced when 15.0 L of air (79% by volume nitrogen) react with an excess of hydrogen? The density of nitrogen at the conditions of the reaction is 1.25 g/L.	
	ANALYSIS	
Information given:	balanced equation: $[N_2(g) + 3H_2(g) \rightarrow 2NH_3(g)]$ V_{air} (15.0 L); % by volume of N_2 in air (79%); density, d, of N_2 (1.25 g/L)	
Information implied:	molar masses of N_2 and NH_3	
Asked for:	mass of NH_3 produced	
	SOLUTION	
mass NH_3	$V_{air} \xrightarrow{\% \ N_2} V_{N_2} \xrightarrow{\text{density}}$ mass $N_2 \rightarrow$ mol $N_2 \rightarrow$ mol $NH_3 \rightarrow$ mass NH_3 $15.0 \text{ L air} \times \dfrac{79 \text{ L } N_2}{100 \text{ L air}} \times \dfrac{1.25 \text{ g } N_2}{1 \text{ L } N_2} \times \dfrac{1 \text{ mol } N_2}{28.02 \text{ g } N_2} \times \dfrac{2 \text{ mol } NH_3}{1 \text{ mol } N_2}$ $\times \dfrac{17.03 \text{ g } NH_3}{1 \text{ mol } NH_3} = 18 \text{ g } NH_3$	

Limiting Reactant and Theoretical Yield

When the two elements antimony and iodine are heated in contact with one another (Figure 3.13, p. 68), they react to form antimony(III) iodide.

$$2Sb(s) + 3I_2(s) \longrightarrow 2SbI_3(s)$$

The coefficients in this equation show that two moles of Sb (243.6 g) react with exactly three moles of I_2 (761.4 g) to form two moles of SbI_3 (1005.0 g). Put another way, the maximum quantity of SbI_3 that can be obtained under these conditions, assuming the reaction goes to completion and no product is lost, is 1005.0 g. This quantity is referred to as the **theoretical yield** of SbI_3.

Ordinarily, in the laboratory, reactants are not mixed in exactly the ratio required for reaction. Instead, an excess of one reactant, usually the cheaper one, is used. For example, 3.00 mol of Sb could be mixed with 3.00 mol of I_2. In that case, after the reaction is over, 1.00 mol of Sb remains unreacted.

$$\text{excess Sb} = 3.00 \text{ mol Sb originally} - 2.00 \text{ mol Sb consumed}$$
$$= 1.00 \text{ mol Sb}$$

The 3.00 mol of I_2 should be completely consumed in forming the 2.00 mol of SbI_3:

$$n_{SbI_3} \text{ formed} = 3.00 \text{ mol } I_2 \times \dfrac{2 \text{ mol } SbI_3}{3 \text{ mol } I_2} = 2.00 \text{ mol } SbI_3$$

Figure 3.12 Fertilizer. The label 5-10-5 on the bag means that the fertilizer has 5% nitrogen, 10% phosphorus, and 5% potassium.

© Cengage Learning/George Semple

The Reactants

1. When antimony (dark gray powder) comes in contact with iodine (violet vapor)...

The Reaction

2. ...a vigorous reaction takes place.

The Product

3. The product is SbI_3, a red solid.

© Cengage Learning/Charles D. Winters

Figure 3.13 **Reaction of antimony with iodine.**

After the reaction is over, the solid obtained would be a mixture of product, 2.00 mol of SbI_3, with 1.00 mol of unreacted Sb.

In situations such as this, a distinction is made between the *excess reactant* (Sb) and the **limiting reactant,** I_2. The amount of product formed is determined (limited) by the amount of limiting reactant. With 3.00 mol of I_2, only 2.00 mol of SbI_3 is obtained, regardless of how large an excess of Sb is used.

Under these conditions, the theoretical yield of product is the amount produced if the limiting reactant is completely consumed. In the case just cited, the theoretical yield of SbI_3 is 2.00 mol, the amount formed from the limiting reactant, I_2.

Often you will be given the amounts of two different reactants and asked to determine which is the limiting reactant, to calculate the theoretical yield of the product and to find how much of the excess reactant is unused. To do so, it helps to follow a systematic, four-step procedure.

1. *Calculate the amount of product that would be formed if the first reactant were completely consumed.*
2. *Repeat this calculation for the second reactant; that is, calculate how much product would be formed if all of that reactant were consumed.*
3. *Choose the smaller of the two amounts calculated in (1) and (2). This is the theoretical yield of product; the reactant that produces the smaller amount is the limiting reactant. The other reactant is in excess; only part of it is consumed.*
4. *Take the theoretical yield of the product and determine how much of the reactant in excess is used up in the reaction. Subtract that from the starting amount to find the amount left.*

To illustrate how this procedure works, suppose you want to make grilled cheese sandwiches ◀ 🛈 from 6 slices of cheese and 18 pieces of bread. The available cheese is enough for 6 grilled cheese sandwiches; the bread is enough for 9. Clearly, the cheese is the limiting reactant; there is an excess of bread. The theoretical yield is 6 sandwiches. Six grilled cheese sandwiches use up 12 slices of bread. Since there are 18 pieces available, 6 pieces of bread are left over.

🛈

Grilled cheese sandwiches are the favorite comfort food of WLM (William L. Masterton) and CNH.

EXAMPLE 3.10

Consider the reaction

$$2Sb(s) + 3I_2(s) \longrightarrow 2SbI_3(s)$$

a What is the limiting reactant and theoretical yield when 1.20 mol of Sb and 2.40 mol of I_2 are mixed?

ANALYSIS

Information given:	moles of each reactant: Sb (1.20), I_2 (2.40) balanced equation: $[2Sb(s) + 3I_2(s) \rightarrow 2SbI_3(s)]$
Asked for:	limiting reactant theoretical yield

STRATEGY

1. Find mol SbI_3 produced by first assuming Sb is limiting, and then assuming I_2 is limiting.

   mol Sb → mol SbI_3;   mol I_2 → mol SbI_3

2. The reactant that gives the smaller amount of SbI_3 is limiting, and the smaller amount of SbI_3 is the theoretical yield.

SOLUTION

1. mol SbI_3	$1.20 \text{ mol Sb} \times \dfrac{2 \text{ mol } SbI_3}{2 \text{ mol Sb}} = 1.20 \text{ mol}$ $2.40 \text{ mol } I_2 \times \dfrac{2 \text{ mol } SbI_3}{3 \text{ mol } I_2} = 1.60 \text{ mol}$
2. limiting reactant	1.20 mol (Sb limiting) < 1.60 mol (I_2 limiting) The limiting reactant is Sb.
theoretical yield	1.20 mol < 1.60 mol The theoretical yield is 1.20 mol SbI_3

b What is the limiting reactant and theoretical yield when 1.20 g of Sb and 2.40 g of I_2 are mixed? What mass of excess reactant is left when the reaction is complete?

ANALYSIS

Information given:	mass of each reactant: Sb (1.20 g), I_2 (2.40 g) balanced equation: $[2Sb(s) + 3I_2(s) \rightarrow 2SbI_3(s)]$
Information implied:	molar masses (MM) of SbI_3 and I_2
Asked for:	limiting reactant theoretical yield mass of excess reactant not used up

STRATEGY

1. Follow the plan outlined in Figure 3.11 and convert mass of Sb and mass of I_2 to mol SbI_3.

2. The smaller number of moles SbI_3 obtained is the theoretical yield. The reactant that yields the smaller amount is the limiting reactant.

3. Convert moles limiting reactant to mass of excess reactant. That is the mass of excess reactant consumed in the reaction.

4. Mass of excess reactant not used up = mass of excess reactant initally − mass excess reactant consumed

SOLUTION

1. mol SbI_3	$1.20 \text{ g Sb} \times \dfrac{1 \text{ mol Sb}}{121.8 \text{ g Sb}} \times \dfrac{2 \text{ mol } SbI_3}{2 \text{ mol Sb}} = 0.00985 \text{ mol } SbI_3$
	$2.40 \text{ g } I_2 \times \dfrac{1 \text{ mol } I_2}{253.8 \text{ g } I_2} \times \dfrac{2 \text{ mol } SbI_3}{3 \text{ mol } I_2} = 0.006304 \text{ mol } SbI_3$

continued

2. limiting reactant	0.006304 mol (I_2 limiting) < 0.00985 mol (Sb limiting); thus I_2 is the limiting reactant.
theoretical yield	0.006304 mol < 0.00985 mol The theoretical yield is 0.006304 mol (3.17 g) SbI_3.
	The reactant in excess is Sb.
3. mass Sb used up	$2.40 \text{ g } I_2 \times \dfrac{1 \text{ mol } I_2}{253.8 \text{ g } I_2} \times \dfrac{2 \text{ mol Sb}}{3 \text{ mol } I_2} \times \dfrac{121.8 \text{ g Sb}}{1 \text{ mol Sb}} = 0.768 \text{ g Sb}$
4. mass unreacted	mass unreacted = mass present initially − mass used up = 1.20 g − 0.768 g = 0.43 g

Figure 3.14 Percent yield. If you started with 20 popcorn kernels, but only 16 of them popped, the percent yield of popcorn from this "reaction" would be 16/20 × 100% = 80%.

Remember that in deciding on the theoretical yield of product, you *choose the smaller of the two calculated amounts*. To see why this must be the case, refer back to Example 3.10b. There, 1.20 g of Sb was mixed with 2.40 g of I_2. Calculations show that the theoretical yield of SbI_3 is 3.17 g, and 0.43 g of Sb is left over. Thus

$$1.20 \text{ g Sb} + 2.40 \text{ g } I_2 \longrightarrow 3.17 \text{ g } SbI_3 + 0.43 \text{ g Sb}$$

This makes sense: 3.60 g of reactants yield a total of 3.60 g of products, including the unreacted antimony. Suppose, however, that 4.95 g of SbI_3 were chosen as the theoretical yield. The following nonsensical situation would arise.

$$1.20 \text{ g Sb} + 2.40 \text{ g } I_2 \longrightarrow 4.95 \text{ g } SbI_3$$

This violates the law of conservation of mass; 3.60 g of reactants cannot form 4.95 g of product.

Experimental Yield; Percent Yield

The theoretical yield is the maximum amount of product that can be obtained (Figure 3.14). In calculating the theoretical yield, it is assumed that the limiting reactant is 100% converted to product. In the real world, that is unlikely to happen. Some of the limiting reactant may be consumed in competing reactions. Some of the product may be lost in separating it from the reaction mixture. For these and other reasons, the **experimental yield** is ordinarily less than the theoretical yield. Put another way, the **percent yield** is expected to be less than 100%:

$$\text{percent yield} = \frac{\text{experimental yield}}{\text{theoretical yield}} \times 100\% \tag{3.3}$$

i You never do quite as well as you hoped, so you never get 100%.

i Remember the 6 grilled cheese sandwiches on page 68? If your dog ate one of them while you weren't looking, your yield would be 83%.

EXAMPLE 3.11

Consider the reaction between magnesium and iodine to produce magnesium iodide. A student determines by calculation that with the amount of reagents he has on hand, he should produce 1.71 mol of MgI_2. After the reaction is complete, he tells his TA that he got a yield of 84.5%. How many grams of MgI_2 did he make?

ANALYSIS	
Information given:	theoretical yield (1.71 mol) percent yield (84.5%)
Asked for:	mass MgI_2 actually obtained

continued

1. Substitute into Equation 3.3.

$$\% \text{ yield} = \frac{\text{actual yield}}{\text{theoretical yield}} \times 100\%$$

2. Your answer will be the actual yield in moles. Convert to grams.

SOLUTION

1. actual yield

$$84.5\% = \frac{\text{actual yield}}{1.71 \text{ mol}} \times 100\%; \text{ actual yield} = 1.44 \text{ mol MgI}_2$$

2. mass MgI_2

$$1.44 \text{ mol MgI}_2 \times \frac{278.1 \text{ g MgI}_2}{1 \text{ mol MgI}_2} = 402 \text{ g}$$

END POINT

If your actual yield is larger than your theoretical yield, something's wrong!

CHEMISTRY BEYOND THE CLASSROOM

Hydrates

Ionic compounds often separate from water solution with molecules of water incorporated into the solid. Such compounds are referred to as **hydrates**. An example is hydrated copper sulfate, which contains five moles of H_2O for every mole of $CuSO_4$. Its formula is $CuSO_4 \cdot 5H_2O$; a dot is used to separate the formulas of the two compounds $CuSO_4$ and H_2O. A Greek prefix is used to show the number of moles of water; the systematic name of $CuSO_4 \cdot 5H_2O$ is copper(II) sulfate pentahydrate.

Certain hydrates, notably $Na_2CO_3 \cdot 10H_2O$ and $FeSO_4 \cdot 7H_2O$, lose all or part of their water of hydration when exposed to dry air. This process is referred to as **efflorescence**; the glassy (vitreous) hydrate crystals crumble to a powder. Frequently, dehydration is accompanied by a color change (Figure A). When $CoCl_2 \cdot 6H_2O$ is exposed to dry air or is heated, it loses water and changes color from red to purple or blue. Crystals of this compound are used as humidity indicators and as an ingredient of invisible ink. Writing becomes visible only when the paper is heated, driving off water and leaving a blue residue.

Molecular as well as ionic substances can form hydrates, but of an entirely different nature. In these crystals, sometimes referred to as *clathrates*, a molecule (such as CH_4, $CHCl_3$) is quite literally trapped in an ice-like cage of water molecules. Perhaps the best-known molecular hydrate is that of chlorine, which has the approximate composition $Cl_2 \cdot 7.3H_2O$. This compound was discovered by the great English physicist and electrochemist Michael Faraday in 1823. You can make it by bubbling chlorine gas through calcium chloride solution at 0°C; the hydrate comes down as feathery white crystals. In the winter of 1914, the German army used chlorine in chemical warfare on the Russian front against the soldiers of the Tsar. They were puzzled by its ineffectiveness; not until spring was deadly chlorine gas liberated from the hydrate, which is stable at cold temperatures.

In the 1930s when high-pressure natural gas (95% methane) pipelines were being built in the United States, it was found that the lines often became plugged in cold weather by a white,

Marna G. Clarke

Figure A Hydrates of copper(II) sulfate and cobalt(II) chloride. $CuSO_4 \cdot 5H_2O$ (*upper left*) is blue; the anhydrous solid—that is, copper sulfate without the water—is white (*lower left*). $CoCl_2 \cdot 5H_2O$ (*upper right*) is reddish pink. Lower hydrates of cobalt such as $CoCl_2 \cdot 4H_2O$ are purple (*lower right*) or blue.

Figure B In one of the cages within which gas molecules are trapped in methane hydrate, water molecules form a pentagonal dodecahedron, a three-dimensional figure in which each of the 12 sides is a regular pentagon.

waxy solid that contained both water and methane (CH_4) molecules. Twenty years later, Walter Claussen at the University of Illinois deduced the structure of that solid, a hydrate of methane. Notice (Figure B) that CH_4 molecules are trapped within a three-dimensional cage of H_2O molecules.

In 1970 a huge deposit of methane hydrate was discovered at the bottom of the Atlantic Ocean, 330 km off the coast of North Carolina. The white solid was stable at the high pressure and low temperature (slightly above 0°C) that prevail at the ocean floor. When raised to the surface, the solid decomposed to give off copious amounts of methane gas (Figure C).

We now know that methane hydrate is widely distributed through the earth's oceans and the permafrost of Alaska and Siberia. The methane in these deposits could, upon combustion, produce twice as much energy as all the world's known resources of petroleum, natural gas, and coal. However, extracting the methane from undersea deposits has proved to be an engineering nightmare. Separating the hydrate from the mud, silt, and rocks with which it is mixed is almost as difficult as controlling its decomposition. Then there is the problem of transporting the methane to the shore, which may be 100 miles or more away.

Figure C Methane hydrate is sometimes referred to as "the ice that burns."

John Pinkston and Laura Stern/Science News/U.S. Geological Survey

Key Concepts

1. Use molar mass to relate
   - moles to mass of a substance
   (**Example 3.1; Problems 1–14**)
   - moles in solution; molarity
   (**Examples 3.2, 3.3; Problems 15–24**)
   - molecular formula to simplest formula
   (**Example 3.7; Problems 43, 44**)
2. Use the formula of a compound to find percent composition or its equivalent.
   (**Example 3.4; Problems 25–34**)
3. Find the simplest formula of a compound by chemical analysis.
   (**Examples 3.5, 3.6; Problems 35–46**)
4. Balance chemical equations by inspection.
   (**Example 3.8; Problems 47–52**)
5. Use a balanced equation to
   - relate masses of products and reactants
   (**Example 3.9; Problems 53–64**)
   - find the limiting reactant, theoretical yield, and percent yield
   (**Examples 3.10, 3.11; Problems 65–72**)

Key Equations

Molar mass $\text{mass} = \text{MM} \times n$

$$\text{Molarity } (M) = \frac{\text{moles of solute}}{\text{liters of solution}}$$

$$\text{Percent yield} = \frac{\text{experimental yield}}{\text{theoretical yield}} \times 100\%$$

Key Terms

limiting reactant	mole	simplest formula
molar mass	percent composition	theoretical yield
molarity	percent yield	

Summary Problem

Consider oxygen gas, which is essential to our survival.

(a) How many moles are there in 19.00 g of oxygen gas?

(b) What is the mass in grams of 0.138 moles of oxygen gas?

(c) When iron is burned in oxygen (a process called "roasting"), iron(III) is formed. Write a balanced equation for this reaction.

(d) What is the molarity of a 485-mL solution made up of 0.384 g of iron(III) oxide and water?

(e) How many grams of oxygen will react with 17.28 g of iron?

(f) If 15.00 g of iron react with an excess of oxygen, how many grams of product are formed?

(g) How many grams of oxygen are required to react with an excess of iron to produce enough iron(III) oxide to make a 300.0 mL solution 0.0744 M?

(h) If 28.00 g of iron react with 28.00 g of oxygen, how many grams of product are formed? How much (in grams) of the reactant in excess is left?

(i) When 10.00 g of oxygen react with iron, the amount of iron(III) oxide actually obtained is 28.14 g. What is the percent yield?

(j) When oxygen is bubbled through a solution of sodium polysulfide, photographic hypo is produced. If "hypo" is made up of sodium, sulfur, and oxygen and contains 29.08% Na and 40.57% S, what is its simplest formula?

Answers

(a) 0.5938 mol
(b) 4.42 g
(c) $4\ Fe(s) + 3O_2(g) \rightarrow 2Fe_2O_3(s)$
(d) 0.00496 M
(e) 7.426 g
(f) 21.45 g
(g) 1.07 g
(h) 40.04 g Fe_2O_3; 15.97 g O_2 left
(i) 84.58%
(j) $Na_2S_2O_3$

Questions and Problems

Even-numbered questions and Challenge Problems have answers in Appendix 5 and fully worked solutions in the *Student Solutions Manual*.

3-1 The Mole

The Mole Molar Mass, and Mole-Gram Conversions

1. One chocolate chip used in making chocolate chip cookies has a mass of 0.324 g.
(a) How many chocolate chips are there in one mole of chocolate chips?
(b) If a cookie needs 15 chocolate chips, how many cookies can one make with a billionth (1×10^{-9}) of a mole of chocolate chips? (A billionth of a mole is scientifically known as a *nanomole*.)

2. The meat from one hazelnut has a mass of 0.985 g.
(a) What is the mass of a millionth of a mole (10^{-6}) of hazelnut meats? (A millionth of a mole is also called a *micromole*.)
(b) How many moles are in a pound of hazelnut meats?

3. Determine
(a) the mass of 0.429 mol of gold.
(b) the number of atoms in 0.715 g of gold.
(c) the number of moles of electrons in 0.336 g of gold.

4. How many electrons are in
(a) an ion of Sc^{3+}?
(b) a mol of Sc^{3+}?
(c) a gram of Sc^{3+}?

5. A cube of sodium has length 1.25 in. How many atoms are in that cube? (Note: $d_{Na} = 0.968$ g/cm^3.)

6. A solid circular cone made of pure platinum ($d = 21.45$ g/cm^3) has a diameter of 2.75 cm and a height of 3.00 inches.
(a) How many moles of platinum are there in the cone?
(b) How many electrons are there in the cone?

7. Calculate the molar masses (in grams per mole) of
(a) cane sugar, $C_{12}H_{22}O_{11}$.
(b) laughing gas, N_2O.
(c) vitamin A, $C_{20}H_{30}O$.

8. Calculate the molar mass (in grams/mol) of
(a) osmium metal, the densest naturally occurring element.
(b) baking soda, $NaHCO_3$.
(c) vitamin D, $C_{28}H_{44}O$, required for healthy bones and teeth.

9. Convert the following to moles.
(a) 4.00×10^3 g of hydrazine, a rocket propellant
(b) 12.5 g of tin(II) fluoride, the active ingredient in fluoride toothpaste
(c) 13 g of caffeine, $C_4H_5N_2O$

10. Convert to moles.
(a) 128.3 g of sucralose, $C_{12}H_{19}O_8Cl_3$, the active ingredient of the artificial sweetener Splenda™
(b) 0.3066 g of uric acid, $C_5H_4N_4O_3$, the compound that can cause gout and arthritis
(c) 2.664 g of cadmium(II) telluride used to coat solar panels

11. Calculate the mass in grams of 3.839 moles of
    (a) hydrazine, a rocket propellant.
    (b) caffeine, $C_4H_5N_2O$.
    (c) theobromine, $C_7H_8N_4O_2$, the alkaloid present in chocolate and cocoa.
12. Calculate the mass in grams of 1.35 mol of
    (a) titanium white, TiO_2, used as a paint pigment.
    (b) sucralose, $C_{12}H_{19}O_8Cl_3$, the active ingredient in the artificial sweetener, Splenda™.
    (c) strychnine, $C_{21}H_{22}N_2O_2$, present in rat poison.
13. Complete the following table for TNT (trinitrotoluene), $C_7H_5(NO_2)_3$.

	Number of Grams	Number of Moles	Number of Molecules	Number of N Atoms
(a)	127.2			
(b)		0.9254		
(c)			1.24×10^{28}	
(d)				7.5×10^{22}

14. Complete the following table for citric acid, $C_6H_8O_7$, the acid found in many citrus fruits.

	Number of Grams	Number of Moles	Number of Molecules	Number of O Atoms
(a)	0.1364			
(b)		1.248		
(c)			4.32×10^{22}	
(d)				5.55×10^{19}

Moles in Solution

15. Household ammonia used for cleaning contains about 10 g (two significant figures) of NH_3 in 100 mL (two significant figures) of solution. What is the molarity of the NH_3 in solution?
16. Household bleach contains about 5.2 g of sodium hypochlorite in 100 g (two significant figures) of water. What is the molarity of sodium hypochlorite in bleach? (Assume no volume change.)
17. What is the molarity of each ion present in aqueous solutions of the following compounds prepared by dissolving 20.00 g of each compound in water to make 4.50 L of solution?
    (a) cobalt(III) chloride
    (b) nickel(III) sulfate
    (c) sodium permanganate
    (d) iron(II) bromide
18. What is the molarity of each ion present in aqueous solutions of the following compounds prepared by dissolving 28.0 g of each compound in water to make 785 mL of solution?
    (a) potassium oxide
    (b) sodium hydrogen carbonate
    (c) scandium(III) iodite
    (d) magnesium phosphate

19. How would you prepare from the solid and pure water
    (a) 0.400 L of 0.155 M $Sr(OH)_2$?
    (b) 1.75 L of 0.333 M $(NH_4)_2CO_3$?
20. Starting with the solid and adding water, how would you prepare 2.00 L of 0.685 M
    (a) $Ni(NO_3)_2$?   (b) $CuCl_2$?   (c) $C_6H_8O_6$ (vitamin C)?
21. You are asked to prepare a 0.8500 M solution of aluminum nitrate. You find that you have only 50.00 g of the solid.
    (a) What is the maximum volume of solution that you can prepare?
    (b) How many milliliters of this prepared solution are required to furnish 0.5000 mol of aluminum nitrate to a reaction?
    (c) If 2.500 L of the prepared solution are required, how much more aluminum nitrate would you need?
    (d) Fifty milliliters of a 0.450 M solution of aluminum nitrate are needed. How would you prepare the required solution from the solution prepared in (a)?
22. An experiment calls for a 0.4500 M solution of aluminum carbonate. The storeroom has only 32.00 g of the solid.
    (a) What is the maximum volume of solution that can be prepared?
    (b) The reaction in the experiment requires 0.1450 mol of carbonate ion. How many milliliters of this solution are required to provide 0.1450 mol?
23. A student combines two solutions of KOH and determines the molarity of the resulting solution. He records the following data:

Solution I:	30.00 mL of 0.125 M KOH
Solution II:	40.00 mL of KOH
Solution I + Solution II:	70.00 mL of 0.203 M KOH

What is the molarity of KOH in Solution II?
24. Twenty-five mL of a 0.388 M solution of Na_2SO_4 is mixed with 35.3 mL of 0.229 M Na_2SO_4. What is the molarity of the resulting solution? Assume that the volumes are additive.

3-2 Mass Relations in Chemical Formulas

25. Turquoise has the following chemical formula: $CuAl_6(PO_4)_4(OH)_8 \cdot 4H_2O$. Calculate the mass percent of each element in turquoise.
26. Diazepam is the addictive tranquilizer also known as Valium®. Its simplest formula is $C_{16}H_{13}N_2OCl$. Calculate the mass percent of each element in this compound.
27. Small amounts of tungsten (W) are usually added to steel to strengthen and harden the steel. Two ores of tungsten are ferberite ($FeWO_4$) and scheelite ($CaWO_4$). How many grams of $FeWO_4$ would contain the same mass of tungsten that is present in 725 g of $CaWO_4$?
28. Allicin is responsible for the distinctive taste and odor of garlic. Its simple formula is $C_6H_{10}O_2S$. How many grams of sulfur can be obtained from 25.0 g of allicin?
29. The active ingredient in Pepto-Bismol® (an over-the-counter remedy for an upset stomach) is bismuth subsalicylate, $C_7H_5BiO_4$. Analysis of a 1.7500-g sample of Pepto-Bismol yields 346 mg of bismuth. What percent by mass is bismuth subsalicylate in the sample? (Assume that there are no other bismuth-containing compounds in Pepto-Bismol.)

30. The active ingredient in some antiperspirants is aluminum chlorohydrate, $Al_2(OH)_5Cl$. Analysis of a 2.000-g sample of antiperspirant yields 0.334 g of aluminum. What percent (by mass) of aluminum chlorohydrate is present in the antiperspirant? (Assume that there are no other compounds containing aluminum in the antiperspirant.)

31. Combustion analysis of 1.00 g of the male sex hormone, testosterone, yields 2.90 g of CO_2 and 0.875 g H_2O. What are the mass percents of carbon, hydrogen, and oxygen in testosterone?

32. Hexachlorophene, a compound made up of atoms of carbon, hydrogen, chlorine, and oxygen, is an ingredient in germicidal soaps. Combustion of a 1.000-g sample yields 1.407 g of carbon dioxide, 0.134 g of water, and 0.523 g of chlorine gas. What are the mass percents of carbon, hydrogen, oxygen, and chlorine in hexachlorophene?

33. A compound NiX_3 is 19.67% (by mass) nickel. What is the molar mass of the compound? What is the name of the compound?

34. A compound R_2O_3 is 32.0% oxygen. What is the molar mass of R_2O_3? What is the element represented by R?

35. Manganese reacts with fluorine to form a fluoride. If 1.873 g of manganese reacts with enough fluorine to form 3.813 g of manganese fluoride, what is the simplest formula of the fluoride? Name the fluoride.

36. Nickel reacts with sulfur to form a sulfide. If 2.986 g of nickel reacts with enough sulfur to form 5.433 g of nickel sulfide, what is the simplest formula of the sulfide? Name the sulfide.

37. Determine the simplest formulas of the following compounds:

(a) the food enhancer monosodium glutamate (MSG), which has the composition 35.51% C, 4.77% H, 37.85% O, 8.29% N, and 13.60% Na

(b) zircon, a diamond-like mineral, which has the composition 34.91% O, 15.32% Si, and 49.76% Zr

(c) nicotine, which has the composition 74.0% C, 8.65% H, and 17.4% N

38. Determine the simplest formulas of the following compounds:

(a) tetraethyl lead, the banned gasoline anti-knock additive, which is composed of 29.71% C, 6.234% H, and 64.07% Pb

(b) citric acid, present in most sour fruit, which is composed of 37.51% C, 4.20% H, and 58.29% O

(c) cisplatin, a drug used in chemotherapy, which is composed of 9.34% N, 2.02% H, 23.36% Cl, and 65.50% Pt

39. Nicotine is found in tobacco leaf and is mainly responsible for the addictive property of cigarette smoking. Nicotine is made up of carbon, hydrogen, and nitrogen atoms. When a 2.500-g sample of nicotine is burned, 6.782 g of CO_2 and 1.944 g of H_2O are obtained. What is the simplest formula for nicotine?

40. Methyl salicylate is a common "active ingredient" in liniments such as Ben-Gay™. It is also known as oil of wintergreen. It is made up of carbon, hydrogen, and oxygen atoms. When a sample of methyl salicylate weighing 5.287 g is burned in excess oxygen, 12.24 g of carbon dioxide and 2.505 g of water are formed. What is the simplest formula for oil of wintergreen?

41. Beta-blockers are a class of drug widely used to manage hypertension. Atenolol, a beta-blocker, is made up of carbon, hydrogen, oxygen, and nitrogen atoms. When a 5.000-g sample is burned in oxygen, 11.57 g of CO_2 and 3.721 g of H_2O are obtained. A separate experiment using the same mass of sample (5.000 g) shows that atenolol has 10.52% nitrogen. What is the simplest formula of atenolol?

42. Saccharin is the active ingredient in many sweeteners used today. It is made up of carbon, hydrogen, oxygen, sulfur, and nitrogen. When 7.500 g of saccharin are burned in oxygen, 12.6 g CO_2, 1.84 g H_2O, and 2.62 g SO_2 are obtained. Another experiment using the same mass of sample (7.500 g) shows that saccharin has 7.65% N. What is the simplest formula for saccharin?

43. Hexamethylenediamine (MM = 116.2 g/mol), a compound made up of carbon, hydrogen, and nitrogen atoms, is used in the production of nylon. When 6.315 g of hexamethylenediamine is burned in oxygen, 14.36 g of carbon dioxide and 7.832 g of water are obtained. What are the simplest and molecular formulas of this compound?

44. Dimethylhydrazine, the fuel used in the Apollo lunar descent module, has a molar mass of 60.10 g/mol. It is made up of carbon, hydrogen, and nitrogen atoms. The combustion of 2.859 g of the fuel in excess oxygen yields 4.190 g of carbon dioxide and 3.428 g of water. What are the simplest and molecular formulas for dimethylhydrazine?

45. Epsom salts are hydrated crystals of magnesium sulfate, $MgSO_4 \cdot xH_2O$. When 6.499 g of Epsom salts are dehydrated by heating, 3.173 g remain. What is the mass percent of water in the hydrated sample? What is x?

46. Sodium borate decahydrate, $Na_2B_4O_7 \cdot 10H_2O$ is commonly known as borax. It is used as a deodorizer and mold inhibitor. A sample weighing 15.86 g is heated until a constant mass is obtained indicating that all the water has been evaporated off.

(a) What percent, by mass of $Na_2B_4O_7 \cdot 10 \ H_2O$ is water?

(b) What is the mass of the anhydrous sodium borate, $Na_2B_4O_7$?

3-3 Mass Relations in Reactions

Balancing Equations

47. Balance the following equations:

(a) $CaC_2(s) + H_2O(l) \longrightarrow Ca(OH)_2(s) + C_2H_2(g)$

(b) $(NH_4)_2Cr_2O_7(s) \longrightarrow Cr_2O_3(s) + N_2(g) + H_2O(g)$

(c) $CH_3NH_2(g) + O_2(g) \longrightarrow CO_2(g) + N_2(g) + H_2O(g)$

48. Balance the following equations:

(a) $C_6H_{12}O_6(s) + O_2(g) \longrightarrow CO_2(g) + H_2O(l)$

(b) $XeF_4(g) + H_2O(l) \longrightarrow Xe(g) + O_2(g) + HF(g)$

(c) $NaCl(s) + H_2O(g) + SO_2(g) + O_2(g) \longrightarrow$
$Na_2SO_4(s) + HCl(g)$

49. Write balanced equations for the reaction of sulfur with the following metals to form solids that you can take to be ionic when the anion is S^{2-}.

(a) potassium   (b) magnesium   (c) aluminum

(d) calcium   (e) iron (forming Fe^{2+} ions)

50. Write balanced equations for the reaction of scandium metal to produce the scandium(III) salt with the following nonmetals:
(a) sulfur   (b) chlorine
(c) nitrogen   (d) oxygen (forming the oxide)

51. Write a balanced equation for
(a) the combustion (reaction with oxygen gas) of glucose, $C_6H_{12}O_6$, to give carbon dioxide and water.
(b) the reaction between xenon tetrafluoride gas and water to give xenon, oxygen, and hydrogen fluoride gases.
(c) the reaction between aluminum and iron(III) oxide to give aluminum oxide and iron.
(d) the formation of ammonia gas from its elements.
(e) the reaction between sodium chloride, sulfur dioxide gas, steam, and oxygen to give sodium sulfate and hydrogen chloride gas.

52. Write a balanced equation for the reaction between
(a) dihydrogen sulfide and sulfur dioxide gases to form sulfur solid and steam.
(b) methane, ammonia, and oxygen gases to form hydrogen cyanide gas and steam.
(c) iron(III) oxide and hydrogen gas to form molten iron and steam.
(d) uranium(IV) oxide and hydrogen fluoride gas to form uranium(IV) fluoride and steam.
(e) the combustion of ethyl alcohol (C_2H_5OH) to give carbon dioxide and water.

Mole-Mass Relations in Reactions

53. Cyanogen gas, C_2N_2, has been found in the gases of outer space. It can react with fluorine to form carbon tetrafluoride and nitrogen trifluoride.

$$C_2N_2(g) + 7F_2(g) \longrightarrow 2CF_4(g) + 2NF_3(g)$$

(a) How many moles of fluorine react with 1.37 mol of cyanogen?
(b) How many moles of CF_4 are obtained from 13.75 mol of fluorine?
(c) How many moles of cyanogen are required to produce 0.8974 mol of NF_3?
(d) How many moles of fluorine will yield 4.981 mol of nitrogen trifluoride?

54. Ammonia reacts with a limited amount of oxygen according to the equation

$$4\,NH_3(g) + 5\,O_2(g) \longrightarrow 4\,NO(g) + 6H_2O(l)$$

(a) How many moles of NO are obtained when 3.914 moles of oxygen are used?
(b) How many moles of oxygen are required to react with 2.611 moles of ammonia?
(c) How many moles of water are obtained when 0.8144 moles of ammonia are used?
(d) How many moles of oxygen are required to produce 0.2179 mol of water?

55. One way to remove nitrogen oxide (NO) from smokestack emissions is to react it with ammonia.

$$4NH_3(g) + 6NO(g) \longrightarrow 5N_2(g) + 6H_2O(l)$$

Calculate
(a) the mass of water produced from 0.839 mol of ammonia.

(b) the mass of NO required to react with 3.402 mol of ammonia.
(c) the mass of ammonia required to produce 12.0 g of nitrogen gas.
(d) the mass of ammonia required to react with 115 g of NO.

56. Phosphine gas reacts with oxygen according to the following equation:

$$4PH_3(g) + 8\,O_2(g) \longrightarrow P_4O_{10}(s) + 6H_2O(g)$$

Calculate
(a) the mass of tetraphosphorus decaoxide produced from 12.43 mol of phosphine.
(b) the mass of PH_3 required to form 0.739 mol of steam.
(c) the mass of oxygen gas that yields 1.000 g of steam.
(d) the mass of oxygen required to react with 20.50 g of phosphine.

57. The combustion of liquid chloroethylene, C_2H_3Cl, yields carbon dioxide, steam, and hydrogen chloride gas.
(a) Write a balanced equation for the reaction.
(b) How many moles of oxygen are required to react with 35.00 g of chloroethylene?
(c) If 25.00 g of chloroethylene react with an excess of oxygen, how many grams of each product are formed?

58. Sand is mainly silicon dioxide. When sand is heated with an excess of coke (carbon), pure silicon and carbon monoxide are produced.
(a) Write a balanced equation for the reaction.
(b) How many moles of silicon dioxide are required to form 20.00 g of silicon?
(c) How many grams of carbon monoxide are formed when 98.76 g of silicon are produced?

59. When copper(II) oxide is heated in hydrogen gas, the following reaction takes place.

$$CuO(s) + H_2(g) \longrightarrow Cu(s) + H_2O(g)$$

A copper rod coated with copper(II) oxide has a mass of 38.72 g. The rod is heated and made to react with 5.67 L of hydrogen gas, whose density at the conditions of the experiment is 0.0519 g/L.
(a) How many grams of CuO were converted to Cu?
(b) What is the mass of the copper after all the hydrogen is consumed? (Assume that CuO is converted only to Cu.)

60. When tin comes in contact with the oxygen in the air, tin(IV) oxide, SnO_2, is formed.

$$Sn(s) + O_2(g) \longrightarrow SnO_2(s)$$

A piece of tin foil, 8.25 cm $\times$ 21.5 cm $\times$ 0.600 mm ($d = 7.28$ g/cm^3), is exposed to oxygen.
(a) Assuming that all the tin has reacted, what is the mass of the oxidized tin foil?
(b) Air is about 21% oxygen by volume ($d = 1.309$ g/L at 25°C, 1 atm). How many liters of air are required to completely react with the tin foil?

61. A crude oil burned in electrical generating plants contains about 1.2% sulfur by mass. When the oil burns, the sulfur forms sulfur dioxide gas:

$$S(s) + O_2(g) \longrightarrow SO_2(g)$$

How many liters of SO_2 ($d = 2.60$ g/L) are produced when 1.00×10^4 kg of oil burns at the same temperature and pressure?

62. When corn is allowed to ferment, the fructose in the corn is converted to ethyl alcohol according to the following reaction

$$C_6H_{12}O_6(aq) \longrightarrow 2C_2H_5OH(l) + 2CO_2(g)$$

(a) What volume of ethyl alcohol ($d = 0.789$ g/mL) is produced from one pound of fructose?

(b) Gasohol can be a mixture of 10 mL ethyl alcohol and 90 mL of gasoline. How many grams of fructose are required to produce the ethyl alcohol in one gallon of gasohol?

63. Consider the hypothetical reaction

$$8A_2B_3(s) + 3X_4(g) \longrightarrow 4A_4X_3(s) + 12B_2(g)$$

When 10.0 g of A_2B_3 (MM = 255 g/mol) react with an excess of X_4, 4.00 g of A_4X_3 are produced.

(a) How many moles of A_4X_3 are produced?

(b) What is the molar mass of A_4X_3?

64. Consider the hypothetical reaction

$$2R_2X_5 + 2\,Z_8 \longrightarrow 5X_2 + 4RZ_4$$

When 25.00 g of Z_8 (MM = 197.4 g/mol) react with an excess of R_2X_5, 21.72 g of X_2 are produced.

(a) How many moles of X_2 are produced?

(b) What is the molar mass of X_2?

65. When solid phosphorus (P_4) reacts with oxygen gas, diphosphorus pentoxide is formed. Initially 2.87 mol of phosphorus and 3.86 mol of oxygen are combined. (Assume 100% yield.)

(a) Write a balanced equation for the reaction.

(b) What is the limiting reactant?

(c) How many moles of excess reactant remain after the reaction is complete?

66. Chlorine and fluorine react to form gaseous chlorine trifluoride. Initially, 1.75 mol of chlorine and 3.68 mol of fluorine are combined. (Assume 100% yield for the reaction.)

(a) Write a balanced equation for the reaction.

(b) What is the limiting reactant?

(c) What is the theoretical yield of chlorine trifluoride in moles?

(d) How many moles of excess reactant remain after reaction is complete.

67. When potassium chlorate is subjected to high temperatures, it decomposes into potassium chloride and oxygen.

(a) Write a balanced equation for the decomposition.

(b) In this decomposition, the actual yield is 83.2%. If 198.5 g of oxygen are produced, how much potassium chlorate decomposed?

68. When iron and steam react at high temperatures, the following reaction takes place.

$$3Fe(s) + 4H_2O(g) \longrightarrow Fe_3O_4(s) + 4H_2(g)$$

How much iron must react with excess steam to form 897 g of Fe_3O_4 if the reaction yield is 69%?

69. When solid silicon tetrachloride reacts with water, solid silicon dioxide and hydrogen chloride gas are formed.

(a) Write a balanced equation for the reaction.

(b) In an experiment, 45.00 g of silicon tetrachloride are treated with 45.00 mL of water ($d = 1.00$ g/mL). What is the theoretical yield of HCl (in grams)?

(c) When the reaction is complete, 17.8 L of HCl gas ($d = 1.49$ g/L at the conditions of the experiment) are obtained. What is the percent yield?

(d) How much of the reactant in excess is unused?

70. The first step in the manufacture of nitric acid by the Ostwald process is the reaction of ammonia gas with oxygen, producing nitrogen oxide and steam. The reaction mixture contains 7.60 g of ammonia and 10.00 g of oxygen. After the reaction is complete, 6.22 g of nitrogen oxide are obtained.

(a) Write a balanced equation for the reaction.

(b) How many grams of nitrogen oxide can be theoretically obtained?

(c) How many grams of excess reactant are theoretically unused?

(d) What is the percent yield of the reaction?

71. Tin(IV) chloride is used as an external coating on glass to toughen glass containers. It is prepared by reacting tin with chlorine gas.

$$Sn\,(s) + 2Cl_2(g) \longrightarrow SnCl_4\,(l)$$

The process requires a 25.0% excess of tin and has a yield of 73.7%. Calculate the mass of tin and the volume of chlorine gas ($d = 2.898$ g/L at 25°C, 1 atm) needed to produce 0.500 L of $SnCl_4$ ($d = 2.226$ g/mL).

72. A student prepares phosphorous acid, H_3PO_3, by reacting solid phosphorus triiodide with water.

$$PI_3(s) + 3H_2O(l) \longrightarrow H_3PO_3(s) + 3HI(g)$$

The student needs to obtain 0.250 L of H_3PO_3 ($d = 1.651$ g/cm^3). The procedure calls for a 45.0% excess of water and a yield of 75.0%. How much phosphorus triiodide should be weighed out? What volume of water ($d = 1.00$ g/cm^3) should be used?

Unclassified

73. Cisplatin, $Pt(NH_3)_2Cl_2$, is a chemotherapeutic agent that disrupts the growth of DNA. If the current cost of Pt is $1118.0/troy ounce (1 troy oz = 31.10 g), how many grams of cisplatin can you make with three thousand dollars worth of platinum? How many pounds?

74. Magnesium ribbon reacts with acid to produce hydrogen gas and magnesium ions. Different masses of magnesium ribbon are added to 10 mL of the acid. The volume of the hydrogen gas obtained is a measure of the number of moles of hydrogen produced by the reaction. Various measurements are given in the table below.

Experiment	Mass of Mg Ribbon (g)	Volume of Acid Used (mL)	Volume of H₂ Gas (mL)
1	0.020	10.0	21
2	0.040	10.0	42
3	0.080	10.0	82
4	0.120	10.0	122
5	0.160	10.0	122
6	0.200	10.0	122

(a) Draw a graph of the results by plotting the mass of Mg versus the volume of the hydrogen gas.
(b) What is the limiting reactant in experiment 1?
(c) What is the limiting reactant in experiment 3?
(d) What is the limiting reactant in experiment 6?
(e) Which experiment uses stoichiometric amounts of each reactant?
(f) What volume of gas would be obtained if 0.300 g of Mg ribbon were used? If 0.010 g were used?

75. Iron reacts with oxygen. Different masses of iron are burned in a constant amount of oxygen. The product, an oxide of iron, is weighed. The graph below is obtained when the mass of product obtained is plotted against the mass of iron used.

(a) How many grams of product are obtained when 0.50 g of iron are used?
(b) What is the limiting reactant when 2.00 g of iron are used?
(c) What is the limiting reactant when 5.00 g of iron are used?
(d) How many grams of iron react exactly with the amount of oxygen supplied?
(e) What is the simplest formula of the product?

76. Acetic acid ($HC_2H_3O_2$) can be prepared by the action of the *acetobacter* organism on dilute solutions of ethanol (C_2H_5OH). The equation for the reaction is

$$C_2H_5OH\,(aq) + O_2(g) \longrightarrow HC_2H_3O_2\,(aq) + H_2O$$

How many milliliters of a 12.5% (by volume) solution of ethanol are required to produce 175 mL of 0.664 *M* acetic acid? (Density of pure ethanol = 0.789 g/mL.)

Conceptual Problems

77. Given a pair of elements and their mass relation, answer the following questions.
(a) The mass of 4 atoms of A = the mass of 6 atoms of B. Which element has the smaller molar mass?
(b) The mass of 6 atoms of C is less than the mass of 3 atoms of the element D. Which element has more atoms/gram?
(c) Six atoms of E have a larger mass than six atoms of F. Which has more atoms/gram?
(d) Six atoms of F have the same mass as 8 atoms of G. Which has more atoms/mole?

78. The reaction between compounds made up of A (squares), B (circles), and C (triangles) is shown pictorially. Using smallest whole-number coefficients, write a balanced equation to represent the picture shown.

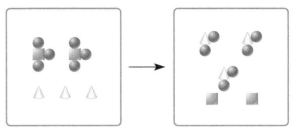

79. Represent the following equation pictorially (see Problem 78), using squares to represent A, circles to represent B, and triangles to represent C.

$$A_2B_3 + C_2 \longrightarrow C_2B_3 + A_2$$

After you have "drawn" the equation, use your drawing as a guide to balance it.

80. Nitrogen reacts with hydrogen to form ammonia. Represent each nitrogen atom by a square and each hydrogen atom with a circle. Starting with five molecules of both hydrogen and nitrogen, show pictorially what you have after the reaction is complete.

81. Consider the following diagram, where atom X is represented by a square and atom Y is represented by a circle.

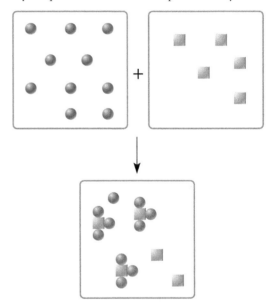

(a) Write the equation for the reaction represented by the diagram.
(b) If each circle stands for a mole of Y and each square a mole of X, how many moles of X did one start with? How many moles of Y?
(c) Using the same representation described in part (b), how many moles of product are formed? How many moles of X and Y are left unreacted?

82. Box A contains 36 atoms of arsenic (As) and 27 molecules of O_2. Box B contains 18 molecules of As_2O_3. Without using your calculator, compare Box A to Box B with respect to
(a) the number of atoms of arsenic and oxygen.
(b) the number of discrete particles.
(c) mass.

83. One mol of ammonia reacts with 1.00 mol of oxygen to form nitrogen oxide and water according to the reaction

$$4NH_3\,(g) + 5O_2\,(g) \longrightarrow 4NO\,(g) + 6H_2O\,(l)$$

State which statements are true about the reaction and make the false statements true.

(a) All the oxygen is consumed.

(b) 4.00 mol NO are produced.

(c) 1.50 mol H_2O are produced.

(d) The description of the experiment does not provide enough information to determine percent yield.

(e) Three moles of water are produced for every two moles of NO obtained.

84. Suppose that the atomic mass of C-12 is taken to be 5.000 amu and that a mole is defined as the number of atoms in 5.000 kg of carbon-12. How many atoms would there be in one mole under these conditions? (*Hint:* There are 6.022×10^{23} C atoms in 12.00 g of C-12.)

85. Suppose that Si-28 ($_{14}^{28}$Si) is taken as the standard for expressing atomic masses and assigned an atomic mass of 10.00 amu. Estimate the molar mass of sodium nitride.

86. Answer the questions below, using **LT** (for *is less than*), **GT** (for *is greater than*), **EQ** (for *is equal to*), or **MI** (for *more information required*) in the blanks provided.

(a) The mass (to three significant figures) of 6.022×10^{23} atoms of Na _____ 23.0 g.

(b) Boron has two isotopes, B-10 (10.01 amu) and B-11 (11.01 amu). The abundance of B-10 _____ the abundance of B-11.

(c) If S-32 were assigned as the standard for expressing relative atomic masses and assigned an atomic mass of 10.00 amu, the atomic mass for H would be _____ 1.00 amu.

(d) When phosphine gas, PH_3, is burned in oxygen, tetraphosphorus decaoxide and steam are formed. In the balanced equation (using smallest whole-number coefficients) for the reaction, the sum of the coefficients on the reactant side is _____ 7.

(e) The mass (in grams) of one mole of bromine molecules is _____ 79.90.

87. Determine whether the statements given below are true or false.

(a) The mass of an atom can have the unit mole.

(b) In N_2O_4, the mass of the oxygen is twice that of the nitrogen.

(c) One mole of chlorine atoms has a mass of 35.45 g.

(d) Boron has an average atomic mass of 10.81 amu. It has two isotopes, B-10 (10.01 amu) and B-11 (11.01 amu). There is more naturally occurring B-10 than B-11.

(e) The compound $C_6H_{12}O_2N$ has for its simplest formula $C_3H_6ON_{1/2}$.

(f) A 558.5-g sample of iron contains ten times as many atoms as 0.5200 g of chromium.

(g) If 1.00 mol of ammonia is mixed with 1.00 mol of oxygen the following reaction occurs,

$$4NH_3(g) + 5O_2(g) \longrightarrow 4NO(g) + 6H_2O(l)$$

All the oxygen is consumed.

(h) When balancing an equation, the total number of moles of reactant molecules must equal the total number of moles of product molecules.

Challenge Problems

88. Chlorophyll, the substance responsible for the green color of leaves, has one magnesium atom per chlorophyll molecule and contains 2.72% magnesium by mass. What is the molar mass of chlorophyll?

89. By x-ray diffraction it is possible to determine the geometric pattern in which atoms are arranged in a crystal and the distances between atoms. In a crystal of silver, four atoms effectively occupy the volume of a cube 0.409 nm on an edge. Taking the density of silver to be 10.5 g/cm^3, calculate the number of atoms in one mole of silver.

90. A 5.025-g sample of calcium is burned in air to produce a mixture of two ionic compounds, calcium oxide and calcium nitride. Water is added to this mixture. It reacts with calcium oxide to form 4.832 g of calcium hydroxide. How many grams of calcium oxide are formed? How many grams of calcium nitride?

91. A mixture of potassium chloride and potassium bromide weighing 3.595 g is heated with chlorine, which converts the mixture completely to potassium chloride. The total mass of potassium chloride after the reaction is 3.129 g. What percentage of the original mixture was potassium bromide?

92. A sample of an oxide of vanadium weighing 4.589 g was heated with hydrogen gas to form water and another oxide of vanadium weighing 3.782 g. The second oxide was treated further with hydrogen until only 2.573 g of vanadium metal remained.

(a) What are the simplest formulas of the two oxides?

(b) What is the total mass of water formed in the successive reactions?

93. A sample of cocaine, $C_{17}H_{21}O_4N$, is diluted with sugar, $C_{12}H_{22}O_{11}$. When a 1.00-mg sample of this mixture is burned, 1.00 mL of carbon dioxide ($d = 1.80$ g/L) is formed. What is the percentage of cocaine in this mixture?

94. A 100.0-g mixture made up of $NaClO_3$, Na_2CO_3, NaCl, and $NaHCO_3$ is heated, producing 5.95 g of oxygen, 1.67 g of water, and 14.5 g of carbon dioxide. NaCl does not react under the conditions of the experiment. The equations for the reactions that take place are:

$$2\,NaClO_3(s) \longrightarrow 2\,NaCl(s) + 3O_2(g)$$
$$Na_2CO_3(s) \longrightarrow Na_2O(s) + CO_2(g)$$
$$2NaHCO_3(s) \longrightarrow Na_2O(s) + 2CO_2(g) + H_2O$$

Assuming 100% decomposition of $NaClO_3$, Na_2CO_3, and $NaHCO_3$, what is the composition of the mixture in grams?

95. An alloy made up of iron (52.6%), nickel (38.0%), cobalt (8.06%), and molybdenum (1.34%) has a density of 7.68 g/cm^3. How many molybdenum atoms are there in a block of the alloy measuring 13.0 cm $\times$ 22.0 cm $\times$ 17.5 cm?

4

Reactions in Aqueous Solution

The three-phase firing process ancient Greek potters used to create this vase utilized both oxidation and reduction processes.

Araldo de Luca/Fine Art/Corbis

He takes up the waters of the sea in his hand, leaving the salt;
He disperses it in mist through the skies;
He recollects and sprinkles it like grain in six-rayed snowy stars over the earth,
There to lie till he dissolves the bonds again.

—HENRY DAVID THOREAU
Journal (January 5, 1856)

Most of the reactions considered in Chapter 3 involved pure substances reacting with each other. However, most of the reactions you will carry out in the laboratory or hear about in lecture take place in water (aqueous) solution. Beyond that, most of the reactions that occur in the world around you involve ions or molecules dissolved in water. For these reasons, among others, you need to become familiar with some of the more important types of aqueous reactions. These include

- precipitation reactions (Section 4-1).
- acid-base reactions (Section 4-2).
- oxidation-reduction reactions (Section 4-3).

All of these reactions involve ions in solution. The corresponding equations are given a special name: net ionic equations. They can be used to do stoichiometric calculations similar to those discussed in Chapter 3.

To carry out these calculations for solution reactions, recall the concentration unit called molarity (Section 3-1), which tells you how many moles of a species are in a given volume of solution.

4-1   Precipitation Reactions

When water (aqueous) solutions of two different ionic compounds are mixed, we often find that an insoluble solid precipitates; this is called a **precipitation reaction**. To identify the solid, we must know which ionic compounds are soluble in water and which are not.

Solubility of Ionic Compounds

When an ionic solid dissolves in water two competing forces come into play:

- the attractive forces between the oppositely charged ions making up the solid
- the attractive forces between water and the ions.

For reasons discussed later in this text (Chapter 7), water has two partially positively charged (δ^+) hydrogen atoms (attractive to anions) and a partially negatively charged (δ^-) oxygen atom (attractive to cations). The extent to which solution occurs depends upon a balance between two forces, which are both electrical in nature:

1. The force of atraction between H_2O molecules and the ions of the solid, which tends to bring the solid into solution. If this factor predominates, we expect the compound to be very soluble in water, as in the case of KCl (Figure 4.1).
2. The force of attraction between oppositely charged ions, which tends to keep them in the solid state. If this is the major factor, we expect water solubility to be low. $SrSO_4$ is almost insoluble, which implies that the interionic forces between Sr^{2+} ions and SO_4^{2-} ions predominate.

Unfortunately, we cannot determine from first principles the relative strengths of these two forces for a given solid. For this reason, among others, we cannot predict in advance the water solubilities of ionic solids, which cover an enormous range of possibilities.

At one extreme, we have the white solid lithium chlorate, $LiClO_3$, which dissolves to the extent of 35 mol/L at room temperature. Mercury(II) sulfide, HgS, found in nature in the red mineral cinnabar, is at the other extreme. Its calculated solubility at 25°C is 10^{-26} mol/L. This means that, in principle at least, about 200 L of a solution of HgS would be required to contain a single pair of Hg^{2+} and S^{2-} ions.

Information on the solubility of common ionic solids in the form of a solubility diagram is given in Figure 4.2.

These rules are quite simple to interpret. For example, the following facts should be evident from Figure 4.2:

- $Ni(NO_3)_2$ is soluble. (All nitrates are soluble.)
- $CoCl_2$ is soluble. (It is not one of the three insoluble chlorides listed.)
- $PbCO_3$ is insoluble.

Figure 4.1 Competing forces when a solute is added to water. When KCl is added to water, the attraction between K^+ and Cl^- and the water molecules competes with the attraction of the ions for each other. In this case, the water to ion attraction is stronger than the ion to ion attraction, so KCl is soluble in water.

	NO_3^-	Cl^-	SO_4^{2-}	OH^-	CO_3^{2-}	PO_4^{3-}
Group 1 cations and NH_4^+						
Group 2 cations			$BaSO_4$	$Mg(OH)_2$		
Transition metal cations, Pb^{2+} and Hg_2^{2+}		AgCl* $PbCl_2$* Hg_2Cl_2*	$PbSO_4$ Ag_2SO_4			

* The bromides and iodides of these cations are also insoluble.

Figure 4.2 Solubility chart for 0.1 *M* solutions of selected anions and cations. Choose the cation (row) and read across for the anion (column). If the block is white, no precipitate will form. If the block is shaded in green, a precipitate will form from dilute solution. Where a formula is shown, this is a cation-anion combination that will precipitate.

Figure 4.3 Partner-exchange reactions. The cation from a soluble compound joins with the anion from another soluble compound. The result may be no reaction, one precipitate, or two precipitates. In this figure, the possible precipitates are AX and/or BR.

Figure 4.4 Precipitation of nickel hydroxide (Ni(OH)$_2$). The precipitate forms when solutions of nickel chloride (NiCl$_2$) and sodium hydroxide (NaOH) are mixed.

Sometimes when water solutions of two different ionic compounds are mixed, an insoluble solid separates out of solution. The **precipitate** (abbreviation: ppt) that forms is itself ionic; the cation comes from one solution, the anion from the other. To predict the occurrence of reactions of this type, you must know which ionic substances are insoluble in water.

The precipitation diagram shown in Figure 4.2 and the schematic diagram (Figure 4.3) enable you to determine whether or not a precipitate will form when dilute solutions of two ionic solutes are mixed. If a cation in solution 1 mixes with an anion in solution 2 to form an insoluble compound (colored squares), that compound will precipitate. Cation-anion combinations that lead to the formation of a soluble compound (white squares) will not give a precipitate. For example, if solutions of NiCl$_2$ (Ni^{2+}, Cl$^-$ ions) and NaOH (Na$^+$, OH$^-$ ions) are mixed (Figure 4.4)

■ a precipitate of Ni(OH)$_2$, an insoluble compound, will form.
■ NaCl, a soluble compound, will not precipitate.

EXAMPLE 4.1

Predict what will happen when the following pairs of dilute aqueous solutions are mixed.
(a) Cu(NO$_3$)$_2$ and (NH$_4$)$_2$SO$_4$ (b) FeCl$_3$ and AgNO$_3$

STRATEGY

1. Follow the schematic diagram in Figure 4.3.
2. Use the precipitation diagram (Figure 4.2) to determine whether or not the possible precipitates are soluble.

SOLUTION

(a) Ions in solution	Cu^{2+} and NO$_3^-$ from Cu(NO$_3$)$_2$; NH$_4^+$ and SO$_4^{2-}$ from (NH$_4$)$_2$SO$_4$
Possible precipitates	CuSO$_4$ and NH$_4$NO$_3$
Solubility	Both are soluble, no precipitate forms
(b) Ions in solution	Fe^{3+} and Cl$^-$ from FeCl$_3$; Ag$^+$ and NO$_3^-$ from AgNO$_3$
Possible precipitates	AgCl and Fe(NO$_3$)$_3$
Solubility	Fe(NO$_3$)$_3$ is soluble, AgCl is insoluble. AgCl precipitates.

Net Ionic Equations

The precipitation reaction that occurs when solutions of Na$_2$CO$_3$ and CaCl$_2$ are mixed (Figure 4.5) can be represented by a simple equation. To obtain that equation, consider the identity of the reactants and products: ◀ ⓘ

Reactants: Na$^+$, CO$_3^{2-}$, Ca^{2+}, and Cl$^-$ ions in water solution:

$$2Na^+(aq) + CO_3^{2-}(aq) + Ca^{2+}(aq) + 2Cl^-(aq)$$

To write a net ionic equation, you first have to identify the ions.

Products: solid $CaCO_3$, Na^+ and Cl^- ions remaining in solution:

$$2Na^+(aq) + 2Cl^-(aq) + CaCO_3(s)$$

The total ionic equation for the reaction is

$$2Na^+(aq) + CO_3{}^{2-}(aq) + Ca^{2+}(aq) + 2Cl^-(aq) \longrightarrow$$

$$2Na^+(aq) + 2Cl^-(aq) + CaCO_3(s)$$

Canceling out the ions that appear on both sides of the equation ($2Na^+$, $2Cl^-$), we obtain the final equation:

$$Ca^{2+}(aq) + CO_3{}^{2-}(aq) \longrightarrow CaCO_3(s)$$

An equation such as this that excludes **spectator ions**, which take no part in the reaction, is referred to as a **net ionic equation**. We will use net ionic equations throughout this chapter and indeed the entire text to represent a wide variety of reactions in water solution. Like all equations, net ionic equations must show

■ *atom balance.* There must be the same number of atoms of each element on both sides. In the preceding equation, the atoms present on both sides are one calcium atom, one carbon atom, and three oxygen atoms.

■ *charge balance.* There must be the same total charge on both sides. In this equation, the total charge is zero on both sides.

$$Ca^{2+}(aq) + CO_3{}^{2-}(aq) \longrightarrow CaCO_3(s)$$

Figure 4.5 Precipitation of calcium carbonate ($CaCO_3$). The precipitate forms when solutions of sodium carbonate (Na_2CO_3) and calcium chloride ($CaCl_2$) are mixed.

© Cengage Learning/Charles D. Winters

Spectator ions are in solution before, during, and after reaction.

EXAMPLE 4.2

Write a net ionic equation for any precipitation reaction that occurs when dilute solutions of the following ionic compounds are mixed.

(a) NaOH and $Cu(NO_3)_2$ (b) $Ba(OH)_2$ and $MgSO_4$ (c) $(NH_4)_3PO_4$ and K_2CO_3

STRATEGY

1. Follow the plan:

   Figure 4.3: compound → ions → possible precipitates

   possible precipitates → (Figure 4.2) → insoluble compound → net ionic equation

2. In writing the net ionic equation, start with the insoluble compound on the right, then write the component ions on the left. Do not forget the physical states: ions (*aq*), product (*s*).

SOLUTION

(a) Ions in solution	Na^+ and OH^- from NaOH; Cu^{2+} and $NO_3{}^-$ from $Cu(NO_3)_2$
Possible precipitates	$NaNO_3$ and $Cu(OH)_2$
Solubility	$NaNO_3$ is soluble; $Cu(OH)_2$ is insoluble.
Net ionic equation	$Cu^{2+}(aq) + 2OH^-(aq) \rightarrow Cu(OH)_2(s)$.
(b) Ions in solution	Ba^{2+} and OH^- from $Ba(OH)_2$; Mg^{2+} and $SO_4{}^{2-}$ from $MgSO_4$
Possible precipitates	$Mg(OH)_2$ and $BaSO_4$
Solubility	Both $BaSO_4$ and $Mg(OH)_2$ are insoluble.
Net ionic equation	$Mg^{2+}(aq) + 2OH^-(aq) \rightarrow Mg(OH)_2(s)$
	$Ba^{2+}(aq) + SO_4{}^{2-}(aq) \rightarrow BaSO_4(s)$
(c) Ions in solution	$NH_4{}^+$ and $PO_4{}^{3-}$ from $(NH_4)_3PO_4$; K^+ and $CO_3{}^{2-}$ from K_2CO_3
Possible precipitates	$(NH_4)_2CO_3$ and K_3PO_4
Solubility	Both $(NH_4)_2CO_3$ and K_3PO_4 are soluble.
Net ionic equation	no reaction

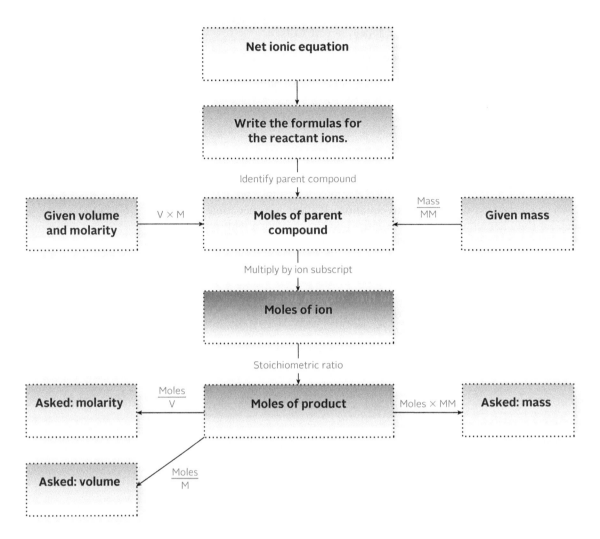

Figure 4.6 Flowchart for solution stoichiometry where the amount of product is asked. If the amount of reactant is needed, reverse the arrows and the operations. *V* is volume, *M* is molarity, and MM is molar mass.

© Cengage Learning/Charles D. Winters

Figure 4.7 Precipitation of iron(III) hydroxide (Fe(OH)₃). The red, gelatinous precipitate forms when aqueous solutions of sodium hydroxide (NaOH) and iron(III) nitrate (Fe(NO₃)₃) are mixed.

Although we have introduced net ionic equations to represent precipitation reactions, they have a much wider application. Indeed, we will use them for all kinds of reactions in water solution. In particular *all of the chemical equations written throughout this chapter are net ionic equations.*

Stoichiometry

The approach followed in Chapter 3, with minor modifications, readily applies to the stoichiometry of solution reactions represented by net ionic equations.

An important consideration when solving these problems is that data are given about the parent compounds and not about the particular ions in the net ionic equation. After all, you do not have reagent bottles with labels that say $3M$ OH^- but rather $3M$ NaOH. Figure 4.6 shows how the modifications fit into the flowchart for the stoichiometry of solution reactions.

EXAMPLE 4.3 GRADED

When aqueous solutions of sodium hydroxide and iron(III) nitrate are mixed, a red precipitate forms (see Figure 4.7).

a Write a net ionic equation for the reaction.

ANALYSIS

Information given:	reactant compounds [NaOH and $Fe(NO_3)_3$]
Asked for:	net ionic equation

STRATEGY

1. Follow the schematic diagram in Figure 4.3 to determine possible precipitates.
2. Use the precipitation diagram (Figure 4.2) to determine whether the possible precipitates are soluble or insoluble.
3. Write the net ionic equation. Start with the product.

SOLUTION

1. Ions in solution	Na^+ and OH^- from NaOH; Fe^{3+} and NO_3^- from $Fe(NO_3)_3$
Possible precipitates	$Fe(OH)_3$ and $NaNO_3$
2. Solubility	$Fe(OH)_3$ is insoluble and forms a precipitate.
3. Net ionic equation	$Fe^{3+}(aq) + 3OH^-(aq) \rightarrow Fe(OH)_3(s)$

b What volume of 0.136 M iron(III) nitrate is required to produce 0.886 g of precipitate?

ANALYSIS

Information given:	net ionic equation from (a): $[Fe^{3+}(aq) + 3OH^-(aq) \rightarrow Fe(OH)_3(s)]$ mass of precipitate (0.886 g); molarity of $Fe(NO_3)_3$ (0.136 M)
Information implied:	molarity of reacting ion, Fe^{3+} molar mass of precipitate
Asked for:	volume of $Fe(NO_3)_3$ used in the reaction

STRATEGY

Reverse the pathway shown in Figure 4.6.

$$\text{mass } Fe(OH)_3 \rightarrow \text{mol ppt} \rightarrow \text{mol ion} \rightarrow \text{mol of parent compound} \rightarrow V \text{ of parent compound}$$

SOLUTION

mol $Fe(NO_3)_3$	$0.886 \text{ g } Fe(OH)_3 \times \dfrac{1 \text{ mol } Fe(OH)_3}{106.87 \text{ g } Fe(OH)_3} \times \dfrac{1 \text{ mol } Fe^{3+}}{1 \text{ mol } Fe(OH)_3} \times \dfrac{1 \text{ mol } Fe(NO_3)_3}{1 \text{ mol } Fe^{3+}}$ $= 0.00829$
$V_{Fe(NO_3)_3}$	$V = \dfrac{\text{mol}}{M} = \dfrac{0.00829 \text{ mol}}{0.136 \text{ mol/L}} = 0.0610 \text{ L}$

c How many grams of precipitate are formed when 50.00 mL of 0.200 M NaOH and 30.00 mL of 0.125 M $Fe(NO_3)_3$ are mixed?

ANALYSIS

Information given:	net ionic equation from (a): $[Fe^{3+}(aq) + 3OH^-(aq) \rightarrow Fe(OH)_3(s)]$ volume (50.00 mL) and molarity (0.200 M) of NaOH volume (30.00 mL) and molarity (0.125 M) of $Fe(NO_3)_3$
Information implied:	number of moles of reacting ions, Fe^{3+} and OH^- Data for moles of both reactants is given, making this a limiting reactant problem.
Asked for:	mass of precipitate formed

continued

STRATEGY

1. Follow the pathway in Figure 4.6 for both NaOH and $Fe(NO_3)_3$ to obtain moles of precipitate formed.

$$\text{mol NaOH } (V \times M) \rightarrow \text{mol OH}^- \rightarrow \text{mol ppt} \quad \text{mol Fe(NO}_3)_3(V \times M) \rightarrow \text{mol Fe}^{3+} \rightarrow \text{mol ppt}$$

2. Choose the smaller number of moles and convert moles to mass.

SOLUTION	
1. mol ppt if NaOH limiting	$0.0500 \text{ L} \times 0.200 \dfrac{\text{mol NaOH}}{\text{L}} \times \dfrac{1 \text{ mol OH}^-}{1 \text{ mol NaOH}} \times \dfrac{1 \text{ mol Fe(OH)}_3}{3 \text{ mol OH}^-}$ $= 0.00333 \text{ mol Fe(OH)}_3$
2. mol ppt if $Fe(NO_3)_3$ limiting	$0.0300 \text{ L} \times 0.125 \dfrac{\text{mol Fe(NO}_3)_3}{\text{L}} \times \dfrac{1 \text{ mol Fe}^{3+}}{1 \text{ mol Fe(NO}_3)_3} \times \dfrac{1 \text{ mol Fe(OH)}_3}{1 \text{ mol Fe}^{3+}}$ $= 0.00375 \text{ mol Fe(OH)}_3$
3. Theoretical yield $Fe(OH)_3$	$0.00333 \text{ mol} < 0.00375 \text{ mol}; 0.00333 \text{ mol Fe(OH)}_3$ is obtained $0.00333 \text{ mol} \times \dfrac{106.87 \text{ g}}{1 \text{ mol}} = 0.356 \text{ g}$

4-2 Acid-Base Reactions

You are probably familiar with a variety of aqueous solutions that are either acidic or basic (Figure 4.8). An **acidic solution** has a sour taste and affects the color of certain organic dyes known as acid-base indicators. For example, litmus turns from blue to red in acidic solution. A **basic solution** has a slippery feeling and changes the colors of indicators (e.g., red to blue for litmus).

The species that give these solutions their characteristic properties are called acids and bases. In this chapter, we use the definitions first proposed by Svante Arrhenius more than a century ago.

> *An **Arrhenius acid** is a species that produces H⁺ ions in water solution.*
> *An **Arrhenius base** is a species that produces OH⁻ ions in water solution.*

We will consider more general definitions of acids and bases in Chapter 13.

Strong and Weak Acids and Bases

There are two types of acids, strong and weak, which differ in the extent of their ionization in water. A **strong acid** ionizes completely, forming H^+ ions and anions. A typical strong acid is HCl. It undergoes the following reaction on addition to water:

$$\text{HCl}(aq) \longrightarrow \text{H}^+(aq) + \text{Cl}^-(aq)$$

In a solution prepared by adding 0.1 mol of HCl to water, there is 0.1 mol of H^+ ions, 0.1 mol of Cl^- ions, and no HCl molecules. There are six common strong acids, whose names and formulas are listed in Table 4.1.

All acids other than those listed in Table 4.1 can be taken to be weak. A **weak acid** is only partially ionized to H^+ ions in water. All of the weak acids considered in this chapter are molecules containing an ionizable hydrogen atom. Their general formula can be represented as HB; the general ionization reaction in water is

$$\text{HB}(aq) \rightleftharpoons \text{H}^+(aq) + \text{B}^-(aq)$$

Figure 4.8 Acidic and basic household solutions. Many common household items, including vinegar, orange juice, and cola drinks, are acidic. In contrast, baking soda and most detergents and cleaning agents are basic.

Table 4.1 Common Strong Acids and Bases

Acid	Name of Acid	Base	Name of Base
HCl	Hydrochloric acid	LiOH	Lithium hydroxide
HBr	Hydrobromic acid	NaOH	Sodium hydroxide
HI	Hydriodic acid	KOH	Potassium hydroxide
HNO_3	Nitric acid	$Ca(OH)_2$	Calcium hydroxide
$HClO_4$	Perchloric acid	$Sr(OH)_2$	Strontium hydroxide
H_2SO_4	Sulfuric acid ◀ⓘ	$Ba(OH)_2$	Barium hydroxide

ⓘ Sulfuric acid ionization:
$H_2SO_4(aq) \longrightarrow H^+(aq) + HSO_4^-(aq)$.

The double arrow implies that this reaction does not go to completion. Instead, a mixture is formed containing significant amounts of both products and reactants. With the weak acid hydrogen fluoride ◀ⓘ

$$HF(aq) \rightleftharpoons H^+(aq) + F^-(aq)$$

a solution prepared by adding 0.1 mol of HF to a liter of water contains about 0.01 mol of H^+ ions, 0.01 mol of F^- ions, and 0.09 mol of HF molecules.

ⓘ Like HCl, HBr, and HI, HF has a halogen bonded to a hydrogen atom. Unlike the three other acids, HF is a weak acid. The others are strong acids.

Bases, like acids, are classified as strong or weak. A **strong base** in water solution is completely ionized to OH^- ions and cations. As you can see from Table 4.1, the strong bases are the hydroxides of the Group 1 and Group 2 metals. These are typical ionic solids, completely ionized both in the solid state and in water solution. The equations written to represent the processes by which NaOH and $Ca(OH)_2$ dissolve in water are

$$NaOH(s) \longrightarrow Na^+(aq) + OH^-(aq)$$
$$Ca(OH)_2(s) \longrightarrow Ca^{2+}(aq) + 2OH^-(aq)$$

In a solution prepared by adding 0.1 mol of NaOH to water, there is 0.1 mol of Na^+ ions, 0.1 mol of OH^- ions, and no NaOH molecules.

A **weak base** produces OH^- ions in a quite different manner. It reacts with H_2O molecules, acquiring H^+ ions and leaving OH^- ions behind. The reaction of ammonia, NH_3, is typical:

$$NH_3(aq) + H_2O \rightleftharpoons NH_4^+(aq) + OH^-(aq)$$

As with all weak bases, this reaction does not go to completion. In a solution prepared by adding 0.1 mol of ammonia to a liter of water, there is about 0.001 mol of NH_4^+, 0.001 mol of OH^-, and nearly 0.099 mol of NH_3.

A common class of weak bases consists of the organic molecules known as *amines*. An **amine** can be considered to be a derivative of ammonia in which one or more hydrogen atoms have been replaced by hydrocarbon groups. ◀ⓘ

In the simplest case, methylamine (Figure 4.9), a hydrogen atom is replaced by a $-CH_3$ group to give the CH_3NH_2 molecule, which reacts with water in a manner very similar to NH_3:

$$CH_3NH_2(aq) + H_2O \rightleftharpoons CH_3NH_3^+(aq) + OH^-(aq)$$

As we have pointed out, strong acids and bases are completely ionized in water. As a result, compounds such as HCl and NaOH are strong electrolytes like NaCl. In contrast, molecular weak acids and weak bases are poor conductors because their water solutions contain relatively few ions. Hydrofluoric acid and ammonia are commonly described as *weak electrolytes*.

Figure 4.9 Methylamine, CH_3NH_2

ⓘ A hydrocarbon group has a string of C and H atoms.

Equations for Acid-Base Reactions

When an acidic water solution is mixed with a basic water solution, an acid-base reaction takes place. The nature of the reaction and hence the equation written for it depend on whether the acid and base involved are strong or weak. ◄ ⓘ

1. *Strong acid–strong base.* Consider what happens when a solution of a strong acid such as HNO_3 is added to a solution of a strong base such as $NaOH$. Because HNO_3 is a strong acid, it is completely converted to H^+ and NO_3^- ions in solution. Similarly, with the strong base $NaOH$, the solution species are the Na^+ and OH^- ions. When the solutions are mixed, the H^+ and OH^- ions react with each other to form H_2O molecules. This reaction, referred to as **neutralization,** is represented by the net ionic equation

You need to know the strong acids and bases to work with acid-base reactions.

$$H^+(aq) + OH^-(aq) \longrightarrow H_2O$$

The Na^+ and NO_3^- ions take no part in the reaction and so do not appear in the equation. Here again, we are dealing with "spectator ions."

There is considerable evidence to indicate that the neutralization reaction occurs when any strong base reacts with any strong acid in water solution. It follows that the neutralization equation written above applies to any strong acid–strong base reaction. ◄ ⓘ

2. *Weak acid–strong base.* When a strong base such as $NaOH$ is added to a solution of a weak acid, HB, a two-step reaction occurs. The first step is the ionization of the HB molecule to H^+ and B^- ions; the second is the neutralization of the H^+ ions produced in the first step by the OH^- ions of the $NaOH$ solution.

When $HClO_4$ reacts with $Ca(OH)_2$, the equation is $H^+(aq) + OH^-(aq) \longrightarrow H_2O$.

$$(1)\ HB(aq) \rightleftharpoons H^+(aq) + B^-(aq)$$
$$(2)\ H^+(aq) + OH^-(aq) \longrightarrow H_2O$$

The equation for the overall reaction is obtained by adding the two equations just written and canceling H^+ ions:

$$HB(aq) + OH^-(aq) \longrightarrow B^-(aq) + H_2O$$

For the reaction between solutions of sodium hydroxide and hydrogen fluoride, ◄ ⓘ the net ionic equation is

When an acid is weak, like HF, its formula appears in the equation.

$$HF(aq) + OH^-(aq) \longrightarrow F^-(aq) + H_2O$$

Here, as always, spectator ions such as Na^+ are not included in the net ionic equation.

3. *Strong acid–weak base.* As an example of a reaction of a strong acid with a weak base, consider what happens when an aqueous solution of a strong acid like HCl is added to an aqueous solution of ammonia, NH_3. Again, we consider the reaction to take place in two steps. The first step is the reaction of NH_3 with H_2O to form NH_4^+ and OH^- ions. Then, in the second step, the H^+ ions of the strong acid neutralize the OH^- ions formed in the first step.

$$(1)\ NH_3(aq) + H_2O \rightleftharpoons NH_4^+(aq) + OH^-(aq)$$
$$(2)\ H^+(aq) + OH^-(aq) \longrightarrow H_2O$$

The overall equation is obtained by summing those for the individual steps. Canceling species (OH^-, H_2O) that appear on both sides, we obtain the net ionic equation ◄ ⓘ

When a base is weak, like NH_3, its formula appears in the equation.

$$H^+(aq) + NH_3(aq) \longrightarrow NH_4^+(aq)$$

In another case, for the reaction of a strong acid such as HNO_3 with methylamine, CH_3NH_2, the net ionic equation is

$$H^+(aq) + CH_3NH_2(aq) \longrightarrow CH_3NH_3^+(aq)$$

Table 4.2 Types of Acid-Base Reactions

Reactants	Reacting Species	Net Ionic Equation
Strong acid–strong base	H^+-OH^-	$H^+(aq) + OH^-(aq) \longrightarrow H_2O$
Weak acid–strong base	$HB-OH^-$	$HB(aq) + OH^-(aq) \longrightarrow$ $H_2O + B^-(aq)$
Strong acid–weak base	H^+-B	$H^+(aq) + B(aq) \longrightarrow BH^+(aq)$

Table 4.2 summarizes the equations written for the three types of acid-base reactions just discussed. Figure 4.10 visually illustrates the process of determining whether a compound is an acid (strong or weak) or a base (strong or weak) and the nature of the **reacting species**. You should find both Table 4.2 and Figure 4.10 useful in writing the equations called for in Example 4.4.

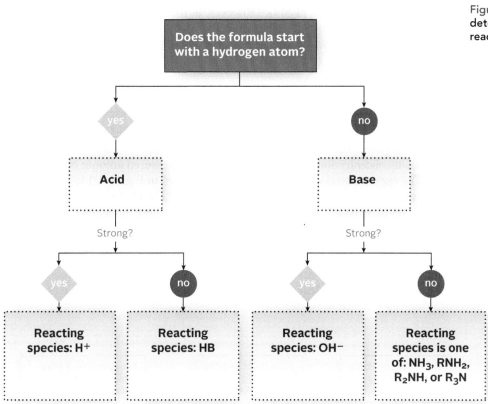

Figure 4.10 Flowchart for determining net ionic equations for reactions between acids and bases.

EXAMPLE 4.4

Write a net ionic equation for each of the following reactions in dilute water solution.

(a) Hypochlorous acid (HClO) and barium hydroxide.

(b) Ammonia with sulfuric acid (H_2SO_4).

(c) Hydrobromic acid (HBr) with potassium hydroxide.

STRATEGY

1. Determine the nature of the compound (acid or base; strong or weak) and its reacting species. (Table 4.1 and Figure 4.10 are helpful.)

2. Recall Table 4.2 and write a net ionic equation for the acid-base reaction.

continued

	SOLUTION
(a) Nature of the compounds	HClO: weak acid; Ba(OH)$_2$: strong base
reacting species	For HClO: HClO; for Ca(OH)$_2$: OH$^-$
net ionic equation	HClO(aq) + OH$^-$(aq) → ClO$^-$(aq) + H$_2$O
(b) Nature of the compounds	HClO$_4$: strong acid; NH$_3$: weak base
reacting species	For HClO$_4$: H$^+$; for NH$_3$: NH$_3$
net ionic equation	H$^+$(aq) + NH$_3$(aq) → NH$_4$$^+$($aq$)
(c) Nature of the compounds	HBr: strong acid; KOH: strong base
reacting species	For HBr: H$^+$; for KOH: OH$^-$
net ionic equation	H$^+$(aq) + OH$^-$(aq) → H$_2$O
	END POINT

Note that there are 2OH$^-$ ions in Ba(OH)$_2$, and 2H$^+$ ions in H$_2$SO$_4$. However, in the balanced equation, you use only one H$^+$ and one OH$^-$ ion.

Acid-Base Titrations

Acid-base reactions in water solution are commonly used to determine the concentration of a dissolved species or its percentage in a solid mixture. This is done by carrying out a **titration,** measuring the volume of a *standard solution* (a solution of known concentration) required to react with a measured amount of sample.

The experimental setup for a titration is shown in Figure 4.11. The flask contains vinegar, a water solution of a weak organic acid called acetic acid. A solution of sodium hydroxide of known concentration is added from a buret. The net ionic equation for the acid-base reaction that occurs is

$$HC_2H_3O_2(aq) + OH^-(aq) \longrightarrow C_2H_3O_2{}^-(aq) + H_2O$$

The objective of the titration is to determine the point at which reaction is complete, called the **equivalence point.** This is reached when the number of moles of

The buret contains a sodium hydroxide solution of known concentration.

The sodium hydroxide solution is slowly added...

...and reacts with the acetic acid in the vinegar solution.

The equivalence point is reached and the volume of NaOH used is noted. The volume of NaOH together with its concentration is used in determining the concentration of acetic acid in the vinegar.

The flask contains vinegar (a dilute solution of acetic acid) and an acid-base indicator (phenolphthalein) that is colorless in acid solution and pink in basic solution.

Figure 4.11 Titration of vinegar with sodium hydroxide (NaOH).

Figure 4.12 **A molecular view of an acid-base titration.**

OH⁻ added is exactly equal to the number of moles of acetic acid, $HC_2H_3O_2$, originally present. To determine this point, a single drop of an *acid-base indicator* such as phenolphthalein is used. It should change color (colorless to pink) at the equivalence point.

A molecular diagram for the titration of NaOH and HCl is shown in Figure 4.12. For simplicity, only the reacting species (H^+ and OH^-) are shown. The spectator ions (Na^+ and Cl^-) and water molecules in solution are not.

EXAMPLE 4.5	GRADED

Consider the weak acid H_2X. It is titrated with NaOH.

a	Write a net ionic equation for the reaction.

STRATEGY

1. Determine the nature of the compound (acid or base; strong or weak) and its reacting species. (Table 4.1 and Figure 4.10 are helpful.)
2. Recall Table 4.2 and write a net ionic equation for the acid-base reaction.

SOLUTION	
1. Nature of the compounds	H_2X—weak acid; NaOH—strong base
2. reacting species	For H_2X—H_2X; for NaOH—OH^-
3. net ionic equation	$H_2X(aq) + 2OH^-(aq) \rightarrow X^{2-}(aq) + 2H_2O$

continued

b	What is the molarity of H_2X if 29.45 mL of 0.187 M NaOH are required to neutralize 15.00 mL of H_2X?

ANALYSIS

Information given:	Volume (29.45 mL) and molarity (0.187 M) of NaOH Volume (15.00 mL) of H_2X
Information implied:	stoichiometric ratio; reacting species Net ionic equation [$H_2X(aq) + 2OH^-(aq) \rightarrow 2H_2O + X^{2-}(aq)$]
Asked for:	Molarity of H_2X

STRATEGY

1. Use the stoichiometric ratio: 2 mol OH^-/1 mol H_2X

2. Follow the flowchart in Figure 4.6.

   H_2X does not break up into ions. Skip the moles parent compound → moles ion step for H_2X.

$$\text{mol NaOH} \xrightarrow{V \times M} \text{mol } OH^- \xrightarrow[\text{ratio}]{\text{stoichiometric}} \text{mol } H_2X \xrightarrow{n \div V} \text{molarity of } H_2X$$

SOLUTION

mol H_2X	$0.02945 \text{ L} \times 0.187 \dfrac{\text{mol NaOH}}{\text{L}} \times \dfrac{1 \text{ mol } OH^-}{1 \text{ mol NaOH}} \times \dfrac{1 \text{ mol } H_2X}{2 \text{ mol } OH^-} = 0.002754$
molarity of H_2X used	$M = n \div V = \dfrac{0.002754 \text{ mol}}{0.01500 \text{ L}} = 0.1836 \ M$

c	What is the molar mass of H_2X if 29.45 mL of 0.187 M NaOH are required to neutralize a solution prepared by adding 0.242 g of H_2X to enough water to make 25.00 mL of solution?

ANALYSIS

Information given:	Volume (29.45 mL) and molarity (0.187 M) of NaOH Mass (0.242 g) of H_2X Volume (25.00 mL) of water
Information implied:	stoichiometric ratio Net ionic equation [$H_2X(aq) + 2OH^-(aq) \rightarrow 2H_2O + X^{2-}(aq)$]
Asked for:	Molar mass of H_2X

STRATEGY

1. Use the stoichiometric ratio: 2 mol OH^-/1 mol H_2X

2. Follow the flowchart in Figure 4.6.

   H_2X does not break up into ions. Skip the moles parent compound → moles ion step for H_2X.

$$\text{mol NaOH} \xrightarrow{V \times M} \text{mol } OH^- \longrightarrow \text{mol NaOH} \xrightarrow[\text{ratio}]{\text{stoichiometric}} \text{mol } H_2X \xrightarrow{\text{mass} \div n} \text{MM of } H_2X$$

SOLUTION

mol H_2X	$0.02945 \text{ L} \times 0.187 \dfrac{\text{mol NaOH}}{\text{L}} \times \dfrac{1 \text{ mol } OH^-}{1 \text{ mol NaOH}} \times \dfrac{1 \text{ mol } H_2X}{2 \text{ mol } OH^-} = 0.002754$
molar mass of H_2X	$MM = \text{mass} \div n = \dfrac{0.242 \text{ g}}{0.002754 \text{ mol}} = 87.9 \text{ g/mol}$

END POINTS

1. You need to figure out the number of moles before you can calculate mass, molar mass, volume, or molarity.

2. The amount of water added to the solid H_2X is irrelevant to the solution of the problem.

CHEMISTRY THE **HUMAN** SIDE

For reasons that are by no means obvious, Sweden produced a disproportionate number of outstanding chemists in the eighteenth and nineteenth centuries. Jöns Jakob Berzelius (1779–1848) determined with amazing accuracy the atomic masses of virtually all the elements known in his time. In his spare time, he invented such modern laboratory tools as the beaker, the flask, the pipet, and the ringstand.

Svante Arrhenius, like Berzelius, was born in Sweden and spent his entire professional career there. According to Arrhenius, the concept of strong and weak acids and bases came to him on May 13, 1883, when he was 24 years old. He added, "I could not sleep that night until I had worked through the entire problem."

Almost exactly one year later, Arrhenius submitted his Ph.D. thesis at the University of Uppsala. He proposed that salts, strong acids, and strong bases are completely ionized in dilute water solution. Today, it seems quite reasonable that solutions of NaCl, HCl, and NaOH contain, respectively, Na^+ and Cl^- ions, H^+ and Cl^- ions, and Na^+ and OH^- ions. It did not seem nearly so obvious to the chemistry faculty at Uppsala in 1884. Arrhenius's dissertation received the lowest passing grade "approved without praise."

Arrhenius sent copies of his Ph.D. thesis to several well-known chemists in Europe and America. Most ignored his ideas; a few were openly hostile. A pair of young chemists gave positive responses: Jacobus van't Hoff (1852–1911) (age 32) at Amsterdam (Holland) and Wilhelm Ostwald (1853–1932) (also 32) at Riga (Latvia). For some years, these three young men were referred to, somewhat disparagingly, as "ionists" or "ionians." As time passed, the situation changed. The first Nobel Prize in chemistry was awarded to van't Hoff in 1901. Two years later, in 1903, Arrhenius became a Nobel laureate; Ostwald followed in 1909.

Among other contributions of Arrhenius, the most important were probably in chemical kinetics (Chapter 11). In 1889 he derived the relation for the temperature dependence of reaction rate. In quite a different area, in 1896 Arrhenius published an article, "On the Influence of Carbon Dioxide in the Air on the Temperature of the Ground." He presented the basic idea of the greenhouse effect, discussed in Chapter 16.

In his later years, Arrhenius turned his attention to popularizing chemistry. He wrote several different textbooks that were well received. In 1925, under pressure from his publisher to submit a manuscript (publishers are like that), Arrhenius started getting up at 4 A.M. to write. As might be expected, rising at such an early hour had an adverse effect on his health. Arrhenius suffered a physical breakdown in 1925, from which he never really recovered, dying two years later.

Svante August Arrhenius
(1859–1927)

4-3 Oxidation-Reduction Reactions

Another common type of reaction in aqueous solution involves a transfer of electrons between two species. Such a reaction is called an oxidation-reduction or **redox reaction.** Many familiar reactions fit into this category, including the reaction of metals with acid.

In a redox reaction, one species *loses* (i.e., donates) electrons and is said to be *oxidized*. The other species, which *gains* (or receives) electrons, is *reduced*. To illustrate, consider the redox reaction that takes place when zinc pellets are added to hydrochloric acid (Figure 4.13). The net ionic equation for the reaction is

$$Zn(s) + 2H^+(aq) \longrightarrow Zn^{2+}(aq) + H_2(g)$$

This equation can be split into two **half-equations,** one of oxidation and the other of reduction. Zinc atoms are oxidized to Zn^{2+} ions by losing electrons. The oxidation half-equation is

$$oxidation: \quad Zn(s) \longrightarrow Zn^{2+}(aq) + 2e^-$$

At the same time, H^+ ions are reduced to H_2 molecules by gaining electrons; the reduction half-equation is

$$reduction: \quad 2H^+(aq) + 2e^- \longrightarrow H_2(g)$$

Figure 4.13 Redox reaction of zinc with a strong acid. The zinc atoms are oxidized to Zn^{2+} ions in solution; the H^+ ions are reduced to H_2 molecules.

Na→Na⁺
Na loses e^-
Na is oxidized
Na is a reducing agent
Na increases its oxidation number ($0 \rightarrow +1$)
OXIDATION HALF-REACTION

S→S²⁻
S gains e^-
S is reduced
S is an oxidizing agent
S decreases its oxidation number ($0 \rightarrow -2$)
REDUCTION HALF-REACTION

Figure 4.14 Summary of occurrences in oxidation and reduction half-reactions.

From this example, it should be clear that—

- *oxidation and reduction occur together* in the same reaction; you can't have one without the other.
- *there is no net change in the number of electrons in a redox reaction.* Those given off in the oxidation half-reaction are taken on by another species in the reduction half-reaction.

In earlier sections of this chapter, we showed how to write and balance equations for precipitation reactions (Section 4.1) and acid-base reactions (Section 4.2). We will show you how to balance redox reactions in Chapter 17. In this section we will identify the species oxidized or reduced, oxidizing or reducing agents, and oxidation and reduction half-reactions. Figure 4.14 summarizes this new terminology. To do that, it is convenient to introduce a new concept, oxidation number.

Oxidation Number

The concept of **oxidation number** is used to simplify the electron bookkeeping in redox reactions. For a monatomic ion (e.g., Na^+, S^{2-}), the oxidation number is, quite simply, the charge of the ion ($+1$, -2). In a molecule or polyatomic ion, the oxidation number of an element is a "pseudo-charge" obtained in a rather arbitrary way, assigning bonding electrons to the atom with the greater attraction for electrons.

In practice, oxidation numbers in all kinds of species are assigned according to a set of arbitrary rules:

1. *The oxidation number of an element in an elementary substance is 0* (Figure 4.15).
   Example: The oxidation number for chlorine in Cl_2 and for phosphorus in P_4 is 0.
2. *The oxidation number of an element in a monatomic ion is equal to the charge of that ion.*
   Example: The oxidation number for chlorine in Cl^- is -1; for sodium in Na^+ it is $+1$.
3. *Certain elements (we will call them "leading elements") have the same oxidation number in all their compounds.*
   Group 1 elements always have an oxidation number of $+1$.
   Group 2 elements always have an oxidation number of $+2$.
   Fluorine (F) always has an oxidation number of -1.
4. *Hydrogen in a compound has an oxidation number of $+1$, unless it is combined with a metal, in which case it is -1.*
   Example: The oxidation number for hydrogen in HCl is $+1$; for hydrogen in NaH it is -1.
5. *The sum of the oxidation numbers in a neutral species is 0 and in a polyatomic ion is equal to the charge of the ion.*
   Examples:
   (a) To determine the oxidation number of P in PH_3, use the fact that H has an oxidation number of $+1$ (since it is combined with P, a nonmetal) and solve algebraically using the above rule. Note that there are 3 H atoms, each with an oxidation number of $+1$. We will call the oxidation number of P, x. $3(+1) + x = 0$; $x = -3$. The oxidation number for P in PH_3 is -3.

Figure 4.15 Rusting and oxidation number. As iron rusts, its oxidation number changes from 0 in $Fe(s)$ to $+3$ in Fe^{3+}.

© Cengage Learning/Charles D. Winters

(b) To determine the oxidation number of N in NH_4^+, use the fact that H has an oxidation number of $+1$ (since it is combined with N, a nonmetal) and solve algebraically using the above rule. Note that there are 4 H atoms, each with an oxidation number of $+1$. We will call the oxidation number of N, y.

$4(+1) + y = -1; y = -3$. The oxidation number for N in NH_4^+ is -3.

6. *Oxygen in a compound has an oxidation number of -2, unless it is combined with a Group 1 metal (always $+1$) or Group 2 metal (always $+2$). Solve algebraically* *for the oxidation number of oxygen.*

Examples:

(a) The oxidation number (O.N.) for oxygen in Na_2O is

$$2(+1) + O.N.\ O = 0;\ O.N.\ O = -2$$

(b) The oxidation number (O.N.) for oxygen in Na_2O_2 is

$$2(+1) + 2(O.N.\ O) = 0;\ O.N.\ O = -1$$

(c) The oxidation number (O.N.) for oxygen in NaO_2 is

$$+1 + 2(O.N.\ O) = 0;\ O.N.\ O = -1/2$$

Oxidation numbers are calculated, not determined experimentally.

Figure 4.16 (page 90) shows you how to use these rules in the form of a flowchart. The application of these rules is illustrated in Example 4.6.

EXAMPLE 4.6

Assign an oxidation number (O.N.) to each element in the following species:

(a) N_2 (b) N^{3-} (c) NO_3^- (d) BaO (e) K_2O_2

SOLUTION

(a) N_2 is in its elementary state. (Rule 1)	O.N. N = 0
(b) N^{3-} is a monoatomic ion. (Rule 2)	O.N. N = -3
(c) There are no Group 1 or Group 2 metals. (Rule 6)	O.N. O = -2
NO_3^- is a polyatomic ion. (Rule 5)	$3(-2) + x = -1$; O.N. N = $+5$
(d) Ba is a Group 2 metal. (Rule 3)	O.N. Ba = $+2$
The sum of the oxidation numbers is 0. (Rule 5)	$+2 + x = 0$; O.N. O = -2
(e) K is a Group 1 metal. (Rule 3)	O.N. K = $+1$
The sum of the oxidation numbers is 0. (Rule 5)	$2(1) + 2x = 0$; O.N. O = -1

END POINT

Always look for the "leading elements" (Group 1 and Group 2 metals and F) in a compound when you start. These elements will lead you to the oxidation numbers of the other elements in the compound. If these leading elements are not present, then look for H and O ($+1$ and -2, respectively, when not combined with Group 1 or 2 metals).

The concept of oxidation number leads directly to a working definition of the terms oxidation and reduction. **Oxidation** is defined as *an increase in oxidation number* and **reduction** as *a decrease in oxidation number.* Consider once again the reaction of zinc with a strong acid:

$$Zn(s) + 2H^+(aq) \longrightarrow Zn^{2+}(aq) + H_2(g)$$

Zn is oxidized (O.N.: $0 \longrightarrow +2$)
H^+ is reduced (O.N.: $+1 \longrightarrow 0$)

These definitions are of course compatible with the interpretation of oxidation and reduction in terms of loss and gain of electrons. An element that loses electrons

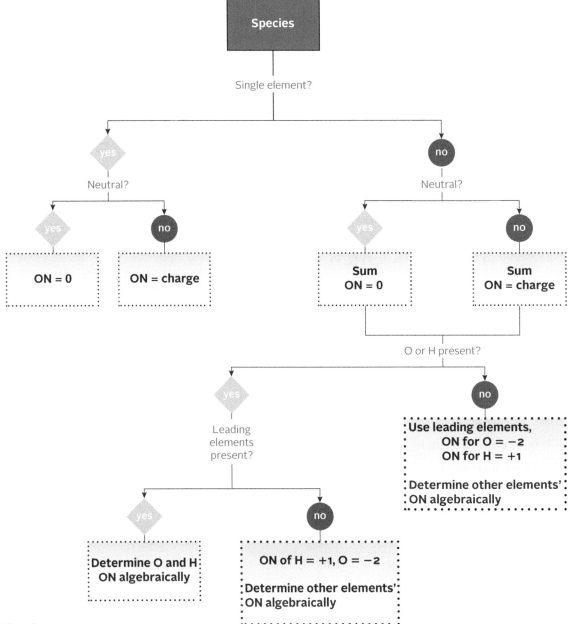

Figure 4.16 Flowchart to determine oxidation number. Leading elements ON are Group 1 = +1, Group 2 = +2, F = −1. (ON in the figure stands for oxidation number.)

must increase in oxidation number. The gain of electrons always results in a decrease in oxidation number.

An easy way to recognize a redox equation is to note changes in oxidation number of two different elements. The net ionic equation

$$2Al(s) + 3Cu^{2+}(aq) \longrightarrow 2Al^{3+}(aq) + 3Cu(s)$$

must represent a redox reaction because aluminum increases in oxidation number, from 0 to +3, and copper decreases from +2 to 0. In contrast, the reaction

$$CO_3^{2-}(aq) + 2H^+(aq) \longrightarrow CO_2(g) + H_2O$$

is not of the redox type because each element has the same oxidation number in both reactants and products: for O it is −2, for H it is +1, and for C it is +4.

The two species that exchange electrons in a redox reaction are given special names. The ion or molecule that accepts electrons is called the **oxidizing agent;** by accepting electrons it brings about the oxidation of another species. Conversely, the species that donates electrons is called the **reducing agent;** when reaction occurs it reduces the other species. ◄ⓘ

ⓘ The oxidizing agent is reduced; the reducing agent is oxidized.

To illustrate these concepts consider the reaction

$$Zn(s) + 2H^+(aq) \longrightarrow Zn^{2+}(aq) + H_2(g)$$

The H^+ ion is the oxidizing agent; it brings about the oxidation of zinc. By the same token, zinc acts as a reducing agent; it furnishes the electrons required to reduce H^+ ions.

Metallic elements taking part in redox reactions, such as zinc in the reaction above, commonly act as reducing agents; they are oxidized to cations such as Zn^{2+}. Other reducing agents include hydrogen gas, which can be oxidized to H^+ ions:

$$H_2(g) \longrightarrow 2H^+(aq) + 2e^-$$

and a few cations such as Fe^{2+} that can be oxidized to a higher state:

$$Fe^{2+}(aq) \longrightarrow Fe^{3+}(aq) + e^-$$

Nonmetallic elements frequently act as oxidizing agents, being reduced to the corresponding anions:

$$Cl_2(g) + 2e^- \longrightarrow 2Cl^-(aq)$$
$$S(s) + 2e^- \longrightarrow S^{2-}(aq)$$

The equations shown above are called half-reactions. They are *not* complete redox reactions.

EXAMPLE 4.7

Consider the following redox equation:

$$Cr^{3+}(aq) + H_2O_2(aq) \rightarrow 2H_2O + Cr_2O_7{}^{2-}(aq)$$

(a) Identify the element oxidized and the element reduced.

(b) What are the oxidizing and reducing agents?

STRATEGY

1. Determine the oxidation number of each element.

2. Find elements whose oxidation numbers change.

SOLUTION

Oxidation numbers	Cr: +3; H: +1; O: −1 → H: +1; O: −2; Cr: +6
Change	Cr: +3 → +6 (increase)
	O: −1 → −2 (decrease)
Element reduced	O (decrease in oxidation number)
Element oxidized	Cr (increase in oxidation number)
Oxidizing agent	H_2O_2 (It is the species that contains the element that is reduced.)
Reducing agent	$Cr_2O_7{}^{2-}$ (It is the species that contains the element that is oxidized.)

Stoichiometric calculations for redox reactions in water solution are carried out in much the same way as those for precipitation reactions (Example 4.3) or acid-base reactions (Example 4.5).

EXAMPLE 4.8

Consider the balanced equation for the reaction between iron(II) and permanganate ions in acidic solution:

$$MnO_4^-(aq) + 5Fe^{2+}(aq) + 8H^+(aq) \rightarrow 5Fe^{3+}(aq) + Mn^{2+}(aq) + 4H_2O$$

What volume of 0.684 M KMnO$_4$ solution is required to completely react with 27.50 mL of 0.250 M Fe(NO$_3$)$_2$ (Figure 4.17)?

ANALYSIS

Information given:	V (27.50 mL) and M (0.250) of Fe(NO$_3$)$_2$ M (0.684) of KMnO$_4$
Information implied:	reacting species; stoichiometric ratios
Asked for:	volume of KMnO$_4$

STRATEGY

Follow the flowchart shown in Figure 4.6.

$$V \times M \rightarrow \text{mol parent} \rightarrow \text{mol ion} \rightarrow \text{mol ion} \rightarrow \text{mol parent} \rightarrow V \times M$$

SOLUTION

1. Parent → ion	Fe(NO$_3$)$_2$ (parent) → Fe^{2+} (ion) KMnO$_4$ (parent) → MnO$_4^-$ (ion)
2. mol Fe(NO$_3$)$_2$	$V \times M = (0.02750 \text{ L})(0.250 \text{ mol/L}) = 0.00688$
3. mol Fe^{2+}	$0.00688 \text{ mol Fe(NO}_3)_2 \times \dfrac{1 \text{ mol Fe}^{2+}}{1 \text{ mol Fe(NO}_3)_2} = 0.00688$
4. mol MnO$_4^-$	$0.00688 \text{ mol Fe}^{2+} \times \dfrac{1 \text{ mol MnO}_4^-}{5 \text{ mol Fe}^{2+}} = 0.00138$
5. mol KMnO$_4$	$0.00138 \text{ mol MnO}_4^- \times \dfrac{1 \text{ mol KMnO}_4}{1 \text{ mol MnO}_4^-} = 0.00138$
6. V KMnO$_4$	moles $= V \times M$; $V = \dfrac{0.00138 \text{ mol}}{0.684 \text{ mol/L}} = 0.00202 \text{ L} = 2.02 \text{ mL}$

A solution of Fe^{2+} in an acidic solution ready to be titrated...

...with a solution of potassium permanganate (KMnO$_4$)

When the potassium permanganate is added, a redox reaction occurs (the equation for the reaction is derived and balanced in the text).

As the reaction takes place, the purple color characteristic of MnO$_4^-$ fades; the Fe^{3+} formed is pale yellow.

Just past the equivalence point a small excess of MnO$_4^-$ gives a light purple color to the solution.

(a)   (b)   (c)

Figure 4.17 A redox titration.

Antacids

As you eat your evening meal, tiny glands in the mucous lining of your stomach discharge hydrochloric acid (HCl) to help you digest your food. If all goes well, the concentration of HCl in your stomach builds up to about 0.1 M, digestion takes place smoothly, and you're ready for a long night of studying chemistry (or whatever). Sometimes, though, too much acid is released and you suffer from a condition called "acid indigestion," "sour stomach," or "heartburn." This is particularly likely to happen if you eat too much spicy or high-fat food.

One way to relieve this discomfort is to take an *antacid*, which may be in the form of solid tablets or a water solution (Figure A). The antacid reduces stomach acidity by a chemical reaction that consumes excess H^+ ions. Several antacids are listed in Table A, along with the brand names of the commercial product that contains them.

Notice that in almost every case, the antacid is an *insoluble carbonate or hydroxide*. Typical reactions of these species with excess H^+ ions can be represented by the following equations:

$$CaCO_3(s) + 2H^+(aq) \rightarrow Ca^{2+}(aq) + CO_2(g) + H_2O$$
$$Mg(OH)_2(s) + 2H^+(aq) \rightarrow Mg^{2+}(aq) + H_2O$$

Using equations such as these, we can readily determine the amount of H^+ ion consumed by a given mass of antacid; let us say one gram. For $CaCO_3$,

$$1.00 \text{ g CaCO}_3 \times \frac{1 \text{ mol CaCO}_3}{100.0 \text{ g CaCO}_3} \times \frac{2 \text{ mol H}^+}{1 \text{ mol CaCO}_3}$$

$$= 0.0200 \text{ mol H}^+$$

Results of similar calculations give

0.0237 mol H^+/g $MgCO_3$, 0.0343 mol H^+/g $Mg(OH)_2$, 0.0385 mol H^+/g $Al(OH)_3$

We conclude that of these four antacids, $Al(OH)_3$ is the most "efficient," so far as getting rid of excess H^+ ions is concerned.

There is another, quite different way to solve the problem caused by acid indigestion. That is to use certain drugs known collectively as "H-2 blockers," scientifically known as H_2 receptor antagonists. They inhibit the release of H^+ ions into the stomach. Prescription drugs of this type have been used for some time to treat disorders of the stomach and esophagus. In 1994, pharmaceutical companies started selling several of these drugs over-the-counter (Figure B).

Another group of drugs, called proton pump inhibitors, have surpassed H-2 blockers in popularity. They are longer lasting and reduce acid secretion in gastric juices more effectively. They do this by an entirely different mechanism from H-2 receptor antagonists. These drugs can now also be found over-the-counter and go by the trade names Prilosec, Prevacid, and Zegerid, to name a few (Figure C). Omeprazole (a generic name) is also available.

Table A Antacids

Product	Antacids(s) Present	Product	Antacids(s) Present
Brioschi	$NaHCO_3$	Phillips	$Mg(OH)_2$
Gaviscon	$MgCO_3$, $Al(OH)_3$	Rolaids	$CaCO_3$, $Mg(OH)_2$
Gelusil	$Mg(OH)_2$, $Al(OH)_3$	Surpass	$CaCO_3$
Maalox	$Mg(OH)_2$, $Al(OH)_3$	Top Care	$MgCO_3$, $Al(OH)_3$
Mylanta	$Mg(OH)_2$, $Al(OH)_3$	Tums	$CaCO_3$

Figure A Some commercially available antacids.

Figure B Over-the-counter "H-2 blockers."

Figure C Over-the-counter proton pump inhibitors.

Key Concepts

1. Apply the precipitation diagram (Figures 4.2 and 4.3) to
   ■ predict solubility and precipitation reactions.
   (**Example 4.1; Problems 1–10**)
   ■ write net ionic equations for precipitation reactions.
   (**Examples 4.2, 4.3; Problems 5–10**)
2. Carry out stoichiometric calculations for reactions.
   (**Examples 4.3, 4.5, 4.8; Problems 11–16, 27–38, 59–70**)

3. With the aid of Tables 4.1 and 4.2 and Figure 4.10, write net ionic equations for acid-base reactions.
   (**Example 4.4; Problems 19–26**)
4. Determine oxidation numbers.
   (**Example 4.6, 4.7; Problems 39–42**)

Key Terms

amine
Arrhenius acid
Arrhenius base
equivalence point
half-equations

net ionic equation
neutralization
oxidation
oxidation number
oxidizing agent

precipitate
redox reaction
reducing agent
reduction
strong acid

strong base
titration
weak acid
weak base

Summary Problem

An aqueous solution of hydrogen chloride is called hydrochloric acid (Figure 4.18). It is widely used for a host of industrial purposes. It reacts with a wide variety of compounds.

(a) Write net ionic equations for the reaction between aqueous solutions of hydrochloric acid with

   1. an aqueous solution of strontium hydroxide.

   2. an aqueous solution of silver nitrate.

   3. an aqueous solution of methylamine (CH_3NH_2).

   4. iron(II) hydroxide. (The chloride ions react with iron(II) hydroxide to form metallic iron. Chlorate ions are also formed.)

(b) When 25.00 mL of 0.695 M HCl reacts with an excess of silver nitrate, a precipitate forms. How many grams of precipitate can be theoretically obtained?

(c) What volume of 0.2500 M strontium hydroxide is required to completely react with 75.00 mL of 0.07942 M HCl?

(d) When 37.5 mL of 0.439 M HCl reacts with 22.0 mL of 0.573 M ammonia, ammonium ions are formed. What is the concentration of each species in solution after reaction is complete? (Assume that volumes are additive.)

(e) An alloy containing aluminum is analyzed. All the aluminum in a 2.500-g sample of the alloy reacts with 212 mL of 0.493 M HCl. The equation for the reaction is:

$$2Al(s) + 6H^+(aq) \longrightarrow 2Al^{3+}(aq) + 3H_2(g)$$

   1. What is the oxidizing agent?

2. Which element is oxidized?

3. Calculate the mass percent of aluminum in the alloy. (Assume that the only component of the alloy that reacts with HCl is aluminum.)

Answers

(a) 1. $H^+(aq) + OH^-(aq) \longrightarrow H_2O$
    2. $Cl^-(aq) + Ag^+(aq) \longrightarrow AgCl(s)$
    3. $CH_3NH_2(aq) + H^+(aq) \longrightarrow CH_3NH_3^+(aq)$
    4. $3Fe(OH)_2(s) + Cl^-(aq) \longrightarrow$
       $3Fe(s) + ClO_3^-(aq) + 3H_2O$
(b) 2.49 g
(c) 11.91 mL
(d) $NH_3 = 0$;
    $H^+ = 0.0648$ M;
    $Cl^- = 0.277$ M;
    $NH_4^+ = 0.212$ M
(e) 1. H^+
    2. Al
    3. 37.6%

Figure 4.18 Hydrochloric acid can be obtained from hardware stores, where it is called muriatic acid. This strong acid is used to clean metal and stone surfaces.

Questions and Problems

Even-numbered questions and Challenge Problems have answers in Appendix 5 and fully worked solutions in the *Student Solutions Manual*.

4-1 Precipitation Reactions

1. Write the formulas of the following compounds and decide which are soluble in water.
(a) sodium sulfate   (b) iron(III) nitrate
(c) silver chloride   (d) chromium(III) hydroxide

2. Follow the instructions for Question 1 for the following compounds:
(a) barium chloride
(b) magnesium hydroxide
(c) chromium(III) carbonate
(d) potassium phosphate

3. Describe how you would precipitate
(a) zinc carbonate from a solution of zinc nitrate.
(b) copper(II) phosphate from a solution of copper(II) chloride.
(c) barium sulfate from a solution of barium hydroxide.

4. Name the reagent, if any, that you would add to a solution of iron(III) chloride to precipitate
(a) iron(III) hydroxide.
(b) iron(III) carbonate.
(c) iron(III) phosphate.

5. Write net ionic equations for the formation of
(a) a precipitate when solutions of magnesium nitrate and potassium hydroxide are mixed.
(b) two different precipitates when solutions of silver(I) sulfate and barium chloride are mixed.

6. Write net ionic equations to explain the formation of
(a) a red precipitate when solutions of iron(III) nitrate and sodium hydroxide are mixed.
(b) two different precipitates when magnesium sulfate and barium hydroxide are mixed.

7. Decide whether a precipitate will form when the following solutions are mixed. If a precipitate forms, write a net ionic equation for the reaction.
(a) potassium nitrate and magnesium sulfate
(b) silver nitrate and potassium carbonate
(c) ammonium carbonate and cobalt(III) chloride
(d) sodium phosphate and barium hydroxide
(e) barium nitrate and potassium hydroxide

8. Follow the directions of Question 7 for solutions of the following:
(a) silver nitrate and sodium chloride
(b) cobalt(II) nitrate and sodium hydroxide
(c) ammonium phosphate and potassium hydroxide
(d) copper(II) sulfate and sodium carbonate
(e) lithium sulfate and barium hydroxide

9. Write a net ionic equation for any precipitation reaction that occurs when 1 M solutions of the following are mixed.
(a) copper(II) sulfate and sodium chloride
(b) manganese(II) nitrate and ammonium hydroxide
(c) silver nitrate and hydrochloric acid

(d) nickel(II) sulfate and potassium hydroxide
(e) ammonium carbonate and sodium nitrate

10. Follow the directions for Question 9 for the following pairs of solutions.
(a) sodium phosphate and barium chloride
(b) zinc sulfate and potassium hydroxide
(c) ammonium sulfate and sodium chloride
(d) cobalt(III) nitrate and sodium phosphate

11. What volume of 0.2500 M cobalt(III) sulfate is required to react completely with
(a) 25.00 mL of 0.0315 M calcium hydroxide?
(b) 5.00 g of sodium carbonate?
(c) 12.50 mL of 0.1249 M potassium phosphate?

12. What is the molarity of a 25.00 mL solution of iron(III) nitrate that reacts completely (100% yield) with
(a) 12.54 mL of 0.1488 M sodium carbonate?
(b) 7.58 g of potassium phosphate?
(c) 10.00 mL of 0.1573 M Sr(OH)$_2$?

13. A 50.00-mL sample of 0.0250 M silver nitrate is mixed with 0.0400 M chromium(III) chloride.
(a) What is the minimum volume of chromium(III) chloride required to completely precipitate silver chloride?
(b) How many grams of silver chloride are produced from (a)?

14. Aluminum ions react with carbonate ions to form an insoluble compound, aluminum carbonate.
(a) Write the net ionic equation for this reaction.
(b) What is the molarity of a solution of aluminum chloride if 30.0 mL is required to react with 35.5 mL of 0.137 M sodium carbonate?
(c) How many grams of aluminum carbonate are formed in (b)?

15. When Na$_3$PO$_4$ and Ca(NO$_3$)$_2$ are combined, the following reaction occurs:

$$2PO_4^{3-}(aq) + 3Ca^{2+}(aq) \longrightarrow Ca_3(PO_4)_2(s)$$

How many grams of Ca$_3$(PO$_4$)$_2$(s) (MM = 310.18 g/mol) are obtained when 15.00 mL of 0.1386 M Na$_3$PO$_4$ are mixed with 20.00 mL of 0.2118 M Ca(NO$_3$)$_2$?

16. When solutions of aluminum sulfate and sodium hydroxide are mixed, a white gelatinous precipitate forms.
(a) Write a balanced net ionic equation for the reaction.
(b) What is the mass of the precipitate when 2.76 g of aluminum sulfate in 125 mL of solution is combined with 85.0 mL of 0.2500 M NaOH?
(c) What is the molarity of the ion in excess? (Ignore spectator ions and assume that volumes are additive.)

4-2 Acid-Base Reactions

17. Classify the following compounds as acids or bases, weak or strong.
(a) perchloric acid   (b) cesium hydroxide
(c) carbonic acid, H$_2$CO$_3$ (d) ethylamine, C$_2$H$_5$NH$_2$

18. Follow the directions of Question 17 for
(a) hydrogen sulfide.   (b) sulfuric acid.
(c) pyridine, C$_5$H$_5$N.   (d) aluminum hydroxide.

19. For an acid-base reaction, what is the reacting species, that is, the ion or molecule that appears in the chemical equation, in the following acids?

(a) perchloric acid   (b) hydriodic acid
(c) nitrous acid   (d) nitric acid
(e) lactic acid, $HC_3H_5O_3$

20. Follow the directions of Question 19 for the following acids:

(a) hypochlorous acid   (b) formic acid, $HCHO_2$
(c) acetic acid, $HC_2H_3O_2$ (d) hydrobromic acid
(e) sulfurous acid

21. For an acid-base reaction, what is the reacting species (the ion or molecule that appears in the chemical equation) in the following bases?

(a) barium hydroxide   (b) trimethylamine $(CH_3)_3N$
(c) aniline, $C_6H_5NH_2$ (d) sodium hydroxide

22. Follow the directions of Question 21 for the following bases:

(a) toluidine, C_7H_9N (b) strontium hydroxide
(c) indol, C_8H_6NH (d) aqueous ammonia

23. Write a balanced net ionic equation for each of the following acid-base reactions in water.

(a) nitrous acid and barium hydroxide
(b) potassium hydroxide and hydrofluoric acid
(c) aniline ($C_6H_5NH_2$) and perchloric acid

24. Write a balanced net ionic equation for each of the following reactions in water.

(a) acetic acid, $HC_2H_3O_2$, with strontium hydroxide
(b) diethylamine, $(C_2H_5)NH$ with sulfuric acid
(c) aqueous hydrofluoric acid with sodium hydroxide

25. Consider the following generic equation:

$$H^+(aq) + B^-(aq) \longrightarrow HB(aq)$$

For which of the following pairs would this be the correct prototype equation for the acid-base reaction in solution? If it is not correct, write the proper equation for the acid-base reaction between the pair.

(a) nitric acid and calcium hydroxide
(b) hydrochloric acid and CH_3NH_2
(c) hydrobromic acid and aqueous ammonia
(d) perchloric acid and barium hydroxide
(e) sodium hydroxide and nitrous acid

26. Consider the following generic equation

$$OH^-(aq) + HB(aq) \longrightarrow B^-(aq) + H_2O$$

For which of the following pairs would this be the correct prototype equation for the acid-base reaction in solution? If it is not correct, write the proper equation for the acid-base reaction between the pair.

(a) hydrochloric acid and pyridine, C_5H_5N
(b) sulfuric acid and rubidium hydroxide
(c) potassium hydroxide and hydrofluoric acid
(d) ammonia and hydriodic acid
(e) strontium hydroxide and hydrocyanic acid

27. What is the molarity of a solution of nitric acid if 0.216 g of barium hydroxide is required to neutralize 20.00 mL of nitric acid?

28. How many milliliters of 0.2315 M sulfuric acid are required to neutralize 38.00 mL of 0.189 M ammonia?

29. What is the volume of 1.222 M sodium hydroxide required to react with

(a) 32.5 mL of 0.569 M sulfurous acid? (One mole of sulfurous acid reacts with two moles of hydroxide ion.)
(b) 5.00 g of oxalic acid, $H_2C_2O_4$? (One mole of oxalic acid reacts with two moles of hydroxide ion.)
(c) 15.0 g of concentrated acetic acid, $HC_2H_3O_2$, that is 88% by mass pure?

30. Consider several 25.00-mL solutions of perchloric acid. What is the molarity of the acid solution neutralized by

(a) 17.25 mL of 0.3471 M ethylamine ($C_2H_5NH_2$).
(b) 14.17 g of strontium hydroxide.
(c) 41.73 mL of an 18% (by mass) solution of ammonia ($d = 0.9295$ g/mL).

31. Analysis shows that a sample of H_2X (MM = 100.0 g/mol) reacts completely with 330.0 mL of 0.2000 M KOH.

$$H_2X(aq) + 2OH^-(aq) \longrightarrow X^{2-}(aq) + 2H_2O$$

What is the volume of the sample? (Density of H_2X = 1.200 g/mL.)

32. A student tries to determine experimentally the molar mass of aspirin (HAsp). She takes 1.00 g of aspirin, dissolves it in water, and neutralizes it with 17.6 mL of 0.315 M KOH. The equation for the reaction is

$$HAsp(aq) + OH^-(aq) \longrightarrow Asp^-(aq) + H_2O$$

What is the molar mass of aspirin?

33. Sulfuric acid is needed in a lead storage battery. A 25.00-mL sample of "battery acid" taken from a functioning battery requires 53.7 mL of 4.552 M NaOH for complete neutralization. What is the molarity of the battery acid?

34. For a product to be called "vinegar," it must contain at least 5.0% acetic acid, $HC_2H_3O_2$, by mass. A 10.00-g sample of a "raspberry vinegar" is titrated with 0.1250 M Ba(OH)$_2$ and required 37.50 mL for complete neutralization. Can the product be called a "vinegar"?

35. The percentage of sodium hydrogen carbonate, $NaHCO_3$, in a powder for stomach upsets is found by titrating with 0.275 M hydrochloric acid. If 15.5 mL of hydrochloric acid is required to react with 0.500 g of the sample, what is the percentage of sodium hydrogen carbonate in the sample?

The balanced equation for the reaction that takes place is

$$NaHCO_3(s) + H^+(aq) \longrightarrow Na^+(aq) + CO_2(g) + H_2O$$

36. A capsule of vitamin C, a weak acid, is analyzed by titration with 0.450 M potassium hydroxide. It is found that 5.94 mL of the base are required to react with the contents of a capsule. If the contents have a mass of 0.634 g, what is the mass percent of vitamin C ($C_6H_8O_6$) in the capsule? (One mol of vitamin C reacts with one mol of OH^-.)

37. An artificial fruit beverage contains 12.0 g of tartaric acid, $H_2C_4H_4O_6$, to achieve tartness. It is titrated with a basic solution that has a density of 1.045 g/cm^3 and contains 5.00 mass percent KOH. What volume of the basic solution is required? (One mole of tartaric acid reacts with two moles of hydroxide ion.)

38. Lactic acid, $C_3H_6O_3$, is the acid present in sour milk. A 0.100-g sample of pure lactic acid requires 12.95 mL of 0.0857 M sodium hydroxide for complete reaction. How many moles of hydroxide ion are required to neutralize one mole of lactic acid?

4-3 Oxidation-Reduction Reactions

39. Assign oxidation numbers to each element in
   (a) nitrogen oxide
   (b) ammonia
   (c) potassium peroxide
   (d) chlorate ion (ClO_3^-)

40. Assign oxidation numbers to each element in
   (a) methane (b) carbonate ion (CO_3^{2-})
   (c) iodate ion (IO_4^-) (d) hydrazine (N_2H_4)

41. Assign oxidation numbers to each element in
   (a) ClO_3^- (b) H_2SO_3 (c) K_2O_2 (d) Na_3N

42. Assign oxidation numbers to each element in
   (a) HIO_3 (b) $NaMnO_4$ (c) SnO_2
   (d) NOF (e) NaO_2

43. Classify each of the following half-reactions as oxidation or reduction.
   (a) $O_2(g) \longrightarrow O^{2-}(aq)$
   (b) $MnO_4^-(aq) \longrightarrow MnO_2(s)$
   (c) $Cr_2O_7^{2-}(aq) \longrightarrow Cr^{3+}(aq)$
   (d) $Cl^-(aq) \longrightarrow Cl_2(g)$

44. Classify each of the following half-equations as oxidation or reduction.
   (a) $CH_3OH(aq) \longrightarrow CO_2(g)$
   (b) $NO_3^-(aq) \longrightarrow NH_4^+(aq)$
   (c) $Fe^{3+}(aq) \longrightarrow Fe(s)$
   (d) $V^{2+}(aq) \longrightarrow VO_3^-(aq)$

45. Classify each of the following half-equations as oxidation or reduction.
   (a) $Mn^{2+}(aq) \longrightarrow MnO_4^-(aq)$
   (b) $CrO_4^{2-}(aq) \longrightarrow Cr^{3+}(aq)$
   (c) $PbO_2(s) \longrightarrow Pb^{2+}(aq)$
   (d) $ClO_2^-(aq) \longrightarrow ClO^-(aq)$

46. Classify each of the following half-reactions as oxidation or reduction.
   (a) $TiO_2(s) \longrightarrow Ti^{3+}(aq)$
   (b) $Zn^{2+}(aq) \longrightarrow Zn(s)$
   (c) $NH_4^+(aq) \longrightarrow N_2(g)$
   (d) $CH_3OH(aq) \longrightarrow CH_2O(aq)$

47. For each unbalanced equation given below
   ■ write unbalanced half-reactions.
   ■ identify the species oxidized and the species reduced.
   ■ identify the oxidizing and reducing agents.
   (a) $Ag(s) + NO_3^-(aq) \longrightarrow Ag^+(aq) + NO(g)$
   (b) $CO_2(g) + H_2O(l) \longrightarrow C_2H_4(g) + O_2(g)$

48. Follow the directions of Question 47 for the following unbalanced equations:
   (a) $As_2O_3(s) + MnO_4^-(aq) \longrightarrow H_3AsO_4(aq) + Mn^{2+}(aq)$
   (b) $N_2H_4(l) + CO_3^{2-}(aq) \longrightarrow N_2(g) + CO(g)$

49. A solution of potassium permanganate reacts with oxalic acid, $H_2C_2O_4$, to form carbon dioxide and solid manganese(IV) oxide (MnO_2) according to the equation

$$3H_2C_2O_4(aq) + 2MnO_4^-(aq) + 2H^+(aq) \longrightarrow$$
$$6CO_2(g) + 2MnO_2(s) + 4H_2O$$

(a) If 20.0 mL of 0.300 *M* potassium permanganate are required to react with 13.7 mL of oxalic acid, what is the molarity of the oxalic acid?
(b) What is the mass of manganese(IV) oxide formed?

50. The vanadium (V) ion in a 0.5000-g sample of ore is converted to VO^{2+} ions. The amount of VO^{2+} in solution can be determined by reaction with an acid solution of $KMnO_4$. The balanced equation for the reaction is

$$5VO^{2+}(aq) + MnO_4^-(aq) + 11H_2O \longrightarrow$$
$$Mn^{2+}(aq) + 5V(OH)_4^+(aq) + 2H^+(aq)$$

What is the mass percent of vanadium in the ore if 26.45 mL of 0.02250 *M* permanganate solution is required for complete reaction?

51. Hydrogen gas is bubbled into a solution of barium hydroxide that has sulfur in it. The equation for the reaction that takes place is

$$H_2(g) + S(s) + 2OH^-(aq) \longrightarrow S^{2-}(aq) + 2H_2O$$

What volume of 0.349 *M* $Ba(OH)_2$ is required to react completely with 3.00 g of sulfur?

52. Consider the reaction between silver and nitric acid for which the equation is

$$Ag(s) + 2H^+(aq) + NO_3^-(aq) \longrightarrow$$
$$Ag^+(aq) + NO_2(g) + H_2O$$

If 42.50 mL of 12.0 *M* nitric acid furnishes enough H^+ to react with silver, how many grams of silver react?

53. The molarity of iodine in solution can be determined by titration with arsenious acid, H_3AsO_4. The unbalanced equation for the reaction is

$$H_3AsO_3(aq) + I_2(aq) + H_2O \longrightarrow$$
$$2I^-(aq) + H_3AsO_4(aq) + 2H^+(aq)$$

A 243-mL solution of aqueous iodine is prepared by dissolving iodine crystals in water. A fifty-mL portion of the solution requires 15.42 mL of 0.134 *M* H_3AsO_3 for complete reaction. What is the molarity of the solution? How many grams of iodine were added to the solution?

54. A wire weighing 0.250 g and containing 92.50% Fe is dissolved in HCl. The iron is completely oxidized to Fe^{3+} by bromine water. The solution is then treated with tin(II) chloride to bring about the reaction

$$Sn^{2+}(aq) + 2Fe^{3+}(aq) \longrightarrow 2Fe^{2+}(aq) + Sn^{4+}(aq) + H_2O$$

If 22.0 mL of tin(II) chloride solution is required for complete reaction, what is the molarity of the tin(II) chloride solution?

55. Glycerol ($C_3H_8O_3$) is commonly used as an additive to moisturizers. The amount of glycerol added can be determined by titration with permanganate in basic solution.

$$14MnO_4^-(aq) + C_3H_8O_3(aq) + 20OH^-(aq) \longrightarrow$$
$$14MnO_4^{2-}(aq) + 3CO_3^{2-}(aq) + 14H_2O$$

How many milliliters of 6.15 *M* $KMnO_4$ are required to completely react with 7.25 mL of an aqueous solution of glycerol ($d = 1.144$ g/mL) that is 51.6% by mass?

56. Laws passed in some states define a drunk driver as one who drives with a blood alcohol level of 0.10% by mass or higher. The level of alcohol can be determined by titrating blood plasma with potassium dichromate according to the following equation

$$16H^+(aq) + 2Cr_2O_7^{2-}(aq) + C_2H_5OH(aq) \longrightarrow$$
$$4Cr^{3+}(aq) + 2CO_2(g) + 11H_2O$$

Assuming that the only substance that reacts with dichromate in blood plasma is alcohol, is a person legally drunk if 38.94 mL of 0.0723 M potassium dichromate is required to titrate a 50.0-g sample of blood plasma?

Unclassified

57. A sample of limestone weighing 1.005 g is dissolved in 75.00 mL of 0.2500 M hydrochloric acid. The following reaction occurs:

$$CaCO_3(s) + 2H^+(aq) \longrightarrow Ca^{2+}(aq) + CO_2(g) + H_2O$$

It is found that 19.26 mL of 0.150 M NaOH is required to titrate the excess HCl left after reaction with the limestone. What is the mass percent of $CaCO_3$ in the limestone?

58. The iron content of hemoglobin is determined by destroying the hemoglobin molecule and producing small water-soluble ions and molecules. The iron in the aqueous solution is reduced to iron(II) ion and then titrated against potassium permanganate. In the titration, iron(II) is oxidized to iron(III) and permanganate is reduced to manganese(II) ion. A 5.00-g sample of hemoglobin requires 32.3 mL of a 0.002100 M solution of potassium permanganate. The reaction with permanganate ion is

$$MnO_4^-(aq) + 8H^+(aq) + 5Fe^{2+}(aq) \longrightarrow$$
$$Mn^{2+}(aq) + 5Fe^{3+}(aq) + 4H_2O$$

What is the mass percent of iron in hemoglobin?

59. In order to determine the purity of ammonium sulfate, a sample with a mass of 0.850 g is dissolved in KOH. The equation for the reaction that takes place is

$$NH_4^+(aq) + OH^-(aq) \longrightarrow NH_3(aq) + H_2O$$

The ammonia liberated is distilled into a flask that contains 50.00 mL of 0.250 M HCl. Not all the HCl is consumed. The excess HCl reacts with 17.3 mL of 0.120 M NaOH. What is the mass percent of $(NH_4)_2SO_4$ in the sample?

60. Gold metal will dissolve only in *aqua regia*, a mixture of concentrated hydrochloric acid and concentrated nitric acid in a 3:1 volume ratio. The products of the reaction between gold and the concentrated acids are $AuCl_4^-(aq)$, $NO(g)$, and H_2O. The equation for this reaction where HNO_3 and HCl are strong acids is

$$Au(s) + 4Cl^-(aq) + 4H^+(aq) + NO_3^-(aq) \longrightarrow$$
$$AuCl_4^-(aq) + NO(g) + 2H_2O$$

(a) What stoichiometric ratio of hydrochloric acid to nitric acid should be used?
(b) What volumes of 12 M HCl and 16 M HNO_3 are required to furnish the Cl^- and NO_3^- ions to react with 25.0 g of gold?

61. Cisplatin, $Pt(NH_3)_2Cl_2$, is a drug widely used in chemotherapy. It can react with the weak base pyridine, C_6H_5N. Suppose 3.11 g of cisplatin are treated with 2.00 mL of pyridine

(d = 0.980 g/mL). The unreacted pyridine is then titrated with HCl according to the following reaction:

$$C_6H_5N(l) + H^+(aq) \longrightarrow C_6H_5NH^+(aq)$$

The complete reaction requires 31.2 mL of 0.0245 M HCl.
(a) How many moles of pyridine were unused in the cisplatin reaction?
(b) How many moles of pyridine would react with one mole of cisplatin?

62. Ten mL of concentrated H_3PO_4 (91.7% by mass, d = 1.69 g/mL) was accidentally poured into a beaker containing 20.0 g of NaOH. Not all the NaOH was consumed. How many grams of NaOH were left unreacted? The equation for the reaction is

$$H_3PO_4(aq) + 3OH^-(aq) \longrightarrow 3H_2O + PO_4^{3-}(aq)$$

Conceptual Questions

63. Consider the following balanced redox reaction in basic medium.

$$3Sn^{2+}(aq) + Cr_2O_7^{2-}(aq) + 4H_2O \longrightarrow$$
$$3Sn^{4+}(aq) + Cr_2O_3(s) + 8OH^-(aq)$$

(a) What is the oxidizing agent?
(b) What species has the element that increases its oxidation number?
(c) What species contains the element with the highest oxidation number?
(d) If the reaction were to take place in acidic medium, what species would not be included in the reaction?

64. Identify the type of aqueous reaction using the symbols **PPT** for precipitation, **SA/SB** for strong acid–strong base, **SA/WB** for strong acid–weak base, **WA/SB** for weak acid–strong base, and **NR** for no reaction.

(a) $CH_3CH_2NH_2$ + HCl
(b) $Ca(OH)_2$ + HF
(c) $Ca(OH)_2$ + Na_3PO_4
(d) Ag_2SO_4 + $BaCl_2$
(e) $Mg(NO_3)_2$ + NaCl

65. Using circles to represent cations and squares to represent anions, show pictorially the reactions that occur between aqueous solutions of

(a) Fe^{3+} and OH^-.
(b) Na^+ and PO_4^{3-}.

66. Assuming that circles represent cations and squares represent anions, match the incomplete net ionic equations to their pictorial representations.

(c)

(**1**) $2Na^+ + SO_4^{2-} \longrightarrow$ _____

(**2**) $Mg^{2+} + 2OH^- \longrightarrow$ _____

(**3**) $Ba^{2+} + CO_3^{2-} \longrightarrow$ _____

67. Consider four beakers. Beaker A has an aqueous solution of NaOH in which the OH^- ions are represented by blue circles. Beaker B has a weak acid; HX is represented by red circles. Beaker C has a weak acid; H_2X is represented by green circles. Beaker D has a weak acid; H_3X is represented by yellow circles. X^- ions are represented by triangles. Match the pictorial representations with the reactions given below.

(**a**) $HX(aq) + OH^-(aq) \longrightarrow X^-(aq) + H_2O$

(**b**) $H_2X(aq) + 2OH^-(aq) \longrightarrow X^-(aq) + 2H_2O$

(**c**) $H_3X(aq) + 3OH^-(aq) \longrightarrow X^-(aq) + 3H_2O$

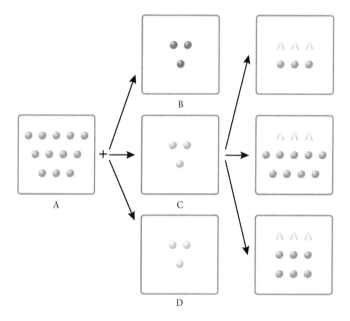

68. The following figures represent species before and after they are dissolved in water. Classify each species as weak electrolyte, strong electrolyte, or nonelectrolyte. You may assume that species that dissociate during solution break up as ions.

(a)

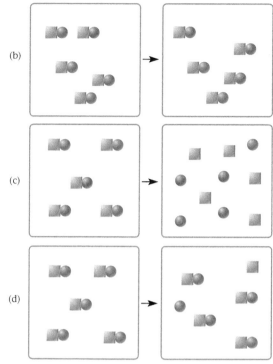

(b)

(c)

(d)

69. A student is asked to identify the metal nitrate present in an aqueous solution. The cation in the solution can be either Na^+, Ba^{2+}, Ag^+, or Ni^{2+}. Results of solubility experiments are as follows:

unknown + chloride ions—no precipitate

unknown + carbonate ions—precipitate

unknown + sulfate ions—precipitate

What is the cation in the solution?

70. Three students titrate different samples of the same solution of HCl to obtain its molarity. Below are their data.

Student A: 20.00 mL HCl + 20.00 mL H_2O
0.100 M NaOH used to titrate to the equivalence point

Student B: 20.00 mL HCl + 40.00 mL H_2O
0.100 M NaOH used to titrate to the equivalence point

Student C: 20.00 mL HCl + 20.00 mL H_2O
0.100 M Ba(OH)$_2$ used to titrate to the equivalence point.

All the students calculated the molarities correctly. Which (if any) of the following statements are true?

(**a**) The molarity calculated by A is half that calculated by B.

(**b**) The molarity calculated by A is equal to that calculated by C.

(**c**) The molarity calculated by B is twice that calculated by C.

(**d**) The molarity calculated by A is twice that calculated by B.

(**e**) The molarity calculated by A is equal to that calculated by B.

Challenge Problems

71. Calcium in blood or urine can be determined by precipitation as calcium oxalate, CaC_2O_4. The precipitate is dissolved in strong acid and titrated with potassium permanganate. The equation for reaction is

$$2MnO_4^-(aq) + 6H^+(aq) + 5H_2C_2O_4(aq) \longrightarrow$$
$$2Mn^{2+}(aq) + 10CO_2(g) + 8H_2O$$

A 24-hour urine sample is collected from an adult patient, reduced to a small volume, and titrated with 26.2 mL of 0.0946 M $KMnO_4$. How many grams of calcium oxalate are in the sample? Normal range for Ca^{2+} output for an adult is 100 to 300 mg per 24 hour. Is the sample within the normal range?

72. Stomach acid is approximately 0.020 M HCl. What volume of this acid is neutralized by an antacid tablet that weighs 330 mg and contains 41.0% $Mg(OH)_2$, 36.2% $NaHCO_3$, and 22.8% NaCl? The reactions involved are

$$Mg(OH)_2(s) + 2H^+(aq) \longrightarrow Mg^{2+}(aq) + 2H_2O$$
$$HCO_3^-(aq) + H^+(aq) \longrightarrow CO_2(g) + H_2O$$

73. Copper metal can reduce silver ions to metallic silver. The copper is oxidized to copper ions according to the reaction

$$2Ag^+(aq) + Cu(s) \longrightarrow Cu^{2+}(aq) + 2Ag(s)$$

A copper strip with a mass of 2.00 g is dipped into a solution of $AgNO_3$. After some time has elapsed, the copper strip is coated with silver. The strip is removed from the solution, dried, and weighed. The coated strip has a mass of 4.18 g. What are the masses of copper and silver metals in the strip? (*Hint:* Remember that the copper metal is being used up as silver metal forms.)

74. A solution contains both iron(II) and iron(III) ions. A 50.00-mL sample of the solution is titrated with 35.0 mL of 0.0280 M $KMnO_4$, which oxidizes Fe^{2+} to Fe^{3+}. The permanganate ion is reduced to manganese(II) ion. The equation for this reaction is

$$MnO_4^-(aq) + 8H^+(aq) + 5Fe^{2+}(aq) \longrightarrow$$
$$Mn^{2+}(aq) + 5Fe^{3+}(aq) + 4H_2O$$

Another 50.00-mL sample of the solution is treated with zinc, which reduces all the Fe^{3+} to Fe^{2+}. The equation for this reaction is

$$2Fe^{3+}(aq) + Zn(s) \longrightarrow 2Fe^{2+}(aq) + Zn^{2+}(aq)$$

The resulting solution is again titrated with 0.0280 M $KMnO_4$; this time 48.0 mL is required. What are the concentrations of Fe^{2+} and Fe^{3+} in the solution?

75. A student is given 0.930 g of an unknown acid, which can be either oxalic acid, $H_2C_2O_4$, or citric acid, $H_3C_6H_5O_7$. To determine which acid she has, she titrates the unknown acid with 0.615 M NaOH. The equivalence point is reached when 33.6 mL are added. What is the unknown acid?

76. Solid iron(III) hydroxide is added to 625 mL of 0.280 M HCl. The resulting solution is acidic and titrated with 238.2 mL of 0.113 M NaOH. What mass of iron(III) hydroxide was added to the HCl?

77. A 300.0-g sample of a solid is made up of a uniform mixture of $NaNO_3$, $MgCl_2$, and $BaCl_2$. A 100.0-g sample of the mixture is dissolved in water and treated with an excess of KOH. The precipitate from the reaction has a mass of 13.47 g. The remaining 200.0-g sample is also dissolved in water and treated with an aqueous solution of $AgNO_3$. The resulting precipitate has a mass of 195.8 g. What are the masses of $NaNO_3$, $MgCl_2$, and $BaCl_2$ in the 300.0-g sample?

78. When 85.0 mL of 0.250 M $Ba(OH)_2$ solution is added to 85.00 mL of 0.250 M $Al(NO_3)_3$ solution, a white gelatinous precipitate of $Al(OH)_3$ is formed. Assuming 100% yield,

(a) what mass (in grams) of $Al(OH)_3$ is formed?
(b) what is the molarity of each of the ions Ba^{2+}, OH^-, Al^{3+}, NO_3^- in the resulting solution?

Gases

Easter Market at the Moscow Kremlin, 1917 (oil on canvas), Kustodiev, Boris Mikhailovich (1878–1927)/State Art Museum, Nizhny Novgorod, Russia/Bridgeman Art Library

The gas in the balloons is most likely helium. Balloons are buoyant because the density of helium is less than that of air.

At 1.4 million atmospheres xenon, a gas, goes metallic.

—ROALD HOFFMANN
Giving In

By far the most familiar gas to all of us is the air we breathe. The Greeks considered air to be one of the four fundamental elements of nature, along with earth, water, and fire. Late in the eighteenth century, Cavendish, Priestley, and Lavoisier studied the composition of air, which is primarily a mixture of nitrogen and oxygen with smaller amounts of argon, carbon dioxide, and water vapor. Today it appears that the concentrations of some of the minor components of the atmosphere may be changing, with adverse effects on the environment. The depletion of the ozone layer and increases in the amounts of "greenhouse" gases are topics for the evening news, television dramas, and movies.

All gases resemble one another closely in their physical behavior. Their volumes respond in almost exactly the same way to changes in pressure, temperature, or amount of gas. In fact, it is possible to write a simple equation relating these four variables that is valid for all gases. This equation, known as the ideal gas law, is the central theme of this chapter; it is introduced in Section 5-2. The law is applied to

- pure gases in Section 5-3.
- gases in chemical reactions in Section 5-4.
- gas mixtures in Section 5-5.

Section 5-6 considers the kinetic theory of gases, the molecular model on which the ideal gas law is based. Finally, in Section 5-7 we describe the extent to which real gases deviate from the law.

Chapter Outline

5-1   Measurements on Gases

To completely describe the state of a gaseous substance, its volume, amount, temperature, and pressure are specified. The first three of these quantities were discussed in earlier chapters and will be reviewed briefly in this section. Pressure, a somewhat more abstract quantity, will be examined in more detail.

Volume, Amount, and Temperature

A gas expands uniformly to fill any container in which it is placed. This means that the volume of a gas is the volume of its container. Volumes of gases can be expressed in liters, cubic centimeters, or cubic meters:

$$1 \text{ L} = 10^3 \text{ cm}^3 = 10^{-3} \text{ m}^3$$

Most commonly, the amount of matter in a gaseous sample is expressed in terms of the number of moles (n). In some cases, the mass in grams is given instead. These two quantities are related through the molar mass, MM:

$$n = \frac{\text{mass}}{\text{MM}}$$

The temperature of a gas is ordinarily measured using a thermometer marked in degrees Celsius. However, as we will see in Section 5-2, *in any calculation involving the physical behavior of gases, temperatures must be expressed on the Kelvin scale.* To convert between °C and K, use the relation introduced in Chapter 1:

$$T_K = t_{°C} + 273.15$$

Typically, in gas law calculations, temperatures are expressed only to the nearest degree. In that case, the Kelvin temperature can be found by simply adding 273 to the Celsius temperature.

Pressure

Pressure is defined as force per unit area. You are probably familiar with the English unit "pounds per square inch," often abbreviated psi. When we say that a gas exerts a pressure of 15 psi, we mean that the pressure on the walls of the gas container is 15 pounds (of force) per square inch of wall area.

A device commonly used to measure atmospheric pressure is the mercury *barometer* (Figure 5.1), first constructed by Evangelista Torricelli in the seventeenth century. This consists of a closed gas tube filled with mercury inverted over a pool of mercury. The pressure exerted by the mercury column exactly equals that of the atmosphere. Hence the height of the column is a measure of the atmospheric pressure. At or near sea level, it typically varies from 740 to 760 mm, depending on weather conditions.

Most barometers contain mercury rather than some other liquid. Its high density allows the barometer to be a convenient size. A water barometer needs to be 10,340 mm (34 ft) high to register atmospheric pressure!

The pressure of a confined gas can be measured by a *manometer* of the type shown in Figure 5.2. Here again the fluid used is mercury. If the level in the inner tube (A) is lower than that in the outer tube (B), the pressure of the gas is greater than that of the atmosphere. If the reverse is true, the gas pressure is less than atmospheric pressure.

Figure 5.1 A mercury barometer. This is the type of barometer first constructed by Torricelli. The pressure of the atmosphere pushes the mercury in the dish to rise into the glass tube. The height of the column of mercury is a measure of the atmospheric pressure.

Because of the way in which gas pressure is measured, it is often expressed in **millimeters of mercury (mm Hg)**. Thus we might say that the atmospheric pressure on a certain day is 772 mm Hg. This means that the pressure of the air is equal to that exerted by a column of mercury 772 mm (30.4 in) high.

The pressure exerted by a column of mercury depends on its density, which varies slightly with temperature. To get around this ambiguity, the *torr* was defined to be the pressure exerted by 1 mm of mercury at certain specified conditions, notably 0°C. Over time, the unit torr has become a synonym for millimeter of mercury. Throughout this text, we will use millimeter of mercury rather than torr because the former has a clearer physical meaning.

Another unit commonly used to express gas pressure is the standard atmosphere, or simply **atmosphere (atm)**. This is the pressure exerted by a column of mercury 760 mm high with the mercury at 0°C. If we say that a gas has a pressure of 0.98 atm, we mean that the pressure is 98% of that exerted by a mercury column 760 mm high.

In the International System (Appendix 1), the standard unit of pressure is the **pascal** (Pa). A pascal is a very small unit; it is approximately the pressure exerted by a film of water 0.1 mm high on the surface beneath it. A related unit is the **bar** (10^5 Pa). A bar is nearly, but not quite, equal to an atmosphere:

$$1.013 \text{ bar} = 1 \text{ atm} = 760 \text{ mm Hg} = 14.7 \text{ psi} = 101.3 \text{ k Pa}$$
$$1 \text{ bar} = 10^5 \text{ Pa}$$

Figure 5.2 A manometer open to the atmosphere, used to measure gas pressure in a closed system. The pressure of the gas is given by $P_{gas} = P_{atm} + P_h$. In the figure, the gas pressure is greater than the atmospheric pressure.

EXAMPLE 5.1	

At room temperature, dry ice (solid CO_2) becomes a gas. At 77°F, 13.6 oz of dry ice are put into a steel tank with a volume of 10.00 ft³. The tank's pressure gauge registers 11.2 psi. Express the volume (V) of the tank in liters, the amount of CO_2 in grams and moles (n), the temperature (T) in °C and K and the pressure (P) in bars, mm Hg, and atmospheres.

ANALYSIS	
Information given:	volume (10.00 ft³); pressure (11.2 psi); temperature (77°F); mass of CO_2 (13.6 oz)
Information implied:	molar mass of CO_2 Table 1.3: conversion factors for volume and mass formulas for temperature conversion from °F to °C and from °C to K
Asked for:	volume in L pressure in atm, mm Hg, and bar temperature in °C and K moles of CO_2

STRATEGY	

1. Find the necessary conversion factors.

2. Use the temperature conversion formula.

3. Convert oz to grams and use the molar mass of CO_2 as a conversion factor.

oz → g → mol

continued

	SOLUTION
volume in L	$10.00 \text{ ft}^3 \times \dfrac{28.32 \text{ L}}{1 \text{ ft}^3} = 283.2 \text{ L}$
pressure in atm	$11.2 \text{ psi} \times \dfrac{1 \text{ atm}}{14.7 \text{ psi}} = 0.762 \text{ atm}$
pressure in mm Hg	$11.2 \text{ psi} \times \dfrac{1 \text{ atm}}{14.7 \text{ psi}} \times \dfrac{760 \text{ mm Hg}}{1 \text{ atm}} = 579 \text{ mm Hg}$
pressure in bar	$11.2 \text{ psi} \times \dfrac{1.013 \text{ bar}}{14.7 \text{ psi}} = 0.772 \text{ bar}$
temperature in °C	$°F = 1.8(°C) + 32;\ 77° = 1.8(°C) + 32°C;\ °C = 25°C$
temperature in K	$K = (°C) + 273.15;\ K = 25°C + 273.15 = 298 \text{ K}$
mol CO_2	$13.6 \text{ oz} \times \dfrac{1 \text{ g}}{0.03527 \text{ oz}} \times \dfrac{1 \text{ mol}}{44.01 \text{ g}} = 8.77 \text{ mol}$

5-2 The Ideal Gas Law

All gases closely resemble each other in the dependence of volume on amount, temperature, and pressure.

1. Volume is directly proportional to amount. Figure 5.3a shows a typical plot of volume (V) versus number of moles (n) for a gas. Notice that the graph is a straight line passing through the origin. The general equation for such a plot is

$$V = k_1 n \qquad \text{(constant } T, P)$$

where k_1 is a constant (the slope of the line in Figure 5.3a). That means k_1 is independent of individual values of V, n, and the nature of the gas. This is the equation of a direct proportionality.

2. Volume is directly proportional to absolute temperature. The dependence of volume (V) on the Kelvin temperature (T) is shown in Figure 5.3b. The graph is a straight line through the origin. The equation of the line is

$$V = k_2 T \qquad \text{(constant } n, P)$$

where k_2 is again the slope of the line in Figure 5.3b. This relationship was first suggested in a different form, by two French scientists, Jacques Charles (1746–1823) and Joseph Gay-Lussac (1778–1850), ◀ⓘ both of whom were balloonists.

ⓘ Gay-Lussac isolated boron and was the first to prepare HF.

Figure 5.3 Relation of gas volume (V) to number of moles (n) and temperature (T) at constant pressure (P). The volume of a gas at constant pressure is directly proportional to (a) the number of moles of gas and (b) the absolute temperature. The volume-temperature plot must be extrapolated to reach zero because most gases liquefy at low temperatures well above 0 K.

© Cengage Learning/Charles D. Winters

Liquid nitrogen is poured into a beaker.

Inflated balloons are pushed into the very cold liquid (T = 77 K).

The volume of the air in the balloons, as predicted by Charles's law, decreases; the balloons all fit into the beaker.

When removed from the liquid nitrogen, the volume of the air in the balloons expands as the air warms up.

Figure 5.4 **An illustration of Charles's law.**

Hence it is often referred to as **Charles's and Gay-Lussac's law** or, more simply, as *Charles's law* (Figure 5.4).

3. *Volume is inversely proportional to pressure.* Figure 5.5 shows a typical plot of volume (V) versus pressure (P). Notice that V decreases as P increases. The graph is a hyperbola. The general relation between the two variables is

$$V = \frac{k_3}{P} \qquad \text{(constant } n, T)$$

The quantity k_3, like k_1 and k_2, is a constant. This is the equation of an inverse proportionality. The fact that volume is inversely proportional to pressure was first established in 1660 by Robert Boyle (1627–1691), an Irish experimental scientist. The equation above is one form of **Boyle's law**.

The three equations relating the volume, pressure, temperature, and amount of a gas can be combined into a single equation. Because V is directly proportional to both n and T,

$$V = k_1 n \qquad V = k_2 T$$

and inversely proportional to P,

$$V = \frac{k_3}{P}$$

it follows that

$$V = \text{constant} \times \frac{n \times T}{P}$$

We can evaluate the constant ($k_1 k_2 k_3$) in this equation by taking advantage of *Avogadro's law*, which states that equal volumes of all gases at the same temperature and pressure contain the same number of moles. For this law to hold, the constant must be the same for all gases. Ordinarily it is represented by the symbol R. Both sides of the equation are multiplied by P to give the **ideal gas law**

$$PV = nRT \tag{5.1}$$

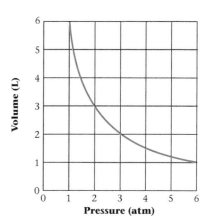

Figure 5.5 **Relation of gas volume (V) to pressure (P) at constant temperature (T).** The volume of a fixed quantity of gas at constant temperature is inversely proportional to the pressure. In this case, the volume decreases from 6 L to 1 L when the pressure increases from 1 atm to 6 atm.

R is a constant, independent of *P*, *V*, *n*, and *T*, but its numerical value depends on the units used.

Table 5.1 Values of *R* in Different Units

Value	Where Used	How Obtained
$0.0821 \dfrac{L \cdot atm}{mol \cdot K}$	Gas law problems with *V* in liters, *P* in atm	From known values of *P*, *V*, *T*, *n*
$8.31 \dfrac{J}{mol \cdot K}$	Equations involving energy in joules	$1 \; L \cdot atm = 101.3 \; J$
$8.31 \times 10^3 \dfrac{g \cdot m^2}{s^2 \cdot mol \cdot K}$	Calculation of molecular speed (page 115)	$1 \; J = 10^3 \dfrac{g \cdot m^2}{s^2}$

18.9 L   5.8 L   2 L

22.4 LITERS

© Cengage Learning/James Maynard

Figure 5.6 Molar volume. The cube has a volume of 22.4 L, which is the volume of one mole of an ideal gas at STP.

where *P* is the pressure, *V* the volume, *n* the number of moles, and *T* the Kelvin temperature. Experimentally, it is found that the ideal gas law predicts remarkably well the experimental behavior of real gases (e.g., H_2, N_2, O_2 ...) at ordinary temperatures and pressures.

The value of the **gas constant** (*R*) can be calculated from experimental values of *P*, *V*, *n*, and *T*. Consider, for example, the situation that applies at 0°C and 1 atm. These conditions are often referred to as **standard temperature and pressure** **(STP)** for a gas. At STP, one mole of any gas occupies a volume of 22.4 L (Figure 5.6). Solving the ideal gas law for *R*,

$$R = \frac{PV}{nT}$$

Substituting *P* = 1.00 atm, *V* = 22.4 L, *n* = 1.00 mol, and *T* = 0 + 273 = 273 K,

$$R = \frac{1.00 \; atm \times 22.4 \; L}{1.00 \; mol \times 273 \; K} = 0.0821 \; L \cdot atm/(mol \cdot K)$$

Notice that *R* has the units of atmospheres, liters, moles, and K. These units must be used for pressure, volume, amount, and temperature in any problem in which this value of *R* is used.

Throughout most of this chapter, we will use 0.0821 L · atm/(mol · K) as the value of *R*. For certain purposes, however, *R* must be expressed in different units (Table 5.1).

5-3 Gas Law Calculations

The ideal gas law can be used to solve a variety of problems. We will show how you can use it to find

- the final state of a gas, knowing its initial state and the changes in *P*, *V*, *n*, or *T* that occur.
- one of the four variables, *P*, *V*, *n*, or *T*, given the values of the other three.
- the molar mass or density of a gas.

Final and Initial State Problems

A gas commonly undergoes a change from an initial to a final state. Typically, you are asked to determine the effect on *V*, *P*, *n*, or *T* of a change in one or more of these variables. For example, starting with a sample of gas at 25°C and 1.00 atm, you might be asked to calculate the pressure developed when the sample is heated to 95°C at constant volume.

The ideal gas law is readily applied to problems of this type. A relationship between the variables involved is derived from this law. In this case, pressure and temperature change, while n and V remain constant.

initial state: $\quad P_1 V = nRT_1$

final state: $\quad P_2 V = nRT_2$

Dividing the second equation by the first cancels V, n, and R, leaving the relation

$$\frac{P_2}{P_1} = \frac{T_2}{T_1} \quad \text{(constant } n, V)$$

Applying this general relation to the problem just described,

$$P_2 = P_1 \times \frac{T_2}{T_1} = 1.00 \text{ atm} \times \frac{368 \text{ K}}{298 \text{ K}} = 1.23 \text{ atm}$$

Similar "two-point" ◄ *i* equations can be derived from the ideal gas law to solve any problem of this type.

> *i* To obtain a two-point equation, write the gas law twice and divide to eliminate constants.

EXAMPLE 5.2

A sealed 15.0-L steel tank is used to deliver propane (C_3H_8) gas. It is fitted with a pressure gauge and a valve. The valve automatically opens to release gas when the pressure in the tank goes above 1.200 atm. The tank is filled with 24.7 g of propane at 25°C. The pressure gauge registers 0.915 atm.

a If the tank is heated to 55°C, will the valve open to release propane? (Assume that the expansion of steel from an increase in temperature is negligible.)

ANALYSIS

Information given:	V(15.0 L); $\quad P$(0.915 atm); $\quad T$(25°C); $\quad$ mass of propane (24.7 g) $\quad T$(55°C)
Information implied:	2 sets of conditions for temperature
Asked for:	Will the valve release gas?

STRATEGY

1. Given two sets of conditions, you need to use the formula for initial state–final state conditions.
2. A sealed steel tank implies that the number of moles and the volume are kept constant.
3. Make sure all temperatures are in K.
4. Find P and compare with 1.200 atm.

SOLUTION

P_2	$\dfrac{V_1 P_1}{n_1 T_1} = \dfrac{V_2 P_2}{n_2 T_2} \rightarrow \dfrac{P_1}{T_1} = \dfrac{P_2}{T_2} \rightarrow \dfrac{0.915}{25 + 273} = \dfrac{P_2}{55 + 273} \rightarrow P_2 = 1.01 \text{ atm}$
Will the valve open?	No. 1.01 atm < 1.200 atm

b The temperature in the tank is increased to 200°C. The valve opens and releases propane. How many grams of propane are released?

ANALYSIS

Information given:	Before T increase: V (15.0 L); P (0.915 atm); T (25°C); mass of propane (24.7 g) After T increase: V (15.0 L); P (1.200 atm); T (200°C); mass of propane (24.7 g)
Information implied:	molar mass of propane
Asked for:	mass of propane released

continued

STRATEGY

1. Convert °C to K and mass of propane to mol propane.

2. Given two sets of conditions, you need to use the formula for initial state–final state conditions.

3. Find n_2.

4. Convert n_2 to mass of propane in the tank at 200°C and find the mass released by difference.

SOLUTION

1. T	°C → K; 25 + 273 = 298; 200 + 273 = 473 K
mol C_3H_8 initially (n_1)	$24.7 \text{ g} \times \dfrac{1 \text{ mol}}{44.1 \text{ g}} = 0.560 \text{ mol}$
2. Initial conditions	$P_1 = 0.915 \text{ atm}; T_1 = 298 \text{ K}; n_1 = 0.560 \text{ mol}; V_1 = 15.0 \text{ L}$
Final conditions	$P_2 = 1.200 \text{ atm}; T_2 = 473 \text{ K}; n_2 = ?; V_2 = 15.0 \text{ L}$
3. n_2	$\dfrac{V_1 P_1}{n_1 T_1} = \dfrac{V_2 P_2}{n_2 T_2} \rightarrow \dfrac{0.915}{0.560 \times 298} = \dfrac{1.200}{n_2 \times 473};\quad n_2 = 0.463 \text{ mol}$
mass C_3H_8 in the tank	$(0.463 \text{ mol})(44.1 \text{ g/mol}) = 20.4 \text{ g}$
mass to be released	$24.7 - 20.5 = 4.3 \text{ g}$

END POINT

Note that the volume of the tank is never used in the calculations. It is important not only to read the problem carefully but also to visualize the description of the gas container. If the gas were in a balloon, instead of in a steel tank, the calculations would be different.

Calculation of P, V, n, or T

Frequently, values are known for three of these quantities (perhaps V, n, and T); the other one (P) must be calculated. This is readily done by direct substitution into the ideal gas law.

EXAMPLE 5.3

Sulfur hexafluoride is a gas used as a long-term tamponade (plug) for a retinal hole to repair detached retinas in the eye. If 2.50 g of this compound is introduced into an evacuated 500.0-mL container at 83°C, what pressure in atmospheres is developed?

ANALYSIS

Information given:	V(500.0 mL); T(83°C); mass of SF_6 (2.50 g)
Information implied:	molar mass of SF_6 ideal gas law (one state) value for R
Asked for:	pressure (P) in atm

STRATEGY

1. Change the given units to conform with the units for R (mL → L; °C → K).

2. You need to find n before you can use the ideal gas law to find P.

3. Substitute into the ideal gas law: $PV = nRT$.

continued

	SOLUTION
1. Change units.	500.0 mL = 0.5000 L 83°C + 273 K = 356 K
2. Find n.	$2.50 \text{ g} \times \dfrac{1 \text{ mol}}{146.07 \text{ g}} = 0.0171 \text{ mol}$
3. P	$P = \dfrac{nRT}{V} = \dfrac{0.0171 \text{ mol} \times 0.0821 \text{ L} \cdot \text{atm/(mol} \cdot \text{K}) \times 356 \text{ K}}{0.5000 \text{ L}} = 1.00 \text{ atm}$

Molar Mass and Density

The ideal gas law offers a simple approach to the experimental determination of the molar mass of a gas. Indeed, this approach can be applied to volatile liquids like acetone (Example 5.4). All you need to know is the mass of a sample confined to a container of fixed volume at a particular temperature and pressure.

EXAMPLE 5.4 GRADED

Acetone has a density of 1.96 g/L at 95°C and 1.02 atm.

a How many moles of acetone are there in a 1.00-L flask under these conditions?

	ANALYSIS
Information given:	volume of the flask (1.00 L); $T = 95°C + 273 = 368$ K; density = 1.96 g/L
Information implied:	mass of acetone (1.96 g)
Asked for:	moles of acetone

STRATEGY
1. A gas occupies the volume of the flask; volume of acetone = volume of flask
2. Find n using the ideal gas law.

	SOLUTION
n	$n = \dfrac{PV}{RT} = \dfrac{1.02 \times 1.00}{0.0821 \times 368} = 0.0338 \text{ mol}$

b What is the molar mass of the acetone?

	ANALYSIS
Information given:	from part (a): $n = 0.0338$ mol
Information implied:	mass of acetone (1.96 g)
Asked for:	molar mass

	SOLUTION
molar mass	$MM = \dfrac{\text{mass}}{\text{moles}} = \dfrac{1.96 \text{ g}}{0.0338 \text{ mol}} = 58.0 \text{ g/mol}$

continued

c   Acetone contains the three elements, C, H, and O. When 1.000 g of acetone is burned, 2.27 g of CO_2 and 0.932 g of H_2O are formed. What is the molecular formula of the acetone?

ANALYSIS

Information given:	from part (b), molar mass of acetone (58.1 g/mol) The combustion of 1.00 g of sample yields 2.27 g CO_2 and 0.932 g H_2O.
Information implied:	mass of C, H, and O in 1.00-g sample
Asked for:	molecular formula of acetone

STRATEGY

1. Recall from Section 3.2 how to convert the mass of the product of combustion to the mass of the element.

2. Follow Figure 3.7 to obtain the simplest formula for the compound.

3. Compare the simplest formula's molar mass to the molar mass obtained in part (b).

SOLUTION

Mass of each element	mass C: $2.27 \text{ g } CO_2 \times \dfrac{12.01 \text{ g C}}{44.01 \text{ g } CO_2} = 0.619$ g
	mass H: $0.932 \text{ g } H_2O \times \dfrac{2(1.008) \text{ g H}}{18.02 \text{ g } H_2O} = 0.104$ g
	mass O = mass sample − (mass C + mass H) = 1.000 g − (0.619 + 0.104) g = 0.277 g
moles of each element	C: $\dfrac{0.619 \text{ g}}{12.01 \text{ g/mol}} = 0.0515$ mol; H: $\dfrac{0.104 \text{ g}}{1.008 \text{ g/mol}} = 0.103$ mol;
	O: $\dfrac{0.277 \text{ g}}{16.00 \text{ g/mol}} = 0.0173$ mol
Atomic ratios	C: $\dfrac{0.0515}{0.0173} = 3$; H: $\dfrac{0.104}{0.0173} = 6$; O: $\dfrac{0.0173}{0.0173} = 1$
Simplest formula	C_3H_6O
MM of simplest formula	$3(12.01) + 6(1.008) + 16.00 = 58.08$ g/mol
MM of vapor (from part (b))	58.1 g/mol
Molecular formula	C_3H_6O (simplest formula = molecular formula)

One way to calculate gas density is to use the ideal gas law where $\dfrac{\text{mass}}{\text{MM}}$ is substituted for n.

$$PV = \frac{\text{mass}}{\text{MM}} RT$$

Since density $= \dfrac{\text{mass}}{V}$ the general relation

$$\text{density} = MM \left(\frac{P}{RT} \right) \tag{5.2}$$

results.

From this equation we see that the density of a gas is dependent (Figure 5.7) on

- *pressure.* Compressing a gas increases its density.
- *temperature.* Hot air rises because a gas becomes less dense when its temperature is increased. Hot air balloons are filled with air at a temperature higher

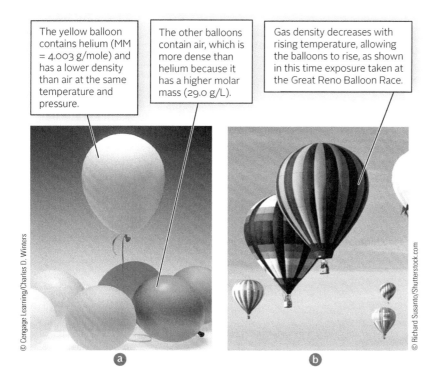

The yellow balloon contains helium (MM = 4.003 g/mole) and has a lower density than air at the same temperature and pressure.

The other balloons contain air, which is more dense than helium because it has a higher molar mass (29.0 g/L).

Gas density decreases with rising temperature, allowing the balloons to rise, as shown in this time exposure taken at the Great Reno Balloon Race.

© Cengage Learning/Charles D. Winters

© Richard Susanto/Shutterstock.com

(a)   **(b)**

Figure 5.7 **Density of a gas.**

than that of the atmosphere, making the air inside less dense than that outside. First used in France in the eighteenth century, they are now seen in balloon races and other sporting events. Heat is supplied on demand by a propane burner.

■ *molar mass.* Hydrogen (MM = 2.016 g/mol) has the lowest molar mass and the lowest density (at a given P and T) of all gases. Hence it has the greatest lifting power in "lighter-than-air" balloons. However, hydrogen has not been used in balloons carrying passengers since 1937, when the *Hindenburg,* a hydrogen-filled airship, exploded and burned. Helium (MM = 4.003 g/mol) is slightly less effective than hydrogen but a lot safer to work with because it is nonflammable. It is used in a variety of balloons, ranging from the small ones used at parties to meteorological balloons with volumes of a billion liters.

Very light gases, notably hydrogen and helium, tend to escape from the earth's atmosphere. The hydrogen you generate in the laboratory today is well on its way into outer space tomorrow. A similar situation holds with helium, which is found in very limited quantities mixed with natural gas in wells below the earth's surface. If helium is allowed to escape, it is gone forever, and our supply of this very useful gaseous element is depleted.

5-4 Stoichiometry of Gaseous Reactions

As pointed out in Chapter 3, a balanced equation can be used to relate moles or grams of substances taking part in a reaction. Where gases are involved, these relations can be extended to include volumes. To do this, we use the ideal gas law and modify Figures 3.11 and 4.6 as shown in Figure 5.8 (page 106). Note that the ideal gas law equation can be used only for gaseous reactants and products.

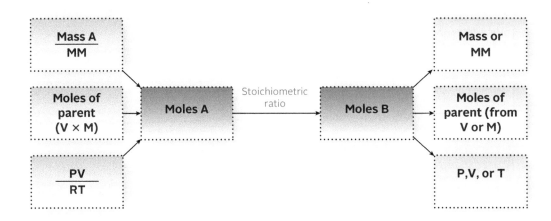

Figure 5.8 Flowchart for stoichiometry calculations involving gases.

EXAMPLE 5.5

Hydrogen peroxide, H_2O_2, is a common bleaching agent. It decomposes quickly to water and oxygen gas at high temperatures.

$$2H_2O_2(l) \longrightarrow 2H_2O(l) + O_2(g)$$

How many liters of oxygen are produced at 78°C and 0.934 atm when 1.27 L of H_2O_2 ($d = 1.00$ g/mL) decompose?

ANALYSIS

Information given:	temperature (78°C); pressure (0.934 atm) H_2O_2: volume (1.27 L); density (1.00 g/mL)
Information implied:	mass and molar mass of H_2O_2 stoichiometric ratio of O_2 to H_2O_2 (2 H_2O_2/1 O_2)
Asked for:	volume of oxygen

STRATEGY

1. Change °C to K and L of H_2O_2 to mL. (Density is given in g/mL.)

2. Find the mass of H_2O_2. Note that you cannot directly use the volume of H_2O_2 to calculate the volume of O_2 because H_2O_2 is NOT a gas.

3. Follow Figure 5.8.

$$\text{mass } H_2O_2 \xrightarrow{\text{MM}} \text{mol } H_2O_2 \xrightarrow[\text{ratio}]{\text{stoichiometric}} \text{moles } O_2 \xrightarrow{nRT/P} \text{volume of oxygen}$$

SOLUTION

Mass H_2O_2	mass = (density)(volume) = (1.00 g/mL)(1.27 × 10³ mL) = 1.27 × 10³ g
Mol O_2	$1.27 \times 10^3 \text{ g} \times \dfrac{1 \text{ mol } H_2O_2}{34.02 \text{ g}} \times \dfrac{1 \text{ mol } O_2}{2 \text{ mol } H_2O_2} = 18.7 \text{ mol}$
Volume O_2	$V = \dfrac{nRT}{P} = \dfrac{(18.7 \text{ mol})(0.0821 \text{ L} \cdot \text{atm/mol} \cdot \text{K})(78 + 273)\text{K}}{0.934 \text{ atm}} = 576 \text{ L}$

EXAMPLE 5.6   GRADED

Sodium bicarbonate (baking soda) is widely used to absorb odors inside refrigerators. When acid is added to baking soda, the following reaction occurs:

$$NaHCO_3(s) + H^+(aq) \longrightarrow Na^+(aq) + CO_2(g) + H_2O$$

All experiments here are performed with 2.45 M HCl and 12.75 g of $NaHCO_3$ at 732 mm Hg and 38°C.

---

a   If an excess of HCl is used, what volume of CO_2 is obtained?

ANALYSIS

Information given:	pressure (732 mm Hg); temperature (38°C); mass of $NaHCO_3$ (12.75 g)
Information implied:	molar mass of $NaHCO_3$ stoichiometric ratio: 1 $NaHCO_3$ /1 CO_2
Asked for:	volume of CO_2 produced

STRATEGY

1. Follow the flowchart in Figure 5.8.

2. Convert to appropriate units of pressure and temperature.

$$\text{mass}_{NaHCO_3} \xrightarrow{MM} n_{NaHCO_3} \xrightarrow{\substack{\text{stoichiometric} \\ \text{ratio}}} n_{CO_2} \xrightarrow{PV = nRT} V_{CO_2}$$

SOLUTION

mol CO_2 (n)	$12.75 \text{ g NaHCO}_3 \times \dfrac{1 \text{ mol}}{84.01 \text{ g}} \times \dfrac{1 \text{ mol CO}_2}{1 \text{ mol NaHCO}_3} = 0.1518$
volume CO_2 (V)	$V = \dfrac{0.1518 \text{ mol} \times 0.0821 \text{ L} \cdot \text{atm/mol} \cdot \text{K} \times (273 + 38)\text{K}}{(732/760)\text{atm}} = 4.02 \text{ L}$

---

b   If $NaHCO_3$ is in excess, what volume of HCl is required to produce 2.65 L of CO_2?

ANALYSIS

Information given:	pressure (732 mm Hg); temperature (38°C); volume of CO_2 produced (2.65 L); molarity of HCl (2.45 M)
Information implied:	H^+ is the reacting species. HCl is the parent compound. stoichiometric ratio: 1 H^+/1 CO_2

STRATEGY

Follow the flowchart in Figure 5.8.

$$V_{CO_2} \xrightarrow{PV = nRT} n_{CO_2} \xrightarrow{\substack{\text{stoichiometric} \\ \text{ratio}}} n_{H^+} \xrightarrow{\substack{\text{atom} \\ \text{ratio}}} n_{HCl} \xrightarrow{M} V_{HCl}$$

SOLUTION

mol CO_2	$n = \dfrac{2.65 \text{ L} \times (732/760)\text{atm}}{0.0821 \text{ L} \cdot \text{atm/mol} \cdot \text{K} \times (273 + 38)\text{K}} = 0.100$
mol HCl	$0.100 \text{ mol CO}_2 \times \dfrac{1 \text{ mol H}^+}{1 \text{ mol CO}_2} \times \dfrac{1 \text{ mol HCl}}{1 \text{ mol H}^+} = 0.100$
Volume HCl	$\dfrac{0.100 \text{ mol HCl}}{2.45 \text{ mol/L}} = 0.0408 \text{ L} = 40.8 \text{ mL}$

continued

| c | What volume of CO_2 is produced when all the $NaHCO_3$ is made to react with 50.0 mL of HCl? |

ANALYSIS

Information given:	molarity of HCl (2.45 M); volume of HCl (50.0 mL); pressure (732 mm Hg); temperature (38°C)
Information implied:	H^+ is the reacting species. HCl is the parent compound. stoichiometric ratios: 1 H^+/1 CO_2; 1 $NaHCO_3$/1 CO_2 from part (a): mol $NaHCO_3$

STRATEGY

1. The presence of enough given data to calculate the number of moles of each reactant tells you that part (c) is a limiting reactant problem.

2. Follow the flowchart in Figure 5.8 to determine the number of moles of CO_2 obtained if HCl is limiting. You can obtain the moles of CO_2 if $NaHCO_3$ is limiting from part (a).

3. Compare the moles of CO_2 obtained using H^+ as the limiting reactant to the moles of CO_2 obtained using $NaHCO_3$ as the limiting reactant. Choose the smaller number of moles of CO_2.

4. Use the ideal gas law to convert mol CO_2 to the volume of CO_2.

SOLUTION

mol CO_2: $NaHCO_3$ limiting	from part (a): 0.1518 mol $NaHCO_3 \times \dfrac{1 \text{ mol } CO_2}{1 \text{ mol } NaHCO_3} = 0.1518$
mol CO_2: HCl limiting	$(0.0500 \text{ L} \times 2.45 \text{ mol/L})$ mol HCl $\times \dfrac{1 \text{ mol } H^+}{1 \text{ mol HCl}} \times \dfrac{1 \text{ mol } CO_2}{1 \text{ mol } H^+} = 0.122$ mol
Theoretical yield of CO_2	$0.122 < 0.1518$; 0.122 mol CO_2 obtained
Volume CO_2	$V = \dfrac{0.122 \text{ mol} \times 0.0821 \text{ L} \cdot \text{atm/mol} \cdot \text{K} \times (273 + 38)\text{K}}{(732/760)\text{atm}} = 3.25 \text{ L}$

END POINTS

1. When a problem comes in several parts, you may not need to use all the given information for each part.

2. You should also check to see whether you can use information that you obtained from the preceding parts for subsequent questions.

Figure 5.9 Electrolysis of water (H_2O). The volume of hydrogen (H_2) formed in the tube at the right is twice the volume of oxygen (O_2) formed in the tube at the left, in accordance with the equation $2H_2O(l) \longrightarrow 2H_2(g) + O_2(g)$.

© Cengage Learning/Charles D. Winters

Perhaps the first stoichiometric relationship to be discovered was the *law of combining volumes,* proposed by Gay-Lussac in 1808: ***The volume ratio of any two gases in a reaction at constant temperature and pressure is the same as the reacting mole ratio.***

To illustrate the law, consider the reaction

$$2H_2O(l) \longrightarrow 2H_2(g) + O_2(g)$$

As you can see from Figure 5.9, the volume of hydrogen produced is twice that of the other gaseous product, oxygen.

The law of combining volumes, like so many relationships involving gases, is readily explained by the ideal gas law. At constant temperature and pressure, volume is directly proportional to number of moles ($V = k_1n$). It follows that for gaseous species involved in reactions, the volume ratio must be the same as the mole ratio given by the coefficients of the balanced equation.

CHEMISTRY THE **HUMAN** SIDE

Avogadro's law (page 99) was proposed in 1811 by an Italian physicist at the University of Turin with the improbable name of Lorenzo Romano Amadeo Carlo Avogadro di Quarequa e di Cerreto (1776–1856).

Avogadro suggested this relationship to explain the law of combining volumes. Today it seems obvious. For example, in the reaction

$$2H_2O(l) \longrightarrow 2H_2(g) + O_2(g)$$

the volume of hydrogen, like the number of moles, is twice that of oxygen. Hence, equal volumes of these gases must contain the same number of moles or molecules. This was by no means obvious to Avogadro's contemporaries. Berzelius, among others, dismissed Avogadro's ideas because he did not believe diatomic molecules composed of identical atoms (H_2, O_2) could exist. Dalton went a step further; he refused to accept the law of combining volumes because he thought it implied splitting atoms.

As a result of arguments like these, Avogadro's ideas lay dormant for nearly half a century. They were revived by another Italian scientist, Stanislao Cannizzaro, professor of chemistry at the University of Genoa. At a conference held in Karlsruhe in 1860, he persuaded the chemistry community of the validity of Avogadro's law and showed how it could be used to determine molar and atomic masses.

The quantity now called "Avogadro's number" (6.02×10^{23}/mol) was first estimated in 1865, nine years after Avogadro died. Not until well into the twentieth century did it acquire its present name. It seems appropriate to honor Avogadro in this way for the contributions he made to chemical theory.

Amadeo Avogadro
(1776–1856)

EXAMPLE	5.7	GRADED

Consider the reaction between hydrogen and oxygen to form water.

$$2H_2(g) + O_2(g) \longrightarrow 2H_2O(l)$$

One liter of oxygen is used and the reactant gases are maintained at 25°C and 1.00 atm. The density of water under the conditions of the experiment is 0.997 g/mL.

a   What volume of $H_2(g)$ at 25°C and 1.00 atm is required to react with 1.00 L of $O_2(g)$?

ANALYSIS	
Information given:	volume O_2 (1.00 L); pressure (1.00 atm); temperature (25°C)
Information implied:	stoichiometric ratio: 1 mol O_2/2 mol H_2 Conditions of temperature and pressure are constant.
Asked for:	Volume of H_2 at constant T and P that reacts with O_2

STRATEGY	

Use the law of combining volumes (T and P are constant).

$$V_{O_2} \xrightarrow{\text{stoichiometric ratio}} V_{H_2}$$

SOLUTION	
volume of H_2	$1.00 \text{ L } O_2 \times \dfrac{2 \text{ L } H_2}{1 \text{ L } O_2} = 2.00 \text{ L } H_2$

b   What volume of $H_2O(l)$ is formed from the reaction?

ANALYSIS	
Information given:	volume of $O_2(g)$ (1.00 L); pressure (1.00 atm); temperature (25°C); density of water (0.997 g/mL)
Asked for:	Volume of water obtained

continued

STRATEGY

The law of combining volumes cannot be used here because water is a liquid.

1. Use the ideal gas law to calculate the number of moles of O_2.
2. Follow the flowchart in Figure 5.8 to calculate the mass of water obtained.
3. Use the density of water to calculate its volume.

SOLUTION

1. Moles of O_2	$n_{O_2} = \dfrac{PV}{RT} = \dfrac{(1.00 \text{ L})(1.00 \text{ atm})}{(0.0821 \text{ L} \cdot \text{atm/mol} \cdot \text{K})(273 + 25)\text{K}} = 0.0409$
2. Mass of water	$0.0409 \text{ mol } O_2 \times \dfrac{2 \text{ mol } H_2O}{1 \text{ mol } O_2} \times \dfrac{18.02 \text{ g } H_2O}{1 \text{ mol}} = 1.47 \text{ g}$
3. Volume of water	$V = \dfrac{\text{mass}}{\text{density}} = \dfrac{1.47 \text{ g}}{0.997 \text{ g/mL}} = 1.48 \text{ mL}$

c What mass of $H_2O(l)$ is formed, assuming a yield of 85.2%?

ANALYSIS

Information given:	% yield (85.2%)
Information implied:	theoretical yield
Asked for:	mass of water obtained (actual yield)

STRATEGY

1. The mass obtained in part (b) is the theoretical yield.
2. Calculate actual yield from percent yield.

$$\% \text{ yield} = \frac{\text{actual yield}}{\text{theoretical yield}} \times 100\%$$

SOLUTION

Actual yield	$\text{actual yield} = \dfrac{\% \text{ yield}}{100\%} \times \text{theoretical yield} = \dfrac{85.2\%}{100\%} \times 1.47 \text{ g} = 1.25 \text{ g}$

5-5 Gas Mixtures: Partial Pressures and Mole Fractions

Because the ideal gas law applies to all gases, you might expect it to apply to gas mixtures. Indeed it does. For a mixture of two gases A and B, the total pressure is given by the expression

$$P_{\text{tot}} = n_{\text{tot}} \frac{RT}{V} = (n_A + n_B) \frac{RT}{V}$$

Separating the two terms on the right,

$$P_{\text{tot}} = n_A \frac{RT}{V} + n_B \frac{RT}{V}$$

The terms $n_A RT/V$ and $n_B RT/V$ are, according to the ideal gas law, the pressures that gases A and B would exert if they were alone. Each quantity is referred to as having a **partial pressure,** ◂ⓘ P_A and P_B.

$$P_A = \text{partial pressure A} = n_A RT/V$$
$$P_B = \text{partial pressure B} = n_B RT/V$$

ⓘ Partial pressure is the pressure a gas would exert if it occupied the entire volume by itself.

Figure 5.10 **Collecting a gas by water displacement.** (a) Hydrogen gas is being generated in the flask by an acidic solution dripping onto a metal. (b) When a gas is collected by displacing water, it becomes saturated with water vapor. The partial pressure of $H_2O(g)$ in the collecting flask is equal to the vapor pressure of liquid water at the temperature of the system.

Substituting P_A and P_B for $n_A RT/V$ and $n_B RT/V$ in the equation for P_{tot},

$$P_{tot} = P_A + P_B \qquad (5.3)$$

The relation just derived was first proposed by John Dalton in 1801; it is often referred to as **Dalton's law** of partial pressures:

The total pressure of a gas mixture is the sum of the partial pressures of the components of the mixture.

To illustrate Dalton's law, consider a gaseous mixture of hydrogen and helium in which

$$P_{H_2} = 2.46 \text{ atm} \qquad P_{He} = 3.69 \text{ atm}$$

It follows from Dalton's law that

$$P_{tot} = 2.46 \text{ atm} + 3.69 \text{ atm} = 6.15 \text{ atm}$$

Wet Gases; Partial Pressure of Water

When a gas such as hydrogen is collected by bubbling through water (Figure 5.10), it picks up water vapor; molecules of H_2O escape from the liquid and enter the gas phase. Dalton's law can be applied to the resulting gas mixture:

$$P_{tot} = P_{H_2O} + P_{H_2}$$

In this case, P_{tot} is the measured pressure. The partial pressure of water vapor, P_{H_2O}, is equal to the **vapor pressure** of liquid water. It has a fixed value at a given temperature (see Appendix 1). The partial pressure of hydrogen, P_{H_2}, can be calculated by subtraction. The number of moles of hydrogen in the wet gas, n_{H_2}, can then be determined using the ideal gas law.

EXAMPLE 5.8

A student prepares a sample of hydrogen gas by electrolyzing water at 25°C. She collects 152 mL of H_2 at a total pressure of 758 mm Hg. Using Appendix 1 to find the vapor pressure of water at 25°C, calculate

(a) the partial pressure of hydrogen.

(b) the number of moles of hydrogen collected.

continued

ANALYSIS	
Information given:	V_{H_2} (152 mL); pressure (758 mm Hg); temperature (25°C)
Information implied:	vapor pressure of water at 25°C (Appendix 1) Volume and temperature are constant.
Asked for:	(a) P_{H_2} (b) n_{H_2}

STRATEGY	
(a) Recall that H_2 and $H_2O(g)$ contribute to the total pressure P_{tot}. Use Dalton's law: $P_{tot} = P_1 + P_2 + \ldots$ (b) Use the ideal gas law to calculate n_{H_2} at P_{H_2}.	

SOLUTION	
(a) P_{H_2}	$P_{tot} = P_{H_2} + P_{H_2O}$ $P_{H_2} = 758 \text{ mm Hg} - 23.76 \text{ mm Hg} = 734 \text{ mm Hg}$
(b) n_{H_2}	$n_{H_2} = \dfrac{P_{H_2}V}{RT} = \dfrac{[(734/760)\text{atm}](0.152 \text{ L})}{(298 \text{ K})(0.0821 \text{ L} \cdot \text{atm/mol} \cdot \text{K})} = 0.00600 \text{ mol}$

Vapor pressure, like density and solubility, is an intensive physical property that is characteristic of a particular substance. The vapor pressure of water at 25°C is 23.76 mm Hg, independent of volume or the presence of another gas. Like density and solubility, vapor pressure varies with temperature; for water it is 55.3 mm Hg at 40°C, 233.7 mm Hg at 70°C, and 760.0 mm Hg at 100°C. We will have more to say in Chapter 9 ◀ ⓘ about the temperature dependence of vapor pressure.

ⓘ Vapor pressure is further discussed in Chapter 9.

Partial Pressure and Mole Fraction

As pointed out earlier, the following relationship applies to a mixture containing gas A (and gas B):

$$P_A = \frac{n_A RT}{V} \qquad P_{tot} = \frac{n_{tot} RT}{V}$$

Dividing P_A by P_{tot} gives

$$\frac{P_A}{P_{tot}} = \frac{n_A}{n_{tot}}$$

The fraction n_A/n_{tot} is referred to as the **mole fraction** of A in the mixture. It is the fraction of the total number of moles that is accounted for by gas A. Using X_A to represent the mole fraction of A (i.e., $X_A = n_A/n_{tot}$),

$$P_A = X_A P_{tot} \qquad (5.4)$$

ⓘ If a mixture contains equal numbers of A and B molecules, $X_A = X_B = 0.50$ and $P_A = P_B = \frac{1}{2}P_{tot}$.

In other words, *the partial pressure of a gas in a mixture is equal to its mole fraction multiplied by the total pressure.* ◀ ⓘ This relation is commonly used to calculate partial pressures of gases in a mixture when the total pressure and the composition of the mixture are known (Example 5.9).

EXAMPLE 5.9

When one mole of methane, CH_4, is heated with four moles of oxygen, the following reaction occurs:

$$CH_4(g) + 2O_2(g) \longrightarrow CO_2(g) + 2H_2O(g)$$

Assuming all of the methane is converted to CO_2 and H_2O, what are the mole fractions of O_2, CO_2, and H_2O in the resulting mixture? If the total pressure of the mixture is 1.26 atm, what are the partial pressures?

ANALYSIS

Information given:	P_{tot} (1.26 atm) Initial amounts of reactants (1.000 mol CH_4 and 4.000 mol O_2)
Information implied:	stoichiometric ratios: 2 mol O_2/1 mol CH_4/1 mol CO_2/2 mol H_2O limiting reactant (CH_4); reactant in excess (O_2)
Asked for:	mol fraction of each gas after reaction partial pressure of each gas after reaction

STRATEGY

1. Find the moles of reactants left after reaction. (Recall that CH_4 is limiting and thus is completely used up.)

$$n_{CH_4} \xrightarrow[\text{ratio}]{\text{stoichiometric}} n_{O_2}$$

n_{O_2} after reaction = n_{O_2} initially − n_{O_2} used

2. Find the moles of products.

$$n_{CH_4} \xrightarrow[\text{ratio}]{\text{stoichiometric}} n_{CO_2} \quad \text{and} \quad n_{CH_4} \xrightarrow[\text{ratio}]{\text{stoichiometric}} n_{H_2O}$$

3. Find n_{tot}.

$$n_{tot} = n_{CH_4} + n_{O_2} + n_{CO_2} + n_{H_2O}$$

4. Find the mol fraction of each gas.

$$X_A = \frac{n_A}{n_{tot}}$$

5. Find the partial pressure of each gas.

$$P_A = (X_A)(P_{tot})$$

SOLUTION

1. mol CH_4	mol CH_4 = 0 (Problem states that CH_4 is completely used up.)
mol O_2 reacted	$1.000 \text{ mol } CH_4 \times \dfrac{2 \text{ mol } O_2}{1 \text{ mol } CH_4} = 2.000$
mol O_2 unreacted	4.000 mol O_2 initially − 2.000 mol reacted = 2.000 mol
2. mol CO_2	$1.000 \text{ mol } CH_4 \times \dfrac{1 \text{ mol } CO_2}{1 \text{ mol } CH_4} = 1.000 \text{ mol } CO_2$ are produced
mol H_2O	$1.000 \text{ mol } CH_4 \times \dfrac{2 \text{ mol } H_2O}{1 \text{ mol } CH_4} = 2.000 \text{ mol } H_2O$ are produced
3. n_{tot}	$n_{tot} = n_{CH_4} + n_{O_2} + n_{CO_2} + n_{H_2O} = 0 + 2.000 + 1.000 + 2.000 = 5.000$
4. X_{CH_4}	$X_{CH_4} = \dfrac{n_{CH_4}}{n_{tot}} = \dfrac{0}{5.000} = 0$
X_{O_2}	$X_{O_2} = \dfrac{n_{O_2}}{n_{tot}} = \dfrac{2.000}{5.000} = 0.4000$

continued

X_{CO_2}	$X_{CO_2} = \dfrac{n_{CO_2}}{n_{tot}} = \dfrac{1.000}{5.000} = 0.2000$
X_{H_2O}	$X_{H_2O} = \dfrac{n_{H_2O}}{n_{tot}} = \dfrac{2.000}{5.000} = 0.4000$
5. P_{CH_4}	$(X_{CH_4})(P_{tot}) = (0)(1.26) = 0$
P_{O_2}	$(X_{O_2})(P_{tot}) = (0.4000)(1.26) = 0.504$ atm
P_{CO_2}	$(X_{CO_2})(P_{tot}) = (0.2000)(1.26) = 0.252$ atm
P_{H_2O}	$(X_{H_2O})(P_{tot}) = (0.4000)(1.26) = 0.504$ atm

END POINT

The mole fractions of all the gases should add up to 1. All the partial pressures should add up to 1.26 atm. They do!

5-6 Kinetic Theory of Gases

The fact that the ideal gas law applies to all gases indicates that the gaseous state is a relatively simple one from a molecular standpoint. Gases must have certain common properties that cause them to follow the same natural law. Between about 1850 and 1880, James Maxwell (1831–1879), Rudolf Clausius (1822–1888), Ludwig Boltzmann (1844–1906), and others developed the **kinetic theory** of gases. They based it on the idea that all gases behave similarly as far as particle motion is concerned.

Molecular Model

By the use of the kinetic theory, it is possible to derive or explain the experimental behavior of gases. To do this we start with a simple molecular model, which assumes that

- *gases are mostly empty space.* The total volume of the molecules is negligibly small compared with that of the container to which they are confined.
- *gas molecules are in constant, chaotic motion.* They collide frequently with one another and with the container walls. ◄ⓘ As a result, their velocities are constantly changing.
- *collisions are elastic.* There are no attractive forces that would tend to make molecules "stick" to one another or to the container walls.
- *gas pressure is caused by collisions of molecules with the walls of the container* (Figure 5.11). As a result, pressure increases with the energy and frequency of these collisions.

Expression for Pressure, *P*

Applying the laws of physics to this simple model, it can be shown that the pressure (*P*) exerted by a gas in a container of volume *V* is

$$P = \frac{N(\text{mass})u^2}{3V}$$

where *N* is the number of molecules, and *u* is the average speed.* We will not attempt to derive this equation. However, it makes sense, at least qualitatively. In particular

- the ratio *N/V* expresses the concentration of gas molecules in the container. The more molecules there are in a given volume, the greater the collision frequency and so the greater the pressure.

ⓘ
In air, a molecule undergoes about 10 billion collisions per second.

Figure 5.11 The kinetic molecular model of a gas. Gas molecules are in constant motion and their collisions are elastic. Collisions with the walls cause gas pressure.

*More rigorously, u^2 is the average of the squares of the speeds of all molecules.

■ the product (mass)(u^2) is a measure of the energy of collision. (When a Cadillac traveling at 100 mph collides with a brick wall, the energy transferred is much greater than would be obtained with a bicycle at 5 mph.) Hence, as this equation predicts, pressure is directly related to (mass)(u^2).

Average Kinetic Energy of Translational Motion, E_t

The kinetic energy, E_t, of a gas molecule of a given mass moving at speed u is

$$E_t = \frac{(\text{mass})u^2}{2}$$

From the equation written above for P, we see that (mass) $u^2 = 3PV/N$. Hence

$$E_t = \frac{3PV}{2N}$$

But the ideal gas law tells us that $PV = nRT$, so

$$E_t = \frac{3nRT}{2N}$$

This equation can be simplified by noting that the number of molecules, N, in a gas sample is equal to the number of moles, n, multiplied by Avogadro's number, N_A, that is, $N = n \times N_A$. Making this substitution in the above equation and simplifying, we obtain the final expression for the *average translational kinetic energy* of a gas molecule:

$$E_t = \frac{3RT}{2N_A} \tag{5.5}$$

This equation contains three constants (3/2, R, N_A) and only one variable, the temperature T. It follows that

■ *at a given temperature, molecules of different gases (e.g., H_2, O_2, ...) must all have the same average kinetic energy of translational motion.* ◀ⓘ
■ *the average translational kinetic energy of a gas molecule is directly proportional to the Kelvin temperature, T.*

ⓘ
A baseball in motion has translational energy.

Average Speed, u

By equating the first and last expressions written above for E_t, we see that

$$\frac{(\text{mass})u^2}{2} = \frac{3RT}{2N_A}$$

Solving this equation for u^2:

$$u^2 = \frac{3RT}{(\text{mass})N_A}$$

But the product (mass of a molecule) times N_A (the number of molecules in a mole) is simply the molar mass MM, so we can write

$$u^2 = \frac{3RT}{\text{MM}}$$

Taking the square root of both sides of this equation, we arrive at the final expression for the **average speed of gas molecules,** ◀ⓘ u,

$$u = \left(\frac{3RT}{\text{MM}}\right)^{1/2} \tag{5.6}$$

From this relation you can see that the average speed u is

ⓘ
E_t depends only on T; u depends on both T and MM.

Figure 5.12 Relation of molecular speed to molar mass. When ammonia gas, which is injected into the left arm of the tube, comes in contact with hydrogen chloride, which is injected into the right arm of the tube, they react to form solid ammonium chloride: $NH_3(g) + HCl(g) \longrightarrow NH_4Cl(s)$. Because NH_3 (MM = 17 g/mol) moves faster than HCl (MM = 36.5 g/mol), the ammonium chloride forms closer to the HCl end of the tube.

- *directly proportional to the square root of the absolute temperature.* For a given gas at two different temperatures, T_2 and T_1, the quantity MM is constant, and we can write

$$\frac{u_2}{u_1} = \left(\frac{T_2}{T_1}\right)^{1/2}$$

- *inversely proportional to the square root of molar mass* (**MM**) (Figure 5.12). For two different gases A and B at the same temperature (T constant):

$$\frac{u_B}{u_A} = \left(\frac{MM_A}{MM_B}\right)^{1/2} \tag{5.7}$$

EXAMPLE 5.10	
Calculate the average speed, *u*, of an N_2 molecule at 25°C.	
ANALYSIS	
Information given:	temperature (25°C)
Information implied:	$R = 8.31 \times 10^3$ g · m²/s² · mol · K MM of N_2
Asked for:	average speed, *u*, of N_2 at 25°C
STRATEGY	

1. Change T to the appropriate units.

2. Note the units of the constant R.

3. Substitute into the equation: $u = \left(\dfrac{3RT}{MM}\right)^{1/2}$

SOLUTION	
Average speed	$u = \left(\dfrac{3 \times 8.31 \times 10^3 \dfrac{\text{g} \cdot \text{m}^2}{\text{s}^2 \cdot \text{mol} \cdot \text{K}} \times 298 \text{ K}}{28.02 \dfrac{\text{g}}{\text{mol}}}\right)^{1/2} = 515 \text{ m/s}$

continued

END POINT

Converting the speed of the nitrogen molecule obtained above to miles per hour:

$$515 \, \frac{m}{s} \times \frac{1 \, mi}{1.609 \times 10^3 \, m} \times \frac{3600 \, s}{1 \, h} = 1.15 \times 10^3 \, mi/h$$

Note that the average cruising speed of a Boeing 777 is about 580 mi/h and the speed of sound is 768 mi/h.

Effusion of Gases; Graham's Law

All of us are familiar with the process of gaseous diffusion, in which gas molecules move through space from a region of high concentration to one of low concentration. If your instructor momentarily opens a cylinder of chlorine gas at the lecture table, you will soon recognize the sharp odor of chlorine, particularly if you have a front-row seat in the classroom. On a more pleasant note, the odor associated with a freshly baked apple pie also reaches you via gaseous diffusion.

Diffusion is a relatively slow process; a sample of gas introduced at one location may take an hour or more to distribute itself uniformly throughout a room. ◀ⓘ At first glance this seems surprising, since as we saw in Example 5.10 gas molecules are moving very rapidly. However, molecules are constantly colliding with one another; at 25°C and 1 atm, an N_2 molecule undergoes more than a billion collisions per second with its neighbors. This slows down the net movement of gas molecules in any given direction.

As we have implied, diffusion is a rather complex process so far as molecular motion is concerned. **Effusion,** the flow of gas molecules at low pressures through tiny pores or pinholes, is easier to analyze using kinetic theory.

The relative rates of effusion of different gases depend on two factors: the pressures of the gases and the relative speeds of their particles. If two different gases A and B are compared at the same pressure, only their speeds are of concern, and

$$\frac{\text{rate of effusion B}}{\text{rate of effusion A}} = \frac{u_B}{u_A}$$

where u_A and u_B are average speeds. As pointed out earlier, at a given temperature,

$$\frac{u_B}{u_A} = \left(\frac{MM_A}{MM_B}\right)^{1/2}$$

It follows that, at constant pressure and temperature,

$$\frac{\text{rate of effusion B}}{\text{rate of effusion A}} = \left(\frac{MM_A}{MM_B}\right)^{1/2}$$

The time that a gas effuses can also be related to its molar mass. Since rate is inversely proportional to time, it follows that

$$\frac{\text{time}_A}{\text{time}_B} = \left(\frac{MM_A}{MM_B}\right)^{1/2}$$

This relation in a somewhat different form was discovered experimentally by the Scottish chemist Thomas Graham (1748–1843) in 1829. Graham was interested in a wide variety of chemical and physical problems, among them the separation of the components of air. **Graham's law** can be stated as

At a given temperature and pressure, the rate of effusion of a gas, in moles per unit time, is inversely proportional to the square root of its molar mass.

Graham's law tells us qualitatively that light molecules effuse more rapidly than heavier ones (Figure 5.13, page 119). In quantitative form, it allows us to determine molar masses of gases (Example 5.11).

ⓘ Diffusion is slower than effusion for the same reason that boarding a subway car is slower at 5 P.M. than at 5 A.M.

EXAMPLE 5.11

In an effusion experiment, argon gas is allowed to expand through a tiny opening into an evacuated flask of volume 120 mL for 32.0 s, at which point the pressure in the flask is found to be 12.5 mm Hg. This experiment is repeated with a gas X of unknown molar mass at the same T and P. It is found that the pressure in the flask builds up to 12.5 mm Hg after 48.0 s. Calculate the molar mass of X.

ANALYSIS

Information given:	volume of both flasks (120 mL); pressure in both flasks (12.5 mm Hg); time for Ar effusion (32.0 s); time for gas (X) effusion (48.0 s)
Information implied:	Temperature, pressure, and volume are the same for both flasks. rate of effusion for each gas MM of argon
Asked for:	MM of X

STRATEGY

1. Since T, P, and V are the same for both gases, the number of moles of gas in both flasks is the same.

$$n_{Ar} = n_X = n$$

2. The rate of effusion is in mol/time.

$$\text{rate} = \frac{n_{Ar}}{\text{time}} = \frac{n_X}{\text{time}} = \frac{n}{\text{time}}$$

3. Substitute into Graham's law of effusion, where A = gas X and B = Ar.

$$\frac{\text{rate B}}{\text{rate A}} = \left(\frac{MM_A}{MM_B}\right)^{1/2} \longrightarrow \frac{\text{rate Ar}}{\text{rate X}} = \left(\frac{MM_X}{MM_{Ar}}\right)^{1/2}$$

SOLUTION

rates	$\text{rate X} = \dfrac{n}{48.0 \text{ s}} \qquad \text{rate Ar} = \dfrac{n}{32.0 \text{ s}}$
MM_X	$\dfrac{\dfrac{n}{32.0 \text{ s}}}{\dfrac{n}{48.0 \text{ s}}} = \left(\dfrac{MM_X}{39.95 \text{ g/mol}}\right)^{1/2} \longrightarrow 1.50 = \left(\dfrac{MM_X}{39.95 \text{ g/mol}}\right)^{1/2}$
	$(1.50)^2 = \left(\left(\dfrac{MM_X}{39.95 \text{ g/mol}}\right)^{1/2}\right)^2 \longrightarrow 2.25 = \dfrac{MM_X}{39.95 \text{ g/mol}} \rightarrow MM_X = 89.9 \text{ g/mol}$

END POINT

Since the unknown gas takes longer to effuse, it should have a larger molar mass than argon. It does!

A practical application of Graham's law arose during World War II, when scientists were studying the fission of uranium atoms as a source of energy. It became necessary to separate $^{235}_{92}U$, which is fissionable, from the more abundant isotope of uranium, $^{238}_{92}U$, which is not fissionable. Because the two isotopes have almost identical chemical properties, chemical separation was not feasible. Instead, an effusion process was worked out using uranium hexafluoride, UF_6. This compound is a gas at room temperature and low pressures. Preliminary experiments indicated that $^{235}_{92}UF_6$ could indeed be separated from $^{238}_{92}UF_6$ by effusion.

The separation factor is very small because the rates of effusion of these two species are nearly equal:

$$\frac{\text{rate of effusion of } ^{235}_{92}UF_6}{\text{rate of effusion of } ^{238}_{92}UF_6} = \left(\frac{352.0}{349.0}\right)^{1/2} = 1.004$$

so a great many repetitive separations are necessary. An enormous plant was built for this purpose in Oak Ridge, Tennessee. In this process, UF_6 effuses many thousands of times through porous barriers. The lighter fractions move on to the next stage, while heavier fractions are recycled through earlier stages. Eventually, a nearly complete separation of the two isotopes is achieved.

Distribution of Molecular Speeds

As shown in Example 5.10, the average speed of an N_2 molecule at 25°C is 515 m/s; that of H_2 is even higher, 1920 m/s. However, not all molecules in these gases have these speeds. The motion of particles in a gas is utterly chaotic. ◄ ⓘ In the course of a second, a particle undergoes millions of collisions with other particles. As a result, the speed and direction of motion of a particle are constantly changing. Over a period of time, the speed will vary from almost zero to some very high value, considerably above the average.

In 1860 James Clerk Maxwell, a Scottish physicist and one of the greatest theoreticians the world has ever known, showed that different possible speeds are distributed among particles in a definite way. Indeed, he developed a mathematical expression for this distribution (referred to today as the **Maxwell distribution**). His results are shown graphically in Figure 5.14 for O_2 at 25°C and 1000°C. On the graph, the relative number of molecules having a certain speed is plotted against that speed. At 25°C, this number increases rapidly with the speed, up to a maximum of about 400 m/s. This is the most probable speed of an oxygen molecule at 25°C. Above about 400 m/s, the number of molecules moving at any particular speed decreases. For speeds in excess of about 1600 m/s, the fraction of molecules drops off to nearly zero. In general, most molecules have speeds rather close to the average value. ◄ ⓘ

As temperature increases, the speed of the molecules increases. The distribution curve for molecular speeds (Figure 5.14) shifts to the right and becomes broader. The chance of a molecule having a very high speed is much greater at 1000°C than at 25°C. Note, for example, that a large number of molecules has speeds greater than 1600 m/s at 1000°C; almost none have that speed at 25°C. The general

Figure 5.13 Effusion of gases. A gas with a higher molar mass (red molecules) effuses into a vacuum more slowly than a gas with a lower molar mass (gray molecules).

ⓘ In a gas sample at any instant, gas molecules are moving at a variety of speeds.

ⓘ Sort of like people—most of them go along with the crowd.

Figure 5.14 Distribution of molecular speeds of oxygen molecules at 25°C and 1000°C. At the higher temperature the fraction of molecules moving at very high speeds is greater.

principle here is one that we will find very useful when we look at the effect of temperature on reaction rate in Chapter 11.

EXAMPLE **5.12** **CONCEPTUAL**

Consider the two boxes A and B shown below. Box B has a volume exactly twice that of box A. The circles and ● represent one mole of HCl and He, respectively. The two boxes are at the same temperature.

(a) Compare the pressures of the gases in the two containers.

(b) Compare the densities of the two gases.

(c) Compare the number of atoms in the two boxes.

(d) If the HCl in box A were transferred to box B, what would be the mole fraction of HCl in the mixture?

(e) Which of the two gases effuses faster?

SOLUTION

(a) Since n/V and T are the same in both cases, $P = nRT/V$ is the same for the two gases.

(b) The mass of HCl is $2(36.5 \text{ g}) = 73.0 \text{ g}$; that of He is $4(4.00 \text{ g}) = 16.0 \text{ g}$. Since $73.0 \text{ g}/V > 16.0 \text{ g}/2V$, HCl has the higher density.

(c) Two moles of diatomic HCl contain the same number of atoms as four moles of He.

(d) $X_{HCl} = 2/6 = 1/3$

(e) Because HCl and He are at the same pressure, the lighter gas, He, effuses faster.

5-7 Real Gases

In this chapter, the ideal gas law has been used in all calculations, with the assumption that it applies exactly. Under ordinary conditions, this assumption is a good one; however, all real gases deviate at least slightly from the ideal gas law. Table 5.2 shows the extent to which two gases, O_2 and CO_2, deviate from ideality at different temperatures and pressures. The data compare the experimentally observed molar volume, V_m

The molar volume is V when $n = 1$.

$$\text{molar volume} = V_m = V/n$$

with the molar volume calculated from the ideal gas law V_m°:

$$V_m^\circ = RT/P$$

Table 5.2 Real Versus Ideal Gases, Percent Deviation* in Molar Volume

P(atm)	O_2			CO_2		
	50°C	0°C	−50°C	50°C	0°C	−50°C
1	−0.0%	−0.1%	−0.2%	−0.4%	−0.7%	−1.4%
10	−0.4%	−1.0%	−2.1%	−4.0%	−7.1%	
40	−1.4%	−3.7%	−8.5%	−17.9%		
70	−2.2%	−6.0%	−14.4%	−34.2%	Condenses to liquid	
100	−2.8%	−7.7%	−19.1%	−59.0%		

*Percent deviation $= \dfrac{(V_m - V_m^\circ)}{V_m^\circ} \times 100\%$

It should be obvious from Table 5.2 that deviations from ideality become larger at *high pressures and low temperatures*. Moreover, the deviations are larger for CO_2 than for O_2. All of these effects can be correlated in terms of a simple, common-sense observation:

> *In general, the closer a gas is to the liquid state, the more it will deviate from the ideal gas law.*

A gas is liquefied by going to low temperatures and/or high pressures. Moreover, as you can see from Table 5.2, carbon dioxide is much easier to liquefy than oxygen.

From a molecular standpoint, deviations from the ideal gas law arise because it neglects two factors:

1. attractive forces between gas particles.
2. the finite volume of gas particles.

We will now consider in turn the effect of these two factors on the molar volumes of real gases.

Attractive Forces

Notice that in Table 5.2 all the deviations are negative; the observed molar volume is less than that predicted by the ideal gas law. This effect can be attributed to attractive forces between gas particles. These forces tend to pull the particles toward one another, reducing the space between them. ◂ⓘ As a result, the particles are crowded into a smaller volume, just as if an additional external pressure were applied. The observed molar volume, V_m, becomes less than V_m°, and the deviation from ideality is *negative:*

ⓘ Attractive forces make the molar volume smaller than expected.

$$\frac{V_m - V_m^\circ}{V_m^\circ} < 0$$

The magnitude of this effect depends on the strength of the attractive forces and hence on the nature of the gas. Intermolecular attractive forces are stronger in CO_2 than they are in O_2, which explains why the deviation from ideality of V_m is greater with carbon dioxide and why carbon dioxide is more readily condensed to a liquid than is oxygen.

Particle Volume

Figure 5.15 shows a plot of V_m/V_m° versus pressure for methane at 25°C. Up to about 150 atm, methane shows a steadily increasing negative deviation from ideality, as might be expected on the basis of attractive forces. At 150 atm, V_m is only about 70% of V_m°.

At very high pressures, methane behaves quite differently. Above 150 atm, the ratio V_m/V_m° *increases,* becoming 1 at about 350 atm. Above that pressure, methane shows a *positive* deviation from the ideal gas law:

$$\frac{V_m - V_m^\circ}{V_m^\circ} > 0$$

Figure 5.15 Deviation of methane gas from ideal gas behavior. Below about 350 atm, attractive forces between methane (CH_4) molecules cause the observed molar volume at 25°C to be less than that calculated from the ideal gas law. At 350 atm, the effect of the attractive forces is just balanced by that of the finite volume of CH_4 molecules, and the gas appears to behave ideally. Above 350 atm, the effect of finite molecular volume predominates and $V_m > V_m^\circ$.

This effect is by no means unique to methane; it is observed with all gases. If the data in Table 5.2 are extended to very high pressures, oxygen and carbon dioxide behave like methane; V_m becomes larger than V_m°.

An increase in molar volume above that predicted by the ideal gas law is related to the finite volume of gas particles. These particles contribute to the observed volume, making V_m greater than V_m°. Ordinarily, this effect becomes evident only at high pressures, where the particles are quite close to one another.

CHEMISTRY **BEYOND** THE CLASSROOM

Measurement of Blood Pressure

One of the first things that a nurse does when you visit your doctor is take your blood pressure with a *sphygmomanometer*. A condition of elevated blood pressure is called *hypertension* and can be a major factor in heart attacks and strokes. How does a sphygmomanometer work?

When the heart contracts and relaxes, it pumps blood through the arteries in the body. The pressure exerted by the blood on the walls of the artery can be measured in the same way as the pressure of a column of air or water.

Blood pressure is measured by a special type of manometer called a sphygmomanometer. It consists of a cuff and a device that inflates an air bladder inside the cuff, which in turn restricts the flow of blood through an artery, together with a device that displays the pressure measured. Commonly, the cuff is placed at about heart level on the upper arm, just above the elbow, with the arm held in a relaxed position against a supporting table or bench. The cuff is inflated to a pressure above that of the blood being pumped by the heart, which cuts off the flow of blood through the artery. The pressure in the cuff is then slowly reduced.

As the cuff is deflated, a "whooshing" sound, called a Korotkoff sound, is detected by using a stethoscope pressed against the brachial artery (the artery that runs from your elbow to your shoulder). The sound results from the restoration of blood flow through the artery as the pressure opposing it in the cuff is relaxed. When two consecutive sounds are heard, the pressure reading is taken from the measuring device. This is called the systolic pressure reading. This reading corresponds to the cycle where the heart is contracting in response to an electrical stimulus.

Further deflation of the cuff leads to several more Korotkoff sounds, ultimately ending in silence as the pressure in the cuff drops below the diastolic blood pressure.

Figure A Aneroid sphygmomanometer.

The disappearance of sound determines the lower of the two readings that comprise a blood pressure measurement—the diastolic reading. The diastolic reading corresponds to the cycle where the heart is relaxing following its contraction.

The standard in sphygmomanometers uses a column of mercury, exactly as a mercury barometer does. It requires a separate stethoscope and a trained ear for proper operation. Another type of device is called an aneroid sphygmomanometer, which uses an analog gauge that is calibrated against a mercury column.

Even more common today are digital sphygmomanometers, which require no external listening device for the Korotkoff sounds. The cuff includes a microphone that is connected to a microprocessor and has an automatic inflator with a pressure sensor connected to an air pump. At the press of a button, the cuff inflates and deflates under computer control, and the reading is taken and processed by the computer chip. These devices measure the mean arterial pressure, which is then processed mathematically to give the systolic and diastolic readings. These are then displayed on a digital panel. Blood pressure readings are commonly reported as even numbers, and in units of millimeters of mercury. Although they are easy to use, digital sphygmomanometers are not suitable for all blood pressure measurements. Some conditions can render their readings significantly inaccurate.

Other types of sphygmomanometers include those that can be placed around a finger. They are easier to use but are less accurate than other types of blood pressure measuring devices.

Figure B Digital sphygmomanometer.

Key Concepts

1. Convert between units of *P*, *V*, *T*, and amount of gas.
   (**Example 5.1; Problems 1–4**)
2. Use the ideal gas law to
   - solve initial and final state problems.
   (**Example 5.2; Problems 5–14**)
   - calculate *P*, *V*, *T* or *n*.
   (**Example 5.3; Problems 15–20**)
   - calculate density or molar mass.
   (**Example 5.4; Problems 21–30**)

- relate amounts and volumes of gases in reactions.
  (**Examples 5.5–5.7; Problems 31–40**)
3. Use Dalton's law.
   (**Examples 5.8, 5.9; Problems 41–52**)
4. Calculate the speeds of gas molecules.
   (**Example 5.10; Problems 53–56, 61, 62**)
5. Use Graham's law to relate rate of effusion to molar mass.
   (**Example 5.11; Problems 57–60**)

Key Equations

Ideal gas law	$PV = nRT$
Gas density	$d = (MM) \times P/RT$
Dalton's law	$P_{tot} = P_A + P_B; P_A = X_A \times P_{tot}$
Average translational kinetic energy	$E_t = 3RT/2N_A$
Average speed of gas molecules	$u = (3RT/MM)^{1/2}$
Graham's law	$rate_B/rate_A = (MM_A/MM_B)^{1/2}$

Key Terms

atmosphere (atm)	kinetic theory	mole fraction	standard temperature and
bar	millimeters of mercury	partial pressure	pressure (STP)
effusion	(mm Hg)	gas constant	vapor pressure

Summary Problem

When aluminum foil is added to a solution of hydrochloric acid, the following reaction occurs:

$$2Al(s) + 6H^+(aq) \longrightarrow 2Al^{3+}(aq) + 3H_2(g)$$

(a) Dry hydrogen gas is collected in a 250.0-mL flask at 22°C and 745 mm Hg. How many grams of hydrogen are in the flask?

(b) The hydrogen gas is transferred without loss from the 250.0-mL flask to a 100.0-mL flask at the same temperature (22°C). What is the pressure in the smaller flask?

(c) What is the density (in g/L) of the gas collected in part (a)?

(d) Suppose the hydrogen is collected over water in a 235-mL flask at 25°C. The total pressure in the flask is 756 mm Hg.

   1. What is the partial pressure of the hydrogen gas?

   2. What is the mass of the hydrogen gas?

   3. What is the mole fraction of hydrogen in the wet gas?

(e) What mass of aluminum is required to generate 1.00 L of $H_2(g)$ at 25°C and 1.00 atm?

(f) How many liters of hydrogen at 25°C and 1.00 atm can be generated by adding 250.0 mL of 6.0 *M* HCl to an excess of aluminum foil?

(g) Compare the rate of effusion of H_2 to that of He.

(h) Compare the time required for an equal number of moles of H_2 and CO_2 to effuse out of identical pinholes.

(i) The average speed of a hydrogen molecule at 25°C is 1.92×10^3 m/s. What is the average speed of an ammonia molecule at that same temperature?

Answers

(a) 0.0204 g
(b) 1.86×10^3 mm Hg
(c) 0.0816 g/L
(d) (1) 732 mm Hg   (2) 0.0187 g   (3) 0.969
(e) 0.735 g
(f) 18 L
(g) 1.41:1
(h) H_2 effuses 4.7 faster than CO_2
(i) 661 m/s

Questions and Problems

Even-numbered questions and Challenge Problems have answers in Appendix 5 and fully worked solutions in the *Student Solutions Manual.*

5-1 Measurements on Gases

1. A ten-gallon methane tank contains 1.243 mol of methane (CH_4) at 74°F. Express the volume of the tank in liters, the amount of methane in the tank in grams, and the temperature of the tank in Kelvin.

2. A 6.00-ft cylinder has a radius of 26 in. It contains 189 lb of helium at 25°C. Express the volume of the cylinder ($V = \pi r^2 h$) in liters, the amount of helium in moles, and the temperature in Kelvin.

3. Complete the following table of pressure conversions.

	mm Hg	atm	kPa	bar
(a)	396	____	____	____
(b)	____	1.15	____	____
(c)	____	____	97.1	____
(d)	____	____	____	1.00

4. Complete the following table of pressure conversions.

	mm Hg	atm	psi	kPa
(a)	____	____	____	42.7
(b)	____	____	29.1	____
(c)	____	0.788	____	____
(d)	1216	____	____	____

5-2 The Ideal Gas Law
5-3 Gas Law Calculations

5. A cylinder with a movable piston records a volume of 12.6 L when 3.0 mol of oxygen is added. The gas in the cylinder has a pressure of 5.83 atm. The cylinder develops a leak and the volume of the gas is now recorded to be 12.1 L at the same pressure. How many moles of oxygen are lost?

6. A tank is filled with gas to a pressure of 875 mm Hg at 25°C. The gas is transferred without loss to a tank twice the size of the original tank. If the pressure is to remain constant, at what temperature (in °C) should the tank be kept?

7. A sample of CO_2 gas at 22°C and 1.00 atm has a volume of 2.00 L. Determine the ratio of the original volume to the final volume when

(a) the pressure and amount of gas remain unchanged and the Celsius temperature is doubled.

(b) the pressure and amount of gas remain unchanged and the Kelvin temperature is doubled.

8. A sample of nitrogen gas has a pressure of 1.22 atm. If the amount of gas and the temperature are kept constant, what is the pressure if

(a) its volume is decreased by 38%?

(b) its volume is decreased to 38% of its original volume?

9. A bicycle owner's manual recommends that a tire have a gauge pressure of 73.0 psi at 75°F for a rider weighing 140 lbs. The bicycle is left overnight in the driveway, where the temperature in the tire is 27°F. What is the actual pressure in the tire? What is the gauge pressure? Gauge pressure is the pressure above atmospheric pressure (14.7 psi).

10. A tire is inflated to a gauge pressure of 28.0 psi at 71°F. Gauge pressure is the pressure above atmospheric pressure, which is 14.7 psi. After several hours of driving, the air in the tire has a temperature of 115°F. What is the gauge pressure of the air in the tire? What is the actual pressure of the air in the tire? Assume that the tire volume changes are negligible.

11. A 38.0-L gas tank at 35°C has nitrogen at a pressure of 4.65 atm. The contents of the tank are transferred without loss to an evacuated 55.0-L tank in a cold room where the temperature is 4°C. What is the pressure in the tank?

12. A sealed tank at room temperature, 25°C, has 22.0 g of CO_2. The gas has a pressure of 732 mm Hg. The tank is moved to a room kept at 12°C and an additional 10.0 g of CO_2 are added to the tank. What is the pressure in the tank? Assume no loss of gas when more CO_2 is added.

13. A balloon filled with helium has a volume of 1.28×10^3 L at sea level where the pressure is 0.998 atm and the temperature is 31°C. The balloon is taken to the top of a mountain where the pressure is 0.753 atm and the temperature is −25°C. What is the volume of the balloon at the top of the mountain?

14. A flask has 1.35 mol of hydrogen gas at 25°C and a pressure of 1.05 atm. Nitrogen gas is added to the flask at the same temperature until the pressure rises to 1.64 atm. How many moles of nitrogen gas are added?

15. A two-liter plastic bottle can withstand a pressure of 5 atm. Approximately 125 mL (about half a cup) of water ($d = 1.00$ g/mL) is poured into the flask. The bottle with water is placed in a water bath maintained at 101°C. Will the bottle explode? (Ignore the volume occupied by the water.)

16. A drum used to transport crude oil has a volume of 162 L. How many grams of water, as steam, are required to fill the drum at 1.00 atm and 108°C? When the temperature in the drum is decreased to 25°C, all the steam condenses. How many mL of water ($d = 1.00$ g/mL) can be collected?

17. A piece of dry ice ($CO_2(s)$) has a mass of 22.50 g. It is dropped into an evacuated 2.50-L flask. What is the pressure in the flask at −4°C?

18. A four-liter tank is filled with propane gas, C_3H_8. The mass of the tank filled with gas is 1236 g. The pressure in the tank is 2.68 atm. The temperature in the room is 37°C. The propane in the tank is used up under the same conditions of temperature and pressure. What is the mass of the empty tank?

19. Complete the following table for dinitrogen tetroxide gas.

Pressure	Volume	Temperature	Moles	Grams
(a) 1.77 atm	4.98 L	43.1°C	_____	_____
(b) 673 mm Hg	488 mL	_____	0.783	_____
(c) 0.899 bar	_____	912°C	_____	6.25
(d) _____	1.15 L	39°F	0.166	_____

20. Use the ideal gas law to complete the following table for propane (C_3H_8) gas.

Pressure	Volume	Temperature	Moles	Grams
(a) 18.9 psi	0.886 L	22°C	_____	_____
(b) 633 mm Hg	1.993 L	_____	0.0844	_____
(c) 1.876 atm	_____	75°F	2.842	_____
(d) _____	2244 mL	13°C	_____	47.25

21. Calculate the densities (in g/L) of the following gases at 78°F and 13.6 psi.
   (a) xenon   (b) methane   (c) acetylene, C_2H_2

22. Calculate the densities (in grams per liter) of the following gases at 97°C and 755 mm Hg.
   (a) hydrogen chloride   (b) sulfur dioxide
   (c) butane (C_4H_{10})

23. Helium-filled balloons rise in the air because the density of helium is less than the density of air.
   (a) If air has an average molar mass of 29.0 g/mol, what is the density of air at 25°C and 770 mm Hg?
   (b) What is the density of helium at the same temperature and pressure?
   (c) Would a balloon filled with carbon dioxide at the same temperature and pressure rise?

24. Space probes to Mars have shown that its atmosphere consists mostly of carbon dioxide. The average temperature on the surface of Mars is −55°C with an average pressure of 0.00592 atm. Compare the density of CO_2 on Mars's surface with that on the earth's surface at 25°C and one atmosphere.

25. Cyclopropane mixed in the proper ratio with oxygen can be used as an anesthetic. At 755 mm Hg and 25°C, it has a density of 1.71 g/L.
   (a) What is the molar mass of cyclopropane?
   (b) Cyclopropane is made up of 85.7% C and 14.3% H. What is the molecular formula of cyclopropane?

26. Phosgene is a highly toxic gas made up of carbon, oxygen, and chlorine atoms. Its density at 1.05 atm and 25°C is 4.24 g/L.
   (a) What is the molar mass of phosgene?
   (b) Phosgene is made up of 12.1% C, 16.2% O, and 71.7% Cl. What is the molecular formula of phosgene?

27. The gas in the discharge cell of a laser contains (in mole percent) 11% CO_2, 5.3% N_2, and 84% He.
   (a) What is the molar mass of this mixture?
   (b) Calculate the density of this gas mixture at 32°C and 758 mm Hg.
   (c) What is the ratio of the density of this gas to that of air (MM = 29.0 g/mol) at the same conditions?

28. Exhaled air contains 74.5% N_2, 15.7% O_2, 3.6% CO_2, and 6.2% H_2O (mole percent).
   (a) Calculate the molar mass of exhaled air.
   (b) Calculate the density of exhaled air at 37°C and 757 mm Hg and compare the value you obtained with that of ordinary air (MM = 29.0 g/mol) under the same conditions.

29. A 1.58-g sample of $C_2H_3X_3(g)$ has a volume of 297 mL at 769 mm Hg and 35°C. Identify the element X.

30. A 0.750-g sample of the gas PX_3 is in a sealed 542-mL bulb at 26°C and 1.00 atm pressure. Identify the element X and name the gas.

5-4 Stoichiometry of Gaseous Reactions

31. Nitrogen oxide is a pollutant commonly found in smokestack emissions. One way to remove it is to react it with ammonia.

$$4NH_3(g) + 6NO(g) \longrightarrow 5N_2(g) + 6H_2O(l)$$

How many liters of ammonia are required to change 12.8 L of nitrogen oxide to nitrogen gas? Assume 100% yield and that all gases are measured at the same temperature and pressure.

32. Nitrogen trifluoride gas reacts with steam to produce the gases HF, NO, and NO_2.
   (a) Write a balanced equation for the reaction.
   (b) What volume of nitrogen oxide is formed when 5.22 L of nitrogen trifluoride are made to react with 5.22 L of steam? Assume 100% yield and constant temperature and pressure conditions throughout the reaction.

33. Dichlorine oxide is used as bactericide to purify water. It is produced by the chlorination of sulfur dioxide gas.

$$SO_2(g) + 2Cl_2(g) \longrightarrow SOCl_2(l) + Cl_2O(g)$$

How many liters of Cl_2O can be produced by mixing 5.85 L of SO_2 and 9.00 L of Cl_2? How many liters of the reactant in excess are present after reaction is complete? Assume 100% yield and that all the gases are measured at the same temperature and pressure.

34. Titanium(III) chloride is used in the manufacture of polyethylene. It is produced by the reaction (at high temperatures) between $TiCl_4$ gas and H_2.

$$2TiCl_4(g) + H_2(g) \longrightarrow 2TiCl_3(s) + 2HCl(g)$$

Assume 100% yield and constant temperature and pressure.
   (a) How many liters of HCl gas can be produced by mixing 3.72 L of $TiCl_4$ and 4.50 L of H_2?
   (b) How many liters of reactant in excess are present after the reaction is complete?

35. Nitric acid can be prepared by bubbling dinitrogen pentoxide into water.

$$N_2O_5(g) + H_2O \longrightarrow 2H^+(aq) + 2NO_3^-(aq)$$

   (a) How many moles of H^+ are obtained when 1.50 L of N_2O_5 at 25°C and 1.00 atm pressure is bubbled into water?
   (b) The solution obtained in (a) after reaction is complete has a volume of 437 mL. What is the molarity of the nitric acid obtained?

36. Potassium peroxide is used to absorb the CO_2 produced by the people in a space vehicle.

$$2K_2O_2(s) + 2CO_2(g) \longrightarrow 2K_2CO_3(s) + O_2(g)$$

If a person at rest exhales 3.0 L of air per minute and CO_2 is 3.4% (by volume) of exhaled air, how many grams of K_2O_2 are needed per person for a five-day trip. Assume a temperature of 25° and 728 mm Hg pressure.

37. Hydrogen cyanide (HCN) is a poisonous gas. It can be formed by the following reaction:

$$H^+(aq) + NaCN(s) \longrightarrow HCN(g) + Na^+(aq)$$

What volume of 6.00 M HCl is required to react with an excess of NaCN to produce enough HCN to fill a room $12 \times 11 \times 9$ feet at a pressure of 0.987 atm and 72°F?

38. When hydrogen peroxide decomposes, oxygen is produced:

$$2H_2O_2(aq) \longrightarrow 2H_2O + O_2(g)$$

What volume of oxygen gas at 25°C and 1.00 atm is produced from the decomposition of 25.00 mL of a 30.0% (by mass) solution of hydrogen peroxide ($d = 1.05$ g/mL)?

39. Ammonium nitrate can be used as an effective explosive because it decomposes into a large number of gaseous products. At a sufficiently high temperature, ammonium nitrate decomposes into nitrogen, oxygen, and steam.

(a) Write a balanced net ionic equation for the decomposition of ammonium nitrate.

(b) If 2.00 kg of ammonium nitrate are sealed in a 50.0-L steel drum and heated to 745°C, what is the resulting pressure in the drum after decomposition? (Assume 100% decomposition.)

40. Acetone peroxide, $C_9H_{18}O_6(s)$, is a powerful but highly unstable explosive that does not contain nitrogen. It can pass undetected through scanners designed to detect the presence of nitrogen in explosives like TNT (trinitrotoluene, $C_7H_5N_3O_6$), or ammonium nitrate.

(a) Write a balanced equation for the combustion (burning in oxygen) of acetone peroxide producing steam and carbon dioxide.

(b) What pressure is generated in a 2.00-L bottle when 5.00 g of acetone peroxide is ignited to 555°C and burned in air? Assume 100% combustion.

5-5 Gas Mixtures: Partial Pressures and Mole Fractions

41. A gaseous mixture is made up of 10.6 g of cyclopropane (C_3H_6), 12.2 g of chloroform ($CHCl_3$), and 5.23 g of methane. What pressure is exerted by the mixture in a 78.0-L tank at 20°C (two significant figures)? Which gas has the highest partial pressure?

42. A certain laser uses a gas mixture consisting of 9.00 g HCl, 2.00 g H_2, and 165.0 g of Ne. What pressure is exerted by the mixture in a 75.0-L tank at 22°C? Which gas has the smallest partial pressure?

43. A sample of a smoke stack emission was collected into a 1.25-L tank at 752 mm Hg and analyzed. The analysis showed 92% CO_2, 3.6% NO, 1.2% SO_2, and 4.1% H_2O by mass. What is the partial pressure exerted by each gas?

44. The contents of a tank of natural gas at 1.20 atm is analyzed. The analysis showed the following mole percents: 88.6% CH_4, 8.9% C_2H_6, and 2.5% C_3H_8. What is the partial pressure of each gas in the tank?

45. A sample of oxygen collected over water at 25°C (vapor pressure H_2O = 23.8 mm Hg). The wet gas occupies a volume of 7.28 L at a total pressure of 1.25 bar. If all the water is removed, what volume will the dry oxygen occupy at a pressure of 1.10 atm and a temperature of 27°C?

46. Hydrogen gas generated in laboratory experiments is usually collected over water. It is called a "wet gas" when collected in this manner because it contains water vapor. A sample of "wet" hydrogen at 25°C fills a 125-mL flask at a pressure of 769 mm Hg. If all the water is removed by heating, what volume will the dry hydrogen occupy at a pressure of 722 mm Hg and a temperature of 37°C? (The vapor pressure of water at 25°C is 23.8 mm Hg.)

47. Consider two bulbs separated by a valve. Both bulbs are maintained at the same temperature. Assume that when the valve between the two bulbs is closed, the gases are sealed in their respective bulbs. When the valve is closed, the following data apply:

Gas	Bulb A	Bulb B
	Ne	CO
V	2.50 L	2.00 L
P	1.09 atm	0.773 atm

Assuming no temperature change, determine the final pressure inside the system after the valve connecting the two bulbs is opened. Ignore the volume of the tube connecting the two bulbs.

48. Follow the instructions of Problem 47 for the following set of bulbs.

Gas	Bulb X	Bulb Y
	CO_2	CH_4
V	3.00 L	6.00 L
P	1.25 atm	2.66 atm

49. When acetylene, C_2H_2, is burned in oxygen, carbon dioxide and steam are formed. A sample of acetylene with a volume of 7.50 L and a pressure of 1.00 atm is burned in excess oxygen at 225°C. The products are transferred without loss to a 10.0-L flask at the same temperature.

(a) Write a balanced equation for the reaction.

(b) What is the total pressure of the products in the 10.0-L flask?

(c) What is the partial pressure of each of the products in the flask?

50. When ammonium nitrate decomposes at 722°C, nitrogen, oxygen, and steam are produced. A 25.0-g sample of ammonium nitrate decomposes, and the products are collected at 125°C into an evacuated flask with a volume of 15.0 L.

(a) Write a balanced equation for the reaction.

(b) What is the total pressure in the collecting flask after decomposition is complete?

(c) What is the partial pressure of each product in the flask?

51. A sample of oxygen is collected over water at 22°C and 752 mm Hg in a 125-mL flask. The vapor pressure of water at 22°C is 19.8 mm Hg.

    (a) What is the partial pressure of oxygen?

    (b) How many moles of dry gas are collected?

    (c) How many moles of wet gas are in the flask?

    (d) If 0.0250 g of $N_2(g)$ are added to the flask at the same temperature, what is the partial pressure of nitrogen in the flask?

    (e) What is the total pressure in the flask after nitrogen is added?

52. Nitrogen gas can be obtained by decomposing ammonium nitrate at high temperatures. The nitrogen gas is collected over water in a 500-mL (three significant figures) flask at 19°C. The ambient pressure is 745 mm Hg. (Vapor pressure of water at 19°C is 16.48 mm Hg.)

    (a) What is the partial pressure of nitrogen?

    (b) How many moles of water are there in the wet gas?

    (c) How many moles of dry gas are collected?

    (d) If 0.128 g of Ne are added to the flask at the same temperature, what is the partial pressure of neon in the flask?

    (e) What is the total pressure after Ne is added?

5-6 Kinetic Theory of Gases

53. Rank the following gases

$$NO \quad Ar \quad N_2 \quad N_2O_5$$

in order of

    (a) increasing speed of effusion through a tiny opening.

    (b) increasing time of effusion.

54. Rank the gases Xe, CH_4, F_2, and CH_2F_2 in order of

    (a) increasing speed of effusion through a pinhole.

    (b) increasing time of effusion.

55. What is the ratio of the rate of effusion of the most abundant gas, nitrogen, to the lightest gas, hydrogen?

56. A balloon filled with nitrogen gas has a small leak. Another balloon filled with hydrogen gas has an identical leak. How much faster will the hydrogen balloon deflate?

57. A gas effuses 1.55 times faster than propane (C_3H_8) at the same temperature and pressure.

    (a) Is the gas heavier or lighter than propane?

    (b) What is the molar mass of the gas?

58. A gas effuses through an opening one-fifth as fast as helium gas effuses through the same opening.

    (a) Is the gas heavier than helium?

    (b) What is the molar mass of the gas?

59. If 0.0129 mol of N_2O_4 effuses through a pinhole in a certain amount of time, how much NO would effuse in that same amount of time under the same conditions?

60. At a certain temperature it takes 11.2 s for 1.78×10^{-3} mol of NH_3 gas to effuse through a pinhole. Under the same conditions, how long will it take for the same amount of phosphine gas, PH_3, to effuse through the same pinhole?

61. At what temperature will a molecule of uranium hexafluoride, the densest gas known, have the same average speed as a molecule of the lightest gas, hydrogen, at 37°C?

62. Calculate the average speed of a

    (a) chlorine molecule at −32°C.

    (b) UF_6 molecule at room temperature (25°C).

5-7 Real Gases

63. The normal boiling points of acetylene (C_2H_2) and NO_2 are −84°C and 21°C, respectively.

    (a) At 25°C and 1 atm, which gas would you expect to have a molar volume closer to the ideal value?

    (b) If you want to reduce the deviations from ideal behavior, in what direction would you change the temperature? The pressure?

64. A sample of methane gas (CH_4) is at 50°C and 20 atm. Would you expect it to behave more or less ideally if

    (a) the pressure were reduced to 1 atm?

    (b) the temperature were reduced to −50°C?

65. Using Figure 5.15

    (a) estimate the density of methane gas at 200 atm.

    (b) compare the value obtained in (a) with that calculated from the ideal gas law.

66. Using Figure 5.15

    (a) estimate the density of methane gas at 100 atm.

    (b) compare the value obtained in (a) with that calculated from the ideal gas law.

    (c) determine the pressure at which the calculated density and the actual density of methane will be the same.

Unclassified

67. When air pollution is high, ozone (O_3) contents can reach 0.60 ppm (i.e., 0.60 mol ozone per million mol air). How many molecules of ozone are present per liter of polluted air if the barometric pressure is 755 mm Hg and the temperature is 79°F?

68. Assume that an automobile burns octane, C_8H_{18} ($d = 0.692$ g/mL).

    (a) Write a balanced equation for the combustion of octane to carbon dioxide and water.

    (b) A car has a fuel efficiency of 22 mi/gal of octane. What volume of carbon dioxide at 25°C and one atmosphere pressure is generated by the combustion when that car goes on a 75-mile trip.

69. A mixture of 3.5 mol of Kr and 3.9 mol of He occupies a 10.00-L container at 300 K. Which gas has the larger

    (a) average translational energy?

    (b) partial pressure?

    (c) mole fraction?

    (d) effusion rate?

70. Given that 1.00 mol of neon and 1.00 mol of hydrogen chloride gas are in separate containers at the same temperature and pressure, calculate each of the following ratios.

    (a) volume Ne/volume HCl   (b) density Ne/density HCl

    (c) average translational energy Ne/average translational energy HCl

    (d) number of Ne atoms/number of HCl molecules

71. An intermediate reaction used in the production of nitrogen-containing fertilizers is that between ammonia and oxygen:

$$4NH_3(g) + 5O_2(g) \longrightarrow 4NO(g) + 6H_2O(g)$$

A 150.0-L reaction chamber is charged with reactants to the following partial pressures at 500°C: $P_{NH_3} = 1.3$ atm, $P_{O_2} = 1.5$ atm. What is the limiting reactant?

72. A student prepared 118.9 mL of CO_2 at a pressure of 758 mm Hg and a temperature of 22°C. He did this by adding 35.47 mL of a 0.1380 M solution of a strong acid (H^+) to Na_2CO_3.

$$Na_2CO_3(s) + 2H^+(aq) \longrightarrow 2Na^+(aq) + CO_2(g) + H_2O$$

Did the student use HCl or H_2SO_4 as the strong acid?

73. At 25°C and 380 mm Hg, the density of sulfur dioxide is 1.31 g/L. The rate of effusion of sulfur dioxide through an orifice is 4.48 mL/s. What is the density of a sample of gas that effuses through an identical orifice at the rate of 6.78 mL/s under the same conditions? What is the molar mass of the gas?

74. Glycine is an amino acid made up of carbon, hydrogen, oxygen, and nitrogen atoms. Combustion of a 0.2036-g sample gives 132.9 mL of CO_2 at 25°C and 1.00 atm and 0.122 g of water. What are the percentages of carbon and hydrogen in glycine? Another sample of glycine weighing 0.2500 g is treated in such a way that all the nitrogen atoms are converted to $N_2(g)$. This gas has a volume of 40.8 mL at 25°C and 1.00 atm. What is the percentage of nitrogen in glycine? What is the percentage of oxygen? What is the empirical formula of glycine?

Conceptual Questions

75. A 250.0-mL flask filled with nitrogen gas at 25°C and 1.10 atm is inverted into cold acetone. As the gas contracts, acetone is forced into the flask to maintain the same pressure of nitrogen. This required 88.7 mL of acetone. What is the final temperature of the nitrogen?

76. The Lamborghini Aventador engine has a 12-cylinder engine in which each cylinder has a volume of 542 cm³. Each cylinder is full of air at 85°C and 1.00 atm.

(a) How many moles of oxygen are in each cylinder? (mole percent of O_2 in air = 21.0)

(b) Assume that the hydrocarbons in gasoline have an average molar mass of 1.0×10^2 g/mol and react with oxygen in a 1:12 mole ratio. How many grams of gasoline should be injected into the cylinder to react with the oxygen?

77. Consider two independent identical bulbs (A and B), each containing a gas. Bulb A has 2.00 moles of CH_4 and bulb B has 2.00 moles of SO_2. These bulbs have a valve that can open into a long tube that has no gas (a vacuum). The tubes for each bulb are identical in length.

(a) If both valves to bulbs A and B are opened simultaneously, which gas will reach the end of the tube first?

(b) If both gases are to reach the end of the tube simultaneously, how would you alter the contents of each bulb? (You may not alter the bulbs or the length of the tube.)

78. Consider two identical sealed steel tanks in a room maintained at a constant temperature. One tank (A) is filled with CO_2, and the other (B) is filled with H_2 until the pressure gauges on both tanks register the same pressure.

(a) Which tank has the greater number of moles?

(b) Which gas has the higher density (g/L)?

(c) Which gas will take longer to effuse out of its tank?

(d) Which gas has a larger average translational energy?

(e) If one mole of helium is added to each tank, which gas (CO_2 or H_2) will have the larger partial pressure?

79. Consider three sealed tanks all at the same temperature, pressure, and volume.

Tank A contains SO_2 gas.
Tank B contains O_2 gas.
Tank C contains CH_4 gas.

Use **LT** (for "*is less than*"), **GT** (for "*is greater than*"), **EQ** (for "*is equal to*"), or **MI** (for "*more information required*") as answers to the blanks below.

(a) The mass of SO_2 in tank A _____ the mass of O_2 in tank B.

(b) The average translational energy of CH_4 in tank C _____ the average translational energy of SO_2 in tank A.

(c) It takes 20 s for all of the O_2 gas in tank B to effuse out of a pinhole in the tank. The time it takes for all of the SO_2 to effuse out of tank A from an identical pinhole _____ 40 s.

(d) The density of O_2 in tank B _____ the density of CH_4 in tank C.

(e) The temperature in tank A is increased from 150 K to 300 K. The temperature in tank B is kept at 150 K. The pressure in tank A is _____ half the pressure in tank B.

80. Consider two bulbs A and B, identical in volume and temperature. Bulb A contains 1.0 mol of N_2 and bulb B has 1.0 mol of NH_3. Both bulbs are connected by a tube with a valve that is closed.

(a) Which bulb has the higher pressure?

(b) Which bulb has the gas with the higher density?

(c) Which bulb has molecules with a higher average kinetic energy?

(d) Which bulb has a gas whose molecules move with a faster molecular speed?

(e) If the valve between the two bulbs is opened, how will the pressure change?

(f) If 2.0 mol of He are added while the valve is opened, what fraction of the total pressure will be due to helium?

81. Sketch a cylinder with ten molecules of helium (He) gas. The cylinder has a movable piston. Label this sketch *before*. Make an *after* sketch to represent

(a) a decrease in temperature at constant pressure.

(b) a decrease in pressure from 1000 mm Hg to 500 mm Hg at constant temperature.

(c) five molecules of H_2 gas added at constant temperature and pressure.

82. Tank A has SO_2 at 2 atm, whereas tank B has O_2 at 1 atm. Tanks A and B have the same volume. Compare the temperature (in K) in both tanks if

(a) tank A has twice as many moles of SO_2 as tank B has of O_2.

(b) tank A has the same number of moles of SO_2 as tank B has of O_2.

(c) tank A has twice as many grams of SO_2 as tank B has of O_2.

83. Two tanks (A and B) have the same volume and are kept at the same pressure. Compare the temperature in both tanks if

(a) tank A has 2.00 mol of methane and tank B has 2.00 mol of neon.

(b) tank A has 2.00 g of methane and tank B has 2.00 g of neon. (Try to do this without a calculator.)

84. The graph below shows the distribution of molecular speeds for helium and carbon dioxide at the same temperature.

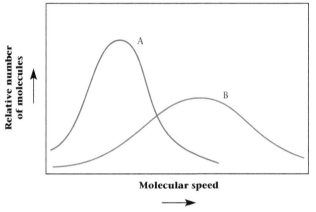

Molecular speed

(a) Which curve could represent the behavior of carbon dioxide?

(b) Which curve represents the gas that would effuse more quickly?

(c) Which curve could represent the behavior of helium gas?

85. Consider the following sketch. Each square in bulb A represents a mole of atoms X. Each circle in bulb B represents a mole of atoms Y. The bulbs have the same volume, and the temperature is kept constant. When the valve is opened, atoms of X react with atoms of Y according to the following equation:

$$2X(g) + Y(g) \longrightarrow X_2Y(g)$$

The gaseous product is represented as □○□, and each □○□ represents one mole of product.

(a) If $P_A = 2.0$ atm, what is P_B before the valve is opened and the reaction is allowed to occur? What is $P_A + P_B$?

(b) Redraw the sketch to represent what happens after the valve is opened.

(c) What is P_A? What is P_B? What is $P_A + P_B$? Compare your answer with the answer in part (a).

86. The following figure shows three 1.00-L bulbs connected by valves. Each bulb contains argon gas with amounts proportional to the number of circles pictorially represented in the chamber. All three bulbs are maintained at the same temperature. Unless stated otherwise, assume that the valves connecting the bulbs are closed and seal the gases in their respective chambers. Assume also that the volume between each bulb is negligible.

(a) Which bulb has the highest pressure?

(b) If the pressure in bulb A is 0.500 atm, what is the pressure in bulb C?

(c) If the pressure in bulb A is 0.500 atm, what is the total pressure?

(d) If the pressure in bulb A is 0.500 atm, and the valve between bulbs A and B is opened, redraw the figure shown above to accurately represent the gas atoms in all three bulbs. What is $P_A + P_B + P_C$? Compare your answer in part (d) to that in part (c).

(e) Follow the instructions of part (d) but now open only the valve between bulbs B and C.

87. Consider three sealed steel tanks, labeled X, Y, and Z. Each tank has the same volume and the same temperature. In each tank, one mole of CH_4 is represented by a circle, one mole of oxygen by a square, and one mole of SO_2 by a triangle. Assume that no reaction takes place between these molecules.

Tank X · Tank Y · Tank Z

(a) In which tank is the total pressure highest?

(b) In which tank is the partial pressure of SO_2 highest?

(c) In which tank is the mass of all three gases the same?

(d) Which tank has the heaviest contents?

Challenge Problems

88. Consider an ideal gas that exerts a pressure of 23.76 mm Hg at 25°C. Assuming n and V are held constant, what would its pressure be at 40°C? 70°C? 100°C? Compare the numbers you have just calculated with the vapor pressures of water at these temperatures. Can you suggest a reason why the two sets of numbers are so different?

89. Two 750-mL (three significant figures) tanks contain identical amounts of a gaseous mixture containing O_2, N_2, and CO_2. Each tank has a pressure of 1.38 atm and is kept at 25°C. The gas in the first tank is passed through a CO_2 absorber. After all the CO_2 is absorbed, the mass of the absorber increases by 0.114 g. All the nitrogen in the second tank is removed. The pressure in the tank without the nitrogen at 25°C is 1.11 atm. What is the composition (in mass percent) of the gaseous mixture in the tanks?

90. A tube 5.0 ft long is evacuated. Samples of NH_3 and HCl, at the same temperature and pressure, are introduced simultaneously through tiny openings at opposite ends of the tube. When the two gases meet, a white ring of $NH_4Cl(s)$ forms. How far from the end at which ammonia was introduced will the ring form?

91. Sulfur trioxide can be prepared by burning sulfur in oxygen.

$$2S(s) + 3O_2(g) \longrightarrow 2SO_3(g)$$

A 5.00-L flask containing 5.00 g of sulfur and oxygen at a pressure of 995 mm Hg and 25°C is heated. When the reaction is complete, the temperature in the flask is 138°C.

(a) What is the pressure of SO_3 in the flask?

(b) What is the total pressure in the flask?

(c) When water is added to SO_3, H_2SO_4 is formed. What is the molarity of the H_2SO_4 formed if 250.0 mL of water are added to the flask? (Assume that there is enough water to convert all the SO_3 to H_2SO_4.)

92. A 0.2500-g sample of an Al-Zn alloy reacts with HCl to form hydrogen gas:

$$Al(s) + 3H^+(aq) \longrightarrow Al^{3+}(aq) + \tfrac{3}{2}H_2(g)$$
$$Zn(s) + 2H^+(aq) \longrightarrow Zn^{2+}(aq) + H_2(g)$$

The hydrogen produced has a volume of 0.147 L at 25°C and 755 mm Hg. What is the percentage of zinc in the alloy?

93. The buoyant force on a balloon is equal to the mass of air it displaces. The gravitational force on the balloon is equal to the sum of the masses of the balloon, the gas it contains, and the balloonist. If the balloon and the balloonist together weigh 168 kg, what would the diameter of a spherical hydrogen-filled balloon have to be (in meters) if the rig is to get off the ground at 22°C and 758 mm Hg? (Take $MM_{air} = 29.0$ g/mol.)

94. A mixture in which the mole ratio of hydrogen to oxygen is 2:1 is used to prepare water by the reaction

$$2H_2(g) + O_2(g) \longrightarrow 2H_2O(g)$$

The total pressure in the container is 0.950 atm at 25°C before the reaction. What is the final pressure in the container at 125°C after the reaction, assuming an 88.0% yield and no volume change?

6
Electronic Structure and the Periodic Table

The tubes of glass in Dale Chihuly's *Glass Forest #3* are filled with neon gas. Neon light is obtained by electrifying the tubes or by placing them over an energized field.

Not chaos-like crush'd and bruis'd
But, as the world, harmoniously
  confus'd,
Where order in variety we see,
And where, though all things
  differ, all agree.

—ALEXANDER POPE
Windsor Forest

In Chapter 2 we briefly considered the structure of the atom. You will recall that every atom has a tiny, positively charged nucleus, consisting of protons and neutrons. The nucleus is surrounded by negatively charged electrons. The number of protons in the nucleus is characteristic of the atoms of a particular element and is referred to as the atomic number. In a neutral atom, the number of electrons is equal to the number of protons and hence to the atomic number.

In this chapter, we focus on electron arrangements in atoms, paying particular attention to the relative energies of different electrons (*energy levels*) and their spatial locations (*orbitals*). Specifically, we consider the nature of the energy levels and orbitals available to

- the single electron in the hydrogen atom (Section 6-2).
- the several electrons in more complex atoms (Sections 6-3, 6-4).

With this background, we show how electron arrangements in multielectron atoms and the monatomic ions derived from them can be described in terms of

- *electron configurations*, which show the number of electrons in each energy level (Sections 6-5, 6-7).
- *orbital diagrams*, which show the arrangement of electrons within orbitals (Sections 6-6, 6-7).

The electron configuration or orbital diagram of an atom of an element can be deduced from its position in the periodic table. Beyond that, position in the table can predict (Section 6-8) the relative sizes of atoms and ions (*atomic radius, ionic radius*) and the relative tendencies of atoms to give up or acquire electrons (*ionization energy, electronegativity*).

Before dealing with electronic structures as such, it is helpful to examine briefly the experimental evidence on which such structures are based (Section 6-1). In particular, we need to look at the phenomenon of *atomic spectra*.

Figure 6.1 Fireworks. The different colors are created by the atomic spectra of different elements.

6-1 Light, Photon Energies, and Atomic Spectra

Fireworks displays are fascinating to watch (Figure 6.1). Neon lights and sodium vapor lamps can transform the skyline of a city with their brilliant colors. The eerie phenomenon of the aurora borealis is an unforgettable experience when you see it for the first time. All of these events relate to the generation of light and its transmission through space.

ℹ️ λ is the Greek letter lambda; ν is the Greek letter nu.

The Wave Nature of Light: Wavelength and Frequency

Light travels through space as a wave, consisting of successive crests, which rise above the midline, and troughs, which sink below it. Waves have three primary characteristics (Figure 6.2), two of which are of particular interest at this point:

1. **Wavelength** (λ), ◀ℹ️ the distance between two consecutive crests or troughs, most often measured in meters or nanometers (1 nm = 10^{-9} m).
2. **Frequency** (ν), the number of wave cycles (successive crests or troughs) that pass a given point in unit time. If 10^8 cycles pass a particular point in one second,

$$\nu = 10^8/s = 10^8 \text{ Hz}$$

The frequency unit **hertz** (Hz) represents one cycle per second.

The speed at which a wave moves through space can be found by multiplying the length of a wave cycle (λ) by the number of cycles passing a point in unit time (ν). For light,

$$\lambda\nu = c \qquad (6.1)$$

where c, the speed of light in a vacuum, is 2.998×10^8 m/s. To use this equation with this value of c—

- λ should be expressed in meters.
- ν should be expressed in reciprocal seconds (hertz).

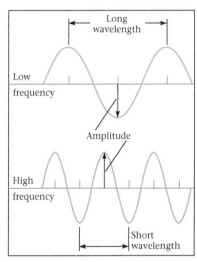

Figure 6.2 Characteristics of waves. The *amplitude* (Ψ) is the height of a crest or the depth of a trough. The *wavelength* (λ) is the distance between successive crests or troughs. The *frequency* (ν) is the number of wave cycles (successive crests or troughs) that pass a given point in a given time.

EXAMPLE 6.1

In a grocery store, a scanner uses a laser with a frequency of 4.62×10^{14} s^{-1} to scan the bar code on a power drink. The store pipes in music from an FM station with a frequency of 286 cm.

a What is the frequency of the FM station?

ANALYSIS	
Information given:	wavelength of the station (286 cm)
Information implied:	speed of light (2.998×10^8 m/s) meter to centimeter conversion factor
Asked for:	frequency of the radio station

continued

1. Recall the Greek letters used as symbols for frequency (ν) and wavelength (λ).
2. Use Equation 6.1 to relate frequency and wavelength.
3. Convert cm to m.

SOLUTION

Wavelength in meters	$286 \text{ cm} \times \dfrac{1 \text{ m}}{100 \text{ cm}} = 2.86 \text{ m}$
Frequency	$\nu = \dfrac{c}{\lambda} = \dfrac{2.998 \times 10^8 \text{ m/s}}{2.86 \text{ m}} = 1.05 \times 10^8 \text{ s}^{-1}$

b What is the wavelength of the laser in nm?

ANALYSIS

Information given:	frequency of the laser ($4.62 \times 10^{14} \text{ s}^{-1}$)
Information implied:	speed of light (2.998×10^8 m/s) meter to nanometer conversion factor
Asked for:	wavelength of the laser

STRATEGY

1. Use Equation 6.1 to relate frequency and wavelength.
2. Convert m to nm.

SOLUTION

Wavelength	$\lambda = \dfrac{c}{\nu} = \dfrac{2.998 \times 10^8 \text{ m/s}}{4.62 \times 10^{14} \text{ s}^{-1}} = 6.49 \times 10^{-7} \text{ m}$
Wavelength in nm	$6.49 \times 10^{-7} \text{ m} \times \dfrac{1 \text{ nm}}{1 \times 10^{-9} \text{ m}} = 649 \text{ nm}$

Light visible to the eye is only a tiny portion of the entire electromagnetic spectrum (Figure 6.3) covering only the narrow wavelength region from 400 to 700 nm. For a substance to be colored, it must absorb somewhere within this region. Ozone in the upper atmosphere absorbs harmful, high-energy **ultraviolet (UV)**

Figure 6.3 The electromagnetic spectrum. Note that only a small fraction is visible to the human eye.

Table 6.1 Relation Between Color and Wavelength

Wavelength (nanometers)	Color Absorbed	Color Transmitted
<400 nm	Ultraviolet	Colorless
400–450 nm	Violet	Red, orange, yellow
450–500 nm	Blue	
500–550 nm	Green	Purple
550–580 nm	Yellow	
580–650 nm	Orange	Blue, green
650–700 nm	Red	
>700 nm	Infrared	Colorless

Figure 6.4 **Crystals of potassium permanganate falling into water.** The purple color of the solution results from absorption at approximately 550 nm.

radiation from the sun. Carbon dioxide absorbs **infrared (IR) radiation** given off by the earth's surface, preventing it from escaping into the outer atmosphere, thereby contributing to global warming. Microwave ovens produce radiation at wavelengths longer than infrared radiation, whereas x-rays have wavelengths shorter than UV radiation (Figure 6.3).

Some of the substances you work with in general chemistry can be identified at least tentatively by their color. Gaseous nitrogen dioxide has a brown color; vapors of bromine and iodine are red and violet, respectively. A water solution of copper sulfate is blue, and a solution of potassium permanganate is purple (Figure 6.4).

The colors of gases and liquids are due to the selective absorption of certain components of visible light. Bromine, for example, absorbs in the violet and blue regions of the spectrum (Table 6.1). The subtraction of these components from visible light accounts for the red color of bromine liquid or vapor. The purple (blue-red) color of a potassium permanganate solution results from absorption in the green region (Figure 6.5).

Figure 6.5 **Absorption spectrum of potassium permanganate.**

The Particle Nature of Light; Photon Energies

A hundred years ago it was generally supposed that all the properties of light could be explained in terms of its wave nature. A series of investigations carried out between 1900 and 1910 by Max Planck (1858–1947) (blackbody radiation) and Albert Einstein (1879–1955) (photoelectric effect) discredited that notion. Today we consider light to be generated as a stream of particles called **photons,** whose energy E (Figure 6.6) is given by the equation

$$E = h\nu = hc/\lambda \tag{6.2}$$

Throughout this text, we will use the SI unit **joule** (J), $1 \text{ kg} \cdot \text{m}^2/\text{s}^2$, to express energy. A joule is a rather small quantity. One joule of electrical energy would keep a 10-W lightbulb burning for only a tenth of a second. For that reason, we will often express energies in *kilojoules* ($1 \text{ kJ} = 10^3 \text{ J}$). The quantity h appearing in Planck's equation is referred to as *Planck's constant.*

$$h = 6.626 \times 10^{-34} \text{ J} \cdot \text{s}$$

Notice from this equation that energy is *inversely* related to wavelength. This explains why you put on sunscreen to protect yourself from UV solar radiation (<400 nm) and a "lead apron" when dental x-rays (<10 nm) are being taken. Conversely, IR (>700 nm) and microwave photons (>80,000 nm) are of relatively low energy (but don't try walking on hot coals).

Figure 6.6 **Energy and wavelength.** A copper wire held in a flame colors the flame green. The energy of the photons of this light can be calculated from its wavelength.

EXAMPLE 6.2 **GRADED**

Sodium vapor lamps are commonly used to illuminate highways because of their intense yellow-orange emissions at 589 nm.

a Calculate the energy, in joules, of one photon of this light.

ANALYSIS

Information given:	wavelength of sodium vapor (589 nm)
Information implied:	speed of light (2.998×10^8 m/s); Planck's constant (6.626×10^{-34} J · s)
Asked for:	energy of one photon in J

STRATEGY

Use Equation 6.2 to relate energy to wavelength.

$$E = \frac{hc}{\lambda}$$

SOLUTION

Energy for one photon	$E = \dfrac{hc}{\lambda} = \dfrac{(6.626 \times 10^{-34} \text{ J · s})(2.998 \times 10^8 \text{ m/s})}{589 \times 10^{-9} \text{ m}} = 3.37 \times 10^{-19}$ J

b Calculate the energy, in kilojoules, of one mole of such photons.

ANALYSIS

Information given:	From part (a), the energy of one photon (3.37×10^{-19} J)
Information implied:	Avogadro's number (6.022×10^{23} units/mol)
Asked for:	energy of one mole of photons in kJ

STRATEGY

Use the appropriate conversion factors to change nm to m, J to kJ, and one photon to one mole of photons.

SOLUTION

E/mol of photons	$E = 1 \text{ mol photons} \times \dfrac{3.37 \times 10^{-19} \text{ J}}{1 \text{ photon}} \times \dfrac{6.022 \times 10^{23} \text{ photons}}{1 \text{ mol photons}} \times \dfrac{1 \text{ kJ}}{1000 \text{ J}} = 203$ kJ

c To sense visible light, the optic nerve needs at least 2.0×10^{-17} J of energy to trigger impulses that reach the brain. How many photons of the sodium lamp emissions are needed to "see" the yellow light?

ANALYSIS

Information given:	Energy required by the optic nerve (2.0×10^{-17} J) From part (a), the energy of one photon (3.37×10^{-19} J)
Asked for:	number of photons needed to "see" yellow light

STRATEGY

Use the energy per photon for yellow light found in part (a) as a conversion factor.

$$\frac{3.37 \times 10^{-19} \text{ J}}{1 \text{ photon}}$$

SOLUTION

Photons needed	$2.0 \times 10^{-17} \text{ J} \times \dfrac{1 \text{ photon}}{3.37 \times 10^{-19} \text{ J}} = 59$ photons

continued

END POINTS

1. In part (a), 3.37×10^{-19} J may seem like a tiny amount of energy, but bear in mind that it comes from a single photon.

2. In part (b), the energy calculated for one mole of photons, 203 kJ, is roughly comparable to the energy effects in chemical reactions. About 240 kJ of heat is evolved when a mole of hydrogen gas burns (more on this in Chapter 8).

3. In part (c), note that not too many photons are needed to sense light.

Atomic Spectra

In the seventeenth century, Sir Isaac Newton showed that visible (white) light from the Sun can be broken down into its various color components by a prism. The **atomic spectrum** obtained is continuous; it contains essentially all wavelengths between 400 and 700 nm.

The situation with high-energy atoms of gaseous elements is quite different (Figure 6.7). Here the spectrum consists of discrete lines given off at specific wavelengths. Each element has a characteristic spectrum that can be used to identify it. In the case of sodium, there are two strong lines in the yellow region at 589.0 nm and 589.6 nm. These lines account for the yellow color of sodium vapor lamps used to illuminate highways.

The fact that the photons making up atomic spectra have only certain discrete wavelengths implies that they can have only certain discrete energies, because

$$E = h\nu = hc/\lambda$$

Since these photons are produced when an electron moves from one energy level to another, the electronic energy levels in an atom must be *quantized*, that is, limited to particular values. Moreover, it would seem that by measuring the spectrum of an element ◄ *i* it should be possible to unravel its electronic energy levels. This is indeed possible, but it isn't easy. Gaseous atoms typically give off hundreds, even thousands, of spectral lines.

One of the simplest of atomic spectra, and the most important from a theoretical standpoint, is that of hydrogen. When energized by a high-voltage discharge, gaseous hydrogen atoms emit radiation at wavelengths that can be grouped into several different series (Table 6.2). The first of these to be discovered, the Balmer ◄ *i* series, lies partly in the visible region. It consists of a strong line at 656.28 nm followed by successively weaker lines, closer and closer together, at lower wavelengths.

i Atomic spectroscopy can identify metals at concentrations as low as 10^{-7} mol/L.

i Balmer was a Swiss high-school teacher.

Figure 6.7 Continuous and line emission spectra. From the top down: The continuous visible spectrum; the line emission spectra for sodium (Na), hydrogen (H), and mercury (Hg).

Table 6.2 Wavelengths (nm) of Lines in the Atomic Spectrum of Hydrogen

Ultraviolet (Lyman Series)	Visible (Balmer Series)	Infrared (Paschen Series)
121.53	656.28	1875.09
102.54	486.13	1281.80
97.23	434.05	1093.80
94.95	410.18	1004.93
93.75	397.01	
93.05		

6-2   The Hydrogen Atom

The hydrogen atom, containing a single electron, has played a major role in the development of models of electronic structure. In 1913 Niels Bohr (1885–1962), a Danish physicist, offered a theoretical explanation of the atomic spectrum of hydrogen. His model was based largely on classical mechanics. In 1922 this model earned him the Nobel Prize in physics. By that time, Bohr had become director of the Institute of Theoretical Physics at Copenhagen. ◄ℹ There, he helped develop the new discipline of quantum mechanics, used by other scientists to construct a more sophisticated model for the hydrogen atom.

Bohr, like all the other individuals mentioned in this chapter, was not a chemist. His only real contact with chemistry came as an undergraduate at the University of Copenhagen. His chemistry teacher, Niels Bjerrum, who later became his close friend and sailing companion, recalled that Bohr set a record for broken glassware that lasted half a century.

ℹ Bohr, a giant of twentieth-century physics, was respected by scientists and politicians alike.

Bohr Model

The **Bohr model** assumed that a hydrogen atom consists of a central proton about which an electron moves in a circular orbit. He related the electrostatic force of attraction of the proton for the electron to the centrifugal force due to the circular motion of the electron. In this way, Bohr was able to express the energy of the atom in terms of the radius of the electron's orbit. To this point, his analysis was purely classical, based on Coulomb's law of electrostatic attraction and Newton's laws of motion. To progress beyond this point, Bohr boldly and arbitrarily assumed, in effect, that the electron in the hydrogen atom can have only certain definite energies. Using arguments that we will not go into, Bohr obtained the following equation for the energy of the hydrogen electron:

$$E_n = -R_H/n^2 \triangleleft ℹ \qquad (6.3)$$

where E_n is the energy of the electron, R_H is a quantity called the *Rydberg constant* (modern value = 2.180×10^{-18} J), and **n** is an integer called the principal quantum number. Depending on the state of the electron, **n** can have any positive, integral value, that is,

$$n = 1, 2, 3, \ldots$$

 $E_n = -2.180 \times 10^{-18}$ J/n²

Before proceeding with the Bohr model, let us make three points:

1. In setting up his model, Bohr designated zero energy as the point at which the proton and electron are completely separated. Energy has to be absorbed to reach that point. This means that the electron, in all its allowed energy states within the atom, must have an energy below zero; that is, it must be negative, hence the minus sign in the equation:

$$E_n = -R_H/n^2$$

2. Ordinarily the hydrogen electron is in its lowest energy state, referred to as the **ground state** or ground level, for which **n** = 1. When an electron absorbs enough energy, it moves to a higher, **excited state.** In a hydrogen atom, the first excited state has **n** = 2, the second **n** = 3, and so on.

3. When an excited electron gives off energy as a photon of light, it drops back to a lower energy state. The electron can return to the ground state (from **n** = 2 to **n** = 1, for example) or to a lower excited state (from **n** = 3 to **n** = 2). In every case, the energy of the photon ($h\nu$) evolved is equal to the difference in energy between the two states:

$$\Delta E = h\nu = E_{hi} - E_{lo}$$

where E_{hi} and E_{lo} are the energies of the higher and lower states, respectively.

Using this expression for ΔE and the equation $E_n = -R_H/n^2$, it is possible to relate the frequency of the light emitted to the quantum numbers, $\mathbf{n}_{hi}$ and $\mathbf{n}_{lo}$, of the two states:

$$h\nu = -R_H\left[\frac{1}{(\mathbf{n}_{hi})^2} - \frac{1}{(\mathbf{n}_{lo})^2}\right]$$

$$\nu = \frac{R_H}{h}\left[\frac{1}{(\mathbf{n}_{lo})^2} - \frac{1}{(\mathbf{n}_{hi})^2}\right] \tag{6.4}$$

The last equation written is the one Bohr derived in applying his model to the hydrogen atom. Given

$$R_H = 2.180 \times 10^{-18}\,\text{J} \qquad h = 6.626 \times 10^{-34}\,\text{J}\cdot\text{s}$$

you can use the equation to find the frequency or wavelength of any of the lines in the hydrogen spectrum.

EXAMPLE 6.3

Calculate the wavelength in nanometers of the line in the Balmer series that results from the transition **n** = 4 to **n** = 2.

ANALYSIS

Information given:	**n** = 2; **n** = 4
Information implied:	speed of light (2.998×10^8 m/s) Rydberg constant (2.180×10^{-18} J) Planck constant (6.626×10^{-34} J $\cdot$ s)
Asked for:	wavelength in nm

STRATEGY

1. Substitute into Equation 6.4 to find the frequency due to the transition.

$$\nu = \frac{R_H}{h}\left(\frac{1}{\mathbf{n}_{lo}{}^2} - \frac{1}{\mathbf{n}_{hi}{}^2}\right)$$

Use the lower value for **n** as $\mathbf{n}_{lo}$ and the higher value for $\mathbf{n}_{hi}$.

2. Use Equation 6.1 to find the wavelength in meters and then convert to nanometers.

continued

SOLUTION

1. Frequency

$$\nu = \frac{2.180 \times 10^{-18} \text{ J}}{6.626 \times 10^{-34} \text{ J} \cdot \text{s}} \left(\frac{1}{(2)^2} - \frac{1}{(4)^2} \right) = 6.169 \times 10^{14} \text{ s}^{-1}$$

2. Wavelength

$$\lambda = \frac{2.998 \times 10^8 \text{ m/s}}{6.169 \times 10^{14} \text{ s}^{-1}} \times \frac{1 \text{ nm}}{1 \times 10^{-9} \text{ m}} = 486.0 \text{ nm}$$

END POINT

Compare this value with that listed in Table 6.2 for the second line of the Balmer series.

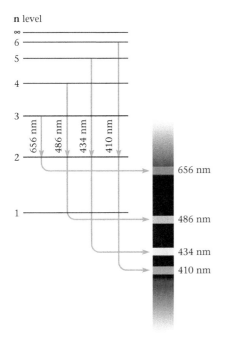

Figure 6.8 Some Balmer series lines for hydrogen. The lines in the visible region result from transitions from levels with values of **n** greater than 2 to the **n** = 2 level.

All of the lines (Figure 6.8) in the Balmer series (Table 6.2) come from transitions to the level n = 2 from higher levels (n = 3, 4, 5, . . .). Similarly, lines in the Lyman series arise when electrons fall to the n = 1 level from higher levels (n = 2, 3, 4, . . .). For the Paschen series, which lies in the infrared, the lower level is always n = 3.

Quantum Mechanical Model

Bohr's theory for the structure of the hydrogen atom was highly successful. Scientists of the day must have thought they were on the verge of being able to predict the allowed energy levels of all atoms. However, the extension of Bohr's ideas to atoms with two or more electrons gave, at best, only qualitative agreement with experiment. Consider, for example, what happens when Bohr's theory is applied to the helium atom. For helium, the errors in calculated energies and wavelengths are of the order of 5% instead of the 0.1% error with hydrogen. There appeared to be no way the theory could be modified to make it work well with helium or other atoms. Indeed, it soon became apparent that there was a fundamental problem with the Bohr model. The idea of an electron moving about the nucleus in a well-defined orbit at a fixed distance from the nucleus had to be abandoned.

Scientists in the 1920s, speculating on this problem, became convinced that an entirely new approach was required to treat electrons in atoms and molecules. In 1924 a young French scientist, Louis de Broglie (1892–1987), in his doctoral dissertation at the Sorbonne, made a revolutionary suggestion. He reasoned that if light could show the behavior of particles (photons) as well as waves, then perhaps an electron, which Bohr had treated as a particle, could behave like a wave. In a few years, de Broglie's postulate was confirmed experimentally. This led to the development of a whole new discipline, first called wave mechanics, more commonly known today as *quantum mechanics*.

The quantum mechanical atom differs from the Bohr model in several ways. In particular, according to quantum mechanics—

- *the kinetic energy of an electron is inversely related to the volume of the region to which it is confined.* This phenomenon has no analog in classical mechanics, but it helps to explain the stability of the hydrogen atom. Consider what happens when an electron moves closer and closer to the nucleus. The electrostatic energy decreases; that is, it becomes more negative. If this were the only factor, the electron should radiate energy and "fall" into the nucleus. However, the kinetic energy is increasing at the same time, because the electron is moving within a smaller and smaller volume. The two effects oppose each other; at some point a balance is reached and the atom is stable.

- *it is impossible to specify the precise position of an electron in an atom at a given instant.* Neither can we describe in detail the path that an electron takes about the nucleus. (After all, if we can't say where the electron is, we certainly don't know how it got there.) The best we can do is to estimate the *probability* of finding the electron within a particular region.

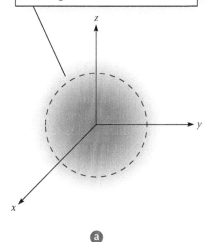

In this rendering the depth of color is proportional to the probability of finding the electron at a given point. The dotted line encloses the volume within which there is a 90% probability of finding the electron.

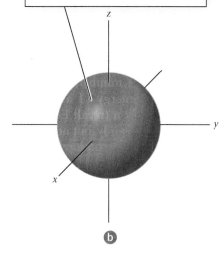

In this rendering the orbital encloses the volume within which there is a 90% probability of finding the electron; the electron density within the orbital is not shown.

Figure 6.9 Two different ways of showing the electron distribution in the ground state of the hydrogen atom.

In 1926 Erwin Schrödinger (1887–1961), an Austrian physicist, made a major contribution to quantum mechanics. He wrote down a rather complex differential equation to express the wave properties of an electron in an atom. This equation can be solved, at least in principle, to find the **amplitude** (height) Ψ of the electron wave at various points in space. The quantity ψ (psi) is known as the *wave function*. Although we will not use the Schrödinger wave equation in any calculations, you should realize that much of our discussion of electronic structure is based on solutions to that equation for the electron in the hydrogen atom. ◄ⓘ

For the hydrogen electron, the square of the wave function, ψ^2, is directly proportional to the probability of finding the electron at a particular point. If ψ^2 at point A is twice as large as at point B, then we are twice as likely to find the electron at A as at B. Putting it another way, over time the electron will turn up at A twice as often as at B.

Figure 6.9a, an *electron cloud* diagram, shows how ψ^2 for the hydrogen electron in its ground state ($n = 1$) varies moving out from the nucleus. The depth of the color is supposed to be directly proportional to ψ^2 and hence to the probability of finding the electron at a point. As you can see, the color fades moving out from the nucleus in any direction; the value of ψ^2 drops accordingly. ◄ⓘ

Another, more common way of showing the electron distribution in the ground state of the hydrogen atom is to draw the **orbital** (Figure 6.9b) within which there is a 90% chance of finding the electron. Notice that the orbital is spherical, which means that the probability is independent of direction; the electron is equally likely to be found north, south, east, or west of the nucleus.

ⓘ In case you're curious, the equation is $\dfrac{d^2\Psi}{dx^2} + \dfrac{8\pi^2 m(E - V)}{h^2}\Psi = 0$ (and that's just in one dimension).

ⓘ Seems strange, but an electron is more likely to be found at the nucleus than at any other point.

6-3 Quantum Numbers

The Schrödinger equation can be solved approximately for atoms with two or more electrons. There are many solutions for the wave function, ψ, each associated with a set of numbers called **quantum numbers.** Three such numbers are given the symbols $\mathbf{n}$, $\boldsymbol{\ell}$, and $\mathbf{m}_\ell$. A wave function corresponding to a particular set of three quantum numbers (e.g., $n = 2$, $\ell = 1$, $m_\ell = 0$) is associated with an electron occupying an atomic orbital. From the expression for ψ, we can deduce the relative energy of that orbital, its shape, and its orientation in space.

For reasons we will discuss later, a fourth quantum number is required to completely describe a specific electron in a multielectron atom. The fourth quantum number is given the symbol m_s. Each electron in an atom has a set of four quantum numbers: n, ℓ, m_ℓ, and m_s. We will now discuss the quantum numbers of electrons as they are used in atoms beyond hydrogen.

First Quantum Number, n; Principal Energy Levels

The first quantum number, given the symbol **n,** is of primary importance in determining the energy of an electron. For the hydrogen atom, the energy depends upon only **n** (recall Equation 6.3). In other atoms, the energy of each electron depends mainly, but not completely, upon the value of **n.** As **n** increases, the energy of the electron increases and, on the average, it is found farther out from the nucleus. The quantum number **n** can take on only integral values, starting with 1:

$$n = 1, 2, 3, 4, \ldots \qquad (6.5)$$

An electron for which $n = 1$ is said to be in the first **principal energy level.** If $n = 2$, we are dealing with the second principal level, and so on.

Second Quantum Number, ℓ; Sublevels (s, p, d, f)

Each principal energy level includes one or more **sublevels.** The sublevels are denoted by the second quantum number, ℓ. As we will see later, the general shape of the electron cloud associated with an electron is determined by ℓ. Larger values of ℓ produce more complex shapes. The quantum numbers **n** and ℓ are related; ℓ can take on any integral value starting with 0 and going up to a maximum of $(n - 1)$. ◄ *ⓘ* That is,

$$\ell = 0, 1, 2, \ldots, (n - 1) \qquad (6.6)$$

*This relation between **n** and ℓ comes from the Schrödinger equation.*

If $n = 1$, there is only one possible value of ℓ—namely 0. This means that, in the first principal level, there is only one sublevel, for which $\ell = 0$. If $n = 2$, two values of ℓ are possible, 0 and 1. In other words, there are two sublevels ($\ell = 0$ and $\ell = 1$) within the second principal energy level. In the same way,

if $n = 3$:	$\ell = 0, 1,$ or 2	(three sublevels)
if $n = 4$:	$\ell = 0, 1, 2,$ or 3	(four sublevels)

In general, *in the nth principal level, there are **n** different sublevels.*

Another method is commonly used to designate sublevels. Instead of giving the quantum number ℓ, the letters s, p, d, or f* indicate the sublevels $\ell = 0, 1, 2,$ or 3, respectively. That is,

quantum number, ℓ	0	1	2	3
type of sublevel	s	p	d	f

Usually, in designating a sublevel, a number is included (Table 6.3) to indicate the principal level as well. Thus reference is made to a 1s sublevel ($n = 1$, $\ell = 0$), a 2s sublevel ($n = 2$, $\ell = 0$), a 2p sublevel ($n = 2$, $\ell = 1$), and so on.

For atoms containing more than one electron, the energy is dependent on ℓ as well as **n.** Within a given principal level (same value of **n**), sublevels increase in energy in the order

$$ns < np < nd < nf$$

---

*These letters come from the adjectives used by spectroscopists to describe spectral lines: *s*harp, *p*rincipal, *d*iffuse, and *f*undamental.

Table 6.3 Sublevel Designations for the First Four Principal Levels

n	1	2		3			4			
ℓ	0	0	1	0	1	2	0	1	2	3
Sublevel	1s	2s	2p	3s	3p	3d	4s	4p	4d	4f

Thus a 2p sublevel has a slightly higher energy than a 2s sublevel. By the same token, when **n** = 3, the 3s sublevel has the lowest energy, the 3p is intermediate, and the 3d has the highest energy. ◀ ⓘ

ⓘ Which would have the higher energy, 4p or 3s?

Third Quantum Number, m_ℓ; Orbitals

Each sublevel contains one or more orbitals, which differ from one another in the value assigned to the third quantum number, m_ℓ. This quantum number determines the direction in space of the electron cloud surrounding the nucleus. The value of m_ℓ is related to that of ℓ. For a given value of ℓ, m_ℓ can have any integral value, including 0, between ℓ and −ℓ, that is

$$m_\ell = \ell, \ldots, +1, 0, -1, \ldots, -\ell \qquad (6.7)$$

To illustrate how this rule works, consider an s sublevel (ℓ = 0). Here m_ℓ can have only one value, 0. This means that an s sublevel contains only one orbital, referred to as an s orbital. ◀ ⓘ For a p orbital (ℓ = 1) m_ℓ = 1, 0, or −1. Within a given p sublevel there are three different orbitals described by the quantum numbers m_ℓ = 1, 0, and −1. All three of these orbitals have the same energy.

ⓘ An electron occupies an orbital within a sublevel of a principal energy level.

For the d and f sublevels

d sublevel:	ℓ = 2	m_ℓ = 2, 1, 0, −1, −2	5 orbitals
f sublevel:	ℓ = 3	m_ℓ = 3, 2, 1, 0, −1, −2, −3	7 orbitals

Here again all the orbitals in a given d or f sublevel have the same energy.

Fourth Quantum Number, m_s; Electron Spin

The fourth quantum number, m_s, is associated with **electron spin.** An electron has magnetic properties that correspond to those of a charged particle spinning on its axis. Either of two spins is possible, clockwise or counterclockwise (Figure 6.10).

The quantum number m_s was introduced to make theory consistent with experiment. In that sense, it differs from the first three quantum numbers, which came from the solution to the Schrödinger wave equation for the hydrogen atom. This quantum number is not related to **n**, ℓ, or m_ℓ. It can have either of two possible values:

$$m_s = +\tfrac{1}{2} \quad \text{or} \quad -\tfrac{1}{2} \qquad (6.8)$$

Electrons that have the same value of m_s (i.e., both $+\tfrac{1}{2}$ or both $-\tfrac{1}{2}$) are said to have *parallel* spins. Electrons that have different m_s values (i.e., one $+\tfrac{1}{2}$ and the other $-\tfrac{1}{2}$) are said to have *opposed* spins.

Figure 6.10 **Electron spin.** The spins can be represented as clockwise and counterclockwise, with the different values of m_s of $+\tfrac{1}{2}$ and $-\tfrac{1}{2}$.

Pauli Exclusion Principle

The four quantum numbers that characterize an electron in an atom have now been considered. There is an important rule, called the **Pauli exclusion principle,** that relates to these numbers. It requires that *no two electrons in an atom can have the same set of four quantum numbers.* This principle was first stated in 1925 by Wolfgang Pauli (1900–1958), a colleague of Bohr, again to make theory consistent with the properties of atoms. ◀ ⓘ

ⓘ Our model for electronic structure is a pragmatic blend of theory and experiment.

Table 6.4 Permissible Values of the Quantum Numbers Through n = 4

n	ℓ	m_ℓ	m_s
1	0 (1s)	0	$+\frac{1}{2}, -\frac{1}{2}$
2	0 (2s)	0	$+\frac{1}{2}, -\frac{1}{2}$
	1 (2p)	$-1, 0, +1$	$\pm\frac{1}{2}$ for each value of m_ℓ
3	0 (3s)	0	$+\frac{1}{2}, -\frac{1}{2}$
	1 (3p)	$-1, 0, +1$	$\pm\frac{1}{2}$ for each value of m_ℓ
	2 (3d)	$-2, -1, 0, +1, +2$	$\pm\frac{1}{2}$ for each value of m_ℓ
4	0 (4s)	0	$+\frac{1}{2}, -\frac{1}{2}$
	1 (4p)	$-1, 0, +1$	$\pm\frac{1}{2}$ for each value of m_ℓ
	2 (4d)	$-2, -1, 0, +1, +2$	$\pm\frac{1}{2}$ for each value of m_ℓ
	3 (4f)	$-3, -2, -1, 0, +1, +2, +3$	$\pm\frac{1}{2}$ for each value of m_ℓ

The Pauli exclusion principle has an implication that is not obvious at first glance. It requires that only two electrons can fit into an orbital, since there are only two possible values of m_s. Moreover, if two electrons occupy the same orbital, they must have opposed spins. Otherwise they would have the same set of four quantum numbers.

The rules for assigning quantum numbers are summarized in Table 6.4 and applied in Examples 6.4 and 6.5.

EXAMPLE 6.4

Consider the following sets of quantum numbers (n, ℓ, m_ℓ, m_s). Which ones could not occur? For the valid sets, identify the orbital involved.

(a) $3, 1, 0, +\frac{1}{2}$ (b) $1, 1, 0, -\frac{1}{2}$ (c) $2, 0, 0, +\frac{1}{2}$

(d) $4, 3, 2, +\frac{1}{2}$ (e) $2, 1, 0, 0$

STRATEGY

1. Use the selection rules to identify quantum numbers that are not valid.

2. Recall the letter and number designations for ℓ.

   $\ell = 0 = $ s; $\ell = 1 = $ p; $\ell = 2 = $ d; $\ell = 3 = $ f

SOLUTION

(a) $3, 1, 0, +\frac{1}{2}$	valid; n = 3, ℓ = 1 = p; 3p
(b) $1, 1, 0, -\frac{1}{2}$	not valid; n = 1, ℓ = 1, ℓ cannot equal n
(c) $2, 0, 0, +\frac{1}{2}$	valid; n = 2, ℓ = 0 = s; 2s
(d) $4, 3, 2, +\frac{1}{2}$	valid; n = 4, ℓ = 3 = f; 4f
(e) $2, 1, 0, 0$	not valid; m_s can only be $+\frac{1}{2}$ or $-\frac{1}{2}$.

EXAMPLE 6.5

(a) What is the capacity for electrons of an s sublevel? A p sublevel? A d sublevel? An f sublevel?

(b) What is the total capacity for electrons of the fourth principal level?

ANALYSIS

Information given:	sublevels
Information implied:	capacity of an orbital ($2e^-$) number of orbitals in a sublevel
Asked for:	(a) number of electrons in each sublevel (b) number of electrons in $n = 4$

STRATEGY

1. Recall the number designations that correspond to the letter designation of sublevels.

2. Use the rule that tells you how many orbitals there are to a particular sublevel.

SOLUTION

(a) s sublevel	$\ell = s = 0$; $m_\ell = 0$; 1 orbital $\times$ $2e^-$/orbital $= 2e^-$
p sublevel	$\ell = p = 1$; $m_\ell = -1, 0, +1$; 3 orbitals $\times$ $2e^-$/orbital $= 6e^-$
d sublevel	$\ell = d = 2$; $m_\ell = -2, -1, 0, +1, +2$; 5 orbitals $\times$ $2e^-$/orbital $= 10e^-$
f sublevel	$\ell = f = 3$; $m_\ell = -3, -2, -1, 0, +1, +2, +3$; 7 orbitals $\times$ $2e^-$/orbital $= 14e^-$
(b) Number of e^- in $n = 4$	$n = 4$; $\ell = 0, 1, 2, 3$ From (a): $(\ell = 0 = 2e^-) + (\ell = 1 = 6e^-) + (\ell = 2 = 10e^-) + (\ell = 3 = 14e^-)$ $= 32e^-$

Using the approach in Example 6.5, you can readily obtain the electron capacities of any principal level or sublevel. In Table 6.5, these capacities are listed through $n = 4$. Notice that

- a principal level of quantum number n has a capacity of $2n^2$. For example, the total capacity of the fourth principal energy level is $2(4)^2 = 32e^-$.
- a sublevel of quantum number (ℓ) has a capacity of $2(2\ell + 1)$. For example, a 3d sublevel ($\ell = 2$) has a capacity of $2(2 \times 2 + 1) = 10e^-$.

Table 6.5 Capacities of Electronic Levels and Sublevels in Atoms

Level n	Total Number of Electrons in Principal Level n: $2n^2$	Maximum Number of Electrons in Sublevels: $2(2\ell + 1)$			
		$\ell = 1$	$\ell = 2$	$\ell = 3$	$\ell = 4$
		s	p	d	f
1	2	2	——	——	——
2	8	2	6	——	——
3	18	2	6	10	——
4	32	2	6	10	14

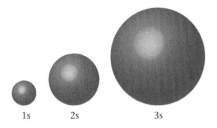

Figure 6.11 s orbitals. The relative sizes of the 90% contours (see Figure 6.9b) are shown for the 1s, 2s, and 3s orbitals.

p_x orbital

p_z orbital

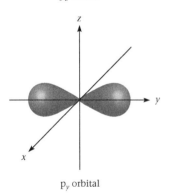

p_y orbital

Figure 6.12 p orbitals. The electron density of the three p orbitals is directed along the *x-*, *y-*, or *z*-axis. The three p orbitals are located at 90° angles to each other.

6-4 Atomic Orbitals; Shapes and Sizes

You will recall (Section 6-2, page 133) that an orbital occupied by an electron in an atom can be represented physically by showing the region of space in which there is a 90% probability of finding the electron. Orbitals are commonly designated by citing the corresponding sublevels. Thus we refer to 1s, 2s, 2p, 3s, 3p, 3d, . . . orbitals.

All s sublevels are spherical; they differ from one another only in size. As **n** increases, the radius of the orbital becomes larger (Figure 6.11). This means that an electron in a 2s orbital is more likely to be found far out from the nucleus than is a 1s electron.

The shapes and orientations of p orbitals are shown in Figure 6.12. Notice that

- a p orbital consists of two lobes along an axis (*x, y,* or *z*). Among other things, this means that, in a p orbital, there is zero probability of finding an electron at the origin, that is, at the nucleus of the atom.
- the three p orbitals in a given sublevel are oriented at right angles to one another along the *x-*, *y-*, and *z*-axis. For that reason, the three orbitals are often designated as p_x, p_y, and p_z.

Although it is not shown in Figure 6.12, p orbitals, like s orbitals, increase in size as the principal quantum number **n** increases. Also not shown are the shapes and sizes of d and f orbitals. We will say more about the nature of d orbitals in Chapter 19.

6-5 Electron Configurations in Atoms

Given the rules referred to in Section 6-3, it is possible to assign quantum numbers to each electron in an atom. Beyond that, electrons can be assigned to specific principal levels, sublevels, and orbitals. There are several ways to do this. Perhaps the simplest way to describe the arrangement of electrons in an atom is to give its **electron configuration,** which shows the number of electrons, indicated by a superscript, in each sublevel. For example, a species with the electron configuration

$$1s^2 2s^2 2p^5 \blacktriangleleft \textit{i}$$

has two electrons in the 1s sublevel, two electrons in the 2s sublevel, and five electrons in the 2p sublevel.

In this section, you will learn how to predict the electron configurations of atoms of elements. There are a couple of different ways of doing this, which we consider in turn. It should be emphasized that, throughout this discussion, we *refer to isolated gaseous atoms in the ground state.* (In *excited* states, one or more electrons are promoted to a higher energy level.)

Electron Configuration from Sublevel Energies

Electron configurations are readily obtained if the order of filling sublevels is known. Electrons enter the available sublevels in order of increasing sublevel energy. Ordinarily, a sublevel is filled to capacity before the next one starts to fill. The relative energies of different sublevels can be obtained from experiment. Figure 6.13 is a plot of these energies for atoms through the **n** = 4 principal level.

From Figure 6.13 it is possible to predict the electron configurations of atoms of elements with atomic numbers 1 through 36. Because an s sublevel can hold only two electrons, the 1s is filled at helium ($1s^2$). With lithium ($Z = 3$), the third electron has to enter a new sublevel: This is the 2s, the lowest sublevel of the second principal energy level. Lithium has one electron in this sublevel ($1s^22s^1$). 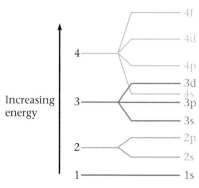 With beryllium ($Z = 4$), the 2s sublevel is filled ($1s^22s^2$). The next six elements fill the 2p sublevel. Their electron configurations are

$_5$B	$1s^22s^22p^1$	$_8$O	$1s^22s^22p^4$
$_6$C	$1s^22s^22p^2$	$_9$F	$1s^22s^22p^5$
$_7$N	$1s^22s^22p^3$	$_{10}$Ne	$1s^22s^22p^6$

Beyond neon, electrons enter the third principal level. The 3s sublevel is filled at magnesium:

$$_{12}\text{Mg} \qquad 1s^22s^22p^63s^2$$

Six more electrons are required to fill the 3p sublevel with argon:

$$_{18}\text{Ar} \qquad 1s^22s^22p^63s^23p^6$$

After argon, an "overlap" of principal energy levels occurs. The next electron enters the *lowest* sublevel of the fourth principal level (4s) instead of the *highest* sublevel of the third principal level (3d). Potassium ($Z = 19$) has one electron in the 4s sublevel; calcium ($Z = 20$) fills it with two electrons:

$$_{20}\text{Ca} \qquad 1s^22s^22p^63s^23p^64s^2$$

Now the 3d sublevel starts to fill with scandium ($Z = 21$). Recall that a d sublevel has a capacity of ten electrons. Hence the 3d sublevel becomes filled at zinc ($Z = 30$):

$$_{30}\text{Zn} \qquad 1s^22s^22p^63s^23p^64s^23d^{10}$$

The next sublevel, 4p, is filled at krypton ($Z = 36$):

$$_{36}\text{Kr} \qquad 1s^22s^22p^63s^23p^64s^23d^{10}4p^6$$

Say one s two not one s squared.

These are ground-state configurations; $1s^22s^12p^2$ would be an excited state for boron.

Figure 6.13 **Electron energy sublevels in the order of increasing energy.** The order shown is the order of sublevel filling as atomic number increases, starting at the bottom with 1s.

The order of filling, through $Z = 36$, is 1s, 2s, 2p, 3s, 3p, 4s, 3d, 4p.

EXAMPLE 6.6

Find the electron configurations of the sulfur and iron atoms.

ANALYSIS	
Information given:	identity of the atoms
Information implied:	atomic number of the atoms Figure 6.13; energy diagram
Asked for:	electron configurations for (a) S and (b) Fe

STRATEGY

1. Find the atomic numbers of S and Fe in the periodic table.

   S: atomic number = 16; Fe: atomic number = 26

2. Use Figure 6.13 and fill the appropriate sublevels.

   Remember 4s fills before 3d.

SOLUTION	
S	$1s^22s^22p^63s^23p^4$
Fe	$1s^22s^22p^63s^23p^64s^23d^6$

END POINT

Use the periodic table to check your answer. See Figure 6.14 (page 141) and the accompanying discussion.

Often, to save space, electron configurations are shortened; the **abbreviated electron configuration** starts with the preceding noble gas. For the elements sulfur and nickel,

	Electron Configuration	Abbreviated Electron Configuration
$_{16}S$	$1s^2 2s^2 2p^6 3s^2 3p^4$	$[Ne]\ 3s^2 3p^4$
$_{28}Ni$	$1s^2 2s^2 2p^6 3s^2 3p^6 4s^2 3d^8$	$[Ar]\ 4s^2 3d^8$

The symbol [Ne] indicates that the first 10 electrons in the sulfur atom have the neon configuration $1s^2 2s^2 2p^6$; similarly, [Ar] represents the first 18 electrons in the nickel atom.

Filling of Sublevels and the Periodic Table

In principle, a diagram such as Figure 6.13 can be extended to include all sublevels occupied by electrons in any element. As a matter of fact, that is a relatively simple thing to do; such a diagram is in effect incorporated into the periodic table introduced in Chapter 2.

To understand how position in the periodic table relates to the filling of sublevels, consider the metals in the first two groups. Atoms of the Group 1 elements all have one s electron in the outermost principal energy level (Table 6.6). In each Group 2 atom, there are two s electrons in the outermost level. A similar relationship applies to the elements in any group:◀ⓘ

> *The atoms of elements in a group of the periodic table have the same distribution of electrons in the outermost principal energy level.*

This means that the order in which electron sublevels are filled is determined by position in the periodic table. Figure 6.14 shows how this works. Notice the following points:

1. *The elements in Groups 1 and 2 are filling an s sublevel.* Thus Li and Be in the second period fill the 2s sublevel. Na and Mg in the third period fill the 3s sublevel, and so on.

2. *The elements in Groups 13 through 18 (six elements in each period) fill p sublevels,* which have a capacity of six electrons. In the second period, the 2p sublevel starts to fill with B $(Z = 5)$ and is completed with Ne $(Z = 10)$. In the third period, the elements Al $(Z = 13)$ through Ar $(Z = 18)$ fill the 3p sublevel.

ⓘ The periodic table works because an element's chemical properties depend on the number of outer electrons.

Table 6.6 Abbreviated Electron Configurations of Group 1 and 2 Elements

Group 1		Group 2	
$_3$Li	$[He]\ 2s^1$	$_4$Be	$[He]\ 2s^2$
$_{11}$Na	$[Ne]\ 3s^1$	$_{12}$Mg	$[Ne]\ 3s^2$
$_{19}$K	$[Ar]\ 4s^1$	$_{20}$Ca	$[Ar]\ 4s^2$
$_{37}$Rb	$[Kr]\ 5s^1$	$_{38}$Sr	$[Kr]\ 5s^2$
$_{55}$Cs	$[Xe]\ 6s^1$	$_{56}$Ba	$[Xe]\ 6s^2$

Figure 6.14 The periodic table and electron configurations. The periodic table can be used to deduce the electron configurations of atoms. The color code in the figure shows the energy sublevels being filled across each period. Elements marked with asterisks have electron configurations slightly different from those predicted by the table.

3. The transition metals, in the center of the periodic table, fill d sublevels. Remember that a d sublevel can hold ten electrons. In the fourth period, the ten elements Sc ($Z = 21$) through Zn ($Z = 30$) fill the 3d sublevel. In the fifth period, the 4d sublevel is filled by the elements Y ($Z = 39$) through Cd ($Z = 48$). The ten transition metals in the sixth period fill the 5d sublevel. Elements 103 to 112 in the seventh period are believed to be filling the 6d sublevel.

4. The two sets of 14 elements listed separately at the bottom of the table are filling f sublevels with a principal quantum number two less than the period number. ◀ 🛈 That is,

- 14 elements in the sixth period ($Z = 57$ to 70) are filling the 4f sublevel. These elements are sometimes called rare earths or, more commonly, **lanthanides,** after the name of the first element in the series, lanthanum (La). Modern separation techniques, notably chromatography, have greatly increased the availability of compounds of these elements. A brilliant red phosphor used in color TV receivers contains a small amount of europium oxide, Eu_2O_3. This is added to yttrium oxide, Y_2O_3, or gadolinium oxide, Gd_2O_3. Cerium(IV) oxide is used to coat interior surfaces of "self-cleaning" ovens, where it prevents the buildup of tar deposits.

🛈 Mendeleev developed the periodic table before the discovery of protons and electrons. Amazing!

CHEMISTRY THE **HUMAN** SIDE

One name, more than any other, is associated with the actinide elements: Glenn Seaborg (1912–1999). Between 1940 and 1957, Seaborg and his team at the University of California, Berkeley, prepared nine of these elements (at. no. 94–102) for the first time. Moreover, in 1945 Seaborg made the revolutionary suggestion that the actinides, like the lanthanides, were filling an f sublevel. For these accomplishments, he received the 1951 Nobel Prize in Chemistry.

Glenn Seaborg was born in a small town in middle America, Ishpeming, Michigan. After obtaining a bachelor's degree in chemistry from UCLA, he spent the rest of his scientific career at Berkeley, first as a Ph.D. student and then as a faculty member. During World War II, he worked on the Manhattan Project. Along with other scientists, Seaborg recommended that the devastating power of the atomic bomb be demonstrated by dropping it on a barren island before United Nations observers. Later he headed the U.S. Atomic Energy Commission under Presidents Kennedy, Johnson, and Nixon.

Sometimes referred to as the "gentle giant" (he was 6 ft 4 in. tall), Seaborg had a charming, self-deprecating sense of humor. He recalled that friends advised him *not* to publish his theory about the position of the actinides in the periodic table, lest it ruin his scientific reputation. Seaborg went on to say that, "I had a great advantage. I didn't have any scientific reputation, so I went ahead and published it." Late in his life there was considerable controversy as to whether a transuranium element should be named for him;

Glenn Theodore Seaborg
(1912–1999)

that honor had always been bestowed posthumously. Seaborg commented wryly that, "They don't want to do it because I'm still alive and they can prove it." Element 106 was named seaborgium (Sg) in 1997; he considered this his greatest honor, even above the Nobel Prize.

■ 14 elements in the seventh period ($Z = 89$ to 102) are filling the 5f sublevel. The first element in this series is actinium (Ac); collectively, these elements are referred to as **actinides.** All these elements are radioactive; only thorium and uranium occur in nature. The other actinides have been synthesized in the laboratory by nuclear reactions. Their stability decreases rapidly with increasing atomic number. The longest lived isotope of nobelium ($_{102}$No) has a half-life of about 3 minutes; that is, in 3 minutes half of the sample decomposes. Nobelium and the preceding element, mendelevium ($_{101}$Md), were identified in samples containing one to three atoms of No or Md.

Electron Configuration from the Periodic Table

Figure 6.14 (or any periodic table) can be used to deduce the electron configuration of any element. ◄ⓘ It is particularly useful for heavier elements such as iodine (Example 6.7).

ⓘ To obtain electron configurations from the periodic table, consider what sublevels are filled going across each period.

EXAMPLE	6.7

For the iodine atom, write its electron configuration and its abbreviated electron configuration.

ANALYSIS	
Information given:	Identity of the atom (I)
Information implied:	atomic number of I periodic table or Figure 6.14
Asked for:	electron configuration abbreviated electron cofiguration

continued

As you can see from Figure 6.14, the electron configurations of several elements (marked*) differ slightly from those predicted. In every case, the difference involves a shift of one or, at the most, two electrons from one sublevel to another of very similar energy. For example, in the first transition series, two elements, chromium and copper, have an extra electron in the 3d as compared with the 4s orbital.

	Predicted	Observed
$_{24}Cr$	$[Ar]\ 4s^2 3d^4$	$[Ar]\ 4s^1 3d^5$
$_{29}Cu$	$[Ar]\ 4s^2 3d^9$	$[Ar]\ 4s^1 3d^{10}$

These anomalies reflect the fact that the 3d and 4s orbitals have very similar energies. Beyond that, it has been suggested that there is a slight increase in stability with a half-filled (Cr) or completely filled (Cu) 3d sublevel.

6-6 Orbital Diagrams of Atoms

For many purposes, electron configurations are sufficient to describe the arrangements of electrons in atoms. Sometimes, however, it is useful to go a step further and show how electrons are distributed among orbitals. In such cases, **orbital diagrams** are used. Each orbital is represented by parentheses ( ), and electrons are shown by arrows written ↑ or ↓, depending on spin.

To show how orbital diagrams are obtained from electron configurations, consider the boron atom ($Z = 5$). Its electron configuration is $1s^2 2s^2 2p^1$. The pair of electrons in the 1s orbital must have opposed spins ($+\frac{1}{2}$, $-\frac{1}{2}$, or ↑↓). The same is true of the two electrons in the 2s orbital. There are three orbitals in the 2p sublevel. The single 2p electron in boron could be in any one of these orbitals.

Its spin could be either "up" or "down." The orbital diagram is ordinarily written

$$\begin{array}{cccc} & 1s & 2s & 2p \\ _5B & (\uparrow\downarrow) & (\uparrow\downarrow) & (\uparrow)(\ )(\ ) \end{array}$$

with the first electron in an orbital arbitrarily designated by an up arrow, $\uparrow$.

With the next element, carbon, a complication arises. In which orbital should the sixth electron go? It could go in the same orbital as the other 2p electron, in which case it would have to have the opposite spin, $\downarrow$. It could go into one of the other two orbitals, either with a parallel spin, $\uparrow$, or an opposed spin, $\downarrow$. Experiment shows that there is an energy difference among these arrangements. The most stable is the one in which the two electrons are in different orbitals with parallel spins. The orbital diagram of the carbon atom is

$$\begin{array}{cccc} & 1s & 2s & 2p \\ _6C & (\uparrow\downarrow) & (\uparrow\downarrow) & (\uparrow)(\uparrow)(\ ) \end{array}$$

Similar situations arise frequently. There is a general principle that applies in all such cases; **Hund's rule** (Friedrich Hund, 1896–1997) ◄ 🛈 predicts that, ordinarily,

> *when several orbitals of equal energy are available, as in a given sublevel, electrons enter singly with parallel spins.*

Only after all the orbitals are half-filled do electrons pair up in orbitals.

Following this principle, the orbital diagrams for the elements boron through neon are shown in Figure 6.15. Notice that

- *in all filled orbitals, the two electrons have opposed spins.* Such electrons are often referred to as being *paired*. There are four paired electrons in the B, C, and N atoms, six in the oxygen atom, eight in the fluorine atom, and ten in the neon atom.
- *in accordance with Hund's rule, within a given sublevel there are as many half-filled orbitals as possible.* Electrons in such orbitals are said to be *unpaired*. There is one unpaired electron in atoms of B and F, two unpaired electrons in C and O atoms, and three unpaired electrons in the N atom. When there are two or more unpaired electrons, as in C, N, and O, those electrons have parallel spins.

Hund's rule, like the Pauli exclusion principle, is based on experiment. It is possible to determine the number of unpaired electrons in an atom. With solids, this is done by studying their behavior in a magnetic field. If there are unpaired electrons present, the solid will be attracted into the field. Such a substance is said to be **paramagnetic**. If the atoms in the solid contain only paired electrons, it is slightly repelled by the field. Substances of this type are called **diamagnetic**. With gaseous atoms, the atomic spectrum can also be used to establish the presence and number of unpaired electrons.

Hund was still lecturing, colorfully and coherently, in his nineties.

Figure 6.15 Orbital diagrams for atoms with five to ten electrons. Orbitals of equal energy are all occupied by unpaired electrons before pairing begins.

Atom	Orbital diagram					Electron configuration
B	$(\uparrow\downarrow)$	$(\uparrow\downarrow)$	$(\uparrow\ )$	$(\ )$	$(\ )$	$1s^22s^22p^1$
C	$(\uparrow\downarrow)$	$(\uparrow\downarrow)$	$(\uparrow\ )$	$(\uparrow\ )$	$(\ )$	$1s^22s^22p^2$
N	$(\uparrow\downarrow)$	$(\uparrow\downarrow)$	$(\uparrow\ )$	$(\uparrow\ )$	$(\uparrow\ )$	$1s^22s^22p^3$
O	$(\uparrow\downarrow)$	$(\uparrow\downarrow)$	$(\uparrow\downarrow)$	$(\uparrow\ )$	$(\uparrow\ )$	$1s^22s^22p^4$
F	$(\uparrow\downarrow)$	$(\uparrow\downarrow)$	$(\uparrow\downarrow)$	$(\uparrow\downarrow)$	$(\uparrow\ )$	$1s^22s^22p^5$
Ne	$(\uparrow\downarrow)$	$(\uparrow\downarrow)$	$(\uparrow\downarrow)$	$(\uparrow\downarrow)$	$(\uparrow\downarrow)$	$1s^22s^22p^6$
	1s	2s	2p			

EXAMPLE 6.8

Construct orbital diagrams for atoms of sulfur and iron.

ANALYSIS

Information given:	identity of the atoms (S and Fe)
Information implied:	periodic table number designations for ℓ number of orbitals in each sublevel
Asked for:	orbital diagram for S and Fe

STRATEGY

1. Write the electron configurations for S and Fe.

   See Example 6.6 where the electron configuration for these atoms is obtained.

2. Recall the number of orbitals per sublevel and the number of electrons allowed in each orbital.

   $m_\ell = 2\ell + 1$; $2e^-$ per orbital

3. Apply Hund's rule.

   Electrons enter singly in parallel spins when several orbitals of equal energy are available.

SOLUTION

For S:	1. Electron configuration	$1s^2 2s^2 2p^6 3s^2 3p^4$
	2. Number of orbitals	$s = 0$; $2(0) + 1 = 1$ orbital for s sublevels
		$p = 1$; $2(1) + 1 = 3$ orbitals for p sublevels

3. Orbital diagram

1s	2s	2p	3s	3p
$(\uparrow\downarrow)$	$(\uparrow\downarrow)$	$(\uparrow\downarrow)(\uparrow\downarrow)(\uparrow\downarrow)$	$(\uparrow\downarrow)$	$(\uparrow\downarrow)(\uparrow\;)(\uparrow\;)$

For Fe:	1. Electron configuration	$1s^2 2s^2 2p^6 3s^2 3p^6 4s^2 3d^6$
	2. Number of orbitals	$s = 0$; $2(0) + 1 = 1$ orbital for s sublevels
		$p = 1$; $2(1) + 1 = 3$ orbitals for p sublevels
		$d = 2$; $2(2) + 1 = 5$ orbitals for p sublevels

3. Orbital diagram

1s	2s	2p	3s	3p	4s	3d
$(\uparrow\downarrow)$	$(\uparrow\downarrow)$	$(\uparrow\downarrow)(\uparrow\downarrow)(\uparrow\downarrow)$	$(\uparrow\downarrow)$	$(\uparrow\downarrow)(\uparrow\downarrow)(\uparrow\downarrow)$	$(\uparrow\downarrow)$	$(\uparrow\downarrow)(\uparrow\;)(\uparrow\;)(\uparrow\;)(\uparrow\;)$

END POINT

You can't write an orbital diagram without knowing: (a) the number designations for ℓ, (b) the number of orbitals in each sublevel, (c) the electron configuration, and (d) Hund's rule.

6-7 Electron Arrangements in Monatomic Ions

The discussion so far in this chapter has focused on electron configurations and orbital diagrams of neutral atoms. It is also possible to assign electronic structures to monatomic ions, formed from atoms by gaining or losing electrons. In general, when a monatomic ion is formed from an atom, *electrons are added to or removed from sublevels in the highest principal energy level.*

Ions with Noble-Gas Structures

As pointed out in Chapter 2, elements close to a noble gas in the periodic table form ions that have the same number of electrons as the noble-gas atom. This means that these ions have noble-gas electron configurations. Thus the three elements preceding neon (N, O, and F) and the three elements following neon (Na, Mg, and Al) all form ions with the neon configuration, $1s^2 2s^2 2p^6$. The three nonmetal atoms achieve this structure by gaining electrons to form anions:

$$_7N\ (1s^2 2s^2 2p^3) + 3e^- \longrightarrow {_7}N^{3-}\ (1s^2 2s^2 2p^6)$$

$$_8O\ (1s^2 2s^2 2p^4) + 2e^- \longrightarrow {_8}O^{2-}\ (1s^2 2s^2 2p^6)$$

$$_9F\ (1s^2 2s^2 2p^5) + e^- \longrightarrow {_9}F^-\ (1s^2 2s^2 2p^6)$$

The three metal atoms acquire the neon structure by losing electrons to form cations:

$$_{11}Na\ (1s^2 2s^2 2p^6 3s^1) \longrightarrow {_{11}}Na^+\ (1s^2 2s^2 2p^6) + e^-$$

$$_{12}Mg\ (1s^2 2s^2 2p^6 3s^2) \longrightarrow {_{12}}Mg^{2+}\ (1s^2 2s^2 2p^6) + 2e^-$$

$$_{13}Al\ (1s^2 2s^2 2p^6 3s^2 3p^1) \longrightarrow {_{13}}Al^{3+}\ (1s^2 2s^2 2p^6) + 3e^-$$

The species N^{3-}, O^{2-}, F^-, Ne, Na^+, Mg^{2+}, and Al^{3+} are said to be *isoelectronic*; that is, they have the same electron configuration.

There are a great many monatomic ions that have noble-gas configurations; Figure 6.16 shows 24 ions of this type. Note, once again, that ions in a given main group have the same charge (+1 for Group 1, +2 for Group 2, −2 for Group 16, −1 for Group 17). This explains, in part, the chemical similarity among elements in the same main group. In particular, ionic compounds formed by such elements have similar chemical formulas. For example,

- halides of the alkali metals have the general formula MX, where M = Li, Na, K, . . . and X = F, Cl, Br, ...
- halides of the alkaline earth metals have the general formula MX_2, where M = Mg, Ca, Sr, . . . and X = F, Cl, Br, ...
- oxides of the alkaline earth metals have the general formula MO, where M = Mg, Ca, Sr, ...

Transition Metal Cations

The transition metals to the right of the scandium subgroup do not form ions with noble-gas configurations. To do so, they would have to lose four or more electrons. The energy requirement is too high for that to happen. However, as pointed out in Chapter 2, these metals do form cations with charges of +1, +2, or +3. Applying the principle that, in forming cations, electrons are removed from the sublevel of highest **n**, you can predict correctly that *when transition metal atoms*

Figure 6.16 Cations, anions, and atoms with ground state noble-gas electron configurations. Atoms and ions shown in the same color are *isoelectronic*; that is, they have the same electron configurations.

									H^-	He
Li^+	Be^{2+}					N^{3-}	O^{2-}	F^-	Ne
Na^+	Mg^{2+}	Al^{3+}					S^{2-}	Cl^-	Ar
K^+	Ca^{2+}	Sc^{3+}					Se^{2-}	Br^-	Kr
Rb^+	Sr^{2+}	Y^{3+}					Te^{2-}	I^-	Xe
Cs^+	Ba^{2+}	La^{3+}							

Bottom row (*left to right*): iron(III) chloride, copper(II) sulfate, manganese(II) chloride, cobalt(II) chloride. Top row (*left to right*): chromium(III) nitrate, iron(II) sulfate, nickel(II) sulfate, potassium dichromate.

Solutions of the compounds in the order listed in (a).

Ⓐ Ⓑ

Figure 6.17 Transition metal ions. Transition metal ions impart color to many of their compounds and solutions.

form positive ions, the outer s electrons are lost first. Consider, for example, the formation of the Mn^{2+} ion from the Mn atom:

$$_{25}Mn \quad [Ar]\ 4s^2 3d^5 \qquad _{25}Mn^{2+} \quad [Ar]3d^5$$

Notice that it is the 4s ◂ ⓘ electrons that are lost rather than the 3d electrons. This is known to be the case because the Mn^{2+} ion has been shown to have five unpaired electrons (the five 3d electrons). If two 3d electrons had been lost, the Mn^{2+} ion would have had only three unpaired electrons.

ⓘ The s electrons are "first in" with the atoms and "first out" with the cations.

All the transition metals (Figure 6.17) form cations by a similar process, that is, loss of outer s electrons. Only after those electrons are lost are electrons removed from the inner d sublevel. Consider, for example, what happens with iron, which, you will recall, forms two different cations. First the 4s electrons are lost to give the Fe^{2+} ion:

$$_{26}Fe(Ar\ 4s^2 3d^6) \longrightarrow {}_{26}Fe^{2+}(Ar\ 3d^6) + 2e^-$$

Then an electron is removed from the 3d level to form the Fe^{3+} ion:

$$_{26}Fe^{2+}(Ar\ 3d^6) \longrightarrow {}_{26}Fe^{3+}(Ar\ 3d^5) + e^-$$

In the Fe^{2+} and Fe^{3+} ions, as in all transition metal ions, there are no outer s electrons.

You will recall that for fourth period *atoms*, the 4s sublevel fills before the 3d. In the corresponding *ions*, the electrons come out of the 4s sublevel before the 3d. This is sometimes referred to as the "first in, first out" ◂ ⓘ rule.

ⓘ Seniority rules don't apply to electrons.

EXAMPLE 6.9	
Give the electron configuration of Fe^{2+} and Br^-.	
ANALYSIS	
Information given:	Identity of the ions and their charge: (Fe^{2+}, Br^-)
Information implied:	atomic number of the atoms; electron configuration of the atoms
Asked for:	electron configuration of the ions
STRATEGY	
1. Write the electron configuration of each atom.	
2. Add electrons (for anions) or subtract electrons (for cations) from sublevels of the highest **n**. If there is more than one sublevel in the highest **n**, add or subtract electrons in the highest ℓ of that **n**.	

continued

	SOLUTION
Fe electron configuration	$1s^2 2s^2 2p^6 3s^2 3p^6 4s^2 3d^6$
Cation with $+2$ charge	subtract 2 electrons
Highest **n**	4 with only one sublevel
Fe^{2+} electron configuration	$1s^2 2s^2 2p^6 3s^2 3p^6 4s^{2-2} 3d^6 = 1s^2 2s^2 2p^6 3s^2 3p^6 3d^6$
Br electron configuration	$1s^2 2s^2 2p^6 3s^2 3p^6 4s^2 3d^{10} 4p^5$
Anion with -1 charge	add 1 electron
Highest **n**	4 with 2 sublevels (s and p)
Highest ℓ in **n**	p
Br^- electron configuration	$1s^2 2s^2 2p^6 3s^2 3p^6 4s^2 3d^{10} 4p^{5+1} = 1s^2 2s^2 2p^6 3s^2 3p^6 4s^2 3d^{10} 4p^6$

END POINT

The electron configuration for Br^- is the same as that for the noble gas closest to it, krypton.

6-8 Periodic Trends in the Properties of Atoms

One of the most fundamental principles of chemistry is the periodic law, which states that

> *The chemical and physical properties of elements are a **periodic function** of atomic number.*

This is, of course, the principle behind the structure of the periodic table. Elements within a given vertical group resemble one another chemically because chemical properties repeat themselves at regular intervals of 2, 8, 18, or 32 elements.

In this section we will consider how the periodic table can be used to correlate properties on an atomic scale. In particular, we will see how atomic radius, ionic radius, ionization energy, and electronegativity vary horizontally and vertically in the periodic table.

Atomic Radius

Strictly speaking, the "size" of an atom is a rather nebulous concept. The electron cloud surrounding the nucleus does not have a sharp boundary. However, a quantity called the **atomic radius** can be defined and measured, assuming a spherical atom. Ordinarily, the atomic radius is taken to be one half the distance of closest approach between atoms in an elemental substance (Figure 6.18).

The atomic radii of the main-group elements are shown at the top of Figure 6.19. Notice that, in general, atomic radii

- decrease across a period from left to right in the periodic table.
- increase down a group in the periodic table.

It is possible to explain these trends in terms of the electron configurations of the corresponding atoms. Consider first the increase in radius observed as we move down the table, let us say among the alkali metals (Group 1). All these elements have a single s electron outside a filled level or filled p sublevel. Electrons in these inner levels are much closer to the nucleus than the outer s electron and hence effectively shield it from the positive charge of the nucleus. To a first approximation, each inner electron cancels the charge of one proton in the nucleus, so the outer s electron is attracted by a net positive charge of $+1$. In this sense, it has the properties of an electron in the hydrogen atom. Because the average distance of the electron from the hydrogen nucleus increases with the principal

$$\text{Atomic radius} = \frac{0.256 \text{ nm}}{2} = 0.128 \text{ nm}$$

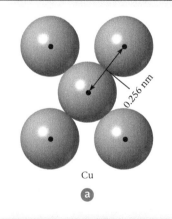

Cu

(a)

$$\text{Atomic radius} = \frac{0.198 \text{ nm}}{2} = 0.099 \text{ nm}$$

0.198 nm

Cl_2

(b)

Figure 6.18 Atomic radii. The radii are determined by assuming that atoms in closest contact in an element touch one another. The atomic radius is taken to be one half of the closest internuclear distance.

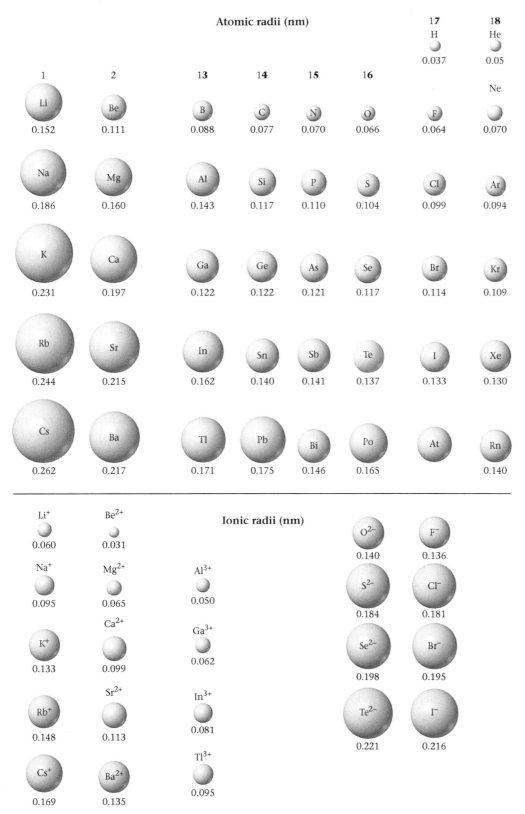

Figure 6.19 Atomic and ionic radii of the main-group elements. Negative ions are always larger than atoms of the same element, whereas positive ions are always smaller than atoms of the same element.

quantum number, **n**, the radius increases moving from Li (2s electron) to Na (3s electron) and so on down the group.

The decrease in atomic radius moving across the periodic table can be explained in a similar manner. Consider, for example, the third period, where electrons are being added to the third principal energy level. The added electrons should be relatively poor shields for each other because they are all at about the same distance from the nucleus. Only the ten **core electrons** in inner, filled levels (n = 1, n = 2) are expected to shield the outer electrons from the nucleus. This means that the charge felt by an outer electron, called the **effective nuclear charge**, should increase steadily with atomic number as we move across the period. As effective nuclear charge increases, the outermost electrons are pulled in more tightly, and atomic radius decreases.

Ionic Radius

The radii of cations and anions derived from atoms of the main-group elements are shown at the bottom of Figure 6.19. The trends referred to previously for atomic radii are clearly visible with ionic radius as well. Notice, for example, that **ionic radius** increases moving down a group in the periodic table. Moreover, the radii of both cations (left) and anions (right) decrease from left to right across a period.

Comparing the radii of cations and anions with those of the atoms from which they are derived

■ *positive ions are smaller than the metal atoms from which they are formed* (Figure 6.20). The Na⁺ ion has a radius, 0.095 nm, only a little more than half that of the Na atom, 0.186 nm.
■ *negative ions are larger than the nonmetal atoms from which they are formed* (Figure 6.20). The radius of the Cl⁻ ion, 0.181 nm, is nearly twice that of the Cl atom, 0.099 nm.

As a result of these effects, anions in general are larger than cations. Compare, for example, the Cl⁻ ion (radius = 0.181 nm) with the Na⁺ ion (radius = 0.095 nm). This means that in sodium chloride, and indeed in the vast majority of all ionic compounds, most of the space in the crystal lattice is taken up by anions.

The differences in radii between atoms and ions can be explained quite simply. A cation is smaller than the corresponding metal atom because the excess of protons in the ion draws the outer electrons in closer to the nucleus. In contrast, an extra electron in an anion adds to the repulsion between outer electrons, making a negative ion larger than the corresponding nonmetal atom.

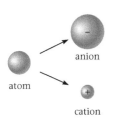

Figure 6.20 Relative size of the parent atom and its ions.

atom → anion
atom → cation

EXAMPLE 6.10

Using only the periodic table, arrange each of the following sets of atoms and ions in order of increasing size.
(a) Mg, Al, Ca   (b) S, Cl, S²⁻   (c) Fe, Fe²⁺, Fe³⁺

STRATEGY

Recall the following:
■ The definition of a period and a group in the periodic table.
■ The radius decreases across a period and increases going down a group.
■ An atom is larger than its cation but smaller than its anion.

SOLUTION

(a) Mg, Al, Ca	Mg and Al belong to the same period: Al < Mg. Mg and Ca belong to the same group: Mg < Ca. Thus Al < Mg < Ca.

continued

(b) S, Cl, S^{2-}	S and Cl belong to the same period: Cl < S. An atom is smaller than its anion: S < S^{2-}. Thus Cl < S < S^{2-}.
(c) Fe, Fe^{2+}, Fe^{3+}	A cation is smaller than its atom: Fe^{2+} < Fe, Fe^{3+} < Fe Increasing the charge of cations of the same atom decreases size: $\qquad Fe^{3+} < Fe^{2+}$. Thus $Fe^{3+} < Fe^{2+} < Fe$ (Figure 6.21).

Ionization Energy

Ionization energy is a measure of how difficult it is to remove an electron from a gaseous atom. Energy must always be *absorbed* to bring about ionization, so ionization energies are always *positive* quantities.

The (first) ionization energy is the energy change for the removal of the outermost electron from a gaseous atom to form a +1 ion:

$$M(g) \longrightarrow M^+(g) + e^- \qquad \Delta E_1 = \text{first ionization energy}$$

The more difficult it is to remove electrons, the larger the ionization energy.

Ionization energies of the main-group elements are listed in Figure 6.22. Notice that ionization energy

■ increases across the periodic table from left to right.
■ decreases moving down the periodic table.

Figures 6.19 and 6.22 show an inverse correlation between ionization energy and atomic radius. The smaller the atom, the more tightly its electrons are held to the positively charged nucleus and the more difficult they are to remove. Conversely, in a large atom such as that of a Group 1 metal, the electron is relatively far from the nucleus, so less energy has to be supplied to remove it from the atom.

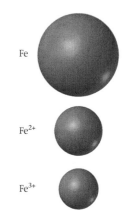

Figure 6.21 Relative sizes of the iron atom and its ions.

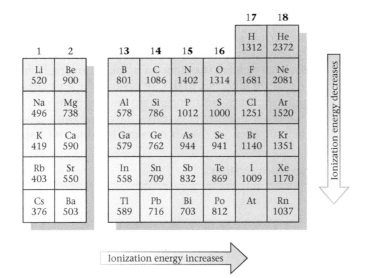

Figure 6.22 First ionization energies of the main-group elements, in kilojoules per mole. In general, ionization energy decreases moving down the periodic table groups and increases across the periods, although there are several exceptions.

EXAMPLE 6.11

Consider the three elements C, N, and Si. Using only the periodic table, predict which of the three elements has

(a) the largest atomic radius; the smallest atomic radius.

(b) the largest ionization energy; the smallest ionization energy.

continued

STRATEGY

1. Because these three elements form a block (N next to C and Si below C), in the periodic table, it is convenient to compare both silicon and nitrogen to carbon.

2. Recall the trends for atomic radius and ionization energy.

SOLUTION

(a) atomic radius	C > N; C < Si
	Si is the largest atom, N is the smallest.
(b) ionization energy	C < N; C > Si
	Si has the smallest first ionization energy.
	N has the largest first ionization energy.

END POINT

Check your answers against Figures 6.19 and 6.22.

If you look carefully at Figure 6.22, you will note a few exceptions to the general trends referred to above and illustrated in Example 6.11. For example, the ionization energy of B (801 kJ/mol) is *less* than that of Be (900 kJ/mol). This happens because the electron removed from the boron atom comes from the 2p as opposed to the 2s sublevel for beryllium. Because 2p is higher in energy than 2s, it is not too surprising that less energy is required to remove an electron from that sublevel.

Electronegativity

The ionization energy of an atom is a measure of its tendency to lose electrons; the larger the ionization energy, the more difficult it is to remove an electron. There are several different ways of comparing the tendencies of different atoms to gain electrons. The most useful of these for our purposes is the **electronegativity,** which measures the ability of an atom to attract to itself the electron pair forming a covalent bond.◂*i*

Electronegativity is a positive quantity.

The greater the electronegativity of an atom, the greater its attraction for electrons. Table 6.7 shows a scale of electronegativities first proposed by Linus Pauling (1901–1994). Each element is assigned a number, ranging from 4.0 for the most electronegative element, fluorine, to 0.8 for cesium, the least electronegative. Among the main-group elements, electronegativity increases moving from left to right in the periodic table. Ordinarily, it decreases, moving down a group. You will find Table 6.7 very helpful when we discuss covalent bonding in Chapter 7.

Table 6.7 Electronegativity Values

H 2.2							—*
Li 1.0	Be 1.6	B 2.0	C 2.5	N 3.0	O 3.5	F 4.0	—*
Na 0.9	Mg 1.3	Al 1.6	Si 1.9	P 2.2	S 2.6	Cl 3.2	—*
K 0.8	Ca 1.0	Sc 1.4	Ge 2.0	As 2.2	Se 2.5	Br 3.0	Kr 3.3
Rb 0.8	Sr 0.9	Y 1.2	Sn 1.9	Sb 2.0	Te 2.1	I 2.7	Xe 3.0
Cs 0.8	Ba 0.9						

*The noble gases He, Ne, and Ar are not listed because they form no stable compounds.

CHEMISTRY **BEYOND** THE CLASSROOM

Why Do Lobsters Turn Red When Cooked?

Harry A. Frank, *University of Connecticut*

Many biological organisms exhibit visible coloration. This occurs when specific molecules become bound in proteins or membranes of an organism and absorb light in the visible region of the electromagnetic spectrum. As described in this chapter, absorption occurs when an atom or molecule is excited by light from the ground electronic state to a higher excited state. Wavelengths of light not absorbed are either scattered or reflected. This may result in coloration or patterns that either blend with the environment or stand out from it. Why animal species select specific molecules for coloration has much to do with the pressures of survival in the wild. However, not all animals are able to synthesize what they need for this purpose. In many cases, dietary intake plays a role.

Animals ingest plant material in the course of their normal dietary intake, and the plant pigments then find their way into the outer covering of the organism, where they may be exhibited as coloration. Many complex biochemical and metabolic pathways are involved after ingestion in modifying and transporting the molecules that ultimately become pigments.

An excellent case in point is the coloration of the American lobster, *Homarus americanus*. The pigment associated with the typical greenish-brown outer layer of the lobster shell is the carotenoid, astaxanthin (Figure A), an oxygenated derivative of β-carotene, also known as the molecule that imparts the orange color to carrots.

Both astaxanthin and β-carotene belong to a class of pigments known as carotenoids. Carotenoids are not synthesized by any member of the animal kingdom, but are taken in through the diet. The lobster diet consists primarily of fish, mollusks, and other crustaceans. Therefore, the color of a live lobster can vary due to differences in available food sources and environmental conditions. This is why some lobsters may be orange, speckled, or even bright blue. Indeed, blue coloration is frequently observed in the joints of a live lobster's appendages (Figure B). When cooked, the color of the lobster turns red. How does this happen?

After being ingested, astaxanthin molecules are assembled in a large protein called crustacyanin and then bound in the calcified outer layer of the lobster shell. When an astaxanthin molecule is bound in the crustacyanin protein, its structure becomes twisted and pairs up with another astaxanthin. This combination of twisting and pairing of astaxanthin molecules produces a pigment-protein complex that absorbs the long wavelengths of visible light (Figure C) and hence appears blue. Free astaxanthin dissolved in organic solvents, such as methanol, absorbs the shorter wavelengths of blue-green light (Figure C) and therefore appears the complement of this color, which is red.

Because the lobster accumulates a combination of protein-bound (blue) and free (red) astaxanthin in its shell, the live animal appears greenish-brown, which apparently serves it well as a disguise against predators on the ocean floor. The color change upon cooking occurs because high temperatures denature the crustacyanin protein and release the bound astaxanthin. No longer constrained by the protein, but still intact, astaxanthin imparts the same red color to the shell that it exhibits as a free molecule in organic solvent. Our world is colored by an abundance of these carotenoid pigments in mammals, fish, reptiles, birds, and other living creatures.

Figure A The molecular structure of astaxanthin. β-Carotene is the same molecule where the ═O and —OH groups are replaced with hydrogens.

Figure B *Homarus americanus* before and after cooking.

© Harry A. Frank. Used with permission.

Figure C Absorption spectra of astaxanthin (red line) in methanol and astaxanthin bound in crustacyanin protein (blue line).

Key Concepts

1. Relate wavelength, frequency, and energy.
   (**Examples 6.1, 6.2; Problems 1–8, 61–63**)
2. Use the Bohr model to identify lines in the hydrogen spectrum.
   (**Example 6.3; Problems 9–16**)
3. Identify quantum numbers of electrons in atoms.
   (**Example 6.4; Problems 17–22, 27, 28**)
4. Derive the electron capacities of energy levels.
   (**Example 6.5; Problems 23–26**)

5. Write electron configurations, full or abbreviated, for atoms or ions.
   (**Examples 6.6, 6.7, 6.9; Problems 29–38, 49, 50**)
6. Draw orbital diagrams for atoms and ions.
   (**Example 6.8; Problems 39–52**)
7. Identify periodic trends in radii, ionization energy, and electronegativity.
   (**Examples 6.10, 6.11; Problems 53–60**)

Key Equations

Frequency-wavelength	$\lambda\nu = c = 2.998 \times 10^8$ m/s	Quantum numbers	$n = 1, 2, 3, 4, \ldots$
Energy-frequency	$E = h\nu = hc/\lambda$;		$\ell = 0, 1, 2, \ldots, (n-1)$
	$h = 6.626 \times 10^{-34}$ J $\cdot$ s		$m_\ell = \ell, \ldots, +1, 0, -1, \ldots, -\ell$
Bohr model	$E_n = -R_H/n^2$;		$m_s = +\frac{1}{2}, -\frac{1}{2}$
	$R_H = 2.180 \times 10^{-18}$ J		
	$\nu = \dfrac{R_H}{h}\left[\dfrac{1}{(n_{lo})^2} - \dfrac{1}{(n_{hi})^2}\right]$		

Key Terms

actinides
atomic radius
atomic spectrum
electron configuration
—abbreviated electron
   configuration

electron spin
electronegativity
excited state
frequency
ground state
Hund's rule

ionic radius
ionization energy
lanthanides
orbital
orbital diagrams
Pauli exclusion principle

photons
principal energy level
quantum numbers
sublevels
wavelength

Summary Problem

Consider the element vanadium ($Z = 23$).

(a) There is a line in the vanadium spectrum at 318.5 nm.

   (1) In what region of the spectrum (ultraviolet, visible, or infrared) is this line found?

   (2) What is the frequency of the line?

   (3) What is the energy difference between the two levels responsible for this line in kilojoules per mole?

   (4) The ionization energy of vanadium from the ground state is 650.2 kJ/mol. Assume that the transition in (3) is from the ground state to an excited state. If that is the case, calculate the ionization energy from the excited state.

(b) Give the electron configuration of the V atom; the V^{3+} ion.

(c) Give the orbital diagram (beyond argon) for V and V^{3+}.

(d) How many unpaired electrons are in the V atom? In the V^{3+} ion?

(e) How many electrons in V have $\ell = 1$ quantum number?

(f) How many electrons in V^{3+} have $\ell = 2$ quantum number?

(g) Rank V, V^{2+}, and V^{3+} in order of increasing size.

Answers

(a) (1) **ultraviolet**; (2) 9.413×10^{14} s^{-1}; (3) **375.6 kJ/mol**; (4) **274.6 kJ/mol**

(b) $1s^2 2s^2 2p^6 3s^2 3p^6 4s^2 3d^3$; $1s^2 2s^2 2p^6 3s^2 3p^6 3d^2$

(c)

	4s	3d
V:	(↑↓)	(↑)(↑)(↑)( )( )
V^{3+}:	( )	(↑)(↑)( )( )( )

(d) 3; 2

(e) 12

(f) 2

(g) $V^{3+} < V^{2+} < V$

Questions and Problems

Even-numbered questions and Challenge Problems have answers in Appendix 5 and fully worked solutions in the *Student Solutions Manual*.

6-1 Light, Photon Energies, and Atomic Spectra

1. A photon of violet light has a wavelength of 423 nm. Calculate
   (a) the frequency.
   (b) the energy in joules per photon.
   (c) the energy in kilojoules per mole.

2. Most retinal tears and detachments are treated by photocoagulation with a laser. A commonly used laser is one with a wavelength of 514 nm. Calculate
   (a) the frequency.
   (b) the energy in joules/photon.
   (c) the energy in kJ/mol.

3. A line in the spectrum of neon has a wavelength of 837.8 nm.
   (a) In what spectral range does the absorption occur?
   (b) Calculate the frequency of this absorption.
   (c) What is the energy in kilojoules per mole?

4. Ozone absorbs energy with a frequency of $1.29 \times 10^{15} \, s^{-1}$.
   (a) What is the wavelength (in nm) of the absorption?
   (b) In what spectral range does the absorption occur?
   (c) What is the energy absorbed by one photon?

5. The ionization energy of rubidium is 403 kJ/mol. Do x-rays with a wavelength of 85 nm have sufficient energy to ionize rubidium?

6. Energy from radiation can cause chemical bonds to break. To break the nitrogen-nitrogen bond in N_2 gas, 941 kJ/mol is required.
   (a) Calculate the wavelength of the radiation that could break the bond.
   (b) In what spectral range does this radiation occur?

7. Lasers are now used for the total or partial removal of skin tattoos. A laser with a wavelenth of 755 nm is used to remove tattoos in green or blue inks.
   (a) How much energy (in kJ/mol) is given off by the laser?
   (b) Compare the energy obtained in (a) with that given off by an x-ray ($\lambda = 1 \times 10^{-9}$ m).

8. Your instructor may use a laser pointer while giving a lecture. The pointer uses a red-orange diode with a wavelength of 635 nm. If 0.255 mol of photons are emitted by the pointer, how much energy (in kJ) does the pointer give off?

6-2 The Hydrogen Atom

9. Consider the transition from the energy levels $n = 5$ to $n = 3$.
   (a) What is the frequency of light associated with this transition?
   (b) In what spectral region does the transition occur?
   (c) Is energy absorbed?

10. Consider the transition from the energy levels $n = 2$ to $n = 5$.
   (a) What is the wavelength associated with this transition?
   (b) In what spectral region does the transition occur?
   (c) Is energy absorbed?

11. According to the Bohr model, the radius of a circular orbit is given by the equation

$$r(\text{in nm}) = 0.0529 \, n^2$$

Draw successive orbits for the hydrogen electron at $n = 1, 2, 3,$ and 4. Indicate by arrows transitions between orbits that lead to lines in the
   (a) Lyman series ($n_{lo} = 1$).
   (b) Balmer series ($n_{lo} = 2$).

12. Calculate E_n for $n = 1, 2, 3,$ and 4 ($R_H = 2.180 \times 10^{-18}$ J). Make a one-dimensional graph showing energy, at different values of n, increasing vertically. On this graph, indicate by vertical arrows transitions in the
   (a) Lyman series ($n_{lo} = 1$).
   (b) Balmer series ($n_{lo} = 2$).

13. For the Pfund series, $n_{lo} = 5$.
   (a) Calculate the wavelength in nanometers of a transition from $n = 7$ to $n = 5$.
   (b) In what region of the spectrum are these lines formed?

14. In the Brackett series, $n_{lo} = 4$.
   (a) Calculate the wavelength in nanometers of a transition from $n = 6$ to $n = 4$.
   (b) In what spectral region are these lines formed?

15. A line in the Lyman series ($n_{lo} = 1$) occurs at 97.23 nm. Calculate n_{hi} for the transition associated with this line.

16. In the Pfund series, $n_{lo} = 5$. Calculate the longest wavelength (in nanometers) possible for a transition in this series.

6-3 Quantum Numbers

17. What are the possible values for m_ℓ for
   (a) the d sublevel?
   (b) the s sublevel?
   (c) all sublevels where $n = 2$?

18. What are the possible values for m_ℓ for
   (a) the d sublevel?
   (b) the s sublevel?
   (c) all sublevels where $n = 5$?

19. For the following pairs of orbitals, indicate which is lower in energy in a many-electron atom.
   (a) 3d or 4s
   (b) 4f or 3d
   (c) 2s or 2p
   (d) 4f or 4d

20. For the following pairs of orbitals, indicate which is higher in energy in a many-electron atom.
   (a) 3s or 2p
   (b) 4s or 4d
   (c) 4f or 6s
   (d) 1s or 2s

21. What type of electron orbital (i.e., s, p, d, or f) is designated by an electron with quantum numbers
    (a) $n = 1, \ell = 0, m_\ell = 0$?
    (b) $n = 3, \ell = 2, m_\ell = -1$?
    (c) $n = 4, \ell = 3, m_\ell = 3$?

22. What type of electron orbital (i.e., s, p, d, or f) is designated by
    (a) $n = 3, \ell = 1, m_\ell = 1$?
    (b) $n = 5, \ell = 0, m_\ell = 0$?
    (c) $n = 6, \ell = 4, m_\ell = -4$?

23. What is the total electron capacity for
    (a) $n = 3$?
    (b) a 5s orbital?
    (c) an f subshell?
    (d) a 7p orbital?

24. Give the number of orbitals in
    (a) $n = 3$.
    (b) a 4p sublevel.
    (c) an f sublevel.
    (d) a d sublevel.

25. How many electrons in an atom can have each of the following quantum number designations?
    (a) $n = 2, \ell = 1, m_\ell = 0$
    (b) $n = 2, \ell = 1, m_\ell = -1$
    (c) $n = 3, \ell = 1, m_\ell = 0, m_s = +\frac{1}{2}$

26. How many electrons in an atom can have the following quantum designation?
    (a) 1s
    (b) 4d, $m_\ell = 0$
    (c) $n = 5, \ell = 2$

27. Given the following sets of quantum numbers, indicate those that could not occur and explain your answer.
    (a) $1, 2, 0, +\frac{1}{2}$
    (b) $2, 1, 2, +\frac{1}{2}$
    (c) $3, 0, 0, -\frac{1}{2}$
    (d) $3, 2, 1, +\frac{1}{2}$
    (e) $4, 2, -2, 0$

28. Given the following sets of electron quantum numbers, indicate those that could not occur, and explain your answer.
    (a) $1, 0, 0, -\frac{1}{2}$
    (b) $1, 1, 0, +\frac{1}{2}$
    (c) $3, 2, -2, +\frac{1}{2}$
    (d) $2, 1, 2, +\frac{1}{2}$
    (e) $4, 0, 2, +\frac{1}{2}$

29. Write the ground state electron configuration for
    (a) C   (b) Cl   (c) Co   (d) Cs   (e) Cd

30. Write the ground state electron configuration for
    (a) S   (b) Sc   (c) Si   (d) Sr   (e) Sb

31. Write an abbreviated ground state electron configuration for
    (a) N   (b) Nb   (c) Na   (d) Ni   (e) Nd

32. Write the abbreviated ground state electron configuration for
    (a) Mg   (b) Os   (c) Ge   (d) V   (e) At

33. Give the symbol of the element of lowest atomic number whose ground state has
    (a) a completed f subshell.
    (b) twenty p electrons.
    (c) two 4d electrons.
    (d) five 5p electrons.

34. Give the symbol of the element of lowest atomic number that has
    (a) an f subshell with 7 electrons.
    (b) twelve d electrons.
    (c) three 3p electrons.
    (d) a completed p subshell.

35. What fraction of the total number of electrons is in p sublevels for the following atoms?
    (a) Al   (b) Au   (c) Ag

36. What fraction of the total number of electrons is in p sublevels in
    (a) Mg   (b) Mn   (c) Mo

37. Which of the following electron configurations (a–f) are for atoms in the ground state? In the excited state? Which are impossible?
    (a) $1s^2 2s^2 3s^2$
    (b) $1s^2 2p^3$
    (c) $1s^2 2s^3 2p^5$
    (d) $1s^2 2s^2 2p^7$
    (e) $1s^2 2s^2 2p^6 3s^1$
    (f) $1s^2 2s^2 2p^6 3s^2 3d^1$

38. Which of the following electron configurations (a–e) are for atoms in the ground state? In the excited state? Which are impossible?
    (a) $1s^2 2s^2 1d^1$
    (b) $1s^2 2s^2 2p^6 3s^2 3p^4$
    (c) $1s^2 2s^1 2p^7 3s^2$
    (d) $1s^2 2s^2 3s^1 3p^4$
    (e) $1s^2 2s^2 3s^2 3p^6 4s^2$

6-6 Orbital Diagrams of Atoms

39. Write the orbital diagram for
    (a) Li   (b) P   (c) F   (d) Fe

40. Write the orbital diagram for an atom of
    (a) Na   (b) O   (c) Co   (d) Cl

41. Give the symbol of the atom with the following orbital diagram.

	1s	2s	2p	3s	3p
(a)	(↑↓)	(↑↓)	(↑↓)(↑↓)(↑↓)	(↑↓)	( )( )( )
(b)	(↑↓)	(↑↓)	(↑↓)(↑↓)(↑↓)	(↑↓)	(↑)(↑)(↑)
(c)	(↑↓)	(↑↓)	(↑↓)(↑)(↑)	( )	( )( )( )

42. What is the symbol of the atom with the following orbital diagram?

	1s	2s	2p	3s	3p
(a)	(↑↓)	(↑)	( )( )( )	( )	( )( )( )
(b)	(↑↓)	(↑↓)	(↑↓)(↑↓)(↑↓)	(↑↓)	(↑↓)(↑↓)(↑)
(c)	(↑↓)	(↑↓)	(↑↓)(↑↓)(↑↓)	(↑)	( )( )( )

43. Write the symbol of
    (a) all the elements in which all the 4d orbitals are half-full.
    (b) all the nonmetals in period 2 that have 2 or more unpaired electrons.
    (c) all the elements in group 2 in which the 4p sublevel is full.
    (d) all the metalloids that have unpaired 5p electrons.

44. Write the symbols of
(a) all the elements in period 5 that have at least two half-filled 5p orbitals.
(b) all the elements in Group 1 that have full 3p orbitals.
(c) all the metalloids that have paired 3p electrons.
(d) all the nonmetals that have full 3d orbitals and 3 half-filled 3p orbitals.

45. How many unpaired electrons are there in an atom of
(a) phosphorus?
(b) potassium?
(c) plutonium (Pu)?

46. How many unpaired electrons are there in the following atoms?
(a) aluminum
(b) argon
(c) arsenic

47. In what main group(s) of the periodic table do elements have the following number of half-filled p-orbitals in the outermost principal energy level?
(a) 0   (b) 1   (c) 2   (d) 3

48. Give the symbol of the main-group metals in period 4 with the following number of unpaired electrons per atom. (Transition metals are not included.)
(a) 0   (b) 1   (c) 2   (d) 3

6-7 Electron Arrangements in Monatomic Atoms

49. Write the ground state electron configuration for
(a) Mg, Mg^{2+}
(b) N, N^{3-}
(c) Ti, Ti^{4+}
(d) Sn^{2+}, Sn^{4+}

50. Write the ground state electron configuration for the following atoms and ions.
(a) F, F^-
(b) Sc, Sc^{3+}
(c) Mn^{2+}, Mn^{5+}
(d) O^-, O^{2-}

51. How many unpaired electrons are there in the following ions?
(a) V^{3+} (b) Sn^{4+} (c) I^- (d) W^{4+}

52. How many unpaired electrons are there in the following ions?
(a) Al^{3+} (b) Cl^- (c) Sr^{2+} (d) Zr^{4+}

6-8 Periodic Trends in the Properties of Atoms

53. Arrange the elements Na, Si, and S in the order of
(a) decreasing atomic radius.
(b) increasing first ionization energy.
(c) decreasing electronegativity.

54. Arrange the elements Mg, S, and Cl in order of
(a) increasing atomic radius.
(b) increasing first ionization energy.
(c) decreasing electronegativity.

55. Which of the four atoms Rb, Sr, Sb, or Cs
(a) has the smallest atomic radius?
(b) has the lowest ionization energy?
(c) is the least electronegative?

56. Which of the four atoms Na, P, Cl, or K
(a) has the largest atomic radius?
(b) has the highest ionization energy?
(c) is the most electronegative?

57. Select the larger member of each pair.
(a) Ca and Ca^{2+}
(b) Sr and Sr^{2+}
(c) F and F^-
(d) Cu^+ and Cu^{2+}

58. Select the smaller member of each pair.
(a) P and P^{3-}
(b) V^{2+} and V^{4+}
(c) K and K^+
(d) Co and Co^{3+}

59. Arrange the following species in order of decreasing radius.
(a) C, Mg, Ca, Si
(b) Sr, Cl, Br, I

60. Arrange the following species in order of increasing radius.
(a) Rb, K, Cs, Kr
(b) Ar, Cs, Si, Al

Unclassified

61. A lightbulb radiates 8.5% of the energy supplied to it as visible light. If the wavelength of the visible light is assumed to be 565 nm, how many photons per second are emitted by a 75-W lightbulb? (1 W = 1 J/s)

62. The speed of a computer chip is measured by its frequency. Advanced Micro Devices (ADM) has developed a microprocessor, "The Piledriver," that operates at 8.429 GHz (1 GHz = 1×10^9 Hz).
(a) What is the wavelength (in nm) of the frequency?
(b) In what spectral range does it belong?
(c) What is the energy (in kJ/mol) associated with this frequency?

63. A carbon dioxide laser produces radiation of wavelength 10.6 micrometers (1 micrometer = 10^{-6} meter). If the laser produces about one joule of energy per pulse, how many photons are produced per pulse?

64. Name and give the symbol of the element that has the characteristic given below.
(a) Its electron configuration in the excited state can be $1s^2 2s^2 2p^6 3s^1\ 3p^3$.
(b) It is the least electronegative element in period 3.
(c) Its +3 ion has the configuration $[_{36}Kr]$.
(d) It is the halogen with the largest atomic radius.
(e) It has the largest ionization energy in Group 16.

Conceptual Questions

65. Compare the energies and frequencies of two photons, one with a short wavelength and the other with a long wavelength.

66. Consider the following transitions
1. n = 3 to n = 1
2. n = 2 to n = 3
3. n = 4 to n = 3
4. n = 3 to n = 5
(a) For which of the transitions is energy absorbed?
(b) For which of the transitions is energy emitted?

(c) Which transitions involve the ground state?

(d) Which transition absorbs the most energy?

(e) Which transition emits the most energy?

67. Write the symbol of each element described below.

(a) largest atomic radius in Group 1

(b) smallest atomic radius in period 3

(c) largest first ionization energy in Group 2

(d) most electronegative in Group 16

(e) element(s) in period 2 with no unpaired p electron

(f) abbreviated electron configuration is [Ar] $4s^2 3d^3$

(g) A +2 ion with abbreviated electron configuration [Ar] $3d^5$

(h) A transition metal in period 4 forming a +2 ion with no unpaired electrons

68. Answer the following questions.

(a) What characteristic of an atomic orbital does the quantum number m_s describe?

(b) Does a photon with a frequency of 35.8 Hz have more or less energy than one with a frequency of 125 Hz?

(c) How many orbitals can be associated with the following set of quantum numbers: $n = 3$, $\ell = 2$?

(d) Is a sample of $ZnCl_2$ containing the Zn^{2+} ion diamagnetic?

69. Explain in your own words what is meant by

(a) the Pauli exclusion principle.

(b) Hund's rule.

(c) a line in an atomic spectrum.

(d) the principal quantum number.

70. Explain the difference between

(a) the Bohr model of the atom and the quantum mechanical model.

(b) wavelength and frequency.

(c) the geometries of the three different p orbitals.

71. Indicate whether each of the following statements is true or false. If false, correct the statement.

(a) An electron transition from $n = 3$ to $n = 1$ gives off energy.

(b) Light emitted by an $n = 4$ to $n = 2$ transition will have a longer wavelength than that from an $n = 5$ to $n = 2$ transition.

(c) A sublevel of $\ell = 3$ has a capacity of ten electrons.

(d) An atom of Group 13 has three unpaired electrons.

72. Criticize or comment on the following statements:

(a) The energy of a photon is inversely proportional to its wavelength.

(b) The energy of the hydrogen electron is inversely proportional to the quantum number ℓ.

(c) Electrons start to enter the fifth principal level as soon as the fourth is full.

73. No currently known elements contain electrons in g ($\ell = 4$) orbitals in the ground state. If an element is discovered that has electrons in the g orbital, what is the lowest value for n in which these g orbitals could exist? What are the possible values of m_ℓ? How many electrons could a set of g orbitals hold?

74. Indicate whether each of the following is true or false.

(a) Effective nuclear charge stays about the same when one goes down a group.

(b) Group 17 elements have seven electrons in their outer level.

(c) Energy is given off when an electron is removed from an atom.

75. Explain why

(a) negative ions are larger than their corresponding atoms.

(b) scandium, a transition metal, forms an ion with a noble-gas structure.

(c) electronegativity decreases down a group in the periodic table.

Challenge Problems

76. The energy of any one-electron species in its nth state (n = principal quantum number) is given by $E = -BZ^2/n^2$, where Z is the charge on the nucleus and B is 2.180×10^{-18} J. Find the ionization energy of the Li^{2+} ion in its first excited state in kilojoules per mole.

77. In 1885, Johann Balmer, a mathematician, derived the following relation for the wavelength of lines in the visible spectrum of hydrogen

$$\lambda = \frac{364.5\, n^2}{(n^2 - 4)}$$

where λ is in nanometers and n is an integer that can be 3, 4, 5, . . . Show that this relation follows from the Bohr equation and the equation using the Rydberg constant. Note that in the Balmer series, the electron is returning to the $n = 2$ level.

78. Suppose the rules for assigning quantum numbers were as follows

$$n = 1, 2, 3, \ldots$$
$$\ell = 0, 1, 2, \ldots, n$$
$$m_\ell = 0, 1, 2, \ldots, \ell + 1$$
$$m_s = +\tfrac{1}{2} \text{ or } -\tfrac{1}{2}$$

Prepare a table similar to Table 6.3 based on these rules for $n = 1$ and $n = 2$. Give the electron configuration for an atom with eight electrons.

79. Suppose that the spin quantum number could have the values $\tfrac{1}{2}$, 0, and $-\tfrac{1}{2}$. Assuming that the rules governing the values of the other quantum numbers and the order of filling sublevels were unchanged,

(a) what would be the electron capacity of an s sublevel? a p sublevel? a d sublevel?

(b) how many electrons could fit in the $n = 3$ level?

(c) what would be the electron configuration of the element with atomic number 8? 17?

80. In the photoelectric effect, electrons are ejected from a metal surface when light strikes it. A certain minimum energy, E_{min}, is required to eject an electron. Any energy absorbed beyond that minimum gives kinetic energy to the electron. It is found that when light at a wavelength of 540 nm falls on a cesium surface, an electron is ejected with a kinetic energy of 2.60×10^{-20} J. When the wavelength is 400 nm, the kinetic energy is 1.54×10^{-19} J.

(a) Calculate E_{min} for cesium in joules.

(b) Calculate the longest wavelength, in nanometers, that will eject electrons from cesium.

Covalent Bonding

The Pyramids at dusk (w/c on paper), Lamplough, Augustus Osborne (1877–1930) / Private Collection / © Christopher Wood Gallery, London, UK / The Bridgeman Art Library

These pyramids of ancient Egypt look like tetrahedra but are not. Tetrahedra have three triangular sides and a triangular base. These pyramids have four triangular sides and a square base.

> What immortal hand
> or eye
> Could frame thy
> fearful symmetry?
>
> —WILLIAM BLAKE
> *The Tiger*

Earlier we referred to the forces that hold nonmetal atoms to one another, covalent bonds. These bonds consist of an electron pair shared between two atoms. To represent the covalent bond in the H_2 molecule, two structures can be written:

$$H:H \quad \text{or} \quad H—H$$

These structures can be misleading if they are taken to mean that the two electrons are fixed in position between the two nuclei. A more accurate picture of the electron density in H_2 is shown in Figure 7.1. At a given instant, the two electrons may be located at any of various points about the two nuclei. However, they are more likely to be found between the nuclei than at the far ends of the molecule.

To understand the stability of the electron-pair bond, consider the graph shown in Figure 7.2, where we plot the energy of interaction between two hydrogen atoms as a function of distance. At large distances of separation (far right) the system consists of two isolated H atoms that do not interact with each other. As the atoms come closer together (moving to the left in Figure 7.2), they experience an attraction that leads gradually to an energy minimum. At an internuclear distance of 0.074 nm and an attractive energy of 436 kJ, the system is in its most stable state; we refer to that state as the H_2 molecule. If the atoms are brought closer together, forces of repulsion become increasingly important and the energy curve rises steeply.

The existence of the energy minimum shown in Figure 7.2 is directly responsible for the stability of the H_2 molecule. The attractive forces that bring about this minimum result from two factors:

1. Locating two electrons between the two protons of the H_2 molecule lowers the electrostatic energy of the system. Attractive energies between oppositely charged particles (electron-proton) slightly exceed the repulsive energies between particles of like charge (electron-electron, proton-proton).

2. When two hydrogen atoms come together to form a molecule, the electrons are spread over the entire volume of the molecule instead of being confined to a particular atom. As pointed out in Chapter 6, quantum mechanics tells us that increasing the volume available to an electron decreases its kinetic energy. We often describe this situation by saying that the two 1s orbitals of the hydrogen atom "overlap" to form a c At internuclear distances less than 0.074 nm, the energy of interaction rises rapidly because of repulsion between the hydrogen nuclei.

b The minimum in the curve, which occurs at the observed internuclear distance of 0.074 nm, corresponds to the most stable state of the H_2 molecule.

a At zero energy, the H atoms are separated.

Increasing energy →

0

−436 kJ

H_2 molecule

0.074 nm

Internuclear distance →

Figure 7.1 Electron density in H_2. The depth of color is proportional to the probability of finding an electron in a particular region.

Figure 7.2 Energy of two hydrogen atoms as a function of the distance between their nuclei.

new bonding orbital. At any rate, calculations suggest that this is the principal factor accounting for the stability of the H_2 molecule.

This chapter is devoted to the covalent bond as it exists in molecules and poly-atomic ions. We consider

- the distribution of outer level *(valence)* electrons in species in which atoms are joined by covalent bonds. These distributions are most simply described by *Lewis structures* (Section 7-1).
- molecular geometries. The so-called *valence shell electron-pair repulsion (VSEPR) model* can be used to predict the angles between covalent bonds formed by a central atom (Section 7-2).
- the polarity of covalent bonds and the molecules they form (Section 7-3). Most bonds and many molecules are polar in the sense that they have a positive and a negative pole.
- the distribution of valence electrons among *atomic orbitals,* using the valence bond approach (Section 7-4).

7-1   Lewis Structures; The Octet Rule

The idea of the covalent bond was first suggested by the American physical chemist Gilbert Newton Lewis (1875–1946) in 1916. He pointed out that the electron configuration of the noble gases appears to be a particularly stable one. Noble-gas atoms are themselves extremely unreactive. Moreover, as pointed out in Chapter 6, a great many monatomic ions have noble-gas structures. Lewis suggested that *non-metal atoms, by sharing electrons to form an electron-pair bond, can acquire a stable noble-gas structure.* Consider, for example, two hydrogen atoms, each with one electron. The process by which they combine to form an H_2 molecule can be shown as

Noble-gas structures are stable in molecules, as they are in atoms and ions.

$$H \cdot + H \cdot \longrightarrow \quad H \; (\!:\!) \; H$$

using dots to represent electrons; the circles emphasize that the pair of electrons in the covalent bond can be considered to occupy the 1s orbital of either hydrogen atom. In that sense, each atom in the H_2 molecule has the electronic structure of the noble gas helium, with the electron configuration $1s^2$.

This idea is readily extended to simple molecules of compounds formed by nonmetal atoms. An example is the HF molecule. You will recall that a fluorine atom has the electron configuration $1s^2 2s^2 2p^5$. It has seven electrons in its outermost principal energy level ($n = 2$). These are referred to as **valence electrons,** in contrast to the core electrons filling the principal level, $n = 1$. If the valence electrons are shown as dots around the symbol of the element, the fluorine atom can be represented as

$$:\ddot{\text{F}}\cdot$$

The combination of a hydrogen with a fluorine atom leads to

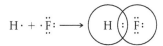

As you can see, the fluorine atom "owns" six valence electrons outright and shares two others. Putting it another way, the F atom is surrounded by eight valence electrons; its electron configuration has become $1s^2 2s^2 2p^6$, which is that of the noble gas neon. This, according to Lewis, explains why the HF molecule is stable in contrast to species such as H_2F, H_3F, . . . none of which exist.

Any structure like this (without the circles) is referred to as a **Lewis structure.** In writing Lewis structures, only the valence electrons written above are shown, because they are the ones that participate in covalent bonding. For the main-group elements, the only ones dealt with here, the number of valence electrons is equal to the last digit of the group number in the periodic table (Table 7.1). Notice that elements in a given main group all have the same number of valence electrons. This explains why such elements behave similarly when they react to form covalently bonded species.

In the Lewis structure of a molecule or polyatomic ion, valence electrons ordinarily occur in pairs. There are two kinds of electron pairs.

1. A pair of electrons shared between two atoms is a **covalent bond,** ordinarily shown as a straight line between bonded atoms.
2. An **unshared electron pair,** owned entirely by one atom, is shown as a pair of dots on that atom. (An unshared pair is often referred to, more picturesquely, as a *lone pair*.)

The Lewis structures for the species OH^-, H_2O, NH_3, and NH_4^+ are

$$\left[:\ddot{\text{O}}{-}\text{H}\right]^- \qquad \text{H}{\overset{\ddot{\text{O}}}{\diagdown}}\text{H} \qquad \text{H}{\overset{\ddot{\text{N}}}{\diagdown}}\text{H} \qquad \left[\begin{array}{c} \text{H} \\ | \\ \text{H}{-}\text{N}{-}\text{H} \\ | \\ \text{H} \end{array}\right]^+$$

Valence electrons are the ones involved in bonding.

Shared electrons are counted for both atoms.

This is why we put the second digit of the group number in bold type.

Table 7.1 Lewis Structures of Atoms Commonly Forming Covalent Bonds

Group:	1	2	13	14	15	16	17	18
No. of valence e^-:	1	2	3	4	5	6	7	8
	H·							
		·Be·	·Ḃ·	·Ċ·	·N̈·	·Ö·	:F̈·	
				·S̈i·	·P̈·	·S̈·	:C̈l·	
				·Ge·	·Äs·	·S̈e·	:B̈r·	·K̈r:
					·S̈b·	·T̈e·	:Ï·	:Ẍe:

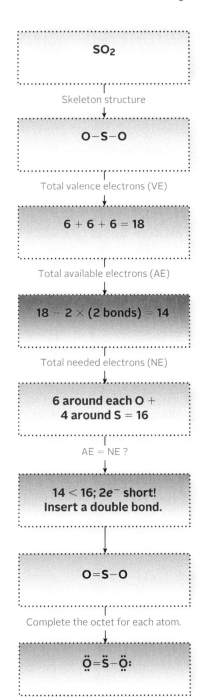

Figure 7.3 Flowchart for writing Lewis structures.

Forming a multiple bond "saves" electrons because bonding pairs are counted for both atoms.

Notice that in each case the oxygen or nitrogen atom is surrounded by eight valence electrons. In each species, a single electron pair is shared between two bonded atoms. Each bond is called a **single bond.** There is one single bond in the OH^- ion, two in the H_2O molecule, three in NH_3, and four in NH_4^+. There are three unshared pairs in the hydroxide ion, two in the water molecule, one in the ammonia molecule, and none in the ammonium ion.

Bonded atoms can share more than one electron pair. A **double bond** occurs when bonded atoms share two electron pairs; in a **triple bond**, three pairs of electrons are shared. In ethylene (C_2H_4) and acetylene (C_2H_2), the carbon atoms are linked by a double bond and triple bond, respectively. Using two parallel lines to represent a double bond and three for a triple bond, we write the structures of these molecules as

$$\underset{\text{ethylene, } C_2H_4}{\overset{H}{\underset{H}{>}}C=C\overset{H}{\underset{H}{<}}} \qquad \underset{\text{acetylene, } C_2H_2}{H-C\equiv C-H}$$

Note that each carbon is surrounded by eight valence electrons and each hydrogen by two.

These examples illustrate the principle that atoms in covalently bonded species tend to have noble-gas electronic structures. This generalization is often referred to as the **octet rule.** Nonmetals, except for hydrogen, achieve a noble-gas structure by sharing in an *octet* of electrons (eight). Hydrogen atoms, in molecules or polyatomic ions, are surrounded by a *duet* of electrons (two).

Writing Lewis Structures

For very simple species, Lewis structures can often be written by inspection. Usually, though, you will save time by following these steps:

1. *Draw a skeleton of the species joining atoms by single bonds*
   Most of the species appearing in this chapter consist of a **central atom** bonded to two or more *terminal atoms.*
   — The central atom is usually written first in the formula.
   — The terminal atoms are most often hydrogen, oxygen, and the halogens.
2. *Count the number of valence electrons (VE).*
   — For a molecule, add the number of valence electrons of all the atoms present.
   — For a polyatomic anion, add the number of valence electrons of each atom plus one electron for each unit of negative charge (e.g., for SO_4^{2-}, add 2 electrons)
   — For a polyatomic cation, add the number of valence electrons of each atom and subtract one electron for each unit of positive charge (e.g., for NH_4^+, subtract 1 electron)
3. *Count the number of valence electrons available for distribution (AE).*
   AE = VE − 2(number of bonds in the skeleton)
4. *Count the number of electrons required to fill out an octet for each atom (except H) in the skeleton (NE).*
   Remember that shared atoms are counted for both atoms.
   (a) If AE = NE, your skeleton is correct. Distribute the available electrons as unshared pairs satisfying the octet rule.
   (b) If AE < NE, modify your skeleton by changing single bonds to double or triple bonds. ◄ ⓘ
      — 2 electrons short: convert one single bond to a double bond.
      — 4 electrons short: convert one single bond to a triple bond, or two single bonds to double bonds.
   Hydrogen and the halogens never form double bonds.

Figure 7.3 shows how to follow these steps for SO_2.

EXAMPLE 7.1

Draw Lewis structures of the hypochlorite ion, OCl⁻ (Figure 7.4), and the ethane, C_2H_6, molecule.

STRATEGY

1. Follow the steps outlined in Figure 7.3.

2. For ethane, hydrogen must be a terminal atom since it cannot form double bonds. Carbon ordinarily forms four bonds.

SOLUTION

OCl⁻ skeleton	$[O–Cl]^-$
VE	6 (for O) + 7 (for Cl) + 1(−1 charge) = 14
AE	AE = VE − 2(bonds) = 14 − 2(1 bond) = 12
NE	6 (for O to have an octet) + 6 (for Cl to have an octet) = 12
AE = NE ?	Yes; distribute electrons.
Lewis structure	$[:\ddot{O}–\ddot{C}l:]^-$

C_2H_6 skeleton	H H
	  | |
	H—C—C—H
	  | |
	H H

VE	2×4 (for C) + 6×1 (for H) = 14
AE	AE = VE − 2(bonds) = 14 − 2(7 bonds) = 0
NE	0 : All the H atoms have duets and both C atoms have octets.
AE = NE ?	Yes; distribute electrons.

Lewis structure	H H
	  | |
	H—C—C—H
	  | |
	H H

END POINT

After you have written the Lewis structure, it is a good idea to add the number of unshared electron pairs and bonding electrons. This sum must equal the number of valence electrons (VE).

EXAMPLE 7.2

Draw the Lewis structures of NO_2^- and N_2.

STRATEGY

Follow the steps outlined in Figure 7.3.

SOLUTION

Skeleton for NO_2^-	$[O–N–O]^-$
VE	2(6 (for O)) + 5 (for N) + 1(−1 charge) = 18
AE	AE = VE − 2(bonds) = 18 − 2(2 bonds) = 14
NE	2(6 (for each O)) + 4 (for N) = 16

continued

AE = NE ?	No; 2 electrons short
	Convert a single bond to a double bond.
Lewis structure	$[:\ddot{O}-\ddot{N}=\ddot{O}:]^-$
Skeleton for N_2	N–N
VE	2(5 (for each N)) = 10
AE	AE = VE − 2(bonds) = 10 − 2(1 bond) = 8
NE	2 × 6 (for each N to have an octet) = 12
AE = NE ?	No; 4 electrons short
	Convert a single bond to a triple bond.
Lewis structure	$:N\equiv N:$

END POINT

For the Lewis structure of NO_2^-, it does not matter which single bond you convert to a double bond. We will talk about this in more detail when we discuss resonance forms.

Figure 7.4 Hypochlorite ions in action. The OCl^- ion is the active bleaching agent in Clorox™ (see Example 7.1).

© Cengage Learning/Charles D. Winters

i
The double-headed arrow is used to separate resonance structures.

Resonance Forms

In certain cases, the Lewis structure does not adequately describe the properties of the ion or molecule that it represents. Consider, for example, the SO_2 structure in Figure 7.3. This structure implies that there are two different kinds of sulfur-to-oxygen bonds in SO_2. One of these appears to be a single bond, the other a double bond. Yet experiment shows that there is only one kind of bond in the molecule.

One way to explain this situation is to assume that each of the bonds in SO_2 is intermediate between a single and a double bond. To express this concept, two structures, separated by a double-headed arrow, ◄ *i* are written

$$\ddot{O}=\overset{\ddots}{S}-\ddot{O}: \longleftrightarrow :\ddot{O}-\overset{\ddots}{S}=\ddot{O}$$

with the understanding that the true structure is intermediate between them. These are referred to as *resonance forms*. The actual structure is intermediate between the two resonance forms and is called a *resonance hybrid*. It is the only structure that actually exists (Figure 7.5). The individual resonance forms do not exist and are merely a convenient way to describe the real structures. The concept of **resonance** is invoked whenever a single Lewis structure does not adequately reflect the properties of a substance.

Another species for which it is necessary to invoke the idea of resonance is the nitrate ion. Here three equivalent structures can be written to explain the experimental observation that the three nitrogen-to-oxygen bonds in the NO_3^- ion are identical in all respects.

$$\left[:\overset{\ddots}{O}\underset{:\overset{\ddots}{O}:}{\overset{}{N}}=\ddot{O}:\right]^- \longleftrightarrow \left[:\overset{\ddots}{O}\underset{:\overset{\ddots}{O}:}{\overset{}{N}}\ddot{O}:\right]^- \longleftrightarrow \left[:\ddot{O}=\underset{:\overset{\ddots}{O}:}{\overset{}{N}}\overset{\ddots}{O}:\right]^-$$

Figure 7.5 The labradoodle (c) is a hybrid of its parents, the labrador (a) and the poodle (b). In similar fashion, a pair of sp hybrid orbitals forms from the parents—an s and a p orbital.

Resonance can also occur with many organic molecules, including benzene, C_6H_6, which is known to have a hexagonal ring structure (Figure 7.6). Benzene can be considered a resonance hybrid of the two forms

Figure 7.6 Benzene ball-and-stick model, showing double bonds.

These structures are commonly abbreviated as

with the understanding that, at each corner of the hexagon, a carbon is attached to a hydrogen atom.

We will encounter other examples of molecules and ions whose properties can be interpreted in terms of resonance. In all such species:

1. Resonance forms do not imply different kinds of molecules with electrons shifting eternally between them. There is only one type of SO_2 molecule; its structure is intermediate between those of the two resonance forms drawn for sulfur dioxide.

2. Resonance can be anticipated when it is possible to write two or more Lewis structures that are about equally plausible. In the case of the nitrate ion, the three structures we have written are equivalent. One could, in principle, write many other structures, but none of them would put eight electrons around each atom.

3. Resonance forms differ only in the distribution of electrons, not in the arrangement of atoms. The molecule

$$H-\overset{\overset{\displaystyle H}{|}}{C}=\overset{\overset{\displaystyle H}{|}}{C}-C\equiv C-\overset{\overset{\displaystyle H}{|}}{C}=\overset{\overset{\displaystyle H}{|}}{C}-H$$

is not a resonance structure of benzene, even though it has the same molecular formula, C_6H_6. Indeed, it is an entirely different substance with different chemical and physical properties.

EXAMPLE 7.3

Write two resonance structures for the NO_2^- ion.

STRATEGY

1. The Lewis structure of NO_2^- is derived in Example 7.2.

$$[:\ddot{O}-\ddot{N}=\ddot{O}:]^-$$

2. Change the position of the multiple bond and one of the unshared electron pairs.

$$[:\ddot{O}-\ddot{N}=\ddot{O}:]^-$$

3. Do not change the skeleton.

SOLUTION

The Lewis structures of the two resonance forms are

$$\left[\begin{array}{c}\ddot{N}\\ :\ddot{O}\quad\ddot{O}:\end{array}\right] \longleftrightarrow \left[\begin{array}{c}\ddot{N}\\ :\ddot{O}\quad\ddot{O}:\end{array}\right]$$

Formal Charge

Often it is possible to write two different Lewis structures for a molecule differing in the arrangement of atoms, that is,

$$A—A—B \quad \text{or} \quad A—B—A$$

Isomers have the same formula but different properties.

Sometimes both structures represent real compounds that are *isomers* of each other. More often, only one structure exists in nature. For example, methyl alcohol (CH_4O) has the structure

$$\begin{array}{c}H\\ |\\ H—C—\ddot{O}—H\\ |\\ H\end{array}$$

In contrast, the structure

$$H—\ddot{C}—\ddot{O}—H\\ \quad |\quad |\\ \quad H\quad H$$

does not correspond to any real compound even though it obeys the octet rule.

There are several ways to choose the more plausible of two structures differing in their arrangement of atoms. As pointed out in Example 7.1, the fact that carbon almost always forms four bonds leads to the correct structure for ethane. Another approach involves a concept called **formal charge,** which can be applied to any atom within a Lewis structure. The formal charge is the difference between the number of valence electrons in the free atom and the number assigned to that atom in the Lewis structure. The assigned electrons include

Formal charge is the charge an atom would have if valence electrons in bonds were distributed evenly.

- all the unshared electrons owned by that atom.
- one half of the bonding electrons shared by that atom.

Thus the formal charge can be determined by counting the electrons "owned" by the atom, its valence electrons (VE), the unshared pairs around the atom, and the bonding electrons around the atom. We arrive at the following equation:

$$C_f = VE - \text{unshared electrons} - \tfrac{1}{2}(\text{bonding electrons})$$

Since a bond is always made up of two electrons, we can find the formal charge by using the modified equation below:

$$C_f = VE - \text{unshared electrons} - \text{number of bonds}$$

To show how this works, let's calculate the formal charges of carbon and oxygen in the two structures written above for methyl alcohol:

(1) $H-\overset{\overset{\displaystyle H}{|}}{\underset{\underset{\displaystyle H}{|}}{C}}-\ddot{\underset{\cdot\cdot}{O}}-H$

(2) $H-\overset{|}{\underset{\underset{\displaystyle H}{|}}{C}}-\overset{\cdot\cdot}{\underset{\underset{\displaystyle H}{|}}{O}}-H$

For C: VE = 4, unshared $e^- = 0$,
  bonds = 4
  $C_f = 4 - 0 - 4 = 0$

For C: VE = 4, unshared $e^- = 2$,
  bonds = 3
  $C_f = 4 - 2 - 3 = -1$

For O: VE = 6, unshared $e^- = 4$,
  bonds = 2
  $C_f = 6 - 4 - 2 = 0$

For O: VE = 6, unshared $e^- = 2$,
  bonds = 3
  $C_f = 6 - 2 - 3 = +1$

Ordinarily, the more likely Lewis structure is the one in which

- the formal charges are as close to zero as possible.
- any negative formal charge is located on the most strongly electronegative atom.

Applying these rules, we can see that structure (1) for methyl alcohol is preferred over structure (2). In (1), both carbon and oxygen have formal charges of zero. In (2), a negative charge is assigned to carbon, which is actually less electronegative than oxygen (2.5 versus 3.5).

The concept of formal charge has a much wider applicability than this short discussion might imply. In particular, it can be used to predict situations in which conventional Lewis structures, written in accordance with the octet rule, may be incorrect (Table 7.2).

Formal charge is not an infallible guide to predicting Lewis structures.

Exceptions to the Octet Rule: Electron-Deficient Molecules

Although most of the molecules and polyatomic ions referred to in general chemistry follow the octet rule, there are some familiar species that do not. Among these are molecules containing an odd number of valence electrons. Nitric oxide, NO, and nitrogen dioxide, NO_2, fall in this category:

NO   no. of valence electrons = 5 + 6 = 11

NO_2 no. of valence electrons = 5 + 6(2) = 17

With an odd number of valence electrons, there's no way they could all be paired.

For such *odd electron* species (sometimes called free radicals) it is impossible to write Lewis structures in which each atom obeys the octet rule. In the NO

Table 7.2 Possible Structures for BeF_2 and BF_3

Structure I	C_f	Structure II	C_f
$:\ddot{F}=Be=\ddot{F}:$	Be = −2 F = +1	$:\ddot{F}-Be-\ddot{F}:$	Be = 0 F = 0
$:\ddot{F}\diagdown_{\!\!B}\diagup\ddot{F}: \\ \;\;\overset{\|}{:\ddot{F}:}$	B = −1 F = +1,0,0	$:\ddot{F}\diagdown_{\!\!B}\diagup\ddot{F}: \\ \;\;\overset{\|}{:\ddot{F}:}$	B = 0 F = 0

Figure 7.7 Nitrogen dioxide. NO_2, a red-brown gas, has an unpaired electron on the N atom.

Figure 7.8 Oxygen (O_2) in a magnetic field. The liquid oxygen, which is blue, is attracted into a magnetic field between the poles of an electromagnet. Both the paramagnetism and the blue color are due to the unpaired electrons in the O_2 molecule.

In most molecules, the central atom is surrounded by 8 electrons. Rarely, it is surrounded by 4 (BeF_2) or 6 (BF_3). Occasionally, the number is 10 (PCl_5) or 12 (SF_6).

molecule, the unpaired electron is put on the nitrogen atom, giving both atoms a formal charge of zero:

$$\cdot \ddot{N}=\ddot{O}:$$

In NO_2 (Figure 7.7), the best structure one can write again puts the unpaired electron on the nitrogen atom:

Elementary oxygen, like NO and NO_2, is paramagnetic (Figure 7.8). Experimental evidence suggests that the O_2 molecule contains two unpaired electrons *and* a double bond. It is impossible to write a conventional Lewis structure for O_2 that has these two characteristics. A more sophisticated model of bonding, using molecular orbitals (Appendix 4), is required to explain the properties of oxygen.

There are a few species in which the central atom violates the octet rule in the sense that it is surrounded by two or three electron pairs rather than four. Examples include the fluorides of beryllium and boron, BeF_2 and BF_3. Although one could write multiple bonded structures for these molecules in accordance with the octet rule (Table 7.2), experimental evidence suggests the structures

in which the central atom is surrounded by four and six valence electrons, respectively, rather than eight. Another familiar substance in which boron is surrounded by only three pairs of electrons rather than four is boric acid, H_3BO_3, used as an insecticide and fungicide.

Exceptions to the Octet Rule: Expanded Octets

The largest class of molecules to violate the octet rule consists of species in which the central atom is surrounded by more than four pairs of valence electrons. Typical molecules of this type are phosphorus pentachloride, PCl_5, and sulfur hexafluoride, SF_6. The Lewis structures of these molecules are

As you can see, the central atoms in these molecules have **expanded octets.** In PCl_5, the phosphorus atom is surrounded by 10 valence electrons (5 shared pairs); in SF_6, there are 12 valence electrons (6 shared pairs) around the sulfur atom.

In molecules of this type, the terminal atoms are most often halogens (F, Cl, Br, I); in a few molecules, oxygen is a terminal atom. The central atom is a nonmetal in the third, fourth, or fifth period of the periodic table. Most frequently, it is one of the following elements:

	Group 15	Group 16	Group 17	Group 18
3rd period	P	S	Cl	
4th period	As	Se	Br	Kr
5th period	Sb	Te	I	Xe

CHEMISTRY THE **HUMAN** SIDE

Born in Massachusetts, G. N. Lewis grew up in Nebraska, then came back East to obtain his B.S. (1896) and Ph.D. (1899) at Harvard. Although he stayed on for a few years as an instructor, Lewis seems never to have been happy at Harvard. A precocious student and an intellectual rebel, he was repelled by the highly traditional atmosphere that prevailed in the chemistry department there in his time. Many years later, he refused an honorary degree from his alma mater.

After leaving Harvard, Lewis made his reputation at MIT, where he was promoted to full professor in only four years. In 1912, he moved across the country to the University of California, Berkeley, as dean of the College of Chemistry and department head. He remained there for the rest of his life. Under his guidance, the chemistry department at Berkeley became perhaps the most prestigious in the country. Among the faculty and graduate students that he attracted were five future Nobel Prize winners: Harold Urey in 1934, William Giauque in 1949, Glenn Seaborg in 1951, Willard Libby in 1960, and Melvin Calvin in 1961.

In administering the chemistry department at Berkeley, Lewis demanded excellence in both research and teaching. Virtually the entire staff was involved in the general chemistry program; at one time eight full professors carried freshman sections.

Lewis's interest in chemical bonding and structure dated from 1902. In attempting to explain "valence" to a class at Harvard, he devised an atomic model to rationalize the octet rule. His model was deficient in many respects; for one thing, Lewis visualized cubic atoms with electrons located at the corners. Perhaps this explains why his ideas of atomic structure were not published until 1916. In that year, Lewis conceived of the electron-pair bond, perhaps his greatest single contribution to chemistry. At that time, it was widely believed that all bonds were ionic; Lewis's ideas were rejected by many well-known organic chemists.

In 1923, Lewis published a classic book (later reprinted by Dover Publications) titled *Valence and the Structure of Atoms and Molecules*. Here, in Lewis's characteristically lucid style, we find many of the basic principles of covalent bonding discussed in this chapter. Included are electron-dot structures, the octet rule, and the concept of electronegativity. Here too is the Lewis definition of acids and bases (Chapter 13). That same year, Lewis published with Merle Randall a text called *Thermodynamics and the Free Energy of Chemical Substances*. Today, a revised edition of that text is still used in graduate courses in chemistry. (Lewis certainly deserved a Nobel Prize, but he never received one.)

The years from 1923 to 1938 were relatively unproductive for Lewis insofar as his own research was concerned. The applications of the electron-pair bond came largely in the areas of organic and quantum chemistry; in neither of these fields did Lewis feel at home. In the early 1930s, he published a series of relatively minor papers dealing with the properties of deuterium. Then in 1939 he began to publish in the field of photochemistry. Of approximately 20 papers in this area, several were of fundamental importance, comparable in quality to the best work of his early years. Retired officially in 1945, Lewis died a year later while carrying out an experiment on fluorescence.

Gilbert Newton Lewis
(1875–1946)

All these atoms have d orbitals available for bonding (3d, 4d, 5d). These are the orbitals in which the extra pairs of electrons are located in such species as PCl_5 and SF_6. Because there is no 2d sublevel, C, N, and O never form expanded octets.

Sometimes, as with PCl_5 and SF_6, it is clear from the formula that the central atom has an expanded octet. Often, however, it is by no means obvious that this is the case. At first glance, formulas such as ClF_3 or XeF_4 look completely straightforward. However, when you try to draw the Lewis structure it becomes clear that an expanded octet is involved. The number of electrons available after the skeleton is drawn is *greater* than the number required to give each atom an octet. When that happens, ***distribute the extra electrons (two or four) around the central atom as unshared pairs.***

The presence of expanded octets requires modification of the process we delineated in Figure 7.3. Figure 7.9 (page 167) is a modification that includes the possibility of an expanded octet. We use ClF_3 as an example.

EXAMPLE 7.4

Draw Lewis structures of XeF_4.

STRATEGY

If AE < NE, follow the process described in Figure 7.3.

If AE = NE, your skeleton is correct; add electrons as unshared pairs to form octets around the atoms.

If AE > NE, follow the process described in Figure 7.9.

SOLUTION

Skeleton	F | F—Xe—F | F
VE	4(7 (for each F)) + 8 (for Xe) = 36
AE	AE = VE − 2(bonds) = 36 − 2(4 bonds) = 28
NE	4(6 (for each F to have an octet)) + 0 (Xe has an octet) = 24
AE = NE ?	No; AE > NE. There are 4 extra electrons.
Satisfy the octet rule	:F: | :F—Xe—F: | :F:
Lewis structure	Add extra electrons (4) to the central atom. :F: | :F—Xe—F: | :F:

7-2 Molecular Geometry

The geometry of a diatomic molecule such as Cl_2 or HCl can be described very simply. Because two points define a straight line, the molecule must be linear.

$$Cl—Cl \qquad H—Cl$$

With molecules containing three or more atoms, the geometry is not so obvious. Here, the angle between bonds, called the **bond angle,** must be considered. For example, a molecule of the type YX_2, where Y represents the central atom and X an atom bonded to it, can be

■ a **linear molecule,** with a bond angle of 180°: $\quad X—Y—X$

■ a **bent molecule,** with a bond angle less than 180°: $\quad X{\diagup}^{Y}{\diagdown}X$

The major features of **molecular geometry** can be predicted on the basis of a quite simple principle—**electron-pair repulsion.** This principle is the essence of the *valence-shell electron-pair repulsion (VSEPR) model,* first suggested by N. V. Sidgwick (1873–1952) and H. M. Powell (1906–1991) in 1940. It was developed and expanded later by R. J. Gillespie (1924–) and R. S. Nyholm (1917–1971). According to the **VSEPR model**, *the valence electron pairs surrounding an atom repel one another. Consequently, the orbitals containing those electron pairs are oriented to be as far apart as possible.*

In this section we apply this model to predict the geometry of some rather simple molecules and polyatomic ions. In all these species, a central atom is surrounded by from two to six pairs of electrons.

Ideal Geometries with Two to Six Electron Pairs on the Central Atom

We begin by considering species in which a central atom, A, is surrounded by from two to six electron pairs, all of which are used to form single bonds with terminal atoms, X. These species have the general formulas AX_2, $AX_3, \ldots, AX_6$. It is understood that there are no unshared pairs around atom A.

To see how the bonding orbitals surrounding the central atom are oriented with respect to one another, consider Figure 7.10, which shows the **ideal geometry** positions taken naturally by two to six balloons tied together at the center. The balloons, like the orbitals they represent, arrange themselves to be as far from one another as possible.

Figure 7.11 shows the geometries predicted by the VSEPR model for molecules of the types AX_2 to AX_6. The geometries for two and three electron pairs are those associated with species in which the central atom has less than an octet of electrons. Molecules of this type include BeF_2 (in the gas state) and BF_3, which have the Lewis structures shown below:

Two electron pairs are as far apart as possible when they are directed at 180° to one another. This gives BeF_2 a linear structure. The three electron pairs around the boron atom in BF_3 are directed toward the corners of an equilateral triangle; the bond angles are 120°. We describe this geometry as **trigonal planar.**

In species that follow the octet rule, the central atom is surrounded by four electron pairs. If each of these pairs forms a single bond with a terminal atom, a molecule of the type AX_4 results. The four bonds are directed

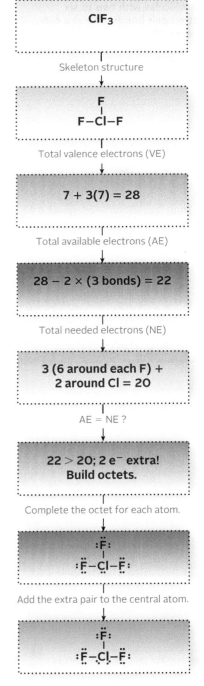

Figure 7.9 Flowchart for Lewis structures with expanded octets.

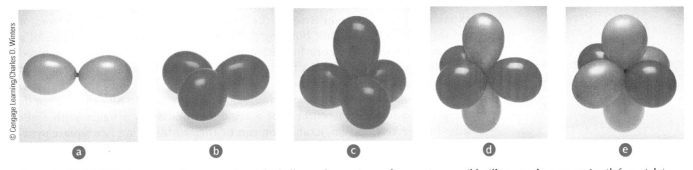

(a)   (b)   (c)   (d)   (e)

© Cengage Learning/Charles D. Winters

Figure 7.10 VSEPR electron-pair geometries. The balloons, by staying as far apart as possible, illustrate the geometries (*left to right*) for two to six electron pairs.

Figure 7.11 **Molecular geometries for molecules with two to six electron-pair bonds around a central atom (A).**

Species type	Orientation of electron pairs	Predicted bond angles	Example	Ball-and-stick model
AX_2	Linear	180°	BeF_2	180°
AX_3	Trigonal planar	120°	BF_3	120°
AX_4	Tetrahedron	109.5°	CH_4	109.5°
AX_5	Trigonal bipyramid	90° 120° 180°	PF_5	90° 120°
AX_6	Octahedron	90° 180°	SF_6	90°

This puts the four electron pairs as far apart as possible.

toward the corners of a regular **tetrahedron.** All the bond angles are 109.5°, the tetrahedral angle. This geometry is found in many polyatomic ions such as NH_4^+ and SO_4^{2-} and in a wide variety of organic molecules, the simplest of which is methane, CH_4.

Molecules of the type AX_5 and AX_6 require the central atom to have an expanded octet. The geometries of these molecules are shown at the bottom of Figure 7.11. In PF_5, the five bonding pairs are directed toward the corners of a **trigonal bipyramid,** a figure formed when two triangular pyramids are fused together, base to base. Three of the fluorine atoms are located at the corners of an equilateral triangle with the phosphorus atom at the center; the other two fluorine atoms are directly above and below the P atom. In SF_6, the six bonds are directed toward the corners of a regular **octahedron,** a figure with eight sides but *six* vertices. An octahedron can be formed by fusing two **square pyramids** base to base. Four of the fluorine atoms in SF_6 are located at the corners of a square with the S atom at the center; one fluorine atom is directly above the S atom, another directly below it.

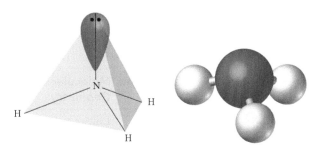

Figure 7.12 Two ways of showing the geometry of the NH_3 molecule. The orientation of the electron pairs, including the unshared pair (black dots), is shown at left. The orientation of the atoms is shown at the right. The nitrogen atom is located directly above the center of the equilateral triangle formed by the three hydrogen atoms. The NH_3 molecule is described as a trigonal pyramid.

Effect of Unshared Pairs on Molecular Geometry

In many molecules and polyatomic ions, one or more of the electron pairs around the central atom are unshared. The VSEPR model is readily extended to predict the geometries of these species. In general

1. The *electron-pair geometry* is approximately the same as that observed when only single bonds are involved. The bond angles are ordinarily a little smaller than the ideal values listed in Figure 7.11.
2. The *molecular geometry* is quite different when one or more unshared pairs are present. In describing molecular geometry, we refer only to the positions of the bonded atoms. These positions can be determined experimentally; positions of unshared pairs cannot be established by experiment. Hence, the locations of unshared pairs are not specified in describing molecular geometry.

With these principles in mind, consider the NH_3 molecule:

$$\overset{\displaystyle \ddot{N}}{\underset{\displaystyle H}{H \diagup | \diagdown H}}$$

The apparent orientation of the four electron pairs around the N atom in NH_3 is shown at the left of Figure 7.12. Notice that, as in CH_4, the four pairs are directed toward the corners of a regular tetrahedron. The diagram at the right of Figure 7.12 shows the positions of the atoms in NH_3. The nitrogen atom is located above the center of an equilateral triangle formed by the three hydrogen atoms. The molecular geometry of the NH_3 molecule is described as a **trigonal pyramid**. The nitrogen atom is at the apex of the pyramid, and the three hydrogen atoms form its triangular base. The molecule is three-dimensional, as the word "pyramid" implies.

The development we have just gone through for NH_3 is readily extended to the water molecule, H_2O. Here the Lewis structure shows that the central oxygen atom is surrounded by two single bonds and two unshared pairs:

$$H \diagup \overset{\displaystyle \ddot{O}}{} \diagdown H$$

The diagram at the left of Figure 7.13 emphasizes that the four electron pairs are oriented tetrahedrally. At the right, the positions of the atoms are shown. Clearly they are not in a straight line; the H_2O molecule is bent.

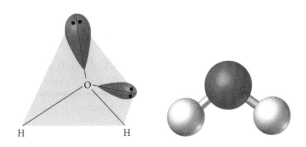

Figure 7.13 Two ways of showing the geometry of the H_2O molecule. At the left, the two unshared pairs are shown. As you can see from the drawing at the right, H_2O is a bent molecule. The bond angle, $105°$, is a little smaller than the tetrahedral angle, $109.5°$. The unshared pairs spread out over a larger volume than that occupied by the bonding pairs.

Table 7.3 Geometries with Two, Three, or Four Electron Pairs Around a Central Atom

No. of Terminal Atoms (X) + Unshared Pairs (E)	Species Type	Ideal Bond Angles*	Molecular Geometry	Examples
2	AX_2	180°	Linear	BeF_2, CO_2
3	AX_3	120°	Trigonal planar	BF_3, SO_3
	AX_2E	120°*	Bent	GeF_2, SO_2
4	AX_4	109.5°	Tetrahedron	CH_4
	AX_3E	109.5°*	Trigonal pyramid	NH_3
	AX_2E_2	109.5°*	Bent	H_2O

*In these species, the observed bond angle is ordinarily somewhat less than the ideal value.

Experiments show that the bond angles in NH_3 and H_2O are slightly less than the ideal value of 109.5°. In NH_3 (three single bonds, one unshared pair around N), the bond angle is 107°. In H_2O (two single bonds, two unshared pairs around O), the bond angle is about 105°. ◀ *i*

These effects can be explained in a rather simple way. An unshared pair is attracted by one nucleus, that of the atom to which it belongs. In contrast, a bonding pair is attracted by two nuclei, those of the two atoms it joins. Hence the electron cloud of an unshared pair is expected to spread out over a larger volume than that of a bonding pair. In NH_3, this tends to force the bonding pairs closer to one another, thereby reducing the bond angle. Where there are two unshared pairs, as in H_2O, this effect is more pronounced. In general, the VSEPR model predicts that unshared electron pairs will occupy slightly more space than bonding pairs.

Table 7.3 summarizes the molecular geometries of species in which a central atom is surrounded by two, three, or four electron pairs. The table is organized in terms of the number of terminal atoms, X, and unshared pairs, E, surrounding the central atom, A.

i Unshared pairs reduce bond angles below ideal values.

EXAMPLE 7.5

Predict the geometry of NH_4^+, BF_3, and PCl_3.

STRATEGY

1. Start by writing Lewis structures for each species.

2. Focus on the central atom, then decide what species type (AX_2, AX_3, . . .) the molecule or ion is.

   A represents the central atom.

   X represents the terminal atoms.

   E represents the unshared electron pairs.

3. Recall Table 7.3, which matches the species type with the molecular geometry and ideal bond angles for the species.

continued

	SOLUTION
Lewis structure for NH_4^+	$$\left[\begin{array}{c} H \\ | \\ H-N-H \\ | \\ H \end{array}\right]^+$$
Species type	A = N, X = H (4), no E $\rightarrow$ AX$_4$
Geometry	tetrahedral, 109.5° bond angles
Lewis structure for BF_3	$:\ddot{F}-B-\ddot{F}:$ $\quad$ $:\ddot{F}:$
Species type	A = B, X = F (3), no E $\rightarrow$ AX$_3$
Geometry	trigonal planar, 120° bond angles
Lewis structure for PCl_3	$\ddot{P}$ $:\ddot{C}l \quad \ddot{C}l:$ $\quad :\ddot{C}l:$
Species type	A = P, X = Cl (3), E = 1 $\rightarrow$ AX$_3$E
Geometry	trigonal pyramid (The ideal bond angles are 109.5° but actually are 104°.)

In many expanded-octet molecules, one or more of the electron pairs around the central atom are unshared. Recall, for example, the Lewis structure of xenon tetra-fluoride, XeF_4 ◄ ⓘ (Example 7.4).

$$:\ddot{F} \diagdown \underset{Xe}{} \diagup \ddot{F}:$$
$$:\ddot{F} \diagup \diagdown \ddot{F}:$$

The way the Lewis structure is written does not necessarily imply geometry.

There are six electron pairs around the xenon atom; four of these are covalent bonds to fluorine and the other two pairs are unshared. This molecule is classified as AX$_4$E$_2$ (Figure 7.14).

Geometries of molecules such as these can be predicted by the VSEPR model. The results are shown in Figure 7.15. The structures listed include those of all types of molecules having five or six electron pairs around the central atom, one or more of which may be unshared. Note that

■ in molecules of the type AX$_4$E$_2$, the two lone pairs occupy opposite rather than adjacent vertices of the octahedron.
■ in the molecules AX$_4$E, AX$_3$E$_2$, and AX$_2$E$_3$ the lone pairs occupy successive positions in the equilateral triangle at the center of the trigonal bipyramid.

Multiple Bonds

The VSEPR model is readily extended to species in which double or triple bonds are present. A simple principle applies: *Insofar as molecular geometry is concerned, a multiple bond behaves like a single bond.* This makes sense. The four electrons in a double bond, or the six electrons in a triple bond, must be located between the two atoms, as are the two electrons in a single bond. This means that the electron pairs in a **multiple bond** must occupy the same region of space as those in a single bond. Hence the "extra" electron pairs in a multiple bond have no effect on geometry.

To illustrate this principle, consider the CO_2 molecule. Its Lewis structure is

$$:\ddot{O}=C=\ddot{O}:$$

Figure 7.14 Xenon tetrafluoride, an AX$_4$E$_2$ molecule.

Figure 7.15 Molecular geometries for molecules with expanded octets and unshared electron pairs. The gray spheres represent terminal atoms (X), and the open ellipses represent unshared electron pairs (E). For example, AX_4E represents a molecule in which the central atom is surrounded by four covalent bonds and one unshared electron pair.

5 ELECTRON PAIRS

Species type	Structure	Description	Example	Bond angles
AX_5		Trigonal bipyramidal	PF_5	90°, 120°, 180°
AX_4E		See-saw	SF_4	90°, 120°, 180°
AX_3E_2		T-shaped	ClF_3	90°, 180°
AX_2E_3		Linear	XeF_2	180°

6 ELECTRON PAIRS

Species type	Structure	Description	Example	Bond angles
AX_6		Octahedron	SF_6	90°, 180°
AX_5E		Square pyramidal	ClF_5	90°, 180°
AX_4E_2		Square planar	XeF_4	90°, 180°

The central atom, carbon, has two double bonds and no unshared pairs. For purposes of determining molecular geometry, we pretend that the double bonds are single bonds, ignoring the extra bonding pairs. The bonds are directed to be as far apart as possible, giving a 180° O—C—O bond angle. The CO_2 molecule, like BeF_2, is linear:

$$\underset{180°}{F—Be—F} \qquad \underset{180°}{O=C=O}$$

> CO_2, like BeF_2, is an AX_2 molecule with no unshared pairs.

This principle can be restated in a somewhat different way for molecules in which there is a single central atom. The geometry of such a molecule depends only on

- *the number of terminal atoms, X, bonded to the central atom, irrespective of whether the bonds are single, double, or triple.*
- *the number of unshared pairs, E, around the central atom.*

This means that Table 7.3 can be used in the usual way to predict the geometry of a species containing multiple bonds.

EXAMPLE 7.6

Predict the geometries of the ClO_3^- ion, the NO_3^- ion, and the N_2O molecule, which have the following Lewis structures, respectively:

$$\left[\ddot{O}—\ddot{Cl}—\ddot{O} \atop \ddot{O} \right]^- \qquad \left[\ddot{O}—N—\ddot{O} \atop \ddot{O} \right]^- \qquad \ddot{N}=N=\ddot{O}$$

STRATEGY

1. Classify each species as AX_mE_n and use Table 7.3.

2. Multiple bonds count as single bonds. It is the number of terminal atoms (X) that are counted, not the number of bonds.

SOLUTION

For ClO_3^-:	
Species type	A = Cl, X = O = 3, E = 1 → AX_3E
Geometry	trigonal pyramid, ideal bond angles are 109.5°
For NO_3^-:	
Species type	A = N, X = O = 3, E = 0 → AX_3
Geometry	trigonal planar, ideal bond angles are 120°
For N_2O:	
Species type	A = N, X = N and O = 2, E = 0 → AX_2
Geometry	linear, ideal bond angles are 180°

The VSEPR model applies equally well to molecules in which there is no single central atom. Consider the acetylene molecule, C_2H_2. Recall that here the two carbon atoms are joined by a triple bond:

$$H—C\equiv C—H$$

Each carbon atom behaves as if it were surrounded by two electron pairs. Both of the bond angles (H—C≡C and C≡C—H) are 180°. The molecule is linear; the four atoms are in a straight line. The two extra electron pairs in the triple bond do not affect the geometry of the molecule.

In ethylene, C_2H_4, there is a double bond between the two carbon atoms. The molecule has the geometry to be expected if each carbon atom had only three pairs of electrons around it.

The six atoms are located in a plane with bond angles of approximately 120°. Actually, the double bond between the carbon atoms occupies slightly more space than a single bond joining carbon to hydrogen. As a result, the H—C=C angles are slightly larger than 120°, and the H—C—H angles are slightly less.

$$
\begin{array}{ccc}
\text{H} & 122° & \text{H} \\
116° & \text{C}=\text{C} & 116° \\
\text{H} & 122° & \text{H}
\end{array}
$$

7-3 Polarity of Molecules

Covalent bonds and molecules held together by such bonds may be—

- *polar.* As a result of an unsymmetrical distribution of electrons, the bond or molecule contains a positive and a negative pole and is therefore a *dipole.*
- *nonpolar.* A symmetrical distribution of electrons leads to a bond or molecule with no positive or negative poles.

Polar and Nonpolar Covalent Bonds

The two electrons in the H_2 molecule are shared equally by the two nuclei. Stated another way, a bonding electron is as likely to be found in the vicinity of one nucleus as another. Bonds of this type are described as nonpolar. A **nonpolar bond** is formed whenever the two atoms joined are identical, as in H_2 and F_2.

In the HF molecule, the distribution of the bonding electrons is somewhat different from that in H_2 or F_2. Here the density of the electron cloud is greater about the fluorine atom. The bonding electrons, on the average, are shifted toward fluorine and away from the hydrogen (atom Y in Figure 7.16). A bond in which the electron density is unsymmetrical is referred to as a **polar bond.**

Atoms of two different elements always differ at least slightly in their electronegativity (recall Table 6.7). Hence covalent bonds between unlike atoms are always polar. ◂ⓘ Consider, for example, the H—F bond. Because fluorine has a higher electronegativity (4.0) than does hydrogen (2.2), bonding electrons are displaced toward the fluorine atom. The H—F bond is polar, with a partial negative charge at the fluorine atom and a partial positive charge at the hydrogen atom.

The extent of polarity of a covalent bond is related to the difference in electronegativities of the bonded atoms. If this difference is large, as in H—F ($\Delta EN = 1.8$), the bond is strongly polar. Where the difference is small, as in H—C ($\Delta EN = 0.3$), the bond is only slightly polar.

Polar and Nonpolar Molecules

A **polar molecule** is one that contains positive and negative poles. There is a partial positive charge (positive pole) at one point in the molecule and a partial negative charge (negative pole) at a different point. As shown in Figure 7.17, polar molecules orient themselves in the presence of an electric field. The positive pole in the molecule tends to align with the external negative charge, and the negative pole with the

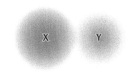

Figure 7.16 Bond polarity. If atom X is more electronegative than atom Y, the electron cloud of the bonding electrons will be concentrated around atom X. Thus the bond is polar.

ⓘ

All molecules, except those of elements, have polar bonds.

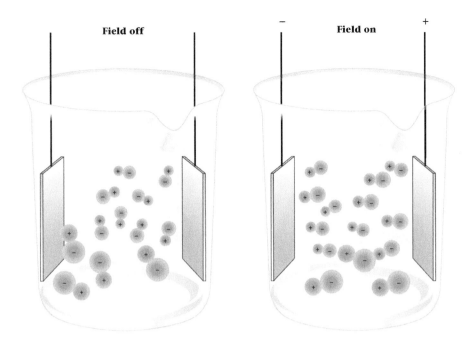

Field off   **Field on**

Figure 7.17 Orientation of polar molecules in an electric field. With the field off, polar molecules are randomly oriented. With the field on, polar molecules such as HF align their positive and negative ends toward the negative and positive poles of the field, respectively. Nonpolar molecules such as H_2 do not line up.

external positive charge. In contrast, there are no positive and negative poles in a **nonpolar molecule.** In an electric field, nonpolar molecules, such as H_2, show no preferred orientation.

The extent to which molecules tend to orient themselves in an electrical field is a measure of their *dipole moment*. A polar molecule such as HF has a dipole moment; a nonpolar molecule such as H_2 or F_2 has a dipole moment of zero.

If a molecule is diatomic, it is easy to decide whether it is polar or nonpolar. A diatomic molecule has only one kind of bond; hence the polarity of the molecule is the same as the polarity of the bond. Hydrogen and fluorine (H_2, F_2) are nonpolar because the bonded atoms are identical and the bond is nonpolar. Hydrogen fluoride, HF, on the other hand, has a polar bond, so the molecule is polar. The bonding electrons spend more time near the fluorine atom so that there is a negative pole at that end and a positive pole at the hydrogen end. This is sometimes indicated by writing

$$H \xrightarrow{\ +\ } F$$

The arrow points toward the negative end of the polar bond (F atom); the plus sign is at the positive end (H atom). The HF molecule is called a **dipole;** it contains positive and negative poles.

If a molecule contains more than two atoms it is not so easy to decide whether it is polar or nonpolar. In this case, not only bond polarity but also molecular geometry determines the polarity of the molecule. To illustrate what is involved, consider the molecules shown in Figure 7.18.

1. In BeF_2 there are two polar Be —F bonds; in both bonds, the electron density is concentrated around the more electronegative fluorine atom. However, because the BeF_2 molecule is linear, the two Be $\xrightarrow{+}$ F dipoles are in opposite directions and cancel one another. The molecule has no net dipole and hence is nonpolar. From a slightly different point of view, in BeF_2 the centers of positive and negative charge coincide with each other at the Be atom. There is no way that a BeF_2 molecule can line up in an electric field.

2. Because oxygen is more electronegative than hydrogen (3.5 versus 2.2) an O—H bond is polar, with the electron density higher around the oxygen atom. In the bent H_2O molecule, the two H $\xrightarrow{+}$ O dipoles do not cancel each

Figure 7.18 Polarity of molecules.
All bonds in these molecules are polar, as shown by the $\longmapsto$ symbol, in which the arrow points to the more negative end of the bond and the $+$ indicates the more positive end. In BeF_2 and CCl_4 the bond dipoles cancel and the molecules are nonpolar. In H_2O and $CHCl_3$ the molecules are polar, with net dipoles shown by the broad arrows pointing toward the negative poles.

other. Instead, they add to give the H_2O molecule a net dipole. The center of negative charge is located at the O atom; this is the negative pole of the molecule. The center of positive charge is located midway between the two H atoms; the positive pole of the molecule is at that point. The H_2O molecule is polar. It tends to line up in an electric field with the oxygen atom oriented toward the positive electrode.

3. Carbon tetrachloride, CCl_4, is another molecule that, like BeF_2, is nonpolar despite the presence of polar bonds. Each of its four bonds is a dipole, C $\longmapsto$ Cl. However, because the four bonds are arranged symmetrically around the carbon atom, they cancel. As a result, the molecule has no net dipole; it is nonpolar. If one of the Cl atoms in CCl_4 is replaced by hydrogen, the situation changes. In the $CHCl_3$ molecule, the H $\longmapsto$ C dipole does not cancel with the three C $\longmapsto$ Cl dipoles. Hence $CHCl_3$ is polar.

There are two criteria for determining the polarity of a molecule: bond polarity and molecular geometry. *If the polar A—X bonds in a molecule AX_mE_n are arranged symmetrically around the central atom A, the molecule is nonpolar.*

EXAMPLE 7.7

Determine which of the following is polar: SO_2, CO_2, and $CHCl_3$.

STRATEGY

1. Write the Lewis structure.

2. Classify the molecule or ion as AX_mE_n.

3. Decide on the geometry (Table 7.3 or Figure 7.15).

4. Consider the A—X bonds and answer the following questions:

(a) Are the terminal atoms identical?

Yes; possibly nonpolar (depends on symmetry). No; polar

(b) Are the A—X bonds arranged symmetrically around the central atom?

No; polar, Yes; nonpolar if the answer to (a) is also yes.

continued

SOLUTION	
Lewis structure for SO_2	$:\ddot{O}-\ddot{S}=\ddot{O}:$
Species type	AX_2E
Geometry	bent
Identical terminal atoms?	yes
Symmetric A–X bonds?	no } polar
Lewis structure for CO_2	$:\ddot{O}=C=\ddot{O}:$
Species type	AX_2
Geometry	linear
Identical terminal atoms?	yes
Symmetric A–X bonds?	yes } nonpolar
Lewis structure for $CHCl_3$	$:\overset{\displaystyle :\ddot{C}l:}{\underset{\displaystyle :\ddot{C}l:}{\ddot{C}l-\overset{\displaystyle |}{\underset{\displaystyle |}{C}}-H}}$
Species type	AX_4
Geometry	tetrahedral
Identical terminal atoms?	no → polar

Generalizing from Example 7.7 and the preceding discussion, we can establish the following rules

■ Molecules of the type AX_2 (linear), AX_3 (trigonal planar), and AX_4 (tetrahedral) are nonpolar if the terminal atoms are identical. Examples: CO_2, BF_3, and CCl_4 are nonpolar; $CHCl_3$ is polar.
■ Molecules of the type AX_2E (bent), AX_2E_2 (bent), and AX_3E (trigonal pyramid) are polar. Examples: SO_2, H_2O, NH_3

EXAMPLE 7.8 CONCEPTUAL

For each of the species in column A, choose the description in column B that best applies.

A	B
CO_2	(a) polar, bent
CH_2Cl_2	(b) nonpolar, trigonal planar
XeF_2	(c) nonpolar, linear
BF_3	(d) nonpolar, trigonal pyramid
	(e) polar, tetrahedral
	(f) polar, trigonal pyramid

continued

STRATEGY
1. Note that the descriptions in column B are about geometry and polarity.
2. Draw the Lewis structures of the compounds in column A (Figure 7.3 or 7.9).
3. Determine series type and geometry (Table 7.3 or Figure 7.15).
4. Determine polarity.
5. Match your description with those given in column B.

SOLUTION	
2. Lewis structure for CO_2	$:\ddot{O}=C=\ddot{O}:$
3. species type → geometry	AX_2 → linear
4. polarity	nonpolar
5. match	c
2. Lewis structure for CH_2Cl_2	$:\ddot{Cl}-\overset{\overset{\textstyle H}{\textstyle \mid}}{\underset{\underset{\textstyle H}{\textstyle \mid}}{C}}-\ddot{Cl}:$
3. species type → geometry	AX_4 → tetrahedral
4. polarity	polar
5. match	e
2. Lewis structure for XeF_2	$:\ddot{F}-\ddot{Xe}-\ddot{F}:$
3. species type → geometry	AX_2E_3 → linear
4. polarity	nonpolar
5. match	c
2. Lewis structure for BF_3	$:\ddot{F}-\overset{\textstyle B}{\underset{\underset{\textstyle :\ddot{F}:}{\textstyle \mid}}{}}-\ddot{F}:$
3. species type → geometry	AX_3 → trigonal planar
4. polarity	nonpolar
5. match	b

7-4 Atomic Orbitals; Hybridization

In the 1930s a theoretical treatment of the covalent bond was developed by, among others, Linus Pauling (1901–1994), then at the California Institute of Technology. The *atomic orbital* or **valence bond model** won him the Nobel Prize in chemistry in 1954. Eight years later, Pauling won the Nobel Peace Prize for his efforts to stop nuclear testing.

According to this model, a covalent bond consists of a pair of electrons of opposed spin within an orbital. For example, a hydrogen atom forms a covalent bond by accepting an electron from another atom to complete its 1s orbital. Using orbital diagrams, we could write

$$1s$$

isolated H atom $(\uparrow)$

H atom in a stable molecule $(\uparrow\downarrow)$

The second electron, shown in blue, is contributed by another atom. This could be another H atom in H_2, an F atom in HF, a C atom in CH_4, and so on.

This simple model is readily extended to other atoms. The fluorine atom (electron configuration $1s^2 2s^2 2p^5$) has a half-filled p orbital:

$$\begin{array}{cccc} & \text{1s} & \text{2s} & \text{2p} \\ \text{isolated F atom} & (\uparrow\downarrow) & (\uparrow\downarrow) & (\uparrow\downarrow)(\uparrow\downarrow)(\uparrow\ ) \end{array}$$

By accepting an electron from another atom, F can complete this 2p orbital:

$$\begin{array}{cccc} & \text{1s} & \text{2s} & \text{2p} \\ \text{F atom in HF, F}_2, \ldots & (\uparrow\downarrow) & (\uparrow\downarrow) & (\uparrow\downarrow)(\uparrow\downarrow)(\uparrow\downarrow) \end{array}$$

According to this model, it would seem that for an atom to form a covalent bond, it must have an unpaired electron. Indeed, the number of bonds formed by an atom should be determined by its number of unpaired electrons. Because hydrogen has an unpaired electron, an H atom should form one covalent bond, as indeed it does. The same holds for the F atom, which forms only one bond. The noble-gas atoms He and Ne, which have no unpaired electrons, should not form bonds at all; they don't.

When this simple idea is extended beyond hydrogen, the halogens, and the noble gases, problems arise. Consider, for example, the three atoms Be ($Z = 4$), B ($Z = 5$), and C ($Z = 6$):

$$\begin{array}{cccc} & \text{1s} & \text{2s} & \text{2p} \\ \text{Be atom} & (\uparrow\downarrow) & (\uparrow\downarrow) & (\ \ )(\ \ )(\ \ ) \\ \text{B atom} & (\uparrow\downarrow) & (\uparrow\downarrow) & (\uparrow\ )(\ \ )(\ \ ) \\ \text{C atom} & (\uparrow\downarrow) & (\uparrow\downarrow) & (\uparrow\ )(\uparrow\ )(\ \ ) \end{array}$$

Notice that the beryllium atom has no unpaired electrons, the boron atom has one, and the carbon atom two. Simple valence bond theory would predict that Be, like He, should not form covalent bonds. A boron atom should form one bond, carbon two. Experience tells us that these predictions are wrong. Beryllium forms two bonds in BeF_2; boron forms three bonds in BF_3. Carbon ordinarily forms four bonds, not two.

To explain these and other discrepancies, simple valence bond theory must be modified. It is necessary to invoke a new kind of orbital, called a **hybrid orbital.**

Hybrid Orbitals: sp, sp², sp³, sp³d, sp³d²

The formation of the BeF_2 molecule can be explained by assuming that, as two fluorine atoms approach Be, the atomic orbitals of the beryllium atom undergo a significant change. **Hybridization** occurs when the 2s orbital is mixed or *hybridized* with a 2p orbital to form two new sp hybrid orbitals (Figure 7.19).

one s atomic orbital + one p atomic orbital $\longrightarrow$ two sp hybrid orbitals

In the BeF_2 molecule, there are two electron-pair bonds. These electron pairs are located in the two sp hybrid orbitals. In each orbital, one electron is a valence electron contributed by beryllium; the other electron comes from the fluorine atom.

A similar argument can be used to explain why boron forms three bonds and carbon forms four. In the case of boron:

one s atomic orbital + two p atomic orbitals $\longrightarrow$ three sp² ◄**ⓘ** hybrid orbitals

ⓘ The number of orbitals is shown by the superscript.

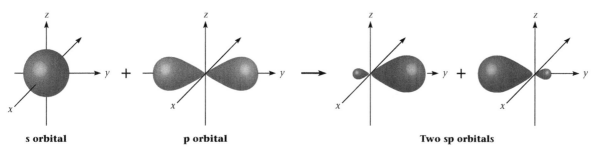

s orbital   **p orbital**   **Two sp orbitals**

Figure 7.19 Formation of sp hybrid orbitals. The mixing of one s orbital and one p orbital gives two sp hybrid orbitals.

Table 7.4 Hybrid Orbitals and Their Geometries

Number of Electron Pairs	Atomic Orbitals	Hybrid Orbitals	Orientation	Examples
2	s, p	sp	Linear	BeF_2, CO_2
3	s, two p	sp^2	Trigonal planar	BF_3, SO_3
4	s, three p	sp^3	Tetrahedron	CH_4, NH_3, H_2O
5	s, three p, d	sp^3d	Trigonal bipyramid	PCl_5, SF_4, ClF_3
6	s, three p, two d	sp^3d^2	Octahedron	SF_6, ClF_5, XeF_4

With carbon:

one s atomic orbital + three p atomic orbitals $\longrightarrow$ four sp^3 hybrid orbitals

You will recall that the bond angles in NH_3 and H_2O are very close to that in CH_4. This suggests that the four electron pairs surrounding the central atom in NH_3 and H_2O, like those in CH_4, occupy sp^3 hybrid orbitals. In NH_3, three of these orbitals are filled by bonding electrons, the other by the unshared pair on the nitrogen atom. In H_2O, two of the sp^3 orbitals of the oxygen atom contain bonding electron pairs; the other two contain unshared pairs. The situation in NH_3 and H_2O is not unique. In general, we find that *unshared as well as shared electron pairs can be located in hybrid orbitals.*

The extra electron pairs in an expanded octet are accommodated by using d orbitals. The phosphorus atom (five valence electrons) in PCl_5 and the sulfur atom (six valence electrons) in SF_6 make use of 3d as well as 3s and 3p orbitals:

With phosphorus:

one s orbital + three p orbitals + one d orbital $\longrightarrow$ five sp^3d hybrid orbitals

With sulfur:

one s orbital + three p orbitals + two d orbitals $\longrightarrow$ six sp^3d^2 hybrid orbitals

Table 7.4 summarizes all we have said about hybrid orbitals ◀ⓘ and also describes their geometry. Note that

ⓘ
Hybridization: sp (AX_2), sp^2 (AX_3, AX_2E), sp^3 (AX_4, AX_3E, AX_2E_2).

■ *the number of hybrid orbitals formed is always equal to the number of atomic orbitals mixed.*
■ the geometries, as found mathematically by quantum mechanics, are exactly as predicted by VSEPR theory. In each case, the hybrid orbitals are directed to be as far apart as possible.

EXAMPLE 7.9

Give the hybridization of carbon in CH_3Cl, phosphorus in PH_3, and sulfur in SF_4.

STRATEGY

1. Draw the Lewis structure of the molecules.
2. Determine the species type: AX_mE_n.
3. Count bonds (m) and unshared pairs (n) around the atom in question.
4. Hybridization:

m + n = 2 = sp; m + n = 3 = sp^2; m + n = 4 = sp^3; m + n = 5 = sp^3d; m + n = 6 = sp^3d^2

continued

	SOLUTION
1. Lewis structure for CH_3Cl	$\begin{array}{c} :\ddot{C}l: \\ \vert \\ H-C-H \\ \vert \\ H \end{array}$
2. species type	AX_4
3. $m + n$	$4 + 0 = 4$
4. hybridization	$m + n = 4 = sp^3$
1. Lewis structure for PH_3	$\begin{array}{c} H-\ddot{P}-H \\ \vert \\ H \end{array}$
2. species type	AX_3E
3. $m + n$	$3 + 1 = 4$
4. hybridization	$m + n = 4 = sp^3$
1. Lewis structure for SF_4	$\begin{array}{c} :\ddot{F}: \quad \ddot{F}: \\ \diagdown \diagup \\ S \\ \diagup \diagdown \\ :\ddot{F} \quad \ddot{F}: \end{array}$
2. species type	AX_4E
3. $m + n$	$4 + 1 = 5$
4. hybridization	$m + n = 5 = sp^3d$

Multiple Bonds

In Section 7-2, we saw that insofar as geometry is concerned, a multiple bond acts as if it were a single bond. In other words, the extra electron pairs in a double or triple bond have no effect on the geometry of the molecule. This behavior is related to hybridization. *The extra electron pairs in a multiple bond (one pair in a double bond, two pairs in a triple bond) are not located in hybrid orbitals.*

To illustrate this rule, consider the ethylene (C_2H_4) and acetylene (C_2H_2) molecules. You will recall that the bond angles in these molecules are 120° for ethylene and 180° for acetylene. This implies sp^2 hybridization in C_2H_4 and sp hybridization in C_2H_2 (see Table 7.4). Using blue lines to represent hybridized electron pairs,

$$\begin{array}{ccc} H & & H \\ \diagdown & & \diagup \\ & C=C & \\ \diagup & & \diagdown \\ H & & H \end{array} \qquad H-C\equiv C-H$$

ethylene   acetylene

In both cases, only one of the electron pairs in the multiple bond occupies a hybrid orbital.

EXAMPLE 7.10

State the hybridization of nitrogen in NH_3, NO_2^-, and N_2.

STRATEGY

1. Start by writing a Lewis structure for each species.

2. Determine the species type.

continued

3. The extra electron pairs in a multiple bond are not located in a hybrid orbital. Thus the number of terminal atoms (m) equals the number of bonds in hybrid orbitals.

4. Add m + n.

5. See Example 7.9 for hybridization based on the (m + n) count.

SOLUTION

1. Lewis structure for NH_3	H—N̈—H
	    |
	   H

2. Species type	AX_3E
3. m + n	3 + 1 = 4
4. hybridization	m + n = 4 = sp^3
1. Lewis structure for NO_2^-	$\left[\ddot{O} = \overset{\ddot{}}{N} \diagdown \ddot{O} \right]^-$
2. Species type	AX_2E
3. m + n	2 + 1 = 3
4. hybridization	m + n = 3 = sp^2
1. Lewis structure for N_2	:N≡N:
2. Species type	AXE
3. m + n	1 + 1 = 2
4. hybridization	m + n = 2 = sp

Sigma and Pi Bonds

We have noted that the extra electron pairs in a multiple bond are not hybridized and have no effect on molecular geometry. At this point, you may well wonder what happened to those electrons. Where are they in molecules like C_2H_4 and C_2H_2?

To answer this question, it is necessary to consider the shape or spatial distribution of the orbitals filled by bonding electrons in molecules. From this point of view, we can distinguish between two types of bonding orbitals. The first of these, and by far the more common, is called a *sigma* bonding orbital. It consists of a single lobe:

in which the electron density is concentrated in the region directly between the two bonded atoms, A and B. A **sigma bond** (σ) consists of an electron pair occupying a sigma bonding orbital.

The unhybridized electron pairs associated with multiple bonds occupy orbitals of a quite different shape, called *pi bonding orbitals*. Such an orbital consists of two lobes, ◄𝑖 one above the bond axis, the other below it.

Sigma and pi orbitals, like s and p, differ in shape; each orbital can hold $2e^-$.

Along the bond axis itself, the electron density is zero. The electron pair of a **pi bond** (π) occupies a pi bonding orbital. There is one π bond in the C_2H_4 molecule, two in

Figure 7.20 **Bonding orbitals in ethylene ($CH_2=CH_2$) and acetylene ($CH\equiv CH$).** The sigma bond backbones are shown in blue. The pi bonds (one in ethylene and two in acetylene) are shown in red. Note that a pi bonding orbital consists of two lobes.

C_2H_2. The geometries of the bonding orbitals in ethylene and acetylene are shown in Figure 7.20.

In general, to find the number of σ and π bonds in a species, remember that

- all single bonds are sigma bonds.
- one of the electron pairs in a multiple bond is a sigma bond; the others are pi bonds.

EXAMPLE 7.11

Give the number of pi and sigma bonds in NH_3, NO_2^-, and N_2.

STRATEGY

1. The Lewis structures for these species are given in Example 7.10.
2. Determine the species type. The number of sigma bonds is m.
3. Count total bonds in the Lewis structure.
4. pi bonds = total bonds − m

SOLUTION

For NH_3

2. species type	AX_3; m = 3 → 3 sigma (σ) bonds
3. number of bonds	3
4. number of pi (π) bonds	3 − m = 3 − 3 = 0 → no pi (π) bonds

For NO_2^-

2. species type	AX_2E; m = 2 → 2 sigma (σ) bonds
3. number of bonds	3
4. number of pi (π) bonds	3 − m = 3 − 2 = 1 → 1 pi (π) bond

For N_2

2. species type	AXE; m = 1 → 1 sigma (σ) bond
3. number of bonds	3
4. number of pi (π) bonds	3 − m = 3 − 1 = 2 → 2 pi (π) bonds

continued

END POINT

If the molecule does not have a defined central atom but instead has a long chain as in

$$
\begin{array}{c}
\text{H} \quad\ \ \text{O} \\
|\qquad\ \ \| \\
\text{H}-\text{C}-\text{C}-\text{C}\equiv\text{N} \\
| \\
\text{H}
\end{array}
$$

Count all single bonds (in this case: 5), and all multiple bonds (here, 1 double bond and 1 triple bond).

The multiple bonds each contribute a sigma bond. The rest of the multiple bonds are pi bonds. Thus this molecule has

sigma bonds: 5 + 1 (from the double bond) + 1 (from the triple bond) = 7

pi bonds: 1 (from the double bond) + 2 (from the triple bond) = 3

CHEMISTRY **BEYOND** THE CLASSROOM

The Noble Gases

The modern periodic table contains six relatively unreactive gases that were unknown to Mendeleev: the noble gases that make up Group **18** at the far right of the table. The first of these elements to be isolated was argon, which makes up about 0.9% of air. The physicist Lord Rayleigh (1842–1919) found that the density of atmospheric nitrogen, obtained by removing O_2, CO_2, and H_2O from air, was slightly greater than that of chemically pure N_2 (MM = 28.02 g/mol). Following up on that observation, Sir William Ramsay (1852–1916) separated argon (MM = 39.95 g/mol) from air. Over a three-year period between 1895 and 1898, this remarkable Scotsman, who never took a formal course in chemistry, isolated three more noble gases: Ne, Kr, and Xe. In effect, Ramsay added a whole new group to the periodic table.

Helium, the first member of the group, was detected in the spectrum of the Sun in 1868. Because of its low density ($\frac{1}{7}$ that of air), helium is used in all kinds of balloons and in synthetic atmospheres to make breathing easier for people suffering from emphysema.

Research laboratories use helium as a liquid coolant to achieve very low temperatures (bp He = −269°C). Argon and, more recently, krypton are used to provide an inert atmosphere in lightbulbs, thereby extending the life of the tungsten filament. In neon signs, a high voltage is passed through a glass tube containing neon at very low pressures. The red glow emitted corresponds to an intense line at 640 nm in the neon spectrum.

Until about 40 years ago, these elements were referred to as "inert gases"; they were believed to be entirely unreactive toward other substances. In 1962 Neil Bartlett (1932–2008), a 29-year-old chemist at the University of British Columbia, shook up the world of chemistry by preparing the first noble-gas compound. In the course of his research on platinum-fluorine compounds, he isolated a reddish solid that he showed to be $O_2{}^+(PtF_6{}^-)$. Bartlett realized that the ionization energy of Xe (1170 kJ/mol) is virtually identical to that of the O_2 molecule (1165 kJ/mol). This encouraged him to attempt to make the analogous compound $XePtF_6$. His success opened up a new era in noble-gas chemistry.

Figure A Crystals of xenon tetrafluoride (XeF_4).

Argonne National Laboratory

The most stable binary compounds of xenon are the three fluorides, XeF_2, XeF_4, and XeF_6. Xenon difluoride can be prepared quite simply by exposing a 1:1 mol mixture of xenon and fluorine to ultraviolet light; colorless crystals of XeF_2 (mp = 129°C) form slowly.

$$Xe(g) + F_2(g) \longrightarrow XeF_2(s)$$

The higher fluorides are prepared using excess fluorine (Figure A). All these compounds are stable in dry air at room temperature. However, they react with water to form compounds in which one or more of the fluorine atoms has been replaced by oxygen. Thus xenon hexafluoride reacts rapidly with water to give the trioxide

$$XeF_6(s) + 3H_2O(l) \longrightarrow XeO_3(s) + 6HF(g)$$

Xenon trioxide is highly unstable; it detonates if warmed above room temperature.

In the past 40 years, compounds have been isolated in which xenon is bonded to several nonmetals (N, C, and Cl) in addition to fluorine and oxygen. In the year 2000, it was reported (*Science*, Volume 290, page 117) that a compound had been isolated in which a metal atom was bonded to xenon. This compound is a dark red solid stable at temperatures below −40°C; it is believed to contain the $[AuXe_4]^{2+}$ cation.

The chemistry of xenon is much more extensive than that of any other noble gas. A group in Finland synthesized HArF, the first known compound of argon. Only one binary compound of krypton, KrF_2, has been prepared. It is a colorless solid that decomposes at room temperature. The chemistry of radon is difficult to study because all its isotopes are radioactive. Indeed, the radiation given off is so intense that it decomposes any reagent added to radon in an attempt to bring about a reaction.

Key Concepts

1. Draw Lewis structures for molecules and polyatomic ions.
   (**Examples 7.1, 7.2, 7.4; Problems 1–20, 67, 68**)
2. Write resonance forms.
   (**Example 7.3; Problems 21–26, 29, 30**)
3. Use Table 7.3 and Figure 7.15, applying the VSEPR, to predict molecular geometry.
   (**Examples 7.5, 7.6, 7.8; Problems 31–42**)
4. Knowing the geometry of a species, predict whether it will be polar.
   (**Examples 7.7, 7.8; Problems 43–48**)
5. State the hybridization of an atom in a bonded species.
   (**Examples 7.9, 7.10; Problems 49–62**)
6. State the number of sigma and pi bonds in a species.
   (**Example 7.11; Problems 63–66**)

Key Terms

bent molecule
bond
—bond angle
—covalent bond
—double bond
—nonpolar bond
—pi bond
—polar bond

—sigma bond
—single bond
—triple bond
dipole
electron-pair repulsion
expanded octets
formal charge
hybrid orbital

Lewis structure
linear molecule
molecular geometry
nonpolar molecule
octahedron
octet rule
polar molecule
resonance

tetrahedron
trigonal bipyramid
trigonal planar
trigonal pyramid
unshared electron pair
valence electrons
VSEPR model

Summary Problem

Consider the species SO_2, XeO_3, RnO_4, and XeF_2.

(a) Draw the Lewis structures of these species.

(b) Draw the resonance structures for SO_2.

(c) Describe the geometries of these species, including bond angles.

(d) Which species are nonpolar?

(e) What is the hybridization of the central atom for each species?

(f) How many sigma and pi bonds are in each species?

(g) SO_2 can have two Lewis structures. One has an expanded octet and two double bonds. The other has an octet with one double bond. Draw both Lewis structures, calculate the formal charges, and state which is the more likely structure based on formal charge alone.

Answers

(a) :Ö—S̈=Ö: :Ö—Xe—Ö:
                                      :Ö:

         :Ö:
          |
    :Ö—Rn—Ö: :F̈—Xe—F̈:
          |
         :Ö:

(b) :Ö—S̈=Ö: ⟷ :Ö=S̈—Ö:

185

(c) SO_2: bent, 120°
　　XeO_3: trigonal pyramid, 109.5°
　　RnO_4: tetrahedron, 109.5°
　　XeF_2: linear, 180°
(d) RnO_4, XeF_2
(e) sp^2 in SO_2;　　sp^3 in XeO_3;　　sp^3 in RnO_4;
　　sp^3d^2 in XeF_2

(f) SO_2: 2σ, 1π　　XeO_3: 3σ,
　　RnO_4: 4σ　　XeF_2: 4σ
(g) $:\ddot{O}=\ddot{S}=\ddot{O}:$　　$:\ddot{O}-\ddot{S}=\ddot{O}:$
　　The structure on the left is more likely based on formal charge alone.

Questions and Problems

Even-numbered questions and Challenge Problems have answers in Appendix 5 and fully worked solutions in the *Student Solutions Manual*.

7-1 Lewis Structures; The Octet Rule

Lewis Structures

1. Write the Lewis structures for the following molecules and polyatomic ions. In each case, the first atom is the central atom.
　　(a) CCl_4　(b) NCl_3　(c) $COCl_2$　(d) SO_3^{2-}
2. Follow the directions of Question 1 for
　　(a) NH_3　(b) KrF_2　(c) NO^+　(d) BrO_2^-
3. Follow the directions of Question 1 for
　　(a) IO_2^-　(b) SiF_4　(c) BrI_3　(d) CN^-
4. Follow the directions of Question 1 for
　　(a) ClF_4^-　(b) PF_6^-　(c) CNS^-　(d) $SnCl_5^-$
5. Follow the directions of Question 1 for
　　(a) OCl_2　(b) PF_3　(c) $SbCl_6^-$　(d) ICl_4^-
6. Follow the directions of Question 1 for
　　(a) C_2^{2-}　(b) NFO　(c) BrF_4^+　(d) NI_3
7. Oxalic acid, $H_2C_2O_4$, is a poisonous compound found in rhubarb leaves. Draw the Lewis structure for oxalic acid. There is a single bond between the two carbon atoms, each hydrogen atom is bonded to an oxygen atom, and each carbon is bonded to two oxygen atoms.
8. Radio astronomers have detected the isoformyl ion, HOC^+, in outer space. Write the Lewis structure for this ion.
9. Draw Lewis structures for the following species. (The skeleton is indicated by the way the molecule is written.)
　　(a) Cl_2CO　　(b) H_3C-CN　　(c) H_2C-CH_2
10. Follow the directions of Question 9 for the following species:
　　(a)
　　　　　　　O
　　　　　　　‖
　　　H_3C-C
　　　　　　　|
　　　　　　　O—H
　　　(b) Br_2C-CF_2　　(c) H_3C-NH_2
11. Dinitrogen pentoxide, N_2O_5, when bubbled into water can form nitric acid. Its skeleton structure has no N—N or O—O bonds. Write its Lewis structure.
12. Formic acid is the irritating substance that gets on your skin when an ant bites. Its formula is HCOOH. Carbon is the central atom, and it has no O—O bonds. Draw its Lewis structure.
13. Two different molecules have the formula $C_2H_2Cl_2$. Draw a Lewis structure for each molecule. (All the H and Cl atoms are bonded to carbon. The two carbon atoms are bonded to each other.)

14. Two different molecules have the formula C_2H_6O. One of the molecules has the oxygen atom bonded to both carbon atoms. The other molecule has the oxygen atom bonded to only one carbon atom while both carbon atoms are bonded to each other. Write Lewis structures for both of these compounds.
15. Give the formula of a polyatomic ion that you would expect to have the same Lewis structure as
　　(a) Cl_2　　(b) H_2SO_4　　(c) CH_4　　(d) $GeCl_4$
16. Give the formula for a molecule that you would expect to have the same Lewis structure as
　　(a) BrO^-　　(b) NH_4^+　　(c) CN^-　　(d) SO_4^{2-}
17. Write a Lewis structure for
　　(a) XeF_3^+　　(b) PCl_4^+
　　(c) BrF_5　　(d) HPO_4^{2-} (no P—H or O—O bonds)
18. Write a Lewis structure for
　　(a) BCl_4　　(b) ClO^-　　(c) $S_2O_3^{2-}$　　(d) $NFCl_2$
19. Write reasonable Lewis structures for the following species, none of which follow the octet rule.
　　(a) BF_3　　(b) NO　　(c) CO^+　　(d) ClO_3
20. Write reasonable Lewis structures for the following species, none of which follow the octet rule.
　　(a) $BeCl_2$　　(b) SeO_2^-　　(c) ClO_3　　(d) CH_3

Resonance Forms and Formal Charge

21. Draw possible resonance structures for
　　(a) $Cl-NO_2$　　(b) H_2C-N-N　　(c) SO_3
22. Draw resonance structures for
　　(a) SeO_3　　(b) CS_3^{2-}　　(c) CNO^-
23. The Lewis structure for hydrazoic acid may be written as

　　　　　　　H
　　　　　　　|
　　　　$:N=N=\ddot{N}:$

(a) Draw two other possible resonance forms for this molecule.
(b) Is

　　　　　　　　$\ddot{N}$
　　　　　　　⟋　⟍
　　　$:N$————$\ddot{N}-H$

another form of hydrazoic acid? Explain.
24. The oxalate ion, $C_2O_4^{2-}$, has the skeleton structure

　　　O　　　　　O
　　　　⟍　　　⟋
　　　　　C—C
　　　⟋　　　⟍
　　　O　　　　　O

(a) Complete the Lewis structure of this ion.
(b) Draw three possible resonance forms for $C_2O_4^{2-}$, equivalent to the Lewis structure drawn in (a).

(c) Is

$$\left[\ddot{O}=\ddot{C}-\ddot{O}-C-\ddot{O}: \atop \quad\quad\quad \underset{:O:}{\|} \right]^{2-}$$

a resonance form of the oxalate ion?

25. The skeleton structure for disulfur dinitride, S_2N_2, is

Draw possible resonance forms of this molecule.

26. Borazine, $B_3N_3H_6$, has the skeleton

Draw the resonance forms of the molecule.

27. What is the formal charge on the indicated atom in each of the following species?
  (a) sulfur in SO_2
  (b) nitrogen in N_2H_4
  (c) each oxygen atom in ozone, O_3

28. Follow the directions in Question 27 for
  (a) oxygen in HOF   **(b)** nitrogen in NO_2^-
  (c) phosphorus in PCl_3

29. Below are two different Lewis structures for nitrous acid (HNO_2). Which is the better Lewis structure based only on formal charge?

Structure I   Structure II

$$H-\ddot{O}-\ddot{N}=\ddot{O}: \qquad H-N=\ddot{O}: \atop \qquad\qquad\qquad\qquad \underset{:\ddot{O}:}{|}$$

30. Below are two different Lewis structures for the thiosulfate ion ($S_2O_3^{2-}$). Which is the better Lewis structure based only on formal charge?

Structure I   Structure II

$$\left(:\ddot{O}-\ddot{S}-\ddot{S}-\ddot{O}: \atop \quad\quad\underset{:\ddot{O}:}{|}\right)^{2-} \qquad \left(\overset{:\ddot{O}:}{\underset{:\ddot{O}:}{\overset{|}{:\ddot{S}-S-\ddot{O}:}}}\right)^{2-}$$

7-2 Molecular Geometry

31. Predict the geometry of the following species:
  (a) SO_2 **(b)** BeF_2 **(c)** $SeCl_4$ **(d)** PCl_5

32. Predict the geometry of the following species:
  (a) O_3 **(b)** OCl_2 **(c)** $SnCl_3^-$ **(d)** CS_2

33. Predict the geometry of the following species:
  (a) $SeOF_2$ **(b)** TeF_5^- **(c)** ICl_4^- **(d)** RnF_2

34. Predict the geometry of the following species:
  (a) NNO   **(b)** ONCl   **(c)** NH_4^+ **(d)** O_3

35. Predict the geometry of the following species:
  (a) ClF_2^- **(b)** SeF_5Br **(c)** SO_3^{2-} **(d)** BrO_2^-

36. Predict the geometry of the following species:
  (a) ClF_5 **(b)** XeF_4 **(c)** SiF_6^{2-} **(d)** PCl_5

37. Give all the ideal bond angles (109.5°, 120°, or 180°) in the following molecules and ions. (The skeleton does not imply geometry.)
  (a) Cl—S—Cl
  (b) F—Xe—F

  (d)

38. Follow the instructions in Question 37 for
  (a) NO_2^- **(b)** BF_3

39. Peroxypropionyl nitrate (PPN) is an eye irritant found in smog. Its skeleton structure is

  (a) Draw the Lewis structure of PPN.
  (b) Indicate all the bond angles.

40. An objectionable component of smog is acetyl peroxide, which has the following skeleton structure.

  (a) Draw the Lewis structure of this compound.
  (b) Write the bond angles indicated by the numbered angles.

41. The uracil molecule is one of the bases in DNA. Estimate the approximate values of the indicated bond angles. Its skeleton (not its Lewis structure) is given below.

42. Niacin is one of the B vitamins (B_3). Estimate the approximate values of the indicated bond angles. Its skeleton (not its Lewis structure) is given below.

7-3 Polarity of Molecules

43. Which of the species with octets in Question 31 are dipoles?

44. Which of the species with octets in Question 32 are dipoles?

45. Which of the species with octets in Question 33 are dipoles?

46. Which of the species with octets in Question 34 are dipoles?

47. There are three compounds with the formula $C_2H_2Cl_2$:

Which of these molecules are polar?

48. There are two different molecules with the formula N_2F_2:

Is either molecule polar? Explain.

7-4 Atomic Orbitals; Hybridization

Hybridization

49. Give the hybridization of the central atom in each species in Question 31.

50. Give the hybridization of the central atom in each species in Question 32.

51. Give the hybridization of the central atom in each species in Question 33.

52. Give the hybridization of the central atom in each species in Question 34.

53. Give the hybridization of the central atom in each species in Question 35.

54. Give the hybridization of the central atom in each species in Question 36.

55. In each of the following polyatomic ions, the central atom has an expanded octet. Determine the number of electron pairs around the central atom and the hybridization in

   (a) SF_2^{2-} (b) $AsCl_6^-$ (c) SCl_4^{2-}

56. Follow the directions of Question 55 for the following polyatomic ions.

   (a) ClF_4^- (b) $GeCl_6^{2-}$ (c) $SbCl_4^-$

57. Give the hybridization of each atom (except H) in the solvent dimethylsulfoxide. (Unshared electron pairs are not shown.)

$$H_3C-S-CH_3$$
$$\overset{\|}{O}$$

58. Acrylonitrile, C_3H_3N, is the building block of the polymer Orlon. Its Lewis structure is

$$H-\underset{①}{C}=\underset{②}{C}-\underset{③}{C}\equiv N:$$

(with H H above the first two carbons)

What is the hybridization of nitrogen and of the three numbered carbon atoms?

59. What is the hybridization of nitrogen in

   (a) $\left[O-\underset{\overset{\|}{O}}{N}-O\right]^-$ (b) $H-\overset{..}{N}-H$ (with Cl below)

   (c) $\overset{..}{N}\equiv\overset{..}{N}$ (d) $\overset{..}{N}\equiv N-O$

60. What is the hybridization of carbon in

   (a) CH_3Cl (b) $\left[O-\underset{\overset{\|}{O}}{C}-O\right]^{2-}$

   (c) $O=C=O$ (d) $H-\underset{\overset{\|}{O}}{C}-OH$

61. Give the hybridization of the central atom (underlined in red).

   (a) $\underline{C}OCl_2$ (b) $H\underline{N}O_2$ (c) $(CH_3)_2\underline{C}HCH_3$

62. Give the hybridization of the central atom (underlined in red).

   (a) $HO\underline{Br}O_3$ (b) $Cl_2\underline{O}$ (c) $O\underline{P}Br_3$

Sigma and Pi Bonds

63. Give the number of sigma and pi bonds in the molecule in Question 57.

64. Give the number of sigma and pi bonds in the molecule in Question 58.

65. Give the number of sigma and pi bonds in each species in Question 59.

66. Give the number of sigma and pi bonds in each species in Question 60.

Unclassified

67. In which of the following molecules does the sulfur have an expanded octet? For those that do, write the Lewis structure.

   (a) SO_2 (b) SF_4 (c) SO_2Cl_2 (d) SF_6

68. Consider the pyrosulfate ion, $S_2O_7^{2-}$. It has no sulfur–sulfur nor oxygen–oxygen bonds.

   (a) Write a Lewis structure for the pyrosulfate ion using only single bonds.

   (b) What is the formal charge on the sulfur atoms for the Lewis structure you drew in part (a)?

   (c) Write another Lewis structure using six $S=O$ bonds and two $O—S$ bonds.

   (d) What is the formal charge on each atom for the structure you drew in part (c)?

69. Consider acetyl salicylic acid, better known as aspirin. Its structure is

(structure diagram of aspirin)

   (a) How many sigma and pi bonds are there in aspirin?

   (b) What are the approximate values of the angles marked (in blue) A, B, and C?

   (c) What is the hybridization of each atom marked (in red) 1, 2, and 3?

70. Complete the table on next page.

Species	Atoms Around Central Atom A	Unshared Pairs Around A	Geometry	Hybridization	Polarity (Assume all X atoms the same)
AX_2E_2	_____	_____	_____	_____	_____
_____	3	0	_____	_____	_____
AX_4E_2	_____	_____	_____	_____	_____
_____	_____	_____	trigonal bipyramid	_____	_____

Conceptual Questions

71. Given the following electronegativities

$$C = 2.5 \qquad N = 3.0 \qquad S = 2.6$$

what is the central atom in CNS^-?

72. Based on the concept of formal charge, what is the central atom in

(a) HCN (do not include H as a possibility)?

(b) NOCl (Cl is always a terminal atom)?

73. Describe the geometry of the species in which there are, around the central atom,

(a) four single bonds, two unshared pairs of electrons.

(b) five single bonds.

(c) two single bonds, one unshared pair of electrons.

(d) three single bonds, two unshared pairs of electrons.

(e) two single bonds, two unshared pairs of electrons.

(f) five single bonds, one unshared pair of electrons.

74. Consider the following molecules: SiH_4, PH_3, H_2S. In each case, a central atom is surrounded by four electron pairs. In which of these molecules would you expect the bond angle to be less than $109.5°$? Explain your reasoning.

75. Give the formula of an ion or molecule in which an atom of

(a) N forms three bonds using sp^3 hybrid orbitals.

(b) N forms two pi bonds and one sigma bond.

(c) O forms one sigma and one pi bond.

(d) C forms four bonds in three of which it uses sp^2 hybrid orbitals.

(e) Xe forms two bonds using sp^3d^2 hybrid orbitals.

76. In each of the following molecules, a central atom is surrounded by a total of three atoms or unshared electron pairs: $SnCl_2$, BCl_3, SO_2. In which of these molecules would you expect the bond angle to be less than $120°$? Explain your reasoning.

77. Explain the meaning of the following terms.

(a) expanded octet   (b) resonance

(c) unshared electron pair   (d) odd-electron species

Challenge Problems

78. A compound of chlorine and fluorine, ClF_x, reacts at about 75°C with uranium to produce uranium hexafluoride and chlorine fluoride, ClF. A certain amount of uranium produced 5.63 g of uranium hexafluoride and 457 mL of chlorine fluoride at 75°C and 3.00 atm. What is x? Describe the geometry, polarity, and bond angles of the compound and the hybridization of chlorine. How many sigma and pi bonds are there?

79. Draw the Lewis structure and describe the geometry of the hydrazine molecule, N_2H_4. Would you expect this molecule to be polar?

80. Consider the polyatomic ion IO_6^{5-}. How many pairs of electrons are around the central iodine atom? What is its hybridization? Describe the geometry of the ion.

81. It is possible to write a simple Lewis structure for the SO_4^{2-} ion, involving only single bonds, which follows the octet rule. However, Linus Pauling and others have suggested an alternative structure, involving double bonds, in which the sulfur atom is surrounded by six electron pairs.

(a) Draw the two Lewis structures.

(b) What geometries are predicted for the two structures?

(c) What is the hybridization of sulfur in each case?

(d) What are the formal charges of the atoms in the two structures?

82. Phosphoryl chloride, $POCl_3$, has the skeleton structure

$$\begin{array}{c} O \\ | \\ Cl-P-Cl \\ | \\ Cl \end{array}$$

Write

(a) a Lewis structure for $POCl_3$ following the octet rule. Calculate the formal charges in this structure.

(b) a Lewis structure in which all the formal charges are zero. (The octet rule need not be followed.)

Thermochemistry

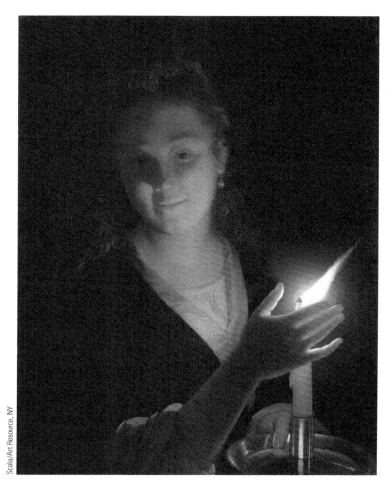

The candle flame gives off heat, melting the candle wax. Wax melting is a phase change from solid to liquid and an endothermic reaction.

Scala/Art Resource, NY

Some say the world will end in fire,
Some say in ice.
From what I've tasted of desire
I hold with those who favor fire.

—ROBERT FROST
Fire and Ice

This chapter deals with energy and heat, two terms used widely by both the general public and scientists. Energy, in the vernacular, is equated with pep and vitality. Heat conjures images of blast furnaces and sweltering summer days. Scientifically, these terms have quite different meanings. **Energy** can be defined as the capacity to do work. **Heat** is a particular form of energy that is transferred from a body at a high temperature to one at a lower temperature when they are brought into contact with each other. Two centuries ago, heat was believed to be a material fluid (caloric); we still use the phrase "heat flow" to refer to heat transfer or to heat effects in general.

Thermochemistry refers to the study of the heat flow that accompanies chemical reactions. Our discussion of this subject will focus on

- the basic principles of heat flow (Section 8-1).
- the experimental measurement of the magnitude and direction of heat flow, known as *calorimetry* (Section 8-2).
- the concept of enthalpy, H (heat content) and *enthalpy change*, ΔH (Section 8-3).
- the calculation of ΔH for reactions, using *thermochemical equations* (Section 8-4) and *enthalpies of formation* (Section 8-5).
- heat effects in the breaking and formation of covalent bonds (Section 8-6).
- the relation between heat and other forms of energy, as expressed by the first law of thermodynamics (Section 8-7).

Principles of Heat Flow

In any discussion of heat flow, it is important to distinguish between system and surroundings. The **system** is that part of the universe on which attention is focused. In a very simple case (Figure 8.1), it might be a sample of water in contact with a hot plate. The **surroundings,** which exchange energy with the system, make up in principle the rest of the universe. For all practical purposes, however, they include only those materials in close contact with the system. In Figure 8.1, the surroundings would consist of the hot plate, the beaker holding the water sample, and the air around it.

When a chemical reaction takes place, we consider the substances involved, reactants and products, to be the system. The surroundings include the vessel in which the reaction takes place (test tube, beaker, and so on) and the air or other material in thermal contact with the reaction system.

State Properties

The **state** of a system is described by giving its composition, temperature, and pressure. The system at the left of Figure 8.1 consists of

$$50.0 \text{ g of } H_2O(l) \text{ at } 50.0°C \text{ and } 1 \text{ atm}$$

When this system is heated, its state changes, perhaps to one described as

$$50.0 \text{ g of } H_2O(l) \text{ at } 80.0°C \text{ and } 1 \text{ atm}$$

Certain quantities, called **state properties,** depend only on the state of the system, not on the way the system reached that state. Putting it another way, if X is a state property, then

$$\Delta X = X_{final} - X_{initial}$$

That is, the change in X is the difference between its values in final and initial states. Most of the quantities that you are familiar with are state properties; volume is a common example. You may be surprised to learn, however, that heat flow is *not* a state property; its magnitude depends on how a process is carried out (Section 8-7).

The universe is a big place.

The distance between two cities depends on path, so it isn't a state property.

Figure 8.1 A system and its surroundings.

When the system (50.0 g of H_2O) absorbs heat from the surroundings (hot plate), its temperature increases from 50.0°C to 80.0°C.

50.0 g H_2O
50.0°C → 80.0°C

OFF ON

When the hot plate is turned off, the system gives off heat to the surrounding air, and its temperature drops.

50.0 g H_2O
80.0°C → 50.0°C

OFF ON

Direction and Sign of Heat Flow

Consider again the setup in Figure 8.1. If the hot plate is turned on, there is a flow of heat from the surroundings into the system, 50.0 g of water. This situation is described by stating that the **heat flow,** q, for the system is a positive quantity. ◄*i*

q is positive when heat flows into the system from the surroundings.

Usually, when heat flows into a system, its temperature rises. In this case, the temperature of the 50.0-g water sample might increase from 50.0°C to 80.0°C. When the hot plate in Figure 8.1 is shut off, the hot water gives off heat to the surrounding air. In this case, q for the system is a negative quantity. ◄*i*

q is negative when heat flows out of the system into the surroundings.

As is usually the case, the temperature of the system drops when heat flows out of it into the surroundings. The 50.0-g water sample might cool from 80.0°C back to 50.0°C.

 This same reasoning can be applied to a reaction in which the system consists of the reaction mixture (products and reactants). We can distinguish between

- *an* **endothermic** *process* ($q > 0$), *in which heat flows from the surroundings into the reaction system.* An example is the melting of ice:

$$H_2O(s) \longrightarrow H_2O(l) \qquad q > 0$$

The melting of ice *absorbs* heat from the surroundings (Figure 8.2a), which might be the water in a glass of iced tea. The temperature of the surroundings drops, perhaps from 25°C to 3°C, as they give up heat to the system.

- *an* **exothermic** *process* ($q < 0$), *in which heat flows from the reaction system into the surroundings.* A familiar example is the combustion of methane, the major component of natural gas.

$$CH_4(g) + 2O_2(g) \longrightarrow CO_2(g) + 2H_2O(l) \qquad q < 0$$

This reaction *evolves* heat to the surroundings, which might be the air around a Bunsen burner in the laboratory (Figure 8.2b) or a potato being baked in a gas oven. In either case, the effect of the heat transfer is to raise the temperature of the surroundings. ◄*i*

Magnitude of Heat Flow

In any process, we are interested not only in the direction of heat flow but also in its magnitude. We will express q in the units introduced in Chapter 6, the **joule** ◄*i* and the *kilojoule*. The joule is named for James Joule (1818–1889), who carried out very precise thermometric measurements that established the first law of thermodynamics (Section 8-7).

 In the past, chemists used the calorie* as an energy unit. This is the amount of heat required to raise the temperature of one gram of water one degree Celsius. The calorie is a larger energy unit than the joule:

$$1 \text{ cal} = 4.184 \text{ J} \qquad 1 \text{ kcal} = 4.184 \text{ kJ}$$

 Most of the remainder of this chapter is devoted to a discussion of the magnitude of the heat flow in chemical reactions or phase changes. However, we will focus on a simpler process in which the only effect of the heat flow is to change the temperature of a system. In general, the relationship between the magnitude of the heat flow, q, and the temperature change, Δt, is given by the equation

$$q = C \times \Delta t \qquad (\Delta t = t_{final} - t_{initial})$$

i When money flows into your checking account, it is a positive quantity. When money flows out, it is a negative quantity.

i endo = into system; exo = out of system.

i 1 J = 1 kg · m²/s² in SI units

---

*The "calorie" referred to by nutritionists is actually a kilocalorie (1 kcal = 10^3 cal). On a 2000-calorie per day diet, you eat food capable of producing 2000 kcal = 2×10^3 kcal = 2×10^6 cal of energy.

Figure 8.2 (a) Endothermic and (b) exothermic processes.

Figure 8.3 **Substances with varying specific heats**. The metals shown (Cu, Al, and Fe) have specific heats ranging from 0.38 J/g · °C (Cu) to 0.90 J/g · °C (Al). This explains why water warms up and cools down more slowly than a container made of aluminum, copper, or iron.

The quantity C appearing in this equation is known as the **heat capacity** of the system. It represents the amount of heat required to raise the temperature of the system 1°C and has the units J/°C.

For a pure substance of a certain mass, the expression for q can be written

$$q = \text{mass} \times c \times \Delta t \tag{8.1}$$

The quantity c is called the **specific heat** or *specific heat capacity*. (In this text, we will use the term *specific heat*.) Specific heat is defined as the amount of heat required to raise the temperature of one gram of a substance one degree Celsius. When the mass of that substance is equal to its molar mass, then c is called the *molar heat capacity*.

Specific heat, like density or melting point, is an intensive property that can be used to identify a substance or determine its purity. Water has an unusually large specific heat, 4.18 J/g·°C. This explains why swimming is not a popular pastime in northern Minnesota in May. Even if the air temperature rises to 90°F, the water temperature will remain below 60°F. Metals have a relatively low specific heat (Table 8.1). When you heat water in a stainless steel saucepan, for example, nearly all of the heat is absorbed by the water, very little by the steel (Figure 8.3).

Table 8.1 Specific Heats of a Few Common Substances

	c (J/g · °C)		c (J/g · °C)
$Br_2(l)$	0.474	$Cu(s)$	0.382
$Cl_2(g)$	0.478	$Fe(s)$	0.446
$C_2H_5OH(l)$	2.43	$H_2O(g)$	1.87
$C_6H_6(l)$	1.72	$H_2O(l)$	4.18
$CO_2(g)$	0.843	$NaCl(s)$	0.866

EXAMPLE 8.1

Compare the amount of heat given off by 1.40 mol of liquid water when it cools from 100.0°C to 30.0°C to that given off when 1.40 mol of steam cools from 200.0°C to 110.0°C.

ANALYSIS

Information given:	$H_2O(l)$: mols (1.40), t_{final} (30.0°C), $t_{initial}$ (100.0°C) $H_2O(g)$: mols (1.40), t_{final} (110.0°C), $t_{initial}$ (200.0°C)
Information implied:	molar mass of water and steam specific heats of water and steam
Asked for:	q for both water and steam

STRATEGY

1. Recall that $\Delta t = t_{final} - t_{initial}$.

2. Convert mols to mass (in grams).

3. Use Table 8.1 to obtain the specific heats of water and steam.

4. Substitute into Equation 8.1

SOLUTION

For $H_2O(l)$:

1. Δt	$\Delta t = t_{final} - t_{initial} = 30.0°C - 100.0°C = -70.0°C$
2. mass	$1.40 \text{ mol} \times \dfrac{18.02 \text{ g}}{1 \text{ mol}} = 25.2 \text{ g}$
3. c	From Table 8.1, $c = 4.18$ J/g · °C.
4. q	$q = \text{mass} \times \Delta t \times c = (25.2 \text{ g})(4.18 \text{ J/g} \cdot °C)(-70.0°C) = -7.37 \times 10^3 \text{ J}$

For $H_2O(g)$:

1. Δt	$\Delta t = t_{final} - t_{initial} = 110.0°C - 200.0°C = -90.0°C$
2. mass	$1.40 \text{ mol} \times \dfrac{18.02 \text{ g}}{1 \text{ mol}} = 25.2 \text{ g}$
3. c	From Table 8.1, $c = 1.87$ J/g · °C.
4. q	$q = \text{mass} \times \Delta t \times c = (25.2 \text{ g})(1.87 \text{ J/g} \cdot °C)(-90.0°C) = -4.24 \times 10^3 \text{ J}$

END POINTS

1. The negative sign indicates that heat flows from the system (water and steam) to the surroundings.

2. Be careful when deciding on initial and final temperatures. The higher temperature is not necessarily the final temperature.

Measurement of Heat Flow; Calorimetry

To measure the heat flow in a reaction, a device known as a **calorimeter** is used. The apparatus contains water and/or other materials of known heat capacity. The walls of the calorimeter are insulated so that there is no exchange of heat with the surrounding air. It follows that the only heat flow is between the reaction system and the calorimeter. The heat flow for the reaction system is equal in magnitude but opposite in sign to that of the calorimeter:

$$q_{reaction} = -q_{calorimeter}$$

Notice that if the reaction is exothermic ($q_{reaction} < 0$), $q_{calorimeter}$ must be positive; that is, heat flows from the reaction mixture into the calorimeter. Conversely, if the reaction is endothermic, the calorimeter gives up heat to the reaction mixture.

The equation just written is basic to calorimetric measurements. It allows you to calculate the amount of heat absorbed or evolved in a reaction if you know the heat capacity, C_{cal}, and the temperature change, Δt, of the calorimeter.

$$q_{cal} = C_{cal} \times \Delta t$$

Coffee-Cup Calorimeter

Figure 8.4 shows a simple **coffee-cup calorimeter** used in the general chemistry laboratory. It consists of two nested polystyrene foam cups partially filled with water. The cups have a tightly fitting cover through which an accurate thermometer is inserted. Because polystyrene foam is a good insulator, there is very little heat flow through the walls of the cups. Essentially all the heat evolved by a reaction taking place within the calorimeter is absorbed by the water. This means that, to a good degree of approximation, the heat capacity of the coffee-cup calorimeter is that of the water:

$$C_{cal} = mass_{H_2O} \times c_{water} = mass_{water} \times 4.18 \ \frac{J}{g \cdot °C}$$

and hence

$$q_{reaction} = -mass_{H_2O} \times 4.18 \ \frac{J}{g \cdot °C} \times \Delta t \qquad (8.2)$$

Figure 8.4 Coffee-cup calorimeter. The heat given off by the sample is absorbed by the water. If you know the mass of the water, its specific heat (4.18 J/g · °C), and the temperature change as read on the thermometer, you can calculate the heat flow, q, for the sample.

EXAMPLE	8.2	GRADED

Calcium chloride, $CaCl_2$, is added to canned vegetables to maintain the vegetables' firmness. When added to water, it dissolves:

$$CaCl_2(s) \longrightarrow Ca^{2+}(aq) + 2Cl^-(aq)$$

A calorimeter contains 50.0 g of water at 25.00°C. When 1.00 g of calcium chloride is added to the calorimeter, the temperature rises to 28.51°C. Assume that all the heat given off by the reaction is transferred to the water.

a	Calculate q for the reaction system.

ANALYSIS	
Information given:	mass of water (50.0 g); mass of $CaCl_2$ (1.00 g) initial temperature (25.00°C); final temperature (28.51°C)
Information implied:	specific heat of water
Asked for:	$q_{reaction}$

continued

STRATEGY	
1. Find Δt.	
2. Substitute into Equation 8.1 to find q_{H_2O}	
3. Recall that q reaction $= -q_{H_2O}$.	

SOLUTION	
1. Δt	$\Delta t = t_{final} - t_{initial} = 28.51°C - 25.00°C = 3.51°C$
2. q_{H_2O}	$q_{H_2O} = \text{mass} \times \Delta t \times c = (50.0\ g)(4.18\ J/g \cdot °C)(3.51°C) = 734\ J$
3. $q_{reaction}$	$q_{reaction} = -q_{H_2O} = -734\ J$

b	How much $CaCl_2$ must be added to raise the temperature of the solution 9.00°C?

ANALYSIS	
Information given:	mass of water (50.0 g) From part (a), $q_{reaction}$ for 1.00 g of $CaCl_2$ used (-734 J). Δt (9.00°C)
Information implied:	specific heat of water
Asked for:	mass $CaCl_2$ to be added

STRATEGY	
1. Find q_{H_2O} by substituting into Equation 8.1.	
2. Recall $q_{reaction} = -q_{H_2O}$	
3. Use -734 J/g $CaCl_2$ obtained in part (a) as a conversion factor.	

SOLUTION	
1. q_{H_2O}	$q_{H_2O} = \text{mass} \times \Delta t \times c = (50.0\ g)(4.18\ J/g \cdot °C)(9.00°C) = 1.88 \times 10^3\ J$
2. $q_{reaction}$	$q_{reaction} = -q_{H_2O} = -1.88 \times 10^3\ J$
3. Mass $CaCl_2$ needed	$-1.88 \times 10^3\ J \times \dfrac{1.00\ g\ CaCl_2}{-734\ J} = 2.56\ g$
Mass $CaCl_2$ to be added	$2.56 - 1.00 = 1.56\ g$

END POINT
Since the final temperature is larger than the initial temperature after the addition of $CaCl_2$, the reaction must be exothermic. Thus $q_{reaction}$ must be negative. It is!

Bomb Calorimeter

A coffee-cup calorimeter is suitable for measuring heat flow for reactions in solution. However, it cannot be used for reactions involving gases, which would escape from the cup. Neither would it be appropriate for reactions in which the products reach high temperatures. The **bomb calorimeter,** shown in Figure 8.5, is more versatile. To use it, a weighed sample of the reactant(s) is added to the heavy-walled metal vessel called a "bomb." This is then sealed and lowered into the insulated outer container. An amount of water sufficient to cover the bomb is added, and the entire apparatus is closed. The initial temperature is measured precisely. The reaction is then started, perhaps by electrical ignition. In an exothermic reaction, the hot products give off heat to the walls of the bomb and to the water. The final temperature is taken to be the highest value read on the thermometer.

Figure 8.5 **Bomb calorimeter.** The heat flow, q, for the reaction is calculated from the temperature change multiplied by the heat capacity of the calorimeter, which is determined in a preliminary experiment.

All of the heat given off by the reaction is absorbed by the calorimeter, which consists of the metal bomb and the water that surrounds it. In other words,

$$q_{reaction} = -(q_{calorimeter} + q_{H_2O})$$

As discussed earlier, the heat absorbed by the water can be determined by substituting into Equation 8.1:

$$q_{H_2O} = c \times mass_{H_2O} \times \Delta t$$

The heat absorbed by the calorimeter is equal to the product of its heat capacity (unique to each calorimeter), C_{cal}, and the temperature change, Δt:

$$q_{calorimeter} = C_{cal} \times \Delta t \qquad (8.3)$$

To find the heat capacity, C_{cal}, the experiment is repeated using the same bomb and the same amount of water. This time, though, we carry out a reaction for which the amount of heat evolved is known, 93.3 kJ. The temperature increase is again measured carefully. Suppose that the water in the calorimeter can has a mass of 1.00 kg (1000 g) and the temperature rises from 20.00°C to 30.00°C, then q for the water would be 41.8 kJ and q for the calorimeter would then be

$$93.3 \text{ kJ} - 41.8 \text{ kJ} = 51.5 \text{ kJ}$$

Substituting into Equation 8.3 we obtain the heat capacity of the calorimeter.

$$C_{cal} = \frac{51.5 \text{ kJ}}{10.00°C} = 5.15 \text{ kJ/°C}$$

Knowing the heat capacity of the calorimeter, the heat flow for any reaction taking place in that calorimeter can be calculated (Example 8.3).

EXAMPLE 8.3

Hydrogen chloride is used in etching semiconductors. It can be prepared by reacting hydrogen and chlorine gases.

$$H_2(g) + Cl_2(g) \longrightarrow 2HCl(g)$$

It is found that when 1.00 g of H_2 is made to react completely with Cl_2 in a bomb calorimeter with a heat capacity of 5.15 kJ/°C, the temperature in the bomb rises from 20.00°C to 29.82°C. The calorimeter can hold 1.000 kg of water. How much heat is evolved by the reaction?

ANALYSIS

Information given:	mass of H_2 (1.00 g); mass of water (1.000 kg = 1.000×10^3 g)
	C_{cal} (5.15 kJ/°C)
	t_{final} (29.82°C), $t_{initial}$ (20.00°C)
Information implied:	specific heat (c) of water
Asked for:	q for the reaction

STRATEGY

1. Find q_{H_2O} by substituting into Equation 8.2.

2. Find q_{cal} by substituting into Equation 8.3.

3. Recall: $q_{reaction} = -(q_{H_2O} + q_{cal})$.

SOLUTION

1. q_{H_2O}	$q_{H_2O} = c_{H_2O} \times mass_{H_2O} \times \Delta t$
	$= 4.18 \dfrac{J}{g \cdot °C} \times 1.000 \times 10^3 \text{ g} \times (29.82 - 20.00)°C$
	$= 4.10 \times 10^4 \text{ J} = 41.0 \text{ kJ}$
2. q_{cal}	$q_{cal} = C_{cal} \times \Delta t = (5.15 \text{ kJ/°C})(29.82 - 20.00)°C = 50.6 \text{ kJ}$
3. $q_{reaction}$	$q_{reaction} = -(q_{cal} + q_{H_2O}) = -(41.0 + 50.6)\text{kJ} = -91.6 \text{ kJ}$

END POINT

The amount of hydrogen gas (1.00 g) that reacted is not relevant to the solution of this problem.

8-3 Enthalpy

We have referred several times to the "heat flow for the reaction system," symbolized as $q_{reaction}$. At this point, you may well find this concept a bit nebulous and wonder if it could be made more concrete by relating $q_{reaction}$ to some property of reactants and products. This can indeed be done; the situation is particularly simple for reactions taking place at constant pressure. Under that condition, the heat flow for the reaction system is equal to the difference in **enthalpy** (H) between products and reactants. The **enthalpy change** is designated as ΔH. That is,

$$q_{reaction} \text{ at constant pressure} = \Delta H = H_{products} - H_{reactants}$$

Enthalpy is a type of chemical energy, sometimes referred to as "heat content." Reactions that occur in the laboratory in an open container or in the world around us take place at a constant pressure, that of the atmosphere. For such reactions, the equation just written is valid, making enthalpy a very useful quantity.

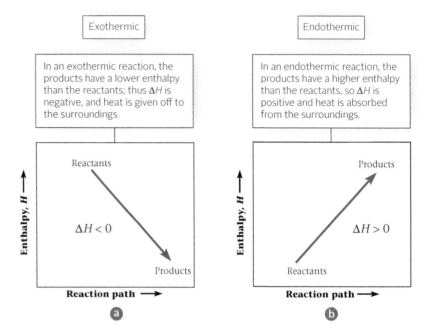

Figure 8.6 **Energy diagram showing ΔH for a reaction.**

When CH_4 burns, ΔH is negative; when ice melts, ΔH is positive.

Figure 8.6a shows the enthalpy relationship between reactants and products for an exothermic reaction such as

$$CH_4(g) + 2O_2(g) \longrightarrow CO_2(g) + 2H_2O(l) \qquad \Delta H < 0$$

Here, the products, 1 mol of $CO_2(g)$ and 2 mol of $H_2O(l)$, have a lower enthalpy than the reactants, 1 mol of $CH_4(g)$ and 2 mol of $O_2(g)$. The decrease in enthalpy is the source of the heat evolved to the surroundings. Figure 8.6b shows the situation for an endothermic process such as

$$H_2O(s) \longrightarrow H_2O(l) \qquad \Delta H > 0$$

Liquid water has a higher enthalpy than ice, so heat must be transferred from the surroundings to melt the ice.

In general, the following relations apply for reactions taking place at constant pressure.

exothermic reaction: $q = \Delta H < 0$ $\qquad H_{products} < H_{reactants}$

endothermic reaction: $q = \Delta H > 0$ $\qquad H_{products} > H_{reactants}$

The enthalpy of a substance, like its volume, is a state property. A sample of one gram of liquid water at 25.00°C and 1 atm has a fixed enthalpy, H. In practice, no attempt is made to determine absolute values of enthalpy. Instead, scientists deal with changes in enthalpy, which are readily determined. For the process

$$1.00 \text{ g } H_2O \ (l, 25.00°C, 1 \text{ atm}) \longrightarrow 1.00 \text{ g } H_2O \ (l, 26.00°C, 1 \text{ atm})$$

ΔH is 4.18 J because the specific heat of water is 4.18 J/g · °C.

8-4 Thermochemical Equations

A chemical equation that shows the enthalpy relation between products and reactants is called a **thermochemical equation**. This type of equation contains, at the right of the balanced chemical equation, the appropriate value and sign for ΔH.

The cold pack contains two separate compartments; one with ammonium nitrate and one with water.

When the seal separating the compartments is broken the following endothermic reaction occurs:

$$NH_4NO_3(s) \longrightarrow NH_4^+(aq) + NO_3^-(aq) \quad \Delta H = +28.1 \text{ kJ}$$

Water

Solid NH_4NO_3

As a result, the temperature, as read on the thermometer, drops.

Figure 8.7 Endothermic reaction in a cold pack.

To see where a thermochemical equation comes from, consider the process by which ammonium nitrate dissolves in water:

$$NH_4NO_3(s) \longrightarrow NH_4^+(aq) + NO_3^-(aq)$$

A simple experiment with a coffee-cup calorimeter shows that when one gram of NH_4NO_3 dissolves, $q_{reaction} = 351$ J. The calorimeter is open to the atmosphere, the pressure is constant, and

$$\Delta H \text{ for dissolving } 1.00 \text{ g of } NH_4NO_3 = 351 \text{ J} = 0.351 \text{ kJ}$$

When one mole (80.05 g) of NH_4NO_3 dissolves, ΔH should be 80 times as great:

$$\Delta H \text{ for dissolving } 1.00 \text{ mol of } NH_4NO_3 = 0.351 \frac{\text{kJ}}{\text{g}} \times 80.05 \text{ g} = 28.1 \text{ kJ}$$

The thermochemical equation for this reaction (Figure 8.5) must then be

$$NH_4NO_3(s) \longrightarrow NH_4^+(aq) + NO_3^-(aq) \quad \Delta H = +28.1 \text{ kJ}$$

By an entirely analogous procedure, the thermochemical equation for the formation of HCl from the elements (Example 8.3) is found to be

$$H_2(g) + Cl_2(g) \longrightarrow 2HCl(g) \quad \Delta H = -185 \text{ kJ}$$

In other words, 185 kJ of heat is evolved when two moles of HCl are formed from H_2 and Cl_2 (Figure 8.8).

These thermochemical equations are typical of those used throughout this text. It is important to realize that

- the sign of ΔH indicates whether the reaction, when carried out at constant pressure, is endothermic (positive ΔH) or exothermic (negative ΔH).
- in interpreting a thermochemical equation, the coefficients represent numbers of moles (ΔH is -185 kJ when *1 mol* H_2 + *1 mol* Cl_2 → *2 mol* HCl).
- the phases (physical states) of all species must be specified, using the symbols (s), (l), (g), or (aq). The enthalpy of one mole of $H_2O(g)$ at 25°C is 44 kJ larger than that of one mole of $H_2O(l)$; the difference, which represents the heat of vaporization of water, is clearly significant. ◀ⓘ
- the value quoted for ΔH applies when products and reactants are at the same temperature, ordinarily taken to be 25°C unless specified otherwise.

Figure 8.8 An exothermic reaction. Hydrogen, a colorless gas, reacts with chlorine, a pale yellow gas, to form colorless hydrogen chloride gas.

ⓘ This explains why burns from steam are more painful than those from boiling water.

Rules of Thermochemistry

To make effective use of thermochemical equations, three basic rules of thermochemistry are applied.

1. *The magnitude of ΔH is directly proportional to the amount of reactant or product.* This is a common-sense rule, consistent with experience. The amount of heat that must be absorbed to boil a sample of water is directly proportional to its mass. ◀ⓘ In another case, the more gasoline you burn in your car's engine, the more energy you produce.

This rule allows you to find ΔH corresponding to any desired amount of reactant or product. To do this, you follow the conversion-factor approach used in Chapter 3 with ordinary chemical equations. Consider, for example,

$$H_2(g) + Cl_2(g) \longrightarrow 2HCl(g) \qquad \Delta H = -185 \text{ kJ}$$

The thermochemical equation allows us to relate the enthalpy change to amounts of reactants and products, leading to conversion factors such as

$$\frac{-185 \text{ kJ}}{1 \text{ mol Cl}_2} \qquad \frac{1 \text{ mol H}_2}{-185 \text{ kJ}} \qquad \frac{-185 \text{ kJ}}{2 \text{ mol HCl}}$$

ⓘ The magnitude of ΔH is dependent not only on the amount of the reactants and products but also on their physical states.

EXAMPLE	8.4	GRADED

The Bunsen burners in your labs are fueled by natural gas, which is mostly methane, CH_4. The thermochemical equation for the combustion (burning in oxygen) of methane is

$$CH_4(g) + 2O_2(g) \longrightarrow CO_2(g) + 2H_2O(l) \qquad \Delta H = -890.3 \text{ kJ}$$

a   Calculate ΔH when 5.00 g of CH_4 react with an excess of oxygen.

ANALYSIS

Information given:	mass of CH_4 (5.00 g), excess O_2
Information implied:	molar mass of CH_4 ΔH for the reaction
Asked for:	ΔH

STRATEGY

1. Use the relationship obtained from the thermochemical equation: -890.3 kJ/mol CH_4 as a conversion factor.

2. Follow the following schematic plan.

$$\text{mass}_{CH_4} \xrightarrow{\text{ MM }} \text{mol}_{CH_4} \xrightarrow{-890.3 \text{ kJ/mol } CH_4} \Delta H$$

SOLUTION

ΔH	$5.00 \text{ g CH}_4 \times \dfrac{1 \text{ mol CH}_4}{16.04 \text{ g CH}_4} \times \dfrac{-890.3 \text{ kJ}}{1 \text{ mol CH}_4} = -278 \text{ kJ}$

b   Calculate ΔH when 2.00 L of CH_4 at 49.0°C and 782 mm Hg react with an excess of oxygen.

ANALYSIS

Information given:	CH_4 data: V (2.00 L), T (49.0°C), P (782 mm Hg). Excess O_2
Information implied:	ΔH for the reaction gas constant, R
Asked for:	ΔH

continued

Follow the following schematic plan.

$$V_{CH_4} \xrightarrow{\;PV/RT\;} mol_{CH_4} \xrightarrow{\;-890.3 \text{ kJ}/1 \text{ mol CH}_4\;} \Delta H$$

SOLUTION

mol CH_4	$n = \dfrac{PV}{RT} = \dfrac{(782/760)\text{atm} \times 2.00 \text{ L}}{(0.0821 \text{ L} \cdot \text{atm/mol} \cdot \text{K}) \times 322 \text{ K}} = 0.0778 \text{ mol}$
ΔH	$0.0778 \text{ mol CH}_4 \times \dfrac{-890.3 \text{ kJ}}{1 \text{ mol CH}_4} = -69.3 \text{ kJ}$

c Calculate ΔH when 2.00 L of CH_4 react with 5.00 L of O_2 in a reaction vessel kept at 49°C and 782 mm Hg.

ANALYSIS

Information given:	V_{CH_4} (2.00 L), V_{O_2} (5.00 L), T and P are constant.
Information implied:	stoichiometric ratios ΔH for the reaction gas constant, R
Asked for:	ΔH

STRATEGY

1. Note that data about both reactants is given (limiting reactant problem) and that the reaction is at constant temperature and pressure.

2. Find V_{CO_2} obtained if CH_4 is limiting and again if O_2 is limiting. Choose the smaller value, then use the following schematic plan:

$$V_{CO_2} \xrightarrow{\;PV/RT\;} mol_{CO_2} \xrightarrow{\;-890.3 \text{ kJ}/1 \text{ mol CO}_2\;} \Delta H$$

SOLUTION

V_{CO_2} produced	If CH_4 is limiting: $2.00 \text{ L CH}_4 \times \dfrac{1 \text{ L CO}_2}{1 \text{ L CH}_4} = 2.00 \text{ L}$
	If O_2 is limiting: $5.00 \text{ L O}_2 \times \dfrac{1 \text{ L CO}_2}{2 \text{ L O}_2} = 2.50 \text{ L}$
	CH_4 is limiting and 2.00 L of CO_2 are produced.
mol CO_2	$n = \dfrac{PV}{RT} = \dfrac{782/760 \text{ atm} \times 2.00 \text{ L}}{(0.0821 \text{ L} \cdot \text{atm/mol} \cdot \text{K}) \times 322 \text{ K}} = 0.0778 \text{ mol}$
ΔH	$0.0778 \text{ mol CO}_2 \times \dfrac{-890.3 \text{ kJ}}{1 \text{ mol CO}_2} = -69.3 \text{ kJ}$

The heat absorbed when a solid melts ($s \rightarrow l$) is referred to as the **heat of fusion;** that absorbed when a liquid vaporizes ($l \rightarrow g$) is called the **heat of vaporization.** Heats of fusion (ΔH_{fus}) and vaporization (ΔH_{vap}) are most often expressed in kilojoules per mole (kJ/mol). Values for several different substances are given in Table 8.2.

2. ΔH for a reaction is equal in magnitude but opposite in sign to ΔH for the reverse reaction (Figure 8.9). Another way to state this rule is to say that the amount of heat evolved in a reaction is exactly equal to the amount of heat

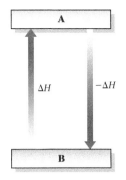

Figure 8.9 Thermochemistry rule 2.

Table 8.2 ΔH(kJ/mol) for Phase Changes

Substance		mp(°C)	ΔH_{fus}*	bp(°C)	ΔH_{vap}*
Benzene	C_6H_6	5	9.84	80	30.8
Bromine	Br_2	−7	10.8	59	29.6
Mercury	Hg	−39	2.33	357	59.4
Naphthalene	$C_{10}H_8$	80	19.3	218	43.3
Water	H_2O	0	6.00	100	40.7

*Values of ΔH_{fus} are given at the melting point, values of ΔH_{vap} at the boiling point. The heat of vaporization of water decreases from 44.9 kJ/mol at 0°C to 44.0 kJ/mol at 25°C to 40.7 kJ/mol at 100°C.

absorbed in the reverse reaction. This again is a common-sense rule. If 6.00 kJ of heat is absorbed when a mole of ice melts,

$$H_2O(s) \longrightarrow H_2O(l) \qquad \Delta H = +6.00 \text{ kJ}$$

then 6.00 kJ of heat should be evolved when a mole of liquid water freezes.

$$H_2O(l) \longrightarrow H_2O(s) \qquad \Delta H = -6.00 \text{ kJ}$$

EXAMPLE 8.5

Given

$$2H_2(g) + O_2(g) \longrightarrow 2H_2O(l) \qquad \Delta H = -571.6 \text{ kJ}$$

calculate ΔH for the equation

$$H_2O(l) \longrightarrow H_2(g) + \tfrac{1}{2}O_2(g)$$

STRATEGY

1. Note that the second equation is the reverse of the first. Apply Rule 2.
2. The coefficients of the second equation are half those of the given equation. Apply Rule 1.

SOLUTION

Apply Rule 2.	$2H_2O(l) \longrightarrow 2H_2(g) + O_2(g)$	$\Delta H = -(-571.6 \text{ kJ}) = 571.6 \text{ kJ}$
Apply Rule 1.	$H_2O(l) \longrightarrow H_2(g) + \tfrac{1}{2}O_2(g)$	$\Delta H = \tfrac{1}{2}(+571.6 \text{ kJ}) = 285.8 \text{ kJ}$

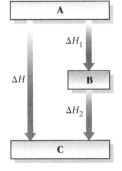

Figure 8.10 Thermochemistry rule 3, Hess's law, $\Delta H = \Delta H_1 + \Delta H_2$ because H is a state property.

ΔH must be independent of path, since H is a state property.

3. *The value of ΔH for a reaction is the same whether it occurs in one step or in a series of steps* (Figure 8.10). If a thermochemical equation can be expressed as the sum of two or more equations,

$$\text{equation} = \text{equation (1)} + \text{equation (2)} + \cdots$$

then ΔH for the overall equation is the sum of the ΔH's for the individual equations:

$$\Delta H = \Delta H_1 + \Delta H_2 + \cdots$$

This relationship is referred to as **Hess's law,** after Germain Hess (1802–1850), professor of chemistry at the University of St. Petersburg, who deduced it in 1840. Hess's law is a direct consequence of the fact that enthalpy is a state property, dependent only on initial and final states. This means that, in Figure 8.10, ΔH *must* equal the sum of ΔH_1 and ΔH_2, because the final and initial states are the same for the two processes.

Hess's law is very convenient for obtaining values of ΔH for reactions that are difficult to carry out in a calorimeter. Consider, for example, the formation of the toxic gas carbon monoxide from the elements

$$C(s) + \tfrac{1}{2}O_2(g) \longrightarrow CO(g)$$

It is difficult, essentially impossible, to measure ΔH for this reaction because when carbon burns, the major product is always carbon dioxide, CO_2. It is possible, however, to calculate ΔH using thermochemical data for two other reactions that are readily carried out in the laboratory (Example 8.6).

EXAMPLE 8.6

Carbon monoxide, CO, is a poisonous gas. It can be obtained by burning carbon in a limited amount of oxygen. Given

(1) $C(s) + O_2(g) \longrightarrow CO_2(g)$ $\Delta H = -393.5$ kJ

(2) $2CO(g) + O_2(g) \longrightarrow 2CO_2(g)$ $\Delta H = -566.0$ kJ

calculate ΔH for the reaction

$$C(s) + \tfrac{1}{2}O_2(g) \longrightarrow CO(g)   \Delta H = ?$$

STRATEGY

1. Work with the given equations until you arrive at two equations that will add to give the equation
$C(s) + \tfrac{1}{2}O_2(g) \longrightarrow CO(g)$

2. Focus on species that appear in only one of the thermochemical equations. In this case C appears only in Equation (1) and CO appears only in Equation (2).

3. For CO

Compare position and number of moles of CO in Equation 2 and in the desired equation.

   Position: Equation (2)—on the left   Desired equation—on the right

   Moles: Equation (2)—2 moles   Desired equation—1 mole

Apply Rule 2 to reverse Equation (2) (to put CO on the right).

Apply Rule 1 and divide Equation (2) by 2 (to get a coefficient of 1 for CO).

4. For C

Compare position and number of moles of C in Equation (1) and in the desired equation.

   Position: Equation (1)—on the left   Desired equation—on the left

   Moles: Equation (2)—1 mole   Desired equation—1 mole

Both the position and coefficient for C are okay, so use Equation (1) "as is."

5. Add the "revised" Equation (2c) to Equation (1).

SOLUTION

For Equation (2)

   Apply Rule 2   (2a) $2CO_2(g) \longrightarrow 2CO(g) + O_2(g)$ $\Delta H = -(-566.0 \text{ kJ}) = 566.0$ kJ

   Apply Rule 1   (2b) $CO_2(g) \longrightarrow CO(g) + \tfrac{1}{2}O_2(g)$ $\Delta H = \tfrac{1}{2}(566.0 \text{ kJ}) = 283.0$ kJ

   "Revised" Equation 2   (2c) $CO_2(g) \longrightarrow CO(g) + \tfrac{1}{2}O_2(g)$ $\Delta H_2 = \tfrac{1}{2}(566.0 \text{ kJ}) = 283.0$ kJ

Equation (1) "as is"   (1) $C(s) + O_2(g) \longrightarrow CO_2(g)$ $\Delta H_1 = -393.5$ kJ

Apply Hess's law   Add "revised" Equation (2c) to Equation (1)

   (2c) $CO_2(g) \longrightarrow CO(g) + \tfrac{1}{2}O_2(g)$ $\Delta H_2 = 283.0$ kJ

   (1) $C(s) + O_2(g) \longrightarrow CO_2(g)$ $\Delta H_1 = -393.5$ kJ

$\tfrac{1}{2}O_2(g)$

$\Delta H = \Delta H_1 + \Delta H_2$ $C(s) + \tfrac{1}{2}O_2(g) \longrightarrow CO(g)$ $\Delta H = 283.0 \text{ kJ} + (-393.5) \text{ kJ} = -110.5$ kJ

Summarizing the rules of thermochemistry:

1. ΔH is directly proportional to the amount of reactant or product.
2. ΔH changes sign when a reaction is reversed.
3. ΔH for a reaction has the same value regardless of the number of steps.

8-5   Enthalpies of Formation

We have now written several thermochemical equations. In each case, we have cited the corresponding value of ΔH. Literally thousands of such equations would be needed to list the ΔH values for all the reactions that have been studied. Clearly, there has to be some more concise way of recording data of this sort. These data should be in a form that can easily be used to calculate ΔH for any reaction. It turns out that there is a simple way to do this, using quantities known as enthalpies of formation.

Meaning of $\Delta H_f°$

Often called "heat of formation." **(i)**

The standard molar **enthalpy of formation** ◄**(i)** of a compound, $\Delta H_f°$, is equal to the enthalpy change when one mole of the compound is formed at a constant pressure of 1 atm and a fixed temperature, ordinarily 25°C, from the elements in their stable states at that pressure and temperature (Figure 8.11). From the equations

$$Ag(s, 25°C) + \tfrac{1}{2}Cl_2(g, 25°C, 1\ atm) \longrightarrow AgCl(s, 25°C) \qquad \Delta H = -127.1\ kJ$$
$$\tfrac{1}{2}N_2(g, 1\ atm, 25°C) + O_2(g, 1\ atm, 25°C) \longrightarrow NO_2(g, 1\ atm, 25°C) \quad \Delta H = +33.2\ kJ$$

it follows that

$$\Delta H_f°\ AgCl(s) = -127.1\ kJ/mol \qquad \Delta H_f°\ NO_2(g) = +33.2\ kJ/mol$$

Enthalpies of formation for a variety of compounds are listed in Table 8.3. Notice that, with a few exceptions, enthalpies of formation are negative quantities. This means that the formation of a compound from the elements is ordinarily exothermic. Conversely, when a compound decomposes to the elements, heat usually must be absorbed.

However, $\Delta H_f°\ Br_2(g)$ is +29.6 kJ/mol. Explain. **(i)**

You will note from Table 8.3 that there are no entries for elemental species such as $Br_2(l)$ ◄**(i)** and $O_2(g)$. This is a consequence of the way in which enthalpies of formation are defined. In effect, the enthalpy of formation of an element in its stable state at 25°C and 1 atm is taken to be zero. That is,

$$\Delta H_f°\ Br_2(l) = \Delta H_f°\ O_2(g) = 0$$

The standard enthalpies of formation of ions in aqueous solution listed at the bottom of Table 8.3 are relative values, established by taking

$$\Delta H_f°\ H^+(aq) = 0$$

From the values listed we see that the Cu^{2+} ion has a heat of formation *greater* than that of H^+ by about 65 kJ/mol, and Cd^{2+} has a heat of formation about 76 kJ/mol *less* than that of H^+.

Calculation of $\Delta H°$

Enthalpies of formation can be used to calculate $\Delta H°$ for a reaction. To do this, apply this general rule:

The standard enthalpy change, $\Delta H°$, for a given thermochemical equation is equal to the sum of the standard enthalpies of formation of the product compounds minus the sum of the standard enthalpies of formation of the reactant compounds.

© Cengage Learning/Charles D. Winters

Figure 8.11 Enthalpy of formation. Magnesium ribbon reacts with oxygen to give MgO, a white solid, and 601.7 kJ of heat per mole of MgO formed. Hence $\Delta H_f° MgO(s) = -601.7\ kJ/mol$.

Table 8.3 Standard Enthalpies of Formation at 25°C (kJ/mol) of Compounds at 1 atm, Aqueous Ions at 1 M

Compounds							
$AgBr(s)$	−100.4	$CaCl_2(s)$	−795.8	$H_2O(g)$	−241.8	$NH_4NO_3(s)$	−365.6
$AgCl(s)$	−127.1	$CaCO_3(s)$	−1206.9	$H_2O(l)$	−285.8	$NO(g)$	+90.2
$AgI(s)$	−61.8	$CaO(s)$	−635.1	$H_2O_2(l)$	−187.8	$NO_2(g)$	+33.2
$AgNO_3(s)$	−124.4	$Ca(OH)_2(s)$	−986.1	$H_2S(g)$	−20.6	$N_2O_4(g)$	+9.2
$Ag_2O(s)$	−31.0	$CaSO_4(s)$	−1434.1	$H_2SO_4(l)$	−814.0	$NaCl(s)$	−411.2
$Al_2O_3(s)$	−1675.7	$CdCl_2(s)$	−391.5	$HgO(s)$	−90.8	$NaF(s)$	−573.6
$BaCl_2(s)$	−858.6	$CdO(s)$	−258.2	$KBr(s)$	−393.8	$NaOH(s)$	−425.6
$BaCO_3(s)$	−1216.3	$Cr_2O_3(s)$	−1139.7	$KCl(s)$	−436.7	$NiO(s)$	−239.7
$BaO(s)$	−553.5	$CuO(s)$	−157.3	$KClO_3(s)$	−397.7	$PbBr_2(s)$	−278.7
$BaSO_4(s)$	−1473.2	$Cu_2O(s)$	−168.6	$KClO_4(s)$	−432.8	$PbCl_2(s)$	−359.4
$CCl_4(l)$	−135.4	$CuS(s)$	−53.1	$KNO_3(s)$	−494.6	$PbO(s)$	−219.0
$CHCl_3(l)$	−134.5	$Cu_2S(s)$	−79.5	$MgCl_2(s)$	−641.3	$PbO_2(s)$	−277.4
$CH_4(g)$	−74.8	$CuSO_4(s)$	−771.4	$MgCO_3(s)$	−1095.8	$PCl_3(g)$	−287.0
$C_2H_2(g)$	+226.7	$Fe(OH)_3(s)$	−823.0	$MgO(s)$	−601.7	$PCl_5(g)$	−374.9
$C_2H_4(g)$	+52.3	$Fe_2O_3(s)$	−824.2	$Mg(OH)_2(s)$	−924.5	$SiO_2(s)$	−910.9
$C_2H_6(g)$	−84.7	$Fe_3O_4(s)$	−1118.4	$MgSO_4(s)$	−1284.9	$SnO_2(s)$	−580.7
$C_3H_8(g)$	−103.8	$HBr(g)$	−36.4	$MnO(s)$	−385.2	$SO_2(g)$	−296.8
$CH_3OH(l)$	−238.7	$HCl(g)$	−92.3	$MnO_2(s)$	−520.0	$SO_3(g)$	−395.7
$C_2H_5OH(l)$	−277.7	$HF(g)$	−271.1	$NH_3(g)$	−46.1	$ZnI_2(s)$	−208.0
$CO(g)$	−110.5	$HI(g)$	+26.5	$N_2H_4(l)$	+50.6	$ZnO(s)$	−348.3
$CO_2(g)$	−393.5	$HNO_3(l)$	−174.1	$NH_4Cl(s)$	−314.4	$ZnS(s)$	−206.0

Cations				Anions			
$Ag^+(aq)$	+105.6	$Hg^{2+}(aq)$	+171.1	$Br^-(aq)$	−121.6	$HPO_4^{2-}(aq)$	−1292.1
$Al^{3+}(aq)$	−531.0	$K^+(aq)$	−252.4	$CO_3^{2-}(aq)$	−677.1	$HSO_4^-(aq)$	−887.3
$Ba^{2+}(aq)$	−537.6	$Mg^{2+}(aq)$	−466.8	$Cl^-(aq)$	−167.2	$I^-(aq)$	−55.2
$Ca^{2+}(aq)$	−542.8	$Mn^{2+}(aq)$	−220.8	$ClO_3^-(aq)$	−104.0	$MnO_4^-(aq)$	−541.4
$Cd^{2+}(aq)$	−75.9	$Na^+(aq)$	−240.1	$ClO_4^-(aq)$	−129.3	$NO_2^-(aq)$	−104.6
$Cu^+(aq)$	+71.7	$NH_4^+(aq)$	−132.5	$CrO_4^{2-}(aq)$	−881.2	$NO_3^-(aq)$	−205.0
$Cu^{2+}(aq)$	+64.8	$Ni^{2+}(aq)$	−54.0	$Cr_2O_7^{2-}(aq)$	−1490.3	$OH^-(aq)$	−230.0
$Fe^{2+}(aq)$	−89.1	$Pb^{2+}(aq)$	−1.7	$F^-(aq)$	−332.6	$PO_4^{3-}(aq)$	−1277.4
$Fe^{3+}(aq)$	−48.5	$Sn^{2+}(aq)$	−8.8	$HCO_3^-(aq)$	−692.0	$S^{2-}(aq)$	+33.1
$H^+(aq)$	0.0	$Zn^{2+}(aq)$	−153.9	$H_2PO_4^-(aq)$	−1296.3	$SO_4^{2-}(aq)$	−909.3

A piece of burning magnesium acts as a fuse when inserted into a pot containing a finely divided mixture of aluminum powder and iron(III) oxide.

The burning magnesium initiates the exothermic reaction:

$2Al(s) + Fe_2O_3(s) \longrightarrow 2Fe(s) + Al_2O_3(s) \; \Delta H° = -851.5 \text{ kJ}$

Enough heat is generated to produce molten iron, which can be seen flowing out of the broken pot onto the protective mat at the bottom of the stand.

© Cengage Learning/Charles D. Winters

Figure 8.12 A thermite reaction.

Using the symbol Σ to represent "the sum of,"

$$\Delta H° = \Sigma \, \Delta H_f° \text{ products} - \Sigma \, \Delta H_f° \text{ reactants} \qquad (8.4)$$

This is a very useful equation, one we will use again and again.

In applying this equation ◂ *i*

■ *elements in their standard states can be omitted,* because their heats of formation are zero. To illustrate, consider the thermite reaction once used to weld rails (Figure 8.12).

$$2Al(s) + Fe_2O_3(s) \longrightarrow 2Fe(s) + Al_2O_3(s)$$

We can write, quite simply,

$$\Delta H° = \Sigma \, \Delta H_f° \text{ products} - \Sigma \, \Delta H_f° \text{ reactants} = \Delta H_f° \, Al_2O_3(s) - \Delta H_f° \, Fe_2O_3(s)$$

■ *the coefficients of products and reactants in the thermochemical equation must be taken into account.* For example, consider the reaction:

$$2Al(s) + 3Cu^{2+}(aq) \longrightarrow 2Al^{3+}(aq) + 3Cu(s)$$

for which:

$$\Delta H° = 2\Delta H_f° \, Al^{3+}(aq) - 3\Delta H_f° \, Cu^{2+}(aq)$$

Strictly speaking, $\Delta H°$ calculated from enthalpies of formation listed in Table 8.3 represents the enthalpy change at 25°C and 1 atm. Actually, ΔH is independent of pressure and varies relatively little with temperature, changing by perhaps 1 to 10 kJ per 100°C.

EXAMPLE 8.7 GRADED

Benzene, C_6H_6, used in the manufacture of plastics, is a carcinogen affecting the bone marrow. Long-term exposure has been shown to cause leukemia and other blood disorders. The combustion of benzene is given by the following equation:

$$C_6H_6(l) + \tfrac{15}{2} O_2(g) \longrightarrow 6CO_2(g) + 3H_2O(l) \qquad \Delta H° = -3267.4 \text{ kJ}$$

continued

a	Calculate the heat of formation of benzene.

ANALYSIS

Information given:	thermochemical equation for the combustion of benzene
Information implied:	ΔH_f° for all species except benzene (Table 8.3)
Asked for:	ΔH_f° for benzene

STRATEGY

1. Find ΔH° for all the species (besides benzene) in Table 8.3 and substitute into Equation 8.4.
2. Recall that ΔH° for $O_2(g)$ is zero.

SOLUTION

Equation 8.4	-3267.4 kJ $= 6(\Delta H_f^\circ\ CO_2) + 3(\Delta H_f^\circ\ H_2O) - (\Delta H_f^\circ\ C_6H_6)$
ΔH_f° for $C_6H_6(l)$	$\Delta H_f^\circ = 3267.4$ kJ $+ 6$ mol $\left(-393.5\ \dfrac{kJ}{mol}\right) + 3$ mol $\left(-285.8\ \dfrac{kJ}{mol}\right)$
	$= +49.0$ kJ/mol

b	Calculate ΔH° for the reaction

$$12CO_2(g) + 6H_2O(l) \longrightarrow 2C_6H_6(l) + 15O_2(g)$$

ANALYSIS

Information given:	thermochemical equation for the combustion of benzene
Asked for:	ΔH° for the reaction: $12CO_2(g) + 6H_2O(l) \longrightarrow 2C_6H_6(l) + 15O_2(g)$

STRATEGY

Note that the given equation is the reverse of the combustion equation and that the coefficients have been doubled. Apply Rules 1 and 2.

SOLUTION

Rule 2	$\Delta H^\circ = -(-3267.4)$ kJ $= 3267.4$ kJ
Rule 1	$\Delta H^\circ = 2(3267.4$ kJ$) = 6534.8$ kJ

c	Calculate ΔH° for the reaction

$$C_6H_6(g) + \tfrac{15}{2}O_2(g) \longrightarrow 6CO_2(g) + 3H_2O(g)$$

ANALYSIS

Information given:	thermochemical equation for the combustion of benzene
Information implied:	ΔH_{vap} for water and benzene (Table 8.2)
Asked for:	$C_6H_6(g) + \tfrac{15}{2}O_2(g) \longrightarrow 6CO_2(g) + 3H_2O(g)$

STRATEGY

1. Notice that the given equation is identical to the combustion equation except for the physical states of benzene and water.
2. a. Write the thermochemical equation for the vaporization of water.
   b. Multiply the equation by 3 since there are three moles of H_2O in the combustion reaction.
3. a. Write the thermochemical equation for the vaporization of benzene.
   b. Note that $C_6H_6(g)$ is a reactant in the equation where ΔH° is needed.
   c. Reverse the vaporization equation for benzene and change the sign of its ΔH°.
4. Apply Hess's law by adding all the equations so you come up with ΔH° for the overall given equation.

continued

SOLUTION

ΔH_1°: $(3 \times \Delta H_{vap} \text{ H}_2\text{O})$	$3(\text{H}_2\text{O}(l) \longrightarrow \text{H}_2\text{O}(g)) = 3(40.7 \text{ kJ/mol})$ $\Delta H_1^\circ = 122.1 \text{ kJ}$
ΔH_2°: (reverse ΔH_{vap} C$_6$H$_6$)	$\text{C}_6\text{H}_6(g) \longrightarrow \text{C}_6\text{H}_6(l) = -(30.8) \text{ kJ/mol}$ $\Delta H_2^\circ = -30.8 \text{ kJ}$
Apply Hess's law	Equation (1) + Equation (2) + combustion equation
	(1) $3\text{H}_2\text{O}(l) \longrightarrow 3\text{H}_2\text{O}(g)$ $\Delta H_1^\circ = 122.1 \text{ kJ}$
	(2) $\text{C}_6\text{H}_6(g) \longrightarrow \text{C}_6\text{H}_6(l)$ $\Delta H_2^\circ = -30.8 \text{ kJ}$
	$\text{C}_6\text{H}_6(l) + \frac{15}{2}\text{O}_2(g) \longrightarrow 6\text{CO}_2(g) + 3\text{H}_2\text{O}(l)$ $\Delta H^\circ = -3267.4 \text{ kJ}$
Overall equation	$\text{C}_6\text{H}_6(g) + \frac{15}{2}\text{O}_2(g) \longrightarrow 6\text{CO}_2(g) + 3\text{H}_2\text{O}(g)$
ΔH°	$122.1 \text{ kJ} + (-30.8 \text{ kJ}) + (-3267.4 \text{ kJ}) = -3176.1 \text{ kJ}$

The relation between ΔH° and enthalpies of formation is perhaps used more often than any other in thermochemistry. Its validity depends on the fact that enthalpy is a state property. For any reaction, ΔH° can be obtained by imagining that the reaction takes place in two steps. First, the reactants (compounds or ions) are converted to the elements:

$$\text{reactants} \longrightarrow \text{elements} \qquad \Delta H_1^\circ = -\Sigma \Delta H_f^\circ \text{ reactants}$$

Then the elements are converted to products:

$$\text{elements} \longrightarrow \text{products} \qquad \Delta H_2^\circ = \Sigma \Delta H_f^\circ \text{ products}$$

Applying a basic rule of thermochemistry, Hess's law,

$$\Delta H^\circ = \Delta H_1^\circ + \Delta H_2^\circ = \Sigma \Delta H_f^\circ \text{ products} + (-\Sigma \Delta H_f^\circ \text{ reactants})$$

Enthalpy changes for reactions in solution can be determined using standard enthalpies of formation of aqueous ions, applying the general relation

$$\Delta H^\circ = \Sigma \Delta H_f^\circ \text{ products} - \Sigma \Delta H_f^\circ \text{ reactants}$$

and taking account of the fact that ΔH_f° H$^+$(aq) = 0 (Example 8.8).

EXAMPLE 8.8 GRADED

Copper carbonate is used to control algae in farm ponds. When hydrochloric acid is added to a solution of copper carbonate, carbon dioxide gas is formed (Figure 8.13). The equation for the reaction is

$$2\text{H}^+(aq) + \text{CO}_3^{2-}(aq) \longrightarrow \text{CO}_2(g) + \text{H}_2\text{O}(l)$$

a   Calculate ΔH° for the thermochemical equation.

ANALYSIS

Information given:	Equation for the reaction: [$2\text{H}^+(aq) + \text{CO}_3^{2-}(aq) \longrightarrow \text{CO}_2(g) + \text{H}_2\text{O}(l)$]
Information implied:	ΔH_f° for all the species in the reaction (Table 8.3)
Asked for:	ΔH° for the reaction

STRATEGY

Use Table 8.3 and recall that ΔH_f° for H$^+$ is zero.

SOLUTION

ΔH°	$\Delta H^\circ = \Delta H_f^\circ \text{ CO}_2 + \Delta H_f^\circ \text{ H}_2\text{O} - [2 \Delta H_f^\circ \text{ H}^+ + \Delta H_f^\circ \text{ CO}_3^{2-}]$
	$= 1 \text{ mol}\left(-393.5 \dfrac{\text{kJ}}{\text{mol}}\right) + 1 \text{ mol}\left(-285.8 \dfrac{\text{kJ}}{\text{mol}}\right) - \left[0 + 1 \text{ mol}\left(-677.1 \dfrac{\text{kJ}}{\text{mol}}\right)\right]$
	$= -2.2 \text{ kJ}$

continued

b	Calculate $\Delta H°$ when 25.00 mL of 0.186 M HCl is added to copper carbonate.

ANALYSIS

Information given:	V_{HCl} (25.00 mL); M_{HCl} (0.186) From part (a): $\Delta H°$ for the reaction (-2.2 kJ)
Asked for:	ΔH when given amounts of HCl are used.

STRATEGY

1. Find moles of H^+ actually used and convert to kJ by using the following plan:

$$(V \times M) \longrightarrow \text{mol HCl} \xrightarrow{\text{1 mol HCl /1 mol } H^+} \text{mol } H^+ \xrightarrow{\Delta H°/2 \text{ mol } H^+} \text{kJ}$$

2. Find $\Delta H°$ for the amount of H^+ used. $\Delta H°$ given is for 2 mol H^+.

SOLUTION

mol H^+	$V \times M = (0.02500 \text{ L})(0.186 \text{ mol/L}) = 0.00465 \text{ mol HCl} = \text{mol } H^+$
$\Delta H°$	$0.00465 \text{ mol } H^+ \times \dfrac{-2.2 \text{ kJ}}{2 \text{ mol } H^+} = -5.1 \times 10^{-3} \text{ kJ}$

END POINT

When you use Table 8.3 to figure out $\Delta H°$ for a reaction, make sure you consider the physical state of the species. Water, for example, has two different $\Delta H_f°$ values given: one for liquid and the other for gas. By the same token, do not forget to write the physical state of each species when you write an equation to represent a reaction.

8-6 Bond Enthalpy

For many reactions, ΔH is a large negative number; the reaction gives off a lot of heat. In other cases, ΔH is positive; heat must be absorbed for the reaction to occur. You may well wonder why the enthalpy change should vary so widely from one reaction to another. Is there some basic property of the molecules involved in the reaction that determines the sign and magnitude of ΔH?

These questions can be answered on a molecular level in terms of a quantity known as bond enthalpy. (More commonly but less properly it is called bond energy.) The **bond enthalpy** is defined as ΔH when one mole of bonds is broken in the gaseous state. From the equations (see also Figure 8.14, page 208).

$$H_2(g) \longrightarrow 2H(g) \quad \Delta H = +436 \text{ kJ}$$

$$Cl_2(g) \longrightarrow 2Cl(g) \quad \Delta H = +243 \text{ kJ}$$

it follows that the H—H bond enthalpy is +436 kJ/mol and that for Cl—Cl is 243 kJ/mol. ◀ⓘ In both reactions, one mole of bonds (H—H and Cl—Cl) is broken.

Bond enthalpies for a variety of single and multiple bonds are listed in Table 8.4. Note that bond enthalpy is always a positive quantity; heat is always

Figure 8.13 **Reaction of an acid with a carbonate.** Hydrochloric acid added to copper carbonate solution produces CO_2 gas.

ⓘ
What is $\Delta H_f°$ for H(g)? Cl(g)? Answer: 218 and 122 kJ/mol, respectively.

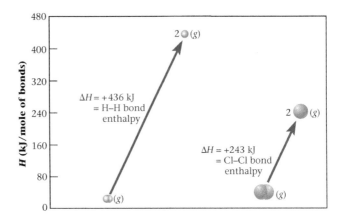

Figure 8.14 Bond enthalpies for H₂ and Cl₂. The H—H bond enthalpy is greater than the Cl—Cl bond enthalpy (+436 kJ/mol versus +243 kJ/mol). This means that the bond in H₂ is stronger than that in Cl₂.

The higher the bond enthalpy, the stronger the bond.

absorbed when chemical bonds are broken. ◄ⓘ Conversely, heat is given off when bonds are formed from gaseous atoms. Thus

$$H(g) + Cl(g) \longrightarrow HCl(g) \quad \Delta H = -431 \text{ kJ}$$

$$H(g) + F(g) \longrightarrow HF(g) \quad \Delta H = -565 \text{ kJ}$$

Using bond enthalpies, it is possible to explain why certain gas phase reactions are endothermic and others are exothermic. In general, a reaction is expected to be endothermic (i.e., heat must be absorbed) if

Table 8.4 Bond Enthalpies

Single Bond Enthalpy (kJ/mol)									
	H	C	N	O	S	F	Cl	Br	I
H	436	414	389	464	339	565	431	368	297
C		347	293	351	259	485	331	276	218
N			159	222	—	272	201	243	—
O				138	—	184	205	201	201
S					226	285	255	213	—
F						153	255	255	277
Cl							243	218	209
Br								193	180
I									151

Multiple Bond Enthalpy (kJ/mol)					
C=C	612	N=N	418	C≡C	820
C=N	615	N=O	607	C≡N	890
C=O	715	O=O	498	C≡O	1075
C=S	477	S=O	498	N≡N	941

■ *the bonds in the reactants are stronger than those in the products.* Consider, for example, the decomposition of hydrogen fluoride to the elements, which can be shown to be endothermic from the data in Table 8.3.

$$2HF(g) \longrightarrow H_2(g) + F_2(g) \qquad \Delta H = -2\Delta H_f^\circ \text{ HF} = +542.2 \text{ kJ}$$

The bonds in HF (565 kJ/mol) are stronger than the average of those in H_2 and F_2: (436 kJ/mol + 153 kJ/mol)/2 = 295 kJ/mol.

■ *there are more bonds in the reactants than in the products.* For the reaction

$$2H_2O(g) \longrightarrow 2H_2(g) + O_2(g) \qquad \Delta H = -2\Delta H_f^\circ \text{ H}_2\text{O}(g) = +483.6 \text{ kJ}$$

there are four moles of O—H bonds in two moles of H_2O as compared with only three moles of bonds in the products (two moles of H—H bonds, one mole of O=O bonds).

You will note from Table 8.4 that the bond enthalpy is larger for a multiple bond than for a single bond between the same two atoms. ◀ ⓘ Thus

$$\text{bond enthalpy C—C} = 347 \text{ kJ/mol}$$
$$\text{bond enthalpy C=C} = 612 \text{ kJ/mol}$$
$$\text{bond enthalpy C≡C} = 820 \text{ kJ/mol}$$

ⓘ But, the double bond enthalpy is less than twice that for a single bond.

This effect is a reasonable one; the greater the number of bonding electrons, the more difficult it should be to break the bond between two atoms.

We should point out a serious limitation of the bond enthalpies listed in Table 8.4. *Whenever the bond involves two different atoms* (e.g., O—H) *the value listed is approximate rather than exact,* because it represents an average taken over two or more different species. Consider, for example, the O—H bond where we find

$$H—O—H(g) \longrightarrow H(g) + OH(g) \qquad \Delta H = +499 \text{ kJ}$$
$$H—O(g) \longrightarrow H(g) + O(g) \qquad \Delta H = +428 \text{ kJ}$$

Both of these reactions involve breaking a mole of O—H bonds, yet the experimental values of ΔH are quite different. The bond enthalpy listed in Table 8.4, 464 kJ/mol, is an average of these two values.

This limitation explains why, whenever possible, we use enthalpies of formation (ΔH_f°) rather than bond enthalpies to calculate the value of ΔH for a reaction. Calculations involving enthalpies of formation are expected to be accurate within ±0.1 kJ; the use of bond enthalpies can result in an error of 10 kJ or more.

8-7 The First Law of Thermodynamics

So far in this chapter our discussion has focused on thermochemistry, the study of the heat effects in chemical reactions. Thermochemistry is a branch of *thermodynamics,* which deals with all kinds of energy effects in all kinds of processes. Thermodynamics distinguishes between two types of energy. One of these is heat (q); the other is **work,** represented by the symbol w. The thermodynamic definition of work is quite different from its colloquial meaning. Quite simply, *work includes all forms of energy except heat.*

The **law of conservation of energy** states that energy (E) can be neither created nor destroyed; it can only be transferred between system and surroundings. That is,

$$\Delta E_{\text{system}} = -\Delta E_{\text{surroundings}}$$

Table 8.5 Heat, Work, and Energy

	q	w	$\Delta E = q + w$	Remarks
I	+ heat absorbed by the system	+ work done on the system	+ energy absorbed by the system	
II	+ heat absorbed by the system	− work done by the system	+ or − 	depends on the magnitude of q and w
III	− heat given off by the system	− work done by the system	− energy given off by the system	
IV	− heat given off by the system	+ work done on the system	+ or − 	depends on the magnitude of q and w

The **first law of thermodynamics** goes a step further. Taking account of the fact that there are two kinds of energy, heat and work, the first law states:

In any process, the total change in energy of a system, ΔE, is equal to the sum of the heat, q, and the work, w, transferred between the system and the surroundings.

$$\Delta E = q + w \qquad (8.5)$$

In applying the first law, note (Table 8.5) that q and w are positive when heat or work enters the system from the surroundings. ◀ⓘ If the transfer is in the opposite direction, from system to surroundings, q and w are negative.

If work is done on the system or if heat is added to it, its energy increases.

EXAMPLE 8.9

Calculate ΔE of a gas for a process in which the gas

(a) absorbs 20 J of heat and does 12 J of work by expanding.

(b) evolves 30 J of heat and has 52 J of work done on it as it contracts.

ANALYSIS

Information given:	(a) heat absorbed (20 kJ), work done by the system (12 kJ) (b) heat evolved (30 kJ), work done on the system (52 kJ)
Asked for:	ΔE for both (a) and (b)

STRATEGY

1. Decide on the signs for work (w) and heat (q).

   Recall that quantities are positive when they enter the system (absorbed, work done on the system) and negative when they leave the system (evolved, work done by the system).

2. Substitute into Equation 8.5.

   $\Delta E = w + q$

SOLUTION

(a) Signs for w and q	w is negative (work is done by the system) q is positive (heat is absorbed)
ΔE	$\Delta E = q + w = 20 \text{ J} + (-12 \text{ J}) = 8 \text{ J}$
(b) Signs for w and q	w is positive (work is done on the system; enters the system) q is negative (heat is evolved)
ΔE	$\Delta E = q + w = -30 \text{ J} + 52 \text{ J} = 22 \text{ J}$

Ordinarily, when a chemical reaction is carried out in the laboratory, any energy evolved is in the form of heat. Consider, for example, the reaction of oxygen with methane, the principal constituent of natural gas.

$$CH_4(g) + 2O_2(g) \longrightarrow CO_2(g) + 2H_2O(l) \qquad \Delta E = -885 \text{ kJ}$$

When you ignite methane in a Bunsen burner, the amount of heat evolved is very close to 885 kJ/mol. There is a small work effect, due to the decrease in volume that occurs when the reaction takes place (Figure 8.15), but this amounts to less than 1% of the energy change.

The situation changes if methane is used as a substitute for gasoline in an internal combustion engine. Here a significant fraction of the energy evolved in combustion is converted to useful work, propelling your car uphill, overcoming friction, charging the battery, or whatever. Depending on the efficiency of the engine, as much as 25% of the available energy might be converted to work; the amount of heat evolved through the tail pipe or radiator drops accordingly.

Finally, the energy available from the above reaction might be used to operate a fuel cell such as those involved in the space program. In that case, as much as 818 kJ/mol of useful electrical work could be obtained; relatively little heat is evolved. Summarizing this discussion in terms of an energy balance (per mole of methane reacting):

$P = 1$ atm

$P = 1$ atm

1 mol $CH_4(g)$
2 mol $O_2(g)$

1 mol $CO_2(g)$

Figure 8.15 Pressure-volume work. When the reaction

$$CH_4(g) + 2O_2(g) \longrightarrow CO_2(g) + 2H_2O(l)$$

is carried out in a cylinder fitted with a piston that exerts a pressure of 1 atm, a contraction occurs (3 mol gas $\rightarrow$ 1 mol gas). The piston falls, and a small amount of work is done on the reaction system.

	ΔE	q	w
Bunsen burner	-885 kJ	-890 kJ	$+5$ kJ
Automobile engine	-885 kJ	-665 kJ	-220 kJ
Fuel cell	-885 kJ	-67 kJ	-818 kJ

Notice that ΔE, like ΔH, is a *state property;* it has the same value regardless of how or where or why the reaction is carried out. In contrast, q and w are path-dependent; their values vary depending on whether the reaction is carried out in the atmosphere, an engine, or an electrical cell.

ΔH Versus ΔE

As noted earlier, for a reaction at constant pressure, such as that taking place in an open coffee-cup calorimeter, the heat flow is equal to the change in enthalpy. If a reaction is carried out at constant volume (as is the case in a sealed bomb calorimeter) and there is no mechanical or electrical work involved, no work is done. Under these conditions, with $w = 0$, the heat flow is equal to the change in energy, ΔE. Hence we have

$$\Delta H = q_p \qquad \Delta E = q_v$$

(q_p = heat flow at constant pressure, q_v = heat flow at constant volume).

Almost always, when you carry out a reaction in the laboratory, you do so at constant pressure, that of the atmosphere. That is why we devoted 90% of the space in this chapter to ΔH rather than ΔE.

We can obtain a relation between ΔH and ΔE for a chemical reaction at constant temperature by starting with the defining equation relating enthalpy, H, to energy, E:

$$H = E + PV$$

where V is the volume at pressure P. It follows that

$$\Delta H = \Delta E + \Delta(PV)$$

To evaluate $\Delta(PV)$ for a reaction note that

- the PV product for a liquid or solid can be ignored, since their molar volume is typically only about 0.1% of that for a gas.
- gases taking part in the reaction can be assumed to obey the ideal gas law, $PV = nRT$.

It follows that

$$\Delta(PV) = \Delta n_g RT$$

where Δn_g is the change in the number of moles of gas when the reaction takes place. Finally, we obtain the relation

$$\Delta H = \Delta E + \Delta n_g RT \qquad (8.6)$$

For the reaction at 25°C

$$CH_4(g) + 2O_2(g) \longrightarrow CO_2(g) + 2H_2O(l)$$

$\Delta n_g = n$ gaseous products $- n$ gaseous reactants

$$= 1 \text{ mol} - 3 \text{ mol} = -2 \text{ mol}$$

$$R = 8.31 \text{ J/mol} \cdot \text{K} \qquad \text{(recall Table 5.1)}$$

$$T = (273 + 25) \text{ K} = 298 \text{ K}$$

Hence:

$$\Delta H = \Delta E - 2 \text{ mol} (8.31 \text{ J/mol} \cdot \text{K})(298 \text{ K})$$

$$= \Delta E - 5.0 \times 10^3 \text{ J} = \Delta E - 5.0 \text{ kJ}$$

Energy can also be expressed in liter-atmospheres: 1 L · atm = 0.1013 kJ.

Recall that ΔE for this reaction is -885 kJ; it follows that ΔH is -890 kJ. The difference between ΔE and ΔH is less than 1%, which is typical for most reactions. Figure 8.15 illustrates the volume change that represents work being done on a system.

EXAMPLE 8.10

Calculate ΔH and ΔE at 25°C for the reaction that takes place when an oxyacetylene torch is used.

$$C_2H_2(g) + \tfrac{5}{2}O_2(g) \longrightarrow 2CO_2(g) + H_2O(g)$$

ANALYSIS

Information given:	chemical equation; T (25°C)
Information implied:	ΔH_f° from Table 8.3 moles of products and reactants R with energy units
Asked for:	ΔH and ΔE

STRATEGY

1. Use Table 8.3 to calculate ΔH for the reaction. Remember ΔH_f° for $O_2(g)$ is zero.
2. Find Δn: $\Delta n = n_{products} - n_{reactants}$
3. Recall that R is 8.31 J/mol · K when energy units are involved (Table 5.1).
4. Substitute into Equation 8.6 to find ΔE.

$$\Delta H = \Delta E + \Delta n_g RT$$

continued

SOLUTION	
1. ΔH	$\Delta H = 2\Delta H_f^\circ\ CO_2(g) + \Delta H_f^\circ\ H_2O(g) - \Delta H_f^\circ\ C_2H_2(g)$
	$\quad = 2\ (-393.5\ \text{kJ}) + (-241.8\ \text{kJ}) - (226.7\ \text{kJ}) = -1255.5\ \text{kJ}$
2. Δn_g	$(2\ \text{mol}\ CO_2 + 1\ \text{mol}\ H_2O) - (1\ \text{mol}\ C_2H_2 + \frac{5}{2}\ \text{mol}\ O_2) = -\frac{1}{2}\ \text{mol}$
3. ΔE	$\Delta H = \Delta E + \Delta n_g RT$
	$-1255.5\ \text{kJ} = \Delta E + (-0.5\ \text{mol})\left(8.31 \times 10^{-3}\ \dfrac{\text{kJ}}{\text{mol}\cdot\text{K}}\right)(298\ \text{K})$
	$\Delta E = -1255.5\ \text{kJ} + 1.24\ \text{kJ} = -1254.3\ \text{kJ}$
END POINT	

Note that in this case ΔH and ΔE differ from one another only by 1.2 kJ (about 0.1%), a very small difference indeed.

CHEMISTRY **BEYOND** THE CLASSROOM

Energy Balance in the Human Body

All of us require energy to maintain life processes and do the muscular work associated with such activities as studying, writing, walking, or jogging. The fuel used to produce that energy is the food we eat. In this discussion, we focus on energy input or energy output and the balance between them.

Energy Input

Energy values of food can be estimated on the basis of the content of carbohydrate, protein, and fat:

carbohydrate:	17 kJ (4.0 kcal) of energy per gram*
protein:	17 kJ (4.0 kcal) of energy per gram
fat:	38 kJ (9.0 kcal) of energy per gram

Alcohol, which doesn't fit into any of these categories, furnishes about 29 kJ (7.0 kcal) per gram.

Using these guidelines, it is possible to come up with the data given in Table A, which lists approximate energy values for standard portions of different types of foods. With a little practice, you can use the table to estimate your energy input within ±10%.

Energy Output

Energy output can be divided, more or less arbitrarily, into two categories.

1. Metabolic energy. This is the largest item in most people's energy budget; it comprises the energy necessary to

Table A Energy Values of Food Portions*

		kJ	kcal
Skim milk	8-oz glass (250 g)	347	83
Whole milk	8-oz glass (250 g)	610	146
Beer	12-oz glass (375 g)	493	118
Mixed vegetables	1/2 cup (125 g)	247	59
Broccoli	1/2 cup (125 g)	63	15
Yellow corn	1 ear	447	107
Fruit cocktail	1/2 cup (125 g)	230	55
Whole wheat bread	1 slice	272	65
Baked potato	1 item	920	220
Black beans	1/2 cup (125 g)	477	114
Lean ground beef	3 oz (84 g)	966	231
Ground turkey	3 oz (84 g)	836	200
Butter	1 tablespoon (15 g)	451	108
Olive oil	1 tablespoon (15 g)	497	119
Canola oil	1 tablespoon (15 g)	502	120
Sugar	1 teaspoon (5 g)	63	15

*Adapted from Michelle McGuire and Kathy A. Beerman: *Table of Food Composition for Nutritional Sciences: From Fundamentals to Food*, Thomson Wadsworth, Belmont, CA, 2007.

*In most nutrition texts, energy is expressed in kilocalories rather than kilojoules, although that situation is changing. Remember that 1 kcal = 4.18 kJ; 1 kg = 2.20 lb.

maintain life. Metabolic rates are ordinarily estimated based on a person's sex, weight, and age. For a young adult female, the metabolic rate is about 100 kJ per day per kilogram of body weight. For a 20-year-old woman weighing 120 lb

$$\text{metabolic rate} = 120 \text{ lb} \times \frac{1 \text{ kg}}{2.20 \text{ lb}} \times 100 \frac{\text{kJ}}{\text{d} \cdot \text{kg}}$$

$$= 5500 \text{ kJ/d} \ (1300 \text{ kcal/day})$$

For a young adult male of the same weight, the figure is about 15% higher.

2. Muscular activity. Energy is consumed whenever your muscles contract, whether in writing, walking, playing tennis, or even sitting in class. Generally speaking, energy expenditure for muscular activity ranges from 50% to 100% of metabolic energy, depending on lifestyle. If the 20-year-old woman referred to earlier is a student who does nothing more strenuous than lift textbooks, the energy spent on muscular activity might be

$$0.50 \times 5500 \text{ kJ} = 2800 \frac{\text{kJ}}{\text{d}}$$

Her total energy output per day would be 5500 kJ + 2800 kJ = 8300 kJ (2000 kcal).

Energy Balance

To maintain constant weight, your daily energy input, as calculated from the foods you eat, should be about 700 kJ (170 kcal) greater than output. The difference allows for the fact that about 40 g of protein is required to maintain body tissues and fluids. If the excess of input over output is greater than 700 kJ/day, the unused food (carbohydrate, protein, or fat) is converted to fatty tissue and stored as such in the body.

Fatty tissue consists of about 85% fat and 15% water; its energy value is

$$0.85 \times 9.0 \frac{\text{kcal}}{\text{g}} \times \frac{454 \text{ g}}{1 \text{ lb}} = 3500 \text{ kcal/lb}$$

This means that to lose one pound of fat, energy input from foods must be decreased by about 3500 kcal. To lose weight at a sensible rate of one pound per week, it is necessary to cut down by 500 kcal (2100 kJ) per day.

It's also possible, of course, to lose weight by increasing muscular activity. Table B shows the amount of energy consumed per hour with various types of exercise. In principle, a person weighing 150 lbs can lose approximately a pound a week by playing a Wii game of tennis for 2.5 hours.

The values listed in Table B reflect both the energy spent in physical activity and the amount used for basal metabolic rate (BMR). To calculate kcalories spent per minute of activity for your own body weight, multiply kcal/lb/hr by your exact weight and then multiply that number by the number of hours spent in the activity.

Table B Energy Spent on Various Activities*

Activity	kcal/lb/hr
Aerobic dance (vigorous)	3.7
Basketball (vigorous, full court)	5.8
Bicycling	
13 mph	2.7
15 mph	2.9
Cross-country skiing	
8 mph	6.2
Running	
5 mph	3.7
6 mph	4.4
Studying	0.66
Swimming	
20 yd/min	1.9
45 yd/min	3.5
50 yd/min	4.2
Walking (brisk pace)	
3.5 mph	2.1
4.5 mph	2.9
Wii games	
Bowling	1.3
Tennis	1.3

*Adapted from Linda Kelly DeBruyne and Kathryn Pinna, *Nutrition for Health and Healthcare, 5th Edition*, Table 6.2, pp. 148–149, Cengage, 2014.

Key Concepts

1. Relate heat flow to specific heat, mass, and Δt.
   (**Example 8.1; Problems 1–6**)
2. Calculate q for a reaction from calorimetric data.
   (**Examples 8.2, 8.3; Problems 7–20**)
3. Apply the rules of thermochemistry
   (**Examples 8.4, 8.5; Problems 21–30**)
4. Apply Hess's law to calculate ΔH.
   (**Example 8.6; Problems 31–36**)
5. Relate $\Delta H°$ to enthalpies of formation.
   (**Examples 8.7, 8.8; Problems 37–52**)
6. Relate ΔE, q, and w.
   (**Example 8.9; Problems 53–58**)
7. Relate ΔH and ΔE.
   (**Example 8.10; Problems 59–62**)

Key Equations

Heat flow	$q = mass \times c \times \Delta t$
Coffee-cup calorimeter	$q_{reaction} = -mass_{water} \times 4.18 \text{ J/g} \cdot {}^{\circ}\text{C} \times \Delta t$
Bomb calorimeter	$q_{cal} = C_{cal} \times \Delta t$
	$q_{reaction} = -(q_{cal} + qH_2O)$
Standard enthalpy change	$\Delta H^{\circ} = \Sigma \, \Delta H_f^{\circ} \text{ products} - \Sigma \, \Delta H_f^{\circ} \text{ reactants}$
First law of thermodynamics	$\Delta E = q + w$
ΔH versus ΔE	$\Delta H = \Delta E + \Delta n_g RT$

Key Terms

bond enthalpy	enthalpy	Hess's law	surroundings
calorimeter	—enthalpy of formation	joule	system
—bomb calorimeter	heat	specific heat	thermochemical equation
—coffee-cup calorimeter	—heat capacity	state properties	work
endothermic	—heat of fusion		
exothermic	—heat of vaporization		

Summary Problem

Ethane, C_2H_6, is a by-product of petroleum refining. The thermochemical equation for the combustion of ethane is:

$$C_2H_6(g) + \tfrac{7}{2}O_2(g) \longrightarrow 2CO_2(g) + 3H_2O(l)$$

$$\Delta H = -1559.7 \text{ kJ}$$

(a) Find ΔH for the thermochemical equations

$$2C_2H_6(g) + 7O_2(g) \longrightarrow 4CO_2(g) + 6H_2O(l)$$

$$2CO_2(g) + 3H_2O(l) \longrightarrow C_2H_6(g) + \tfrac{7}{2}O_2(g)$$

(b) The heat of vaporization for $H_2O(l)$ is $+44.0$ kJ/mol. Calculate ΔH for the equation

$$C_2H_6(g) + \tfrac{7}{2}O_2(g) \longrightarrow 2CO_2(g) + 3H_2O(g)$$

(c) The heats of formation of $CO_2(g)$ and $H_2O(l)$ are -393.5 and -285.8 kJ/mol, respectively. What is ΔH_f° for $C_2H_6(g)$?

(d) How much heat is evolved when 1.00 g of $C_2H_6(g)$ is burned in an open container to give $CO_2(g)$ and $H_2O(l)$?

(e) When 1.00 g of C_2H_6 is burned in a bomb calorimeter, the temperature rises from 22.00°C to 33.13°C. In a separate experiment, it is found that absorption of 10.0 kJ of heat by the calorimeter raises the temperature of the calorimeter by 2.15°C. Find $C_{calorimeter}$ and $q_{reaction}$.

(f) Find $\Delta E = q_v$ at 25°C for the combustion of one mole of ethane where $\Delta H = q_p = -1559.7$ kJ.

(g) Using bond enthalpies estimate ΔH for the formation of one mole of ethane gas from its elements.

Answers

(a) -3119.4 kJ; $+1559.7$ kJ
(b) -1427.7 kJ
(c) -84.7 kJ
(d) 51.9 kJ
(e) 4.65 kJ/°C; -51.8 kJ
(f) -1553.5 kJ
(g) -1523 kJ

Questions and Problems

Even-numbered questions and Challenge Problems have answers in Appendix 5 and fully worked solutions in the *Student Solutions Manual.*

8-1 Principles of Heat Flow

1. Titanium is a metal used in jet engines. Its specific heat is 0.523 J/g · °C. If 5.88 g of titanium absorb 4.78 J, what is the change in temperature?

2. Gold has a specific heat of 0.129 J/g · °C. When a 5.00-g piece of gold absorbs 1.33 J of heat, what is the change in temperature?

3. Copper is used in building the integrated circuits, chips, and printed circuit boards for computers. When 228 J of heat are absorbed by 125 g of copper at 22.38°C, the temperature rises to 27.12°C. What is the specific heat of copper?

4. Mercury was once used in thermometers and barometers. When 46.9 J of heat are absorbed by 100.0 g of mercury at 25.00°C, the temperature increases to 28.35°C. What is the specific heat of mercury?

5. The specific heat of aluminum is 0.902 J/g · °C. How much heat is absorbed by an aluminum pie tin with a mass of 473 g to raise its temperature from room temperature (23.00°C) to oven temperature (375°F)?

6. Chromium has a specific heat of 0.450 J/g · °C. How much heat is absorbed by 35.0 g of chromium if the temperature increases from 45°F to 88°F?

8-2 Measurement of Heat Flow; Calorimetry

7. Magnesium sulfate is often used in first-aid hot packs, giving off heat when dissolved in water. A coffee-cup calorimeter at 25°C contains 15.0 mL of water at 25°C. A 2.00-g sample of $MgSO_4$ is dissolved in the water and 1.51 kJ of heat are evolved. (You can make the following assumptions about the solution: volume = 15.0 mL, density = 1.00 g/mL, specific heat = 4.18 J/g · °C.)
   (a) Write a balanced equation for the solution process.
   (b) Is the process exothermic?
   (c) What is q_{H_2O}?
   (d) What is the final temperature of the solution?
   (e) What are the initial and final temperatures in °F?

8. Sodium chloride is added in cooking to enhance the flavor of food. When 10.00 g of NaCl are dissolved in 200.0 mL of water at 25.0°C in a coffee-cup calorimeter, 669 J of heat are absorbed. (You can make the following assumptions about the solution: volume = 200.0 mL, density = 1.00 g/mL, specific heat = 4.18 J/g · °C)
   (a) Is the solution process exothermic?
   (b) What is q_{H_2O}?
   (c) What is the final temperature of the solution?

9. When 375 mL of water (d = 1.00 g/mL) at 32°C are mixed with 125 mL of water at 92°C, what is the final temperature of the mixture? (Assume no heat is lost to the surroundings.)

10. How many mL of water at 10°C (2 significant figures) must be added to 75 mL of water at 35°C to obtain a final temperature of 19°C? (Make the same assumptions as in Question 9.)

11. When one mol of KOH is neutralized by sulfuric acid, q = −56 kJ. (This is called the heat of neutralization.) At 23.7°C, 25.0 mL of 0.475 M H_2SO_4 is neutralized by 0.613 M KOH in a coffee-cup calorimeter. Assume that the specific heat of all solutions is 4.18 J/g · °C, that the density of all solutions is 1.00 g/mL, and that volumes are additive.
   (a) How many mL of KOH is required to neutralize H_2SO_4?
   (b) What is the final temperature of the solution?

12. The heat of neutralization, ΔH_{neut}, can be defined as the amount of heat released (or absorbed), q, per mole of acid (or base) neutralized. ΔH_{neut} for nitric acid is −52 kJ/mol HNO_3. At 27.3°C, 50.00 mL of 0.743M HNO_3 is neutralized by 1.00 M $Sr(OH)_2$ in a coffee-cup calorimeter.
   (a) How many mL of $Sr(OH)_2$ were used in the neutralization?
   (b) What is the final temperature of the resulting solution? (Use the assumptions in Question 11.)

13. Fructose is a sugar commonly found in fruit. A sample of fructose, $C_6H_{12}O_6$, weighing 4.50 g is burned in a bomb calorimeter that contains 1.00 L of water (d = 1.00 g/mL). The heat capacity of the calorimeter is 16.97 kJ/°C. The temperature of the calorimeter and water rise from 23.49°C to 27.72°C.
   (a) What is q for the calorimeter?
   (b) What is q for water in the calorimeter?
   (c) What is q when 4.50 g of fructose are burned in the calorimeter?
   (d) What is q for the combustion of one mole of fructose?

14. In earlier times, ethyl ether was commonly used as an anesthetic. It is, however, highly flammable. When five milliliters of ethyl ether, $C_4H_{10}O(l)$ (d = 0.714 g/mL), are burned in a bomb calorimeter, the temperature rises from 23.5°C to 39.7°C. The calorimeter contains 1.200 kg of water and has a heat capacity of 5.32 kJ/°C.
   (a) What is q_{H_2O}?
   (b) What is q_{cal}?
   (c) What is q for the combustion of 5.00 mL of ethyl ether?
   (d) What is q for the combustion of one mole of ethyl ether?

15. Isooctane is a primary component of gasoline and gives gasoline its octane rating. Burning 1.00 mL of isooctane (d = 0.688 g/mL) releases 33.0 kJ of heat. When 10.00 mL of isooctane are burned in a bomb calorimeter, the temperature in the bomb and water rises from 23.2°C to 66.5°C. The bomb contains 1.00 kg of water. What is the heat capacity of the calorimeter?

16. Ethyl alcohol, C_2H_5OH, is the intoxicating agent in liquor. Burning 1.00 g of ethyl alcohol in an excess of oxygen at 23.28°C and constant volume releases 29.52 kJ of heat. When 7.40 g of ethyl alcohol is burned in oxygen under the same conditions in a bomb calorimeter, the temperature of the bomb and water rises from 23.28°C to 48.04°C. The bomb holds 0.750 kg of water.
   (a) What is q for the combustion of the ethyl alcohol in the bomb calorimeter?
   (b) What is q_{H_2O}?
   (c) What is q_{cal}?
   (d) What is the heat capacity of the calorimeter?

17. Acetic acid, $HC_2H_3O_2$, is responsible for the sour taste of vinegar. Combustion of acetic acid gives off 14.52 kJ/g of heat. When 15.00 g of acetic acid is burned in a bomb calorimeter (heat capacity = 2.166 kJ/°C) containing 0.800 kg of water, the final temperature in the calorimeter is 64.14°C. What was the initial temperature of the calorimeter?

18. Acetylene, C_2H_2, is used in welding torches. It releases a lot of energy when burned in oxygen. The combustion of one gram of acetylene releases 48.2 kJ. A 0.750-g sample of acetylene is burned in a bomb calorimeter (heat capacity = 1.117 kJ/°C) that contains 800.0 g of water. The final temperature of the bomb and water after combustion is 35.2°C. What is the initial temperature of the bomb and water?

19. Salicylic acid, $C_7H_6O_3$, is one of the starting materials in the manufacture of aspirin. When 1.00 g of salicylic acid burns in a bomb calorimeter, the temperature of the bomb and water goes from 23.11°C to 28.91°C. The calorimeter and water absorb 21.9 kJ of heat. How much heat is given off when one mole of salicylic acid burns?

20. Methanol (CH_3OH) is also known as wood alcohol and can be used as a fuel. When one mole of methanol is burned, 1453 kJ of heat are evolved. When methanol is burned in a bomb calorimeter, 71.8 kJ of heat are evolved by the methanol. How many mL of methanol ($d = 0.791$ g/mL) were burned?

8-4 Thermochemical Equations

21. Nitrogen oxide (NO) has been found to be a key component in many biological processes. It also can react with oxygen to give the brown gas NO_2. When one mole of NO reacts with oxygen, 57.0 kJ of heat are evolved.

(a) Write the thermochemical equation for the reaction between one mole of nitrogen oxide and oxygen.

(b) Is the reaction exothermic or endothermic?

(c) Draw an energy diagram showing the path of this reaction. (Figure 8.6 is an example of such an energy diagram.)

(d) What is ΔH when 5.00 g of nitrogen oxide react?

(e) How many grams of nitrogen oxide must react with an excess of oxygen to liberate ten kilojoules of heat?

22. Small amounts of oxygen gas can be prepared in the laboratory by decomposing potassium chlorate with heat. A by-product of the decomposition is potassium chloride. When one mole of potassium chlorate decomposes, 44.7 kJ are evolved.

(a) Write a balanced thermochemical equation for the decomposition of one mole of potassium chlorate.

(b) Is the reaction exothermic?

(c) Draw an energy diagram showing the path of this reaction. (Figure 8.6 is an example of such an energy diagram.)

(d) What is ΔH when 3.00 g of potassium chlorate decompose?

(e) How many grams of potassium chlorate need to be decomposed to liberate fifteen kilojoules of heat?

23. In the late eighteenth century Priestley prepared ammonia by reacting $HNO_3(g)$ with hydrogen gas. The thermodynamic equation for the reaction is

$$HNO_3(g) + 4H_2(g) \longrightarrow NH_3(g) + 3H_2O(g) \quad \Delta H = -637\,kJ$$

(a) Calculate ΔH when one mole of hydrogen gas reacts.

(b) What is ΔH when 10.00 g of $NH_3(g)$ are made to react with an excess of steam to form HNO_3 and H_2 gases?

24. Barium chloride is sometimes used in fireworks to give a bright green color. It gives off heat when dissolved in water.

$$BaCl_2(s) \longrightarrow Ba^{2+}(aq) + 2Cl^-(aq)$$

(a) What is ΔH when one mol of $BaCl_2$ precipitates from solution?

(b) What is ΔH when 15.00 g of $BaCl_2$ precipitate?

25. Strontium metal is responsible for the red color in fireworks. Fireworks manufacturers use strontium carbonate, which can be produced by combining strontium metal, graphite (C), and oxygen gas. The formation of one mole of $SrCO_3$ releases 1.220×10^3 kJ of energy.

(a) Write a balanced thermochemical equation for the reaction.

(b) What is ΔH when 10.00 L of oxygen at 25°C and 1.00 atm are used by the reaction?

26. Nitroglycerin, $C_3H_5(NO_3)_3(l)$, is an explosive most often used in mine or quarry blasting. It is a powerful explosive because four gases (N_2, O_2, CO_2, and steam) are formed when nitroglycerin is detonated. In addition, 6.26 kJ of heat are given off per gram of nitroglycerin detonated.

(a) Write a balanced thermochemical equation for the reaction.

(b) What is ΔH when 4.65 mol of products are formed?

27. A typical fat in the body is glyceryl trioleate, $C_{57}H_{104}O_6$. When it is metabolized in the body, it combines with oxygen to produce carbon dioxide, water, and 3.022×10^4 kJ of heat per mole of fat.

(a) Write a balanced thermochemical equation for the metabolism of fat.

(b) How many kilojoules of energy must be evolved in the form of heat if you want to get rid of five pounds of this fat by combustion?

(c) How many nutritional calories is this? (1 nutritional calorie = 1×10^3 calories)

28. Use the same fat described in Question 27.

(a) Write a thermochemical equation for the metabolism of fat.

(b) A dieter reduces his food intake by 425 nutritional calories a day (1 nutritional calorie = 1000 calories). How many kJ can he lose in one day?

(c) How many pounds of fat are "lost" in one week with this diet?

29. Which requires the absorption of a greater amount of heat—vaporizing 100.0 g of benzene or boiling 20.0 g of water? (Use Table 8.2.)

30. Which evolves more heat—freezing 100.0 g of benzene or 100.0 g of bromine?

31. A student is asked to calculate the amount of heat involved in changing 10.0 g of liquid bromine at room temperature (22.5°C) to vapor at 59.0°C. To do this, one must use Tables 8.1 and 8.2 for information on the specific heat, boiling point, and heat of vaporization of bromine. In addition, the following step-wise process must be followed.

(a) Calculate ΔH for: $Br_2(l, 22.5°C) \longrightarrow Br_2(l, 59.0°C)$

(b) Calculate ΔH for: $Br_2(l, 59.0°C) \longrightarrow Br_2(g, 59.0°C)$

(c) Using Hess's law, calculate ΔH for:
$Br_2(l, 22.5°C) \longrightarrow Br_2(g, 59.0°C)$

32. Follow the step-wise process outlined in Problem 31 to calculate the amount of heat involved in condensing 100.00 g of benzene gas (C_6H_6) at 80.00°C to liquid benzene at 25.00°C. Use Tables 8.1 and 8.2 for the specific heat, boiling point, and heat of vaporization of benzene.

33. A lead ore, galena, consisting mainly of lead(II) sulfide, is the principal source of lead. To obtain the lead, the ore is first heated in the air to form lead oxide.

$$PbS(s) + \tfrac{3}{2}O_2(g) \longrightarrow PbO(s) + SO_2(g) \quad \Delta H = -415.4 \text{ kJ}$$

The oxide is then reduced to metal with carbon.

$$PbO(s) + C(s) \longrightarrow Pb(s) + CO(g) \qquad \Delta H = +108.5 \text{ kJ}$$

Calculate ΔH for the reaction of one mole of lead(II) sulfide with oxygen and carbon, forming lead, sulfur dioxide, and carbon monoxide.

34. A reaction used to produce the silicon for semiconductors from sand (SiO_2), can be broken up into three steps:

$$SiO_2(s) + 2C(s) \longrightarrow Si(s) + 2CO(g) \qquad \Delta H = 689.9 \text{ kJ}$$
$$Si(s) + 2Cl_2(g) \longrightarrow SiCl_4(g) \qquad \Delta H = -657.0 \text{ kJ}$$
$$SiCl_4(g) + 2Mg(s) \longrightarrow 2MgCl_2(s) + Si(s) \quad \Delta H = -625.6 \text{ kJ}$$

(a) Write a thermochemical equation for the overall reaction where silicon is obtained from silicon dioxide and CO and $MgCl_2$ are by-products.
(b) What is ΔH for the formation of one mole of silicon?
(c) Is the overall reaction exothermic?

35. Given the following thermochemical equations,

$$C_2H_2(g) + \tfrac{5}{2}O_2(g) \longrightarrow 2CO_2(g) + H_2O(l) \quad \Delta H = -1299.5 \text{ kJ}$$
$$C(s) + O_2(g) \longrightarrow CO_2(g) \qquad \Delta H = -393.5 \text{ kJ}$$
$$H_2(g) + \tfrac{1}{2}O_2(g) \longrightarrow H_2O(l) \qquad \Delta H = -285.8 \text{ kJ}$$

calculate ΔH for the decomposition of one mole of acetylene, $C_2H_2(g)$, to its elements in their stable state at 25°C and 1 atm.

36. Given the following thermochemical equations:

$$4B(s) + 3O_2(g) \longrightarrow 2B_2O_3(s) \qquad \Delta H° = -2543.8 \text{ kJ}$$
$$H_2(g) + \tfrac{1}{2}O_2(g) \longrightarrow H_2O(g) \qquad \Delta H° = -241.8 \text{ kJ}$$
$$B_2H_6(s) + 3O_2 \longrightarrow B_2O_3(s) + 3H_2O(g) \quad \Delta H° = -2032.9 \text{ kJ}$$

Calculate $\Delta H°$ for the decomposition of B_2H_6 into its elements.

8-5 Enthalpies of Formation

37. Write thermochemical equations for the decomposition of one mole of the following compounds into the elements in their stable states at 25°C and 1 atm.
(a) ethyl alcohol, C_2H_5OH (l)
(b) sodium fluoride (s)
(c) magnesium sulfate (s)
(d) ammonium nitrate (s)

38. Write thermochemical equations for the formation of one mole of the following compounds from the elements in their native states at 25°C and 1 atm.
(a) solid potassium chlorate
(b) liquid carbon tetrachloride
(c) gaseous hydrogen iodide
(d) solid silver(I) oxide

39. Given

$$2Al_2O_3(s) \longrightarrow 4Al(s) + 3O_2(g) \quad \Delta H° = 3351.4 \text{ kJ}$$

(a) What is the heat of formation of aluminum oxide?
(b) What is $\Delta H°$ for the formation of 12.50 g of aluminum oxide?

40. Given

$$2Cr_2O_3(s) \longrightarrow 4Cr(s) + 3O_2(g) \quad \Delta H° = +2269.4 \text{ kJ}$$

(a) What is the heat of formation of chromium(III) oxide?
(b) What is $\Delta H°$ for the formation of 13.65 g of chromium (III) oxide?

41. Limestone, $CaCO_3$, when subjected to a temperature of 900°C in a kiln, decomposes to calcium oxide and carbon dioxide. How much heat is evolved or absorbed when one gram of limestone decomposes? (Use Table 8.3.)

42. When hydrazine reacts with oxygen, nitrogen gas and steam are formed.
(a) Write a thermochemical equation for the reaction.
(b) How much heat is evolved or absorbed if 1.683 L of steam at 125°C and 772 mm Hg are obtained?

43. Use Table 8.3 to obtain $\Delta H°$ for the following thermochemical equations:
(a) $Mg(OH)_2(s) + 2NH_4^+(aq) \longrightarrow Mg^{2+}(aq) + 2NH_3(g) + 2H_2O(l)$
(b) $PbO(s) + C(s) \longrightarrow CO(g) + Pb(s)$
(c) $Mn(s) + 4H^+(aq) + SO_4^{2-}(aq) \longrightarrow Mn^{2+}(aq) + SO_2(g) + 2H_2O(l)$

44. Use Table 8.3 to obtain $\Delta H°$ for the following thermochemical equations:
(a) $Zn(s) + 2H^+(aq) \longrightarrow Zn^{2+}(aq) + H_2(g)$
(b) $2H_2S(g) + 3O_2(g) \longrightarrow 2SO_2(g) + 2H_2O(g)$
(c) $3Ni(s) + 2NO_3^-(aq) + 8H^+(aq) \longrightarrow$
$$3Ni^{2+}(aq) + 2NO(g) + 4H_2O(l)$$

45. Use the appropriate tables to calculate $\Delta H°$ for
(a) the reaction between copper(II) oxide and carbon monoxide to give copper metal and carbon dioxide.
(b) the decomposition of one mole of methyl alcohol (CH_3OH) to methane and oxygen gases.

46. Use the appropriate tables to calculate $\Delta H°$ for
(a) the reaction between $MgCO_3(s)$ and a strong acid to give $Mg^{2+}(aq)$, $CO_2(g)$, and water.
(b) the precipitation of iron(III) hydroxide from the reaction between iron(III) and hydroxide ions.

47. When one mole of nitroglycerine, $C_3H_5(NO_3)_3(l)$ decomposes, nitrogen, oxygen, carbon dioxide, and liquid water are formed. The decomposition liberates 5725 kJ of heat.
(a) Write a balanced thermochemical equation for the reaction.
(b) Using Table 8.3, calculate $\Delta H_f°$ for nitroglycerine.

48. When one mole of calcium carbonate reacts with ammonia, solid calcium cyanamide, $CaCN_2$, and liquid water are formed. The reaction absorbs 90.1 kJ of heat.
(a) Write a balanced thermochemical equation for the reaction.
(b) Using Table 8.3, calculate $\Delta H_f°$ for calcium cyanamide.

49. Chlorine trifluoride is a toxic, intensely reactive gas. It was used in World War II to make incendiary bombs. It reacts with ammonia and forms nitrogen, chlorine, and hydrogen fluoride gases. When two moles of chlorine trifluoride react, 1196 kJ of heat are evolved.

(a) Write a thermochemical equation for the reaction.
(b) What is $\Delta H_f°$ for ClF_3?

50. When one mole of ethylene gas, C_2H_4, reacts with fluorine gas, hydrogen fluoride and carbon tetrafluoride gases are formed and 2496.7 kJ of heat are given off. What is $\Delta H_f°$ for $CF_4(g)$?

51. Glucose, $C_6H_{12}O_6(s)$, ($\Delta H_f° = -1275.2$ kJ/mol) is converted to ethyl alcohol, $C_2H_5OH(l)$, and carbon dioxide in the fermentation of grape juice. What quantity of heat is liberated when 750.0 mL of wine containing 12.0% ethyl alcohol by volume ($d = 0.789$ g/cm^3) are produced by the fermentation of grape juice?

52. When ammonia reacts with dinitrogen oxide gas ($\Delta H_f° = 82.05$ kJ/mol), liquid water and nitrogen gas are formed. How much heat is liberated or absorbed by the reaction that produces 345 mL of nitrogen gas at 25°C and 717 mm Hg?

8-7 First Law of Thermodynamics

53. How many kJ are equal to 3.27 L · atm of work?

54. How many L · atm are equal to 12.2 kJ of work?

55. Find
(a) ΔE when a gas absorbs 18 J of heat and has 13 J of work done on it.
(b) q when 72 J of work are done on a system and its energy is increased by 61 J.

56. Calculate
(a) q when a system does 54 J of work and its energy decreases by 72 J.
(b) ΔE for a gas that releases 38 J of heat and has 102 J of work done on it.

57. Consider the following reaction in a vessel with a movable piston.

$$R(g) + T(g) \longrightarrow X(g)$$

As the reaction takes place, the piston loses 1072 J of heat. The piston moves down and the surroundings do 549 J of work on the system. What is ΔE?

58. Consider the following reaction in the vessel described in Question 57.

$$A(g) + B(g) \longrightarrow C(s)$$

For this reaction, $\Delta E = 286$ J, the piston moves up and the system absorbs 388 J of heat from its surroundings.
(a) Is work done by the system?
(b) How much work?

59. Determine the difference between ΔH and ΔE at 25°C for

$$2CO(g) + O_2(g) \longrightarrow 2CO_2(g)$$

60. For the vaporization of one mole of bromine at 59°C, determine
(a) ΔH (Table 8.2)
(b) ΔPV
(c) ΔE

61. Consider the combustion of propane, C_3H_8, the fuel that is commonly used in portable gas barbeque grills. The products of combustion are carbon dioxide and liquid water.
(a) Write a thermochemical equation for the combustion of one mole of propane.
(b) Calculate ΔE for the combustion of propane at 25°C.

62. Consider the combustion of one mole of methyl alcohol, $CH_3OH(l)$, which yields carbon dioxide gas and steam.
(a) Write a thermochemical equation for the reaction. (Use Table 8.2.)
(b) Calculate ΔE at 25°C.

Unclassified

63. Butane gas, C_4H_{10}, is sold to campers as bottled fuel. Its density at 25°C and 1.00 atm is 2.38 g/L. What volume of butane gas at 25°C and 1.00 atm is required to heat one gallon of water ($d = 1.00$ g/mL) from 25°C to 98°C? The reaction for the combustion of butane ($\Delta H_f° = -125.6$ kJ/mol) is

$$C_4H_{10}(g) + \tfrac{13}{2}O_2(g) \longrightarrow 4CO_2(g) + 5H_2O(g)$$

64. The BTU (British thermal unit) is the unit of energy most commonly used in the United States. One joule = 9.48×10^{-4} BTU. What is the specific heat of water in BTU/lb · °F? (Specific heat of water is 4.18 J/g · °C.)

65. Natural gas companies in the United States use the "therm" as a unit of energy. One therm is 1×10^5 BTU.
(a) How many joules are in one therm? (1 J = 9.48×10^{-4} BTU)
(b) When propane gas, C_3H_8, is burned in oxygen, CO_2 and steam are produced. How many therms of energy are given off by 1.00 mol of propane gas?

66. A 12-oz can of most colas has about 120 nutritional calories (1 nutritional calorie = 1 kcal). Approximately how many minutes of walking are required to burn up as energy the calories taken in after drinking a can of cola? (Walking uses up about 250 kcal/h.)

67. Given the following reactions,

$$N_2H_4(l) + O_2(g) \longrightarrow N_2(g) + 2H_2O(g) \quad \Delta H° = -534.2 \text{ kJ}$$
$$H_2(g) + \tfrac{1}{2}O_2(g) \longrightarrow H_2O(g) \quad \Delta H° = -241.8 \text{ kJ}$$

Calculate the heat of formation of hydrazine.

68. In World War II, the Germans made use of otherwise unusable airplane parts by grinding them up into powdered aluminum. This was made to react with ammonium nitrate to produce powerful bombs. The products of this reaction were nitrogen gas, steam, and aluminum oxide. If 10.00 kg of ammonium nitrate are mixed with 10.00 kg of powdered aluminum, how much heat is generated?

69. One way to lose weight is to exercise. Playing tennis for half an hour consumes about 225 kcal of energy. How long would you have to play tennis to lose one pound of body fat? (One gram of body fat is equivalent to 32 kJ of energy.)

70. Consider the reaction between methane and oxygen producing carbon dioxide and water. Suppose that the reaction is carried out in a furnace used to heat a house. If $q = -890$ kJ and $w = +5$ kJ, what is ΔE? ΔH at 25°C?

71. Consider burning ethane gas, C_2H_6 in oxygen (combustion) forming CO_2 and water.
(a) How much energy (in J) is produced in the combustion of one molecule of ethane?
(b) What is the energy of a photon of ultraviolet light with a wavelength of 12.6 nm?
(c) Compare your answers for (a) and (b).

72. On complete combustion at constant pressure, a 1.00-L sample of a gaseous mixture at 0°C and 1.00 atm (STP) evolves 75.65 kJ of heat. If the gas is a mixture of ethane (C_2H_6) and propane (C_3H_8), what is the mole fraction of ethane in the mixture?

73. Microwave ovens convert radiation to energy. A microwave oven uses radiation with a wavelength of 12.5 cm. Assuming that all the energy from the radiation is converted to heat without loss, how many moles of photons are required to raise the temperature of a cup of water (350.0 g, specific heat = 4.18 J/g · °C) from 23.0°C to 99.0°C?

74. In 2010, 3.30×10^9 gallons of gasoline were consumed in the United States. The following assumptions can be made:

- Gasoline is mainly n-octane, C_8H_{18} (d = 0.7028 g/mL).
- Burning one mole of n-octane in oxygen releases 5564.2 kJ of heat.
- The heat capacity C of the surface region of the earth is 2.6×10^{23} J/K.

What is the increase in temperature of the surface region of the earth due to gasoline consumption in the United States?

75. Some solar-heated homes use large beds of rocks to store heat.

(a) How much heat is absorbed by 100.0 kg of rocks if their temperature increases by 12°C? (Assume that c = 0.82 J/g · °C.)

(b) Assume that the rock pile has total surface area 2 m². At maximum intensity near the earth's surface, solar power is about 170 watts/m². (1 watt = 1 J/s.) How many minutes will it take for solar power to produce the 12°C increase in part (a)?

Conceptual Questions

76. Consider a solution prepared by dissolving 10.00 g of NaOH in 1.00 L of water.

(a) When the solid dissolves, will the temperature of the solution increase?

(b) What is the sign of H for the process?

(c) Will dissolving 5.00 g of NaOH increase Δt?

(d) Will dissolving one mole of NaOH in 1.00 L of water increase t_f?

77. Draw a cylinder with a movable piston containing six molecules of a liquid. A pressure of 1 atm is exerted on the piston. Next draw the same cylinder after the liquid has been vaporized. A pressure of one atmosphere is still exerted on the piston. Is work done *on* the system or *by* the system?

78. Redraw the cylinder in Question 77 after work has been done on the system.

79. Which statement(s) is/are true about bond enthalpy?

(a) The bond energy for a triple bond between A and B is three times that of a single bond between A and B.

(b) ΔH for the breaking of a bond is always a negative number.

(c) Energy is required to make a bond.

(d) Bond enthalpy is defined only for bonds broken or formed in the gaseous state.

(e) The presence of π bonds does not influence the geometry of a molecule. However, the presence of π bonds affects the value of the bond enthalpy between two atoms.

80. Equal masses of liquid A, initially at 100°C, and liquid B, initially at 50°C, are combined in an insulated container. The final temperature of the mixture is 80°C. All the heat flow occurs between the two liquids. The two liquids do not react with each other. Is the specific heat of liquid A larger than, equal to, or smaller than the specific heat of liquid B?

81. Determine whether the statements given below are true or false. Consider an endothermic process taking place in a beaker at room temperature.

(a) Heat flows from the surroundings to the system.

(b) The beaker is cold to the touch.

(c) The pressure of the system decreases.

(d) The value of q for the system is positive.

82. An exothermic reaction is carried out in a coffee-cup calorimeter. Which of the following statements is/are NOT true for the process?

(a) The temperature of the water increases.

(b) Heat is absorbed by the water.

(c) The enthalpy of the products is higher than the enthalpy of the reactants.

(d) $q_{H_2O} = q_{rxn}$

(e) $q_{rxn} > 0$

(f) $q_{rxn} + q_{H_2O} = 0$

83. Determine whether the statements given below are true or false. Consider enthalpy (H).

(a) It is a state property.

(b) $q_{reaction}$ (at constant P) = $\Delta H = H_{products} - H_{reactants}$

(c) The magnitude of ΔH is independent of the amount of reactant.

(d) In an exothermic process, the enthalpy of the system remains unchanged.

Challenge Problems

84. Microwave ovens emit microwave radiation that is absorbed by water. The absorbed radiation is converted to heat that is transferred to other components of the food. Suppose the microwave radiation has wavelength 12.5 cm. How many photons are required to increase the temperature of 1.00×10^2 mL of water (d = 1.0 g/mL) from 20°C to 100°C if all the energy of the photons is converted to heat?

85. On a hot day, you take a six-pack of soda on a picnic, cooling it with ice. Each empty (aluminum) can weighs 12.5 g. A can contains 12.0 oz of soda. The specific heat of aluminum is 0.902 J/g · °C; take that of soda to be 4.10 J/g · °C.

(a) How much heat must be absorbed from the six-pack to lower the temperature from 25.0° to 5.0°C?

(b) How much ice must be melted to absorb this amount of heat? (ΔH_{fus} of ice is given in Table 8.2.)

86. A cafeteria sets out glasses of tea at room temperature. The customer adds ice. Assuming that the customer wants to have some ice left when the tea cools to 0°C, what fraction of the total volume of the glass should be left empty for adding ice? Make any reasonable assumptions needed to work this problem.

87. The thermite reaction was once used to weld rails:

$$2Al(s) + Fe_2O_3(s) \longrightarrow Al_2O_3(s) + 2Fe(s)$$

(a) Using heat of formation data, calculate ΔH for this reaction.

(b) Take the specific heats of Al_2O_3 and Fe to be 0.77 and 0.45 J/g · °C, respectively. Calculate the temperature to which the products of this reaction will be raised, starting at room temperature, by the heat given off in the reaction.

(c) Will the reaction produce molten iron (mp Fe = 1535°C, ΔH_{fus} = 270 J/g)?

88. A sample of sucrose, $C_{12}H_{22}O_{11}$, is contaminated by sodium chloride. When the contaminated sample is burned in a bomb calorimeter, sodium chloride does not burn. What is the percentage of sucrose in the sample if a temperature increase of 1.67°C is observed when 3.000 g of the sample are burned in the calorimeter? Sucrose gives off 5.64×10^3 kJ/mol when burned. The heat capacity of the calorimeter and water is 22.51 kJ/°C.

89. A wad of steel wool (specific heat = 0.45 J/g · °C) at 25°C is misplaced in a microwave oven, which later is accidentally turned on high. The oven generates microwaves of 13.5 cm wavelength at the rate of 925 moles of those photons per second. All of the photons are converted to high-heat energy, raising the temperature of the steel wool. Steel wool reacts explosively with the oxygen in the oven when the steel wool reaches 400.0°C. If the steel wool explodes after 1.55 seconds, how many grams of steel wool were accidentally put into the oven?

90. Consider a metal ion A^{2+} and its nitrate salt, $A(NO_3)_2$. In an experiment, 35.00 mL of a 0.217 M solution of $A(NO_3)_2$ is made to react with 25.00 mL of 0.195 M NaOH. A precipitate, $A(OH)_2$, forms. Along with the precipitation, the temperature increases from 24.8°C to 28.2°C. What is ΔH for the precipitation of $A(OH)_2$? The following assumptions can be made.

- The density of the solution is 1.00 g/mL.
- Volumes are additive.
- The specific heat of the solution is 4.18 J/g · °C.

9

Liquids and Solids

The painting shows water in two of its physical states: liquid and solid.

See plastic Nature working
    to this end,
The single atoms each to
    other tend,
Attract, attracted to, the
    next in place
Form'd and impell'd its
    neighbour to embrace.

—ALEXANDER POPE
From *An Essay on Man:*
Epistle III

Bridgeman Art Library, London/SuperStock

In Chapter 5, we pointed out that at ordinary temperatures and pressures, all gases follow the ideal gas law. In this chapter we will examine liquids and solids. Unfortunately, there are no simple relations analogous to the ideal gas law that correlate the properties of these two physical states.

- In Section 9-1 we will compare some of the properties of the three different phases and give two reasons for this difference in behavior.
- Section 9-2 will focus on the different phases of a pure substance. We will consider different phenomena related to gas-liquid equilibria, including *vapor pressure, boiling point behavior,* and *critical properties.*
- Section 9-3 deals with phase diagrams, which describe all these types of phase equilibria (gas-liquid, gas-solid, and liquid-solid).
- Section 9-4 explores the relationship between particle structure, interparticle forces, and physical properties in molecular substances.
- Section 9-5 extends the discussion to nonmolecular solids (network covalent, ionic, and metallic).
- Section 9-6 is devoted to the crystal structure of ionic and metallic solids.

9-1 Comparing Solids, Liquids, and Gases

Why are liquids and solids so different from gases? There are two reasons for this difference in behavior.

1. *Molecules are much closer to one another in liquids and solids.* In the gas state, particles are typically separated by ten molecular diameters or more; in liquids and solids, they touch one another. This explains why liquids and solids have densities so much larger than those of gases. At 100°C, 1 atm, water ($H_2O(l)$) has a density of 0.95 g/mL; that of steam ($H_2O(g)$) under the same condition is only 0.00059 g/mL (Table 9.1). Because of such small numbers for gas densities, they are usually given in g/L, whereas those of liquids and solids are in g/mL.

Table 9.1 Comparison of the Three Phases of Water

Phase	Molecular Spacing	Density	Volume of One Mole
Steam ($H_2O(g)$)		at 100°C, 1 atm 0.00059 g/mL	at 100°C, 1 atm 31 L
Water ($H_2O(l)$)		at 100°C, 1 atm 0.95 g/mL	19 mL
Ice ($H_2O(s)$)		at 0°C, 1 atm 0.92 g/mL	20 mL

Table 9.1 also shows that the volume of one mole of water and one mole of ice are similar, whereas that of steam is about 1500 times greater. The low density and large molar volume of steam result from the large separation between the water molecules.

For the same reason, liquids and solids are much less compressible than gases. When the pressure on liquid water is increased from 1 to 2 atm, the volume decreases by about 0.0045%. The same change in pressure reduces the volume of an equal amount of ideal gas by 50%.

2. *Intermolecular forces, which are essentially negligible with gases, play a much more important role in liquids and solids.* Among the effects of these forces is the phenomenon of *surface tension,* shown by liquids. Molecules at the surface are attracted inward by unbalanced intermolecular forces; as a result, liquids tend to form spherical drops, for which the ratio of surface area to volume is as small as possible (Figure 9.1). Water, in which there are relatively strong intermolecular forces, has a high surface tension. That of most organic liquids is lower, which explains why they tend to "wet" or spread out on solid surfaces more readily than water. The wetting ability of water can be increased by adding a soap or detergent, which drastically lowers the surface tension.

Figure 9.1 **Surface tension causes water to bead on the polished surface of a car.**

9-2 Liquid-Vapor Equilibrium

All of us are familiar with the process of vaporization, in which a liquid is converted to a gas, commonly referred to as a *vapor.* In an open container, *evaporation* continues until all the liquid is gone. If the container is closed, the situation is quite different. At first, the movement of molecules is primarily in one direction, from liquid to vapor. Here, however, the vapor molecules cannot escape from the container. Some of them collide with the surface and re-enter the liquid. As time passes and the concentration of molecules in the vapor increases, so does the rate of condensation. When the rate of condensation becomes equal to the rate of vaporization, the liquid and vapor are in a state of dynamic **equilibrium**:

$$\text{liquid} \rightleftharpoons \text{vapor}$$

The double arrow implies that the forward and reverse processes are occurring at the same rate, which is characteristic of a dynamic equilibrium (Figure 9.2). The vapor, like any gas, can be assumed to obey the ideal gas law.

Figure 9.2 **Liquid-vapor equilibrium.** Under the conditions shown, the bromine liquid and vapor have established a dynamic equilibrium.

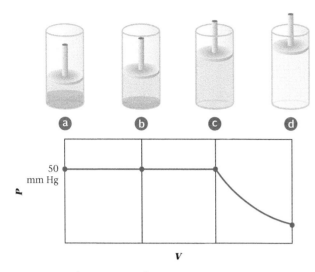

Figure 9.3 The pressure of a vapor in equilibrium with a liquid is independent of the volume of the container. The volumes of the cylinders in the figure are determined by the movable piston. In A and B, some liquid remains. In C all the liquid is just vaporized. A further increase in volume in D decreases the pressure in accordance with Boyle's law.

Vapor Pressure

Once equilibrium between liquid and vapor is reached, the number of molecules per unit volume in the vapor does not change with time. This means that *the pressure exerted by the vapor over the liquid remains constant*. The pressure of vapor in equilibrium with a liquid is called the **vapor pressure**. This quantity is a characteristic property of a given liquid at a particular temperature. It varies from one liquid to another, depending on the strength of the intermolecular forces. At 25°C, the vapor pressure of water is 24 mm Hg; that of ether, in which intermolecular forces are weaker, is 537 mm Hg.

It is important to realize that *so long as both liquid and vapor are present, the pressure exerted by the vapor is independent of the volume of the container*. If a small amount of liquid is introduced into a closed container, some of it will vaporize, establishing its equilibrium vapor pressure. The greater the volume of the container, the greater will be the amount of liquid that vaporizes to establish that pressure. The ratio n/V stays constant, so $P = nRT/V$ does not change. Only if all the liquid vaporizes will the pressure drop below the equilibrium value (Figure 9.3).

EXAMPLE 9.1 GRADED

A "cool-mist" vaporizer with capacity 2.00 L is used to add moisture to dry air in a room at 25°C. The room has dimensions 12 ft by 12 ft by 8 ft. The vapor pressure of water at 25°C is 24 mm Hg. Take the density of water at 25°C to be 1.00 g/mL.

a | If the vaporizer runs until it is empty, what is the vapor pressure of water in the room?

ANALYSIS

Information given:	volume of vaporizer (2.00 L), T (25°C) room dimensions (12 ft × 12 ft × 8 ft) vapor pressure of water at 25°C (24 mm Hg) density of water (1.00 g/mL)
Information implied:	volume of water to be "vaporized" molar mass of H_2O ft^3 to L conversion factor R value
Asked for:	vapor pressure in the room when all the water is vaporized

STRATEGY

1. Assume that all the water in the vaporizer has been converted to vapor.

2. Find the volume of the room = volume of vapor (V).

3. Find the moles of water, n, from the vaporizer that will vaporize:

$$\text{Volume of water in vaporizer} \xrightarrow{\text{density}} \text{mass of water} \xrightarrow{\text{MM}} \text{mol water } (n)$$

4. Substitute into the ideal gas law and find P.

5. Check whether your assumption in (1) is correct.

   Calculated P from (3) > vapor pressure at 25°C: assumption wrong;

      Vapor pressure is vapor pressure of water at 25°C.

   Calculated P from (3) < vapor pressure at 25°C: assumption correct;

      Vapor pressure is calculated pressure from (3)

continued

SOLUTION

2. $V_{room} = V_{vapor}$	$(12 \times 12 \times 8) \text{ ft}^3 \times \dfrac{28.32 \text{ L}}{1 \text{ ft}^3} = 3.3 \times 10^4 \text{ L}$
3. n_{vapor}	$2.00 \text{ L} \times \dfrac{1000 \text{ mL}}{1 \text{ L}} \times \dfrac{1.00 \text{ g}}{1 \text{ mL}} \times \dfrac{1 \text{ mol}}{18.02 \text{ g}} = 111 \text{ mol}$
4. P_{calc}	$P_{calc} = \dfrac{nRT}{V} = \dfrac{(111 \text{ mol})(0.0821 \text{ L} \cdot \text{atm/mol} \cdot \text{K})(298 \text{ K})}{3.3 \times 10^4 \text{ L}} = 0.082 \text{ atm} = 62 \text{ mm Hg}$
5. Check assumption	vapor pressure of water at 25°C = 24 mm Hg; P_{calc} = 62 mm Hg P_{calc} > 24 mm Hg; the assumption is wrong. The vapor pressure of water in the room is 24 mm Hg.

b   How much water is required to completely saturate the air at 25°C?

ANALYSIS

Information given:	From part (a): P_{vapor} (24 mm Hg), V_{vapor} (3.3 × 10⁴ L) T(25°C)
Information implied:	molar mass of H_2O R value
Asked for:	volume of water required to saturate the room

STRATEGY

1. Substitute into the ideal gas law to find n_{steam}.
2. Moles of vapor = moles of water. Convert to mass of water.

SOLUTION

1. n_{H_2O}	$n_{H_2O} = \dfrac{PV}{RT} = \dfrac{(24/760 \text{ atm})(3.3 \times 10^4 \text{ L})}{(0.0821 \text{ L} \cdot \text{atm/mol} \cdot \text{K})(298 \text{ K})} = 43 \text{ mol}$
2. $mass_{H_2O}$	$(43 \text{ mol})(18.02 \text{ g/mol}) = 7.7 \times 10^2 \text{ g}$

END POINTS

1. The volume of the room is the volume of the water vapor obtained from the vaporizer.
2. The volume of the water in the vaporizer cannot be used as V in the ideal gas law because it is the volume of a liquid, not a gas.
3. The calculations in part (b) show that about 770 grams (0.77 L) are required for saturation (100% relative humidity).

Vapor Pressure Versus Temperature

The vapor pressure of a liquid always increases as temperature rises. Water evaporates more readily on a hot, dry day. Stoppers in bottles of volatile liquids such as ether or gasoline pop out when the temperature rises.

The vapor pressure of water, which is 24 mm Hg at 25°C, becomes 92 mm Hg at 50°C and 1 atm (760 mm Hg) at 100°C. The data for water are plotted in Figure 9.4a. As you can see, the graph of vapor pressure versus temperature is not a straight line, as it would be if pressure were plotted versus temperature for an ideal gas. Instead, the slope increases steadily as temperature rises, reflecting the fact that more molecules vaporize at higher temperatures. At 100°C, the concentration of H_2O molecules in the vapor in equilibrium with liquid is 25 times as great as at 25°C.

In working with the relationship between two variables, such as vapor pressure and temperature, scientists prefer to deal with linear (straight-line) functions. Straight-line graphs are easier to construct and to interpret. In this case, it is possible to obtain a linear function by making a simple shift in variables. Instead of plotting vapor pressure (P) versus temperature (T), we plot the *natural logarithm** of the

---

*Many natural laws are most simply expressed in terms of natural logarithms, which are based on the number $e = 2.71828\ldots$. If $y = e^x$, then the natural logarithm of y, ln y, is equal to x (see Appendix 3).

A plot of vapor pressure vs. temperature (in °C) for a liquid is an exponentially increasing curve.

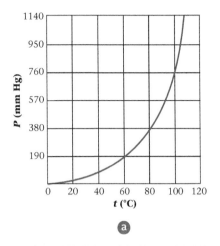

a

A plot of the logarithm of the vapor pressure vs. $1/T$ (in K) is a straight line. (To provide larger numbers, we have plotted $1000/T$.)

b

Figure 9.4 Vapor pressure versus temperature.

We'll use this value of R frequently in future chapters.

vapor pressure ($\ln P$) versus the reciprocal of the absolute temperature ($1/T$). Such a plot for water is shown in Figure 9.4b. As with all other liquids, a plot of $\ln P$ versus $1/T$ is a straight line.

The general equation of a straight line is

$$y = mx + b \qquad (m = \text{slope}, b = y\text{-intercept})$$

The y-coordinate in Figure 9.4b is $\ln P$, and the x-coordinate is $1/T$. The slope, which is a negative quantity, turns out to be $-\Delta H_{vap}/R$, where ΔH_{vap} is the molar heat of vaporization and R is the gas constant, in the proper units. Hence the equation of the straight line in Figure 9.4b is

$$\ln P = \frac{-\Delta H_{vap}}{RT} + b$$

For many purposes, it is convenient to have a two-point relation between the vapor pressures (P_2, P_1) at two different temperatures (T_2, T_1). Such a relation is obtained by first writing separate equations at each temperature:

$$\text{at } T_2: \qquad \ln P_2 = b - \frac{\Delta H_{vap}}{RT_2}$$

$$\text{at } T_1: \qquad \ln P_1 = b - \frac{\Delta H_{vap}}{RT_1}$$

On subtraction, the constant b is eliminated, and we obtain

$$\ln P_2 - \ln P_1 = \ln \frac{P_2}{P_1} = -\frac{\Delta H_{vap}}{R}\left[\frac{1}{T_2} - \frac{1}{T_1}\right] = \frac{\Delta H_{vap}}{R}\left[\frac{1}{T_1} - \frac{1}{T_2}\right] \qquad (9.1)$$

This equation is known as the **Clausius-Clapeyron equation.** Rudolph Clausius (1822–1888) was a prestigious nineteenth-century German scientist. B. P. E. Clapeyron (1799–1864), a French engineer, first proposed a modified version of the equation in 1834.

In using the Clausius-Clapeyron equation, the units of ΔH_{vap} and R must be consistent. If ΔH is expressed in joules, then R must be expressed in joules per mole per kelvin. ◂ *i* Recall (Table 5.1) that

$$R = 8.31 \text{ J/mol} \cdot \text{K}$$

This equation can be used to calculate any one of the five variables (P_2, P_1, T_2, T_1, and ΔH_{vap}), knowing the values of the other four. For example, we can use it to find the vapor pressure (P_1) at temperature T_1, knowing P_2 at T_2 and the value of the heat of vaporization (Example 9.2).

EXAMPLE 9.2

Benzene has a vapor pressure of 183 mm Hg at 40°C. Taking its heat of vaporization to be 30.8 kJ/mol, calculate its vapor pressure at 25°C.

ANALYSIS	
Information given:	vapor pressure at 40°C (183 mm Hg) temperature (25°C) ΔH_{vap} (30.8 kJ/mol)
Information implied:	R value with energy units
Asked for:	pressure at 25°C

continued

STRATEGY

1. Use subscript 2 for the higher temperature, pressure pair: $P_2 = 183$ mm Hg; $T_2 = 40°C$

2. Substitute into Equation 9.1 using the appropriate R value and T in K.

$$\ln P_2 - \ln P_1 = \frac{\Delta H_{vap}}{R}\left[\frac{1}{T_1} - \frac{1}{T_2}\right]$$

SOLUTION

Substitute into Equation 9.1	$\ln 183 - \ln P_1 = \dfrac{30.8 \text{ kJ/mol}}{8.31 \times 10^{-3} \text{ kJ/mol} \cdot \text{K}}\left[\dfrac{1}{298 \text{ K}} - \dfrac{1}{313 \text{ K}}\right]$
	$\ln 183 = 5.209$
	$\dfrac{30.8 \text{ kJ/mol}}{8.31 \times 10^{-3} \text{ kJ/mol} \cdot \text{K}}\left[\dfrac{1}{298 \text{ K}} - \dfrac{1}{313 \text{ K}}\right] = 0.596$
P_1	$5.209 - \ln P_1 = 0.596; \ln P_1 = 4.613; P_1 = 101$ mm Hg

END POINT

This value is reasonable. Lowering the temperature (40°C to 25°C) should decrease the pressure. The answer shows that it does (183 mm Hg to 101 mm Hg)!

Boiling Point

When a liquid is heated in an open container, bubbles form, usually at the bottom, where heat is applied. The first small bubbles are air, driven out of solution by the increase in temperature. Eventually, at a certain temperature, large vapor bubbles form throughout the liquid. These vapor bubbles rise to the surface, where they break. When this happens, the liquid is said to be boiling. For a pure liquid, the temperature remains constant throughout the boiling process.

The temperature at which a liquid boils depends on the pressure above it. To understand why this is the case, consider Figure 9.5. This shows vapor bubbles rising in a boiling liquid. For a vapor bubble to form, the pressure within it, P_1, must be at least equal to the pressure above it, P_2. Because P_1 is simply the vapor pressure of the liquid, it follows that *a liquid boils at a temperature at which its vapor pressure is equal to the pressure above its surface.* If this pressure is 1 atm (760 mm Hg), the temperature is referred to as the **normal boiling point.** (When the term **"boiling point"** is used without qualification, normal boiling point is implied.) The normal boiling point of water is 100°C; its vapor pressure is 760 mm Hg at that temperature.

As you might expect, the boiling point of a liquid can be reduced by lowering the pressure above it. Water can be made to boil at 25°C by evacuating the space above it. When a pressure of 24 mm Hg, the equilibrium vapor pressure at 25°C, is reached, the water starts to boil. Chemists often take advantage of this effect in purifying a high-boiling compound that might decompose or oxidize at its normal boiling point. They distill it at a reduced temperature under vacuum and condense the vapor.

If you have been fortunate enough to camp in the high Sierras or the Rockies, you may have noticed that it takes longer at high altitudes to cook foods in boiling water. The reduced pressure lowers the temperature at which water boils in an open container and thus slows down the physical and chemical changes that take place when foods like potatoes or eggs are cooked. In principle, this problem can be solved by using a pressure cooker. In that device, the pressure that develops is high enough to raise the boiling point of water

Figure 9.5 Boiling and vapor pressure. A liquid boils when it reaches the temperature at which the vapor pressure in its vapor bubbles (P_1) exceeds the pressure above the liquid (P_2).

© Cengage Learning/Charles D. Winters

The vapor pressure of a substance at its normal bp is 760 mm Hg.

Or in the Presidential Range of New Hampshire (William L. Masterton), or in the Swiss Alps (CNH).

above 100°C. Pressure cookers are indeed used in places like Salt Lake City, Utah (elevation 1340 m, bp $H_2O(l)$ = 95°C), but not by mountain climbers, who have to carry all their equipment on their backs.

Critical Temperature and Pressure

Consider an experiment in which liquid carbon dioxide is introduced into an otherwise evacuated glass tube, which is then sealed (Figure 9.6). At 0°C, the pressure above the liquid is 34 atm, the equilibrium vapor pressure of $CO_2(l)$ at that temperature. As the tube is heated, some of the liquid is converted to vapor, and the pressure rises, to 44 atm at 10°C and 56 atm at 20°C. Nothing spectacular happens (unless there happens to be a weak spot in the tube) until 31°C is reached, where the vapor pressure is 73 atm. Suddenly, as the temperature goes above 31°C, the meniscus between the liquid and vapor disappears! The tube now contains only one phase.

It is impossible to have liquid carbon dioxide at temperatures above 31°C, no matter how much pressure is applied. Even at pressures as high as 1000 atm, carbon dioxide gas does not liquefy at 35 or 40°C. This behavior is typical of all substances. There is a temperature, called the **critical temperature,** above which the liquid phase of a pure substance cannot exist. The pressure that must be applied to cause condensation at that temperature is called the **critical pressure.** Quite simply, the critical pressure is the vapor pressure of the liquid at the critical temperature.

Table 9.2 lists the critical temperatures of several common substances. The species in the column at the left all have critical temperatures below 25°C. They are often referred to as "permanent gases." Applying pressure at room temperature will not condense a permanent gas. It must be cooled as well. When you see a truck labeled "liquid nitrogen" on the highway, you can be sure that the cargo trailer is refrigerated to at least −147°C, the critical temperature of N_2.

"Permanent gases" are most often stored and sold in steel cylinders under high pressures, often 150 atm or greater. When the valve on a cylinder of N_2 or O_2 is opened, gas escapes, and the pressure drops accordingly. The substances listed

Liquid carbon dioxide under pressure is sealed in an evacuated glass tube; some vaporizes.

When the tube is heated, vapor bubbles form in the liquid.

Suddenly, at 31°C, which is the critical temperature of CO_2, the meniscus (the surface of the liquid in the tube) disappears. Above this temperature, only one phase is present, no matter what the applied pressure is.

Marna G. Clarke

(a)   (b)   (c)

Figure 9.6 Critical temperature.

Table 9.2 Critical Temperatures (°C)

Permanent Gases		Condensable Gases		Liquids	
Helium	−268	Carbon dioxide	31	Ethyl ether	194
Hydrogen	−240	Ethane	32	Ethyl alcohol	243
Nitrogen	−147	Propane	97	Benzene	289
Argon	−122	Ammonia	132	Bromine	311
Oxygen	−119	Chlorine	144	Water	374
Methane	−82	Sulfur dioxide	158		

in the center column of Table 9.2, all of which have critical temperatures above 25°C, are handled quite differently. They are available commercially as liquids in high-pressure cylinders. When the valve on a cylinder of propane is opened, the gas that escapes is replaced by vaporization of liquid. The pressure quickly returns to its original value. Only when the liquid is completely vaporized does the pressure drop as gas is withdrawn. This indicates that almost all of the propane is gone, and it is time to recharge the tank. ◄ⓘ

Above the critical temperature and pressure, a substance is referred to as a *supercritical fluid*. Such fluids have unusual characteristics. They can diffuse through a solid like a gas and dissolve materials like a liquid. Carbon dioxide and water are the most commonly used supercritical fluids and hold the promise of many practical new applications (see Beyond the Classroom at the end of this chapter).

ⓘ A pressure gauge on a propane tank doesn't indicate how much gas you have left.

9-3 Phase Diagrams

In the preceding section, we discussed several features of the equilibrium between a liquid and its vapor. For a pure substance, at least two other types of phase equilibria need to be considered. One is the equilibrium between a solid and its vapor, the other between solid and liquid at the **melting point** (also called the **freezing point**). Many of the important relations in all these equilibria can be shown in a **phase diagram.** A phase diagram is a graph that shows the pressures and temperatures at which different phases are in equilibrium with each other. The phase diagram of water is shown in Figure 9.7 (page 224). This figure, which covers a wide range of temperatures and pressures, is not drawn to scale.

To understand what a phase diagram implies, consider first curves **b** (in green) and **c** (in red) and line **a** (in blue) in Figure 9.7. Each of these shows the pressures and temperatures at which two adjacent phases are in equilibrium.

1. Curve **b** is a portion of the pressure-temperature curve of liquid water. At any temperature and pressure along this curve, liquid water is in equilibrium with water vapor. At point **X** on the curve, these two phases are in equilibrium at 0°C and about 5 mm Hg (more exactly, 0.01°C and 4.56 mm Hg). At point **Y** corresponding to 100°C, the pressure exerted by the vapor in equilibrium with liquid water is 1 atm; this is the normal boiling point of water. The extension of the curve **b** beyond point **Y** gives the equilibrium vapor pressure of water above the normal boiling point. The extension ends at 374°C, the critical temperature of water, where the pressure is 218 atm.
2. Curve **c** represents the vapor pressure curve of ice. At any point along this curve, such as point **X** (0°C, 5 mm Hg) or point **Z**, which might represent −3°C and 3 mm Hg, ice and water vapor are in equilibrium with each other.
3. Line **a** gives the temperatures and applied pressures at which liquid water is in equilibrium with ice.

Figure 9.7 **Phase diagram of water (not to scale).** The curves and line represent the temperatures and pressures at which phases are in equilibrium. The triple point is at 0.01°C, 4.56 mm Hg; the critical point is at 374°C, 1.66×10^5 mm Hg (218 atm).

The equivalent of point **X** on any phase diagram is the only one at which all three phases, liquid, solid, and vapor, are in equilibrium with each other. It is called the **triple point.** For water, the triple point temperature is 0.01°C. At this temperature, liquid water and ice have the same vapor pressure, 4.56 mm Hg.

In the three areas of the phase diagram labeled solid, liquid, and vapor, only one phase is present. To understand this, consider what happens to an equilibrium mixture of two phases when the pressure or temperature is changed. Suppose we start at the point on the curve **b** indicated by an open circle. Here liquid water and vapor are in equilibrium with each other, let us say at 70°C and 234 mm Hg. If the pressure on this mixture is increased, condensation occurs. The phase diagram confirms this; increasing the pressure at 70°C *(vertical arrow)* puts us in the liquid region. In another experiment, the temperature might be increased at a constant pressure. This should cause the liquid to vaporize. The phase diagram shows that this is indeed what happens. An increase in temperature *(horizontal arrow)* shifts us to the vapor region.

EXAMPLE 9.3 CONCEPTUAL

Consider a sample of H_2O at point X in Figure 9.7.

(a) What phase(s) is (are) present?

(b) If the temperature of the sample were reduced at constant pressure, what would happen?

(c) How would you convert the sample to vapor without changing the temperature?

STRATEGY

1. Use the phase diagram in Figure 9.7.

2. Note that P increases moving up vertically; T increases moving to the right.

SOLUTION

(a) **X** is the triple point. Ice, liquid water, and water vapor are present.

(b) Move to the left to reduce T. This penetrates the solid area, which implies that the sample freezes completely.

(c) Reduce the pressure to below the triple point value, perhaps to 4 mm Hg.

Sublimation

The process by which a solid changes directly to vapor without passing through the liquid phase is called **sublimation.** The opposite of sublimation, the phase transition from the vapor phase to solid without passing through the liquid phase, is called **deposition.** The pressure of the solid in equilibrium with the gas is called the vapor pressure of the solid (analogous to the vapor pressure of the liquid). A solid can sublime only at temperatures below the triple point; above that temperature it will melt to liquid (Figure 9.7). At temperatures below the triple point, a solid can be made to sublime by reducing the pressure of the vapor above it to less than the equilibrium value. To illustrate what this means, consider the conditions under which ice sublimes. This happens on a cold, dry, winter day when the temperature is below 0°C and the pressure of water vapor in the air is less than the equilibrium value (4.5 mm Hg at 0°C). The rate of sublimation can be increased by evacuating the space above the ice. This is how foods are freeze-dried. The food is frozen, put into a vacuum chamber, and evacuated to a pressure of 1 mm Hg or less. The ice crystals formed on freezing sublime, which leaves a product whose mass is only a fraction of that of the original food.

Your freezer can have the conditions necessary for the sublimation of ice or the deposition of water vapor. Ice cubes left in the freezer after a period of time shrink. Ice crystals formed on meat (even in air-tight containers) provide an example of deposition. The crystals are a product of the deposition of the water vapor in the meat to ice, leaving the meat dehydrated.

Iodine sublimes more readily than ice because its triple-point pressure, 90 mm Hg, is much higher. Sublimation occurs on heating (Figure 9.8) below the triple-point temperature, 114°C. If the triple point is exceeded, the solid melts. Solid carbon dioxide (dry ice) has a triple-point pressure above 1 atm (5.2 atm at −57°C). Liquid carbon dioxide cannot exist at 1 atm pressure regardless of temperature. Solid CO_2 always passes directly to vapor if allowed to warm up in an open container. ◄ *i*

Figure 9.8 **Sublimation.** Solid iodine passes directly to the vapor at any temperature below the triple point, 114°C. The vapor deposes back to a solid on a cold surface like that of the upper flask, which is filled with ice.

i The white "smoke" around a piece of dry ice is $CO_2(g)$, formed by sublimation.

Melting Point

For a pure substance, the melting point is identical to the freezing point. It represents the temperature at which solid and liquid phases are in equilibrium. Melting points are usually measured in an open container, that is, at atmospheric pressure. For most substances, the melting point at 1 atm (the "normal" melting point) is virtually identical with the triple-point temperature. For water, the difference is only 0.01°C.

Although the effect of pressure on melting point is very small, its direction is still important. To decide whether the melting point will be increased or decreased by compression, a simple principle is applied. *An increase in pressure favors the formation of the more dense phase.*

Two types of behavior are shown in Figure 9.9 (page 226).

1. *The solid is the more dense phase* (Figure 9.9a). The solid-liquid equilibrium line is inclined to the right, shifting away from the *y*-axis as it rises. At higher pressures, the solid becomes stable at temperatures above the normal melting point. In other words, the melting point is raised by an increase in pressure. This behavior is shown by most substances.

2. *The liquid is the more dense phase* (Figure 9.9b). The liquid-solid line is inclined to the left, toward the *y*-axis. An increase in pressure favors the formation of liquid; that is, the melting point is decreased by raising the pressure. Water is one of the few substances that behave this way; ice is less dense than liquid water. The effect is exaggerated for emphasis in Figure 9.9b. Actually, an increase in pressure of 134 atm is required to lower the melting point of ice by 1°C.

When the solid is the more dense phase, an increase in pressure converts liquid to solid; the melting point increases.

If the liquid is the more dense phase, an increase in pressure converts solid to liquid and the melting point decreases.

a

b

Figure 9.9 Effect of pressure on the melting point of a solid.

Figure 9.10 Molecular liquids. The bottom layer, carbon tetrachloride (CCl_4), and the top layer, octane (C_8H_{18}), are nonpolar molecular liquids that are not soluble in water. The middle layer is a water solution of blue copper sulfate.

© Cengage Learning/Charles D. Winters

It's much easier to separate two molecules than to separate two atoms within a molecule.

9-4   Molecular Substances; Intermolecular Forces

Molecules are the characteristic structural units of gases, most liquids, and many solids. As a class, molecular substances tend to have the following characteristics. They are:

1. *Nonconductors of electricity when pure.* Molecules are uncharged, so they cannot carry an electric current. In most cases (e.g., iodine, I_2, and ethyl alcohol, C_2H_5OH), water solutions of molecular substances are also nonconductors. A few polar molecules, including HCl, react with water to form ions:

$$HCl(g) \longrightarrow H^+(aq) + Cl^-(aq)$$

and hence produce a conducting water solution.

2. *Insoluble in water but soluble in nonpolar solvents such as CCl_4 or benzene.* Iodine is typical of most molecular substances; it is only slightly soluble in water (0.0013 mol/L at 25°C), much more soluble in benzene (0.48 mol/L). A few molecular substances, including ethyl alcohol, are very soluble in water. As you will see later in this section, such substances have intermolecular forces similar to those in water.

3. *Low melting and boiling.* Many molecular substances are gases at 25°C and 1 atm (e.g., N_2, O_2, and CO_2), which means that they have boiling points below 25°C. Others (such as octane and CCl_4) are liquids (Figure 9.10) with melting (freezing) points below room temperature. Of the molecular substances that are solids at ordinary temperatures, most are low-melting. For example, iodine melts at 114°C; the melting point of naphthalene, used in mothballs, is 80°C. The upper limit for melting and boiling points of most molecular substances is about 300°C.

The generally low melting and boiling points of molecular substances reflect the fact that the forces between molecules (**intermolecular forces**) are weak. To melt or boil a molecular substance, the molecules must be set free from one another. This requires only that enough energy be supplied to overcome the weak attractive forces between molecules. The strong covalent bonds within molecules remain intact when a molecular substance melts or boils. ◂ⓘ

The boiling points of different molecular substances are directly related to the strength of the intermolecular forces involved. *The stronger the intermolecular forces, the higher the boiling point of the substance.* In the remainder of this section,

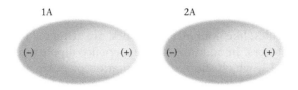

1A   2A

(−)   (+)   (−)   (+)

Figure 9.11 **Dispersion force.** Temporary dipoles in adjacent molecules line up to create an electrical attraction force known as the dispersion force. Deeply shaded areas indicate regions where the electron cloud is momentarily concentrated and creates partial charges, indicated by (+) and (−).

we examine the nature of the three different types of intermolecular forces: *dispersion forces, dipole forces,* and *hydrogen bonds.*

Dispersion (London) Forces

The most common type of intermolecular force, found in all molecular substances, is referred to as a **dispersion force.** It is basically electrical in nature, involving an attraction between temporary or *induced* dipoles in adjacent molecules. To understand the origin of dispersion forces, consider Figure 9.11.

On the average, electrons in a nonpolar molecule, ◀ *ⓘ* such as H_2, are as close to one nucleus as to the other. However, at a given instant, the electron cloud may be concentrated at one end of the molecule (position 1A in Figure 9.11). This momentary concentration of the electron cloud on one side of the molecule creates a temporary dipole in H_2. One side of the molecule, shown in deeper color in Figure 9.11, acquires a partial negative charge; the other side has a partial positive charge of equal magnitude.

ⓘ It is time to review the material in Chapter 7.

This temporary dipole induces a similar dipole (an induced dipole) in an adjacent molecule. When the electron cloud in the first molecule is at 1A, the electrons in the second molecule are attracted to 2A. These temporary dipoles, both in the same direction, lead to an attractive force between the molecules. This is the dispersion force.

All molecules have dispersion forces ◀ *ⓘ*. The strength of these forces depends on two factors

ⓘ In nonpolar molecules, dispersion is the only intermolecular force.

- the number of electrons in the atoms that make up the molecule.
- the ease with which electrons are *dispersed* to form temporary dipoles.

Both these factors increase with increasing molecular size. Large molecules are made up of more and/or larger atoms. The outer electrons in larger atoms are relatively far from the nucleus and are easier to disperse than the electrons in small atoms. In general, molecular size and molar mass parallel one another. Thus within a given class of substances (Table 9.3) we can say that *as molar mass increases, dispersion forces become stronger and the boiling point of nonpolar molecular substances increases.* ◀ *ⓘ*

ⓘ Dispersion forces are weak in H_2, much stronger in I_2.

Table 9.3 Effect of Molar Mass on Boiling Points of Molecular Substances

Noble Gases*			Halogens			Hydrocarbons		
	MM (g/mol)	bp (°C)		MM (g/mol)	bp (°C)		MM (g/mol)	bp (°C)
He	4	−269	F_2	38	−188	CH_4	16	−161
Ne	20	−246	Cl_2	71	−34	C_2H_6	30	−88
Ar	40	−186	Br_2	160	59	C_3H_8	44	−42
Kr	84	−152	I_2	254	184	*n*-C_4H_{10}	58	0

*Strictly speaking, the noble gases are "atomic" rather than molecular. However, like molecules, the noble-gas atoms are attracted to one another by dispersion forces.

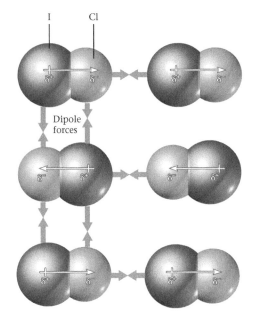

I Cl

Dipole forces

Figure 9.12 Dipole forces in the ICl crystal. The δ^+ and δ^- indicate partial charges on the I and Cl atoms in the polar molecules. The existence of these partial charges causes the molecules to line up in the pattern shown. Adjacent molecules are attracted to each other by the dipole forces between the δ^+ of one molecule and the δ^- of another molecule.

ⓘ In most molecules, dispersion forces are stronger than dipole forces.

Dipole Forces

Polar molecules, like nonpolar molecules, are attracted to one another by dispersion forces. In addition, they experience **dipole forces** as illustrated in Figure 9.12, which shows the orientation of polar molecules, such as ICl, in a crystal. Adjacent molecules line up so that the negative pole of one molecule (small Cl atom) is as close as possible to the positive pole (large I atom) of its neighbor. Under these conditions, there is an electrical attractive force, referred to as a dipole force, between adjacent polar molecules.

When iodine chloride is heated to 27°C, the weak intermolecular forces are unable to keep the molecules rigidly aligned, and the solid melts. Dipole forces are still important in the liquid state, because the polar molecules remain close to one another. Only in the gas, where the molecules are far apart, do the effects of dipole forces become negligible. ◂ⓘ Hence boiling points as well as melting points of polar compounds such as ICl are somewhat higher than those of nonpolar substances of comparable molar mass. This effect is shown in Table 9.4.

Table 9.4 Boiling Points of Nonpolar Versus Polar Substances

	Nonpolar			Polar	
Formula	MM (g/mol)	bp (°C)	Formula	MM (g/mol)	bp (°C)
N_2	28	−196	CO	28	−192
SiH_4	32	−112	PH_3	34	−88
GeH_4	77	−90	AsH_3	78	−62
Br_2	160	59	ICl	162	97

EXAMPLE 9.4 CONCEPTUAL

Explain, in terms of intermolecular forces, why the boiling point of O_2 (−183°C) is higher than that of N_2 (−196°C) and the boiling point of NO (−151°C) is higher than that of either O_2 or N_2.

STRATEGY

1. Draw the Lewis structure of the molecule.

2. Determine its polarity.

3. Identify the intermolecular forces present.

4. Remember that dispersion forces are always present and increase with molar mass.

SOLUTION

O_2 vs. N_2	
1. Lewis structures	:Ö=Ö: vs :N≡N:
2. Polarity	Both are nonpolar.
3. Intermolecular forces	Only dispersion forces for both O_2 and N_2
4. Strength of forces	MM_{O_2} = 32 g/mol; MM_{N_2} = 28 g/mol
	Dispersion forces of O_2 are larger.

continued

NO vs. O_2 and N_2	
1. Lewis structures	$\cdot\ddot{N}{=}\ddot{O}\colon$; see above for O_2 and N_2
2. Polarity	NO is polar, O_2 and N_2 are nonpolar.
3. Intermolecular forces	NO: dispersion and dipole forces;
	O_2 and N_2 only have dispersion forces.
4. Strength of forces	$MM_{O_2} = 32$ g/mol; $MM_{N_2} = 28$ g/mol; $MM_{NO} = 30$ g/mol
	All have similar dispersion force strength.
	Only NO has dipole forces in addition to the dispersion forces.

Hydrogen Bonds

Ordinarily, polarity has a relatively small effect on boiling point. In the series HCl → HBr → HI, boiling point increases steadily with molar mass, even though polarity decreases moving from HCl to HI. However, when hydrogen is bonded to a small, highly electronegative atom (N, O, F), polarity has a much greater effect on boiling point. Hydrogen fluoride, HF, despite its low molar mass (20 g/mol), has the highest boiling point of all the hydrogen halides. Water (MM = 18 g/mol) and ammonia (MM = 17 g/mol) also have abnormally high boiling points (Table 9.5). In these cases, the effect of polarity reverses the normal trend expected from molar mass alone.

The unusually high boiling points of HF, H_2O, and NH_3 result from an unusually strong type of dipole force called a **hydrogen bond.** The hydrogen bond is a force exerted between an H atom bonded to an F, O, or N atom in one molecule and an unshared pair on the F, O, or N atom of a neighboring molecule:

$$:X{-}H\text{---}:X{-}H \qquad X{=}N,\ O,\ \text{or}\ F$$

hydrogen bond

There are two reasons why hydrogen bonds are stronger than ordinary dipole forces:

1. The difference in electronegativity between hydrogen (2.2) and fluorine (4.0), oxygen (3.5), or nitrogen (3.0) is quite large. It causes the bonding electrons in molecules such as HF, H_2O, and NH_3 to be primarily associated with the more electronegative atom (F, O, or N). So the hydrogen atom, insofar as its interaction with a neighboring molecule is concerned, behaves almost like a bare proton.

2. The small size of the hydrogen atom allows the unshared pair of an F, O, or N atom of one molecule to approach the H atom in another very closely. It is significant that hydrogen bonding occurs only with these three nonmetals, all of which have small atomic radii.

Table 9.5 Effect of Hydrogen Bonding on Boiling Point

	bp (°C)		bp (°C)		bp (°C)
NH_3	−33	H_2O	100	HF	19
PH_3	−88	H_2S	−60	HCl	−85
AsH_3	−63	H_2Se	−42	HBr	−67
SbH_3	−18	H_2Te	−2	HI	−35

Note: Molecules in blue show hydrogen bonding.

Hydrogen bonding is important in biochemistry, particularly with proteins.

Hydrogen bonds can exist in many molecules other than HF, H_2O, and NH_3. The basic requirement is simply that hydrogen be bonded to a fluorine, oxygen, or nitrogen atom with at least one unshared pair. Consider, for example, the two compounds whose condensed structural formulas are

$$CH_3CH_2CH_2-\underset{\underset{propylamine}{}}{\overset{\overset{H}{|}}{N}}-H \qquad CH_3-\underset{\underset{trimethylamine}{}}{\overset{\overset{CH_3}{|}}{N}}-CH_3$$

Propylamine, in which two hydrogen atoms are bonded to nitrogen, can show hydrogen bonding; it is a liquid with a normal boiling point of 49°C. Trimethylamine, which like propylamine has the molecular formula C_3H_9N, cannot hydrogen bond to itself; it boils at 3°C.

EXAMPLE 9.5

Would you expect to find hydrogen bonds in acetic acid

$$H-\underset{\underset{H}{|}}{\overset{\overset{H}{|}}{C}}-\underset{\underset{O}{\|}}{C}-O-H$$

, diethyl ether

$$H-\underset{\underset{H}{|}}{\overset{\overset{H}{|}}{C}}-\underset{\underset{H}{|}}{\overset{\overset{H}{|}}{C}}-O-\underset{\underset{H}{|}}{\overset{\overset{H}{|}}{C}}-\underset{\underset{H}{|}}{\overset{\overset{H}{|}}{C}}-H$$
,

and hydrazine, N_2H_4?

STRATEGY

1. Skeleton structures for acetic acid and diethyl ether are given; draw the Lewis structure for all three compounds.

2. For H-bonding to occur, one of the following bonds has to be present in the molecule: H—F, H—O, or H—N.

SOLUTION

Acetic acid	
1. Lewis structure:	Given: $H-\underset{\underset{H}{\|}}{\overset{\overset{N}{\|}}{C}}-\underset{\underset{:O:}{\|}}{C}-\overset{..}{O}-H$
2. H—F, H—O, or H—N?	Yes; hydrogen bonding is present.
Diethyl ether	
1. Lewis structure:	Given: $H-\underset{\underset{H}{\|}}{\overset{\overset{N\ H}{\|}}{C}}-\underset{\underset{H}{\|}}{\overset{\overset{H}{\|}}{C}}-\overset{..}{O}-\underset{\underset{H}{\|}}{\overset{\overset{H}{\|}}{C}}-\underset{\underset{H}{\|}}{\overset{\overset{H}{\|}}{C}}-H$
2. H—F, H—O, or H—N?	No. The presence of O and H atoms in the molecule does not mean that H-bonding can occur.
Hydrazine	
1. Lewis structure:	$H-\underset{\underset{H}{\|}}{\overset{\overset{..}{N}}{}}-\underset{\underset{H}{\|}}{\overset{\overset{..}{N}}{}}-H$
2. H—F, H—O, or H—N?	Yes; H-bonding can occur.

END POINT

In acetic acid, the H atom bonded to oxygen in one molecule forms a hydrogen bond with an oxygen in an adjacent molecule. The same situation applies in hydrazine if you substitute nitrogen for oxygen.

Water has many unusual properties in addition to its high boiling point. As pointed out in Chapter 8, it has a very high specific heat, 4.18 J/g · °C. Its heat of vaporization per gram, 2.26 kJ/g, is the highest of all molecular substances. Both of these properties reflect the hydrogen-bonded structure of the liquid. Many of these bonds have to be broken when the liquid is heated; all of them disappear on boiling.

Figure 9.13 The structure of ice. In ice, the water molecules are arranged in an open pattern that gives ice its low density. Each oxygen atom (red) is bonded covalently to two hydrogen atoms (gray) and forms hydrogen bonds with two other hydrogen atoms.

In contrast to most substances, water expands on freezing. Ice is one of the very few solids that has a density less than that of the liquid from which it is formed (d ice at 0°C = 0.917 g/cm^3; d water at 0°C = 1.000 g/cm^3). This behavior is an indirect result of hydrogen bonding. When water freezes to ice, an open hexagonal pattern of molecules results (Figure 9.13). Each oxygen atom in an ice crystal is bonded to four hydrogens. Two of these are attached by ordinary covalent bonds at a distance of 0.099 nm. The other two form hydrogen bonds 0.177 nm in length. The large proportion of empty space in the ice structure explains why ice is less dense than water. ◀ 𝒊

𝒊 If ice were more dense than water, skating would be a lot less popular.

EXAMPLE 9.6	CONCEPTUAL

What types of intermolecular forces are present in nitrogen, N_2, chloroform, $CHCl_3$, carbon dioxide, CO_2, and ammonia, NH_3?

STRATEGY

1. Write the Lewis structure for each molecule.

2. Determine polarity. Polar molecules have dipole forces.

3. Check for the presence of H—N, H—O, and H—F bonds. The presence of these bonds indicates hydrogen bonding.

4. All molecules have dispersion forces.

SOLUTION

N_2

1. Lewis structure:

2. Polarity:

3. H—N, H—O, and H—F bonds?

:N≡N:

Nonpolar—no dipole forces present

No. Hydrogen bonding not possible.
Intermolecular forces: dispersion forces

$CHCl_3$

1. Lewis structure:

```
        :Cl:
         |
  :Cl—C—H
         |
        :Cl:
```

2. Polarity:

Polar—dipole forces present

continued

3. H—N, H—O, and H—F bonds?	No. Hydrogen bonding not possible.
	Intermolecular forces: dispersion and dipole forces
CO_2	
1. Lewis structure:	$:\ddot{O}=C=\ddot{O}:$
2. Polarity:	Nonpolar—no dipole forces present
3. H—N, H—O, and H—F bonds?	No. Hydrogen bonding not possible
	Intermolecular forces: dispersion forces
NH_3	
1. Lewis structure:	H—$\ddot{N}$—H 　　| 　　H
2. Polarity:	Polar—dipole forces present
3. H—N, H—O, and H—F bonds?	Yes. Hydrogen bonding possible.
	Intermolecular forces: dispersion, hydrogen bonds, and dipole forces

We have now discussed three types of intermolecular forces: dispersion forces, dipole forces, and hydrogen bonds. You should bear in mind that **all these forces are relatively weak compared with ordinary covalent bonds.** Consider, for example, the situation in H_2O. The total intermolecular attractive energy in ice is about 50 kJ/mol. In contrast, to dissociate one mole of water vapor into atoms requires the absorption of 928 kJ of energy, that is, 2(OH bond energy). This explains why it is a lot easier to boil water than to decompose it into the elements. Even at a temperature of 1000°C and 1 atm, only about one H_2O molecule in a billion decomposes to hydrogen and oxygen atoms.

9-5 Network Covalent, Ionic, and Metallic Solids

Virtually all substances that are gases or liquids at 25°C and 1 atm are molecular. In contrast, there are three types of nonmolecular solids (Figure 9.14). These are

- **network covalent** solids, in which atoms are joined by a continuous network of covalent bonds. The entire crystal, in effect, consists of one huge molecule.
- **ionic solids,** held together by strong electrical forces (ionic bonds) between oppositely charged ions adjacent to one another.
- **metals,** in which the structural units are electrons (e^-) and cations, which may have charges of +1, +2, or +3.

Figure 9.14 Diagrams of four types of substances (see text discussion). X represents a nonmetal atom, — represents a covalent bond, M^+ a cation, X^- an anion, and e^- an electron.

Figure 9.15 **The structure of a diamond.** Diamond has a three-dimensional structure in which each carbon atom is surrounded tetrahedrally by four other carbon atoms.

Network Covalent Solids

As a class, network covalent solids

- *have high melting points, often about 1000°C.* To melt the solid, covalent bonds between atoms must be broken. In this respect, solids of this type differ markedly from molecular solids, which have much lower melting points.
- *are insoluble in all common solvents.* For solution to occur, covalent bonds throughout the solid have to be broken.
- *are poor electrical conductors.* In most network covalent substances (graphite is an exception), there are no mobile electrons to carry a current.

Graphite and Diamond

Several nonmetallic elements and metalloids have a network covalent structure. The most important of these is carbon, which has two different crystalline forms of the network covalent type. Both graphite and diamond have high melting points, above 3500°C. However, the bonding patterns in the two solids are quite different.

In diamond, each carbon atom forms single bonds with four other carbon atoms arranged tetrahedrally around it. The hybridization in diamond is sp³. ◄ 𝑖 The three-dimensional covalent bonding contributes to diamond's unusual hardness. Diamond is one of the hardest substances known; it is used in cutting tools and quality grindstones (Figure 9.15).

𝑖 Hybridization is discussed in Section 7-4.

Graphite is planar, with the carbon atoms arranged in a hexagonal pattern. Each carbon atom is bonded to three others, two by single bonds, one by a double bond. The hybridization is sp². The forces between adjacent layers in graphite are of the dispersion type and are quite weak. A "lead" pencil really contains a graphite rod, thin layers of which rub off onto the paper as you write (Figure 9.16).

At 25°C and 1 atm, graphite is the stable form of carbon. Diamond, in principle, should slowly transform to graphite under ordinary conditions. Fortunately

Figure 9.16 **The structure of graphite.** Graphite has a two-dimensional layer structure with weak dispersion forces between the layers.

for the owners of diamond rings, this transition occurs at zero rate unless the diamond is heated to about 1500°C, at which temperature the conversion occurs rapidly. For understandable reasons, no one has ever become very excited over the commercial possibilities of this process. The more difficult task of converting graphite to diamond has aroused much greater enthusiasm.

At high pressures, diamond is the stable form of carbon, since it has a higher density than graphite (3.51 vs 2.26 g/cm³). The industrial synthesis of diamond from graphite or other forms of carbon is carried out at about 100,000 atm and 2000°C.

Compounds of Silicon

Perhaps the simplest compound with a network covalent structure is quartz, the most common form of SiO_2 and the major component of sand. In quartz, each silicon atom bonds tetrahedrally to four oxygen atoms. Each oxygen atom bonds to two silicons, thus linking adjacent tetrahedra to one another (Figure 9.17). Notice that the network of covalent bonds extends throughout the entire crystal. Unlike most pure solids, quartz does not melt sharply to a liquid. Instead, it turns to a viscous mass over a wide temperature range, first softening at about 1400°C. The viscous fluid probably contains long —Si—O—Si—O— chains, with enough bonds broken to allow flow.

More than 90% of the rocks and minerals found in the earth's crust are silicates, which are essentially ionic. Typically the anion has a network covalent structure in which SiO_4^{4-} tetrahedra are bonded to one another in one, two, or three dimensions. The structure shown at the left of Figure 9.18, where the anion is a one-dimensional infinite chain, is typical of fibrous minerals such as diopside, $CaSiO_3 \cdot MgSiO_3$. Asbestos has a related structure in which two chains are linked together to form a double strand.

The structure shown at the right of Figure 9.18 is typical of layer minerals such as talc, $Mg_3(OH)_2Si_4O_{10}$. Here SiO_4^{4-} tetrahedra are linked together to form an infinite sheet. The layers are held loosely together by weak dispersion forces, so they easily slide past one another. As a result, talcum powder, like graphite, has a slippery feeling.

Among the three-dimensional silicates are the *zeolites*, which contain cavities or tunnels in which Na^+ or Ca^{2+} ions may be trapped. Synthetic zeolites with made-to-order holes are used in home water softeners. When hard water containing Ca^{2+} ions flows through a zeolite column, an exchange reaction occurs. If we represent the formula of the zeolite as NaZ, where Z^- represents a complex, three-dimensional anion, the water-softening reaction can be represented by the equation

$$Ca^{2+}(aq) + 2NaZ(s) \longrightarrow CaZ_2(s) + 2Na^+(aq)$$

Sodium ions migrate out of the cavities; ◀ⓘ Ca^{2+} ions from the hard water move in to replace them.

This type of water softener shouldn't be used if you're trying to reduce sodium intake.

© Martin Novak/Shutterstock.com

Figure 9.17 Crystal structure of quartz. The Si (gray) and O (red) atoms form six-membered rings. Each Si atom is bonded tetrahedrally to four O atoms.

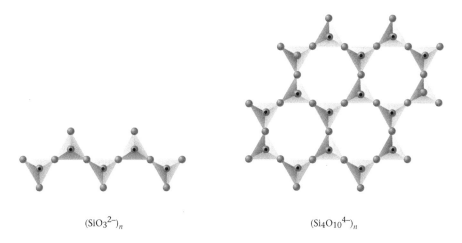

$(SiO_3^{2-})_n$

$(Si_4O_{10}^{4-})_n$

Figure 9.18 Silicate lattices. The red circles represent oxygen atoms. The black dot in the center of the red circle represents the Si atom, which is at the center of a tetrahedron. *(Left)* Diopside has a one-dimensional infinite chain. *(Right)* A portion of the talc structure, which is composed of infinite sheets.

Ionic Solids

An ionic solid consists of cations and anions (e.g., Na^+, Cl^-). No simple, discrete molecules are present in NaCl or other ionic compounds; rather, the ions are held in a regular, repeating arrangement by strong ionic bonds, electrostatic interactions between oppositely charged ions. Because of this structure, shown in Figure 9.14, ionic solids have the following properties:

1. *Ionic solids are nonvolatile and have high-melting points* (typically from 600°C to 2000°C). Ionic bonds must be broken to melt the solid, separating oppositely charged ions from each other. Only at high temperatures do the ions acquire enough kinetic energy for this to happen.

2. *Ionic solids do not conduct electricity because the charged ions are fixed in position.* ◄ **ⓘ** They become good conductors, however, when melted or dissolved in water. In both cases, in the melt or solution, the ions (such as Na^+ and Cl^-) are free to move through the liquid and thus can conduct an electric current.

3. *Many, but not all, ionic compounds* (e.g., NaCl but not $CaCO_3$) *are soluble in water, a polar solvent.* In contrast, ionic compounds are insoluble in nonpolar solvents such as benzene (C_6H_6) or carbon tetrachloride (CCl_4).

The relative strengths of different ionic bonds can be estimated from Coulomb's law, which gives the electrical energy of interaction between a cation and anion in contact with one another:

$$E = \frac{k \times Q_1 \times Q_2}{d}$$

Here, Q_1 and Q_2 are the charges of anion and cation, and d, the distance between the centers of the two ions, is the sum of the ionic radii (Appendix 2):

$$d = r_{cation} + r_{anion}$$

The quantity k is a constant whose magnitude need not concern us. Because the cation and anion have opposite charges, E is a negative quantity. This makes sense; energy is evolved when two oppositely charged ions, originally far apart with $E = 0$, approach one another closely. Conversely, energy has to be absorbed to separate the ions from each other.

From Coulomb's law, the strength of the ionic bond should depend on two factors:

1. *The charges of the ions.* The bond in CaO ($+2$, -2 ions) is considerably stronger than that in NaCl ($+1$, -1 ions). This explains why the melting point of calcium oxide (2927°C) is so much higher than that of sodium chloride (801°C).

2. *The size of the ions.* The ionic bond in NaCl (mp = 801°C) is somewhat stronger than that in KBr (mp = 734°C) because the internuclear distance is smaller in NaCl:

$$d_{NaCl} = r_{Na^+} + r_{Cl^-} = 0.095 \text{ nm} + 0.181 \text{ nm} = 0.276 \text{ nm}$$
$$d_{KBr} = r_{K^+} + r_{Br^-} = 0.133 \text{ nm} + 0.195 \text{ nm} = 0.328 \text{ nm}$$

ⓘ Charged particles must move to carry a current.

A more sophisticated model of metals is described in Appendix 4.

Metals

Figure 9.14d illustrates a simple model of bonding in metals known as the **electron-sea model.** The metallic crystal is pictured as an array of positive ions, for example, Na^+, Mg^{2+}. These are anchored in position, like buoys in a mobile "sea" of electrons. These electrons are not attached to any particular positive ion but rather can wander through the crystal. The electron-sea model explains many of the characteristic properties of metals:

1. *High electrical* conductivity. The presence of large numbers of relatively mobile electrons explains why metals have electrical conductivities several hundred times greater than those of typical nonmetals. Silver is the best electrical conductor but is too expensive for general use. Copper, with a conductivity close to that of silver, is the metal most commonly used for electrical wiring. Although a much poorer conductor than copper, mercury is used in many electrical devices, such as silent light switches, in which a liquid conductor is required.

2. *High* thermal conductivity. Heat is carried through metals by collisions between electrons, which occur frequently. Saucepans used for cooking commonly contain aluminum, copper, or stainless steel; their handles are made of a nonmetallic material that is a good thermal insulator.

3. *Ductility and* malleability. Most metals are ductile (capable of being drawn out into a wire) and malleable (capable of being hammered into thin sheets). In a metal, the electrons act like a flexible glue holding the atomic nuclei together. As a result, metal crystals can be deformed without shattering.

4. Luster. Polished metal surfaces reflect light. Most metals have a silvery white metallic color because they reflect light of all wavelengths. Because electrons are not restricted to a particular bond, they can absorb and re-emit light over a wide wavelength range. Gold and copper absorb some light in the blue region of the visible spectrum and so appear yellow (gold) or red (copper).

5. *Insolubility in water and other common solvents*. No metals dissolve in water; electrons cannot go into solution, and cations cannot dissolve by themselves. The only liquid metal, mercury, dissolves many metals, forming solutions called *amalgams*. An Ag-Sn-Hg amalgam is still used in filling teeth.

In general, the melting points of metals cover a wide range, from $-39°C$ for mercury to $3410°C$ for tungsten. This variation in melting point corresponds to a similar variation in the strength of the metallic bond. Generally speaking, the lowest melting metals are those that form $+1$ cations, like sodium (mp = 98°C) and potassium (mp = 64°C).

Much of what has been said about the four structural types of solids in Sections 9-4 and 9-5 is summarized in Table 9.6.

Diamond, a network covalent solid

Hydrated nickel nitrate, $Ni(NO_3)_2 \cdot 6H_2O$ an ionic solid.

Manganese, a metallic solid

(a)   (b)   (c)

Solids with different structures.

EXAMPLE	9.7	CONCEPTUAL

For each species in column A, choose the description in column B that best applies.

A	B
CO_2	(a) ionic, high-melting
$CuSO_4$	(b) liquid metal, good conductor
SiO_2	(c) polar molecule, soluble in water
Hg	(d) ionic, insoluble in water
	(e) network covalent, high-melting
	(f) nonpolar molecule, gas at 25°C

STRATEGY

1. Characterize each species with respect to type, forces within and between particles, and if necessary, physical properties.
2. Find the appropriate matches.

SOLUTION

(a) CO_2	molecule, nonpolar
	Only match is (f) even if you did not know that CO_2 is a gas at 25°C.
(b) $CuSO_4$	ionic, water soluble
	Only match is (a) even if you did not know that $CuSO_4$ has a high melting point.
(c) SiO_2	network covalent
	Only match is (e).
(d) Hg	metal, liquid at room temperature
	Only match is (b).

Table 9.6 Structures and Properties of Types of Substances

Type	Structural Particles	Forces Within Particles	Forces Between Particles	Properties	Examples
Molecular	Molecules (a) nonpolar	Covalent bond	Dispersion	Low mp, bp; often gas or liquid at 25°C; nonconductors; insoluble in water, soluble in organic solvents	H_2 CCl_4
	(b) polar	Covalent bond	Dispersion, dipole, H bond	Similar to nonpolar but generally higher mp and bp, more likely to be water-soluble	$HCl(g)$ NH_3
Network covalent	Atoms	—	Covalent bond	Hard solids with very high melting points; nonconductors; insoluble in common solvents	C SiO_2
Ionic	Ions	—	Ionic bond	High mp; conductors in molten state or water solution; often soluble in water, insoluble in organic solvents	NaCl MgO $CaCO_3$
Metallic	Cations, mobile electrons	—	Metallic bond	Variable mp; good conductors in solid; insoluble in common solvents	Na Fe

9-6 Crystal Structures

Solids tend to crystallize in definite geometric forms that often can be seen by the naked eye. In ordinary table salt, cubic crystals of NaCl are clearly visible. Large, beautifully formed crystals of such minerals as fluorite, CaF_2, are found in nature. It is possible to observe distinct crystal forms of many metals under a microscope.

Crystals have definite geometric forms because the atoms or ions present are arranged in a definite, three-dimensional pattern. The nature of this pattern can be deduced by a technique known as x-ray diffraction. The basic information that comes out of such studies has to do with the dimensions and geometric form of the **unit cell,** the smallest structural unit that, repeated over and over again in three dimensions, generates the crystal. In all, there are 14 different kinds of unit cells. Our discussion will be limited to a few of the simpler unit cells found in metals and ionic solids.

Metals

Three of the simpler unit cells found in metals, shown in Figure 9.19, are the following:

1. Simple cubic cell (SC). This is a cube that consists of eight atoms whose centers are located at the corners of the cell. Atoms at adjacent corners of the cube touch one another.

2. Face-centered cubic cell (FCC). Here, there is an atom at each corner of the cube and one in the center of each of the six faces of the cube. In this structure, atoms at the corners of the cube do not touch one another; they are forced slightly apart. Instead, contact occurs along a face diagonal. The atom at the center of each face touches atoms at opposite corners of the face.

3. Body-centered cubic cell (BCC). This is a cube with atoms at each corner and one in the center of the cube. Here again, corner atoms do not touch each

Figure 9.19 Three types of unit cells. In each case, there is an atom at each of the eight corners of the cube. In the body-centered cubic unit cell, there is an additional atom in the center of the cube. In the face-centered cubic unit cell, there is an atom in the center of each of the six faces.

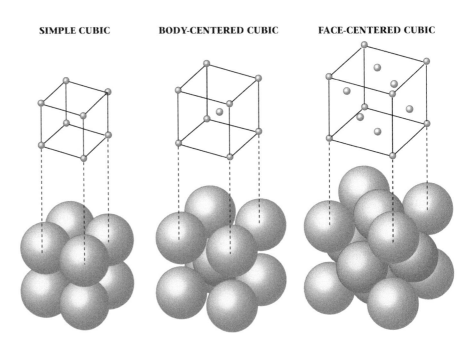

SIMPLE CUBIC   **BODY-CENTERED CUBIC**   **FACE-CENTERED CUBIC**

Table 9.7 Properties of Cubic Unit Cells

	Simple	BCC	FCC
Number of atoms per unit cell	1	2	4
Relation between side of cell, s, and atomic radius, r	$2r = s$	$4r = s\sqrt{3}$	$4r = s\sqrt{2}$
% of empty space	47.6	32.0	26.0

other. Instead, contact occurs along the body diagonal; the atom at the center of the cube touches atoms at opposite corners.

Table 9.7 lists three other ways in which these types of cubic cells differ from one another.

1. *Number of atoms per unit cell.* Keep in mind that a huge number of unit cells are in contact with each other, interlocking to form a three-dimensional crystal. This means that several of the atoms in a unit cell do not belong exclusively to that cell. Specifically

- an atom at the corner of a cube forms a part of eight different cubes that touch at that point. (To convince yourself of this, look back at Figure 2.15 focus on the small sphere in the center.) In this sense, only $\frac{1}{8}$ of a corner atom belongs to a particular cell.
- an atom at the center of the face of a cube is shared by another cube that touches that face. In effect, only $\frac{1}{2}$ of that atom can be assigned to a given cell. This means, for example, that the number of atoms per FCC unit cell is

$$(8 \text{ corner atoms} \times \tfrac{1}{8}) + (6 \text{ face atoms} \times \tfrac{1}{2}) = 4 \text{ atoms per cube}$$

2. *Relation between side of cell (s) and atomic radius (r).* To see how these two quantities are related, consider an FCC cell in which atoms touch along a face diagonal. As you can see from Figure 9.20.

- the distance along the face diagonal, d, is equal to four atomic radii

$$d = 4r$$

- d can be related to the length of a side of the cell by the Pythagorean theorem, $d^2 = s^2 + s^2 = 2s^2$. Taking the square root of both sides,

$$d = s\sqrt{2}$$

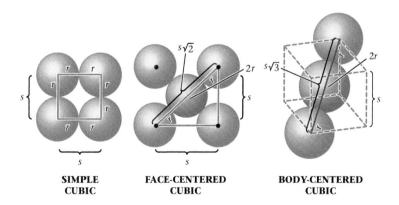

SIMPLE
CUBIC

FACE-CENTERED
CUBIC

BODY-CENTERED
CUBIC

Figure 9.20 **Relation between atomic radius (*r*) and length of edge (*s*) for cubic cells.** In the simple cubic cell, $2r = s$. In the face-centered cubic cell, the face diagonal is equal to $s\sqrt{2}$ (hypotenuse of a right triangle) and to $4r$. Thus, $s\sqrt{2} = 4r$. In the body-centered cubic cell, the body diagonal is equal to $s\sqrt{3}$ (diagonal of a cube) and to $4r$. Thus, $s\sqrt{3} = 4r$.

Equating the two expressions for d, we have

$$4r = s\sqrt{2}$$

This relation offers an experimental way of determining the atomic radius of a metal, if the nature and dimensions of the unit cell are known.

EXAMPLE 9.8 GRADED

Silver is a metal commonly used in jewelry and photography. It crystallizes with a face-centered cubic (FCC) unit cell 0.407 nm on an edge.

a What is the atomic radius of silver in cm? (1 nm = 10^{-7} cm)

ANALYSIS

Information given:	type of cubic cell (face-centered) length of side, s(0.407 nm) nm to cm conversion (1 nm = 1×10^{-7} cm)
Information implied:	side and atomic radius relationship in a face-centered cubic cell
Asked for:	atomic radius of silver in cm

STRATEGY

1. Relate the atomic radius, r, to the side of the cube, s, in a face-centered cubic cell (FCC). See Table 9.7.
2. Substitute into the equation $4r = s\sqrt{2}$.
3. Convert nm to cm.

SOLUTION

$4r = s\sqrt{2}$	$r = \dfrac{0.407 \text{ nm } (\sqrt{2})}{4} = 0.144 \text{ nm} \times \dfrac{1 \times 10^{-7} \text{ cm}}{1 \text{ nm}} = 1.44 \times 10^{-8} \text{ cm}$

b What is the volume of a single silver atom? (The volume of a spherical ball of radius r is $V = \frac{4}{3}\pi r^3$.)

ANALYSIS

Information given:	from part (a); atomic radius, r (1.44 $\times 10^{-8}$ cm) formula for the volume of a sphere ($V = \frac{4}{3}\pi r^3$)
Asked for:	volume of a single Ag atom

STRATEGY

Assume that the atom is a perfect sphere and substitute into the formula for the volume of a sphere.

SOLUTION

V	$V = \frac{4}{3}\pi r^3 = \frac{4}{3}\pi(1.44 \times 10^{-8} \text{ cm})^3 = 1.25 \times 10^{-23} \text{ cm}^3$

c What is the density of a single silver atom?

ANALYSIS

Information given:	from part (b): atomic volume, V(1.25 $\times 10^{-23}$ cm^3) formula for the volume of a sphere ($V = \frac{4}{3}\pi r^3$)
Information implied:	molar mass of Ag Avogadro's number
Asked for:	density of a single Ag atom

STRATEGY

1. Find the mass of a single Ag atom. Recall that there are 6.022 $\times 10^{23}$ atoms of silver in one molar mass of silver (107.9 g/mol). Use that as a conversion factor.
2. Recall that density = mass/volume.

continued

SOLUTION	
1. mass of 1 Ag atom	$1 \text{ Ag atom} \times \dfrac{107.9 \text{ g}}{6.022 \times 10^{23} \text{ atoms}} = 1.792 \times 10^{-22} \text{ g}$
2. density	$\text{density} = \dfrac{\text{mass}}{\text{volume}} = \dfrac{1.792 \times 10^{-22} \text{ g}}{1.25 \times 10^{-23} \text{ cm}^3} = 14.3 \text{ g/cm}^3$

END POINTS
1. In face-centered cubic cells, the fraction of empty space is 0.26.
2. The calculated density in part (c) assumes no empty space. If empty space is factored in, [(0.26)(14.3) = 3.7], then 3.7 g/cm³ has to be subtracted from the density obtained in part (c). The calculated density is therefore [14.3 − 3.7] = 10.6 g/cm³. The experimentally obtained value is 10.5 g/cm³.

3. *Percentage of empty space.* Metal atoms in a crystal, like marbles in a box, tend to pack closely together. As you can see from Table 9.7, nearly half of a simple cubic unit cell is empty space. This makes the SC structure very unstable; only one metal (polonium, $Z = 84$) has this type of unit cell. The body-centered cubic structure has less waste space; about 20 metals, including all those in Group 1, have a BCC unit cell. A still more efficient way of packing spheres of the same size is the face-centered cubic structure, where the fraction of empty space is only 0.26. ◀ⓘ About 40 different metals have a structure based on a face-centered cubic cell or a close relative in which the packing is equally efficient (hexagonal closest packed structure).

ⓘ
Golf balls and oranges pack naturally in an FCC structure.

CHEMISTRY THE **HUMAN** SIDE

The crystal structures discussed in this section were determined by a powerful technique known as x-ray diffraction (Figure A, page 242). By studying the pattern produced when the scattered rays strike a target, it is possible to deduce the geometry of the unit cell. With molecular crystals, one can go a step further, identifying the geometry and composition of the molecule. A father and son team of two English physicists, William H. (1862–1942) and William L. Bragg (1890–1971), won the Nobel Prize in 1915 for pioneer work in this area.

The Braggs and their successors strongly sought out talented women scientists to develop the new field of x-ray crystallography. Foremost among these women was Dorothy Crowfoot Hodgkin, who spent almost all of her professional career at Oxford University in England. As the years passed she unraveled the structures of successively more complex natural products. These included penicillin, which she studied between 1942 and 1949, vitamin B-12 (1948–1957), and her greatest triumph, insulin, which she worked on for more than 30 years. Among her students at Oxford was Margaret Thatcher, the future prime minister. In 1964, Dorothy Hodgkin became the third woman to win the Nobel Prize in Chemistry. The first two women Nobel laureates in Chemistry were Marie Curie and Irène Joliot-Curie. Forty-five years later (in 2009), a fourth woman, Ada Yonath, an Israeli crystallographer, won the Nobel Prize in Chemistry.

Hodgkin's accomplishments were all the more remarkable when you consider some of the obstacles she had to overcome. At the age of 24, she developed rheumatoid arthritis, a crippling disease of the immune system. Gradually she lost the use of her hands; ultimately she was confined to a wheelchair. Dorothy Hodgkin succeeded because

Bettmann/Corbis

Dorothy Crowfoot Hodgkin
(1910–1994)

she combined a first-rate intellect with an almost infinite capacity for hard work. Beyond that, she was one of those rare individuals who inspire loyalty by taking genuine pleasure in the successes of other people. A colleague referred to her as "the gentle genius."

Dorothy's husband, Thomas Hodgkin, was a scholar in his own right with an interest in the history of Africa. Apparently realizing that hers was the greater talent, he acted as a "house-husband" for their three children so she would have more time to devote to research. One wonders how Dorothy and Thomas Hodgkin reacted to the 1964 headline in a London tabloid, "British Wife Wins Nobel Prize."

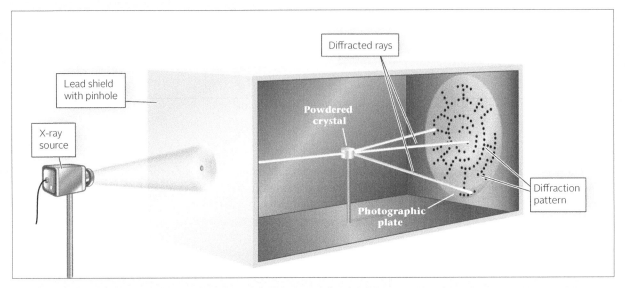

Figure A **X-ray diffraction.** Knowing the angles and intensities at which x-rays are diffracted by a crystal, it is possible to calculate the distances between layers of atoms.

Figure 9.21 Three types of lattices in ionic crystals. In LiCl, the Cl^- ions are in contact with each other, forming a face-centered cubic lattice. In NaCl, the Cl^- ions are forced slightly apart by the larger Na^+ ions. In CsCl, the large Cs^+ ion at the center touches the Cl^- ions at each corner of the cube.

LiCl   NaCl   CsCl

Ionic Crystals

The geometry of ionic crystals, in which there are two different kinds of ions, is more difficult to describe than that of metals. However, in many cases the packing can be visualized in terms of the unit cells described in Figure 9.21. Lithium chloride, LiCl, is a case in point. Here, the larger Cl^- ions form a face-centered cubic lattice (Figure 9.21). The smaller Li^+ ions fit into "holes" between the Cl^- ions. This puts a Li^+ ion at the center of each edge of the cube.

In the sodium chloride crystal, the Na^+ ion is slightly too large to fit into holes in a face-centered lattice of Cl^- ions (Figure 9.21). As a result, the Cl^- ions are pushed slightly apart so that they are no longer touching, and only Na^+ ions are in contact with Cl^- ions. However, the relative positions of positive and negative ions remain the same as in LiCl: ◄ ⓘ Each anion is surrounded by six cations and each cation by six anions.

The structures of LiCl and NaCl are typical of all the alkali halides (Group 1 cation, Group 17 anion) except those of cesium. Because of the large size of the Cs^+ ion, CsCl crystallizes in a quite different structure. Here, each Cs^+ ion is located at the center of a simple cube outlined by Cl^- ions. The Cs^+ ion at the center touches all the Cl^- ions at the corners; the Cl^- ions do not touch each other. As you can see, each Cs^+ ion is surrounded by eight Cl^- ions, and each Cl^- ion is surrounded by eight Cs^+ ions. ◄ ⓘ

NaCl is FCC in both Na^+ and Cl^- ions; CsCl is BCC in both Cs^+ and Cl^- ions.

EXAMPLE 9.9

Consider Figure 9.21. The length of an edge of a cubic cell, *s*, is the distance between the center of an atom or ion at the "top" of the cell and the center of the atom or ion at the "bottom." Taking the ionic radii of Li⁺, Na⁺, and Cl⁻ to be 0.060 nm, 0.095 nm, and 0.181 nm, respectively, determine *s* for NaCl and LiCl.

STRATEGY

Use Figure 9.21 to determine along which lines the ions touch.

SOLUTION

NaCl	The atoms touch along a side.
	$s = 1\ r$ of Cl^- + $2\ r$ of Na^+ + $1\ r$ of Cl^-
	$= 0.181$ nm + $2(0.095$ nm$)$ + 0.181 nm $= 0.552$ nm
LiCl	The chloride atoms touch along a face diagonal.
	$s = 1\ r$ of Cl^- + $2\ r$ of Cl^- + $1\ r$ of Cl^- = $4\ r$ of Cl^-
	$= 4(0.181$ nm$) = 0.724$ nm
	length of face diagonal $= s\sqrt{2} = (0.724$ nm$)\ (\sqrt{2}) = 0.512$ nm

CHEMISTRY **BEYOND** THE CLASSROOM

Supercritical Carbon Dioxide

Carbon dioxide has been in the news recently because of its possible contribution to global climate change. However, supercritical carbon dioxide has found many practical applications. It is the supercritical fluid most extensively studied because it is cheap, nontoxic, and relatively easy to bring to supercritical status.

Carbon dioxide usually behaves as a gas in air or as a solid called dry ice when it is frozen. It behaves as a supercritical fluid above its critical temperature (31°C) and critical pressure (73 atm). As a supercritical fluid it expands to fill its container like a gas but it has a density like that of a liquid (Figure A).

Caffeine used to be extracted using the solvent dichloromethane (CH_2Cl_2). Unfortunately, dichloromethane was suspected to be a carcinogen so a different solvent had to be found. More than 20 years ago, supercritical CO_2 was tried as an alternative method for the extraction of caffeine from coffee beans. It was forced through the green coffee beans and the infused beans were then sprayed under high pressure with water to remove the dissolved caffeine. The caffeine was isolated by passing the water through activated charcoal filters, or by distillation (Chapter 1), crystallization, or reverse osmosis (Chapter 10). The recovered caffeine was then sold mostly to beverage manufacturers and pharmaceutical companies. This process was so effective that most of the coffee sold today is decaffeinated by this process (Figure B).

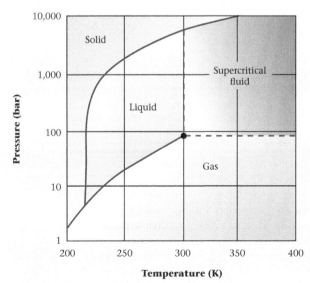

Figure A **Phase diagram for carbon dioxide.** Supercritical carbon dioxide is shown in the violet-shaded region.

Figure B **Decaffeinated coffee is one product that employs supercritical carbon dioxide in the production process.**

The dry cleaning industry now has several pilot plants using supercritical CO_2 as a solvent for removing oils, stains, and grime from clothes that cannot be washed in water. The industry is trying to replace the more commonly used solvent, perchloroethylene (PERC), which has been linked to liver and kidney damage in people working in dry cleaning plants and in people in urban areas who live near these plants. Supercritical CO_2 has one drawback, however. It does not dissolve many of the proteins in food stains, waxes, and salts because of its nonpolar nature. Research is now ongoing to find detergents that will enhance carbon dioxide's effectivity in stain removal. Several polymer detergents look promising.

A decade ago, a study by a team of researchers from the Massachusetts Institute of Technology reported the use of supercritical CO_2 to deactivate live bacteria. Medical research is now engaged in the study of supercritical CO_2 as an alternative option for tissue transplant sterilization. It is particularly attractive for sterilizing bone and tendon fragments used for torn ACLs (anterior cruciate ligaments) or spinal fusions. These fragments are typically sterilized by irradiation or by treatment with steam. These treatments are often found to compromise the strength of the fragments by degrading the collagen, making them unusable for grafts.

Other projects involving supercritical CO_2, to name a few, are the extraction of cholesterol from meat, fat from potato chips, and nicotine from tobacco. Supercritical CO_2 heat pumps are now marketed in Asia using CO_2 as a natural refrigerant.

Many scientists are very excited at the prospects of finding more uses for this novel compound. Ironically, its supercritical phase may redeem it from the bad press that gaseous CO_2 has received lately.

Key Concepts

1. Use the ideal gas law to determine whether a liquid will completely vaporize.
   (**Example 9.1; Problems 1–6**)
2. Use the Clausius-Clapeyron equation to relate vapor pressure to temperature.
   (**Example 9.2; Problems 7–14**)
3. Use a phase diagram to determine the phase(s) present at a given T and P.
   (**Example 9.3; Problems 15–24**)
4. Identify the type of intermolecular forces in different substances.
   (**Examples 9.4–9.6; Problems 25–38**)
5. Classify substances as molecular, network covalent, ionic, or metallic.
   (**Example 9.7; Problems 39–48**)
6. Relate unit cell dimensions to atomic or ionic radii.
   (**Examples 9.8, 9.9; Problems 49–58**)

Key Equations

Clausius-Clapeyron

$$\ln \frac{P_2}{P_1} = \frac{\Delta H_{vap}}{R}\left[\frac{1}{T_1} - \frac{1}{T_2}\right]$$

Atomic radius (r) versus side (s) of unit cell: SC: $2r = s$ FCC: $4r = s\sqrt{2}$ BCC: $4r = s\sqrt{3}$

Key Terms

body-centered cubic cell (BCC)	face-centered cubic cell (FCC)	hydrogen bond	simple cubic cell (SC)
boiling point	deposition	intermolecular forces	sublimation
critical pressure	dipole forces	network covalent	triple point
critical temperature	dispersion force	normal boiling point	unit cell
	electron-sea model	phase diagram	vapor pressure

Summary Problem

Consider propylamine, $C_3H_7NH_2$. It is used as a component of some herbicides, a regulating agent for plastics, and is used as an additive for the petroleum industry. Its structural formula is

$$H_3C—CH_2—CH_2—NH_2$$

Some of its physical properties are:
density: 0.719 g/mL
normal boiling point: 48.7°C
normal freezing point: −83°C
critical temperature: 234°C
heat of vaporization: 30.7 kJ/mol

(a) What intermolecular forces would you expect to find in propylamine?

(b) Compare the boiling point of propylamine to nitrogen, methylamine (CH_3NH_2), ammonium chloride, and silicon dioxide (SiO_2).

(c) Would you expect propylamine to conduct electricity?

(d) Compare its solubility in water to that of nitrogen gas and silicon dioxide.

(e) Give the symbol of a nonmetal that is a better conductor of electricity than propylamine.

(f) Give the formula of a solid compound that, unlike propylamine, has a network covalent structure.

(g) Draw a rough sketch of the phase diagram of propylamine. Assume that the solid phase is more dense than the liquid phase.

(h) If you want to purify propylamine by sublimation, at what approximate temperature would you operate?

(i) A sample of propylamine vapor at 250°C and 1 atm is cooled at constant pressure. At what temperature does the liquid first appear?

(j) Estimate the vapor pressure of propylamine at 25°C.

(k) Estimate the temperature at which the vapor pressure of propylamine is 0.500 atm.

(l) A sample of 1.00 mL of propylamine is placed in a one-liter flask at 22°C where the vapor pressure is 269 mm Hg. Show by calculation whether there is any liquid left in the flask when equilibrium is established.

Answers

(a) dispersion, dipole, H bonds
(b) higher than N_2 and methylamine; lower than NH_4Cl and SiO_2
(c) no
(d) more soluble than N_2 or SiO_2
(e) C (graphite) or Si
(f) SiO_2
(g) See Figure 9.9 for the general configuration. Put in the appropriate temperatures and pressures.
(h) below about −83°C
(i) 48.7°C
(j) 305 mm Hg
(k) 30.5°C
(l) no liquid left

Questions and Problems

Even-numbered questions and Challenge Problems have answers in Appendix 5 and fully worked solutions in the *Student Solutions Manual*.

9-2 Liquid-Vapor Equilibrium

1. Methyl alcohol can be used as a fuel instead of, or combined with, gasoline. A sample of methyl alcohol, CH_3OH, in a flask of constant volume exerts a pressure of 254 mm Hg at 57°C. The flask is slowly cooled.
    (a) Assuming no condensation, use the ideal gas law to calculate the pressure of the vapor at 35°C; at 45°C.
    (b) Compare your answers in (a) with the equilibrium vapor pressures of methyl alcohol: 203 mm Hg at 35°C; 325 mm Hg at 45°C.
    (c) On the basis of your answers to (a) and (b), predict the pressure exerted by the methyl alcohol in the flask at 35°C; at 45°C.
    (d) What physical states of methyl alcohol are present in the flask at 35°C? at 45°C?

2. Benzene, a known carcinogen, was once widely used as a solvent. A sample of benzene vapor in a flask of constant volume exerts a pressure of 325 mm Hg at 80°C. The flask is slowly cooled.
    (a) Assuming no condensation, use the ideal gas law to calculate the pressure of the vapor at 50°C; at 60°C.
    (b) Compare your answers in (a) to the equilibrium vapor pressures of benzene: 269 mm Hg at 50°C, 389 mm Hg at 60°C.
    (c) On the basis of your answers to (a) and (b), predict the pressure exerted by the benzene at 50°C; at 60°C.

3. The vapor pressure of $I_2(s)$ at 30°C is 0.466 mm Hg.
    (a) How many milligrams of iodine will sublime into an evacuated 750.0-mL flask?
    (b) If 3.00 mg of I_2 are used, what will the final pressure in the flask be?
    (c) If 7.85 mg of I_2 are used, what will the final pressure in the flask be?

4. Camphor, $C_{10}H_{16}O$, is the active ingredient in vapor-steam products like Vicks VapoRub®. Its vapor pressure at 20°C is 0.18 mm Hg.

(a) How many milligrams of camphor will sublime into an evacuated 0.500-L flask?

(b) A 125-mL flask contains 0.15 mg of camphor at 20°C. What is the pressure in the flask?

5. Trichloroethane, $C_2H_3Cl_3$ is used as a degreaser (solvent for waxes and oils). Its density is 1.435 g/mL and its vapor pressure at 20°C is 124 mm Hg.

(a) How many mL will vaporize in an evacuated 1.50-L flask at 20°C?

(b) A 3.00-mL sample is poured into an evacuated 1.5-L flask at 20°C. Will all the liquid vaporize? If not, what is the pressure in the flask?

(c) A similar 3.00-mL sample is poured into an evacuated 20.00-L flask at 20°C. What physical state(s) is/are in the flask?

6. p-Dichlorobenzene, $C_6H_4Cl_2$, can be one of the ingredients in mothballs. Its vapor pressure at 20°C is 0.40 mm Hg.

(a) How many milligrams of $C_6H_4Cl_2$ will sublime into an evacuated 750-mL flask at 20°C?

(b) If 5.0 mg of p-dichlorobenzene were put into an evacuated 750-mL flask, how many milligrams would remain in the solid phase?

(c) What is the final pressure in an evacuated 500-mL flask at 20°C that contains 2.00 mg of p-dichlorobenzene? Will there be any solid in the flask?

7. Chloroform, $CHCl_3$, was once used as an anesthetic. In spy movies it is the liquid put in handkerchiefs to render victims unconscious. Its vapor pressure is 197 mm Hg at 23°C and 448 mm Hg at 45°C. Estimate its

(a) heat of vaporization.

(b) normal boiling point.

8. Dichloromethane, CH_2Cl_2, is widely used as a "degreaser" and paint stripper. Its vapor pressure is 381.0 mm Hg at 21.9°C and 465.8 mm Hg at 26.9°C. Estimate

(a) its heat of vaporization (ΔH_{vap}).

(b) its normal boiling point.

9. Winter Park, a town in the Rocky Mountains of Colorado, has an altitude of 9100 (2 significant figures) feet. Water ($\Delta H_{vap} = 40.7$ kJ/mol) boils at 91°C in Winter Park. What is the normal atmospheric pressure in Winter Park?

10. Glacier National Park in Montana is a favorite vacation spot for backpackers. It is about 4100 ft above sea level with an atmospheric pressure of 681 mm Hg. At what temperature does water ($\Delta H_{vap} = 40.7$ kJ/mol) boil in Glacier National Park?

11. At a resort in Sante Fe, New Mexico, the barometric pressure is 625 mm Hg. Water boils in an open pot at 94.5°C. A pressure cooker is set for 1.75 atm.

(a) At what temperature will water boil in that pressure cooker? (For water, $\Delta H_{vap} = 40.7$ kJ/mol.)

(b) What is the difference between the boiling point in the open pot and in the pressure cooker?

12. At sea level, the barometric pressure is 760 mm Hg. Calculate the difference in the normal boiling point for water boiled in an open pot and water boiled in a pressure cooker set for 1.75 atm. (ΔH_{vap} for water = 40.7 kJ/mol)

13. The data below give the vapor pressure of octane, a major component of gasoline.

vp (mm Hg)	10	40	100	400
$t(°C)$	19.2	45.1	65.7	104.0

Plot ln(vp) versus $1/T$. Use your graph to estimate the heat of vaporization of octane. ($\ln P = A - \dfrac{\Delta H_{vap}}{R}\left(\dfrac{1}{T}\right)$, where A is the y-intercept and ΔH_{vap} is the slope.)

14. Consider the following data for the vapor pressure of diethyl ether, a widely used anesthetic in the early days of surgery.

vp (mm Hg)	146	231	355	531
$t(°C)$	−5	5	15	25

Follow the instructions in Question 13 to estimate the heat of vaporization of diethyl ether.

9-3 Phase Diagrams

15. Referring to Figure 9.7, state what phase(s) is/are present at

(a) 1 atm, 100°C.

(b) 0.5 atm, 100°C.

(c) 0.8 atm, 50°C.

16. Referring to Figure 9.7, state what phase(s) is (are) present at

(a) 1 atm, 10°C.

(b) 3 mm Hg, 20°C.

(c) 1000 mm Hg, 75°C.

17. Consider the phase diagram of the compound X below. Use the phase diagram to answer the following questions.

_____ (a) What is the physical state of the compound at 35 mm Hg and −50°C?

_____ (b) What is the normal boiling point of the compound?

_____ (c) Where is the critical point?

_____ (d) Where is the triple point?

_____ (e) What change occurs when at a constant pressure of 30 mm Hg, the temperature is decreased from 120°C to 20°C?

_____ (f) Which phase is the densest phase?

18. Consider the phase diagram of the compound in Problem 17 to answer the following questions.

_____ (a) What is the physical state of the compound at 35 mm Hg and 120°C?

_____ (b) What is the normal freezing point of the compound?

_____ (c) What is the point _A_ called?

_____ (d) What is the point _B_ called?

_____ (e) What is the point _C_ called?

_____ (f) What change occurs when at a constant pressure of 33 mm Hg, the temperature is decreased from 40°C to −20°C?

_____ (g) Will the solid float on the liquid?

_____ (h) Can the compound exist as a liquid at 180°C and 2 atm pressure?

19. Argon gas has its triple point at −189.3°C and 516 mm Hg. It has a critical point at −122°C and 48 atm. The density of the solid is 1.65 g/cm^3, whereas that of the liquid is 1.40 g/cm^3. Sketch the phase diagram for argon and use it to fill in the blanks below with the words "boils," "melts," "sublimes," or "condenses."

(a) Solid argon at 500 mm Hg _____ when the temperature is increased.

(b) Solid argon at 2 atm _____ when the temperature is increased.

(c) Argon gas at −150°C _____ when the pressure is increased.

(d) Argon gas at −165°C _____ when the pressure is increased.

20. Iodine has a triple point at 114°C, 90 mm Hg. Its critical temperature is 535°C. The density of the solid is 4.93 g/cm^3, and that of the liquid is 4.00 g/cm^3. Sketch the phase diagram for iodine and use it to fill in the blanks using either "liquid" or "solid."

(a) Iodine vapor at 80 mm Hg condenses to the _____ when cooled sufficiently.

(b) Iodine vapor at 125°C condenses to the _____ when enough pressure is applied.

(c) Iodine vapor at 700 mm Hg condenses to the _____ when cooled above the triple point temperature.

21. Given the following data about xenon,

normal boiling point = −108°C

normal melting point = −112°C

triple point = −121°C at 281 mm Hg

critical point = 16.6°C at 58 atm

(a) Construct an approximate phase diagram for xenon.

(b) Estimate the vapor pressure of xenon at −115°C.

(c) Is the density of solid Xe larger than that for liquid Xe?

22. Given the following data about CO_2:

■ triple point: −57°C, 5.11 atm

■ critical point: 31°C, 73 atm

■ sublimes at −78.5°C, 1 atm

■ The solid is the denser phase.

(a) Construct a simplified (not to scale) phase diagram for CO_2.

(b) Estimate the vapor pressure of CO_2 at 0°C.

(c) At 10 atm pressure and −50°C, is CO_2 a liquid?

23. A pure substance X has the following properties: mp = 90°C, increasing slightly as pressure increases; normal bp = 120°C; liquid vp = 65 mm Hg at 100°C, 20 mm Hg at the triple point.

(a) Draw a phase diagram for X.

(b) Label solid, liquid, and vapor regions of the diagram.

(c) What changes occur if, at a constant pressure of 100 mm Hg, the temperature is raised from 100°C to 150°C?

24. A pure substance A has a liquid vapor pressure of 6 atm at 550°C. Its triple point is at 5 atm, 350°C. The solid phase is denser than the liquid phase.

(a) Sketch the phase diagram for compound A.

(b) From the phase diagram, estimate the temperature at which the compound boils when 6.3 atm of pressure are applied.

(c) What changes occur when, at a constant pressure of 4 atm, the temperature drops from 400°C to 100°C?

9-4 Molecular Substances; Intermolecular Forces

25. Arrange the following in order of decreasing boiling point.

(a) I_2 (b) F_2 (c) Cl_2 (d) Br_2

26. Arrange the following compounds in order of increasing boiling point.

(a) Ar (b) He (c) Ne (d) Xe

27. Which of the following would you expect to show dispersion forces? dipole forces?

(a) $GeBr_4$ (b) C_2H_2 (c) HF (_g_) (d) $TeCl_2$

28. Which of the following compounds would you expect to show dispersion forces? Dipole forces?

(a) F_2 (b) CO (c) CO_2 (d) H_2CO

29. Which of the following would show hydrogen bonding?

(a) CH_3OH (b) $CH_3—\underset{\underset{H}{|}}{N}—CH_3$

(c) CH_3NH_2 (d) HF

30. Which of the following compounds show hydrogen bonding?

(a) CH_3F (b) HO—OH

(c) NH_3F (d) $H_3C—O—CH_3$

31. Explain in terms of forces between structural units why

(a) HI has a higher boiling point than HBr.

(b) GeH_4 has a higher boiling point than SiH_4.

(c) H_2O_2 has a higher melting point than C_3H_8.

(d) NaCl has a higher boiling point than CH_3OH.

32. Explain in terms of structural units why

(a) CO_2 has a lower boiling point than Na_2CO_3.

(b) N_2H_4 has a higher boiling point than C_2H_6.

(c) formic acid, $H—\underset{\underset{O}{\|}}{C}—OH$, has a lower boiling point

than benzoic acid, $C_6H_5—\underset{\underset{O}{\|}}{C}—OH$.

(d) CO has a higher boiling point than N_2.

33. In which of the following processes is it necessary to break covalent bonds as opposed to simply overcoming intermolecular forces?

(a) melting mothballs made of naphthalene

(b) dissolving HBr gas in water to form hydrobromic acid

(c) vaporizing ethyl alcohol, C_2H_5OH

(d) changing ozone, O_3, to oxygen gas, O_2

34. In which of the following processes is it necessary to break covalent bonds as opposed to simply overcoming intermolecular forces?
  (a) Decomposing HCl to H_2 and Cl_2
  (b) Dissolving NaCl in water
  (c) Freezing ethyl alcohol
  (d) Subliming iodine

35. For each of the following pairs, choose the member with the lower boiling point. Explain your reason in each case.
  (a) NaCl or PCl_3 (b) NH_3 or AsH_3
  (c) C_3H_7OH or $C_2H_5OCH_3$ (d) HI(g) or HCl(g)

36. Follow the directions for Question 35 for the following substances.
  (a) LiCl or CCl_4 (b) CH_3OH or CH_3F
  (c) H_2O or SO_2 (d) N_2 or Cl_2

37. What are the strongest attractive forces that must be overcome to
  (a) boil silicon hydride, SiH_4?
  (b) vaporize calcium chloride?
  (c) dissolve Cl_2 in carbon tetrachloride, CCl_4?
  (d) melt iodine?

38. What are the strongest attractive forces that must be overcome to
  (a) melt ice?   (b) sublime bromine?
  (c) boil chloroform ($CHCl_3$)?   (d) vaporize benzene (C_6H_6)?

9-5 Network Covalent, Ionic, and Metallic Solids

39. Classify each of the following solids as metallic, network covalent, ionic, or molecular.
  (a) It is insoluble in water, melts above 500°C, and does not conduct electricity either as a solid, dissolved in water, or molten.
  (b) It dissolves in water but does not conduct electricity as an aqueous solution, as a solid, or when molten.
  (c) It dissolves in water, melts above 100°C, and conducts electricity when present in an aqueous solution.

40. Classify each of the following solids as metallic, network covalent, ionic, or molecular.
  (a) It melts below 100°C and is insoluble in water.
  (b) It conducts electricity only when melted.
  (c) Mostly insoluble in water and conducts electricity.

41. Of the four general types of solids, which one(s)
  (a) are generally low-boiling?
  (b) are ductile and malleable?
  (c) are generally soluble in nonpolar solvents?

42. Of the four general types of solids, which one(s)
  (a) are generally insoluble in water?
  (b) have very high melting points?
  (c) conduct electricity as solids?

43. Classify each of the following species as molecular, network covalent, ionic, or metallic.
  (a) Na   (b) Na_2SO_4 (c) C_6H_6
  (d) C_{60} (e) HCl(aq)

44. Classify each of the following species as metallic, network covalent, ionic, or molecular.
  (a) W   (b) NO_2 (c) C (diamond)
  (d) $FeCl_2$ (e) C_2H_2

45. Give the formula of a solid containing Si that is
  (a) molecular
  (b) ionic
  (c) network covalent

46. Give the formula of a solid containing oxygen that is
  (a) a polar molecule   (b) ionic
  (c) network covalent   (d) a nonpolar molecule

47. Describe the structural units in
  (a) NaI   (b) N_2 (c) KO_2 (d) Au

48. Describe the structural units in
  (a) C (graphite)
  (b) SiC
  (c) $FeCl_2$
  (d) C_2H_2

9-6 Crystal Structures

49. Molybdenum has an atomic radius of 0.145 nm. The volume of its cubic unit cell is 0.0375 nm^3. What is the geometry of the molybdenum unit cell?

50. Nickel has an atomic radius of 0.162 nm. The edge of its cubic unit cell is 0.458 nm. What is the geometry of the nickel unit cell?

51. Lead (atomic radius = 0.181 nm) crystallizes with a face-centered cubic unit cell. What is the length of a side of the cell?

52. Iridium crystallizes in a face-centered unit cell. The side of the cell is 0.3833 nm. Calculate the volume of the cell and the radius of an iridium atom.

53. In the LiCl structure shown in Figure 9.21, the chloride ions form a face-centered cubic unit cell 0.513 nm on an edge. The ionic radius of Cl^- is 0.181 nm.
  (a) Along a cell edge, how much space is between the Cl^- ions?
  (b) Would an Na^+ ion ($r = 0.095$ nm) fit into this space? a K^+ ion ($r = 0.133$ nm)?

54. Potassium iodide has a unit cell similar to that of sodium chloride (Figure 9.21). The ionic radii of K^+ and I^- are 0.133 nm and 0.216 nm, respectively. How long is
  (a) one side of the cube?
  (b) the face diagonal of the cube?

55. For a cell of the CsCl type (Figure 9.21), how is the length of one side of the cell, s, related to the sum of the radii of the ions, $r_{cation} + r_{anion}$?

56. Consider the CsCl cell (Figure 9.21). The ionic radii of Cs^+ and Cl^- are 0.169 and 0.181 nm, respectively. What is the length of
  (a) the body diagonal?
  (b) the side of the cell?

57. Consider the sodium chloride unit cell shown in Figure 9.21. Looking only at the front face (five large Cl^- ions, four small Na^+ ions),
  (a) how many cubes share each of the Na^+ ions in this face?
  (b) how many cubes share each of the Cl^- ions in this face?

58. Consider the CsCl unit shown in Figure 9.21. How many Cs^+ ions are there per unit cell? How many Cl^- ions? (Note that each Cl^- ion is shared by eight cubes.)

Unclassified

59. A 1.25-L clean and dry flask is sealed. The air in the flask is at 27°C and 38% relative humidity. The flask is put in a cooler at 5°C. How many grams of water will condense in the flask? (Use the table in Appendix 1 for the vapor pressure of water at various temperatures.)

60. Aluminum metal crystallizes with a face-centered cubic unit cell. The volume of the cell is 0.0662 nm³.
(a) What is the atomic radius of aluminum in cm?
(b) What is the volume of a single aluminum atom?
(c) What is the density of a single aluminum atom?
(d) In face-centered cubic cell packing, the fraction of empty space is 26.0%. When this is factored in, what is the calculated density of aluminum?

61. Consider a sealed flask with a movable piston that contains 5.25 L of O_2 saturated with water vapor at 25°C. The piston is depressed at constant temperature so that the gas is compressed to a volume of 2.00 L. (Use the table in Appendix 1 for the vapor pressure of water at various temperatures.)
(a) What is the vapor pressure of water in the compressed gas mixture?
(b) How many grams of water condense when the gas mixture is compressed?

62. Packing efficiency is defined as the percent of the total volume of a solid occupied by (spherical) atoms. The formula is

$$\text{packing efficiency} = \frac{\text{volume of the atom(s) in the cell}}{\text{volume of the cell}} \times 100$$

The volume of one atom is $\frac{4}{3}\pi r^3$ and the volume of the cell is s^3. Calculate the packing efficiency of
(a) a simple cubic cell (1 atom/cell).
(b) a face-centered cubic cell (4 atoms/cell).
(c) a body-centered cubic cell (2 atoms/cell).
Use Table 9.6 to relate r to s.

63. Mercury is an extremely toxic substance. Inhalation of the vapor is just as dangerous as swallowing the liquid. How many milliliters of mercury will saturate a room that is $15 \times 12 \times 8.0$ ft with mercury vapor at 25°C? The vapor pressure of Hg at 25°C is 0.00163 mm Hg and its density is 13 g/mL.

64. An experiment is performed to determine the vapor pressure of formic acid. A 30.0-L volume of helium gas at 20.0°C is passed through 10.00 g of liquid formic acid (HCOOH) at 20.0°C. After the experiment, 7.50 g of liquid formic acid remains. Assume that the helium gas becomes saturated with formic acid vapor and the total gas volume and temperature remain constant. What is the vapor pressure of formic acid at 20.0°C?

65. Chloroform, $CHCl_3$, has a normal boiling point of 61°C. Its vapor pressure at 43°C is 0.526 atm. What is the concentration (in g/L) of $CHCl_3$ when it saturates the air at 27°C?

Conceptual Problems

66. Which of the following statements are true?
(a) The critical temperature must be reached to change liquid to gas.
(b) To melt a solid at constant pressure, the temperature must be above the triple point.
(c) CHF_3 can be expected to have a higher boiling point than $CHCl_3$ because CHF_3 has hydrogen bonding.

(d) One metal crystallizes in a body-centered cubic cell and another in a face-centered cubic cell of the same volume. The two atomic radii are related by the factor $\sqrt{1.5}$.

67. Write the letter(s) corresponding to the bond(s) and/or force(s) associated with each species below.
   A – dispersion forces
   B – dipole-dipole interaction
   C – ionic bond
   D – metallic bond
   E – covalent bond within a network covalent solid
   F – hydrogen bond.
(a) C (diamond)   (b) CH_3 (c) Au
(d) SO_3 (e) NH_3 (f) KCl

68. In the blanks provided, answer the questions below, using **LT** (for *is less than*), **GT** (for *is greater than*), **EQ** (for *is equal to*), or **MI** (for *more information required*).
(a) The boiling point of C_3H_7OH (MM = 60.0 g/mol) _____ the boiling point of $C_2H_6C{=}O$ (MM = 58.0 g/mol).
(b) The vapor pressure of X is 250 mm Hg at 57°C. Given a sealed flask at 57°C that contains only gas, the pressure in the flask _____ 245 mm Hg.
(c) The melting-point curve for Y tilts to the right of a straight line. The density of Y(*l*) _____ the density of Y(*s*).
(d) The normal boiling point of A is 85°C, while the normal boiling point of B is 45°C. The vapor pressure of A at 85°C _____ the vapor pressure of B at 45°C.
(e) The triple point of A is 25 mm Hg and 5°C. The melting point of A _____ 5°C.

69. Answer the questions below, by filling in the blanks with **LT** for *is less than,* **GT** for *is greater than,* **EQ** for *is equal to,* or **MI** for *more information required.*
(a) At 50°C, benzene has a vapor pressure of 269 mm Hg. A flask that contains both benzene liquid and vapor at 50°C has a pressure _____ 269 mm Hg.
(b) Ether has a vapor pressure of 537 mm Hg at 25°C. A flask that contains only ether vapor at 37°C has a pressure _____ 537 mm Hg.
(c) The boiling point of H_2O _____ the boiling point of C_3H_8.
(d) The energy required to vaporize liquid bromine _____ the energy required to decompose Br_2 into Br atoms.
(e) The dispersion forces present in naphthalene, $C_{10}H_8$, _____ the dispersion forces present in butane, C_4H_{10}.

70. A liquid has a vapor pressure of 159 mm Hg at 20°C and 165 mm Hg at 30°C. Different amounts of the liquid are added to three identical evacuated steel tanks kept at 20°C. The tanks are all fitted with pressure gauges. For each part, write

L/G if both liquid and gas are present.
G if only gas is present.
I if the situation is impossible.

(a) The pressure gauge in Flask I registers a pressure of 256 mm Hg.
(b) The pressure gauge in Flask II registers a pressure of 135 mm Hg.
(c) The pressure gauge in Flask III registers a pressure of 165 mm Hg at 30°C. The temperature is lowered to 20°C, and the gauge registers a pressure of 159 mm Hg.

71. Indicate whether the following statements are true or false.

(a) The triple point in the phase diagram of a pure substance is the temperature and pressure at which the substance can boil, freeze, and sublime at the same time.

(b) CHF_3 can be expected to have a higher boiling point than $CHCl_3$ because $CHCl_3$ has hydrogen bonds.

(c) The strength of covalent bonds within a molecule has no effect on the melting or boiling points of the compound made up of that molecule.

(d) As the gas from a condensable gas (e.g., propane) in a tank is used up, the pressure gradually decreases.

(e) The critical temperature is the temperature above which only the gaseous phase can exist.

72. What is the difference between

(a) a covalent bond and an ionic bond?

(b) boiling point and critical point?

(c) deposition and sublimation?

(d) freezing point and triple point?

73. Four shiny solids are labeled A, B, C, and D. Given the following information about the solids, deduce the identity of A, B, C, and D.

(1) The solids are a graphite rod, a silver bar, a lump of "fool's gold" (iron sulfide), and iodine crystals.

(2) B, C, and D are insoluble in water. A is slightly soluble.

(3) Only C can be hammered into a sheet.

(4) C and D conduct electricity as solids; B conducts when melted; A does not conduct as a solid, melted, or dissolved in water.

74. Consider the vapor pressure curves of molecules A, B, and C shown below.

(a) Which compound (A, B, or C) has the weakest forces between molecules?

(b) Which compound (A, B, or C) has a normal boiling point at about 15°C?

(c) At what temperature will B boil if the atmospheric pressure is 500 mm Hg?

(d) At 25°C and 400 mm Hg, what is the physical state of A?

(e) At what pressure will C boil at 40°C?

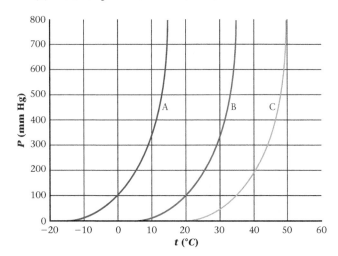

Challenge Problems

75. The following data are given for CCl_4:

normal melting point = −23°C
normal boiling point = 77°C
density of liquid = 1.59 g/mL
vapor pressure at 25°C = 110 mm Hg

How much heat is required to vaporize 20.0 L of CCl_4 at its normal boiling point?

76. Iron crystallizes in a body-centered unit cell. Its atomic radius is 0.124 nm. Its density is 7.86 g/cm³. Using this information, estimate Avogadro's number.

77. A flask with a volume of 10.0 L contains 0.400 g of hydrogen gas and 3.20 g of oxygen gas. The mixture is ignited and the reaction

$$2H_2(g) + O_2(g) \longrightarrow 2H_2O$$

goes to completion. The mixture is cooled to 27°C. Assuming 100% yield,

(a) What physical state(s) of water is (are) present in the flask?

(b) What is the final pressure in the flask?

(c) What is the pressure in the flask if 3.2 g of each gas is used?

78. Trichloroethane, $C_2H_3Cl_3$, is the active ingredient in aerosols that claim to stain-proof men's ties. Trichloroethane has a vapor pressure of 100.0 mm Hg at 20.0°C and boils at 74.1°C. An uncovered cup ($\frac{1}{2}$ pint) of trichloroethane ($d = 1.325$ g/mL) is kept in an 18-ft³ refrigerator at 39°F. What percentage (by mass) of the trichloroethane is left as a liquid when equilibrium is established?

79. It has been suggested that the pressure exerted on a skate blade is sufficient to melt the ice beneath it and form a thin film of water, which makes it easier for the blade to slide over the ice. Assume that a skater weighs 120 lb and the blade has an area of 0.10 in². Calculate the pressure exerted on the blade (1 atm = 15 lb/in²). From information in the text, calculate the decrease in melting point at this pressure. Comment on the plausibility of this explanation and suggest another mechanism by which the water film might be formed.

80. As shown in Figure 9.20, Li^+ ions fit into a closely packed array of Cl^- ions, but Na^+ ions do not. What is the value of the r_{cation}/r_{anion} ratio at which a cation just fits into a structure of this type?

81. When the temperature drops from 20°C to 10°C, the pressure of a cylinder of compressed N_2 drops by 3.4%. The same temperature change decreases the pressure of a propane (C_3H_8) cylinder by 42%. Explain the difference in behavior.

10 Solutions

Collecting maple sap and converting it to maple syrup utilize many of the chemical principles found in this chapter.

Water, water, everywhere,
And all the boards did shrink;
Water, water, everywhere,
Nor any drop to drink.

—SAMUEL TAYLOR COLERIDGE

From *The Rime of the Ancient Mariner*

Sugaring Off, c.1865 (oil on canvas), Johnson, Eastman (1824–1906) / Huntington Library and Art Gallery, San Marino, CA, USA / ©The Huntington Library, Art Collections & Botanical Gardens/The Bridgeman Art Library

In the course of a day, you use or make solutions many times. Your morning cup of coffee is a solution of solids (sugar and coffee) in a liquid (water). The gasoline you fill your gas tank with is a solution of several different liquid hydrocarbons. The soda you drink at a study break is a solution containing a gas (carbon dioxide) in a liquid (water).

A solution is a homogeneous mixture of a **solute** (substance being dissolved) distributed through a **solvent** (substance doing the dissolving). Solutions exist in any of the three physical states: gas, liquid, or solid. Air, the most common gaseous solution, is a mixture of nitrogen, oxygen, and lesser amounts of other gases. Many metal alloys are solid solutions. An example is the U.S. "nickel" coin (25% Ni, 75% Cu). The most familiar solutions are those in the liquid state, especially ones in which water is the solvent. Aqueous solutions are most important for our purposes in chemistry and will be emphasized in this chapter.

This chapter covers several of the physical aspects of solutions, including

- methods of expressing solution concentrations by specifying the relative amounts of solute and solvent (Section 10-1).
- factors affecting solubility, including the nature of the solute and the solvent, the temperature, and the pressure (Section 10-2).
- the effect of solutes on such solvent properties as vapor pressure, freezing point, and boiling point (Sections 10-3 and 10-4).

10-1 Concentration Units

Several different methods are used to express relative amounts of solute and solvent in a solution. Two concentration units, *molarity* and *mole fraction*, were referred to in previous chapters. Two others, *mass percent* and *molality*, are considered for the first time.

Molarity (*M*)

In Chapter 4, **molarity** was the concentration unit of choice in dealing with solution stoichiometry. You will recall that molarity is defined as

$$\text{molarity } (M) = \frac{\text{moles solute}}{\text{liters solution}}$$

A solution can be prepared to a specified molarity by weighing out the calculated mass of solute and dissolving in enough solvent to form the desired volume of solution. Alternatively, you can start with a more concentrated solution and dilute with water to give a solution of the desired molarity ◀ *i* (Figure 10.1). The calculations are straightforward if you keep a simple point in mind: Adding solvent cannot change the number of moles of solute. That is,

$$n_{\text{solute}} \text{ (concentrated solution)} = n_{\text{solute}} \text{ (dilute solution)}$$

In both solutions, *n* can be found by multiplying the molarity, *M*, by the volume in liters, *V*. Hence

$$M_cV_c = M_dV_d \tag{10.1}$$

where the subscripts c and d stand for concentrated and dilute solutions, respectively.

The advantage of preparing solutions by the method illustrated in Figure 10.1 is that only volume measurements are necessary. If you wander into the general chemistry storeroom, you're likely to find concentrated "stock" solutions of various chemicals. Storeroom personnel prepare the more dilute solutions that you use in the laboratory on the basis of calculations like those in the following example.

i It's easier to dilute a concentrated solution than to start from "scratch."

A volumetric flask with 50.0 mL of 2.00 *M* CuSO$_4$ that was accurately measured using the pipet.

Water is added to just below the narrow neck of the flask. The flask is then shaken to ensure complete mixing.

Finally, water is carefully added to bring the level exactly to the mark on the neck of the flask corresponding to 1.00 L.

Figure 10.1 Preparation of one liter of 0.100 *M* CuSO$_4$ by dilution.

EXAMPLE 10.1

Copper sulfate is widely used as a dietary supplement for animal feed. A lab technician prepares a "stock" solution of $CuSO_4$ by adding 40.00 g of $CuSO_4$ to enough water to make 500.0 mL of solution. An experiment requires a 0.1000 M solution of $CuSO_4$.

a What is the molarity of the $CuSO_4$ "stock" solution prepared by the technician?

ANALYSIS

Information given:	mass $CuSO_4$ (40.00 g); $V_{solution}$ (500.0 mL)
Information implied:	molar mass of $CuSO_4$
Asked for:	molarity of stock solution

STRATEGY

Recall (from Chapter 3) the definition of molarity:

$$M = \frac{\text{moles solute}}{\text{volume of solution (L)}}$$

SOLUTION

n_{CuSO_4}	40.00 g $CuSO_4 \times \dfrac{1 \text{ mol}}{159.6 \text{ g}} = 0.2506$ mol
$M_{\text{stock solution}}$	$\dfrac{0.2506 \text{ mol}}{0.5000 \text{ L}} = 0.5013 \ M$

b How would you prepare 1.500 L of 0.1000 M solution from the stock solution?

ANALYSIS

Information given:	for stock solution from part (a): M_c (0.5013 M) for diluted solution: M_d (0.1000 M); V_d (1.500 L)
Asked for:	how to prepare 1.500 L of 0.1000 M $CuSO_4$ from the stock solution

STRATEGY

1. The question really is: What volume of "stock" solution needs to be diluted to give the desired volume and molarity?

2. Substitute into Equation 10.1 to calculate V_c, the volume of the concentrated or stock solution.

$$M_c V_c = M_d V_d$$

SOLUTION

V_c	$V_c = \dfrac{M_d V_d}{M_c} = \dfrac{(0.1000 \ M)(1.500 \text{ L})}{0.5013 \ M} = 0.2992$ L
Directions	Measure out 299.2 mL of the stock solution and dilute with enough water to make 1.500 L of solution.

END POINT

The molarity of the $CuSO_4$ stock solution (0.5013 M) is about five times what you need (0.1000 M), so it is reasonable to use only one fifth of the volume of the stock solution (≈ 0.3 L $\rightarrow$ 1.500 L).

Mole Fraction (X)

Recall from Chapter 5 the defining equation for **mole fraction** (X) of a component A:

$$X_A = \frac{\text{moles A}}{\text{total moles}} = \frac{n_A}{n_{tot}}$$

The mole fractions of all components of a solution (A, B, . . .) must add to unity:

$$X_A + X_B + \cdots = 1$$

EXAMPLE 10.2

Hydrogen peroxide is used by some water treatment systems to remove the disagreeable odor of sulfides in drinking water. It is available commercially in a 30.0% by mass aqueous solution. What is the mole fraction of H_2O_2?

ANALYSIS

Information given:	mass percent of H_2O_2 (30.0%)
Information implied:	molar masses of H_2O_2 and H_2O
Asked for:	mol fraction (X) of H_2O_2

STRATEGY

1. Start with a fixed mass of solution such as one hundred grams.
2. Calculate moles H_2O_2 ($n_{H_2O_2}$), moles H_2O (n_{H_2O}), and find total moles (n_{tot}).
3. Substitute into the equation:

$$X_A = \frac{n_A}{n_{tot}}$$

SOLUTION

1. Assume 100.0 g of solution — mass H_2O_2 = 30.0 g; mass H_2O = 70.0 g

2. n_{H_2O}

$$n_{H_2O} = \frac{70.0 \text{ g}}{18.02 \text{ g/mol}} = 3.88 \text{ mol}$$

$n_{H_2O_2}$

$$n_{H_2O_2} = \frac{30.0 \text{ g}}{34.02 \text{ g/mol}} = 0.882 \text{ mol}$$

n_{tot}

$$n_{tot} = n_{H_2O_2} + n_{H_2O} = 0.882 \text{ mol} + 3.88 \text{ mol} = 4.76 \text{ mol}$$

3. $X_{H_2O_2}$

$$X_{H_2O_2} = \frac{n_{H_2O_2}}{n_{tot}} = \frac{0.882 \text{ mol}}{4.76 \text{ mol}} = 0.185$$

END POINTS

1. Multiplying the mol fraction of H_2O_2 by 100% gives the mol percent of H_2O_2 in the solution (18.5%).
2. Notice that the mole percent (18.5) is considerably less than the mass percent (30.0) of H_2O_2 in solution. That is because the molar mass of H_2O_2 is larger than that of H_2O.

Mass Percent; Parts per Million; Parts per Billion

The **mass percent** of solute in solution is expressed quite simply:

$$\text{mass percent of solute} = \frac{\text{mass solute}}{\text{total mass solution}} \times 100\%$$

In a solution prepared by dissolving 24 g of NaCl in 152 g of water,

$$\text{mass percent of NaCl} = \frac{24\text{ g}}{24\text{ g} + 152\text{ g}} \times 100\% = \frac{24}{176} \times 100\% = 14\%$$

When the amount of solute is very small, as with trace impurities in water, concentration is often expressed in **parts per million** (ppm) or **parts per billion** (ppb).

In the United States, by law, drinking water cannot contain more than 5×10^{-8} g of arsenic per gram of water:

$$\text{ppm As} = \frac{5 \times 10^{-8}\text{ g}}{\text{g sample}} \times 10^6 = 0.05; \qquad \text{ppb As} = \frac{5 \times 10^{-8}\text{ g}}{\text{g sample}} \times 10^9 = 50$$

For very dilute aqueous solutions, 1 ppm is approximately equivalent to 1 mg/L. This is because the density of water at 25°C is 1.0 g/mL, and very dilute solutions have a density almost equal to that of pure water (1 g = 1 mL). Thus

$$1\text{ ppm} = \frac{1\text{ g solute}}{1 \times 10^6\text{ g solution}} = \frac{1\text{ g solute}}{1 \times 10^6\text{ mL solution}}$$

Converting g to mg and mL to L we get

$$\frac{1\text{ g solute}}{1 \times 10^6\text{ mL solution}} \times \frac{1000\text{ mg}}{1\text{ g}} \times \frac{1000\text{ mL}}{1\text{ L}} = 1\text{ mg/L}$$

Molality (*m*)

The concentration unit **molality,** symbol *m,* is the number of moles of solute per kilogram (1000 g) of solvent.

Molality and molarity are concentration units; morality is something else.

$$\text{molality } (m) = \frac{\text{moles solute}}{\text{kilograms solvent}}$$

You can readily calculate the molality of a solution if you know the masses of solute and solvent (Example 10.3).

EXAMPLE 10.3

Glucose, $C_6H_{12}O_6$, in water is often used for intravenous feeding. Sometimes sodium ions are added to the solution. A pharmacist prepares a solution by adding 2.0 mg of sodium ions (in the form of NaCl), 6.00 g of glucose, and 112 g of water.

a What is the molality of the glucose in solution?

ANALYSIS

Information given:	mass of glucose (6.00 g) mass of water (112 g)
Information implied:	molar mass of glucose
Asked for:	*m* of glucose

STRATEGY

1. Recall the definition of molality and identify the solute and solvent.

$$m = \frac{\text{moles of solute}}{\text{mass of solvent (kg)}}$$

2. Find the numerator (moles of solute). Find the denominator (kg of solvent).

3. Substitute into the definition of *m*.

continued

	SOLUTION
numerator	moles solute (glucose) = $\dfrac{6.00 \text{ g glucose}}{180.16 \text{ g/mol}}$ = 0.0333
denominator	mass of solvent (H_2O) in kg = 112 g/1000 = 0.112 kg
m	$m = \dfrac{\text{moles glucose}}{\text{kg } H_2O} = \dfrac{0.0333 \text{ mol}}{0.112 \text{ kg}} = $ 0.297 m

b How many ppm of Na^+ does the solution contain?

	ANALYSIS
Information given:	mass of Na^+ ions (2.0 mg), mass of glucose (6.00 g), mass of water (112 g)
Asked for:	ppm Na^+

STRATEGY
Substitute into the equation: ppm of $Na^+ = \dfrac{\text{mass } Na^+ \text{ (in grams)}}{\text{mass solution}} \times 10^6$

	SOLUTION
mass of solution	mass of solution = mass Na^+ + mass glucose + mass water
	mass of solution = $\left(2.00 \text{ mg} \times \dfrac{1 \text{ g}}{1000 \text{ mg}}\right) + 6.00 \text{ g} + 112 \text{ g} = 118 \text{ g}$
ppm Na^+	ppm $= \dfrac{\text{mass } Na^+}{\text{mass solution}} \times 10^6 = \dfrac{0.00200 \text{ g}}{118 \text{ g}} \times 10^6 = $ 17

END POINT
Theoretically, you would also need to calculate the mass of Cl^- ions and add that to the mass of Na^+, glucose, and water. Practically, the mass of Cl^- ions (like the mass of Na^+ ions) is negligible when compared with the mass of the solution.

Conversions Between Concentration Units

It is frequently necessary to convert from one concentration unit to another. This problem arises, for example, in making up solutions of hydrochloric acid. Typically, the analysis or assay that appears on the label (Figure 10.2) does not give the molarity or molality of the acid. Instead, it lists the mass percent of solute and the density of the solution.

Conversions between concentration units are relatively straightforward provided you *first decide on a fixed amount of solution*. The amount chosen depends on the unit in which concentration is originally expressed. Some suggested starting quantities are listed below.

When the Original Concentration Is	Start With
Mass percent	100 g solution
Molarity (*M*)	1.00 L solution
Molality (*m*)	1000 g solvent
Mole fraction (*X*)	1 mol (solute + solvent)

ACTUAL ANALYSIS. LOT **320037**		MEETS A.C.S. SPECIFICATIONS	
* Assay (HCl)(by acidimetry)		37.7	%
Appearance		Passes Test	
Color (APHA)		< 5	
Specific Gravity at 60°/60°F		1.1906	
Residue after Ignition		0.00005	%
Free Chlorine (Cl)		Passes Test	
Bromide (Br)		< 0.005	%
	Trace Impurities (in ppm):		
Ammonium (NH₄)		< 3	
Sulfate (SO₄)		0.25	
Sulfite (SO₃)		0.2	
Arsenic (As)		< 0.004	
Copper (Cu)		0.0004	
Iron (Fe)		0.002	
Heavy Metals (as Pb)		< 0.05	
Nickel (Ni)		0.0004	

*Assay value tends to be less than reported due to vapor loss, especially when opening container.

Marna G. Clarke

Figure 10.2 The label on a bottle of concentrated hydrochloric acid. The label gives the mass percent of HCl in the solution (known as the *assay*) and the density (or *specific gravity*) of the solution. The molality, molarity, and mole fraction of HCl in the solution can be calculated from this information.

In these kinds of calculations, there can be many different types of concentration units that you need to obtain from one given concentration unit. To keep your data organized, it is best to prepare and fill in a table ◄ ⓘ like the one below.

Complex problems can usually be solved if you know where to start.

	moles	←── MM ──→	mass	density ──→	volume
Solute					
Solvent					
Solution					

Here is an illustration of how you can fill in and use the table. Suppose that you want to find the molality (*m*) of a solution X (MM = 100.0 g/mol) that is 1.35 *M* and has a density of 1.28 g/mL. The given concentration unit is molarity. Recall the definition of molarity (1.35 moles solute in 1000 mL of solution) and fill in the table accordingly. The concentration unit asked for is molality (moles solute in 1 kg of solvent). The table shown below has the given data filled in the appropriate spaces and the spots needed for the molality are shaded. Your table now looks like this:

Molarity and molality. On the left is a 0.10 *M* solution of potassium chromate. On the right is a 0.10 *m* solution that contains the same amount of potassium chromate (19.4 g, in dish). The 0.10 *M* solution was made by placing the solid in the flask and adding water to give 1 L of solution. The 0.10 *m* solution was prepared by placing the solid in the flask and adding 1000 g of water. You can see that the 0.10 *m* solution has a slightly larger volume.

	moles	←── MM ──→	mass	density ──→	volume
Solute	1.35				
Solvent					
Solution					1000 mL

You can calculate the mass of the solvent if you know the mass of the solution and the mass of the solute. Since the density and the volume of the solution are known, the mass of the solution is 1280 g [(1.28 g/mL)(1000 mL)]. The mass of the solute is 135 g [(1.35 mol)(100.0 g/mol)]. Thus the mass of the solvent is 1145 g (1280 g − 135 g) or 1.145 kg.

© Cengage Learning/Charles D. Winters

Your table now looks like this:

	moles	← MM	mass	density →	volume
Solute	1.35		135 g		
Solvent			1145 g		
Solution			1280 g		1000 mL

Finding the molality is now simply a matter of using the data in the shaded spaces to fit the defintion of molality.

$$\text{molality } (m) = \frac{n_{\text{solute}}}{\text{mass solvent (kg)}} = \frac{1.35 \text{ mol}}{1.145 \text{ kg}} = 1.18 \; m$$

EXAMPLE 10.4 GRADED

Consider the information given on the label of the bottle of HCl shown in Figure 10.2 (page 252).

a What is the mass percent of water in the bottle?

ANALYSIS	
Information given:	from the label: mass % of HCl (37.7%)
Asked for:	mass % of H_2O

SOLUTION	
mass % H_2O	100% = mass % HCl + mass % H_2O mass % H_2O = 100% − 37.7% = 62.3%

b What is the molality of HCl?

ANALYSIS	
Information given:	from label: mass % of HCl (37.7%), from part (a): mass % of H_2O (62.3%)
Asked for:	molality m

STRATEGY

1. Assume 100.0 g of solution.

2. Draw the table, fill in the mass of solute and solvent, and shade in the spots needed for molality: moles of solute and mass of solvent in kg.

3. Find the moles of solute and add to the table.

4. Substitute values into the defining equation for molality.

$$m = \frac{\text{moles solute}}{\text{mass solvent (kg)}}$$

SOLUTION

2. Table

	moles	← MM	mass	density →	volume
Solute			37.7 g		
Solvent			62.3 g		
Solution			100.0 g		

continued

3. moles solute

Table

$$37.7 \text{ g HCl} \times \frac{1 \text{ mol}}{36.46 \text{ g}} = 1.03 \text{ g}$$

	moles	$\xleftarrow{\text{MM}}$	mass	$\xrightarrow{\text{density}}$	volume
Solute	1.03		37.7 g		
Solvent			0.0623 kg		
Solution			100.0 g		

4. m

$$\frac{1.03 \text{ mol}}{0.0623 \text{ kg}} = 16.5 \, m$$

c What is the molarity of the HCl in the bottle?

ANALYSIS

Information given:	from label: mass % of HCl (37.7%) from part (a): mass % of H_2O (62.3%) from part (b): moles of solute (1.03)
Asked for:	M

STRATEGY

1. Draw the table as in part (b). Shade in the spots needed for molarity: moles of solute and volume of solution.

2. Find the volume of solution and fill in the table with that information.

3. Substitute values into the defining equation for molarity.

$$M = \frac{\text{moles solute}}{\text{volume solution (L)}}$$

SOLUTION

1. Table

	moles	$\xleftarrow{\text{MM}}$	mass	$\xrightarrow{\text{density}}$	volume
Solute	1.03		37.7 g		
Solvent			0.0623 kg		
Solution			100.0 g		

2. V_{solution}

$$V_{\text{solution}} = \frac{\text{mass}}{\text{density}} = \frac{100.0 \text{ g}}{1.19 \text{ g/mL}} = 84.0 \text{ mL} = 0.0840 \text{ L}$$

Table

	moles	$\xleftarrow{\text{MM}}$	mass	$\xrightarrow{\text{density}}$	volume
Solute	1.03		37.7 g		
Solvent			0.0623 kg		
Solution			100.0 g		0.0840 L

3. M

$$M = \frac{\text{mol solute}}{V_{\text{solution}} \text{ (L)}} = \frac{1.03 \text{ mol}}{0.0840 \text{ L}} = 12.3 \, M$$

10-2 Principles of Solubility

The extent to which a solute dissolves in a particular solvent depends on several factors. The most important of these are

■ the nature of solvent and solute particles and the interactions between them.
■ the temperature at which the solution is formed.
■ the pressure of a gaseous solute.

In this section we consider in turn the effect of each of these factors on **solubility**.

Solute-Solvent Interactions

In discussing solubility, it is sometimes stated that "like dissolves like." A more meaningful way to express this idea is to say that two substances with intermolecular forces of about the same type and magnitude are likely to be very soluble in one another (Figure 10.3). To illustrate, consider the hydrocarbons pentane, C_5H_{12}, and hexane, C_6H_{14}, which are completely soluble in each other. Molecules of these nonpolar substances are held together by dispersion forces of about the same magnitude. A pentane molecule experiences little or no change in intermolecular forces when it goes into solution in hexane.

Most nonpolar substances have very small water solubilities. ◄ ⓘ Petroleum, a mixture of hydrocarbons, spreads out in a thin film on the surface of a body of water rather than dissolving. The mole fraction of pentane, C_5H_{12}, in a saturated water solution is only 0.0001. These low solubilities are readily understood in terms of the structure of liquid water, which you will recall (Chapter 9) is strongly hydrogen-bonded. Dissimilar intermolecular forces between C_5H_{12} (dispersion) and H_2O (H bonds) lead to low solubility.

Of the relatively few organic compounds that dissolve readily in water, many contain —OH groups. Three familiar examples are methyl alcohol, ethyl alcohol, and ethylene glycol, all of which are infinitely soluble in water.

$$
\begin{array}{ccc}
\overset{\displaystyle H}{\underset{\displaystyle H}{H-C-OH}} & \overset{\displaystyle H \quad H}{\underset{\displaystyle H \quad H}{H-C-C-OH}} & \overset{\displaystyle H \quad H}{\underset{\displaystyle HO \quad OH}{H-C-C-H}} \\[2ex]
\text{methyl alcohol} & \text{ethyl alcohol} & \text{ethylene glycol}
\end{array}
$$

> You can't remove the "hot" taste of chili peppers by drinking water because the compound responsible is nonpolar and water-insoluble.

In these compounds, as in water, the principal intermolecular forces are hydrogen bonds. When a substance like methyl alcohol dissolves in water, it forms hydrogen bonds with H_2O molecules. These hydrogen bonds, joining a CH_3OH molecule to an H_2O molecule, are about as strong as those in the pure substances.

Not all organic compounds that contain —OH groups are soluble in water (Table 10.1). As molar mass increases, the polar —OH group represents an increasingly smaller portion of the molecule. At the same time, the nonpolar hydrocarbon portion becomes larger. As a result, solubility decreases with increasing molar mass. Butanol, $CH_3CH_2CH_2CH_2OH$, is much less soluble in water than methyl alcohol, CH_3OH. The hydrocarbon portion, shaded in gold, is much larger in butanol.

The solubility (or insolubility) of different vitamins is of concern in nutrition. Molecules of vitamins B and C contain several —OH groups that can form hydrogen bonds with water (Figure 10.4). As a result, they are water-soluble, readily excreted by the body, and must be consumed daily. In contrast, vitamins A, D, E, and K, whose molecules are relatively nonpolar, are water-insoluble. These vitamins are not so readily excreted; they tend to stay behind in fatty tissues. This means that the body can draw on its reservoir of vitamins A, D, E, and K to deal with sporadic deficiencies. Conversely, megadoses of these vitamins can lead to very high, possibly toxic, concentrations in the body.

© Cengage Learning/Charles D. Winters

Figure 10.3 Solubility and intermolecular forces. Oil *(left)* is made up of nonpolar molecules. It does not mix well with water because water is polar. Ethylene glycol *(right)*, commonly used as antifreeze, mixes with water in all proportions because both form hydrogen bonds.

Table 10.1 Solubilities of Alcohols in Water

Substance	Formula	Solubility (g solute/L H_2O)
Methyl alcohol	CH_3OH	Completely soluble
Ethyl alcohol	CH_3CH_2OH	Completely soluble
Propanol	$CH_3CH_2CH_2OH$	Completely soluble
Butanol	$CH_3CH_2CH_2CH_2OH$	74
Pentanol	$CH_3CH_2CH_2CH_2CH_2OH$	27
Hexanol	$CH_3CH_2CH_2CH_2CH_2CH_2OH$	6.0
Heptanol	$CH_3CH_2CH_2CH_2CH_2CH_2CH_2OH$	1.7

Figure 10.4 Molecular structures of vitamin D_2 and vitamin B_6. Polar groups are shown in color. Vitamin D_2 is water-insoluble and vitamin B_6 is water-soluble.

As we noted in Chapter 4, the solubility of ionic compounds in water varies tremendously from one solid to another. The extent to which solution occurs depends on a balance between two forces, both electrical in nature:

We can't predict which factor will predominate, so we can't predict solubility.

1. The force of attraction between H_2O molecules and the ions, which tends to bring the solid into solution. If this factor predominates, the compound is very soluble in water, as is the case with NaCl, NaOH, and many other ionic solids.
2. The force of attraction between oppositely charged ions, which tends to keep them in the solid state. If this is the major factor, the water solubility is very low. The fact that $CaCO_3$ and $BaSO_4$ are almost insoluble in water implies that interionic attractive forces predominate with these ionic solids.

Effect of Temperature on Solubility

When an excess of a solid such as sodium nitrate, $NaNO_3$, is shaken with water, an equilibrium is established between ions in the solid state and in solution:

$$NaNO_3(s) \longrightarrow Na^+(aq) + NO_3^-(aq)$$

At 20°C, the saturated solution contains 87.6 g of $NaNO_3$ per one hundred grams of water.

A similar type of equilibrium is established when a gas such as oxygen is bubbled through water.

$$O_2(g) \rightleftharpoons O_2(aq)$$

At 20°C and 1 atm, 0.00138 mol of O_2 dissolves per liter of water.

The effect of a temperature change on solubility equilibria such as these can be predicted by applying a simple principle. ***An increase in temperature always shifts the position of an equilibrium to favor an endothermic process.*** This means that if the solution process absorbs heat ($\Delta H_{soln.} > 0$), an increase in temperature increases the solubility. Conversely, if the solution process is exothermic ($\Delta H_{soln.} < 0$), an increase in temperature decreases the solubility.

Dissolving a solid in a liquid is usually an endothermic process; heat must be absorbed to break down the crystal lattice.

$$\text{solid} + \text{liquid} \rightleftharpoons \text{solution} \qquad \Delta H_{soln.} > 0$$

For example, for sodium nitrate,

$$\Delta H_{soln.} = \Delta H_f° \, Na^+(aq) + \Delta H_f° \, NO_3^-(aq) - \Delta H_f° \, NaNO_3(s) = +22.8 \text{ kJ}$$

Consistent with this effect, the solubility of $NaNO_3$ and most (but not all) other solids increases with temperature (Figure 10.5).

Gases behave quite differently from solids. When a gas condenses to a liquid, heat is always evolved. By the same token, heat is usually evolved when a gas dissolves in a liquid:

$$\text{gas} + \text{liquid} \rightleftharpoons \text{solution} \qquad \Delta H_{soln.} < 0$$

This means that the reverse process (gas coming out of solution) is endothermic. Hence, it is favored by an increase in temperature; typically, gases become less soluble as the temperature rises. This rule is followed by all gases in water. You have probably noticed this effect when opening carbonated drinks (Figure 10.6). Cold drinks produce fewer bubbles than warm drinks because carbon dioxide is less soluble at higher temperature. For the same reason, warm carbonated beverages go flat faster than cold ones. You have probably noticed this effect when heating water in an open pan or beaker. Bubbles of air are driven out of the water by an increase in temperature. The reduced solubility of oxygen in water at high temperatures (Figure 10.7a) may explain why trout congregate at the bottom of deep pools on hot summer days when the surface water is depleted of dissolved oxygen.

Effect of Pressure on Solubility

Pressure has a major effect on solubility only for gas-liquid systems. At a given temperature, raising the pressure increases the solubility C_g, of a gas. Indeed, at low to moderate pressures, gas solubility is directly proportional to pressure (Figure 10.7b).

$$C_g = kP_g \tag{10.2}$$

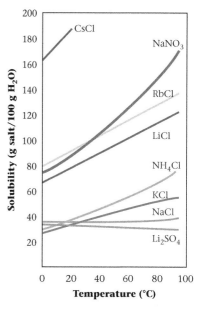
Figure 10.5 Solubility of ionic compounds vs. temperature.

Figure 10.6 Gas solubility and temperature. The two bottles of carbonated water show the dramatic difference between the solubility of carbon dioxide at low temperature *(left)* and room temperature *(right)*.

Figure 10.7 Oxygen solubility. The solubility of $O_2(g)$ in water decreases as temperature rises (a) and increases as pressure increases (b). In (a), the pressure is held constant at 1 atm; in (b), the temperature is held constant at 25°C.

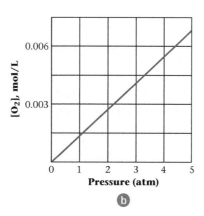

where P_g is the partial pressure of the gas over the solution, C_g is its concentration in the solution, and k is a constant characteristic of the particular gas-liquid system. This relation is called **Henry's law** after its discoverer, William Henry (1775–1836), a friend of John Dalton.

Henry's law arises because increasing the pressure raises the concentration of molecules in the gas phase. To balance this change and maintain equilibrium, more gas molecules enter the solution, increasing their concentration in the liquid phase (Figure 10.8).

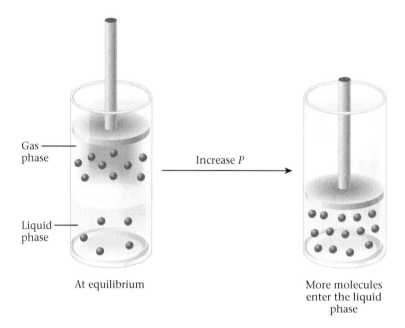

Figure 10.8 Henry's law at work. An increase in pressure, caused by a decrease in volume for the gas phase, causes the O_2 molecules to enter the liquid phase, resulting in more O_2 dissolved.

Gas phase

Liquid phase

Increase P

At equilibrium

More molecules enter the liquid phase

EXAMPLE 10.5

The solubility of pure nitrogen in blood at body temperature, 37°C, and one atmosphere is 6.2×10^{-4} M. If a diver breathes air ($X_{N_2} = 0.78$) at a depth where the total pressure is 2.5 atm, calculate the concentration of nitrogen in his blood.

ANALYSIS

Information given:	solubility of N_2 (C_{N_2}) (6.2×10^{-4} M) at 1.00 atm X_{N_2} (0.78) P_{tot} (2.5 atm)
Information implied:	k for N_2 in blood P_{N_2}
Asked for:	solubility of N_2 (C_{N_2}) at 2.50 atm

STRATEGY

1. At a given temperature, k is dependent only on the nature of the gas-liquid system. Find k using Henry's law and the solubility for pure N_2 at 1.00 atm.

$$C_{N_2} = kP_{N_2}$$

2. Find P_{N_2} when P_{tot} is 2.5 atm using the relationship between mol fraction and partial pressure (Chapter 5).

$$P_{N_2} = X_{N_2} P_{tot}$$

3. Substitute into Henry's law to get C_{N_2} at the higher pressure.

continued

SOLUTION	
1. k	$k = \dfrac{C_{N_2}}{P_{N_2}} = \dfrac{6.2 \times 10^{-4}\ M}{1.00\ \text{atm}} = 6.2 \times 10^{-4}\ M/\text{atm}$
2. P_{N_2}	$P_{N_2} = X_{N_2}P_{\text{tot}} = (0.78)(2.5\ \text{atm}) = 2.0\ \text{atm}$
3. C_{N_2}	$C_{N_2} = kP_{N_2} = 6.2 \times 10^{-4}\ \dfrac{M}{\text{atm}} \times 2.0\ \text{atm} = \boxed{1.2 \times 10^{-3}\ M}$

The influence of partial pressure on gas solubility is used in making carbonated beverages such as beer, sparkling wines, and many soft drinks. These beverages are bottled under pressures of CO_2 as high as 4 atm. When the bottle or can is opened, the pressure above the liquid drops to 1 atm, and the carbon dioxide rapidly bubbles out of solution (Figure 10.9). Pressurized containers for shaving cream, whipped cream, and cheese spreads work on a similar principle. Pressing a valve reduces the pressure on the dissolved gas, causing it to rush from solution, carrying liquid with it as a foam.

Another consequence of the effect of pressure on gas solubility is the painful, sometimes fatal, affliction known as the "bends." *ⓘ* This occurs when a person goes rapidly from deep water (high pressure) to the surface (lower pressure), where gases are less soluble. The rapid decompression causes air, dissolved in blood and other body fluids, to bubble out of solution. These bubbles impair blood circulation and affect nerve impulses (Figure 10.10). To minimize these effects, deep-sea divers and aquanauts breathe a helium-oxygen mixture rather than compressed air (nitrogen-oxygen). Helium is only about one-third as soluble as nitrogen, and hence much less gas comes out of solution on decompression.

Figure 10.9 Henry's law. As soon as the pressure is released, carbon dioxide begins to bubble out of solution in a carbonated beverage.

ⓘ

SCUBA divers have to worry about this.

SCUBA divers must pay attention to the solubility of gases in the blood and the fact that solubility increases with pressure.

A hyperbaric chamber. People who have problems breathing can be placed in a hyperbaric chamber where they are exposed to a higher partial pressure of oxygen.

a

b

Figure 10.10 **Solubility affected by pressure.**

10-3   Colligative Properties of Nonelectrolytes

The properties of a solution differ considerably from those of the pure solvent. Those solution properties that depend primarily on the *concentration of solute particles* rather than their nature are called **colligative properties.** Such properties include vapor pressure lowering, osmotic pressure, boiling point elevation, and freezing point depression. This section considers the relations between colligative properties and solute concentration, with nonelectrolytes that exist in solution as molecules.

The relationships among colligative properties and solute concentration are best regarded as limiting laws. They are approached more closely as the solution becomes more dilute. In practice, the relationships discussed in this section are valid, for nonelectrolytes, to within a few percent at concentrations as high as 1 M. At higher concentrations, solute-solute interactions lead to larger deviations.

Vapor Pressure Lowering

You may have noticed that concentrated aqueous solutions evaporate more slowly than does pure water. This reflects the fact that the vapor pressure of water over the solution is less than that of pure water (Figure 10.11).

Vapor pressure lowering is a true colligative property; that is, it is independent of the nature of the solute but directly proportional to its concentration. For example, the vapor pressure of water above a 0.10 M solution of either glucose or sucrose at 0°C is the same, about 0.008 mm Hg less than that of pure water. In a 0.30 M solution, the vapor pressure lowering is almost exactly three times as great, 0.025 mm Hg.

The relationship between solvent vapor pressure and concentration is ordinarily expressed as

$$P_1 = X_1 P_1^{\circ} \tag{10.3}$$

In this equation, P_1 is the vapor pressure of solvent over the solution, P_1° is the vapor pressure of the pure solvent at the same temperature, and X_1 is the mole fraction of solvent. Note that because X_1 in a solution must be less than 1, P_1 must be less than P_1°. This relationship is called **Raoult's law;** François Raoult (1830–1901) carried out a large number of careful experiments on vapor pressures and freezing point lowering.

Vapor pressure lowering is directly proportional to solute mole fraction.

$X_{solvent} = 1.0$ $X_{solvent} = 0.5$

Figure 10.11 Raoult's law. Adding a solute lowers the concentration of solvent molecules in the liquid phase. To maintain equilibrium, the concentration of solvent molecules in the gas phase must decrease, thereby lowering the solvent vapor pressure.

To obtain a direct expression for vapor pressure lowering, note that $X_1 = 1 - X_2$, where X_2 is the mole fraction of solute. Substituting $1 - X_2$ for X_1 in Raoult's law,

$$P_1 = (1 - X_2) \, P_1{}^\circ$$

Rearranging,

$$P_1{}^\circ - P_1 = X_2 P_1{}^\circ$$

The quantity $(P_1{}^\circ - P_1)$ is the vapor pressure lowering (ΔP). It is the difference between the solvent vapor pressure in the pure solvent and in solution.

$$\Delta P = X_2 P_1{}^\circ \qquad\qquad (10.4)$$

EXAMPLE 10.6

A solution contains 82.0 g of glucose, $C_6H_{12}O_6$, in 322 g of water. Calculate the vapor pressure of the solution at 25°C (vapor pressure of pure water at 25°C = 23.76 mm Hg).

ANALYSIS

Information given:	mass of solute, glucose (82.0 g) mass of solvent, H_2O (322 g) vapor pressure of pure water at 25°C (23.76 mm Hg)
Information implied:	molar masses of glucose and water
Asked for:	vapor pressure of the solution at 25°C

STRATEGY

1. Find moles of solute, moles of solvent, and mole fraction of solvent.

2. Substitute into Equation 10.3, where the subscript 1 refers to the solvent (in this case, water).
   $$P_1 = X_1 P_1{}^\circ$$

SOLUTION

1. moles solute	$n_{glucose} = 82.0 \text{ g} \times \dfrac{1 \text{ mol glucose}}{180.2 \text{ g}} = 0.455 \text{ mol}$
moles solvent	$n_{H_2O} = 322 \text{ g} \times \dfrac{1 \text{ mol } H_2O}{18.02 \text{ g}} = 17.9 \text{ mol}$
mole fraction of solvent	$X_{H_2O} = \dfrac{n_{H_2O}}{n_{H_2O} + n_{glucose}} = \dfrac{17.9 \text{ mol}}{(17.9 + 0.455) \text{ mol}} = 0.975$
2. P_{H_2O}	$P_{H_2O} = (X_{H_2O})(P^\circ{}_{H_2O}) = (0.975)(23.76 \text{ mm Hg}) = \boxed{23.17 \text{ mm Hg}}$

END POINT

The vapor pressure of water in the solution decreases only by 0.589 mm Hg. This is because there is a relatively small amount of solute (glucose) in the solution.

Boiling Point Elevation and Freezing Point Lowering

When a solution of a nonvolatile solute is heated, it does not begin to boil until the temperature exceeds the boiling point of the solvent. The difference in temperature is called the **boiling point elevation,** ΔT_b.

$$\Delta T_b = T_b - T_b{}^\circ$$

where T_b and $T_b{}^\circ$ are the boiling points of the solution and the pure solvent, respectively. As boiling continues, pure solvent distills off, the concentration of solute increases, and the boiling point continues to rise (Figure 10.12).

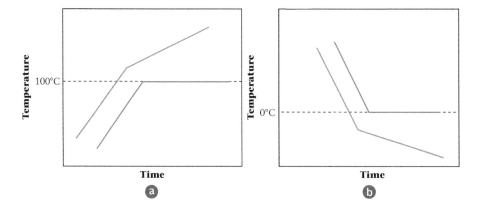

Figure 10.12 Boiling (a) and freezing (b) curves for pure water (red) and an aqueous solution (green). For pure water, the temperature remains constant during boiling or freezing. For the solution, the temperature changes steadily during the phase change because water is being removed, increasing the concentration of solute.

When a solution is cooled, it does not begin to freeze until a temperature below the freezing point of the pure solvent is reached. The **freezing point lowering**, ΔT_f, is defined to be a positive quantity:

$$\Delta T_f = T_f^\circ - T_f$$

where T_f°, the freezing point of the solvent, lies above T_f, the freezing point of the solution. As freezing takes place, pure solvent freezes out, the concentration of solute increases, and the freezing point continues to drop. This is what happens with "ice beer." When beer is cooled below 0°C, pure ice separates and the percentage of ethyl alcohol increases.

Boiling point elevation is a direct result of vapor pressure lowering. At any given temperature, a solution of a nonvolatile solute* has a vapor pressure *lower* than that of the pure solvent. Hence a *higher* temperature must be reached before the solution boils, that is, before its vapor pressure becomes equal to the external pressure. Figure 10.13 illustrates this reasoning graphically.

The freezing point lowering, like the boiling point elevation, is a direct result of the lowering of the solvent vapor pressure by the solute. Notice from Figure 10.13 that the freezing point of the solution is the temperature at which the solvent in solution has the same vapor pressure as the pure solid solvent. This implies that it is pure solvent (e.g., ice) that separates when the solution freezes.

Solutes raise the boiling point and lower the freezing point.

Figure 10.13 Effects of vapor pressure lowering. Because a nonvolatile solute lowers the vapor pressure of a solvent, the boiling point of a solution will be higher and the freezing point lower than the corresponding values for the pure solvent. Water solutions freeze *below* 0°C at point A and boil *above* 100°C at point B.

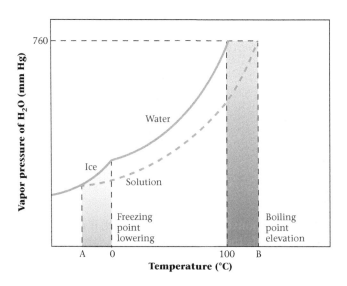

*Volatile solutes ordinarily lower the boiling point because they contribute to the total vapor pressure of the solution.

Boiling point elevation and freezing point lowering, like vapor pressure lowering, are colligative properties. They are directly proportional to solute concentration, generally expressed as molality, m. The relevant equations are

$$\Delta T_b = k_b \, (\text{molality})$$

$$\Delta T_f = k_f \, (\text{molality})$$

The proportionality constants in these equations, k_b and k_f, are called the *molal boiling point constant* and the *molal freezing point constant,* respectively. Their magnitudes depend on the nature of the solvent (Table 10.2). Note that when the solvent is water,

$$k_b = 0.52°C/m \qquad k_f = 1.86°C/m$$

EXAMPLE 10.7

An antifreeze solution is prepared containing 50.0 cm³ of ethylene glycol, $C_2H_6O_2$ ($d = 1.12$ g/cm³), in 50.0 g of water. Calculate the freezing point of this 50-50 mixture.

ANALYSIS

Information given:	volume of ethylene glycol (50.0 cm³) density of ethylene glycol (1.12 g/cm³) mass of water (50.0 g)
Information implied:	mass of ethylene glycol molar mass of ethylene glycol k_f for water (Table 10.2) freezing point of water
Asked for:	freezing point of the solution

STRATEGY

1. Determine the number of moles of ethylene glycol in solution. Use the following plan:

$$\text{Volume} \xrightarrow{\text{density}} \text{mass} \xrightarrow{\text{MM}} \text{moles}$$

2. Find the molality m of the solution by using the defining equation for molality.

$$m = \frac{\text{mol solute}}{\text{mass solvent (kg)}}$$

3. Find ΔT_f.

$$\Delta T_f = k_f(m)$$

4. Find T_{solution}.

$$\Delta T_f = T°_{\text{solvent}} - T_{\text{solution}}$$

SOLUTION

1. mol ethylene glycol	$50.0 \text{ cm}^3 \times \dfrac{1.12 \text{ g}}{1 \text{ cm}^3} \times \dfrac{1 \text{ mol}}{62.04 \text{ g}} = 0.903 \text{ mol}$
2. m	$m = \dfrac{\text{mol solute}}{\text{mass solvent (kg)}} = \dfrac{0.903 \text{ mol}}{0.05000 \text{ kg}} = 18.1 \, m$
3. ΔT_f	$\Delta T_f = k_f(m) = (1.86°C/m)(18.1 \, m) = 33.7°C$
4. T_{solution}	$T_{\text{solution}} = T°_{\text{solvent}} - \Delta T_f = 0°C - 33.7°C = -33.7°C$

END POINT

Actually, the freezing point is somewhat lower, about $-37°C$ ($-35°F$), which reminds us that the equation used, $\Delta T_f = k_f(m)$, is a limiting law, strictly valid only in very dilute solutions.

Table 10.2 Molal Freezing Point and Boiling Point Constants

Solvent	fp (°C)	k_f (°C/m)	bp (°C)	k_b (°C/m)
Water	0.00	1.86	100.00	0.52
Acetic acid	16.66	3.90	117.90	2.53
Benzene	5.50	5.10	80.10	2.53
Cyclohexane	6.50	20.2	80.72	2.75
Camphor	178.40	40.0	207.42	5.61
p-Dichlorobenzene	53.1	7.1	174.1	6.2
Naphthalene	80.29	6.94	217.96	5.80

You take advantage of freezing point lowering when you add antifreeze to your automobile radiator in winter. Ethylene glycol, $HO(CH_2)_2OH$, is the solute commonly used.◀ⓘ It has a high boiling point (197°C), is virtually nonvolatile at 100°C, and raises the boiling point of water. Hence antifreeze that contains ethylene glycol does not boil away in summer driving.

Propylene glycol, $HO—(CH_2)_3—OH$, is much less toxic.

Osmotic Pressure

One interesting effect of vapor pressure lowering is shown at the left of Figure 10.14. We start with two beakers, one containing pure water and the other containing a sugar solution. These are placed next to each other under a bell jar (Figure 10.14a). As time passes, the liquid level in the beaker containing the solution rises. The level of pure water in the other beaker falls. Eventually, by evaporation and condensation, all the water is transferred to the solution (Figure 10.14b). At the end of the experiment, the beaker that contained pure water is empty. The driving force behind this process is the difference in vapor pressure of water in the two beakers. *Water moves from a region where its vapor pressure or mole fraction is high ($X_1 = 1$ in pure water) to one in which its vapor pressure or mole fraction is lower ($X_1 < 1$ in sugar solution).*

The apparatus shown in Figure 10.14c and d can be used to achieve a result similar to that found in the bell jar experiment. In this case, a sugar solution is

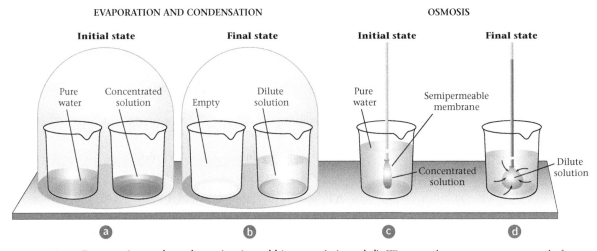

Figure 10.14 **Evaporation and condensation (a and b); osmosis (c and d).** Water tends to move spontaneously from a region where its vapor pressure is high to a region where it is low. In a → b, movement of water molecules occurs through the air trapped under the bell jar. In c → d, water molecules move by osmosis through a semipermeable membrane. The driving force is the same in the two cases, although the mechanism differs.

separated from water by a semipermeable membrane. This may be an animal bladder, a slice of vegetable tissue, or a piece of parchment. The membrane, by a mechanism that is not well understood, allows water molecules to pass through it, but not sugar molecules. *i* As before, water moves from a region where its mole fraction is high (pure water) to a region where it is lower (sugar solution). This process, taking place through a membrane permeable only to the solvent, is called **osmosis.** As a result of osmosis, the water level rises in the tube and drops in the beaker (Figure 10.14d).

The osmotic pressure, π, is equal to the external pressure, P, just sufficient to prevent osmosis (Figure 10.15). If P is less than π, osmosis takes place in the normal way, and water moves through the membrane into the solution (Figure 10.15a). By making the external pressure large enough, it is possible to reverse this process (Figure 10.15b). When $P > \pi$, water molecules move through the membrane from the solution to pure water. This process, called *reverse osmosis*, is used to obtain fresh water from seawater in arid regions of the world, including Saudi Arabia. *i* It is also used to concentrate the sap from which maple syrup is made.

Osmotic pressure, like vapor pressure lowering, is a colligative property. For any nonelectrolyte, π is directly proportional to molarity, M. The equation relating these two quantities is very similar to the ideal gas law:

$$\pi = \frac{nRT}{V} = MRT \qquad (10.5)$$

where R is the gas law constant, $0.0821 \text{ L} \cdot \text{atm/mol} \cdot \text{K}$, and T is the Kelvin temperature. Even in dilute solution, the osmotic pressure is quite large. Consider, for example, a $0.10\ M$ solution at 25°C:

$$\pi = (0.10 \text{ mol/L}) \left(0.0821 \frac{\text{L} \cdot \text{atm}}{\text{mol} \cdot \text{K}}\right) (298 \text{ K}) = 2.4 \text{ atm}$$

A pressure of 2.4 atm is equivalent to that of a column of water 25 m (more than 80 ft) high.

i The semipermeable membranes act like molecular sieves.

i Reverse osmosis absorbs energy, but there's plenty of that in Saudi Arabia.

When the external pressure P is less than the osmotic pressure, π ($P < \pi$), normal osmosis occurs.

When the external pressure exceeds the osmotic pressure, water flows in the opposite direction, producing reverse osmosis. Reverse osmosis can be used to obtain fresh water from seawater.

a $P < \pi$

Solution

Pure water

Normal osmotic system

b $P > \pi$

P

Pure water

Solution

Reverse osmotic system

Figure 10.15 Reverse osmosis.

EXAMPLE 10.8

Calculate the osmotic pressure at 15°C of a solution prepared by dissolving 50.0 g of sugar, $C_{12}H_{22}O_{11}$, in enough water to form one liter of solution.

ANALYSIS

Information given:	T (15°C) mass of sugar (50.0 g) volume of solution (1.00 L)
Information implied:	molar mass of sugar R value
Asked for:	Osmotic pressure of the solution (π)

STRATEGY

1. Determine the molarity M of the solution using the following plan:

$$\text{mass of solute} \xrightarrow{\text{MM}} \text{moles of solute} \xrightarrow{V \text{ of solution}} (M)$$

2. Substitute into Equation 10.5 to determine the osmotic pressure, π. Use the R value 0.0821 L · atm/mol · K.

SOLUTION

1. mol solute	$50.0 \text{ g} \times \dfrac{1 \text{ mol}}{342.3 \text{ g}} = 0.146 \text{ mol}$
M	$M = \text{mol solute/volume solution (L)} = 0.146 \text{ mol}/1.000 \text{ L} = 0.146 \ M$
2. π	$\pi = MRT = 0.146 \dfrac{\text{mol}}{\text{L}} \times 0.0821 \dfrac{\text{L · atm}}{\text{mol · K}} \times (273 + 15)\text{K} = 3.45 \text{ atm}$

Marna G. Clarke

Figure 10.16 Effect of osmosis on cucumbers and prunes. When a cucumber is pickled, water moves out of the cucumber by osmosis into the concentrated brine solution. A prune placed in pure water swells as water moves into the prune, again by osmosis.

Dishwashers' hands get wrinkled too. **ⓘ**

If a cucumber is placed in a concentrated brine solution, it shrinks and assumes the wrinkled skin of a pickle. ◀ⓘ The skin of the cucumber acts as a semipermeable membrane. The water solution inside the cucumber is more dilute than the solution surrounding it. As a result, water flows out of the cucumber into the brine (Figure 10.16).

When a dried prune is placed in water, the skin also acts as a semipermeable membrane. This time the solution inside the prune is more concentrated than the water, so that water flows into the prune, making the prune less wrinkled.

Nutrient solutions used in intravenous feeding must be *isotonic* with blood; that is, they must have the same osmotic pressure as blood. If the solution is too dilute, its osmotic pressure will be less than that of the fluids inside blood cells; in that case, water will flow into the cell until it bursts. Conversely, if the nutrient solution has too high a concentration of solutes, water will flow out of the cell until it shrivels and dies.

Determination of Molar Masses from Colligative Properties

Colligative properties, particularly freezing point depression, can be used to determine molar masses of a wide variety of nonelectrolytes. The approach used is illustrated in Examples 10.9 and 10.10.

EXAMPLE 10.9

A student is directed to determine the molar mass of 1.00 g of an unknown solid dissolved in cyclohexane. She is given the following pertinent information about the solvent:

    volume of solvent in solution: 225 mL
    freezing point: 6.50°C
    freezing point constant: 20.2°C/m
    density: 0.779 g/mL

She does a freezing point determination and finds the freezing point of the solution to be 6.18°C. What is the molar mass of the solute?

ANALYSIS	
Information given:	mass of solute (1.00 g) volume of solvent (225 mL) freezing point of solution, T_f (6.18°C) solvent—cyclohexane: freezing point, T_f° (6.50°C), k_f (20.2°C/m), density (0.779 g/mL)
Information implied:	mass of solvent ΔT_f
Asked for:	molar mass of solute

STRATEGY

1. Determine the freezing point depression.

$$\Delta T_f = T_f^\circ - T_f$$

2. Find the molality of the solution.

$$\Delta T_f = mk_f$$

3. Find the mass of the solvent in kg using the density.
4. Using the defining equation for molality, find the moles of solute.
5. Find the molar mass using the mass and number of moles of solute.

SOLUTION

1. ΔT_f	$\Delta T_f = T_f^\circ - T_f = 6.50°C - 6.18°C = 0.32°C$
2. m	$m = \dfrac{\Delta T_f}{k_f} = \dfrac{0.32°C}{20.2°C/m} = 0.016$
3. mass of solvent	$225 \text{ mL} \times 0.779 \dfrac{g}{mL} \times \dfrac{1 \text{ kg}}{1000 \text{ g}} = 0.175 \text{ kg}$
4. moles of solute	moles solute $= (m)(\text{mass of solvent}) = (0.016)(0.175) = 2.8 \times 10^{-3}$
5. molar mass	molar mass $= \dfrac{\text{mass}}{\text{moles}} = \dfrac{1.00 \text{ g}}{2.8 \times 10^{-3} \text{ mol}} = 3.6 \times 10^2 \text{ g/mol}$

In carrying out a molar mass determination by freezing point depression, we must choose a solvent in which the solute is readily soluble. Usually, several such solvents are available. Of these, we tend to pick one that has the largest k_f. This makes ΔT_f large and thus reduces the percent error in the freezing point measurement. From this point of view, cyclohexane or other organic solvents are better choices than water, because their k_f values are larger.

Molar masses can also be determined using other colligative properties. Osmotic pressure measurements are often used, particularly for solutes of high molar mass, where the concentration is likely to be quite low. The advantage of using osmotic pressure is that the effect is relatively large. Consider, for example, a 0.0010 M aqueous solution, for which

$$\pi \text{ at } 25°C = 0.024 \text{ atm} = 18 \text{ mm Hg}$$
$$\Delta T_f \approx 1.86 \times 10^{-3} \text{ °C}$$
$$\Delta T_b \approx 5.2 \times 10^{-4} \text{ °C}$$

A pressure of 18 mm Hg can be measured relatively accurately; temperature differences of the order of 0.001°C are essentially impossible to measure accurately.

EXAMPLE 10.10

A solution made up of 2.00 g of an unknown solute and 275 mL of water ($d = 1.00$ g/mL) has an osmotic pressure of 0.872 atm at 25°C. What is the molar mass of the unknown solute?

ANALYSIS

Information given:	mass of solute (2.00 g) volume of solvent (275 mL) π (0.872 atm); T (25°C) density of solvent (1.00 g/mL); density of solution (1.00 g/mL)
Information implied:	mass of solution volume of solution R value
Asked for:	molar mass of solute

STRATEGY

1. Find the molarity M, by substituting into Equation 10.5.

    $\pi = MRT$

2. Determine the volume of the solution by first finding its mass.

    mass of solution = mass of solute + mass of water

    volume of solution = mass/density

3. Find the moles of solute using the defining equation for molarity.

    M = moles solute/volume of solution (L)

4. Find the molar mass; molar mass = mass/moles

SOLUTION

1. M	$M = \dfrac{\pi}{RT} = \dfrac{0.872 \text{ atm}}{(0.0821 \text{ L} \cdot \text{atm/mol} \cdot \text{K})(298 \text{ K})} = 0.0356 \text{ mol/L}$
2. Volume of solution	mass of solution = mass of solute + mass of solvent $\quad\quad\quad\quad\quad$ = 2.00 g + (275 mL × 1.00 g/mL) = 277 g volume of solution = $\dfrac{277 \text{ g}}{1.00 \text{ g/mL}}$ = 277 mL = 0.277 L

continued

3. Moles of solute	mol solute = MV = (0.0356 mol/L)(0.277 L) = 0.00987
4. Molar mass	molar mass = mass/moles = 2.00 g/0.00987 mol = 203 g/mol

10-4 Colligative Properties of Electrolytes

As noted earlier, colligative properties of solutions are directly proportional to the concentration of solute *particles*. On this basis, it is reasonable to suppose that, at a given concentration, an electrolyte should have a greater effect on these properties than does a nonelectrolyte. When one mole of a nonelectrolyte such as glucose dissolves in water, one mole of solute molecules is obtained. On the other hand, one mole of the electrolyte NaCl yields two moles of ions (1 mol of Na^+, 1 mol of Cl^-). With $CaCl_2$, three moles of ions are produced per mole of solute (1 mol of Ca^{2+}, 2 mol of Cl^-).

This reasoning is confirmed experimentally. Compare, for example, the vapor pressure lowerings for 1.0 M solutions of glucose, sodium chloride, and calcium chloride at 25°C.

	Glucose	NaCl	CaCl$_2$
ΔP	0.42 mm Hg	0.77 mm Hg	1.3 mm Hg

With many electrolytes, ΔP is so large that the solid, when exposed to moist air, picks up water *(deliquesces)*. This occurs with calcium chloride, whose saturated solution has a vapor pressure only 30% that of pure water. If dry $CaCl_2$ is exposed to air in which the relative humidity is greater than 30%, it absorbs water and forms a saturated solution. Deliquescence continues until the vapor pressure of the solution becomes equal to that of the water in the air.

The freezing points of electrolyte solutions, like their vapor pressures, are lower than those of nonelectrolytes at the same concentration. Sodium chloride and calcium chloride are used to lower the melting point of ice on highways ◂ 🛈; their aqueous solutions can have freezing points as low as −21 and −55°C, respectively (Figure 10.17).

To calculate the freezing point lowering of an electrolyte in water, we use the general equation

$$\Delta T_f = i \times 1.86°C/m \times \text{molality} \qquad (10.6)$$

the multiplier i in this equation is known as the *Van't Hoff factor*. It tells us the number of moles of particles in solution (molecules or ions) per mole of solute. For sugar or other nonelectrolytes, i is 1:

$$C_{12}H_{22}O_{11}(s) \longrightarrow C_{12}H_{22}O_{11}(aq); \qquad \text{1 mol sugar} \longrightarrow \text{1 mol molecules}$$

For NaCl and $CaCl_2$, i should be 2 and 3, respectively.

$$NaCl(s) \longrightarrow Na^+(aq) + Cl^-(aq); \qquad \text{1 mol NaCl} \longrightarrow \text{2 mol ions}$$
$$CaCl_2(s) \longrightarrow Ca^{2+}(aq) + 2Cl^-(aq); \qquad \text{1 mol CaCl}_2 \longrightarrow \text{3 mol ions}$$

Similar equations apply for other colligative properties.

$$\Delta T_b = i \times 0.52°C/m \times \text{molality} \qquad (10.7)$$
$$\pi = i \times \text{molarity} \times RT$$

🛈 Potassium chloride is sold for home use because it's kinder to the environment.

Figure 10.17 Truck applying salt, NaCl, on a snow-packed road.

EXAMPLE 10.11	

Estimate the freezing points of 0.20 m aqueous solutions of KNO_3 and $Cr(NO_3)_3$.

Assume that i is the number of moles of ions formed per mole of electrolyte.

ANALYSIS	
Information given:	molality of solutions
Information implied:	i; k_f; T_f°
Asked for:	T_f

STRATEGY

1. Determine i by counting the moles of ions present after the solute dissociates.

2. Apply Equation 10.6 to find ΔT; then find the freezing point of the solution, T_f.

SOLUTION	
KNO_3: i	$KNO_3(s) \longrightarrow K^+(aq) + NO_3^-(aq)$
	2 ions: 1 K^+ and NO_3^-; $i = 2$
T_f	$\Delta T = ik_f m = 2(1.86°C/m)(0.20\ m) = 0.74°C$
	$T_f = T_f^\circ - \Delta T = 0°C - 0.74°C = -0.74°C$
$Cr(NO_3)_3$: i	$Cr(NO_3)_3(s) \longrightarrow Cr^{3+}(aq) + 3NO_3^-(aq)$
	4 ions: 1 Cr^{3+} and 3 NO_3^-; $i = 4$
T_f	$\Delta T = ik_f m = 4(1.86°C/m)(0.20\ m) = 1.5°C$
	$T_f = T_f^\circ - \Delta T = 0°C - 1.5°C = -1.5°C$

The data in Table 10.3 suggest that the situation is not as simple as this discussion implies. The observed freezing point lowerings of NaCl and $MgSO_4$ are smaller than would be predicted with $i = 2$. For example, 0.50 m solutions of NaCl and $MgSO_4$ freeze at −1.68 and −0.995°C, respectively; the predicted freezing point is −1.86°C. Only in very dilute solution does the multiplier i approach the predicted value of 2.

Table 10.3 Freezing Point Lowerings of Solutions

	ΔT_f Observed (°C)		i (Calc from ΔT_f)	
Molality	NaCl	MgSO₄	NaCl	MgSO₄
0.00500	0.0182	0.0160	1.96	1.72
0.0100	0.0360	0.0285	1.94	1.53
0.0200	0.0714	0.0534	1.92	1.44
0.0500	0.176	0.121	1.89	1.30
0.100	0.348	0.225	1.87	1.21
0.200	0.685	0.418	1.84	1.12
0.500	1.68	0.995	1.81	1.07

This behavior is generally typical of electrolytes. Their colligative properties deviate considerably from ideal values, even at concentrations below 1 m. There are at least a couple of reasons for this effect.

1. Because of electrostatic attraction, an ion in solution tends to surround itself with more ions of opposite than of like charge (Figure 10.18). The existence of this *ionic atmosphere,* first proposed in 1923 by Peter Debye (1884–1966), a Dutch physical chemist, prevents ions from acting as completely independent solute particles. The result is to make an ion somewhat less effective than a nonelectrolyte molecule in its influence on colligative properties.

2. Oppositely charged ions may interact strongly enough to form a discrete species called an *ion pair.* This effect is essentially nonexistent with electrolytes such as NaCl, in which the ions have low charges ($+1, -1$). However, with $MgSO_4$ ($+2, -2$ ions), ion pairing plays a major role. Even at concentrations as low as 0.1 m, there are more $MgSO_4$ ion pairs than free Mg^{2+} and SO_4^{2-} ions.

Freezing point lowering (or other colligative properties) can be used to determine the extent of dissociation of a weak electrolyte in water. The procedure followed is illustrated in Example 10.12.

Figure 10.18 Ionic atmosphere. An ion, on the average, is surrounded by more ions of opposite charge than of like charge.

EXAMPLE 10.12	CONCEPTUAL

The freezing point of a 0.50 m solution of oxalic acid, $H_2C_2O_4$, in water is $-1.12°C$. Which of the following equations best represents what happens when oxalic acid dissolves in water?

(1) $H_2C_2O_4(s) \longrightarrow H_2C_2O_4(aq)$

(2) $H_2C_2O_4(s) \longrightarrow H^+(aq) + HC_2O_4^-(aq)$

(3) $H_2C_2O_4(s) \longrightarrow 2H^+(aq) + C_2O_4^{2-}(aq)$

ANALYSIS	
Information given:	m (0.50) $T_f = -1.12°C$
Information implied:	k_f; $T_f°$
Asked for:	Which equation best represents the dissociation of oxalic acid?

STRATEGY

1. Find i for all 3 equations using Equation 10.6.

2. The calculated i values should be 1 for equation (1), 2 for equation (2), and 3 for equation (3).

SOLUTION	
i	$i = \dfrac{\Delta T}{k_f m} = \dfrac{0°C - (-1.12°C)}{(1.86°C/m)(0.50\ m)} = 1.2$ Equation (1) gives the value closest to 1.2.

END POINT

Actually about 20% of the oxalic acid is ionized via Equation (2).

CHEMISTRY BEYOND THE CLASSROOM

Maple Syrup

The collection of maple sap and its conversion to syrup or sugar illustrate many of the principles covered in this chapter. Moreover, in northern New England, making maple syrup is an interesting way to spend the month of March (Figure A), which separates midwinter from "mud season."

The driving force behind the flow of maple sap is by no means obvious. The calculated osmotic pressure of sap, a 2% solution of sucrose (MM = 342 g/mol), is

$$\pi = \frac{20/342 \text{ mol}}{1 \text{ L}} \times 0.0821 \frac{\text{L} \cdot \text{atm}}{\text{mol} \cdot \text{K}} \times 280 \text{ K} = 1.3 \text{ atm}$$

This is sufficient to push water to a height of about 45 ft; many maple trees are taller than that. Besides, a maple tree continues to bleed sap for several days after it has been cut down. An alternative theory suggests that sap is forced out of the tree by bubbles of $CO_2(g)$, produced by respiration. When the temperature drops at night, the carbon dioxide goes into solution, and the flow of sap ceases. This would explain the high sensitivity of sap flow to temperature; the aqueous solubility of carbon dioxide doubles when the temperature falls by 15°C.

If you want to make your own maple syrup on a small scale, perhaps 10 to 20 L per season, there are a few principles to keep in mind.

1. Make sure the trees you tap are *maples;* hemlocks would be a particularly poor choice. Identify the trees to be tapped in the fall, before the leaves fall. If possible, select sugar maples, which produce about one liter of maple syrup per tree per season. Other types of maples are less productive.

2. It takes 20 to 40 L of sap to yield one liter of maple syrup. To remove the water, you could freeze the sap, as the Native Americans did 300 years ago. The ice that forms is pure water; by discarding it, you increase the concentration of sugar in the remaining solution. Large-scale operators today use reverse osmosis to remove about half of the water. The remainder must be boiled off. The characteristic flavor of maple syrup is caused by compounds formed on heating, such as

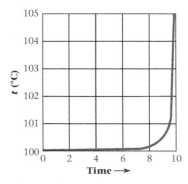

acetol   cyclotene   vanillin

It's best to boil off the water outdoors. If you do this in the kitchen, you may not be able to open the doors and windows for a couple of weeks. Wood tends to swell when it absorbs a few hundred liters of water.

3. When the concentration of sugar reaches 66%, you are at the maple syrup stage. The calculated boiling point elevation (660 g of sugar, 340 g of water)

$$\Delta T_b = 0.52°C \times \frac{660/342}{0.340} = 3.0°C$$

is somewhat less than the observed value, about 4°C. The temperature rises very rapidly around 104°C (Figure B), which makes thermometry the method of choice for detecting the end point. Shortly before that point, add a drop or two of vegetable oil to prevent foaming; perhaps this is the source of the phrase "spreading oil on troubled waters."

Figure A William L. Masterton collecting sap from red maple trees. (WLM made delicious maple syrup — CNH)

Figure B Boiling maple sap. It seems to take forever to boil down maple sap until you reach 101°C. Then the temperature rises rapidly to the end point, 104°C. Further *careful* heating produces maple sugar.

Key Concepts

1. Make dilution calculations.
   (**Example 10.1; Problems 11–14, 16**)
2. Calculate a concentration (*M*, *X*, mass %, *m*, ppm).
   (**Examples 10.2, 10.3; Problems 1–8**)
3. Convert from one concentration unit to another.
   (**Example 10.4; Problems 9, 10, 15–20**)
4. Apply Henry's law to relate gas solubility to partial pressure.
   (**Example 10.5; Problems 27–30**)

5. Apply Raoult's law to calculate vapor pressure lowering. (**Example 10.6; Problems 33–36**)
6. Relate freezing point, boiling point, osmotic pressure to solute concentration. (**Examples 10.7, 10.8, 10.10; Problems 31, 32, 37–44, 53–56**)
7. Use colligative properties to determine molar mass of a solute. (**Example 10.9; Problems 45–52**)
8. Use colligative properties to determine extent of ionization. (**Example 10.11; Problems 57, 58**)

Key Equations

Dilution of solution	$M_c V_c = M_d V_d$
Henry's law	$C_g = kP_g$
Raoult's law	$P_1 = X_1 P_1°; \Delta P = X_2 P_1°$
Osmotic pressure	$\pi = MRT \times i$

Boiling point	$\Delta T_b = k_b \times m \times i$ ($k_b = 0.52°C/m$ for water) (i = no. of moles of particles per mole of solute)
Freezing point	$\Delta T_f = k_f \times m \times i$ ($k_f = 1.86°C/m$ for water)

Key Terms

boiling point elevation
colligative properties
freezing point lowering

mass percent
molality
molarity

mole fraction
osmosis
osmotic pressure

parts per billion
parts per million

Summary Problem

Consider naphthalene, $C_{10}H_8$, the active ingredient of moth balls. A solution of naphthalene is prepared by mixing 25.0 g of naphthalene with 0.750 L of carbon disulfide, CS_2 (d = 1.263 g/mL). Assume that the volume of the solution is the same as that of the solvent after the solution is prepared.

(a) What is the mass percent of naphthalene in the solution?

(b) What is the concentration of naphthalene in parts per million?

(c) What is the density of the solution?

(d) What is the molarity of the solution?

(e) What is the molality of the solution?

(f) The vapor pressure of pure CS_2 at 25°C is 358 mm Hg. Assume that the vapor pressure exerted by naphthalene is negligible. (It is 1 mm Hg at 50°C.) What is the vapor pressure of the solution at this temperature?

(g) What is the osmotic pressure of the solution at 25°C?

(h) The normal boiling point of CS_2 is 46.13°C (k_b = 2.34°C/m). What is the normal boiling point of the solution?

(i) Vasopressin is a hormone secreted by the pituitary gland to regulate the rate at which water is reabsorbed by the kidneys. To determine its molar mass, 10.00 g of vasopressin is dissolved in 50.00 g of naphthalene (k_f = 6.94°C/m). The freezing point of the mixture is determined to be 79.01°C; that of pure naphthalene is 80.29°C. What is the molar mass of vasopressin?

Answers

(a) 2.57%
(b) 2.57 × 10⁴ ppm
(c) 1.30 g/mL
(d) 0.260 M
(e) 0.206 m
(f) 352 mm Hg
(g) 6.36 atm
(h) 46.61°C
(i) 1.08 × 10³ g/mol

Questions and Problems

Even-numbered questions and Challenge Problems have answers in Appendix 5 and fully worked solutions in the *Student Solutions Manual.*

10-1 Concentration Units

1. A solution is prepared by dissolving 12.15 g of nickel(II) nitrate in 175 mL of water ($d = 1.00$ g/mL). Calculate
   (a) the mass percent of nickel(II) nitrate in the solution.
   (b) the mole fraction of nickel(II) ions in the solution.

2. Acetone, C_3H_6O, is the main ingredient of nail polish remover. A solution is made up by adding 35.0 mL of acetone ($d = 0.790$ g/mL) to 50.0 mL of ethyl alcohol, C_2H_6O ($d = 0.789$ g/mL). Assuming volumes are additive, calculate
   (a) the mass percent of acetone in the solution.
   (b) the volume percent of ethyl alcohol in the solution.
   (c) the mole fraction of acetone in the solution.

3. For a solution of acetic acid (CH_3COOH) to be called "vinegar," it must contain 5.00% acetic acid by mass. If a vinegar is made up only of acetic acid and water, what is the molarity of acetic acid in the vinegar? The density of the vinegar is 1.006 g/mL.

4. Solutions introduced directly into the bloodstream have to be "isotonic" with blood; that is, they must have the same osmotic pressure as blood. An aqueous NaCl solution has to be 0.90% by mass to be isotonic with blood. What is the molarity of the sodium ions in solution? Take the density of the solution to be 1.00 g/mL.

5. Silver ions can be found in some of the city water piped into homes. The average concentration of silver ions in city water is 0.028 ppm.
   (a) How many milligrams of silver ions would you ingest daily if you drank eight glasses (eight oz/glass) of city water daily?
   (b) How many liters of city water are required to recover 1.00 g of silver chemically?

6. Lead is a poisonous metal that especially affects children because children retain a larger fraction of lead than adults do. To date, there is no "safe" concentration of lead in blood. Research shows that 100 μg/L (2 significant figures) of lead in the blood of young children can cause delayed cognitive development. How many moles of lead per liter in a child's blood does 100 μg Pb/L represent? How many ppm? Assume the density of blood is 1.00 g/mL.

7. Complete the following table for aqueous solutions of copper(II) sulfate.

	Mass of Solute	Volume of Solution	Molarity
(a)	12.50 g	478 mL	_____
(b)	_____	283 mL	0.299 M
(c)	4.163 g	_____	0.8415 M

8. Complete the following table for aqueous solutions of aluminum nitrate.

	Mass of Solute	Volume of Solution	Molarity
(a)	1.672 g	145.0 mL	_____
(b)	2.544 g	_____	1.688 M
(c)	_____	894 mL	0.729 M

9. Complete the following table for aqueous solutions of caffeine, $C_8H_{10}O_2N_4$.

	Molality	Mass Percent Solvent	Ppm Solute	Mole Fraction Solvent
(a)	_____	_____	_____	0.900
(b)	_____	_____	1269	_____
(c)	_____	85.5	_____	_____
(d)	0.2560	_____	_____	_____

10. Complete the following table for aqueous solutions of theobromine, $C_7H_8N_4O_2$, the alkaloid present in chocolate and cocoa.

	Molality	Mass Percent Solute	Ppm Solvent	Mole Fraction Solute
(a)	_____	_____	_____	0.0542
(b)	_____	_____	557,169	_____
(c)	_____	17.4	_____	_____
(d)	0.8341	_____	_____	_____

11. Describe how you would prepare 465 mL of 0.3550 M potassium dichromate solution starting with
   (a) solid potassium dichromate.
   (b) 0.750 M potassium dichromate solution.

12. Describe how you would prepare 750.0 mL of 0.362 M barium hydroxide solution starting with
   (a) solid barium hydroxide.
   (b) 4.93 M barium hydroxide solution.

13. A solution is prepared by diluting 225 mL of 0.1885 M aluminum sulfate solution with water to a final volume of 1.450 L. Calculate
   (a) the number of moles of aluminum sulfate before dilution.
   (b) the molarities of the aluminum sulfate, aluminum ions, and sulfate ions in the diluted solution.

14. A solution is prepared by diluting 0.7850 L of 1.262 M potassium sulfide solution with water to a final volume of 2.000 L.

    (a) How many grams of potassium sulfide were dissolved to give the original solution?

    (b) What are the molarities of the potassium sulfide, potassium ions, and sulfide ions in the diluted solution?

15. A bottle of phosphoric acid is labeled "85.0% H_3PO_4 by mass; density = 1.689 g/cm^3." Calculate the molarity, molality, and mole fraction of the phosphoric acid in solution.

16. A bottle of concentrated aqueous ammonia is labelled "29.8% NH_3 by mass; density = 0.8960 g/mL."

    (a) What is the molarity of the ammonia solution?

    (b) If 300.0 mL of the commercial ammonia is diluted with water to make 2.50 L of solution, what is the molarity of the diluted solution?

17. Complete the following table for aqueous solutions of potassium hydroxide.

	Density (g/mL)	Molarity	Molality	Mass Percent of Solute
(a)	1.05	1.13	————	————
(b)	1.29	————	————	30.0
(c)	1.43	————	14.2	————

18. Complete the following table for aqueous solutions of potassium carbonate.

	Density	Molarity	Molality	Mass Percent of Solute
(a)	1.02 g/mL	0.632	————	————
(b)	1.16 g/mL	————	————	28.5
(c)	1.20 g/mL	————	5.11	————

19. Assume that 30 L of maple sap yields one kilogram of maple syrup (66% sucrose, $C_{12}H_{22}O_{11}$). What is the molality of the sucrose solution after one fourth of the water content of the sap has been removed?

20. Juice (d = 1.0 g/mL) from freshly harvested grapes has about 24% sucrose by mass. What is the molality of sucrose, $C_{12}H_{22}O_{11}$, in the grape juice after 25% (by mass) of the water content has been removed? Assume a volume of 15.0 L.

10-2 Principles of Solubility

21. Which of the following is more likely to be soluble in benzene (C_6H_6)? In each case, explain your answer.

    (a) CCl_4 or NaCl

    (b) hexane (C_6H_{14}) or glycerol ($CH_2OHCHOHCH_2OH$)

    (c) acetic acid (CH_3COOH) or heptanoic acid ($C_6H_{13}COOH$)

    (d) HCl or propylchloride ($CH_3CH_2CH_2Cl$)

22. Which of the following is more likely to be soluble in CCl_4? In each case explain your answer.

    (a) benzene or KCl   **(b)** octane or glycerol

    (c) $CHCl_3$ or $C_6H_{11}Cl_3$ **(d)** CBr_4 or $CHBr_3$

23. Choose the member of each set that you would expect to be more soluble in water. Explain your answer.

    (a) naphthalene, $C_{10}H_8$, or hydrogen peroxide, H—O—O—H

    (b) silicon dioxide or sodium hydroxide

    (c) chloroform, $CHCl_3$, or hydrogen chloride

    (d) methyl alcohol, CH_3OH, or methyl ether, H_3C—O—CH_3

24. Choose the member of each set that you would expect to be more soluble in water.

    (a) ethyl chloride (C_2H_5Cl) or ethanol (C_2H_5OH)

    (b) ammonia (NH_3) or phosphine (PH_3)

    (c) heptanoic acid ($CH_3(CH_2)_5COOH$) or propionic acid (CH_3CH_2COOH)

    (d) iodine trichloride or sodium iodide

25. Consider the process by which lead chloride dissolves in water:

$$PbCl_2(s) \longrightarrow Pb^{2+}(aq) + 2Cl^-(aq)$$

    (a) Using data from tables in Chapter 8, calculate ΔH for this reaction.

    (b) Based only on thermodynamic data, would you expect the solubility of $PbCl_2$ to increase if the temperature is increased?

26. Consider the process by which calcium carbonate dissolves in water:

$$CaCO_3(s) \longrightarrow Ca^{2+}(aq) + CO_3^{2-}(aq)$$

    (a) Using data from tables in Chapter 8, calculate ΔH for this reaction.

    (b) Based only on thermodynamic data, would you expect the solubility of $CaCO_3$ to increase when the temperature is increased?

27. The Henry's law constant for the solubility of helium gas in water is 3.8×10^{-4} M/atm at 25°C.

    (a) Express the constant for the solubility of helium gas in M/mm Hg.

    (b) If the partial pressure of He at 25°C is 293 mm Hg, what is the concentration of dissolved He in mol/L at 25°C?

    (c) What volume of helium gas can be dissolved in 10.00 L of water at 293 mm Hg and 25°C? (Ignore the partial pressure of water.)

28. The Henry's law constant for the solubility of argon gas in water is 1.0×10^{-3} M/atm at 30°C.

    (a) Express the constant in M/mm Hg.

    (b) If the partial pressure of argon gas at 30°C is 693 mm Hg, what is the concentration (in M) of the dissolved argon gas at 30°C?

    (c) How many grams of argon gas can be dissolved in 29 L of water at 693 mm Hg and 30°C? (Ignore the partial pressure of water.)

29. A carbonated beverage is made by saturating water with carbon dioxide at 0°C and a pressure of 3.0 atm. The bottle is then opened at room temperature (25°C), and comes to equilibrium with air in the room containing CO_2 (P_{CO_2} = 3.4×10^{-4} atm). The Henry's law constant for the solubility of CO_2 in water is 0.0769 M/atm at 0°C and 0.0313 M/atm at 25°C.

(a) What is the concentration of carbon dioxide in the bottle before it is opened?

(b) What is the concentration of carbon dioxide in the bottle after it has been opened and come to equilibrium with the air?

30. Air contains 78% nitrogen. At 25°C, Henry's law constant for nitrogen is 6.8×10^{-4} M/atm, while at 37°C, it is 6.2×10^{-4} M/atm.

(a) How many milligrams of the nitrogen in the air will dissolve in 3.0 L of water at 25°C? The air in the head space above the water has a pressure of 1.38 atm.

(b) How many milligrams of the nitrogen in the air will dissolve in 3.0 L of water at 37°C? The air in the head space above the water has a pressure of 1.38 atm.

(c) How much more nitrogen is dissolved at the lower temperature in mass percent?

10-3 Colligative Properties of Nonelectrolytes

31. Vodka is advertised to be *80 proof.* That means that the ethanol (C_2H_5OH) concentration is 40% (two significant figures) by volume. Assuming the density of the solution to be 1.0 g/mL, what is the freezing point of vodka? The density of ethanol is 0.789 g/mL.

32. What is the freezing point of maple syrup (66% sucrose)? Sucrose is $C_{12}H_{22}O_{11}$.

33. Calculate the vapor pressure of water over each of the following ethylene glycol ($C_2H_6O_2$) solutions at 22°C (vp pure water = 19.83 mm Hg). Ethylene glycol can be assumed to be nonvolatile.

(a) $X_{\text{ethylene glycol}}$ = 0.288

(b) percent ethylene glycol by mass = 39.0%

(c) 2.42 m ethylene glycol

34. Calculate the vapor pressure of water over each of the following solutions of glycerol, $C_3H_8O_3$, at 28°C (vapor pressure of pure water = 28.35 mm Hg). Glycerol can be assumed to be nonvolatile.

(a) percent glycerol by mass = 30.6%

(b) 2.74 m

(c) mole fraction of glycerol = 0.188

35. The vapor pressure of pure CCl_4 at 65°C is 504 mm Hg. How many grams of naphthalene ($C_{10}H_8$) must be added to 25.00 g of CCl_4 so that the vapor pressure of CCl_4 over the solution is 483 mm Hg? Assume that the vapor pressure of naphthalene at 65°C is negligible.

36. Consider an aqueous solution of urea, ($CO(NH_2)_2$) at 26°C. Its vapor pressure is 21.15 mm Hg (vapor pressure of pure H_2O = 25.21 mm Hg). How would you prepare 250.0 mL of this solution (d = 1.06 g/mL)?

37. Calculate the osmotic pressure of the following solutions of urea, $(NH_2)_2CO$, at 22°C.

(a) 0.217 M urea

(b) 25.0 g urea dissolved in enough water to make 685 mL of solution.

(c) 15.0% urea by mass (density of the solution = 1.12 g/mL)

38. Pepsin is an enzyme involved in the process of digestion. Its molar mass is about 3.50×10^4 g/mol. What is the osmotic pressure in mm Hg at 30°C of a 0.250-g sample of pepsin in 55.0 mL of an aqueous solution?

39. Calculate the freezing point and normal boiling points of each of the following aqueous solutions.

(a) 2.63 m acetic acid

(b) 33.0% by mass lactose, $C_{12}H_{22}O_{11}$

(c) 32.15 mL of ethylene glycol, $C_2H_6O_2$ (d = 1.113 g/mL) in 624 mL of water (d = 1.00 g/mL)

40. How many grams of the following nonelectrolytes would have to be mixed with 100.0 g of *p*-dichlorobenzene to increase the boiling point by 3.0°C? To decrease the freezing point by 2.0°C? (Use Table 10.2.)

(a) succinic acid ($C_4H_6O_4$)

(b) caffeine ($C_8H_{10}N_4O_2$)

41. What is the freezing point and normal boiling point of a solution made by adding 39 mL of acetone, C_3H_6O, to 225 mL of water? The densities of acetone and water are 0.790 g/cm³ and 1.00 g/cm³, respectively.

42. Antifreeze solutions are aqueous solutions of ethylene glycol, $C_2H_6O_2$ (d = 1.12 g/mL). In Connecticut, cars are "winterized" by filling radiators with an antifreeze solution that will protect the engine for temperatures as low as −20°F.

(a) What is the minimum molality of antifreeze solution required?

(b) How many milliliters of ethylene glycol need to be added to 250 mL of water to prepare the solution called for in (a)?

43. When 13.66 g of lactic acid, $C_3H_6O_3$, are mixed with 115 g of stearic acid, the mixture freezes at 62.7°C. The freezing point of pure stearic acid is 69.4°C. What is the freezing point constant of stearic acid?

44. A solution consisting of 4.50 g of propylene glycol, $C_3H_8O_2$, in 40.5 mL of t-butyl alcohol (d = 0.780 g/mL) freezes at 8.5°C. The freezing point of t-butyl alcohol is 25.5°C. What is its freezing point constant?

45. Insulin is a hormone responsible for the regulation of glucose levels in the blood. An aqueous solution of insulin has an osmotic pressure of 2.5 mm Hg at 25°C. It is prepared by dissolving 0.100 g of insulin in enough water to make 125 mL of solution. What is the molar mass of insulin?

46. Epinephrine (or adrenaline) is a hormone and neurotransmitter. It participates in the fight-or-flight response of the sympathetic nervous system. An aqueous solution of epinephrine is prepared by dissolving 0.250 g of the hormone in enough water to make 0.350 L of solution. The solution has an osmotic pressure of 71.8 mm Hg at 22°C. What is the molar mass of epinephrine?

47. Lauryl alcohol is obtained from the coconut and is an ingredient in many hair shampoos. Its empirical formula is $C_{12}H_{26}O$. A solution of 5.00 g of lauryl alcohol in 100.0 g of benzene boils at 80.78°C. Using Table 10.2, find the molecular formula of lauryl alcohol.

48. The Rast method uses camphor ($C_{10}H_{16}O$) as a solvent for determining the molar mass of a compound. When 2.50 g of cortisone acetate are dissolved in 50.00 g of camphor (k_f = 40.0°C/m), the freezing point of the mixture is 173.44°C; that of pure camphor is 178.40°C. What is the molar mass of cortisone acetate?

49. Caffeine is made up of 49.5% C, 5.2% H, 16.5% O, and 28.9% N. A solution made up of 8.25 g of caffeine and 100.0 mL of benzene ($d = 0.877$ g/mL) freezes at 3.03°C. Pure benzene ($k_f = 5.10°C/m$) freezes at 5.50°C. What are the simplest and molecular formulas for caffeine?

50. A compound contains 42.9% C, 2.4% H, 16.6% N, and 38.1% O. The addition of 3.16 g of this compound to 75.0 mL of cyclohexane ($d = 0.779$ g/cm^3) gives a solution with a freezing point at 0.0°C. Using Table 10.2, determine the molecular formula of the compound.

51. A biochemist isolates a new protein and determines its molar mass by osmotic pressure measurements. A 50.0-mL solution is prepared by dissolving 225 mg of the protein in water. The solution has an osmotic pressure of 4.18 mm Hg at 25°C. What is the molar mass of the new protein?

52. A new enzyme is isolated and purified. Its molar mass is determined by measuring the osmotic pressure of 0.850 L of an aqueous solution containing 7.89 g of the enzyme. The osmotic pressure of the solution is 1.14 mm Hg at 22°C. What is the molar mass of the enzyme?

10-4 Colligative Properties of Electrolytes

53. Estimate the freezing and normal boiling points of 0.25 m aqueous solutions of
    (a) NH_4NO_3 (b) $NiCl_3$ (c) $Al_2(SO_4)_3$

54. Arrange 0.10 m aqueous solutions of the following solutes in order of decreasing freezing point and boiling point.
    (a) $Al(ClO_3)_3$ (b) CH_3OH
    (c) $(NH_4)_2Cr_2O_7$ (d) $MgSO_4$

55. Aqueous solutions introduced into the bloodstream by injection must have the same osmotic pressure as blood; that is, they must be "isotonic" with blood. At 25°C, the average osmotic pressure of blood is 7.7 atm. What is the molarity of an isotonic saline solution (NaCl in H_2O)? Recall that NaCl is an electrolyte; assume complete conversion to Na$^+$ and Cl$^-$ ions.

56. What is the osmotic pressure of a 0.135 M solution of Na_2SO_4 at 20°C? (Assume complete dissociation.)

57. The freezing point of a 0.11 m solution of HNO_2 is $-0.20°C$.
    (a) What is i for the solution?
    (b) Is the solution made
       (i) of HNO_2 molecules only?
       (ii) of H^+ and NO_2^- only?
       (iii) of more HNO_2 molecules than H^+ ions?
       (iv) primarily of H^+ and NO_2^- ions with some HNO_2 molecules?

58. The freezing point of a 0.21 m aqueous solution of H_2SO_4 is $-0.796°C$.
    (a) What is i?
    (b) Is the solution made up primarily of
       (i) H_2SO_4 molecules only?
       (ii) H^+ and HSO_4^- ions?
       (iii) 2 H^+ and 1 SO_4^{2-} ions?

59. An aqueous solution of LiX is prepared by dissolving 3.58 g of the electrolyte in 283 mL of H_2O ($d = 1.00$ g/mL). The solution freezes at $-1.81°C$. What is X$^-$? (Assume complete dissociation of LiX to Li$^+$ and X$^-$.)

60. An aqueous solution of M_2O is prepared by dissolving 10.91 g of the electrolyte in 435 mL of H_2O ($d = 1.00$ g/mL).

The solution freezes at $-4.68°C$. What is M$^+$? (Assume complete dissociation of M_2O to M$^+$ and O^{2-}.)

Unclassified

61. A sucrose ($C_{12}H_{22}O_{11}$) solution that is 45.0% sucrose by mass has a density of 1.203 g/mL at 25°C. Calculate its
    (a) molarity.
    (b) molality.
    (c) vapor pressure (vp H_2O at 25°C = 23.76 mm Hg).
    (d) normal boiling point.

62. An aqueous solution made up of 32.47 g of iron(III) chloride in 100.0 mL of solution has a density of 1.249 g/mL at 25°C. Calculate its
    (a) molarity.
    (b) molality.
    (c) osmotic pressure at 25°C (assume $i = 4$).
    (d) freezing point.

63. How would you prepare 5.00 L of a solution that is 43.0% H_2SO_4 by mass if the resulting solution has a density of 1.329 g/mL? What is the molarity of the resulting solution?

64. Carbon tetrachloride (CCl_4) boils at 76.8°C and has a density of 1.59 g/mL.
    (a) A solution prepared by dissolving 0.287 mol of a nonelectrolyte in 255 mL of CCl_4 boils at 80.3°C. What is the boiling point constant (k_b) for CCl_4?
    (b) Another solution is prepared by dissolving 37.1 g of an electrolyte (MM = 167 g/mol) in 244 mL of CCl_4. The resulting solution boils at 85.2°C. What is i for the electrolyte?

65. Twenty-five milliliters of a solution ($d = 1.107$ g/mL) containing 15.25% by mass of sulfuric acid is added to 50.0 mL of 2.45 M barium chloride.
    (a) What is the expected precipitate?
    (b) How many grams of precipitate are obtained?
    (c) What is the chloride concentration after precipitation is complete?

66. The Henry's law constant for the solubility of radon in water at 30°C is 9.57×10^{-6} M/mm Hg. Radon is present with other gases in a sample taken from an aquifer at 30°C. Radon has a mole fraction of 2.7×10^{-6} in the gaseous mixture. The gaseous mixture is shaken with water at a total pressure of 28 atm. Calculate the concentration of radon in the water. Express your answers using the following concentration units.
    (a) molarity
    (b) ppm (Assume that the water sample has a density of 1.00 g/mL.)

67. Pure bromobenzene, C_6H_5Br, boils at 156.0°C and has a boiling point constant of 6.26°C/m. A sample of bromobenzene is contaminated by anthracene, $C_{14}H_{10}$. The boiling point of the impure sample is 159.2°C. How pure is the sample? (Express your answer as mass percent bromobenzene.)

68. Consider two solutions at a certain temperature. Solution X has a nonelectrolyte as a solute and an osmotic pressure of 1.8 atm. Solution Y also has a nonelectrolyte as a solute and an osmotic pressure of 4.2 atm. What is the osmotic pressure of a solution made up of equal volumes of solutions X and Y at the same temperature? Assume that the volumes are additive.

Conceptual Problems

69. A pharmacist prepares an isotonic saline solution for intravenous infusion. Instead of preparing a 0.15 *M* solution, a 1.5 *M* solution is prepared. What would happen to the red blood cells if this erroneously prepared solution is infused?

70. One mole of $CaCl_2$ is represented as □◯ where □ represents Ca and ◯ represents Cl. Complete the picture showing only the calcium and chloride ions. The water molecules need not be shown.

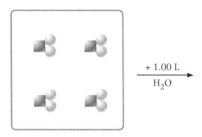

What is the molarity of Ca^{2+}? of Cl^-?

71. One mole of Na_2S is represented as ▯◯ where □ represents Na and ◯ represents S. Complete the picture showing only the sodium and sulfide ions. The water molecules need not be shown.

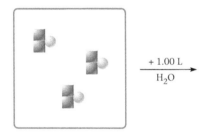

What is the molarity of Na^+? of S^{2-}?

72. Consider two nonelectrolytes X and Y. X has a higher molar mass than Y. Twenty-five grams of X are dissolved in 100 g of solvent C and labeled solution 1. Twenty-five grams of Y are dissolved in 100 g of solvent C and labeled solution 2. Both solutions have the same density. Which solution has

(a) a higher molarity?

(b) a higher mass percent?

(c) a higher molality?

(d) a larger multiplier *i*?

(e) a larger mole fraction of solvent?

73. Consider three test tubes. Tube A has pure water. Tube B has an aqueous 1.0 *m* solution of ethanol, C_2H_5OH. Tube C has an aqueous 1.0 *m* solution of NaCl. Which of the following statements are true? (Assume that for these solutions 1.0 *m* = 1.0 *M*.)

(a) The vapor pressure of the solvent over tube A is greater than the solvent pressure over tube B.

(b) The freezing point of the solution in tube B is higher than the freezing point of the solution in tube A.

(c) The freezing point of the solution in tube B is higher than the freezing point of the solution in tube C.

(d) The boiling point of the solution in tube B is higher than the boiling point of the solution in tube C.

(e) The osmotic pressure of the solution in tube B is greater than the osmotic pressure of the solution in tube C.

74. The freezing point of 0.20 *m* HF is −0.38°C. Is HF primarily nonionized in this solution (HF molecules), or is it dissociated to H^+ and F^- ions?

75. A certain gaseous solute dissolves in water, evolving 12.0 kJ of heat. Its solubility at 25°C and 4.00 atm is 0.0200 *M*. Would you expect the solubility to be greater or less than 0.0200 *M* at

(a) 5°C and 6 atm? (b) 50°C and 2 atm?

(c) 20°C and 4 atm? (d) 25°C and 1 atm?

76. The freezing point of 0.10 *M* $KHSO_3$ is −0.38°C. Which of the following equations best represents what happens when $KHSO_3$ dissolves in water?

(a) $KHSO_3(s) \longrightarrow KHSO_3(aq)$

(b) $KHSO_3(s) \longrightarrow K^+(aq) + HSO_3^-(aq)$

(c) $KHSO_3(s) \longrightarrow K^+(aq) + SO_3^{2-}(aq) + H^+(aq)$

77. Consider 2 vapor pressure curves A and B. They are the vapor pressure curves of pure benzene and a contaminated solution of benzene with a nonvolatile solute.

(a) Which graph shows the vapor pressure curve for pure benzene?

(b) What is the boiling point of the contaminated benzene solution?

(c) Estimate the molality of the contaminated solution ($k_b = 2.53°C/m$).

78. A gaseous solute dissolves in water. The solution process has $\Delta H = -15$ kJ. Its solubility at 22°C and 6.00 atm is 0.0300 *M*. Would you expect the solubility to be greater or less at

(a) 22°C and 1 atm?

(b) 18°C and 6 atm?

(c) 15°C and 10 atm?

(d) 35°C and 3 atm?

79. In your own words, explain

(a) why seawater has a lower freezing point than fresh water.

(b) why one often obtains a "grainy" product when making fudge (a supersaturated sugar solution).

(c) why the concentrations of solutions used for intravenous feeding must be controlled carefully.

(d) why fish in a lake (and fishermen) seek deep, shaded places during summer afternoons.

(e) why champagne "fizzes" in a glass.

80. Explain, in your own words,
    (a) how to determine experimentally whether a pure substance is an electrolyte or a nonelectrolyte.
    (b) why a cold glass of beer goes "flat" upon warming.
    (c) why the molality of a solute is ordinarily larger than its mole fraction.
    (d) why the boiling point is raised by the presence of a solute.

81. Beaker A has 1.00 mol of chloroform, $CHCl_3$, at 27°C. Beaker B has 1.00 mol of carbon tetrachloride, CCl_4, also at 27°C. Equal masses of a nonvolatile, nonreactive solute are added to both beakers. In answering the questions below, the following data may be helpful.

	CHCl₃ (A)	CCl₄ (B)
Vapor pressure at 27°C	0.276 atm	0.164 atm
Boiling point	61.26°C	76.5°C
k_b (°C/m)	3.63	5.03

Write <, >, =, or *more information needed* in the blanks provided.
    (a) Vapor pressure of solvent over beaker B _____ vapor pressure of solvent over beaker A.
    (b) Boiling point of solution in beaker A _____ boiling point of solution in beaker B.
    (c) Vapor pressure of pure $CHCl_3$ _____ vapor pressure of solvent over beaker A.
    (d) Vapor pressure lowering of solvent in beaker A _____ vapor pressure lowering of solvent in beaker B.
    (e) Mole fraction of solute in beaker A _____ mole fraction of solute in beaker B.

Challenge Problems

82. What is the density of an aqueous solution of potassium nitrate that has a normal boiling point of 103.0°C and an osmotic pressure of 122 atm at 25°C?

83. A solution contains 158.2 g of KOH per liter; its density is 1.13 g/mL. A lab technician wants to prepare 0.250 m KOH, starting with 100.0 mL of this solution. How much water or solid KOH should be added to the 100.0-mL portion?

84. Consider a mixture of methyl alcohol (CH_3OH) and n-propyl alcohol ($CH_3CH_2CH_2OH$). A mixture of both alcohols with a mass of 14.2 g is stirred into 100.0 g of water. The freezing point of the new solution is depressed by 5.5°C. What is the mol percent of methyl alcohol in the mixture of both alcohols?

85. The water-soluble nonelectrolyte X has a molar mass of 410 g/mol. A 0.100-g mixture containing this substance and sugar (MM = 342 g/mol) is added to 1.00 g of water to give a solution whose freezing point is −0.500°C. Estimate the mass percent of X in the mixture.

86. A martini, weighing about 5.0 oz (142 g), contains 30.0% by mass of alcohol. About 15% of the alcohol in the martini passes directly into the bloodstream (7.0 L for an adult). Estimate the concentration of alcohol in the blood (g/cm³) of a person who drinks two martinis before dinner. (A concentration of 0.00080 g/cm³ or more is frequently considered indicative of intoxication in a "normal" adult.)

87. When water is added to a mixture of aluminum metal and sodium hydroxide, hydrogen gas is produced. This is the reaction used in commercial drain cleaners:

$$2Al(s) + 6H_2O(l) + 2OH^-(aq) \longrightarrow 2Al(OH)_4^-(aq) + 3H_2(g)$$

A sufficient amount of water is added to 49.92 g of NaOH to make 0.600 L of solution; 41.28 g of Al is added to this solution and hydrogen gas is formed.
    (a) Calculate the molarity of the initial NaOH solution.
    (b) How many moles of hydrogen were formed?
    (c) The hydrogen was collected over water at 25°C and 758.6 mm Hg. The vapor pressure of water at this temperature is 23.8 mm Hg. What volume of hydrogen was generated?

88. Consider an aqueous solution with a boiling point of 105.0°C. The solute in the solution is an ionic compound M_2X. What is the mol fraction of the solute in the mixture?

89. It is found experimentally that the volume of a gas that dissolves in a given amount of water is independent of the pressure of the gas; that is, if 5 cm³ of a gas dissolves in 100 g of water at 1 atm pressure, 5 cm³ will dissolve at a pressure of 2 atm, 5 atm, 10 atm, ... Show that this relationship follows logically from Henry's law and the ideal gas law.

11 Rate of Reaction

The rate at which the speeding train is traveling can be expressed in km/h. Similarly, the rate of a reaction can be expressed in M/h.

Not every collision,
not every punctilious trajectory
by which billiard-ball complexes
arrive at their calculable meeting
places leads to reaction.

Men (and women) are not as
different from molecules as they
think.

—ROALD HOFFMANN
Excerpt from *Men and Molecules*

© 2014 Artists Rights Society (ARS), New York/VISCOPY, Australia

For a chemical reaction to be feasible, it must occur at a reasonable rate. Consequently, it is important to be able to control the rate of reaction. Most often, this means making it occur more rapidly. When you carry out a reaction in the general chemistry laboratory, you want it to take place quickly. A research chemist trying to synthesize a new drug has the same objective. Sometimes, though, it is desirable to reduce the rate of reaction. The aging process, a complex series of biological oxidations, believed to involve "free radicals" with unpaired electrons such as

$$\cdot \ddot{O} - H \quad \text{and} \quad \cdot \ddot{O} - \ddot{O} \colon^{-}$$

is one we would all like to slow down.

This chapter sets forth the principles of **chemical kinetics**, the study of reaction rates. The main emphasis is on those factors that influence rate. These include

- the concentrations of reactants (Sections 11-2 and 11-3).
- the process by which the reaction takes place (Section 11-4).
- the temperature (Section 11-5).
- the presence of a catalyst (Section 11-6).
- the reaction mechanism (Section 11-7).

11-1   Meaning of Reaction Rate

To discuss **reaction rate** meaningfully, it must be defined precisely. *The rate of reaction is a positive quantity that expresses how the concentration of a reactant or product changes with time.* To illustrate what this means, consider the reaction

$$N_2O_5(g) \longrightarrow 2NO_2(g) + \tfrac{1}{2}O_2(g)$$

Figure 11.1 **Changes in reactant and product concentrations with time.** For the reaction $N_2O_5(g) \longrightarrow 2NO_2(g) + \frac{1}{2}O_2(g)$, the concentrations of NO_2 and O_2 increase with time, whereas that of N_2O_5 decreases. The reaction rate is defined as $-\Delta[N_2O_5]/\Delta t = \Delta[NO_2]/2\Delta t = \Delta[O_2]/\frac{1}{2}\Delta t$.

As you can see from Figure 11.1, the concentration of N_2O_5 decreases with time; the concentrations of NO_2 and O_2 increase. ◄ *i* Because these species have different coefficients in the balanced equation, their concentrations do not change at the same rate. When *one* mole of N_2O_5 decomposes, *two* moles of NO_2 and *one-half* mole of O_2 are formed. This means that

i Δ [reactants] < 0.
Δ [products] > 0.

$$-\Delta[N_2O_5] = \frac{\Delta[NO_2]}{2} = \frac{\Delta[O_2]}{\frac{1}{2}}$$

where $\Delta[\ ]$ refers to the change in concentration in moles per liter. The minus sign in front of the N_2O_5 term is necessary because $[N_2O_5]$ decreases as the reaction takes place; the numbers in the denominator of the terms on the right $(2, \frac{1}{2})$ are the coefficients of these species in the balanced equation. The rate of reaction can now be defined by dividing by the change in time, Δt:

$$\text{rate} = \frac{-\Delta[N_2O_5]}{\Delta t} = \frac{\Delta[NO_2]}{2\Delta t} = \frac{\Delta[O_2]}{\frac{1}{2}\Delta t}$$

More generally, for the reaction

$$aA + bB \longrightarrow cC + dD$$

where A, B, C, and D represent substances in the gas phase (*g*) or in aqueous solution (*aq*), and *a, b, c, d* are their coefficients in the balanced equation,

$$\text{rate} = \frac{-\Delta[A]}{a\,\Delta t} = \frac{-\Delta[B]}{b\,\Delta t} = \frac{\Delta[C]}{c\,\Delta t} = \frac{\Delta[D]}{d\,\Delta t} \quad ◄ i \qquad (11.1)$$

i This is the defining equation for rate.

To illustrate the use of this expression, suppose that for the formation of ammonia,

$$N_2(g) + 3H_2(g) \longrightarrow 2NH_3(g)$$

molecular nitrogen is disappearing at the rate of 0.10 mol/L per minute, that is, $\Delta[N_2]/\Delta t = -0.10 \text{ mol/L} \cdot \text{min}$. From the coefficients of the balanced equation, we see that the concentration of H_2 must be decreasing three times as fast: $\Delta[H_2]/\Delta t = -0.30 \text{ mol/L} \cdot \text{min}$. By the same token, the concentration of NH_3 must be increasing at the rate of $2 \times 0.10 \text{ mol/L} \cdot \text{min}$: $\Delta[NH_3]/\Delta t = 0.20 \text{ mol/L} \cdot \text{min}$. It follows that

$$\text{rate} = \frac{-\Delta[N_2]}{\Delta t} = \frac{-\Delta[H_2]}{3\Delta t} = \frac{\Delta[NH_3]}{2\Delta t} = \frac{0.10 \text{ mol}}{L \cdot \text{min}}$$

By defining rate this way, it is independent of which species we focus on: N_2, H_2, or NH_3.

Notice that reaction rate has the units of concentration divided by time. We will always express concentration in moles per liter. Time, on the other hand, can be expressed in seconds, minutes, hours, ... A rate of 0.10 mol/L · min corresponds to

$$0.10 \frac{mol}{L \cdot min} \times \frac{1\ min}{60\ s} = 1.7 \times 10^{-3} \frac{mol}{L \cdot s}$$

or

$$0.10 \frac{mol}{L \cdot min} \times \frac{60\ min}{1\ h} = 6.0 \frac{mol}{L \cdot h}$$

EXAMPLE 11.1

Consider the following balanced hypothetical equation.

$$A(g) + 3B(g) \longrightarrow C(g) + 2D(g)$$

a | Express the average rate of the reaction with respect to each of the products and reactants.

SOLUTION

$$rate = \frac{-\Delta[A]}{\Delta t} = \frac{-\Delta[B]}{3\Delta t} = \frac{\Delta[C]}{\Delta t} = \frac{\Delta[D]}{2\Delta t}$$

b | In the first 20 seconds of the reaction, the concentration of B dropped from 0.100 M to 0.0357 M. What is the average rate of the reaction in the first 20 seconds of the reaction?

ANALYSIS

Information given:	time, t (20 s) $[B]_o$ (0.100 M); $[B]$ after 20 seconds (0.0357 M) from part (a): the reaction rate for B $\left(\dfrac{-\Delta[B]}{3\Delta t} \right)$
Asked for:	average rate of the reaction

STRATEGY

Substitute into the rate equation obtained in part (a).

SOLUTION

average rate	$rate = \left(\dfrac{-\Delta[B]}{3\Delta t} \right) = -\dfrac{(0.0357\ M - 0.100\ M)}{3(20\ s)} = 1.07 \times 10^{-3}\ M/s$

c | Predict the change in the concentration of D during the first 20 seconds of the reaction.

ANALYSIS

Information given:	from part (a): the rate equation for D $\left(\dfrac{-\Delta[D]}{2\Delta t} \right)$; from part (b): the rate of reaction (1.07 $\times$ 10^{-3} M/s)
Asked for:	change in the concentration of D after 20 s, ($\Delta[D]$)

STRATEGY

Substitute into the rate equation obtained in part (a).

SOLUTION

$\Delta[D]$	$rate = \dfrac{\Delta[D]}{2\Delta t}$; $\Delta[D] = (1.07 \times 10^{-3}\ M/s)(2)(20\ s) = 0.0428\ M$

END POINT

The answer obtained in part (c) is not the concentration of D after 20 s but rather the change in the concentration of D. If you were given an initial concentration of D ($[D]_o$), then you would be able to obtain $[D]$ after 20 s.

Bleach is added to a solution of a blue dye to give an initial solution with known concentrations of bleach and dye. The timer is started.

With time, the dye fades as it reacts with the bleach.

The changing concentration of the dye could be measured by following the color change with a spectrophotometer.

© Cengage Learning/Charles D. Winters

(a)　　　　　　　(b)　　　　　　　(c)

Figure 11.2 **Measurement of reaction rate by observing color change.**

Measurement of Rate

Rate is often measured by following the color change that occurs in a reaction (Figure 11.2).

For the reaction

$$N_2O_5(g) \longrightarrow 2NO_2(g) + \tfrac{1}{2}O_2(g)$$

the rate could be determined by measuring

- the absorption of visible light by the NO_2 formed; this species has a reddish-brown color, whereas N_2O_5 and O_2 are colorless.
- the change in pressure that results from the increase in the number of moles of gas (1 mol reactant $\rightarrow 2\tfrac{1}{2}$ mol product).

The graphs in Figure 11.1 show plots of data from measurements of this type.

To find the rate of decomposition of N_2O_5, it is convenient to use Figure 11.3, which is a magnified version of a portion of Figure 11.1. If a tangent is drawn to the curve of concentration versus time, its slope at that point must equal $\Delta[N_2O_5]/\Delta t$. But because the reaction rate is $-\Delta[N_2O_5]/\Delta t$, it follows that

$$\text{rate} = -\text{slope of tangent}$$

From Figure 11.3 it appears that the slope of the tangent at $t = 2$ min is -0.028 mol/L · min. Hence

$$\text{rate at 2 min} = -(-0.028 \text{ mol/L} \cdot \text{min}) = 0.028 \text{ mol/L} \cdot \text{min}$$

Slope tangent $= \dfrac{-0.056}{2.0}$

$= -0.028 \dfrac{\text{mole}}{\text{L} \cdot \text{min}}$

$\Delta y = -0.056$

$\Delta x = 2.0$

Time (minutes)

[N₂O₅] (vertical axis)

Figure 11.3 **Determination of the instantaneous rate at a particular concentration.** To determine the rate of reaction, plot concentration versus time and take the tangent to the curve at the desired point. For the reaction $N_2O_5(g) \longrightarrow 2NO_2(g) + \tfrac{1}{2}O_2(g)$, it appears that the reaction rate at $[N_2O_5] = 0.080\ M$ is 0.028 mol/L · min.

11-2　Reaction Rate and Concentration

Ordinarily, reaction rate is directly related to reactant concentration (Figure 11.4). The higher the concentration of starting materials, the more rapidly a reaction takes place. Pure hydrogen peroxide, in which the concentration of H_2O_2 molecules is about 40 mol/L, is an extremely dangerous substance. In the presence of trace impurities, it decomposes explosively

$$H_2O_2(l) \longrightarrow H_2O(g) + \tfrac{1}{2}O_2(g)$$

Figure 11.4 Reaction rate and concentration. The rate of the reaction of potassium permanganate with hydrogen peroxide depends on the concentration of the permanganate. With dilute KMnO₄ the reaction is slow (a), but it is more rapid in more concentrated KMnO₄ (b).

Rate depends on concentration, but the rate constant does not.

at a rate too rapid to measure. The hydrogen peroxide you buy in a drugstore is a dilute aqueous solution in which $[H_2O_2] \approx 1$ *M*. At this relatively low concentration, decomposition is so slow that the solution is stable for several months.

The dependence of reaction rate on concentration is readily explained. Ordinarily, ***reactions occur as the result of collisions between reactant molecules.*** The higher the concentration of molecules, the greater the number of collisions in unit time and hence the faster the reaction. As reactants are consumed, their concentrations drop, collisions occur less frequently, and reaction rate decreases. This explains the common observation that reaction rate drops off with time, eventually going to zero when the limiting reactant is consumed.

Rate Expression and Rate Constant

The dependence of reaction rate on concentration is readily determined for the decomposition of N_2O_5. Figure 11.5 shows what happens when reaction rate is plotted versus $[N_2O_5]$. As you would expect, rate increases as concentration increases, going from zero when $[N_2O_5] = 0$ to about 0.06 mol/L · min when $[N_2O_5] = 0.16$ *M*. Moreover, as you can see from the figure, the plot of rate versus concentration is a straight line through the origin, which means that rate must be directly proportional to the concentration.

$$\text{rate} = k[N_2O_5]$$

This equation is referred to as the **rate expression** for the decomposition of N_2O_5. It tells how the rate of the reaction

$$N_2O_5(g) \longrightarrow 2NO_2(g) + \tfrac{1}{2}O_2(g)$$

depends on the concentration of reactant. The proportionality constant k is called a **rate constant.** It is independent of the other quantities in the equation. ◂*i*

As we will see shortly, the rate expression can take various forms, depending on the nature of the reaction. It can be quite simple, as in the N_2O_5 decomposition, or exceedingly complex.

Order of Reaction Involving a Single Reactant

Rate expressions have been determined by experiment for a large number of reactions. For the process

$$A \longrightarrow \text{products}$$

the rate expression has the general form

$$\text{rate} = k[A]^m \text{ ◂} i$$

Remember, [A] means the concentration of A in moles per liter.

The power to which the concentration of reactant A is raised in the rate expression is called the **order of reaction,** m. If m is 0, the reaction is said to be "zero-order." If $m = 1$, the reaction is "first-order"; if $m = 2$, it is "second-order"; and

Figure 11.5 A plot of rate versus concentration for the decomposition of N₂O₅ is a straight line. The line, if extrapolated, passes through the origin. This means that rate is directly proportional to concentration; that is, rate = k [N₂O₅].

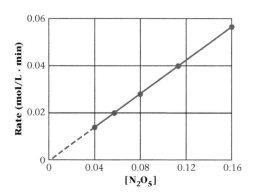

so on. Ordinarily, the reaction order is integral (0, 1, 2, . . .), but fractional orders such as $\frac{3}{2}$ are possible.

The order of a reaction must be determined experimentally; *it cannot be deduced from the coefficients in the balanced equation.* This must be true because there is only one reaction order, but there are many different ways in which the equation for the reaction can be balanced. For example, although we wrote

$$N_2O_5(g) \longrightarrow 2NO_2(g) + \tfrac{1}{2}O_2(g)$$

to describe the decomposition of N_2O_5, it could have been written

$$2N_2O_5(g) \longrightarrow 4NO_2(g) + O_2(g)$$

The reaction is still first-order no matter how the equation is written.

One way to find the order of a reaction is to measure the initial rate (i.e., the rate at $t = 0$) as a function of the concentration of reactant. Suppose, for example, that we make up two different reaction mixtures differing only in the concentration of reactant A. We now measure the rates at the beginning of reaction, before the concentration of A has decreased appreciably. This gives two different initial rates (rate$_1$, rate$_2$) corresponding to two different starting concentrations of A, $[A]_1$ and $[A]_2$. From the rate expression,

$$\text{rate}_2 = k[A]_2{}^m \qquad \text{rate}_1 = k[A]_1{}^m$$

Dividing the second rate by the first,

$$\frac{\text{rate}_2}{\text{rate}_1} = \frac{[A]_2{}^m}{[A]_1{}^m} = \left(\frac{[A]_2}{[A]_1}\right)^m$$

Because all the quantities in this equation are known except m, the reaction order can be calculated (Example 11.2).

EXAMPLE 11.2

Acetaldehyle, CH_3CHO, occurs naturally in oak and tobacco leaves, and also is present in automobile and diesel exhaust. The initial rate of decomposition of acetaldehyde at 600°C

$$CH_3CHO(g) \longrightarrow CH_4(g) + CO(g)$$

was measured at a series of concentrations with the following results:

$[CH_3CHO]$	0.20 M	0.30 M	0.40 M	0.50 M
Rate (mol/L·s)	0.34	0.76	1.4	2.1

Using these data, determine the reaction order; that is, determine the value of m in the equation

$$\text{rate} = k[CH_3CHO]^m$$

ANALYSIS	
Information given:	experiments with initial concentrations and rates
Asked for:	order of the reaction

STRATEGY

1. Choose two initial concentrations and their corresponding rates.

2. Calculate the rate ratio.

3. Calculate the concentration ratio.

4. Substitute into the following equation to obtain the order of the reaction, m.

$$\frac{\text{rate}_2}{\text{rate}_1} = \left(\frac{[A]_2}{[A]_1}\right)^m$$

continued

SOLUTION

1. Choose the first two experiments	$\text{rate}_1 = 0.34$ mol/L·s; $\text{rate}_2 = 0.76$ mol/L·s; $[CH_3CHO]_1 = 0.20$ M; $[CH_3CHO]_2 = 0.30$ M
2. Rate ratio	$\dfrac{\text{rate}_2}{\text{rate}_1} = \dfrac{0.76}{0.34} = 2.2$
3. Concentration ratio	$\dfrac{[CH_3CHO]_2}{[CH_3CHO]_1} = \dfrac{0.30}{0.20} = 1.5$
4. m	$\dfrac{\text{rate}_2}{\text{rate}_1} = \left(\dfrac{[CH_3CHO]_2}{[CH_3CHO]_1}\right)^m \longrightarrow 2.2 = (1.5)^m \longrightarrow m = 2$ The reaction is second-order.

END POINTS

1. If m is not obvious, then solve for m algebraically by taking the log of both sides:

$$2.2 = (1.5)^m \text{ becomes } \log 2.2 = m(\log 1.5); \quad m = \frac{\log 2.2}{\log 1.5} = \frac{0.34}{0.18} = 1.9 \longrightarrow 2$$

2. You would get the same result ($m = 2$) if you used any two experiments. Try it!

Once the order of the reaction is known, the rate constant is readily calculated. Consider, for example, the decomposition of acetaldehyde, where we have shown that the rate expression is

$$\text{rate} = k[CH_3CHO]^2$$

There are two variables in this equation, rate and concentration, and two constants, k and reaction order.

The data in Example 11.2 show that the rate at 600°C is 0.34 mol/L·s when the concentration is 0.20 mol/L. It follows that

$$k = \frac{\text{rate}}{[CH_3CHO]^2} = \frac{0.34 \text{ mol/L·s}}{(0.20 \text{ mol/L})^2} = 8.5 \text{ L/mol·s}$$

The same value of k would be obtained, within experimental error, using any other data pair.

Having established the value of k and the reaction order, the rate is readily calculated at any concentration. Again, using the decomposition of acetaldehyde as an example, we have established that

$$\text{rate} = 8.5 \frac{L}{\text{mol·s}} [CH_3CHO]^2$$

If the concentration of acetaldehyde were 0.60 M,

$$\text{rate} = 8.5 \frac{L}{\text{mol·s}} (0.60 \text{ mol/L})^2 = 3.1 \text{ mol/L·s}$$

Order of Reaction with More Than One Reactant

Many (indeed, most) reactions involve more than one reactant. For a reaction between two species A and B,

$$a\text{A} + b\text{B} \longrightarrow \text{products}$$

the general form of the rate expression is

$$\text{rate} = k[\text{A}]^m \times [\text{B}]^n$$

In this equation m is referred to as "the order of the reaction with respect to A." Similarly, n is "the order of the reaction with respect to B." The *overall order* of the reaction is the sum of the exponents, $m + n$. If $m = 1$, $n = 2$, then the reaction is first-order in A, second-order in B, and third-order overall.

When more than one reactant is involved, the order can be determined by holding the initial concentration of one reactant constant while varying that of the other reactant. From rates measured under these conditions, it is possible to deduce the order of the reaction with respect to the reactant whose initial concentration is varied.

To see how to do this, consider the reaction between A and B referred to above. Suppose we run two different experiments in which the initial concentrations of A differ ($[A]_1$, $[A]_2$) but that of B is held constant at $[B]$. Then

$$\text{rate}_1 = k[A]_1{}^m \times [B]^n \qquad \text{rate}_2 = k[A]_2{}^m \times [B]^n$$

Dividing the second equation by the first ◄ ⓘ

$$\frac{\text{rate}_2}{\text{rate}_1} = \frac{k[A]_2{}^m \times [B]^n}{k[A]_1{}^m \times [B]^n} = \frac{[A]_2{}^m}{[A]_1{}^m} = \left(\frac{[A]_2}{[A]_1}\right)^m$$

> ⓘ This way, the concentration terms cancel for that reactant.

Knowing the two rates and the ratio of the two concentrations, we can readily find the value of m.

EXAMPLE 11.3

Consider the reaction between t-butylbromide and a base at 55°C:

$$(CH_3)_3CBr(aq) + OH^-(aq) \longrightarrow (CH_3)_3COH(aq) + Br^-(aq)$$

A series of experiments is carried out with the following results:

	Expt. 1	Expt. 2	Expt. 3	Expt. 4	Expt. 5
$[(CH_3)_3CBr]$	0.50	1.0	1.5	1.0	1.0
$[OH^-]$	0.050	0.050	0.050	0.10	0.20
Rate (mol/L·s)	0.0050	0.010	0.015	0.010	0.010

a   Find the order of the reaction with respect to both $(CH_3)_3CBr$ and OH^-.

ANALYSIS

Information given:	results of initial state experiments
Asked for:	order of the reaction with respect to $(CH_3)_3CBr$ and OH^-

STRATEGY

For m: **1.** Choose two experiments where $[OH^-]$ is constant.

        2. Find the rate and concentration ratios for $(CH_3)_3CBr$.

        3. Substitute into the rate equation

            rate ratio $= ([(CH_3)_3CBr]$ ratio$)^m$

For n: **1.** Choose two experiments where $[(CH_3)_3CBr]$ is constant.

        2. Find the rate and concentration ratios for OH^-.

        3. Substitute into the rate equation

            rate ratio $= ([OH^-]$ ratio$)^n$

continued

SOLUTION

For m:	1. Experiments	Experiments 1 and 3
	2. ratios	rate ratio: $\dfrac{\text{rate}_3}{\text{rate}_1} = \dfrac{0.015}{0.005} = 3$
		concentration ratio: $\dfrac{[(CH_3)_3CBr]_3}{[(CH_3)_3CBr]_1} = \dfrac{1.5}{0.50} = 3$
	3. m	$3 = (3)^m$; $m = 1$
For n:	1. Experiments	Experiments 2 and 5
	2. ratios	rate ratio: $\dfrac{\text{rate}_5}{\text{rate}_2} = \dfrac{0.010}{0.010} = 1$
		concentration ratio: $\dfrac{[OH^-]_5}{[OH^-]_2} = \dfrac{0.20}{0.050} = 4$
	3. n	$1 = (4)^n$; $n = 0$
reaction order		The reaction is first-order with respect to $(CH_3)_3CBr$ and zero-order with respect to OH^-.

b Write the rate expression for the reaction.

ANALYSIS

Information given:	from part (a): m (1), n (0)
Asked for:	rate expression for the reaction

SOLUTION

rate expression	rate $= k[(CH_3)_3CBr]^1 [OH^-]^0 = k[(CH_3)_3CBr]$

c Calculate the rate constant at 55°C.

ANALYSIS

Information given:	from part (b): rate expression (rate $= k[(CH_3)_3CBr]$) experiments with rates and concentrations at initial states
Asked for:	k

STRATEGY

1. Substitute a rate and a concentration for $(CH_3)_3CBr$ and OH^- into the rate expression.

2. Use any rate/concentration pair from any experiment. (We choose experiment 3.)

SOLUTION

k	rate $= k[(CH_3)_3CBr] \longrightarrow 0.015 \dfrac{\text{mol}}{\text{L} \cdot \text{s}} = k\left(1.5 \dfrac{\text{mol}}{\text{L}}\right) \longrightarrow k = 0.010 \text{ s}^{-1}$

d What effect does doubling the concentration of OH^- have on the reaction if $[(CH_3)_3CBr]$ is kept at 1.0 M?

SOLUTION

Changing $[OH^-]$ has no effect on the rate of the reaction. The reaction is zero-order ($n = 0$) with respect to OH^-, which means that the rate is independent of its concentration.

11-3   Reactant Concentration and Time

The rate expression

$$\text{rate} = k[N_2O_5]$$

shows how the rate of decomposition of N_2O_5 changes with concentration. From a practical standpoint, however, it is more important to know the relation between concentration and *time* ◄ 🛈 rather than between concentration and rate. Suppose, for example, you are studying the decomposition of dinitrogen pentoxide. Most likely, you would want to know how much N_2O_5 is left after 5 min, 1 h, or several days. An equation relating *rate* to concentration does not provide that information.

There are no speedometers for reactions, but every lab has a clock.

Using calculus, it is possible to develop integrated rate equations relating reactant concentration to time. We now examine several such equations, starting with first-order reactions.

First-Order Reactions

For the decomposition of N_2O_5 and any other **first-order reaction** of the type

$$A \longrightarrow \text{products} \qquad \text{rate} = k[A]$$

it can be shown by using calculus that the relationship between concentration and time is*

$$\ln \frac{[A]_o}{[A]} = kt \qquad (11.2a)$$

where $[A]_o$ is the original concentration of reactant, $[A]$ is its concentration at time t, k is the first-order rate constant, and the abbreviation "ln" refers to the natural logarithm.

Because $\ln a/b = \ln a - \ln b$, the first-order equation can be written in the form

$$\ln [A]_o - \ln [A] = kt \qquad (11.2b)$$

Solving for $\ln [A]$,

$$\ln [A] = \ln [A]_o - kt$$

Comparing this equation with the general equation of a straight line,

$$y = b + mx \qquad (b = y\text{-intercept}, m = \text{slope})$$

it is clear that a plot of $\ln [A]$ versus t should be a straight line with a y-intercept of $\ln [A]_o$ and a slope of $-k$. This is indeed the case, as you can see from Figure 11.6, in which we have plotted $\ln [N_2O_5]$ versus t for the decomposition of N_2O_5. Drawing the best straight line through the points and taking the slope based on a point on the x-axis ($y_2 = -3.5, x_2 = 4.9$) and another on the y-axis ($y_1 = -1.8, x_1 = 0$)

$$\text{slope} = \frac{y_2 - y_1}{x_2 - x_1} = \frac{-3.5 + 1.8}{4.9 - 0} = \frac{-1.7}{4.9} = -0.35$$

It follows that the rate constant is 0.35/min; ◄ 🛈 the integrated first-order equation for the decomposition of N_2O_5 is

The slope is negative but k is always positive.

$$\ln \frac{[N_2O_5]_o}{[N_2O_5]} = \frac{0.35}{\text{min}} t \qquad (\text{at } 67°C)$$

---

*Throughout Section 11-3, balanced chemical equations are written in such a way that the coefficient of the reactant is 1. In general, if the coefficient of the reactant is a, where a may be 2 or 3 or . . . , then k in each integrated rate equation must be replaced by the product ak. (See Problem 101.)

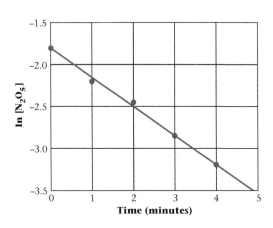

Figure 11.6 A first-order reaction plot. The rate constant for a first-order reaction can be determined from the slope of a plot of ln[A] versus time. The reaction illustrated, at 67°C, is $N_2O_5(g) \longrightarrow 2NO_2(g) + \frac{1}{2}O_2(g)$. Using two points on the line, one can find the slope of the line, which is the rate constant. In this case, k is found to be about 0.35/min.

EXAMPLE 11.4

Consider the first-order decomposition of N_2O_5 at 67°C, where $k = 0.35$/min.

a | What is the concentration after six minutes, starting at 0.200 M?

	ANALYSIS
Information given:	$k(0.35/\text{min})$; $t(6.00 \text{ min})$; $[N_2O_5]_o$ (0.200 M) reaction order (first-order)
Asked for:	$[N_2O_5]$ after 6 minutes

STRATEGY

Substitute into Equation 11.2a or 11.2b.

$$\ln \frac{[N_2O_5]_o}{[N_2O_5]} = kt \quad \text{or} \quad \ln [N_2O_5]_o - \ln [N_2O_5] = kt$$

SOLUTION

$[N_2O_5]$	$\ln(0.200 \text{ mol/L} \cdot \text{min}) - \ln [N_2O_5] = -1.609 - \ln(N_2O_5) = 0.35 \frac{1}{\text{min}} \times 6.00 \text{ min}$ $\ln [N_2O_5] = -1.609 - 2.1 = -3.7 \longrightarrow [N_2O_5] = e^{-3.7} = 0.024 \text{ mol/L}$

b | How long does it take for the concentration to drop from 0.200 M to 0.150 M?

	ANALYSIS
Information given:	$k(0.35/\text{min})$; $[N_2O_5]_o$ (0.200 M); $[N_2O_5]_t$ (0.150 M)
Asked for:	t

STRATEGY

Substitute into Equation 11.2a or 11.2b.

SOLUTION

t	$\ln(0.200 \, M) - \ln(0.150 \, M) = \dfrac{0.35}{\text{min}} \times t \longrightarrow -1.61 - (-1.90) = 0.35 \, t$ $t = 0.82 \text{ min}$

c | How long does it take for half a sample of N_2O_5 to decompose?

	ANALYSIS
Information given:	$k(0.35/\text{min})$; $[N_2O_5] = (\frac{1}{2}) [N_2O_5]_o$
Asked for:	t

continued

SOLUTION

t $\ln \dfrac{[N_2O_5]_o}{\frac{1}{2}[N_2O_5]_o} = (0.35/\text{min})\, t \longrightarrow \ln 2 = (0.35/\text{min})t \longrightarrow t = 0.693/0.35$

$t = 2.0 \text{ min}$

The analysis of Example 11.4c reveals an important feature of a first-order reaction: *The time required for one half of a reactant to decompose via a first-order reaction has a fixed value, independent of concentration.* ◄ ⓘ This quantity, called the **half-life**, is given by the expression

$$t_{1/2} = \frac{\ln 2}{k} = \frac{0.693}{k} \qquad \text{first-order reaction} \qquad (11.3)$$

where k is the rate constant. For the decomposition of N_2O_5, where $k = 0.35/\text{min}$, $t_{1/2} = 2.0$ min. Thus every two minutes, one half of a sample of N_2O_5 decomposes (Table 11.1).

Notice that for a first-order reaction the rate constant has the units of reciprocal time, for example, min^{-1}. This suggests a simple physical interpretation of k (at least where k is small); it is the fraction of reactant decomposing in unit time. For a first-order reaction in which

$$k = 0.010/\text{min} = 0.010 \text{ min}^{-1}$$

we can say that 0.010 (i.e., 1.0%) of the sample decomposes per minute.

Perhaps the most important first-order reaction is that of radioactive decay, in which an unstable nucleus decomposes (Chapter 2). Letting X be the amount of a radioactive isotope present at time t,

$$\text{rate} = kX$$

and

$$\ln \frac{X_o}{X} = kt$$

where X_o is the initial amount. The amount of a radioactive isotope can be expressed in terms of moles, grams, or number of atoms.

ⓘ This relation holds only for a first-order reaction.

Table 11.1 Decomposition of N_2O_5 at 67°C ($t_{1/2}$ = 2.0 min)

t (min)	0.0	2.0	4.0	6.0	8.0
$[N_2O_5]$	0.160	0.080	0.040	0.020	0.010
Fraction of N_2O_5 decomposed	0	$\frac{1}{2}$	$\frac{3}{4}$	$\frac{7}{8}$	$\frac{15}{16}$
Fraction of N_2O_5 left	1	$\frac{1}{2}$	$\frac{1}{4}$	$\frac{1}{8}$	$\frac{1}{16}$
Number of half-lives	0	1	2	3	4

EXAMPLE 11.5 GRADED

Plutonium-240 (Pu-240) is a by-product of the nuclear reaction that takes place in a reactor. It takes one thousand years for 10.0% of a 4.60-g sample to decay.

a   What is the half-life of Pu-240?

ANALYSIS

Information given:	time, t (1000 y); $[\text{Pu-240}]_0$ (4.60 g); rate of decay (10%/1000 years)
Information implied:	reaction order; k; $[\text{Pu-240}]$ after 1000 years
Asked for:	$t_{1/2}$

STRATEGY

1. Find $[\text{Pu-240}]$ after 1000 years.

    $[\text{Pu-240}] = [\text{Pu-240}]_0 - 0.10([\text{Pu-240}]_0)$

2. All nuclear reactions are first-order. Find k by substituting into Equation 11.2a or 11.2b.

3. Find $t_{1/2}$ by substituting into Equation 11.3.

    $t_{1/2} = 0.693/k$

SOLUTION

$[\text{Pu-240}]$	$[\text{Pu-240}] = 4.60 \text{ g} - (0.10)(4.60 \text{ g}) = 4.14 \text{ g}$
k	$\ln 4.60 - \ln 4.14 = k(1000 \text{ y}) \longrightarrow k = 1.05 \times 10^{-4} \text{ y}^{-1}$
$t_{1/2}$	$t_{1/2} = \dfrac{0.693}{1.05 \times 10^{-4} \text{ y}^{-1}} = 6.60 \times 10^3 \text{ y}$

b   How long will it take to reduce a 2.00-g sample to 15% of its original amount?

ANALYSIS

Information given:	$[\text{Pu-240}]_0$ (2.00 g); $[\text{Pu-240}]$ (15% of 2.00 g) from part (a): $k(1.05 \times 10^{-4} \text{ y}^{-1})$
Asked for:	t

STRATEGY

1. Find $[\text{Pu-240}]$.

2. Find t by substituting into Equation 11.2a or 11.2b.

SOLUTION

$[\text{Pu-240}]$	$[\text{Pu-240}] = 0.15(2.00 \text{ g}) = 0.30 \text{ g}$
t	$\ln 2.00 - \ln 0.30 = (1.05 \times 10^{-4} \text{ y}^{-1})t \longrightarrow t = 1.8 \times 10^4 \text{ y}$

c   What is the rate of decay of a 5.00-g sample in g/year?

ANALYSIS

Information given:	$[\text{Pu-240}]_0$ (5.00 g); from part (a): $k(1.05 \times 10^{-4} \text{ y}^{-1})$
Information implied:	reaction order ($m = 1$)
Asked for:	rate of decay

STRATEGY

Since the question is now to relate concentration and rate, you must substitute into the general rate expression for Pu-240 decay.

rate $= k[\text{Pu-240}]_0{}^1$

continued

SOLUTION	
Rate	rate $= (1.05 \times 10^{-4}\ y^{-1})(5.00\ g) = 5.25 \times 10^{-4}\ g/y$
END POINT	
In part (c) the rate is dependent on the initial mass unlike in part (a), where the half-life is independent of the original amount.	

Half-lives can be interpreted in terms of the level of radiation of the corresponding isotopes. Uranium has a very long half-life (4.5×10^9 y), so it gives off radiation very slowly. At the opposite extreme is fermium-258, which decays with a half-life of 3.8×10^{-4} s. You would expect the rate of decay to be quite high. Within a second virtually all the radiation from fermium-258 is gone. Species such as this produce very high radiation during their brief existence.

Zero- and Second-Order Reactions

A **zero-order reaction** is a reaction whose rate is independent of reactant concentration. This is because any non-zero quantity raised to the zero power, including [A], is equal to 1.

$$A \longrightarrow \text{products} \qquad \text{rate} = k[A]^0 = k$$

In other words, the rate of a zero-order reaction is constant, independent of concentration. As you might expect from our earlier discussion, zero-order reactions are relatively rare. Most of them take place at solid surfaces, where the rate is independent of concentration in the gas phase. A typical example is the thermal decomposition of hydrogen iodide on gold:

$$HI(g) \xrightarrow{\text{Au}} \tfrac{1}{2}H_2(g) + \tfrac{1}{2}I_2(g)$$

When the gold surface is completely covered with HI molecules, increasing the concentration of HI(g) has no effect on reaction rate.

It can be shown using the integrated rate law that the concentration-time relation for a zero-order reaction is

$$[A] = [A]_o - kt \qquad \textbf{(11.4)}$$

Comparing this equation with that for a straight line,

$$y = b + mx \qquad (b = y\text{-intercept},\ m = \text{slope})$$

it should be clear that a plot of [A] versus t should be a straight line with a slope of $-k$. Putting it another way, if a plot of concentration versus time is linear, the reaction must be zero-order; the rate constant k is numerically equal to the slope of that line but has the opposite sign.

For a **second-order reaction** (a reaction whose rate depends on the second power of the reactant concentration) involving a single reactant, such as acetaldehyde (recall Example 11.2),

$$A \longrightarrow \text{products} \qquad \text{rate} = k[A]^2$$

it is again necessary to resort to calculus* to obtain the concentration-time relationship

$$\frac{1}{[A]} - \frac{1}{[A]_o} = kt \qquad \textbf{(11.5)}$$

In a zero-order reaction, all the reactant is consumed in a finite time, which is $2t_{1/2}$.

---

*If you are taking a course in calculus, you may be surprised to learn how useful it can be in the real world (e.g., chemistry). The general rate expressions for zero-, first-, and second-order reactions are

$$-d[A]/dt = k \qquad -d[A]/dt = k[A] \qquad -d[A]/dt = k[A]^2$$

Integrating these equations from 0 to t and from $[A]_o$ to [A], you should be able to derive the equations for zero-, first-, and second-order reactions.

Table 11.2 Characteristics of Zero-, First-, and Second-Order Reactions for A $(g) \longrightarrow$ products, where [A] = conc. A at t, $[A]_o$ = conc. A at $t = 0$

Order	Rate Expression	Conc.-Time Relation	Half-Life	Linear Plot
0	rate = k	$[A]_o - [A] = kt$	$[A]_o/2k$	[A] vs. t
1	rate = $k[A]$	$\ln\dfrac{[A]_o}{[A]} = kt$	$0.693/k$	ln [A] vs. t
2	rate = $k[A]^2$	$\dfrac{1}{[A]} - \dfrac{1}{[A]_o} = kt$	$1/k[A]_o$	$\dfrac{1}{[A]}$ vs. t

where the symbols [A], $[A]_o$, t, and k have their usual meanings. For a second-order reaction, a plot of 1/[A] versus t should be linear.

The characteristics of zero-, first-, and second-order reactions are summarized in Table 11.2. To determine reaction order, the properties in either of the last two columns at the right of the table can be used (Example 11.6).

EXAMPLE 11.6

The following data were obtained for the gas-phase decomposition of hydrogen iodide:

Time (h)	0	2	4	6
[HI]	1.00	0.50	0.33	0.25

Is this reaction zero-, first-, or second-order in HI?

STRATEGY

1. Prepare a table listing [HI], ln[HI], and 1/[HI] as a function of time from the experimental data given.

2. Make a plot of each concentration-time relationship.

3. See Table 11.2 for the order of the reaction based on the linear plot that you obtain.

SOLUTION

concentration-time table

t	[HI]	ln [HI]	1/[HI]
0	1.00	0	1.0
2	0.50	−0.69	2.0
4	0.33	−1.10	3.0
6	0.25	−1.39	4.0

plots

See Figure 11.7 (page 289).

order of the reaction

1/[HI] vs time gives a linear plot. The reaction is second-order.

END POINTS

1. Note that the concentration drops from 1.00 M to 0.500 M in 2 hours. If the reaction were zero-order, it would be all over in 2 hours.

2. If the reaction is first-order, the concentration would be 0.25 M after 4 hours.

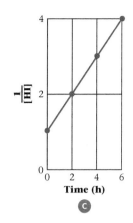

Figure 11.7 **Determination of reaction order.** One way to determine reaction order is to search for a linear relation between some function of concentration and time (Table 11.2). Because a plot of 1/[HI] versus t (plot c) is linear, the decomposition of HI must be second-order.

EXAMPLE **11.7** CONCEPTUAL

A certain reaction is first-order in A and second-order in B. In the box shown below, which is assumed to have a volume of one liter, a mole of A is represented by ●, a mole of B by ●.

In which of the three boxes shown below is the rate of reaction the same as that in the box shown above?

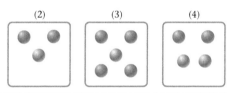

ANALYSIS	
Information given:	order of the reaction with respect to A (first-order) and to B (second-order) concentrations of A and B
Asked for:	Which box has the same reaction rate as box (1)?

STRATEGY	
Write the reaction rate for each box and compare.	

SOLUTION	
Box (1)	rate = $k(1)(2)^2 = 4k$
Box (2)	rate = $k(2)(1)^2 = 2k$
Box (3)	rate = $k(4)(1)^2 = 4k$
Box (4)	rate = $k(2)(2)^2 = 8k$
	The rates in boxes (1) and (3) are the same.

▼
11-4   **Models for Reaction Rate**

So far in this chapter we have approached reaction rate from an experimental point of view, describing what happens in the laboratory or the world around us. Now we change emphasis and try to explain why certain reactions occur rapidly

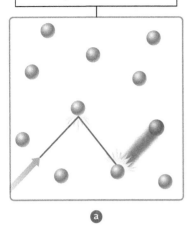

A single red molecule moves among 10 blue molecules and collides with 2 of them per second.

a

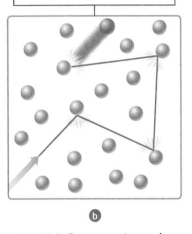

With double the number of blue molecules, the number of collisions doubles to 4 per second.

b

Figure 11.8 Concentration and molecular collisions. A red molecule must collide with a blue molecule for a reaction to take place.

i There would be an explosion.

while others take place slowly. To do this, we look at a couple of models that chemists have developed to predict rate constants for reactions.

Collision Model; Activation Energy

Consider the reaction

$$CO(g) + NO_2(g) \longrightarrow CO_2(g) + NO(g)$$

Above 600 K, this reaction takes place as a direct result of collisions between CO and NO_2 molecules. When the concentration of CO doubles (Figure 11.8), the number of these collisions in a given time increases by a factor of 2; doubling the concentration of NO_2 has the same effect. Assuming that reaction rate is directly proportional to the collision rate, the following relation should hold:

$$\text{reaction rate} = k[CO] \times [NO_2]$$

Experimentally, this prediction is confirmed; the reaction is first-order in both carbon monoxide and nitrogen dioxide.

There is a restriction on this simple model for the CO-NO_2 reaction. According to the kinetic theory of gases, for a reaction mixture at 700 K and concentrations of 0.10 M, every CO molecule should collide with about 10^9 NO_2 molecules in one second. If every collision were effective, the reaction should be over in a fraction of a second. ◂ *i* In reality, this does not happen; under these conditions, the half-life is about 10 s. This implies that not every CO-NO_2 collision leads to reaction.

There are a couple of reasons why collision between reactant molecules does not always lead to reaction. For one thing, the molecules have to be properly oriented with respect to one another when they collide. Suppose, for example, that the carbon atom of a CO molecule strikes the nitrogen atom of an NO_2 molecule (Figure 11.9). This is very unlikely to result in the transfer of an oxygen atom from NO_2 to CO, which is required for reaction to occur.

A more important factor that reduces the number of effective collisions has to do with the kinetic energy of reactant molecules. These molecules are held together by strong chemical bonds. Only if the colliding molecules are moving very rapidly will the kinetic energy be large enough to supply the energy required to break these bonds. Molecules with small kinetic energies bounce off one another without reacting. As a result, only a small fraction of collisions are effective.

For every reaction, there is a certain minimum energy that molecules must possess for collision to be effective. This is referred to as the **activation energy**. It has the symbol E_a and is expressed in kilojoules/mol. For the reaction between one mole of CO and one mole of NO_2, E_a is 134 kJ/mol. The colliding molecules (CO and NO_2) must have a total kinetic energy of at least 134 kJ/mol if they are to react. The activation energy for a reaction is a positive quantity ($E_a > 0$) whose value depends on the nature of the reaction.

Figure 11.9 Effective and ineffective molecular collisions. For a collision to result in reaction, the molecules must be properly oriented. For the reaction $CO(g) + NO_2(g) \longrightarrow CO_2(g) + NO(g)$, the carbon atom of the CO molecule must strike an oxygen atom of the NO_2 molecule, forming CO_2 as one product, NO as the other.

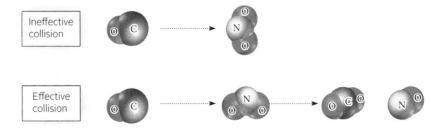

Ineffective collision

Effective collision

The collision model of reaction rates just developed can be made quantitative. We can say that the rate constant for a reaction, k, is a product of three factors:

$$k = p \times Z \times f$$

where

- p, called a *steric factor*, takes into account the fact that only certain orientations of colliding molecules are likely to lead to reaction (recall Figure 11.9). Unfortunately, the collision model cannot predict the value of p; about all we can say is that it is less than 1, sometimes much less.
- Z, the *collision frequency*, which gives the number of molecular collisions occurring in unit time at unit concentrations of reactants. This quantity can be calculated quite accurately from the kinetic theory of gases, but we will not describe that calculation.
- f, the *fraction of collisions in which the energy of the colliding molecules is equal to or greater than E_a*. It can be shown that

$$f = e^{-E_a/RT}$$

where e is the base of natural logarithms, R is the gas constant, and T is the absolute temperature in K.

Substituting this expression for f into the equation written above for k, we obtain the basic equation for the collision model:

$$k = p \times Z \times e^{-E_a/RT}$$

This equation tells us, among other things, that *the larger the value of E_a, the smaller the rate constant.* Thus

$$\text{if } E_a = 0, e^{-E_a/RT} = e^0 = 1$$
$$E_a = RT, e^{-E_a/RT} = e^{-1} = 0.37$$
$$E_a = 2RT, e^{-E_a/RT} = e^{-2} = 0.14$$

and so on. This makes sense; the larger the activation energy, the smaller the fraction of molecules having enough energy to react on collision, so the slower the rate of reaction. ◀ *i*

High-energy molecules get the job done.

Table 11.3 compares observed rate constants for several reactions with those predicted by collision theory, arbitrarily taking $p = 1$. As you might expect, the calculated k's are too high, suggesting that the steric factor is indeed less than 1.

Transition-State Model; Activation Energy Diagrams

Figure 11.10 is an energy diagram for the CO-NO_2 reaction. Reactants CO and NO_2 are shown at the left. Products CO_2 and NO are at the right; they have an energy 226 kJ less than that of the reactants. The enthalpy change, ΔH, for the reaction is -226 kJ. In the center of the figure is an intermediate called an **activated complex**. This is an unstable, high-energy species that must be formed before the reaction can occur. ◀ *i* It has an energy 134 kJ greater than that of the

In the activated complex, electrons are often in excited states.

Table 11.3 Observed and Calculated Rate Constants for Second-Order Gas-Phase Reactions

Reaction	k Observed	Collision Model	Transition-State Model
$NO + O_3 \longrightarrow NO_2 + O_2$	6.3×10^7	4.0×10^9	3.2×10^7
$NO + Cl_2 \longrightarrow NOCl + Cl$	5.2	130	1.6
$NO_2 + CO \longrightarrow NO + CO_2$	1.2×10^{-4}	6.4×10^{-4}	1.0×10^{-4}
$2NO_2 \longrightarrow 2NO + O_2$	5.0×10^{-3}	1.0×10^{-1}	1.2×10^{-2}

Figure 11.10 Reaction energy diagram. During the reaction initiation, 134 kJ—the activation energy E_a—must be furnished to the reactants for every mole of CO that reacts. This energy activates each CO—NO_2 complex to the point at which reaction can occur.

reactants and 360 kJ greater than that of the products. The state of the system at this point is often referred to as a *transition state,* intermediate between reactants and products.

The exact nature of the activated complex is difficult to determine. For this reaction, the activated complex might be a "pseudomolecule" made up of CO and NO_2 molecules in close contact. The path of the reaction might be more or less as follows:

$$O{\equiv}C + O{-}N{\diagdown}_O \longrightarrow O{\equiv}C{\cdots}O{\cdots}N{\diagdown}_O \longrightarrow O{=}C{=}O + N{=}O$$

$$\text{reactants} \qquad\qquad \text{activated complex} \qquad\qquad \text{products}$$

The dotted lines stand for "partial bonds" in the activated complex. The N—O bond in the NO_2 molecule has been partially broken. A new bond between carbon and oxygen has started to form.

The idea of the activated complex was developed by, among others, Henry Eyring at Princeton in the 1930s. It forms the basis of the *transition-state model* for reaction rate, which assumes that the activated complex

- is in equilibrium, at low concentrations, with the reactants.
- may either decompose to products by "climbing" over the energy barrier or, alternatively, revert back to reactants.

With these assumptions, it is possible to develop an expression for the rate constant k. Although the expression is too complex to discuss here, we show some of the results in Table 11.3.

The transition-state model is generally somewhat more accurate than the collision model (at least with $p = 1$). Another advantage is that it explains why

CHEMISTRY THE **HUMAN** SIDE

Chemical kinetics, the subject of this chapter, evolved from an art into a science in the first half of the twentieth century. One person more than any other was responsible for that change: Henry Eyring. Born in a Mormon settlement in Mexico, Henry immigrated to Arizona with his parents in 1912, when the Mexican revolution broke out. Many years later, he became a U.S. citizen, appearing before a judge who had just naturalized Albert Einstein.

Eyring's first exposure to kinetics came as a young chemistry instructor at the University of Wisconsin in the 1920s. There he worked on the thermal decomposition of N_2O_5 (Section 11-2). In 1931 Eyring came to Princeton, where he developed the transition-state model of reaction rates. Using this model and making reasonable guesses about the nature of the activated complex, Eyring calculated rate constants that agreed well with experiment. After 15 productive years at Princeton, Eyring returned to his Mormon roots at the University of Utah in Salt Lake City. While serving as dean of the graduate school, he found time to do ground-breaking research on the kinetics of life processes, including the mechanism of aging.

Henry Eyring's approach to research is best described in his own words:

I perceive myself as rather uninhibited, with a certain mathematical facility and more interest in the broad aspects of a problem than the delicate nuances. I am more interested in discovering what is over the next rise than in assiduously cultivating the beautiful garden close at hand.

Henry Eyring
(1901–1981)

Henry's enthusiasm for chemistry spilled over into just about every other aspect of his life. Each year he challenged his students to a 50-yard dash held at the stadium of the University of Utah. He continued these races until he was in his mid-seventies. Henry never won, but on one occasion he did make the *CBS Evening News.*

the activation energy is ordinarily much smaller than the bond enthalpies in the reactant molecules. Consider, for example, the reaction

$$CO(g) + NO_2(g) \longrightarrow CO_2(g) + NO(g)$$

where E_a = 134 kJ/mol. This is considerably smaller than the amount of energy required to break any of the bonds in the reactants.

$$C\equiv O \quad 1075 \text{ kJ} \qquad N=O \quad 607 \text{ kJ} \qquad N-O \quad 222 \text{ kJ}$$

Forming the activated complex shown in Figure 11.10 requires the absorption of relatively little energy, because it requires only the weakening of reactant bonds rather than their rupture.

11-5   Reaction Rate and Temperature

The rates of most reactions increase as the temperature rises (Figure 11.11). A person in a hurry to prepare dinner applies this principle by turning the dial on the oven to the highest possible setting. By storing the leftovers in a refrigerator, the chemical reactions responsible for food spoilage are slowed down. ◄ⓘ As a general and very approximate rule, it is often stated that an increase in temperature of 10°C doubles the reaction rate. If this rule holds, foods should deteriorate four times as rapidly at room temperature (25°C) as they do in a refrigerator at 5°C.

The effect of temperature on reaction rate ◄ⓘ can be explained in terms of kinetic theory. Recall from Chapter 5 that raising the temperature greatly increases the fraction of molecules having very high speeds and hence high kinetic energies (kinetic energy = $mu^2/2$). These are the molecules that are most likely to react when they collide. The higher the temperature, the larger the fraction of molecules that can provide the activation energy required for reaction. This effect is apparent from Figure 11.12 where the distribution of kinetic energies is shown at two different temperatures. Notice that the fraction of molecules having a kinetic energy equal to or greater than the activation energy E_a (shaded area) is considerably larger at the

ⓘ Hibernating animals lower their body temperature, slowing down life processes.

ⓘ
$T\uparrow \rightarrow k\uparrow \rightarrow$ rate ↑

The light sticks outside the beaker at room temperature are brighter than the two sticks cooled to 0.2°C because the reaction rate is larger at the higher temperature.

With the light sticks in the beaker heated well above room temperature, to 59°C, these sticks are now much brighter than those at room temperature.

ⓐ ⓑ

© Cengage Learning/Charles D. Winters

Figure 11.11 Temperature and reaction rate. A chemical reaction in the Cyalume® light sticks produces light as it takes place.

Figure 11.12 Temperature and activation energy. When T is increased to $T + 10$, the number of molecules with high energies increases. Hence, many more molecules possess sufficient energy to react and the reaction occurs more rapidly. If E_a is 50 kJ/mol, any molecule with this or a higher energy can react. The number of molecules that react at T is represented by the red area under the T curve, and the number that react at $T + 10$ is represented by the sum of the red and blue areas under the $T + 10$ curve.

higher temperature. Hence the fraction of effective collisions increases; this is the major factor causing reaction rate to increase with temperature.

The Arrhenius Equation

You will recall from Section 11-4 that the collision model yields the following expression for the rate constant:

$$k = p \times Z \times e^{-E_a/RT} \qquad (11.6)$$

The steric factor, p, is presumably temperature-independent. The collision number, Z, is relatively insensitive to temperature. For example, when the temperature increases from 500 to 600 K, Z increases by less than 10%. Hence to a good degree of approximation we can write, as far as temperature dependence is concerned,

$$k = Ae^{-E_a/RT}$$

where A is a constant. Taking the natural logarithm of both sides of this equation,

$$\ln k = \ln A - E_a/RT$$

This equation was first shown to be valid by Svante Arrhenius (Chapter 4) in 1889. It is commonly referred to as the **Arrhenius equation**. This equation expresses the temperature dependence of the rate constant.

Comparing the Arrhenius equation with the general equation for a straight line,

$$y = b + mx$$

it should be clear that a plot of $\ln k$ ("y") versus $1/T$ ("x") should be linear. The slope of the straight line, m, is equal to $-E_a/R$. Figure 11.13 shows such a plot for

Figure 11.13 Arrhenius plot. A plot of $\ln k$ versus $1/T$ (k = rate constant, T = Kelvin temperature) is a straight line. From the slope of this line, the activation energy can be determined: $E_a = -R \times$ slope.

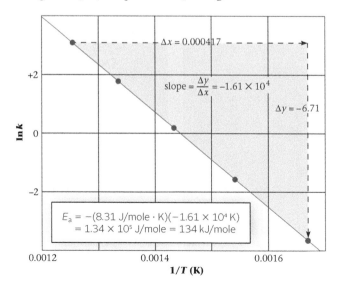

the CO-NO_2 reaction. The slope appears to be about -1.61×10^4 K. Hence

$$\frac{-E_a}{R} = -1.61 \times 10^4 \text{ K}$$

$$E_a = 8.31\frac{\text{J}}{\text{mol} \cdot \text{K}}(1.61 \times 10^4 \text{ K}) = 1.34 \times 10^5 \text{ J/mol} = 134 \text{ kJ/mol}$$

Two-Point Equation Relating k and T

The Arrhenius equation can be expressed in a different form by following the procedure used with the Clausius-Clapeyron equation in Chapter 9. At two different temperatures, T_2 and T_1,

$$\ln k_2 = \ln A - \frac{E_a}{RT_2}$$

$$\ln k_1 = \ln A - \frac{E_a}{RT_1}$$

Subtracting the second equation from the first eliminates $\ln A$, and we obtain

$$\ln k_2 - \ln k_1 = \frac{-E_a}{R}\left[\frac{1}{T_2} - \frac{1}{T_1}\right] \tag{11.7a}$$

$$\ln \frac{k_2}{k_1} = \frac{E_a}{R}\left[\frac{1}{T_1} - \frac{1}{T_2}\right] \tag{11.7b}$$

Taking $R = 8.31$ J/mol·K, the activation energy is expressed in joules per mole.

EXAMPLE 11.8 GRADED

Consider the first-order decomposition of A. The following is known about it:

- the rate constant doubles when the temperature increases from 15°C to 25°C.
- the rate constant for the decomposition at 40°C is 0.0125 s^{-1}.

a What is the activation energy for the decomposition?

ANALYSIS	
Information given:	k at 15°C(k_1); k at 25°C($k_2 = 2k_1$)
Information implied:	R value
Asked for:	E_a

STRATEGY
1. Take 15°C (288 K) as T_1 where the rate constant is k_1.
2. Take 25°C (298 K) as T_2 where the rate constant is $k_2 = 2k_1$.
3. Substitute into Equation 11.7b.
$\ln \dfrac{k_2}{k_1} = \dfrac{E_a}{R}\left[\dfrac{1}{T_1} - \dfrac{1}{T_2}\right]$

SOLUTION	
E_a	$\ln \dfrac{2k_1}{k_1} = \dfrac{E_a}{8.31 \text{ J/mol}\cdot\text{K}}\left[\dfrac{1}{288} - \dfrac{1}{298}\right] \longrightarrow (0.693)(8.31) = E_a\left[\dfrac{1}{288} - \dfrac{1}{298}\right]$
	$E_a = 4.9 \times 10^4 \text{ J/mol} = 49 \text{ kJ/mol}$

continued

b	What is the half-life of A at 78°C?	
	ANALYSIS	
Information given:	from part (a): E_a (4.9 × 10^4 J/mol) k_1 (0.0125 s^{-1}) at T_1 (40°C); T_2 (78°C)	
Information implied:	R value	
Asked for:	$t_{1/2}$ at 78°C	
	STRATEGY	
1. Find k_2 at 78°C (T_2) by substituting into Equation 11.7b. Recall that k_1 and T_1 are given.		
2. Find $t_{1/2}$ at 78°C by substituting into Equation 11.3.		
	SOLUTION	
k_2 at 78°C	$\ln k_2 - \ln(0.0125) = \dfrac{4.9 \times 10^4 \text{ J/mol}}{8.31 \text{ J/mol} \cdot \text{K}}\left(\dfrac{1}{313 \text{ K}} - \dfrac{1}{351 \text{ K}}\right)$	
	$\ln k_2 = 2.04 + (-4.38) = -2.34 \longrightarrow k_2 = e^{-2.34} = 0.0963 \text{ s}^{-1}$	
$t_{1/2}$	$t_{1/2} = \dfrac{0.693}{0.0963 \text{ s}^{-1}} = 7.20 \text{ s}$	

c	At what temperature will the rate of the decomposition of 0.165 M be 0.124 mol/L·s?	
	ANALYSIS	
Information given:	from part (a): E_a (4.9 × 10^4 J/mol); from part (b): k_1 (0.0125 s^{-1}) at T_1 (40°C) rate (0.124 mol/L·s) for [A] (0.165 M)	
Information implied:	R value	
Asked for:	T at the given rate for A.	
	STRATEGY	
1. Find k (k_2) for the decomposition at T_2 by substituting into the rate expression.		
2. Substitute into the Arrhenius equation (11.7b).		
	SOLUTION	
k	0.124 mol/L·s $= k$(0.165 mol/L) $\longrightarrow k = 0.752 \text{ s}^{-1}$	
T_2	$\ln(0.752 \text{ s}^{-1}) - \ln(0.0125 \text{ s}^{-1}) = \dfrac{4.9 \times 10^4 \text{ J/mol}}{8.31 \text{ J/mol} \cdot \text{K}}\left(\dfrac{1}{313 \text{ K}} - \dfrac{1}{T_2}\right)$	
	$4.10 = 5.9 \times 10^3 \left(\dfrac{1}{313 \text{ K}} - \dfrac{1}{T_2}\right) \longrightarrow T_2 = 4.0 \times 10^2 \text{ K} = 1.3 \times 10^2 °\text{C}$	

11-6 Catalysis

A catalyst *is a substance that increases the rate of a reaction without being consumed by it.* It does this by changing the reaction path to one with a lower activation energy. Frequently the catalyzed path consists of two or more steps. In this case, the activation energy for the uncatalyzed reaction exceeds that for any of the steps in the catalyzed reaction (Figure 11.14).

Heterogeneous Catalysis

A *heterogeneous catalyst* is one that is in a different phase from the reaction mixture. Most commonly, the catalyst is a solid that increases the rate of a gas-phase or liquid-phase reaction. An example is the decomposition of N_2O on gold:

$$N_2O(g) \xrightarrow{\text{Au}} N_2(g) + \tfrac{1}{2}O_2(g)$$

In the catalyzed decomposition, N_2O is chemically adsorbed on the surface of the solid. A chemical bond is formed between the oxygen atom of an N_2O molecule and a gold atom on the surface. This weakens the bond joining nitrogen to oxygen, making it easier for the N_2O molecule to break apart. Symbolically, this process can be shown as

$$N\equiv N-O(g) + Au(s) \longrightarrow N\equiv N\text{-}\text{-}\text{-}O\text{-}\text{-}\text{-}Au(s) \longrightarrow N\equiv N(g) + O(g) + Au(s)$$

where the broken lines represent weak covalent bonds.

Perhaps the most familiar example of heterogeneous catalysis is the series of reactions that occur in the catalytic converter of an automobile (Figure 11.15). Typically this device contains 1 to 3 g of platinum metal mixed with rhodium. The platinum catalyzes the oxidation of carbon monoxide and unburned hydrocarbons such as benzene, C_6H_6:

$$2CO(g) + O_2(g) \xrightarrow{\text{Pt}} 2CO_2(g)$$

$$C_6H_6(g) + \tfrac{15}{2}O_2(g) \xrightarrow{\text{Pt}} 6CO_2(g) + 3H_2O(g)$$

The rhodium acts as a catalyst to destroy nitrogen oxide by the reaction

$$2NO(g) \longrightarrow N_2(g) + O_2(g)$$

One problem with heterogeneous catalysis is that the solid catalyst is easily "poisoned." Foreign materials deposited on the catalytic surface during the reaction reduce or even destroy its effectiveness. A major reason for using unleaded gasoline is that lead metal poisons the Pt-Rh mixture in the catalytic converter. ◄ *i*

Homogeneous Catalysis

A *homogeneous catalyst* is one that is present in the same phase as the reactants. It speeds up the reaction by forming a reactive intermediate that decomposes to give products. In this way, the catalyst provides an alternative process of lower activation energy.

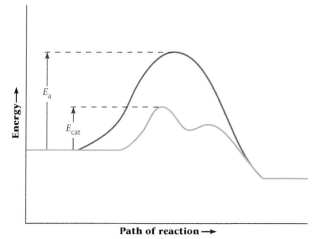

Figure 11.14 Catalysis and activation energy. By changing the path by which a reaction occurs, a catalyst can lower the activation energy that is required and so speed up a reaction.

i Many industrial processes use heterogeneous catalysts.

Figure 11.15 Automobile catalytic converter. Catalytic converters contain a "three-way" catalyst designed to convert CO to CO_2, unburned hydrocarbons to CO_2 and H_2O, and NO to N_2. The active components of the catalysts are the precious metals platinum and rhodium; palladium is sometimes used as well.

Figure 11.16 Enzyme catalysis. An enzyme in the potato is catalyzing the decomposition of a hydrogen peroxide solution, as shown by the bubbles of oxygen.

An example of a reaction that is subject to homogeneous catalysis is the decomposition of hydrogen peroxide in aqueous solution:

$$2H_2O_2(aq) \longrightarrow 2H_2O + O_2(g)$$

Under ordinary conditions, this reaction occurs very slowly. However, if a solution of sodium iodide, NaI, is added, reaction occurs almost immediately; you can see the bubbles of oxygen forming.

The catalyzed decomposition of hydrogen peroxide is believed to take place by a two-step path:

Step 1: $H_2O_2(aq) + I^-(aq) \longrightarrow H_2O + IO^-(aq)$

Step 2: $\underline{H_2O_2(aq) + IO^-(aq) \longrightarrow H_2O + O_2(g) + I^-(aq)}$

$$2H_2O_2(aq) \longrightarrow 2H_2O + O_2(g)$$

Notice that the end result is the same as in the direct reaction.

The I^- ions are not consumed in the reaction. For every I^- ion used up in the first step, one is produced in the second step. The activation energy for this two-step process is much smaller than for the uncatalyzed reaction.

Enzymes

Many reactions that take place slowly under ordinary conditions occur readily in living organisms in the presence of catalysts called *enzymes*. Enzymes are protein molecules of high molar mass. An example of an enzyme-catalyzed reaction is the decomposition of hydrogen peroxide:

$$2H_2O_2(aq) \longrightarrow 2H_2O + O_2(g)$$

In blood or tissues, this reaction is catalyzed by an enzyme called catalase (Figure 11.16). When 3% hydrogen peroxide is used to treat a fresh cut or wound, oxygen gas is given off rapidly. The function of catalase in the body is to prevent the build-up of hydrogen peroxide, a powerful oxidizing agent.

Many enzymes are extremely specific. For example, the enzyme maltase catalyzes the hydrolysis of maltose:

$$\underset{\text{maltose}}{C_{12}H_{22}O_{11}(aq)} + H_2O \xrightarrow{\text{maltase}} \underset{\text{glucose}}{2C_6H_{12}O_6(aq)}$$

This is the only function of maltase, but it is one that no other enzyme can perform. Many such digestive enzymes are required for the metabolism of carbohydrates, proteins, and fats. It has been estimated that without enzymes, it would take upward of 50 years to digest a meal. ◀ *i*

That would make dieting a lot easier.

Enzymes, like all other catalysts, lower the activation energy for reaction. They can be enormously effective; it is not uncommon for the rate constant to increase by a factor of 10^{12} or more. However, from a commercial standpoint, enzymes have some drawbacks. A particular enzyme operates best over a narrow range of temperature. An increase in temperature frequently deactivates an enzyme by causing the molecule to "unfold," changing its characteristic shape. Recently, chemists have discovered that this effect can be prevented if the enzyme is immobilized by bonding to a solid support. Among the solids that have been used are synthetic polymers, porous glass, and even stainless steel. The development of immobilized enzymes has led to a host of new products. The sweetener aspartame, used in many diet soft drinks, is made using the enzyme aspartase in immobilized form.

11-7 Reaction Mechanisms

A **reaction mechanism** is a description of a path, or a sequence of steps, by which a reaction occurs at the molecular level. In the simplest case, only a single step is involved. This is a collision between two reactant molecules. This is the mechanism for the reaction of CO with NO_2 at high temperatures, above about 600 K:

$$CO(g) + NO_2(g) \longrightarrow NO(g) + CO_2(g)$$

At lower temperatures the reaction between carbon monoxide and nitrogen dioxide takes place by a quite different mechanism. ◄*i* Two steps are involved:

$$NO_2(g) + NO_2(g) \longrightarrow NO_3(g) + NO(g)$$
$$\underline{CO(g) + NO_3(g) \longrightarrow CO_2(g) + NO_2(g)}$$
$$CO(g) + NO_2(g) \longrightarrow NO(g) + CO_2(g)$$

Notice that the overall reaction, obtained by summing the individual steps, is identical with that for the one-step process. The rate expressions are quite different, however:

high temperatures: rate = $k[CO][NO_2]$
low temperatures: rate = $k[NO_2]^2$

In general, the nature of the rate expression and hence *the reaction order depends on the mechanism by which the reaction takes place.*

Reaction mechanisms frequently change with T, sometimes with P.

Elementary Steps

The individual steps that constitute a reaction mechanism are referred to as **elementary steps.** These steps may be *unimolecular*

$$A \longrightarrow B + C \qquad rate = k[A]$$

bimolecular

$$A + B \longrightarrow C + D \qquad rate = k[A][B]$$

or, in rare cases, *termolecular*

$$A + B + C \longrightarrow D + E \qquad rate = k[A][B][C]$$

Notice from the rate expressions just written that *the rate of an elementary step is equal to a rate constant k multiplied by the concentration of each reactant molecule.* This rule is readily explained. Consider, for example, a step in which two molecules, A and B, collide effectively with each other to form C and D. As pointed out earlier, the rate of collision and hence the rate of reaction will be directly proportional to the concentration of each reactant.

Slow Steps

Often, one step in a mechanism is much slower than any other. If this is the case, *the slowest step is a* **rate-determining step.** That is, the rate of the overall reaction can be taken to be that of the slow step. Consider, for example, a three-step reaction:

Step 1: A $\longrightarrow$ B   (fast)
Step 2: B $\longrightarrow$ C   (slow)
Step 3: $\underline{C \longrightarrow D}$ (fast)
A $\longrightarrow$ D

The rate at which A is converted to D (the overall reaction) is approximately equal to the rate of conversion of B to C (the slow step).

To understand the rationale behind this rule, and its limitations, consider an analogous situation. Suppose three people (A, B, and C) are assigned to grade general chemistry examinations that contain three questions. On the average, A spends 10 s grading question 1 and B spends 15 s grading question 2. In contrast,

Figure 11.17 Reaction plot for a reaction with a three-step mechanism. The second of the three steps is rate-determining.

C, the ultimate procrastinator, takes 5 min to grade question 3. The rate at which exams are graded is

$$\frac{1 \text{ exam}}{10 \text{ s} + 15 \text{ s} + 300 \text{ s}} = \frac{1 \text{ exam}}{325 \text{ s}} = 0.00308 \text{ exam/s}$$

This is approximately equal to the rate of the slower grader:

$$\frac{1 \text{ exam}}{300 \text{ s}} = 0.00333 \text{ exam/s}$$

Extrapolating from general chemistry exams to chemical reactions, we can say that

- the overall rate of the reaction cannot exceed that of the slowest step.
- if that step is by far the slowest, its rate will be approximately equal to that of the overall reaction.
- the slowest step in a mechanism is ordinarily the one with the highest activation energy (Figure 11.17).

Deducing a Rate Expression from a Proposed Mechanism

As we have seen, rate expressions for reactions must be determined experimentally. Once this has been done, it is possible to derive a plausible mechanism compatible with the observed rate expression. This, however, is a rather complex process and we will not attempt it here. Instead, we will consider the reverse process, which is much more straightforward. *Given a mechanism for a several-step reaction, how can you deduce the rate expression corresponding to that mechanism?*

In principle, at least, all you have to do is to apply the rules just cited.

1. *Find the slowest step and equate the rate of the overall reaction to the rate of that step.*
2. *Find the rate expression for the slowest step.*

To illustrate this process, consider the two-step mechanism for the low-temperature reaction between CO and NO_2.

$$NO_2(g) + NO_2(g) \longrightarrow NO_3(g) + NO(g) \quad \text{(slow)}$$
$$\underline{CO(g) + NO_3(g) \longrightarrow CO_2(g) + NO_2(g) \quad \text{(fast)}}$$
$$CO(g) + NO_2(g) \longrightarrow CO_2(g) + NO(g)$$

Applying the above rules in order,

$$\text{rate of overall reaction} = \text{rate of first step} = k[NO_2]^2$$

This analysis explains why the rate expression for the two-step mechanism is different from that for the direct, one-step reaction.

Elimination of Intermediates

Sometimes the rate expression obtained by the process just described involves a reactive intermediate, that is, a species produced in one step of the mechanism and consumed in a later step. Ordinarily, concentrations of such species are too small to be determined experimentally. Hence they must be eliminated from the rate expression if it is to be compared with experiment. The final rate expression usually includes only those species that appear in the balanced equation for the reaction. Sometimes, the concentration of a catalyst is included, but never that of a reactive intermediate.

To illustrate this situation, consider the reaction between nitric oxide and chlorine, which is believed to proceed by a two-step mechanism:

Step 1: $\quad NO(g) + Cl_2(g) \underset{k_{-1}}{\overset{k_1}{\rightleftharpoons}} NOCl_2(g) \qquad$ (fast)

Step 2: $\quad \dfrac{NOCl_2(g) + NO(g) \overset{k_2}{\longrightarrow} 2NOCl(g) \quad \text{(slow)}}{2NO(g) + Cl_2(g) \longrightarrow 2NOCl(g)}$

The first step occurs rapidly and reversibly; a dynamic equilibrium is set up in which the rates of forward and reverse reactions are equal. Because the second step is slow and rate-determining, it follows that

$$\text{rate of overall reaction} = \text{rate of step 2} = k_2[NOCl_2][NO]$$

The rate expression just written is unsatisfactory in that it cannot be checked against experiment. The species $NOCl_2$ is a reactive intermediate whose concentration is too small to be measured accurately, if at all. To eliminate the $[NOCl_2]$ term from the rate expression, recall that the rates of forward and reverse reactions in step 1 are equal, which means that

$$k_1[NO] \times [Cl_2] = k_{-1}[NOCl_2]$$

Solving for $[NOCl_2]$ and substituting in this rate expression,

$$[NOCl_2] = \frac{k_1[NO][Cl_2]}{k_{-1}}$$

$$\text{rate of overall reaction} = k_2[NOCl_2] \times [NO] = \frac{k_2k_1[NO]^2[Cl_2]}{k_{-1}}$$

The quotient k_2k_1/k_{-1} is the experimentally observed rate constant for the reaction, which is found to be second-order in NO and first-order in Cl_2, as predicted by this mechanism.

> If the concentration of a species can't be measured, it had better not appear in the rate expression.

EXAMPLE 11.9

The decomposition of ozone, O_3, to diatomic oxygen, O_2, is believed to occur by a two-step mechanism:

Step 1: $\qquad O_3(g) \underset{k_{-1}}{\overset{k_1}{\rightleftharpoons}} O_2(g) + O(g) \quad$ (fast)

Step 2: $\quad \dfrac{O_3(g) + O(g) \overset{k_2}{\longrightarrow} 2O_2(g) \qquad \text{(slow)}}{2O_3(g) \longrightarrow 3O_2(g)}$

Obtain the rate expression corresponding to this mechanism.

STRATEGY

1. The rate-limiting step is the slow step (step 2). Write its rate expression.

2. Write the rate expressions for the forward and reverse reactions of step 1. Since step 1 is in equilibrium, rate forward reaction = rate backward reaction.

3. Express the rates of step 1 in terms of [O] and substitute into the rate expression for step 2.

4. Combine all constants into a single constant k.

continued

SOLUTION

1. Rate expression for step 2	rate $= k_2[O_3][O]$
2. Rate of forward reaction	rate $= k_1[O_3]$
Rate of reverse reaction	rate $= k_{-1}[O_2][O]$
Rate forward reaction = rate reverse reaction	$k_1[O_3] = k_{-1}[O_2][O]$
3. [O]	$[O] = \dfrac{k_1[O_3]}{k_{-1}[O_2]}$
Overall rate	rate $= k_2[O_3]\left(\dfrac{k_1[O_3]}{k_{-1}[O_2]}\right)$
4. Combine all constants	$k = \dfrac{k_2 k_1}{k_{-1}}$
	rate $= k\,\dfrac{[O_3]^2}{[O_2]}$

It is important to point out one of the limitations of mechanism studies. Usually more than one mechanism is compatible with the same experimentally obtained rate expression. To make a choice between alternative mechanisms, other evidence must be considered. A classic example of this situation is the reaction between hydrogen and iodine

$$H_2(g) + I_2(g) \longrightarrow 2HI(g)$$

for which the observed rate expression is

$$\text{rate} = k[H_2][I_2]$$

For many years, it was assumed that the H_2-I_2 reaction occurs in a single step, a collision between an H_2 molecule and an I_2 molecule. That would, of course, be compatible with the rate expression above. However, there is now evidence to indicate that a quite different and more complex mechanism is involved (see Problem 77).

CHEMISTRY **BEYOND** THE CLASSROOM

The Ozone Story

In recent years, a minor component of the atmosphere, ozone, has received a great deal of attention. Ozone, molecular formula O_3, is a pale blue gas with a characteristic odor that can be detected after lightning activity, in the vicinity of electric motors, or near a subway train.

Depending on its location in the atmosphere, ozone can be a villain or a beleaguered hero. In the lower atmosphere (the *troposphere*), ozone is bad news; it is a major component of photochemical smog, formed by the reaction sequence

$$NO_2(g) \longrightarrow NO(g) + O(g)$$
$$O_2(g) + O(g) \longrightarrow O_3(g)$$
$$\overline{NO_2(g) + O_2(g) \longrightarrow NO(g) + O_3(g)}$$

Ozone is an extremely powerful oxidizing agent, which explains its toxicity to animals and humans. At partial pressures as low as 10^{-7} atm, it can cut in half the rate of photosynthesis by plants.

In the upper atmosphere (the *stratosphere*), the situation is quite different. There the partial pressure of ozone goes through a maximum of about 10^{-5} atm at an altitude of 30 km. From 95% to 99% of sunlight in the ultraviolet region between 200 and 300 nm is absorbed by ozone in this region, commonly referred to as the "ozone layer." The mechanism by which this occurs can be represented by the following pair of equations:

$$O_3(g) + \text{UV radiation} \longrightarrow O_2(g) + O(g)$$
$$O_2(g) + O(g) \longrightarrow O_3(g) + \text{heat}$$
$$\overline{\text{UV radiation} \longrightarrow \text{heat}}$$

The net effect is simply the conversion of ultraviolet to thermal energy.

If the UV radiation were to reach the surface of the earth, it could have several adverse effects. A 5% decrease in ozone concentration could increase the incidence of skin cancer by 10% to 20%. Ultraviolet radiation is also a factor in diseases of the eye, including cataract formation.

Ozone molecules in the stratosphere can decompose by the reaction

$$O_3(g) + O(g) \longrightarrow 2O_2(g)$$

This reaction takes place rather slowly by direct collision between an O_3 molecule and an O atom. It can occur more rapidly by a two-step process in which a chlorine atom acts as a catalyst:

$$\frac{\begin{aligned} O_3(g) + Cl(g) &\longrightarrow O_2(g) + ClO(g) \quad k = 5.2 \times 10^9 \text{ L/mol} \cdot \text{s at 220 K} \\ ClO(g) + O(g) &\longrightarrow Cl(g) + O_2(g) \quad k = 2.6 \times 10^{10} \text{ L/mol} \cdot \text{s at 220 K} \end{aligned}}{O_3(g) + O(g) \longrightarrow 2O_2(g)}$$

A single chlorine atom can bring about the decomposition of tens of thousands of ozone molecules. Bromine atoms can substitute for chlorine; indeed the rate constant for the Br-catalyzed reaction is larger than that for the reaction just cited.

The chlorine atoms that catalyze the decomposition of ozone come from chlorofluorocarbons (CFCs) used in many refrigerators and air conditioners. A major culprit is CF_2Cl_2, Freon, which forms Cl atoms when exposed to ultraviolet radiation at 200 nm:

$$CF_2Cl_2(g) \longrightarrow CF_2Cl(g) + Cl(g)$$

The two scientists who first suggested (in 1974) that CFCs could deplete the ozone layer, F. Sherwood Rowland (1927–2012) and Mario Molina (1943–), won the 1995 Nobel Prize in chemistry, along with Paul Crutzen (1933–), who first suggested that oxides of nitrogen in the atmosphere could catalyze the decomposition of ozone.

Recent research suggests that ozone depletion has affected Antarctica's climate, cooling the interior and warming the extremities. The catalyzed decomposition of ozone is known to be responsible for the ozone hole (Figure A) that develops in Antarctica each year in September and October, at the end of winter in the Southern Hemisphere. No ozone is generated during the long, dark Antarctic winter. Meanwhile, a heterogeneous reaction occurring on clouds of ice crystals at −85°C produces species such as

Cl_2, Br_2, and HOCl. When the Sun reappears in September, these molecules decompose photochemically to form Cl or Br atoms, which catalyze the decomposition of ozone, lowering its concentration.

Ozone depletion is by no means restricted to the Southern Hemisphere. In the extremely cold winter of 1994–1995, a similar "ozone hole" was found in the Arctic. Beyond that, the concentration of ozone in the atmosphere over parts of Siberia dropped by 40%.

In 1987 an international treaty was signed in Montreal to cut back on the use of CFCs. Production of Freon in the United States ended in 1996. It has been replaced in automobile air conditioners by a related compound with no chlorine atoms, $C_2H_2F_4$:

Figure A **The ozone (O_3) layer over the southern hemisphere stratosphere August 30, 2013.** The thickness is measured in Dobsons (= 0.01 mm thick). The normal ozone layer for the stratosphere is 360 Dobsons. The ozone "hole" is 200–220 Dobsons.

Key Concepts

1. Determine the rate expression (reaction order) from
   - initial rate data.
   (**Examples 11.1, 11.2, 11.3; Problems 19–28**)
   - concentration-time data, using Table 11.2.
   (**Example 11.6; Problems 29, 30**)
   - reaction mechanism.
   (**Example 11.9; Problems 77, 78**)

2. Relate concentration to time for a first-order reaction.
   (**Examples 11.4, 11.5; Problems 35–48**)
3. Use the Arrhenius equation to relate rate constant to temperature.
   (**Example 11.8; Problems 57–70**)

Key Equations

Rate of reaction	$aA + bB \longrightarrow cC + dD$
	$$\text{rate} = \frac{-\Delta[A]}{a\,\Delta t} = \frac{-\Delta[B]}{b\,\Delta t} = \frac{\Delta[C]}{c\,\Delta t} = \frac{\Delta[D]}{d\,\Delta t}$$
Zero-order reaction	$\text{rate} = k \qquad [A] = [A]_o - kt$
First-order reaction	$\text{rate} = k[A] \qquad \ln[A]_o/[A] = kt \qquad t_{1/2} = 0.693/k$
Second-order reaction	$\text{rate} = k[A]^2 \qquad 1/[A] - 1/[A]_o = kt$
Rate constant	$\text{rate} = p \times Z \times e^{-E_a/RT}$
Arrhenius equation	$$\ln \frac{k_2}{k_1} = \frac{E_a}{R}\left[\frac{1}{T_1} - \frac{1}{T_2}\right]$$

Key Terms

activation energy	half-life	rate expression	zero-order reaction
catalyst	order of reaction	reaction mechanism	
elementary steps	rate constant	reaction rate	
first-order reaction	rate-determining step	second-order reaction	

Summary Problem

Hydrogen peroxide decomposes to water and oxygen according to the following reaction:

$$H_2O_2(aq) \longrightarrow H_2O + \tfrac{1}{2}O_2(g)$$

Its rate of decomposition is measured by titrating samples of the solution with potassium permanganate ($KMnO_4$) at certain timed intervals.

(a) At what rate is H_2O_2 decomposing if O_2 is being formed at the rate of 0.0215 mol/L · min? At what rate is H_2O being produced?

(b) Initial rate experiments at 25°C for the decomposition give the following data:

[H$_2$O$_2$]	Initial rate (mol/L · min)
0.200	1.04×10^{-4}
0.300	1.55×10^{-4}
0.500	2.59×10^{-4}

1. What is the order of the reaction?
2. Write the rate equation for the decomposition.
3. Calculate the rate constant and the half-life for the reaction at 25°C.

(c) Hydrogen peroxide is sold commercially as a 30.0% solution. If the solution is kept at room temperature (25°C),

how long will it take for the solution to become 10.0% H_2O_2?

(d) It has been determined that at 40°C, the rate constant for the reaction is 1.93×10^{-3} min^{-1}. Calculate the activation energy for the decomposition of H_2O_2.

(e) Manufacturers recommend that solutions of hydrogen peroxide be kept in a refrigerator at 4°C. How much longer will it take a 30.0% solution to decompose to 10.0% if one follows the manufacturer's recommendations?

(f) The rate constant for the catalyzed decomposition at 25°C is 2.95×10^8 min^{-1}.

1. What is the half-life of the catalyzed reaction?
2. What is the rate of the catalyzed decomposition of 0.500 M H_2O_2?
3. How much faster is the rate of the catalyzed reaction compared to that of the uncatalyzed decomposition?

(g) Hydrogen peroxide in basic solution oxidizes iodide ions to iodine. The proposed mechanism for this reaction is

$$H_2O_2(aq) + I^-(aq) \longrightarrow HOI(aq) + OH^-(aq) \quad \text{slow}$$
$$HOI(aq) + I^-(aq) \rightleftharpoons I_2(aq) + OH^-(aq) \quad \text{fast}$$

Write the overall redox reaction. Write a rate law consistent with this proposed mechanism.

Answers

(a) rate H_2O_2 = rate H_2O = 0.0430 mol/L·min
(b) 1. first-order
2. rate = $k[H_2O_2]$
3. $k = 5.17 \times 10^{-4}$ min^{-1}; $t_{1/2}$ = 22.3 hours
(c) 35.4 hours
(d) 68 kJ/mol
(e) 249 hours
(f) 1. 2.35×10^{-9} min
2. 1.48×10^{8} mol/L·min
3. 5.71×10^{11} times faster, or about 500 trillion times faster
(g) $H_2O_2(aq) + 2I^-(aq) \longrightarrow I_2(aq) + 2OH^-(aq)$;
rate = $k[H_2O_2] \times [I^-]$

Questions and Problems

Even-numbered questions and Challenge Problems have answers in Appendix 5 and fully worked solutions in the *Student Solutions Manual*.

11-1 Meaning of Reaction Rate

1. Express the rate of the reaction

$$2N_2H_4\ (l) + N_2O_4\ (l) \longrightarrow 3N_2\ (g) + 4H_2O(g)$$

in terms of
(a) $\Delta[N_2O_4]$ (b) $\Delta[N_2]$

2. Express the rate of the reaction

$$2N_2O(g) \longrightarrow 2N_2(g) + O_2(g)$$

in terms of
(a) $\Delta[N_2O]$ (b) $\Delta[O_2]$

3. Consider the following hypothetical reaction:

$$X(g) \longrightarrow Y(g)$$

A 200.0-mL flask is filled with 0.120 moles of X. The disappearance of X is monitored at timed intervals. Assume that temperature and volume are kept constant. The data obtained are shown in the table below.

Time (min)	0	20	40	60	80
moles of X	0.120	0.103	0.085	0.071	0.066

(a) Make a similar table for the appearance of Y.
(b) Calculate the average disappearance of X in *M*/s in the first two 20-minute intervals.
(c) What is the average rate of appearance of Y between the 20- and 60-minute intervals?

4. Consider the following hypothetical reaction:

$$2AB_2(g) \longrightarrow A_2(g) + 2B_2(g)$$

A 500.0-mL flask is filled with 0.384 mol of AB_2. The appearance of A_2 is monitored at timed intervals. Assume that temperature and volume are kept constant. The data obtained are shown in the table below.

Time (min)	0	10	20	30	40	50
moles of A_2	0	0.0541	0.0833	0.1221	0.1432	0.1567

(a) Make a similar table for the disappearance of AB_2.
(b) What is the average rate of disappearance of AB_2 over the second and third 10-minute intervals?
(c) What is the average rate of appearance of A_2 between $t = 30$ and $t = 50$?

5. Consider the combustion of ethane:

$$2C_2H_6(g) + 7O_2(g) \longrightarrow 4CO_2(g) + 6H_2O(g)$$

If the ethane is burning at the rate of 0.20 mol/L·s, at what rates are CO_2 and H_2O being produced?

6. For the reaction

$$5Br^-(aq) + BrO_3^-(aq) + 6H^+(aq) \longrightarrow 3Br_2(aq) + 3H_2O$$

It was found that at a particular instant, bromine was being formed at the rate of 0.029 mol/L·s. At that instant, at what rate
(a) are the hydrogen ions being consumed?
(b) is water being formed?
(c) are the bromide ions being consumed?

7. Nitrosyl chloride (NOCl) decomposes to nitrogen oxide and chlorine gases.
(a) Write a balanced equation using smallest whole-number coefficients for the decomposition.
(b) Write an expression for the reaction rate in terms of $\Delta[NOCl]$.
(c) The concentration of NOCl drops from 0.580 *M* to 0.238 *M* in 8.00 min. Calculate the average rate of reaction over this time interval.

8. Ammonia is produced by the reaction between nitrogen and hydrogen gases.
(a) Write a balanced equation using smallest whole-number coefficients for the reaction.
(b) Write an expression for the rate of reaction in terms of $\Delta[NH_3]$.
(c) The concentration of ammonia increases from 0.257 *M* to 0.815 *M* in 15.0 min. Calculate the average rate of reaction over this time interval.

9. Experimental data are listed for the following hypothetical reaction:

$$A + B \longrightarrow products$$

Time (s)	0	20	40	60	80	100
[B]	0.51	0.43	0.39	0.35	0.33	0.31

(a) Plot these data as in Figure 11.3.
(b) Draw a tangent to the curve to find the instantaneous rate at 60 seconds.
(c) Find the average rate over the 20 to 80 second interval.
(d) Compare the instantaneous rate at 60 seconds with the average rate over the 60-second interval.

10. Experimental data are listed for the hypothetical reaction

$$X \longrightarrow Y + Z$$

Time (s)	0	10	20	30	40	50
[X]	0.0038	0.0028	0.0021	0.0016	0.0012	0.00087

(a) Plot these data as in Figure 11.3.
(b) Draw a tangent to the curve to find the instantaneous rate at 40 s.
(c) Find the average rate over the 10 to 50 s interval.
(d) Compare the instantaneous rate at 40 s with the average rate over the 40-s interval.

11-2 Reaction Rate and Concentration

11. A reaction has two reactants X and Y. What is the order with respect to each reactant and the overall order of the reaction described by the following rate expressions?
    (a) rate = $k_1[X][Y]^2$ (b) rate = $k_2[X]^2$
    (c) rate = $k_3[X][Y]$ (d) rate = k_4

12. A reaction has two reactants Q and P. What is the order with respect to each reactant and the overall order of the reaction described by the following rate expressions?
    (a) rate = k_1 (b) rate = $k_2[P]^2[Q]$
    (c) rate = $k_3[Q]^2$ (d) rate = $k_4[P][Q]$

13. What will the units of the rate constants in Question 11 be if the rate is expressed in mol/L · min?

14. What will the units of the rate constants in Question 12 be if the rate is expressed in mol/L · min?

15. Consider the reaction

$$Z \longrightarrow products$$

The data below give the rate of the reaction and the concentration of Z (as M).

rate (M/min)	[Z]
0.02	0.1
0.08	0.2
0.32	0.4
1.28	0.8

(a) What is the order of the reaction with respect to Z?
(b) What is the rate expression for the reaction?
(c) Estimate the value of k.

16. Consider the reaction

$$Y \longrightarrow products$$

The graph below plots the rate of the reaction versus the concentration of Y.
(a) What is the order of the reaction with respect to Y?
(b) Write the rate expression for the reaction.
(c) Estimate the value of k.

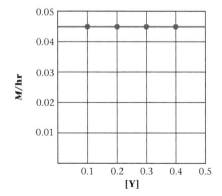

17. Complete the following table for the reaction

$$2R(g) + 3S(g) \longrightarrow products$$

that is first-order in R and second-order in S.

	[R]	[S]	k (L²/mol²·min)	Rate (mol/L·min)
(a)	0.200	0.200	1.49	————
(b)	————	0.633	0.42	0.833
(c)	0.100	————	0.298	0.162
(d)	0.0500	0.0911	————	0.00624

18. Complete the following table for the reaction below. It is first-order in both X and Y.

$$2X(g) + Y(g) \longrightarrow products$$

	[X]	[Y]	k (L/mol·h)	rate (mol/L·h)
(a)	0.100	0.400	1.89	————
(b)	0.600	————	0.884	0.159
(c)	————	0.250	13.4	0.0479
(d)	0.600	0.233	————	0.00112

19. The decomposition of nitrogen dioxide is a second-order reaction. At 550 K, a 0.250 M sample decomposes at the rate of 1.17 mol/L · min.
(a) Write the rate expression.
(b) What is the rate constant at 550 K?
(c) What is the rate of decomposition when $[NO_2] = 0.800\ M$?

20. The decomposition of ammonia on tungsten at 1100°C is zero-order with a rate constant of 2.5×10^{-4} mol/L · min.
(a) Write the rate expression.
(b) Calculate the rate when $[NH_3] = 0.075\ M$.
(c) At what concentration of ammonia is the rate equal to the rate constant?

21. The reaction

$$ICl(g) + \tfrac{1}{2}H_2(g) \longrightarrow \tfrac{1}{2}I_2(g) + HCl(g)$$

is first-order in both reactants. At a certain temperature, the rate of the reaction is 2.86×10^{-5} mol/L·s when [ICl] = 0.0500 M and $[H_2]$ = 0.0150 M.
(a) What is the value of k at that temperature?
(b) At what concentration of hydrogen is the rate 2.66×10^{-4} mol/L · s and [ICl] = 0.100 M?
(c) What is [ICl] when the rate is 0.0715 mol/L·s and H_2 = 3[ICl]?

22. The hypothetical reaction

$$X(g) + \tfrac{1}{2}Y(g) \longrightarrow products$$

is first-order in X and second-order in Y. The rate of the reaction is 0.00389 mol/L · min when [X] is 0.150 M and [Y] is 0.0800 M.
(a) What is the value for k?
(b) At what concentration of [Y] is the rate 0.00948 mol/L · min and [X] is 0.0441 M?
(c) At what concentration of [X] is the rate 0.0124 mol/L · min and [Y] = 2[X]?

11-3 Reaction Concentration and Time

Determination of Reaction Order

23. For a reaction involving the decomposition of Z at a certain temperature, the following data are obtained:

Rate (mol/ L · min)	6.27×10^{-3}	5.33×10^{-3}	4.58×10^{-3}	3.54×10^{-3}
[Z]	0.700	0.645	0.598	0.526

(a) What is the order of the reaction?
(b) Write the rate expression for the decomposition of Z.
(c) Calculate k for the decomposition at that temperature.

24. For a reaction involving the decomposition of Y, the following data are obtained:

Rate (mol/L · min)	0.288	0.245	0.202	0.158
[Y]	0.200	0.170	0.140	0.110

(a) Determine the order of the reaction.
(b) Write the rate expression for the decomposition of Y.
(c) Calculate k for the experiment above.

25. When boron trifluoride reacts with ammonia, the following reaction occurs:

$$BF_3(g) + NH_3(g) \longrightarrow BF_3NH_3(g)$$

The following data are obtained at a particular temperature:

Expt.	[BF$_3$]	[NH$_3$]	Initial Rate (mol/L · s)
1	0.100	0.100	0.0341
2	0.200	0.233	0.159
3	0.200	0.0750	0.0512
4	0.300	0.100	0.102

(a) What is the order of the reaction with respect to BF_3, NH_3, and overall?
(b) Write the rate expression for the reaction.
(c) Calculate k for the reaction.
(d) When [BF$_3$] = 0.533 M and NH$_3$ = 0.300 M, what is the rate of the reaction at the temperature of the experiment?

26. When nitrogen dioxide reacts with carbon monoxide, the following reaction occurs.

$$NO_2(g) + CO(g) \longrightarrow NO(g) + CO_2(g)$$

The following data are obtained at a certain temperature:

Expt.	[NO$_2$]	[CO]	Initial Rate (mol/L · s)
1	0.138	0.100	0.00565
2	0.189	0.200	0.0106
3	0.276	0.100	0.0226
4	0.276	0.300	0.0226

(a) What is the order of the reaction with respect to NO$_2$, CO, and overall?
(b) Write the rate expression of the reaction.
(c) Calculate k for the reaction.

(d) When [NO$_2$] = 0.421 M and [CO] = 0.816 M, what is the rate of the reaction at the temperature of the experiments?

27. Hydrogen bromide is a highly reactive and corrosive gas used mainly as a catalyst for organic reactions. It is produced by reacting hydrogen and bromine gases together.

$$H_2(g) + Br_2(g) \longrightarrow 2HBr(g)$$

The rate is followed by measuring the intensity of the orange color of the bromine gas. The following data are obtained:

Expt.	[H$_2$]	[Br$_2$]	Initial Rate (mol/L · s)
1	0.100	0.100	4.74×10^{-3}
2	0.100	0.200	6.71×10^{-3}
3	0.250	0.200	1.68×10^{-2}

(a) What is the order of the reaction with respect to hydrogen, bromine, and overall?
(b) Write the rate expression for the reaction.
(c) Calculate k for the reaction. What are the units for k?
(d) When [H$_2$] = 0.455 M and [Br$_2$] = 0.215 M, what is the rate of the reaction?

28. Diethylhydrazine reacts with iodine according to the following equation:

$$(C_2H_5)_2(NH)_2(l) + I_2(aq) \longrightarrow (C_2H_5)_2 N_2(l) + 2HI(aq)$$

The rate of the reaction is followed by monitoring the disappearance of the purple color due to iodine. The following data are obtained at a certain temperature.

Expt.	[(C$_2$H$_5$)$_2$(NH)$_2$]	[I$_2$]	Initial Rate (mol/L · h)
1	0.150	0.250	1.08×10^{-4}
2	0.150	0.3620	1.56×10^{-4}
3	0.200	0.400	2.30×10^{-4}
4	0.300	0.400	3.44×10^{-4}

(a) What is the order of the reaction with respect to diethylhydrazine, iodine, and overall?
(b) Write the rate expression for the reaction.
(c) Calculate k for the reaction.
(d) What must [(C$_2$H$_5$)$_2$(NH)$_2$] be so that the rate of the reaction is 5.00×10^{-4} mol/L · h when [I$_2$] = 0.500 M?

29. The equation for the reaction between iodide and bromate ions in acidic solution is

$$6I^-(aq) + BrO_3^-(aq) + 6H^+(aq) \longrightarrow 3I_2(aq) + Br^-(aq) + 3H_2O$$

The rate of the reaction is followed by measuring the appearance of I$_2$. The following data are obtained:

[I$^-$]	[BrO$_3^-$]	[H$^+$]	Initial Rate (mol/L · s)
0.0020	0.0080	0.020	8.89×10^{-5}
0.0040	0.0080	0.020	1.78×10^{-4}
0.0020	0.0160	0.020	1.78×10^{-4}
0.0020	0.0080	0.040	3.56×10^{-4}
0.0015	0.0040	0.030	7.51×10^{-5}

(a) What is the order of the reaction with respect to each reactant?
(b) Write the rate expression for the reaction.
(c) Calculate k.
(d) What is the hydrogen ion concentration when the rate is 5.00×10^{-4} mol/L·s and $[I^-] = [BrO_3^-] = 0.0075 \ M$?

30. Consider the reaction between nitrogen oxide and oxygen.

$$2NO(g) + O_2(g) \longrightarrow 2NO_2(g)$$

The rate of the reaction is followed by measuring the appearance of NO_2. The following data are obtained:

Expt.	[NO]	[O₂]	Initial Rate (mol/L·min)
1	0.0244	0.0372	2.20×10^{-3}
2	0.0244	0.0122	7.23×10^{-4}
3	0.0244	0.0262	1.55×10^{-3}
4	0.0732	0.0372	1.98×10^{-2}

(a) What is the order of the reaction with respect to each reactant?
(b) Write the rate expression for the reaction.
(c) Calculate k for the reaction. What are the units for k?
(d) What is the rate of the reaction at the temperature of the experiment when [NO] = 0.0100 M and $[O_2]$ = 0.0462 M?

31. In a solution at a constant H^+ concentration, iodide ions react with hydrogen peroxide to produce iodine.

$$H^+(aq) + I^-(aq) + \tfrac{1}{2}H_2O_2(aq) \longrightarrow \tfrac{1}{2}I_2(aq) + H_2O$$

The reaction rate can be followed by monitoring the appearance of I_2. The following data are obtained:

[I⁻]	[H₂O₂]	Initial Rate (mol/L·min)
0.015	0.030	0.0022
0.035	0.030	0.0052
0.055	0.030	0.0082
0.035	0.050	0.0087

(a) Write the rate expression for the reaction.
(b) Calculate k.
(c) What is the rate of the reaction when 25.0 mL of a 0.100 M solution of KI is added to 25.0 mL of a 10.0% by mass solution of H_2O_2 (d = 1.00 g/mL)? Assume volumes are additive.

32. Consider the reaction

$$CH_3CO_2CH_3(aq) + OH^-(aq) \longrightarrow CH_3CO_2^-(aq) + CH_3OH(aq)$$

The following data are obtained at a certain temperature:

Expt.	[CH₃CO₂CH₃]	[OH⁻]	Initial Rate (mol/L·s)
1	0.050	0.120	0.00158
2	0.050	0.154	0.00203
3	0.084	0.154	0.00340
4	0.084	0.200	0.00442

(a) Write the rate expression for the reaction.
(b) Calculate k.

(c) What is the rate of the reaction when 10.00 mL of CH_3COOCH_3 (d = 0.932 g/mL) is added to 75.00 mL of 1.50 M NaOH? (Assume that volumes are additive.)

33. Nitrosyl bromide decomposes to nitrogen oxide and bromine. Use the following data to determine the order of the decomposition of nitrosyl bromide.

Time (s)	0	6	12	18	24
[NOBr]	0.0286	0.0253	0.0229	0.0208	0.0190

34. Consider the decomposition of Q. Use the following data to determine the order of the decomposition.

Time (min)	0	4	8	12	16
[Q]	0.334	0.25	0.20	0.167	0.143

First-Order Reactions

35. Azomethane decomposes into nitrogen and ethane at high temperatures according to the following equation:

$$(CH_3)_2N_2(g) \longrightarrow N_2(g) + C_2H_6(g)$$

The following data are obtained in an experiment:

Time (h)	[(CH₃)₂N₂]
1.00	0.905
2.00	0.741
3.00	0.607
4.00	0.497

(a) By plotting the data, show that the reaction is first-order.
(b) From the graph, determine k.
(c) Using k, find the time (in hours) that it takes to decrease the concentration to 0.100 M.
(d) Calculate the rate of the reaction when $[(CH_3)_2N_2]$ = 0.415 M.

36. The decomposition of sulfuryl chloride, SO_2Cl_2, to sulfur dioxide and chlorine gases is a first-order reaction. The following kinetic data for the decomposition at a certain temperature are given below:

Time (min)	0	10	30	50	70
[SO₂Cl₂]	0.0100	0.00987	0.00962	0.00937	0.00913

(a) Plot the data to show that the reaction is first-order.
(b) From the graph, determine k.
(c) What is the half-life of the decomposition?
(d) Using k from (b), find the time it takes to decrease the concentration to 0.00427 M.
(e) What is the rate of decomposition when $[SO_2Cl_2]$ = 0.0153 M?

37. The first-order rate constant for the decomposition of a certain hormone in water at 25°C is 3.42×10^{-4} day^{-1}.
(a) If a 0.0200 M solution of the hormone is stored at 25°C for two months, what will its concentration be at the end of that period?
(b) How long will it take for the concentration of the solution to drop from 0.0200 M to 0.00350 M?
(c) What is the half-life of the hormone?

38. Consider the first-order decomposition of phosgene at a certain temperature.

$$COCl_2(g) \longrightarrow products$$

It is found that the concentration of phosgene is 0.0450 *M* after 300 seconds and 0.0200 *M* after 500 seconds. Calculate the following:
  (a) the rate constant at the temperature of the decomposition
  (b) the half-life of the decomposition
  (c) the intial concentration of phosgene

39. The decomposition of azomethane, $(CH_3)_2N_2$, to nitrogen and ethane gases is a first-order reaction,

$$(CH_3)_2N_2\ (g) \rightarrow N_2\ (g) + C_2H_6\ (g).$$

At a certain temperature, a 29-mg sample of azomethane is reduced to 12 mg in 1.4 s.
  (a) What is the rate constant *k* for the decomposition at that temperature?
  (b) What is the half-life of the decomposition?
  (c) How long will it take to decompose 78% of the azomethane?

40. The first-order rate constant for the decomposition of a certain drug at 25°C is 0.215 month^{-1}.
  (a) If 10.0 g of the drug is stored at 25°C for one year, how many grams of the drug will remain at the end of the year?
  (b) What is the half-life of the drug?
  (c) How long will it take to decompose 65% of the drug?

41. In the first-order decomposition of acetone at 500°C

$$CH_3COCH_3(g) \longrightarrow products$$

It is found that it takes 12.7 h to decompose 0.0622 *M* acetone to 0.0133 *M*.
  (a) Find *k* at 500°C.
  (b) What is the rate of decomposition when $[CH_3COCH_3] = 0.0500\ M$?
  (c) How many minutes will it take to decompose enough acetone so that 68% remains?

42. The decomposition of sulfuryl chloride, SO_2Cl_2, to sulfur dioxide and chlorine gases is a first-order reaction. It is found that at a certain temperature, it takes 1.43 hours to decompose 0.0714 *M* to 0.0681 *M*.
  (a) What is the rate constant for the decomposition?
  (b) What is the rate of decomposition when $[SO_2Cl_2] = 0.0462\ M$?
  (c) How long will it take to decompose SO_2Cl_2 so that 45% remains?

43. Dinitrogen pentoxide gas decomposes to form nitrogen dioxide and oxygen. The reaction is first-order and has a rate constant of 0.247 h^{-1} at 25°C. If a 2.50-L flask originally contains N_2O_5 at a pressure of 756 mm Hg at 25°C, then how many moles of O_2 are formed after 135 minutes? (*Hint*: First write a balanced equation for the decomposition.)

44. Sucrose ($C_{12}H_{22}O_{11}$) hydrolyzes into glucose and fructose. The hydrolysis is a first-order reaction. The half-life for the hydrolysis of sucrose is 64.2 min at 25°C. How many grams of sucrose in 1.25 L of a 0.389 *M* solution are hydrolyzed in 1.73 hours?

45. Iodine-131 is used to treat tumors in the thyroid. Its decomposition is first-order with a half-life of 8.1 days. If a patient is given a sample containing 5.00 mg of I-131, how long will it take for 32% of the isotope to remain in her system?

46. Cesium-131 is the latest tool of nuclear medicine. It is used to treat malignant tumors by implanting Cs-131 directly into the tumor site. Its first-order half-life is 9.7 days. If a patient is implanted with 20.0 mg of Cs-131, how long will it take for 33% of the isotope to remain in his system?

47. Bromine-82 is used to study the flow and distribution of waste water. A sample contains 0.500 mg of Br-82. After 24 hours, 0.306 mg of Br-82 remains. What is the half-life of Br-82?

48. A sample of sodium-24 chloride contains 0.050 mg of Na-24 to study the sodium balance of an animal. After 24.9 h, 0.016 mg of Na-24 is left. What is the half-life of Na-24?

Zero- and Second-Order Reactions

49. The decomposition of A at 85°C is a zero-order reaction. It takes 35 minutes to decompose 37% of an initial mass of 282 mg.
  (a) What is *k* at 85°C?
  (b) What is the half-life of 282 mg at 85°C?
  (c) What is the rate of decomposition for 282 mg at 85°C?
  (d) If one starts with 464 mg, what is the rate of its decomposition at 85°C?

50. The decomposition of R at 33°C is a zero-order reaction. It takes 128 minutes to decompose 41.0% of an initial mass of 739 mg at 33°C. At 33°C,
  (a) what is *k*?
  (b) what is the half-life of 739 mg?
  (c) what is the rate of decomposition for 739 mg?
  (d) what is the rate of decomposition if one starts with an initial amount of 1.25 g?

51. For the zero-order decomposition of HI on a gold surface

$$HI(g) \xrightarrow{\ Au\ } \tfrac{1}{2}H_2(g) + \tfrac{1}{2}I_2(g)$$

it takes 16.0 s for the pressure of HI to drop from 1.00 atm to 0.200 atm.
  (a) What is the rate constant for the reaction?
  (b) How long will it take for the pressure to drop from 0.150 atm to 0.0432 atm?
  (c) What is the half-life of HI at a pressure of 0.500 atm?

52. For the zero-order decomposition of ammonia on tungsten

$$NH_3(g) \xrightarrow{\ W\ } \tfrac{1}{2}N_2(g) + \tfrac{3}{2}H_2(g)$$

the rate constant is 2.08×10^{-4} mol/L·s.
  (a) What is the half-life of a 0.250 *M* solution of ammonia?
  (b) How long will it take for the concentration of ammonia to drop from 1.25 *M* to 0.388 *M*?

53. Ammonium cyanate, NH_4NCO, in water rearranges to produce urea, a common fertilizer, $(NH_2)_2CO$:

$$NH_4NCO(aq) \longrightarrow (NH_2)_2CO(aq)$$

The rearrangement is a second-order reaction. It takes 11.6 h for the concentration of NH_4NCO to go from 0.250 *M* to 0.0841 *M*.
  (a) What is *k* for the reaction?
  (b) What is the half-life of the reaction when NH_4NCO is 0.100 *M*?

(c) How long will it take to rearrange 39% of a 0.450 M solution?

(d) How fast is a 0.839 M solution being changed to urea?

54. Butadiene, C_4H_6, dimerizes according to the following reaction:

$$C_4H_6(g) \longrightarrow C_8H_{12}(g)$$

The dimerization is a second-order reaction. It takes 145 s for the concentration of C_4H_6 to go from 0.350 M to 0.197 M.

(a) What is k for the dimerization?

(b) What is the half-life of the reaction when butadiene is 0.200 M?

(c) How long will it take to dimerize 28.9% of a 0.558 M sample?

(d) How fast is 0.128 M butadiene dimerizing?

55. The rate constant for the second-order reaction

$$NOBr(g) \longrightarrow NO(g) + \tfrac{1}{2}Br_2(g)$$

is 48 L/mol · min at a certain temperature. How long will it take to decompose 90.0% of a 0.0200 M solution of nitrosyl bromide?

56. The decomposition of nitrosyl chloride

$$NOCl(g) \longrightarrow NO(g) + \tfrac{1}{2}Cl_2(g)$$

is a second-order reaction. If it takes 0.20 min to decompose 15% of a 0.300 M solution of nitrosyl chloride, what is k for the reaction?

11-4 Models for Reaction Rate

11-5 Reaction Rate and Temperature

57. An increase in temperature from 23°C to 36°C increases the rate constant of a certain reaction by 50.0%. What is the activation energy of the reaction?

58. If the activation energy of a reaction is 9.13 kJ, then what is the percent increase in the rate constant when the temperature is increased from 27°C to 69°C?

59. The following data are obtained for the gas-phase decomposition of acetaldehyde:

k (L/mol·s)	0.0105	0.101	0.60	2.92
T (K)	700	750	800	850

Plot these data (ln k versus 1/T) and find the activation energy for the reaction.

60. The following data are obtained for the reaction

$$SiH_4(g) \longrightarrow Si(s) + 2H_2(g)$$

k (s^{-1})	0.048	2.3	49	590
t (°C)	500	600	700	800

Plot these data (ln k versus 1/T) and find the activation energy for the reaction.

61. Consider the following hypothetical reaction:

$$X + Y \longrightarrow \text{Products} \qquad \Delta H = -75 \text{ kJ}$$

Draw a reaction-energy diagram for the reaction if the activation energy is 32 kJ.

62. For the reaction

$$Q + R \longrightarrow Y + Z \qquad \Delta H = 128 \text{ kJ}$$

Draw a reaction-energy diagram for the reaction if its activation energy is 284 kJ.

63. The uncoiling of deoxyribonucleic acid (DNA) is a first-order reaction. Its activation energy is 420 kJ. At 37°C, the rate constant is 4.90×10^{-4} min^{-1}.

(a) What is the half-life of the uncoiling at 37°C (normal body temperature)?

(b) What is the half-life of the uncoiling if the organism has a temperature of 40°C (≈ 104°F)?

(c) By what factor does the rate of uncoiling increase (per °C) over this temperature interval?

64. The precipitation of egg albumin in water at 100°C has an activation energy of 52.0 kJ/mol. By what percent does the rate of precipitation decrease if the water is at 92°C?

65. Consider the hypothetical decomposition

$$Z \longrightarrow \text{products}$$

The rate of the reaction as a function of temperature in M/min is

$$\text{rate} = 2.7t - 19$$

where t is the temperature in °C.

(a) Calculate the rate of decomposition at 17°C and at 27°C.

(b) Estimate the activation energy of the reaction.

(c) What is the percent increase in rate for a 10°C increase in temperature?

66. The chirping rate of a cricket, X, in chirps per minute, near room temperature is

$$X = 7.2t - 32$$

where t is the temperature in °C.

(a) Calculate the chirping rates at 25°C and 35°C.

(b) Use your answers in (a) to estimate the activation energy for the chirping.

(c) What is the percentage increase for a 10°C rise in temperature?

67. For the reaction

$$2N_2O(g) \longrightarrow 2N_2(g) + O_2(g)$$

the rate constant is 0.066 L/mol · min at 565°C and 22.8 L/mol · min at 728°C.

(a) What is the activation energy of the reaction?

(b) What is k at 485°C?

(c) At what temperature is k, the rate constant, equal to 11.6 L/mol · min?

68. For the decomposition of a peroxide, the activation energy is 17.4 kJ/mol. The rate constant at 25°C is 0.027 s^{-1}.

(a) What is the rate constant at 65°C?

(b) At what temperature will the rate constant be 25% greater than the rate constant at 25°C?

69. Consider a 5.000 M solution of the hypothetical compound R. Its decomposition is a second-order reaction and its activation energy is 57.6 kJ/mol. At 759°C, the half-life of the 5.000 M solution is 8.49 h. What is the half-life of the same solution at 700°C (3 significant figures)?

70. The decomposition of N_2O_5 to NO_2 and NO_3 is a first-order gas-phase reaction. At 25°C, the reaction has a half-life of 2.81 s. At 45°C, the reaction has a half-life of 0.313 s. What is the activation energy of the reaction?

11-6 Catalysis

71. For a certain reaction, E_a is 135 kJ and $\Delta H = 45$ kJ. In the presence of a catalyst, the activation energy is 39% of that for the uncatalyzed reaction. Draw a diagram similar to Figure 11.14 but instead of showing two activated complexes (two humps) show only one activated complex (i.e., only one hump) for the reaction. What is the activation energy of the uncatalyzed reverse reaction?

72. Consider a reaction in which $E_a = 129$ kJ and $\Delta H = -29$ kJ. In the presence of a catalyst, the activation energy is 48% of the uncatalyzed reaction. Follow the directions in Question 71 in drawing an energy diagram.

73. A catalyst lowers the activation energy of a reaction from 215 kJ to 206 kJ. By what factor would you expect the reaction-rate constant to increase at 25°C? Assume that the frequency factors (A) are the same for both reactions. (*Hint:* Use the formula $\ln k = \ln A - E_a/RT$.)

74. A reaction has an activation energy of 363 kJ at 25°C. If the rate constant has to increase ten-fold, what should the activation energy of the catalyzed reaction be? (See Question 73 for assumptions and a hint.)

11-7 Reaction Mechanisms

75. Write the rate expression for each of the following elementary steps:
  (a) $NO_3 + CO \longrightarrow NO_2 + CO_2$
  (b) $I_2 \longrightarrow 2I$
  (c) $NO + O_2 \longrightarrow NO_3$

76. Write the rate expression for each of the following elementary steps:
  (a) $Cl_2 \longleftrightarrow 2Cl$
  (b) $N_2O_2 + O_2 \longrightarrow 2NO_2$
  (c) $I^- + HClO \longrightarrow HIO + Cl^-$

77. For the reaction between hydrogen and iodine,

$$H_2(g) + I_2(g) \longrightarrow 2HI(g)$$

the experimental rate expression is rate $= k[H_2][I_2]$. Show that this expression is consistent with the mechanism

$$I_2(g) \rightleftharpoons 2I(g) \qquad \text{(fast)}$$
$$H_2(g) + I(g) + I(g) \longrightarrow 2HI(g) \qquad \text{(slow)}$$

78. For the reaction

$$2H_2(g) + 2NO(g) \longrightarrow N_2(g) + 2H_2O(g)$$

the experimental rate expression is rate $= k[NO]^2[H_2]$. The following mechanism is proposed:

$$2NO \rightleftharpoons N_2O_2 \qquad \text{(fast)}$$
$$N_2O_2 + H_2 \longrightarrow H_2O + N_2O \qquad \text{(slow)}$$
$$N_2O + H_2 \longrightarrow N_2 + H_2O \qquad \text{(fast)}$$

Is this mechanism consistent with the rate expression?

79. At low temperatures, the rate law for the reaction

$$CO(g) + NO_2(g) \longrightarrow CO_2(g) + NO(g)$$

is as follows: rate $=$ (constant) $[NO_2]^2$. Which of the following mechanisms is consistent with the rate law?

  (a) $CO + NO_2 \longrightarrow CO_2 + NO$
  (b) $2NO_2 \rightleftharpoons N_2O_4$ (fast)
       $N_2O_4 + 2CO \longrightarrow 2CO_2 + 2NO$ (slow)
  (c) $2NO_2 \longrightarrow NO_3 + NO$ (slow)
       $NO_3 + CO \longrightarrow NO_2 + CO_2$ (fast)
  (d) $2NO_2 \longrightarrow 2NO + O_2$ (slow)
       $O_2 + 2CO \longrightarrow 2CO_2$ (fast)

80. Two mechanisms are proposed for the reaction

$$2NO(g) + O_2(g) \longrightarrow 2NO_2(g)$$

Mechanism 1: $NO + O_2 \rightleftharpoons NO_3$ (fast)

             $NO_3 + NO \longrightarrow 2NO_2$ (slow)

Mechanism 2: $NO + NO \rightleftharpoons N_2O_2$ (fast)

             $N_2O_2 + O_2 \longrightarrow 2NO_2$ (slow)

Show that each of these mechanisms is consistent with the observed rate law: rate $= k[NO]^2[O_2]$.

Unclassified

81. The hypothetical reaction

$$QR + X \longrightarrow \text{products}$$

was monitored at 27°C as a function of time. The following plots are obtained:

 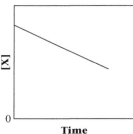

  (a) Write the rate expression for the reaction.
  (b) What is the rate constant if the rate of the reaction is 0.29 mol/L · min when $[QR] = [X] = 1.000$ *M*?
  (c) What is the half-life for the consumption of QR when $[QR] = 0.0866$ *M*?
  (d) If the initial concentration of $[X] = 5.00$ *M*, what is $[X]$ after 4.7 min?

82. When a base is added to an aqueous solution of chlorine dioxide gas, the following reaction occurs:

$$2ClO_2(aq) + 2OH^-(aq) \longrightarrow ClO_3^-(aq) + ClO_2^-(aq) + H_2O$$

The reaction is first-order in OH^- and second-order for ClO_2. Initially, when $[ClO_2] = 0.010$ *M* and $[OH^-] = 0.030$ *M*, the rate of the reaction is 6.00×10^{-4} mol/L · s. What is the rate of the reaction when 50.0 mL of 0.200 *M* ClO_2 and 95.0 mL of 0.155 *M* NaOH are added?

83. The decomposition of sulfuryl chloride, SO_2Cl_2, to sulfur dioxide and chlorine gases is a first-order reaction.

$$SO_2Cl_2(g) \longrightarrow SO_2(g) + Cl_2(g)$$

At a certain temperature, the half-life of SO_2Cl_2 is 7.5×10^2 min. Consider a sealed flask with 122.0 g of SO_2Cl_2.
  (a) How long will it take to reduce the amount of SO_2Cl_2 in the sealed flask to 45.0 g?
  (b) If the decomposition is stopped after 29.0 h, what volume of Cl_2 at 27°C and 1.00 atm is produced?

84. How much faster would a reaction proceed at 46°C than at 28°C if the activation energy of the reaction is 121 kJ/mol?

85. A reaction has an activation energy of 85 kJ. What percentage increase would one get for k for every 5°C increase in temperature around room temperature, 25°C?

86. For the first-order thermal decomposition of ozone

$$O_3(g) \longrightarrow O_2(g) + O(g)$$

$k = 3 \times 10^{-26} \, s^{-1}$ at 25°C. What is the half-life for this reaction in years? Comment on the likelihood that this reaction contributes to the depletion of the ozone layer.

87. A drug decomposes in the blood by a first-order process. A pill containing 0.500 g of the active ingredient reaches its maximum concentration of 2.5 mg/100 mL of blood. If the half-life of the active ingredient is 75 min, what is its concentration in the blood 2.0 h after the maximum concentration has been reached?

88. Page 295 has a two-point equation relating k (the rate constant) and T (temperature). Derive a two-point equation relating k and activation energy for a catalyzed and an uncatalyzed reaction at the same temperature. Assume that A is the same for both reactions.

Conceptual Problems

89. The greatest increase in the reaction rate for the reaction between A and C, where rate = $k[A]^{1/2} [C]$, is caused by
  (a) doubling [A]   (b) halving [C]
  (c) halving [A]   (d) doubling [A] and [C]

90. Consider the reaction between methyl acetate and water:

$$CH_3COOCH_3 \, (aq) + H_2O \longrightarrow CH_3COOH(aq) + CH_3OH(aq)$$

The rate law for the reaction at 25°C is:

$$\text{rate} = k[CH_3COOCH_3][H^+]$$

The reaction is carried out in one liter of a solution that is 0.48 M CH_3COOCH_3 and 0.122 M H^+. Fill in the blanks with increases, decreases, or remains the same.
  (a) The rate of the reaction _____ when 125 mL of water is added.
  (b) The value of k _____ when the temperature is increased to 38°C.
  (c) The value of E_a _____ when HCl is added.
  (d) The rate of the reaction _____ when NaOH is added.
  (e) The value of k _____ when a catalyst is added.

91. Consider the decomposition of A represented by circles, where each circle represents a molecule of A.

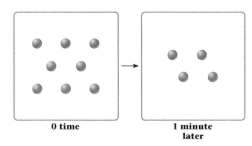

0 time   1 minute later

Is the reaction zero-order, first-order, or second-order?

92. Consider the decomposition reaction

$$2X \longrightarrow 2Y + Z$$

The following graph shows the change in concentration with respect to time for the reaction. What does each of the curves labeled 1, 2, and 3 represent?

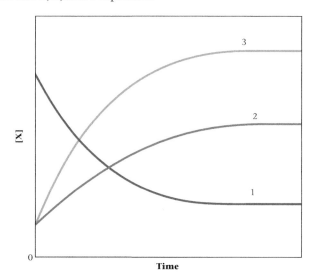

93. Consider the following activation energy diagram.

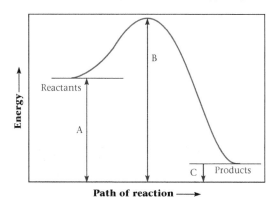

Which of the following statements about the diagram are true?
  (a) E_a (forward) < E_a (reverse)
  (b) B represents the activation energy of the forward reaction.
  (c) Energy (ΔH) for the reaction is A − C.
  (d) E_a (forward) = B − A
  (e) E_a (forward) = E_a (reverse) − C

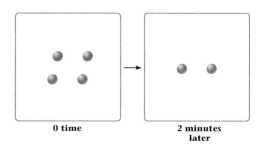

0 time   2 minutes later

94. Three first-order reactions have the following activation energies:

Reaction	A	B	C
E_a(kJ)	75	136	292

(a) Which reaction is the fastest?
(b) Which reaction has the largest half-life?
(c) Which reaction has the largest rate?

95. Consider the first-order decomposition reaction

$$A \longrightarrow B + C$$

where A (circles) decomposes to B (triangles) and C (squares). Given the following boxes representing the reaction at $t = 2$ minutes, fill in the boxes with products at the indicated time. Estimate the half-life of the reaction.

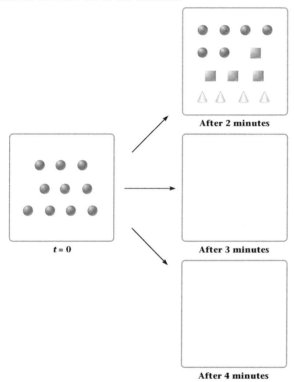

96. Consider the following energy diagram (not to scale) for the reaction.

$$2CH_3(g) \longrightarrow C_2H_6(g)$$

(a) What is the activation energy of the forward reaction?
(b) At what point would $2CH_3$ (g) appear?
(c) What is ΔH for the reaction?
(d) What is the activation energy of the reverse reaction?
(e) At what point would C_2H_6 certainly be found?
(f) What is any species at point C called?

97. For the reaction

$$A + B \longrightarrow C$$

the rate expression is rate = $k[A][B]$
(a) Given three test tubes, with different concentrations of A and B, which test tube has the smallest rate?
  (1) 0.10 M A; 0.10 M B
  (2) 0.15 M A; 0.15 M B
  (3) 0.06 M A; 1.0 M B
(b) If the temperature is increased, describe (using the words *increases, decreases,* or *remains the same*) what happens to the rate, the value of k, and E_a.

98. The following experiments are performed for the first-order reaction:

$$A \longrightarrow \text{products}$$

Fill in the blanks. If the answer cannot be calculated with the given information, write NC on the blank(s) provided.

	Expt. 1	Expt. 2	Expt. 3	Expt. 4
[A]	0.100	0.200	0.100	0.100
Catalyst	No	No	Yes	No
Temperature	25°C	25°C	25°C	30°C
k (min^{-1})	0.5	_____	1	0.6
Rate (mol/L · min)	_____	_____	_____	_____
E_a (kJ)	32	_____	_____	_____

Challenge Problems

99. The gas-phase reaction between hydrogen and iodine

$$H_2(g) + I_2(g) \rightleftharpoons 2HI(g)$$

proceeds with a rate constant for the forward reaction at 700°C of 138 L/mol·s and an activation energy of 165 kJ/mol.
(a) Calculate the activation energy of the reverse reaction given that ΔH_f° for HI is 26.48 kJ/mol and ΔH_f° for I_2(g) is 62.44 kJ/mol.
(b) Calculate the rate constant for the reverse reaction at 700°C. (Assume A in the equation $k = Ae^{-E_a/RT}$ is the same for both forward and reverse reactions.)
(c) Calculate the rate of the reverse reaction if the concentration of HI is 0.200 M. The reverse reaction is second-order in HI.

100. Consider the coagulation of a protein at 100°C. The first-order reaction has an activation energy of 69 kJ/mol. If the protein takes 5.4 minutes to coagulate in boiling water at 100°C, then how long will it take to coagulate the protein at an altitude where water boils at 87°C?

101. For a first-order reaction $aA \longrightarrow$ products, where $a \neq 1$, the rate is $-\Delta[A]/a\,\Delta t$, or in derivative notation, $-\dfrac{1}{a}\dfrac{d\,[A]}{dt}$. Derive the integrated rate law for the first-order decomposition of a moles of reactant.

102. The following data apply to the reaction

$$A(g) + 3B(g) + 2C(g) \longrightarrow products$$

[A]	[B]	[C]	Rate
0.20	0.40	0.10	X
0.40	0.40	0.20	$8X$
0.20	0.20	0.20	X
0.40	0.40	0.10	$4X$

Determine the rate law for the reaction.

103. Using calculus, derive the equation for
(a) the concentration-time relation for a second-order reaction (see Table 11.2).
(b) the concentration-time relation for a third-order reaction, A $\longrightarrow$ product.

104. In a first-order reaction, suppose that a quantity X of a reactant is added at regular intervals of time, Δt. At first the amount of reactant in the system builds up; eventually, however, it levels off at a saturation value given by the expression

$$saturation\ value = \frac{X}{1 - 10^{-a}} \qquad where\ a = 0.30\ \frac{\Delta t}{t_{1/2}}$$

This analysis applies to prescription drugs, of which you take a certain amount each day. Suppose that you take 0.100 g of a drug three times a day and that the half-life for elimination is 2.0 days. Using this equation, calculate the mass of the drug in the body at saturation. Suppose further that side effects show up when 0.500 g of the drug accumulates in the body. As a pharmacist, what is the maximum dosage you could assign to a patient for an 8-h period without causing side effects?

105. Consider the hypothetical first-order reaction

$$2A(g) \rightarrow X(g) + \tfrac{1}{2}Y(g)$$

At a certain temperature, the half-life of the reaction is 19.0 min. A 1.00-L flask contains A with a partial pressure of 622 mm Hg. If the temperature is kept constant, what are the partial pressures of A, X, and Y after 42 minutes?

12

Gaseous Chemical Equilibrium

The painting shows children on a seesaw in a state of dynamic equilibrium. The boy at the center can change the position of equilibrium by moving either to the left or to the right of the seesaw.

Order is not pressure which is imposed on society from without, but an equilibrium which is set up from within.

—JOSÉ ORTEGA Y GASSET

Reactions can often be reversed by changing the temperature or pressure.

Among the topics covered in Chapter 9 was the equilibrium between liquid and gaseous water:

$$H_2O(l) \rightleftharpoons H_2O(g)$$

The state of this equilibrium system at a given temperature can be described in a simple way by citing the equilibrium pressure of water vapor: 0.034 atm at 25°C, 1.00 atm at 100°C, and so on.

Chemical reactions involving gases carried out in closed containers resemble in many ways the $H_2O(l)$–$H_2O(g)$ system. The reactions are reversible; reactants are not completely consumed. Instead, an equilibrium mixture containing both products and reactants is obtained. At equilibrium, forward and reverse reactions take place at the same rate. As a result, the amounts of all species at equilibrium remain constant with time.

Ordinarily, where gaseous chemical equilibria are involved, more than one gaseous substance is present. We might, for example, have a system,

$$a\text{A}(g) + b\text{B}(g) \rightleftharpoons c\text{C}(g) + d\text{D}(g)$$

in which the equilibrium mixture contains two gaseous products (C and D) and two gaseous reactants (A and B). To describe the position of this equilibrium, we must cite the partial pressure of each species, that is, P_C, P_D, P_A, and P_B.

Recall from Chapter 5 that the partial pressure of a gas, i, in a mixture is given by the expression

$$P_i = n_i RT/V$$

It follows that, for an equilibrium mixture confined in a closed container of volume V at temperature T, *the partial pressure (P_i) of each species is directly proportional to the number of moles (n_i) of that species.*

It turns out that there is a relatively simple relationship between the partial pressures of different gases present at equilibrium in a reaction system. This relationship is expressed in terms of a quantity called the *equilibrium constant*, symbol K.* In this chapter, you will learn how to

- write the expression for K ◀ *i* corresponding to any chemical equilibrium (Section 12-2).
- calculate the value for K from experimental data for the equilibrium system (Section 12-3).
- use the value of K to predict the extent to which a reaction will take place (Section 12-4).
- use K, along with Le Châtelier's principle, to predict the result of disturbing an equilibrium system (Section 12-5).

i Don't confuse the *equilibrium* constant K with the *rate* constant k.

12-1 The N₂O₄–NO₂ Equilibrium System

Consider what happens when a sample of N_2O_4, a colorless gas, is placed in a closed, evacuated container at 100°C. Instantly, a reddish-brown color develops. This color is due to nitrogen dioxide, NO_2, formed by decomposition of part of the N_2O_4 (Figure 12.1):

$$N_2O_4(g) \longrightarrow 2NO_2(g)$$

At first, this is the only reaction taking place. As soon as some NO_2 is formed, however, the reverse reaction can occur:

$$2NO_2(g) \longrightarrow N_2O_4(g)$$

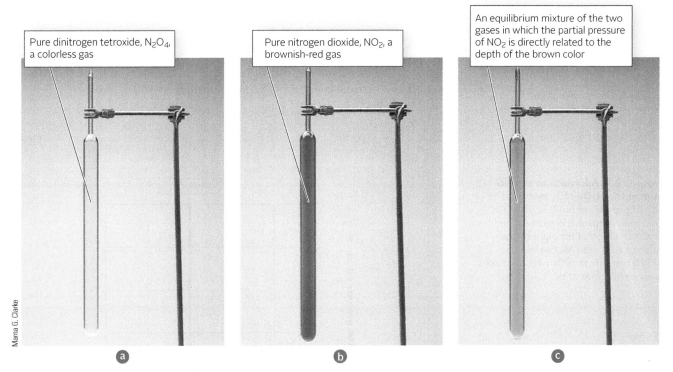

Pure dinitrogen tetroxide, N_2O_4, a colorless gas

Pure nitrogen dioxide, NO_2, a brownish-red gas

An equilibrium mixture of the two gases in which the partial pressure of NO_2 is directly related to the depth of the brown color

Marna G. Clarke

a **b** **c**

Figure 12.1 The N₂O₄–NO₂ system.

---

*Thermodynamically, the quantity that should appear for each species in the expression for K is the *activity* rather than the partial pressure. Activity is defined as the ratio of the equilibrium partial pressure to the standard pressure, 1 atm. This means that the activity and hence the equilibrium constant K are dimensionless, that is, pure numbers without units.

Table 12.1 Establishment of Equilibrium in the System $N_2O_4(g) \rightleftharpoons 2NO_2(g)$ (at 100°C)

Time	0	20	40	60	80	100
$P_{N_2O_4}$ (atm)	1.00	0.60	0.35	0.22*	0.22	0.22
P_{NO_2} (atm)	0.00	0.80	1.30	1.56	1.56	1.56

*Boldface numbers are equilibrium pressures.

Overall, the partial pressure of N_2O_4 drops, and the forward reaction slows down. Conversely, the partial pressure of NO_2 increases, so the rate of the reverse reaction increases. Soon these rates become equal. A dynamic equilibrium has been established.

$$N_2O_4(g) \rightleftharpoons 2NO_2(g)$$

At equilibrium, appreciable amounts of both gases are present. From that point on, the amounts of both NO_2 and N_2O_4 and their partial pressures remain constant, so long as the volume of the container and the temperature remain unchanged.

The characteristics just described are typical of all systems at equilibrium. First, *the forward and reverse reactions are taking place at the same rate.* This explains why *the concentrations of species present remain constant with time.* Moreover, these concentrations are independent of the direction from which equilibrium is approached.

The approach to equilibrium in the N_2O_4–NO_2 system is illustrated by the data in Table 12.1 and by Figure 12.2 (time is in arbitrary units). Originally, only N_2O_4 is present; its pressure is 1.00 atm. Because no NO_2 is around, its original pressure is zero. As equilibrium is approached, the overall reaction is

$$N_2O_4(g) \longrightarrow 2NO_2(g)$$

The amount and hence the partial pressure of N_2O_4 drop, rapidly at first, then more slowly. The partial pressure of NO_2 increases. Finally, both partial pressures level off and become constant. ◀ⓘ At equilibrium, at 100°C,

$$P_{N_2O_4} = 0.22 \text{ atm} \qquad P_{NO_2} = 1.56 \text{ atm}$$

These pressures don't change because forward and reverse reactions occur at the same rate.

Figure 12.2 Approach to equilibrium in the N_2O_4–NO_2 system. The partial pressure of N_2O_4 starts off at 1.00 atm, drops sharply at first, and finally levels off at the equilibrium value of 0.22 atm. Meanwhile, the partial pressure of NO_2 rises from zero to its equilibrium value, 1.56 atm.

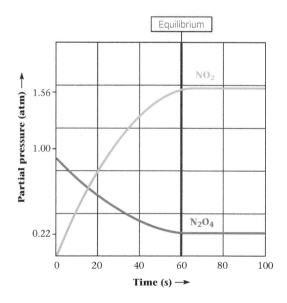

Table 12.2 Pressure Measurements at Equilibrium in the N$_2$O$_4$–NO$_2$ System at 100°C

		Original Pressure (atm)	Equilibrium Pressure (atm)
Expt. 1	N$_2$O$_4$	1.00	0.22
	NO$_2$	0.00	1.56
Expt. 2	N$_2$O$_4$	0.00	0.07
	NO$_2$	1.00	0.86
Expt. 3	N$_2$O$_4$	1.00	0.42
	NO$_2$	1.00	2.16

There are many ways to approach equilibrium in the N$_2$O$_4$–NO$_2$ system. Table 12.2 gives data for three experiments in which the original conditions are quite different. Experiment 1 is that just described, starting with pure N$_2$O$_4$. Experiment 2 starts with pure NO$_2$ at a partial pressure of 1.00 atm. As equilibrium is approached, some of the NO$_2$ reacts to form N$_2$O$_4$:

$$2NO_2(g) \longrightarrow N_2O_4(g)$$

Finally, in Experiment 3, both N$_2$O$_4$ and NO$_2$ are present originally, each at a partial pressure of 1.00 atm.

Looking at the data in Table 12.2, you might wonder whether these three experiments have anything in common. Specifically, is there any relationship between the equilibrium partial pressures of NO$_2$ and N$_2$O$_4$ that is valid for all the experiments? It turns out that there is, although it is not an obvious one. The value of the quotient $(P_{NO_2})^2/P_{N_2O_4}$ is the same, about 11, ◄ *i* in each case:

i Note that this relationship holds only at equilibrium; initial pressures can have any value.

$$\text{Expt. 1} \quad \frac{(P_{NO_2})^2}{P_{N_2O_4}} = \frac{(1.56)^2}{0.22} = 11$$

$$\text{Expt. 2} \quad \frac{(P_{NO_2})^2}{P_{N_2O_4}} = \frac{(0.86)^2}{0.07} = 11$$

$$\text{Expt. 3} \quad \frac{(P_{NO_2})^2}{P_{N_2O_4}} = \frac{(2.16)^2}{0.42} = 11$$

You may wonder why the equilibrium constant, 11, has no units. The reason is that each term in the reaction quotient represents the ratio of the measured pressure of the gas to the thermodynamic standard state of one atmosphere. Thus the quotient $(P_{NO_2})^2/P_{N_2O_4}$ in Experiment 1 becomes

$$K = \frac{\left(\dfrac{1.56 \text{ atm}}{1 \text{ atm}}\right)^2}{\left(\dfrac{0.22 \text{ atm}}{1 \text{ atm}}\right)} = 11$$

This relationship holds for any equilibrium mixture containing N$_2$O$_4$ and NO$_2$ at 100°C. More generally, it is found that, at any temperature, the quantity

$$\frac{(P_{NO_2})^2}{P_{N_2O_4}}$$

where P_{NO_2} and $P_{N_2O_4}$ are equilibrium partial pressures in atmospheres, is a constant, independent of the original composition, the volume of the container, or the total pressure. This constant is referred to as the **equilibrium constant** (K) for the system

$$N_2O_4(g) \rightleftharpoons 2NO_2(g)$$

The equilibrium constant for this system, like all equilibrium constants, changes with temperature. At 100°C, K for the $N_2O_4-NO_2$ system is 11; at 150°C, it has a different value, about 110. Any mixture of NO_2 and N_2O_4 at 100°C will react in such a way that the ratio $(P_{NO_2})^2/P_{N_2O_4}$ becomes equal to 11. At 150°C, reaction occurs until this ratio becomes 110.

12-2 The Equilibrium Constant Expression

For every gaseous chemical system, an *equilibrium constant expression* can be written stating the condition that must be attained at equilibrium. For the general system involving only gases,

$$aA(g) + bB(g) \rightleftharpoons cC(g) + dD(g)$$

where A, B, C, and D represent different substances and *a*, *b*, *c*, and *d* are their coefficients in the balanced equation,

$$K = \frac{(P_C)^c \times (P_D)^d}{(P_A)^a \times (P_B)^b} \qquad (12.1)$$

where P_C, P_D, P_A, and P_B are the partial pressures of the four gases at equilibrium. *These partial pressures must be expressed in atmospheres.* Notice that in the expression for K

- the equilibrium partial pressures of *products* (right side of equation) appear in the *numerator*.
- the equilibrium partial pressures of *reactants* (left side of equation) appear in the *denominator*.
- each partial pressure is raised to a *power* equal to its *coefficient* in the balanced equation.

This equilibrium constant is often given the symbol K_p to emphasize that it involves partial pressures. Other equilibrium expressions for gases are sometimes used, including K_c:

$$K_c = \frac{[C]^c \times [D]^d}{[A]^a \times [B]^b}$$

where the brackets represent equilibrium concentrations in moles per liter. The two constants, K_c and K_p, are simply related (see Problem 78):

$$K_p = K_c(RT)^{\Delta n_g}$$

where Δn_g is the change in the number of moles of gas in the equation, R is the gas law constant, 0.0821 L · atm/mol · K, and T is the Kelvin temperature. Throughout this chapter, we deal only with K_p, referring to it simply as K.

Changing the Chemical Equation

It is important to realize that **the expression for K depends on the form of the chemical equation written to describe the equilibrium system.** ◀ ⓘ To illustrate what this statement means, consider the $N_2O_4-NO_2$ system:

$$N_2O_4(g) \rightleftharpoons 2NO_2(g) \qquad K = \frac{(P_{NO_2})^2}{P_{N_2O_4}}$$

K is meaningless unless accompanied by a chemical equation.

Many other equations could be written for this system, for example,

$$\tfrac{1}{2}N_2O_4(g) \rightleftharpoons NO_2(g)$$

In this case, the expression for the equilibrium constant would be

$$K' = \frac{P_{NO_2}}{(P_{N_2O_4})^{1/2}}$$

Comparing the expressions for K and K', it is clear that K' is the square root of K. This illustrates a general rule, sometimes referred to as the **coefficient rule**, which states,

If the coefficients in a balanced equation are multiplied by a factor n, the equilibrium constant is raised to the nth power:

$$K' = K^n \tag{12.2}$$

In this particular case, $n = \tfrac{1}{2}$ because the coefficients were divided by 2. Hence K' is the square root of K. If $K = 11$, then $K' = (11)^{1/2} = 3.3$.

Another equation that might be written to describe the N_2O_4–NO_2 system is

$$2NO_2(g) \rightleftharpoons N_2O_4(g)$$

for which the equilibrium constant expression is

$$K'' = \frac{P_{N_2O_4}}{(P_{NO_2})^2}$$

This chemical equation is simply the reverse of that written originally; N_2O_4 and NO_2 switch sides of the equation. Notice that K'' is the reciprocal of K; the numerator and denominator have been inverted. This illustrates the **reciprocal rule**:

The equilibrium constants for forward and reverse reactions are the reciprocals of each other,

$$K'' = 1/K \tag{12.3}$$

If, for example, $K = 11$, then $K'' = 1/11 = 0.091$.

Adding Chemical Equations

A property of K that you will find very useful in this and succeeding chapters is expressed by the **rule of multiple equilibria**, ◂ⓘ which states

If a reaction can be expressed as the sum of two or more reactions, K for the overall reaction is the product of the equilibrium constants of the individual reactions.

We will use this relationship frequently in later chapters.

That is, if

$$\text{reaction 3} = \text{reaction 1} + \text{reaction 2}$$

then

$$K \text{ (reaction 3)} = K \text{ (reaction 1)} \times K \text{ (reaction 2)} \tag{12.4}$$

To illustrate the application of this rule, consider the following reactions at 700°C:

$$SO_2(g) + \tfrac{1}{2}O_2(g) \rightleftharpoons SO_3(g) \qquad K = 2.2$$
$$NO_2(g) \rightleftharpoons NO(g) + \tfrac{1}{2}O_2(g) \qquad K = 4.0$$

Adding these equations eliminates $\tfrac{1}{2}O_2$; the result is

$$SO_2(g) + NO_2(g) \rightleftharpoons SO_3(g) + NO(g)$$

For this overall reaction, $K = 2.2 \times 4.0 = 8.8$.

Table 12.3 Dependence of K on the Form of the Chemical Reaction

$$R(g) \Longrightarrow Y(g) \qquad K = \frac{P_Y}{P_R}$$

Form of Equation	K Expression	Relation to K	Rule
$Y(g) \Longrightarrow R(g)$	$K' = \dfrac{P_R}{P_Y}$	$K' = \dfrac{1}{K}$	Reciprocal Rule
$nR(g) \Longrightarrow nY(g)$	$K'' = \dfrac{(P_Y)^n}{(P_R)^n}$	$K'' = K^n$	Coefficient Rule
$R(g) \Longrightarrow A(g)$	$K_1 = \dfrac{P_A}{P_R}$		
$A(g) \Longrightarrow Y(g)$	$K_2 = \dfrac{P_Y}{P_A}$		
$R(g) \Longrightarrow Y(g)$		$K = K_1 \times K_2$	Rule of Multiple Equilibria

The validity of this rule can be demonstrated by writing the expression for K for the individual reactions and multiplying:

$$\frac{P_{SO_3}}{(P_{SO_2})(P_{O_2})^{1/2}} \times \frac{(P_{NO})(P_{O_2})^{1/2}}{P_{NO_2}} = \frac{(P_{SO_3})(P_{NO})}{(P_{SO_2})(P_{NO_2})}$$

The product expression, at the right, is clearly the equilibrium constant for the overall reaction, as it should be according to the rule of multiple equilibria.

Table 12.3 summarizes the rules for writing the expression for the equilibrium constant.

EXAMPLE 12.1

Consider the air pollutants NO and NO_2, contributors to photochemical smog. Both can be formed by the reaction between atmospheric nitrogen and oxygen. At 25°C,

$$N_2(g) + O_2(g) \Longrightarrow 2NO(g) \qquad K_1 = 4.2 \times 10^{-31} \qquad (1)$$
$$N_2(g) + 2O_2(g) \Longrightarrow 2NO_2(g) \qquad K_2 = 1.0 \times 10^{-18} \qquad (2)$$

a   Write the equilibrium constant expression for the formation of two moles of NO at 25°C.

SOLUTION

K expression	$K_1 = \dfrac{(P_{NO})^2}{(P_{N_2})(P_{O_2})}$

b   Calculate K for the formation of one mole of NO at 25°C.

STRATEGY

1. Write the reaction for the formation of one mole of NO. Note that all the coefficients of the original equation are divided by two.

2. Find K. Apply the coefficient rule.

SOLUTION

1. Equation for the reaction	$\frac{1}{2}N_2(g) + \frac{1}{2}O_2(g) \Longrightarrow NO(g)$
2. K	$K = (K_1)^{1/2} = (4.2 \times 10^{-31})^{1/2} = 6.5 \times 10^{-16}$

continued

| c | Calculate K for the decomposition of one mole of NO_2 at 25°C. |

STRATEGY

1. Write the reaction for the decomposition of NO_2 and note that you are asked for the decomposition of *one* mole.
2. To arrive at the equation for the decomposition of one mole of NO_2, you switch sides (reciprocal rule).
3. To arrive at the equation for the decomposition of *one* mole of NO_2, multiply the coefficients by 1/2 (coefficient rule).
4. Apply both rules.

SOLUTION

1. Equation for the reaction	$N_2(g) + 2O_2(g) \rightleftharpoons 2NO_2(g) \longrightarrow 2NO_2(g) \rightleftharpoons N_2(g) + 2O_2(g) \longrightarrow$ $NO_2(g) \rightleftharpoons \frac{1}{2}N_2(g) + O_2(g)$
2. Reciprocal rule	$K' = 1/K_2$
3. Coefficient rule	$K = (K')^{1/2} = (1/K_2)^{1/2}$
4. K	$\left(\dfrac{1}{1.0 \times 10^{-18}}\right)^{1/2} = 1.0 \times 10^9$

| d | Calculate K for the following reaction at 25°C: |

$$2NO(g) + O_2(g) \rightleftharpoons 2NO_2(g)$$

STRATEGY

1. Start with two moles of NO, which is a product in Equation (1) but a reactant in the desired equation. Reverse Equation (1) and apply the reciprocal rule.
2. Focus on two moles of NO_2, which is the product in Equation (2) and in the desired equation. Change is unnecessary.
3. Add the two equations to get the overall equation.
4. Apply the rule of multiple equilibria.

SOLUTION

1. Reverse Equation (1).	$2NO(g) \rightleftharpoons N_2(g) + O_2(g)$	K'
Apply reciprocal rule.	$K' = \dfrac{1}{K_1} = \dfrac{1}{4.2 \times 10^{-31}}$	
2. Keep Equation (2).	$N_2(g) + 2O_2(g) \rightleftharpoons 2NO_2(g)$	$K_2 = 1.0 \times 10^{-18}$
3. Overall equation	$2NO(g) \rightleftharpoons N_2(g) + O_2(g)$	$K' = \dfrac{1}{4.2 \times 10^{-31}}$
	$\dfrac{N_2(g) + 2O_2(g) \rightleftharpoons 2NO_2(g)}{2NO(g) + O_2(g) \rightleftharpoons 2NO_2(g)}$	$\dfrac{K_2 = 1.0 \times 10^{-18}}{K = (K')(K_2)}$
4. K	$K = \left(\dfrac{1}{4.2 \times 10^{-31}}\right)(1.0 \times 10^{-18}) = 2.4 \times 10^{12}$	

Heterogeneous Equilibria

In the reactions described so far, all the reactants and products have been gaseous; the equilibrium systems are *homogeneous*. In certain reactions, at least one of the substances involved is a pure liquid or solid; the others are gases (Figure 12.3). Such a system is *heterogeneous,* because more than one phase is present. Examples include

$$CO_2(g) + H_2(g) \rightleftharpoons CO(g) + H_2O(l)$$
$$I_2(s) \rightleftharpoons I_2(g)$$

Figure 12.3 **A heterogeneous equilibrium system: solid I₂–gaseous I₂.**

Table 12.4 Equilibrium Constant Expressions for the Reaction
$$CO_2(g) + H_2(g) \rightleftharpoons CO(g) + H_2O(l)$$

	Expt. 1	Expt. 2	Expt. 3	Expt. 4
Mass $H_2O(l)$	8 g	6 g	4 g	2 g
P_{H_2O} (atm)	3×10^{-2}	3×10^{-2}	3×10^{-2}	3×10^{-2}
K_I	9×10^{-6}	9×10^{-6}	9×10^{-6}	9×10^{-6}
K_{II}	3×10^{-4}	3×10^{-4}	3×10^{-4}	3×10^{-4}

$$K_I = \frac{P_{CO} \times P_{H_2O}}{P_{CO_2} \times P_{H_2}} \qquad K_{II} = \frac{P_{CO}}{P_{CO_2} \times P_{H_2}}$$

Experimentally, it is found that for such systems

- *the position of the equilibrium is independent of the amount of solid or liquid, as long as some is present.*
- *terms for pure liquids or solids need not appear in the expression for K.*

To see why this is the case, consider the data in Table 12.4 for the system

$$CO_2(g) + H_2(g) \rightleftharpoons CO(g) + H_2O(l)$$

at 25°C. In Experiments 1 through 4, liquid water is present, so the partial pressure of water in the gas phase, P_{H_2O}, is constant at 0.03 atm, the equilibrium vapor pressure of water at 25°C.

Notice that

1. In all cases, the quantity

$$K_I = \frac{P_{CO} \times P_{H_2O}}{P_{CO_2} \times P_{H_2}}$$

   is constant at 9×10^{-6}.

2. In all cases (Experiments 1 through 4), the quantity

$$K_{II} = \frac{P_{CO}}{P_{CO_2} \times P_{H_2}}$$

   is constant at 3×10^{-4}. This must be true because

$$K_{II} = K_I/P_{H_2O}$$

and P_{H_2O} is constant at 3×10^{-2} atm. It follows that K_{II} is a valid equilibrium constant for the system

$$CO_2(g) + H_2(g) \rightleftharpoons CO(g) + H_2O(l)$$

Because the expression for K_{II} is simpler than that for K_I, it is the equilibrium constant of choice for the heterogeneous system.

A similar argument can be applied to the equilibrium involved in the sublimation of iodine:

$$I_2(s) \rightleftharpoons I_2(g)$$

At a given temperature, the pressure of iodine vapor is constant, independent of the amount of solid iodine or any other factor. The equilibrium constant expression is

$$K = P_{I_2}$$

Applying the reciprocal rule, we can deduce the equilibrium constant expression for the reverse reaction

$$I_2(g) \rightleftharpoons I_2(s) \qquad K = 1/P_{I_2}$$

© Cengage Learning/Charles D. Winters

EXAMPLE 12.2

Write the expression for K for

(a) the reduction of black solid copper(II) oxide (1 mol) with hydrogen to form copper metal and steam.

(b) the reaction of one mole of steam with red-hot coke (carbon) to form a mixture of hydrogen and carbon monoxide, called water gas.

STRATEGY

1. Write a balanced chemical equation for the equilibrium system. Do not forget to include physical states.

2. Write the expression for K, leaving out pure solids and liquids.

3. Recall that gases are represented by their partial pressures.

SOLUTION

(a) Reaction	$CuO(s) + H_2(g) \rightleftharpoons Cu(s) + H_2O(g)$	
K expression	$K = \dfrac{P_{H_2O}}{P_{H_2}}$ ◀ⓘ	ⓘ In part (a) of Example 12.2, if the product were $H_2O(l)$, K would equal $1/P_{H_2}$.
(b) Reaction	$H_2O(g) + C(s) \rightleftharpoons CO(g) + H_2(g)$	
K expression	$K = \dfrac{(P_{H_2})(P_{CO})}{P_{H_2O}}$	

In this and succeeding chapters, a wide variety of different types of equilibria will be covered. They may involve gases, pure liquids or solids, and species in aqueous solution. It will always be true that in the expression for the equilibrium constant

- *gases enter as their partial pressures in atmospheres.*
- *pure liquids or solids do not appear; neither does the solvent for a reaction in dilute solution.*
- *species (ions or molecules) in water solution enter as their molar concentrations.*

Thus for the equilibrium attained when zinc metal reacts with acid,

$$Zn(s) + 2H^+(aq) \rightleftharpoons Zn^{2+}(aq) + H_2(g)$$

$$K = \frac{P_{H_2} \times [Zn^{2+}]}{[H^+]^2}$$

We will have more to say in later chapters about equilibria involving species in aqueous solution.

12-3 Determination of K

Numerical values of equilibrium constants can be calculated if the partial pressures of products and reactants at equilibrium are known. Sometimes you will be given equilibrium partial pressures directly (Example 12.3). At other times you will be given the original partial pressures and the equilibrium partial pressure of one species (Example 12.4). In that case, the calculation of K is a bit more difficult, because you have to calculate the equilibrium partial pressures of all the species.

EXAMPLE 12.3

Ammonium chloride is sometimes used as a flux in soldering because it decomposes on heating:

$$NH_4Cl(s) \rightleftharpoons NH_3(g) + HCl(g)$$

The HCl formed removes oxide films from metals to be soldered. In a certain equilibrium system at 400°C, 22.6 g of NH_4Cl is present; the partial pressures of NH_3 and HCl are 2.5 atm and 4.8 atm, respectively. Calculate K at 400°C.

ANALYSIS

Information given:	equation for the reaction at $T = 400°C$ mass of NH_4Cl (22.6 g) partial pressures of NH_3 (2.5 atm) and HCl (4.8 atm)
Asked for:	K at 400°C

STRATEGY

1. Write the expression for K. Recall that pure liquids and solids are not included in the expression.
2. Substitute the partial pressures of the gases into the expression for K.

SOLUTION

1. K expression	$K = (P_{NH_3})(P_{HCl})$
2. K	$K = (2.5 \text{ atm})(4.8 \text{ atm}) = $ 12

END POINT

Note that the mass of NH_4Cl is not relevant to the calculation because NH_4Cl is a solid and hence is not included in the expression for K.

EXAMPLE 12.4

Consider the equilibrium system

$$2HI(g) \rightleftharpoons H_2(g) + I_2(g)$$

Originally, an evacuated flask contains only HI at a pressure of 1.00 atm at 520°C. At equilibrium the partial pressure of H_2 is found to be 0.10 atm.

a	What are P_{I_2} and P_{HI} at equilibrium?

ANALYSIS

Information given:	initial pressure (P_o) for HI (1.00 atm) equilibrium pressure (P_{eq}) for H_2 (0.10 atm)
Information implied:	initial pressures (P_o) for H_2 and I_2
Asked for:	equilibrium partial pressures (P_{eq}) for I_2 and HI

STRATEGY AND SOLUTION

1. Create a table (shown below) and enter the pertinent information (both explicit and implied) from the statement of the problem. Note that the problem states that *only* HI is initially present. Your table looks like this:

	2HI(g)	$\rightleftharpoons$	H₂(g)	+	I₂(g)
P_o (atm)	1.00		0		0
ΔP (atm)					
P_{eq} (atm)			0.10		

continued

2. Find a column where only one entry is missing. In this case it is the $H_2(g)$ column. The missing entry can be determined algebraically. By noting that initially the pressure is 0 atm and at equilibrium, the pressure is 0.10 atm. The change, ΔP, must be +0.10 atm. All changes will be shaded in green.

	2HI(g)	$\rightleftharpoons$	H₂(g)	+	I₂(g)
P_o (atm)	1.00		0		0
ΔP (atm)			+0.10		
P_{eq} (atm)			0.10		

3. Since the system is at constant volume and temperature, the number of moles must be directly proportional to the pressure. Use the coefficients of the reaction to determine stoichiometric ratios. (Recall Chapter 3.) Fill in the rest of the row labeled ΔP with 0.10 atm multiplied by the stoichiometric ratios:

$$1 \text{ mol } I_2/1 \text{ mol } H_2 \qquad 2 \text{ mol HI}/1 \text{ mol } H_2$$

Use + signs for species produced and − signs for species used up.

	2HI(g)	$\rightleftharpoons$	H₂(g)	+	I₂(g)
P_o (atm)	1.00		0		0
ΔP (atm)	$-(\frac{2}{1})(0.10)$		+0.10		$+(\frac{1}{1})(0.10)$
P_{eq} (atm)			0.10		

4. Fill in the row P_{eq} by performing the indicated calculation for each column.

	2HI(g)	$\rightleftharpoons$	H₂(g)	+	I₂(g)
P_o (atm)	1.00		0		0
ΔP (atm)	$-(\frac{2}{1})(0.10)$		+0.10		$+(\frac{1}{1})(0.10)$
P_{eq} (atm)	0.80		0.10		0.10

5. Thus from the row P_{eq}, the equilibrium partial pressure for HI is 0.80 atm, and that for I_2 is 0.10 atm.

b | What is K for the system?

ANALYSIS

Information given:	from part (a): equilibrium pressures for HI (0.80 atm) and for I_2 (0.10 atm); equilibrium pressure (P_{eq}) for H_2 (0.10 atm)
Information implied:	K expression
Asked for:	equilibrium constant K at 520°C

STRATEGY

1. Write the equilibrium expression.
2. Substitute into the expression for K.

SOLUTION

1. equilibrium expression	$K = \dfrac{(P_{H_2})(P_{I_2})}{(P_{HI})^2}$
2. K	$K = \dfrac{(0.10)(0.10)}{(0.80)^2} = 0.016$

END POINT

You should expect K to be rather small (0.016). Recall that at equilibrium, P_{H_2} is 0.10 atm, only one tenth of the original pressure of HI (1.00 atm).

Example 12.4 illustrates a principle that you will find very useful in solving equilibrium problems throughout this (and later) chapters. *As a system approaches equilibrium, changes in partial pressures of reactants and products—like changes in molar amounts—are related to one another through the coefficients of the balanced equation.*

12-4 Applications of the Equilibrium Constant

Sometimes, knowing only the magnitude of the equilibrium constant, it is possible to decide on the feasibility of a reaction. Consider, for example, a possible method for "fixing" atmospheric nitrogen—converting it to a compound—by reaction with oxygen:

$$N_2(g) + O_2(g) \rightleftharpoons 2NO(g)$$

$$K = \frac{(P_{NO})^2}{(P_{N_2})(P_{O_2})} = 1 \times 10^{-30} \text{ at } 25°C$$

Because K is so small, the partial pressure of NO in equilibrium with N_2 and O_2, and hence the amount of NO, must be extremely small, approaching zero. Clearly, this would not be a suitable way to fix nitrogen, at least at 25°C.

An alternative approach to nitrogen fixation involves reacting it with hydrogen:

$$N_2(g) + 3H_2(g) \rightleftharpoons 2NH_3(g)$$

$$K = \frac{(P_{NH_3})^2}{(P_{N_2})(P_{H_2})^3} = 6 \times 10^5 \text{ at } 25°C$$

In this case, the equilibrium system must contain mostly ammonia. A mixture of N_2 and H_2 should be almost completely converted to NH_3 at equilibrium.

In general, if K is a very small number, the equilibrium mixture will contain mostly unreacted starting materials; for all practical purposes, the forward reaction does not go. Conversely, a large K implies a reaction that, at least in principle, is feasible; products should be formed in high yield. ◀ *i* Frequently, K has an intermediate value, in which case you must make quantitative calculations concerning the *direction* or *extent* of reaction.

> *i* The concept of equilibrium is most useful when K is neither very large nor very small.

Direction of Reaction; the Reaction Quotient (Q)

Consider the general gas-phase reaction

$$aA(g) + bB(g) \rightleftharpoons cC(g) + dD(g)$$

for which

$$K = \frac{(P_C)^c(P_D)^d}{(P_A)^a(P_B)^b}$$

As we have seen, for a given system at a particular temperature, the value of K is fixed.

In contrast, the actual pressure ratio, Q,

$$Q = \frac{(P_C)^c(P_D)^d}{(P_A)^a(P_B)^b}$$

can have any value at all. ◀ *i* If you start with pure reactants (A and B), P_C and P_D are zero, and the value of Q is zero:

> *i* Q can have any value; K is fixed.

$$Q = \frac{(0)(0)}{(P_A)^a(P_B)^b} = 0$$

Conversely, starting with only C and D, the value of Q becomes infinite as the values of P_A and P_B approach zero:

$$Q = \frac{(P_C)^c(P_D)^d}{(P_A)^a(P_B)^b \to 0} \longrightarrow \infty$$

If all four substances are present, Q can have any value between zero and infinity.

The form of the expression for Q, known as the **reaction quotient,** is the same as that for the equilibrium constant, K. The difference is that the partial pressures that appear in Q are those that apply at a particular moment, not necessarily when the system is at equilibrium. By comparing the numerical value of Q with that of K, it is possible to decide in which direction the system will move to achieve equilibrium.

1. If $Q < K$ ◀ⓘ the reaction proceeds from left to right:

$$aA(g) + bB(g) \longrightarrow cC(g) + dD(g)$$

In this way, the partial pressures of products increase, while those of reactants decrease. As this happens, the reaction quotient Q increases and eventually at equilibrium becomes equal to K.

2. If $Q > K$ ◀ⓘ the partial pressures of products are "too high" and those of the reactants "too low" to meet the equilibrium condition. Reaction proceeds in the reverse direction:

$$aA(g) + bB(g) \longleftarrow cC(g) + dD(g)$$

increasing the partial pressures of A and B while reducing those of C and D. This lowers Q to its equilibrium value, K.

3. If perchance ◀ⓘ $Q = K$, the system is already at equilibrium, so the amounts of products and reactants remain unchanged.

To illustrate these statements, consider the simple system shown in Table 12.5.

ⓘ
$Q < K; \longrightarrow$

ⓘ
$Q > K; \longleftarrow$

ⓘ
As equilibrium is approached, Q approaches K.

ⓘ
The direction of the reaction may change, but the expression for the equilibrium constant does not.

Table 12.5 Approach to Equilibrium in the System $A \rightleftharpoons B$ ◀ⓘ for which $K = 1.00$

	Experiment 1*				Experiment 2*			
t	0	20	40	60	0	20	40	60
[B]	1.00	1.35	1.50	1.50	2.00	1.65	1.50	1.50
[A]	2.00	1.65	1.50	1.50	1.00	1.35	1.50	1.50
$Q = [B]/[A]$	0.500	0.818	1.00	1.00	2.00	1.22	1.00	1.00
	$Q < K$		$Q = K$		$Q > K$		$Q = K$	

*In both experiments, systems to the right of the broken line have reached equilibrium.

EXAMPLE 12.5

Consider the following system at 100°C:

$$N_2O_4(g) \rightleftharpoons 2NO_2(g) \qquad K = 11$$

Predict the direction in which reaction will occur to reach equilibrium, starting with 0.10 mol of N_2O_4 and 0.20 mol of NO_2 in a 2.0-L container.

ANALYSIS

Information given:	reaction and K at 100°C (11) initial amounts of N_2O_4 (0.10 mol) and NO_2 (0.20 mol) volume of the container (2.0 L)
Information implied:	R value
Asked for:	direction of the reaction

continued

STRATEGY

1. Calculate the partial pressures of all species using the ideal gas law and stoichiometric ratios.

2. Write the expression for Q and find its value.

3. Compare Q with K to predict the direction of the reaction.

SOLUTION

1. $P_{N_2O_4}$	$P = \dfrac{nRT}{V} = \dfrac{(0.10)(0.0821)(373)}{2.0} = 1.5$ atm
P_{NO_2}	Since V and T are constant, n is proportional to P, and P can be obtained using the stoichioimetric ratio of NO_2 and N_2O_4.
	1.5 atm for $N_2O_4 \times \dfrac{2 \text{ mol } NO_2}{1 \text{ mol } N_2O_4} = 3.0$ atm
2. Q	$Q = \dfrac{(P_{NO_2})^2}{P_{N_2O_4}} = \dfrac{(3.0)^2}{1.5} = 6.0$
3. Direction	$Q(6.0) < K(11)$ The reaction proceeds from left to right $(\rightarrow)$.

END POINT

When we say that the reaction proceeds from left to right or in the forward direction, we refer to the reaction as written. In this case, the partial pressure of NO_2 increases while that of N_2O_4 decreases until the system reaches equilibrium.

Extent of Reaction; Equilibrium Partial Pressures

The equilibrium constant for a chemical system can be used to calculate the partial pressures of the species present at equilibrium. In the simplest case, one equilibrium pressure can be calculated, knowing all the others. Consider, for example, the system

$$N_2(g) + O_2(g) \rightleftharpoons 2NO(g)$$

$$K = \frac{(P_{NO})^2}{(P_{N_2})(P_{O_2})} = 1 \times 10^{-30} \text{ at } 25°C$$

In air, the partial pressures of N_2 and O_2 are about 0.78 and 0.21 atm, respectively. The expression for K can be used to calculate the equilibrium partial pressure of NO under these conditions:

$$(P_{NO})^2 = (P_{N_2})(P_{O_2})(K)$$
$$= (0.78)(0.21)(1 \times 10^{-30}) = 1.6 \times 10^{-31}$$
$$P_{NO} = (1.6 \times 10^{-31})^{1/2} = 4 \times 10^{-16} \text{ atm}$$

This is an extremely small pressure, as you might have guessed from the small magnitude of the equilibrium constant. ◀ⓘ

More commonly K is used to determine the equilibrium partial pressures of all species, reactants, and products, knowing their original pressures. To do this, it helps to follow this path.

1. *Create a table to keep track of all information.* This was illustrated in Example 12.4.

2. *Choose ΔP for one of the species to be represented by x.* A good choice would be a species that has a coefficient of 1 in its balanced equation and/or a

ⓘ That's just as well; if N_2 and O_2 in the atmosphere were converted to NO, we'd be in big trouble.

species that is not present initially. Relate ΔP of the chosen species to the other species by using the stoichiometric ratios provided by the coefficients in the balanced equation.

3. *Determine the direction of the reaction and assign $+$ or $-$ signs to the entries in the row for ΔP.*

   —The reaction proceeds in the direction where initial pressure is 0.

   —If all the species have initial partial pressures, find Q.

4. *Fill in the row for equilibrium partial pressures.*

5. *Write the expression for K and substitute the entries for equilibrium partial pressure into the expression.* Solve for x.

6. *Having found x, refer back to the table and calculate the equilibrium partial pressures of all species.*

EXAMPLE 12.6 GRADED

For the system

$$CO_2(g) + H_2(g) \rightleftharpoons CO(g) + H_2O(g)$$

K is 0.64 at 900 K.

a Calculate the equilibrium partial pressures of all species, starting with

$$P_{CO_2} = P_{H_2} = 1.00 \text{ atm}; P_{CO} = P_{H_2O} = 0$$

ANALYSIS

Information given:	initial partial pressures for CO_2 (1.00 atm), H_2 (1.00 atm), CO (0 atm), and H_2O (0 atm); K (0.64) at 900 K
Information implied:	direction of the reaction
Asked for:	equilibrium partial pressures of all species

STRATEGY AND SOLUTION

1. Create a table.

	$CO_2(g)$	$+$	$H_2(g)$	$\rightleftharpoons$	$CO(g)$	$+$	$H_2O(g)$
P_o (atm)	1.00		1.00		0		0
ΔP (atm)							
P_{eq} (atm)							

2. Choose ΔP for CO as x. Since all the coefficients of the reaction are 1, all the species have x for ΔP.

	$CO_2(g)$	$+$	$H_2(g)$	$\rightleftharpoons$	$CO(g)$	$+$	$H_2O(g)$
P_o (atm)	1.00		1.00		0		0
ΔP (atm)	x		x		x		x
P_{eq} (atm)							

3. Direction of the reaction ($\rightarrow$)

	$CO_2(g)$	$+$	$H_2(g)$	$\rightleftharpoons$	$CO(g)$	$+$	$H_2O(g)$
P_o (atm)	1.00		1.00		0		0
ΔP (atm)	$-x$		$-x$		$+x$		$+x$
P_{eq} (atm)							

continued

4. Equilibrium partial pressures

	$CO_2(g)$	$+$	$H_2(g)$	$\rightleftharpoons$	$CO(g)$	$+$	$H_2O(g)$
P_o (atm)	1.00		1.00		0		0
ΔP (atm)	$-x$		$-x$		$+x$		$+x$
P_{eq} (atm)	$1.00 - x$		$1.00 - x$		x		x

5. K expression

$$K = \frac{(P_{CO})(P_{H_2O})}{(P_{CO_2})(P_{H_2})} = \frac{(x)(x)}{(1.00 - x)(1.00 - x)} = 0.64$$

Solve for x Take the square root of both sides: $0.80 = \dfrac{x}{1.00 - x} \longrightarrow x = 0.44$

6. P_{eq} $P_{CO} = P_{H_2O} = x = 0.44$ atm; $P_{CO_2} = P_{H_2} = 1.00 - x = (1.00 - 0.44)$ atm $= 0.56$ atm

---

b   Calculate the equilibrium partial pressures of all species, starting with

$$P_{CO_2} = 2.00 \text{ atm}, P_{H_2} = 1.00 \text{ atm}; P_{CO} = P_{H_2O} = 0$$

ANALYSIS

Information given:	initial partial pressures for CO_2 (2.00 atm), H_2 (1.00 atm), CO (0.500 atm), and H_2O (0.500 atm); K (0.64) at 900 K
Information implied:	direction of the reaction
Asked for:	equilibrium partial pressures of all species

STRATEGY

1. Determine Q so you know the direction of the reaction.
2. Make a table. Follow the first four steps (shown in detail in part (a)).
3. Substitute into the K expression.
4. Solve for the equilibrium pressure of all species.

ANALYSIS

1. Q

$$Q = \frac{(2.00)(1.00)}{(0.500)(0.500)} = 0.125$$

$0.125 < 0.64$; direction of reaction $\longrightarrow$

2. Table

	$CO_2(g)$	$+$	$H_2(g)$	$\rightleftharpoons$	$CO(g)$	$+$	$H_2O(g)$
P_o (atm)	2.00		1.00		0.500		0.500
ΔP (atm)	$-x$		$-x$		$+x$		$+x$
P_{eq} (atm)	$2.00 - x$		$1.00 - x$		$0.500 + x$		$0.500 + x$

3. K expression $K = \dfrac{(P_{CO})(P_{H_2O})}{(P_{CO_2})(P_{H_2})} = \dfrac{(0.500 + x)(0.500 + x)}{(2.00 - x)(1.00 - x)} = 0.64$ *ⓘ For the most part, we'll avoid tedious calculations of this type (Example 12.6b).*

It is not useful to take the square root of both sides, so rearrange the equation to the standard form for using the quadratic equation: $ax^2 + bx + c = 0 \longrightarrow 0.36\,x^2 + 2.92\,x - 1.03 = 0$

Apply the **quadratic formula**: $x = \dfrac{-b \pm \sqrt{b^2 - 4ac}}{2a}$ ◄ⓘ $\longrightarrow x = 0.34$ or -8.40

0.34 atm is plausible; -8.40 atm would imply a negative partial pressure.

continued

4. P_{eq} $\quad P_{CO} = P_{H_2O} = 0.500 + x = 0.500 + 0.34 = 0.84$ atm

$\qquad P_{CO_2} = 2.00 - 0.34 = 1.66$ atm $\qquad P_{H_2} = 1.00 - 0.34 = 0.66$ atm

END POINT

Check your algebra by substituting the equilibrium partial pressures back into the expression for K.

$$K = \frac{(P_{CO})(P_{H_2O})}{(P_{CO_2})(P_{H_2})} = \frac{(0.84)(0.84)}{(1.66)(0.66)} = 0.64 \qquad \text{Voila!}$$

We should point out that the calculations in Example 12.6 assume ideal gas behavior. At the conditions specified (1 atm, relatively high temperatures), this assumption is a good one. However, many industrial gas-phase reactions are carried out at very high pressures. In that case, intermolecular forces become important, and calculated yields based on ideal gas behavior may be seriously in error.

12-5   Effect of Changes in Conditions on an Equilibrium System

Once a system has attained equilibrium, it is possible to change the ratio of products to reactants by changing the external conditions. We will consider three ways in which a chemical equilibrium can be disturbed:

1. Adding or removing a gaseous reactant or product.
2. Compressing or expanding the system.
3. Changing the temperature.

You can deduce the direction in which an equilibrium will shift when one of these changes is made by applying the following principle:

> *If a system at equilibrium is disturbed by a change in concentration, pressure, or temperature, the system will, if possible, shift to partially counteract the change.* ◄ ⓘ

Equilibrium systems, like people, resist change.

This law was first stated, in a more complex form, in 1884 by Henri Le Châtelier (1850–1936), a French chemist who had studied a variety of industrial equilibria, including those involved in the production of iron in the blast furnace. The statement is commonly referred to as **Le Châtelier's principle.**

Adding or Removing a Gaseous Species

According to Le Châtelier's principle, *if a chemical system at equilibrium is disturbed by adding a gaseous species (reactant or product), the reaction will proceed in such a direction as to consume part of the added species. Conversely, if a gaseous species is removed, the system shifts to restore part of that species.* In that way, the change in concentration brought about by adding or removing a gaseous species is partially counteracted when equilibrium is restored.

To apply this general rule, consider the reaction

$$N_2O_4(g) \rightleftharpoons 2NO_2(g)$$

Suppose this system has reached equilibrium at a certain temperature. This equilibrium could be disturbed by

- *adding N_2O_4*. Reaction will occur in the forward direction (left to right). In this way, part of the N_2O_4 will be consumed.
- *adding NO_2*, which causes the reverse reaction (right to left) to occur, using up part of the NO_2 added.

■ *removing* N_2O_4. Reaction occurs in the reverse direction to restore part of the N_2O_4.

■ *removing* NO_2, which causes the forward reaction to occur, restoring part of the NO_2 removed.

It is possible to use K to calculate the extent to which reaction occurs when an equilibrium is disturbed by adding or removing a product or reactant. To see how this is done, consider the effect of adding hydrogen iodide to the HI–H_2–I_2 system (Example 12.7).

$$2HI(g) \rightleftharpoons H_2(g) + I_2(g)$$

The system does *not* return to its original equilibrium state.

EXAMPLE 12.7

In Example 12.4 you found that the HI–H_2–I_2 system is in equilibrium at 520°C when $P_{HI} = 0.80$ atm and $P_{H_2} = P_{I_2} = 0.10$ atm. Suppose enough HI is added to raise its pressure temporarily to 1.00 atm. When equilibrium is restored, what are P_{HI}, P_{H_2}, and P_{I_2}?

ANALYSIS

Information given:	equilibrium pressures for HI (0.80 atm), I_2 (0.10 atm), and H_2 (0.10 atm) from Example 12.4; K (0.016) equilibrium disturbed by adding HI (now 1.00 atm)
Information implied:	direction of the reaction
Asked for:	equilibrium pressures when equilibrium is reestablished

STRATEGY

1. Create a table. Note that the equilibrium partial pressures now become initial pressures for HI and I_2. P_o for HI is now 1.00 atm.

2. Since HI is added, the reaction goes in the direction of using up the HI, thus to the right.

3. Write the K expression and solve for x.

4. Substitute the value for x into the equilibrium pressures for all species.

SOLUTION

Table

	2HI(g)	$\rightleftharpoons$	H₂(g)	+	I₂(g)
P_o (atm)	1.00		0.10		0.10
ΔP (atm)	$-2x$		$+x$		$+x$
P_{eq} (atm)	$1.00 - 2x$		$0.10 + x$		$0.10 + x$

K expression

$$K = 0.016 = \frac{(P_{H_2})(P_{I_2})}{(P_{HI})^2} = \frac{(0.10 + x)(0.10 + x)}{(1.00 - 2x)^2} = \frac{(0.10 + x)^2}{(1.00 - 2x)^2}$$

x

Take the square root of both sides: $0.13 = \dfrac{(0.10 + x)}{(1.00 - 2x)} \longrightarrow x = 0.024$ atm

Equilibrium pressures

$P_{I_2} = P_{H_2} = 0.10 + 0.024 = 0.12$ atm; $\quad P_{HI} = 1.00 - 2(0.024) = 0.95$ atm

END POINT

Note that the equilibrium partial pressure of HI is intermediate between its value before equilibrium was established (0.80 atm) and that immediately afterward (1.00 atm). This is exactly what LeChâtelier's principle predicts: part of the added HI is consumed to reestablish equilibrium!

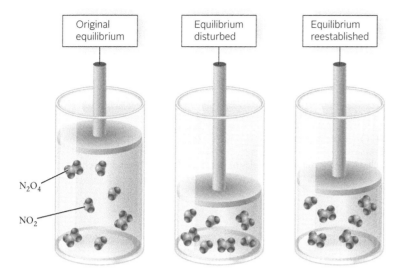

Figure 12.4 Effect of compression on the $N_2O_4(g) \rightleftharpoons 2NO_2(g)$ system at equilibrium. The immediate effect (middle cylinder) is to crowd the same number of moles of gas into a smaller volume and thus increase the total pressure. This is partially compensated for by the conversion of some of the NO_2 to N_2O_4, thereby reducing the total number of moles of gas.

We should emphasize that adding a pure liquid or solid has no effect on a system at equilibrium. The rule is a simple one: *For a species to shift the position of an equilibrium, it must appear in the expression for K.*

Compression or Expansion

To understand how a change in pressure can change the position of an equilibrium, consider again the N_2O_4–NO_2 system:

$$N_2O_4(g) \rightleftharpoons 2NO_2(g)$$

Suppose the system is compressed by pushing down the piston shown in Figure 12.4. The immediate effect is to increase the gas pressure because the same number of molecules are crowded into a smaller volume ($P = nRT/V$). According to Le Châtelier's principle, the system will shift to partially counteract this change. There is a simple way in which the gas pressure can be reduced. ◄ ⓘ Some of the NO_2 molecules combine with each other to form N_2O_4 (cylinder at right of Figure 12.4). That is, reaction occurs in the reverse direction:

ⓘ When V decreases, NO_2 is converted to N_2O_4.

$$N_2O_4(g) \longleftarrow 2NO_2(g)$$

This reduces the number of moles, n, and hence the pressure, P.

It is possible (but not easy) to calculate from K the extent to which NO_2 is converted to N_2O_4 when the system is compressed. The results of such calculations are given in Table 12.6. As the pressure is increased from 1.0 to 10.0 atm, more and more of the NO_2 is converted to N_2O_4. Notice that the total number of moles of gas decreases steadily as a result of this conversion.

Table 12.6 Effect of Compression on the Equilibrium System
$N_2O_4(g) \rightleftharpoons 2NO_2(g)$; $K = 11$ at 100°C

P_{tot} (atm)	n_{NO_2}	$n_{N_2O_4}$	n_{tot}
1.0	0.92	0.08	1.00
2.0	0.82	0.13	0.95
5.0	0.64	0.22	0.86
10.0	0.50	0.29	0.79

Table 12.7 Effect of Pressure on the Position of Gaseous Equilibria

System	Δn_{gas}*	P_{tot} Increases	P_{tot} Decreases
1. $N_2O_4(g) \rightleftharpoons 2NO_2(g)$	$+1$	$\longleftarrow$	$\longrightarrow$
2. $SO_2(g) + \frac{1}{2}O_2(g) \rightleftharpoons SO_3(g)$	$-\frac{1}{2}$	$\longrightarrow$	$\longleftarrow$
3. $N_2(g) + 3H_2(g) \rightleftharpoons 2NH_3(g)$	-2	$\longrightarrow$	$\longleftarrow$
4. $C(s) + H_2O(g) \rightleftharpoons CO(g) + H_2(g)$	$+1$	$\longleftarrow$	$\longrightarrow$
5. $N_2(g) + O_2(g) \rightleftharpoons 2NO(g)$	0	0	0

*Δn_{gas} is the change in the number of moles of gas as the forward reaction occurs.

The analysis we have just gone through for the N_2O_4–NO_2 system can be applied to any equilibrium system involving gases

- *When the system is compressed, thereby increasing the total pressure, reaction takes place in the direction that decreases the total number of moles of gas.*
- *When the system is expanded, thereby decreasing the total pressure, reaction takes place in the direction that increases the total number of moles of gas.*

The application of this principle to several different systems is shown in Table 12.7. In system 2, the number of moles of gas decreases from $\frac{3}{2}$ to 1 as the reaction goes to the right. Hence increasing the pressure causes the forward reaction to occur; a decrease in pressure has the reverse effect. Notice that it is the change in the number of moles of *gas* that determines which way the equilibrium shifts (system 4). When there is no change in the number of moles of gas (system 5), a change in pressure has no effect on the position of the equilibrium.

We should emphasize that in applying this principle it is important to realize that an "increase in pressure" means that the system is compressed; that is, the volume is decreased. Similarly, a "decrease in pressure" corresponds to expanding the system by increasing its volume. There are other ways in which pressure can be changed. One way is to add an unreactive gas such as helium at constant volume. This increases the total number of moles and hence the total pressure. It has no effect, however, on the position of the equilibrium, because it does not change the partial pressures of any of the gases taking part in the reaction. Remember that the partial pressure of a gas A is given by the expression

$$P_A = n_A RT/V \qquad (n_A = \text{no. of moles of A})$$

Adding a different gas at constant volume does not change any of the quantities on the right side of this equation, so P_A stays the same.

Change in Temperature

Le Châtelier's principle can be used to predict the effect of a change in temperature on the position of an equilibrium. In general, *an increase in temperature causes the endothermic reaction to occur.* This absorbs heat and so tends to reduce the temperature of the system, partially compensating for the original temperature increase.

Applying this principle to the N_2O_4–NO_2 system,

$$N_2O_4(g) \rightleftharpoons 2NO_2(g) \qquad \Delta H° = +57.2 \text{ kJ}$$

it follows that raising the temperature causes the forward reaction to occur, because that reaction absorbs heat. This is confirmed by experiment (Figure 12.5). At higher temperatures, more NO_2 is produced, and the reddish-brown color of that gas becomes more intense.

Marna G. Clarke

Figure 12.5 Effect of temperature on the N_2O_4–NO_2 system at equilibrium. At 0°C *(tube at right)*, N_2O_4, which is colorless, predominates. At 50°C *(tube at left)*, some of the N_2O_4 has dissociated to give the deep brown color of NO_2.

For the synthesis of ammonia,

$$N_2(g) + 3H_2(g) \rightleftharpoons 2NH_3(g) \qquad \Delta H° = -92.2 \text{ kJ}$$

an increase in temperature shifts the equilibrium to the left; some ammonia decomposes to the elements. This reflects the fact that the reverse reaction is endothermic:

$$2NH_3(g) \longrightarrow N_2(g) + 3H_2(g) \qquad \Delta H° = +92.2 \text{ kJ}$$

As pointed out earlier, the equilibrium constant of a system changes with temperature. The form of the equation relating K to T is a familiar one, similar to the Clausius-Clapeyron equation (Chapter 9) and the Arrhenius equation (Chapter 11). This one is called the *van't Hoff equation*, honoring Jacobus van't Hoff (1852–1911), who was the first to use the equilibrium constant, K. Coincidentally, van't Hoff was a good friend of Arrhenius. The equation is

$$\ln \frac{K_2}{K_1} = \frac{\Delta H°}{R}\left[\frac{1}{T_1} - \frac{1}{T_2}\right] \qquad (12.5)$$

where K_2 and K_1 are the equilibrium constants at T_2 and T_1, respectively, $\Delta H°$ is the standard enthalpy change for the forward reaction, and R is the gas constant, 8.31 J/mol·K.

To illustrate how this equation is used, let us apply it to calculate the equilibrium constant at 100°C for the system

$$N_2(g) + 3H_2(g) \rightleftharpoons 2NH_3(g) \qquad \Delta H° = -92.2 \text{ kJ}$$

given that $K = 6 \times 10^5$ at 25°C. The relation is

$$\ln \frac{K \text{ at } 100°C}{6 \times 10^5} = \frac{-92,200 \text{ J/mol}}{8.31 \text{ J/mol} \cdot \text{K}}\left[\frac{1}{298 \text{ K}} - \frac{1}{373 \text{ K}}\right] = -7.5$$

Taking inverse logarithms,

$$\frac{K \text{ at } 100°C}{6 \times 10^5} = 6 \times 10^{-4}$$

$$K \text{ at } 100°C = 4 \times 10^2$$

Notice that the equilibrium constant becomes smaller as the temperature increases. In general,

- if the forward reaction is exothermic, as is the case here ($\Delta H° = -92.2$ kJ), K decreases as T increases.
- if the forward reaction is endothermic as in the decomposition of N_2O_4,

$$N_2O_4(g) \rightleftharpoons 2NO_2(g) \qquad \Delta H° = +57.2 \text{ kJ}$$

K increases as T increases. For this reaction K is 0.11 at 25°C and 11 at 100°C.

EXAMPLE 12.8

Consider the following systems:

(a) $2CO(g) + O_2(g) \rightleftharpoons 2CO_2(g) \qquad \Delta H = -566 \text{ kJ}$

(b) $H_2(g) + I_2(g) \rightleftharpoons 2HI(g) \qquad \Delta H = -2.7 \text{ kJ}$

(c) $H_2(g) + I_2(s) \rightleftharpoons 2HI(g) \qquad \Delta H = +53.0 \text{ kJ}$

(d) $I_2(g) \rightleftharpoons 2I(g) \qquad \Delta H = +36.2 \text{ kJ}$

What will happen to the position of the equilibrium if the system is compressed (at constant temperature)? Heated at constant pressure?

continued

STRATEGY		
1. An increase in pressure tends to drive the equilibrium in a direction where there are fewer moles of gas.		
2. An increase in temperature favors an endothermic reaction ($\Delta H > 0$), i.e., creates more products.		

SOLUTION		
(a) Increase in P Increase in T	(3 mol of gas on the left; 2 mol of gas on the right) $\longrightarrow$ exothermic reaction $\longleftarrow$	
(b) Increase in P Increase in T	(2 mol of gas on the left; 2 mol of gas on the right) no effect exothermic reaction $\longleftarrow$ (slightly)	
(c) Increase in P Increase in T	(1 mol of gas on the left; 2 mol of gas on the right) $\longleftarrow$ endothermic reaction $\longrightarrow$	
(d) Increase in P Increase in T	(1 mol of gas on the left; 2 mol of gas on the right) $\longleftarrow$ endothermic reaction $\longrightarrow$	

We should emphasize that of the three changes in conditions described in this section

- adding or removing a gaseous species
- compressing or expanding the system
- changing the temperature

the only one that changes the value of the equilibrium constant is a change in temperature. In the other two cases, K remains constant.

CHEMISTRY **BEYOND** THE CLASSROOM

An Industrial Application of Gaseous Equilibrium

A wide variety of materials, both pure substances and mixtures, are made by processes that involve one or more gas-phase reactions. Among these is one of the most important industrial chemicals, ammonia.

The process used to make this chemical applies many of the principles of chemical equilibrium, discussed in this chapter, and chemical kinetics (Chapter 11).

Haber Process for Ammonia

Combined ("fixed") nitrogen in the form of protein is essential to all forms of life. There is more than enough elementary nitrogen in the air, about 4×10^{18} kg, to meet all our needs. The problem is to convert the element to compounds that can be used by plants to make proteins. At room temperature and atmospheric pressure, N_2 does not react with any nonmetal. However, in 1908 in Germany, Fritz Haber (Figure A) showed that nitrogen does react with hydrogen at high temperatures and pressures to form ammonia in good yield:

$$N_2(g) + 3H_2(g) \rightleftharpoons 2NH_3(g) \qquad \Delta H° = -92.2 \text{ kJ}$$

The **Haber process**, represented by this equation, is now the main source of fixed nitrogen. Its feasibility depends on choosing conditions under which nitrogen and hydrogen react rapidly to give a high yield of ammonia. At 25°C and atmospheric pressure, the position of the equilibrium favors the formation of NH₃ ($K = 6 \times 10^5$). Unfortunately, however, the rate of reaction is virtually zero. Equilibrium is reached more rapidly by raising the temperature. However, because the synthesis of ammonia is exothermic, high temperatures reduce K and hence the yield of ammonia. High pressures, on the other hand, have a favorable effect on both the rate of the reaction and the position of the equilibrium (Table A).

Figure A Fritz Haber
(1868–1934).

The values of the equilibrium constant K listed in Table A are those obtained from data at low pressures, where the gases behave ideally. At higher pressures the mole percent of ammonia observed is generally larger than the calculated value. For example, at 400°C and 300 atm, the observed mole percent of NH₃ is 47; the calculated value is only 41.

Table A Effect of Temperature and Pressure on the Yield of Ammonia in the Haber Process ($P_{H_2} = 3P_{N_2}$)

°C	K	Mole Percent NH₃ in Equilibrium Mixture				
		10 atm	50 atm	100 atm	300 atm	1000 atm
200	0.30	51	74	82	90	98
300	3.7×10^{-3}	15	39	52	71	93
400	1.5×10^{-4}	4	15	25	47	80
500	1.4×10^{-5}	1	6	11	26	57
600	2.1×10^{-6}	0.5	2	5	14	31

The goal of Haber's research was to find a catalyst to synthesize ammonia at a reasonable rate without going to very high temperatures. These days two different catalysts are used. One consists of a mixture of iron, potassium oxide, K_2O, and aluminum oxide, Al_2O_3. The other, which uses finely divided ruthenium, Ru, metal on a graphite surface, is less susceptible to poisoning by impurities. Reaction takes place at 450°C and a pressure of 200 to 600 atm. The ammonia formed is passed through a cooling chamber. Ammonia boils at −33°C, so it is condensed out as a liquid, separating it from unreacted nitrogen and hydrogen. The yield is typically less than 50%, so the reactants are recycled to produce more ammonia.

Fritz Haber

Haber had a distinguished scientific career. He received, among other honors, the 1918 Nobel Prize in chemistry. For more than 20 years, he was director of the prestigious Kaiser Wilhelm Institute for Physical Chemistry at Dahlem, Germany, a suburb of Berlin. Presumably because of Haber's Jewish ancestry, he was forced to resign that position when Adolf Hitler came to power in 1933. A contributing factor may have been his comment after listening to a radio broadcast of one of Hitler's hysterical harangues.

When asked what he thought of it, Haber replied, "Give me a gun so I can shoot him."

Haber, like all of us, suffered disappointment and tragedy in his life. In the years after World War I, he devised a scheme to pay Germany's war debts by extracting gold from seawater. The project, which occupied Haber for several years, was a total failure because the concentration of gold turned out to be only about one thousandth of the literature value.

During World War I, Haber was in charge of the German poison gas program. In April of 1915, the Germans used chlorine for the first time on the Western front, causing 5000 fatalities. Haber's wife, Clara, was aghast; she pleaded with her husband to forsake poison gas. When he adamantly refused to do so, she committed suicide.

On a lighter note, there is an amusing anecdote told by the author Morris Goran (*The Story of Fritz Haber*, 1967). It seems that in a Ph.D. examination, Haber asked the candidate how iodine was prepared. The student made a wild guess that it was obtained from a tree (it isn't). Haber played along with him, asking where the iodine tree grew, what it looked like, and, finally, when it bloomed. "In the fall," was the answer. Haber smiled, put his arm around the student, and said, "Well, my friend, I'll see you again when the iodine blossoms appear once more."

Key Concepts

1. Relate the expression for K to the corresponding chemical equation.
   (**Examples 12.1, 12.2; Problems 5–14**)
2. Calculate K, knowing
   ■ appropriate K's for other reactions.
   (**Example 12.1; Problems 15–20**)
   ■ all the equilibrium partial pressures.
   (**Example 12.3; Problems 21–24**)
   ■ all the original and one equilibrium partial pressure.
   (**Example 12.4; Problems 25, 26**)

3. Use the value of K to determine
   ■ the direction of reaction.
   (**Example 12.5; Problems 27–32**)
   ■ equilibrium partial pressures of all species.
   (**Examples 12.6, 12.7; Problems 33–46**)
4. Use Le Châtelier's principle to predict what will happen when the conditions on an equilibrium system are changed.
   (**Examples 12.7, 12.8; Problems 47–60**)

Key Equations

Expression for K	$aA(g) + bB(g) \rightleftharpoons cC(g) + dD(g)$
	$$K = \frac{(P_C)^c(P_D)^d}{(P_A)^a(P_B)^b}$$
Coefficient rule	$K' = K^n$
Reciprocal rule	$K'' = 1/K$
Multiple equilibrium	$K_3 = K_1 \times K_2$
van't Hoff equation	$$\ln\frac{K_2}{K_1} = \frac{\Delta H°}{R}\left[\frac{1}{T_1} - \frac{1}{T_2}\right]$$

Key Terms

equilibrium constant   Le Châtelier's principle   reaction quotient

Summary Problem

Consider phosgene, $COCl_2(g)$. It is an important intermediate in the manufacture of some plastics. It can decompose into carbon monoxide and chlorine gases. Its molar heat of formation ($\Delta H_f°$) is -220.1 kJ/mol.

(a) Write the chemical equation for the decomposition of phosgene using simplest whole-number coefficients.

(b) Write the equilibrium expression for the decompostion.

(c) At 500°C, after the reaction has reached equilibrium, the partial pressures of CO, Cl_2, and $COCl_2$ are 0.413 atm, 0.237 atm, and 0.217 atm, respectively. Calculate K at 500°C.

(d) Initially, a 5.00-L flask contains $COCl_2$, Cl_2, and CO at partial pressures of 0.689 atm, 0.250 atm, and 0.333 atm, respectively. After equilibrium is established at 500°C, the partial pressure of $COCl_2$ is 0.501 atm. Calculate the equilibrium partial pressures of chlorine and carbon monoxide gases.

(e) Initially, a reaction vessel contains Cl_2, CO, and $COCl_2$ at 500°C, where both Cl_2 and CO gases have partial pressures of 0.750 atm. The partial pressure of phosgene is 0.500 atm.

1. In what direction will the reaction proceed at that temperature?

2. What is the partial pressure of each gas when equilibrium is established?

3. Suppose enough phosgene is added to raise its pressure temporarily to 2.00 atm. In which direction will the equilibrium shift?

4. When equilibrium is restored, what is the partial pressure of each gas?

(f) In which direction will this system shift if at equilibrium

1. the system is expanded?

2. helium gas is added?

3. the temperature is increased?

(g) What is the equilibrium constant at 375°C?

Answers

(a) $COCl_2(g) \rightleftharpoons Cl_2(g) + CO(g)$

(b) $K = \dfrac{P_{CO} \times P_{Cl_2}}{P_{COCl_2}}$

(c) 0.451

(d) $P_{Cl_2} = 0.438$ atm; $P_{CO} = 0.521$ atm

(e) 1. ⟵   2. $P_{Cl_2} = P_{CO} = 0.558$ atm; $P_{COCl_2} = 0.692$ atm

3. ⟶   4. $P_{Cl_2} = P_{CO} = 0.872$ atm; $P_{COCl_2} = 1.686$ atm

(f) 1. ⟶   2. no change   3. ⟶

(g) 0.0168

Questions and Problems

Even-numbered questions and Challenge Problems have answers in Appendix 5 and fully worked solutions in the *Student Solutions Manual*.

12-1 The N_2O_4–NO_2 Equilibrium System

1. The following data are for the system

$$A(g) \rightleftharpoons 2B(g)$$

Time (s)	0	20	40	60	80	100
P_A (atm)	1.00	0.83	0.72	0.65	0.62	0.62
P_B (atm)	0.00	0.34	0.56	0.70	0.76	0.76

(a) How long does it take the system to reach equilibrium?
(b) How does the rate of the forward reaction compare with the rate of the reverse reaction after 30 s? After 90 s?

2. The following data are for the system

$$A(g) \rightleftharpoons 2B(g)$$

Time (s)	0	30	45	60	75	90
P_A (atm)	0.500	0.390	0.360	0.340	0.325	0.325
P_B (atm)	0.000	0.220	0.280	0.320	0.350	0.350

(a) How long does it take the system to reach equilibrium?
(b) How does the rate of the forward reaction compare with the rate of the reverse reaction after 45 s? After 90 s?

3. Complete the table below for the reaction

$$2A(g) + B(g) \longrightarrow 3C(g)$$
$$A + B \longrightarrow \text{products}$$

Time (s)	0	10	20	30	40	50	60
P_A (atm)	1.850	___	1.500	___	1.100	___	0.920
P_B (atm)	0.900	___	___	0.600	___	0.435	___
P_C (atm)	0.000	0.0850	___	___	___	___	___

4. Complete the table below for the reaction:

$$3A(g) + B(g) \rightleftharpoons 2C(g)$$

Time (min)	0	1	2	3	4	5	6
P_A (atm)	1.000	0.778	___	___	___	0.325	___
P_B (atm)	0.400	___	0.260	___	0.185	___	0.175
P_C (atm)	0.000	___	___	0.390	___	___	___

12-2 The Equilibrium Constant Expression

5. Write the equilibrium expressions (K) for the following reactions:
(a) $CH_4(g) + H_2O(l) \rightleftharpoons CO(g) + 3H_2(g)$
(b) $4NH_3(g) + 5O_2(g) \rightleftharpoons 4NO(g) + 6H_2O(g)$
(c) $BaCO_3(s) \rightleftharpoons BaO(s) + CO_2(g)$
(d) $NH_3(g) + HCl(g) \rightleftharpoons NH_4Cl(s)$

6. Write equilibrium constant (K) expressions for the following reactions:
(a) $Na_2CO_3(s) \rightleftharpoons 2NaO(s) + CO_2(g)$
(b) $C_2H_6(g) + 2H_2O(l) \rightleftharpoons 2CO(g) + 5H_2(g)$
(c) $4NO(g) + 6H_2O(g) \rightleftharpoons 4NH_3(g) + 5O_2(g)$
(d) $NH_3(g) + HI(l) \rightleftharpoons NH_4I(s)$

7. Write equilibrium constant expressions (K) for the following reactions:
(a) $2NO_3^-(aq) + 8H^+(aq) + 3Cu(s) \rightleftharpoons$
$\qquad 2NO(g) + 3Cu^{2+}(aq) + 4H_2O(l)$
(b) $2PbS(s) + 3O_2(g) \rightleftharpoons 2PbO(s) + 2SO_2(g)$
(c) $Ca^{2+}(aq) + CO_3^{2-}(aq) \rightleftharpoons CaCO_3(s)$

8. Write equilibrium constant (K) expressions for the following reactions:
(a) $I_2(s) + 2Cl^-(g) \rightleftharpoons Cl_2(g) + 2I^-(aq)$
(b) $CH_3NH_2(aq) + H^+(aq) \rightleftharpoons CH_3NH_3^+(aq)$
(c) $Au^{2+}(aq) + 4CN^-(aq) \rightleftharpoons Au(CN)_4^{2-}(aq)$

9. Given the following descriptions of reversible reactions, write a balanced equation (smallest whole-number coefficients) and the equilibrium constant expression for each.
(a) Nickel metal reacts with carbon monoxide to form nickel tetracarbonyl ($Ni(CO)_4$) gas.
(b) Aqueous nitrous acid in equilibrium with hydrogen and nitrite ions.
(c) Chlorine gas and bromide ions in equilibrium with liquid bromine and chloride ions.

10. Given the following descriptions of reversible reactions, write a balanced net ionic equation (simplest whole-number coefficients) and the equilibrium constant expression (K) for each.
(a) Liquid acetone (C_3H_6O) is in equilibrium with its vapor.
(b) Hydrogen gas reduces nitrogen dioxide gas to form ammonia and steam.
(c) Hydrogen sulfide gas (H_2S) bubbled into an aqueous solution of lead(II) ions produces lead sulfide precipitate and hydrogen ions.

11. Write an equation for an equilibrium system that would lead to the following expressions (a–c) for K.

(a) $K = \dfrac{(P_{CO})^2(P_{H_2})^5}{(P_{C_2H_6})(P_{H_2O})^2}$

(b) $K = \dfrac{(P_{NH_3})^4(P_{O_2})^5}{(P_{NO})^4(P_{H_2O})^6}$

(c) $K = \dfrac{[ClO_3^-]^2[Mn^{2+}]^2}{(P_{Cl_2})[MnO_4^-]^2[H^+]^4}$; liquid water is a product

12. Write a chemical equation for an equilibrium system that would lead to the following expressions (a–d) for K.

(a) $K = \dfrac{(P_{H_2S})^2(P_{O_2})^3}{(P_{SO_2})^2(P_{H_2O})^2}$

(b) $K = \dfrac{(P_{F_2})^{1/2}(P_{I_2})^{1/2}}{P_{IF}}$

(c) $K = \dfrac{[Cl^-]^2}{(P_{Cl_2})[Br^-]^2}$

(d) $K = \dfrac{(P_{NO})^2(P_{H_2O})^4[Cu^{2+}]^3}{[NO_3^-]^2[H^+]^8}$

13. Consider the following reaction at 250°C:

$$A(s) + 2B(g) \rightleftharpoons C(s) + 2D(g)$$

(a) Write an equilibrium constant expression for the reaction. Call the equilibrium constant K_1.
(b) Write an equilibrium constant expression for the formation of one mole of B(g) and call the equilibrium constant K_2.
(c) Relate K_1 and K_2.

14. Consider the following reaction at 100°C:

$$NO(g) + \tfrac{1}{2}Cl_2(g) \rightleftharpoons NOCl(g)$$

(a) Write an equilibrium constant expression for the reaction and call it K'.
(b) Write an equilibrium constant expression for the decomposition of NOCl to produce one mole of chlorine gas. Call the constant K''.
(c) Relate K' and K''.

12-3 Determination of K

15. At 627°C, $K = 0.76$ for the reaction

$$2SO_2(g) + O_2(g) \rightleftharpoons 2SO_3(g)$$

Calculate K at 627°C for
(a) the synthesis of one mole of sulfur trioxide gas.
(b) the decomposition of two moles of SO_3.

16. At 800°C, $K = 2.2 \times 10^{-4}$ for the following reaction

$$2H_2S(g) \rightleftharpoons 2H_2(g) + S_2(g)$$

Calculate K at 800°C for
(a) the synthesis of one mole of H_2S from H_2 and S_2 gases.
(b) the decomposition of one mole of H_2S gas.

17. Given the following reactions and their equilibrium constants,

$$H_2O(g) + CO(g) \rightleftharpoons H_2(g) + CO_2(g) \qquad K = 1.6$$
$$FeO(s) + CO(g) \rightleftharpoons Fe(s) + CO_2(g) \qquad K = 0.67$$

calculate K for the reaction

$$Fe(s) + H_2O(g) \rightleftharpoons FeO(s) + H_2(g)$$

18. Given the following data at 25°C

$$2NO(g) \rightleftharpoons N_2(g) + O_2(g) \qquad K = 1 \times 10^{-30}$$
$$2NO(g) + Br_2(g) \rightleftharpoons 2NOBr(g) \qquad K = 8 \times 10^{1}$$

Calculate K for the formation of one mole of NOBr from its elements in the gaseous state.

19. Given the following data at a certain temperature,

$$2N_2(g) + O_2(g) \rightleftharpoons 2N_2O(g) \qquad K = 1.2 \times 10^{-35}$$
$$N_2O_4(g) \rightleftharpoons 2NO_2(g) \qquad K = 4.6 \times 10^{-3}$$
$$\tfrac{1}{2}N_2(g) + O_2(g) \rightleftharpoons NO_2(g) \qquad K = 4.1 \times 10^{-9}$$

calculate K for the reaction between one mole of dinitrogen oxide gas and oxygen gas to give dinitrogen tetroxide gas.

20. Consider the following hypothetical reactions and their equilibrium constants at 75°C,

$$3A(g) \rightleftharpoons 3B(g) + 2C(g) \qquad K_1 = 0.31$$
$$3D(g) + 2B(g) \rightleftharpoons 2C(g) \qquad K_2 = 2.8$$

Find the equilibrium constant at 75°C for the following reaction

$$A(g) \rightleftharpoons D(g) + \tfrac{5}{3}B(g)$$

21. When one mole of carbon disulfide gas reacts with hydrogen gas, methane and hydrogen sulfide gases are formed. When equilibrium is reached at 900°C, analysis shows that $P_{CH_4} = 0.0833$ atm, $P_{H_2S} = 0.163$ atm, $P_{CS_2} = 1.27$ atm, and $P_{H_2} = 0.873$ atm.

(a) Write a balanced equation (smallest whole-number coefficients) for the reaction.
(b) Find K at 900°C.

22. Calculate K for the formation of methyl alcohol at 100°C:

$$CO(g) + 2H_2(g) \rightleftharpoons CH_3OH(g)$$

given that at equilibrium, the partial pressures of the gases are $P_{CO} = 0.814$ atm, $P_{H_2} = 0.274$ atm, and $P_{CH_3OH} = 0.0512$ atm.

23. Ammonium carbamate solid ($NH_4CO_2NH_2$) decomposes at 313 K into ammonia and carbon dioxide gases. At equilibrium, analysis shows that there are 0.0451 atm of CO_2, 0.0961 atm of ammonia, and 0.159 g of ammonium carbamate.

(a) Write a balanced equation for the decomposition of one mole of $NH_4CO_2NH_2$.
(b) Calculate K at 313 K.

24. Consider the decomposition at 25°C of one mole of NOBr gas into NO and Br_2 gases. At equilibrium, the concentrations of NOBr, NO, and Br_2 gases are 0.0162 M, 0.0011 M, and 0.072 M, respectively.

(a) Write a balanced equation for the reaction.
(b) Calculate K for the reaction at 25°C. (Note that the gases need to be expressed as pressure in atm.)

25. Consider the decomposition of ammonium hydrogen sulfide:

$$NH_4HS(s) \rightleftharpoons NH_3(g) + H_2S(g)$$

In a sealed flask at 25°C are 10.0 g of NH_4HS, ammonia with a partial pressure of 0.692 atm, and H_2S with a partial pressure of 0.0532 atm. When equilibrium is established, it is found that the partial pressure of ammonia has increased by 12.4%. Calculate K for the decomposition of NH_4HS at 25°C.

26. A sealed flask has 0.541 atm of SO_3 at 1000 K. The following equilibrium is established.

$$2SO_3(g) \rightleftharpoons 2SO_2(g) + O_2(g)$$

At equilibrium, the partial pressure of oxygen is measured to be 0.216 atm. Calculate K for the decomposition of SO_3 at 1000 K.

12-4 Applications of the Equilibrium Constant

K; Direction of the Reaction

27. A gaseous reaction mixture contains 0.30 atm SO_2, 0.16 atm Cl_2, and 0.50 atm SO_2Cl_2 in a 2.0-L container. $K = 0.011$ for the equilibrium system

$$SO_2Cl_2(g) \rightleftharpoons SO_2(g) + Cl_2(g)$$

(a) Is the system at equilibrium? Explain.
(b) If it is not at equilibrium, in which direction will the system move to reach equilibrium?

28. For the system

$$PCl_5(g) \rightleftharpoons PCl_3(g) + Cl_2(g)$$

K is 26 at 300°C. In a 10.0-L flask at 300°C, a gaseous mixture consists of all three gases with the following partial pressures: $P_{PCl_5} = 0.026$ atm, $P_{PCl_3} = 0.65$ atm, $P_{Cl_2} = 0.33$ atm

(a) Is the system at equilibrium? Explain.
(b) If the system is not at equilibrium, in which direction will the system move to reach equilibrium?

29. The reversible reaction between hydrogen chloride gas and one mole of oxygen gas produces steam and chlorine gas:

$$4HCl(g) + O_2(g) \rightleftharpoons 2Cl_2(g) + 2H_2O(g) \qquad K = 0.79$$

Predict the direction in which the system will move to reach equilibrium if one starts with

(a) $P_{H_2O} = P_{HCl} = P_{O_2} = 0.20$ atm
(b) $P_{HCl} = 0.30$ atm, $P_{H_2O} = 0.35$ atm, $P_{Cl_2} = 0.2$ atm, $P_{O_2} = 0.15$ atm

30. For the reaction

$$N_2(g) + 3H_2(g) \rightleftharpoons 2NH_3(g)$$

K at a certain temperature is 3.7×10^{-4}. Predict the direction in which the system will move to reach equilibrium if one starts with

(a) $P_{N_2} = P_{H_2} = P_{NH_3} = 0.01$ atm
(b) $P_{NH_3} = 0.0045$ atm
(c) $P_{N_2} = 1.2$ atm, $P_{H_2} = 1.88$ atm, $P_{NH_3} = 0.0058$ atm

31. A compound, X, decomposes at 131°C according to the following equation:

$$2X(g) \rightleftharpoons A(g) + 3C(g) \qquad K = 1.1 \times 10^{-3}$$

If a flask initially contains X, A, and C, all at partial pressures of 0.250 atm, in which direction will the reaction proceed?

32. Consider the following reaction at 75°C:

$$3R(s) + 2Q(g) \rightleftharpoons A(g) + 5B(l) \qquad K = 9.4$$

A 10.0-L sample contains 0.30 mol of R and Q and 0.50 mol of A and B. In which direction will the reaction proceed?

K; Extent of the Reaction

33. Consider the reaction between nitrogen and steam:

$$2N_2(g) + 6H_2O(g) \rightleftharpoons 4NH_3(g) + 3O_2(g)$$

At a certain temperature, $K = 28.6$. Calculate the equilibrium partial pressure of steam if $P_{NH_3} = 1.75$ atm, $P_{O_2} = 0.963$ atm, and $P_{N_2} = 0.996$ atm at equilibrium.

34. At 500°C, K for the formation of ammonia from nitrogen and hydrogen gases is 1.5×10^{-5}.

$$N_2(g) + 3H_2(g) \rightleftharpoons 2NH_3(g)$$

Calculate the equilibrium partial pressure of hydrogen if the equilibrium partial pressures of ammonia and nitrogen are 0.015 atm and 1.2 atm, respectively.

35. At a certain temperature, K is 4.9 for the formation of one mole of bromine chloride gas (BrCl) from its elements. A mixture at equilibrium at this temperature contains all three gases. The partial pressures at equilibrium of bromine and chlorine gas is 0.19 atm. What is the partial pressure of bromine chloride in this mixture at equilibrium?

36. At a certain temperature, $K = 0.29$ for the decomposition of two moles of iodine trichloride, $ICl_3(s)$, to chlorine and iodine gases. The partial pressure of chlorine gas at equilibrium is three times that of iodine gas. What are the partial pressures of iodine and chlorine at equilibrium?

37. For the reaction

$$N_2(g) + 2H_2O(g) \rightleftharpoons 2NO(g) + 2H_2(g)$$

K is 1.54×10^{-3}. When equilibrium is established, the partial pressure of nitrogen is 0.168 atm, and that of NO is 0.225 atm. The total pressure of the system at equilibrium is 1.87 atm. What are the equilibrium partial pressures of hydrogen and steam?

38. Nitrogen dioxide can decompose to nitrogen oxide and oxygen.

$$2NO_2(g) \rightleftharpoons 2NO(g) + O_2(g)$$

K is 0.87 at a certain temperature. A 5.0-L flask at equilibrium is determined to have a total pressure of 1.25 atm and oxygen to have a partial pressure of 0.515 atm. Calculate P_{NO} and P_{NO_2} at equilibrium.

39. Consider the following reaction:

$$N_2(g) + O_2(g) \rightleftharpoons 2NO(g)$$

At a certain temperature, the equilibrium constant for the reaction is 0.0639. What are the partial pressures of all gases at equilibrium if the initial partial pressure of the gases (both products and reactants) is 0.400 atm?

40. Consider the hypothetical reaction at 325°C

$$R(g) + Q(g) \rightleftharpoons 2Z(g) \qquad K = 2.71$$

What are the equilibrium partial pressures of all the gases if all the gases (products and reactants) have an initial partial pressure of 0.228 atm?

41. At a certain temperature, the equilibrium constant for the following reaction is 0.0472.

$$NO(g) + SO_3(g) \rightleftharpoons SO_2(g) + NO_2(g)$$

All gases are at an initial pressure of 0.862 atm.

(a) Calculate the partial pressure of each gas at equilibrium.
(b) Compare the initial total pressure with the total pressure of the gases at equilibrium. Would that relation be true of all gaseous systems?

42. At 460°C, the reaction

$$SO_2(g) + NO_2(g) \rightleftharpoons NO(g) + SO_3(g)$$

has $K = 84.7$. All gases are at an initial pressure of 1.25 atm.

(a) Calculate the partial pressure of each gas at equilibrium.
(b) Compare the total pressure initially with the total pressure at equilibrium. Would that relation be true of all gaseous systems?

43. Solid ammonium iodide decomposes to ammonia and hydrogen gases at sufficiently high temperatures.

$$NH_4I(s) \rightleftharpoons NH_3(g) + HI(g)$$

The equilibrium constant for the decomposition at 400°C is 0.215. Twenty grams of ammonium iodide are sealed in a 7.50-L flask and heated to 400°C.

(a) What is the total pressure in the flask at equilibrium?
(b) How much solid NH_4I is left after the decomposition?

44. Consider the following decomposition at 80°C.

$$PH_3BCl_3(s) \rightleftharpoons PH_3(g) + BCl_3(g) \qquad K = 0.054$$

Twenty grams of PH_3BCl_3 are sealed in a 5.0-L flask and heated to 80°C.
  (a) What is the total pressure in the flask at equilibrium?
  (b) How many grams of PH_3BCl_3 are left in the flask at equilibrium?

45. Hydrogen cyanide, a highly toxic gas, can decompose to cyanogen and hydrogen gases,

$$2HCN(g) \rightleftharpoons C_2N_2(g) + H_2(g)$$

At a certain temperature, K for this decomposition is 0.17. What are the partial pressures of all gases at equilibrium if initially the partial pressures are $P_{C_2N_2} = P_{H_2} = 0.32$ atm, $P_{HCN} = 0.45$ atm?

46. At 800 K, hydrogen iodide can decompose into hydrogen and iodine gases.

$$2HI(g) \rightleftharpoons I_2(g) + H_2(g)$$

At this temperature, $K = 0.0169$. What are the partial pressures at equilibrium of the hydrogen and iodine if initially a sealed flask at 800 K contains only HI at a pressure of 0.200 atm?

12-5 Effect of Changes in Conditions on an Equilibrium System

Changing Reaction Conditions; Le Châtelier's Principle

47. For the following reactions, predict whether the pressure of the reactants or products increases or remains the same when the volume of the reaction vessel is increased.
  (a) $H_2O(l) \rightleftharpoons H_2O(g)$
  (b) $N_2(g) + 3H_2(g) \rightleftharpoons 2NH_3(g)$
  (c) $C_2H_4(g) + H_2O(g) \rightleftharpoons C_2H_5OH(g)$

48. Follow the directions of Question 47 for the following reactions:
  (a) $2H_2O(g) + O_2(g) \rightleftharpoons 2H_2O_2(g)$
  (b) $2CH_4(g) \rightleftharpoons C_2H_2(g) + 3H_2(g)$
  (c) $Br_2(g) + H_2(g) \rightleftharpoons 2HBr(g)$

49. Consider the system

$$SO_3(g) \rightleftharpoons SO_2(g) + \tfrac{1}{2}O_2(g) \qquad \Delta H = 98.9 \text{ kJ}$$

  (a) Predict whether the forward or reverse reaction will occur when the equilibrium is disturbed by
    1. adding oxygen gas.
    2. compressing the system at constant temperature.
    3. adding argon gas.
    4. removing $SO_2(g)$.
    5. decreasing the temperature.
  (b) Which of the above factors will increase the value of K? Which will decrease it?

50. Consider the system

$$4NH_3(g) + 3O_2(g) \rightleftharpoons 2N_2(g) + 6H_2O(l) \qquad \Delta H = -1530.4 \text{ kJ}$$

  (a) How will the amount of ammonia at equilibrium be affected by
    1. removing $O_2(g)$?
    2. adding $N_2(g)$?
    3. adding water?

  4. expanding the container at constant pressure?
  5. increasing the temperature?
  (b) Which of the above factors will increase the value of K? Which will decrease it?

51. Predict the direction in which each of the following equilibria will shift if the pressure on the system is decreased by expansion.
  (a) $Ni(s) + 4CO(g) \rightleftharpoons Ni(CO)_4(g)$
  (b) $2CH_4(g) \rightleftharpoons C_2H_2(g) + 3H_2(g)$
  (c) $Br_2(g) + H_2(g) \rightleftharpoons 2HBr(g)$

52. Predict the direction in which each of the following equilibria will shift if the pressure on the system is decreased by expansion.
  (a) $2H_2O_2(l) \rightleftharpoons O_2(g) + 2H_2O(l)$
  (b) $2CH_4(g) \rightleftharpoons C_2H_2(g) + 3H_2(g)$
  (c) $I_2(s) + H_2(g) \rightleftharpoons 2HI(g)$

53. At a certain temperature, nitrogen and oxygen gases combine to form nitrogen oxide gas.

$$N_2(g) + O_2(g) \rightleftharpoons 2NO(g)$$

When equilibrium is established, the partial pressures of the gases are: $P_{N_2} = 1.200$ atm, $P_{O_2} = 0.800$ atm, $P_{NO} = 0.0220$ atm.
  (a) Calculate K at the temperature of the reaction.
  (b) After equilibrium is reached, more oxygen is added to make its partial pressure 1.200 atm. Calculate the partial pressure of all gases when equilibrium is reestablished.

54. Consider the following hypothetical reaction:

$$X_2(g) + R(s) \rightleftharpoons X_2R(g)$$

R has a molar mass of 73 g/mol. When equilibrium is established, a 2.5-L reaction vessel at 125°C contains 15.0 g of R, 4.3 atm of X_2, and 0.98 atm of X_2R.
  (a) Calculate K for the reaction at 125°C.
  (b) The mass of R is doubled. What are the partial pressures of X_2 and X_2R when equilibrium is reestablished?
  (c) The partial pressure of X_2 is decreased to 2.0 atm. What are the partial pressures of X_2 and X_2R when equilibrium is reestablished?

55. Iodine chloride decomposes at high temperatures to iodine and chlorine gases.

$$2ICl(g) \rightleftharpoons I_2(g) + Cl_2(g)$$

Equilibrium is established at a certain temperature when the partial pressures of ICl, I_2, and Cl_2 are (in atm) 0.43, 0.16, and 0.27, respectively.
  (a) Calculate K.
  (b) If enough iodine condenses to decrease its partial pressure to 0.10 atm, in which direction will the reaction proceed? What is the partial pressure of iodine when equilibrium is reestablished?

56. Sulfur oxychloride, SO_2Cl_2, decomposes to sulfur dioxide and chlorine gases.

$$SO_2Cl_2(g) \rightleftharpoons SO_2(g) + Cl_2(g)$$

At a certain temperature, the equilibrium partial pressures of SO_2, Cl_2, and SO_2Cl_2 are 1.88 atm, 0.84 atm, and 0.27 atm, respectively.
  (a) What is K at that temperature?
  (b) Enough Cl_2 condenses to reduce its partial pressure to 0.68 atm. What are the partial pressures of all gases when equilibrium is reestablished?

57. For the following reaction

$$C(s) + 2H_2(g) \rightleftharpoons CH_4(g)$$

$K = 0.26$ at 1000°C (3 significant figures). What is the equilibrium constant at 750°C (3 significant figures)?

58. For the system

$$2SO_3(g) \rightleftharpoons 2SO_2(g) + O_2(g)$$

$K = 1.32$ at 627°C. What is the equilibrium constant at 555°C?

59. For a certain reaction, $\Delta H°$ is +33 kJ. What is the ratio of the equilibrium constant at 400 K to that at 200 K?

60. What is the value of $\Delta H°$ for a reaction in which K at 37°C is 48% of K at 27°C?

Unclassified

61. Hemoglobin (Hb) binds to both oxygen and carbon monoxide. When the carbon monoxide replaces the oxygen in an organism, the following reaction occurs:

$$HbO_2(aq) + CO(g) \rightleftharpoons HbCO(aq) + O_2(g)$$

At 37°C, K is about 200. When equal concentrations of HbO_2 and HbCO are present, the effect of CO inhalation is fatal. Assuming $P_{O_2} = 0.21$ atm, what is P_{CO} when $[HbO_2] = [HbCO]$?

62. Mustard gas, used in chemical warfare in World War I, has been found to be an effective agent in the chemotherapy of Hodgkin's disease. It can be produced according to the following reaction:

$$SCl_2(g) + 2C_2H_4(g) \rightleftharpoons S(CH_2CH_2Cl)_2(g)$$

An evacuated 5.0-L flask at 20.0°C is filled with 0.258 mol SCl_2 and 0.592 mol C_2H_4. After equilibrium is established, 0.0349 mol mustard gas is present.

(a) What is the partial pressure of each gas at equilibrium?

(b) What is K at 20.0°C?

63. At 1000 K, hydrogen dissociates into gaseous atoms:

$$H_2(g) \rightleftharpoons 2H(g)$$

where K is 5.0×10^{-18}. Ten moles of hydrogen gas are pumped into an evacuated 15.0-L flask and heated to 1000 K.

(a) How many atoms of H are in the flask when equilibrium is reached?

(b) What percent (in moles) of H_2 dissociated?

64. For the decomposition of $CaCO_3$ at 900°C, $K = 1.04$.

$$CaCO_3(s) \rightleftharpoons CaO(s) + O_2(g)$$

Find the smallest mass of $CaCO_3$ needed to reach equilibrium in a 5.00-L vessel at 900°C.

65. Isopropyl alcohol is the main ingredient in rubbing alcohol. It can decompose into acetone (the main ingredient in nail polish remover) and hydrogen gas according to the following reaction:

$$C_3H_7OH(g) \rightleftharpoons C_2H_6CO(g) + H_2(g)$$

At 180°C, the equilibrium constant for the decomposition is 0.45. If 20.0 mL ($d = 0.785$ g/mL) of isopropyl alcohol is placed in a 5.00-L vessel and heated to 180°C, what percentage remains undissociated at equilibrium?

66. Consider the equilibrium

$$H_2(g) + S(s) \rightleftharpoons H_2S(g)$$

When this system is at equilibrium at 25°C in a 2.00-L container, 0.120 mol of H_2, 0.034 mol of H_2S, and 0.4000 mol of S are present. When the temperature is increased to 35°C, the partial pressure of H_2 increases to 1.56 atm.

(a) What is K for the reaction at 25°C?

(b) What is K for the reaction at 35°C?

Conceptual Problems

67. Consider the following graph representing the progression of a reaction with time.

$$A(g) \rightleftharpoons 2B(g)$$

(a) What is the relationship between Q and K at Points 1, 3, and 4?

(b) What caused the disturbance at Point 2?

(c) Draw a similar graph for the same reaction, but increase the pressure in A.

68. The following data apply to the *unbalanced* equation

$$A(g) \rightleftharpoons B(g)$$

Time (s)	0	50	100	150	200	250
P_A (atm)	2.00	1.25	0.95	0.80	0.71	0.68
P_B (atm)	0.10	0.60	0.80	0.90	0.96	0.98

(a) On the basis of the data, balance the equation (simplest whole-number coefficients).

(b) Has the system reached equilibrium? Explain.

69. Consider the reaction:

$$A(g) + 2B(g) + C(s) \rightleftharpoons 2D(g)$$

At 25°C, only A, B, and C are present. The partial pressures of A, B, and D are given as P_A, P_B, and P_D. Equilibrium is established 18 minutes after the reaction starts. Use terms is less than (**LT**), is greater than (**GT**), is equal to (**EQ**), or insufficient information (**X**) to answer the following questions.

(a) P_D at 5 min _____ P_A.

(b) P_A at 21 min _____ P_A at 27 min.

(c) P_B at 7 min _____ P_B at 13 min.

(d) After 20 min, more B is added. When equilibrium is reestablished, K before the addition _____ K after the addition.

(e) After 20 min, the temperature is increased to 35°C. P_A before the temperature increase _____ P_A after the temperature increase after equilibrium is reestablished.

70. For the reaction

$$C(s) + CO_2(g) \rightleftharpoons 2CO(g)$$

$K = 168$ at 1273 K. If one starts with 0.3 atm of CO_2 and 12.0 g of C at 1273 K, will the equilibrium mixture contain
(a) mostly CO_2?
(b) mostly CO?
(c) roughly equal amounts of CO_2 and CO?
(d) only C?

71. Consider the system

$$A(g) + 2B(g) + C(s) \rightleftharpoons 2D(g)$$

at 25°C. At zero time, only A, B, and C are present. The reaction reaches equilibrium 10 min after the reaction is initiated. Partial pressures of A, B, and D are written as P_A, P_B, and P_D.

Answer the questions below, using **LT** (for *is less than*), **GT** (for *is greater than*), **EQ** (for *is equal to*), or **MI** (for *more information required*).
(a) P_D at 11 min _____ P_D at 12 min.
(b) P_A at 5 min _____ P_A at 7 min.
(c) K for the forward reaction _____ K for the reverse reaction.
(d) At equilibrium, K _____ Q.
(e) After the system is at equilibrium, more of gas B is added. After the system returns to equilibrium, K before the addition of B _____ K after the addition of B.
(f) The same reaction is initiated, this time with a catalyst. K for the system without a catalyst _____ K for the system with a catalyst.
(g) K for the formation of one mole of D _____ K for the formation of two moles of D.
(h) The temperature of the system is increased to 35°C. P_B at equilibrium at 25°C _____ P_B at equilibrium at 35°C.
(i) Ten more grams of C are added to the system. P_B before the addition of C _____ P_B after the addition of C.

72. The graph below is similar to that of Figure 12.2.

If after 100 s have elapsed the partial pressure of N_2O_4 is increased to 1.0 atm, what will the graph for N_2O_4 look like beyond 100 s?

73. The system

$$3Z(g) + Q(g) \rightleftharpoons 2R(g)$$

is at equilibrium when the partial pressure of Q is 0.44 atm. Sufficient R is added to increase the partial pressure of Q temporarily to 1.5 atm. When equilibrium is reestablished, the partial pressure of Q could be which of the following?
(a) 1.5 atm   (b) 1.2 atm   (c) 0.80 atm
(d) 0.44 atm   (e) 0.40 atm

74. The figures below represent the following reaction at equilibrium at different temperatures.

$$A_2(g) + 3B_2(g) \rightleftharpoons 2AB_3(g)$$

where squares represent atom A and circles represent atom B. Is the reaction exothermic?

T = 300 K   **T = 400 K**

75. Criticize the statement "The rate of most reactions increases as the temperature rises. Thus, the equilibrium constant of most reactions increases as temperature rises."

76. Consider the statement "The equilibrium constant for a mixture of hydrogen, nitrogen, and ammonia is 3.41." What information is missing from this statement?

Challenge Problems

77. Consider the following reaction at a certain temperature:

$$2NO(g) + O_2(g) \rightleftharpoons 2NO_2(g)$$

A reaction mixture contains 0.70 atm of O_2 and 0.81 atm of NO. When equilibrium is established, the total pressure in the reaction vessel is 1.20 atm. Find K.

78. Derive the relationship

$$K = K_c \times (RT)^{\Delta n_g}$$

where K_c is the equilibrium constant using molarities and Δn_g is the change in the number of moles of gas in the reaction (see page 310). (*Hint:* Recall that $P_A = n_A RT/V$ and $n_A/V = [A]$.)

79. Ammonia can decompose into its constituent elements according to the reaction

$$2NH_3(g) \rightleftharpoons N_2(g) + 3H_2(g)$$

The equilibrium constant for the decomposition at a certain temperature is 2.5. Calculate the partial pressures of all the gases at equilibrium if ammonia with a pressure of 1.00 atm is sealed in a 3.0-L flask.

80. Hydrogen iodide gas decomposes to hydrogen gas and iodine gas:

$$2HI(g) \rightleftharpoons H_2(g) + I_2(g)$$

To determine the equilibrium constant of the system, identical one-liter glass bulbs are filled with 3.20 g of HI and maintained at a certain temperature. Each bulb is periodically opened and analyzed for iodine formation by titration with sodium thiosulfate, $Na_2S_2O_3$.

$$I_2(aq) + 2S_2O_3{}^{2-}(aq) \longrightarrow S_4O_6{}^{2-}(aq) + 2I^-(aq)$$

It is determined that when equilibrium is reached, 37.0 mL of 0.200 M $Na_2S_2O_3$ is required to titrate the iodine. What is K at the temperature of the experiment?

81. For the system

$$SO_3(g) \rightleftharpoons SO_2(g) + \tfrac{1}{2}O_2(g)$$

at 1000 K, $K = 0.45$. Sulfur trioxide, originally at 1.00 atm pressure, partially dissociates to SO_2 and O_2 at 1000 K. What is its partial pressure at equilibrium?

82. A student studies the equilibrium

$$I_2(g) \rightleftharpoons 2I(g)$$

at a high temperature. She finds that the total pressure at equilibrium is 40% greater than it was originally, when only I_2 was present. What is K for this reaction at that temperature?

83. At a certain temperature, the reaction

$$Xe(g) + 2F_2(g) \rightleftharpoons XeF_4(g)$$

gives a 50.0% yield of XeF_4, starting with Xe ($P_{Xe} = 0.20$ atm) and F_2 ($P_{F_2} = 0.40$ atm). Calculate K at this temperature. What must the initial pressure of F_2 be to convert 75.0% of the xenon to XeF_4?

84. Benzaldehyde, a flavoring agent, is obtained by the dehydrogenation of benzyl alcohol.

$$C_6H_5CH_2OH(g) \rightleftharpoons C_6H_5CHO(g) + H_2(g)$$

K for the reaction at 250°C is 0.56. If 1.50 g of benzyl alcohol is placed in a 2.0-L flask and heated to 250°C,

(a) what is the partial pressure of the benzaldehyde when equilibrium is established?

(b) how many grams of benzyl alcohol remain at equilibrium?

85. Hydrogen and nitrogen react to form NH_3

$$3H_2(g) + N_2(g) \rightleftharpoons 2NH_3(g)$$

At a certain temperature, it is determined that an equilibrium mixture of N_2, H_2, and NH_3 has a total pressure of 3.065 atm. The mixture has 20.4% NH_3. What is K at that temperature?

86. At 165°C, a solid chloride (X) reacts with 0.0417 mol of steam to form the solid oxychloride, Y, and HCl gas. Steam and HCl have a 1:2 stoichiometric ratio in the reaction

$$X(s) + H_2O(g) \rightleftharpoons Y(s) + 2HCl(g)$$

When the reaction reaches equilibrium, the gases are transferred without loss to a 2.50-L vessel. The temperature is kept at 165°C. Sufficient silver(I) ion is added to precipitate the chloride as AgCl. If 7.29 g of AgCl are obtained, what is K for the reaction of the chloride with steam at 165°C?

Acids and Bases

The principal acid present in lemons is citric acid, $H_3C_6H_5O_2$. Citric acid is a triprotic acid capable of donating three protons.

There is nothing in the Universe but alkali and acid, From which Nature composes all things.

—OTTO TACHENIUS (1671)

Van Gogh Museum

Among the solution reactions considered in Chapter 4 were those between acids and bases. In this chapter, we take a closer look at the properties of acidic and basic water solutions. In particular, we examine

- the ionization of water and the equilibrium between hydrated H^+ ions (H_3O^+ ions) and OH^- ions in water solution (Section 13-2).
- the quantities pH and pOH, used to describe the acidity or basicity of water solutions (Section 13-3).
- the types of species that act as weak acids and weak bases and the equilibria that apply in their water solutions (Sections 13-4 and 13-5).
- the acid-base properties of salt solutions (Section 13-6).
- the Lewis model, which expands the range of compounds that can be considered acids and bases (Section 13-7).

13-1 Brønsted-Lowry Acid-Base Model

In Chapter 4 we considered an acid to be a substance that produces an excess of H^+ ions in water. A base was similarly defined to be a substance that forms excess OH^- ions in water solution. This approach, first proposed by Svante Arrhenius in

1884, is a very practical one, but it has one disadvantage. It severely limits the number of reactions that qualify as acid-base.

The model of acids and bases in this chapter is a somewhat more general one developed independently by Johannes Brønsted (1879–1947) in Denmark and Thomas Lowry (1874–1936) in England in 1923. The *Brønsted-Lowry model* ◀ ⓘ focuses on the nature of acids and bases and the reactions that take place between them. Specifically, it considers that

Sometimes called the Lowry-Brønsted model, at least in England.

■ *an* acid *is a proton (H⁺ ion) donor.*
■ *a* base *is a proton (H⁺ ion) acceptor.*
■ *in an acid-base reaction, a proton is transferred from an acid to a base.*

A Brønsted-Lowry acid-base reaction might be represented as

$$HB(aq) + A^-(aq) \rightleftharpoons HA(aq) + B^-(aq)$$

The species HB and HA both act as a **Brønsted-Lowry acid** in the forward and reverse reactions, respectively; A^- and B^- both act as a **Brønsted-Lowry base**.

The Brønsted-Lowry model introduces some new terminology.

1. The species formed when a proton is removed from an acid is referred to as the **conjugate base** of that acid; B^- is the conjugate base of HB. The species formed when a proton is added to a base is called the **conjugate acid** of that base: HA is the conjugate acid of A^-. In other words, conjugate acid-base pairs differ only in the presence or absence of a proton, H^+. A conjugate acid has one more proton than its conjugate base. Thus we have

Conjugate Acid	Conjugate Base
HF	F^-
HSO_4^-	SO_4^{2-}
NH_4^+	NH_3

2. A species that can either accept or donate a proton is referred to as an **amphiprotic species.** An example is the H_2O molecule, which can gain a proton to form the **hydronium ion,** H_3O^+, or lose a proton, leaving the hydroxide ion, OH^-.

$$\left[:\ddot{O}-H\right]^- \xleftarrow{-H^+} H-\ddot{O}-H \xrightarrow{+H^+} [H-\overset{\overset{\displaystyle H}{\displaystyle |}}{O}-H]^+$$

hydroxide ion   water molecule   hydronium ion

EXAMPLE 13.1

Consider the following species: HNO_2, F^-, and HCO_3^-. Write the formula of the conjugate base for the weak acid, the conjugate acid for the weak base, and the conjugate acid and base of the amphiprotic ion.

STRATEGY

1. Form the conjugate base by removing one H atom. Decrease the charge by one unit (e.g., −1 to −2).

2. Form the conjugate acid by adding one H atom. Increase the charge by one unit (e.g., −1 to 0).

SOLUTION

HNO_2 conjugate base	$HNO_2 \longrightarrow NO_2^{0-1} \longrightarrow NO_2^-$
F^- conjugate acid	$F^- \longrightarrow HF^{-1+1} \longrightarrow HF$
HCO_3^- conjugate base	$HCO_3^- \longrightarrow CO_3^{-1-1} \longrightarrow CO_3^{2-}$
HCO_3^- conjugate acid	$HCO_3^- \longrightarrow H_2CO_3^{-1+1} \longrightarrow H_2CO_3$

13-2 The Ion Product of Water

The acidic and basic properties of aqueous solutions are dependent on an equilibrium that involves the solvent water. The reaction involved can be regarded as a Brønsted-Lowry acid-base reaction in which the H_2O molecule shows its amphiprotic nature:

$$H_2O + H_2O \rightleftharpoons H_3O^+(aq) + OH^-(aq)$$

Alternatively, and somewhat more simply, the reaction can be viewed as the ionization of a single H_2O molecule:

$$H_2O \rightleftharpoons H^+(aq) + OH^-(aq)$$

Recall (Chapter 12) that in the equilibrium constant expression for reactions in solution

- solutes enter as their molarity, [ ].
- the solvent, H_2O in this case, does not appear. Its concentration is essentially the same in all dilute solutions.

Hence for the ionization of water, the equilibrium constant expression can be written*

$$K_w = [H^+][OH^-] \tag{13.1}$$

where K_w, referred to as the **ion product constant of water**, has a very small value. At 25°C,

$$K_w = 1.0 \times 10^{-14} \tag{13.2}$$

This equation applies to *any* water solution at 25°C.

The concentrations of H^+ ($[H^+] = [H_3O^+]$) ions and OH^- ions in *pure water* at 25°C are readily calculated from K_w. Notice from the equation for ionization that these two ions are formed in equal numbers. Hence in pure H_2O,

$$[H^+] = [OH^-]$$
$$K_w = [H^+][OH^-] = [H^+]^2 = 1.0 \times 10^{-14}$$
$$[H^+] = 1.0 \times 10^{-7} M = [OH^-]$$

An aqueous solution in which $[H^+]$ *equals* $[OH^-]$ is called a **neutral solution**. It has an $[H^+]$ of $1.0 \times 10^{-7} M$ at 25°C.

In most water solutions, the concentrations of H^+ and OH^- are not equal. The equation $[H^+][OH^-] = 1.0 \times 10^{-14}$ indicates that these two quantities are inversely proportional to one another (Figure 13.1). When $[H^+]$ is very high, $[OH^-]$ is very low, and vice versa.

An aqueous solution in which $[H^+]$ is greater than $[OH^-]$ is termed acidic. An aqueous solution in which $[OH^-]$ is greater than $[H^+]$ is basic (alkaline). Therefore,

if $[H^+] > 1.0 \times 10^{-7} M$, $[OH^-] < 1.0 \times 10^{-7} M$, solution is acidic

if $[OH^-] > 1.0 \times 10^{-7} M$, $[H^+] < 1.0 \times 10^{-7} M$, solution is basic

Figure 13.1 **The $[H^+]$-$[OH^-]$ relationship.** A graph of $[OH^-]$ versus $[H^+]$ looks very much like a graph of gas volume versus pressure. In both cases, the two variables are inversely proportional to one another. When $[H^+]$ gets larger, $[OH^-]$ gets smaller.

In any water solution at 25°C $[H^+] \times [OH^-] = 1.0 \times 10^{-14}$

There are very few H^+ and OH^- ions in pure water.

13-3 pH and pOH

As just pointed out, the acidity or basicity of a solution can be described in terms of its H^+ concentration. In 1909, Søren Sørensen (1868–1939), a biochemist working at the Carlsberg Brewery in Copenhagen, proposed an alternative method of

---

*For simplicity, throughout this chapter, we will use H^+ rather than H_3O^+ in equilibrium constant expressions.

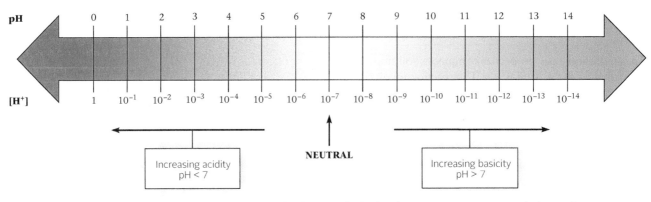

Figure 13.2 pH-[H⁺] relationship. Acidity is inversely related to pH; the higher the H⁺ ion concentration, the lower the pH. In neutral solution, $[H^+] = [OH^-] = 1.0 \times 10^{-7}\ M$; pH = 7.00 at 25°C.

specifying the acidity of a solution. He defined a term called **pH** (for "power of the hydrogen ion"):

$$pH = -\log_{10}[H^+] = -\log_{10}[H_3O^+] \qquad (13.3)$$

or

$$[H^+] = [H_3O^+] = 10^{-pH}$$

Figure 13.2 shows the relationship between pH and [H⁺]. Notice that, as the defining equation implies, pH increases by one unit when the concentration of H⁺ decreases by a power of 10. Moreover, *the higher the pH, the less acidic the solution.* Most aqueous solutions have hydrogen ion concentrations between 1 and $10^{-14}\ M$ and hence have a pH between 0 and 14. ◄ 𝒊

The pH, like [H⁺] or [OH⁻], can be used to differentiate acidic, neutral, and basic solutions. At 25°C,

if pH < 7.0,   solution is acidic
if pH = 7.0,   solution is neutral
if pH > 7.0,   solution is basic

Table 13.1 shows the pH of some common solutions.

A similar approach is used for the hydroxide ion concentration. The **pOH** of a solution is defined as

$$pOH = -\log_{10}[OH^-] \qquad (13.4)$$

Because $[H^+][OH^-] = 1.0 \times 10^{-14}$ at 25°C, it follows that at this temperature

$$pH + pOH = 14.00$$

Thus a solution that has a pH of 6.20 must have a pOH of 7.80, and vice versa.

𝒊
It is simpler to use numbers (pH = 4) than exponents ($[H^+] = 10^{-4}\ M$) to describe acidity.

Table 13.1 pH of Some Common Materials

Lemon juice	2.2–2.4	Urine, human	4.8–8.4
Wine	2.8–3.8	Cow's milk	6.3–6.6
Vinegar	3.0	Saliva, human	6.5–7.5
Tomato juice	4.0	Drinking water	5.5–8.0
Beer	4–5	Blood, human	7.3–7.5
Cheese	4.8–6.4	Seawater	8.3

EXAMPLE 13.2

Consider the following solutions at 25°C.

a   Find the $[H^+]$ and pH of a tap water sample in which $[OH^-] = 2.0 \times 10^{-7}$.

ANALYSIS

Information given:	$[OH^-]$ $(2.0 \times 10^{-7}\ M)$
Information implied:	K_w value
Asked for:	$[H^+]$ and pH

STRATEGY

1. Substitute into Equation 13.1 to obtain $[H^+]$.
2. Substitute into Equation 13.3 to convert $[H^+]$ to pH.

SOLUTION

1. $[H^+]$	$K_w = 1.0 \times 10^{-14} = [H^+][OH^-] = [H^+](2.0 \times 10^{-7})$
	$[H^+] = \dfrac{1.0 \times 10^{-14}}{2.0 \times 10^{-7}} = 5.0 \times 10^{-8}\ M$ ⓘ Remember to enter the minus sign before pressing the 10^x key.
2. pH	$pH = -\log_{10}[H^+] = -\log_{10}(5.0 \times 10^{-8}) = 7.30$ ◂ⓘ

b   Find the $[H^+]$ and $[OH^-]$ of human blood at pH 7.40.

ANALYSIS

Information given:	pH (7.40)
Information implied:	K_w value
Asked for:	$[H^+]$ and $[OH^-]$

STRATEGY

1. Substitute into Equation 13.3 to convert pH to $[H^+]$.
2. Substitute into Equation 13.1 to obtain $[OH^-]$.

SOLUTION

1. $[H^+]$	$pH = -\log_{10}[H^+] \longrightarrow 7.40 = -\log_{10}[H^+] \longrightarrow [H^+] = 10^{-7.40} = 4.0 \times 10^{-8}\ M$
2. $[OH^-]$	$1.0 \times 10^{-14} = [OH^-](4.0 \times 10^{-8}) \longrightarrow [OH^-] = 2.5 \times 10^{-7}\ M$

c   Find the pOH of a solution in which $[H^+] = (5.0)[OH^-]$.

ANALYSIS

Information given:	$[H^+]$ and $[OH^-]$ relation $([H^+] = 5.0\ [OH^-])$
Information implied:	K_w value
Asked for:	pOH

STRATEGY

1. Substitute into Equation 13.1 to obtain $[OH^-]$.
2. Substitute into Equation 13.4 to convert $[OH^-]$ to pOH.

SOLUTION

1. $[OH^-]$	$1.0 \times 10^{-14} = [OH^-][H^+] = [OH^-](5.0[OH^-]) = 5.0[OH^-]^2$
	$[OH^-]^2 = \dfrac{1.0 \times 10^{-14}}{5.0} \longrightarrow [OH^-] = 4.5 \times 10^{-8}\ M$
2. pOH	$pOH = -\log_{10}[OH^-] = -\log_{10}(4.5 \times 10^{-8}) = 7.35$

The number calculated in part (b) for the concentration of H^+ in blood, 4.0×10^{-8} M, is very small. You may wonder what difference it makes whether $[H^+]$ is 4.0×10^{-8} M, 4.0×10^{-7} M, or some other such tiny quantity. In practice, it makes a great deal of difference because a large number of biological processes involve H^+ as a reactant, so the rates of these processes depend on its concentration. If $[H^+]$ increases from 4.0×10^{-8} M to 4.0×10^{-7} M, the rate of a first-order reaction involving H^+ increases by a factor of 10. Indeed, if $[H^+]$ in blood increases by a much smaller amount, from 4.0×10^{-8} M to 5.0×10^{-8} M (pH $7.40 \longrightarrow 7.30$), a condition called *acidosis* develops. The nervous system is depressed; fainting and even coma can result.

pH of Strong Acids and Strong Bases

As pointed out in Chapter 4, the following acids are strong

$$HCl \qquad HBr \qquad HI$$
$$HClO_4 \qquad HNO_3 \qquad H_2SO_4*$$

in the sense that they are completely ionized in water. The reaction of the strong acid HCl with water can be represented by the equation (Figure 13.3)

$$HCl(aq) + H_2O \longrightarrow H_3O^+(aq) + Cl^-(aq)$$

Because this reaction goes to completion, it follows that a 0.10 M solution of HCl is 0.10 M in both H_3O^+ (i.e., H^+) and Cl^- ions. ◄ⓘ

The strong bases

$$LiOH \qquad NaOH \qquad KOH$$
$$Ca(OH)_2 \qquad Sr(OH)_2 \qquad Ba(OH)_2$$

are completely ionized in dilute water solution. A 0.10 M solution of NaOH is 0.10 M◄ⓘ in both Na^+ and OH^- ions.

The fact that strong acids and bases are completely ionized in water makes it relatively easy to calculate the pH of their solutions (Example 13.3).

ⓘ
1 M HCl has a pH of 0,
1 M NaOH a pH of 14.

---

*The *first* ionization of H_2SO_4 is complete:

$$H_2SO_4(aq) \longrightarrow H^+(aq) + HSO_4^-(aq)$$

The *second* ionization is reversible:

$$HSO_4^-(aq) \rightleftharpoons H^+(aq) + SO_4^{2-}(aq)$$

EXAMPLE 13.3 GRADED

Consider barium hydroxide, $Ba(OH)_2$, a white powdery substance used in the treatment of waste water. A student is asked by his TA to prepare 455 mL of a solution containing 4.23 g of $Ba(OH)_2$.

a	What is the calculated pH of the prepared solution?

ANALYSIS	
Information given:	mass $Ba(OH)_2$ (4.23 g); volume of solution (455 mL)
Information implied:	molar mass of $Ba(OH)_2$; K_w
Asked for:	pH of the solution

continued

STRATEGY

1. Start by expressing the concentration in g $Ba(OH)_2$/L of solution.

2. Follow the pathway:

$$\text{mass } Ba(OH)_2/L \xrightarrow{MM} [Ba(OH)_2] \xrightarrow{2[OH^-]/[Ba(OH)_2]} [OH^-] \xrightarrow{K_w} [H^+] \xrightarrow{Eq\ 13.3} pH$$

SOLUTION

$[OH^-]$	$\dfrac{4.23 \text{ g } Ba(OH)_2}{0.455 \text{ L}} \times \dfrac{1 \text{ mol } Ba(OH)_2}{171.3 \text{ g}} \times \dfrac{2 \text{ mol } OH^-}{1 \text{ mol } Ba(OH)_2} = 0.109 \text{ mol/L} = 0.109 \ M$
$[H^+]$	$[H^+] = \dfrac{1.0 \times 10^{-14}}{0.109} = 9.2 \times 10^{-14} \ M$
pH	$pH = -\log_{10}(9.2 \times 10^{-14}) = 13.04$

b His TA measures the pH of the prepared solution with a pH meter and finds it to be 12.91. How much $Ba(OH)_2$ did the student really add?

ANALYSIS

Information given:	pH (12.91) mass $Ba(OH)_2$ to be added (4.23 g); volume of solution (455 mL)
Information implied:	molar mass of $Ba(OH)_2$; K_w
Asked for:	mass $Ba(OH)_2$ added compared to student's calculation

STRATEGY

The pathway to follow is the reverse of that in (a):

$$pH \xrightarrow{Eq\ 13.3} [H^+] \xrightarrow{K_w} [OH^-] \xrightarrow{2[OH^-]/[Ba(OH)_2]} [Ba(OH)_2] \xrightarrow{V} \text{mol } Ba(OH)_2 \xrightarrow{MM} \text{mass}$$

SOLUTION

$[OH^-]$	$[H^+] = 10^{-12.91} = 1.2 \times 10^{-13} \ M; \quad [OH^-] = \dfrac{1.0 \times 10^{-14}}{1.2 \times 10^{-13}} = 0.081 \ M$
mass $Ba(OH)_2$ really added	$\dfrac{0.081 \text{ mol } OH^-}{1 \text{ L}} \times \dfrac{1 \text{ mol } Ba(OH)_2}{2 \text{ mol } OH^-} \times 0.455 \text{ L} \times \dfrac{171.3 \text{ g } Ba(OH)_2}{1 \text{ mol}} = 3.16 \text{ g}$

c The TA added 0.50 g of NaOH to the prepared solution to increase the pH. What is the pH of the prepared solution after the addition of NaOH?

ANALYSIS

Information given:	from (b) $[OH^-]$ due to $Ba(OH)_2$ (0.081 M); volume of solution (455 mL) mass of NaOH added (0.50 g)
Information implied:	molar mass of NaOH; K_w
Asked for:	pH of the solution

STRATEGY

1. Find moles OH^- contributed by $Ba(OH)_2$ from (b).

2. Find moles OH^- contributed by NaOH.

3. Find $[OH^-]$ after NaOH addition.

$$\dfrac{(\text{mol } OH^-)_{NaOH} + (\text{mol } OH^-)_{Ba(OH)_2}}{V}$$

4. Find $[H^+]$ and pH.

continued

	SOLUTION
1. mol OH⁻ from Ba(OH)₂	$\dfrac{0.081 \text{ mol OH}^-}{1 \text{ L}} \times 0.455 \text{ L} = 0.0369$
2. mol OH⁻ from NaOH	$0.50 \text{ g NaOH} \times \dfrac{1 \text{ mol}}{40.0 \text{ g}} \times \dfrac{1 \text{ mol OH}^-}{1 \text{ mol NaOH}} = 0.0125$
3. [OH⁻]	$\dfrac{0.0369 \text{ mol} + 0.0125 \text{ mol}}{0.455 \text{ L}} = 0.108 \ M$
4. [H⁺] and pH	$[\text{H}^+] = \dfrac{1.0 \times 10^{-14}}{0.108} = 9.2 \times 10^{-14} \ M$ $$\text{pH} = -\log_{10}(9.2 \times 10^{-14}) = 13.04$$

Measuring pH

The pH of a solution can be measured by an instrument called a pH meter. A pH meter translates the H⁺ ion concentration of a solution into an electrical signal that is converted into either a digital display or a deflection on a meter that reads pH directly (Figure 13.4). Later, in Chapter 17, we will consider the principle on which the pH meter works.

A less accurate but more colorful way to measure pH uses a universal indicator, which is a mixture of acid-base indicators that shows changes in color at different pH values (Figure 13.5). A similar principle is used with pH paper. ◀ⓘ Strips of this paper are coated with a mixture of pH-sensitive dyes; these strips are widely used to test the pH of biological fluids, groundwater, and foods. Depending on the indicators used, a test strip can measure pH over a wide or narrow range.

Home gardeners frequently measure and adjust soil pH in an effort to improve the yield and quality of grass, vegetables, and flowers. Interestingly enough, the colors of many flowers depend on pH; among these are dahlias, delphiniums, and, in particular, hydrangeas (Figure 13.6). The first research in this area was carried out by Robert Boyle of gas-law fame, who published a paper on the relation between flower color and acidity in 1664.

ⓘ One type of paper, pH paper, is also known as litmus paper.

Figure 13.4 A pH meter with a digital readout. With a pH of 3.12, cola drinks are quite acidic.

Figure 13.5 pH as shown by a universal indicator. Universal indicator is deep red in strongly acidic solution *(upper left)*. It changes to yellow and green at pH 6 to 8, and then to deep violet in strongly basic solution *(lower right)*.

© Cengage Learning/Charles D. Winters

You can assume that all common acids other than HCl, HBr, HI, HNO$_3$, HClO$_4$, and H$_2$SO$_4$ are weak.

13-4 Weak Acids and Their Equilibrium Constants

Many solutes behave as weak acids; that is, they react reversibly with water to form H$_3$O$^+$ ions. Using HB to represent a weak acid, its Brønsted-Lowry reaction with water is

$$HB(aq) + H_2O \longrightarrow H_3O^+(aq) + B^-(aq)$$

Typically, this reaction occurs to a very small extent; usually, fewer than 1% of the HB molecules are converted to ions.

Most weak acids (Figure 13.7) fall into one of two categories:

1. Molecules containing an ionizable hydrogen atom. This type of weak acid was discussed in Chapter 4. There are literally thousands of molecular weak acids, most of them organic in nature. Among the molecular inorganic weak acids is nitrous acid:

$$HNO_2(aq) + H_2O \rightleftharpoons H_3O^+(aq) + NO_2^-(aq)$$

2. Cations (acidic ion). The ammonium ion, NH$_4^+$, behaves as a weak acid in water; a 0.10 M solution of NH$_4$Cl has a pH of about 5. The process by which the NH$_4^+$ ion lowers the pH of water can be represented by the (Brønsted-Lowry) equation:

$$NH_4^+(aq) + H_2O \rightleftharpoons H_3O^+(aq) + NH_3(aq)$$

Comparing this equation with the one above for HNO$_2$, you can see that they are very similar. In both cases, a weak acid (HNO$_2$, NH$_4^+$) is converted to its conjugate base (NO$_2^-$, NH$_3$). The fact that the NH$_4^+$ ion has a +1 charge, whereas the HNO$_2$ molecule is neutral, is really irrelevant here.

You may be surprised to learn that many metal cations act as weak acids in water solution. A 0.10 M solution of Al$_2$(SO$_4$)$_3$ has a pH close to 3; you can change the color of hydrangeas from red to blue by adding aluminum salts to soil. At first glance it is not at all obvious how a cation such as Al^{3+} can make a water solution acidic. However, the aluminum cation in water solution is really a hydrated species, Al(H$_2$O)$_6^{3+}$, in which six water molecules are bonded to the central Al^{3+} ion. This species can transfer a proton to a solvent water molecule to form an H$_3$O$^+$ ion:

$$Al(H_2O)_6^{3+}(aq) + H_2O \rightleftharpoons H_3O^+(aq) + Al(H_2O)_5(OH)^{2+}(aq)$$

Figure 13.8 shows the structures of the Al(H$_2$O)$_6^{3+}$ cation and its conjugate base, Al(H$_2$O)$_5$(OH)$^{2+}$.

A very similar equation can explain why solutions of zinc salts are acidic. Here it appears that the hydrated cation in water solution contains four H$_2$O molecules bonded to a central Zn^{2+} ion. The Brønsted-Lowry equation is

$$Zn(H_2O)_4^{2+}(aq) + H_2O \rightleftharpoons H_3O^+(aq) + Zn(H_2O)_3(OH)^+(aq)$$

The Equilibrium Constant for a Weak Acid

In discussing the equilibrium involved when a weak acid is added to water, it is convenient to represent the proton transfer

$$HB(aq) + H_2O \rightleftharpoons H_3O^+(aq) + B^-(aq)$$

as a simple ionization

$$HB(aq) \rightleftharpoons H^+(aq) + B^-(aq)$$

in which case the expression for the equilibrium constant becomes

$$K_a = \frac{[H^+][B^-]}{[HB]} \tag{13.5}$$

©John Copland/Shutterstock.com

©Denise Kappa/Shutterstock.com

Figure 13.6 Influence of acidic or basic soil. Hydrangeas grown in strongly acidic soil (below pH 5) are blue. When they are grown in neutral or basic soil, the flowers are rosy pink.

The color change of hydrangeas is exactly the opposite of that for litmus.

To get red hydrangeas, add lime (CaO).

All transition metal cations behave this way.

Figure 13.7 Some household weak acids. Vinegar contains acetic acid. Fruit juices and sodas contain citric acid. Baking powder contains the Al^{3+} ion.

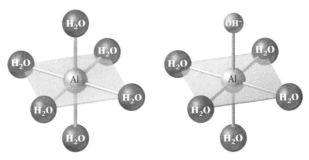

Figure 13.8 A weakly acidic cation. In water solution, the Al^{3+} ion is bonded to six water molecules in the $Al(H_2O)_6^{3+}$ ion *(left)*. In the $Al(H_2O)_5OH^{2+}$ ion *(right)*, one of the H_2O molecules has been replaced by an OH^- ion.

The equilibrium constant K_a is called, logically enough, the **acid equilibrium constant** of the weak acid HB. Table 13.2 lists the K_a values of some weak acids in order of decreasing strength. The weaker the acid, the smaller the value of K_a.

Table 13.2 Equilibrium Constants for Weak Acids and Their Conjugate Bases

	Acid	K_a	Base	K_b
Sulfurous acid	H_2SO_3	1.7×10^{-2}	HSO_3^-	5.9×10^{-13}
Hydrogen sulfate ion	HSO_4^-	1.0×10^{-2}	SO_4^{2-}	1.0×10^{-12}
Phosphoric acid	H_3PO_4	7.1×10^{-3}	$H_2PO_4^-$	1.4×10^{-12}
Hexaaquairon(III) ion	$Fe(H_2O)_6^{3+}$	6.7×10^{-3}	$Fe(H_2O)_5OH^{2+}$	1.5×10^{-12}
Hydrofluoric acid	HF	6.9×10^{-4}	F^-	1.4×10^{-11}
Nitrous acid	HNO_2	6.0×10^{-4}	NO_2^-	1.7×10^{-11}
Formic acid	$HCHO_2$	1.9×10^{-4}	CHO_2^-	5.3×10^{-11}
Lactic acid	$HC_3H_5O_3$	1.4×10^{-4}	$C_3H_5O_3^-$	7.1×10^{-11}
Benzoic acid	$HC_7H_5O_2$	6.6×10^{-5}	$C_7H_5O_2^-$	1.5×10^{-10}
Acetic acid	$HC_2H_3O_2$	1.8×10^{-5}	$C_2H_3O_2^-$	5.6×10^{-10}
Hexaaquaaluminum (III) ion	$Al(H_2O)_6^{3+}$	1.2×10^{-5}	$Al(H_2O)_5OH^{2+}$	8.3×10^{-10}
Carbonic acid	H_2CO_3	4.4×10^{-7}	HCO_3^-	2.3×10^{-8}
Dihydrogen phosphate ion	$H_2PO_4^-$	6.2×10^{-8}	HPO_4^{2-}	1.6×10^{-7}
Hydrogen sulfite ion	HSO_3^-	6.0×10^{-8}	SO_3^{2-}	1.7×10^{-7}
Hypochlorous acid	HClO	2.8×10^{-8}	ClO^-	3.6×10^{-7}
Hydrocyanic acid	HCN	5.8×10^{-10}	CN^-	1.7×10^{-5}
Ammonium ion	NH_4^+	5.6×10^{-10}	NH_3	1.8×10^{-5}
Tetraaquazinc(II) ion	$Zn(H_2O)_4^{2+}$	3.3×10^{-10}	$Zn(H_2O)_3OH^+$	3.0×10^{-5}
Hydrogen carbonate ion	HCO_3^-	4.7×10^{-11}	CO_3^{2-}	2.1×10^{-4}
Hydrogen phosphate ion	HPO_4^{2-}	4.5×10^{-13}	PO_4^{3-}	2.2×10^{-2}

$$HB(aq) \rightleftharpoons H^+(aq) + B^-(aq) \qquad K_a = \frac{[H^+] \times [B^-]}{[HB]}$$

$$B^-(aq) + H_2O \rightleftharpoons HB(aq) + OH^-(aq) \qquad K_b = \frac{[HB] \times [OH^-]}{[B^-]}$$

For example, HCN ($K_a = 5.8 \times 10^{-10}$) is a weaker acid than HNO_2, for which $K_a = 6.0 \times 10^{-4}$.

We sometimes refer to the **pK_a** value of a weak acid

$$pK_a = -\log_{10} K_a \qquad (13.6)$$

HNO_2	$K_a = 6.0 \times 10^{-4}$	$pK_a = 3.22$	
HCN	$K_a = 5.8 \times 10^{-10}$	$pK_a = 9.24$	

EXAMPLE 13.4

Consider acetic acid, $HC_2H_3O_2$, and the hydrated zinc cation, $Zn(H_2O)_4{}^{2+}$.

a Write equations to show why these species are acidic.

STRATEGY AND SOLUTION

1. To prove that a species is acidic, you must produce a hydronium ion (H_3O^+) obtained by transferring an H atom to water.

$$HC_2H_3O_2(aq) + H_2O \rightleftharpoons C_2H_3O_2{}^-(aq) + HH_2O$$
$$(H_3O^+)$$

$$\overline{HC_2H_3O_2(aq) + H_2O \rightleftharpoons C_2H_3O_2{}^-(aq) + H_3O^+}$$

2. For the hydrated cation, one of the water molecules in the ion donates an H atom to an unattached water molecule. Think of $Zn(H_2O)_4{}^{2+}$ as $Zn(H_2O)(H_2O)_3{}^{2+}$.

$$Zn(H_2O)(H_2O)_3{}^{2+}(aq) + H_2O \rightleftharpoons Zn(OH^-)(H_2O)_3{}^{2+}(aq) + HH_2O$$
$$(Zn(OH)(H_2O)_3{}^+) \qquad (H_3O^+)$$

$$\overline{Zn(H_2O)(H_2O)_3{}^{2+}(aq) + H_2O \rightleftharpoons Zn(OH)(H_2O)_3{}^+(aq) + H_3O^+}$$

b Which is the stronger acid?

SOLUTION

K_a values from Table 13.2	$HC_2H_3O_2(K_a = 1.8 \times 10^{-5})$ vs $Zn(H_2O)_4{}^{2+}(K_a = 3.3 \times 10^{-10})$
	$1.8 \times 10^{-5} > 3.3 \times 10^{-10}$ $HC_2H_3O_2$ is the stronger acid.

c What is the pK_a of $Zn(H_2O)_4{}^{2+}$?

SOLUTION

pK_a	$pK_a = -\log_{10} K_a = -\log_{10}(3.3 \times 10^{-10}) = 9.48$

There are several ways to determine K_a of a weak acid. A simple approach involves measuring $[H^+]$ or pH in a solution prepared by dissolving a known amount of the weak acid to form a given volume of solution.

EXAMPLE 13.5

Aspirin, a commonly used pain reliever, is a weak organic acid whose molecular formula may be written as $HC_9H_7O_4$. An aqueous solution of aspirin has total volume 350.0 mL and contains 1.26 g of aspirin. The pH of the solution is found to be 2.60. Calculate K_a for aspirin.

continued

ANALYSIS	
Information given:	molecular formula for aspirin ($HC_9H_7O_4$); mass of aspirin (1.26 g); volume of solution (350.0 mL); pH of solution (2.60)
Information implied:	molar mass of aspirin; $[H^+]$
Asked for:	K_a

STRATEGY

1. Determine the original concentration, $[\ ]_o$, of aspirin.

2. pH = $[H^+]_{eq}$

3. Draw a table as illustrated in Example 12.4. Substitute $[\ ]_o$ for P_o, $\Delta[\ ]$ for ΔP, and $[\ ]_{eq}$ for P_{eq}. Since only one H atom ionizes at a time, $\Delta[\ ]$ for all species is the same. Recall that $[\ ]$ stands for the concentration in molarity.

4. Write the K expression for the ionization and calculate K_a.

SOLUTION

1. $[\ ]_o$ for aspirin	$\dfrac{1.26\ g}{0.3500\ L} \times \dfrac{1\ mol}{180.15\ g} = 0.0200\ M$
2. $[H^+]_{eq}$	$2.60 = -\log_{10}[H^+]$; $[H^+] = 10^{-2.60} = 2.5 \times 10^{-3}\ M$

3. Table

	$HC_9H_7O_4(aq)$	$\rightleftharpoons$	$H^+(aq)$	$+$	$C_9H_7O_4^-(aq)$
$[\ ]_o$	0.0200		0.0000		0.0000
$\Delta[\ ]$	−0.0025		+0.0025		+0.0025
$[\ ]_{eq}$	0.0175		0.0025		0.0025

4. K expression

$$HC_9H_7O_4(aq) \rightleftharpoons H^+(aq) + C_9H_7O_4^-(aq)$$

K_a

$$K_a = \frac{[H^+][C_9H_7O_4^-]}{[HC_9H_7O_4]} = \frac{(0.0025)(0.0025)}{0.0175} = 3.6 \times 10^{-4}$$

END POINT

Aspirin is a relatively *strong* weak acid (Figure 13.9). It would be located near the top of Table 13.2.

For weak acids, it is always true that $\Delta[H^+] = \Delta[B^-] = -\Delta[HB]$.

Figure 13.9 Aspirin is more acidic than vinegar.

In discussing the ionization of a weak acid,

$$HB(aq) \rightleftharpoons H^+(aq) + B^-(aq)$$

we often refer to the **percent ionization**:

$$\% \text{ ionization} = \frac{[H^+]_{eq}}{[HB]_o} \times 100\% \qquad (13.7)$$

For the aspirin solution referred to in Example 13.5,

$$\% \text{ ionization} = \frac{2.5 \times 10^{-3}}{2.0 \times 10^{-2}} \times 100\% = 12\%$$

As you might expect, percent ionization at a given concentration is directly related to K_a. Ibuprofen ($K_a = 2.5 \times 10^{-5}$), a weaker acid than aspirin ($K_a = 3.6 \times 10^{-4}$), should be only 3.6% ionized at 0.020 M compared with 12% for aspirin.

Percent ionization also depends on the concentration of weak acid, increasing as the acid is diluted (Figure 13.10).

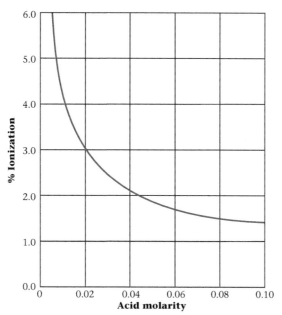

Figure 13.10 Percent ionization of acetic acid ($K_a = 1.8 \times 10^{-5}$).
Like any weak acid, the percent ionization of acetic acid is inversely
related to its molar concentration.

EXAMPLE 13.6 CONCEPTUAL

The box below shows a system at equilibrium. It has a volume of 0.50 L and the symbol ⬤ represents
0.10 mol of a weak acid, HB. The symbol ⬤ represents 0.10 mol of the conjugate base, B^-. Hydronium
ions and water molecules are not shown. What is the percent ionization of the acid?

STRATEGY

1. Since there are 4 red circles (HB) and 1 blue circle (B^-) at equilibrium, there must have been 5 red
   circles to start with.

2. Substitute into Equation 13.7.

$$\% \text{ ionization} = \frac{[\ ]_{eq}}{[\ ]_o} \times 100$$

SOLUTION

% ionization	1 blue circle = $[B^-]_{eq}$; 5 red circles = $[HB]_o$
	$\% \text{ ionization} = \dfrac{1}{5} \times 100\% = 20\%$

Calculation of [H⁺] in a Water Solution of a Weak Acid

Given the ionization constant of a weak acid and its original concentration, the
H^+ concentration in solution is easy to calculate. The approach used is the inverse
of that followed in Example 13.5, where K_a was calculated knowing $[H^+]$; here,
K_a is known and $[H^+]$ must be calculated.

EXAMPLE 13.7

Nicotinic acid, $HC_6H_4O_2N$ ($K_a = 1.4 \times 10^{-5}$), is another name for niacin, an important member of the vitamin B group. Determine [H^+] in a solution prepared by dissolving 3.0 g of nicotinic acid (MM = 123.11 g/mol), HNic, in enough water to form 245 mL of solution.

ANALYSIS

Information given:	molar mass for nicotinic acid, HNic (123.11 g/mol); mass of HNic (3.0 g)
	volume of solution (245 mL)
	K_a (1.4×10^{-5})
Information implied:	[HNic]$_o$
Asked for:	[H^+]$_{eq}$

STRATEGY

1. Determine the original concentration, [ ]$_o$, of HNic.

2. Let $x = \Delta$[H^+].

   Since all the coefficients in the reaction are one, Δ[HNic] and Δ[Nic$^-$] also equal x.

3. Draw a table as illustrated in Example 13.5.

4. Write the K expression for the ionization and substitute the equilibrium concentrations for HNic, Nic$^-$, and H^+ obtained from the table.

5. To avoid the quadratic equation, assume $x << $ [HNic]$_o$. Thus [HNic]$_{eq}$ = 0.10. Then, substitute into the K_a expression and solve for x.

SOLUTION

1. [ ]$_o$ for HNic	$\dfrac{3.0 \text{ g}}{0.245 \text{ L}} \times \dfrac{1 \text{ mol}}{123.11 \text{ g}} = 0.10 \ M$
2. Δ[ ]	Δ[HNic] = Δ[Nic$^-$] = Δ[H^+] = x

3. Table

	HNic(aq)	$\rightleftharpoons$	H$^+$(aq)	+	Nic$^-$(aq)
[ ]$_o$	0.10		0.00		0.00
Δ[ ]	$-x$		$+x$		$+x$
[ ]$_{eq}$	0.10 $- x$		x		x

4. K_a expression	$1.4 \times 10^{-5} = \dfrac{[H^+][Nic^-]}{[HNic]} = \dfrac{(x)(x)}{0.10 - x}$
5. x	$1.4 \times 10^{-5} = \dfrac{(x)(x)}{0.10} \longrightarrow x = 0.0012 \ M$

END POINT

Note that the concentration of H^+, 0.0012 M, is

- much smaller than the original concentration of the weak acid, 0.10 M. In this case, then, the approximation 0.10 $- x \approx 0.10$ is justified. This will usually, but not always, be the case (see Example 13.8).

- much larger than [H^+] in pure water, 1×10^{-7} M, justifying the assumption that the ionization of water can be neglected. *This will always be the case, provided* [H^+] *from the weak acid is* $\geq 10^{-6}$ M.

In general, the value of K_a is seldom known more accurately than $\pm 5\%$. Hence in the expression

$$K_a = \frac{x^2}{a - x}$$

where $x = $ [H^+] and $a = $ *original* concentration of weak acid, you can neglect the x in the denominator if doing so does not introduce an error of more than 5%. In other words,

$$\text{if } \frac{x}{a} \leq 0.05$$

then

$$a - x \approx a$$

Note that $x = [H^+]_{eq}$ and $a = [HB]_o$. Hence,

$$\frac{x}{a} = \frac{[H^+]_{eq}}{[HB]_o}$$

If we multiply by 100%, then

$$\frac{x}{a}\% = \frac{[H^+]_{eq}}{[HB]_o} \times 100\%,$$

which is the equation for determining percent ionization (13.7). In other words, the approximation $x - a \approx a$ is valid if the % ionization is less than 5% (this is known as the **five percent rule**).

In most of the problems you will work, the approximation $a - x \approx a$ is valid, and you can solve for $[H^+]$ quite simply, as in Example 13.7, where $x = 0.012a$. Sometimes, though, you will find that the calculated $[H^+]$ is greater than 5% of the original concentration of weak acid. In that case, you can solve for x by using either the quadratic formula or the method of **successive approximations**.

EXAMPLE 13.8

Calculate $[H^+]$ in a 0.100 M solution of nitrous acid, HNO_2, for which $K_a = 6.0 \times 10^{-4}$.

ANALYSIS

Information given:	$[HNO_2]_o$ (0.100 M); K_a (6.0 × 10⁻⁴)
Asked for:	$[H^+] = [H^+]_{eq}$

STRATEGY

1. The table setup is identical to that of Example 13.7 giving the equilibrium expression

   Substitute values for equilibrium from the table. $K_a = \dfrac{[H^+][NO_2^-]}{[HNO_2]} = \dfrac{(x)(x)}{0.100 - x}$

2. Assume $x \ll 0.100$ and solve for x.

3. Check your assumption by calculating % ionization.

4. If the % ionization is less than 5%, your assumption is valid. If not, solve for x using the quadratic equation.

SOLUTION

1. Equilibrium expression	$6.0 \times 10^{-4} = \dfrac{[H^+][NO_2^-]}{[HNO_2]} = \dfrac{(x)(x)}{0.100 - x}$
2. Assume $x \ll 0.100$.	$6.0 \times 10^{-4} = \dfrac{x^2}{0.100} \longrightarrow x = 0.0077\ M = [H^+]$
3. Check the assumption.	% ionization $= \dfrac{[H^+]_{eq}}{[HB]_o} \times 100\% = \dfrac{0.0077}{0.100} \times 100\% = 7.7\%$
	$7.7\% > 5.0\% \longrightarrow$ The assumption is not valid.

continued

4. Use the quadratic equation.	$x^2 + (6.0 \times 10^{-4})x - (6.0 \times 10^{-5}) = 0$ $x = \dfrac{-b \pm \sqrt{b^2 - 4ac}}{2a} = \dfrac{-(6.0 \times 10^{-4}) \pm \sqrt{(6.0 \times 10^{-4})^2 - 4(-6.0 \times 10^{-5})}}{2}$ $x = 0.0074\ M$ or $-0.0080\ M$ $-0.0080\ M$ is physically impossible, so $x = [H^+] = 0.0074\ M$

Polyprotic Weak Acids

Certain weak acids are polyprotic; a **polyprotic acid** contains more than one ioniz-able hydrogen atom. Such acids ionize in steps, with a separate equilibrium con-stant for each step. Oxalic acid, a weak organic acid sometimes used to remove bloodstains, is *diprotic:*

$$H_2C_2O_4(aq) \rightleftharpoons H^+(aq) + HC_2O_4^-(aq) \qquad K_{a1} = 5.9 \times 10^{-2}$$
$$HC_2O_4^-(aq) \rightleftharpoons H^+(aq) + C_2O_4^{2-}(aq) \qquad K_{a2} = 5.2 \times 10^{-5}$$

Phosphoric acid, a common ingredient of cola drinks, is *triprotic:*

$$H_3PO_4(aq) \rightleftharpoons H^+(aq) + H_2PO_4^-(aq) \qquad K_{a1} = 7.1 \times 10^{-3}$$
$$H_2PO_4^-(aq) \rightleftharpoons H^+(aq) + HPO_4^{2-}(aq) \qquad K_{a2} = 6.2 \times 10^{-8}$$
$$HPO_4^{2-}(aq) \rightleftharpoons H^+(aq) + PO_4^{3-}(aq) \qquad K_{a3} = 4.5 \times 10^{-13}$$

The behavior of these acids is typical of all polyprotic acids in that

- *the anion formed in one step* (e.g., $HC_2O_4^-$, $H_2PO_4^-$) *produces another H^+ ion in the next step.*
- *the acid equilibrium constant becomes smaller with each successive step.*

$$K_{a1} > K_{a2} > K_{a3}$$

Putting it another way, *the acids in successive steps become progressively weaker.* This is reasonable; it should be more difficult to remove a positively charged pro-ton, H^+, from a negatively charged species like $H_2PO_4^-$ than from a neutral mol-ecule like H_3PO_4.

Ordinarily, successive values of K_a for polyprotic acids decrease by a fac-tor of at least 100 (Table 13.3). In that case, essentially all the H^+ ions in the solution come from the first step. This makes it relatively easy to calculate the pH of a solution of a polyprotic acid.

Table 13.3 Equilibrium Constants for Some Weak Polyprotic Acids at 25°C

Acid	Formula	K_{a1}	K_{a2}	K_{a3}
Carbonic acid*	H_2CO_3	4.4×10^{-7}	4.7×10^{-11}	
Oxalic acid	$H_2C_2O_4$	5.9×10^{-2}	5.2×10^{-5}	
Phosphoric acid	H_3PO_4	7.1×10^{-3}	6.2×10^{-8}	4.5×10^{-13}
Sulfurous acid	H_2SO_3	1.7×10^{-2}	6.0×10^{-8}	

*Carbonic acid is a water solution of carbon dioxide:
$$CO_2(g) + H_2O \rightleftharpoons H_2CO_3(aq)$$
The ionization constants listed are calculated assuming that all the carbon dioxide that dissolves is in the form of H_2CO_3.

EXAMPLE 13.9

The distilled water you use in the laboratory is slightly acidic because of dissolved CO_2, which reacts to form carbonic acid, H_2CO_3. Calculate the pH of a 0.0010 M solution of H_2CO_3 and $[CO_3^{2-}]$ at equilibrium.

ANALYSIS

Information given:	$[H_2CO_3]_0$ (0.0010 M)
Information implied:	two-step ionization, K_a value for each ionization (Table 13.3)
Asked for:	pH, $[CO_3^{2-}]_{eq}$

STRATEGY

1. Write the ionization reaction for each ionization. Recall that ionization of polyprotic acids takes place one H^+ at a time.

2. Most of the H^+ is obtained from the first ionization. Find $[H^+]_{eq}$ for the first ionization and convert to pH.

3. Note $[H^+] = [HCO_3^-]$. Substitute into the expression for the second ionization to obtain $[CO_3^{2-}]_{eq}$.

SOLUTION

1. Ionization reactions	$H_2CO_3(aq) \rightleftharpoons H^+(aq) + HCO_3^-(aq)$ $\quad K_{a1} = 4.4 \times 10^{-7}$
	$HCO_3^-(aq) \rightleftharpoons H^+(aq) + CO_3^{2-}(aq)$ $\quad K_{a2} = 4.7 \times 10^{-11}$
2. $[H^+]_{eq}$ from 1st ionization	$4.4 \times 10^{-7} = \dfrac{[H^+][HCO_3^-]}{[H_2CO_3]} = \dfrac{x^2}{0.0010 - x}$
Assume $x \ll 0.0010$	$4.4 \times 10^{-7} = \dfrac{x^2}{0.0010} \longrightarrow x^2 = (0.0010)(4.4 \times 10^{-7})$
	$x = 2.1 \times 10^{-5}\,M$
Check assumption	% ionization = 2.1%; assumption is valid.
pH	$[H^+] = [HCO_3^-] = 2.1 \times 10^{-5}\,M$; pH = 4.68
3. $[CO_3^{2-}]_{eq}$	$4.7 \times 10^{-11} = \dfrac{[H^+][CO_3^{2-}]}{[HCO_3^-]} \longrightarrow [H^+] = [HCO_3^-]$
	$4.7 \times 10^{-11} = [CO_3^{2-}]$

The conclusions in Example 13.9 are ordinarily valid for any weak diprotic acid, H_2B.

$$[H^+] \text{ is calculated from } K_{a1}: [H^+] = (K_{a1} \times [H_2B])^{1/2}$$

$$[HB^-] = [H^+]$$

$$[B^{2-}] = K_{a2}$$

In a water solution of CO_2, there are very, very few CO_3^{2-} ions.

Molecular Structure and Acid Strength

Equilibrium constants, K_a, are a measure of the strength of an acid. The values range from 0.010 to 1×10^{-13}. What is responsible for this variation?

As we saw earlier, a Brønsted-Lowry acid is a proton donor. The strength of the acid depends on how easily the bond between hydrogen and another atom is broken and how many of the H^+ ions are donated. So, as you might suspect, the stronger the bond, the weaker the acid because the more tightly the hydrogen atom is held, the less likely it is to come off.

For oxoacids—where hydrogen is bonded to oxygen, which in turn is bonded to some other atom—two other factors affect the strength of the acid. Those are the electronegativity of the nonmetal to which the oxygen atoms are bonded and the number of oxygen atoms bonded to the nonmetal. The more electronegative the nonmetal bonded to oxygen is, the more it polarizes and therefore weakens the O—H

Figure 13.11 Weak bases.

bond, resulting in a larger equilibrium constant. Increasing the number of oxygen atoms bonded to the nonmetal also weakens the O—H bond, resulting in a stronger acid. HClO is a much weaker acid than $HClO_2$, whereas $HClO_4$ is a strong acid. The acidity of oxoacids is discussed in more detail in Chapter 21.

13-5 Weak Bases and Their Equilibrium Constants

Like the weak acids, a large number of solutes act as weak bases (Figure 13.11). It is convenient to classify weak bases into two groups, molecules and anions.

1. *Molecules.* As pointed out in Chapter 4, there are many molecular weak bases, including the organic compounds known as amines. The simplest weak base is ammonia, whose reversible Brønsted-Lowry reaction with water is represented by the equation

$$NH_3(aq) + H_2O \rightleftharpoons NH_4^+(aq) + OH^-(aq)$$

This reaction proceeds to only a very small extent. In 0.10 M NH_3, the concentrations of NH_4^+ and OH^- are only about 0.0013 M.

2. *Anions* (basic ion). *An anion derived from a weak acid is itself a weak base.* A typical example is the fluoride ion, F^-, which is the conjugate base of the weak acid HF. The reaction of the F^- ion with water is

$$F^-(aq) + H_2O \rightleftharpoons HF(aq) + OH^-(aq)$$

The OH^- ions formed make the solution basic. A 0.10 M solution of sodium fluoride, NaF, has a pH of about 8.1.

Notice the similarity between the equation just written and that for the NH_3 molecule cited above

- both weak bases (NH_3, F^-) accept protons to form the conjugate acid (NH_4^+, HF).
- water acts as a Brønsted-Lowry acid in each case, donating a proton to form the OH^- ion.

EXAMPLE 13.10

Write an equation to explain why NO_2^-, Na_2CO_3, and $KHCO_3$ produce basic water solutions.

STRATEGY

1. React each basic anion with a water molecule.

2. The weak base picks up the proton (H^+) and increases its charge by one unit to create its conjugate acid.

3. OH^- is the other product of the reaction.

SOLUTION

NO_2^-:
$$NO_2^-(aq) + H_2O \rightarrow HNO_2^{-1+1}(aq) + OH^-(aq) \rightarrow HNO_2(aq) + OH^-(aq)$$
$$\quad\quad\quad\quad HOH$$

Na_2CO_3:
$$CO_3^{2-}(aq) + H_2O \rightarrow HCO_3^{-2+1}(aq) + OH^-(aq) \rightarrow HCO_3^-(aq) + OH^-(aq)$$
$$\quad\quad\quad\quad HOH$$

$KHCO_3$:
$$HCO_3^-(aq) + H_2O \rightarrow HHCO_3^{-1+1}(aq) + OH^-(aq) \rightarrow H_2CO_3(aq) + OH^-(aq)$$
$$\quad\quad\quad\quad HOH$$

END POINT

The presence of OH^- as a product is the reason these anions in water are considered to be basic.

The Equilibrium Constant for a Weak Base

For the weak base ammonia

$$NH_3(aq) + H_2O \rightleftharpoons NH_4^+(aq) + OH^-(aq)$$

the **base equilibrium constant** (K_b) is written in the usual way, omitting the term for solvent water.

$$K_b = \frac{[NH_4^+] \times [OH^-]}{[NH_3]}$$

For an anion B$^-$ that acts as a weak base,

$$B^-(aq) + H_2O \rightleftharpoons HB(aq) + OH^-(aq)$$

a very similar expression can be written:

$$K_b = \frac{[HB] \times [OH^-]}{[B^-]} \qquad (13.8)$$

Values of K_b are listed at the right of Table 13.2 (page 340). The larger the value of K_b, the stronger the base. Because K_b for NH_3 (1.8×10^{-5}) is larger than K_b for $C_2H_3O_2^-$ (5.6×10^{-10}), ammonia is a stronger base than the acetate ion. The quantity **pK_b** is defined as

$$pK_b = -\log_{10}K_b \qquad (13.9)$$

The pK_b of ammonia ($K_b = 1.8 \times 10^{-5}$) is 4.74.

Calculation of [OH$^-$] in a Water Solution of a Weak Base

We showed in Section 13-4 how K_a of a weak acid can be used to calculate [H$^+$] in a solution of that acid. In a very similar way, K_b can be used to find [OH$^-$] in a solution of a weak base.

EXAMPLE	13.11	GRADED

Consider sodium hypochlorite, NaOCl, the main component in household bleach. The hypochlorite ion, OCl$^-$, has $K_b = 3.6 \times 10^{-7}$. A solution is prepared by dissolving 12.0 g of NaOCl (MM = 74.45 g/mol) in enough water to make 835 mL of solution.

a	What is the pH of the solution?

ANALYSIS	
Information given:	K_b for OCl$^-$ (3.6×10^{-7}) mass of NaOCl (12.0 g); molar mass of NaOCl (74.45 g/mol) volume of solution (0.835 L)
Information implied:	K_w
Asked for:	pH of the solution

continued

STRATEGY

1. Determine the original concentration, $[\ ]_o$, of NaOCl.

2. Draw a table as illustrated in Example 13.5.

3. Write the K expression and find $[OH^-]$ assuming $[OH^-]_{eq} << [OCl^-]_o$. Check the validity of the assumption by finding % ionization.

4. Find $[H^+]$ using Equation 13.1 and pH using Equation 13.3.

SOLUTION

1. $[\ ]_o$ for NaOCl	$\dfrac{12.0\ g}{0.835\ L} \times \dfrac{1\ mol}{74.45\ g} = 0.193\ M$

2. Table

		$OCl^-(aq)$	$+\ H_2O \rightleftharpoons$	$HOCl(aq)$	$+$	$OH^-(aq)$
	$[\ ]_o$	0.193		0.00		0.00
	$\Delta[\ ]$	$-x$		$+x$		$+x$
	$[\ ]_{eq}$	$0.193 - x$		x		x

3. K expression

$$K_b = \frac{[HOCl][OH^-]}{[OCl^-]} = 3.6 \times 10^{-7} = \frac{(x)(x)}{0.193 - x}$$

Assume $x << 0.193$

$$x^2 = 0.193(3.6 \times 10^{-7}) \longrightarrow x = 2.6 \times 10^{-4}$$

Check assumption

% ionization = 0.14%; the assumption is justified. $[OH^-] = 2.6 \times 10^{-4}\ M$

4. $[H^+]$; pH

$$1.0 \times 10^{-14} = [H^+](2.6 \times 10^{-4}) \longrightarrow [H^+] = 3.9 \times 10^{-11} \longrightarrow pH = 10.41$$

---

b Household bleach is 5.25% NaOCl by mass. Assuming that its density is 1.00 g/mL, is household bleach more alkaline than the prepared solution?

ANALYSIS

Information given:	NaOCl content of household bleach (5.25% by mass) density of bleach (1.00 g/mL) pH of solution in part (a) (10.41)
Asked for:	Compare pH of solution (a) and pH of bleach.

STRATEGY

1. Assume 100.0 g (= 100.0 mL) of bleach. Thus, there are 5.25 g of NaOCl in 100.0 mL of solution.

2. Find $[OH^-]$, $[H^+]$, and pH of bleach as in part (a).

3. Compare the pH of both solutions. The solution with a higher pH is more alkaline.

SOLUTION

$[NaOCl]_o = [OCl^-]_o$	$\dfrac{5.25\ g}{0.100\ L} \times \dfrac{1\ mol}{74.45\ g} = 0.705\ M$
K expression	(as in part (a)): $K_b = \dfrac{[HOCl][OH^-]}{[OCl^-]} = 3.6 \times 10^{-7} = \dfrac{(x)(x)}{0.705 - x}$
Assume $x << 0.705$	$x^2 = 0.705(3.6 \times 10^{-7}) \longrightarrow x = 5.0 \times 10^{-4} = [OH^-]$
Check assumption.	% ionization = 0.071%; the assumption is justified. $[OH^-] = 5.0 \times 10^{-4}\ M$
$[H^+]$; pH	$1.0 \times 10^{-14} = [H^+](5.0 \times 10^{-4}) \longrightarrow [H^+] = 2.0 \times 10^{-11} \longrightarrow pH = 10.70$
Comparison	pH of the solution in part (a) = 10.41; pH of bleach = 10.70
	10.70 > 10.41; bleach is more alkaline than the solution prepared in part (a).

Relation Between K_a and K_b

Equilibrium constants of weak bases can be measured in the laboratory by procedures very much like those used for weak acids. In practice, though, it is simpler to take advantage of a simple mathematical relationship between K_b for a weak base and K_a for its conjugate acid. This relationship can be derived by adding together the equations for the ionization of the weak acid HB and the reaction of the weak base B^- with water:

(1) $\qquad HB(aq) \rightleftharpoons H^+(aq) + B^-(aq) \qquad K_I = K_a$ of HB

(2) $B^-(aq) + H_2O(aq) \rightleftharpoons HB(aq) + OH^-(aq) \qquad K_{II} = K_b$ of B^-

(3) $\qquad H_2O \rightleftharpoons H^+(aq) + OH^-(aq) \qquad K_{III} = K_w$

Because Equation (1) + Equation (2) = Equation (3), we have, according to the rule of multiple equilibria (Chapter 12),

$$K_I \times K_{II} = K_{III}$$

or

$$(K_a \text{ of HB})(K_b \text{ of } B^-) = K_w = 1.0 \times 10^{-14} \qquad (13.10)$$

Checking Table 13.2, you can see that this relation holds in each case. For example,

$$(K_a HNO_2)(K_b NO_2^-) = (6.0 \times 10^{-4})(1.7 \times 10^{-11})$$
$$= 10 \times 10^{-15} = 1.0 \times 10^{-14}$$

From the general relation between K_a of a weak acid and K_b of its conjugate weak base, it should be clear that these two quantities are inversely related to each other. *The larger the value of K_a, the smaller the value of K_b and vice versa.* Table 13.4 shows this effect and some other interesting features. In particular,

1. Brønsted-Lowry acids (left column) can be divided into three categories:
   (a) strong acids ($HClO_4, \ldots$), which are stronger proton donors than the H_3O^+ ion.
   (b) weak acids ($HF, \ldots$), which are weaker proton donors than the H_3O^+ ion, but stronger than the H_2O molecule.
   (c) species such as C_2H_5OH, which are weaker proton donors than the H_2O molecule and hence do not form acidic water solutions.
2. Brønsted-Lowry bases (right column) can be divided similarly:
   (a) strong bases ($H^-, \ldots$) (Figure 13.12), which are stronger proton acceptors than the OH^- ion.
   (b) weak bases ($F^-, \ldots$), which are weaker proton acceptors than the OH^- ion, but stronger than the H_2O molecule.
   (c) the anions of strong acids ($ClO_4^-, \ldots$), which are weaker proton acceptors than the H_2O molecule and hence do not form basic water solutions.

We should point out that, just as the three strong acids at the top of Table 13.4 are completely converted to H_3O^+ ions in aqueous solution:

$$HClO_4(aq) + H_2O \longrightarrow H_3O^+(aq) + ClO_4^-(aq)$$

$$HCl(aq) + H_2O \longrightarrow H_3O^+(aq) + Cl^-(aq)$$

$$HNO_3(aq) + H_2O \longrightarrow H_3O^+(aq) + NO_3^-(aq)$$

Figure 13.12 Hydride ion–water reaction. As water is dropped onto solid calcium hydride, the hydride ion (H^-) reacts immediately and vigorously to form $H_2(g)$ (which is ignited by the heat of the reaction) and OH^-.

Table 13.4 Relative Strengths of Brønsted-Lowry Acids and Bases

K_a	Conjugate Acid	Conjugate Base	K_b
Very large	$HClO_4$	ClO_4^-	Very small
Very large	HCl	Cl^-	Very small
Very large	HNO_3	NO_3^-	Very small
	H_3O^+	H_2O	
6.9×10^{-4}	HF	F^-	1.4×10^{-11}
1.8×10^{-5}	$HC_2H_3O_2$	$C_2H_3O_2^-$	5.6×10^{-10}
1.2×10^{-5}	$Al(H_2O)_6^{3+}$	$Al(H_2O)_5(OH)^{2+}$	8.3×10^{-10}
4.4×10^{-7}	H_2CO_3	HCO_3^-	2.3×10^{-8}
2.8×10^{-8}	$HClO$	ClO^-	3.6×10^{-7}
5.6×10^{-10}	NH_4^+	NH_3	1.8×10^{-5}
4.7×10^{-11}	HCO_3^-	CO_3^{2-}	2.1×10^{-4}
	H_2O	OH^-	
Very small	C_2H_5OH	$C_2H_5O^-$	Very large
Very small	OH^-	O^{2-}	Very large
Very small	H_2	H^-	Very large

Species shown in deep pink or deep blue are unstable because they react with H_2O molecules.

The —OH group in C_2H_5OH does not act as an OH^- ion.

the three strong bases at the bottom of the table are completely converted to OH^- ions:

$$H^-(aq) + H_2O \longrightarrow OH^-(aq) + H_2(g)$$
$$O^{2-}(aq) + H_2O \longrightarrow OH^-(aq) + OH^-(aq)$$
$$C_2H_5O^-(aq) + H_2O \longrightarrow OH^-(aq) + C_2H_5OH(aq)$$

In other words, the species H^-, O^{2-}, and $C_2H_5O^-$ do not exist in water solution.

Figure 13.13 **Sodium chloride solution.** When the NaCl is completely dissolved, the solution will contain only Na^+, Cl^-, and H_2O, and it is neither acidic nor basic.

© Cengage Learning/Charles D. Winters

13-6 Acid-Base Properties of Salt Solutions

*A **salt** is an ionic solid containing a cation other than H^+ and an anion other than OH^- or O^{2-}.* When a salt such as NaCl (Figure 13.13), K_2CO_3, or $Al(NO_3)_3$ dissolves in water, the cation and anion separate from one another.

$$NaCl(s) \longrightarrow Na^+(aq) + Cl^-(aq)$$
$$K_2CO_3(s) \longrightarrow 2K^+(aq) + CO_3^{2-}(aq)$$
$$Al(NO_3)_3(s) \longrightarrow Al^{3+}(aq) + 3NO_3^-(aq)$$

To predict whether a given salt solution will be acidic, basic, or neutral, you consider three factors in turn.

1. *Decide what effect, if any, the cation has on the pH of water.*
2. *Decide what effect, if any, the anion has on the pH of water.*
3. *Combine the two effects to decide on the behavior of the salt.*

Table 13.5 Acid-Base Properties of Ions* in Water Solution

	Spectator		Basic		Acidic	
Anion	Cl^-	NO_3^-	$C_2H_3O_2^-$	CO_3^{2-}		
	Br^-	ClO_4^-	F^-	PO_4^{3-}		
	I^-		Many others			
Cation	Li^+	Ca^{2+}			NH_4^+	Al^{3+}
	Na^+	Sr^{2+}			Mg^{2+}	
	K^+	Ba^{2+}			Transition metal ions	

*For the acid-base properties of amphiprotic anions such as HCO_3^- or $H_2PO_4^-$, see the discussion at the end of this section.

Cations: Weak Acids or Spectator Ions?

As we pointed out in Section 13-4, certain cations act as weak acids in water solution because of reactions such as

$$NH_4^+(aq) + H_2O \rightleftharpoons H_3O^+(aq) + NH_3(aq)$$
$$Zn(H_2O)_4^{2+}(aq) + H_2O \rightleftharpoons H_3O^+(aq) + Zn(H_2O)_3(OH)^+(aq)$$

Essentially all transition metal ions behave like Zn^{2+}, forming a weakly acidic solution. Among the main-group cations, Al^{3+} and, to a lesser extent, Mg^{2+}, act as weak acids. In contrast the cations in Group 1 show little or no tendency to react with water.

If we say that *to classify an ion as acidic or basic in water solution, it must change the pH by more than 0.5 unit in 0.1 M solution*, then the cations derived from strong bases are:

- the alkali metal cations ($Li^+, Na^+, K^+ \ldots$)
- the heavier alkaline earth cations ($Ca^{2+}, Sr^{2+}, Ba^{2+}$)

are spectator ions ◀ⓘ as far as pH is concerned (Table 13.5).

ⓘ Learn the spectator ions, not the players.

Anions: Weak Bases or Spectator Ions?

As pointed out in Section 13-5, anions that are the conjugate bases of weak acids act themselves as weak bases in water. They accept a proton from a water molecule, leaving an OH^- ion that makes the solution basic. The reactions of the fluoride and carbonate ions are typical:

$$F^-(aq) + H_2O \rightleftharpoons HF(aq) + OH^-(aq)$$
$$CO_3^{2-}(aq) + H_2O \rightleftharpoons HCO_3^-(aq) + OH^-(aq)$$

Most anions behave this way, except those derived from strong acids (Cl^-, Br^-, I^-, NO_3^-, ClO_4^-), which show little or no tendency to react with water to form OH^- ions. Like the cations in the left column of Table 13.5, they act as spectator ions as far as pH is concerned. You can also use the flowchart shown in Figure 13.14.

Salts: Acidic, Basic, or Neutral?

If you know how the cation and anion of a salt affect the pH of water, it is a relatively simple matter to decide what the net effect will be (see the flowchart in Figure 13.14).

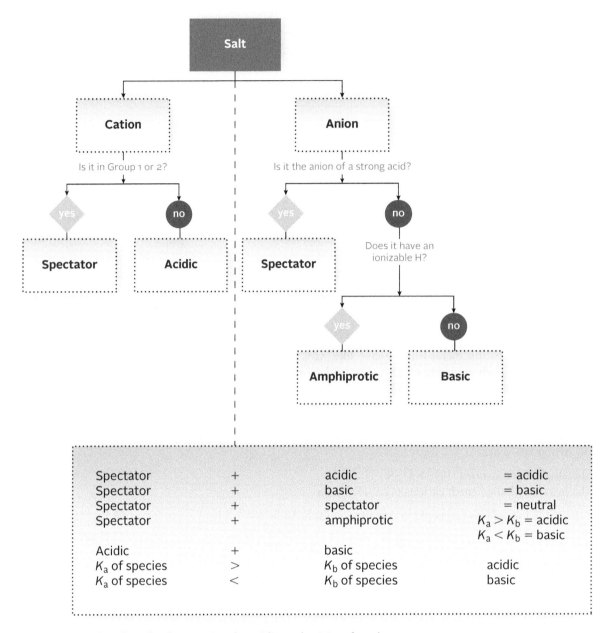

Figure 13.14 Flowchart for determining the acidity or basicity of a salt.

EXAMPLE 13.12

Consider aqueous solutions of the following salts:

$Zn(NO_3)_2$, $KClO_4$, Na_3PO_4, NH_4F, and $NaHCO_3$

Which of theses solutions are acidic? basic? neutral?

STRATEGY

Follow the flowchart in Figure 13.14.

SOLUTION

$Zn(NO_3)_2$:	
cation: Zn^{2+}	not in group 1 or 2 $\longrightarrow$ acidic
anion: NO_3^-	anion of a strong acid $\longrightarrow$ spectator
	$Zn(NO_3)_2$ is acidic

continued

$KClO_4$:	
cation: K^+	group 1 $\longrightarrow$ spectator
anion: ClO_4^-	anion of a strong acid $\longrightarrow$ spectator
	$KClO_4$ is neutral
Na_3PO_4:	
cation: Na^+	group 1 $\longrightarrow$ spectator
anion: PO_4^{3-}	not an anion of a strong acid; no ionizable H $\longrightarrow$ basic
	Na_3PO_4 is basic
NH_4F:	
cation: NH_4^+	not in group 1 or 2 $\longrightarrow$ acidic
anion: F^-	not an anion of a strong acid; no ionizable H $\longrightarrow$ basic
	acidic + basic $\longrightarrow$ Check K_a for NH_4^+ and K_b for F^-
	K_a for $NH_4^+ = 5.6 \times 10^{-10}$; K_b for $F^- = 1.4 \times 10^{-11}$ $K_a > K_b$
	NH_4F is acidic.
$NaHCO_3$:	
cation: Na^+	group 1 $\longrightarrow$ spectator
anion: HCO_3^-	not an anion of a strong acid; has an ionizable H $\longrightarrow$ amphiprotic
	spectator + amphiprotic $\longrightarrow$ Check K_a and K_b for HCO_3^-
	K_a for $HCO_3^- = 4.7 \times 10^{-11}$; K_b for $HCO_3^- = 2.3 \times 10^{-8}$ $K_a < K_b$
	$NaHCO_3$ is basic.

13-7 Extending the Concept of Acids and Bases: The Lewis Model

We have seen that the Brønsted-Lowry model extends the Arrhenius picture of acid-base reactions considerably. However, the Brønsted-Lowry model is restricted in one important respect. It can be applied only to reactions involving a proton transfer. For a species to act as a Brønsted-Lowry acid, it must contain an ionizable hydrogen atom.

The Lewis acid-base model removes that restriction. A **Lewis acid** is a species that in an acid-base reaction, *accepts an electron pair.* In this reaction, a **Lewis base** *donates the electron pair.*

From a structural point of view, the Lewis model of a base does not differ in any essential way from the Brønsted-Lowry model. For a species to accept a proton and thereby act as a Brønsted-Lowry base, it must possess an unshared pair of electrons. Consider, for example, the NH_3 molecule, the H_2O molecule, and the F^- ion, all of which can act as Brønsted-Lowry bases:

Each of these species contains an unshared pair of electrons to form the NH_4^+ ion, the H_3O^+ ion, or the HF molecule:

Clearly, NH_3, H_2O, and F^- can also be Lewis bases because they possess an unshared electron pair that can be donated to an acid. We see then that the Lewis model does not significantly change the number of species that can behave as bases.

On the other hand, the Lewis model greatly increases the number of species that can be considered to be acids. The substance that accepts an electron pair and therefore acts as a Lewis acid can be a proton:

$$\underset{\text{acid}}{H^+(aq)} + \underset{\text{base}}{H_2O} \longrightarrow H_3O^+(aq)$$

$$\underset{\text{acid}}{H^+(aq)} + \underset{\text{base}}{NH_3(aq)} \longrightarrow NH_4^+(aq)$$

It can equally well be a cation such as Zn^{2+}, which can accept electron pairs from a Lewis base:

$$\underset{\text{acid}}{Zn^{2+}(aq)} + \underset{\text{base}}{4H_2O} \longrightarrow Zn(H_2O)_4^{2+}(aq)$$

$$\underset{\text{acid}}{Zn^{2+}(aq)} + \underset{\text{base}}{4NH_3(aq)} \longrightarrow Zn(NH_3)_4^{2+}(aq)$$

We will discuss reactions of this type in more detail in Chapters 15 and 19.

Another important class of Lewis acids are molecules containing an incomplete octet of electrons. A classic example is boric acid, an antiseptic sometimes found in eyewashes.

$$H-\ddot{O}-B-\ddot{O}-H$$
$$|$$
$$:\ddot{O}:$$
$$|$$
$$H$$

When boric acid is added to water, it acts as a Lewis acid, picking up an OH^- to complete the boron octet and at the same time liberating a proton.

$$\underset{\substack{\text{Lewis} \\ \text{acid}}}{B(OH)_3(s)} + \underset{\substack{\text{Lewis} \\ \text{base}}}{H_2O} \longrightarrow B(OH)_4^-(aq) + H^+(aq)$$

The Lewis model is commonly used in chemistry to consider the catalytic behavior of such Lewis acids as $ZnCl_2$ and BF_3. In general, when proton transfer reactions are involved, most chemists use the Arrhenius or Brønsted-Lowry models. Table 13.6 summarizes the acid-base models we have discussed.

Table 13.6 Alternative Definitions of Acids and Bases

Model	Acid	Base
Arrhenius	Supplies H^+ to water	Supplies OH^- to water
Brønsted-Lowry	H^+ donor	H^+ acceptor
Lewis	Electron pair acceptor	Electron pair donor

CHEMISTRY **BEYOND** THE CLASSROOM

Organic Acids and Bases

As pointed out earlier in this chapter, most molecular weak acids are organic in nature; they contain carbon as well as hydrogen atoms. By the same token, most molecular weak bases are organic compounds called amines, which were discussed briefly in Chapter 4.

Carboxylic Acids

Table A lists some of the organic acids found in foods. All of these compounds contain the carboxyl group

$$-C-O-H$$
$$\parallel$$
$$O$$

Table A Some Naturally Occurring Organic Acids

Name		Source
Acetic acid	CH_3-COOH	Vinegar
Citric acid	$HOOC-CH_2-\underset{\underset{COOH}{\vert}}{\overset{\overset{OH}{\vert}}{C}}-CH_2-COOH$	Citrus fruits
Lactic acid	$CH_3-\underset{\underset{OH}{\vert}}{CH}-COOH$	Sour milk
Malic acid	$HOOC-CH_2-\underset{\underset{OH}{\vert}}{CH}-COOH$	Apples, watermelons, grape juice, wine
Oxalic acid	$HOOC-COOH$	Rhubarb, spinach, tomatoes
Quinic acid		Cranberries
Tartaric acid	$HOOC-\underset{\underset{OH}{\vert}}{CH}-\underset{\underset{OH}{\vert}}{CH}-COOH$	Grape juice, wine

The general equation for the reversible dissociation of a carboxylic acid is

$$RCOOH(aq) \rightleftharpoons RCOO^-(aq) + H^+(aq)$$

where R is a hydrocarbon group such as CH_3, C_2H_5, ... The reaction of a carboxylic acid with a strong base can be represented by the equation

$$RCOOH(aq) + OH^-(aq) \longrightarrow RCOO^-(aq) + H_2O$$

This reaction, as is the case with all weak acids, goes essentially to completion.

Certain drugs, both prescription and over-the-counter, contain organic acids. Two of the most popular products of this type are the analgesics aspirin and ibuprofen (Advil®, Nuprin®, and so on).

aspirin

ibuprofen

Because these compounds are acidic, they can cause stomach irritation unless taken with food or water. Beyond that, aspirin inhibits blood clotting, which explains why it is often prescribed to reduce the likelihood of a stroke or heart attack. Indeed, it is now recommended for a person in the throes of a heart attack.

Amines

An **amine** is a derivative of ammonia, NH_3, in which one or more of the hydrogen atoms have been replaced by a hydro-

Table B Types of Amines

Type	General Formula	Example
Primary	RNH_2	$CH_3-\underset{\underset{H}{\vert}}{N}-H$
Secondary	R_2NH	$CH_3-\underset{\underset{H}{\vert}}{N}-CH_3$
Tertiary	R_3N	$CH_3-\underset{\underset{CH_3}{\vert}}{N}-CH_3$

where $R = CH_3$, C_2H_5, ...

carbon group (e.g., CH_3, C_2H_5). Amines can be classified according to the number (1, 2, or 3) of hydrocarbon groups bonded to nitrogen (Table B). Most amines of low molar mass are volatile with distinctly unpleasant odors. For example, $(CH_3)_3N$ is a gas at room temperature (bp = 3°C) with an odor somewhere between those of ammonia and spoiled fish.

The general equation for the reaction of a primary amine with water is

$$RNH_2(aq) + H_2O \rightleftharpoons RNH_3^+(aq) + OH^-(aq)$$

while its reaction with strong acid is

$$RNH_2(aq) + H^+(aq) \longrightarrow RNH_3^+(aq)$$

Secondary and tertiary amines react similarly, forming cations of general formula $R_2NH_2^+$ and R_3NH^+. In each case the added proton bonds to the unshared pair on the nitrogen atom. (Compare RNH_3^+, $R_2NH_2^+$, and R_3NH^+ with the ammonium ion, NH_4^+.)

The reaction of amines with H^+ ions has an interesting practical application. Amines of high molar mass, frequently used as drugs, have very low water solubilities. They can be converted to a water-soluble form by treatment with strong acid. For example,

$$C_9H_{10}NO_2-\underset{\underset{C_2H_5}{|}}{N}-C_2H_5(s) + HCl(aq) \longrightarrow$$

novocaine

$$[C_9H_{10}NO_2-\underset{\underset{C_2H_5}{|}}{\overset{\overset{H}{|}}{N}}-C_2H_5]^+(aq) + Cl^-(aq)$$

novocaine hydrochloride

Novocaine hydrochloride is about 200 times as soluble as novocaine itself. When your dentist injects "novocaine," the liquid in the syringe is a water solution of novocaine hydrochloride.

Alkaloids such as caffeine, coniine, and morphine (Figure A) are amines that are extracted from plants.

caffeine

coniine

morphine

Caffeine occurs in tea leaves, coffee beans, and cola nuts. Morphine is obtained from unripe opium poppy seed pods. Coniine, extracted from hemlock, is the alkaloid that killed Socrates. He was sentenced to death because of unconventional teaching methods; teacher evaluations had teeth in them in ancient Greece.

Figure A Flowers and unripe seed capsules of the opium poppy. Within the capsule is a milky, gummy substance that contains the alkaloid morphine.

© zhuda/Shutterstock.com

Key Concepts

1. Classify a species as a Brønsted-Lowry acid or base and explain by a net ionic equation.
   (**Examples 13.1, 13.4, 13.10; Problems 1–12**)
2. Given $[H^+]$, $[OH^-]$, pH, or pOH, calculate the other three quantities.
   (**Examples 13.2, 13.3; Problems 13–32**)
3. Given the pH and original concentration of a weak acid solution, calculate K_a.
   (**Example 13.5; Problems 43–48**)
4. Given K_a of a weak acid and its original concentration, calculate $[H^+]$.
   (**Examples 13.7–13.9; Problems 49–56**)
5. Given K_b of a weak base and its original concentration, calculate $[OH^-]$.
   (**Example 13.11; Problems 71–76**)
6. Given K_a for a weak acid, calculate K_b for its conjugate base (or vice versa).
   (**Problems 69, 70, 73, 74**)
7. Predict whether a salt solution is acidic, basic, or neutral.
   (**Example 13.12; Problems 77–84**)

Key Equations

Ionization of water	$K_w = [H^+][OH^-] = 1.0 \times 10^{-14}$ at 25°C
pH, pOH	$pH = -\log_{10}[H^+]$ $pOH = -\log_{10}[OH^-]$
Expressions for K_a, K_b	$K_a = \dfrac{[H^+][B^-]}{[HB]}$ $pK_a = -\log_{10}K_a$
	$K_b = \dfrac{[HB][OH^-]}{[B^-]}$ $pK_b = -\log_{10}K_b$
	$(K_a)(K_b) = K_w$
% ionization (weak acid)	% ionization $= \dfrac{[H^+]_{aq}}{[HB]_o} \times 100\%$

Key Terms

acid equilibrium constant
base equilibrium constant
Brønsted-Lowry acid
Brønsted-Lowry base

conjugate acid
conjugate base
ion product constant
    of water

Lewis acid
Lewis base
pH
pK_a

pK_b
pOH
polyprotic acid
salt

Summary Problem

Consider aqueous solutions of nitric acid, sodium hydroxide, ammonia, and hydrogen fluoride.

(a) When aqueous solutions of nitric acid and ammonia are combined, ammonium and nitrate ions are produced. When aqueous solutions of sodium hydroxide and hydrogen fluoride are combined, a solution containing sodium and fluoride ions is obtained. Classify each reactant as a strong or weak acid or base.

(b) Using the Brønsted-Lowry model, write equations to explain the acidity or basicity of HNO_3, NH_3, and HF.

(c) Calculate the pH of 0.45 M solutions of HNO_3, NaOH, NH_3, and HF.

(d) Consider NH_4^+.

   1. What is its conjugate base?

   2. What is the value of K_b for its conjugate base?

(e) Consider F^-.

   1. What is its conjugate acid?

   2. What is the value of K_a for its conjugate acid?

(f) Classify the salts NH_4NO_3 and NaF as acidic, basic, or neutral. Write net ionic equations to explain your answers.

(g) Would ammonium fluoride, NH_4F, be acidic or basic?

Answers

(a) HNO_3, strong acid; NaOH, strong base; NH_3, weak base; HF, weak acid.

(b) $HNO_3(aq) + H_2O \longrightarrow H_3O^+(aq) + NO_3^-(aq)$
   $NH_3(aq) + H_2O \rightleftharpoons NH_4^+(aq) + OH^-(aq)$
   $HF(aq) + H_2O \rightleftharpoons H_3O^+(aq) + F^-(aq)$

(c) HNO_3: pH = 0.35; NaOH: pH = 13.65;
   HF: pH = 1.75; NH_3: pH = 11.45

(d) 1. NH_3 2. $K_b = 1.8 \times 10^{-5}$

(e) 1. HF   2. $K_a = 6.9 \times 10^{-4}$

(f) NH_4NO_3 is acidic;
   $NH_4^+(aq) + H_2O \rightleftharpoons NH_3(aq) + H_3O^+(aq)$
   NaF is basic; $F^-(aq) + H_2O \rightleftharpoons HF(aq) + OH^-(aq)$

(g) ammonium fluoride is acidic.

Questions and Problems

Even-numbered questions and Challenge Problems have answers in Appendix 5 and fully worked solutions in the *Student Solutions Manual*.

13-1 Brønsted-Lowry Acid-Base Model

1. For each of the following reactions, indicate the Brønsted-Lowry acids and bases. What are the conjugate acid/base pairs?
(a) $H_3O^+(aq) + CN^-(aq) \rightleftharpoons HCN(aq) + H_2O$
(b) $HNO_2(aq) + OH^-(aq) \rightleftharpoons NO_2^-(aq) + H_2O$
(c) $HCHO_2(aq) + H_2O \rightleftharpoons CHO_2^-(aq) + H_3O^+(aq)$

2. Follow the directions for Question 1 for the following reactions.
(a) $CN^-(aq) + H_2O \rightleftharpoons HCN(aq) + OH^-(aq)$
(b) $HCO_3^-(aq) + H_3O^+(aq) \rightleftharpoons H_2CO_3(aq) + H_2O$
(c) $HC_2H_3O_2(aq) + HS^-(aq) \rightleftharpoons$
$$C_2H_3O_2^-(aq) + H_2S(aq)$$

3. According to the Brønsted-Lowry theory, which of the following would you expect to act as an acid? Which as a base? Both acid and base?
(a) CHO_2^-
(b) NH_4^+
(c) HSO_3^-

4. According to the Brønsted-Lowry theory, which of the following would you expect to act as an acid? Which as a base?
(a) CH_3O^-
(b) CO_3^{2-}
(c) $HAsO_4^{2-}$

5. Give the formula of the conjugate acid of
(a) OH^- (b) HPO_4^{2-} (c) NH_3
(d) F^- (e) $Zn(H_2O)_2(OH)_2$

6. Give the formula for the conjugate base of
(a) $H_2AsO_4^-$ (b) $Fe(H_2O)_5(OH)^+$
(c) $HClO_3$ (d) NH_4^+ (e) $HC_2H_3O_2$

7. Write a balanced equation showing how the $H_2PO_4^-$ ion can be either a Brønsted-Lowry acid or a Brønsted-Lowry base.

8. Follow the instructions of Question 7 for the hydrogen carbonate ion, HCO_3^-.

9. Using the Brønsted-Lowry model, write equations to show why the following species behave as weak acids in water.
(a) $Ni(H_2O)_5OH^+$ (b) $Al(H_2O)_6^{3+}$
(c) H_2S (d) HPO_4^{2-}
(e) $HClO_2$ (f) $Cr(H_2O)_5(OH)^+$

10. Follow the directions of Question 9 for the following species.
(a) $Zn(H_2O)_3OH^+$ (b) HSO_4^- (c) HNO_2
(d) $Fe(H_2O)_6^{2+}$ (e) $HC_2H_3O_2$ (f) $H_2PO_4^-$

11. Using the Brønsted-Lowry model, write an equation to show why each of the following species produces a basic aqueous solution.
(a) NH_3 (b) NO_2^- (c) $C_6H_5NH_2$
(d) CO_3^{2-} (e) F^- (f) HCO_3^-

12. Follow the directions of Question 11 for the following
(a) $(CH_3)_3N$ (b) $HC_2O_4^-$ (c) CN^-
(d) S^{2-} (e) BrO_3^- (f) $H_2C_6H_5O_7^-$

13-2 The Ion Product of Water
13-3 pH and pOH

[H⁺], [OH⁻], pH, and pOH

13. Find the pH of solutions with the following $[H^+]$. Classify each as acidic or basic.
(a) $6.0\ M$ (b) $0.33\ M$
(c) $4.6 \times 10^{-8}\ M$ (d) $7.2 \times 10^{-14}\ M$

14. Find the pH of solutions with the following $[H^+]$. Classify each as acidic or basic.
(a) $2.7 \times 10^{-3}\ M$ (b) $1.5\ M$
(c) $1.45 \times 10^{-13}\ M$ (d) $6.4 \times 10^{-9}\ M$

15. Calculate H^+ and OH^- in solutions with the following pH.
(a) 9.0 (b) 3.20 (c) -1.05 (d) 7.46

16. Calculate $[H^+]$ and $[OH^-]$ in solutions with the following pH.
(a) -0.76 (b) 9.11 (c) 3.81 (d) 12.08

17. Complete the following table for solutions at 25°C.

	[H⁺]	[OH⁻]	pH	pOH	Acidic?
(a)	2.4×10^{-8}	_____	_____	_____	_____
(b)	_____	1.9×10^{-2}	_____	_____	_____
(c)	_____	_____	8.62	_____	_____
(d)	_____	_____	_____	12.22	_____

18. Complete the table below for solutions at 25°C.

	[H⁺]	[OH⁻]	pH	pOH	Basic?
(a)	_____	_____	_____	8.14	_____
(b)	_____	_____	13.28	_____	_____
(c)	_____	4.3×10^{-3}	_____	_____	_____
(d)	2.75	_____	_____	_____	_____

19. Solution 1 has $[H^+] = 1.7 \times 10^{-2}\ M$. Solution 2 has $[H^+] = 4.3 \times 10^{-4}\ M$. Which solution is more acidic? Which has the higher pH?

20. Solution R has pH 13.42. Solution Q has $[OH^-] = 0.16\ M$. Which solution is more basic? Which has the lower pH?

21. Consider three solutions, R, Z, and Q.
- Solution R has $[H^+] = 0.0134$.
- Solution Z has a pH 6 times as large as solution R.
- Solution Q has $[OH^-]$ half as high as that of solution Z.
(a) Calculate the pH in solutions R, Q, and Z.
(b) What is the ratio of $[OH^-]$ in solutions R and Z and solutions R and Q?
(c) Classify each solution as acidic, basic, or neutral.

22. Solution A has a pH of 12.32. Solution B has $[H^+]$ three times as large as that of solution A. Solution C has a pH half that of solution A.
(a) What is $[H^+]$ for all three solutions?
(b) What is the pH of solutions B and C?
(c) Classify each solution as acidic, basic, or neutral.

23. Unpolluted rain water has a pH of about 5.5. Acid rain has been shown to have a pH as low as 3.0. Calculate the $[H^+]$ ratio of acid rain to unpolluted rain.

24. Most cola soft drinks have a pH of 3.1. Green tea has a pH of 5.8.
    (a) Calculate $[H^+]$ for the cola soft drink and green tea.
    (b) How much more acidic is a cola soft drink than green tea? Express your answer as a ratio of $[H^+]$ (cola) to $[H^+]$ (green tea).

25. Find $[OH^-]$ and the pH of the following solutions.
    (a) 0.25 g of $Ba(OH)_2$ dissolved in enough water to make 0.655 L of solution.
    (b) A 3.00-L solution of KOH is prepared by diluting 300.0 mL of 0.149 M KOH with water. What is the molarity of the diluted solution? What is the effect of a tenfold dilution on the pH?

26. Find $[H^+]$ and the pH of the following solutions.
    (a) A 456-mL sample of a 12.0% (by mass) solution of HNO_3 (d = 1.00 g/mL). What is the pH of 10.0 mL of the same sample?
    (b) A solution made up of 1.0 g of HCl dissolved in enough water to make 1.28 L of solution. What is the pH of the solution? If the same mass of HCl is dissolved in enough water to make 128 mL of solution, what would the pH be?

27. Find $[H^+]$, $[OH^-]$, and the pH of the following solutions.
    (a) 30.0 mL of a 0.216 M solution of HCl diluted with enough water to make 125 mL of solution.
    (b) A solution made by dissolving 275 mL of HBr gas at 25°C and 1.00 atm in enough water to make 475 mL of solution. Assume that all the HBr dissolves in water.

28. Find $[OH^-]$, $[H^+]$, and the pH of the following solutions.
    (a) Thirty-eight mL of a 0.106 M solution of $Sr(OH)_2$, diluted with enough water to make 275 mL of solution.
    (b) A solution prepared by dissolving 5.00 g of KOH in enough water to make 447 mL of solution.

29. How many grams of HI should be added to 265 mL of 0.215 M HCl so that the resulting solution has a pH of 0.38? Assume that the addition of HI does not change the volume of the resulting solution.

30. What is the pH of a solution obtained by adding 145 mL of 0.575 M HCl to 493 mL of a HNO_3 solution with a pH of 1.39? Assume that volumes are additive.

31. What is the pH of a solution obtained by adding 13.0 g of NaOH to 795 mL of a 0.200 M solution of $Sr(OH)_2$? Assume no volume change after NaOH is added.

32. What is the pH of a solution obtained by mixing 235 mL of NaOH with a pH of 11.57 and 316 mL of $Sr(OH)_2$ with a pH of 12.09? Assume that volumes are additive.

13-4 Weak Acids and Their Equilibrium Constants

Ionization Expressions, Weak Acids

33. Write the ionization equation and the K_a for each of the following acids.
    (a) AsH_4^+ (b) $H_2C_3H_5O_7^-$ (c) H_2SO_3

34. Write the ionization equation and the K_a expression for each of the following acids.
    (a) HSO_3^- (b) HPO_4^{2-} (c) HNO_2

35. Calculate K_a for the weak acids that have the following pK_a values.
    (a) 3.9   (b) 10.12   (c) 13.07

36. Calculate pK_a for the weak acids that have the following K_a values:
    (a) 1.8×10^{-4} (b) 6.8×10^{-8} (c) 4.0×10^{-11}

37. Consider these acids

Acid	A	B	C	D
pK_a	3.7	9.2	7.4	1.6

(a) Arrange the acids in order of decreasing acid strength from strongest to weakest.
(b) Which acid has the largest K_a value?

38. Consider these acids

Acid	A	B	C	D
K_a	1.6×10^{-3}	9×10^{-4}	2×10^{-6}	3×10^{-4}

(a) Arrange the acids in order of increasing acid strength from weakest to strongest.
(b) Which acid has the smallest pK_a value?

39. Rank the following solutions in order of increasing $[H^+]$. (Use Table 13.2.)

0.1 M HNO_3, 0.1 M HNO_2, 0.1 M $HC_7H_5O_2$, 0.1 M HClO

40. Rank the following acids (M = 0.10) in order of decreasing $[H^+]$:

HF   HCl   HCN   HClO

41. Rank the solutions in Question 39 in order of increasing pH.

42. Rank the solutions in Question 40 in order of increasing pH.

Equilibrium Calculations, Weak Acids

43. The pH of a 0.129 M solution of a weak acid, HB, is 2.34. What is K_a for the weak acid?

44. The pH of a 2.642 M solution of a weak acid, HB, is 5.32. What is K_a for the weak acid?

45. Paraminobenzene (PABA), $HC_7H_6NO_2$, is used in some sunscreens. A solution is made up of 0.263 mol PABA in enough water to make 750.0 mL of solution. The solution has $[H^+]$ = 2.6×10^{-3} M. What is K_a for PABA?

46. Acetaminophen, $HC_8H_8NO_2$ (MM = 151.17 g/mol), is the active ingredient in Tylenol®, a common pain reliever. A solution is made by dissolving 6.54 g of acetaminophen in enough water to make 250.0 mL of solution. The resulting solution has a pH of 5.24. What is K_a for acetaminophen?

47. Caproic acid, $HC_6H_{11}O_2$, is found in coconut oil and is used in making artificial flavors. A 2.00-L solution of caproic acid has a pH of 2.77 and contains 52.3 g of caproic acid. What is K_a for caproic acid?

48. Barbituric acid, $HC_4H_3N_2O_3$, is used to prepare barbiturates, a class of drugs used as sedatives. A 325-mL aqueous solution of barbituric acid has a pH of 2.34 and contains 9.00 g of the acid. What is K_a for barbituric acid?

49. When aluminum chloride dissolves in water, $Al(H_2O)_6^{3+}$ and Cl^- ions are obtained. Using the K_a in Table 13.2, calculate the pH of a 1.75 M solution of $AlCl_3$.

50. Using the K_a values in Table 13.2, calculate the pH of a 0.47 M solution of sodium hydrogen sulfite, $NaHSO_3$.

51. Barbituric acid, $HC_4H_3N_2O_3$, is used to prepare barbiturates, a class of drugs used as sedatives. Its K_a is 9.8×10^{-5}. Calculate $[H^+]$ in solutions prepared by adding enough water to the following to make 1.45 L.
   (a) 0.344 mol
   (b) 28.9 g

52. Penicillin (MM = 356 g/mol), an antibiotic often used to treat bacterial infections, is a weak acid. Its K_a is 1.7×10^{-3}. Calculate $[H^+]$ in solutions prepared by adding enough water to the following to make 725 mL.
   (a) 0.187 mol   (b) 127 g

53. Gallic acid, $HC_7H_5O_5$, is an ingredient in some ointments used to treat psoriasis. Its K_a is 3.9×10^{-5}. For a 0.168 M solution of gallic acid, calculate
   (a) $[H^+]$ (b) $[OH^-]$
   (c) pH   (d) % ionization

54. Anisic acid ($K_a = 3.38 \times 10^{-5}$) is found in anise seeds and is used as a flavoring agent. For a 0.279 M solution of anisic acid, calculate
   (a) $[H^+]$ (b) $[OH^-]$
   (c) pH   (d) % ionization

55. Phenol, once known as carbolic acid, HC_6H_5O, is a weak acid. It was one of the first antiseptics used by Lister. Its K_a is 1.1×10^{-10}. A solution of phenol is prepared by dissolving 14.5 g of phenol in enough water to make 892 mL of solution. For this solution, calculate
   (a) pH   (b) % ionization

56. Benzoic acid ($K_a = 6.6 \times 10^{-5}$) is present in many berries. Calculate the pH and % ionization of a 726-mL solution that contains 0.288 mol of benzoic acid.

Polyprotic Acids

Use the K_a values listed in Table 13.3 for polyprotic acids.

57. Chromic acid, H_2CrO_4, is commonly obtained by adding concentrated H_2SO_4 to a dichromate. It is used to clean glass. For chromic acid, K_a is 1.8×10^{-1} and 3.2×10^{-7} for its first and second dissociation, respectively. Write the overall equation and calculate K for the complete dissociation of chromic acid.

58. Consider citric acid, $H_3C_6H_5O_7$, added to many soft drinks. The equilibrium constants for its step-wise ionization are $K_{a1} = 7.5 \times 10^{-4}$, $K_{a2} = 1.7 \times 10^{-5}$, and $K_{a3} = 4.0 \times 10^{-7}$. Write the overall net ionic equation and calculate K for the complete ionization of citric acid.

59. Consider a 0.45 M solution of ascorbic acid, $H_2C_6H_6O_6$, one form of Vitamin C.

$H_2C_6H_6O_6(aq) \rightleftharpoons HC_6H_6O_6^-(aq) + H^+(aq)$ $K_a = 7.9 \times 10^{-5}$
$HC_6H_6O_6^-(aq) \rightleftharpoons C_6H_6O_6^{2-}(aq) + H^+(aq)$ $K_a = 1.6 \times 10^{-12}$

Calculate the pH of this solution and estimate $[HC_6H_6O_6^-]$ and $[C_6H_6O_6^{2-}]$.

60. Consider a 0.33 M solution of the diprotic acid H_2X.

$H_2X \rightleftharpoons H^+(aq) + HX^-(aq)$ $K_{a1} = 3.3 \times 10^{-4}$
$HX^- \rightleftharpoons H^+(aq) + X^{2-}(aq)$ $K_{a2} = 9.7 \times 10^{-8}$

Calculate the pH of the solution and estimate $[HX^-]$ and $[X^{2-}]$.

61. Phthalic acid, $H_2C_8H_4O_4$, is a diprotic acid. It is used to make phenolphthalein indicator. $K_{a1} = 0.0012$, and $K_{a2} = 3.9 \times 10^{-6}$. Calculate the pH of a 2.9 M solution of phthalic acid. Estimate $[HC_8H_4O_4^-]$ and $[C_8H_4O_4^{2-}]$.

62. Selenious acid, H_2SeO_3, is primarily used to chemically darken copper, brass, and bronze. It is a diprotic acid with the following K_a values: $K_{a1} = 2.7 \times 10^{-3}$ and $K_{a2} = 5.0 \times 10^{-8}$. What is the pH of a 2.89 M solution of selenious acid? Estimate $[HSeO_3^-]$ and $[SeO_3^{2-}]$.

13-5 Weak Bases and Their Equilibrium Constants

Ionization Expressions; Weak Bases

63. Write the ionization expression and the K_b expression for 0.1 M solutions of the following bases:
   (a) $C_2H_3O_2^-$
   (b) $C_2H_5NH_2$
   (c) $HC_2O_4^-$

64. Follow the instructions for Question 63 for the following bases.
   (a) NH_3 (b) HS^- (c) $(CH_3)_3N$

65. Using the equilibrium constants in Table 13.2, rank the following 0.1 M aqueous solutions in order of increasing K_b
   (a) NO_2^- (b) $H_2PO_4^-$ (c) CO_3^{2-}

66. Follow the directions of Question 65 for the following bases:
   (a) SO_3^{2-} (b) CHO_2^- (c) NO_2^-

67. Using the equilibrium constants listed in Table 13.2, arrange the following 0.1 M aqueous solutions in order of increasing pH (from lowest to highest).
   (a) $NaNO_2$ (b) HCl
   (c) NaF (d) $Zn(H_2O)_3(OH)(NO_3)$

68. Using the equilibrium constants listed in Table 13.2, arrange the following 0.1 M aqueous solutions in order of decreasing pH (from highest to lowest).
   (a) KOH (b) $NaCN$ (c) HCO_3^- (d) $Ba(OH)_2$

Equilibrium Calculations; Weak Bases

69. Find the value of K_b for the conjugate base of the following organic acids.
   (a) picric acid used in the manufacture of explosives; $K_a = 0.16$
   (b) trichloroacetic acid used in the treatment of warts; $K_a = 0.20$

70. Find the values of K_b for the conjugate bases of the following organic acids:
   (a) glycolic acid, used by dermatologists as a chemical peel; $K_a = 1.5 \times 10^{-4}$
   (b) butyric acid, responsible for the odor of rancid butter; $K_a = 1.5 \times 10^{-5}$

71. Determine $[OH^-]$, pOH and pH of a 0.28 M aqueous solution of Na_2CO_3.

72. Determine the $[OH^-]$ and pH of a 0.72 M solution of $NaHCO_3$.

73. Codeine (Cod), a powerful and addictive painkiller, is a weak base.
   (a) Write a reaction to show its basic nature in water. Represent the codeine molecule as Cod.

(b) The K_a for its conjugate acid is 1.2×10^{-8}. What is K_b for the reaction written in (a)?

(c) What is the pH of a 0.0020 M solution of codeine?

74. Consider pyridine, C_5H_5N, a pesticide and deer repellent. Its conjugate acid, $C_5H_5NH^+$, has $K_a = 6.7 \times 10^{-6}$.

(a) Write a balanced net ionic equation for the reaction that shows the basicity of aqueous solutions of pyridine.

(b) Calculate K_b for the reaction in (a).

(c) Find the pH of a solution prepared by mixing 2.74 g of pyridine in enough water to make 685 mL of solution.

75. A solution of baking soda, $NaHCO_3$, has a pH of 10.08. What is the percent (by mass) of $NaHCO_3$ in a 235-mL solution? (Assume a density of 1.00 g/mL.)

76. A solution of sodium cyanide, NaCN, has a pH of 12.10. How many grams of NaCN are in 425 mL of a solution with the same pH?

13-6 Acid-Base Properties of Salt Solutions

77. Write formulas for two salts that
(a) contain Ni^{3+} and are acidic.
(b) contain Na^+ and are basic.
(c) contain ClO_4^- and are neutral.
(d) contain NH_4^+ and are acidic.

78. Write formulas for two salts that
(a) contain NH_4^+ and are basic.
(b) contain CO_3^{2-} and are basic.
(c) contain Br^- and are neutral.
(d) contain CrO_4^{2-} and are acidic.

79. State whether 1 M solutions of the following salts in water are acidic, basic, or neutral.
(a) K_2CO_3 (b) NH_4F (c) LiH_2PO_4
(d) $NaNO_2$ (e) $Ba(ClO_4)_2$

80. State whether 1 M solutions of the following salts in water would be acidic, basic, or neutral.
(a) $FeCl_3$ (b) BaI_2 (c) NH_4NO_2
(d) Na_2HPO_4 (e) K_3PO_4

81. Write net ionic equations to explain the acidity or basicity of the various salts listed in Question 79.

82. Write net ionic equations to explain the acidity or basicity of the various salts listed in Question 80.

83. Arrange the following aqueous 0.1 M solutions in order of decreasing pH (highest to lowest).

$Ca(NO_3)_2$ HCl H_2SO_4 NaClO   KOH

84. Arrange the following aqueous 0.1 M solutions in order of increasing pH (lowest to highest).

$ZnBr_2$ NaOH   LiF $Sr(OH)_2$ $KClO_4$

Unclassified

85. At 25°C, a 0.20 M solution of methylamine, CH_3NH_2, is 5.0% ionized. What is K_b for methylamine?

86. Ammonium chloride is mixed with sodium nitrite. An acid-base reaction could take place. Write a balanced equation and calculate K for the reaction. How likely is the reaction taking place?

87. There are 324 mg of acetylsalicylic acid (MM = 180.15 g/mol) per aspirin tablet. If two tablets are dissolved in water to give two ounces ($\frac{1}{16}$ quart) of solution, estimate the pH. K_a of acetylsalicylic acid is 3.6×10^{-4}.

88. A student is asked to bubble enough ammonia gas through water to make 4.00 L of an aqueous ammonia solution with a pH of 11.55. What volume of ammonia gas at 25°C and 1.00 atm pressure is necessary?

89. Consider an acid HY (MM = 100 g/mol [3 significant figures]). What is the pH of a 33.0% by mass solution of HY ($d = 1.10$ g/mL)? The K_a of HY is 2.0×10^{-8}.

90. A student prepares 455 mL of a KOH solution, but neglects to write down the mass of KOH added. His TA suggests that he take the pH of the solution. The pH is 13.33. How many grams of KOH were added?

91. Consider the process

$$H_2O \rightleftharpoons H^+(aq) + OH^-(aq) \qquad \Delta H° = 55.8 \text{ kJ}$$

(a) Will the pH of pure water at body temperature (37°C) be 7.0?

(b) If not, calculate the pH of pure water at 37°C.

92. Household bleach is prepared by dissolving chlorine in water.

$$Cl_2(g) + H_2O \rightleftharpoons H^+(aq) + Cl^-(aq) + HOCl(aq)$$

K_a for HOCl is 3.2×10^{-8}. How much chlorine must be dissolved in one liter of water so that the pH of the solution is 1.19?

93. A tablet with a mass of 4.08 g contains 71.2% $NaHCO_3$. It is recommended for heartburn relief. If the stomach can hold 2.0×10^2 mL, will four tablets neutralize an "acid stomach" with a pH of 1.5?

94. Consider a weak organic base (nonelectrolyte) with molar mass 281 g/mol. An aqueous solution of the base has a pH of 8.73 and an osmotic pressure of 55 mm Hg at 25°C. What is K_b for the weak base?

95. Is a saline (NaCl) solution at 80°C acidic, basic, or neutral?

Conceptual Problems

96. Which of the following is/are true regarding a 0.1 M solution of $Ba(OH)_2$?
(a) $[OH^-] = 0.1 M$ (b) $[OH^-] = [Ba^{2+}]$
(c) pH = 13.30   (d) $[H^+] = 1 \times 10^{-7} M$

97. Which of the following is/are true about a 0.10 M solution of a strong acid, HY?
(a) $[Y^-] = 0.10 M$ (b) $[HY] = 0.10 M$
(c) $[H^+] = 0.10 M$ (d) pH = 1.0
(e) $[H^+] + [Y^-] = 0.20 M$

98. Consider the following six beakers. All have 100 mL of aqueous 0.1 M solutions of the following compounds:

beaker A has HI
beaker B has HNO_2
beaker C has NaOH
beaker D has $Ba(OH)_2$
beaker E has NH_4Cl
beaker F has $C_2H_5NH_2$

Answer the questions below, using **LT** (for *is less than*), **GT** (for *is greater than*), **EQ** (for *is equal to*), or **MI** (for *more information required*).

(a) The pH in beaker A _____ the pH in beaker B.
(b) The pH in beaker C _____ the pH in beaker D.
(c) The % ionization in beaker A _____ the % ionization in beaker C.
(d) The pH in beaker B _____ the pH in beaker E.
(e) The pH in beaker E _____ the pH in beaker F.
(f) The pH in beaker C _____ the pH in beaker F.

99. Each box represents an acid solution at equilibrium. Squares represent H^+ ions. Circles represent anions. (Although the anions have different identities in each figure, they are all represented as circles.) Water molecules are not shown. Assume that all solutions have the same volume.

(a) Which figure represents the strongest acid?
(b) Which figure represents the acid with the smallest K_a?
(c) Which figure represents the acid with the lowest pH?

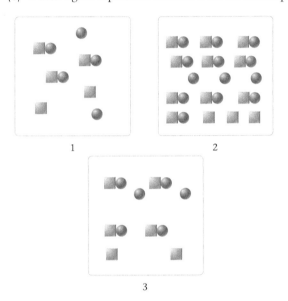

1   2

3

100. Each box represents an acid solution at equilibrium. Squares represent H^+ ions, and circles represent the anion. Water molecules are not shown. Which figure represents a strong acid? Which figure is a weak acid?

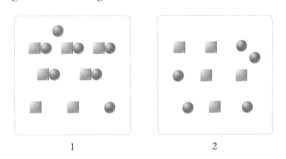

1   2

101. If □○ represents the weak acid HA (□ = H^+, ○ = A^-) in Figure (a), fill in Figure (b) with □○, □, and/or ○ to represent HA as being 10% ionized. (Water molecules are omitted.)

(a)   (b)

102. You are asked to determine whether an unknown white solid is acidic or basic. You also need to say whether the acid or base is weak or strong. You are given the molar mass of the solid and told that it is soluble in water. Describe an experiment that you can perform to obtain the desired characteristics of the white solid.

Challenge Problems

103. What is the pH of a 0.020 M solution of H_2SO_4? You may assume that the first ionization is complete. The second ionization constant is 0.010.

104. Using the Tables in Appendix 1, calculate ΔH for the reaction of the following.
(a) 1.00 L of 0.100 M NaOH with 1.00 L of 0.100 M HCl
(b) 1.00 L of 0.100 M NaOH with 1.00 L of 0.100 M HF, taking the heat of formation of HF(aq) to be -320.1 kJ/mol

105. What is the pH of a solution obtained by mixing 0.30 L of 0.233 M Ba(OH)$_2$ and 0.45 L of 0.12 M HCl? Assume that volumes are additive.

106. A solution is made up of 273 mL of 0.164 M HNO$_3$ and 0.800 M HCl. The resulting pH of the solution is 0.39. How many mL of HCl are used to make up the solution? Assume volumes are additive.

107. What is the freezing point of vinegar, which is an aqueous solution of 5.00% acetic acid, HC$_2$H$_3$O$_2$, by mass ($d = 1.006$ g/cm^3)?

108. The solubility of Ca(OH)$_2$ at 25°C is 0.153 g/100 g H$_2$O. Assuming that the density of a saturated solution is 1.00 g/mL, calculate the maximum pH one can obtain when Ca(OH)$_2$ is dissolved in water.

109. Consider two weak acids, HA (MM = 138 g/mol) and HB (MM = 72.0 g/mol). A solution consisting of 11.0 g of HA in 745 mL has the same pH as a solution made up of 5.00 g of HB in 525 mL. Which of the two acids is stronger? Justify your answer by an appropriate calculation.

110. Consider an aqueous solution of a weak base, NaB (MM = 233 g/mol). It has a pH of 10.54. The freezing point of the solution is -0.89°C and its density is 1.00 g/mol. Find K_b for the weak base B^-.

14 Equilibria in Acid-Base Solutions

Blue delphinium petals contain an anthocyanin called delphinin. It changes from bluish red in acid to blue to violet in base. The delphiniums in the painting must be growing in soil with a basic pH.

The meeting of two personalities is like the contact of two chemical substances: if there is any reaction, both are transformed.

—CARL JUNG
Modern Man in Search of a Soul

The Metropolitan Museum of Art/Gift of Betty G. Newell, 1996

Chapter Outline

14-1   Buffers

14-2   Acid-Base Indicators

14-3   Acid-Base Titrations

In Chapter 13 we dealt with the equilibrium established when a single solute, either a weak acid or a weak base, is added to water. This chapter focuses on the equilibrium established when two different solutes are mixed in water solution. These solutes may be

- **a weak acid, HB, and its conjugate base, B⁻.** Solutions called *buffers* contain roughly equal amounts of these two species. The equilibria involved in buffer solutions are considered in Section 14-1.
- **an acid and a base used in an acid-base titration.** This type of reaction was discussed in Chapter 4. Section 14-3 examines the equilibria involved, the way pH changes during the titration, and the choice of indicator for the titration.

We will also consider the equilibrium involved in using an acid-base indicator to estimate pH (Section 14-2).

14-1 Buffers

Any solution containing appreciable amounts of both a weak acid and its conjugate base

- *is highly resistant to changes in pH brought about by addition of strong acid or strong base.*
- *has a pH close to the pK_a of the weak acid.*

A solution with these properties is called a **buffer** ◂ⓘ because it cushions the "shock" (i.e., the drastic change in pH) that occurs when a strong acid or strong base is added to water.

ⓘ By definition, a buffer is any solution that resists a change in pH.

To prepare a buffer, we can mix solutions of a weak acid HB and the sodium salt of that acid NaB, which consists of Na^+ and B^- ions. This mixture can react with either a strong base

$$HB(aq) + OH^-(aq) \longrightarrow B^-(aq) + H_2O$$

or a strong acid

$$B^-(aq) + H^+(aq) \longrightarrow HB(aq)$$

These reactions have very large equilibrium constants, as we will see in Section 14-3, and so go virtually to completion. As a result, the added H^+ or OH^- ions are consumed and do not directly affect the pH. This is the principle of buffer action, which explains why a buffered solution is much more resistant to a change in pH than one that is unbuffered (Figure 14.1).

Buffers are widely used to maintain nearly constant pH in a variety of commercial products and laboratory procedures (Figure 14.2). For these applications

Both beakers contain distilled water.

Both beakers contain a pH 4 buffer.

After the addition of 20 mL of 0.30 M HCl, the pH of the water has dropped significantly.

Addition of the same volume of 0.30 M HCl to the buffer has barely changed the pH.

© Cengage Learning/Charles D. Winters

Figure 14.1 **The effect of a buffer upon addition of acid**. Note the reading on the pH meter.

Buffered aspirin   Blood plasma   Buffer tablets

Figure 14.2 Some applications of buffers. Many products, including aspirin and blood plasma, are buffered. Buffer tablets are also available in the laboratory *(far right)* to make up a solution to a specified pH.

and others, it is essential to be able to determine

- the pH of a buffer system made by mixing a weak acid with its conjugate base.
- the appropriate buffer system to maintain a desired pH.
- the (small) change in pH that occurs when a strong acid or base is added to a buffer.
- the capacity of a buffer to absorb H^+ or OH^- ions.

Determination of [H⁺] in a Buffer System

The concentration of H^+ ion in a buffer can be calculated if you know the concentrations of the weak acid HB and its conjugate base B^-. ◀ⓘ These three quantities are related through the acid equilibrium constant of HB:

$$HB(aq) \rightleftharpoons H^+(aq) + B^-(aq) \qquad K_a = \frac{[H^+][B^-]}{[HB]}$$

Solving for $[H^+]$,

$$[H^+] = K_a \times \frac{[HB]}{[B^-]} \qquad \qquad \textbf{(14.1)}$$

By taking the logarithm of both sides of Equation 14.1 and multiplying through by -1, you can show that, for any buffer system,

$$pH = pK_a + \log_{10} \frac{[B^-]}{[HB]}$$

This relation, known as the *Henderson-Hasselbalch equation,* is often used in biology and biochemistry to calculate the pH of buffers. Historically, it was Henderson who "discovered" Equation 14.1 in 1908. Hasselbalch put it in logarithmic form eight years later.

Equation 14.1 is a completely general equation, applicable to all buffer systems. The calculation of $[H^+]$—and hence pH—can be simplified if you keep two points in mind:

1. ***You can always assume that equilibrium is established without appreciably changing the original concentrations of either HB or B⁻.*** That is,

$$[HB] = [HB]_o$$
$$[B^-] = [B^-]_o$$

To explain why these relations hold, consider, for example, what happens when the weak acid HB is added to water:

$$HB(aq) \rightleftharpoons H^+(aq) + B^-(aq)$$

B⁻ comes from adding a salt of the weak acid, such as NaB.

We saw in Chapter 13 that under these conditions, it is usually a good approximation to take $[HB] \approx [HB]_o$. The approximation is even more accurate when a considerable amount of B^- is added, as is the case with a buffer. By Le Châtelier's principle, the reverse reaction occurs, the ionization of HB is repressed, and $[HB] = [HB]_o$.

2. *Because the two species HB and B$^-$ are present in the same solution, the ratio of their concentrations is also their mole ratio.* That is,

$$\frac{[HB]}{[B^-]} = \frac{\text{no. moles } HB/V}{\text{no. moles } B^-/V} = \frac{n_{HB}}{n_{B^-}}$$

where n = amount in moles and V = volume of solution. Hence Equation 14.1 can be rewritten as

$$[H^+] = K_a \times \frac{n_{HB}}{n_{B^-}} \blacktriangleleft \textit{i}$$
(14.2)

i If you start with 0.20 mol HB and 0.10 mol B^-, $[H^+] = 2K_a$.

Frequently, Equation 14.2 is easier to work with than Equation 14.1.

EXAMPLE	14.1	GRADED

Lactic acid, $C_3H_6O_3$, is a weak organic acid present in both sour milk and buttermilk. It is also a product of carbohydrate metabolism and is found in the blood after vigorous muscular activity (Figure 14.3). A buffer is prepared by dissolving lactic acid, HLac ($K_a = 1.4 \times 10^{-4}$), and sodium lactate, $NaC_3H_5O_3$, NaLac.

a   Calculate $[H^+]$ and the pH of the buffer made up of 1.00 mol of sodium lactate and 1.00 mol of lactic acid in enough water to form 550.0 mL of solution.

ANALYSIS

Information given:	K_a HLac (1.4×10^{-4}); n_{HLac} (1.00 mol); n_{Lac^-} (1.00 mol) volume of solution (550.0 mL)
Asked for:	$[H^+]$ and pH of the buffer

STRATEGY

1. Substitute into Equation 14.2 to obtain $[H^+]$.

$$[H^+] = K_a \times \frac{n_{HB}}{n_{B^-}}$$

2. Find pH: $pH = -\log_{10}[H^+]$

SOLUTION

1. $[H^+]$	$[H^+] = 1.4 \times 10^{-4} \times \dfrac{1.00}{1.00} = 1.4 \times 10^{-4}\ M$
2. pH	$pH = -\log_{10}(1.4 \times 10^{-4}) = 3.85$

b   Calculate $[H^+]$ and the pH of the buffer made up of 34.6 g of NaLac dissolved in 550.0 mL of a 1.20 *M* aqueous solution of HLac. (Assume no volume change after addition of NaLac.)

ANALYSIS

Information given:	K_a HLac (1.4×10^{-4}); mass of NaLac (34.6 g); M HLac (1.20) volume of solution (550.0 mL)
Information implied:	MM NaLac
Asked for:	$[H^+]$ and pH of the buffer

STRATEGY

1. Find n_{HLac} in solution: $n = V \times M$

2. Find n_{Lac^-} in solution. Recall 1 mol Lac^-/1 mol NaLac. Thus

$$n_{NaLac} = n_{Lac^-} = \text{mass NaLac/MM NaLac}$$

continued

3. Substitute into Equation 14.2 to find $[H^+]$.

4. $pH = -\log_{10}[H^+]$

SOLUTION

1. n_{HLac} $n = 0.5500 \text{ L} \times 1.20 \dfrac{\text{mol}}{\text{L}} = 0.660 \text{ mol}$

2. n_{Lac^-} $n_{Lac^-} = n_{NaLac} = \dfrac{34.6 \text{ g NaLac}}{112.06 \text{ g/mol}} = 0.309 \text{ mol}$

3. $[H^+]$ $[H^+] = 1.4 \times 10^{-4} \times \dfrac{0.660}{0.309} = 3.0 \times 10^{-4} \ M$

4. pH $pH = -\log_{10}(3.0 \times 10^{-4}) = 3.52$

END POINT

Looking back at part (a), note that when equal amounts of a weak acid and its conjugate base are present, $pH = pK_a$.

Figure 14.3 An exhausted distance runner crouches in pain. Exertion beyond the point at which the body can clear lactic acid from the muscles causes the pain.

Notice in Example 14.1(a) that the total volume of solution is irrelevant. All that is required to solve for $[H^+]$ is the number of moles of lactic acid and sodium lactate. The pH of the buffer is 3.85 whether the volume of solution is 550 mL, 1 L, or even 10 L. This explains why *diluting a buffer with water does not change the pH.*

Choosing a Buffer System

Suppose you want to make up a buffer in the laboratory with a specified pH (e.g., 4.0, 7.0, 10.0, . . .). Looking at the equation

$$[H^+] = K_a \times \frac{[HB]}{[B^-]} = K_a \times \frac{n_{HB}}{n_{B^-}}$$

it is evident that the pH of the buffer depends on two factors:

1. *The acid equilibrium constant of the weak acid, K_a.* The value of K_a has the greatest influence on buffer pH. Because HB and B^- are likely to be present in nearly equal amounts,

$$[H^+] \approx K_a \qquad pH \approx pK_a$$

Thus to make up a buffer with a pH close to 7, you should use a conjugate weak acid–weak base pair in which K_a of the weak acid is about 10^{-7}.

2. *The ratio of the concentrations or amounts of HB and B^-.* Small variations in pH can be achieved by adjusting this ratio. To obtain a slightly more acidic buffer, add more weak acid, HB; addition of more weak base, B^-, will make the buffer a bit more basic.

EXAMPLE 14.2   GRADED

Suppose you need to prepare a buffer with a pH of 9.00.

a   Which of the buffer systems in Table 14.1 would you choose?

STRATEGY AND SOLUTION

Check the K_a values in Table 14.1. Find the buffer system with a pK_a value closest to 9.00. The clear choice is the NH_4^+/NH_3 system with a pK_a of 9.25.

continued

b	What should be the ratio of the concentration of weak acid, HB, to its conjugate base, B⁻?

ANALYSIS

Information given:	pH (9.00); from part (a): buffer system (NH_4^+/NH_3)
Information implied:	K_a of NH_4^+
Asked for:	$[NH_4^+]/[NH_3]$

STRATEGY

1. Find $[H^+]$.
2. Substitute into Equation 14.1 where [HB] is $[NH_4^+]$ and B⁻ is $[NH_3]$.

SOLUTION

1. $[H^+]$	pH $= -\log_{10} 9.00 = 1.0 \times 10^{-9}$ M
2. $\dfrac{NH_4^+}{NH_3}$	$[H^+] = K_a \times \dfrac{NH_4^+}{NH_3} \longrightarrow \dfrac{NH_4^+}{NH_3} = \dfrac{[H^+]}{K_a} = \dfrac{1.0 \times 10^{-9}}{5.6 \times 10^{-10}} = 1.8$

c	What mass in grams of B⁻ should be added to 245 mL of 0.880 M HB to give a pH of 9.00?

ANALYSIS

Information given:	from part (a): buffer system (NH_4^+/NH_3) from part (b): $[H^+]$ (1.0×10^{-9} M); NH_4^+/NH_3 (1.8) NH_4^+ solution: V (0.245 L); M (0.880)
Information implied:	molar mass of NH_3
Asked for:	mass of NH_3 required to prepare a buffer with pH 9.0.

STRATEGY

1. Find mol NH_4^+.

   $n = V \times M$

2. Since there is only one solution, $[NH_4^+]/[NH_3] = $ mol NH_4^+/mol $NH_3 = 1.8$. Substitute mol NH_4^+ and find mol NH_3.
3. Find the mass of NH_3 required using its molar mass.

SOLUTION

1. mol NH_4^+	$n = (0.245$ L$)(0.880$ mol/L$) = 0.216$ mol
2. mol NH_3	$\dfrac{0.216}{\text{mol } NH_3} = 1.8 \longrightarrow$ mol $NH_3 = 0.12$
3. mass NH_3	mass $= (0.12$ mol$)(17.03$ g/mol$) = 2.0$ g

As Example 14.2 suggests, the most common way to prepare buffers is by mixing a weak acid and its conjugate base. However, a somewhat different approach is also possible: partial neutralization of a weak acid or a weak base gives a buffer. To illustrate, suppose that 0.18 mol of HCl is added to 0.28 mol of NH_3. The following reaction occurs:

$$H^+(aq) + NH_3(aq) \longrightarrow NH_4^+(aq)$$

Table 14.1 Buffer Systems at Different pH Values

| Desired pH | Buffer System | | K_a (Weak Acid) | pK_a |
	Weak Acid	Weak Base		
4	Lactic acid (HLac)	Lactate ion (Lac$^-$)	1.4×10^{-4}	3.85
5	Acetic acid (HC$_2$H$_3$O$_2$)	Acetate ion (C$_2$H$_3$O$_2{}^-$)	1.8×10^{-5}	4.74
6	Carbonic acid (H$_2$CO$_3$)	Hydrogen carbonate ion (HCO$_3{}^-$)	4.4×10^{-7}	6.36
7	Dihydrogen phosphate ion (H$_2$PO$_4{}^-$)	Hydrogen phosphate ion (HPO$_4{}^{2-}$)	6.2×10^{-8}	7.21
8	Hypochlorous acid (HClO)	Hypochlorite ion (ClO$^-$)	2.8×10^{-8}	7.55
9	Ammonium ion (NH$_4{}^+$)	Ammonia (NH$_3$)	5.6×10^{-10}	9.25
10	Hydrogen carbonate ion (HCO$_3{}^-$)	Carbonate ion (CO$_3{}^{2-}$)	4.7×10^{-11}	10.32

The limiting reactant, H$^+$, is consumed. Hence we have

	n_{H^+}	n_{NH_3}	$n_{NH_4^+}$
Original	0.18	0.28	0.00
Change	-0.18	-0.18	$+0.18$
Final	0.00	0.10	0.18

Note that complete neutralization would not produce a buffer; only NH$_4{}^+$ ions would be present.

The solution formed contains appreciable amounts of both NH$_4{}^+$ and NH$_3$. It is a buffer. ◄ *i* As we saw in Example 14.2, when the ratio [NH$_4{}^+$]/[NH$_3$] is 1.8, the pH of the buffer is 9.00.

In buffers, the limiting reactant is always the reactant with the smaller number of moles because the ratio of reactants to products is always 1.

EXAMPLE 14.3

The food industry uses the acetic acid/sodium acetate buffer to control the pH of some prepared foods. When NaOH is added to acetic acid (HAc) to prepare the buffer, the following reaction occurs:

$$\text{HAc}(aq) + \text{OH}^-(aq) \longrightarrow \text{Ac}^-(aq) + \text{H}_2\text{O}$$

a Show by calculation whether a solution made up of 0.300 mol of NaOH and 0.500 mol of HC$_2$H$_3$O$_2$ is a buffer.

ANALYSIS

Information given:	mol NaOH = mol OH$^-$ (0.300); mol HAc (0.500)
Information implied:	K_a for HAc (Table 13.2)
Asked for:	Is the solution a buffer?

STRATEGY

1. Fill in a table like the one shown in the preceding discussion.
2. Recall that for a solution to be a buffer, the solution must have a weak acid and its conjugate base.

continued

SOLUTION

1. Table

	n_{OH^-}	n_{HAc}	n_{Ac^-}
Original	0.300	0.500	0
Change	−0.300	−0.300	+0.300
Final	0	0.200	0.300

2. Buffer? There are 0.200 mol HAc and 0.300 mol Ac⁻ after reaction. The solution is a buffer.

---

b Is a buffer produced when 25.00 mL of 0.100 M NaOH is added to 35.00 mL of 0.125 M HC₂H₃O₂?

ANALYSIS

Information given:	NaOH: V (25.00 mL); M (0.100) HAc: V (35.00 mL); M (0.125)
Information implied:	K_a for HAc
Asked for:	Is the solution a buffer?

STRATEGY

1. Find mol OH⁻.
2. Find mol HAc.
3. Make a table as in part (a).
4. Check for the presence of the weak acid (HAc) and its conjugate base (Ac⁻) after reaction is complete.

SOLUTION

1. mol OH⁻ $(0.02500 \text{ L})(0.100 \text{ mol/L}) = 2.50 \times 10^{-3}$ mol

2. mol HAc $(0.03500 \text{ L})(0.125 \text{ mol/L}) = 4.38 \times 10^{-3}$ mol

3. Table

	n_{OH^-}	n_{HAc}	n_{Ac^-}
Original	2.50×10^{-3}	4.38×10^{-3}	0
Change	-2.50×10^{-3}	-2.50×10^{-3}	$+2.50 \times 10^{-3}$
Final	0	1.88×10^{-3}	2.50×10^{-3}

4. Buffer? There are 2.50×10^{-3} mol Ac⁻ and 1.88×10^{-3} mol HAc after reaction. The solution is a buffer.

---

c When 5.00 g of NaOH are added to 150.0 mL of 0.500 M HC₂H₃O₂ (without a volume change), is the resulting solution a buffer?

ANALYSIS

Information given:	NaOH: mass (5.00 g) HAc: V (150.0 mL); M (0.500)
Information implied:	K_a for HAc
Asked for:	Is the solution a buffer?

STRATEGY

1. Find mol OH⁻.
2. Find mol HAc.
3. Make a table as in part (a).
4. Check for the presence of the weak acid (HAc) and its conjugate base (Ac⁻) after reaction is complete.

continued

SOLUTION

		n_{OH^-}	n_{HAc}	n_{Ac^-}
1. mol OH^-	mol NaOH = $\dfrac{5.00 \text{ g}}{40.0 \text{ g/mol}}$ = 0.125 mol = mol OH^-			
2. mol HAc	(0.1500 L)(0.500 mol/L) = 0.0750 mol			
3. Table	Original	0.125	0.0750	0
	Change	−0.0750	−0.0750	+0.0750
	Final	0.050	0	0.0750
4. Buffer?	There are 0.0750 mol Ac^- and no mol HAc after reaction. The solution is not a buffer.

END POINT

To make a buffer from a strong base and a weak acid (or a strong acid and a weak base), the strong base (or strong acid) must be the limiting reactant (the reactant with the smaller number of moles).

Effect of Added H^+ or OH^- on Buffer Systems

The pH of a buffer does change slightly on addition of moderate amounts of a strong acid or strong base (Figure 14.4). Addition of H^+ ions converts an equal amount of weak base B^- to its conjugate acid HB:

$$H^+(aq) + B^-(aq) \longrightarrow HB(aq)$$

By the same token, addition of OH^- ions converts an equal amount of weak acid to its conjugate base B^-:

$$HB(aq) + OH^-(aq) \longrightarrow B^-(aq) + H_2O$$

In either case, the ratio n_{HB}/n_{B^-} changes. This in turn changes the H^+ ion concentration and pH of the buffer. ◄ ⓘ The effect ordinarily is small, as Example 14.4 illustrates.

ⓘ A buffer "works" because it contains species (HB and B^-) that react with both H^+ and OH^-.

Figure 14.4 Comparison of pH changes in water and a buffer solution.

These tubes show the effect of adding a few drops of strong acid or strong base to water. The pH changes drastically, giving a pronounced color change with universal indicator.

The experiment is repeated in these tubes using a buffer of pH 7 instead of water. This time the pH changes only very slightly, and there is no change in color of the indicator.

Marna G. Clarke

EXAMPLE 14.4

Consider the buffer described in Example 14.1, where $n_{HLac} = n_{Lac^-} = 1.00$ mol (K_a HLac $= 1.4 \times 10^{-4}$). You will recall that in this buffer the pH is 3.85.

a Calculate the pH after addition of 0.08 mol of HCl.

ANALYSIS

Information given:	mol HLac (1.00); mol Lac$^-$ (1.00); mol HCl = mol H$^+$ (0.08) pH of the buffer (3.85) K_a for HLac (1.4×10^{-4})
Asked for:	pH of the buffer after the addition of acid

STRATEGY

1. Write the reaction between the strong acid H$^+$ and the conjugate base, Lac$^-$.

2. Adding H$^+$ uses up the conjugate base in a 1:1 stoichiometric ratio.

   mol Lac$^-$ after addition = mol Lac$^-$ − mol H$^+$

3. Adding H$^+$ produces more weak acid in a 1:1 stoichiometric ratio.

   mol HLac after addition = mol HLac + mol H$^+$

4. Substitute into Equation 14.2 to find [H$^+$] and pH.

SOLUTION

1. Reaction	Lac$^-$(aq) + H$^+$(aq) $\longrightarrow$ HLac (aq)
2. mol Lac$^-$ after H$^+$ addition	mol Lac$^-$ = 1.00 − 0.08 = 0.92 mol
3. mol HLac after H$^+$ addition	mol HLac = 1.00 + 0.08 = 1.08 mol
4. [H$^+$]; pH	$[H^+] = 1.4 \times 10^{-4} \times \dfrac{1.08}{0.92} = 1.6 \times 10^{-4}$ $pH = -\log_{10}(1.6 \times 10^{-4}) = 3.80$

b Calculate the pH after addition of 0.08 mol of NaOH.

ANALYSIS

Information given:	mol HLac (1.00); mol Lac$^-$ (1.00); mol NaOH = mol OH$^-$ (0.08) pH of the buffer (3.85) K_a for HLac (1.4×10^{-4})
Asked for:	pH of the buffer after the addition of strong base

STRATEGY

1. Write the reaction between the strong base OH$^-$ and the weak acid, HLac.

2. Adding OH$^-$ uses up the weak acid in a 1:1 stoichiometric ratio.

   mol HLac after addition = mol HLac − mol OH$^-$

3. Adding OH$^-$ produces more conjugate base in a 1:1 stoichiometric ratio.

   mol Lac$^-$ after addition = mol Lac$^-$ + mol OH$^-$

4. Substitute into Equation 14.2 to find [H$^+$] and pH.

SOLUTION

1. Reaction	HLac(aq) + OH$^-$(aq) $\longrightarrow$ Lac$^-$(aq) + H$_2$O
2. mol HLac after OH$^-$ addition	mol HLac = 1.00 − 0.08 = 0.92 mol
3. mol Lac$^-$ after OH$^-$ addition	mol Lac$^-$ = 1.00 + 0.08 = 1.08 mol

continued

4. $[H^+]$; pH	$[H^+] = 1.4 \times 10^{-4} \times \dfrac{0.92}{1.08} = 1.2 \times 10^{-4}$
	$pH = -\log_{10}(1.2 \times 10^{-4}) = 3.92$

END POINTS

1. Adding a strong acid to a buffer

   ■ increases the number of moles of the weak acid.

   ■ decreases the number of moles of the conjugate base.

   ■ decreases the pH by a small amount. (In this case: $3.85 \longrightarrow 3.80$)

2. Adding a strong base to a buffer

   ■ increases the number of moles of the conjugate base.

   ■ decreases the number of moles of the weak acid.

   ■ increases the pH by a small amount. (In this case: $3.85 \longrightarrow 3.92$)

3. One cannot add an unlimited amount of strong acid or base. When the weak acid or its conjugate base becomes the limiting reactant, then the buffer is destroyed and only H^+ and the weak acid are present (if H^+ is added) or OH^- and the weak base are left (if OH^- is added).

Example 14.4 shows, among other things, how effectively a buffer "soaks up" H^+ or OH^- ions. That can be important. Suppose you are carrying out a reaction whose rate is first-order in H^+. If the pH increases from 5 to 7, perhaps by the absorption of traces of ammonia from the air, the rate will decrease by a factor of 100. A reaction that should have been complete in three hours will still be going on when you come back ten days later. Small wonder that chemists frequently work with buffered solutions to avoid disasters of that type.

Buffer Capacity and Buffer Range

A buffer has a limited capacity to react with H^+ or OH^- ions without undergoing a drastic change in pH. To see why this is the case, consider Figure 14.5, which applies to the H_2CO_3–HCO_3^- buffer system ($K_a\,H_2CO_3 = 4 \times 10^{-7}$).

To interpret Figure 14.5, it is important to realize that the **buffer capacity** *to absorb added OH^- or H^+ ions is inversely related to the slope of the curve.* Notice that

■ at point A where

$$pH = pK_a\,H_2CO_3 = 6.4$$

the slope is very small and the buffer has its maximum capacity to absorb OH^- or H^+ ions without a drastic change in pH.

■ at point B, where

$$pH = 7.4; \qquad [HCO_3^-] = 10 \times [H_2CO_3]$$

the slope is very large. Very few H_2CO_3 molecules are left, so the buffer has lost its capacity to absorb more OH^- ions.

■ at point C, where

$$pH = 5.4; \qquad [H_2CO_3] = 10 \times [HCO_3^-]$$

the pH is dropping off precipitously. Very few HCO_3^- ions are left, so the buffer has lost its capacity to absorb added H^+ ions.

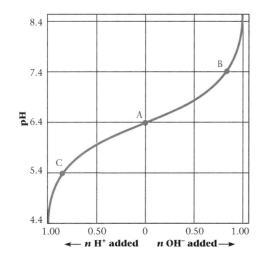

Figure 14.5 Variation of pH of an H_2CO_3–HCO_3^- buffer with addition of strong acid or base. Starting with one mole of both H_2CO_3 and HCO_3^- at point A, the curve to the left shows how pH decreases when H^+ is added. The curve to the right shows how pH increases as OH^- is added.

Blood is buffered at pH 7.4, in part by the H_2CO_3–HCO_3^- system. As you can see from Figure 14.5 and the above discussion, that puts us at point B on the curve. We conclude that blood has very little capacity to absorb OH^- ions (moving to the right of point B) but a large capacity to absorb H^+ ions (moving to the left of point B). This is indeed fortunate because life processes produce many more H^+ ions than OH^- ions.

The *buffer range* is the pH range over which the buffer is effective. It is related to the ratio of concentrations of the weak acid and its conjugate base. The further the ratio is from 1, the less effective the buffering action. Ideally, the acid/base ratio should be between 0.1 and 10. Since $\log_{10} 10 = 1$ and $\log_{10} 0.1 = -1$, buffers are most useful within ± 1 pH unit of the pK_a of the weak acid.

EXAMPLE 14.5 CONCEPTUAL

Consider the buffer system shown in the box below. The symbol ● represents a mole of the weak acid; the symbol ● represents a mole of its conjugate base. The pH of the buffer is 6.0. What is K_a of the weak acid?

ANALYSIS

Information given:	mol HB (3); mol B⁻ (2) pH (6.0)
Asked for:	K_a for HB

STRATEGY

1. Find $[H^+]$.
2. Substitute into Equation 14.2.

SOLUTION

1. $[H^+]$	$6.0 = -\log_{10}[H^+] \longrightarrow [H^+] = 1 \times 10^{-6}\ M$
2. K_a	$1.0 \times 10^{-6} = K_a \times \dfrac{3}{2} \longrightarrow \boxed{K_a = 7 \times 10^{-7}}$

14-2 Acid-Base Indicators

As pointed out in Chapter 4, an *acid-base indicator* is useful in determining the **equivalence point** of an acid-base titration. This is the point at which reaction is complete; equivalent quantities of acid and base have reacted. If the indicator is chosen properly, the point at which it changes color (its **end point**) ◀ ⓘ coincides with the equivalence point. To understand how and why an indicator changes color, we need to understand the equilibrium principle involved.

An acid-base indicator (Figure 14.6) is derived from a weak acid HIn:

$$HIn(aq) \rightleftharpoons H^+(aq) + In^-(aq) \qquad K_a = \frac{[H^+] \times [In^-]}{[HIn]}$$

where the weak acid HIn and its conjugate base In⁻ have different colors (Figure 14.7). The color that you see when a drop of indicator solution is added in an acid-base titration depends on the ratio

$$\frac{[HIn]}{[In^-]}$$

Three cases can be distinguished:

1. If $\dfrac{[HIn]}{[In^-]} \geq 10$

   the principal species is HIn; you see the "acid" color, that of the HIn molecule.

ⓘ It's called an end point because that's when you stop the titration.

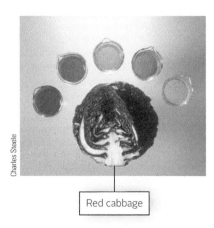

Charles Steele

Red cabbage

Figure 14.6 Red cabbage juice, a natural acid-base indicator. The picture shows (*left to right*) its colors at pH 1, 4, 7, 10, and 13.

Methyl red goes from red at low pH, to orange at about pH 5, to yellow at high pH.

Bromthymol blue is yellow at low pH, blue at high pH, and green at about pH 7.

Phenolphthalein goes from colorless to pink at about pH 9.

Marna G. Clarke

Figure 14.7 Acid-base indicators.

2. If $\dfrac{[\text{HIn}]}{[\text{In}^-]} \leq 0.1$

the principal species is In^-; you see the "base" color, that of the In^- ion. ◀ ⓘ

For litmus, HIn is red; In^- is blue.

3. If $\dfrac{[\text{HIn}]}{[\text{In}^-]} \approx 1$

the color observed is intermediate between those of the two species, HIn and In^-.

The expression for the ionization of the indicator molecule, HIn,

$$K_a = \frac{[\text{H}^+] \times [\text{In}^-]}{[\text{HIn}]}$$

can be rearranged to give

$$\frac{[\text{HIn}]}{[\text{In}^-]} = \frac{[\text{H}^+]}{K_a} \qquad (14.3)$$

From this expression, it follows that the ratio $[\text{HIn}]/[\text{In}^-]$ and hence the color of an indicator depends on two factors.

1. *[H⁺] or the pH of the solution.* ◀ ⓘ At high $[\text{H}^+]$ (low pH), you see the color of the HIn molecule; at low $[\text{H}^+]$ (high pH), the color of the In^- ion dominates.

Because only a drop or two of indicator is used, it does not affect the pH of the solution.

2. *K_a of the indicator.* Because K_a varies from one indicator to another, different indicators change colors at different pHs. A color change occurs when

$$[\text{H}^+] \approx K_a \qquad \text{pH} \approx \text{p}K_a$$

Table 14.2 Colors and End Points of Indicators

	Color [HIn]	Color [In⁻]	K_a	pH at End Point
Methyl red	Red	Yellow	1×10^{-5}	5
Bromthymol blue	Yellow	Blue	1×10^{-7}	7
Phenolphthalein	Colorless	Pink	1×10^{-9}	9

Table 14.2 shows the characteristics of three indicators: methyl red, bromthymol blue, and phenolphthalein. ◀ ⓘ These indicators change colors at pH 5, 7, and 9, respectively.

In practice it is usually possible to detect an indicator color change over a range of about two pH units. Consider, for example, what happens with bromthymol blue as the pH increases:

ⓘ What is the color of phenolphthalein at pH 11? pH 7?

- below pH 6 (e.g., . . . 3, 4, 5) the color is pure yellow.
- between pH 6 and 7, the indicator starts to take on a greenish tinge; we might describe its color as yellow-green.
- at pH 7, the color is a 50–50 mixture of yellow and blue so it looks green.
- between pH 7 and 8, a blue color becomes visible.
- above pH 8 (or 9, 10, 11, . . .) the color is pure blue.

EXAMPLE 14.6

Consider bromthymol blue ($K_a = 1 \times 10^{-7}$). At pH 6.5,

(a) calculate the ratio [In⁻]/[HIn].

(b) what is the color of the indicator at this point?

ANALYSIS

Information given:	K_a for bromthymol blue (1×10^{-7}) pH (6.5)
Information implied:	color of bromthymol blue at different pH values
Asked for:	(a) [In⁻]/[HIn] (b) color of bromthymol blue at pH 6.5

STRATEGY

(a) Convert pH to [H⁺] and substitute into Equation 14.1 to find [In⁻]/[HIn].

(b) Find the color of bromthymol blue at pH 6.5 by using the information in the above discussion of the colors for bromthymol blue at different pH values.

SOLUTION

(a) [In⁻]/[HIn]	$6.5 = -\log_{10}[H^+] \longrightarrow [H^+] = 3 \times 10^{-7}$ $\dfrac{[In^-]}{[HIn]} = \dfrac{K_a}{[H^+]} = \dfrac{1 \times 10^{-7}}{3 \times 10^{-7}} = \dfrac{1}{3}$
(b) bromthymol blue color	At pH 7, bromthymol blue is green. Below pH 6, bromthymol blue is yellow. Since 6.5 is halfway between 6 and 7, bromthymol blue at pH 6.5 is yellow-green.

14-3 Acid-Base Titrations

The discussion of **acid-base titrations** in Chapter 4 focused on stoichiometry. Here, the emphasis is on the equilibrium principles that apply to the acid-base reactions involved. It is convenient to distinguish between titrations involving

■ a strong acid (e.g., HCl) and a strong base (e.g., NaOH, Ba(OH)$_2$).
■ a weak acid (e.g., HC$_2$H$_3$O$_2$) and a strong base (e.g., NaOH).
■ a strong acid (e.g., HCl) and a weak base (e.g., NH$_3$).

Strong Acid–Strong Base

As pointed out in Chapter 13, strong acids ionize completely in water to form H$_3$O$^+$ ions; strong bases dissolve in water to form OH$^-$ ions. The neutralization reaction that takes place when *any* strong acid reacts with *any* strong base can be represented by a net ionic equation of the Brønsted-Lowry type:

$$H_3O^+(aq) + OH^-(aq) \longrightarrow 2H_2O$$

or, more simply, substituting H$^+$ ions for H$_3$O$^+$ ions, as

$$H^+(aq) + OH^-(aq) \longrightarrow H_2O$$

The equation just written is the reverse of that for the ionization of water, so the equilibrium constant can be calculated by using the reciprocal rule (Chapter 12).

$$K = 1/K_w = 1/(1.0 \times 10^{-14}) = 1.0 \times 10^{14}$$

The enormous value of K means that for all practical purposes this reaction goes to completion, consuming the limiting reactant, H$^+$ or OH$^-$.

Consider now what happens when HCl, a typical strong acid, is titrated with NaOH. Figure 14.8 shows how the pH changes during the titration. Two features of this curve are of particular importance:

1. At the equivalence point, when all the HCl has been neutralized by NaOH, a solution of NaCl, a neutral salt, is present. The pH at the equivalence point is 7.
2. Near the equivalence point, the pH rises very rapidly. Indeed, the pH may increase by as much as six units (from 4 to 10) when half a drop, ≈ 0.02 mL, of NaOH is added (Example 14.7).

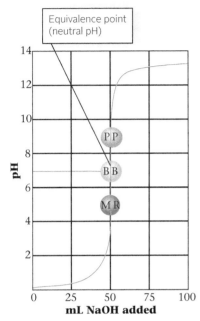

Figure 14.8 A strong acid–strong base titration. The curve represents the titration of 50.00 mL of 1.000 *M* HCl with 1.000 *M* NaOH. The solution at the equivalence point is neutral (pH = 7). The pH rises so rapidly near the equivalence point that any of the three indicators, methyl red (MR, end point pH = 5), bromthymol blue (BB, end point pH = 7), or phenolphthalein (PP, end point pH = 9) can be used.

EXAMPLE 14.7

When 50.00 mL of 1.000 *M* HCl is titrated with 0.7450 *M* NaOH (Figure 14.9), the pH increases.

| a | How many milliliters of NaOH are required to reach the equivalence point and a pH of 7.00? |

ANALYSIS

Information given:	HCl: *V* (50.00 mL); *M* (1.000) NaOH: *M* (0.7450)
Information implied:	acid-base reaction
Asked for:	volume of NaOH required to reach the equivalence point

STRATEGY

1. Recall the stoichiometry of acid-base reactions discussed in Chapter 4 and write the reaction.
2. Find mol HCl.
3. Follow the plan below to find the mol of NaOH and the volume of NaOH needed.

$$\text{mol HCl} \xrightarrow[\text{ratio}]{\text{atomic}} \text{mol H}^+ \xrightarrow[\text{ratio}]{\text{stoichiometric}} \text{mol OH}^- \xrightarrow[\text{ratio}]{\text{atomic}} \text{mol NaOH} \xrightarrow{M} V \text{ NaOH}$$

continued

SOLUTION

1. Reaction	$H^+(aq) + OH^-(aq) \longrightarrow H_2O$
2. mol HCl	mol HCl $= V \times M = (0.05000\ \text{L})(1.000\ \text{mol/L}) = 0.05000$
3. mol NaOH	$0.05000\ \text{mol HCl} \times \dfrac{1\ \text{mol H}^+}{1\ \text{mol HCl}} \times \dfrac{1\ \text{mol OH}^-}{1\ \text{mol H}^+} \times \dfrac{1\ \text{mol NaOH}}{1\ \text{mol OH}^-}$
	$= 0.05000\ \text{mol NaOH}$
Volume of NaOH	$V = \dfrac{0.05000\ \text{mol}}{0.7450\ \text{mol/L}} = 0.06711\ L$

b Find the pH when the volume of NaOH added is 0.02 mL less than the volume required to reach the equivalence point.

ANALYSIS

Information given:	HCl: V (0.05000 L); M (1.000)
	from part (a): mol H^+ (0.05000); V NaOH (67.11 mL)
	volume NaOH in the titration ($67.11 - 0.02 = 67.09$ mL)
Asked for:	pH of the solution after NaOH is added

STRATEGY

1. Find mol OH^-.

2. Fill in the following stoichiometric table.

	H^+	OH^-
Mol before reaction		
Change		
Mol after reaction		
Volume		

This table almost looks like the equilibrium table in Chapter 12.

3. Find [excess reactant]

$$[\text{excess reactant}] = \frac{\text{mol excess reactant}}{(\text{volume H}^+) + (\text{volume OH}^-)}$$

4. Find pH

SOLUTION

1. mol OH^-	$(0.06709\ \text{L})(0.7450\ \text{mol/L}) = 0.04998\ \text{mol}$

2. Table

	H^+	OH^-
Mol before reaction	0.05000	0.04998
Change	-0.04998	-0.04998
Mol after reaction	2×10^{-5}	0
Volume	50.00 mL	67.09 mL

3. [Excess reactant] $=$ [H^+] $[\text{H}^+] = \dfrac{2 \times 10^{-5}}{(0.05000 + 0.06709)\ \text{L}} = 2 \times 10^{-4}\ M$

4. pH $\text{pH} = -\log_{10}(2 \times 10^{-4}) = 3.7$

continued

c Find the pH when the volume of NaOH added is 0.02 mL more than the volume required to reach the equivalence point.

ANALYSIS

Information given:	HCl: V (0.05000 L); M (1.000) from part (a): mol H^+ (0.05000); V NaOH (67.11 mL) volume NaOH in the titration (67.11 + 0.02 = 67.13 mL)
Asked for:	pH of the solution after NaOH is added

STRATEGY

1. Find mol OH^-.

2. Fill in a stoichiometric table as in part (b).

3. Find [excess reactant] as in part (b).

4. Find pH.

SOLUTION

1. mol OH^- (0.06713 L)(0.7450 mol/L) = 0.05001 mol

2. Table

	H^+	OH^-
Mol before reaction	0.05000	0.05001
Change	−0.05000	−0.05000
Mol after reaction	0	1×10^{-5}
Volume	50.00 mL	67.13 mL

3. [Excess reactant] = $[OH^-]$ $[OH^-] = \dfrac{1 \times 10^{-5}}{0.05000 + 0.06713 \text{ L}} = 9 \times 10^{-5} \, M$; $[H^+] = 1 \times 10^{-10} \, M$

4. pH $pH = -\log_{10}(1 \times 10^{-10}) = 10.0$

END POINT

Note that half a drop before the end point (pH = 7) is reached, the pH is 3.8. When half a drop is added after the end point is reached, the end point changes from 7 to 10. This is predicted by Figure 14.8.

Figure 14.9 Titrating a solution of HCl with NaOH using phenolphthalein as the indicator.

© Cengage Learning/Charles D. Winters

For an effective titration, the reaction must go to completion, which requires $K > 10^4$ (approximately).

From Figure 14.8 and Example 14.7, we conclude that any indicator that changes color between pH 4 and 10 should be satisfactory for a strong acid–strong base titration. Bromthymol blue (BB: end point pH = 7) would work very well, but so would methyl red (MR: end point pH = 5) or phenolphthalein (PP: end point pH = 9).

Weak Acid–Strong Base

A typical weak acid–strong base titration is that of acetic acid with sodium hydroxide. The net ionic equation for the reaction is

$$HC_2H_3O_2(aq) + OH^-(aq) \longrightarrow C_2H_3O_2{}^-(aq) + H_2O$$

Notice that this reaction is the reverse of the reaction of the weak base $C_2H_3O_2{}^-$ (the acetate ion) with water (Chapter 13). It follows from the reciprocal rule that for this reaction,

$$K = \frac{1}{(K_b \; C_2H_3O_2{}^-)} = \frac{1}{(5.6 \times 10^{-10})} = 1.8 \times 10^9$$

Here again, K is a very large number; **ⓘ** the reaction of acetic acid with a strong base goes essentially to completion.

Figure 14.10 A weak acid–strong base titration. The curve represents the titration of 50.00 mL of 1.000 M acetic acid, $HC_2H_3O_2$, with 1.000 M NaOH. The solution at the equivalence point is basic (pH = 9.22). Phenolphthalein is a suitable indicator. Methyl red would change color much too early, when only about 33 mL of NaOH had been added. Bromthymol blue would change color slightly too quickly.

Figure 14.10 shows how the pH changes as fifty milliliters of one molar acetic acid is titrated with one molar sodium hydroxide. Notice that

- the pH starts off above 2; we are dealing with a weak acid, $HC_2H_3O_2$.
- there is a region, centered at the *halfway point* of the titration, where pH changes very slowly. The solution contains equal amounts of unreacted $HC_2H_3O_2$ and its conjugate base, $C_2H_3O_2^-$, formed by reaction with NaOH. As pointed out earlier, such a system acts as a buffer. Figure 14.11 shows how $[H^+] = K_a$ at half-neutralization.
- at the equivalence point, there is a solution of sodium acetate, $NaC_2H_3O_2$. The acetate ion is a weak base, so the pH at the equivalence point must be greater than 7.

Figure 14.11 Schematic showing that $[H^+] = K_a$ at half-neutralization.

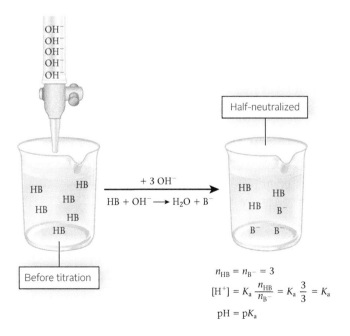

EXAMPLE 14.8

50.00 mL of 1.000 M acetic acid, $HC_2H_3O_2$, is titrated with 0.8000 M NaOH.

a Find the pH of the solution before any base is added.

ANALYSIS

Information given:	$HC_2H_3O_2$ (HAc): V (0.05000 L), M (1.000) NaOH: M (0.8000)
Information implied:	K_a for HAc
Asked for:	pH before titration starts (no base added)

STRATEGY

1. This is simply determining the pH of a weak acid. Recall Example 13.7.
2. Let $x = [H^+] = [Ac^-]$ at equilibrium. HAc at equilibrium $= [HAc]_o - x$. Make the assumption that $x = [H^+] \ll [HAc]_o$.
3. Substitute into Equation 13.5, solve for x, and check the assumption.
4. Find pH.

SOLUTION

2. $x = [H^+]$	$1.8 \times 10^{-5} = \dfrac{(x)(x)}{1.000 - x} = \dfrac{x^2}{1.000} \longrightarrow x = 4.2 \times 10^{-3} \, M$
3. Check assumption	% ionization $= \dfrac{4.2 \times 10^{-3}}{1.00} \times 100\% = 0.42\% < 5\%$ The assumption is valid.
4. pH	$pH = -\log_{10}(4.2 \times 10^{-3}) = 2.38$

b Find the pH of the solution when half the acetic acid has been neutralized.

ANALYSIS

Information given:	$HC_2H_3O_2$ (HAc): V (0.05000 L), M (1.000) NaOH: M (0.8000)
Information implied:	K_a for HAc
Asked for:	pH at half-neutralization.

SOLUTION

1. $[H^+]$	At half-neutralization $[H^+] = K_a$; $[H^+] = 1.8 \times 10^{-5}$
2. pH	$pH = -\log_{10}(1.8 \times 10^{-5}) = 4.74$

c Find the pH of the solution at the equivalence point.

ANALYSIS

Information given:	$HC_2H_3O_2$ (HAc): V (0.05000 L), M (1.000); from part (a): mol HAc (0.05000 mol) NaOH: M (0.8000)
Information implied:	K_a for HAc and K_b for Ac$^-$
Asked for:	pH at the equivalence point

STRATEGY

1. Write the reaction for the titration.
2. Find the volume of NaOH required to reach the equivalence point.

For the titration: mol HAc = mol OH$^-$

continued

3. At the equivalence point, all the acetic acid has been converted to acetate ions and mol $Ac^- = $ mol HAc at the start. Find $[Ac^-]$.

$$[Ac^-] = \frac{mol\ Ac^-}{V_{HAc} + V_{OH^-}}$$

4. Find $[OH^-]$ by substituting into Equation 13.8.
5. Find $[H^+]$ and pH.

	SOLUTION
1. Reaction	$HAc(aq) + OH^-(aq) \longrightarrow Ac^-(aq) + H_2O$
2. Volume NaOH required	mol HAc $= 0.05000$ mol $=$ mol $OH^- =$ mol NaOH
	volume $= \dfrac{0.05000\ mol}{0.800\ mol/L} = 0.0625$ L
3. $[Ac^-]$	mol $Ac^- =$ mol HAc $= 0.05000$ mol
	$[Ac^-] = \dfrac{0.05000\ mol}{(0.05000 + 0.06250)\ L} = 0.4444\ M$
4. $[OH^-]$	$K_b = \dfrac{[OH^-][HAc^-]}{[Ac^-]} \longrightarrow 5.6 \times 10^{-10} = \dfrac{(x)(x)}{0.4444}$
	$x = [OH^-] = 1.6 \times 10^{-5}\ M$
5. $[H^+]$; pH	$[H^+] = \dfrac{1.0 \times 10^{-14}}{1.6 \times 10^{-5}} = 6.2 \times 10^{-10};\ pH = 9.20$

From Figure 14.10 and Example 14.8, it should be clear that the indicator used in this titration must change color at about pH 9. Phenolphthalein (end point pH = 9) is satisfactory. Methyl red (end point pH = 5) is not suitable. If we used methyl red, we would stop the titration much too early, when reaction is only about 65% complete. This situation is typical of *weak acid–strong base titrations.* ◀ⓘ For such a titration, we choose an indicator that *changes color above pH 7.*

ⓘ If either the acid or base is weak, pH changes relatively slowly near the end point.

Strong Acid–Weak Base

The reaction between solutions of hydrochloric acid and ammonia can be represented by the Brønsted-Lowry equation:

$$H_3O^+(aq) + NH_3(aq) \longrightarrow NH_4^+(aq) + H_2O$$

Again, we simplify the equation by replacing H_3O^+ with H^+:

$$H^+(aq) + NH_3(aq) \longrightarrow NH_4^+(aq)$$

To find the equilibrium constant, note that this equation is the reverse of that for the acid dissociation of the NH_4^+ ion. Hence

$$K = \frac{1}{K_a NH_4^+} = \frac{1}{5.6 \times 10^{-10}} = 1.8 \times 10^9$$

Because K is so large, the reaction goes virtually to completion.

Figure 14.12 shows how pH changes when fifty milliliters of one molar NH_3 is titrated with one molar HCl. In many ways, this curve is the inverse of that shown in Figure 14.10 for the weak acid–strong base case. In particular,

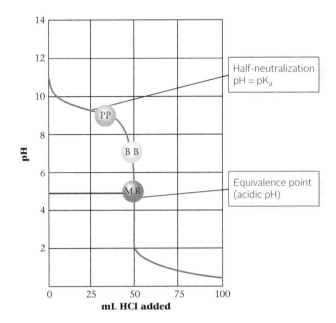

Figure 14.12 A strong acid–weak base titration. The curve represents the titration of 50.00 mL of 1.000 M ammonia, a weak base, with 1.000 M HCl. The solution at the equivalence point is acidic because of the NH_4^+ ion. Methyl red is a suitable indicator; phenolphthalein would change color much too early.

As strong acid is added, the pH drops.

■ the original pH is that of 1.000 M NH_3, a weak base ($K_b = 1.8 \times 10^{-5}$). As you would expect, it lies between 7 and 14:

$$[OH^-]^2 \approx K_b \times [NH_3]_o = (1.8 \times 10^{-5})(1.000)$$
$$[OH^-] = 4.2 \times 10^{-3} \, M; \, pOH = 2.38$$
$$pH = 14.00 - 2.38 = 11.62$$

■ addition of 25 mL of HCl gives a buffer containing equal amounts of NH_3 and NH_4^+ ($K_a = 5.6 \times 10^{-10}$). Hence

$$pH = pK_a \text{ of } NH_4^+ = 9.25$$

■ at the equivalence point, there is a 0.5000 M solution of NH_4^+, a weak acid. As you would expect, the pH is on the acid side. ◀ⓘ

$$[H^+]^2 \approx (0.5000)(5.6 \times 10^{-10})$$
$$[H^+] = 1.7 \times 10^{-5} \, M$$
$$pH = 4.77$$

From Figure 14.12, it should be clear that the only suitable indicator listed is methyl red. The other two indicators would change color too early, before the equivalence point. In general, *for the titration of a weak base with a strong acid, the indicator should change color on the acid side of pH 7.*

Titration of Diprotic Acids

Recall from Chapter 13 that a diprotic acid is one with two ionizable hydrogen ions. The hydrogen ions come off the diprotic acid one at a time and each ionization has its own equilibrium constant.

$$H_2X(aq) \rightleftharpoons H^+(aq) + HX^-(aq) \quad K_{a1}$$
$$HX^-(aq) \rightleftharpoons H^+(aq) + X^{2-}(aq) \quad K_{a2}$$

Because of this stepwise ionization, titration curves of diprotic acids have two equivalence points as shown in Figure 14.13. The first equivalence point in the titration curve represents the titration of the first H^+, and the second equivalence point represents the titration of the second proton.

Notice from Figure 14.13 that the volume of NaOH required to reach the second equivalence point is the same as that required to reach the first equivalence point. That is because the number of moles of H_2X yields HX^- in a 1:1 ratio.

Figure 14.13 A titration curve for a diprotic acid. Two equivalence points are shown, each corresponding to the reaction of one H^+ ion.

Thus the titration to the second equivalence point requires the same number of moles to neutralize H_2X to HX^- as to neutralize HX^- to X^{2-}. This makes sense since the overall neutralization reaction for one mole of H_2X requires two moles of OH^-.

$$H_2X(aq) + 2OH^-(aq) \longrightarrow X^{2-}(aq) + 2H_2O$$

Table 14.3 Characteristics of Acid-Base Titrations

	Strong Acid–Strong Base				
Example	Equation	K	Species at Equivalence Point	pH at Equivalence Point	Indicator*
NaOH–HCl	$H^+(aq) + OH^-(aq) \longrightarrow H_2O$	$K = 1/K_w$ 1.0×10^{14}	Na^+, Cl^-	7.00	MR, BB, PP
$Ba(OH)_2$–HNO_3	$H^+(aq) + OH^-(aq) \longrightarrow H_2O$	1.0×10^{14}	Ba^{2+}, NO_3^-	7.00	MR, BB, PP
	Weak Acid–Strong Base				
$HC_2H_3O_2$–NaOH	$HC_2H_3O_2(aq) + OH^-(aq) \longrightarrow$ $C_2H_3O_2^-(aq) + H_2O$	$K = 1/K_b$ 1.8×10^9	$Na^+, C_2H_3O_2^-$	$9.22^\dagger$	PP
HF–KOH	$HF(aq) + OH^-(aq) \longrightarrow$ $F^-(aq) + H_2O$	6.9×10^{10}	K^+, F^-	$8.42^\dagger$	PP
	Strong Acid–Weak Base				
NH_3–HCl	$NH_3(aq) + H^+(aq) \longrightarrow NH_4^+(aq)$	$K = 1/K_a$ 1.8×10^9	NH_4^+, Cl^-	$4.78^\dagger$	MR
ClO^-–HCl	$ClO^-(aq) + H^+(aq) \longrightarrow HClO(aq)$	3.6×10^7	$HClO, Cl^-$	$3.93^\dagger$	MR

*MR = methyl red (end point pH = 5); BB = bromthymol blue (end point pH = 7); PP = phenolphthalein (end point pH = 9).
†When 1 M acid is titrated with 1 M base.

Summary

Table 14.3 summarizes our discussion of acid-base titrations. Notice that for these three types of titrations,

■ *the equations (second column) that describe the reactions are quite different.* Strong acids and bases are represented by H^+ and OH^- ions, ◂ⓘ respectively; weak acids and weak bases are represented by their chemical formulas.

■ *the equilibrium constants (K) for all these reactions are very large, indicating that the reactions go essentially to completion.*

■ *the pH at the equivalence point is determined by which species are present.* When only "spectator ions" are present, as in a strong acid–strong base titration,

ⓘ H^+ and OH^- ions are the reacting species in strong acids and strong bases, respectively.

the pH at the equivalence point is 7. Basic anions (F^-, $C_2H_3O_2^-$), formed during a weak acid–strong base titration, make the pH at the equivalence point greater than 7. Conversely, acidic species (NH_4^+, $HClO$) formed during a strong acid–weak base titration make the pH at the equivalence point less than 7.

EXAMPLE 14.9

Consider the titration of formic acid, $HCHO_2$, with barium hydroxide.

(a) Write a balanced net ionic equation for the reaction.

(b) Calculate K for the reaction.

(c) Is the solution at the equivalence point acidic, basic, or neutral?

(d) What would be an appropriate indicator for the titration?

SOLUTION

(a) Reaction	$HCH_2O(aq) + OH^-(aq) \longrightarrow CHO_2^-(aq) + H_2O$
(b) K	$CHO_2^-(aq) + H_2O \rightleftharpoons HCH_2O(aq) + OH^-(aq)$ K_b
	$K = 1/K_b = 1/5.3 \times 10^{-11} = 1.9 \times 10^{10}$
(c) Acidic, basic or neutral?	basic, due to the presence of CHO_2^-
(d) Indicator	phenolphthalein

CHEMISTRY BEYOND THE CLASSROOM

Acid Rain

Natural rainfall has a pH of about 5.5. It is slightly acidic because of dissolved carbon dioxide, which reacts with water to form the weak acid H_2CO_3. In contrast, the average pH of rainfall in the eastern United States and southeastern Canada is about 4.4, corresponding to an H^+ ion concentration more than ten times the normal value. In an extreme case, rain with a pH of 1.5 was recorded in Wheeling, West Virginia (distilled vinegar has a pH of 2.4).

Acid rain results from the presence of two strong acids in polluted air: H_2SO_4 and HNO_3. Nitric acid comes largely from a three-step process. The first step involves the formation of NO from the elements by high-temperature combustion, as in an automobile engine. This is then oxidized to NO_2, which in turn undergoes the following reaction:

$$2NO_2(g) + H_2O \longrightarrow HNO_3(aq) + HNO_2(aq)$$

The principal source of sulfur dioxide, the precursor of sulfuric acid, is high-sulfur coal. Coal used in power plants can contain up to 4% sulfur, mostly in the form of minerals such as pyrite, FeS_2. Combustion forms sulfur dioxide:

$$4FeS_2(s) + 11O_2(g) \longrightarrow 2Fe_2O_3(s) + 8SO_2(g)$$

Sulfur dioxide formed by reactions such as this is converted to H_2SO_4 in the air by a two-step process:

$$(1)\ SO_2(g) + \tfrac{1}{2}O_2(g) \longrightarrow SO_3(g)$$
$$(2)\ SO_3(g) + H_2O \longrightarrow H_2SO_4(aq)$$

The sulfuric acid forms as tiny droplets high in the atmosphere. These may be carried by prevailing winds as far as 1500 km. The acid rain that falls in the Adirondacks of New York (where up to 40% of the lakes are acidic) comes from sulfur dioxide produced by power plants in Ohio and Illinois.

The effects of acid rain are particularly severe in areas where the bedrock is granite or other materials incapable of neutralizing H^+ ions. As the concentration of acid builds up in a lake, aquatic life, from algae to brook trout, dies. The end product is a crystal-clear, totally sterile lake.

Acid rain adversely affects trees as well (Figure A). It appears that the damage is largely due to the leaching of

Figure A Trees damaged by acid rain in the Great Smoky Mountains.

metal cations from the soil. In particular, H^+ ions in acid rain can react with insoluble aluminum compounds in the soil, bringing Al^{3+} ions into solution. The following reaction is typical:

$$Al(OH)_3(s) + 3H^+(aq) \longrightarrow Al^{3+}(aq) + 3H_2O$$

Al^{3+} ions in solution can have a direct toxic effect on the roots of trees. Perhaps more important, Al^{3+} can replace Ca^{2+}, an ion that is essential to the growth of all plants.

Strong acids in the atmosphere can also attack building materials such as limestone or marble (calcium carbonate):

$$CaCO_3(s) + 2H^+(aq) \longrightarrow Ca^{2+}(aq) + CO_2(g) + H_2O$$

The relatively soluble calcium compounds formed this way [$CaSO_4$, $Ca(NO_3)_2$] are gradually washed away (Figure B). This process is responsible for the deterioration of the Greek ruins on the Acropolis in Athens. These structures suffered more damage in the twentieth century than in the preceding 2000 years.

Sulfur dioxide emissions from power plants can be reduced by spraying a water solution of calcium hydroxide directly into the smokestack. This "scrubbing" operation

Figure B Effect of acid rain on a marble statue of George Washington in New York City. The photograph on the left was taken in the 1930s; the one on the right was taken in the 1990s.

brings about the reaction

$$Ca^{2+}(aq) + 2OH^-(aq) + SO_2(g) + \tfrac{1}{2}O_2(g) \longrightarrow CaSO_4(s) + H_2O$$

The good news is that processes such as this have reduced SO_2 emissions by 40% over the past 30 years. The bad news is that NO emissions have slowly but steadily increased.

Key Concepts

1. Calculate the pH of a buffer as originally prepared. (**Example 14.1; Problems 9–14**)
2. Choose a buffer to get a specified pH. (**Example 14.2; Problems 23–26**)
3. Determine whether a combination of a strong acid/base and its salt is a buffer. (**Example 14.3; Problems 19–22**)
4. Calculate the pH of a buffer after addition of H^+ or OH^- ions. (**Example 14.4; Problems 29–36**)
5. Determine the color of an indicator at a given pH. (**Example 14.6; Problems 39–40**)
6. Calculate the pH during an acid-base titration. (**Examples 14.7, 14.8; Problems 41–52**)
7. Choose the proper indicator for an acid-base titration. (**Example 14.9; Problems 37–38**)
8. Calculate K for an acid-base reaction. (**Example 14.9; Problems 5–8**)

Key Equations

Buffer $\qquad\qquad [H^+] = K_a \times \dfrac{[HB]}{[B^-]} = K_a \times \dfrac{n_{HB}}{n_{B^-}}$

Indicator $\qquad\quad \dfrac{[HIn]}{[In^-]} = \dfrac{[H^+]}{K_a}$

Key Terms

acid-base titrations
buffer

buffer capacity
end point

equivalence point

Summary Problem

Consider formic acid, $HCHO_2$ ($K_a = 1.9 \times 10^{-4}$), the agent responsible for the sting in ant bites. Its conjugate base is CHO_2^- ($K_b = 5.3 \times 10^{-11}$).

(a) Write net ionic equations and calculate K for the reactions in aqueous solutions between the following:

1. $HCHO_2$ and NaOH
2. CHO_2^- and HCl
3. $HCHO_2$ and NaCN ($K_b = 1.7 \times 10^{-5}$)

(b) A buffer is prepared by dissolving 25.0 g of sodium formate, $NaCHO_2$, in 1.30 L of 0.660 M $HCHO_2$ without a volume change.

1. Calculate the pH of the buffer.
2. Calculate the pH of the buffer after 0.100 mol of HNO_3 is added.
3. Calculate the pH of the buffer after 10.00 g of KOH are added.
4. Calculate the pH of the buffer after 75.0 mL of 0.517 M HNO_3 are added.
5. What is the buffer capacity?

(c) Thirty-five milliliters of 0.739 M $HCHO_2$ is titrated with 0.431 M KOH.

1. What is the pH of the $HCHO_2$ solution before titration?

2. How many milliliters of KOH are required to reach the equivalence point?
3. What is the pH of the solution halfway to neutralization?
4. What is the pH of the solution at the equivalence point?

Answers

(a) 1. $HCHO_2(aq) + OH^-(aq) \rightleftharpoons H_2O +$
$CHO_2^-(aq)$ $K = 1.9 \times 10^{10}$
2. $CHO_2^-(aq) + H^+(aq) \rightleftharpoons HCHO_2(aq)$
$K = 5.3 \times 10^3$
3. $HCHO_2(aq) + CN^-(aq) \rightleftharpoons HCN(aq) +$
$CHO_2^-(aq)$ $K = 3.2 \times 10^5$

(b) 1. 3.34
2. 3.16
3. 3.63
4. 3.29
5. The buffer can theoretically absorb a maximum of 0.368 mol of H^+ and 0.858 mol of OH^-.

(c) 1. 1.93
2. 60.0 mL
3. 3.72
4. 8.58

Questions and Problems

Even-numbered questions and Challenge Problems have answers in Appendix 5 and fully worked solutions in the *Student Solutions Manual*.

Equilibrium constants required to solve these problems can be found in the tables in Chapter 13 or in Appendix 1.

14-1 Buffers

Acid-Base Reactions

1. Write a net ionic equation for the reaction between aqueous solutions of
   (a) ammonia and hydrofluoric acid.
   (b) perchloric acid and rubidium hydroxide.
   (c) sodium sulfite and hydriodic acid.
   (d) nitric acid and calcium hydroxide.
2. Write a net ionic equation for the reaction between aqueous solutions of
   (a) sodium acetate ($NaC_2H_3O_2$) and nitric acid.
   (b) hydrobromic acid and strontium hydroxide.
   (c) hypochlorous acid and sodium cyanide.
   (d) sodium hydroxide and nitrous acid.

3. Write a balanced net ionic equation for the reaction of each of the following aqueous solutions with H^+ ions.
   (a) sodium sulfite
   (b) lithium hydroxide
   (c) potassium benzoate ($KC_7H_5O_2$)
4. Write a balanced net ionic equation for the reaction between the following aqueous solutions and OH^- ions.
   (a) $Fe(H_2O)_6^{3+}$
   (b) sodium hydrogen carbonate
   (c) ammonium chloride
5. Calculate K for the reactions in Question 1.
6. Calculate K for the reactions in Question 2.
7. Calculate K for the reactions in Question 3.
8. Calculate K for the reactions in Question 4.

Buffers

9. Calculate $[H^+]$ and pH in a solution in which lactic acid, $HC_3H_5O_3$, is 0.250 M and the lactate ion, $C_3H_5O_3^-$, is
   (a) 0.250 M (b) 0.125 M
   (c) 0.0800 M (d) 0.0500 M

10. Calculate $[OH^-]$ and pH in a solution in which the hydrogen sulfite ion, HSO_3^-, is 0.429 M and the sulfite ion is
   (a) 0.0249 M (b) 0.247 M
   (c) 0.504 M (d) 0.811 M
   (e) 1.223 M

11. A buffer is prepared by dissolving 0.0250 mol of sodium nitrite, $NaNO_2$, in 250.0 mL of 0.0410 M nitrous acid, HNO_2. Assume no volume change after HNO_2 is dissolved. Calculate the pH of this buffer.

12. A buffer is prepared by dissolving 0.062 mol of sodium fluoride in 127 mL of 0.0399 M hydrofluoric acid. Assume no volume change after NaF is added. Calculate the pH of this buffer.

13. A buffer solution is prepared by adding 15.00 g of sodium acetate ($NaC_2H_3O_2$) and 12.50 g of acetic acid to enough water to make 500 mL (three significant figures) of solution.
   (a) What is the pH of the buffer?
   (b) The buffer is diluted by adding enough water to make 1.50 L of solution. What is the pH of the diluted buffer?

14. A buffer solution is prepared by adding 5.50 g of ammonium chloride and 0.0188 mol of ammonia to enough water to make 155 mL of solution.
   (a) What is the pH of the buffer?
   (b) If enough water is added to double the volume, what is the pH of the solution?

15. A solution with a pH of 9.22 is prepared by adding water to 0.413 mol of KX to make 2.00 L of solution. What is the pH of the solution after 0.368 mol of HX is added?

16. An aqueous solution of 0.057 M weak acid, HX, has a pH of 4.65. What is the pH of the solution if 0.018 mol of KX is dissolved in one liter of the weak acid?

17. Which of the following would form a buffer if added to 250.0 mL of 0.150 M SnF_2?
   (a) 0.100 mol of HCl   (b) 0.060 mol of HCl
   (c) 0.040 mol of HCl   (d) 0.040 mol of NaOH
   (e) 0.040 mol of HF

18. Which of the following would form a buffer if added to 295 mL of 0.380 M HCl?
   (a) 0.100 mol of NH_4Cl (b) 0.033 mol of KF
   (c) 0.167 mol of $Sr(OH)_2$ (d) 0.279 mol of KNO_2
   (e) 0.112 mol of KClO

19. Calculate the pH of a solution prepared by mixing 2.50 g of hypobromous acid (HOBr) and 0.750 g of KOH in water. (K_a HOBr $= 2.5 \times 10^{-9}$).

20. Calculate the pH of a solution prepared by mixing 20.00 mL of aniline, $C_6H_5NH_2$ ($d = 1.022$ g/mL), with 35.0 mL of 1.67 M HCl. K_b for aniline is 4.3×10^{-10}. (Assume volumes are additive.)

21. Calculate the pH of a solution prepared by mixing 49.0 mL of butyric acid, $HC_4H_7O_2$, with 6.15 g of KOH in water. The following data about butyric acid may be helpful:

$$\text{density} = 0.9595 \text{ g/mL}; K_a = 1.54 \times 10^{-5}$$

22. Calculate the pH of a solution prepared by mixing 100.0 mL of 1.20 M ethanolamine, $C_2H_5ONH_2$, with 50.0 mL of 1.0 M HCl. K_a for $C_2H_5ONH_3^+$ is 3.6×10^{-10}.

23. Consider the weak acids in Table 13.2. Which acid-base pair would be best for a buffer at a pH of
   (a) 3.0   (b) 6.5   (c) 12.0

24. Follow the instructions of Question 23 for a pH of
   (a) 4.5   (b) 9.2   (c) 11.0

25. A sodium hydrogen carbonate-sodium carbonate buffer is to be prepared with a pH of 9.40.
   (a) What must the $[HCO_3^-]/[CO_3^{2-}]$ ratio be?
   (b) How many moles of sodium hydrogen carbonate must be added to a liter of 0.225 M Na_2CO_3 to give this pH?
   (c) How many grams of sodium carbonate must be added to 475 mL of 0.336 M $NaHCO_3$ to give this pH? (Assume no volume change.)
   (d) What volume of 0.200 M $NaHCO_3$ must be added to 735 mL of a 0.139 M solution of Na_2CO_3 to give this pH? (Assume that volumes are additive.)

26. You want to make a buffer with a pH of 10.00 from NH_4^+/NH_3.
   (a) What must the $[NH_4^+]/[NH_3]$ ratio be?
   (b) How many moles of NH_4Cl must be added to 465 mL of an aqueous solution of 1.24 M NH_3 to give this pH?
   (c) How many milliliters of 0.236 M NH_3 must be added to 2.08 g of NH_4Cl to give this pH?
   (d) What volume of 0.499 M NH_3 must be added to 395 mL of 0.109 M NH_4Cl to give this pH?

27. The buffer capacity indicates how much OH^- or H^+ ions a buffer can react with. What is the buffer capacity of the buffers in Problem 9?

28. The buffer capacity indicates how much OH^- or H^+ ions a buffer can react with. What is the buffer capacity of the buffers in Problem 10?

29. A buffer is made up of 0.300 L each of 0.500 M KH_2PO_4 and 0.317 M K_2HPO_4. Assuming that volumes are additive, calculate
   (a) the pH of the buffer.
   (b) the pH of the buffer after the addition of 0.0500 mol of HCl to 0.600 L of buffer.
   (c) the pH of the buffer after the addition of 0.0500 mol of NaOH to 0.600 L of buffer.

30. A buffer is made up of 239 mL of 0.187 M potassium hydrogen tartrate ($KHC_4H_4O_6$) and 137 mL of 0.288 M potassium tartrate ($K_2C_4H_4O_6$). K_a for ($H_2C_4H_4O_6$) is 4.55×10^{-5}. Assuming volumes are additive, calculate
   (a) the pH of the buffer.
   (b) the pH of the buffer after adding 0.0250 mol of HCl to 0.376 L of the buffer.
   (c) the pH of the buffer after adding 0.0250 mol of KOH to 0.376 L of the buffer.

31. Enough water is added to the buffer in Question 29 to make the total volume 10.0 L. Calculate
   (a) the pH of the buffer.
   (b) the pH of the buffer after the addition of 0.0500 mol of HCl to 0.600 L of diluted buffer.
   (c) the pH of the buffer after the addition of 0.0500 mol of NaOH to 0.600 L of diluted buffer.
   (d) Compare your answers to Question 29(a)–(c) with your answers to (a)–(c) in this problem.
   (e) Comment on the effect of dilution on the pH of a buffer and on its buffer capacity.

32. Enough water is added to the buffer in Question 30 to make the total volume 5.00 L.
   (a) Calculate the pH of the buffer.
   (b) Calculate the pH of the buffer after adding 0.0250 mol of HCl to 0.376 L of the buffer.

(c) Calculate the pH of the buffer after adding 0.0250 mol of KOH to 0.376 L of the buffer.

(d) Compare your answers to Question 30 (a–c) with your answers to (a–c) of this problem.

(e) Comment on the effect of dilution on the pH of a buffer and on its buffer capacity.

33. A buffer is prepared in which the ratio $[H_2PO_4^-]/[HPO_4^{2-}]$ is 3.0.

(a) What is the pH of this buffer?

(b) Enough strong acid is added to convert 15% of HPO_4^{2-} to $H_2PO_4^-$. What is the pH of the resulting solution?

(c) Enough strong base is added to make the pH 7.00. What is the ratio of $[H_2PO_4^-]$ to $[HPO_4^{2-}]$ at this point?

34. A buffer is prepared using the butyric acid/butyrate $(HC_4H_7O_2/C_4H_7O_2^-)$ acid-base pair. The ratio of acid to base is 2.2 and K_a for butyric acid is 1.54×10^{-5}.

(a) What is the pH of this buffer?

(b) Enough strong base is added to convert 15% of butyric acid to the butyrate ion. What is the pH of the resulting solution?

(c) Strong acid is added to the buffer to increase its pH. What must the acid/base ratio be so that the pH increases by exactly one unit (e.g., from 2 to 3) from the answer in (a)?

35. Blood is buffered mainly by the $HCO_3^- - H_2CO_3$ buffer system. The normal pH of blood is 7.40.

(a) Calculate the $[H_2CO_3]/[HCO_3^-]$ ratio.

(b) What does the pH become if 15% of the HCO_3^- ions are converted to H_2CO_3?

(c) What does the pH become if 15% of the H_2CO_3 molecules are converted to HCO_3^- ions?

36. There is a buffer system in blood $(H_2PO_4^- - HPO_4^{2-})$ that helps keep the blood pH at 7.40. $(K_a\ H_2PO_4^- = 6.2 \times 10^{-8})$.

(a) Calculate the $[H_2PO_4^-]/[HPO_4^{2-}]$ ratio at the normal pH of blood.

(b) What percentage of the HPO_4^{2-} ions are converted to $H_2PO_4^-$ when the pH is 7.00?

(c) What percentage of the $H_2PO_4^-$ ions are converted to HPO_4^{2-} when the pH is 8.00?

14-2 Acid-Base Indicators

37. Given three acid-base indicators—methyl orange (end point at pH 4), bromthymol blue (end point at pH 7), and phenolphthalein (end point at pH 9)—which would you select for the following acid-base titrations?

(a) perchloric acid with an aqueous solution of ammonia

(b) nitrous acid with lithium hydroxide

(c) hydrobromic acid with strontium hydroxide

(d) sodium fluoride with nitric acid

38. Given the acid-base indicators in Question 37, select a suitable indicator for the following titrations.

(a) sodium formate $(NaCHO_2)$ with HNO_3

(b) hypochlorous acid with barium hydroxide

(c) nitric acid with HI

(d) hydrochloric acid with ammonia

39. Metacresol purple is an indicator that changes from yellow to purple at pH 8.2.

(a) What is K_a for this indicator?

(b) What is its pH range?

(c) What is the color of a solution with pH 9.0 and a few drops of metacresol purple?

40. Thymolphthalein is an indicator that changes from colorless to blue at pH 9.5.

(a) What is its K_a?

(b) What is its pH range?

(c) What is the color of a solution with a pH of 10.0 and a few drops of thymolphthalein?

14-3 Acid-Base Titrations

41. When 25.00 mL of HNO_3 are titrated with $Sr(OH)_2$, 58.4 mL of a 0.218 M solution are required.

(a) What is the pH of HNO_3 before titration?

(b) What is the pH at the equivalence point?

(c) Calculate $[NO_3^-]$ and $[Sr^{2+}]$ at the equivalence point. (Assume that volumes are additive.)

42. A solution of KOH has a pH of 13.29. It requires 27.66 mL of 0.2500 M HCl to reach the equivalence point.

(a) What is the volume of the KOH solution?

(b) What is the pH at the equivalence point?

(c) What is $[K^+]$ and $[Cl^-]$ at the equivalence point? Assume volumes are additive.

43. A solution consisting of 25.00 g NH_4Cl in 178 mL of water is titrated with 0.114 M KOH.

(a) How many milliliters of KOH are required to reach the equivalence point?

(b) Calculate $[Cl^-]$, $[K^+]$, $[NH_3]$, and $[OH^-]$ at the equivalence point. (Assume that volumes are additive.)

(c) What is the pH at the equivalence point?

44. A 50.0-mL sample of $NaHSO_3$ is titrated with 22.94 mL of 0.238 M KOH.

(a) Write a balanced net ionic equation for the reaction.

(b) What is $[HSO_3^-]$ before the titration?

(c) Find $[HSO_3^-]$, $[SO_3^{2-}]$, $[OH^-]$, $[K^+]$, and $[Na^+]$ at the equivalence point.

(d) What is the pH at the equivalence point?

45. A 20.00-mL sample of 0.220 M triethylamine, $(CH_3CH_2)_3 N$, is titrated with 0.544 M HCl. $(K_b(CH_3CH_2)_3N = 5.2 \times 10^{-4})$

(a) Write a balanced net ionic equation for the titration.

(b) How many milliliters of HCl are required to reach the equivalence point?

(c) Calculate $[(CH_3CH_2)_3N]$, $[(CH_3CH_2)_3NH^+]$, $[H^+]$, and $[Cl^-]$ at the equivalence point. (Assume that volumes are additive.)

(d) What is the pH at the equivalence point?

46. A 35.00-mL sample of 0.487 M KBrO is titrated with 0.264 M HNO$_3$. $(K_b\ BrO^- = 4.0 \times 10^{-6})$

(a) Write a balanced net ionic equation for the reaction.

(b) How many milliliters of HCl are required to reach the equivalence point?

(c) What is the pH at the equivalence point?

(d) Calculate $[K^+]$, $[NO_3^-]$, $[H^+]$, $[BrO^-]$, and $[HBrO]$ at the equivalence point. (Assume volumes are additive.)

47. A 0.4000 M solution of nitric acid is used to titrate 50.00 mL of 0.237 M barium hydroxide. (Assume that volumes are additive.)

(a) Write a balanced net ionic equation for the reaction that takes place during titration.

(b) What are the species present at the equivalence point?

(c) What volume of nitric acid is required to reach the equivalence point?
(d) What is the pH of the solution before any HNO_3 is added?
(e) What is the pH of the solution halfway to the equivalence point?
(f) What is the pH of the solution at the equivalence point?

48. A 0.2481 M solution of KOH is used to titrate 30.00 mL of 0.269 M hydrobromic acid. Assume that volumes are additive.
(a) Write a balanced net ionic equation for the reaction that takes place during the titration.
(b) What are the species present at the equivalence point?
(c) What volume of KOH is required to reach the equivalence point?
(d) What is the pH of the solution
  1. before any KOH is added?
  2. halfway to the equivalence point?
  3. at the equivalence point?

49. Consider the titration of butyric acid (HBut) with sodium hydroxide. In an experiment, 50.00 mL of 0.350 M butyric acid is titrated with 0.225 M NaOH. K_a HBut = 1.5×10^{-5}.
(a) Write a balanced net ionic equation for the reaction that takes place during titration.
(b) What are the species present at the equivalence point?
(c) What volume of sodium hydroxide is required to reach the equivalence point?
(d) What is the pH of the solution before any NaOH is added?
(e) What is the pH of the solution halfway to the equivalence point?
(f) What is the pH of the solution at the equivalence point?

50. Morphine, $C_{17}H_{19}O_3N$, is a weak base ($K_b = 7.4 \times 10^{-7}$). Consider its titration with hydrochloric acid. In the titration, 50.0 mL of a 0.1500 M solution of morphine is titrated with 0.1045 M HCl.
(a) Write a balanced net ionic equation for the reaction that takes place during titration.
(b) What are the species present at the equivalence point?
(c) What volume of hydrochloric acid is required to reach the equivalence point?
(d) What is the pH of the solution before any HCl is added?
(e) What is the pH of the solution halfway to the equivalence point?
(f) What is the pH of the solution at the equivalence point?

51. Consider a 10.0% (by mass) solution of hypochlorous acid. Assume the density of the solution to be 1.00 g/mL. A 30.0-mL sample of the solution is titrated with 0.419 M KOH. Calculate the pH of the solution
(a) before titration.
(b) halfway to the equivalence point.
(c) at the equivalence point.

52. A solution is prepared by dissolving 0.350 g of benzoic acid, $HC_7H_5O_2$, in water to make 100.0 mL of solution. A 30.00-mL sample of the solution is titrated with 0.272 M KOH. Calculate the pH of the solution
(a) before titration.

(b) halfway to the equivalence point.
(c) at the equivalence point.

Unclassified

53. A student is given 250.0 g of sodium lactate, $NaC_3H_5O_3$, and a bottle of lactic acid marked "73.0% by mass $HC_3H_5O_3$, $d = 1.20$ g/mL." How many milliliters of 73.0% lactic acid should the student add to the sodium lactate to produce a buffer with a pH of 4.50?

54. Ammonia gas is bubbled into 275 mL of water to make an aqueous solution of ammonia. To prepare a buffer with a pH of 9.56, 15.0 g of NH_4Cl are added. How many liters of NH_3 at 25°C and 0.981 atm should be used to prepare the buffer? Assume no volume changes and ignore the vapor pressure of water.

55. For an aqueous solution of acetic acid to be called "distilled white vinegar" it must contain 5.0% acetic acid by mass. A solution with a density of 1.05 g/mL has a pH of 2.95. Can the solution be called "distilled white vinegar"?

56. Consider an unknown base, RNH. One experiment titrates a 50.0-mL aqueous solution containing 2.500 g of the base. This titration requires 59.90 mL of 0.925 M HCl to reach the equivalence point. A second experiment uses a 50.0-mL solution of the unknown base identical to what was used in the first experiment. To this solution is added 29.95 mL of 0.925 M HCl. The pH after the HCl addition is 10.77.
(a) What is the molar mass of the unknown base?
(b) What is K_b for the unknown base?
(c) What is K_a for RNH_2^+?

57. A painful arthritic condition known as gout is caused by an excess of uric acid, HUric, in the blood. An aqueous solution contains 4.00 g of uric acid. A 0.730 M solution of KOH is used for titration. After 12.00 mL of KOH are added, the resulting solution has pH 4.12. The equivalence point is reached after a total of 32.62 mL of KOH are added.

$$HUric(aq) + OH^-(aq) \longrightarrow Uric^-(aq) + H_2O$$

(a) What is the molar mass of uric acid?
(b) What is its K_a?

58. Water is accidentally added to 350.00 mL of a stock solution of 6.00 M HCl. A 75.00-mL sample of the diluted solution is titrated to pH 7.00 with 78.8 mL of 4.85 M NaOH. How much water was accidentally added? (Assume that volumes are additive.)

59. A solution of an unknown weak base (nonelectrolyte) at 25°C has an osmotic pressure of 1.287 atm and a pH of 8.94. (Assume that, in the equation for π [Chapter 10], $i \approx 1$.) What is K_a for its conjugate acid?

60. Consider an aqueous solution of HF. The molar heat of formation for aqueous HF is -320.1 kJ/mol.
(a) What is the pH of a 0.100 M solution of HF at 100°C?
(b) Compare with the pH of a 0.100 M solution of HF at 25°C.

Conceptual Problems

61. Each symbol in the box below represents a mole of a component in one liter of a buffer solution; ◯ represents the anion (X^-), ▢◯ = the weak acid (HX), ▢ = H^+, and

△ = OH⁻. Water molecules and the few H⁺ and OH⁻ ions from the dissociation of HX and X⁻ are not shown. The box contains 10 mol of a weak acid, □○, in a liter of solution. Show what happens upon

(a) the addition of 2 mol of OH⁻ (2 △).
(b) the addition of 5 mol of OH⁻ (5 △).
(c) the addition of 10 mol of OH⁻ (10 △).
(d) the addition of 12 mol of OH⁻ (12 △).

Which addition (a)–(d) represents neutralization halfway to the equivalence point?

62. Use the same symbols as in Question 61 (○ = anion, △ = OH⁻) for the box below.

(a) Fill in a similar box (representing one liter of the same solution) after 2 mol of H⁺ (2 □) have been added. Indicate whether the resulting solution is an acid, base, or buffer.
(b) Follow the directions of part (a) for the resulting solution after 2 mol of OH⁻ (2 △) have been added.
(c) Follow the directions of part (a) for the resulting solution after 5 mol of OH⁻ (5 △) have been added.

(*Hint:* Write the equation for the reaction before you draw the results.)

63. The following is the titration curve for the titration of 50.00 mL of a 0.100 *M* acid with 0.100 *M* KOH.

mL 0.100 *M* KOH

(a) Is the acid strong or weak?
(b) If the acid is weak, what is its K_a?
(c) Estimate the pH at the equivalence point.

64. Consider the following five beakers:

Beaker A has 0.100 mol HA and 0.100 mol NaA in 100 mL of solution.
Beaker B has 0.100 mol HA and 0.100 mol NaA in 200 mL of solution.
Beaker C has 0.100 mol HA, 0.100 mol NaA, and 0.0500 mol HCl in 100 mL of solution.
Beaker D has 0.100 mol HA, 0.100 mol NaA, and 0.100 mol NaOH in 100 mL of solution.
Beaker E has 0.100 mol HCl and 0.100 mol NaOH in 100 mL of solution.

Answer the questions below, using **LT** (for *is less than*), **GT** (for *is greater than*), **EQ** (for *is equal to*), or **MI** (for *more information required*).

(a) The pH in beaker A _____ the pH in beaker B.
(b) The pH in beaker A _____ the pH in beaker C.
(c) The pH in beaker D _____ the pH in beaker E.
(d) The pH in beaker A _____ the pH in beaker E.
(e) The pH in beaker A _____ the pH in beaker D.

65. Follow the directions of Question 64. Consider two beakers:

Beaker A has a weak acid ($K_a = 1 \times 10^{-5}$).
Beaker B has HCl.

The volume and molarity of each acid in the beakers are the same. Both acids are to be titrated with a 0.1 *M* solution of NaOH.

(a) Before titration starts (at zero time), the pH of the solution in Beaker A is _____ the pH of the solution in Beaker B.
(b) At half-neutralization (halfway to the equivalence point), the pH of the solution in Beaker A _____ the pH of the solution in Beaker B.
(c) When each solution has reached its equivalence point, the pH of the solution in Beaker A _____ the pH of the solution in Beaker B.
(d) At the equivalence point, the volume of NaOH used to titrate HCl in Beaker B _____ the volume of NaOH used to titrate the weak acid in Beaker A.

66. Which statements are true about the titration represented by the following equation?

$$H^+(aq) + F^-(aq) \rightleftharpoons HF(aq)$$

(a) The net ionic equation could represent the titration of NaF by H_2SO_4.
(b) The reaction could be considered to be that of a strong acid with a strong base.
(c) At the equivalence point, the resulting solution can act as a buffer.
(d) Diluting the solution obtained at half-neutralization with water does not change the pH of the solution.
(e) Phenolphthalein (pH range: 9–11) would be an effective indicator in this titration.

67. Indicate whether each of the following statements is true or false. If the statement is false, restate it to make it true.

(a) The formate ion (CHO_2^-) concentration in 0.10 M $HCHO_2$ is the same as in 0.10 M $NaCHO_2$.

(b) A buffer can be destroyed by adding too much strong acid.

(c) A buffer can be made up by any combination of weak acid and weak base.

(d) Because K_a for HCO_3^- is 4.7×10^{-11}, K_b for HCO_3^- is 2.1×10^{-4}.

68. The solutions in three test tubes labeled A, B, and C all have the same pH. The test tubes are known to contain 1.0×10^{-3} M HCl, 6.0×10^{-3} M $HCHO_2$, and 4×10^{-2} M $C_6H_5NH_3^+$. Describe a procedure for identifying the solutions.

69. Consider the following titration curves. The solution in the buret is 0.1 M. The solution in the beaker has a volume of 50.0 mL. Answer the following questions.

(a) Is the titrating agent (solution in the buret) an acid or a base?

(b) Which curve shows the titration of the weakest base?

(c) What is the K_a of the conjugate acid of the base titrated in curve B?

(d) What is the molarity of the solution in the beaker for curve C?

(e) What is the pH at the equivalence point for curve A?

Challenge Problems

70. Consider the titration of HF ($K_a = 6.7 \times 10^{-4}$) with NaOH. What is the pH when a third of the acid has been neutralized?

71. The species called *glacial acetic acid* is 98% acetic acid by mass (d = 1.0542 g/mL). What volume of glacial acetic acid must be added to 100.0 mL of 1.25 M NaOH to give a buffer with a pH of 4.20?

72. Four grams of a monoprotic weak acid are dissolved in water to make 250.0 mL of solution with a pH of 2.56. The solution is divided into two equal parts, A and B. Solution A is titrated with strong base to its equivalence point. Solution B is added to solution A after solution A is neutralized. The pH of the resulting solution is 4.26. What is the molar mass of the acid?

73. Explain why it is not possible to prepare a buffer with a pH of 6.50 by mixing NH_3 and NH_4Cl.

74. Fifty cm^3 of 1.000 M nitrous acid is titrated with 0.850 M NaOH. What is the pH of the solution

(a) before any NaOH is added?

(b) at half-neutralization?

(c) at the equivalence point?

(d) when 0.10 mL less than the volume of NaOH to reach the equivalence point is added?

(e) when 0.10 mL more than the volume of NaOH to reach the equivalence point is added?

(f) Use your data to construct a plot similar to that shown in Figure 14.10 (pH versus volume NaOH added).

75. A diprotic acid, H_2B (MM = 126 g/mol), is determined to be a hydrate, $H_2B \cdot xH_2O$. A 10.00-g sample of this hydrate is dissolved in enough water to make 150.0 mL of solution. Twenty-five milliliters of this solution requires 48.5 mL of 0.425 M NaOH to reach the equivalence point. What is x?

76. What is the pH of a 0.1500 M H_2SO_4 solution if

(a) the ionization of HSO_4^- is ignored?

(b) the ionization of HSO_4^- is taken into account? (K_a for HSO_4^- is 1.1×10^{-2}.)

77. Two students were asked to determine the K_b of an unknown base. They were given a bottle with a solution in it. The bottle was labeled "aqueous solution of a monoprotic strong acid." They were also given a pH meter, a buret, and an appropriate indicator. They reported the following data:

volume of acid required for neutralization = 21.0 mL
pH after 7.00 mL of strong acid added = 8.95

Use the students' data to determine the K_b of the unknown base.

78. How many grams of NaOH must be added to 1.00 L of a buffer made from 0.150 M NH_3 and 10.0 g of NH_4Cl so that the pH increases by one unit (e.g., from 5 to 6)? K_a for NH_4^+ is 5.6×10^{-10}.

79. How many grams of NaF must be added to 70.00 mL of 0.150 M HNO_3 to obtain a buffer with a pH of 4.68?

80. A student is asked to determine the molarity of 25.00 mL of a solution of $HClO_4$. He uses 0.731 M KOH. After adding 42.35 mL of KOH, he realizes that he forgot to add an indicator. His TA suggests he take the pH of the solution. The pH is 12.39.

(a) Did the student go beyond the equivalence point?

(b) What is the molarity of the strong acid?

(c) If he did, how many milliliters of KOH did he add in excess? If he did not, how much more (in mL) KOH should he add?

Complex Ion and Precipitation Equilibria

The limestone cliffs of Etretat are rocks of compressed calcium carbonate ($CaCO_3$). This compound is insoluble in water but can be dissolved by a strong acid. Acid rain can erode these cliffs.

Surrounded by beakers, by strange coils,
By ovens and flasks with twisted necks,
The chemist, fathoming the whims of attractions,
Artfully imposes on them their precise meetings.

—SULLY-PRUDHOMME
The Naked World (translated by William Dock)

Thus far, we have looked at the equilibrium established when acids, bases or both of these are added to water.

In this chapter we will discuss the equilibrium of complex ion formation (Section 15-1). Solubility (Section 15-2) and precipitation (Section 15-3), discussed in Chapter 4, are reexamined with a focus on an equilibrium constant K_{sp}. We will also learn how to dissolve precipitates (Section 15-4) by making them react with strong acids or complexing agents forming complex ions.

The equilibria involving complex ions and precipitates have applications in geology, medicine, and agriculture. In chemistry, you are more likely to meet up with these equilibria in the laboratory when you carry out experiments in qualitative analysis.

15-1 Complex Ion Equilibria; Formation Constant (K_f)

Recall from Chapter 13 the Lewis model for the definition of an acid and a base. Many of the electron acceptors for Lewis acids are transition metal ions (Lewis acids). These metal ions combine with molecules (e.g., H_2O, NH_3) or other ions (e.g., Cl^-, OH^-) that are good electron-pair donors (Lewis bases). The product of this combination is a **complex ion**. Chapter 19 will discuss the chemistry of these ions in greater detail.

The equilibrium constant for the formation of a complex ion is called a **formation constant** (or stability constant) and given the symbol K_f. A typical example is

$$Cu^{2+}(aq) + 4NH_3(aq) \rightleftharpoons Cu(NH_3)_4{}^{2+}(aq) \qquad K_f = \frac{[Cu(NH_3)_4{}^{2+}]}{[Cu^{2+}][NH_3]^4}$$

Table 15.1 lists formation constants of complex ions. In each case, K_f applies to the formation of the complex by a reaction of the type just cited. Notice that

Table 15.1 Formation Constants of Complex Ions

Complex Ion	K_f	Complex Ion	K_f
$AgCl_2^-$	1.8×10^5	$CuCl_2^-$	1×10^5
$Ag(CN)_2^-$	2×10^{20}	$Cu(NH_3)_4^{2+}$	2×10^{12}
$Ag(NH_3)_2^+$	1.7×10^7	$FeSCN^{2+}$	9.2×10^2
$Ag(S_2O_3)_2^{3-}$	1×10^{13}	$Fe(CN)_6^{3-}$	4×10^{52}
$Al(OH)_4^-$	1×10^{33}	$Fe(CN)_6^{4-}$	4×10^{45}
$Cd(CN)_4^{2-}$	2×10^{18}	$Hg(CN)_4^{2-}$	2×10^{41}
$Cd(NH_3)_4^{2+}$	2.8×10^7	$Ni(NH_3)_6^{2+}$	9×10^8
$Cd(OH)_4^{2-}$	1.2×10^9	$PtCl_4^{2-}$	1×10^{16}
$Co(NH_3)_6^{2+}$	1×10^5	$trans$-$Pt(NH_3)_2Cl_2$	3×10^{28}
$Co(NH_3)_6^{3+}$	1×10^{23}	cis-$Pt(NH_3)_2Cl_2$	3×10^{29}
$Co(NH_3)_5Cl^{2+}$	2×10^{28}	$Zn(CN)_4^{2-}$	6×10^{16}
$Co(NH_3)_5NO_2^{2+}$	1×10^{24}	$Zn(NH_3)_4^{2+}$	3.6×10^8
		$Zn(OH)_4^{2-}$	3×10^{14}

for most complex ions in Table 15.1, K_f is a large number: 10^5 or greater. This means that equilibrium considerations strongly favor complex formation. Consider, for example, the system

$$Ag^+(aq) + 2NH_3(aq) \rightleftharpoons Ag(NH_3)_2^+(aq)$$

$$K_f = \frac{[Ag(NH_3)_2^+]}{[Ag^+][NH_3]^2} = 1.7 \times 10^7$$

The large K_f value means that the forward reaction goes virtually to completion. Addition of ammonia to a solution of $AgNO_3$ will convert nearly all the Ag^+ ions to $Ag(NH_3)_2^+$ (see Example 15.1).

EXAMPLE 15.1

Consider the equilibrium

$$Ag^+(aq) + 2NH_3(aq) \rightleftharpoons Ag(NH_3)_2^+(aq) \qquad K_f = 1.7 \times 10^7$$

a   What is the ratio $[Ag(NH_3)_2^+]/[Ag^+]$ in 0.10 M NH_3?

ANALYSIS

Information given:	reaction $(Ag^+(aq) + 2NH_3(aq) \rightleftharpoons Ag(NH_3)_2^+(aq))$ $K_f(1.7 \times 10^7)$ $[NH_3]$ (0.10 M)
Asked for:	$\dfrac{[Ag(NH_3)_2^+]}{[Ag^+]}$

STRATEGY

1. Write the K_f expression for the reaction.

2. Substitute the values for $[NH_3]$ and K_f into the expression to get the desired ratio.

continued

	SOLUTION
1. Equilibrium expression	$K_f = \dfrac{[Ag(NH_3)_2^+]}{[Ag^+][NH_3]^2}$
2. $\dfrac{[Ag(NH_3)_2^+]}{[Ag^+]}$	$1.7 \times 10^7 = \dfrac{[Ag(NH_3)_2^+]}{[Ag^+](0.10)^2} \longrightarrow \dfrac{[Ag(NH_3)_2^+]}{[Ag^+]} = 1.7 \times 10^7(0.10)^2 = 1.7 \times 10^5$

b   What concentration of NH_3 is required to convert 99% of Ag^+ to $Ag(NH_3)_2^+$?

ANALYSIS

Information given:	reaction ($Ag^+(aq) + 2NH_3(aq) \rightleftharpoons Ag(NH_3)_2^+(aq)$) $K_f (1.7 \times 10^7)$ 99% Ag^+ is converted to $Ag(NH_3)_2^+$
Asked for:	$[NH_3]$

STRATEGY

1. Converting 99% Ag^+ to $Ag(NH_3)_2^+$ means that for every 1 mol of Ag^+ there are 99 mol of $Ag(NH_3)_2^+$ or $\dfrac{[Ag(NH_3)_2^+]}{[Ag^+]} = \dfrac{99}{1}$.

2. Substitute the K_f value and the value for the ratio to find $[NH_3]$.

SOLUTION

$[NH_3]$	$1.7 \times 10^7 = \left(\dfrac{99}{1}\right) \times \dfrac{1}{[NH_3]^2} \longrightarrow [NH_3] = 2.4 \times 10^{-3}\ M$

c   What is the equilibrium constant for the reaction:

$$Ag(NH_3)_2^+(aq) + 2S_2O_3^{2-}(aq) \rightleftharpoons Ag(S_2O_3)_2^{3-}(aq) + 2NH_3(aq)$$

(Take K_f $Ag(S_2O_3)_2^{3-}$ to be 1×10^{13}.)

ANALYSIS

Information given:	reaction ($Ag(NH_3)_2^+(aq) + 2S_2O_3^{2-}(aq) \rightleftharpoons 2NH_3(aq) + Ag(S_2O_3)_2^{3-}(aq)$) K_f for $Ag(NH_3)_2^+$ (1.7×10^7); K_f for $Ag(S_2O_3)_2^{3-}$ (1×10^{13})
Asked for:	K for the reaction

STRATEGY AND SOLUTION

1. Focus on the reactant, $Ag(NH_3)_2^+$, in the desired equation. It contributes NH_3 as a product. Find an equation that "liberates" NH_3 from $Ag(NH_3)_2^+$. That equation is the reverse of the equation that forms $Ag(NH_3)_2^+$.

    (1) $Ag(NH_3)_2^+(aq) \rightleftharpoons Ag^+(aq) + 2NH_3(aq)$ $K_1 = 1/K_f = 1/(1.7 \times 10^7)$

2. Focus on the second reactant, $S_2O_3^{2-}$, in the desired equation. It is part of the reaction for the formation of $Ag(S_2O_3)_2^{3-}$. Write the equation for that reaction.

    (2) $Ag^+(aq) + 2S_2O_3^{2-}(aq) \rightleftharpoons Ag(S_2O_3)_2^{3-}(aq)$ $K_2 = K_f = 1 \times 10^{13}$

3. Combine both equations to get the final equation. Combine K_1 and K_2 to get K for the reaction using the rule of multiple equilibria.

(1) $Ag(NH_3)_2^+(aq) \rightleftharpoons Ag^+(aq) + 2NH_3(aq)$	$K_1 = 1/K_f = 1/(1.7 \times 10^7)$
(2) $Ag^+(aq) + 2S_2O_3^{2-}(aq) \rightleftharpoons Ag(S_2O_3)_2^{3-}(aq)$	$K_2 = K_f = 1 \times 10^{13}$
$Ag(NH_3)_2^+(aq) + 2S_2O_3^{2-}(aq) \rightleftharpoons Ag(S_2O_3)_2^{3-}(aq) + 2NH_3(aq)$	$K = \dfrac{1 \times 10^{13}}{1.7 \times 10^7} = 6 \times 10^5$

END POINT

Looking back at part (a), a ratio of 1.7×10^5 for $\dfrac{[Ag(NH_3)_2^+]}{[Ag^+]}$ means that in 0.10 M NH_3, there are 170,000 $Ag(NH_3)_2^+$ complex ions for every Ag^+ ion.

Figure 15.1 Two coordination complexes are shown. Left is $Co(NH_3)_6^{3+}$. Right is $Co(NH_3)_5Cl^{2+}$.

Figure 15.2 An aqueous solution of $SrCrO_4$ in equilibrium with solid $SrCrO_4$.

Although the solid doesn't appear in K_{sp}, it must be present for equilibrium.

From Example 15.1 we see that the reaction for the conversion of $Ag(NH_3)_2^+$ to $Ag(S_2O_3)_2^{3-}$ has a large equilibrium constant. Looking at it in a slightly different way, we can say that of these two complexes of Ag^+, the $Ag(S_2O_3)_2^{3-}$ ion is the more stable.

A similar situation applies to two complex ions of Co^{3+}:

$$Co(NH_3)_6^{3+}(aq) + Cl^-(aq) \rightleftharpoons Co(NH_3)_5Cl^{2+}(aq) + NH_3(aq)$$

$$K = \frac{K_f\ Co(NH_3)_5Cl^{2+}}{K_f\ Co(NH_3)_6^{3+}} = \frac{2 \times 10^{28}}{1 \times 10^{23}} = 2 \times 10^5$$

Addition of Cl^- ions to the complex $Co(NH_3)_6^{3+}$ converts it to the more stable complex $Co(NH_3)_5Cl^{2+}$, which has a pink color (Figure 15.1).

15-2 Solubility; Solubility Product Constant (K_{sp})

In Chapter 4, we considered a compound soluble if it dissolved in water and insoluble if it did not. In this section, we look at the dissolution process in the context of an equilibrium existing between the component ions of an ionic solid and the solid itself.

Consider $SrCrO_4$ in water (Figure 15.2). Much like the weak acid HF in water,

$$HF(aq) \rightleftharpoons H^+(aq) + F^-(aq)$$

an equilibrium is established in solution between the compound ($SrCrO_4$) and its corresponding ions (Sr^{2+} and CrO_4^{2-}). The net ionic equation representing this equilibrium is

$$SrCrO_4(s) \rightleftharpoons Sr^{2+}(aq) + CrO_4^{2-}(aq)$$

K_{sp} Expression

The rules in Chapters 12 and 13 make it possible to write the equilibrium constant expression for the dissolving of $SrCrO_4$. In particular, the solid does not appear in the expression; the concentration of each ion is raised to a power equal to its coefficient in the chemical equation.

$$K_{sp} = [Sr^{2+}][CrO_4^{2-}]$$

The symbol K_{sp} represents a particular type of equilibrium constant known as the **solubility product constant**. Like all equilibrium constants, K_{sp} has a fixed value for a given system at a particular temperature. At 25°C, K_{sp} for $SrCrO_4$ is about 3.6×10^{-5}; that is,

$$[Sr^{2+}][CrO_4^{2-}] = 3.6 \times 10^{-5}$$

This relation says that the product of the two ion concentrations at equilibrium must be 3.6×10^{-5}, regardless of how equilibrium is established.

EXAMPLE   15.2

Write expressions for K_{sp} for Ag_2CrO_4 and $Ca_3(PO_4)_2$.

STRATEGY

1. Start by writing the chemical equation for the solution process (solid on the left; ions on the right).

2. Write the K_{sp} expression. Recall that

   ■ solids do not appear.

   ■ concentrations ([ ]) are used.

   ■ the concentration of each ion is raised to a power equal to its coefficient in the chemical equation.

continued

<table>
<tr><td colspan="2" align="center">SOLUTION</td></tr>
<tr><td>Ag$_2$CrO$_4$:</td><td></td></tr>
<tr><td>   **1.** Equation</td><td>Ag$_2$CrO$_4$(s) $\rightleftharpoons$ 2Ag$^+$(aq) + CrO$_4$$^{2-}$(aq)</td></tr>
<tr><td>   **2.** K_{sp} expression</td><td>K_{sp} = [Ag$^+$]^2[CrO$_4$$^{2-}$]</td></tr>
<tr><td>Ca$_3$(PO$_4$)$_2$:</td><td></td></tr>
<tr><td>   **1.** Equation</td><td>Ca$_3$(PO$_4$)$_2$(s) $\rightleftharpoons$ 3Ca^{2+}(aq) + 2PO$_4$$^{3-}$(aq)</td></tr>
<tr><td>   **2.** K_{sp}</td><td>K_{sp} = [Ca^{2+}]^3[PO$_4$$^{3-}$]^2</td></tr>
</table>

K_{sp} and the Equilibrium Concentrations of Ions

The relation

$$K_{sp}\ SrCrO_4 = [Sr^{2+}][CrO_4^{2-}] = 3.6 \times 10^{-5}$$

can be used to calculate the equilibrium concentration of one ion if you know that of the other. Suppose, for example, the concentration of CrO_4^{2-} in a certain solution in equilibrium with $SrCrO_4$ is known to be 2.0×10^{-3} M. It follows that

$$[Sr^{2+}] = \frac{K_{sp}\ SrCrO_4}{[CrO_4^{2-}]} = \frac{3.6 \times 10^{-5}}{2.0 \times 10^{-3}} = 1.8 \times 10^{-2}\ M$$

If in another case, $[Sr^{2+}] = 1.0 \times 10^{-4}$ M,

$$[CrO_4^{2-}] = \frac{K_{sp}\ SrCrO_4}{[Sr^{2+}]} = \frac{3.6 \times 10^{-5}}{1.0 \times 10^{-4}} = 3.6 \times 10^{-1}\ M$$

Example 15.3 illustrates the same kind of calculation for a different compound.

EXAMPLE 15.3

Calcium phosphate, Ca$_3$(PO$_4$)$_2$, is a water-insoluble mineral, large quantities of which are used to make commercial fertilizers. Use the K_{sp} value for Ca$_3$(PO$_4$)$_2$ to determine the concentration of Ca^{2+} in equilibrium with the solid if $[PO_4^{3-}] = 5 \times 10^{-5}$ M.

ANALYSIS	
Information given:	$[PO_4^{3-}]$ (5×10^{-5} M)
Information implied:	K_{sp} for Ca$_3$(PO$_4$)$_2$
Asked for:	$[Ca^{2+}]$

STRATEGY

1. Write the chemical equation for the dissolution of Ca$_3$(PO$_4$)$_2$.

2. Write the K_{sp} expression.

3. Substitute into the K_{sp} expression to find $[Ca^{2+}]$.

SOLUTION	
1. Equation	Ca$_3$(PO$_4$)$_2$(s) $\rightleftharpoons$ 3Ca^{2+}(aq) + 2PO$_4$$^{3-}$(aq)
2. K_{sp} expression	K_{sp} = [Ca^{2+}]^3[PO$_4$$^{3-}$]^2
3. $[Ca^{2+}]$	1×10^{-33} = [Ca^{2+}]^3(5×10^{-5})^2 $\longrightarrow$ [Ca^{2+}]^3 = 4×10^{-25} $\longrightarrow$ [Ca^{2+}] = 7×10^{-9} M

END POINT

To check your answer, use the calculated $[Ca^{2+}]$ (7×10^{-9} M) and the given $[PO_4^{3-}]$ (5×10^{-5} M) to find K_{sp}.

$$K_{sp} = [Ca^{2+}]^3[PO_4^{3-}]^2 = (7 \times 10^{-9})^3(5 \times 10^{-5})^2 = 9 \times 10^{-34}$$

The calculated K_{sp} is very close to the value listed in Table 15.2: 1×10^{-33}.

Table 15.2 Solubility Product Constants at 25°C

		K_{sp}				K_{sp}
Acetates	$AgC_2H_3O_2$	1.9×10^{-3}	Hydroxides	$Al(OH)_3$		2×10^{-31}
				$Ca(OH)_2$		4.0×10^{-6}
Bromides	$AgBr$	5×10^{-13}		$Fe(OH)_2$		5×10^{-17}
	Hg_2Br_2	6×10^{-23}		$Fe(OH)_3$		3×10^{-39}
	$PbBr_2$	6.6×10^{-6}		$Mg(OH)_2$		6×10^{-12}
Carbonates	Ag_2CO_3	8×10^{-12}		$Tl(OH)_3$		2×10^{-44}
	$BaCO_3$	2.6×10^{-9}		$Zn(OH)_2$		4×10^{-17}
	$CaCO_3$	4.9×10^{-9}	Iodides	AgI		1×10^{-16}
	$MgCO_3$	6.8×10^{-6}		Hg_2I_2		5×10^{-29}
	$PbCO_3$	1×10^{-13}		PbI_2		8.4×10^{-9}
	$SrCO_3$	5.6×10^{-10}				
Chlorides	$AgCl$	1.8×10^{-10}	Phosphates	Ag_3PO_4		1×10^{-16}
	Hg_2Cl_2	1×10^{-18}		$AlPO_4$		1×10^{-20}
	$PbCl_2$	1.7×10^{-5}		$Ca_3(PO_4)_2$		1×10^{-33}
				$Mg_3(PO_4)_2$		1×10^{-24}
Chromates	Ag_2CrO_4	1×10^{-12}	Sulfates	$BaSO_4$		1.1×10^{-10}
	$BaCrO_4$	1.2×10^{-10}		$CaSO_4$		7.1×10^{-5}
	$PbCrO_4$	2×10^{-14}		$PbSO_4$		1.8×10^{-8}
	$SrCrO_4$	3.6×10^{-5}		$SrSO_4$		3.4×10^{-7}
Fluorides	BaF_2	1.8×10^{-7}				
	CaF_2	1.5×10^{-10}				
	MgF_2	7×10^{-11}				
	PbF_2	7.1×10^{-7}				

K_{sp} and Water Solubility

One way to establish equilibrium between a slightly soluble solid and its ions in solution is to stir the solid with water to form a saturated solution. As you might expect, the solubility of the solid, s, in moles per liter, is related to the solubility product constant, K_{sp}. In the case of barium sulfate dissolving in water we have

$$\underset{s}{BaSO_4(s)} \rightleftharpoons \underset{s}{Ba^{2+}(aq)} + \underset{s}{SO_4{}^{2-}(aq)}$$

The solubility of $BaSO_4$ is exactly equal to the equilibrium concentration of either Ba^{2+} or $SO_4{}^{2-}$. In other words,

$$[Ba^{2+}] = [SO_4{}^{2-}] = s$$

and

$$K_{sp}\ BaSO_4 = [Ba^{2+}][SO_4{}^{2-}] = s^2$$

Solving,

$$s = (K_{sp})^{\frac{1}{2}} = (1.1 \times 10^{-10})^{\frac{1}{2}} = 1.0 \times 10^{-5}\ mol/L$$

This means that a saturated solution can be prepared by dissolving 2.3 mg of $BaSO_4$ in enough water to make 1.0 L of solution. Putting it another way, dissolving *less* than 2.3 mg results in an unsaturated solution. If more than 2.3 mg are added, an equilibrium is established between the solution containing $1.0 \times 10^{-5}\ M\ Ba^{2+}$ and $SO_4{}^{2-}$ and the solid $BaSO_4$.

When an ionic solid consists of anions and cations of different charges, the relation between K_{sp} and s takes a different form, but the principle is the same (Example 15.4).

s is always the molar solubility.

When you hear the word "dissolve" imagine the solid breaking up into its component ions.

The relationship between K_{sp} and s depends on the type of solid.

EXAMPLE 15.4 GRADED

Barium fluoride, BaF_2, is used in metallurgy as a welding and soldering agent. Its K_{sp} is 1.8×10^{-7}.

a What is its solubility in mol/L (molar solubility)?

ANALYSIS

Information given:	K_{sp} for BaF_2 (1.8×10^{-7})
Asked for:	molar solubility (solubility in mol/L)

STRATEGY

1. Write a net ionic equation to represent dissolving BaF_2.
2. Let s be the molar solubility of BaF_2; $s = [BaF_2]$ that dissolves.
3. Use stoichiometric ratios to relate BaF_2 to its ions.
4. Write the K_{sp} expression. Substitute into it and solve for s.

SOLUTION

1. Equation	$BaF_2(s) \rightleftharpoons Ba^{2+}(aq) + 2F^-(aq)$
2. Stoichiometric ratios	1 mol BaF_2 dissolves/L $\longrightarrow$ 1 mol/L Ba^{2+} $\longrightarrow$ 2 mol/L F^-
	$\qquad\qquad\quad \downarrow \qquad\qquad\qquad\qquad \downarrow \qquad\qquad\quad \downarrow$
	$\qquad\qquad\quad s \qquad\qquad\qquad\qquad\quad s \qquad\qquad\quad 2s$
s	$K_{sp} = [Ba^{2+}][F^-]^2 = (s)(2s)^2 = 4s^3$
	$1.8 \times 10^{-7} = 4s^3; \quad s = \left(\dfrac{1.8 \times 10^{-7}}{4}\right)^{\frac{1}{3}} = 3.6 \times 10^{-3}\ M$

b What is its solubility in g/L?

ANALYSIS

Information given:	K_{sp} for BaF_2 (1.8×10^{-7}) from part (a): molar solubility of BaF_2 ($3.6 \times 10^{-3}\ M$)
Information implied:	MM BaF_2
Asked for:	solubility of BaF_2 in g/L

STRATEGY

Use the molar mass of BaF_2 to change the molar solubility to the solubility in g/L.

SOLUTION

Solubility (g/L)	$3.6 \times 10^{-3}\ \dfrac{\text{mol}}{\text{L}} \times 175.3\ \dfrac{\text{g}}{\text{mol}} = 0.63$ g/L

c How many grams of BaF_2 are required to prepare 298 mL of a saturated solution of BaF_2?

ANALYSIS

Information given:	K_{sp} for BaF_2 (1.8×10^{-7}) from part (b): solubility of BaF_2 in g/L (0.63 g/L) volume of saturated solution (298 mL)
Asked for:	mass of BaF_2 to prepare a saturated solution.

STRATEGY

1. Recall the definition of a saturated solution. It has the maximum number of grams of a solute that can be dissolved (solubility).
2. Use the solubility in g/L as a conversion factor to find the mass needed to prepare 298 mL of a saturated solution.

continued

	SOLUTION
Mass BaF_2	$0.298 \text{ L} \times \dfrac{0.63 \text{ g}}{1 \text{ L}} = 0.19 \text{ g}$

END POINT
You have to know either the molar solubility or the solubility in g/L to prepare a saturated solution.

Yellow orpiment, As_2S_3

Galena, PbS

Iron pyrite, FeS_2

© Cengage Learning/Charles D. Winters

Sulfide minerals and K_{sp}. Sulfides are among the least soluble ionic compounds. Their K_{sp} values are often smaller than 10^{-25}. For this reason, many sulfides are found as minerals.

Frequently we find that the experimentally determined solubility of an ionic solid is larger than that predicted from K_{sp}. Consider, for example, $PbCl_2$, where the solubility calculated from the relation

$$4s^3 = K_{sp} = 1.7 \times 10^{-5}$$

is 0.016 mol/L. The measured solubility is considerably larger, 0.036 mol/L. The explanation for this is that some of the lead in $PbCl_2$ goes into solution in the form of species other than Pb^{2+}. In particular, we can detect the presence of the ions $Pb(OH)^+$ and $PbCl^+$.

The reverse of Example 15.4 involves finding K_{sp} of a compound given its solubility. The solubilities of many ionic compounds are determined experimentally and tabulated in chemical handbooks. Most solubility values are given in grams of solute dissolved in 100 grams of water. To obtain the molar solubility in moles/L, we have to assume that the density of the solution is equal to that of water. Then the number of grams of solute per 100 g water is equal to the number of grams of solute per 100 mL of solution. This assumption is valid because the mass of the compound in solution is small. To solve for K_{sp}, find the molar solubility of the solute and determine the concentration of its component ions. Substitute into the K_{sp} expression.

EXAMPLE 15.5

Consider the pesticide magnesium arsenate, $Mg_3(AsO_4)_2$. Its solubility is determined experimentally to be 1.6×10^{-3} g/100 g H_2O. What is the K_{sp} for $Mg_3(AsO_4)_2$? (Assume that the density of water is equal to the density of the solution.)

ANALYSIS	
Information given:	solubility (1.6×10^{-3} g/100 g H_2O)
Information implied:	density of the solution = density of water MM $Mg_3(AsO_4)_2$
Asked for:	K_{sp}

STRATEGY
1. Assume 1.6×10^{-3} g/100 g H_2O = 1.6×10^{-3} g/100 mL of solution. Convert this solubility data to molar solubility (mol/L). Use the molar mass of $Mg_3(AsO_4)_2$.
2. Write a net ionic equation to represent dissolving $Mg_3(AsO_4)_2$.
3. Use stoichiometric ratios to relate $[Mg_3(AsO_4)_2]$ to $[Mg^{2+}]$ and $[AsO_4^{3-}]$.
4. Substitute into the K_{sp} expression and find K_{sp}.

continued

	SOLUTION
1. Molar solubility	$\dfrac{1.6 \times 10^{-3} \text{ g}}{0.100 \text{ L}} \times \dfrac{1 \text{ mol}}{350.7 \text{ g}} = 4.6 \times 10^{-5} \ M$
2. Reaction	$Mg_3(AsO_4)_2(s) \rightleftharpoons 3Mg^{2+}(aq) + 2AsO_4^{3-}(aq)$
3. Stoichiometric ratios	1 mol $Mg_3(AsO_4)_2$ dissolved $\longrightarrow$ 3 mol Mg^{2+}; $[Mg^{2+}] = 3(4.6 \times 10^{-5}) = 1.4 \times 10^{-4}$
	1 mol $Mg_3(AsO_4)_2$ dissolved $\longrightarrow$ 2 mol AsO_4^{3-}; $[AsO_4^{3-}] = 2(4.6 \times 10^{-5}) = 9.2 \times 10^{-5}$
4. K_{sp}	$K_{sp} = [Mg^{2+}]^3[AsO_4^{3-}]^2 = (1.4 \times 10^{-4})^3(9.2 \times 10^{-5})^2 = 2.3 \times 10^{-20}$

K_{sp} and the Common Ion Effect

How would you expect the solubility of barium sulfate in water

$$BaSO_4(s) \rightleftharpoons Ba^{2+}(aq) + SO_4^{2-}(aq)$$

to compare with that in a 0.10 M solution of Na_2SO_4, which contains the same *(common)* ◄ i anion, SO_4^{2-}? A moment's reflection should convince you that the solubility in 0.10 M Na_2SO_4 must be *less* than that in pure water. Recall that when $BaSO_4$ is dissolved in pure water, $[SO_4^{2-}]$ is $1.0 \times 10^{-5} \ M$. Increasing the concentration of SO_4^{2-} to 0.10 M should, by Le Châtelier's principle, drive the above equilibrium to the *left*, repressing the solubility of barium sulfate. This is, indeed, the case (Example 15.6).

i A "common" ion comes from two sources such as $BaSO_4$ and Na_2SO_4.

EXAMPLE   15.6

Taking K_{sp} of $BaSO_4$ to be 1.1×10^{-10}, estimate its solubility (moles per liter) in 0.10 M Na_2SO_4 solution.

ANALYSIS	
Information given:	K_{sp} of $BaSO_4$ (1.1×10^{-10}) M for Na_2SO_4(0.10)
Asked for:	molar solubility

STRATEGY

1. Let s = mol $BaSO_4$ dissolved in a liter of solution and relate $[Ba^{2+}]$ and $[SO_4^{2-}]$ to $BaSO_4$ dissolved (s).

2. Set up the following table.

	[Ba^{2+}]	[SO$_4^{2-}$]
Original		
Change		
Equilibrium		

Show that at the start, there are no Ba^{2+} ions present and no SO_4^{2-} ions contributed by $BaSO_4$ but contributed by Na_2SO_4.

Let s be in the "change" row to show the amount of Ba^{2+} ions and SO_4^{2-} ions contributed by the dissolved $BaSO_4$.

3. Substitute your entries in the equilibrium row into the K_{sp} expression for $BaSO_4$.

4. Solve for s and assume that $s \ll SO_4^{2-}$ originally present.

continued

SOLUTION

1. Ba^{2+}; SO_4^{2-}

2. Table

amount of $BaSO_4$ dissolved $= s = [Ba^{2+}] = [SO_4^{2-}]$

	$[Ba^{2+}]$	$[SO_4^{2-}]$
Original	0	0.10
Change	$+s$	$+s$
Equilibrium	s	$0.10 + s$

3. K_{sp} expression

4. s

$1.1 \times 10^{-10} = [Ba^{2+}][SO_4^{2-}] = (s)(0.10 + s)$

Assume $s \ll 0.10$.

$1.1 \times 10^{-10} = (s)(0.10) \longrightarrow s = 1.1 \times 10^{-9}\ M$

s is indeed much smaller than 0.10, so the assumption is justified.

END POINT

The solubility in 0.10 M Na_2SO_4 is much less than that in pure water ($1.1 \times 10^{-9} \ll 1.0 \times 10^{-5}$), which is exactly what we predicted.

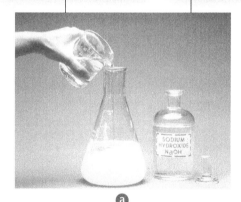

Addition of sodium hydroxide, which contains the Na^+ common ion, causes NaCl to precipitate from the solution.

The flasks in both (a) and (b) contain a saturated solution of NaCl.

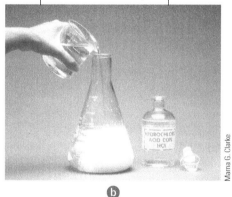

Addition of hydrochloric acid, which contains the Cl^- common ion, also causes NaCl to precipitate from the solution.

a

b

Marna G. Clarke

Figure 15.3 Sodium chloride and the common ion effect.

The **common ion effect** illustrated in Example 15.6 is a general one. *An ionic solid is less soluble in a solution containing a common ion than it is in water* (Figure 15.3).

© Cengage Learning/Charles D. Winters

Figure 15.4 Precipitation of strontium chromate. A solution prepared by mixing solutions of $Sr(NO_3)_2$ and K_2CrO_4 is in equilibrium with yellow $SrCrO_4(s)$. In such a solution $[Sr^{2+}] \times [CrO_4^{2-}] = 3.6 \times 10^{-5}$.

15-3 Precipitate Formation

As we saw in Chapter 4, a precipitate forms when a cation from one solution combines with an anion from another solution to form an insoluble ionic solid. We also considered how to predict whether such a reaction would occur and, if so, how to represent it by a net ionic equation.

Precipitation reactions, like all reactions, reach a position of equilibrium. Putting it another way, even the most "insoluble" electrolyte dissolves to at least a slight extent, thereby establishing equilibrium with its ions in solution. Suppose, for example, solutions of $Sr(NO_3)_2$ and K_2CrO_4 are mixed. In this case, Sr^{2+} ions combine with CrO_4^{2-} ions to form a yellow precipitate of strontium chromate, $SrCrO_4$ (Figure 15.4).

Very quickly, an equilibrium is established between the solid and the corresponding ions in solution:

$$SrCrO_4(s) \rightleftharpoons Sr^{2+}(aq) + CrO_4^{2-}(aq)$$

K_{sp} and Precipitate Formation

K_{sp} values can be used to make predictions as to whether or not a precipitate will form when two solutions are mixed. To do this, we follow an approach very similar to that used in Chapter 12, to determine the direction in which a system will move to reach equilibrium. We work with a quantity known as the **ion product constant**, Q, which has the same mathematical form as K_{sp}. The difference is that the concentrations that appear in Q are those that apply at a particular moment. ◄ⓘ Those that appear in K_{sp} are equilibrium concentrations. That is, the value of Q is expected to change as a precipitation reaction proceeds, approaching K_{sp} and eventually becoming equal to it.

Three cases can be distinguished (Figure 15.5):

1. If $Q > K_{sp}$, the solution contains a higher concentration of ions than it can hold at equilibrium. A *precipitate forms,* decreasing the concentrations until the ion product becomes equal to K_{sp} and equilibrium is established.
2. If $Q < K_{sp}$, the solution contains a lower concentration of ions than is required for equilibrium with the solid. The solution is unsaturated. *No precipitate forms;* equilibrium is not established.
3. If $Q = K_{sp}$, the solution is just saturated with ions and is at the point of precipitation.

Figure 15.5 Equilibrium curve for silver chloride. Silver chloride (*s*) is in contact with Ag^+ and Cl^- ions in aqueous solution. The product Q of the concentration of ions $[Ag^+] \times [Cl^-]$ is equal to K_{sp} *(curved line)* when equilibrium exists. If $Q > K_{sp}$, $AgCl(s)$ tends to precipitate out until equilibrium is reached. If $Q < K_{sp}$, additional solid dissolves.

ⓘ K_{sp} is a constant; Q can have any value.

EXAMPLE 15.7 GRADED

In the laboratory, strontium chromate ($K_{sp} = 3.6 \times 10^{-5}$) can be prepared by mixing solutions containing strontium ions and chromate ions, where the concentrations of these ions are high enough to form a precipitate.

a A solution has the following initial concentrations: $[Sr^{2+}] = 6.0 \times 10^{-3}$ M, $[CrO_4^{2-}] = 3.0 \times 10^{-3}$ M. Will a precipitate form?

ANALYSIS

Information given:	K_{sp} of $SrCrO_4$ (3.6×10^{-5}) $[Sr^{2+}]_{initial}$ (6.0×10^{-3} M); $[CrO_4^{2-}]_{initial}$ (3.0×10^{-3} M)
Asked for:	Will a precipitate form?

STRATEGY

1. Write the equation for the dissolution of $SrCrO_4$.
2. Substitute initial concentrations into the Q expression and find Q.
3. Compare the Q value to the K_{sp} value.

SOLUTION

1. Equation	$SrCrO_4(s) \rightleftharpoons Sr^{2+}(aq) + CrO_4^{2-}(aq)$
2. Q	$Q = [Sr^{2+}]_{initial}[CrO_4^{2-}]_{initial} = (6.0 \times 10^{-3})(3.0 \times 10^{-3}) = 1.8 \times 10^{-5}$
3. Precipitate?	$K_{sp} = 3.6 \times 10^{-5} > Q = 1.8 \times 10^{-5}$; No precipitate forms.

b Is a precipitate obtained when 275 mL of 6.0×10^{-3} M $Sr(NO_3)_2$ solution is mixed with 825 mL of 0.040 M K_2CrO_4?

ANALYSIS

Information given:	K_{sp} of $SrCrO_4$ (3.6×10^{-5}) Sr^{2+} source is $Sr(NO_3)_2$: V (275 mL), M (6.0×10^{-3}) CrO_4^{2-} source is K_2CrO_4: V (825 mL), M (0.040)
Asked for:	Will a precipitate form?

continued

<table>
<tr><td colspan="2" align="center">STRATEGY</td></tr>
</table>

1. Find the initial number of moles n for Sr^{2+} and $CrO_4{}^{2-}$ ions ($V \times M$).
2. Find $[Sr^{2+}]_{initial}$ and $[CrO_4{}^{2-}]_{initial}$ by taking n for each ion and dividing it by the total volume.
3. Substitute initial concentrations into the Q expression and find Q.
4. Compare the Q value to the K_{sp} value.

SOLUTION	
1. $n_{initial}$ for Sr^{2+}	$n = (0.275 \text{ L})(6.0 \times 10^{-3} \text{ mol/L}) = 1.65 \times 10^{-3}$ mol
$n_{initial}$ for $CrO_4{}^{2-}$	$n = (0.825 \text{ L})(0.040 \text{ mol/L}) = 0.033$ mol
2. $[Sr^{2+}]_{initial}$	$\dfrac{1.65 \times 10^{-3} \text{ mol}}{(0.275 + 0.825) \text{ L}} = 1.5 \times 10^{-3}$ M
$[CrO_4{}^{2-}]_{initial}$	$\dfrac{0.033 \text{ mol}}{(0.275 + 0.825) \text{ L}} = 0.030$ M
3. Q	$Q = [Sr^{2+}]_{initial}[CrO_4{}^{2-}]_{initial} = (1.5 \times 10^{-3})(0.030) = 4.5 \times 10^{-5}$
4. Precipitate?	$K_{sp} = 3.6 \times 10^{-5} < Q = 4.5 \times 10^{-5}$; Yes, a precipitate forms (just barely!).

c   Is a precipitate obtained when 12.0 g of Na_2CrO_4 are added to 333 mL of 4.5×10^{-3} M $Sr(NO_3)_2$? Assume no volume change after Na_2CrO_4 is added.

ANALYSIS	
Information given:	K_{sp} of $SrCrO_4$ (3.6×10^{-5}) Sr^{2+} source is $Sr(NO_3)_2$: V (333 mL), M (4.5×10^{-3}) $CrO_4{}^{2-}$ source is Na_2CrO_4: mass (12.0 g)
Information implied:	Molar mass of Na_2CrO_4
Asked for:	Will a precipitate form?

<table>
<tr><td colspan="2" align="center">STRATEGY</td></tr>
</table>

1. Find initial number of moles n for $CrO_4{}^{2-}$ ions. (mass/MM). Since the volume does not change, it is not necessary to find n for Sr^{2+}.
2. Find $[CrO_4{}^{2-}]_{initial}$ by taking n and dividing it by the volume. $[Sr^{2+}]_{initial}$ is given.
3. Substitute initial concentrations into the Q expression and find Q.
4. Compare the Q value to the K_{sp} value.

SOLUTION	
1. $n_{initial}$ for $CrO_4{}^{2-}$	$n = \dfrac{12.0 \text{ g}}{162 \text{ g/mol}} = 0.0741$ mol
2. $[Sr^{2+}]_{initial}$; $[CrO_4{}^{2-}]_{initial}$	$[Sr^{2+}]_{initial} = 4.5 \times 10^{-3}$ M; $[CrO_4{}^{2-}]_{initial} = \dfrac{0.0741 \text{ mol}}{0.333 \text{ L}} = 0.222$ M
3. Q	$Q = [Sr^{2+}]_{initial}[CrO_4{}^{2-}]_{initial} = (4.5 \times 10^{-3})(0.222) = 1.0 \times 10^{-3}$
4. Precipitate?	$K_{sp} = 3.6 \times 10^{-5} < Q = 1.0 \times 10^{-3}$; Yes, a precipitate forms.

Selective Precipitation

One way to separate two cations in water solution is to add an anion that precipitates only one of the cations. This approach is known as **selective precipitation.** To see how it works, consider a simple case, a solution containing Mg^{2+} and Na^+ ions. Referring to Table 15.2, you can see that Mg^{2+} forms a couple of insoluble compounds: $MgCO_3$ ($K_{sp} = 6.8 \times 10^{-6}$) and $Mg(OH)_2$ ($K_{sp} = 6 \times 10^{-12}$). In contrast, all of the common compounds of sodium are soluble, including the carbonate and hydroxide. It follows that you could readily separate Mg^{2+} from Na^+ by adding either CO_3^{2-} or OH^- ions to the solution. In either case, Mg^{2+} will precipitate while Na^+ remains in solution.

Now let us consider a somewhat more complex case. Suppose you have a solution containing Mg^{2+} and Ba^{2+}, both at 0.10 M, which you'd like to separate by adding CO_3^{2-} ions. From Table 15.2, it appears that $BaCO_3$ ($K_{sp} = 2.6 \times 10^{-9}$) is less soluble than $MgCO_3$ ($K_{sp} = 6.8 \times 10^{-6}$). Perhaps, by adding CO_3^{2-} ions carefully, you could precipitate Ba^{2+} selectively, leaving Mg^{2+} in solution.

This is indeed the case. To precipitate $BaCO_3$, add enough CO_3^{2-} ions to make their concentration 2.6×10^{-8} M:

$$[CO_3^{2-}] = \frac{K_{sp} \; BaCO_3}{[Ba^{2+}]} = \frac{2.6 \times 10^{-9}}{0.10} = 2.6 \times 10^{-8} \; M$$

At this very low concentration of CO_3^{2-}, $MgCO_3$ does not precipitate:

$$Q = [Mg^{2+}][CO_3^{2-}] = (0.10)(2.6 \times 10^{-8}) = 2.6 \times 10^{-9} < K_{sp} \; MgCO_3 = 6.8 \times 10^{-6}$$

To bring down $MgCO_3$, add more CO_3^{2-} until its concentration becomes 6.8×10^{-5} M:

$$[CO_3^{2-}] = \frac{K_{sp} \; MgCO_3}{[Mg^{2+}]} = \frac{6.8 \times 10^{-6}}{0.10} = 6.8 \times 10^{-5} \; M$$

At that point, virtually all of the Ba^{2+} ions have been precipitated:

$$[Ba^{2+}] \text{ remaining} = \frac{K_{sp} \; BaCO_3}{[CO_3^{2-}]} = \frac{2.6 \times 10^{-9}}{6.8 \times 10^{-5}} = 3.8 \times 10^{-5} \; M$$

$$\% \; Ba^{2+} \text{ remaining in solution} = \frac{3.8 \times 10^{-5}}{0.10} \times 100\% = 0.038\%$$

$$\% \; Ba^{2+} \text{ precipitated} = 100.000 - 0.038 = 99.962$$

In other words it is possible to precipitate essentially all of the Ba^{2+} ions before Mg^{2+} starts to precipitate.

EXAMPLE 15.8 GRADED

A flask (Figure 15.6) contains a solution 0.10 M in Cl^- and 0.010 M in CrO_4^{2-}. $AgNO_3$ is added to the solution.

a | Which anion, Cl^- or CrO_4^{2-}, precipitates first?

ANALYSIS	
Information given:	$[Cl^-]$ (0.10 M); $[CrO_4^{2-}]$ (0.010 M)
Information implied:	K_{sp} of AgCl; K_{sp} of Ag_2CrO_4
Asked for:	Which anion will precipitate first?

STRATEGY	

1. Substitute into the K_{sp} expression for AgCl and find $[Ag^+]$.

2. Substitute into the K_{sp} expression for Ag_2CrO_4 and find $[Ag^+]$.

3. The anion that requires the smaller amount of Ag^+ precipitates first.

SOLUTION	
1. $[Ag^+]$ for AgCl	$K_{sp} = [Ag^+][Cl^-]$; $1.8 \times 10^{-10} = [Ag^+](0.10)$; $[Ag^+] = 1.8 \times 10^{-9} \; M$
2. $[Ag^+]$ for Ag_2CrO_4	$K_{sp} = [Ag^+]^2[CrO_4^{2-}]$; $1 \times 10^{-12} = [Ag^+]^2(0.010)$; $[Ag^+] = 1 \times 10^{-5} \; M$
3. First anion to precipitate	$[Ag^+]$ required for AgCl precipitation (1.8×10^{-9} M) is *less* than $[Ag^+]$ required for Ag_2CrO_4 precipitation (1×10^{-5} M). AgCl precipitates first.

continued

b	What percentage of the first anion has been precipitated when the second anion starts to precipitate?
	ANALYSIS
Information given:	$[Cl^-]$ (0.10 M); $[CrO_4^{2-}]$ (0.010 M) From part (a): $[Ag^+]$ when Ag_2CrO_4 starts to precipitate (1×10^{-5} M)
Asked for:	% Cl^- in solution when Ag_2CrO_4 starts to precipitate

STRATEGY

1. Find $[Cl^-]$ in solution when $[Ag^+] = 1 \times 10^{-5}$, the concentration of Ag^+ when Ag_2CrO_4 starts to precipitate.

2. Find % Cl^- remaining in solution

$$\frac{[Cl^-] \text{ remaining in solution}}{[Cl^-] \text{ initial}} \times 100\%$$

3. Find % Cl^- that precipitated

% Cl^- precipitated = 100.00% − % Cl^- still in solution

SOLUTION

1. $[Cl^-]$ remaining in solution	$K_{sp} = [Ag^+][Cl^-]$; $1.8 \times 10^{-10} = (1 \times 10^{-5})[Cl^-]$; $[Cl^-] = 2 \times 10^{-5}$ M
2. % Cl^- remaining in solution	$\dfrac{[Cl^-] \text{ remaining in solution}}{[Cl^-] \text{ initial}} \times 100\% = \dfrac{2 \times 10^{-5}}{0.10} \times 100\% = 0.02\%$
3. % Cl^- precipitated	100.00% − % Cl^- still in solution = 100.00% − 0.02% = 99.8%

END POINT

Look at Figure 15.6. *White* AgCl does indeed come down first, followed by *red* Ag_2CrO_4.

Figure 15.6 Titration of Cl^- with Ag^+, using CrO_4^{2-} as an indicator.
AgCl (cream color) comes down first, then Ag_2CrO_4 (red) precipitates.

Dissolving Precipitates

Many different methods can be used to bring water-insoluble ionic solids into solution. Most commonly, this is done by adding a reagent to react with either the anion or the cation. The two most useful reagents for this purpose are

- a strong acid, H^+, used to react with basic anions (Figure 15.7).
- a complexing agent, most often NH_3 or OH^-, added to react with metal cations.

Strong Acid

Water-insoluble metal hydroxides can be brought into solution with a strong acid such as HCl. The reaction with zinc hydroxide is typical:

$$Zn(OH)_2(s) + 2H^+(aq) \longrightarrow Zn^{2+}(aq) + 2H_2O$$

We can imagine that this reaction occurs in two (reversible) steps:

dissolving $Zn(OH)_2$ in water: $Zn(OH)_2(s) \rightleftharpoons Zn^{2+}(aq) + 2\,OH^-(aq)$

neutralizing OH^- ions by H^+: $2H^+(aq) + 2\,OH^-(aq) \rightleftharpoons 2H_2O$

$$Zn(OH)_2(s) + 2H^+(aq) \rightleftharpoons Zn^{2+}(aq) + 2H_2O$$

The equilibrium constant for the neutralization (Example 15.9) is so large that the overall reaction goes essentially to completion.

The tube contains white silver chloride (AgCl) at the bottom and yellow silver phosphate (Ag_3PO_4) on top of the chloride.

Adding a strong acid dissolves the silver phosphate, which has a basic anion, but leaves the silver chloride undissolved.

© Cengage Learning/Charles D. Winters

Figure 15.7 Dissolving a precipitate in acid.

EXAMPLE 15.9 GRADED

Consider the following reaction:

$$Zn(OH)_2(s) + 2H^+(aq) \rightleftharpoons Zn^{2+}(aq) + 2H_2O$$

a Determine K for this system, applying the rule of multiple equilibria to the two-step process referred to above.

STRATEGY AND SOLUTION

1. Write the equation for the dissolution of $Zn(OH)_2$.

    $$Zn(OH)_2(s) \rightleftharpoons Zn^{2+}(aq) + 2OH^-(aq) \qquad K_1$$

    $K_1 = K_{sp}$ for $Zn(OH)_2 = 4 \times 10^{-17}$

2. Note that H^+ must react with OH^- to get H_2O, a product of the reaction. There are 2 mol OH^- produced. Thus

    $$2H^+(aq) + 2OH^-(aq) \rightleftharpoons 2H_2O \qquad\qquad K_2$$

    Note that the equation for K_2 is the *reverse* of the equation for the ionization of water ($H_2O \rightleftharpoons H^+(aq) + OH^-(aq)$). Thus,

    $$K_2 = \frac{1}{K_w}$$

    and since there are 2 mol water,

    $$K_2 = \frac{1}{(K_w)^2} = \frac{1}{(1 \times 10^{-14})^2} = 1 \times 10^{28}$$

continued

3. Adding the two equations and applying the rule of multiple equilibria, we obtain

$$Zn(OH)_2(s) \rightleftharpoons Zn^{2+}(aq) + 2OH^-(aq) \qquad\qquad K_1 = 4 \times 10^{-17}$$

$$\underline{2H^+(aq) + 2OH^-(aq) \rightleftharpoons 2H_2O \qquad\qquad K_2 = 1 \times 10^{28}}$$

$$Zn(OH)_2(s) + 2H^+(aq) \rightleftharpoons Zn^{2+}(aq) + 2H_2O \qquad K = (4 \times 10^{-17})(1 \times 10^{28}) = 4 \times 10^{11}$$

b   Calculate the molar solubility, s, of $Zn(OH)_2$ in acid at pH 5.0.

ANALYSIS

Information given:	From part (a): equilibrium reaction ($Zn(OH)_2(s) + 2H^+(aq) \rightleftharpoons Zn^{2+}(aq) + 2H_2O$) From part (a): K for the reaction (4×10^{11}) pH (5.0)
Asked for:	molar solubility of $Zn(OH)_2$

STRATEGY

1. Find $[H^+]$.

2. Write the K expression.

3. Let $s = [Zn(OH)_2]$ that dissolves (molar solubility). Since $[Zn(OH)_2] = [Zn^{2+}]$, then $[Zn^{2+}] = s$. Substitute into the equilibrium expression.

SOLUTION

1. $[H^+]$	$5.0 = -\log_{10}[H^+]$; $[H^+] = 1.0 \times 10^{-5}$
2. K expression	$K = \dfrac{[Zn^{2+}]}{[H^+]^2}$
3. s	$4 \times 10^{11} = \dfrac{s}{(1.0 \times 10^{-5})^2} \quad\longrightarrow\quad s = 4 \times 10^1\ M$

END POINT

This extremely high concentration, 40 M, means that $Zn(OH)_2$ is completely soluble in acid at pH 5 (or any lower pH).

ⓘ $BaSO_4$ cannot be dissolved by acid. Explain.

Strong acid can also be used to dissolve many water-insoluble salts in which the anion is a weak base. ◀ ⓘ In particular, H^+ ions will dissolve

■ *virtually all carbonates* (CO_3^{2-}). The product is the weak acid H_2CO_3, which then decomposes into CO_2 and H_2O. The equation for the reaction of H^+ ions with $ZnCO_3$ is typical:

$$ZnCO_3(s) + 2H^+(aq) \longrightarrow Zn^{2+}(aq) + H_2CO_3(aq)$$

■ *many sulfides* (S^{2-}). The driving force behind this reaction is the formation of the weak acid H_2S, much of which evolves as a gas. The reaction in the case of zinc sulfide is

$$ZnS(s) + 2H^+(aq) \longrightarrow Zn^{2+}(aq) + H_2S(aq)$$

EXAMPLE 15.10

Write balanced equations to explain why each of the following precipitates dissolves in strong acid: $Al(OH)_3$, $CaCO_3$ and CoS

SOLUTION

$$Al(OH)_3(s) + 3H^+(aq) \longrightarrow Al^{3+}(aq) + 3H_2O$$

$$CaCO_3(s) + 2H^+(aq) \longrightarrow Ca^{2+}(aq) + H_2CO_3(aq)$$

$$CoS(s) + 2H^+(aq) \longrightarrow Co^{2+}(aq) + H_2S(aq)$$

Complex Formation

Ammonia and sodium hydroxide are commonly used to dissolve precipitates containing a cation that forms a stable complex with NH_3 or OH^- (Table 15.3). *i*
The reactions with zinc hydroxide are typical:

$$Zn(OH)_2(s) + 4NH_3(aq) \longrightarrow Zn(NH_3)_4^{2+}(aq) + 2OH^-(aq)$$

$$Zn(OH)_2(s) + 2OH^-(aq) \longrightarrow Zn(OH)_4^{2-}(aq)$$

i $Mg(OH)_2$ cannot be dissolved by NH_3 or NaOH. Explain.

The equilibrium constant for the solubility reaction is readily calculated. Consider, for example, the reaction by which zinc hydroxide dissolves in ammonia. Again, imagine that the reaction occurs in two steps:

$$Zn(OH)_2(s) \rightleftharpoons Zn^{2+}(aq) + 2OH^-(aq) \qquad K_{sp}\ Zn(OH)_2$$

$$Zn^{2+}(aq) + 4NH_3(aq) \rightleftharpoons Zn(NH_3)_4^{2+}(aq) \qquad K_f\ Zn(NH_3)_4^{2+}$$

$$\overline{Zn(OH)_2(s) + 4NH_3(aq) \rightleftharpoons Zn(NH_3)_4^{2+}(aq) + 2OH^-(aq)}$$

Applying the rule of multiple equilibria,

$$K = K_{sp}\ Zn(OH)_2 \times K_f\ Zn(NH_3)_4^{2+} = (4 \times 10^{-17})(3.6 \times 10^8) = 1 \times 10^{-8}$$

In general, for any reaction of this type

$$K = K_{sp} \times K_f$$

where K_{sp} is the solubility product constant of the solid and K_f is the formation constant of the complex.

Table 15.3 Complexes of Cations with NH_3 and OH^-

Cation	NH_3 Complex	OH^- Complex
Ag^+	$Ag(NH_3)_2^+$	—
Cu^{2+}	$Cu(NH_3)_4^{2+}$ (blue)	—
Cd^{2+}	$Cd(NH_3)_4^{2+}$	—
Sn^{4+}	—	$Sn(OH)_6^{2-}$
Sb^{3+}	—	$Sb(OH)_4^-$
Al^{3+}	—	$Al(OH)_4^-$
Ni^{2+}	$Ni(NH_3)_6^{2+}$ (blue)	—
Zn^{2+}	$Zn(NH_3)_4^{2+}$	$Zn(OH)_4^{2-}$

EXAMPLE 15.11

Consider the reaction by which silver chloride dissolves in ammonia (Figure 15.8):

$$AgCl(s) + 2NH_3(aq) \rightleftharpoons Ag(NH_3)_2^+(aq) + Cl^-(aq)$$

a Taking $K_{sp}\ AgCl = 1.8 \times 10^{-10}$ and $K_f\ Ag(NH_3)_2^+ = 1.7 \times 10^7$, calculate K for this reaction.

STRATEGY AND SOLUTION

1. Write the equation for the dissolution of AgCl.

$$AgCl(s) \rightleftharpoons Ag^+(aq) + Cl^-(aq) \qquad K_1 = K_{sp}\ \text{for AgCl} = 1.8 \times 10^{-10}$$

2. Write the equation for the reaction between Ag^+ and NH_3.

$$Ag^+(aq) + 2NH_3(aq) \rightleftharpoons Ag(NH_3)_2^+(aq) \qquad K_2 = K_f = 1.7 \times 10^7$$

continued

3. Adding the two equations and applying the rule of multiple equilibria, we obtain

$$AgCl(s) \rightleftharpoons Ag^+(aq) + Cl^-(aq) \qquad\qquad K_1 = 1.8 \times 10^{-10}$$

$$\underline{Ag^+(aq) + 2NH_3(aq) \rightleftharpoons Ag(NH_3)_2^+(aq) \qquad\qquad K_2 = 1.7 \times 10^7}$$

$$AgCl(s) + 2NH_3(aq) \rightleftharpoons Ag(NH_3)_2^+(aq) + Cl^-(aq) \qquad K = (1.8 \times 10^{-10})(1.7 \times 10^7) = 3.1 \times 10^{-3}$$

b   Calculate the number of moles of AgCl that dissolve in one liter of 6.0 M NH_3.

ANALYSIS

Information given:	from part (a): equilibrium reaction $(AgCl(s) + 2NH_3(aq) \rightleftharpoons Ag(NH_3)_2^+(aq))$ from part (a): K for the reaction (3.1×10^{-3}) M for NH_3 (6.0)
Asked for:	molar solubility of AgCl in 6 M NH_3

STRATEGY

1. Write the K expression.

2. Set up an equilibrium table. Include only the ions. Note that for every mole/L of AgCl that dissolves (x), 2 mol of NH_3 are consumed $(-2x)$, one mole of $Ag(NH_3)_2^+$ and one mole of $Cl^-(+x)$ are formed.

3. Substitute into the K expression and solve for x.

SOLUTION

1. K expression	$K = \dfrac{[Ag(NH_3)_2^+][Cl^-]}{[NH_3]^2}$

2. table

	$[NH_3]$	$[Ag(NH_3)_2^+]$	$[Cl^-]$
Original	6.0	0	0
Change	$-2x$	$+x$	$+x$
Equilibrium	$6.0 - 2x$	x	x

3. x (molar solubility)

$$3.1 \times 10^{-3} = \frac{(x)(x)}{(6.0 - 2x)^2} = \frac{x^2}{(6.0 - 2x)^2}$$

Taking the square root of both sides: $0.056 = \dfrac{x}{6.0 - 2x} \longrightarrow x = 0.30\ M$

0.30 moles of AgCl dissolves in one liter of 6.0 M NH_3.

Figure 15.8 Dissolving a precipitate by complex formation.

Silver chloride (AgCl) is precipitated by adding a solution of sodium chloride (NaCl) to a silver nitrate ($AgNO_3$) solution.

As a solution of ammonia is added, the AgCl precipitate is dissolving to form the complex ion $Ag(NH_3)_2^+$.

$NH_3(aq)$

AgCl(s)

$[Ag(NH_3)_2]^+(aq)$

© Cengage Learning/Charles D. Winters

EXAMPLE 15.12 CONCEPTUAL

Column A lists a series of ionic solids, all of which are insoluble in water. Match each of these compounds with the appropriate description(s) listed in Column B. Note that more than one description can fit a particular compound.

A		B
$BaSO_4$		(a) $K_{sp} = s^2$
$AgCl$		(b) $K_{sp} = 4s^3$
$Zn(OH)_2$		(c) more soluble in water than in Na_2SO_4 solution
$CaSO_4$		(d) soluble in strong acid
		(e) soluble in strong base
		(f) soluble in ammonia

SOLUTION

$BaSO_4$ matches with (a) and (c)

$AgCl$ matches with (a) and (f)

$Zn(OH)_2$ matches with (b), (d), (e), (f)

$CaSO_4$ matches with (a) and (c)

CHEMISTRY **BEYOND** THE CLASSROOM

Qualitative Analysis

In the laboratory you will most likely carry out one or more experiments involving the separation and identification of cations present in an "unknown" solution. A scheme of **qualitative analysis** for 21 different cations is shown in Table A. As you can see, the general approach is to take out each group (I, II, III, IV) in succession, using *selective precipitation*.

Group I consists of the only three common cations that form insoluble chlorides: Ag^+, Pb^{2+}, and Hg_2^{2+}. Addition of hydrochloric acid precipitates $AgCl$, $PbCl_2$, and Hg_2Cl_2. Cations in Groups II, III, and IV remain in solution, since their chlorides are soluble.

Group II consists of six different cations, all of which form very insoluble sulfides (Figure A). These compounds are precipitated selectively by adding hydrogen sulfide, a

Figure A Group II sulfides. *(From left to right)* CuS, Bi_2S_3, and HgS are black; CdS is orange-yellow; Sb_2S_3 is brilliant red-orange; and SnS_2 is yellow.

toxic, foul-smelling gas, at a pH of 0.5. At this rather high H^+ ion concentration, 0.3 M, the equilibrium

$$H_2S(aq) \rightleftharpoons 2H^+(aq) + S^{2-}(aq)$$

Table A Cation Groups of Qualitative Analysis

Group	Cations	Precipitating Reagent/Conditions
I	Ag^+, Pb^{2+}, Hg_2^{2+}	6 M HCl
II	Cu^{2+}, Bi^{3+}, Hg^{2+}, Cd^{2+}, Sn^{4+}, Sb^{3+}	0.1 M H_2S at a pH of 0.5
III	Al^{3+}, Cr^{3+}, Co^{2+}, Fe^{2+}, Mn^{2+}, Ni^{2+}, Zn^{2+}	0.1 M H_2S at a pH of 9
IV	Ba^{2+}, Ca^{2+}, Mg^{2+}; Na^+, K^+	0.2 M $(NH_4)_2CO_3$ at a pH of 9.5. No precipitate with Na^+, K^+

lies far to the left. The concentration of S^{2-} is extremely low, but sufficient to precipitate all the cations in this group, including Cu^{2+}.

$$Cu^{2+}(aq) + H_2S(aq) \longrightarrow CuS(s) + 2H^+(aq) \quad K = 1 \times 10^{16}$$

Group III sulfides are much more difficult to precipitate than those of Group II. Compare, for example, the equilibrium constant for the reaction

$$Mn^{2+}(aq) + H_2S(aq) \rightleftharpoons MnS(s) + 2H^+(aq) \quad K = 2 \times 10^{-7}$$

with that given above for the precipitation of CuS. However, when the solution is made basic (pH = 9), the situation changes. The concentration of H^+ ion drops sharply and, as you would expect from Le Châtelier's principle, the reaction just cited becomes much more spontaneous. Five of the Group III cations precipitate as sulfides. The other two come down as hydroxides, e.g.,

$$Al^{3+}(aq) + 3OH^-(aq) \longrightarrow Al(OH)_3(s)$$

The colors of these precipitates are shown in Figure B.

Group IV cations have soluble chlorides and sulfides, so they are still in solution at this point. The alkaline earth cations (Mg^{2+}, Ca^{2+}, Ba^{2+}) are precipitated as carbonates

$$M^{2+}(aq) + CO_3^{2-}(aq) \longrightarrow MCO_3(s) \quad M = Mg, Ca, Ba$$

The alkali metal cations are identified by flame tests (Figure C).

Once the groups are separated from one another, the individual cations in each group must, of course, be separated and identified. This is where things get complicated. A process that might be referred to as "selective dissolving" is often involved. Consider, for example, the precipitate obtained in Group I: $PbCl_2$, $AgCl$, and Hg_2Cl_2. Of the three solids, lead chloride is by far the most soluble in water. Stirring it with hot water brings it into solution

$$PbCl_2(s) \longrightarrow Pb^{2+}(aq) + 2Cl^-(aq)$$

leaving the other two compounds behind. Further treatment with ammonia dissolves silver chloride

$$AgCl(s) + 2NH_3(aq) \longrightarrow Ag(NH_3)_2^+(aq) + Cl^-(aq)$$

but not Hg_2Cl_2.

Figure B Group III sulfides and hydroxides. *(Left)* Five Group III cations precipitate as sulfides. These are NiS, CoS, and FeS (all of which are black), MnS (salmon), and ZnS (white). *(Right)* Two cations precipitate as hydroxides, $Al(OH)_3$ (white) and $Cr(OH)_3$ (gray-green).

Figure C Flame tests for Na⁺ (yellow) and K⁺ (violet). A drop of solution is picked up on a platinum loop and immersed in the flame. The test for K^+ is best done with a filter that hides the strong Na^+ color.

Key Concepts

1. Relate K_f for a complex ion to the ratio of concentration of complex ion to metal ion.
   (Example 15.1, Problems 1–4)
2. Write the K_{sp} expression for any ionic solid.
   (Example 15.2, Problems 5–8)
3. Use the value of K_{sp} to
   ■ calculate the concentration of one ion knowing that of the other.
   (Example 15.3, Problems 9–14)
   ■ calculate molar solubility or use molar solubility to calculate the value of K_{sp}.
   (Examples 15.4, 15.5, Problems 15–26)
   ■ calculate solubility in a solution containing a common ion.
   (Example 15.6, Problems 19, 20)
   ■ determine whether a precipitate forms.
   (Example 15.7, Problems 27–32)
   ■ determine which ion will precipitate first in a solution.
   (Example 15.8, Problems 33–38)
4. Calculate K for
   ■ dissolving a metal oxide in strong acid.
   (Example 15.9, Problem 46)
   ■ dissolving a precipitate in a complexing agent.
   (Example 15.11, Problems 45, 47–54)
5. Write a balanced net ionic equation to explain why a precipitate dissolves in
   ■ a strong acid.
   (Example 15.10, Problems 39, 40)
   ■ NH_3 or OH^-.
   (Example 15.12, Problems 41–44)

Key Terms

common ion effect
complex ion

formation constant
ion product constant

selective precipitation
solubility product constant

Summary Problem

Consider zinc hydroxide ($K_{sp} = 4 \times 10^{-17}$), commonly used as a protective coating on metals. It can be formed when solutions of zinc nitrate and sodium hydroxide are combined. (Assume volumes are additive in all cases.)

(a) Write the K_{sp} expression for zinc hydroxide.

(b) How many milligrams of NaOH must be added to 125 mL of 0.0120 M Zn(NO$_3$)$_2$ to just start the precipitation of Zn(OH)$_2$? (Ignore OH$^-$ ions from water.)

(c) Will a precipitate form if 10.00 mL of 0.0200 M NaOH are added to 250.0 mL of 0.098 M Zn(NO$_3$)$_2$?

(d) How many milligrams of zinc hydroxide are required to prepare 700.0 mL of a saturated Zn(OH)$_2$ solution?

(e) What is the pH of a saturated solution of Zn(OH)$_2$?

(f) Which would precipitate first, Zn(OH)$_2$ or Al(OH)$_3$, when 0.075 mol of NaOH are added to a 0.500-L solution containing 0.0100 mol Zn^{2+} and 0.0100 mol Al^{3+} ions? (K_{sp} of Al(OH)$_3$ is 2×10^{-31}.)

(g) Write equations and calculate K for the following reactions:

1. Dissolving Zn(OH)$_2$ in strong acid.

2. Dissolving Zn(OH)$_2$ in strong base.

3. Dissolving Zn(OH)$_2$ in aqueous NH$_3$ forming Zn(NH$_3$)$_4$$^{2+}$.

(h) How many grams of Zn(OH)$_2$ can be dissolved in 2.50 L of 0.166 M NaOH? (The complex ion Zn(OH)$_4$$^{2-}$ is formed.)

Answers

(a) $K_{sp} = [Zn^{2+}][OH^-]^2$
(b) 3×10^{-4} mg
(c) Yes
(d) 0.2 mg
(e) 8.6
(f) Al(OH)$_3$
(g) 1. Zn(OH)$_2$(s) + 2H$^+$(aq) $\rightleftharpoons$ 2H$_2$O + Zn^{2+}(aq)
$$K = 4 \times 10^{11}$$
2. Zn(OH)$_2$(s) + 2OH$^-$(aq) $\rightleftharpoons$ Zn(OH)$_4$$^{2-}$(aq)
$$K = 0.01$$
3. Zn(OH)$_2$(s) + 4NH$_3$(aq) $\rightleftharpoons$ Zn(NH$_3$)$_4$$^{2+}$(aq) +
$$2OH^-(aq) \quad K = 1 \times 10^{-8}$$
(h) 0.07 g

Questions and Problems

Even-numbered questions and Challenge Problems have answers in Appendix 5 and fully worked solutions in the *Student Solutions Manual*.

15-1 Complex Ion Equilibria; Formation Constant (K_f)

1. At what concentration of ammonia is
   (a) $[Cd^{2+}] = [Cd(NH_3)_4^{2+}]$?
   (b) $[Co^{2+}] = [Co(NH_3)_6^{2+}]$?
2. At what concentration of cyanide ion is
   (a) $[Cd^{2+}] = 10^{-8} \times [Cd(CN)_4^{2-}]$?
   (b) $[Fe^{2+}] = 10^{-20} \times [Fe(CN)_6^{4-}]$?
3. Consider the complex ion [Ni(en)$_3$]$^{2+}$. Its K_f is 2.1×10^{18}. At what concentration of en is 67% of the Ni^{2+} converted to [Ni(en)$_3$]$^{2+}$?
4. Consider the complex ion Hg(Cl)$_4$$^{2-}$. Its K_f value is 1.2×10^{15}. At what concentration of Cl$^-$ is 58% of the Hg^{2+} converted to Hg(Cl)$_4$$^{2-}$?

15-2 Solubility; Solubility Product Constant (K_{sp})

Expression for K_{sp}

5. Write the equilibrium equation and the K_{sp} expression for each of the following:
   (a) AgBr   (b) Ca(OH)$_2$
   (c) Co$_2$S$_3$ (d) Fe$_3$(PO$_4$)$_2$
6. Write the equilibrium equation and the K_{sp} expression for each of the following.
   (a) AgCl   (b) Al$_2$(CO$_3$)$_3$
   (c) MnS$_2$ (d) Mg(OH)$_2$
7. Write the equilibrium equations on which the following K_{sp} expressions are based.
   (a) $[Hg_2^{2+}][Cl^-]^2$ (b) $[Pb^{2+}][CrO_4^{2-}]$
   (c) $[Mn^{4+}][O^{2-}]^2$ (d) $[Al^{3+}]^2[S^{2-}]^3$
8. Write the equilibrium equations on which the following K_{sp} expressions are based.
   (a) $[Pb^{4+}][O^{2-}]^2$ (b) $[Hg^{2+}]^3[PO_4^{3-}]^2$
   (c) $[Ni^{3+}][OH^-]^3$ (d) $[Ag^+]^2[SO_4^{2-}]$

K_{sp} and the Equilibrium Concentration of Ions

9. Given K_{sp} and the equilibrium concentration of one ion, calculate the equilibrium concentration of the other ion.

(a) barium bromate: $K_{sp} = 2.4 \times 10^{-4}$; $[Ba^{2+}] = 1.9 \times 10^{-2}$

(b) cadmium(II) phosphate: $K_{sp} = 2.5 \times 10^{-33}$; $[Cd^{2+}] = 8.2 \times 10^{-6}$

(c) iron(II) fluoride: $K_{sp} = 2.4 \times 10^{-6}$; $[F^-] = 3.7 \times 10^{-3}$

10. Follow the directions for Question 9 for the following compounds:

(a) lithium phosphate: $K_{sp} = 3.2 \times 10^{-9}$; $[PO_4^{3-}] = 7.5 \times 10^{-4} M$

(b) silver nitrite: $K_{sp} = 6.0 \times 10^{-4}$; $[Ag^+] = 0.025 M$

(c) tin(II) hydroxide: $K_{sp} = 1.4 \times 10^{-28}$; pH = 9.35

11. Calculate the concentration of the following ions in equilibrium with $1.63 \times 10^{-3} M\ Ba^{2+}$.

(a) SO_4^{2-} (b) F^- (c) CO_3^{2-}

12. Calculate the concentration of the following ions in equilibrium with $1.24 \times 10^{-4} M\ Ca^{2+}$.

(a) CO_3^{2-} (b) OH^- (c) PO_4^{3-}

13. Fill in the blanks in the following table.

	Compound	[cation]	[anion]	K_{sp}
(a)	$CoCO_3$	_____	_____	1×10^{-10}
(b)	LaF_3	_____	7×10^{-6}	2×10^{-19}
(c)	$Ba_3(PO_4)_2$	4.2×10^{-8}	_____	6×10^{-39}

14. Fill in the blanks in the following table.

	Compound	[cation]	[anion]	K_{sp}
(a)	BaC_2O_4	_____	_____	1.6×10^{-6}
(b)	$Cr(OH)_3$	2.7×10^{-8}	_____	6.3×10^{-31}
(c)	$Pb_3(PO_4)_2$	_____	8×10^{-6}	1×10^{-54}

K_{sp} and Solubility

15. Calculate the molar solubility of the following compounds:

(a) calcium iodate: $K_{sp} = 6.5 \times 10^{-6}$

(b) cadmium(III) phosphate: $K_{sp} = 2.5 \times 10^{-33}$

(c) silver oxalate: $K_{sp} = 5.4 \times 10^{-12}$

16. Calculate the molar solubility of the following compounds.

(a) MgF_2 (b) $Fe(OH)_3$ (c) $Mg_3(PO_4)_2$

17. Calculate the K_{sp} of the following compounds, given their molar solubilities.

(a) $TlBr$: 1.9×10^{-3}

(b) $Ni(OH)_2$: 5.2×10^{-6}

(c) $Cu_3(AsO_4)_2$: 3.7×10^{-8}

18. Calculate the K_{sp} of the following compounds, given their molar solubilities.

(a) MgC_2O_4, $9.2 \times 10^{-3} M$

(b) $Mn(OH)_2$, $3.5 \times 10^{-5} M$

(c) $Cd_3(PO_4)_2$, $1.5 \times 10^{-7} M$

19. Calculate the solubility (in grams per liter) of silver chloride in the following.

(a) pure water

(b) $0.025 M\ BaCl_2$

(c) $0.17 M\ AgNO_3$

20. Calculate the solubility (in grams per liter) of magnesium hydroxide in the following.

(a) pure water

(b) $0.041 M\ Ba(OH)_2$

(c) $0.0050 M\ MgCl_2$

21. Copper(I) chloride, $CuCl$, is the starting material for the fungicide copper oxychloride. At 25°C, a saturated solution of $CuCl$ is prepared by dissolving 10.1 mg of $CuCl$ in 0.250 L of water. What is the K_{sp} for copper(I) chloride?

22. A saturated solution of $Ni(OH)_2$ can be prepared by dissolving 0.239 mg of $Ni(OH)_2$ in water to make 500.0 mL of solution. What is the K_{sp} for $Ni(OH)_2$?

23. One gram of $PbCl_2$ is dissolved in 1.0 L of hot water. When the solution is cooled to 25°C, will some of the $PbCl_2$ crystallize out? If so, how much?

24. K_{sp} for $CaSO_4$ at 100°C is estimated to be 1.6×10^{-5}. At 25°C, 0.915 g of $CaSO_4$ is added to one liter of water.

(a) Will all the $CaSO_4$ dissolve?

(b) If the solution is heated to 100°C, will all the $CaSO_4$ dissolve?

25. At 25°C, 100.0 mL of a $Ba(OH)_2$ solution is prepared by dissolving $Ba(OH)_2$ in an alkaline solution. At equilibrium, the saturated solution has $0.138 M\ Ba^{2+}$ and a pH of 13.28. Estimate K_{sp} for $Ba(OH)_2$.

26. At 25°C, 10.24 mg of $Cr(OH)_2$ are dissolved in enough water to make 125 mL of solution. When equilibrium is established, the solution has a pH of 8.49. Estimate K_{sp} for $Cr(OH)_2$.

15-3 Precipitate Formation

K_{sp} and Precipitation

27. Calcium nitrate is added to a sodium sulfate solution that is $0.0150 M$.

(a) At what concentration of Ca^{2+} does a precipitate first start to form?

(b) Enough $Ca(NO_3)_2$ is added to make $[Ca^{2+}] = 0.0075 M$. What percentage of the original sulfate ion has precipitated?

28. Cadmium(II) chloride is added to a solution of potassium hydroxide with a pH of 9.62. (K_{sp} $Cd(OH)_2 = 2.5 \times 10^{-14}$)

(a) At what concentration of Cd^{2+} does a precipitate first start to form?

(b) Enough cadmium(II) chloride is added to make $[Cd^{2+}] = 0.0013 M$. What is the pH of the resulting solution?

(c) What percentage of the original hydroxide ion is left in solution?

29. Water from a well is found to contain 3.0 mg of calcium ion per liter. If 0.50 mg of sodium sulfate is added to one liter of the well water without changing its volume, will a precipitate form? What should $[SO_4^{2-}]$ be to just start precipitation?

30. Silver(I) sulfate ($K_{sp} = 1.2 \times 10^{-5}$) is used in the electroplating of silver. A 1.0-L solution is prepared by mixing 15 g of silver nitrate with 20 g (2 significant figures) of potassium sulfate. Will a precipitate form? At what $[Ag^+]$ concentration will precipitation (if any) start?

31. A solution is prepared by mixing 13.00 mL of 0.0021 M aqueous $Hg_2(NO_3)_2$ with 25.0 mL of 0.015 M HCl. Assume that volumes are additive.
   (a) Will precipitation occur?
   (b) Calculate $[Hg_2^{2+}]$, $[Cl^-]$, and $[NO_3^-]$ after equilibrium is established.

32. A solution is prepared by mixing 45.00 mL of 0.022 M $AgNO_3$ with 13.00 mL of 0.0014 M Na_2CO_3. Assume that volumes are additive.
   (a) Will precipitation occur?
   (b) Calculate $[Ag^+]$, $[CO_3^{2-}]$, $[Na^+]$, and $[NO_3^-]$ after equilibrium is established.

Selective Precipitation

33. A solution is 0.047 M in both NaF and Na_2CO_3. Solid strontium nitrate, $Sr(NO_3)_2$, is added without changing the volume of the solution.
   (a) Which salt, $SrCO_3$ or SrF_2 ($K_{sp} = 4.3 \times 10^{-9}$), will precipitate first?
   (b) What is $[Sr^{2+}]$ when the salt in (a) first begins to precipitate?

34. Solid lead nitrate is added to a solution that is 0.020 M in OH^- and SO_4^{2-}. Addition of the lead nitrate does not change the volume of the solution.
   (a) Which compound, $PbSO_4$ or $Pb(OH)_2$ ($K_{sp} = 2.8 \times 10^{-16}$), will precipitate first?
   (b) What is the pH of the solution when $PbSO_4$ first starts to precipitate?

35. A solution is made up by adding 0.632 g of barium nitrate and 0.920 g of lanthanum nitrate, $La(NO_3)_3$, to enough water to make 0.500 L of solution. Solid sodium iodate, $NaIO_3$, is added (without volume change) to the solution.
   (a) Which salt will precipitate first? $La(IO_3)_3$ ($K_{sp} = 7.50 \times 10^{-12}$) or $BaIO_3$ ($K_{sp} = 4.0 \times 10^{-9}$)?
   (b) What is $[IO_3^-]$ when the salt in (a) first begins to precipitate?

36. A solution is made up by adding 0.839 g of silver(I) nitrate and 1.024 g of lead(II) nitrate to enough water to make 492 mL of solution. Solid sodium chromate, Na_2CrO_4, is added without changing the volume of the solution.
   (a) Which salt will precipitate first, Ag_2CrO_4 or $PbCrO_4$?
   (b) What is the concentration of the chromate ion when the first salt starts to precipitate?

37. A solution is made up by mixing 125 mL of 0.100 M $AuNO_3$ and 225 mL of 0.049 M $AgNO_3$. Twenty-five mL of a 0.0100 M solution of HCl is then added. K_{sp} of AuCl = 2.0×10^{-13}. When equilibrium is established, will there be
   ■ no precipitate?
   ■ a precipitate of AuCl only?
   ■ a precipitate of AgCl only?
   ■ a precipitate of both AgCl and AuCl?

38. To a beaker with 500.0 mL of water are added 95 mg of $Ba(NO_3)_2$, 95 mg of $Ca(NO_3)_2$, and 100.0 mg of Na_2CO_3. After equilibrium is established, will there be
   ■ no precipitate?
   ■ a precipitate of $BaCO_3$ only?
   ■ a precipitate of $CaCO_3$ only?
   ■ a precipitate of both $CaCO_3$ and $BaCO_3$?
Assume that the volume of the solution is still 500.0 mL after the addition of the salts.

15-4 Dissolving Precipitates

39. Write net ionic equations for the reaction of H^+ with
   (a) Fe_2S_3
   (b) $Mg(OH)_2$
   (c) $MgCO_3$
   (d) $Pt(NH_3)_4^{2+}$
   (e) Hg_2I_2

40. Write net ionic equations for the reactions of each of the following compounds with a strong acid.
   (a) CaF_2
   (b) $CuCO_3$
   (c) $Ti(OH)_3$
   (d) $Sn(OH)_6^{2-}$
   (e) $Cd(NH_3)_4^{2+}$

41. Write a net ionic equation for the reaction with aqueous NH_3 in which
   (a) Pt^{2+} forms a complex ion.
   (b) Ag^+ forms a precipitate.
   (c) $Ni(OH)_3$ dissolves.

42. Write a net ionic equation for the reaction with ammonia by which
   (a) $Cu(OH)_2$ dissolves.
   (b) Cd^{2+} forms a complex ion.
   (c) Pb^{2+} forms a precipitate.

43. Write a net ionic equation for the reaction with OH^- by which
   (a) Sb^{3+} forms a precipitate.
   (b) antimony(III) hydroxide dissolves when more OH^- is added.
   (c) Sb^{3+} forms a complex ion.

44. Write a net ionic equation for the reaction with Al^{3+} by which
   (a) a complex ion forms when it reacts with OH^-.
   (b) a precipitate forms when it reacts with the phosphate ion.
   (c) the precipitate formed with OH^- is dissolved by a strong acid.

Solution Equilibria

45. Write an overall net ionic equation and calculate K for the reaction where $CoCO_3$ ($K_{sp} = 1.0 \times 10^{-10}$) is dissolved by $Na_2C_2O_4$ (Na_2ox) to form $[Coox_3]^{4-}$ ($K_f = 5 \times 10^9$).

46. Write an overall net ionic equation and calculate K for the reaction where $Co(OH)_2$ ($K_{sp} = 2 \times 10^{-16}$) is dissolved by HCl.

47. Consider the reaction

$$AgC_2H_3O_2(s) + Cl^-(aq) \longrightarrow AgCl(s) + C_2H_3O_2^-(aq)$$

   (a) Calculate K for the reaction.
   (b) Will more AgCl precipitate if H^+ ions are added?

48. Consider the reaction

$$BaF_2(s) + SO_4^{2-}(aq) \rightleftharpoons BaSO_4(s) + 2F^-(aq)$$

   (a) Calculate K for the reaction.
   (b) Will $BaSO_4$ precipitate if Na_2SO_4 is added to a saturated solution of BaF_2?

49. Aluminum hydroxide reacts with an excess of hydroxide ions to form the complex ion $Al(OH)_4^-$.
   (a) Write an equation for this reaction.
   (b) Calculate K.
   (c) Determine the solubility of $Al(OH)_3$ (in mol/L) at pH 12.0.

50. Consider the reaction

$$Cu(OH)_2(s) + 4NH_3(aq) \rightleftharpoons Cu(NH_3)_4^{2+}(aq) + 2OH^-(aq)$$

(a) Calculate K given that for $Cu(OH)_2$ $K_{sp} = 2 \times 10^{-19}$ and for $Cu(NH_3)_4^{2+}$ $K_f = 2 \times 10^{12}$.
(b) Determine the solubility of $Cu(OH)_2$ (in mol/L) in 4.5 M NH_3.

51. Calculate the molar solubility of gold(I) chloride ($K_{sp} = 2.0 \times 10^{-13}$) in 0.10 M NaCN. The complex ion formed is $[Au(CN)_2]^-$ with $K_f = 2 \times 10^{38}$. Ignore any other competing equilibria.

52. Calculate the molar solubility of $PbCl_2$ in 0.2 M NaOH. The complex formed is $Pb(OH)_3^-$ ($K_f = 3.8 \times 10^{14}$). Ignore any other competing equilibria.

53. For the reaction

$$CdC_2O_4(s) + 4NH_3(aq) \rightleftharpoons Cd(NH_3)_4^{2+}(aq) + C_2O_4^{2-}(aq)$$

(a) calculate K. (K_{sp} for CdC_2O_4 is 1.5×10^{-8}.)
(b) calculate $[NH_3]$ at equilibrium when 2.00 g of CdC_2O_4 are dissolved in 1.00 L of solution.

54. For the reaction

$$Zn(OH)_2(s) + 2OH^-(aq) \longrightarrow Zn(OH)_4^{2-}(aq)$$

(a) Calculate K for the reaction.
(b) Find $[OH^-]$ when 10.0 g of $Zn(OH)_2$ are dissolved in 1.0 L of solution.

Unclassified

55. What are the concentrations of Cu^{2+}, NH_3, and $Cu(NH_3)_4^{2+}$ at equilibrium when 18.8 g of $Cu(NO_3)_2$ are added to 1.0 L of a 0.400 M solution of aqueous ammonia? Assume that the reaction goes to completion and forms $Cu(NH_3)_4^{2+}$.

56. For the system

$$hemoglobin \cdot O_2(aq) + CO(g) \rightleftharpoons hemoglobin \cdot CO(aq) + O_2(g)$$

$K = 2.0 \times 10^2$. What must be the ratio of P_{CO}/P_{O_2} if 12.0% of the hemoglobin in the bloodstream is converted to the CO complex?

57. Calcium ions in blood trigger clotting. To prevent that in donated blood, sodium oxalate, $Na_2C_2O_4$, is added to remove calcium ions according to the following equation.

$$C_2O_4^{2-}(aq) + Ca^{2+}(aq) \longrightarrow CaC_2O_4(s)$$

Blood contains about 0.10 mg Ca^{2+}/mL. If a 250.0-mL sample of donated blood is treated with an equal volume of 0.160 M $Na_2C_2O_4$, estimate $[Ca^{2+}]$ after precipitation. (K_{sp} $CaC_2O_4 = 4 \times 10^{-9}$)

58. A saturated solution of lithium carbonate (Li_2CO_3) freezes at $-0.33°C$. Estimate K_{sp} for Li_2CO_3. Assume complete dissociation and a density for the solution of 1.0 g/mL.

59. A town adds 2.0 ppm of F^- ion to fluoridate its water supply. (Fluoridation of water reduces the incidence of dental caries). If the concentration of Ca^{2+} in the water is 3.5×10^{-4} M, will a precipitate of CaF_2 form when the water is fluoridated?

60. Consider the following hypothetical dissociation:

$$AB_3(s) \rightleftharpoons A^{3+}(aq) + 3B^-(aq) \qquad \Delta H < 0$$

What effect will each of the following have on the position of equilibrium?

(a) addition of $A(NO_3)_3$
(b) increase in temperature
(c) adding Na^+, forming NaB

61. When 25.0 mL of 0.500 M iron(II) sulfate is combined with 35.0 mL of 0.332 M barium hydroxide, two different precipitates are formed.

(a) Write a net ionic equation for the reaction that takes place.
(b) Estimate the mass of the precipitates formed.
(c) What are the equilibrium concentrations of the ions in solution?

62. Consider a 2.0-L aqueous solution of 4.17 M NH_3, where 21.0 g of NH_4Cl are dissolved. To this solution, 4.8 g of $CaCl_2$ are added.

(a) What is $[OH^-]$ before $CaCl_2$ is added?
(b) Will a precipitate form?
(c) What is $[Ca^{2+}]$ after equilibrium is established?

63. Marble is almost pure $CaCO_3$. Acid rain has a devastating effect on marble statuary left outdoors. Assume that the reaction which occurs is

$$CaCO_3(s) + H^+(aq) \longrightarrow Ca^{2+}(aq) + HCO_3^-(aq)$$

Neglecting all other competing equilibria and using Tables 15.1 and 13.2, calculate

(a) K for the reaction.
(b) the molar solubility of $CaCO_3$ in pure water.
(c) the molar solubility of $CaCO_3$ in acid rainwater with a pH of 4.00.

64. Consider the following solubility data for calcium oxalate (CaC_2O_4):

$$K_{sp} \text{ at } 25°C = 4 \times 10^{-9}$$
$$K_{sp} \text{ at } 95°C = 1 \times 10^{-8}$$

Five hundred mL of a saturated solution are prepared at 95°C. How many milligrams of CaC_2O_4 will precipitate when the solution is cooled to 25°C? (Assume that supersaturation does not take place.)

Conceptual Problems

65. Consider three complexes of Ag^+ and their formation constants, K_f.

Complex Ion	K_f
$Ag(NH_3)_2^+$	1.6×10^7
$Ag(CN)_2^-$	5.6×10^{18}
$AgBr_2^-$	1.3×10^7

Which statements are true?

(a) $Ag(NH_3)_2^+$ is more stable than $Ag(CN)_2^-$.
(b) Adding a strong acid (HNO_3) to a solution that is 0.010 M in $Ag(NH_3)_2^+$ will tend to dissociate the complex ion into Ag^+ and NH_4^+.
(c) Adding a strong acid (HNO_3) to a solution that is 0.010 M in $AgBr_2^-$ will tend to dissociate the complex ion into Ag^+ and Br^-.
(d) To dissolve AgI, one can add either NaCN or HCN as a source of the cyanide ion. Fewer moles of NaCN would be required.
(e) Solution A is 0.10 M in Br^- and contains the complex ion $AgBr_2^-$. Solution B is 0.10 M in CN^- and contains the complex ion $Ag(CN)_2^-$. Solution B will have more particles of complex ion per particle of Ag^+ than solution A.

66. The box below represents one liter of a saturated solution of the species , where squares represent the cation and circles represent the anion. Water molecules, though present, are not shown.

Complete the next three figures below by filling one-liter boxes to the right of the arrow, showing the state of the ions after water is added to form saturated solutions. The species represented to the left of the arrow is the solid form of the ions represented above. Do not show the water molecules.

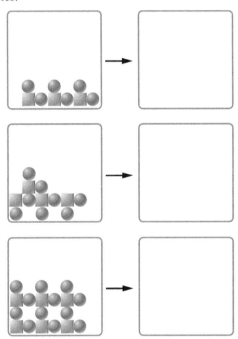

67. Consider a saturated solution of $BaCO_3$ at 75°C. Answer the following questions using **LT** (for *is less than*), **GT** (for *is greater than*), **EQ** (for *is equal to*), or write **MI** if *more information* is required.

    (a) At equilibrium $[Ba^{2+}]$ _____$[CO_3^{2-}]$.

    (b) The temperature is increased to 200°C. The K_{sp} at 75°C _____K_{sp} at 200°C.

    (c) A solution of $NaNO_3$ is added. $[Ba^{2+}]$ before $NaNO_3$ is added _____$[Ba^{2+}]$ after $NaNO_3$ is added.

    (d) Solid Na_2CO_3 is added. After equilibrium is reestablished, $[Ba^{2+}]$ _____$[CO_3^{2-}]$.

68. Which of the following statements are true?

    (a) For an insoluble metallic salt, K_{sp} is always less than 1.

    (b) More $PbCl_2$ can be dissolved at 100°C than at 25°C. One can conclude that dissolving $PbCl_2$ is an exothermic process.

    (c) When strips of copper metal are added to a saturated solution of $Cu(OH)_2$, a precipitate of $Cu(OH)_2$ can be expected to form because of the common ion effect.

69. Consider the insoluble salts JQ, K_2R, L_2S_3, MT_2, and NU_3. They are formed from the metal ions J^+, K^+, L^{3+}, M^{2+}, and N^{3+} and the nonmetal ions Q^-, R^{2-}, S^{2-}, T^-, and U^-. All the salts have the same K_{sp}, 1×10^{-10}, at 25°C.

    (a) Which salt has the highest molar solubility?

    (b) Does the salt with the highest molar solubility have the highest solubility in g salt/100 g water?

    (c) Can the solubility of each salt in g/100 g water be determined from the information given? If yes, calculate the solubility of each salt in g/100 g water. If no, why not?

70. Consider the following graph of the solubility of a compound X versus temperature.

Is dissolving X an exothermic process?

71. Consider the equilibrium curve for AgCl shown below. Which of the following statements about a solution at point A on the curve are true?

    (a) The solution is saturated and at equilibrium.

    (b) Addition of NaCl increases the concentration of Cl^- in solution.

    (c) Addition of NaCl increases the concentration of Ag^+ in solution.

    (d) Addition of Ag^+ results in the precipitation of AgCl.

    (e) Addition of solid $NaNO_3$ to the solution without change in volume does not change $[Ag^+]$ or $[Cl^-]$.

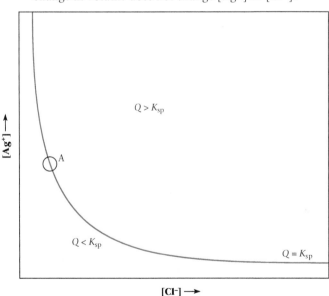

72. Dissolving $CaCO_3$ is an endothermic reaction. The following five graphs represent an experiment done on $CaCO_3$. Match the experiment to the graph.

    (a) HCl is added.

    (b) The temperature is increased.

    (c) $CaCl_2$ is added.

    (d) NaCl is added.

(1)

(2)

(3)

(4)

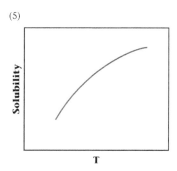

(5)

Challenge Problems

73. Insoluble hydroxides such as $Zn(OH)_2$ are better dissolved in solutions that are slightly acidic, such as aqueous HCN. Find the molar solubility of $Ca(OH)_2$ in 0.15 M HCN and compare it with the solubility in water: (*Hint:* Find K for the following reaction.)

$$Zn(OH)_2(s) + 2HCN(aq) \rightleftharpoons Zn^{2+}(aq) + 2CN^-(aq) + 2H_2O$$

74. What is the solubility of CaF_2 in a buffer solution containing 0.30 M $HCHO_2$ and 0.20 M $NaCHO_2$? *Hint:* Consider the equation

$$CaF_2(s) + 2H^+(aq) \longrightarrow Ca^{2+}(aq) + 2HF(aq)$$

and solve the equilibrium problem.

75. What is $[Br^-]$ just as AgCl begins to precipitate when 1.5 M $AgNO_3$ is slowly added to a solution containing 0.050 M Cl^- and 0.050 M Br^-?

76. The concentrations of various cations in seawater, in moles per liter, are

Ion	Na^+	Mg^{2+}	Ca^{2+}	Al^{3+}	Fe^{3+}
Molarity (M)	0.46	0.056	0.01	4×10^{-7}	2×10^{-7}

(a) At what $[OH^-]$ does $Mg(OH)_2$ start to precipitate?
(b) At this concentration, will any of the other ions precipitate?
(c) If enough OH^- is added to precipitate 50% of the Mg^{2+}, what percentage of each of the other ions will precipitate?
(d) Under the conditions in (c), what mass of precipitate will be obtained from one liter of seawater?

77. Consider the equilibrium

$$Zn(NH_3)_4^{2+}(aq) + 4OH^-(aq) \rightleftharpoons Zn(OH)_4^{2-}(aq) + 4NH_3(aq)$$

(a) Calculate K for this reaction.
(b) What is the ratio $[Zn(NH_3)_4^{2+}]/[Zn(OH)_4^{2-}]$ in a solution 1.0 M in NH_3?

78. Calculate K for the following reaction:

$$Cd(NH_3)_4^{2+}(aq) + 4H^+(aq) + 2F^-(aq) \rightleftharpoons$$
$$CdF_2(s) + 4NH_4^+(aq)$$

K_{sp} for CdF_2 is 6.4×10^{-3}.

16 Spontaneity of Reaction

The ballerinas in Degas's painting *The Dancers* look like they have depleted their glucose by using it for energy to dance.

Things are always at their best in their beginnings.

—BLAISE PASCAL

Geoffrey Clements/Corbis

The goal of this chapter is to answer a basic question: Will a given reaction occur "by itself" at a particular temperature and pressure, without the exertion of any outside force? In that sense, is the reaction *spontaneous*? This is a critical question in just about every area of science and technology. A synthetic organic chemist looking for a source of acetylene, C_2H_2, would like to know whether this compound can be made by heating the elements together. A metallurgist trying to produce titanium metal from TiO_2 would like to know what reaction to use; would hydrogen, carbon, or aluminum be feasible reducing agents?

To develop a general criterion for spontaneity, we will apply the principles of **thermodynamics,** the science that deals with heat and energy effects. Three different thermodynamic functions are of value in analyzing spontaneity.

1. ΔH, the change in enthalpy (Chapter 8); a *negative* value of ΔH tends to make a reaction spontaneous.
2. ΔS, the change in entropy (Section 16-2); a *positive* value of ΔS tends to make a reaction spontaneous.
3. ΔG, the change in free energy (Sections 16-3 and 16-4); a reaction at constant temperature and pressure will be *spontaneous* if ΔG is *negative*, no ifs, ands, or buts.

Besides serving as a general criterion for spontaneity, the free energy change can be used to:

- determine the effect of temperature, pressure, and concentration on reaction spontaneity (Section 16-5).
- calculate the equilibrium constant for a reaction (Section 16-6).
- determine whether coupled reactions will be spontaneous (Section 16-7).

16-1 Spontaneous Processes

All of us are familiar with certain **spontaneous processes.** For example,

■ an ice cube melts when added to a glass of water at room temperature

$$H_2O(s) \longrightarrow H_2O(l)$$

■ a mixture of hydrogen and oxygen burns if ignited by a spark ◀ ⓘ

$$2H_2(g) + O_2(g) \longrightarrow 2H_2O(l)$$

■ an iron (steel) tool exposed to moist air rusts

$$2Fe(s) + \tfrac{3}{2}O_2(g) + 3H_2O(l) \longrightarrow 2Fe(OH)_3(s)$$

ⓘ A spark is OK, but a continuous input of energy isn't.

In other words, these three reactions are spontaneous at 25°C and 1 atm.

The word "spontaneous" does not imply anything about how rapidly a reaction occurs. Some spontaneous reactions, notably the rusting of iron, are quite slow. Often a reaction that is spontaneous does not occur without some sort of stimulus to get the reaction started. A mixture of hydrogen and oxygen shows no sign of reaction in the absence of a spark or match. Once started, though, a spontaneous reaction continues by itself without further input of energy from the outside.

If a reaction is spontaneous under a given set of conditions, the reverse reaction must be nonspontaneous. For example, water does not spontaneously decompose to the elements by the reverse of the reaction referred to above.

$$2H_2O(l) \longrightarrow 2H_2(g) + O_2(g) \qquad \text{nonspontaneous}$$

However, it is often possible to bring about a nonspontaneous reaction by supplying energy in the form of work. Electrolysis can be used to decompose water to the elements. Electrical energy must be furnished for the decomposition, perhaps from a storage battery.

Perhaps the simplest way to define spontaneity is to say that **a spontaneous process is one that moves the reaction system toward equilibrium.** A nonspontaneous process moves the system away from equilibrium.

The Energy Factor

Many spontaneous processes proceed with a decrease of energy. Boulders roll downhill, not uphill. A storage battery discharges when you leave your car's headlights on. Extrapolating to chemical reactions, one might guess that spontaneous reactions would be exothermic ($\Delta H < 0$). Marcellin Berthelot (1827–1907) in Paris and Julius Thomsen (1826–1909) in Copenhagen proposed this as a general principle, applicable to all reactions.

Instead it turns out that in almost all cases the reverse is true. Nearly all exothermic chemical reactions are spontaneous at 25°C and 1 atm. Consider, for example, the formation of water from the elements and the rusting of iron:

$$2H_2(g) + O_2(g) \longrightarrow 2H_2O(l) \qquad \Delta H = -571.6 \text{ kJ}$$
$$2Fe(s) + \tfrac{3}{2}O_2(g) + 3H_2O(l) \longrightarrow 2Fe(OH)_3(s) \qquad \Delta H = -788.6 \text{ kJ}$$

For both of these spontaneous reactions, ΔH is a negative quantity.

On the other hand, this simple rule fails for many familiar phase changes. An example of a spontaneous reaction that is not exothermic is the melting of ice. This takes place spontaneously at 1 atm above 0°C, even though it is endothermic:

$$H_2O(s) \longrightarrow H_2O(l) \qquad \Delta H = +6.0 \text{ kJ}$$

There is still another basic objection to using the sign of ΔH as a general criterion for spontaneity. Endothermic reactions that are nonspontaneous at room temperature often become spontaneous when the temperature is raised. Consider,

Figure 16.1 Certain "states" are more probable than others. For example, when you toss a pair of dice, a 7 is much more likely to come up than a 12.

for example, the decomposition of limestone:

$$CaCO_3(s) \longrightarrow CaO(s) + CO_2(g) \qquad \Delta H = +178.3 \text{ kJ}$$

At 25°C and 1 atm, this reaction does not occur. Witness the existence of the white cliffs of Dover and other limestone deposits over eons of time. However, if the temperature is raised to about 1100 K, limestone decomposes to give off carbon dioxide gas at 1 atm. In other words, this endothermic reaction becomes spontaneous at high temperatures. This is true despite the fact that ΔH remains about 178 kJ, nearly independent of temperature.

The Randomness Factor

Clearly, the direction of a spontaneous change is not always determined by the tendency for a system to go to a state of lower energy. There is another natural tendency that must be taken into account to predict the direction of spontaneity. *Nature tends to move spontaneously from a state of lower probability to one of higher probability.* Or, as G. N. Lewis put it,

> Each system which is left to its own, will over time, change toward a condition of maximum probability.

To illustrate what these statements mean, consider a pastime far removed from chemistry: tossing dice. If you've ever shot craps (and maybe even if you haven't), you know that when a pair of dice is thrown, a 7 is much more likely to come up than a 12. Figure 16.1 shows why this is the case. There are six different ways to throw a 7 and only one way to throw a 12. Over time, dice will come up 7 six times as often as 12. A 7 is a state of "high" probability; a 12 is a state of "low" probability.

Now let's consider a process a bit closer to chemistry (Figure 16.2). Two different gases, let us say H_2 and N_2, are originally contained in different glass bulbs, separated by a stopcock. When the stopcock is opened, the two different kinds of molecules distribute themselves evenly between the two bulbs. Eventually, half of the H_2 molecules will end up in the left bulb and half in the right; the same holds for the N_2 molecules. Each gas achieves its own most probable distribution, independent of the presence of the other gas.

We could explain the results of this experiment the way we did before; the final distribution is clearly much more probable than the initial distribution. There is, however, another useful way of looking at this process. The system has gone from a highly ordered state (all the H_2 molecules on the left, all the N_2 molecules on the right) to a more disordered, or random, state in which the molecules are distributed evenly between the two bulbs. The same situation holds when marbles rather than molecules are mixed (Figure 16.3). *In general, nature tends to move spontaneously from more ordered to more random states.*

Closed

Open

Figure 16.2 Spontaneous change. Different kinds of gas molecules mix spontaneously, going from a more ordered to a more random state.

Charles Steele

Figure 16.3 Probability. Some states are much more probable than others. If you shake red and black marbles with each other, the random distribution at the left is much more probable than the highly ordered distribution at the right.

This statement is quite easy for parents or students to understand. Your room tends to get messy because an ordered room has few options for objects to be moved around (leaving socks on the floor is not an option for an orderly room). The comedian Bill Cosby insists that with an army of 80 two-year-olds he could take over any country in the world, because they have a remarkable ability for disorganization.

16-2 Entropy, S

Entropy is often described as a measure of disorder or randomness. While useful, these terms are subjective and should be used cautiously. It is better to think about entropic changes in terms of the change in the number of *microstates* of the system. Microstates are different ways in which molecules can be distributed. An increase in the number of possible microstates (i.e., disorder) results in an increase of entropy. Entropy treats the randomness factor quantitatively. Rudolf Clausius gave it the symbol S for no particular reason. In general, the more random the state, the larger the number of its possible microstates, the more probable the state, thus the greater its entropy.

Entropy, like enthalpy (Chapter 8), is a state property. That is, the entropy depends only on the state of a system, not on its history. The **entropy change** is determined by the entropies of the final and initial states, not on the path followed from one state to another.

$$\Delta S = S_{final} - S_{initial}$$

Several factors influence the amount of entropy that a system has in a particular state. In general,

- *a liquid has a higher entropy than the solid from which it is formed.* In a solid the atoms or molecules are confined to fixed positions, so the number of microstates is small. In the liquid, these atoms or molecules can occupy many more positions as they move away from the lattice. Thus, the number of microstates increases and there are then many more ways to arrange the particles. The phase transition is from an "ordered" to a "disordered" system.
- *a gas has a higher entropy than the liquid from which it is formed.* When vaporization occurs, the particles acquire greater freedom to move about. They are distributed throughout the entire container instead of being restricted to a small volume. In vaporization, the number of microstates increases.
- *increasing the temperature of a substance increases its entropy.* Raising the temperature increases the kinetic energy of the molecules (or atoms or ions) and hence their freedom of motion. In the solid, the molecules vibrate with a greater amplitude at higher temperatures. In a liquid or a gas, they move about more rapidly. ◂ⓘ

ⓘ One way or another, heating a substance increases its entropy.

These effects are shown in Figure 16.4, where the entropy of ammonia, NH_3, is plotted versus temperature. Note that the entropy of solid ammonia at 0 K is

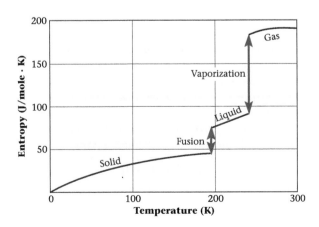

Figure 16.4 Molar entropy of ammonia as a function of temperature. Note the large increases in entropy upon fusion (melting) and vaporization.

zero. This reflects the fact that molecules are completely ordered in the solid state at this temperature; there is no randomness whatsoever. More generally, the **third law of thermodynamics** ◂ 𝓲 tells us that *a completely ordered pure crystalline solid has an entropy of zero at 0 K.*

We'll get to the second law shortly.

Notice from the figure that the effect of temperature on entropy is due almost entirely to phase changes. The slope of the curve is small in regions where only one phase is present. In contrast, there is a large jump in entropy when the solid melts and an even larger one when the liquid vaporizes. This behavior is typical of all substances; melting and vaporization are accompanied by relatively large increases in entropy.

EXAMPLE 16.1 CONCEPTUAL	

Predict whether ΔS is positive or negative for each of the following processes:

(a) taking dry ice from a freezer where its temperature is $-80°C$ and allowing it to warm to room temperature.

(b) dissolving bromine in hexane.

(c) condensing gaseous bromine to liquid bromine.

STRATEGY	

1. Consider the relative disorder of final and initial states.

2. Recall that entropy increases from solid to liquid to gas.

SOLUTION	
(a) Dry ice warming	Increase in temperature and a phase change; $\Delta S > 0$
(b) Dissolving bromine	A solution with two different molecules is more random than a solution with only one kind of molecule; $\Delta S > 0$
(c) $Br_2(g) \longrightarrow Br_2(l)$	A phase change from gas to liquid; $\Delta S < 0$

Decomposition of ammonium nitrate. The value of ΔS for this reaction,

$$NH_4NO_3(s) \longrightarrow N_2O(g) + 2H_2O(g)$$

is positive. That can be predicted because a solid decomposes to gases.

Standard Molar Entropies

The entropy of a substance, unlike its enthalpy, can be evaluated directly. The details of how this is done are beyond the level of this text, but Figure 16.4 shows the results for one substance, ammonia. From such a plot you can read off the **standard molar entropy** at 1 atm pressure and any given temperature, most often 25°C. This quantity is given the symbol $S°$ and has the units of joules per mole per kelvin (J/mol·K). From Figure 16.4, it appears that

$$S° \ NH_3(g) \ at \ 25°C \approx 192 \ J/mol·K$$

Standard molar entropies of elements, compounds, and aqueous ions are listed in Table 16.1. Notice that

- *elements have nonzero standard entropies.* This means that in calculating the standard entropy change for a reaction, $\Delta S°$, elements as well as compounds must be taken into account.
- *standard molar entropies of pure substances* (elements and compounds) *are always positive quantities* ($S° > 0$).
- *aqueous ions may have negative $S°$ values.* This is a consequence of the arbitrary way in which ionic entropies are defined, taking

$$S° \ H^+(aq) = 0$$

The fluoride ion has a standard entropy 13.8 units less than that of H^+; hence $S° \ F^-(aq) = -13.8 \ J/mol·K.$

Table 16.1 Standard Entropies at 25°C (J/mol·K) of Elements and Compounds at 1 atm, Aqueous Ions at 1 M

Elements							
$Ag(s)$	42.6	$Cl_2(g)$	223.0	$I_2(s)$	116.1	$O_2(g)$	205.0
$Al(s)$	28.3	$Cr(s)$	23.8	$K(s)$	64.2	$Pb(s)$	64.8
$Ba(s)$	62.8	$Cu(s)$	33.2	$Mg(s)$	32.7	$P_4(s)$	164.4
$Br_2(l)$	152.2	$F_2(g)$	202.7	$Mn(s)$	32.0	$S(s)$	31.8
$C(s)$	5.7	$Fe(s)$	27.3	$N_2(g)$	191.5	$Si(s)$	18.8
$Ca(s)$	41.4	$H_2(g)$	130.6	$Na(s)$	51.2	$Sn(s)$	51.6
$Cd(s)$	51.8	$Hg(l)$	76.0	$Ni(s)$	29.9	$Zn(s)$	41.6

Compounds							
$AgBr(s)$	107.1	$CaCl_2(s)$	104.6	$H_2O(g)$	188.7	$NH_4NO_3(s)$	151.1
$AgCl(s)$	96.2	$CaCO_3(s)$	92.9	$H_2O(l)$	69.9	$NO(g)$	210.7
$AgI(s)$	115.5	$CaO(s)$	39.8	$H_2O_2(l)$	109.6	$NO_2(g)$	240.0
$AgNO_3(s)$	140.9	$Ca(OH)_2(s)$	83.4	$H_2S(g)$	205.7	$N_2O_4(g)$	304.2
$Ag_2O(s)$	121.3	$CaSO_4(s)$	106.7	$H_2SO_4(l)$	156.9	$NaCl(s)$	72.1
$Al_2O_3(s)$	50.9	$CdCl_2(s)$	115.3	$HgO(s)$	70.3	$NaF(s)$	51.5
$BaCl_2(s)$	123.7	$CdO(s)$	54.8	$KBr(s)$	95.9	$NaOH(s)$	64.5
$BaCO_3(s)$	112.1	$Cr_2O_3(s)$	81.2	$KCl(s)$	82.6	$NiO(s)$	38.0
$BaO(s)$	70.4	$CuO(s)$	42.6	$KClO_3(s)$	143.1	$PbBr_2(s)$	161.5
$BaSO_4(s)$	132.2	$Cu_2O(s)$	93.1	$KClO_4(s)$	151.0	$PbCl_2(s)$	136.0
$CCl_4(l)$	216.4	$CuS(s)$	66.5	$KNO_3(s)$	133.0	$PbO(s)$	66.5
$CHCl_3(l)$	201.7	$Cu_2S(s)$	120.9	$MgCl_2(s)$	89.6	$PbO_2(s)$	68.6
$CH_4(g)$	186.2	$CuSO_4(s)$	107.6	$MgCO_3(s)$	65.7	$PCl_3(g)$	311.7
$C_2H_2(g)$	200.8	$Fe(OH)_3(s)$	106.7	$MgO(s)$	26.9	$PCl_5(g)$	364.5
$C_2H_4(g)$	219.5	$Fe_2O_3(s)$	87.4	$Mg(OH)_2(s)$	63.2	$SiO_2(s)$	41.8
$C_2H_6(g)$	229.5	$Fe_3O_4(s)$	146.4	$MgSO_4(s)$	91.6	$SnO_2(s)$	52.3
$C_3H_8(g)$	269.9	$HBr(g)$	198.6	$MnO(s)$	59.7	$SO_2(g)$	248.1
$CH_3OH(l)$	126.8	$HCl(g)$	186.8	$MnO_2(s)$	53.0	$SO_3(g)$	256.7
$C_2H_5OH(l)$	160.7	$HF(g)$	173.7	$NH_3(g)$	192.3	$ZnI_2(s)$	161.1
$CO(g)$	197.6	$HI(g)$	206.5	$N_2H_4(l)$	121.2	$ZnO(s)$	43.6
$CO_2(g)$	213.6	$HNO_3(l)$	155.6	$NH_4Cl(s)$	94.6	$ZnS(s)$	57.7

Cations				Anions			
$Ag^+(aq)$	72.7	$Hg^{2+}(aq)$	−32.2	$Br^-(aq)$	82.4	$HPO_4^{2-}(aq)$	−33.5
$Al^{3+}(aq)$	−321.7	$K^+(aq)$	102.5	$CO_3^{2-}(aq)$	−56.9	$HSO_4^-(aq)$	131.8
$Ba^{2+}(aq)$	9.6	$Mg^{2+}(aq)$	−138.1	$Cl^-(aq)$	56.5	$I^-(aq)$	111.3
$Ca^{2+}(aq)$	−53.1	$Mn^{2+}(aq)$	−73.6	$ClO_3^-(aq)$	162.3	$MnO_4^-(aq)$	191.2
$Cd^{2+}(aq)$	−73.2	$Na^+(aq)$	59.0	$ClO_4^-(aq)$	182.0	$NO_2^-(aq)$	123.0
$Cu^+(aq)$	40.6	$NH_4^+(aq)$	113.4	$CrO_4^{2-}(aq)$	50.2	$NO_3^-(aq)$	146.4
$Cu^{2+}(aq)$	−99.6	$Ni^{2+}(aq)$	−128.9	$Cr_2O_7^{2-}(aq)$	261.9	$OH^-(aq)$	−10.8
$Fe^{2+}(aq)$	−137.7	$Pb^{2+}(aq)$	10.5	$F^-(aq)$	−13.8	$PO_4^{3-}(aq)$	−222
$Fe^{3+}(aq)$	−315.9	$Sn^{2+}(aq)$	−17.4	$HCO_3^-(aq)$	91.2	$S^{2-}(aq)$	−14.6
$H^+(aq)$	0.0	$Zn^{2+}(aq)$	−112.1	$H_2PO_4^-(aq)$	90.4	$SO_4^{2-}(aq)$	20.1

As a group, gases have higher entropies than liquids or solids. Moreover, among substances of similar structure and physical state, entropy usually increases with molar mass. Compare, for example, the hydrocarbons

$$CH_4(g) \qquad S° = 186.2 \text{ J/mol·K}$$
$$C_2H_6(g) \qquad S° = 229.5 \text{ J/mol·K}$$
$$C_3H_8(g) \qquad S° = 269.9 \text{ J/mol·K}$$

As the molecule becomes more complex, there are more ways for the atoms to move about with respect to one another, leading to a higher entropy.

$\Delta S°$ for Reactions

Table 16.1 can be used to calculate the **standard entropy change**, $\Delta S°$, for reactions, from the relation

$$\Delta S° = \sum S°_{products} - \sum S°_{reactants} \qquad (16.1)$$

In taking these sums, the standard molar entropies are multiplied by the number of moles specified in the balanced chemical equation.

To show how this relation is used, consider the reaction

$$CaCO_3(s) \longrightarrow CaO(s) + CO_2(g)$$
$$\Delta S° = S° \, CaO(s) + S° \, CO_2(g) - S° \, CaCO_3(s)$$
$$\Delta S° = 1 \, mol\left(39.8 \, \frac{J}{mol \cdot K}\right) + 1 \, mol\left(213.6 \, \frac{J}{mol \cdot K}\right) - 1 \, mol\left(92.9 \, \frac{J}{mol \cdot K}\right)$$
$$= 39.8 \, J/K + 213.6 \, J/K - 92.9 \, J/K = +160.5 \, J/K$$

The units of $S°$ are J/mol·K: Those of $\Delta S°$ are J/K. Why the difference?

Observe that $\Delta S°$ for the decomposition of calcium carbonate is a positive quantity. This is reasonable because the gas formed, CO_2, has a much higher molar entropy than either of the solids, CaO or $CaCO_3$. As a matter of fact, *a reaction that results in an increase in the number of moles of gas is almost always accompanied by an increase in entropy. Conversely, if the number of moles of gas decreases, $\Delta S°$ is a negative quantity.* Consider, for example, the reaction

If there is no change to the number of moles of gas, ΔS is usually small.

$$2H_2(g) + O_2(g) \longrightarrow 2H_2O(l)$$
$$\Delta S° = 2S° \, H_2O(l) - 2S° \, H_2(g) - S° \, O_2(g)$$
$$= 139.8 \, J/K - 261.2 \, J/K - 205.0 \, J/K = -326.4 \, J/K$$

EXAMPLE 16.2

Consider calcium hydroxide, commonly called slaked lime. It is used as the lye substitute in "no-lye" hair relaxers. It is also found in most depilatory creams.

a Calculate $\Delta S°$ for dissolving one mole of calcium hydroxide in water.

STRATEGY

1. Write a balanced equation for dissolving $Ca(OH)_2$.
2. Find $\Delta S°$ by substituting $S°$ values found in Table 16.1 into Equation 16.1.

SOLUTION

1. Equation	$Ca(OH)_2(s) \longrightarrow Ca^{2+}(aq) + 2OH^-(aq)$
2. $\Delta S°$	$\Delta S° = S° \, Ca^{2+}(aq) + 2S° \, OH^-(aq) - S° \, Ca(OH)_2(s)$
	$= 1 \, mol\left(\frac{-53.1 \, J}{mol \cdot K}\right) + 2 \, mol\left(\frac{-10.8 \, J}{mol \cdot K}\right) - 1 \, mol\left(\frac{+83.4 \, J}{mol \cdot K}\right) = -158.1 \, J/K$

b Calculate $\Delta S°$ for the formation of one gram of $Ca(OH)_2$ by reacting lime (CaO) with water.

STRATEGY

1. Write a balanced equation for the reaction.
2. Find $\Delta S°$ by substituting $S°$ values from Table 16.1 into Equation 16.1.
   Note that the value you obtained is for the difference in entropy for *one mole*.
3. Use $\Delta S°$ for one mole as a conversion factor to obtain $\Delta S°$ for one gram of $Ca(OH)_2$.

continued

	SOLUTION
1. Equation	$CaO(s) + H_2O \longrightarrow Ca(OH)_2(s)$
2. $\Delta S°$ for one mole	$\Delta S° = S° \, Ca(OH)_2 - [S° \, CaO + S° \, H_2O]$
	$\Delta S° = 1 \text{ mol} \left(\dfrac{+83.4 \text{ J}}{\text{mol} \cdot \text{K}} \right) - \left[1 \text{ mol} \left(\dfrac{+39.8 \text{ J}}{\text{mol} \cdot \text{K}} \right) + 1 \text{ mol} \left(\dfrac{+69.9 \text{ J}}{\text{mol} \cdot \text{K}} \right) \right]$
	$= -26.3 \text{ J/K}$
3. $\Delta S°$ for one gram	$\dfrac{-26.3 \text{ J/K}}{1 \text{ mol } Ca(OH)_2} \times \dfrac{1 \text{ mol } Ca(OH)_2}{74.10 \text{ g}} = -0.355 \text{ J/K}$

The Second Law of Thermodynamics

The relationship between entropy change and spontaneity can be expressed through a basic principle of nature known as the **second law of thermodynamics**. One way to state this law is to say that *in a spontaneous process, there is a net increase in entropy, taking into account both system and surroundings.* ◄ ⓘ That is,

$$\Delta S_{universe} = (\Delta S_{system} + \Delta S_{surroundings}) > 0 \qquad \text{spontaneous process}$$

ⓘ In this sense, the universe is running down.

(Recall from Chapter 8 that the system is that portion of the universe on which attention is focused; the surroundings include everything else.)

Notice that the second law refers to the total entropy change, involving both system and surroundings. For many spontaneous processes, the entropy change for the system is a *negative* quantity. Consider, for example, the rusting of iron, a spontaneous process:

$$2Fe(s) + \tfrac{3}{2}O_2(g) + 3H_2O(l) \longrightarrow 2Fe(OH)_3(s)$$

$\Delta S°$ for this system at 25°C and 1 atm can be calculated from a table of standard entropies; it is found to be -358.4 J/K. The negative sign of $\Delta S°$ is entirely consistent with the second law. All the law requires is that the entropy change of the surroundings be greater than 358.4 J/K, so that $\Delta S_{universe} > 0$.

In principle, the second law can be used to determine whether a reaction is spontaneous. To do that, however, requires calculating the entropy change for the surroundings, which is not easy. We follow a conceptually simpler approach (Section 16-3), which deals only with the thermodynamic properties of chemical *systems*.

16-3 Free Energy, *G*

As pointed out earlier, two thermodynamic quantities affect reaction spontaneity. One of these is the enthalpy, *H;* the other is the entropy, *S.* The problem is to put these two quantities together in such a way as to arrive at a single function whose sign will determine whether a reaction is spontaneous. This problem was first solved more than a century ago by J. Willard Gibbs, who introduced a new quantity, now called the *Gibbs free energy* and given the symbol *G.* Gibbs showed that for a reaction taking place at constant pressure and temperature, the **free energy of formation,** ΔG represents that portion of the total energy change that is available (i.e., "free") to do useful work. ◄ ⓘ If, for example, ΔG for a reaction is -270 kJ, it is possible to obtain 270 kJ of useful work from the reaction. Conversely, if ΔG is $+270$ kJ, at least that much energy in the form of work must be supplied to make the reaction take place.

ⓘ A spontaneous process is capable of producing useful work.

The basic definition of the Gibbs free energy is

$$G = H - TS$$

where *T* is the absolute (Kelvin) temperature. The **free energy** of a substance, like its enthalpy and entropy, is a state property; its value is determined only by the

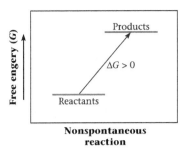

Figure 16.5 Sign of free energy and spontaneity. For a spontaneous reaction, the free energy of the products is less than that of the reactants: $\Delta G < 0$. For a nonspontaneous reaction, the reverse is true, $\Delta G > 0$.

Spontaneous reaction **Nonspontaneous reaction**

state of a system, not by how it got there. Putting it another way, the **free energy change**, ΔG for a reaction depends only on the nature of products and reactants and the conditions (temperature, pressure, and concentration). It does *not* depend on the path by which the reaction is carried out.

The sign of the free energy change can be used to determine the spontaneity of a reaction carried out at constant temperature and pressure.

1. *If ΔG is negative, the reaction is spontaneous.*
2. *If ΔG is positive, the reaction will not take place spontaneously.* Instead, the reverse reaction will be spontaneous.
3. *If ΔG is 0, the system is at equilibrium;* there is no tendency for reaction to occur in either direction.

A system has its minimum free energy at equilibrium.

In other words, ΔG is a measure of the driving force of a reaction. *Reactions, at constant pressure and temperature, go in such a direction as to decrease the free energy of the system.* This means that the direction in which a reaction takes place is determined by the relative free energies of products and reactants. If the products at the specified conditions of temperature, pressure, and concentration have a lower free energy than the reactants ($G_{\text{products}} < G_{\text{reactants}}$), the forward reaction will occur (Figure 16.5). If the reverse is true ($G_{\text{reactants}} < G_{\text{products}}$), the reverse reaction is spontaneous. Finally, if $G_{\text{products}} = G_{\text{reactants}}$, there is no driving force to make the reaction go in either direction.

Relation Among ΔG, ΔH, and ΔS

From the defining equation for free energy, it follows that at constant temperature

$$\Delta G = \Delta H - T\Delta S$$

where ΔG, ΔH, and ΔS are the changes in free energy, enthalpy, and entropy, respectively, for a reaction. This relation, known as the **Gibbs-Helmholtz equation**, is perhaps the most important equation in chemical thermodynamics. As you can see, two factors tend to make ΔG negative and hence lead to a spontaneous reaction:

1. *A negative value of ΔH.* Exothermic reactions ($\Delta H < 0$) tend to be spontaneous, inasmuch as they contribute to a negative value of ΔG. On the molecular level, this means that there will be a tendency to form "strong" bonds at the expense of "weak" ones.
2. *A positive value of ΔS.* If the entropy change is positive ($\Delta S > 0$), the term $-T\Delta S$ will make a negative contribution to ΔG. Hence there will be a tendency for a reaction to be spontaneous if the products are less ordered than the reactants.

In many physical changes, the entropy increase is the major driving force. This situation applies when two liquids with similar intermolecular forces, such as benzene (C_6H_6) and toluene (C_7H_8), are mixed. There is no change in enthalpy, but the entropy increases because the molecules of benzene and toluene are mixed randomly in solution.*

---

*The formation of a *water* solution is often accompanied by a *decrease* in entropy because hydrogen bonding or hydration effects lead to a highly ordered solution structure. Recall from Example 16.2 that when one mole of $Ca(OH)_2$ dissolves in water, $\Delta S° = -158.1$ J/K.

CHEMISTRY THE **HUMAN** SIDE

Two theoreticians working in the latter half of the nineteenth century changed the very nature of chemistry by deriving the mathematical laws that govern the behavior of matter undergoing physical or chemical change. One of these was James Clerk Maxwell, whose contributions to kinetic theory were discussed in Chapter 5. The other was J. Willard Gibbs, Professor of Mathematical Physics at Yale from 1871 until his death in 1903.

In 1876, Gibbs published the first portion of a remarkable paper in the *Transactions of the Connecticut Academy of Sciences* titled, "On the Equilibrium of Heterogeneous Substances." When the paper was completed in 1878 (it was 323 pages long), the foundation was laid for the science of chemical thermodynamics. For the first time, the concept of free energy appeared. Included as well were the basic principles of chemical equilibrium (Chapter 12), phase equilibrium (Chapter 9), and the relations governing energy changes in electrical cells (Chapter 17).

If Gibbs had never published another paper, this single contribution would have placed him among the greatest theoreticians in the history of science. Generations of experimental scientists have established their reputations by demonstrating in the laboratory the validity of the relationships that Gibbs derived at his desk. Many of these relationships were rediscovered by others; an example is the Gibbs-Helmholtz equation developed in 1882 by Hermann von Helmholtz (1821–1894), a prestigious German physiologist and physicist who was completely unaware of Gibbs's work.

Gibbs is often cited as an example of the "prophet without honor in his own country." His colleagues in New Haven and elsewhere in the United States seem not to have realized the significance of his work until late in his life. During his first ten years as a professor at Yale he received no salary. They started paying him when he got an offer from Johns Hopkins. In 1920, when he was first proposed for the Hall of Fame of Distinguished Americans at New York University, he received 9 votes out of a possible 100. Not until 1950 was he elected to that body. Even today the name of Gibbs is generally unknown among educated Americans outside of those interested in the natural sciences.

J. Willard Gibbs
(1839–1903)

Admittedly, Gibbs himself was largely responsible for the fact that for many years his work did not attract the attention it deserved. He made little effort to publicize it; the *Transactions of the Connecticut Academy of Sciences* was hardly the leading scientific journal of its day. Gibbs was one of those rare individuals who seem to have no inner need for recognition by contemporaries. His satisfaction came from solving a problem in his mind; having done so, he was ready to proceed to other problems. His papers are not easy to read; he seldom cites examples to illustrate his abstract reasoning. Frequently, the implications of the laws that he derives are left for the readers to grasp on their own. One of his colleagues at Yale confessed many years later that none of the members of the Connecticut Academy of Sciences understood his paper on thermodynamics; as he put it, "We knew Gibbs and took his contributions on faith."

In certain reactions, ΔS is nearly zero, and ΔH is the only important component of the driving force for spontaneity. An example is the synthesis of hydrogen fluoride from the elements

$$\tfrac{1}{2}H_2(g) + \tfrac{1}{2}F_2(g) \longrightarrow HF(g)$$

For this reaction, ΔH is a large negative number, -271.1 kJ, showing that the bonds in HF are stronger than those in the H_2 and F_2 molecules. As you might expect for a gaseous reaction in which there is no change in the number of moles, ΔS is very small, about 0.0070 kJ/K. The free energy change, ΔG, at 1 atm is -273.2 kJ at 25°C, almost identical to ΔH. Even at very high temperatures, the difference between ΔG and ΔH is small, amounting to only about 14 kJ at 2000 K.

16-4 Standard Free Energy Change, $\Delta G°$

Although the Gibbs-Helmholtz equation is valid under all conditions we will apply it only under *standard conditions,* where

- gases are at one atmosphere partial pressure.
- ions or molecules in solution are at one molar concentration. ◀ ⓘ

In other words, we will use the equation in the form

$$\Delta G° = \Delta H° - T\Delta S° \qquad (16.2)$$

ⓘ "Standard conditions" has quite a different meaning from "STP."

415

Figure 16.6 Precipitation of silver chloride. A spontaneous reaction for which $\Delta G°$ is negative.

where $\Delta G°$ is the **standard free energy change** (1 atm, 1 M); $\Delta H°$ is the standard enthalpy change, which can be calculated from heats of formation, $\Delta H_f°$ (listed in Table 8.3, Chapter 8); and $\Delta S°$ is the standard entropy change (Table 16.1).

Recall that the sign of ΔG correlates with the spontaneity of reaction. We can do the same thing with $\Delta G°$ provided we restrict our attention to standard conditions (1 atm, 1 M).

1. **If $\Delta G°$ is negative, the reaction is spontaneous at standard conditions.** For example, the following reaction is spontaneous at 25°C:

$$CaO(s) + CO_2(g, 1 \text{ atm}) \longrightarrow CaCO_3(s) \quad \Delta G° \text{ at } 25°C = -130.4 \text{ kJ}$$

2. **If $\Delta G°$ is positive, the reaction is nonspontaneous at standard conditions.** The reaction

$$AgCl(s) \longrightarrow Ag^+(aq, 1 M) + Cl^-(aq, 1 M) \quad \Delta G° \text{ at } 25°C = +55.7 \text{ kJ}$$

is nonspontaneous at 25°C. The reverse reaction is spontaneous; when solutions of $AgNO_3$ and HCl are mixed in such a way that $[Ag^+] = [Cl^-] = 1 M$, silver chloride precipitates (Figure 16.6).

3. **If $\Delta G°$ is 0, the system is at equilibrium at standard conditions;** there is no tendency for the reaction to occur in either direction. An example is the vaporization of water at 100°C and 1 atm:

$$H_2O(l) \longrightarrow H_2O(g, 1 \text{ atm}) \qquad \Delta G° = 0$$

under these conditions (i.e., at the normal boiling point), the molar free energies of liquid and gaseous water are identical. Hence $\Delta G° = 0$ and the system is at equilibrium.

EXAMPLE 16.3

Calcium sulfate, $CaSO_4$, is used as a drying agent and sold under the trade name Drierite. The reaction for dissolving calcium sulfate is

$$CaSO_4(s) \longrightarrow Ca^{2+}(aq) + SO_4{}^{2-}(aq)$$

a | Find $\Delta H°$ for the reaction.

ANALYSIS	
Information given:	equation for the reaction ($CaSO_4(s) \longrightarrow Ca^{2+}(aq) + SO_4{}^{2-}(aq)$)
Information implied:	Table 8.3 ($\Delta H_f°$ values)
Asked for:	$\Delta H°$

STRATEGY	

1. Recall Equation 8.4 to determine $\Delta H°$.

$$\Delta H° = \Sigma \Delta H_f°{}_{\text{products}} - \Sigma \Delta H_f°{}_{\text{reactants}}$$

2. Obtain $\Delta H_f°$ values from Table 8.3 and substitute into Equation 8.4.

SOLUTION	
$\Delta H°$	$\Delta H° = \Delta H_f° Ca^{2+}(aq) + \Delta H_f° SO_4{}^{2-}(aq) - \Delta H_f° CaSO_4(s)$
	$= -542.8 \text{ kJ} - 909.3 \text{ kJ} - (-1434.1 \text{ kJ}) = -18.0 \text{ kJ}$

b | Find $\Delta S°$ for the reaction.

ANALYSIS	
Information given:	equation for the reaction ($CaSO_4 \longrightarrow Ca^{2+}(aq) + SO_4{}^{2-}(aq)$)
Information implied:	Table 16.1 ($S°$ values)
Asked for:	$\Delta S°$

continued

	STRATEGY
	Obtain $S°$ values from Table 16.1 and substitute into Equation 16.1.
	SOLUTION
$\Delta S°$	$\Delta S° = S°\ Ca^{2+}(aq) + S°\ SO_4{}^{2-}(aq) - S°\ CaSO_4(s)$
	$= -53.1\ J/K + 20.1\ J/K - 106.7\ J/K = -139.7\ J/K$

c	Find $\Delta G°$ at 25°C for the reaction.
	ANALYSIS
Information given:	From part (a): $\Delta H°(-18.0\ kJ)$
	From part (b): $\Delta S°(-139.7\ J/K)$
Asked for:	$\Delta G°$
	STRATEGY
	1. Convert $\Delta S°$ into kJ and °C to K.
	2. Substitute into the Gibbs-Helmholtz equation (Equation 16.2).
	SOLUTION
1. $\Delta S°$ in kJ; T in K	$\Delta S° = -139.7\ J/K = -0.1397\ kJ/K$; 25°C = 298 K
2. $\Delta G°$	$\Delta G° = \Delta H° - T\Delta S° = -18.0\ kJ - 298\ K(-0.1397\ kJ/K) = 23.6\ kJ$
	END POINT

$\Delta G°$ is positive, so this reaction at standard conditions

$$CaSO_4(s) \longrightarrow Ca^{2+}(aq, 1\ M) + SO_4{}^{2-}(aq, 1\ M)$$

should not be spontaneous. In other words, calcium sulfate should not dissolve in water to give a 1 M solution. This is indeed the case. The solubility of $CaSO_4$ at 25°C is considerably less than 1 mol/L.

Calculation of $\Delta G°$ at 25°C; Free Energies of Formation

To illustrate the use of the Gibbs-Helmholtz equation (16.2), we apply it first to find $\Delta G°$ for a reaction at 25°C (Example 16.3). To do this, the units of $\Delta H°$ and $\Delta S°$ must be consistent. Using $\Delta H°$ in kilojoules, it is necessary to convert $\Delta S°$ from joules per kelvin to *kilojoules per kelvin*.

The Gibbs-Helmholtz equation can be used to calculate the **standard free energy of formation** of a compound. This quantity, $\Delta G_f°$, is analogous to the enthalpy of formation, $\Delta H_f°$. ◀ⓘ It is defined as the free energy change per mole when a compound is formed from the elements in their stable states at 1 atm.

ⓘ $\Delta G_f°$ for an element, like $\Delta H_f°$, is zero.

Tables of standard free energies of formation at 25°C of compounds and ions in solution are given in Appendix 1 (along with standard heats of formation and standard entropies). Notice that, for most compounds, $\Delta G_f°$ is a negative quantity, which means that the compound can be formed spontaneously from the elements. This is true for water:

$$H_2(g) + \tfrac{1}{2}O_2(g) \longrightarrow H_2O(l) \qquad \Delta G_f°\ H_2O(l) = -237.2\ kJ/mol$$

and, at least in principle, for methane:

$$C(s) + 2H_2(g) \longrightarrow CH_4(g) \qquad \Delta G_f°\ CH_4(g) = -50.7\ kJ/mol$$

A few compounds, including acetylene, have positive free energies of formation ($\Delta G_f°\ C_2H_2(g) = +209.2\ kJ/mol$). These compounds cannot be made from the elements at ordinary temperatures and pressures; indeed, they are potentially unstable with respect to the elements. In the case of acetylene, the reaction

$$C_2H_2(g) \longrightarrow 2C(s) + H_2(g) \qquad \Delta G° \text{ at } 25°C = -209.2\ kJ$$

occurs with explosive violence unless special precautions are taken.

Values of ΔG_f° can be used to calculate free energy changes for reactions. The relationship is entirely analogous to that for enthalpies in Chapter 8:

$$\Delta G^\circ{}_{\text{reaction}} = \sum \Delta G_f^\circ{}_{\text{products}} - \sum \Delta G_f^\circ{}_{\text{reactants}} \qquad (16.3)$$

If you calculate ΔG° in this way, you should keep in mind an important limitation. $\Delta G^\circ{}_{\text{reaction}}$ *is valid only at the temperature at which ΔG_f° data are tabulated, in this case 25°C.* ◄ⓘ ΔG° varies considerably with temperature, so this approach is not even approximately valid at other temperatures.

To find ΔG° at temperatures other than 25°C, use the Gibbs-Helmholtz equation.

EXAMPLE 16.4

Using ΔG_f° values from Appendix 1, calculate the standard free energy change at 25°C for the reaction referred to in Example 16.3.

ANALYSIS

Information given:	equation for the reaction ($CaSO_4(s) \longrightarrow Ca^{2+}(aq) + SO_4^{2-}(aq)$)
Information implied:	ΔG_f° values (Appendix 1)
Asked for:	ΔG°

STRATEGY

Obtain ΔG_f° values from Appendix 1 and substitute into Equation 16.3.

SOLUTION

ΔG°	$\Delta G^\circ = \Delta G_f^\circ\, Ca^{2+}(aq) + \Delta G_f^\circ\, SO_4^{2-}(aq) - \Delta G_f^\circ\, CaSO_4(s)$ $= -553.6 \text{ kJ} - 744.5 \text{ kJ} + 1321.8 \text{ kJ} = +23.7 \text{ kJ}$

END POINT

Notice that the value of ΔG° at 25°C is essentially identical to that obtained in Example 16.3, which is reassuring.

Calculation of ΔG° at Other Temperatures

To a good degree of approximation, the temperature variation of ΔH° and ΔS° can be neglected.* This means that to apply the Gibbs-Helmholtz equation

$$\Delta G^\circ = \Delta H^\circ - T\Delta S^\circ$$

at temperatures other than 25°C, you need only change the value of T. The quantities ΔH° and ΔS° can be calculated in the usual way from tables of standard enthalpies and entropies.

---

*As far as the Gibbs-Helmholtz equation is concerned, there is another reason for ignoring the temperature dependence of ΔH and ΔS. These two quantities always change in the same direction as the temperature changes (i.e., if ΔH becomes more positive, so does ΔS). Hence the two effects tend to cancel each other.

EXAMPLE 16.5

Iron, a large component of steel, is obtained by reducing iron(III) oxide (present in hematite ore) with hydrogen in a blast furnace. Steam is a byproduct of the reaction. Calculate ΔG° at 230°C for the reduction of one mole of Fe_2O_3.

ANALYSIS

Information given:	n Fe_2O_3 (one mole); temperature (230°C)
Information implied:	Table 8.3 (ΔH_f° values) Table 16.1 (S° values)
Asked for:	ΔG°

continued

STRATEGY

1. Write a balanced equation for the reaction.

2. Find ΔH_f° values in Table 8.3 (or Appendix 1) and substitute into Equation 8.3 to obtain ΔH°.

3. Find S° values in Table 16.1 (or Appendix 1) and substitute into Equation 16.1 to obtain ΔS°. (Remember to convert J/K to kJ/K.)

4. Change °C to K and substitute the values for ΔH° and ΔS° into the Gibbs-Helmholtz equation (Equation 16.2) to obtain ΔG°.

SOLUTION

1. Equation	$Fe_2O_3(s) + 3H_2(g) \longrightarrow 2Fe(s) + 3H_2O(g)$
2. ΔH°	$\Delta H^\circ = 3\Delta H_f^\circ\ H_2O(g) - \Delta H_f^\circ\ Fe_2O_3(s) = -725.4\ kJ + 824.2\ kJ = +98.8\ kJ$
3. ΔS°	$\Delta S^\circ = 3S^\circ\ H_2O(g) + 2S^\circ\ Fe(s) - 3S^\circ\ H_2(g) - S^\circ\ Fe_2O_3(s)$
	$\quad = 566.1\ J/K + 54.6\ J/K - 391.8\ J/K - 87.4\ J/K = +141.5\ J/K = +0.1415\ kJ/K$
4. ΔG°	$\Delta G^\circ = \Delta H^\circ - T\Delta S^\circ = 98.8\ kJ - (273 + 230)K\ (0.1415\ kJ/K) = +27.6\ kJ$

Figure 16.7 Variation of ΔG° with T.
For the reaction:

$$Fe_2O_3(s) + 3H_2(g) \longrightarrow 2Fe(s) + 3H_2O(g)$$

ΔG° decreases from +98.8 kJ at 0 K (the value of ΔH°), to +27.6 kJ at 503 K, 0.0 kJ at 698 K, and −71.0 kJ at 1200 K.

From Example 16.5 and the preceding discussion, it should be clear that ΔG°, unlike ΔH° and ΔS°, is strongly dependent on temperature. This comes about, of course, because of the T in the Gibbs-Helmholtz equation:

$$\Delta G^\circ = \Delta H^\circ - T\Delta S^\circ$$

Comparing this equation with that of a straight line,

$$y = b + mx$$

it is clear that a plot of ΔG° versus T should be linear, with a slope of $-\Delta S^\circ$ and a y-intercept (at 0 K) of ΔH° (Figure 16.7).

16-5 Effect of Temperature, Pressure, and Concentration on Reaction Spontaneity

A change in reaction conditions can, and often does, change the direction in which a reaction occurs spontaneously. The Gibbs-Helmholtz equation in the form

$$\Delta G^\circ = \Delta H^\circ - T\Delta S^\circ$$

is readily applied to deduce the effect of temperature on reaction spontaneity. To consider the effect of pressure or concentration, we need an analogous expression for the dependence of ΔG on these factors.

Temperature

When the temperature of a reaction system is increased, the sign of $\Delta G°$, and hence the direction in which the reaction proceeds spontaneously, may or may not change. Whether it does or does not depends on the signs of $\Delta H°$ and $\Delta S°$. The four possible situations, deduced from the Gibbs-Helmholtz equation, are summarized in Table 16.2.

If $\Delta H°$ and $\Delta S°$ have opposite signs (Table 16.2, I and II), it is impossible to reverse the direction of spontaneity by a change in temperature alone. The two terms $\Delta H°$ and $-T\Delta S°$ reinforce one another. Hence $\Delta G°$ has the same sign at all temperatures. Reactions of this type are rather uncommon. One such reaction is

$$Cu(s) + H_2O(g) \longrightarrow CuO(s) + H_2(g)$$

Here $\Delta H°$ is $+84.5$ kJ and $\Delta S°$ is -0.0487 kJ/K. Hence

$$\Delta G° = \Delta H° - T\Delta S°$$
$$= +84.5 \text{ kJ} + T(0.0487 \text{ kJ/K})$$

Clearly $\Delta G°$ is positive at all temperatures. The reaction cannot take place spontaneously at 1 atm regardless of temperature. ◄ ⓘ

It is more common to find that $\Delta H°$ and $\Delta S°$ have the same sign (Table 16.2, III and IV). When this happens, the enthalpy and entropy factors oppose each other. $\Delta G°$ changes sign as temperature increases, and the direction of spontaneity reverses. At low temperatures, $\Delta H°$ predominates, and the exothermic reaction, which may be either the forward or the reverse reaction, occurs. As the temperature rises, the quantity $T\Delta S°$ increases in magnitude and eventually exceeds $\Delta H°$. At high temperatures, the reaction that leads to an increase in entropy occurs. In most cases, 25°C is a "low" temperature, at least at a pressure of 1 atm. This explains why exothermic reactions are usually spontaneous at room temperature and atmospheric pressure.

An example of a reaction for which $\Delta H°$ and $\Delta S°$ have the same sign is the decomposition of calcium carbonate: ◄ ⓘ

$$CaCO_3(s) \longrightarrow CaO(s) + CO_2(g)$$

Here $\Delta H° = +178.3$ kJ and $\Delta S° = +160.5$ J/K. Hence

$$\Delta G° = +178.3 \text{ kJ} - T(0.1605 \text{ kJ/K})$$

For this reaction at "low" temperatures, such as 298 K,

$$\Delta G° = +178.3 \text{ kJ} - 47.8 \text{ kJ} = +130.5 \text{ kJ}$$

the $\Delta H°$ term predominates, $\Delta G°$ is positive, and the reaction is nonspontaneous at 1 atm. Conversely, at "high" temperatures, such as 2000 K,

$$\Delta G° = +178.3 \text{ kJ} - 321.0 \text{ kJ} = -142.7 \text{ kJ}$$

the $T\Delta S°$ term predominates, $\Delta G°$ is negative, and the reaction is spontaneous at 1 atm.

ⓘ Copper metal doesn't react with water vapor, period.

ⓘ CaO is made commercially by heating $CaCO_3$.

Table 16.2 Effect of Temperature on Reaction Spontaneity

	$\Delta H°$	$\Delta S°$	$\Delta G° = \Delta H° - T\Delta S°$	Remarks
I	−	+	Always −	Spontaneous at all T; reverse reaction always nonspontaneous
II	+	−	Always +	Nonspontaneous at all T; reverse reaction occurs
III	+	+	+ at low T − at high T	Nonspontaneous at low T; becomes spontaneous as T is raised
IV	−	−	− at low T + at high T	Spontaneous at low T; at high T, reverse reaction becomes spontaneous

EXAMPLE 16.6	
At what temperature does ΔG° become zero for the reaction considered in Example 16.5?	
$$Fe_2O_3(s) + 3H_2(g) \longrightarrow 2Fe(s) + 3H_2O(g)$$	
ANALYSIS	
Information given:	equation for the reaction ($Fe_2O_3(s) + 3H_2(g) \longrightarrow 2Fe(s) + 3H_2O(g)$) ΔG° (0)
Information implied:	from Example 16.5 (ΔH° and ΔS°)
Asked for:	T
STRATEGY	
Substitute ΔH° and ΔS° values from Example 16.5 into the Gibbs-Helmholtz equation.	
SOLUTION	
ΔH°; ΔS°	$\Delta H^\circ = 98.8$ kJ; $\Delta S^\circ = 0.1415$ kJ/K
T	$\Delta G^\circ = \Delta H^\circ - T\Delta S^\circ$; $0 = 98.8$ kJ $- T(0.1415$ kJ/K$)$; $T = \dfrac{98.8 \text{ kJ}}{0.1415 \text{ kJ/K}} = 698$ K

The development presented in Example 16.6 is an important one from a practical standpoint. It tells us the temperature at which the direction of spontaneity changes. In the reduction of Fe_2O_3 by hydrogen, this temperature is approximately 700 K. At lower temperatures, the reaction does not occur at standard conditions; recall from Example 16.5 that ΔG° at 500 K is +27.6 kJ. At temperatures above 700 K, ΔG° has a negative sign and the reaction

$$Fe_2O_3(s) + 3H_2(g) \longrightarrow 2Fe(s) + 3H_2O(g)$$

is spontaneous.

From a slightly different point of view, we can say that the equation $T = \Delta H^\circ/\Delta S^\circ$ allows us to calculate the temperature at which a chemical or physical change is at equilibrium at 1 atm pressure. Consider, for example, the vaporization of water:

P and T affect ΔG but not ΔH; P affects ΔS.

$$H_2O(l) \longrightarrow H_2O(g)$$

where $\Delta H^\circ = +40.7$ kJ, $\Delta S^\circ = +0.109$ kJ/K. The temperature at which ΔG° is zero is

$$T = \frac{40.7}{0.109} \qquad T = 373 \quad K = 100^\circ C$$

This is, of course, the normal boiling point of water (the temperature at which liquid water is at equilibrium with vapor at 1 atm).

Pressure and Concentration

All of the free energy calculations to this point have involved the standard free energy change, ΔG°. It is possible, however, to write a general relation for the free energy change, ΔG, valid under any conditions. This relation is a relatively simple one, but we will not attempt to derive it. It says that

$$\Delta G = \Delta G^\circ + RT \ln Q \qquad (16.4)$$

If $Q < 1$, $\Delta G < \Delta G^\circ$. Explain.

The quantity Q that appears in this equation is the "reaction quotient" referred to in Chapter 12. It has the same mathematical form as the equilibrium constant, K;

the difference is that the terms that appear in Q are arbitrary, initial pressures or concentrations rather than equilibrium values.

To use this general expression for ΔG, you need to express T in K, R in kilojoules per kelvin (8.31×10^{-3} kJ/K). As far as Q is concerned, the general rules are as follows:

- gases enter as their partial pressures in atmospheres.
- species in aqueous solution enter as their molar concentrations.
- pure liquids or solids do not appear; neither does the solvent in a dilute solution.

As an example, consider the reaction of zinc with strong acid:

$$Zn(s) + 2H^+(aq) \longrightarrow Zn^{2+}(aq) + H_2(g)$$

Here Equation 16.4 would take the following form:

$$\Delta G = \Delta G° + RT \ln \frac{[Zn^{2+}](P_{H_2})}{[H^+]^2}$$

EXAMPLE 16.7 GRADED	

When zinc is dissolved in a strong acid at 25°C, zinc ions and hydrogen gas are produced. The equation for the reaction is

$$Zn(s) + 2H^+(aq) \longrightarrow Zn^{2+}(aq) + H_2(g)$$

a	Calculate $\Delta G°$ at the temperature of the reaction.

ANALYSIS

Information given:	equation for the reaction ($Zn(s) + 2H^+(aq) \longrightarrow Zn^{2+}(aq) + H_2(g)$) $T(25°C)$
Information implied:	$\Delta G_f°$ values at 25°C (Appendix 1)
Asked for:	$\Delta G°$

STRATEGY

1. Find $\Delta G°$ by substituting $\Delta G_f°$ values into Equation 16.3.
2. Recall that $\Delta G_f°$ for elements in their native state at 25°C and H$^+$ (aq) is zero.

SOLUTION

$\Delta G°$	$\Delta G° = \Delta G_f° \, Zn^{2+}(aq) + \Delta G_f° \, H_2(g) - [\Delta G_f \, Zn(s) + 2(\Delta G_f° \, H^+(aq))]$
	$= -147.1 \text{ kJ} \quad + \quad 0 \quad - \quad [0 \quad + \quad 2(0)] = -147.1 \text{ kJ}$

b	Calculate ΔG when $P_{H_2} = 750$ mm Hg, $[Zn^{2+}] = 0.10$ M, $[H^+] = 1.0 \times 10^{-4}$ M.

ANALYSIS

Information given:	P_{H_2} (750 mm Hg); $[Zn^{2+}]$(0.10 M); $[H^+]$(1.0 $\times$ 10^{-4} M); T(25°C) from part (a): $\Delta G°(-147.1$ kJ)
Information implied:	R value in energy units
Asked for:	ΔG

STRATEGY

1. Write the expression for Q and find its value. Recall that solids and pure liquids are not included. Make sure that concentrations are expressed in molarity and that pressure is in atm.
2. Substitute into Equation 16.4. Remember that R must be in kJ/K and T must be in K.

continued

	SOLUTION
1. Q	$Q = \dfrac{[Zn^{2+}]P_{H_2}}{[H^+]^2} = \dfrac{(0.10)(750/760)}{(1.0 \times 10^{-4})^2} = 9.9 \times 10^6$
2. ΔG	$\Delta G = \Delta G° + RT \ln Q = -147.1 \text{ kJ} + (0.00831 \text{ kJ/K})(298 \text{ K}) \ln (9.9 \times 10^6)$
	$= -147.1 \text{ kJ} + 39.9 \text{ kJ} = -107.2 \text{ kJ}$

c What is the pH when $\Delta G = -100.0$ kJ, $P_{H_2} = 0.933$ atm, $[Zn^{2+}] = 0.220$ M and the mass of Zn is 155 g?

	ANALYSIS
Information given:	$P_{H_2}(0.933 \text{ atm})$; $[Zn^{2+}](0.220 \text{ M})$; mass Zn(155 g); $T(25°C)$ $\Delta G(-100.0 \text{ kJ})$ From part (a): $\Delta G°(-147.1 \text{ kJ})$
Information implied:	R value in energy units
Asked for:	pH

	STRATEGY
	1. Find Q by substituting into Equation 16.4.
	2. Write the equation for Q.
	3. Solve for $[H^+]$ by substituting into the Q expression. Change $[H^+]$ into pH.

	SOLUTION
1. Q	$-100.0 \text{ kJ} = -147.1 \text{ kJ} + (0.00831 \text{ kJ/K})(298 \text{ K}) \ln Q$ $\ln Q = \dfrac{47.1}{(0.00831)(298)} = 19.0 \longrightarrow Q = e^{19.0} = 1.82 \times 10^8$
2. Q expression	$Q = \dfrac{[Zn^{2+}](P_{H_2})}{[H^+]^2}$; $1.82 \times 10^8 = \dfrac{(0.220)(0.933)}{[H^+]^2}$
3. $[H^+]$; pH	$[H^+] = \left(\dfrac{(0.220)(0.933)}{1.82 \times 10^8}\right)^{1/2} = 3.35 \times 10^{-5}$ M; pH $= -\log_{10} 3.35 \times 10^{-5} = $ **4.47**

As Example 16.7 implies, changes in pressure and/or concentration can have a considerable effect on ΔG. Sometimes, when $\Delta G°$ is close to zero, changing the pressure from 1 atm to some other value can change the direction of spontaneity. Consider, for example, the reaction at 300°C:

$$NH_4Cl(s) \longrightarrow NH_3(g) + HCl(g) \qquad \Delta G° = +13.0 \text{ kJ}$$

The positive sign of $\Delta G°$ implies that ammonium chloride will not decompose at 300°C to give NH_3 and HCl, both at 1 atm pressure. However, notice what happens if $P_{NH_3} = P_{HCl} = 0.10$ atm, in which case $Q = (0.10)^2$:

$$\Delta G = \Delta G° + RT \ln(0.10)^2$$
$$= +13.0 \text{ kJ} + (8.31 \times 10^{-3})(573 \text{ kJ})(\ln 0.010)$$
$$= +13.0 \text{ kJ} - 21.9 \text{ kJ} = -8.9 \text{ kJ}$$

This means that NH_3 and HCl *can* be formed, each at 0.10 atm pressure, by heating ammonium chloride to 300°C.

Another example of how a change in concentration can change the direction of spontaneity involves the reaction

$$SrCrO_4(s) \longrightarrow Sr^{2+}(aq) + CrO_4^{2-}(aq) \qquad \Delta G° = +25.3 \text{ kJ at 25°C}$$

Because $\Delta G°$ is a positive quantity, $SrCrO_4$ does not dissolve spontaneously to form a 1 M solution at 25°C (Figure 16.8). Suppose, however, the concentrations

Figure 16.8 **A saturated solution of strontium chromate.** The solution of $SrCrO_4$ contains 0.006 mol/L of Sr^{2+} and CrO_4^{2-}. For the reaction at 25°C: $SrCrO_4(s) \rightleftharpoons Sr^{2+}(aq, 0.006 \text{ M}) + CrO_4^{2-}(aq, 0.006 \text{ M})$, $\Delta G = 0$.

© Cengage Learning/Charles D. Winters

of Sr^{2+} and CrO_4^{2-} are reduced to 0.0010 M:

$$\Delta G = \Delta G^\circ + RT \ln [Sr^{2+}][CrO_4^{2-}]$$
$$= +25.3 \text{ kJ} + [(8.31 \times 10^{-3})(298)] \text{ kJ} [\ln (1.0 \times 10^{-3})^2]$$
$$= +25.3 \text{ kJ} - 34.2 \text{ kJ} = -8.9 \text{ kJ}$$

Strontium chromate should, and does, dissolve spontaneously to form a 0.0010 M water solution. As you might expect, its solubility at 25°C lies between 1 M and 0.0010 M, at about 0.0060 M.

$$\Delta G \text{ at } 0.0060 \; M = +25.3 \text{ kJ} + RT \ln (0.0060)^2 = 0$$

16-6 The Free Energy Change and the Equilibrium Constant

Throughout this chapter we have stressed the relation between the free energy change and reaction spontaneity. For a reaction to be spontaneous at standard conditions (1 atm, 1 M), ΔG° must be negative. Another indicator of reaction spontaneity is the equilibrium constant, K; if K is greater than 1, the reaction is spontaneous at standard conditions. As you might suppose, the two quantities ΔG° and K are related. The nature of that relationship is readily deduced by starting with the general equation

$$\Delta G = \Delta G^\circ + RT \ln Q$$

Suppose now that reaction takes place until equilibrium is established, at which point $Q = K$ and $\Delta G = 0$. ◄ ⓘ

Remember, when $\Delta G = 0$, the system is at equilibrium. ⓘ

$$0 = \Delta G^\circ + RT \ln K$$

hence

$$\Delta G^\circ = -RT \ln K \qquad \qquad (16.5)$$

Equation 16.5 is generally applicable to any system. The equilibrium constant may be the K referred to in our discussion of gaseous equilibrium (Chapter 12), or any of the solution equilibrium constants (K_w, K_a, K_b, K_{sp}, . . .) discussed in subsequent chapters. Notice that ΔG° is the *standard* free energy change (gases at 1 atm, species in solution at 1 M). That is why, in the expression for K, gases enter as their partial pressures in *atmospheres* and ions or molecules in solution as their *molarities*.

EXAMPLE 16.8

Using ΔG_f° tables in Appendix 1, calculate the solubility product constant, K_{sp}, for $PbCl_2$ at 25°C.

ANALYSIS

Information given:	$T(25°C)$
Information implied:	Equation for dissolving $PbCl_2$ ΔG_f° values (Appendix 1) R value
Asked for:	K_{sp}

STRATEGY

1. Obtain ΔG_f° values from Appendix 1 and substitute into Equation 16.3.

2. Substitute into Equation 16.5 to find K_{sp}, which is K.

$$\Delta G^\circ = -RT \ln K_{sp}$$

continued

	SOLUTION
$\Delta G°$	$\Delta G° = \Delta G_f°\ Pb^{2+}(aq) + 2\Delta G_f°\ Cl^-(aq) - \Delta G_f°\ PbCl_2(s)$
	$= -24.4\ kJ + 2(-131.2\ kJ) + 314.1\ kJ = +27.3\ kJ$
K_{sp}	$\Delta G° = -RT \ln K_{sp};\ 27.3\ kJ = -(0.00831\ kJ/K)(298\ K) \ln K_{sp}$
	$\ln K_{sp} = -11.0 \longrightarrow K_{sp} = e^{-11.0} = 1.7 \times 10^{-5}$
	END POINT

This is the value listed in Chapter 15 for K_{sp} of $PbCl_2$.

The relationship $\Delta G° = -RT \ln K$ allows us to relate the standard free energy change to the extent of reaction. Consider, for example, the simple equilibrium system

$$A(g) \rightleftharpoons B(g)$$

at 25°C. Figure 16.9 shows how the extent of reaction varies with the value of $\Delta G°$ for this system. Notice that

- if $\Delta G°$ is greater than about $+20$ kJ, the equilibrium constant is so small that virtually no A is converted to B. We would say that, for all practical purposes, the reaction does not occur.
- if $\Delta G°$ is less than about -20 kJ, the equilibrium constant is so large that virtually all of A is converted to B. In essence, the reaction goes to completion.
- only if $\Delta G°$ lies between $+20$ kJ and -20 kJ will the equilibrium mixture contain appreciable amounts of both A and B. In particular, if $\Delta G° = 0$, $K = 1$ and the equilibrium system contains equal amounts of A and B.

Figure 16.9 **Dependence of extent of reaction on the value of $\Delta G°$ for the general system A(g) $\rightleftharpoons$ B(g).**

16-7 Additivity of Free Energy Changes; Coupled Reactions

Free energy changes for reactions, like enthalpy or entropy changes, are additive. That is,

$$\text{if Reaction 3 = Reaction 1 + Reaction 2}$$
$$\text{then } \Delta G_3 = \Delta G_1 + \Delta G_2$$

This relation can be regarded as the free energy equivalent of Hess's law (Chapter 8). To illustrate its application, consider the synthesis of $CuCl_2$ from the elements

$$Cu(s) + \tfrac{1}{2}Cl_2(g) \longrightarrow CuCl(s) \qquad \Delta G° \text{ at } 25°C = -119.9\ kJ$$
$$\underline{CuCl(s) + \tfrac{1}{2}Cl_2(g) \longrightarrow CuCl_2(s) \qquad \Delta G° \text{ at } 25°C = -55.8\ kJ}$$
$$Cu(s) + Cl_2(g) \longrightarrow CuCl_2(s)$$

For the overall reaction, $\Delta G°$ at 25°C $= -119.9\ kJ + (-55.8\ kJ) = -175.7\ kJ$.

Because free energy changes are additive, it is often possible to bring about a nonspontaneous reaction by coupling it with a reaction for which $\Delta G°$ is a large negative number. As an example, consider the preparation of iron metal from hematite ore. The reaction

$$Fe_2O_3(s) \longrightarrow 2Fe(s) + \tfrac{3}{2}O_2(g) \qquad \Delta G° \text{ at } 25°C = +742.2\ kJ$$

is clearly nonspontaneous; even at temperatures as high as 2000°C, $\Delta G°$ is a positive quantity. Suppose, though, that this reaction is "coupled" with the spontaneous oxidation of carbon monoxide:

$$CO(g) + \tfrac{1}{2}O_2(g) \longrightarrow CO_2(g) \qquad \Delta G° \text{ at } 25°C = -257.2\ kJ$$

The overall reaction is spontaneous:

$$Fe_2O_3(s) \longrightarrow 2Fe(s) + \tfrac{3}{2}O_2(g)$$
$$3CO(g) + \tfrac{3}{2}O_2(g) \longrightarrow 3CO_2(g)$$
$$\overline{Fe_2O_3(s) + 3CO(g) \longrightarrow 2Fe(s) + 3CO_2(g)}$$

For the coupled reaction at 25°C,

$$\Delta G° = +742.2 \text{ kJ} + 3(-257.2 \text{ kJ}) = -29.4 \text{ kJ}$$

The negative sign of $\Delta G°$ implies that although Fe_2O_3 does not spontaneously decompose, it can be converted to iron by reaction with carbon monoxide. This is in fact the reaction used in the blast furnace when iron ore consisting mainly of Fe_2O_3 is reduced to iron (Chapter 20).

Many industrial processes involve coupled reactions.

Coupled reactions are common in human metabolism. Spontaneous processes, such as the oxidation of glucose,

$$C_6H_{12}O_6(aq) + 6O_2(g) \longrightarrow 6CO_2(g) + 6H_2O \qquad \Delta G° = -2870 \text{ kJ at } 25°C$$

ordinarily do not serve directly as a source of energy. Instead these reactions are used to bring about a nonspontaneous reaction:

$$ADP(aq) + HPO_4{}^{2-}(aq) + 2H^+(aq) \longrightarrow ATP(aq) + H_2O \qquad (16.6)$$
$$\Delta G° = +31 \text{ kJ at } 25°C$$

ADP (adenosine diphosphate) and ATP (adenosine triphosphate) are complex organic molecules (Figure 16.10) that, in essence, differ only by the presence of an extra phosphate group in ATP. In the coupled reaction with glucose, about 38 mol of ATP are synthesized for every mole of glucose consumed. This gives an overall free energy change for the coupled reaction of

$$-2870 \text{ kJ} + 38(+31 \text{ kJ}) \approx -1700 \text{ kJ}$$

Figure 16.10 Structures of ADP and ATP.

adenosine diphosphate (ADP)

adenosine triphosphate (ATP)

In a very real sense, your body "stores" energy available from the metabolism of foods in the form of ATP. This molecule in turn supplies the energy required for all sorts of biochemical reactions taking place in the body. It does this by reverting to ADP, that is, by reversing reaction 16.6. ◀ ❶ The amount of ATP consumed is amazingly large; a competitive sprinter may hydrolyze as much as 500 g (about 1 lb) of ATP per minute.

Conversion of ATP to ADP gives you energy in a hurry.

EXAMPLE 16.9

The lactic acid ($C_3H_6O_3(aq)$, $\Delta G_f° = -559$ kJ) produced in muscle cells by vigorous exercise eventually is absorbed into the bloodstream, where it is metabolized back to glucose ($\Delta G_f° = -919$ kJ) in the liver. The reaction is

$$2C_3H_6O_3(aq) \longrightarrow C_6H_{12}O_6(aq)$$

(a) Calculate $\Delta G°$ at 25°C for this reaction, using free energies of formation.

(b) If the hydrolysis of ATP to ADP is coupled with this reaction, how many moles of ATP must react to make the process spontaneous?

ANALYSIS	
Information given:	equation for the reaction ($2C_3H_6O_3(aq) \longrightarrow C_6H_{12}O_6(aq)$) $\Delta G_f°$ values: $C_3H_6O_3(aq)$ (-559 kJ); $C_6H_{12}O_6(aq)$(-919 kJ) T(25°C) energy from ATP/mol (31 kJ)
Asked for:	(a) $\Delta G°$ (b) mol ATP for spontaneity

STRATEGY	
1. Find $\Delta G°$ using the $\Delta G_f°$ values given in Appendix 1. $\Delta G° = \Sigma\Delta G_f°_{products} - \Sigma\Delta G_f°_{reactants}$	
2. Convert the energy obtained in (1) to moles ATP by using the conversion factor: 31 kJ/mol ATP	

SOLUTION	
1. $\Delta G°$	$\Delta G° = \Delta G_f° \, C_6H_{12}O_6(aq) - 2\Delta G_f° \, C_3H_6O_3(aq)$ $= -919$ kJ $- 2(-559$ kJ$) = $ +199 kJ
2. mol ATP	$199 \text{ kJ} \times \dfrac{1 \text{ mol ATP}}{31 \text{ kJ}} = 6.4 \text{ mol ATP}$

CHEMISTRY **BEYOND** THE CLASSROOM

Rubber Elasticity: An Entropic Phenomenon

Leslie Sperling, Lehigh University

Molecular Structure of Rubbery Materials

Along with such materials as plastics, adhesives, fibers, and coatings, rubber is polymeric in nature. Such materials consist of long chains, with molecular masses generally of the order of 50,000 to 500,000 g/mol. Common rubbery materials—often called elastomers—include automotive tires and rubber bands.

For a polymer to exhibit rubber elasticity, it must have two properties:

1. It must be *crosslinked* or *vulcanized*. Crosslinking is the chemical joining together of polymer chains, usually by sulfur bonds at random positions, to make a three-dimensional network (see Figure A).

2. It must be above its glass transition temperature, which means that the polymer chains have sufficient thermal energy to move freely. Many rubbery materials have glass transition temperatures around 200 K, below which they are glassy, like plastics.

Figure A Schematic of polymer chains randomly placed in space, with crosslinks also placed randomly, but frequently averaging every 5,000–10,000 g/mol along the chains.

Common rubbery materials consist of butadiene and styrene statistical copolymers, written poly(butadiene-*stat*-styrene). The butadiene polymer has the repeating structure $(—CH_2—CH=CH—CH_2^-)$ while styrene has the repeating structure

$$—CH_2 - CH—$$
$$\underset{C_6H_5}{|}$$

How Does Entropy Apply to Elastomeric Materials?

Entropy is a measure of disorder in materials. Relaxed polymer chains with a random conformation (shape in space), like cooked spaghetti or a box of fishing worms, have a high degree of entropy, which is favored by Mother Nature. If the chains are stretched out (stressed), the number of conformations the chains can have in space is limited, and the entropy is reduced (see Figure B). The ratio of final length to initial length is denoted α.

For materials such as rubber bands, the quantity α may be as large as five. At that point, the chains are substantially fully extended. Further stretching may break the rubber band, actually severing the polymer chains at the break point (a chemical reaction!).

There is an interesting demonstration experiment that you can do with a rubber band, preferably a large and/or thick one. Touch the unstretched rubber band to your lips. Then, stretch the rubber band rapidly and immediately, retouch it to your lips. The rubber band will be warmer. This motion creates a thermodynamic cycle in which the system goes through a series of different states and then returns to its original state. The cycle acts as a heat engine.

The major applications of rubbery materials today include automotive tires, rubber bands, tubing of various kinds, electric wire insulation, elastomeric urethane fibers for undergarments, and silicone rubber. Such types of polymers are important materials in our twenty-first-century world.

References

1. A. J. Etzel, S. J. Goldstein, H. J. Panabaker, D. G. Fradkin, and L. H. Sperling, *J. Chem. Ed.*, **63**, 731 (1986).
2. L. H. Sperling, *Introduction to Physical Polymer Science*, 4th Ed., Wiley, New York, 2006.

RELAXED   **STRESSED**

Figure B Pulling down on the elastomer. The chains become oriented in the elongational direction, lowering the entropy of the system.

Key Concepts

1. Deduce the sign of ΔS for a process from randomness considerations.
   (**Example 16.1; Problems 7–14**)
2. Calculate $\Delta S°$ for a reaction, using Table 16.1.
   (**Example 16.2; Problems 17–22**)
3. Calculate $\Delta G°$ at any temperature, knowing $\Delta H°$ and $\Delta S°$.
   (**Examples 16.3, 16.5; Problems 23–26**)
4. Calculate $\Delta G°$ at 25°C from free energies of formation.
   (**Examples 16.4, 16.7–16.9; Problems 27–32**)

5. Calculate the temperature at which $\Delta G° = 0$.
   (**Example 16.6; Problems 39–44, 49–54**)
6. Calculate ΔG from $\Delta G°$, knowing all pressures and/or concentrations.
   (**Example 16.7; Problems 55–60**)

7. Relate $\Delta G°$ to K.
   (**Example 16.8; Problems 61–70**)
8. Calculate $\Delta G°$ for coupled reactions.
   (**Example 16.9; Problems 71–76**)

Key Equations

Entropy change	$\Delta S° = \sum S°_{products} - \sum S°_{reactants}$
Gibbs-Helmholtz equation	$\Delta G° = \Delta H° - T\Delta S°$
Free energy change	$\Delta G° = \sum \Delta G_f° \; products - \sum \Delta G_f° \; reactants$
	$\Delta G = \Delta G° + RT \ln Q$
	$\Delta G° = -RT \ln K$

Key Terms

entropy	spontaneous processes	standard free	standard free energy
free energy	standard entropy change	energy change	of formation

Summary Problem

Consider formic acid, HCOOH. It is given off by many insects when they sting their victims. Its name comes from *formica*, the Latin name for ant. Extensively used by the textile industry, it can be obtained by the exposure of methyl alcohol (an alcohol present in gasohol) to air.

$$CH_3OH(aq) + O_2(g) \longrightarrow HCOOH(aq) + H_2O(l)$$

The following data may be useful:

$CH_3OH(aq)$: $\Delta H_f° = -246$ kJ/mol; $S° = 133.1$ J/mol · K
$HCOOH(aq)$: $\Delta H_f° = -425$ kJ/mol; $S° = 163$ J/mol · K

(a) Calculate $\Delta H°$ and $\Delta S°$ for the process.

(b) Is the reaction spontaneous at 25°C? At 100°C?

(c) The heat of vaporization for formic acid is 23.1 kJ/mol. Its normal boiling point is 101°C. Calculate $\Delta S°$ for the reaction

$$HCOOH(l) \longrightarrow HCOOH(g)$$

(d) What is the standard molar entropy for $HCOOH(g)$ taking $S°$ for $HCOOH(l)$ to be 129 J/mol · K?

(e) Calculate ΔG at 25°C for the formation of formic acid from methyl alcohol (equation given earlier) when $[HCOOH] = 0.250$ M, $P_{O_2} = 0.75$ atm, and $[CH_3OH] = 0.0100$ M.

(f) Calculate $\Delta G°$ for the ionization of formic acid at 25°C. ($K_a = 1.9 \times 10^{-4}$)

Answers

(a) $\Delta H° = -465$ kJ; $\Delta S° = -105$ J/K
(b) $\Delta G° = -434$ kJ; yes; yes
(c) 61.8 J/K
(d) 191 J/mol · K
(e) −425 kJ
(f) 21.2 kJ

Questions and Problems

Even-numbered questions and Challenge Problems have answers in Appendix 5 and fully worked solutions in the *Student Solutions Manual*.

16-1 Spontaneous Processes

1. Which of the following processes are spontaneous?
   (a) building a sand castle
   (b) outlining your chemistry notes
   (c) wind scattering leaves in a pile
2. Which of the following processes are spontaneous?
   (a) ice cream melting at 75°F
   (b) sorting a list of names alphabetically
   (c) gathering leaves in a pile

3. Which of the following processes are spontaneous?
   (a) a ball rolling down a hill
   (b) a drop of ink dispersing in water
   (c) melting wax at 10°C

4. Which of the following processes are spontaneous?
   (a) cleaning up your desk
   (b) building a house of cards
   (c) sugar dissolving in hot water

5. On the basis of your experience, predict which reactions are spontaneous:
   (a) $PbO_2(s) \longrightarrow Pb(s) + O_2(g)$
   (b) $N_2(l) \longrightarrow N_2(g)$ at 25°C
   (c) $C_6H_{12}O_6(s) \longrightarrow C_6H_{12}O_6(l)$ at 25°C
   (d) $Ca^{2+}(aq) + CO_3{}^{2-}(aq) \longrightarrow CaCO_3(s)$

6. On the basis of your experience, predict which of the following reactions are spontaneous.
   (a) $CO_2(s) \longrightarrow CO_2(g)$ at 25°C
   (b) $NaCl(s) \longrightarrow NaCl(l)$ at 25°C
   (c) $2NaCl(s) \longrightarrow 2Na(s) + Cl_2(g)$
   (d) $CO_2(g) \longrightarrow C(s) + O_2(g)$

16-2 Entropy, S

7. In each of the following pairs, choose the substance with a lower entropy.
   (a) $H_2O(l)$ at 10°C, $H_2O(l)$ at 30°C
   (b) C (graphite), C (diamond)
   (c) $Cl_2(l)$, $Cl_2(g)$, both at room temperature

8. In each of the following pairs, choose the substance with a lower entropy.
   (a) One mole of $O_2(g)$ with 758 mm Hg pressure, one mole of $O_2(g)$ with 493 mm Hg pressure, both at room temperature
   (b) glucose (s), glucose (aq)
   (c) $Hg(l)$, $Hg(g)$, both at room temperature

9. Predict the sign of ΔS for the following:
   (a) a lake freezing
   (b) ice cream thawing
   (c) a candle burning
   (d) weeding a garden

10. Predict the sign of ΔS for the following:
   (a) precipitating solid AgCl from a solution containing Ag^+ and Cl^- ions
   (b) dissolving sugar in hot coffee
   (c) glass turning into sand

11. Predict the sign of $\Delta S°$ for each of the following reactions.
   (a) $2Na(s) + Cl_2(g) \longrightarrow 2NaCl(s)$
   (b) $2NO(g) + O_2(g) \longrightarrow 2NO_2(g)$
   (c) $C_2H_4(g) + 3O_2(g) \longrightarrow 2CO_2(g) + 3H_2O(l)$
   (d) $NH_4NO_3(s) + H_2O(l) \longrightarrow 2NH_3(g) + O_2(g)$

12. Predict the sign of $\Delta S°$ for each of the following reactions:
   (a) $CCl_4(l) + 5O_2(g) \longrightarrow CO_2(g) + 4ClO_2(g)$
   (b) $8H_2O(l) + S_8(s) \longrightarrow 8H_2S(g) + 4O_2(g)$
   (c) $Br_2(l) \longrightarrow Br_2(s)$
   (d) $2NH_3(g) \longrightarrow N_2(g) + 3H_2(g)$

13. Predict the sign of $\Delta S°$ for each of the following reactions.
   (a) $O_3(g) \longrightarrow O_2(g) + O(g)$
   (b) $PCl_3(g) + Cl_2(g) \longrightarrow PCl_5(g)$
   (c) $CuSO_4(s) + 5H_2O(l) \longrightarrow CuSO_4 \cdot 5H_2O(s)$

14. Predict the sign of $\Delta S°$ for each of the following reactions.
   (a) $H_2(g) + Ni^{2+}(aq) \longrightarrow 2H^+(aq) + Ni(s)$
   (b) $Cu(s) + 2H^+(aq) \longrightarrow H_2(g) + Cu^{2+}(aq)$
   (c) $N_2O_4(g) \longrightarrow 2NO_2(g)$

15. Predict the order of the following reactions in terms of increasing ΔS:
   (a) $2H_2O_2(l) \longrightarrow 2H_2O(l) + O_2(g)$
   (b) $Si(s) + O_2(g) \longrightarrow SiO_2(s)$
   (c) $Br_2(g) + 3I_2(g) \longrightarrow 2BrI_3(l)$
   (d) $NH_3(g) + HBr(g) \longrightarrow NH_4Br(s)$

16. Predict the order of the following reactions in terms of increasing ΔS:
   (a) $N_2(g) + 3H_2(g) \longrightarrow 2NH_3(g)$
   (b) $2H_2(g) + O_2(g) \longrightarrow 2H_2O(l)$
   (c) $C(g) + O_2(g) \longrightarrow CO_2(s)$
   (d) $I_2(g) + Cl_2(g) \longrightarrow 2ICl(g)$

17. Use Table 16.1 to calculate $\Delta S°$ for each of the following reactions.
   (a) $CO(g) + 2H_2(g) \longrightarrow CH_3OH(l)$
   (b) $N_2(g) + O_2(g) \longrightarrow 2NO(g)$
   (c) $BaCO_3(s) \longrightarrow BaO(s) + CO_2(g)$
   (d) $2NaCl(s) + F_2(g) \longrightarrow 2NaF(s) + Cl_2(g)$

18. Use Table 16.1 to calculate $\Delta S°$ for each of the following reactions:
   (a) $4NH_3(g) + 5O_2(g) \longrightarrow 4NO(g) + 6H_2O(g)$
   (b) $2H_2O_2(l) + N_2H_4(l) \longrightarrow N_2(g) + 4H_2O(g)$
   (c) $C(s) + O_2(g) \longrightarrow CO_2(g)$
   (d) $CH_4(g) + 3Cl_2(g) \longrightarrow CHCl_3(l) + 3HCl(g)$

19. Use Table 16.1 to calculate $\Delta S°$ for each of the following reactions.
   (a) $2Cl^-(aq) + I_2(s) \longrightarrow$
   $$Cl_2(g) + 2I^-(aq)$$
   (b) $SO_4{}^{2-}(aq) + 4H^+(aq) + Cd(s) \longrightarrow$
   $$Cd^{2+}(aq) + SO_2(g) + 2H_2O(l)$$
   (c) $2Br^-(aq) + 2H_2O(l) \longrightarrow$
   $$Br_2(l) + H_2(g) + 2OH^-(aq)$$

20. Use Table 16.1 to calculate $\Delta S°$ for each of the following reactions.
   (a) $5NO(g) + 3MnO_4{}^-(aq) + 4H^+(aq) \longrightarrow$
   $$5NO_3{}^-(aq) + 3Mn^{2+}(aq) + 2H_2O(l)$$
   (b) $6Fe^{2+}(aq) + CrO_4{}^{2-}(aq) + 8H^+(aq) \longrightarrow$
   $$Cr(s) + 6Fe^{3+}(aq) + 4H_2O(l)$$
   (c) $C_2H_2(g) + \frac{5}{2}O_2(g) \longrightarrow 2CO_2(g) + H_2O(g)$

21. Use Table 16.1 to calculate $\Delta S°$ for each of the following reactions:
   (a) $2Cr(s) + 3CO_2(g) \longrightarrow Cr_2O_3(s) + 3CO(g)$
   (b) $4C(s) + O_2(g) + 6H_2(g) \longrightarrow 2C_2H_5OH(l)$
   (c) $N_2O_4(g) + 4H_2(g) \longrightarrow N_2(g) + 4H_2O(g)$
   (d) $2CuO(s) \longrightarrow 2Cu(s) + O_2(g)$

22. Use Table 16.1 to calculate $\Delta S°$ for each of the following reactions.
   (a) $2HNO_3(l) + 3H_2S(g) \longrightarrow$
   $$4H_2O(l) + 2NO(g) + 3S(s)$$
   (b) $PCl_5(g) + 4H_2O(l) \longrightarrow$
   $$5Cl^-(aq) + H_2PO_4{}^-(aq) + 6H^+(aq)$$
   (c) $MnO_4{}^-(aq) + 3Fe^{2+}(aq) + 4H^+(aq) \longrightarrow$
   $$3Fe^{3+}(aq) + MnO_2(s) + 2H_2O(l)$$

16-4 Standard Free Energy Change, $\Delta G°$

23. Calculate $\Delta G°$ at 82°C for reactions in which
  (a) $\Delta H° = 293$ kJ; $\Delta S° = -695$ J/K
  (b) $\Delta H° = -1137$ kJ; $\Delta S° = 0.496$ kJ/K
  (c) $\Delta H° = -86.6$ kJ; $\Delta S° = -382$ J/K

24. Calculate $\Delta G°$ at 72°C for reactions in which
  (a) $\Delta H° = -136$ kJ; $\Delta S° = +457$ J/K
  (b) $\Delta H° = 41.5$ kJ; $\Delta S° = -0.288$ kJ/K
  (c) $\Delta H° = -795$ kJ; $\Delta S° = -861$ J/K

25. Calculate $\Delta G°$ at 355 K for each of the reactions in Question 17. State whether the reactions are spontaneous.

26. Calculate $\Delta G°$ at 415 K for each of the reactions in Question 18. State whether the reactions are spontaneous.

27. From the values for $\Delta G_f°$ given in Appendix 1, calculate $\Delta G°$ at 25°C for each of the reactions in Question 19.

28. Follow the directions of Problem 27 for each of the reactions in Question 20.

29. Use standard entropies and heats of formation to calculate $\Delta G_f°$ at 25°C for
  (a) cadmium(II) chloride (s).
  (b) methyl alcohol, $CH_3OH(l)$.
  (c) copper(I) sulfide (s).

30. Follow the directions of Question 29 for the following compounds:
  (a) solid ammonium nitrate
  (b) liquid methyl alcohol
  (c) solid copper(II) sulfide

31. It has been proposed that wood alcohol, CH_3OH, a relatively inexpensive fuel to produce, be decomposed to produce methane. Methane is a natural gas commonly used for heating homes. Is the decomposition of wood alcohol to methane and oxygen thermodynamically feasible at 25°C and 1 atm?

32. A student warned his friends not to swim in a river close to an electric plant. He claimed that the ozone produced by the plant turned the river water to hydrogen peroxide, which would bleach hair. The reaction is

$$O_3(g) + H_2O(l) \longrightarrow H_2O_2(aq) + O_2(g)$$

Assuming that the river water is at 25°C and all species are at standard concentrations, show by calculation whether his claim is plausible. Take $\Delta G_f°$ $O_3(g)$ at 25°C to be $+163.2$ kJ/mol and $\Delta G_f°$ $H_2O_2(aq) = -134$ kJ/mol.

33. Sodium carbonate, also called "washing soda," can be made by heating sodium hydrogen carbonate:

$$2NaHCO_3(s) \longrightarrow Na_2CO_3(s) + CO_2(g) + H_2O(l)$$
$$\Delta H° = +135.6 \text{ kJ}; \Delta G° = +34.6 \text{ kJ at 25°C}$$

  (a) Calculate $\Delta S°$ for this reaction. Is the sign reasonable?
  (b) Calculate $\Delta G°$ at 0 K; at 1000 K.

34. The reaction between magnesium metal and water (l) produces solid magnesium hydroxide and hydrogen gas. Calculate $\Delta G°$ for the formation of one mole of $Mg(OH)_2$ at 27°C and at 39°C.

35. In the laboratory, $POCl_3$ (phosphorus oxychloride) is used in the manufacture of phosphate esters, which are used in flame retardants and pesticides. It can be prepared by the following reaction at 25°C:

$$2PCl_3(g) + O_2(g) \longrightarrow 2POCl_3(g)$$
$$\Delta H° = -572 \text{ kJ}; \Delta G° = -518.7 \text{ kJ}$$

  (a) Calculate $\Delta S°$. Is the sign reasonable?
  (b) Calculate $S°$ for $POCl_3$.
  (c) Calculate $\Delta H_f°$ for $POCl_3$.

36. Oxygen can be made in the laboratory by reacting sodium peroxide and water.

$$2Na_2O_2(s) + 2H_2O(l) \longrightarrow 4NaOH(s) + O_2(g)$$
$$\Delta H° = -109.0 \text{ kJ}; \Delta G° = -148.4 \text{ kJ at 25°C}$$

  (a) Calculate $\Delta S°$. Is the sign reasonable?
  (b) Calculate $S°$ for $Na_2O_2(s)$.
  (c) Calculate $\Delta H_f°$ for $Na_2O_2(s)$.

37. Phosgene, $COCl_2$, can be formed by the reaction of chloroform, $CHCl_3(l)$, with oxygen:

$$2CHCl_3(l) + O_2(g) \longrightarrow 2COCl_2(g) + 2HCl(g)$$
$$\Delta H° = -353.2 \text{ kJ}; \Delta G° = -452.4 \text{ kJ at 25°C}$$

  (a) Calculate $\Delta S°$ for the reaction. Is the sign reasonable?
  (b) Calculate $S°$ for phosgene.
  (c) Calculate $\Delta H_f°$ for phosgene.

38. When permanganate ions in aqueous solution react with cobalt metal in strong acid, the equation for the reaction that takes place is

$$2MnO_4^-(aq) + 16H^+(aq) + 5Co(s) \longrightarrow$$
$$2Mn^{2+}(aq) + 5Co^{2+}(aq) + 8H_2O(l)$$
$$\Delta H° = -2024.6 \text{ kJ}; \Delta G° \text{ at 25°C} = -1750.9 \text{ kJ}$$

  (a) Calculate $\Delta S°$ for the reaction at 25°C.
  (b) Calculate $S°$ for Co^{2+}, given $S°$ for Co is 30.04 J/mol·K.

16-5 Effect of Temperature, Pressure, and Concentration on Reaction Spontaneity

Temperature Dependence of Spontaneity

39. Discuss the effect of temperature change on the spontaneity of the following reactions at 1 atm:
  (a) $Al_2O_3(s) + 2Fe(s) \longrightarrow 2Al(s) + Fe_2O_3(s)$
         $\Delta H° = 851.4$ kJ; $\Delta S° = 38.5$ J/K
  (b) $N_2H_4(l) \longrightarrow N_2 + 2H_2(g)$
         $\Delta H° = -50.6$ kJ; $\Delta S° = 0.3315$ kJ/K
  (c) $SO_3(g) \longrightarrow SO_2(g) + \frac{1}{2}O_2(g)$
         $\Delta H° = 98.9$ kJ; $\Delta S° = 0.0939$ kJ/K

40. Discuss the effect of temperature on the spontaneity of reactions with the following values for $\Delta H°$ and $\Delta S°$.
  (a) $\Delta H° = 128$ kJ; $\Delta S° = 89.5$ J/K
  (b) $\Delta H° = -20.4$ kJ; $\Delta S° = -156.3$ J/K
  (c) $\Delta H° = -127$ kJ; $\Delta S° = 43.2$ J/K

41. At what temperature does $\Delta G°$ become zero for each of the reactions in Problem 39? Explain the significance of your answers.

42. Over what temperature range are the reactions in Problem 40 spontaneous?

43. For the reaction

$$2Cl^-(aq) + Br_2(l) \longrightarrow Cl_2(g) + 2Br^-(aq)$$

calculate the temperature at which $\Delta G° = 0$.

44. For the reaction

$$SnO_2(s) + 2CO(g) \longrightarrow 2CO_2(g) + Sn(s)$$

calculate the temperature at which $\Delta G° = 0$.

45. For the decomposition of Ag_2O:

$$2\,Ag_2O(s) \longrightarrow 4\,Ag(s) + O_2(g)$$

(a) Obtain an expression for $\Delta G°$ as a function of temperature. Prepare a table of $\Delta G°$ values at 100 K intervals between 100 K and 500 K.
(b) Calculate the temperature at which $\Delta G° = 0$.

46. Consider the following hypothetical equation

$$A(s) + B(s) \longrightarrow C(s) + D(s)$$

where $\Delta H° = 492$ kJ and $\Delta S° = 327$ J/K.

(a) Obtain an expression for $\Delta G°$ as a function of temperature. Prepare a table of $\Delta G°$ values at 100 K intervals between 100 K and 500 K.
(b) Find the temperature at which $\Delta G°$ becomes zero.

47. Two possible ways of producing iron from iron ore are
(a) $Fe_2O_3(s) + \frac{3}{2}C(s) \longrightarrow 2Fe(s) + \frac{3}{2}CO_2(g)$
(b) $Fe_2O_3(s) + 3H_2(g) \longrightarrow 2Fe(s) + 3H_2O(g)$
Which of these reactions proceeds spontaneously at the lower temperature?

48. Manganese is a metal used in making stainless steel alloys. It can be obtained from pyrolusite, an ore containing MnO_2. If the production process calls for as low a temperature as possible, the ore could be
(a) decomposed on heating, producing Mn and O_2.
(b) heated with hydrogen gas, producing Mn and steam.
(c) heated with coke ($C(s)$), producing Mn and CO_2.

49. Red phosphorus is formed by heating white phosphorus. Calculate the temperature at which the two forms are at equilibrium, given

white P: $\Delta H_f° = 0.00$ kJ/mol; $S° = 41.09$ J/mol·K
red P: $\Delta H_f° = -17.6$ kJ/mol; $S° = 22.80$ J/mol·K

50. Organ pipes in unheated churches develop "tin disease," in which white tin is converted to gray tin. Given

white Sn: $\Delta H_f° = 0.00$ kJ/mol; $S° = 51.55$ J/mol·K
gray Sn: $\Delta H_f° = -2.09$ kJ/mol; $S° = 44.14$ J/mol·K

calculate the equilibrium temperature for the transition.

51. Sulfur has about 20 different allotropes. The most common are rhombic sulfur (the stable form at 25°C and 1 atm) and monoclinic sulfur. They differ in their crystal structures. Given

S(s, monoclinic): $\Delta H_f° = 0.30$ kJ/mol, $S° = 0.0326$ kJ/mol·K

at what temperature are the two forms in equilibrium?

52. Pencil "lead" is almost pure graphite. Graphite is the stable elemental form of carbon at 25°C and 1 atm. Diamond is an allotrope of graphite. Given

diamond: $\Delta H_f° = 1.9$ kJ/mol; $S° = 2.4$ J/mol·K

at what temperature are the two forms in equilibrium at 1 atm?

$$C\ (graphite) \rightleftharpoons C\ (diamond)$$

53. Given the following data for sodium

Na(s): $S° = 51.2$ J/mol·K
Na(g): $S° = 153.6$ J/mol·K $\Delta H_f° = 108.7$ kJ/mol

estimate the temperature at which sodium sublimes at 1 atm.

$$Na(s) \rightleftharpoons Na(g)$$

54. Given the following data for bromine:

$Br_2(l)$; $S° = 152.2$ J/mol·K
$Br_2(g)$; $S° = 245.4$ J/mol·K $\Delta H_f° = 30.91$ kJ/mol

Estimate the normal boiling point of bromine.

Effect of Concentration/Pressure on Spontaneity

55. Show by calculation, using Appendix 1, whether dissolving lead(II) chloride

$$PbCl_2(s) \rightleftharpoons Pb^{2+}(aq) + 2Cl^-(aq)$$

is spontaneous at 25°C
(a) when $[Pb^{2+}] = 1.0$ M; $[Cl^-] = 2.0$ M.
(b) when $[Pb^{2+}] = 1.0 \times 10^{-5}$ M; $[Cl^-] = 2.0 \times 10^{-5}$ M.

56. Show by calculation whether the reaction

$$HF(aq) \rightleftharpoons H^+(aq) + F^-(aq) \qquad \Delta G° = 18.0 \text{ kJ}$$

is spontaneous at 25°C when
(a) $[H^+] = [F^-] = 0.78$ M and $[HF] = 0.24$ M
(b) $[H^+] = [F^-] = 0.0030$ M and $[HF] = 1.85$ M

57. For the reaction

$$2H_2O(l) + 2Cl^-(aq) \longrightarrow H_2(g) + Cl_2(g) + 2OH^-(aq)$$

(a) calculate $\Delta G°$ at 25°C.
(b) calculate ΔG at 25°C when $P_{H_2} = P_{Cl_2} = 0.250$ atm, $[Cl^-] = 0.335$ M, and the pH of the solution is 11.98.

58. For the reaction

$$2H_2O(l) + 2Br^-(aq) \longrightarrow H_2(g) + Br_2(l) + 2OH^-(aq)$$

(a) Calculate $\Delta G°$ at 25°C.
(b) Calculate ΔG at 25°C when $P_{H_2} = 0.878$ atm, $[Br^-] = 0.423$ M, and the pH of the solution is 9.72.

59. Consider the reaction

$$2SO_2(g) + O_2(g) \longrightarrow 2SO_3(g)$$

(a) Calculate $\Delta G°$ at 25°C.
(b) If the partial pressures of SO_2 and SO_3 are kept at 0.400 atm, what partial pressure should O_2 have so that the reaction just becomes nonspontaneous (i.e., $\Delta G = +1.0$ kJ)?

60. Consider the reaction

$$AgCl(s) \longrightarrow Ag^+(aq) + Cl^-(aq)$$

(a) Calculate $\Delta G°$ at 25°C.
(b) What should the concentration of Ag^+ and Cl^- be so that $\Delta G = -1.0$ kJ (just spontaneous)? Take $[Ag^+] = [Cl^-]$.
(c) The K_{sp} for AgCl is 1.8×10^{-10}. Is the answer to (b) reasonable?

16-6 The Free Energy Change and the Equilibrium Constant

61. Consider the reaction

$$CO(g) + H_2O(g) \rightleftharpoons CO_2(g) + H_2(g)$$

Use the appropriate tables to calculate
(a) $\Delta G°$ at 552°C   (b) K at 552°C

62. Consider the reaction

$$NH_4^+(aq) \rightleftharpoons H^+(aq) + NH_3(aq)$$

Use ΔG_f° for $NH_3(aq)$ at $25°C = -26.7$ kJ/mol and the appropriate tables to calculate
   (a) $\Delta G°$ at $25°C$ (b) K_a at $25°C$
63. Consider the following reaction at $25°C$:

$$Cl_2(g) \rightleftharpoons 2Cl(g) \qquad K = 1.0 \times 10^{-37}$$

   (a) Calculate $\Delta G°$ for the reaction at $25°C$.
   (b) Calculate ΔG_f° for $Cl(g)$ at $25°C$.
64. Consider the reaction

$$N_2O(g) + NO_2(g) \longrightarrow 3NO(g) \qquad K = 4.4 \times 10^{-19}$$

   (a) Calculate $\Delta G°$ for the reaction at $25°C$.
   (b) Calculate ΔG_f° for N_2O at $25°C$.
65. For the reaction

$$CO(g) + 3H_2(g) \rightleftharpoons CH_4(g) + H_2O(g)$$

$K = 2.2 \times 10^{11}$ at 473 K and 4.6×10^8 at 533 K. Calculate $\Delta G°$ at both temperatures.
66. Consider the decomposition of N_2O_4 at $100°C$.

$$N_2O_4(g) \rightleftharpoons 2NO_2(g) \qquad K = 11$$

   (a) What is $\Delta G°$ at that temperature?
   (b) What is K at $25°C$? (ΔG_f° for $N_2O_4 = 97.9$ kJ/mol)
67. Use the values for ΔG_f° in Appendix 1 to calculate K_{sp} for barium sulfate at $25°C$. Compare with the value given in Chapter 15.
68. Given that ΔH_f° for $HF(aq)$ is -320.1 kJ/mol and $S°$ for $HF(aq)$ is 88.7 J/mol·K, find K_a for HF at $25°C$.
69. At $25°C$, a 0.327 M solution of a weak acid HX has a pH of 5.12. What is $\Delta G°$ for the dissociation of the weak acid?
70. A 0.250 M solution of a weak base R_2NH has a pH of 10.60 at $25°C$. What is $\Delta G°$ for the dissociation of the weak base in water at $25°C$?

$$R_2NH(aq) + H_2O \longrightarrow R_2NH_2^+(aq) + OH^-(aq)$$

16-7 Additivity of Free Energy Changes; Coupled Reactions

71. Given the following standard free energies at $25°C$,

$$SO_2(g) + 3CO(g) \longrightarrow COS(g) + 2CO_2(g) \quad \Delta G° = -246.5 \text{ kJ}$$
$$CO(g) + H_2O(g) \longrightarrow CO_2(g) + H_2(g) \quad \Delta G° = -28.5 \text{ kJ}$$

find $\Delta G°$ at $25°C$ for the following reaction.

$$SO_2(g) + CO(g) + 2H_2(g) \longrightarrow COS(g) + 2H_2O(g)$$

72. Given the following standard free energies at $25°C$ for the following reactions:

$$N_2O_5(g) \longrightarrow 2NO(g) + \tfrac{3}{2}O_2(g) \qquad \Delta G° = -59.2 \text{ kJ}$$
$$NO(g) + \tfrac{1}{2}O_2(g) \longrightarrow NO_2(g) \qquad \Delta G° = -35.6 \text{ kJ}$$

Calculate $\Delta G°$ at $25°C$ for the following reaction:

$$2NO_2(g) + \tfrac{1}{2}O_2(g) \longrightarrow N_2O_5(g)$$

73. Natural gas, which is mostly methane, CH_4, is a resource that the United States has in abundance. In principle, ethane can be obtained from methane by the reaction

$$2CH_4(g) \longrightarrow C_2H_6(g) + H_2(g)$$

   (a) Calculate $\Delta G°$ at $25°C$ for the reaction. Comment on the feasibility of this reaction at $25°C$.

   (b) Couple the reaction above with the formation of steam from the elements:

$$H_2(g) + \tfrac{1}{2}O_2(g) \longrightarrow H_2O(g) \qquad \Delta G° = -228.6 \text{ kJ}$$

What is the equation for the overall reaction? Comment on the feasibility of the overall reaction.
74. Theoretically, one can obtain zinc from an ore containing zinc sulfide, ZnS, by the reaction

$$ZnS(s) \longrightarrow Zn(s) + S(s)$$

   (a) Show by calculation that this reaction is not feasible at $25°C$.
   (b) Show that by coupling the above reaction with the reaction

$$S(s) + O_2(g) \longrightarrow SO_2(g)$$

the overall reaction, in which Zn is obtained by roasting in oxygen, is feasible at $25°C$.
75. When glucose, $C_6H_{12}O_{11}$, is metabolized to CO_2 and H_2O in living systems, the first step is the conversion of glucose to glucose-6-phosphate.

$$\text{glucose} + H_3PO_4 \longrightarrow \text{glucose-6-phosphate} + H_2O$$
$$\Delta G° = 13.4 \text{ kJ}$$

This nonspontaneous reaction can be made to proceed by coupling it with the hydrolysis of ATP.

$$ATP + H_2O \longrightarrow ADP + H_3PO_4 \qquad \Delta G° = -31.0 \text{ kJ}$$

   (a) Write an equation for the coupled reaction and calculate $\Delta G°$.
   (b) Estimate the number of moles of ATP required to convert glucose to glucose-6-phosphate.
76. Consider the following reactions at $25°C$:

$$C_6H_{12}O_6(aq) + 6O_2(g) \longrightarrow 6CO_2(g) + 6H_2O$$
$$\Delta G° = -2870 \text{ kJ}$$
$$ADP(aq) + HPO_4^{2-}(aq) + 2H^+(aq) \longrightarrow ATP(aq) + H_2O$$
$$\Delta G° = 31 \text{ kJ}$$

Write an equation for a coupled reaction between glucose, $C_6H_{12}O_6$, and ADP in which $\Delta G° = -390$ kJ.

Unclassified

77. At 1200 K, an equilibrium mixture of CO and CO_2 gases contains 98.31 mol percent CO and some solid carbon. The total pressure of the mixture is 1.00 atm. For the system

$$CO_2(g) + C(s) \rightleftharpoons 2CO(g)$$

calculate
   (a) P_{CO} and P_{CO_2} (b) K (c) $\Delta G°$ at 1200 K
78. A protein is said to be denatured when it loses some of its unique characteristics because its structure is changed. The enthalpy change for the denaturation of a certain protein is 142 kJ/mol, while its entropy change is 278 J/K. What is the minimum temperature at which the protein would denature spontaneously?
79. A student is asked to prepare a 0.030 M aqueous solution of $PbCl_2$.
   (a) Is this possible at $25°C$? (*Hint:* Is dissolving 0.030 mol of $PbCl_2$ at $25°C$ possible?)
   (b) If the student used water at $100°C$, would this be possible?

80. Some bacteria use light energy to convert carbon dioxide and water to glucose and oxygen:

$$6CO_2(g) + 6H_2O(l) \longrightarrow C_6H_{12}O_6(aq) + 6O_2(g)$$
$$\Delta G° = 2870 \text{ kJ at } 25°C$$

Other bacteria, those that do not have light available to them, couple the reaction

$$H_2S(g) + \tfrac{1}{2}O_2(g) \longrightarrow H_2O(l) + S(s)$$

to the glucose synthesis above. Coupling the two reactions, the overall reaction is

$$24H_2S(g) + 6CO_2(g) + 6O_2(g) \longrightarrow$$
$$C_6H_{12}O_6(aq) + 18H_2O(l) + 24S(s)$$

Show that the reaction is spontaneous at 25°C.

81. Calculate the first and second ionization constants (K_{a1} and K_{a2}) for the ionization of the diprotic acid H_2SO_4 given the following information:

$$H_2SO_4(aq): \Delta H_f° = -902.9 \text{ kJ} \qquad S° = 17.14 \text{ J/K}$$

82. Carbon monoxide poisoning results when carbon monoxide replaces oxygen bound to hemoglobin. The oxygenated form of hemoglobin, $Hb \cdot O_2$ carries O_2 to the lungs.

$$Hb \cdot O_2(aq) + CO(g) \rightleftharpoons Hb \cdot CO(aq) + O_2(g)$$

At 98.6°F (37°C), $\Delta G°$ for the reaction is about -14 kJ. What is the ratio of $[Hb \cdot O_2]$ to $[Hb \cdot CO]$ when the pressure of CO is the same as that of O_2?

83. The formation constant for the following reaction at 25°C

$$Zn^{2+}(aq) + 4NH_3(aq) \rightleftharpoons Zn(NH_3)_4^{2+}(aq)$$

is 3.6×10^8.
(a) What is $\Delta G°$ at this temperature?
(b) If standard state concentrations of the reactants and products are combined, in which direction will the reaction proceed?
(c) What is ΔG when $[Zn(NH_3)_4^{2+}] = 0.010 \, M, [Zn^{2+}] = 0.0010 \, M$ and $[NH_3] = 3.5 \times 10^{-4}$?

Conceptual Problems

84. Determine whether each of the following statements is true or false.
(a) An exothermic reaction is spontaneous.
(b) When $\Delta G°$ is positive, the reaction cannot occur under any conditions.
(c) $\Delta S°$ is positive for a reaction in which there is an increase in the number of moles.
(d) If $\Delta H°$ and $\Delta S°$ are both negative, $\Delta G°$ will be negative.

85. Which of the following quantities can be taken to be independent of temperature? independent of pressure?
(a) ΔH for a reaction   (b) ΔS for a reaction
(c) ΔG for a reaction   (d) S for a substance

86. Fill in the blanks:
(a) $\Delta H°$ and $\Delta G°$ become equal at _____ K.
(b) $\Delta G°$ and ΔG are equal when $Q = $ _____.
(c) $S°$ for steam is _____ than $S°$ for water.

87. Fill in the blanks:
(a) At equilibrium, ΔG is _____.
(b) For $C_6H_6(l) \rightleftharpoons C_6H_6(g)$, $\Delta H°$ is _____. $(+, -, 0)$.
(c) When a pure solid melts, the temperature at which liquid and solid are in equilibrium and $\Delta G° = 0$ is called _____.

88. In your own words, explain why
(a) $\Delta S°$ is negative for a reaction in which the number of moles of gas decreases.
(b) we take $\Delta S°$ to be independent of T, even though entropy increases with T.
(c) a solid has lower entropy than its corresponding liquid.

89. Consider the following reaction with its thermodynamic data:

$$2A(g) + B_2(g) \longrightarrow 2AB(g) \ \Delta H° < 0; \ \Delta S° < 0; \ \Delta G° \text{ at } 60°C = +10 \text{ kJ}$$

Which statements about the reaction are true?
(a) When $\Delta G = 1$, the reaction is at equilibrium.
(b) When $Q = 1$, $\Delta G = \Delta G°$.
(c) At 75°C, the reaction is definitely nonspontaneous.
(d) At 100°C, the reaction has a positive entropy change.
(e) If A and B_2 are elements in their stable states, $S°$ for A and B_2 at 25°C is 0.
(f) K for the reaction at 60°C is less than 1.

90. Consider the graph below:

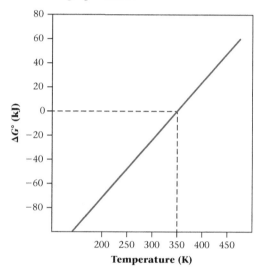

(a) Describe the relationship between the spontaneity of the process and temperature.
(b) Is the reaction exothermic?
(c) Is $\Delta S° > 0$?
(d) At what temperature is the reaction at standard conditions likely to be at equilibrium?
(e) Estimate K for the reaction at 97°C.

91. The relationship between $\Delta G°$ and T is linear. Draw the graph of a reaction with the properties described below. You need only label the point at which $\Delta G° = 0$.

■ The reaction is exothermic.
■ $\Delta n_g < 0$ ($\Delta n_g =$ moles of gas products $-$ moles of gas reactants).
■ At 300 K, the system is at equilibrium and $K = 1$.

92. Answer the questions below by writing **LT** (for *is less than*), **GT** (for *is greater than*), **EQ** (for *is equal to*), or **MI** (for *more information required*) in the blanks.
The reaction given below takes place in a cylinder that feels warm to the touch after the reaction is complete.

$$A_2(g) + B_2(g) \longrightarrow 2AB(s)$$

(a) At all temperatures, $\Delta S°$ _____ 0.
(b) At all temperatures, $\Delta H°$ _____ 0.
(c) At all temperatures, $\Delta G°$ _____ 0.

Challenge Problems

93. Consider the following data about ethyl alcohol:

ΔH_{vap} = 42.9 kJ   vapor pressure = 392 mm Hg at 63°C

What is $S°$ for $C_2H_5OH(g)$?

94. $\Delta H_f°$ for iodine gas is 62.4 kJ/mol, and $S°$ is 260.7 J/mol·K. Calculate the equilibrium partial pressures of $I_2(g)$, $H_2(g)$, and $HI(g)$ for the system

$$2HI(g) \rightleftharpoons H_2(g) + I_2(g)$$

at 500°C if the initial partial pressures are all 0.200 atm.

95. The heat of fusion of ice is 333 J/g. For the process

$$H_2O(s) \longrightarrow H_2O(l)$$

determine

  (a) $\Delta H°$ (b) $\Delta G°$ at 0°C   (c) $\Delta S°$

  (d) $\Delta G°$ at −20°C   (e) $\Delta G°$ at 20°C

96. The overall reaction that occurs when sugar is metabolized is

$$C_{12}H_{22}O_{11}(s) + 12O_2(g) \longrightarrow 12CO_2(g) + 11H_2O(l)$$

For this reaction, $\Delta H°$ is −5650 kJ and $\Delta G°$ is −5790 kJ at 25°C.

  (a) If 25% of the free energy change is actually converted to useful work, how many kilojoules of work are obtained when one gram of sugar is metabolized at body temperature, 37°C?

  (b) How many grams of sugar would a 120-lb woman have to eat to get the energy to climb the Jungfrau in the Alps, which is 4158 m high? ($w = 9.79 \times 10^{-3}$ mh, where w = work in kilojoules, m is body mass in kilograms, and h is height in meters.)

97. Hydrogen has been suggested as the fuel of the future. One way to store it is to convert it to a compound that can be heated to release the hydrogen. One such compound is calcium hydride, CaH_2. This compound has a heat of formation of −186.2 kJ/mol and a standard entropy of 42.0 J/mol·K. What is the minimum temperature to which calcium hydride would have to be heated to produce hydrogen at one atmosphere pressure?

98. When a copper wire is exposed to air at room temperature, it becomes coated with a black oxide, CuO. If the wire is heated above a certain temperature, the black oxide is converted to a red oxide, Cu_2O. At a still higher temperature, the oxide coating disappears. Explain these observations in terms of the thermodynamics of the reactions

$$2CuO(s) \longrightarrow Cu_2O(s) + \tfrac{1}{2}O_2(g)$$
$$Cu_2O(s) \longrightarrow 2Cu(s) + \tfrac{1}{2}O_2(g)$$

and estimate the temperatures at which the changes occur.

99. K_a for acetic acid ($HC_2H_3O_2$) at 25°C is 1.754×10^{-5}. At 50°C, K_a is 1.633×10^{-5}. Assuming that $\Delta H°$ and $\Delta S°$ are not affected by a change in temperature, calculate $\Delta S°$ for the ionization of acetic acid.

100. Consider the reaction

$$2HI(g) \rightleftharpoons H_2(g) + I_2(g)$$

At 500°C, a flask initially has all three gases, each at a partial pressure of 0.200 atm. When equilibrium is established, the partial pressure of HI is determined to be 0.48 atm. What is $\Delta G°$ for the reaction at 500°C?

101. Consider the formation of ammonia from its elements:

$$N_2(g) + 3H_2(g) \longrightarrow 2NH_3(g)$$

At what temperature is $K = 1.00$? (*Hint:* Use the thermodynamic tables in Appendix 1.)

102. Consider the formation of $HI(g)$ from $H_2(g)$ and $I_2(g)$.

$$H_2(g) + I_2(g) \rightleftharpoons 2HI(g)$$

For this reaction, K at 333°C is 98, while K at 527°C is 62.5. Find $S°$ for $I_2(g)$ at 527°C.

17 Electrochemistry

Most "silver" tableware like the ones in the painting are not made of pure silver but rather are silver-plated using an electrolytic cell.

If by fire,
Of sooty coal th' empiric
    Alchymist
Can turn, or holds it possible
    to turn
Metals of drossiest Ore to
    Perfect Gold.

—JOHN MILTON
Paradise Lost (Book V, Lines 439–442)

Suzanne C. Oullette

i Cathode = reduction; anode = oxidation. The consonants go together, as do the vowels.

Electrochemistry is the study of the interconversion of electrical and chemical energy. This conversion takes place in an electrochemical cell that may be a(n)

- **voltaic (galvanic) cell** (Section 17-2), in which a spontaneous reaction generates electrical energy.
- **electrolytic cell** (Section 17-6), in which electrical energy is used to bring about a nonspontaneous reaction.

All of the reactions considered in this chapter are of the oxidation-reduction type. You will recall from Chapter 4 that such a reaction can be split into two half-reactions. In one half-reaction, referred to as **reduction**, electrons are consumed; in the other, called **oxidation**, electrons are produced. There can be no net change in the number of electrons; the number of electrons consumed in reduction must be exactly equal to the number produced in the oxidation half-reaction. These two half-reactions combine to give a balanced redox reaction (Section 17-1).

In an electrochemical cell, these two half-reactions occur at two different **electrodes**, which most often consist of metal plates or wires. Reduction occurs at the **cathode**; ◂*i* a typical half-reaction might be

$$\text{cathode:} \qquad Cu^{2+}(aq) + 2e^- \longrightarrow Cu(s)$$

Oxidation takes place at the **anode**, ◂*i* where a species such as zinc metal produces electrons:

$$\text{anode:} \qquad Zn(s) \longrightarrow Zn^{2+}(aq) + 2e^-$$

It is always true that in an electrochemical cell, *anions* move to the *anode; cations* move to the *cathode*.

One of the most important characteristics of a cell is its *voltage*, which is a measure of reaction spontaneity. Cell voltages depend on the nature of the half-reactions occurring at the electrodes (Section 17-3) and on the concentrations of species involved (Section 17-5). From the voltage measured at standard concentrations, it is possible to calculate the standard free energy change and the equilibrium constant (Section 17-4) of the reaction involved.

The principles discussed in this chapter have a host of practical applications. Whenever you start your car, turn on your cell phone, or use a remote control for your television or other devices, you are making use of a voltaic cell. Many of the most important elements, including hydrogen and chlorine, are prepared in electrolytic cells. These applications, among others, are discussed in Section 17-7.

17-1 Oxidation-Reduction Reactions Revisited

In Chapter 4, redox reactions were introduced as one of the three important types of reactions in aqueous solutions. In this section, we will review some of the terminology introduced in Chapter 4 and then go on to describe how to balance redox equations by first learning how to balance redox half-reactions.

Review of Oxidation-Reduction Terminology and Concepts

- A **redox reaction** involves a transfer of electrons between two species.
- **Oxidation** occurs when a species loses electrons and increases its oxidation number (O. N.). An equation that shows this loss is called an *oxidation half-reaction*. In this half-reaction, the electrons are in the product side.
- **Reduction** occurs when a species gains electrons and decreases its oxidation number (O. N.). An equation that shows this gain is called a *reduction half-reaction*. In this half-reaction, the electrons are in the reactant side.
- The ion or molecule that donates the electrons (i.e., oxidized) is called the **reducing agent**.
- The ion or molecule that accepts the electrons (i.e., reduced) is called the **oxidizing agent**.
- Oxidation and reduction occur together in the same reaction. There is no net change in the number of electrons, just an exchange of electrons.

Balancing Half-Equations (Oxidation or Reduction)

Before you can balance an overall redox equation, you have to be able to balance two **half-equations,** one for oxidation (electron loss) and one for reduction (electron gain). Sometimes that's easy. Given the oxidation half-equation

$$Fe^{2+}(aq) \longrightarrow Fe^{3+}(aq) \qquad (\text{O.N. Fe: } +2 \longrightarrow +3)$$

it is clear that mass and charge balance can be achieved by adding an electron to the right:

$$Fe^{2+}(aq) \longrightarrow Fe^{3+}(aq) + e^-$$

In another case, this time a reduction half-equation,

$$Cl_2(g) \longrightarrow Cl^-(aq) \qquad (\text{O.N. Cl: } 0 \longrightarrow -1)$$

mass balance is obtained by writing a coefficient of 2 for Cl^-; charge is then balanced by adding two electrons to the left. The balanced half-equation is

$$Cl_2(g) + 2e^- \longrightarrow 2Cl^-(aq)$$

Sometimes, though, it is by no means obvious how a given half-equation is to be balanced. This commonly happens when elements other than those being oxidized or reduced take part in the reaction. Most often, these elements are oxygen (O.N. = −2) and hydrogen (O.N. = +1). Consider, for example, the half-equation for the reduction of the permanganate ion,

$$MnO_4^-(aq) \longrightarrow Mn^{2+}(aq) \qquad (\text{O.N. Mn: } +7 \longrightarrow +2)$$

or the oxidation of chromium(III) hydroxide,

$$Cr(OH)_3(s) \longrightarrow CrO_4^{2-}(aq) \qquad (\text{O.N. Cr: } +3 \longrightarrow +6)$$

To balance half-equations such as these, proceed as follows:

(a) *Assign oxidation numbers to each element.*
(b) *Balance the atoms of each element being oxidized or reduced.*
(c) *Multiply the oxidation number by the number of atoms that have that oxidation number. This gives you the "total" oxidation number.*
    The number of atoms is indicated either by the coefficient that you used to balance the atoms or by the subscript of the atom.
(d) *Balance oxidation number by adding electrons.*
    Electrons are added to the left for a reduction half-equation and to the right for an oxidation half-equation.
    The number of electrons added should equal the change in "total" number from part (c).
(e) *Balance charge by adding H⁺ ions in acidic solution and OH⁻ ions in basic solution.*
(f) *Balance hydrogen by adding H₂O molecules.*
(g) *Check to make sure that oxygen is balanced.*
    If it is, the half-equation is almost certainly balanced correctly with respect to mass and charge.

Figure 17.1 is a flowchart schematically showing the steps described above. Example 17.1 shows how these rules are applied to balance half-equations.

Figure 17.1 Flowchart for balancing half-reactions in acid or base medium.

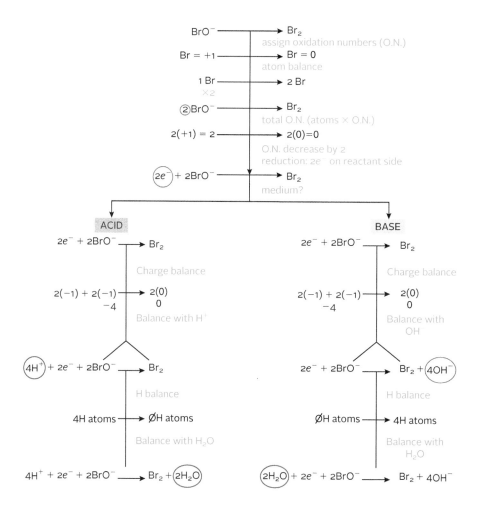

EXAMPLE 17.1

Balance the following half-equations. Follow the steps outlined in the order given.

a $NO_3^-(aq) \rightarrow NO(g)$ (basic solution)

	SOLUTION
1. Oxidation numbers	N: $+5 \rightarrow +2$; O: $-2 \rightarrow -2$; N is reduced.
2. Atom balance	1 N on each side; no adjustment is required.
3. "total oxidation" number	N: $5(1) \rightarrow$ N: $2(1)$
4. Add electrons	The oxidation number for N goes from $+5$ to $+2$. It is reduced by 3. Add 3 electrons to the reactant side. $NO_3^-(aq) + 3e^- \rightarrow NO(g)$
5. Balance charge	reactants: $-1 + 3(-1) = -4$ products: 0 basic medium: add OH^-. To balance, add $4OH^-$ on the right. $NO_3^-(aq) + 3e^- \rightarrow NO(g) + 4OH^-(aq)$ reactants: $-1 + 3(-1) = -4$ products: $4(-1) = -4$
6. Balance H	reactants: 0 H products: 4 H To balance, add $2H_2O$ on the left. $NO_3^-(aq) + 3e^- + 2H_2O \rightarrow NO(g) + 4OH^-(aq)$ reactants: 4 H products: 4 H
7. Check O:	reactants: $3 + 2 = 5$ products: $4 + 1 = 5$
The half-equation is balanced:	$NO_3^-(aq) + 3e^- \rightarrow NO(g) + 4OH^-(aq)$

b $Cl_2(g) \rightarrow ClO_3^-(aq)$ (acidic solution)

	SOLUTION
1. Oxidation numbers	Cl: $0 \rightarrow +5$; O: $-2 \rightarrow -2$; Cl is oxidized.
2. Atom balance	reactant: 2 Cl product: 1 Cl Multiply ClO_3^- by 2. $Cl_2(g) \rightarrow 2ClO_3^-(aq)$
3. "total oxidation" number	Cl: $0(2) = 0 \rightarrow$ Cl: $5(2) = 10$
4. Add electrons	The oxidation number for Cl goes from 0 to 10. The oxidation number increases by 10. Add 10 electrons to the product side. $Cl_2(g) \rightarrow 2ClO_3^-(aq) + 10e^-$
5. Balance charge	reactants: 0 products: $2(-1) + 10(-1) = -12$ acidic medium: add H^+. To balance, add $12 H^+$ on the right. $Cl_2(g) \rightarrow 2ClO_3^-(aq) + 10e^- + 12H^+(aq)$ reactants: 0 products: $2(-1) + 10(-1) + 12(+1) = 0$
6. Balance H	reactants: 0 H products: 12 H To balance, add $6H_2O$ on the left. $Cl_2(g) + 6H_2O \rightarrow 2ClO_3^-(aq) + 10e^- + 12H^+(aq)$ reactants: 12 H products: 12 H
7. Check O:	reactants: 6 products: $2(3) = 6$
The half-equation is balanced:	$Cl_2(g) + 6H_2O \rightarrow 2ClO_3^-(aq) + 10e^- + 12H^+(aq)$

Balancing Redox Equations

The process used to balance an overall redox equation is relatively straightforward, provided you know how to balance half-equations. Follow a systematic, four-step procedure:

1. *Split the equation into two half-equations,* one for reduction, the other for oxidation.
2. *Balance one of the half-equations* with respect to both atoms and charge as described above (steps a through e).
3. *Balance the other half-equation.*
4. *Combine the two half-equations in such a way as to eliminate electrons.* Suppose, for example, the two balanced half-equations are

$$A(s) \longrightarrow A^{2+}(aq) + 2e^-$$
$$B^{3+}(aq) + 3e^- \longrightarrow B(s)$$

Multiplying the first half-equation by 3, the second by 2, and adding gives

$$3A(s) + 2B^{3+}(aq) \longrightarrow 3A^{2+}(aq) + 2B(s)$$

The electrons, six on both sides, cancel.

EXAMPLE 17.2

Balance the following redox reactions by following the four-step process outlined above.

a $Fe_2^+(aq) + NO_3^-(aq) \rightarrow Fe_3^+(aq) + NO(g)$ (basic solution)

SOLUTION

1. Split into two half-equations.	$Fe^{2+}(aq) \rightarrow Fe^{3+}(aq)$ $NO_3^-(aq) \rightarrow NO(g)$
2-3. Balance the half-equations.	Check the text. This has been done earlier. $Fe^{2+}(aq) \rightarrow Fe^{3+}(aq) + e^-$ $NO_3^-(aq) + 3e^- + 2H_2O \rightarrow NO(g) + 4OH^-(aq)$
4. Eliminate electrons.	Multiply the oxidation half-equation by 3. $3[Fe^{2+}(aq) \rightarrow Fe^{3+}(aq) + e^-]$
Combine half-equations.	$NO_3^-(aq) + 3Fe^{2+}(aq) + 2H_2O \rightarrow NO(g) + 4OH^-(aq) + 3Fe^{3+}(aq)$

b $MnO_4^-(aq) + Cl_2(g) \rightarrow Mn_2^+(aq) + ClO_3^-(aq)$ (acidic solution)

SOLUTION

1. Split into two half-equations.	$MnO_4^-(aq) \rightarrow Mn^{2+}(aq)$ $Cl_2(g) \rightarrow ClO_3^-(aq)$
2-3. Balance the half-equations.	The oxidation half-equation is balanced in Example 17.1. $Cl_2(g) + 6H_2O \rightarrow 2ClO_3^-(aq) + 10e^- + 12H^+(aq)$ Try to balance the reduction half-equation. $MnO_4^-(aq) + 8H^+(aq) + 5e^- \rightarrow Mn^{2+}(aq) + 4H_2O$
4. Eliminate electrons.	Multiply the reduction half-equation by 2. $2[MnO_4^-(aq) + 8H^+(aq) + 5e^- \rightarrow Mn^{2+}(aq) + 4H_2O]$
Combine half-equations.	$Cl_2(g) + 6H_2O + 2MnO_4^-(aq) + 16H^+(aq) \rightarrow$ $\quad 2ClO_3^-(aq) + 12H^+(aq) + 2Mn^{2+}(aq) + 8H_2O$
Net ionic equation	$6H_2O \rightarrow 8H_2O = 2H_2O$ (product side) $16H^+ \rightarrow 12H^+ = 4H^+$ (reactant side)
Balanced net ionic equation	$Cl_2(g) + 2MnO_4^-(aq) + 4H^+(aq) \rightarrow 2ClO_3^-(aq) + 2Mn^{2+}(aq) + 2H_2O$

continued

		END POINT			

It is a good idea to check both mass and charge balance in the final balanced net ionic equation. In (2), for example:

	Cl Atoms	Mn Atoms	O Atoms	H Atoms	Charge
Left	2	2	$2(4) = 8$	4	$+0 - 2 + 4 = +2$
Right	2	2	$2(3) + 2 = 8$	$2(2) = 4$	$-2 + 4 + 0 = +2$

17-2 Voltaic Cells

In principle at least, any spontaneous redox reaction can serve as a source of energy in a **voltaic (galvanic) cell.** The cell must be designed in such a way that oxidation occurs at one electrode (anode) with reduction at the other electrode (cathode). The electrons produced at the anode must be transferred to the cathode, where they are consumed. To do this, the electrons move through an external circuit, where they do electrical work.

To understand how a voltaic cell operates, let us start with some simple cells that are readily made in the general chemistry laboratory.

The Zn-Cu^{2+} Cell

When a piece of zinc is added to a water solution containing Cu^{2+} ions, the following redox reaction takes place:

$$Zn(s) + Cu^{2+}(aq) \longrightarrow Zn^{2+}(aq) + Cu(s)$$

In this reaction, copper metal plates out on the surface of the zinc. The blue color of the aqueous Cu^{2+} ion fades as it is replaced by the colorless aqueous Zn^{2+} ion (Figure 17.2). Clearly, this redox reaction is spontaneous; it involves electron transfer from a Zn atom to a Cu^{2+} ion.

To design a voltaic cell using the Zn-Cu^{2+} reaction as a source of electrical energy, the electron transfer must occur indirectly; that is, the electrons given off by zinc atoms must be made to pass through an external electric circuit before

© Cengage Learning/Charles D. Winters

Figure 17.2 Zn-Cu^{2+} spontaneous redox reaction. The blue of the Cu^{2+} ion in solution fades *(from left to right)* as the ion is reduced to metallic Cu that deposits on the zinc strip in the solution.

Figure 17.3 A Zn-Cu²⁺ voltaic cell. In this voltaic cell, a voltmeter *(left)* is connected to a half-cell consisting of a Cu cathode in a solution of blue Cu^{2+} ions and a half-cell consisting of a Zn anode in a solution of colorless Zn^{2+} ions. The following spontaneous reaction takes place in this cell: $Zn(s) + Cu^{2+}(aq) \longrightarrow Zn^{2+}(aq) + Cu(s)$. The salt bridge allows ions to pass from one solution to the other to complete the circuit and prevents direct contact between Zn atoms and the Cu^{2+} ions.

The Zn electrode must not come in contact with Cu^{2+} ions. Why?

they reduce Cu^{2+} ions to copper atoms. ◂ⓘ Figure 17.3 shows one way to do this. The voltaic cell consists of an anode **half-cell** and a cathode **half-cell**:

- a Zn anode dipping into a solution containing Zn^{2+} ions, seen in the beaker at the far right.
- a Cu cathode dipping into a solution containing Cu^{2+} ions (blue), seen in the other beaker in Figure 17.3.

The "external circuit" consists of a voltmeter with leads (red and black) to the anode and cathode.

Let us trace the flow of electric current through the cell.

1. At the zinc *anode* (connected to the black wire in Figure 17.3), electrons are produced by the oxidation half-reaction

$$Zn(s) \longrightarrow Zn^{2+}(aq) + 2e^-$$

This electrode, which "pumps" electrons into the external circuit, is ordinarily marked as the negative pole of the cell.

2. Electrons generated at the anode move through the external circuit (right to left in Figure 17.3) to the copper *cathode,* connected through the red wire to the voltmeter. At the cathode, the electrons are consumed, reducing Cu^{2+} ions present in the solution around the electrode:

$$Cu^{2+}(aq) + 2e^- \longrightarrow Cu(s)$$

The copper electrode, which "pulls" electrons from the external circuit, is considered to be the positive pole of the cell.

3. As the above half-reactions proceed, a surplus of positive ions (Zn^{2+}) tends to build up around the zinc electrode. The region around the copper electrode tends to become deficient in positive ions as Cu^{2+} ions are consumed. To maintain electrical neutrality, cations must move toward the copper cathode or, alternatively, anions must move toward the zinc anode. In practice, both migrations occur.

In the cell in Figure 17.3, movement of ions occurs through a **salt bridge** connecting the two beakers. The salt bridge is an inverted glass U-tube, plugged with glass wool at each end. The tube is filled with a solution of a salt that takes no part in the electrode reactions; potassium nitrate, KNO_3, is frequently used. As current is drawn from the cell, K^+ ions move from the salt bridge into the cathode half-cell. At the same time, NO_3^- ions move into the anode half-cell. In this way, electrical neutrality is maintained without Cu^{2+} ions coming in contact with the Zn electrode, which would short-circuit the cell.

The cell in Figure 17.3 is often abbreviated as

$$Zn \mid Zn^{2+} \parallel Cu^{2+} \mid Cu \blacktriangleleft \textbf{\textit{i}}$$

In this notation,

Anode $\parallel$ cathode.

- the *anode* reaction *(oxidation)* is shown at the left. Zn atoms are oxidized to Zn^{2+} ions.
- the salt bridge (or other means of separating the half cells) is indicated by the symbol $\parallel$.
- the *cathode* reaction *(reduction)* is shown at the right. Cu^{2+} ions are reduced to Cu atoms.
- a single vertical line indicates a phase boundary, such as that between a solid electrode and an aqueous solution.

Notice that the anode half-reaction comes first in the cell notation, just as the letter *a* comes before *c*.

Other Salt Bridge Cells

Cells similar to that in Figure 17.3 can be set up for many different spontaneous redox reactions. Consider, for example, the reaction

$$Ni(s) + Cu^{2+}(aq) \longrightarrow Ni^{2+}(aq) + Cu(s)$$

This reaction, like that between Zn and Cu^{2+}, can serve as a source of electrical energy in a voltaic cell. The cell is similar to the one in Figure 17.3 except that, in the anode compartment, a nickel electrode is surrounded by a solution of a nickel(II) salt, such as $NiCl_2$ or $NiSO_4$. The cell notation is $Ni \mid Ni^{2+} \parallel Cu^{2+} \mid Cu$.

Another spontaneous redox reaction that can serve as a source of electrical energy is that between zinc metal and Co^{3+} ions:

$$Zn(s) + 2Co^{3+}(aq) \longrightarrow Zn^{2+}(aq) + 2Co^{2+}(aq)$$

A voltaic cell using this reaction is similar to the Zn-Cu^{2+} cell; the $Zn \mid Zn^{2+}$ half-cell and the salt bridge are the same. Because no metal is involved in the cathode half-reaction, an *inert* electrode that conducts an electric current is used. Frequently, the cathode is made of platinum $\blacktriangleleft \textbf{\textit{i}}$ (Figure 17.4). In the cathode, Co^{3+} ions are provided by a solution of $Co(NO_3)_3$. The half-reactions occurring in the cell are

Nichrome or graphite can also be used.

anode: $Zn(s) \longrightarrow Zn^{2+}(aq) + 2e^-$ (oxidation)

cathode: $Co^{3+}(aq) + e^- \longrightarrow Co^{2+}(aq)$ (reduction)

The cell notation is $Zn \mid Zn^{2+} \parallel Co^{3+}, Co^{2+} \mid Pt$. Note that a comma separates the half-cell components that are in the *same* phase. The symbol Pt is used to indicate the presence of an inert platinum electrode. A single vertical line separates Pt (a solid) from the components of the half-cell, which are in the liquid phase.

Figure 17.4 A Zn-Co³⁺ voltaic cell. A platinum electrode is immersed in a solution containing Co^{3+} and Co^{2+} ions. The spontaneous cell reaction is

$$Zn(s) + 2Co^{3+}(aq) \longrightarrow Zn^{2+}(aq) + 2Co^{2+}(aq)$$

The platinum electrode does not participate in the reaction. The zinc electrode does.

EXAMPLE 17.3

When chlorine gas is bubbled through an aqueous solution of NaBr, chloride ions and liquid bromine are the products of the spontaneous reaction.

(a) Draw a sketch of the cell, labeling the anode, the cathode, and the direction of electron flow.

(b) Write the half-reaction that takes place at the anode and at the cathode.

(c) Write a balanced equation for the cell reaction.

(d) Write an abbreviated notation for the cell.

STRATEGY

1. Split the equation into two half-reactions.

2. Recall that the anode
   - is where oxidation takes place.
   - is the electrode toward which anions move.
   - is where electrons are produced.

3. The cathode
   - is where reduction takes place.
   - is the electrode toward which cations move.
   - is where electrons are released by the anode through an external circuit.

SOLUTION

(a) Sketch of the cell	See Figure 17.5, where all the appropriate parts are labeled and the direction of electron flow is indicated.
(b) Half-reactions	cathode: $Cl_2(g) + 2e^- \longrightarrow 2Cl^-(aq)$ (reduction)
	anode: $2Br^-(aq) \longrightarrow Br_2(l) + 2e^-$ (oxidation)
(c) Balanced equation	$Cl_2(g) + 2Br^-(aq) \longrightarrow 2Cl^-(aq) + Br_2(l)$
(d) Abbreviated cell notation	$Pt \mid Br_2, Br^- \parallel Cl^- \mid Cl_2 \mid Pt$

Figure 17.5 A Cl$_2$-Br$_2$ voltaic cell. In this cell the spontaneous cell reaction is

$$Cl_2(g) + 2Br^-(aq) \longrightarrow 2Cl^-(aq) + Br_2(l)$$

Both electrodes are platinum, with the anode at the left and the cathode at the right.

To summarize our discussion of the structure of voltaic cells,

- *a voltaic cell consists of two half-cells.* They are joined by an external electric circuit through which electrons move and a salt bridge through which ions move.
- *each half-cell consists of an electrode dipping into a water solution.* If a metal participates in the cell reaction, either as a product or as a reactant, it is ordinarily used as the electrode; otherwise, an inert electrode such as platinum is used.
- *in one half-cell, oxidation occurs at the anode; in the other, reduction takes place at the cathode.* The overall cell reaction is the sum of the half-reactions taking place at the anode and cathode.

17-3 Standard Voltages

The driving force behind the spontaneous reaction in a voltaic cell is measured by the cell voltage, which is an *intensive* property, independent of the number of electrons passing through the cell. **Cell voltage** depends on the nature of the redox reaction and the concentrations of the species involved; for the moment, we'll concentrate on the first of these factors.

The **standard cell voltage** ($E°$) (also referred to as the *standard voltage*) for a given cell is that measured when the current flow is essentially zero, *all ions and molecules in solution are at a concentration of 1 M, and all gases are at a pressure of 1 atm.* To illustrate, consider the Zn-H$^+$ cell. Let us suppose that the half-cells are set up in such a way that the concentrations of Zn^{2+} and H$^+$ are both 1 M and the pressure of H$_2(g)$ is 1 atm. Under these conditions, the cell voltage at very low current flow is +0.762 V. This quantity is referred to as the standard voltage and is given the symbol $E°$.

$$Zn(s) + 2H^+(aq, 1\ M) \longrightarrow Zn^{2+}(aq, 1\ M) + H_2(g, 1\ atm) \qquad E° = +0.762\ V$$

$E°_{red}$ and $E°_{ox}$

Any redox reaction can be split into two half-reactions, an oxidation and a reduction. It is possible to associate the standard voltages, **standard oxidation voltage ($E°_{ox}$)** and **standard reduction voltage ($E°_{red}$)**, with the oxidation and reduction half-reactions. The standard voltage for the overall reaction, $E°$, is the sum of these two quantities

$$E° = E°_{red} + E°_{ox} \blacktriangleleft \textit{i} \tag{17.1}$$

i You need two half-cells to measure a voltage.

To illustrate, consider the reaction between Zn and H$^+$ ions, for which the standard voltage is +0.762 V.

$$+0.762\ V = E°_{red}(H^+ \longrightarrow H_2) + E°_{ox}(Zn \longrightarrow Zn^{2+})$$

There is no way to measure the standard voltage for a half-reaction; only $E°$ can be measured directly. To obtain values for $E°_{ox}$ and $E°_{red}$, the value zero is arbitrarily assigned to the standard voltage for reduction of H$^+$ ions to H$_2$ gas:

$$2H^+(aq, 1\ M) + 2e^- \longrightarrow H_2(g, 1\ atm) \qquad E°_{red}(H^+ \longrightarrow H_2) = 0.000\ V$$

Using this convention, it follows that the standard voltage for the oxidation of zinc must be +0.762 V; that is,

$$Zn(s) \longrightarrow Zn^{2+}(aq, 1\ M) + 2e^- \qquad E°_{ox}(Zn \longrightarrow Zn^{2+}) = +0.762\ V$$

As soon as one half-reaction voltage is established, others can be calculated. For example, the standard voltage for the Zn-Cu^{2+} cell shown in Figure 17.3 is found to be +1.101 V. Knowing that $E°_{ox}$ for zinc is +0.762 V, it follows that

$$E°_{red}(Cu^{2+} \longrightarrow Cu) = E° - E°_{ox}(Zn \longrightarrow Zn^{2+})$$
$$= 1.101\ V - 0.762\ V = +0.339\ V$$

Standard half-cell voltages are ordinarily obtained from a list of standard voltages such as those in Table 17.1. The unit for the electric potential is the **volt**. Each **standard potential** listed is the standard voltage for a given half-reaction, that is

$$\text{standard potential} = E^\circ_{\text{red}}$$

For example, the standard potentials listed in the table for $Zn^{2+} \longrightarrow Zn$ and $Cu^{2+} \longrightarrow Cu$ are -0.762 V and $+0.339$ V, respectively; it follows that at 25°C

$$Zn^{2+}(aq, 1\ M) + 2e^- \longrightarrow Zn(s) \qquad E^\circ_{\text{red}} = -0.762 \text{ V}$$
$$Cu^{2+}(aq, 1\ M) + 2e^- \longrightarrow Cu(s) \qquad E^\circ_{\text{red}} = +0.339 \text{ V}$$

To obtain the standard voltage for an oxidation half-reaction, all you have to do is *change the sign of the standard potential* ◀ ⓘ *listed in Table 17.1.* For example, knowing that

$$Zn^{2+}(aq) + 2e^- \longrightarrow Zn(s) \qquad E^\circ_{\text{red}} = -0.762 \text{ V}$$

it follows that

$$Zn(s) \longrightarrow Zn^{2+}(aq) + 2e^- \qquad E^\circ_{\text{ox}} = +0.762 \text{ V}$$

Again, we have

$$Cu^{2+}(aq) + 2e^- \longrightarrow Cu(s) \qquad E^\circ_{\text{red}} = +0.339 \text{ V}$$

so,

$$Cu(s) \longrightarrow Cu^{2+}(aq) + 2e^- \qquad E^\circ_{\text{ox}} = -0.339 \text{ V}$$

Another way to express this principle is to say that *standard voltages for forward and reverse half-reactions are equal in magnitude but opposite in sign.*

Strength of Oxidizing and Reducing Agents

As pointed out in Section 17-1, an oxidizing agent is a species that can gain electrons; it is a reactant in a reduction half-reaction. Table 17.1 lists reduction half-reactions from left to right; it follows that *oxidizing agents are in the left column of the table.* All of the species appearing in that column ($Li^+, \ldots, F_2$) are, in principle, oxidizing agents. A "strong" oxidizing agent is one that has a strong attraction for electrons and hence *can readily oxidize other species.* In contrast, a "weak" oxidizing agent does not gain electrons readily. It is capable of reacting only with those species that are very easily oxidized.

The strength of an oxidizing agent is directly related to the standard voltage for its reduction, E°_{red}. *The more positive E°_{red} is, the stronger the oxidizing agent.* Looking at Table 17.1, you can see that *oxidizing strength increases moving down the left column.* The Li^+ ion, at the top of that column, is a very weak oxidizing agent with a large negative reduction voltage ($E^\circ_{\text{red}} = -3.040$ V). In practice, cations of the Group 1 metals (Li^+, Na^+, K^+, . . .) and the Group 2 metals (Mg^{2+}, Ca^{2+}, . . .) never act as oxidizing agents in water solution.

Further down the left column of Table 17.1, the H^+ ion ($E^\circ_{\text{red}} = 0.000$ V) is a reasonably strong oxidizing agent. It is capable of oxidizing many metals, including magnesium and zinc:

$$Mg(s) + 2H^+(aq) \longrightarrow Mg^{2+}(aq) + H_2(g) \qquad E^\circ = +2.357 \text{ V}$$
$$Zn(s) + 2H^+(aq) \longrightarrow Zn^{2+}(aq) + H_2(g) \qquad E^\circ = +0.762 \text{ V}$$

The strongest oxidizing agents are those at the bottom of the left column. Species such as $Cr_2O_7^{2-}$ ($E^\circ_{\text{red}} = +1.33$ V) and Cl_2 ($E^\circ_{\text{red}} = +1.360$ V) are commonly used as oxidizing agents in the laboratory. The fluorine molecule, at the bottom of the left column, should be the strongest of all oxidizing agents. In practice, fluorine is seldom used as an oxidizing agent in water solution. The F_2

Standard potential = E°_{red}. ⓘ

Table 17.1 Standard Potentials in Water Solution at 25°C

Acidic Solution, [H⁺] = 1 M

Reaction		$E°_{red}$ (V)
$Li^+(aq) + e^-$	$\longrightarrow Li(s)$ ◄ ⓘ	-3.040
$K^+(aq) + e^-$	$\longrightarrow K(s)$	-2.936
$Ba^{2+}(aq) + 2e^-$	$\longrightarrow Ba(s)$	-2.906
$Ca^{2+}(aq) + 2e^-$	$\longrightarrow Ca(s)$	-2.869
$Na^+(aq) + e^-$	$\longrightarrow Na(s)$	-2.714
$Mg^{2+}(aq) + 2e^-$	$\longrightarrow Mg(s)$	-2.357
$Al^{3+}(aq) + 3e^-$	$\longrightarrow Al(s)$	-1.68
$Mn^{2+}(aq) + 2e^-$	$\longrightarrow Mn(s)$	-1.182
$Zn^{2+}(aq) + 2e^-$	$\longrightarrow Zn(s)$	-0.762
$Cr^{3+}(aq) + 3e^-$	$\longrightarrow Cr(s)$	-0.744
$Fe^{2+}(aq) + 2e^-$	$\longrightarrow Fe(s)$	-0.409
$Cr^{3+}(aq) + e^-$	$\longrightarrow Cr^{2+}(aq)$	-0.408
$Cd^{2+}(aq) + 2e^-$	$\longrightarrow Cd(s)$	-0.402
$PbSO_4(s) + 2e^-$	$\longrightarrow Pb(s) + SO_4^{2-}(aq)$	-0.356
$Tl^+(aq) + e^-$	$\longrightarrow Tl(s)$	-0.336
$Co^{2+}(aq) + 2e^-$	$\longrightarrow Co(s)$	-0.282
$Ni^{2+}(aq) + 2e^-$	$\longrightarrow Ni(s)$	-0.236
$AgI(s) + e^-$	$\longrightarrow Ag(s) + I^-(aq)$	-0.152
$Sn^{2+}(aq) + 2e^-$	$\longrightarrow Sn(s)$	-0.141
$Pb^{2+}(aq) + 2e^-$	$\longrightarrow Pb(s)$	-0.127
$2H^+(aq) + 2e^-$	$\longrightarrow H_2(g)$	0.000
$AgBr(s) + e^-$	$\longrightarrow Ag(s) + Br^-(aq)$ ◄ ⓘ	0.073
$S(s) + 2H^+(aq) + 2e^-$	$\longrightarrow H_2S(aq)$	0.144
$Sn^{4+}(aq) + 2e^-$	$\longrightarrow Sn^{2+}(aq)$	0.154
$SO_4^{2-}(aq) + 4H^+(aq) + 2e^-$	$\longrightarrow SO_2(g) + 2H_2O$	0.155
$Cu^{2+}(aq) + e^-$	$\longrightarrow Cu^+(aq)$	0.161
$Cu^{2+}(aq) + 2e^-$	$\longrightarrow Cu(s)$	0.339
$Cu^+(aq) + e^-$	$\longrightarrow Cu(s)$	0.518
$I_2(s) + 2e^-$	$\longrightarrow 2I^-(aq)$	0.534
$Fe^{3+}(aq) + e^-$	$\longrightarrow Fe^{2+}(aq)$	0.769
$Hg_2^{2+}(aq) + 2e^-$	$\longrightarrow 2Hg(l)$	0.796
$Ag^+(aq) + e^-$	$\longrightarrow Ag(s)$	0.799
$2Hg^{2+}(aq) + 2e^-$	$\longrightarrow Hg_2^{2+}(aq)$	0.908
$NO_3^-(aq) + 4H^+(aq) + 3e^-$	$\longrightarrow NO(g) + 2H_2O$	0.964
$AuCl_4^-(aq) + 3e^-$	$\longrightarrow Au(s) + 4Cl^-(aq)$	1.001
$Br_2(l) + 2e^-$	$\longrightarrow 2Br^-(aq)$	1.077
$O_2(g) + 4H^+(aq) + 4e^-$	$\longrightarrow 2H_2O$	1.229
$MnO_2(s) + 4H^+(aq) + 2e^-$	$\longrightarrow Mn^{2+}(aq) + 2H_2O$	1.229
$Cr_2O_7^{2-}(aq) + 14H^+(aq) + 6e^-$	$\longrightarrow 2Cr^{3+}(aq) + 7H_2O$	1.33
$Cl_2(g) + 2e^-$	$\longrightarrow 2Cl^-(aq)$	1.360
$ClO_3^-(aq) + 6H^+(aq) + 5e^-$	$\longrightarrow \frac{1}{2}Cl_2(g) + 3H_2O$	1.458
$Au^{3+}(aq) + 3e^-$	$\longrightarrow Au(s)$	1.498
$MnO_4^-(aq) + 8H^+(aq) + 5e^-$	$\longrightarrow Mn^{2+}(aq) + 4H_2O$	1.512
$PbO_2(s) + SO_4^{2-}(aq) + 4H^+(aq) + 2e^-$	$\longrightarrow PbSO_4(s) + 2H_2O$	1.687
$H_2O_2(aq) + 2H^+(aq) + 2e^-$	$\longrightarrow 2H_2O$	1.763
$Co^{3+}(aq) + e^-$	$\longrightarrow Co^{2+}(aq)$	1.953
$F_2(g) + 2e^-$ ◄ ⓘ	$\longrightarrow 2F^-(aq)$	2.889

Increasing strength of oxidizing agents (left margin, downward arrow)

Increasing strength of reducing agents (right margin, upward arrow)

Basic Solution, [OH⁻] = 1 M

Reaction		$E°_{red}$ (V)
$Fe(OH)_2(s) + 2e^-$	$\longrightarrow Fe(s) + 2OH^-(aq)$	-0.891
$2H_2O + 2e^-$	$\longrightarrow H_2(g) + 2OH^-(aq)$	-0.828
$Fe(OH)_3(s) + e^-$	$\longrightarrow Fe(OH)_2(s) + OH^-(aq)$	-0.547
$S(s) + 2e^-$	$\longrightarrow S^{2-}(aq)$	-0.445
$NO_3^-(aq) + 2H_2O + 3e^-$	$\longrightarrow NO(g) + 4OH^-(aq)$	-0.140
$NO_3^-(aq) + H_2O + 2e^-$	$\longrightarrow NO_2^-(aq) + 2OH^-(aq)$	0.004
$ClO_4^-(aq) + H_2O + 2e^-$	$\longrightarrow ClO_3^-(aq) + 2OH^-(aq)$	0.398
$O_2(g) + 2H_2O + 4e^-$	$\longrightarrow 4OH^-(aq)$	0.401
$ClO_3^-(aq) + 3H_2O + 6e^-$	$\longrightarrow Cl^-(aq) + 6OH^-(aq)$	0.614
$ClO^-(aq) + H_2O + 2e^-$	$\longrightarrow Cl^-(aq) + 2OH^-(aq)$	0.890

ⓘ Lithium is the strongest reducing agent.

ⓘ Lithium and fluorine are very dangerous materials to work with.

ⓘ $\boxed{R}$ / $\boxed{O}$ O = strongest oxidizing agent; R = strongest reducing agent.

ⓘ Lithium and fluorine are very dangerous materials to work with.

ⓘ Fluorine is the strongest oxidizing agent.

Figure 17.6 Lithium as a reducing agent. Lithium, the strongest reducing agent listed in Table 17.1, is shown reacting with water. The heat of reaction is igniting the hydrogen produced.

molecule takes electrons away from just about anything, including water, often with explosive violence.

The arguments we have just gone through can be applied, in reverse, to *reducing agents, which are in the column to the right of Table 17.1* (Li, . . . , F⁻). In principle, at least, all of them can supply electrons to another species in a redox reaction. Their strength as reducing agents is directly related to their $E°_{ox}$ values. *The more positive $E°_{ox}$ is, the stronger the reducing agent.* Looking at the values, and remembering that $E°_{ox} = -E°_{red}$,

$$Li(s) \longrightarrow Li^+(aq) + e^- \qquad E°_{ox} = +3.040 \text{ V}$$
$$H_2(g) \longrightarrow 2H^+(aq) + 2e^- \qquad E°_{ox} = 0.000 \text{ V}$$
$$2F^-(aq) \longrightarrow F_2(g) + 2e^- \qquad E°_{ox} = -2.889 \text{ V}$$

It should be clear that *reducing strength decreases moving down the table.* Actually, the five metals at the top of the right column (Li, K, Ba, Ca, Na) cannot be used as reducing agents in water solution because they react directly with water, reducing it to hydrogen gas (Figure 17.6):

$$Li(s) + H_2O \longrightarrow Li^+(aq) + OH^-(aq) + \tfrac{1}{2}H_2(g)$$
$$Ba(s) + 2H_2O \longrightarrow Ba^{2+}(aq) + 2OH^-(aq) + H_2(g)$$

EXAMPLE	17.4	CONCEPTUAL

Consider the following species in acidic solution: MnO_4^-, I^-, NO_3^-, H_2S, and Fe^{2+}.

a   Using Table 17.1, classify each of these as an oxidizing and/or reducing agent.

STRATEGY

Recall that the oxidizing agents are located in the left column of Table 17.1 and the reducing agents are in the right column of the same table.

SOLUTION

MnO_4^-	found in the left column, oxidizing agent
I^-	found in the right column, reducing agent
NO_3^-	found in the left column, oxidizing agent
H_2S	found in the right column, reducing agent
Fe^{2+}	found in both the left column and right column, oxidizing and reducing agent

b   Arrange the oxidizing agents and the reducing agents in order of increasing strength.

STRATEGY AND SOLUTION

Going *down* the left column, the oxidizing agents increase in strength.

$$Fe^{2+} < NO_3^- < MnO_4^-$$

Going *up* the right column, the reducing agents increase in strength.

$$Fe^{2+} < I^- < H_2S$$

Calculation of $E°$ from $E°_{red}$ and $E°_{ox}$

As pointed out earlier, the standard voltage for a redox reaction is the sum of the standard voltages of the two half-reactions, reduction and oxidation; that is,

$$E° = E°_{red} + E°_{ox}$$

This simple relation makes it possible, using Table 17.1, to calculate the standard voltages for more than 3000 different redox reactions.

EXAMPLE 17.5 GRADED

Consider the voltaic cell in which the reaction is

$$2Ag^+(aq) + Cd(s) \longrightarrow 2Ag(s) + Cd^{2+}(aq)$$

a Use Table 17.1 to calculate $E°$ for the voltaic cell.

STRATEGY

1. Assign oxidation numbers to each element so you can decide which element is reduced and which one is oxidized.

2. Write the oxidation and reduction half-reactions together with the corresponding $E°_{ox}$ and $E°_{red}$. Recall that $E°_{ox} = -(E°_{red})$.

3. Add both half-reactions (make sure you cancel electrons) and take the sum of $E°_{ox}$ and $E°_{red}$ to obtain $E°$ for the cell.

SOLUTION

1. Oxidation numbers	Ag: $+1 \longrightarrow 0$ (reduction)
	Cd: $0 \longrightarrow +2$ (oxidation)
2. Half-reactions	$2Ag^+(aq) + 2e^- \longrightarrow 2Ag(s)$ $\quad E°_{red} = +0.799$ V
	$Cd(s) \longrightarrow Cd^{2+}(aq) + 2e^-$ $\quad E°_{ox} = -(E°_{red}) = -(-0.402$ V$)$
	$= +0.402$ V
3. $E°$	$Cd(s) + 2Ag^+(aq) \longrightarrow Cd^{2+}(aq) + 2Ag(s)$ $\quad E° = 0.799$ V $+ 0.402$ V
	$= 1.201$ V

b If the value zero is arbitrarily assigned to the standard voltage for the reduction of Ag^+ ions to Ag, what is $E°_{red}$ for the reduction of Cd^{2+} ions to Cd?

STRATEGY AND SOLUTION

$E°$ for the cell does not change. It does not matter what you choose to be $E°_{red}$ of the half-reaction. Naturally, $E°_{ox}$ will also change and you cannot choose to change that.

If you choose $E°_{red}$ to be zero, then

$\quad E°_{red} + E°_{ox} = 1.201$ V

$\quad\quad 0 + E°_{ox} = 1.201$ V; $E°_{ox}$ for the half-reaction $Cd(s) \longrightarrow Cd^{2+}(aq) + 2e^- = 1.201$ V

Since $E°_{red}$ for Cd^{2+} is asked for, then $E°_{red} = -(E°_{ox}) = -1.201$ V

Example 17.5 illustrates two general points concerning cell voltages.

1. The calculated voltage, $E°$, is always a positive quantity for a reaction taking place in a voltaic cell.

2. The quantities $E°$, $E°_{ox}$, and $E°_{red}$ are independent of how the equation for the cell reaction is written. You *never* multiply the voltage by the coefficients of the balanced equation. ◀ ⓘ

ⓘ It would be nice if you could double the voltage by rewriting the equation, but you can't.

Spontaneity of Redox Reactions

To determine whether a given redox reaction is spontaneous, apply a simple principle:

If the calculated voltage for a redox reaction is a positive quantity, the reaction will be spontaneous. If the calculated voltage is negative, the reaction is not spontaneous.

Nickel metal immersed in a solution of copper sulfate ($CuSO_4$).

Nickel metal reacts spontaneously with Cu^{2+} ions, producing Cu metal and Ni^{2+} ions. Copper plates out on the surface of the nickel, and the blue color of Cu^{2+} is replaced by the green color of Ni^{2+}.

Marna G. Clarke

Figure 17.7 Oxidation of Ni by Cu^{2+}.

Ordinarily, this principle is applied at standard concentrations (1 atm for gases, 1 M for species in aqueous solution). Hence it is the sign of $E°$ that serves as the criterion for spontaneity. To show how this works, consider the problem of oxidizing nickel metal to Ni^{2+} ions. This cannot be accomplished by using 1 M Zn^{2+} ions:

$$Ni(s) + Zn^{2+}(aq, 1\ M) \longrightarrow Ni^{2+}(aq, 1\ M) + Zn(s)$$

$$E° = E°_{ox}\ Ni + E°_{red}\ Zn^{2+} = +0.236\ V - 0.762\ V = -0.526\ V$$

Since you add to obtain $E°$, it makes no difference which you write first, $E°_{red}$ or $E°_{ox}$.

Sure enough, if you immerse a bar of nickel in a solution of 1 M $ZnSO_4$, nothing happens. Suppose, however, you add the nickel bar to a solution of 1 M $CuSO_4$ (Figure 17.7; left beaker). As time passes, the nickel is oxidized and the Cu^{2+} ions reduced (Figure 17.7; right beaker) through the spontaneous redox reaction

$$Ni(s) + Cu^{2+}(aq, 1\ M) \longrightarrow Ni^{2+}(aq, 1\ M) + Cu(s)$$

$$E° = E°_{ox}\ Ni + E°_{red}\ Cu^{2+} = +0.236\ V + 0.339\ V = +0.575\ V$$

EXAMPLE 17.6

Consider Fe^{2+} at standard conditions.

a | Will the reaction below occur?

$$2Fe^{3+}(aq) + 2I^-(aq) \longrightarrow 2Fe^{2+}(aq) + I_2(s)$$

ANALYSIS	
Information given:	equation for the reaction ($2Fe^{3+}(aq) + 2I^-(aq) \longrightarrow 2Fe^{2+}(aq) + I_2(s)$)
Information implied:	Table 17.1 (standard reduction potentials)
Asked for:	Will the reaction occur?

continued

STRATEGY

1. Assign oxidation numbers.

2. Write oxidation and reduction half-reactions. Include $E°_{ox}$ and $E°_{red}$.

3. Find $E°$. The reaction will occur if $E° > 0$.

SOLUTION

1. oxidation numbers	Fe: $+3 \longrightarrow +2$ reduction I: $-1 \longrightarrow 0$ oxidation
2. half-reactions	$2Fe^{3+}(aq) + 2e^- \longrightarrow 2Fe^{2+}(aq)$ $E°_{red} = +0.769$ V $2I^-(aq) \longrightarrow I_2(s) + 2e^-$ $E°_{ox} = -0.534$ V
3. $E°$	$E° = 0.769$ V $+ (-0.534$ V$) = +0.235$ V $E° > 0$, the reaction will occur at standard conditions.

---

b   Can Fe(s) be oxidized to Fe^{2+} by treatment with hydrochloric acid?

ANALYSIS

Information given:	oxidation half-reaction $(Fe(s) \longrightarrow Fe^{2+}(aq) + 2e^-)$
Information implied:	Table 17.1 (standard reduction potentials)
Asked for:	Will HCl oxidize Fe?

STRATEGY

1. $HCl(aq)$ is made up of two ions, H^+ and Cl^-. Since an oxidizing agent is needed (to oxidize Fe to Fe^{2+}), find either H^+ or Cl^- (or both) in the left column of Table 17.1.

2. Write the possible half-reactions.

3. Write the redox reaction and find $E°$.

SOLUTION

1. Oxidizing agent	Only H^+ appears in the left column.
2. Half-reactions	$2H^+(aq) + 2e^- \longrightarrow H_2(g)$ $E°_{red} = 0.000$ V $Fe(s) \longrightarrow Fe^{2+}(aq) + 2e^-$ $E°_{ox} = 0.409$ V
3. Redox reaction	$Fe(s) + 2H^+(aq) \longrightarrow Fe^{2+}(aq) + H_2(g)$ $E° = 0.409$ V $E° > 0$, HCl will oxidize Fe at standard conditions (Figure 17.8).

---

c   What redox reactions occur when the following species are mixed in acidic solution: Cl^-, Fe^{2+}, Cr^{2+}, I_2?

ANALYSIS

Information given:	ions in acidic solution $(Cl^-, Fe^{2+}, Cr^{2+}, I_2)$
Information implied:	Table 17.1
Asked for:	Will a redox reaction occur when the ions are mixed?

STRATEGY

1. Check the left column of Table 17.1 to determine which of the ions are oxidizing agents (i.e., they are reduced). Write the reduction half-reactions of the oxidizing agents.

2. Check the right column of Table 17.1 to determine which of the ions are reducing agents (i.e., they are oxidized). Write the reduction half-reactions of the reducing agents.

3. Write all possible combinations of oxidation and reduction half-reactions. The combination(s) that give positive $E°$ values are possible.

4. Write the redox equation(s) for the reaction(s) that occur.

continued

	SOLUTION	
1. Oxidizing agents	$Fe^{2+}(aq) + 2e^- \longrightarrow Fe(s)$ $I_2(s) + 2e^- \longrightarrow 2I^-(aq)$	$E^\circ_{red} = -0.409$ V $E^\circ_{red} = +0.534$ V
2. Reducing agents	$2Cl^-(aq) \longrightarrow Cl_2(g) + 2e^-$ $Cr^{2+}(aq) \longrightarrow Cr^{3+}(aq) + e^-$ $Fe^{2+}(aq) \longrightarrow Fe^{3+}(aq) + e^-$	$E^\circ_{ox} = -1.360$ V $E^\circ_{ox} = +0.408$ V $E^\circ_{ox} = -0.769$ V
3. Possible combinations	$Fe^{2+} + Cl^-: E^\circ < 0$ $I_2 + Cl^-: E^\circ < 0$ $Fe^{2+} + Cr^{2+}: E^\circ < 0$ $I_2 + Cr^{2+}: E^\circ > 0$ ✔ $Fe^{2+} + Fe^{2+}: E^\circ < 0$ $I_2 + Fe^{2+}: E^\circ < 0$	
4. Redox reaction	$I_2(s) + 2Cr^{2+}(aq) \longrightarrow 2I^-(aq) + 2Cr^{3+}(aq)$	$E^\circ = 0.942$ V

Figure 17.8 Oxidation of Fe by H⁺. Finely divided iron in the form of steel wool reacts with hydrochloric acid to evolve hydrogen: $Fe(s) + 2H^+(aq)$ $\longrightarrow Fe^{2+}(aq) + H_2(g)$. The hydrogen gas is creating foam as it is produced along the strands of the steel wool.

$F = 9.648 \times 10^4$ C/mol
$ = 9.648 \times 10^4$ J/mol·V

17-4   Relations Between E°, ΔG°, and K

As pointed out previously, the value of the standard cell voltage, E°, is a measure of the spontaneity of a cell reaction. In Chapter 16, we showed that the standard free energy change, ΔG°, is a general criterion for reaction spontaneity. As you might suppose, these two quantities have a simple relation to one another and to the equilibrium constant, K, for the cell reaction.

E° and ΔG°

It can be shown by a thermodynamic argument that we will not go through here that these two quantities are related by the equation

$$\Delta G^\circ = -nFE^\circ \qquad (17.2)$$

where

- ΔG° is the standard free energy change (gases at 1 atm, species in solution at 1 M) for the reaction.
- E° is the standard voltage for the cell reaction.
- n is the number of moles of electrons transferred in the reaction.
- F, called the **Faraday constant,** ◂ 🛈 has the following value: 9.648×10^4 J/mol·V (J = joule, V = volt).

Notice from this equation that ΔG° and E° have opposite signs. This is reasonable; a spontaneous reaction is one for which ΔG° is *negative* but E° is *positive*.

E° and K

Redox reactions, like all reactions, eventually reach a state of equilibrium. It is possible to calculate the equilibrium constant for a redox reaction from the standard voltage. To do that, we start with the relation obtained in Chapter 16:

$$\Delta G^\circ = -RT \ln K$$

We have just seen that $\Delta G^\circ = -nFE^\circ$. It follows that

$$RT \ln K = nFE^\circ$$

that is,

$$E^\circ = \frac{RT}{nF} \ln K$$

The quantity RT/F is readily evaluated at 25°C, the temperature at which standard potentials are tabulated.

$$\frac{RT}{F} = \frac{8.31 \text{ J/mol} \cdot \text{K} \times 298 \text{ K}}{9.648 \times 10^4 \text{ J/mol} \cdot \text{V}} = 0.0257 \text{ V}$$

Hence

$$E° = \frac{(0.0257 \text{ V})}{n} \ln K \qquad \text{(at 25°C)} \qquad \textbf{(17.3)}$$

Notice that if the standard voltage is positive, $\ln K$ is also positive, and K is greater than 1. Conversely, if the standard voltage is negative, $\ln K$ is also negative, and K is less than 1.

Spontaneous reaction: $E° > 0$, $\Delta G° < 0$, $K > 1$.

EXAMPLE 17.7	

Consider the following reaction at 25°C:

$$3Ag(s) + NO_3^-(aq) + 4H^+(aq) \longrightarrow 3Ag^+(aq) + NO(g) + 2H_2O$$

a	Find $\Delta G°$.

ANALYSIS	
Information given:	reaction: $(3Ag(s) + NO_3^-(aq) + 4H^+(aq) \longrightarrow 3Ag^+(aq) + NO(g) + 2H_2O)$ temperature (25°C)
Information implied:	Table 17.1 (standard reduction potentials) Faraday constant (F)
Asked for:	$\Delta G°$

STRATEGY	

1. Assign oxidation numbers.
2. Split the redox reaction into oxidation and reduction half-reactions and find $E°$.
3. Determine the number of electrons cancelled out when balancing the equation to find n.
4. Substitute into Equation 17.2. Note that $\Delta G°$ will be in joules since the Faraday constant is in joules.

SOLUTION	
1. oxidation numbers	N: $+5 \longrightarrow +2$ reduction Ag: $0 \longrightarrow +1$ oxidation
2. half-reactions	$NO_3^-(aq) + 4H^+(aq) + 3e^- \longrightarrow NO(g) + 2H_2O$ $E°_{\text{red}} = 0.964 \text{ V}$ $Ag(s) + e^- \longrightarrow Ag^+(aq)$ $E°_{\text{ox}} = -0.799 \text{ V}$ $E° = 0.964 \text{ V} + (-0.799 \text{ V}) = 0.165 \text{ V}$
3. n	Multiply the oxidation half-reaction by 3 to cancel out the three electrons in the reduction half-reaction. Three electrons cancel out: $n = 3$
4. $\Delta G°$	$\Delta G° = -nFE = (-3 \text{ mol})(9.648 \times 10^4 \frac{\text{J}}{\text{mol} \cdot \text{V}})(0.165 \text{ V}) = -4.78 \times 10^4 \text{ J}$ $= -47.8 \text{ kJ}$

b	Find K.

ANALYSIS	
Information given:	From part (a): $E°$ (0.165 V); n (3 mol)
Asked for:	K

continued

STRATEGY	
Substitute into Equation 17.3.	
SOLUTION	
K	$E° = \dfrac{0.0257\ V}{n} \ln K;\ 0.165\ V = \dfrac{0.0257\ V}{3} \ln K;\ \ln K = 19.3$ $K = e^{19.3} = 2.4 \times 10^8$

Table 17.2 Relation Between $E°$, K, and $\Delta G°$ ($n = 2$)

$E°$ (V)	K	$\Delta G°$ (kJ)	$E°$ (V)	K	$\Delta G°$ (kJ)
+2.00	4×10^{67}	−400	−2.00	3×10^{-68}	+400
+1.00	6×10^{33}	−200	−1.00	2×10^{-34}	+200
+0.50	8×10^{16}	−100	−0.50	1×10^{-17}	+100
+0.25	3×10^{8}	−50	−0.25	4×10^{-9}	+50
+0.10	2×10^{3}	−20	−0.10	0.0004	+20
+0.05	50	−10	−0.05	0.02	+10
0.00	1	0			

The approach used in Example 17.7 to find the number of moles of electrons transferred, *n*, is generally useful. What you do is to break down the equation for the cell reaction into two half-equations, oxidation and reduction. The quantity *n* is the number of electrons appearing in either half-equation.

Table 17.2 lists values of K and $\Delta G°$ corresponding to various values of $E°$ with $n = 2$. Notice that:

■ if $E°$ is greater than about +0.10 V, K is very large, and the reaction goes virtually to completion.
■ if $E°$ is smaller than about −0.10 V, K is very small, so the reaction does not proceed to any appreciable extent. ◄ *ⓘ*

Only if the standard voltage falls within a rather narrow range, say +0.10 to −0.10 V, will the value of K (and that of $\Delta G°$) be such that the reaction will produce an equilibrium mixture containing appreciable amounts of both reactants and products.

ⓘ
Most redox reactions either go to completion or not at all.

17-5 Effect of Concentration on Voltage

To this point, we have dealt only with "standard" voltages, that is, voltages when all gases are at 1 atm pressure and all species in aqueous solution are at a concentration of 1 *M*. When the concentration of a reactant or product changes, the voltage changes as well. Qualitatively, the direction in which the voltage shifts is readily predicted when you realize that cell voltage is directly related to reaction spontaneity.

1. Voltage will *increase* if the concentration of a reactant is increased or that of a product is decreased. Either of these changes increases the driving force behind the redox reaction, making it more spontaneous.
2. Voltage will *decrease* if the concentration of a reactant is decreased or that of a product is increased. Either of these changes makes the redox reaction less spontaneous.

When a voltaic cell operates, supplying electrical energy, the concentration of reactants decreases and that of the products increases. As time passes, the voltage drops steadily. Eventually it becomes zero, and we say that the cell is "dead." At that point, the redox reaction taking place within the cell is at equilibrium, and there is no driving force to produce a voltage. ◄*i*

That's what happens when you leave your car lights on, unless your car turns them off automatically.

Nernst Equation

To obtain a quantitative relation between cell voltage and concentration, it is convenient to start with the general expression for the free energy change discussed in Chapter 16:

$$\Delta G = \Delta G° + RT \ln Q$$

where Q is the reaction quotient in which products appear in the numerator, reactants in the denominator. But $\Delta G° = -nFE°$ and $\Delta G = -nFE$, so we have

$$-nFE = -nFE° + RT \ln Q$$

Solving for the cell voltage E,

$$E = E° - \frac{RT}{nF} \ln Q$$

This relationship is known as the **Nernst equation**, after Walther Nernst ◄*i* (1864–1941), a brilliant and egocentric colleague of Arrhenius, who first proposed it in 1888. Recalling that, at 25°C, the quantity RT/F is 0.0257 V,

Nernst won the Nobel Prize in Chemistry in 1920.

$$E = E° - \frac{(0.0257 \text{ V})}{n} \ln Q \qquad \text{at } 25°C \qquad (17.4)$$

In this equation, E is the cell voltage, $E°$ is the standard voltage, n is the number of moles of electrons exchanged in the reaction, and Q is the reaction quotient. Notice that:

- if $Q > 1$, which means that the concentrations of products are high relative to those of reactants, $\ln Q$ is positive and $E < E°$.
- if $Q < 1$ (concentrations of products low relative to reactants), $\ln Q$ is negative and $E > E°$.
- if $Q = 1$, as is the case under standard conditions, $\ln Q = 0$ and $E = E°$.

Remember that gases enter Q as their partial pressures in atmospheres. Species in water solution enter as their molar concentrations. Pure liquids and solids do not appear in the expression for Q. For example,

$$a\text{A}(s) + b\text{B}(aq) \longrightarrow c\text{C}(aq) + d\text{D}(g) \qquad Q = \frac{[\text{C}]^c(\text{P}_\text{D})^d}{[\text{B}]^b}$$

EXAMPLE 17.8 GRADED

Consider a voltaic cell in which the following reaction occurs:

$$O_2(g, 0.98 \text{ atm}) + 4H^+(aq, \text{pH} = 1.24) + 4Br^-(aq, 0.15 \text{ M}) \longrightarrow 2H_2O + 2Br_2(l)$$

a	Calculate E for the cell at 25°C.

ANALYSIS	
Information given:	reaction: ($O_2(g) + 4H^+(aq) + 4Br^-(aq) \longrightarrow 2Br_2(l) + 2H_2O$) P_{O_2} (0.98 atm); [H^+] (pH = 1.24); [Br^-] (0.15 M) temperature (25°C)
Information implied:	Table 17.1 (standard reduction potentials)
Asked for:	E

continued

	STRATEGY

1. Change pH to $[H^+]$ and find Q.

2. Assign oxidation numbers, write oxidation and reduction half-reactions, and cancel electrons to find n.

3. Find $E°$. $(E°_{red} + E°_{ox})$

4. Substitute into the Nernst equation (Equation 17.4) for $T = 25°C$

$$E = E° - \frac{0.0257}{n} \ln Q$$

	SOLUTION
1. $[H^+]$	$1.24 = -\log_{10}[H^+]$; $[H^+] = 0.058\ M$
Q	$Q = \dfrac{1}{(P_{O_2})[H^+]^4[Br^-]^4} = \dfrac{1}{(0.98)(0.058)^4(0.15)^4} = 1.8 \times 10^8$
2. Oxidation numbers	O: $0 \longrightarrow -2$ (reduction); Br: $-1 \longrightarrow 0$ (oxidation)
Half-reactions	$O_2(g) + 4H^+(aq) + 4e^-(aq) \longrightarrow 2H_2O$
	$2Br^-(aq) \longrightarrow Br_2(l) + 2e^-$
n	The oxidation half-reaction must be multiplied by 2 to cancel out the four electrons in the reduction half-reaction.
	$n = 4$
3. $E°$	$E°_{red}$ for $O_2 = 1.229$ V; $E°_{ox}$ for $Br^- = -1.077$ V
	$E° = 1.229$ V $+ (-1.077$ V$) = 0.152$ V
4. E	$E = 0.152$ V $- \dfrac{0.0257}{4} \ln (1.8 \times 10^8) = 0.030$ V

b	When the voltaic cell is at 35°C, E is measured to be 0.039 V. What is $E°$ at 35°C?

	ANALYSIS
Information given:	E (0.039 V) at T (35°C)
	From part (a): Q (1.8×10^8); n (4 moles)
Information implied:	R and F values in joules
Asked for:	$E°$ at 35°C

	STRATEGY

Substitute into the Nernst equation.

$$E = E° - \frac{RT}{nF} \ln Q$$

	SOLUTION
$E°$	0.039 V $= E° - \dfrac{(8.31\ \text{J/mol} \cdot \text{K})(308\ \text{K})}{4\,(9.648 \times 10^4\ \text{J/mol} \cdot \text{V})} \ln (1.8 \times 10^8)$
	$E° = 0.039$ V $+ 0.126$ V $= 0.165$ V

The Nernst equation can also be used to determine the effect of changes in concentration on the voltage of an individual half-cell, $E°_{red}$ or $E°_{ox}$. Consider, for example, the half-reaction

$$MnO_4^-(aq) + 8H^+(aq) + 5e^- \longrightarrow Mn^{2+}(aq) + 4H_2O \qquad E°_{red} = +1.512\ \text{V}$$

The Nernst equation takes the form

$$E_{red} = +1.512\ \text{V} - \frac{(0.0257\ \text{V})}{5}\ \ln \frac{[\text{Mn}^{2+}]}{[\text{MnO}_4^-][\text{H}^+]^8}$$

where E_{red} is the observed reduction voltage corresponding to any given concentrations of Mn^{2+}, MnO_4^-, and H^+.

Use of the Nernst Equation to Determine Ion Concentrations

In chemistry, the most important use of the Nernst equation lies in the experimental determination of the concentrations of ions in solution. Suppose you measure the cell voltage E and know the concentration of all but one species in the two half-cells. It should then be possible to calculate the concentrations of that species by using the Nernst equation (Example 17.9).

EXAMPLE 17.9

Consider a voltaic cell at 25°C in which the reaction is

$$\text{Zn}(s) + 2\text{H}^+(aq) \longrightarrow \text{Zn}^{2+}(aq) + \text{H}_2(g)$$

It is found that the voltage is +0.560 V when $[\text{Zn}^{2+}] = 0.85\ M$ and $P_{\text{H}_2} = 0.988$ atm. What is the pH in the H_2-H half-cell?

ANALYSIS

Information given:	reaction: ($\text{Zn}(s) + 2\text{H}^+(aq) \longrightarrow \text{Zn}^{2+}(aq) + \text{H}_2(g)$) E (0.560 V); T (25°C) P_{H_2} (0.988 atm); $[\text{Zn}^{2+}]$(0.85 M) temperature (25°C)
Information implied:	Table 17.1 (standard reduction potentials)
Asked for:	pH

STRATEGY

1. Assign oxidation numbers, write oxidation and reduction half-reactions, and cancel electrons to find n.

2. Find $E°$. ($E°_{red} + E°_{ox}$)

3. Substitute into the Nernst equation (Equation 17.4) for $T = 25°C$ and find Q.

4. Write the Q expression and substitute given concentrations and pressures to find $[\text{H}^+]$. Change $[\text{H}^+]$ to pH.

SOLUTION

1. Oxidation numbers	Zn: $0 \longrightarrow +2$ oxidation; H^+: $+1 \longrightarrow 0$ reduction
Half-reactions	$2\text{H}^+(aq) + 2e^- \longrightarrow \text{H}_2(g)$ $\text{Zn}(s) \longrightarrow 2e^- + \text{Zn}^{2+}(aq)$
n	2 electrons cancel out so $n = 2$.
2. $E°$	$E° = E°_{ox}\ \text{Zn} + E°_{red}\ \text{H}^+ = 0.762\ \text{V} + 0\ \text{V} = 0.762\ \text{V}$
3. Q	$E = E° - \dfrac{0.0257}{n}\ \ln Q;\ 0.560\ \text{V} = 0.762\ \text{V} - \dfrac{0.0257}{2}\ \ln Q$
	$\dfrac{0.0257}{2}\ \ln Q = 0.762\ \text{V} - 0.560\ \text{V};\ \ln Q = 15.7;\ Q = 6.7 \times 10^6$
4. $[\text{H}^+]$	$Q = \dfrac{[\text{Zn}^{2+}](P_{\text{H}_2})}{[\text{H}^+]^2};\ [\text{H}^+] = \left[\dfrac{(0.85)(0.988)}{6.7 \times 10^6}\right]^{\frac{1}{2}} = 3.5 \times 10^{-4}$
pH	$\text{pH} = -\log_{10}(3.5 \times 10^{-4}) = 3.45$

Figure 17.9 Electrode for a pH meter. The pH of a solution can be determined with the aid of a "glass electrode." The voltage between the glass electrode and the reference electrode is directly related to pH. The leads from the electrodes are connected to a pH meter.

As Example 17.9 implies, the $Zn|Zn^{2+}||H^+|H_2|Pt$ cell can be used to measure the concentration of H^+ or pH of a solution. Indeed, cells of this type are commonly used to measure pH; Figure 17.9 shows a schematic diagram of a cell used with a pH meter. The pH meter, referred to in Chapter 13, is actually a high-resistance voltmeter calibrated to read pH rather than voltage. The cell connected to the pH meter consists of two half-cells. One of these is a reference half-cell of known voltage. The other half-cell contains a solution of known pH separated by a thin, fragile *glass electrode* from a solution whose pH is to be determined. The voltage of this cell is a function of the pH of the solution in the beaker.

Specific ion electrodes, similar in design to the glass electrode, have been developed to analyze for a variety of cations and anions. One of the first to be used extensively was a fluoride ion electrode that is sensitive to F^- at concentrations as low as 0.1 part per million and hence is ideal for monitoring fluoridated water supplies. An electrode that is specific for Cl^- ions is used to diagnose cystic fibrosis. Attached directly to the skin, it detects the abnormally high concentrations of sodium chloride in sweat that are a characteristic symptom of this disorder. Diagnoses that used to require an hour or more can now be carried out in a few minutes; as a result, large numbers of children can be screened rapidly and routinely.

The general approach illustrated by Example 17.9 is widely used to determine equilibrium constants for solution reactions. The pH meter in particular can be used to determine acid or base equilibrium constants by measuring the pH of solutions containing known concentrations of weak acids or bases. Specific ion electrodes are readily adapted to the determination of solubility product constants. For example, a chloride ion electrode can be used to find $[Cl^-]$ in equilibrium with $AgCl(s)$ and a known $[Ag^+]$. From that information, K_{sp} of AgCl can be calculated.

17-6 Electrolytic Cells

In an **electrolytic cell,** a nonspontaneous redox reaction is made to occur by pumping electrical energy into the system. A general diagram for such a cell is shown in Figure 17.10. The storage battery at the left provides a source of direct electric current. From the terminals of the battery, two wires lead to the electrolytic cell. This consists of two electrodes, A and C, dipping into a solution containing ions M^+ and X^-.

The battery acts as an electron pump, pushing electrons into the *cathode*, C, and removing them from the *anode*, A. ◄ ⓘ To maintain electrical neutrality, some process within the cell must consume electrons at C and liberate them at A. This process is an oxidation-reduction reaction; when carried out in an electrolytic cell, it is called **electrolysis.** At the cathode, an ion or molecule undergoes reduction by accepting electrons. At the anode, electrons are produced by the oxidation of an ion or molecule.

Oxidation occurs at the anode, reduction at the cathode, in both voltaic and electrolytic cells.

Quantitative Relationships

There is a simple relationship between the amount of electricity passed through an electrolytic cell and the amounts of substances produced by oxidation or reduction at the electrodes. From the balanced half-equations

$$Ag^+(aq) + e^- \longrightarrow Ag(s)$$
$$Cu^{2+}(aq) + 2e^- \longrightarrow Cu(s)$$
$$Au^{3+}(aq) + 3e^- \longrightarrow Au(s)$$

you can deduce that

$$1 \text{ mol of } e^- \longrightarrow 1 \text{ mol of Ag (107.9 g of Ag)}$$
$$2 \text{ mol of } e^- \longrightarrow 1 \text{ mol of Cu (63.55 g of Cu)}$$
$$3 \text{ mol of } e^- \longrightarrow 1 \text{ mol of Au (197.0 g of Au)}$$

Relations of this type, obtained from balanced half-equations, can be used in many practical calculations involving electrolytic cells. You will also need to become familiar with certain electrical units, including those of

quantity of electrical charge. The common unit here is the **coulomb** (C). The coulomb is related to the charge carried by a mole of electrons through the Faraday constant:

$$1 \text{ mol electrons} = 9.648 \times 10^4 \text{ C}$$

rate of current flow. Here the common unit is the **ampere** (A). When a current of one ampere passes through an electrical circuit, one coulomb passes a given point in the circuit in one second. That is,

$$1 \text{ A} = 1 \text{ C/s}$$

amount of electrical energy. This can be expressed in joules, J. When a charge of one coulomb (C) moves through a potential difference of one volt (V), it acquires an energy of one joule:

$$1 \text{ J} = 1 \text{ C} \cdot \text{V}$$

Your electric bill deals with a different unit of electrical energy called the kilowatt-hour (kWh). The joule and the kilowatt-hour are related through a simple conversion factor:

$$1 \text{ kWh} = 3.600 \times 10^6 \text{ J} = 3.600 \times 10^3 \text{ kJ}$$

These relations and others appear in Table 17.3.

Figure 17.10 Diagram of an electrolytic cell. Electrons enter the cathode from an external source. Cations move to the cathode, where they are reduced, and anions move to the anode, where they are oxidized.

Table 17.3 Electrical Units

Quantity	Unit	Defining Relation	Conversion Factors
Charge	coulomb (C)	$1 \text{ C} = 1 \text{ A} \cdot \text{s} = 1 \text{ J/V}$	$1 \text{ mol } e^- = 9.648 \times 10^4 \text{ C}$
Current	ampere (A)	$1 \text{ A} = 1 \text{ C/s}$	
Potential	volt (V)	$1 \text{ V} = 1 \text{ J/C}$	
Power	watt (W)	$1 \text{ W} = 1 \text{ J/s}$	
Energy	joule (J)	$1 \text{ J} = 1 \text{ V} \cdot \text{C}$	$1 \text{ kWh} = 3.600 \times 10^6 \text{ J}$

EXAMPLE 17.10

Chromium metal can be electroplated from an aqueous solution of potassium dichromate. The reduction half-reaction is

$$\text{Cr}_2\text{O}_7{}^{2-}(aq) + 14\text{H}^+(aq) + 12e^- \longrightarrow 2\text{Cr}(s) + 7\text{H}_2\text{O}$$

A current of 6.00 A and a voltage of 4.5 V are used in the electroplating.

continued

a	How many grams of chromium can be plated if the current is run for 48 minutes?
ANALYSIS	
Information given:	reduction half-reaction: $(Cr_2O_7^{2-}(aq) + 14H^+(aq) + 12e^- \longrightarrow 2Cr(s) + 7H_2O)$ current (6.00 A); voltage (4.5 V); time in s (48 × 60)
Information implied:	1 C = 1 A · s; MM Cr
Asked for:	mass Cr plated

STRATEGY

1. Since mass is asked for, you may assume that you have all the information to convert the amount of electricity to moles of electrons. Moles of electrons provide the bridge that connects the amount of electricity to the stoichiometry of the chemical reaction. Use the following plan:

$$\text{amperes (A)} \xrightarrow{\times \text{ time (s)}} \text{coulomb (C)} \xrightarrow{1 \text{ mol } e^- = 9.648 \times 10^4 \text{ C}} \text{mol } e^-$$

2. Convert mol e^- to mass of Cr using the stoichiometry of the reaction.

$$\text{mol } e^- \xrightarrow{12 \text{ mol } e^-/2 \text{ mol Cr}} \text{mol Cr} \xrightarrow{\text{MM}} \text{mass Cr}$$

SOLUTION

1. mol e^-	$6.00 \text{ A} = 6.00 \text{ C/s}$ $6.00 \dfrac{\text{C}}{\text{s}} \times (48 \times 60)\text{s} \times \dfrac{1 \text{ mol } e^-}{9.648 \times 10^4 \text{ C}} = 0.179 \text{ mol } e^-$
2. Mass Cr	$0.179 \text{ mol } e^- \times \dfrac{2 \text{ mol Cr}}{12 \text{ mol } e^-} \times \dfrac{52.00 \text{ g Cr}}{1 \text{ mol Cr}} = 1.55 \text{ g}$

b	How long will it take to completely convert 215 mL of 1.25 M $K_2Cr_2O_7$ to elemental chromium?
ANALYSIS	
Information given:	reduction half-reaction: $(Cr_2O_7^{2-}(aq) + 14H^+(aq) + 12e^- \longrightarrow 2Cr(s) + 7H_2O)$ current (6.00 A); voltage (4.5 V) $K_2Cr_2O_7$: V (0.215 L); M (1.25)
Information implied:	1 C = 1 A · s
Asked for:	time

STRATEGY

1. Since V and M are given for $K_2Cr_2O_7$ (thus for $Cr_2O_7^{2-}$), you have enough information to convert to mole e^- using the stoichiometry of the reaction.

$$V \times M \longrightarrow \text{mol } Cr_2O_7^{2-} \xrightarrow{12 \text{ mol } e^-/1 \text{ mol } Cr_2O_7^{2-}} \text{mol } e^-$$

2. Convert mol e^- to coulombs.

$$\text{mol } e^- \xrightarrow{9.648 \times 10^4 \text{ C/1 mol } e^-} \text{coulombs}$$

3. Convert coulomb to time. Recall 1 A = 1 C/s.

SOLUTION

1. mol e^-	$(0.215 \text{ L})(1.25 \text{ mol/L})(12 \text{ mol } e^-/\text{mol } Cr_2O_7^{2-}) = 3.225 \text{ mol } e^-$
2. C	$3.225 \text{ mol } e^- \times \dfrac{9.648 \times 10^4 \text{ C}}{1 \text{ mol } e^-} = 3.11 \times 10^5 \text{ C}$
3. Time	$\text{time} = \dfrac{3.11 \times 10^5 \text{ C}}{6.00 \text{ C/s}} = 5.16 \times 10^4 \text{ s} = 14.4 \text{ h}$

continued

c	How many kilowatt-hours of electrical energy are required to plate 1.00 g of chromium?

ANALYSIS

Information given:	reduction half-reaction: $(Cr_2O_7^{2-}(aq) + 14H^+(aq) + 12e^- \longrightarrow 2Cr(s) + 7H_2O)$ current (6.00 A); voltage (4.5 V) mass Cr (1.00 g)
Information implied:	1 A = 1 C/s; 1 kWh = 3.600×10^6 J; 1 J = 1 C · V MM for Cr
Asked for:	kilowatt-hours

STRATEGY

1. Convert the mass of Cr to mol e^- using stoichiometry

$$\text{mass Cr} \xrightarrow{\text{MM}} \text{mol Cr} \xrightarrow{\text{2 mol Cr/12 mol } e^-} \text{mol } e^-$$

2. Find kWh by finding the energy in joules (C × V) and then convert to kWh (3.600×10^6 J = 1 kWh)

$$\text{mol } e^- \xrightarrow{\text{1 mol } e^- = 9.648 \times 10^4 \text{ C}} \text{coulomb (C)} \xrightarrow{\text{V}} \text{C · V} \longrightarrow \text{J} \longrightarrow \text{kWh}$$

SOLUTION

1. mol e^-	$1.00 \text{ g Cr} \times \dfrac{1 \text{ mol Cr}}{52.00 \text{ g}} \times \dfrac{12 \text{ mol } e^-}{2 \text{ mol Cr}} = 0.115 \text{ mol } e^-$
2. kWh	$0.115 \text{ mol } e^- \times \dfrac{9.468 \times 10^4 \text{ C}}{1 \text{ mol } e^-} \times 4.5 \text{ V} \times \dfrac{1 \text{ J}}{1 \text{ C · V}} \times \dfrac{1 \text{ kWh}}{3.600 \times 10^6 \text{ J}} = 0.014 \text{ kWh}$

END POINTS

1. Note that whether you are given data to determine the moles of a species in a reaction or the coulombs of electricity, you can get to moles of electrons.

2. The value given for the voltage used is irrelevant for parts (a) and (b). You only need it to find the number of kilowatt hours.

In working Example 17.11, we have in effect assumed that the electrolyses were 100% efficient in converting electrical energy into chemical energy. In practice, this is almost never the case. Some electrical energy is wasted in side reactions at the electrodes and in the form of heat. This means that the actual yield of products is less than the theoretical yield. ◄ *i*

i
Nobody is 100% efficient.

Cell Reactions (Water Solution)

As is always the case, a reduction half-reaction occurs at the cathode of an electrolytic cell. This half-reaction in aqueous solution may be

■ *the reduction of a cation to the corresponding metal.* This commonly occurs with transition metal cations, which are relatively easy to reduce. Examples include

$$Ag^+(aq) + e^- \longrightarrow Ag(s) \qquad E°_{red} = +0.799 \text{ V}$$
$$Cu^{2+}(aq) + 2e^- \longrightarrow Cu(s) \qquad E°_{red} = +0.339 \text{ V}$$

This type of half-reaction is characteristic of electroplating processes, in which a metal object serves as the cathode (Figure 17.11).

■ *the reduction of a water molecule to hydrogen gas*

$$2H_2O + 2e^- \longrightarrow H_2(g) + 2OH^-(aq) \qquad E°_{red} = -0.828 \text{ V}$$

This half-reaction commonly occurs when the cation in solution is very difficult to reduce. For example, electrolysis of a solution containing K^+ ions

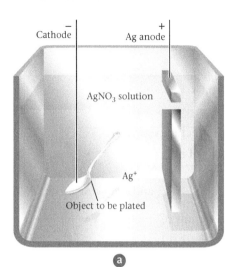

Silver ions are reduced to metallic silver by electrolysis. The silver coats the spoon to be plated.

The Oscar statue is gold plated.

Figure 17.11 Electroplating.

Table 17.4 Electrolysis of Water Solutions

Solution	Cathode Product	Anode Product
$CuBr_2(aq)$	$Cu(s)$	$Br_2(l)$
$AgNO_3(aq)$	$Ag(s)$	$O_2(g)$
$KI(aq)$	$H_2(g)$	$I_2(s)$
$Na_2SO_4(aq)$	$H_2(g)$	$O_2(g)$

Figure 17.12 Electrolysis of potassium iodide (KI) solution. The electrolysis of aqueous KI is similar to that of aqueous NaCl. The cathode reaction *(left)* is the reduction of water to $H_2(g)$ and OH^-, as shown by the pink color of phenolphthalein indicator in the water. The anode reaction *(right)* is the oxidation of $I^-(aq)$ to $I_2(aq)$, as shown by the brown color of the solution.

($E^\circ_{red} = -2.936$ V) or Na^+ ions ($E^\circ_{red} = -2.714$ V) yields hydrogen gas at the cathode (Table 17.4).

At the anode of an electrolytic cell, the half-reaction may be

- *the oxidation of an anion to the corresponding nonmetal*

$$2I^-(aq) \longrightarrow I_2(s) + 2e^- \qquad E^\circ_{ox} = -0.534 \text{ V}$$

Electrolysis of a water solution of KI gives a saturated solution of iodine at the anode (Figure 17.12).

- *the oxidation of a water molecule to oxygen gas*

$$2H_2O \longrightarrow O_2(g) + 4H^+(aq) + 4e^- \qquad E^\circ_{ox} = -1.229 \text{ V}$$

This half-reaction occurs when the anion cannot be oxidized. Examples include nitrate and sulfate anions, where the nonmetal present is already in its highest oxidation state (+5 for N, +6 for S).

17-7 Commercial Cells

To a chemist, electrochemical cells are of interest primarily for the information they yield concerning the spontaneity of redox reactions, the strengths of oxidizing and reducing agents, and the concentrations of trace species in solution. The viewpoint of an engineer is somewhat different; applications of electrolytic cells in electroplating and electrosynthesis are of particular importance. To the nonscientist, electrochemistry is important primarily because of commercial voltaic cells, which supply the electrical energy for instruments ranging in size from pacemakers to automobiles.

Electrolysis of Aqueous NaCl

From a commercial standpoint, the most important electrolysis carried out in water solution is that of sodium chloride (Figure 17.13). At the anode, Cl^- ions are oxidized to chlorine gas:

$$\text{anode:} \qquad 2Cl^-(aq) \longrightarrow Cl_2(g) + 2e^-$$

At the cathode, the half-reaction involves H_2O molecules, which are easier to reduce ($E^\circ_{red} = -0.828$ V) than Na^+ ions ($E^\circ_{red} = -2.714$ V).

$$\text{cathode:} \qquad 2H_2O + 2e^- \longrightarrow H_2(g) + 2OH^-(aq)$$

To obtain the overall cell reaction, add the half-reactions:

$$2Cl^-(aq) + 2H_2O \longrightarrow Cl_2(g) + H_2(g) + 2OH^-(aq)$$

Chlorine gas bubbles out of solution at the anode. At the cathode, hydrogen gas is formed, and the solution around the electrode becomes strongly basic.

The products of this electrolysis have a variety of uses. Chlorine is used to purify drinking water; large quantities of it are consumed in making plastics such as polyvinyl chloride (PVC). Hydrogen, prepared in this and many other industrial processes, is used chiefly in the synthesis of ammonia (Chapter 12). Sodium hydroxide (lye), obtained on evaporation of the electrolyte, is used in processing pulp and paper, in the purification of aluminum ore, in the manufacture of glass and textiles, and for many other purposes.

Figure 17.13 Electrolysis of aqueous NaCl (brine). The anode and cathode half-cells are separated by a membrane that is permeable to ions. Thus the flow of ions maintains charge balance. The products, NaOH(aq), Cl_2, and H_2, are all valuable industrial chemicals.

Figure 17.14 Zn-MnO₂ dry cell. This cell produces 1.5 V and will deliver a current of about half an ampere for six hours.

Primary (Nonrechargeable) Voltaic Cells

The construction of the ordinary dry cell (Leclanché cell) used in flashlights is shown in Figure 17.14. The zinc wall of the cell is the anode. The graphite rod through the center of the cell is the cathode. The space between the electrodes is filled with a moist paste. This contains MnO_2, $ZnCl_2$, and NH_4Cl. When the cell operates, the half-reaction at the anode is

$$Zn(s) \longrightarrow Zn^{2+}(aq) + 2e^-$$

At the cathode, manganese dioxide is reduced to species in which Mn is in the +3 oxidation state, such as Mn_2O_3:

$$2MnO_2(s) + 2NH_4^+(aq) + 2e^- \longrightarrow Mn_2O_3(s) + 2NH_3(aq) + H_2O$$

The overall reaction occurring in this voltaic cell is

$$Zn(s) + 2MnO_2(s) + 2NH_4^+(aq) \longrightarrow$$
$$Zn^{2+}(aq) + Mn_2O_3(s) + 2NH_3(aq) + H_2O$$

If too large a current is drawn from a Leclanché cell, the ammonia forms a gaseous insulating layer around the carbon cathode. When this

CHEMISTRY THE **HUMAN** SIDE

The laws of electrolysis were discovered by Michael Faraday, perhaps the most talented experimental scientist of the nineteenth century. Faraday lived his entire life in what is now greater London. The son of a blacksmith, he had no formal education beyond the rudiments of reading, writing, and arithmetic. Apprenticed to a bookbinder at the age of 13, Faraday educated himself by reading virtually every book that came into the shop. One that particularly impressed him was a textbook, *Conversations in Chemistry,* written by Mrs. Jane Marcet. Anxious to escape a life of drudgery as a tradesman, Faraday wrote to Sir Humphry Davy at the Royal Institution, requesting employment. Shortly afterward, a vacancy arose, and Faraday was hired as a laboratory assistant.

Davy quickly recognized Faraday's talents and as time passed allowed him to work more and more independently. In his years with Davy, Faraday published papers covering almost every field of chemistry. They included studies on the condensation of gases (he was the first to liquefy ammonia), the reaction of silver compounds with ammonia, and the isolation of several organic compounds, the most important of which was benzene. In 1825, Faraday began a series of lectures at the Royal Institution that were brilliantly successful. That same year he succeeded Davy as director of the laboratory. As Faraday's reputation grew, it was said that "Humphrey Davy's greatest discovery was Michael Faraday." Perhaps it was witticisms of this sort that led to an estrangement between master and protégé. Late in his life, Davy opposed Faraday's nomination as a Fellow of the Royal Society and is reputed to have cast the only vote against him.

To Michael Faraday, science was an obsession; one of his biographers described him as a "work maniac." An observer (Faraday had no students) said of him,

> ... *if he had to cross the laboratory for anything, he did not walk, he ran; the quickness of his perception was equalled by the calm rapidity of his movements.*

Michael Faraday (1791–1867)

In 1839, he suffered a nervous breakdown, the result of overwork. For much of the rest of his life, Faraday was in poor health. He gradually gave up more and more of his social engagements but continued to do research at the same pace as before.

Faraday developed the laws of electrolysis between 1831 and 1834. In mid-December of 1833, he began a quantitative study of the electrolysis of several metal cations, including Sn^{2+}, Pb^{2+}, and Zn^{2+}. Despite taking a whole day off for Christmas, he managed to complete these experiments, write up the results of three years' work, and get his paper published in the *Philosophic Transactions of the Royal Society* on January 9, 1834. In this paper, Faraday introduced the basic vocabulary of electrochemistry, using for the first time the terms "anode," "cathode," "ion," "electrolyte," and "electrolysis."

happens, the voltage drops sharply and then returns slowly to its normal value of 1.5 V. ◀ⓘ This problem can be avoided by using an alkaline dry cell, in which the paste between the electrodes contains KOH rather than NH_4Cl. In this case the overall cell reaction is simply

$$Zn(s) + 2MnO_2(s) \longrightarrow ZnO(s) + Mn_2O_3(s)$$

No gas is produced. The alkaline dry cell, although more expensive than the Leclanché cell, has a longer shelf life and provides more current.

Another important primary battery is the mercury cell. ◀ⓘ It usually comes in very small sizes and is used in hearing aids, watches, cameras, and some calculators. The anode of this cell is a zinc-mercury amalgam; the reacting species is zinc. The cathode is a plate made up of mercury(II) oxide, HgO. The electrolyte is a paste containing HgO and sodium or potassium hydroxide. The electrode reactions are

$$\begin{aligned}
\text{anode:} \quad & Zn(s) + 2OH^-(aq) \longrightarrow Zn(OH)_2(s) + 2e^- \\
\text{cathode:} \quad & HgO(s) + H_2O + 2e^- \longrightarrow Hg(l) + 2\,OH^-(aq) \\
\hline
& Zn(s) + HgO(s) + H_2O \longrightarrow Zn(OH)_2(s) + Hg(l)
\end{aligned}$$

Notice that the overall reaction does not involve any ions in solution, so there are no concentration changes when current is drawn. As a result, the battery maintains a constant voltage of about 1.3 V throughout its life.

Storage (Rechargeable) Voltaic Cells

A storage cell, unlike an ordinary dry cell, can be recharged repeatedly. This can be accomplished because the products of the reaction are deposited directly on the electrodes. By passing a current through a storage cell, it is possible to reverse the electrode reactions and restore the cell to its original condition.

Lead Storage Battery

The rechargeable 12-V lead storage battery ◀ⓘ used in automobiles consists of six voltaic cells of the type shown in Figure 17.15. A group of lead plates, the grids of which are filled with spongy gray lead, forms the anode of the cell. The multiple cathode consists of another group of plates of similar design filled with lead(IV) oxide, PbO_2. These two sets of plates alternate through the cell. They are immersed in a water solution of sulfuric acid, H_2SO_4, which acts as the electrolyte.

When a lead storage battery is supplying current, the lead in the anode grids is oxidized to Pb^{2+} ions, which precipitate as $PbSO_4$. At the cathode, lead dioxide is reduced to Pb^{2+} ions, which also precipitate as $PbSO_4$.

$$\begin{aligned}
\text{Anode:} \quad & Pb(s) + HSO_4^-(aq) \longrightarrow PbSO_4(s) + H^+(aq) + 2e^- \\
\text{Cathode:} \quad & PbO_2(s) + 3H^+(aq) + HSO_4^-(aq) + 2e^- \longrightarrow PbSO_4(s) + 2H_2O \\
\hline
& Pb(s) + PbO_2(s) + 2H^+(aq) + 2HSO_4^-(aq) \longrightarrow 2PbSO_4(s) + 2H_2O \\
& \Delta G° = -371.4 \text{ kJ (25°C)}
\end{aligned}$$

Deposits of lead sulfate slowly build up on the plates, partially covering and replacing the lead and lead dioxide.

ⓘ A flashlight draws about 1 A and runs for about an hour before "dying."

ⓘ Mercury cells are no longer sold, because of mercury's toxicity.

ⓘ A 12-V storage battery can deliver 300 A for a minute or so.

Figure 17.15 Lead storage battery. Three advantages of the lead storage battery are its ability to deliver large amounts of energy for a short time, the ease of recharging, and a nearly constant voltage from full charge to discharge. A disadvantage is its high mass-to-energy ratio.

To recharge a lead storage battery, a direct current is passed through it in the proper direction so as to reverse the above reaction. In an automobile, the energy required to carry out the recharging comes from the engine.

When the battery is charged, some water may be electrolyzed:

$$2H_2O \longrightarrow 2H_2(g) + O_2(g) \qquad \Delta G° \text{ at } 25°C = 474.4 \text{ kJ}$$

The hydrogen and oxygen produced create a safety hazard. Beyond that, they can cause Pb, PbO_2, and $PbSO_4$ to flake off the plates. In a modern "maintenance-free" battery, the lead plates are alloyed with small amounts of calcium, which inhibits the electrolysis reaction.

One of the advantages of the lead storage battery is that its voltage stays constant at 2 V per cell over a wide range of sulfuric acid concentrations. Only when the battery is nearly completely discharged does the voltage drop. It is also true that the cell voltage is virtually independent of temperature. You have trouble starting your car on a cold morning because the conductivity of the electrolyte drops off sharply with temperature; the voltage is still 2 V per cell at $-40°C$.

Nickel-Based Batteries

Nickel-cadmium (Nicad) batteries have an anode of solid cadmium and a cathode of solid nickeloxy hydroxide, NiO(OH). Aqueous potassium hydroxide is often used as an electrolyte. The reaction taking place in the cell is

anode: $\qquad$ $Cd(s) + 2OH^-(aq) \longrightarrow Cd(OH)_2(s) + 2e^-$

cathode: $\quad 2NiO(OH)(s) + 2H_2O + 2e^- \longrightarrow 2Ni(OH)_2(s) + 2OH^-(aq)$

The battery produces about 1.3 V. As the battery is used, solid hydroxides of nickel and cadmium are deposited on the electrode. When the battery is recharged by a current it runs in the opposite direction, where the hydroxides of nickel and cadmium regenerate Cd and NiO(OH). Nicad batteries are a popular choice for emergency medical equipment and power tools.

Nicad batteries have been popular for many years but are being replaced by the nickel-metal **hydride**, NiMH, battery. This is because of the toxicity of cadmium and the difficulty of disposal for Nicad batteries.

In nickel-metal hydride batteries, the cathode is also nickeloxy hydroxide. The anode is a metal hydride, MH, where the "metal" is not a single element but an alloy of several metals. The half-reactions for this battery are

anode: $\qquad$ $MH(s) + OH^-(aq) \longrightarrow M(s) + H_2O + e^-$

cathode: $\quad NiO(OH)(s) + H_2O + e^- \longrightarrow Ni(OH)_2(s) + OH^-(aq)$

Besides being environmentally friendlier than Nicad batteries, nickel-metal hydride batteries have a higher energy content per unit mass. This is called the battery's energy density with units expressed in watt-hour/kg. These batteries are commonly used in hybrid electric cars and buses.

Lithium-Ion (Li-Ion) Batteries

The latest type of rechargeable battery is the lithium-ion battery. The reaction taking place in the cell is more complicated than in either the lead storage battery or the nickel-based batteries. Suffice it to say that lithium ions move from the anode to the cathode. The anode is made up of carbon layers in which lithium ions are embedded. The cathode is a lithium metal oxide like $LiCoO_2$. The electrolyte is a lithium salt in an organic solvent.

These batteries have a number of advantages. Lithium is the lightest metal ($d = 0.53 \text{ g/cm}^3$) and extremely reactive. These characteristics translate into a high-energy density. A typical Li-ion battery can store 150 watt-hours in 1 kg of battery. In comparison, NiMH batteries can store 80 Wh/kg while a lead storage battery can store 2.5 Wh/kg. Li-ion batteries hold their charge, losing only about 5% a month, whereas NiMH batteries lose about 20% per month. Li-ion batteries also have no memory effect, which means that you do not have to completely discharge them before recharging. However, if you completely discharge a Li-ion battery, it is ruined. The biggest disadvantage to the consumer is its cost and limited life span.

Li-ion batteries are used when light weight and high energy density are required. They are commonly used in cell phones, laptops, and digital cameras.

Fuel Cells

A fuel cell (Figure 17.16) is a voltaic cell in which a fuel, usually hydrogen, is oxidized at the anode. At the cathode, oxygen is reduced. The reaction taking place in the alkaline fuel cells used in the space program since the 1960s is

anode: $2H_2(g) + 4OH^-(aq) \longrightarrow 4H_2O + 4e^-$

cathode: $O_2(g) + 2H_2O + 4e^- \longrightarrow 4OH^-(aq)$

$2H_2(g) + O_2(g) \longrightarrow 2H_2O$ $\Delta G°$ at 25°C = −474.4 kJ

Using a platinum catalyst, this reaction can now be carried out at temperatures as low as 40°C. The hydrogen used must be very pure; traces of carbon monoxide can poison the catalyst.

Current research on fuel cells ◂ⓘ is directed toward the replacement of the internal combustion engine. To do this, hydrogen must be stored in the vehicle and replenished from time to time at "filling stations." Three kilograms of hydrogen should be enough to drive a small car 500 km (300 miles) between fill-ups.

The rationale for using a hydrogen fuel cell is that the cell reaction produces only water. In contrast, your car's engine produces small amounts of air pollutants such as NO and copious amounts of CO_2, which contributes to global warming. On the other hand, there are a couple of deterrents to the use of fuel cells, i.e.,

■ the cost per kilojoule of energy is higher than with a gasoline- or diesel-powered engine.

■ the storage of hydrogen in a vehicle is a serious problem. To carry three kilograms of hydrogen as a compressed gas at 200 atm (∼3000 psi) requires a heavy tank with a volume of 200 L (50 gallons). Liquid hydrogen requires a smaller tank, but its extremely low critical temperature (−240°C) creates all sorts of problems. Another possibility is to convert hydrogen reversibly to a transition metal hydride (e.g., TiH_2). It's hard to say at this point how well that would work. All in all, it is almost certain that your next car will *not* use hydrogen as a fuel.

Figure 17.16 Fuel cell under Honda Clarity hood.

ⓘ Fuel cells have been around for about 50 years.

CHEMISTRY **BEYOND** THE CLASSROOM

Fuel Cells: The Next Step in Chemical-to-Electrical-Energy Conversion?

Steven R. Shaw, Montana State University

Fuel cells have attracted considerable interest because of their potential for efficient conversion of the energy (ΔG) from a chemical reaction to electrical energy (ΔE). This efficiency is achieved by directly converting chemical energy to electricity. Conventional systems burn fuel in an engine and convert the resulting mechanical output to electrical power. Potential applications include stationary multi-megawatt power plants, battery replacements for personal electronics, and even fuel-cell-powered unmanned autonomous vehicles (UAVs).

The different types of fuel cells depend on the electrolytes used and the temperatures at which the electrolytes operate. One popular fuel cell that operates at low temperatures (80°C) used a polymer-electrolyte membrane, also known as a proton-exchange membrane (PEM). This membrane is made of a fluorinated polymer, like Teflon, that is highly conductive to electrons. It allows an energetically favorable reaction ($\Delta G < 0$) between the fuel gas and oxidant to be spatially confined, so that the electrons evolving from and consumed by the half-cell reactions can do work in an external circuit. Figure A shows a schematic of a hydrogen/oxygen fuel cell with a proton-exchange membrane.

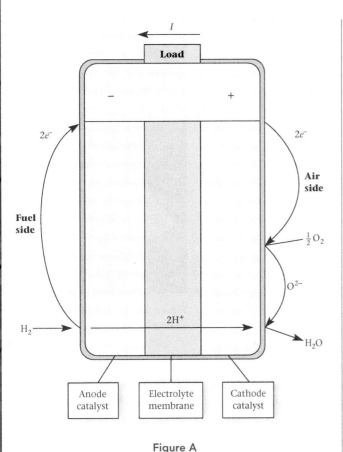

Figure A

The half-cell reactions supported by this structure are

$$H_2 \longrightarrow 2e^- + 2H^+$$

$$2H^+ + 2e^- + \tfrac{1}{2}O_2 \longrightarrow H_2O$$

Taken together, the overall reaction is simply the energetically favorable oxidation of hydrogen.

$$H_2(g) + \tfrac{1}{2}O_2(g) \longrightarrow H_2O$$

Unlike burning hydrogen in air, in a fuel cell the electrolyte partitions the overall reaction into half-cell reactions on either side of the cell. Hydrogen ions flow through the membrane, but since the electrolyte is nonconductive to electrons, it forces the electrons to flow through an external electrical circuit to complete the reaction. This flow of electrons can be harnessed to do useful work, such as illuminating a light bulb or running a laptop. The voltage at the terminals of a single cell is rather small, on the order of a volt, so in applications a number of cells are usually connected in series to form a fuel cell "stack."

Although it is attractive to directly convert chemical energy to electricity, PEM fuel cells face significant practical obstacles. Expensive heavy metals like platinum are typically used as catalysts to reduce energy barriers associated with the half-cell reactions. PEM fuel cells also cannot use practical hydrocarbon fuels like diesel without complicated preprocessing steps. Those significantly increase the complexity of the overall system. At this time, it appears likely that PEM fuel cells will be confined to niche applications where high cost and special fuel requirements are tolerable.

On the other hand, high-temperature fuel cells, which work at 800°C, offer some significant advantages over low-temperature cells like PEMs. One of their principal advantages is the potential to use more practical hydrocarbon fuels such as methane, or even mixtures of gasoline and diesel. This feature avoids the need for the "new energy infrastructure" sometimes associated with hydrogen fuel cells. Another advantage is the lower cost of the catalysts needed for the operation of a high-temperature fuel cell. As an added benefit, the exhaust gases are hot enough for secondary energy recovery. For example, "waste heat" from a high-temperature fuel cell can supply energy needed to convert conventional hydrocarbon fuels into simpler components (like carbon monoxide and hydrogen) that work efficiently in the fuel cell. The exhaust of a high-temperature fuel cell can also be used to drive a conventional turbine and generator in what is called a combined cycle or hybrid system. Some high-temperature fuel-cell test installations use the exhaust to heat nearby buildings.

One leading prototype of a high-temperature fuel cell is the solid oxide fuel cell, or SOFC. The basic principle of the SOFC, like the PEM, is to use an electrolyte layer with high ionic conductivity but very small electronic conductivity. Figure B shows a schematic illustration of a SOFC fuel cell using carbon monoxide as fuel.

Unlike the PEM, the ionic conduction occurs for the oxygen ion instead of the hydrogen ion. SOFCs are made of ceramic materials like zirconium ($Z = 40$) stabilized by yttrium ($Z = 39$). High-temperature oxygen conductivity is achieved by creating oxygen vacancies in the lattice

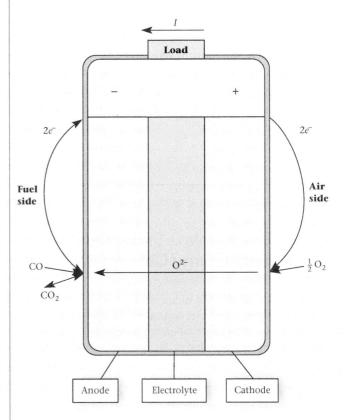

Figure B

structure of the electrolyte material. The half-cell reactions in this case are

$$\tfrac{1}{2}O_2 + 2e^- \longrightarrow O^{2-}$$

$$O^{2-} + CO \longrightarrow CO_2 + 2e^-$$

Note that the overall reaction

$$\tfrac{1}{2}O_2(g) + CO(g) \longrightarrow CO_2(g)$$

is the oxidation of CO, but the fuel cell's materials partition the reaction so that the electrons flow through an external circuit and perform useful work.

Figure C shows an electron photomicrograph of a broken planar SOFC. The thick portion on the left is the porous anode structure. This is an anode-supported cell, meaning that in addition to collecting current and supporting the anode reaction, the anode layer stiffens the whole cell. The layer on the right is the cathode, and the interface between the two is the thin electrolyte. One of the challenges of this design is to ensure that the rates of expansion of the cathode and the anode match. If the anode expands faster than the cathode, the planar cell tends to curl like a potato chip when the temperature changes.

Note that fuel cells do not provide a new source of energy. They are designed to use conventional, currently available fuels. Their main advantage lies in the efficiency of their operation, which creates fewer byproducts to threaten the environment.

Figure C

Key Concepts

1. Balance redox half-equations and overall equation.
   (**Examples 17.1, 17.2, Problems 1–10**)
2. Draw a diagram for a voltaic cell, labeling electrodes and direction of current flow.
   (**Example 17.3; Problems 11–18**)
3. Use standard potentials (Table 17.1) to
   ■ compare the relative strengths of different oxidizing agents; different reducing agents.
   (**Example 17.4; Problems 19–28**)
   ■ calculate $E°$ and/or reaction spontaneity.
   (**Examples 17.5, 17.6; Problems 29–48**)
4. Relate $E°$ to $\Delta G°$ and K.
   (**Example 17.7; Problems 49–62**)
5. Use the Nernst equation to relate voltage to concentration.
   (**Examples 17.8, 17.9; Problems 63–76**)
6. Relate mass of product to amount of electricity (coulombs) or amount of energy (joules) used in electrolysis reactions.
   (**Example 17.10; Problems 77–84**)

Key Equations

Standard voltage	$E° = E°_{red} + E°_{ox}$
$E°, \Delta G°, K$	$E° = \dfrac{-\Delta G°}{nF} = \dfrac{RT \ln K}{nF} = \dfrac{0.0257 \text{ V}}{n} \ln K$ (at 25°C)
Nernst equation	$E = E° - \dfrac{RT}{nF} \ln Q = E° - \dfrac{0.0257 \text{ V}}{n} \ln Q$ (at 25°C)

Key Terms

ampere	electrolytic cell	reducing agent	standard reduction
anode	Faraday constant	reduction	voltage ($E°_{red}$)
cathode	half-cell	salt bridge	voltaic (galvanic) cell
coulomb	half-equations	standard oxidation	
electrodes	oxidation	voltage ($E°_{ox}$)	
electrolysis	oxidizing agent	standard potential ($E°$)	

463

Summary Problem

A voltaic cell consists of two half-cells. One of the half-cells contains a platinum electrode dipped in an aqueous solution of permanganate and manganese(II) ions. The other half-cell has a platinum electrode dipped in an aqueous solution of chlorate ions. Chlorine gas is bubbled into the half-cell solution. Assume that the cell reaction, which produces a positive voltage, involves permanganate ions and chlorine gas in acidic solution. The temperature of the cells is kept constant at 25°C.

(a) Write a balanced oxidation half-reaction. Where does it occur?

(b) Write a balanced reduction half-reaction. Where does it occur?

(c) Write a balanced overall equation for the redox reaction taking place.

(d) Write the cell notation.

(e) Calculate $E°$ for the cell.

(f) For the redox reaction in (c), calculate K and $\Delta G°$.

(g) Calculate the pH of the cell, when the cell voltage is 0.023 V, all ionic species except H^+ are at 0.200 M, and P_{Cl_2} is 0.873 atm.

An electrolytic cell contains a solution of manganese(II) nitrate. Assume that in this cell, manganese plates out at one electrode and that oxygen gas is evolved at the other electrode.

(h) Write the anode half-reaction, the cathode half-reaction, and a balanced overall equation for the electrolysis.

(i) How long will it take to deposit 5.00 g of manganese using a current of 6.20 A?

(j) A current of 5.00 A is passed through the cell for 1.00 h. Starting out with 175 mL of 1.25 M $Mn(NO_3)_2$, what is $[Mn^{2+}]$ after electrolysis? What is the pH of the solution? (Assume that the H^+ present before electrolysis is negligible.) What is the volume of oxygen gas produced at 757 mm Hg and 25°C? Assume 100% efficiency and no change in volume during electrolysis.

Answers

(a) oxidation half-reaction at the anode:
$Cl_2(g) + 6H_2O \longrightarrow 2ClO_3^-(aq) + 12H^+(aq) + 10e^-$

(b) reduction half-reaction at the cathode:
$MnO_4^-(aq) + 8H^+(aq) + 5e^- \longrightarrow Mn^{2+}(aq) + 4H_2O$

(c) $Cl_2(g) + 2MnO_4^-(aq) + 4H^+(aq) \longrightarrow$
$2ClO_3^-(aq) + 2Mn^{2+}(aq) + 2H_2O$

(d) $Pt \mid Cl_2 \mid ClO_3^- \parallel MnO_4^-, Mn^{2+} \mid Pt$

(e) 0.054 V

(f) $\Delta G° = -52$ kJ; $K = 1 \times 10^9$

(g) 1.65

(h) anode: $2H_2O \longrightarrow O_2(g) + 4H^+(aq) + 4e^-$
cathode: $Mn^{2+}(aq) + 2e^- \longrightarrow Mn(s)$
overall: $2Mn^{2+}(aq) + 2H_2O \longrightarrow$
$2Mn(s) + O_2(g) + 4H^+(aq)$

(i) 47.2 min

(j) 0.717 M; −0.028; 1.15 L

Questions and Problems

Even-numbered questions and Challenge Problems have answers in Appendix 5 and fully worked solutions in the *Student Solutions Manual*.

17-1 Oxidation-Reduction Reactions Revisited

1. Balance the following half-equations. Balance (a) and (b) in basic medium, (c) and (d) in acidic medium.
(a) $O_2(g) \longrightarrow O^{2-}(aq)$
(b) $MnO_4^-(aq) \longrightarrow MnO_2(s)$
(c) $Cr_2O_7^{2-}(aq) \longrightarrow Cr^{3+}(aq)$
(d) $Cl^-(aq) \longrightarrow Cl_2(g)$

2. Balance the following half-equations. Balance (a) and (b) in acidic medium, (c) and (d) in basic medium.
(a) $CH_3OH(aq) \longrightarrow CO_2(g)$
(b) $NO_3^-(aq) \longrightarrow NH_4^+(aq)$
(c) $Fe^{3+}(aq) \longrightarrow Fe(s)$
(d) $V^{2+}(aq) \longrightarrow VO_3^-(aq)$

3. Balance the following reactions in base:
(a) $Ag(s) + NO_3^-(aq) \longrightarrow Ag^+(aq) + NO(g)$
(b) $CO_2(g) + H_2O(l) \longrightarrow C_2H_4(g) + O_2(g)$

4. Balance the following reactions in acid:
(a) $H_2O_2(aq) + Ni^{2+}(aq) \longrightarrow Ni^{3+}(aq) + H_2O$
(b) $Cr_2O_7^{2-}(aq) + Sn^{2+}(aq) \longrightarrow Cr^{3+}(aq) + Sn^{4+}(aq)$

5. Write balanced equations for the following reactions in acid solution.
(a) $Ni^{2+}(aq) + IO_4^-(aq) \longrightarrow Ni^{3+}(aq) + I^-(aq)$
(b) $O_2(g) + Br^-(aq) \longrightarrow H_2O + Br_2(l)$
(c) $Ca(s) + Cr_2O_7^{2-}(aq) \longrightarrow Ca^{2+}(aq) + Cr^{3+}(aq)$
(d) $IO_3^-(aq) + Mn^{2+}(aq) \longrightarrow I^-(aq) + MnO_2(s)$

6. Write balanced equations for the following reactions in acid solution.
(a) $P_4(s) + Cl^-(aq) \longrightarrow PH_3(g) + Cl_2(g)$
(b) $MnO_4^-(aq) + NO_2^-(aq) \longrightarrow Mn^{2+}(aq) + NO_3^-(aq)$
(c) $HBrO_3(aq) + Bi(s) \longrightarrow HBrO_2(aq) + Bi_2O_3(s)$
(d) $CrO_4^{2-}(aq) + SO_3^{2-}(aq) \longrightarrow Cr^{3+}(aq) + SO_4^{2-}(aq)$

7. Write balanced equations for the following reactions in basic solution.
   (a) $SO_2(g) + I_2(aq) \longrightarrow SO_3(g) + I^-(aq)$
   (b) $Zn(s) + NO_3^-(aq) \longrightarrow NH_3(aq) + Zn^{2+}(aq)$
   (c) $ClO^-(aq) + CrO_2^-(aq) \longrightarrow Cl^-(aq) + CrO_4^{2-}(aq)$
   (d) $K(s) + H_2O \longrightarrow K^+(aq) + H_2(g)$

8. Write balanced net ionic equations for the following reactions in basic medium.
   (a) $Ca(s) + VO_4^{3-}(aq) \longrightarrow Ca^{2+}(aq) + V^{2+}(aq)$
   (b) $C_2H_4(g) + BiO_3^-(aq) \longrightarrow CO_2(g) + Bi^{3+}(aq)$
   (c) $PbO_2(s) + H_2O \longrightarrow O_2(g) + Pb^{2+}$
   (b) $IO_3^-(aq) + Cl^-(aq) \longrightarrow Cl_2(g) + I_3^-(aq)$

9. Write balanced net ionic equations for the following reactions in acid solution.
   (a) Liquid hydrazine reacts with an aqueous solution of sodium bromate. Nitrogen gas and bromide ions are formed.
   (b) Solid phosphorus (P_4) reacts with an aqueous solution of nitrate to form nitrogen oxide gas and dihydrogen phosphate ($H_2PO_4^-$) ions.
   (c) Aqueous solutions of potassium sulfite and potassium permanganate react. Sulfate and manganese(II) ions are formed.

10. Write balanced net ionic equations for the following reactions in acid solution.
    (a) Nitrogen oxide and hydrogen gases react to form ammonia gas and steam.
    (b) Hydrogen peroxide reacts with an aqueous solution of sodium hypochlorite to form oxygen and chlorine gases.
    (c) Zinc metal reduces the vanadyl ion (VO^{2+}) to vanadium(III) ions. Zinc ions are also formed.

17-2 Voltaic Cells

11. Write a balanced chemical equation for the overall cell reaction represented as
    (a) $Ag \mid Ag^+ \parallel Sn^{4+}, Sn^{2+} \mid Pt.$
    (b) $Al \mid Al^{3+} \parallel Cu^{2+} \mid Cu.$
    (c) $Pt \mid Fe^{2+}, Fe^{3+} \parallel MnO_4^-, Mn^{2+} \mid Pt.$

12. Write a balanced net ionic equation for the overall cell reaction represented by
    (a) $Cd \mid Cd^{2+} \parallel Sb^{3+} \mid Sb$
    (b) $Pt \mid Cu^+, Cu^{2+} \parallel Mg^{2+} \mid Mg$
    (c) $Pt \mid Cr^{3+}, Cr_2O_7^{2-} \parallel ClO_3^-, Cl^- \mid Pt$ (acid)

13. Draw a diagram for a salt bridge cell for each of the following reactions. Label the anode and cathode, and indicate the direction of current flow throughout the circuit.
    (a) $Zn(s) + Cd^{2+}(aq) \longrightarrow Zn^{2+}(aq) + Cd(s)$
    (b) $2AuCl_4^-(aq) + 3Cu(s) \longrightarrow$
    $\qquad\qquad\qquad 2Au(s) + 8Cl^-(aq) + 3Cu^{2+}(aq)$
    (c) $Fe(s) + Cu(OH)_2(s) \longrightarrow Cu(s) + Fe(OH)_2(s)$

14. Follow the directions in Question 13 for the following reactions:
    (a) $Sn(s) + 2Ag^+(aq) \longrightarrow Sn^{2+}(aq) + 2Ag(s)$
    (b) $H_2(g) + Hg_2Cl_2(s) \longrightarrow 2H^+(aq) + 2Cl^-(aq) + 2Hg(l)$
    (c) $Pb(s) + PbO_2(s) + 4H^+(aq) + 2SO_4^{2-}(aq) \longrightarrow$
    $\qquad\qquad\qquad 2PbSO_4(s) + 2H_2O$

15. Consider a voltaic salt bridge cell represented by the following reaction:
    $$Ca(s) + 2H^+(aq) \longrightarrow Ca^{2+}(aq) + H_2(g)$$
    (a) What is the direction of the electrons in the external circuit?

    (b) What electrode can be used at the cathode?
    (c) What is the reaction occurring at the anode?

16. Consider a salt bridge voltaic cell represented by the following reaction
    $$MnO_4^-(aq) + 8H^+(aq) + 5Fe^{2+}(aq) \longrightarrow$$
    $$Mn^{2+}(aq) + 5Fe^{3+}(aq) + 4H_2O$$
    (a) What is the direction of the electrons in the external circuit?
    (b) What electrode can be used at the anode?
    (c) What is the reaction occurring at the cathode?

17. Consider a salt bridge cell in which the anode is a manganese rod immersed in an aqueous solution of manganese(II) sulfate. The cathode is a chromium strip immersed in an aqueous solution of chromium(III) sulfate. Sketch a diagram of the cell, indicating the flow of the current throughout. Write the half-equations for the electrode reactions, the overall equation, and the abbreviated notation for the cell.

18. Follow the directions in Question 17 for a salt bridge cell in which the anode is a platinum rod immersed in an aqueous solution of sodium iodide containing solid iodine crystals. The cathode is another platinum rod immersed in an aqueous solution of sodium bromide with bromine liquid.

17-3 Standard Voltages

Strength of Oxidizing and Reducing Species

19. Which species in the pair is the stronger oxidizing agent?
    (a) Cr or Cd
    (b) I^- or Br^-
    (c) OH^- or NO_2^- (basic solution)
    (d) NO in acidic solution or NO in basic solution

20. Which species in each pair is the stronger reducing agent?
    (a) Cl^- or Br^- (b) Cu or Ni   (c) Hg_2^{2+} or $NO(g)$

21. Using Table 17.1, arrange the following reducing agents in order of increasing strength.
    $$Br^- \qquad Zn \qquad Co \qquad PbSO_4 \qquad H_2S \text{ (acidic)}$$

22. Use Table 17.1 to arrange the following oxidizing agents in order of increasing strength:
    $$Mn^{2+} \qquad S \qquad Co^{3+} \qquad Cl_2 \qquad K^+$$

23. Consider the following species.
    $$Cr^{3+} \qquad Hg(l) \qquad H_2 \text{ (acidic)} \qquad Sn^{2+} \qquad Br_2 \text{ (acidic)}$$
    Classify each species as oxidizing agent, reducing agent, or both. Arrange the oxidizing agents in order of increasing strength. Do the same for the reducing agents.

24. Follow the directions of Question 23 for the following species:
    $$Cu^+ \qquad Zn \qquad Ni^{2+} \qquad Fe^{2+} \qquad H^+ \text{(acidic)}$$

25. For the following half-reactions, answer these questions.
$Ce^{4+}(aq) + e^- \longrightarrow Ce^{3+}(aq)$	$E° = +1.61\ V$
$Ag^+(aq) + e^- \longrightarrow Ag(s)$	$E° = +0.80\ V$
$Hg_2^{2+}(aq) + 2e^- \longrightarrow 2Hg(l)$	$E° = +0.80\ V$

$$Sn^{2+}(aq) + 2e^- \longrightarrow Sn(s) \qquad E° = -0.14 \text{ V}$$
$$Ni^{2+}(aq) + 2e^- \longrightarrow Ni(s) \qquad E° = -0.24 \text{ V}$$
$$Al^{3+}(aq) + 3e^- \longrightarrow Al(s) \qquad E° = -1.68 \text{ V}$$

(a) Which is the weakest oxidizing agent?
(b) Which is the strongest oxidizing agent?
(c) Which is the strongest reducing agent?
(d) Which is the weakest reducing agent?
(e) Will $Sn(s)$ reduce $Ag^+(aq)$ to $Ag(s)$?
(f) Will $Hg(l)$ reduce $Sn^{2+}(aq)$ to $Sn(s)$?
(g) Which ion(s) can be reduced by $Sn(s)$?
(h) Which metal(s) can be oxidized by $Ag^+(aq)$?

26. For the following half-reactions, answer the questions below.

$$Co^{3+}(aq) + e^- \longrightarrow Co^{2+}(aq) \qquad E° = +1.953 \text{ V}$$
$$Fe^{3+}(aq) + e^- \longrightarrow Fe^{2+}(aq) \qquad E° = +0.769 \text{ V}$$
$$I_2(aq) + 2e^- \longrightarrow 2I^-(aq) \qquad E° = +0.534 \text{ V}$$
$$Pb^{2+}(aq) + 2e^- \longrightarrow Pb(s) \qquad E° = -0.127 \text{ V}$$
$$Cd^{2+}(aq) + 2e^- \longrightarrow Cd(s) \qquad E° = -0.402 \text{ V}$$
$$Mn^{2+}(aq) + 2e^- \longrightarrow Mn(s) \qquad E° = -1.182 \text{ V}$$

(a) Which is the weakest reducing agent?
(b) Which is the strongest reducing agent?
(c) Which is the strongest oxidizing agent?
(d) Which is the weakest oxidizing agent?
(e) Will $Pb(s)$ reduce $Fe^{3+}(aq)$ to $Fe^{2+}(aq)$?
(f) Will $I^-(aq)$ reduce $Pb^{2+}(aq)$ to $Pb(s)$?
(g) Which ion(s) can be reduced by $Pb(s)$?
(h) Which if any metal(s) can be oxidized by $Fe^{3+}(aq)$?

27. Use Table 17.1 to select
(a) a reducing agent in acidic solution that converts ClO_3^- to Cl_2 but not $Cr_2O_7^{2-}$ to Cr^{3+}.
(b) an oxidizing agent that converts Mg to Mg^{2+} but not Mn to Mn^{2+}.
(c) a reducing agent in basic solution that converts NO_2^- to NO_3^- but not OH^- to O_2.

28. Use Table 17.1 to select
(a) an oxidizing agent in basic solution that converts ClO_3^- to ClO_4^- but not Cl^- to ClO_3^-.
(b) a reducing agent that converts Mg^{2+} to Mg but not Ba^{2+} to Ba.
(c) a reducing agent that converts Na^+ to Na but not Li^+ to Li.

Calculation of E°

29. Calculate $E°$ for the following voltaic cells:
(a) $MnO_2(s) + 4H^+(aq) + 2I^-(aq) \longrightarrow$
$$Mn^{2+}(aq) + 2H_2O + I_2(s)$$
(b) $H_2(g) + 2OH^-(aq) + S(s) \longrightarrow 2H_2O + S^{2-}(aq)$
(c) an Ag-Ag^+ half-cell and an Au-$AuCl_4^-$ half-cell

30. Calculate $E°$ for the following voltaic cells:
(a) $2Na(s) + Fe^{2+}(aq) \longrightarrow 2Na^+(aq) + Fe(s)$
(b) $3SO_4^{2-}(aq) + 12H^+(aq) + 2Al(s) \longrightarrow$
$$3SO_2(g) + 2Al^{3+}(aq) + 6H_2O$$
(c) $2S(s) + 4OH^-(aq) \longrightarrow$
$$O_2(g) + 2S^{2-}(aq) + 2H_2O \qquad \text{(basic)}$$

31. Using Table 17.1, calculate $E°$ for the reaction between
(a) lead and silver nitrate ion to produce lead(II) nitrate.

(b) oxygen and iron(II) ions to produce water and iron(III) ions.
(c) sulfur and nitrogen oxide gas in basic solution to form nitrate and sulfide ions.

32. Using Table 17.1, calculate $E°$ for the reaction between
(a) chromium(II) ions and tin(IV) ions to produce chromium(III) ions and tin(II) ions.
(b) manganese(II) ions and hydrogen peroxide to produce solid manganese dioxide (MnO_2).

33. Calculate $E°$ for the following cells:
(a) $Pb \mid PbSO_4 \parallel Pb^{2+} \mid Pb$
(b) $Pt \mid Cl_2 \mid ClO_3^- \parallel O_2 \mid H_2O \mid Pt$
(c) $Pt \mid OH^- \mid O_2 \parallel ClO_3^-, Cl^- \mid Pt$ (basic medium)

34. Calculate $E°$ for the following cells:
(a) $Ag \mid Ag^+ \parallel Sn^{4+}, Sn^{2+} \mid Pt$
(b) $Al \mid Al^{3+} \parallel Cu^{2+} \mid Cu$
(c) $Pt \mid Fe^{2+}, Fe^{3+} \parallel MnO_4^-, Mn^{2+} \mid Pt$

35. Suppose $E°_{red}$ for $Ag^+ \longrightarrow Ag$ were set equal to zero instead of that of $H^+ \longrightarrow H_2$. What would be
(a) $E°_{red}$ for $H^+ \longrightarrow H_2$?
(b) $E°_{ox}$ for $Ca \longrightarrow Ca^{2+}$?
(c) $E°$ for the cell in 33(c)? Compare your answer with that obtained in 33(c).

36. Suppose $E°_{red}$ for $H^+ \longrightarrow H_2$ were taken to be 0.300 V instead of 0.000 V. What would be
(a) $E°_{ox}$ for $H_2 \longrightarrow H^+$?
(b) $E°_{red}$ for $Br_2 \longrightarrow Br^-$?
(c) $E°$ for the cell in 34(c)? Compare your answer with that obtained in 34(c).

Spontaneity and E°

37. Which of the following reactions is/are spontaneous at standard conditions?
(a) $Zn(s) + 2Fe^{3+} \longrightarrow Zn^{2+}(aq) + 2Fe^{2+}(aq)$
(b) $Cu(s) + 2H^+(aq) \longrightarrow Cu^{2+}(aq) + H_2(g)$
(c) $2Br^-(aq) + I_2(s) \longrightarrow Br_2(l) + 2I^-(aq)$

38. Which of the following reactions is(are) spontaneous at standard conditions?
(a) $Ba(s) + 2Cr^{3+}(aq) \longrightarrow 2Cr^{2+}(aq) + Ba^{2+}(aq)$
(b) $Co^{2+}(aq) + Cu(s) \longrightarrow Co(s) + Cu^{2+}(aq)$
(c) $2S(s) + 4OH^-(aq) \longrightarrow$
$$O_2(g) + 2S^{2-}(aq) + 2H_2O \qquad \text{(basic)}$$

39. Use the following half-equations to write three spontaneous reactions. Justify your answers by calculating $E°$ for the cells.
1. $MnO_4^-(aq) + 8H^+(aq) + 5e^- \longrightarrow$
$$Mn^{2+}(aq) + 4H_2O \quad E° = +1.512 \text{ V}$$
2. $O_2(g) + 4H^+(aq) + 4e^- \longrightarrow 2H_2O$
$$E° = +1.229 \text{ V}$$
3. $Co^{2+}(aq) + 2e^- \longrightarrow Co(s) \qquad E° = -0.282 \text{ V}$

40. Follow the directions of Question 39 for the following half-reactions:
1. $I_2(s) + 2e^- \longrightarrow 2I^-(aq)$
2. $Co^{3+}(aq) + e^- \longrightarrow Co^{2+}(aq)$
3. $Cr_2O_7^{2-}(aq) + 14H^+(aq) + 6e^- \longrightarrow$
$$2Cr^{3+}(aq) + 7H_2O$$

41. Use Table 17.1 to answer the following questions:
(a) Will chlorate ions in acidic solution oxidize Cl^- to Cl_2 or F^- to F_2?
(b) What is the product when Fe reduces Cr^{3+}?

(c) Oxygen is reduced to water in acidic medium and to OH^- in basic medium. If ClO_3^- (oxidized to ClO_4^-) is the reducing agent, in which medium will oxygen be reduced spontaneously? ($E°_{red}$ for ClO_4^- in acidic medium is 1.19 V.)

42. Use Table 17.1 to answer the following questions.
(a) Will nitrate ions in acidic solution oxidize copper to copper(II) ions or gold to gold(III) ions?
(b) What is the product of the oxidation of copper by iron(II) ions? copper(I) ions or copper(II) ions?
(c) Oxygen is reduced to water in acidic medium and to OH^- in basic medium. If ClO_3^- (oxidized to ClO_4^-) is the reducing agent, in which medium will oxygen be reduced spontaneously? (The standard reduction potential for ClO_4^- in acidic medium is 1.19 V.)

43. Write the equation for the reaction, if any, that occurs when each of the following experiments is performed under standard conditions.
(a) Crystals of iodine are added to an aqueous solution of potassium bromide.
(b) Liquid bromine is added to an aqueous solution of sodium chloride.
(c) A chromium wire is dipped into a solution of nickel(II) chloride.

44. Write the equation for the reaction, if any, that occurs when each of the following experiments is performed under standard conditions.
(a) Sulfur is added to mercury.
(b) Manganese dioxide in acidic solution is added to liquid mercury.
(c) Aluminum metal is added to a solution of potassium ions.

45. Which of the following species will be oxidized by 1 M HCl?
(a) Au   (b) Mg   (c) Cu   (d) F^-

46. Which of the following species will be oxidized by 1 M HBr?
(a) Na   (b) Hg   (c) Pb   (d) Mn^{2+}

47. Use Table 17.1 to predict what reaction, if any, will occur when the following species are mixed in acidic solution at standard conditions.
(a) Cr, Ni^{2+}, Cl^- (b) F_2, Na^+, Br^- (c) SO_2, Fe^{3+}, NO_3^-

48. Use Table 17.1 to predict what reaction, if any, will occur if sulfur is added to acidic aqueous solutions of the following species at standard conditions.
(a) $MgBr_2$ (b) $Sn(NO_3)_2$ (c) $Cr(ClO_3)_2$

17-4 Relations between $E°$, $\Delta G°$, and K

49. Consider a cell reaction at 25°C where $n = 2$. Fill in the following table.

	$\Delta G°$	$E°$	K
(a)	19 kJ	_____	_____
(b)	_____	0.035 V	_____
(c)	_____	_____	0.095

50. Consider a cell reaction at 25°C where $n = 4$. Fill in the following table.

	$\Delta G°$	$E°$	K
(a)	_____	_____	1.6×10^{-3}
(b)	_____	0.117 V	_____
(c)	−5.8 kJ	_____	_____

51. For a certain cell, $\Delta G° = 25.0$ kJ. Calculate $E°$ if n is
(a) 1   (b) 2   (c) 4
Comment on the effect that the number of electrons exchanged has on the voltage of a cell.

52. For a certain cell, $E° = 1.08$ V. Calculate $\Delta G°$ if n is
(a) 1   (b) 2   (c) 3

53. Calculate $E°$, $\Delta G°$, and K at 25°C for the reaction
$$3Mn^{2+}(aq) + 2MnO_4^-(aq) + 2H_2O \longrightarrow$$
$$5MnO_2(s) + 4H^+(aq)$$

54. Calculate $E°$, $\Delta G°$, and K at 25°C for the reaction
$$2MnO_4^-(aq) + 4H^+(aq) + Cl_2(g) \longrightarrow$$
$$2Mn^{2+}(aq) + 2ClO_3^-(aq) + 2H_2O$$

55. Calculate $\Delta G°$ at 25°C for each of the reactions referred to in Question 29. Assume smallest whole-number coefficients.

56. Calculate $\Delta G°$ at 25°C for each of the reactions referred to in Question 30. Assume smallest whole-number coefficients.

57. Calculate K at 25°C for each of the reactions referred to in Question 31.

58. Calculate K at 25°C for each of the reactions referred to in Question 32. Assume smallest whole-number coefficients.

59. Consider the reaction for the formation of the complex ion $CuCl_2^-$.
$$Cu^+(s) + 2Cl^-(aq) \rightarrow CuCl_2^-(aq)$$
The formation constant K_f for this reaction is 3.0×10^5.
(a) Write the two half-reactions that give rise to the formation of the complex ion.
(b) Find the standard reduction potential at 25°C for the reaction.

60. Use Table 17.1 to find K_f for $AuCl_4^-(aq)$ at 25°C.

61. Given the following information:
$$Ag_2CrO_4(s) \rightleftharpoons 2Ag^+(aq) + CrO_4^{2-}(aq) \qquad K_{sp} = 1 \times 10^{-12}$$
$$Ag^+(aq) + e^- \longrightarrow Ag(s) \qquad E° = +0.799 \text{ V}$$
find the standard reduction potential at 25°C for the half-reaction
$$Ag_2CrO_4(s) + 2e^- \longrightarrow 2Ag(s) + CrO_4^{2-}(aq)$$

62. What is $E°$ at 25°C for the following reaction?
$$Ba^{2+}(aq) + SO_4^{2-}(aq) \longrightarrow BaSO_4(s)$$
K_{sp} for $BaSO_4$ is 1.1×10^{-10}.

17-5 Effect of Concentration on Voltage

63. Consider a voltaic cell at 25°C in which the following reaction takes place.
$$3H_2O_2(aq) + 6H^+(aq) + 2Au(s) \longrightarrow 2Au^{3+}(aq) + 6H_2O$$

(a) Calculate $E°$.

(b) Write the Nernst equation for the cell.

(c) Calculate E when $[Au^{3+}] = 0.250\ M$, $[H^+] = 1.25\ M$, $[H_2O_2] = 1.50\ M$.

64. Consider the voltaic cell at 25°C in which the following reaction takes place:

$$MnO_2(s) + 4H^+(aq) + 2Br^-(aq) \longrightarrow$$
$$Br_2(l) + Mn^{2+}(aq) + 2H_2O$$

(a) Calculate $E°$.

(b) Write the Nernst equation for the cell.

(c) Find E under the following conditions: $[Mn^{2+}] = 0.60\ M$, $[Br^-] = 0.83\ M$, pH = 3.17

65. Consider a voltaic cell in which the following reaction takes place.

$$2Fe^{2+}(aq) + H_2O_2(aq) + 2H^+(aq) \longrightarrow 2Fe^{3+}(aq) + 2H_2O$$

(a) Calculate $E°$.

(b) Write the Nernst equation for the cell.

(c) Calculate E at 25°C under the following conditions: $[Fe^{2+}] = 0.00813\ M$, $[H_2O_2] = 0.914\ M$, $[Fe^{3+}] = 0.199\ M$, pH = 2.88.

66. Consider a voltaic cell in which the following reaction takes place in basic medium at 25°C.

$$2NO_3^-(aq) + 3S^{2-}(aq) + 4H_2O \longrightarrow$$
$$3S(s) + 2NO(g) + 8OH^-(aq)$$

(a) Calculate $E°$.

(b) Write the Nernst equation for the cell voltage E.

(c) Calculate E under the following conditions: $P_{NO} = 0.994$ atm, pH = 13.7, $[S^{2-}] = 0.154\ M$, $[NO_3^-] = 0.472\ M$

67. Calculate the voltages of the following cells at 25°C and under the following conditions:

(a) $Cu\ |\ Cu^+(0.80\ M)\ ||\ Hg_2^{2+}\ (0.10\ M)\ |\ Hg\ |\ Pt$

(b) $Cr\ |\ Cr^{3+}(0.615\ M)\ ||\ Ni^{2+}\ (0.228\ M)\ |\ Ni$

68. Calculate voltages of the following cells at 25°C and under the following conditions.

(a) $Zn\ |\ Zn^{2+}\ (0.50\ M)\ ||\ Cd^{2+}\ (0.020\ M)\ |\ Cd$

(b) $Cu\ |\ Cu^{2+}\ (0.0010\ M)\ ||\ H^+\ (0.010\ M)\ |\ H_2\ (1.00\ atm)\ |\ Pt$

69. Consider the reaction

$$2Cu^{2+}(aq) + Sn^{2+}(aq) \longrightarrow Sn^{4+}(aq) + 2Cu^+(aq)$$

At what concentration of Cu^{2+} is the voltage zero, if all other species are at 0.200 M?

70. Consider the reaction at 25°C:

$$S(s) + 2H^+(aq) + 2Ag(s) + 2Br^-(aq) \longrightarrow 2AgBr(s) + H_2S(aq)$$

At what pH is the voltage zero if all the species are at standard conditions?

71. Complete the following cell notation.

$$Ag\ |\ Br^-\ (3.73\ M)\ |\ AgBr\ ||\ H^+\ (?)\ |\ H_2\ (1.0\ atm)\ |\ Pt$$
$$E = -0.030\ V$$

72. Complete the following cell notation.

$$Zn\ |\ Zn^{2+}(1.00\ M)\ ||\ H^+\ (?)\ |\ H_2(1.0\ atm)\ |\ Pt \qquad E = +0.40\ V$$

73. Consider the reaction below at 25°C:

$$2MnO_4^-(aq) + 16H^+(aq) + 10Br^-(aq) \longrightarrow$$
$$2Mn^{2+}(aq) + 5Br_2(l) + 8H_2O$$

Use Table 17.1 to answer the following questions. Support your answers with calculations.

(a) Is the reaction spontaneous at standard conditions?

(b) Is the reaction spontaneous at a pH of 2.00 with all other ionic species at 0.100 M?

(c) Is the reaction spontaneous at a pH of 5.00 with all other ionic species at 0.100 M?

(d) At what pH is the reaction at equilibrium with all other ionic species at 0.100 M?

74. Consider the reaction below at 25°C:

$$3SO_4^{2-}(aq) + 12H^+(aq) + 2Cr(s) \longrightarrow$$
$$3SO_2(g) + 2Cr^{3+}(aq) + 6H_2O$$

Use Table 17.1 to answer the following questions. Support your answers with calculations.

(a) Is the reaction spontaneous at standard conditions?

(b) Is the reaction spontaneous at a pH of 3.00 with all other ionic species at 0.100 M and all gases at 1.00 atm?

(c) Is the reaction spontaneous at a pH of 8.00 with all other ionic species at 0.100 M and all gases at 1.00 atm?

(d) At what pH is the reaction at equilibrium with all other ionic species at 0.100 M and all gases at 1.00 atm?

75. Consider a cell in which the reaction is

$$2Ag(s) + Cu^{2+}(aq) \longrightarrow 2Ag^+(aq) + Cu(s)$$

(a) Calculate $E°$ for this cell.

(b) Chloride ions are added to the $Ag\ |\ Ag^+$ half-cell to precipitate AgCl. The measured voltage is +0.060 V. Taking $[Cu^{2+}] = 1.0\ M$, calculate $[Ag^+]$.

(c) Taking $[Cl^-]$ in (b) to be 0.10 M, calculate K_{sp} of AgCl.

76. Consider a cell in which the reaction is

$$Pb(s) + 2H^+(aq) \longrightarrow Pb^{2+}(aq) + H_2(g)$$

(a) Calculate $E°$ for this cell.

(b) Chloride ions are added to the $Pb\ |\ Pb^{2+}$ half-cell to precipitate $PbCl_2$. The voltage is measured to be +0.210 V. Taking $[H^+] = 1.0\ M$ and $P_{H_2} = 1.0$ atm, calculate $[Pb^{2+}]$.

(c) Taking $[Cl^-]$ in (b) to be 0.10 M, calculate K_{sp} of $PbCl_2$.

17-6 Electrolytic Cells

77. An electrolytic cell produces aluminum from Al_2O_3 at the rate of ten kilograms a day. Assuming a yield of 100%,

(a) how many moles of electrons must pass through the cell in one day?

(b) how many amperes are passing through the cell?

(c) how many moles of oxygen (O_2) are being produced simultaneously?

78. The electrolysis of an aqueous solution of NaCl has the overall equation

$$2H_2O + 2Cl^- \longrightarrow H_2(g) + Cl_2(g) + 2OH^-(ag)$$

During the electrolysis, 0.228 mol of electrons pass through the cell.

(a) How many electrons does this represent?

(b) How many coulombs does this represent?

(c) Assuming 100% yield, what masses of H_2 and Cl_2 are produced?

79. A solution containing a metal ion ($M^{2+}(aq)$) is electrolyzed by a current of 5.00 A. After 15.3 minutes, 2.83 g of the metal plated out.
  (a) How many coulombs are supplied by the battery?
  (b) What is the metal? (Assume 100% efficiency.)

80. A solution containing a metal ion ($M^{2+}(aq)$) is electrolyzed by a current of 7.8 A. After 15.5 minutes, 2.39 g of the metal is plated out.
  (a) How many coulombs are supplied by the battery?
  (b) What is the metal? (Assume 100% efficiency.)

81. A baby's spoon with an area of 6.25 cm^2 is plated with silver from $AgNO_3$ using a current of 2.00 A for two hours and 25 minutes.
  (a) If the current efficiency is 82.0%, how many grams of silver are plated?
  (b) What is the thickness of the silver plate formed ($d = 10.5$ g/cm^3)?

82. A metallurgist wants to gold-plate an object with a surface area of 17.21 in^2. The gold plating must be 0.00200 in. thick (assume uniform thickness).
  (a) How many grams of gold ($d = 10.5$ g/cm^3) are required?
  (b) How many minutes will it take to plate the object from a solution of AuCN using a current of 7.00 A? Assume 100% efficiency.

83. A lead storage battery delivers a current of 6.00 A for one hour and 22 minutes at a voltage of 12.0 V.
  (a) How many grams of lead are converted to $PbSO_4$?
  (b) How much electrical energy is produced in kilowatt hours?

84. Calcium metal can be obtained by the direct electrolysis of molten $CaCl_2$, at a voltage of 3.2 V.
  (a) How many joules of electrical energy are required to obtain 12.0 lb of calcium?
  (b) What is the cost of the electrical energy obtained in (a) if electrical energy is sold at the rate of nine cents per kilowatt hour?

Unclassified

85. Given the following data:

$$PtCl_4{}^{2-}(aq) + 2e^- \longrightarrow Pt(s) + 4Cl^-(aq)$$
$$E^\circ_{red} = 0.73 \text{ V}$$

$$Pt^{2+}(aq) + 2e^- \longrightarrow Pt(s) \qquad E^\circ_{red} = 1.20 \text{ V}$$

Find K_f for $PtCl_4{}^{2-}$ at 25°C.

86. In a nickel-cadmium battery (Nicad), cadmium is oxidized to $Cd(OH)_2$ at the anode, while Ni_2O_3 is reduced to $Ni(OH)_2$ at the cathode. A portable CD player uses 0.175 amp of current. How many grams of Cd and Ni_2O_3 are consumed when the CD player is used for an hour and a half?

87. Hydrogen gas is produced when water is electrolyzed.

$$2H_2O(g) \longrightarrow 2H_2(g) + O_2(g)$$

A balloonist wants to fill a balloon with hydrogen gas. How long must a current of 12.0 A be used in the electrolysis of water to fill the balloon to a volume of 10.00 L and a pressure of 0.924 atm at 22°C?

88. Consider the electrolysis of $NiCl_2$ to $Ni(s)$ and $Cl_2(g)$.
  (a) What is the minimum voltage required to carry out this reaction at standard conditions?
  (b) If a voltage of 3.0 V is actually used, and 10.00 kJ of energy are consumed, how many grams of $Ni(s)$ are obtained? (1 J = 1 volt-coulomb)

89. An electrolysis experiment is performed to determine the value of the Faraday constant (number of coulombs per mole of electrons). In this experiment, 28.8 g of gold is plated out from a AuCN solution by running an electrolytic cell for two hours with a current of 2.00 A. What is the experimental value obtained for the Faraday constant?

90. An electrolytic cell consists of a 100.0-g strip of copper in 0.200 M $Cu(NO_3)_2$ and a 100.0-g strip of Cr in 0.200 M $Cr(NO_3)_3$. The overall reaction is:

$$3Cu(s) + 2Cr^{3+}(aq) \longrightarrow 3Cu^{2+}(aq) + 2Cr(s) \quad E^\circ = -1.083 \text{ V}$$

An external battery provides 3 amperes for 70 minutes and 20 seconds with 100% efficiency. What is the mass of the copper strip after the battery has been disconnected?

91. After use, a nickel-cadmium battery (Nicad) has 0.133 g of $Cd(OH)_2$ deposited on the anode. The battery is inserted into a recharger to convert all the $Cd(OH)_2$ back to Cd. It takes the recharger 18.3 minutes to complete the process. How many amperes did the recharger supply? Assume 100% efficiency. (See Problem 86 for a description of a Nicad battery.)

92. Consider the following reaction carried out at 1000°C.

$$CO(g) + \tfrac{1}{2}O_2(g) \longrightarrow CO_2(g)$$

Assuming that all gases are at 1.00 atm, calculate the voltage produced at the given conditions. (Use Appendix 1 and assume that ΔH° and ΔS° do not change with an increase in temperature.)

93. Atomic masses can be determined by electrolysis. In one hour, a current of 0.600 A deposits 2.42 g of a certain metal, M, which is present in solution as M^+ ions. What is the atomic mass of the metal?

94. Consider the following reaction at 25°C:

$$O_2(g) + 4H^+(aq) + 4Cl^-(aq) \longrightarrow 2H_2O + 2Cl_2(g)$$

The [H^+] is adjusted by adding a buffer that is 0.125 M in lactic acid (HLac) and 0.125 M in sodium lactate (NaLac). The pressure of both gases is 1.00 atm and [Cl^-] is 0.200 M. (K_a for HLac is 1.4×10^{-4}.) Will the cell function as a voltaic cell?

95. Given the standard reduction potential for $Zn(OH)_4{}^{2-}$:

$$Zn(OH)_4{}^{2-}(aq) + 2e^- \longrightarrow Zn(s) + 4OH^-(aq)$$
$$E^\circ_{red} = -1.19 \text{ V}$$

Calculate the formation constant (K_f) for the reaction

$$Zn^{2+}(aq) + 4OH^-(aq) \rightleftharpoons Zn(OH)_4{}^{2-}(aq)$$

Conceptual Problems

96. Choose the figure that best represents the results after the electrolysis of water. (Circles represent hydrogen atoms and squares represent oxygen atoms.)

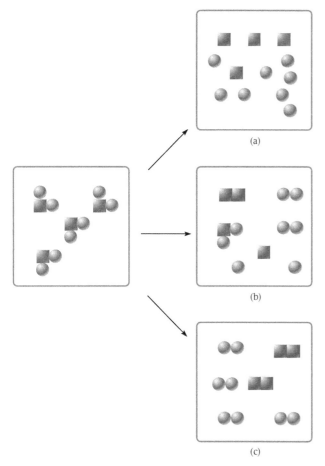

(a)

(b)

(c)

97. For the cell:

$$Cr \mid Cr^{3+} \parallel Co^{2+} \mid Co$$

$E°$ is 0.46 V. The same cell was prepared in the laboratory at standard conditions. The voltage obtained was 0.40 V. A possible explanation for the difference is

(a) the surface area of the chromium electrode was smaller than the cobalt electrode.
(b) the mass of the chromium electrode was larger than the mass of the cobalt electrode.
(c) the concentration of $Co(NO_3)_2$ solution used was less than 1.0 M.
(d) the concentration of $Cr(NO_3)_2$ solution used was less than 1.0 M.
(e) the volume of $Co(NO_3)_2$ solution used was larger than the volume of $Cr(NO_3)_2$ solution used.

98. Which of the changes below will increase the voltage of the following cell?

$$Co \mid Co^{2+} \ (0.010 \ M) \parallel H^+ \ (0.010 \ M) \mid H_2 \ (0.500 \ atm) \mid Pt$$

(a) Increase the volume of $CoCl_2$ solution from 100 mL to 300 mL.
(b) Increase $[H^+]$ from 0.010 M to 0.500 M.
(c) Increase the pressure of H_2 from 0.500 atm to 1 atm.
(d) Increase the mass of the Co electrode from 15 g to 25 g.
(e) Increase $[Co^{2+}]$ from 0.010 M to 0.500 M.

99. The standard potential for the reduction of AgSCN is 0.0895 V.

$$AgSCN(s) + e^- \longrightarrow Ag(s) + SCN^-(aq)$$

Find another electrode potential to use together with the above value and calculate K_{sp} for AgSCN.

100. Consider the following standard reduction potentials:

$$Tl^+(aq) + e^- \longrightarrow Tl(s) \qquad E°_{red} = -0.34 \ V$$
$$Tl^{3+}(aq) + 3e^- \longrightarrow Tl(s) \qquad E°_{red} = 0.74 \ V$$
$$Tl^{3+}(aq) + 2e^- \longrightarrow Tl^+(aq) \qquad E°_{red} = 1.28 \ V$$

and the following abbreviated cell notations:

(1) $Tl \mid Tl^+ \parallel Tl^{3+}, Tl^+ \mid Pt$

(2) $Tl \mid Tl^{3+} \parallel Tl^{3+}, Tl^+ \mid Pt$

(3) $Tl \mid Tl^+ \parallel Tl^{3+} \mid Tl$

(a) Write the overall equation for each cell.
(b) Calculate $E°$ for each cell.
(c) Calculate $\Delta G°$ for each overall equation.
(d) Comment on whether $\Delta G°$ and/or $E°$ are state properties. (*Hint:* A state property is path-independent.)

101. Use Table 17.1 to answer the following questions. Use **LT** (for *is less than*), **GT** (for *is greater than*), **EQ** (for *is equal to*), or **MI** (for *more information required*).

(a) For the half reaction

$$\tfrac{1}{2}Br_2(l) + e^- \longrightarrow Br^-(aq)$$

$E°_{red}$ _____ 1.077 V.

(b) For the reaction

$$2Br^-(aq) + Co^{2+}(aq) \longrightarrow Br_2(l) + Co(s)$$

$E°$ _____ 0.

(c) If the half reaction

$$Co^{2+}(aq) + 2e^- \longrightarrow Co(s)$$

is designated as the new standard where $E°_{red}$ is 0.00, then $E°_{red}$ for

$$2H^+(aq) + 2e^- \longrightarrow H_2(g)$$

is _____ 0.

(d) For the reaction

$$2Cr^{3+}(aq) + 3Co(s) \longrightarrow 2Cr(s) + 3Co^{2+}(aq)$$

the number of electrons exchanged is _____ 6.

(e) For the reaction described in (d), the number of coulombs that passes through the cell is _____ 9.648×10^4.

102. Consider three metals, X, Y, and Z, and their salts, XA, YA, and ZA. Three experiments take place with the following results:

- $X + hot \ H_2O \longrightarrow H_2$ bubbles
- $X + YA \longrightarrow$ no reaction
- $X + ZA \longrightarrow X$ discolored $+ Z$

Rank metals X, Y, and Z, in order of decreasing strength as reducing agents.

Challenge Problems

103. An alloy made up of tin and copper is prepared by simultaneously electroplating the two metals from a solution containing $Sn(NO_3)_2$ and $Cu(NO_3)_2$. If 20.0% of the total current is used to plate tin, while 80.0% is used to plate copper, what is the percent composition of the alloy?

104. In a fully charged lead storage battery, the electrolyte consists of 38% sulfuric acid by mass. The solution has a density of 1.286 g/cm³. Calculate E for the cell. Assume all the H^+ ions come from the first dissociation of H_2SO_4, which is complete.

105. Consider a voltaic cell in which the following reaction occurs.

$$Zn(s) + Sn^{2+}(aq) \longrightarrow Zn^{2+}(aq) + Sn(s)$$

(a) Calculate $E°$ for the cell.

(b) When the cell operates, what happens to the concentration of Zn^{2+}? The concentration of Sn^{2+}?

(c) When the cell voltage drops to zero, what is the ratio of the concentration of Zn^{2+} to that of Sn^{2+}?

(d) If the concentration of both cations is 1.0 M originally, what are the concentrations when the voltage drops to zero?

106. In biological systems, acetate ion is converted to ethyl alcohol in a two-step process:

$$CH_3COO^-(aq) + 3H^+(aq) + 2e^- \longrightarrow CH_3CHO(aq) + H_2O$$

$$E°' = -0.581 \text{ V}$$

$$CH_3CHO(aq) + 2H^+(aq) + 2e^- \longrightarrow C_2H_5OH(aq)$$

$$E°' = -0.197 \text{ V}$$

($E°'$ is the standard reduction voltage at 25°C and pH of 7.00.)

(a) Calculate $\Delta G°'$ for each step and for the overall conversion.

(b) Calculate $E°'$ for the overall conversion.

107. Consider the cell

$$Pt \mid H_2 \mid H^+ \parallel H^+ \mid H_2 \mid Pt$$

In the anode half-cell, hydrogen gas at 1.0 atm is bubbled over a platinum electrode dipping into a solution that has a pH of 7.0. The other half-cell is identical to the first except that the solution around the platinum electrode has a pH of 0.0. What is the cell voltage?

108. A hydrogen-oxygen fuel cell operates on the reaction:

$$H_2(g) + \tfrac{1}{2}O_2(g) \longrightarrow H_2O(l)$$

If the cell is designed to produce 1.5 amp of current and if the hydrogen is contained in a 1.0-L tank at 200 atm pressure and 25°C, how long can the fuel cell operate before the hydrogen runs out? Assume that oxygen gas is in excess.

109. An alloy is made up of 68% zinc and 32% tin. It is prepared by simultaneously electroplating the two metals from a solution containing both $Zn(NO_3)_2$ and $Sn(NO_3)_2$. What percent of the total current is used to plate each metal?

Nuclear Reactions

Pedro Saura via Science/AAAS

Hand stencils on the walls of the El Castillo caves are estimated to be done by Neanderthals at least 40,800 years ago. They are now the oldest known cave art in the world, 5,000 to 10,000 years older than those in the Chauvet Caves in France. These stencils are dated by measuring the uranium decay, since there is no organic matter in them.

The changing of Bodies into Light, and Light into Bodies, is very conformable to the Course of Nature, which seems delighted with Transmutations.

—SIR ISAAC NEWTON
Opticks, 2nd edition (1718)
Book 3, Query 30, 349

The "ordinary chemical reactions" discussed to this point involve changes in the outer electronic structures of atoms or molecules. In contrast, nuclear reactions result from changes taking place within atomic nuclei. You will recall (Chapter 2) that atomic nuclei are represented by symbols such as

$$^{12}_{6}C \qquad ^{14}_{6}C$$

The atomic number Z (number of protons in the nucleus) is shown as a left subscript. The mass number A (number of protons + number of neutrons in the nucleus) appears as a left superscript.

The reactions that we discuss in this chapter will be represented by **nuclear equations**. An equation of this type uses nuclear symbols such as those above; in other respects it resembles an ordinary chemical equation. A nuclear equation must be balanced with respect to nuclear charge (atomic number) and nuclear mass (mass number). To see what that means, consider an equation that we will have a lot more to say about later in this chapter:

$$^{14}_{7}N + ^{1}_{0}n \longrightarrow ^{14}_{6}C + ^{1}_{1}H$$

The reactants are an N-14 nucleus and a neutron; the products are a C-14 nucleus and an H-1 nucleus. ◄ⓘ The atomic numbers add to 7 on both sides:

$$\text{reactants: } 7 + 0 = 7 \qquad \text{products: } 6 + 1 = 7$$

whereas the mass numbers add to 15:

$$\text{reactants: } 14 + 1 = 15 \qquad \text{products: } 14 + 1 = 15$$

Chapter Outline

18-1 Nuclear Stability

18-2 Radioactivity

18-3 Rate of Radioactive Decay

18-4 Mass-Energy Relations

18-5 Nuclear Fission

18-6 Nuclear Fusion

ⓘ

Nuclear reactions usually involve the transmutation of elements.

18-1 Nuclear Stability

As we have learned from Chapter 2, atomic nuclei consist of protons, which are positively charged, and neutrons, which have zero charge. According to classical electrostatics, we should expect the protons to repel one another and the nucleus to fly apart. It turns out, however, that at the very short distances of separation characteristic of atomic nuclei (about 10^{-15} m), there are strong attractive forces between nuclear particles. The stability of a nucleus depends upon the balance between these forces and electrostatic repulsion.

Despite extensive research, we still do not have a clear understanding of the way in which those forces lead to nuclear stability. All we have is a series of empirical rules for predicting which nuclei will be most stable. The most important of these rules are listed below.

- The neutron-to-proton ratio (n/p^+) required for stability varies with atomic number. The 264 known stable nuclei, shown as red dots in Figure 18.1, fall within a relatively narrow *belt of stability*. For light elements ($Z < 20$), this ratio is close to one. For example, the isotopes $^{12}_{6}C$, $^{14}_{7}N$, and $^{16}_{8}O$ are stable. As atomic number increases, the ratio increases; the belt of stability shifts to higher numbers of neutrons. With very heavy isotopes such as $^{206}_{82}Pb$, the stable neutron-to-proton ratio is about 1.5.

$$(206 - 82)/82 = 1.51$$

- Nuclei containing more than 83 protons are unstable. Putting it another way, no element beyond bismuth ($Z = 83$) has a stable isotope. All the isotopes of such elements are radioactive.
- Nuclei with an even number of **nucleons**, either protons or neutrons, tend to be more stable than those with an odd number of nuclear particles.
- Certain numbers of protons and neutrons appear to be particularly stable: 2, 8, 20, 28, 50, 82, and 126. These *magic numbers* were first suggested by Maria Mayer (1906–1972), who developed a nuclear shell model similar to, but much less regular than, the electron shell model discussed in Chapter 6. For her work,

Figure 18.1 Neutron-to-proton ratios of stable isotopes. The ratios of stable isotopes (red dots) fall within a narrow range, referred to as the *belt of stability*. For light isotopes of small atomic number the stable ratio is 1:1. For heavier isotopes the ratio gradually increases to about 1.5:1. Isotopes outside the belt of stability are unstable and radioactive. There are no stable isotopes for elements of atomic number greater than 83 (Bi).

Maria Mayer won the Nobel Prize in Physics in 1963. The Nobel Prize in Physics has been awarded yearly (except between 1940 and 1942) for 106 years, and the only women to have won this prize so far are Maria Mayer and Marie Curie.

EXAMPLE 18.1

For each pair of nuclei, list which one is more stable.

a $^{6}_{3}Li$ or $^{9}_{3}Li$

STRATEGY

1. Determine the neutron-to-proton ratio.
2. Check the belt of stability.

SOLUTION

| 1. n/p^+ | $^{6}_{3}Li$: $n/p^+ = 3/3 = 1$; $^{9}_{3}Li$: $n/p^+ = 6/3 = 2$ |
| 2. Stability | $^{9}_{3}Li$: is unstable. A $n/p^+ = 2$ lies well below the belt of stability. |

b $^{204}_{82}Pb$ or $^{209}_{85}At$

STRATEGY AND SOLUTION

For Pb, $Z = 82$, whereas Z for At is 85. Pb is the more stable element.

18-2   Radioactivity

Unstable isotopes decompose (*decay*) by a process referred to as **radioactivity**. A few such nuclei occur in nature, accounting for *natural radioactivity*. Many more can be made ("induced") artificially by bombarding stable nuclei with high-energy particles. The radiation given off (Figure 18.2) may be in the form of

beta (β) particles identical in their properties to electrons

alpha (α) particles, which are $^{4}_{2}He$ nuclei, carrying a +2 charge because the two extranuclear electrons of the helium atom are missing

gamma (γ) rays, which consist of high-energy radiation

Mode of Decay

Naturally occurring radioactive nuclei commonly decompose by

- *alpha particle emission,* in which an ordinary helium nucleus, $^{4}_{2}He$, is given off. Uranium-238 behaves this way:

$$^{238}_{92}U \longrightarrow {}^{4}_{2}He + {}^{234}_{90}Th$$

  Notice that when alpha emission occurs, the atomic number decreases by two units; the mass number decreases by four units. Here, as in all nuclear equations, there is a balance of both atomic number ($90 + 2 = 92$) and mass number ($234 + 4 = 238$) on both sides of the arrow.

  The **alpha particle** is by far the most massive particle emitted by radioactive nuclei. It has the highest *ionizing power*, i.e., the ability to ionize other molecules or ions. Because of their size, alpha particles have difficulty penetrating cells. Thus they can be stopped by clothing, skin, or even a sheet of paper. Alpha-emitting nuclei kept outside the body are relatively safe. Inhaling or ingesting these particles is much more dangerous, because cell membranes cannot stop them.

- *beta particle emission,* which produces an electron, given the symbol $^{0}_{-1}e$. An example of beta emission is

$$^{234}_{90}Th \longrightarrow {}^{0}_{-1}e + {}^{234}_{91}Pa$$

Figure 18.2 Nuclear radiation. In the presence of an electric field, alpha particles are deflected toward the negative pole, showing that they are positively charged, and beta particles are deflected toward the positive pole, showing that they are negatively charged. Because they are lighter, beta particles are deflected more than alpha particles. Gamma rays pass straight through the field and so must have no charge.

Notice that the product nucleus, $^{234}_{91}Pa$, has the same mass number as the reactant, $^{234}_{90}Th$, but its atomic number is one unit larger. Putting it another way, beta emission converts a neutron to a proton. It occurs with nuclei that contain "too many neutrons" for stability.

Beta particles are smaller and lighter than alpha particles. Thus although they can penetrate the cell more easily, they have a lower ionizing power. A sheet of metal can stop their entry into cells, and even when ingested or inhaled, they are less destructive than alpha particles.

■ **gamma radiation,** which consists of high-energy photons. Because gamma emission changes neither the mass number nor the atomic number, it is ordinarily omitted in writing nuclear equations.

Gamma rays, because of their size, have the lowest ionizing power. However, several inches of lead or a thick cement block is necessary to stop gamma ray penetration.

Radioactive nuclei produced "artificially" in the laboratory can show α-, β-, or γ-emission. They can also decompose by

■ *positron emission.* A **positron** is identical to an electron except that it has a charge of $+1$ rather than -1. A positron has the symbol $^{0}_{1}e$.

$$^{40}_{19}K \longrightarrow {}^{0}_{1}e + {}^{40}_{18}Ar$$

Here, a proton in K-40 (19 p^+, 21 n) is converted to a neutron in Ar-40 (18 p^+, 22 n). Positron emission is characteristic of nuclei that have too many protons for stability.

■ **K-electron capture,** in which an electron in the innermost energy level ($n = 1$) "falls" into the nucleus.

$$^{82}_{37}Rb + {}^{0}_{-1}e \longrightarrow {}^{82}_{36}Kr$$

Notice that the result of K-electron capture is the same as positron emission; mass number remains unchanged, whereas atomic number decreases by one unit. Electron capture is more common with heavy nuclei, presumably because the $n = 1$ level is closer to the nucleus.

Table 18.1 is a summary of the different modes of radioactive decay described here.

Table 18.1 Modes of Radioactive Decay

Mode	Emission	Example	Change in A	Change in Z	Penetrating Power
α-decay	^4_2He	$^{238}_{92}U \longrightarrow\ ^4_2He +\ ^{234}_{90}Th$	-4	-2	≈ 0.03 mm
β-decay	$^0_{-1}e^-$	$^{234}_{90}Th \longrightarrow e +\ ^{234}_{91}Pa$	0	$+1$	≈ 2 mm
γ-decay	photons		0	0	≈ 10 cm
positron emission	^0_1e	$^{40}_{19}K \longrightarrow\ ^0_1e +\ ^{40}_{18}Ar$	0	-1	
K-electron capture	none	$^{82}_{37}Rb +\ ^0_{-1}e^- \longrightarrow\ ^{82}_{36}Kr$	0	-1	

EXAMPLE 18.2

Promethium ($Z = 61$) is essentially nonexistent in nature; *i* all of its isotopes are radioactive. Write balanced nuclear equations for the decomposition of various isotopes of promethium:

Pm-142 by positron emission, Pm-147 by beta emission, and Pm-150 by alpha emission.

STRATEGY

1. Recall the symbol of the particle emitted for the specified decay mode.
2. Balance mass number and atomic number.
3. Find the symbol of the product isotope in the periodic table by using its atomic number.

SOLUTION

Pm-142 decomposition by positron emission.

1. particle emitted — positron: ^0_1e
2. mass and atomic number balance — $^{142}_{61}Pm \longrightarrow\ ^0_1e +\ ^{142}_{60}$
3. reaction — $^{142}_{61}Pm \longrightarrow\ ^0_1e +\ ^{142}_{60}Nd$

Pm-147 decomposition by beta emission.

1. particle emitted — β-particle: $^0_{-1}e$
2. mass and atomic number balance — $^{147}_{61}Pm \longrightarrow\ ^0_{-1}e +\ ^{147}_{62}$
3. reaction — $^{147}_{61}Pm \longrightarrow\ ^0_{-1}e +\ ^{147}_{62}Sm$

Pm-150 decomposition by alpha emission.

1. particle emitted — α-particle: ^4_2He
2. mass and atomic number balance — $^{150}_{61}Pm \longrightarrow\ ^4_2He +\ ^{146}_{59}$
3. reaction — $^{150}_{61}Pm \longrightarrow\ ^4_2He +\ ^{146}_{59}Pr$

Bombardment Reactions

More than 1500 radioactive isotopes have been prepared in the laboratory. The number of such isotopes per element ranges from 1 (hydrogen and boron) to 34 (indium). They are all prepared by bombardment reactions in which a stable nucleus is converted to one that is radioactive, which in turn decays to stable products. The bombarding particle may be

- a *neutron,* usually of rather low energy, produced in a fission reactor (Section 18-5). A typical reaction of this type is

$$^{27}_{13}Al +\ ^1_0n \longrightarrow\ ^{28}_{13}Al$$

The product nucleus, Al-28, is radioactive, decaying by beta emission:

$$^{28}_{13}Al \longrightarrow\ ^{28}_{14}Si +\ ^0_{-1}e$$

i Promethium is by far the rarest of the "rare earths" (Example 18.2).

■ a *charged particle* (electron, positron, α-particle, . . .), which can be accelerated to very high velocities in electric and/or magnetic fields. In this way, the charged particle acquires enough energy to bring about a nuclear reaction, despite electrostatic repulsion with the components of the atom.

The first radioactive isotopes to be made in the laboratory were prepared in 1934 by Irene Curie and her husband, Frederic Joliot. They achieved this by bombarding certain stable isotopes with high-energy alpha particles. One reaction was

$$^{27}_{13}\text{Al} + ^{4}_{2}\text{He} \longrightarrow ^{30}_{15}\text{P} + ^{1}_{0}n$$

The product, phosphorus-30, is radioactive, decaying by positron emission:

$$^{30}_{15}\text{P} \longrightarrow ^{30}_{14}\text{Si} + ^{0}_{1}e$$

An interesting application of bombardment reactions is the preparation of the so-called transuranium elements. Beginning with neptunium (Np, Z = 93) in 1940, a series of elements with atomic numbers greater than that of uranium have been synthesized. Much of this work was done by a group at the University of California, Berkeley, under the direction of Glenn Seaborg (1912–1999) and later Albert Ghiorso (1915–2010). A Russian group at Dubna near Moscow, led by G. N. Flerov (1913–1999), made significant contributions. The heavy elements (197 through 212) were prepared in the late twentieth century by a German group at Darmstadt under the direction of Peter Armbruster (1931–). Most of the nuclei produced have half-lives of a few microseconds. Experiments reporting the formation of elements 113, 115, and 117 have been reported in peer-reviewed journals, but their identification is yet to be confirmed by IUPAC (International Union of Pure and Applied Chemists).

In October 2006, a research team of scientists from the Lawrence Livermore National Laboratory in California, USA, and the Joint Institute of Nuclear Research in Dubna, Russia, reported the indirect detection of Uuo-294 (Element 118). It is reported to be produced by the following collisions.

$$^{249}_{98}\text{Cf} + ^{48}_{20}\text{Ca} \longrightarrow ^{294}_{118}\text{Uuo} + 3^{1}_{0}n$$

The decay products of Uuo-294, not the atoms themselves, were observed.

Some of the reactions used to prepare transuranium elements are listed in Table 18.2. Neutron bombardment is effective for the lower members of the series (elements 93 through 95), but the yield of product decreases sharply with increasing atomic number. To form the heavier transuranium elements, it is necessary to bombard appropriate targets with high-energy positive ions. By using relatively

ⓘ
The names of these elements reflect the names of their discoverers and other nuclear scientists.

Table 18.2 Synthesis of Transuranium Elements ◀ⓘ

	Neutron Bombardment
Neptunium, plutonium	$^{238}_{92}\text{U} + ^{1}_{0}n \longrightarrow ^{239}_{93}\text{Np} + ^{0}_{-1}e$
	$^{239}_{93}\text{Np} \longrightarrow ^{239}_{94}\text{Pu} + ^{0}_{-1}e$
Americium	$^{239}_{94}\text{Pu} + 2^{1}_{0}n \longrightarrow ^{241}_{95}\text{Am} + ^{0}_{-1}e$
	Positive Ion Bombardment
Curium	$^{239}_{94}\text{Pu} + ^{4}_{2}\text{He} \longrightarrow ^{242}_{96}\text{Cm} + ^{1}_{0}n$
Californium	$^{242}_{96}\text{Cm} + ^{4}_{2}\text{He} \longrightarrow ^{245}_{98}\text{Cf} + ^{1}_{0}n$
Rutherfordium	$^{249}_{98}\text{Cf} + ^{12}_{6}\text{C} \longrightarrow ^{257}_{104}\text{Rf} + 4^{1}_{0}n$
Dubnium	$^{249}_{98}\text{Cf} + ^{15}_{7}\text{N} \longrightarrow ^{260}_{105}\text{Db} + 4^{1}_{0}n$
Seaborgium	$^{249}_{98}\text{Cf} + ^{18}_{8}\text{O} \longrightarrow ^{263}_{106}\text{Sg} + 4^{1}_{0}n$

Figure 18.3 Cobalt-60 cancer therapy. Gamma rays from the rotating radiation source are concentrated at the location of the diseased tissue.

heavy bombarding particles such as carbon-12, one can achieve a considerable increase in atomic number.

Applications

A large number of radioactive nuclei have been used both in industry and in many areas of basic and applied research. A few of these are discussed below.

Medicine

Radioactive isotopes are commonly used in cancer therapy, usually to eliminate any malignant cells left after surgery. Cobalt-60 is most often used; γ-rays from this source are focused at small areas where cancer is suspected (Figure 18.3). Certain types of cancer can be treated internally with radioactive isotopes. If a patient suffering from cancer of the thyroid drinks a solution of NaI containing radioactive iodide ions (^{131}I or ^{123}I), the iodine moves preferentially to the thyroid gland. There the radiation destroys malignant cells without affecting the rest of the body.

More commonly, radioactive nuclei are used for diagnosis (Table 18.3). Positron emission tomography (PET) is a technique used to study brain disorders. The patient is given a dose of glucose ($C_6H_{12}O_6$) which has a radioactive isotope of fluorine attached to it. The fluorine isotope is a positron emitter. The brain is then scanned to detect positron emission from the radioactive "labeled" glucose. In this way, differences in glucose uptake and metabolism in patients with normal and abnormal brains are established. For example, PET scans have determined that the brain of a schizophrenic metabolizes only about 20% as much glucose as that of most people.

SPECT scans (single proton emission computer tomography) also use radioactive tracers and a scanner to give two- or three-dimensional images. The scans give information about blood flow to tissues and chemical reactions (metabolism) in the body.

Chemistry

Radioactive nuclei are used extensively in chemical analysis. One technique of particular importance is *neutron activation analysis*. This procedure depends on the phenomenon of induced radioactivity. A sample is bombarded by neutrons, bringing about such reactions as

$$^{84}_{38}\text{Sr} + ^{1}_{0}n \longrightarrow ^{85}_{38}\text{Sr}$$

Table 18.3 Diagnostic Uses of Radioactive Isotopes

Isotope	Use
$^{18}_{9}$F	Brain scan (PET); see text
$^{24}_{11}$Na	Circulatory disorders
$^{32}_{15}$P	Detection of eye tumors
$^{59}_{26}$Fe	Anemia
$^{67}_{31}$Ga	Scan for lung tumors, abscesses
$^{75}_{34}$Se	Pancreas scan
$^{99}_{43}$Tc	Imaging of brain, liver, kidneys, bone marrow
$^{133}_{54}$Xe	Lung imaging
$^{201}_{81}$Tl	Heart disorders

Ionization chamber

Alarm triggering circuit

Alarm

Radioactive source

Ions

© magicoven/Shutterstock.com

Chris Priest/Science Source

(a)

(b)

(c)

Figure 18.4 Smoke detector. Most smoke detectors use a tiny amount of a radioactive isotope to produce a current flow that drops off sharply in the presence of smoke particles, emitting an alarm in the process.

Ordinarily the element retains its chemical identity, but the isotope formed is radioactive, decaying by gamma emission. The magnitude of the energy change and hence the wavelength of the gamma ray vary from one element to another and so can serve for the qualitative analysis of the sample. The intensity of the radiation depends on the amount of the element present in the sample; this permits quantitative analysis of the sample. Neutron activation analysis can be used to analyze for 50 different elements in amounts as small as one picogram (10^{-12} g).

One application of neutron activation analysis is in the field of archaeology. By measuring the amount of strontium in the bones of prehistoric humans, it is possible to get some idea of their diet. Plants contain considerably more strontium than animals do, so a high strontium content suggests a largely vegetarian diet. Strontium analyses of bones taken from ancient farming communities consistently show a difference by sex; women have higher strontium levels than men. Apparently, in those days, women did most of the farming; men spent a lot of time away from home hunting and eating their kill.

Commercial Applications

Most smoke alarms (Figure 18.4) use a radioactive species, typically americium-241. A tiny amount of this isotope is placed in a small ionization chamber; decay of Am-241 ionizes air molecules within the chamber. Under the influence of a potential applied by a battery, these ions move across the chamber, producing an electric current. If smoke particles get into the chamber, the flow of ions is impeded and the current drops. This is detected by electronic circuitry, and an alarm sounds. The alarm also goes off if the battery voltage drops, indicating that it needs to be replaced.

Another potential application of radioactive species is in food preservation (Figure 18.5). It is well known that gamma rays can kill insects, larvae, and parasites such as trichina that cause trichinosis in pork. Radiation can also inhibit the sprouting of onions and potatoes. Perhaps most important from a commercial standpoint, it can extend the shelf lives of many foods for weeks or even months. Many chemicals used to preserve foods have later been shown to have adverse health effects, so irradiation is an attractive alternative. Finally, irradiation can destroy microorganisms such as *E. coli* (which explains its use in treating beef) and anthrax (which explains its use in "sterilizing" suspected mail).

Nordion International

Figure 18.5 Food preservation. Strawberries irradiated with gamma rays from radioactive isotopes to keep them fresh.

CHEMISTRY THE **HUMAN** SIDE

The history of radiochemistry is in no small measure the story of two remarkable women, Marie and Irene Curie, and their husbands, Pierre Curie and Frederic Joliot. Marie Curie (1867–1934) was born Maria Sklodowska in Warsaw, Poland, then a part of the Russian empire. In 1891, she emigrated to Paris to study at the Sorbonne, where she met and married a French physicist, Pierre Curie (1859–1906). The Curies were associates of Henri Becquerel (1852–1928), the man who discovered that uranium salts are radioactive. They showed that thorium, like uranium, is radioactive and that the amount of radiation emitted is directly proportional to the amount of uranium or thorium in the sample.

In 1898, Marie and Pierre Curie isolated two new radioactive elements, which they named radium and polonium. To obtain a few milligrams of these elements, they started with several tons of pitchblende ore and carried out a long series of tedious separations. Their work was done in a poorly equipped, unheated shed where the temperature reached 6°C (43°F) in winter. Four years later, in 1902, Marie determined the atomic mass of radium to within 0.5%, working with a tiny sample.

In 1903, the Curies received the Nobel Prize in Physics (with Becquerel) for the discovery of radioactivity. Three years later, Pierre Curie died at the age of 46, the victim of a tragic accident. He stepped from behind a carriage in a busy Paris street and was run down by a horse-driven truck. That same year, Marie became the first woman instructor at the Sorbonne. In 1911, she won the Nobel Prize in Chemistry for the discovery of radium and polonium, thereby becoming the first person to win two Nobel Prizes.

When Europe exploded into war in 1914, scientists largely abandoned their studies to go to the front. Marie Curie, with her daughter Irene, then 17 years old, organized

medical units equipped with x-ray machinery. These were used to locate foreign metallic objects in wounded soldiers. Many of the wounds were to the head; French soldiers came out of the trenches without head protection because their government had decided that helmets looked too German. In November of 1918, the

Marie and Pierre Curie with daughter Irene, at their home near Paris

Curies celebrated the end of World War I; France was victorious, and Marie's beloved Poland was free again.

In 1921, Irene Curie (1897–1956) began research at the Radium Institute. Five years later she married Frederic Joliot (1900–1958), a brilliant young physicist who was also an assistant at the Institute. In 1931, they began a research program in nuclear chemistry that led to several important discoveries and at least one near miss. The Joliot-Curies were the first to demonstrate induced radioactivity. They also discovered the positron, a particle that scientists had been seeking for many years. They narrowly missed finding another, more fundamental particle, the neutron. That honor went to James Chadwick in England (Chadwick, a student of Rutherford, discovered the neutron in 1932). In 1935, Irene Curie and Frederic Joliot received the Nobel Prize in Physics. The award came too late for Irene's mother, who had died of leukemia in 1934. Twenty-two years later, Irene Curie-Joliot died of the same disease. Both women acquired leukemia through prolonged exposure to radiation.

18-3 Rate of Radioactive Decay

As pointed out in Chapter 11, radioactive decay is a first-order process. This means that the following equations, discussed on pages 285–287, apply:

$$\text{rate} = kX$$

$$\ln \frac{X_0}{X} = kt$$

$$k = \frac{0.693}{t_{1/2}}$$

where k ◀ 🛈 is the first-order rate constant, $t_{1/2}$ is the half-life, X is the amount of radioactive species present at time t, and X_0 is the amount at $t = 0$.

Because of the way in which rate of decay is measured (Figure 18.6), it is often described by the **activity** (A) of the sample, which expresses the number of atoms decaying in unit time. The first equation above can be written

$$A = kN \qquad (18.1)$$

where A is the activity, k the first-order rate constant, and N the number of radioactive nuclei present.

🛈 k = rate constant = fraction of atoms decaying in unit time.

Figure 18.6 A liquid scintillation counter. This instrument is used to detect radiation and measure disintegrations per minute quickly and accurately.

Activity can be expressed in terms of the number of atoms decaying per second, or becquerels (Bq).

$$1 \text{ Bq} = 1 \text{ atoms/s}$$

Alternatively, activity may be cited in disintegrations per minute or, perhaps most commonly, in **curies** (Ci).

$$1 \text{ Ci} = 3.700 \times 10^{10} \text{ atoms/s}$$

EXAMPLE 18.3 GRADED

The half-life of radium-226 is 1.60×10^3 y $= 5.05 \times 10^{10}$ s.

a   Calculate k in s^{-1}.

ANALYSIS	
Information given:	$t_{1/2}$ for Ra-226 (5.05×10^{10} s)
Asked for:	k in s^{-1}

STRATEGY

Substitute into the formula relating half-life and rate constant in a first-order reaction.

$$k = \frac{0.693}{t_{1/2}}$$

SOLUTION	
k	$k = \dfrac{0.693}{5.05 \times 10^{10} \text{ s}} = 1.37 \times 10^{-11} \text{ s}^{-1}$

b   What is the activity in curies of a 1.00-g sample of Ra-226?

ANALYSIS	
Information given:	mass of sample (1.00 g) from part (a): k(1.37×10^{-11} s^{-1})
Information implied:	atoms/s to Ci conversion factor Avogadro's number
Asked for:	activity (A) in Ci

STRATEGY

1. Find the number of nuclei N in 1.00 g of Ra-226 using Avogadro's number and 226 g/mol as the molar mass of Ra-226.

2. Substitute into Equation 18.1 to find activity in atoms/s.

   $A = kN$

3. Use the conversion factor

   $1 \text{ Ci} = 3.700 \times 10^{10}$ atoms/s

to find the activity in Ci.

SOLUTION	
1. N	$1.00 \text{ g} \times \dfrac{6.022 \times 10^{23} \text{ atoms}}{226 \text{ g}} = 2.66 \times 10^{21} \text{ atoms}$
2. A (atoms/s)	$A = (1.37 \times 10^{-11} \text{ s}^{-1})(2.66 \times 10^{21} \text{ atom}) = 3.64 \times 10^{10} \text{ atoms/s}$
3. A (Ci)	$3.64 \times 10^{10} \text{ atoms/s} \times \dfrac{1 \text{ Ci}}{3.700 \times 10^{10} \text{ atoms/s}} = 0.985 \text{ Ci}$

continued

| **C** | What is the mass in grams of a sample of Ra-226 that has an activity of 1.00×10^9 atoms/min? |

ANALYSIS

Information given:	activity ($A = 1.00 \times 10^9$ atoms/min) from part (a): $k(1.37 \times 10^{-11} \text{ s}^{-1})$
Information implied:	Avogadro's number
Asked for:	mass of Ra-226 with the given activity

STRATEGY

1. Since k is in s^{-1}, convert the given activity in atoms/min to atoms/s.
2. Find N by substituting into Equation 18.1.
3. Determine the mass of Ra-226 by using Avogadro's number as the conversion factor.

SOLUTION

1. A in atoms/s	$\left(1.00 \times 10^9 \dfrac{\text{atoms}}{\text{min}}\right)\left(\dfrac{1 \text{ min}}{60 \text{ s}}\right) = 1.67 \times 10^7$ atoms/s
2. N	$N = \dfrac{1.67 \times 10^7 \text{ atoms/s}}{1.37 \times 10^{-11} \text{ s}^{-1}} = 1.22 \times 10^{18}$ atoms
3. Mass	1.22×10^{18} atoms $\times \dfrac{226 \text{ g}}{6.022 \times 10^{23} \text{ atoms}} = 4.58 \times 10^{-4}$ g

END POINT

The curie (Ci) was supposed to be the activity of a one-gram sample of radium, the element discovered by the Curies. Part (b) shows it isn't quite.

Age of Organic Material

During the 1950s, Professor W. F. Libby (1908–1980) of the University of Chicago and others worked out a method for determining the age of organic material. It is based on the decay rate of carbon-14. The method can be applied to objects from a few hundred up to 50,000 years old. It has been used to determine the authenticity of canvases of Renaissance painters and to check the ages of relics left by prehistoric cave dwellers.

Carbon-14 is produced in the atmosphere by the interaction of neutrons from cosmic radiation with ordinary nitrogen atoms:

$$^{14}_{7}\text{N} + ^{1}_{0}n \longrightarrow ^{14}_{6}\text{C} + ^{1}_{1}\text{H}$$

The carbon-14 formed by this nuclear reaction is eventually incorporated into the carbon dioxide of the air. A steady-state concentration, amounting to about one atom of carbon-14 for every 10^{12} atoms of carbon-12, is established in atmospheric CO_2. More specifically, the concentration of C-14 is such that a sample containing one gram of carbon has an activity of 15.3 atoms/min. A living plant, taking in carbon dioxide, has this same activity, as do plant-eating animals or human beings.

When a plant or animal dies, the intake of radioactive carbon stops. Consequently, the radioactive decay of carbon-14

$$^{14}_{6}\text{C} \longrightarrow ^{14}_{7}\text{N} + ^{0}_{-1}e \qquad t_{1/2} = 5730 \text{ y}$$

takes over, and the C-14 activity drops. The activity of a sample is directly proportional to the amount of C-14, ◄ *i* so we can write:

$$\ln \frac{A_o}{A} = kt \qquad (18.2)$$

Since activity is directly proportional to amount, $X_o/X = A_o/A$.

where A_o is the original activity, assumed to be 15.3 atoms/min, and A is the measured activity today; t is the age of the sample.

EXAMPLE 18.4

A tiny piece of paper taken from the Dead Sea Scrolls, believed to date back to the first century A.D., was found to have an activity per gram of carbon of 12.1 atoms/min. Taking A_o to be 15.3 atoms/min, estimate the age of the scrolls.

ANALYSIS

Information given:	A (12.1 atoms/min); A_o (15.3 atoms/min)
Information implied:	$t_{1/2}$ for C-14 (5730 y)
Asked for:	Age of the scrolls

STRATEGY

1. Find k by substituting into the equation relating half-life and rate constant for a first-order reaction.

$$k = \frac{0.693}{t_{1/2}}$$

2. Substitute into Equation 18.2 to find t.

$$\ln \frac{A_o}{A} = kt$$

SOLUTION

1. k	$k = \dfrac{0.693}{5730 \text{ y}} = 1.21 \times 10^{-4} \text{ y}^{-1}$
2. t	$\ln \dfrac{15.3 \text{ atoms/min}}{12.1 \text{ atoms/min}} = (1.21 \times 10^{-4} \text{ y}^{-1})(t) \longrightarrow 0.235 = (1.21 \times 10^{-4} \text{ y}^{-1})(t)$ $t = 1.94 \times 10^3 \text{ y}$ The scrolls do date back to the first century A.D.

Figure 18.7 The Shroud of Turin. This piece of linen first appeared in about 1350 A.D. It bears a faint image (enhanced here) of a man who appears to have been crucified.

Mary Evans Picture Library/Alamy

Such a reaction is exothermic; ΔH, like ΔE, is negative.

As you can imagine, it is not easy to determine accurately activities of the order of 10 atoms decaying per minute, about one "event" every six seconds. Elaborate precautions have to be taken to exclude background radiation. Moreover, relatively large samples must be used to increase the counting rate. Recently a technique has been developed whereby C-14 atoms can be counted very accurately in a specially designed mass spectrometer. This method was used to date the Shroud of Turin (Figure 18.7), using six samples with a total mass of about 0.1 g. Analysis by an international team of scientists in 1988 showed that the flax used to make the linen of which the Shroud is composed grew in the fourteenth century A.D. That means this remarkable burial garment could not have been used for the body of Christ.

18-4 Mass-Energy Relations

The energy change accompanying a nuclear reaction can be calculated from the relation

$$\Delta E = c^2 \Delta m$$

where

Δm = change in mass = mass products − mass reactants

ΔE = energy products − energy reactants

c = speed of light

In a spontaneous nuclear reaction, the products weigh less than the reactants (Δm negative). In this case, the energy of the products is less than that of the reactants (ΔE negative) and energy is evolved to the surroundings.

In an "ordinary chemical reaction," Δm is immeasurably small. In a nuclear reaction, on the other hand, Δm is appreciable, amounting to 0.002% or more of the mass of reactants. The change in mass can readily be calculated from a table of nuclear masses (Table 18.4).

Table 18.4 Nuclear Masses on the ^{12}C Scale*

	At. No.	Mass No.	Mass (amu)		At. No.	Mass No.	Mass (amu)
e	0	0	0.00055	Br	35	79	78.8992
n	0	1	1.00867		35	81	80.8971
H	1	1	1.00728		35	87	86.9028
	1	2	2.01355	Rb	37	89	88.8913
	1	3	3.01550	Sr	38	90	89.8869
He	2	3	3.01493	Mo	42	99	98.8846
	2	4	4.00150	Ru	44	106	105.8832
Li	3	6	6.01348	Ag	47	109	108.8790
	3	7	7.01436	Cd	48	109	108.8786
Be	4	9	9.00999		48	115	114.8791
	4	10	10.01134	Sn	50	120	119.8748
B	5	10	10.01019	Ce	58	144	143.8817
	5	11	11.00656		58	146	145.8868
C	6	11	11.00814	Pr	59	144	143.8809
	6	12	11.99671	Sm	62	152	151.8857
	6	13	13.00006	Eu	63	157	156.8908
	6	14	13.99995	Er	68	168	167.8951
O	8	16	15.99052	Hf	72	179	178.9065
	8	17	16.99474	W	74	186	185.9138
	8	18	17.99477	Os	76	192	191.9197
F	9	18	17.99601	Au	79	196	195.9231
	9	19	18.99346	Hg	80	196	195.9219
Na	11	23	22.98373	Pb	82	206	205.9295
Mg	12	24	23.97845		82	207	206.9309
	12	25	24.97925		82	208	207.9316
	12	26	25.97600	Po	84	210	209.9368
Al	13	26	25.97977		84	218	217.9628
	13	27	26.97439	Rn	86	222	221.9703
	13	28	27.97477	Ra	88	226	225.9771
Si	14	28	27.96924	Th	90	230	229.9837
S	16	32	31.96329	Pa	91	234	233.9934
Cl	17	35	34.95952	U	92	233	232.9890
	17	37	36.95657		92	235	234.9934
Ar	18	40	39.95250		92	238	238.0003
K	19	39	38.95328		92	239	239.0038
	19	40	39.95358	Np	93	239	239.0019
Ca	20	40	39.95162	Pu	94	239	239.0006
Ti	22	48	47.93588		94	241	241.0051
Cr	24	52	51.92734	Am	95	241	241.0045
Fe	26	56	55.92066	Cm	96	242	242.0061
Co	27	59	58.91837	Bk	97	245	245.0129
Ni	28	59	58.91897	Cf	98	248	248.0186
Zn	30	64	63.91268	Es	99	251	251.0255
	30	72	71.91128	Fm	100	252	252.0278
Ge	32	76	75.90380		100	254	254.0331
As	33	79	78.90288				

*Note that these are *nuclear masses*. The masses of the corresponding atoms can be calculated by adding the masses of each extranuclear electron (0.000549). For example, for an *atom* of $^{4}_{2}He$ we have 4.00150 + 2(0.000549) = 4.00260. Similarly, for an atom of $^{12}_{6}C$, 11.99671 + 6(0.000549) = 12.00000.

To obtain a more useful form of the equation $\Delta E = c^2\,\Delta m$, we substitute for c its value in meters per second.

$$c = 3.00 \times 10^8 \text{ m/s}$$

Thus

$$\Delta E = 9.00 \times 10^{16}\,\frac{\text{m}^2}{\text{s}^2} \times \Delta m$$

But

$$1\text{ J} = 1\text{ kg} \cdot \frac{\text{m}^2}{\text{s}^2}; \qquad 1\,\frac{\text{m}^2}{\text{s}^2} = 1\,\frac{\text{J}}{\text{kg}}$$

So

$$\Delta E = 9.00 \times 10^{16}\,\frac{\text{J}}{\text{kg}} \times \Delta m$$

This equation can be used to calculate the energy change in *joules*, if you know Δm in *kilograms*. Ordinarily, Δm is expressed in *grams*; ΔE is calculated in *kilojoules*. The relationship between ΔE and Δm in these units can be found by using conversion factors:

$$\Delta E = 9.00 \times 10^{16}\,\frac{\text{J}}{\text{kg}} \times \frac{1\text{ kg}}{10^3\text{ g}} \times \frac{1\text{ kJ}}{10^3\text{ J}} \times \Delta m$$

ⓘ

This is the most useful form of the mass-energy relation.

$$\Delta E = 9.00 \times 10^{10}\,\frac{\text{kJ}}{\text{g}} \times \Delta m \; ◀ⓘ \tag{18.3}$$

EXAMPLE 18.5

For the radioactive decay of radium, $^{226}_{88}\text{Ra} \longrightarrow {}^{222}_{86}\text{Rn} + {}^{4}_{2}\text{He}$, calculate ΔE in kilojoules when 10.2 g of radium decays.

ANALYSIS

Information given:	nuclear reaction ($^{226}_{88}\text{Ra} \longrightarrow {}^{222}_{86}\text{Rn} + {}^{4}_{2}\text{He}$) mass of Ra-226 (10.2 g)
Information implied:	nuclear masses (Table 18.4)
Asked for:	ΔE

STRATEGY

1. Use Table 18.4 to find nuclear masses and Δm for the decay of one mole of Ra-226. Do not round off.

   Δm = nuclear masses of products − nuclear masses of reactants.

2. Find Δm for the decay of 10.2 g of Ra-226.

3. Determine ΔE by substituting into Equation 18.3.

SOLUTION

1. Δm for one mole	Δm = (mass of 1 mol He-4 + mass of 1 mol Rn-222) − (mass of 1 mol Ra-226) = 4.0015 g + 221.9703 g − 225.9771 g = −0.0053 g/mol Ra-226
2. Δm for 10.2 g	$\Delta m = 10.2\text{ g Ra} \times \dfrac{1\text{ mol Ra}}{226.0\text{ g Ra}} \times \dfrac{-0.0053\text{ g}}{1\text{ mol Ra}} = -2.4 \times 10^{-4}\text{ g}$
3. ΔE	$\Delta E = \left(9.00 \times 10^{10}\,\dfrac{\text{kJ}}{\text{g}}\right)(\Delta m) = \left(9.00 \times 10^{10}\,\dfrac{\text{kJ}}{\text{g}}\right)(-2.4 \times 10^{-4}\text{ kJ}) = -2.2 \times 10^7\text{ kJ}$

END POINT

In ordinary chemical reactions, the energy change is of the order of 50 kJ/g or less. In this nuclear reaction, the energy change is much, much greater.

Nuclear Binding Energy

It is always true that *a nucleus weighs less than the individual protons and neutrons of which it is composed.* Consider, for example, the $^{6}_{3}\text{Li}$ nucleus, which contains three protons and three neutrons. According to Table 18.4, one mole of

Li-6 nuclei weighs 6.01348 g. In contrast, the total mass of three moles of neutrons and three moles of protons is

$$3(1.00867 \text{ g}) + 3(1.00728 \text{ g}) = 6.04785 \text{ g}$$

Clearly, one mole of Li-6 weighs *less* than the corresponding protons and neutrons. For the process

$$^{6}_{3}\text{Li} \longrightarrow 3^{1}_{1}\text{H} + 3^{1}_{0}n$$

$\Delta m = 6.04785 \text{ g} - 6.01348 \text{ g} = 0.03437 \text{ g/mol Li}.$

The quantity just calculated is referred to as the **mass defect**. The corresponding energy difference is

$$\Delta E = 9.00 \times 10^{10} \frac{\text{kJ}}{\text{g}} \times 0.03437 \frac{\text{g}}{\text{mol Li}} = 3.09 \times 10^{9} \frac{\text{kJ}}{\text{mol Li}}$$

This energy is referred to as the **binding energy.** It follows that 3.09×10^{9} kJ of energy would have to be absorbed to decompose one mole of Li-6 nuclei into protons and neutrons.

$$^{6}_{3}\text{Li} \longrightarrow 3^{1}_{0}n + 3^{1}_{1}\text{H} \qquad \Delta E = 3.09 \times 10^{9} \frac{\text{kJ}}{\text{mol Li}}$$

It takes a lot of energy to blow a nucleus apart.

By the same token, 3.09×10^{9} kJ of energy would be evolved when one mole of Li-6 is formed from protons and neutrons.

EXAMPLE 18.6

Calculate the binding energy of C-14, in kilojoules per mole.

STRATEGY

1. Recall that the C-14 nucleus is made up of 6 protons ($^{1}_{1}\text{H}$) and 8 neutrons ($^{1}_{0}n$). Thus

$$^{14}_{6}\text{C} \longrightarrow 6^{1}_{1}\text{H} + 8^{1}_{0}n$$

2. Use Table 18.4 to find Δm.

3. Substitute into Equation 18.3 to find ΔE in kJ/mol.

SOLUTION

1. Δm	$\Delta m = 6(1.00728 \text{ g}) + 8(1.00867 \text{ g}) - 13.99995 \text{ g} = 0.11309 \text{ g/mol C-14}$
2. ΔE	$\Delta E = 9.00 \times 10^{10} \dfrac{\text{kJ}}{\text{g}} \times 0.11309 \dfrac{\text{g}}{\text{mol}} = 1.02 \times 10^{10} \text{ kJ/mol}$

The binding energy of a nucleus is, in a sense, a measure of its stability. The greater the binding energy, the more difficult it would be to decompose the nucleus into protons and neutrons. As you might expect, the binding energy as calculated above increases steadily as the nucleus gets heavier, containing more protons and neutrons. A better measure of the relative stabilities of different nuclei is the *binding energy per mole of nuclear particles (nucleons)*. This quantity is calculated by dividing the binding energy per mole of nuclei by the number of particles per nucleus. Thus

$$\text{Li-6:} \quad 3.09 \times 10^{9} \frac{\text{kJ}}{\text{mol Li-6}} \times \frac{1 \text{ mol Li-6}}{6 \text{ mol nucleons}} = 5.15 \times 10^{8} \frac{\text{kJ}}{\text{mol}}$$

$$\text{C-14:} \quad 1.02 \times 10^{10} \frac{\text{kJ}}{\text{mol}} \times \frac{1 \text{ mol C-14}}{14 \text{ mol nucleons}} = 7.29 \times 10^{8} \frac{\text{kJ}}{\text{mol}}$$

Figure 18.8 shows a plot of this quantity, binding energy/mole of nucleons, versus mass number. Notice that the curve has a broad maximum in the vicinity of mass numbers 50 to 80. Consider what would happen if a heavy nucleus such as $^{235}_{92}\text{U}$

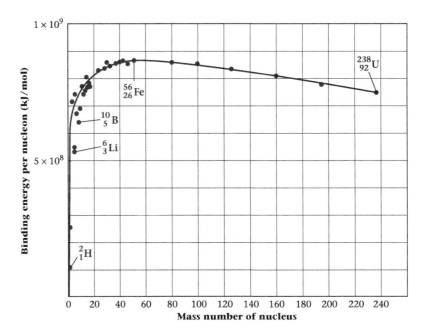

Figure 18.8 Binding energy per nucleon, a measure of nuclear stability. The binding energy has its maximum value for nuclei of intermediate mass, falling off for very heavy or very light nuclei. The form of this curve accounts for the fact that both fission and fusion give off large amounts of energy.

were to split into smaller nuclei with mass numbers near the maximum. This process, referred to as nuclear **fission**, should result in an evolution of energy. The same result is obtained if very light nuclei such as ^2_1H combine with one another to form a heavier nucleus. ◀ ⓘ Indeed, this process, called nuclear **fusion**, should evolve even more energy, because the binding energy per nucleon increases very sharply at the beginning of the curve.

Both very heavy and very light nuclei are relatively unstable; they decompose with the evolution of energy. ⓘ

▼
| 18-5 | **Nuclear Fission** |

The process of nuclear fission was discovered in Germany more than half a century ago in 1938 by Lise Meitner (1878–1968) and Otto Hahn (1879–1968). With the outbreak of World War II a year later, interest focused on the enormous amount of energy released in the process (Figure 18.9). At Los Alamos, in the mountains of New Mexico, a group of scientists led by J. Robert Oppenheimer (1904–1967) worked feverishly to produce the fission, or "atomic," bomb. Many of the members of this group were exiles from Nazi Germany. They were spurred on by the fear that Hitler would obtain the bomb first. Their work led to the explosion of the first atomic bomb in the New Mexico desert at 5:30 A.M. on July 16, 1945. Less than a month later (August 6, 1945), the world learned of this new weapon when another bomb was exploded over Hiroshima. This bomb killed 70,000 people and completely devastated an area of 10 square kilometers. ◀ ⓘ Three days later, Nagasaki and its inhabitants met a similar fate. On August 14, Japan surrendered, and World War II was over.

Figure 18.9 A nuclear fission explosion. Such a dramatic and destructive release of energy had never been seen before the development of the "atomic bomb" during World War II.

The Hiroshima bomb was equivalent to 20,000 tons of TNT. ⓘ

The Fission Process ($^{235}_{92}U$)

Several isotopes of the heavy elements undergo fission if bombarded by neutrons of high enough energy. In practice, attention has centered on two particular isotopes, $^{235}_{92}U$ and $^{239}_{94}Pu$. Both of these can be split into fragments by relatively low-energy neutrons.

Our discussion concentrates on the uranium-235 isotope. It makes up only about 0.7% of naturally occurring uranium. The more abundant isotope, uranium-238, does not undergo fission. The first process used to separate these isotopes, and until recently the only one available, was that of gaseous effusion (Chapter 5). The volatile compound uranium hexafluoride, UF_6, which sublimes at 56°C, is used for this purpose.

When a uranium-235 atom undergoes fission, it splits into two unequal fragments and a number of neutrons and beta particles. The fission process is complicated by the fact that different uranium-235 atoms split up in many different ways. For example, while one atom of $^{235}_{92}U$ is splitting to give isotopes of rubidium ($Z = 37$) and cesium ($Z = 55$), another may break up to give isotopes of bromine ($Z = 35$) and lanthanum ($Z = 57$), while still another atom yields isotopes of zinc ($Z = 30$) and samarium ($Z = 62$):

$$^{1}_{0}n + {}^{235}_{92}U \begin{cases} \longrightarrow {}^{90}_{37}Rb + {}^{144}_{55}Cs + 2\,{}^{1}_{0}n \\ \longrightarrow {}^{87}_{35}Br + {}^{146}_{57}La + 3\,{}^{1}_{0}n \\ \longrightarrow {}^{72}_{30}Zn + {}^{160}_{62}Sm + 4\,{}^{1}_{0}n \end{cases}$$

More than 200 isotopes of 35 different elements have been identified among the fission products of uranium-235.

The stable neutron-to-proton ratio near the middle of the periodic table, where the fission products are located, is considerably smaller ($\sim$1.2) than that of uranium-235 (1.6). Hence the immediate products of the fission process contain too many neutrons for stability; they decompose by beta emission. In the case of rubidium-90, three steps are required to reach a stable nucleus:

$$^{90}_{37}Rb \longrightarrow {}^{90}_{38}Sr + {}^{0}_{-1}e \qquad t_{1/2} = 2.8 \text{ min}$$
$$^{90}_{38}Sr \longrightarrow {}^{90}_{39}Y + {}^{0}_{-1}e \qquad t_{1/2} = 29 \text{ y}$$
$$^{90}_{39}Y \longrightarrow {}^{90}_{40}Zr + {}^{0}_{-1}e \qquad t_{1/2} = 64 \text{ h}$$

The radiation hazard associated with fallout from nuclear weapons testing arises from radioactive isotopes such as these. One of the most dangerous is strontium-90. In the form of strontium carbonate, $SrCO_3$, it is incorporated into the bones of animals and human beings, where it remains for a lifetime.

Notice from the fission equations above that two to four neutrons are produced by fission for every one consumed. Once a few atoms of uranium-235 split, the neutrons produced can bring about the fission of many more uranium-235 atoms. This creates the possibility of a *chain reaction*, whose rate increases exponentially with time. This is precisely what happens in the atomic bomb. The energy evolved in successive fissions escalates to give a tremendous explosion within a few seconds.

For nuclear fission to result in a chain reaction, the sample must be large enough so that most of the neutrons are captured internally. If the sample is too small, most of the neutrons escape, breaking the chain. The *critical mass* of uranium-235 required to maintain a chain reaction in a bomb appears to be about 1 to 10 kg. In the bomb dropped on Hiroshima, the critical mass was achieved by using a conventional explosive to fire one piece of uranium-235 into another.

The evolution of energy in nuclear fission is directly related to the decrease in mass that takes place. About 80,000,000 kJ of energy is given off for every gram of $^{235}_{92}U$ that reacts. This is about 40 times as great as the energy change for simple nuclear reactions such as radioactive decay. The heat of combustion of coal is only about 30 kJ/g; the energy given off when TNT explodes is still smaller, about 2.8 kJ/g. Putting it another way, the fission of one gram of $^{235}_{92}U$ produces as much energy as the combustion of 2700 kg of coal or the explosion of 30 metric tons (3×10^4 kg) of TNT.

Nuclear Reactors

About 20% of the electricity generated in the United States come from nuclear reactors, which use the fission of U-235 to generate heat. That heat boils water and turns a turbine. A light-water reactor is shown schematically in Figure 18.10. The *fuel rods* contain cylindrical pellets of uranium dioxide (UO_2) in zirconium alloy tubes. The uranium in these reactors is "enriched" so that it contains about 3% U-235, the fissionable isotope. Natural uranium is less than 1% U-235. The

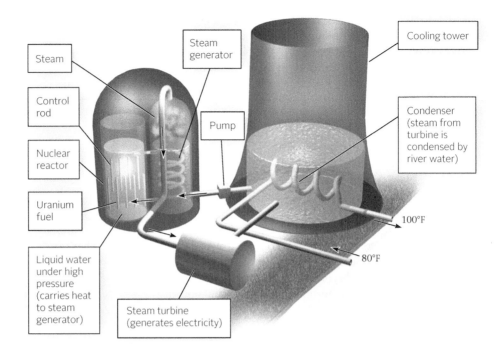

Figure 18.10 A pressurized water nuclear reactor. The control rods are made of a material such as cadmium or boron, which absorbs neutrons effectively. The fuel rods contain uranium oxide enriched in U-235.

control rods are cylinders that contain substances such as boron and cadmium, which absorb neutrons. By varying the depth to which these are inserted into the reactor core, the speed of the chain reactions can be controlled. In a pressurized water reactor, water at 140 atm absorbs the heat released by the chain reaction and is heated to about 320°C. This heated water passes through the steam generators, which also contain water under pressure. The water turns to steam at 270°C and drives a turbogenerator that produces electricity.

In a light water reactor, the circulating water in the core serves a purpose beyond cooling. It slows down, or moderates, the neutrons produced in the fission reactions. This is necessary if the chain reaction is to continue. Fast neutrons, unmoderated, are not readily absorbed by the U-235 nuclei. Reactors designed in Canada use heavy water, D_2O, which has had an important advantage over ordinary water. Its moderating properties are such that naturally occurring uranium can be used as a fuel. Energy-intensive enrichment has not been necessary. The next generation of Canadian reactors, called Advanced CANDU, include the use of lightly enriched uranium, just as light water reactors do.

By the 1970s it was generally supposed that nuclear fission would replace fossil fuels (oil, natural gas, coal) as an energy source. That hasn't happened. The most evident reasons are

- *Nuclear accidents* at Three Mile Island, Pennsylvania, in 1979 and Chernobyl, Ukraine, in 1986 had a devastating effect on public opinion in the United States and, to a smaller degree, elsewhere in the world. At Three Mile Island only about 50 curies of radiation were released to the environment and there were no casualties. The explosion at Chernobyl was a very different story. About 100 million curies were released, leading to at least 31 fatalities. Moreover, 135,000 people were permanently evacuated from the region surrounding the reactor. Since then, all the other reactors at Chernobyl—three, in addition to the one that exploded—have been permanently shut down.

- *Disposal of radioactive wastes* from nuclear reactors has proved to be a serious political issue. The NIMBY (not in my backyard) attitude applies here. The U.S. government has spent billions of dollars to develop a permanent nuclear waste storage repository at Yucca Mountain, Nevada, as the site for burying some 70,000 metric tons of nuclear waste. Current plans (mostly as a result of political pressure), however, include closing the Yucca Mountain repository permanently, precluding waste storage there. The problem of radioactive waste disposal is not a uniquely American problem. Other nations have struggled with the disposal of wastes as well.

Currently, nuclear energy is being reevaluated in light of its carbon neutrality. That is, nuclear reactors do not directly contribute the gases, believed by many scientists to contribute to global warming, that the combustion of carbon-based fuels do. Furthermore, reactor designs now under construction or on the drawing boards (called Generation III+ reactors) are addressing some of the safety and cost issues that have plagued past designs. Generation III+ reactors are now under construction in China, Finland, Korea, and Japan, and one U.S. utility has made a firm commitment to construct one such reactor (the Westinghouse AP1000), adding two additional units to a pair of currently operating reactors in the state of Georgia. These will be the first new reactors constructed since the 1970s. Older designs are also being revived. The last reactor to go online in the United States, at Watts Bar in Tennessee, opened in 1996. Construction of what was to be a two-unit plant began in 1973. The second unit was never completed. Now, the Tennessee Valley Authority is actively working to restart construction on the unfinished reactor, expecting to have it online by December of 2015.

18-6 Nuclear Fusion

Recall (Figure 18.8) that very light nuclei, such as those of hydrogen, are unstable with respect to fusion into heavier isotopes. Indeed, the energy available from nuclear fusion is considerably greater than that given off in the fission of an equal mass of a heavy element (Example 18.7).

EXAMPLE 18.7

Calculate ΔE, in kilojoules per gram of reactants, in a fusion reaction,

$$^2_1\text{H} + {}^2_1\text{H} \longrightarrow {}^4_2\text{He},$$

and in a fission reaction,

$$^{235}_{92}\text{U} \longrightarrow {}^{90}_{38}\text{Sr} + {}^{144}_{58}\text{Ce} + {}^1_0n + 4\,{}^{\,0}_{-1}e.$$

STRATEGY

1. Find Δm for the reaction as written by using the nuclear masses in Table 18.4.

2. Find Δm for one gram of reactant.

3. Substitute into Equation 18.3 to determine ΔE.

SOLUTION

fusion reaction

1. $\Delta m/2$ mol H-2

$$\Delta m = (\text{mass He-4}) - 2(\text{mass H-2})$$
$$= 4.00150\text{ g} - 2(2.01355\text{ g}) = -0.02560\text{ g}$$

2. $\Delta m/$g H-2

$$\Delta m = \frac{-0.02560\text{ g}}{2\text{ mol H-2}} \times \frac{2\text{ mol H-2}}{2(2.01355\text{ g})} = -0.006357\text{ g/g H-2}$$

3. ΔE

$$\Delta E = 9.00 \times 10^{10}\,\frac{\text{kJ}}{\text{g}} \times \frac{-6.357 \times 10^{-3}\text{ g}}{1\text{ g H-2}} = -5.72 \times 10^8\text{ kJ/g H-2}$$

fission reaction

1. $\Delta m/$mol U-235

$$\Delta m = \text{mass(Sr-90)} + \text{mass(Ce-144)} + \text{mass}({}^1_0n) + 4[\text{mass }({}^{\,0}_{-1}e)] - \text{mass(U-235)}$$
$$= [89.8869 + 143.8817 + 1.00867 + 4(0.00055)]\text{ g} - 234.9934\text{ g} = -0.21393\text{ g}$$

continued

2. Δm/g U-235	$\Delta m = \dfrac{-0.2139 \text{ g}}{\text{mol U-235}} \times \dfrac{1 \text{ mol U-235}}{235.0 \text{ g}} = -9.102 \times 10^{-4}$ g/g U-235
3. ΔE	$\Delta E = 9.00 \times 10^{10} \dfrac{\text{kJ}}{\text{g}} \times \dfrac{-9.102 \times 10^{-4} \text{ g}}{1 \text{ g U-235}} = -8.19 \times 10^{7}$ kJ/g U-235

END POINT

Comparing the answers to (a) and (b), it appears that the fusion reaction produces about seven times as much energy per gram of reactant (57.2×10^{7} versus 8.19×10^{7} kJ) as does the fission reaction. This factor varies from about 3 to 10, depending on the particular reaction chosen to represent the fission and fusion processes.

As an energy source, nuclear fusion possesses several additional advantages over nuclear fission. In particular, light isotopes suitable for fusion are far more abundant than the heavy isotopes required for fission. You can calculate, for example (Problem 79), that the fusion of only $2 \times 10^{-9}\%$ of the deuterium ($^2_1$H) in seawater would meet the total annual energy requirements of the world.

Unfortunately, fusion processes, unlike neutron-induced fission, have very high activation energies. ◀ⓘ To overcome the electrostatic repulsion between two deuterium nuclei and cause them to react, they have to be accelerated to velocities of about 10^6 m/s, about 10,000 times greater than ordinary molecular velocities at room temperature. The corresponding temperature for fusion, as calculated from kinetic theory, is of the order of 10^{9}°C. In the hydrogen bomb, temperatures of this magnitude were achieved by using a fission reaction to trigger nuclear fusion. If fusion reactions are to be used to generate electricity, it will be necessary to develop equipment in which very high temperatures can be maintained long enough to allow fusion to occur and give off energy. In any conventional container, the reactant nuclei would quickly lose their high kinetic energies by collisions with the walls.

ⓘ In April 1997, the Tokamak fusion reactor at Princeton shut down when government funding was withdrawn.

One fusion reaction currently under study is a two-step process involving deuterium and lithium as the basic starting materials:

$$^2_1\text{H} + ^3_1\text{H} \longrightarrow {}^4_2\text{He} + {}^1_0n$$
$$^6_3\text{Li} + ^1_0n \longrightarrow {}^4_2\text{He} + {}^3_1\text{H}$$
$$\overline{^2_1\text{H} + ^6_3\text{Li} \longrightarrow 2\,{}^4_2\text{He}}$$

This process is attractive because it has a lower activation energy than other fusion reactions.

One possible way to achieve nuclear fusion is to use magnetic fields to confine the reactant nuclei and prevent them from touching the walls of the container, where they would quickly slow down below the velocity required for fusion. Using 400-ton magnets, it is possible to sustain the reaction for a fraction of a second. To achieve a net evolution of energy, this time must be extended to about one second. A practical fusion reactor would have to produce 20 times as much energy as it consumes. Optimists predict that this goal may be reached in 50 years.

Another approach to nuclear fusion is shown in Figure 18.11. Tiny glass pellets (about 0.1 mm in diameter) filled with frozen deuterium and tritium serve as a target. The pellets are illuminated by a powerful laser beam, which delivers 10^{12} kilowatts of power in one nanosecond (10^{-9} s). The reaction is the same as with magnetic confinement; unfortunately, at this point energy breakeven seems many years away.

A mixture of deuterium and tritium is sealed inside the tiny capsule (1 mm in diameter) at the tip of the laser target.

Exposure of the capsule to the energy of a powerful laser beam produces a 0.5-picosecond burst of energy from the fusion of the deuterium and tritium.

(a)   (b)

Figure 18.11 Laser fusion.

CHEMISTRY **BEYOND** THE CLASSROOM

Biological Effects of Radiation

The harmful effects of radiation result from its high energy, sufficient to form unstable free radicals (species containing unpaired electrons) such as

$$\text{H·} \qquad \text{H—}\ddot{\text{O}}\text{·} \qquad \text{:}\ddot{\text{O}}\text{—}\ddot{\text{O}}\text{·}$$

These free radicals can react with and in that sense destroy organic molecules essential to life.

The extent of damage from radiation depends mainly on two factors. These are the amount of radiation absorbed and the type of radiation. The former is commonly expressed in *rads* (radiation *a*bsorbed *d*ose). A rad corresponds to the absorption of 10^{-2} J of energy per kilogram of tissue:

$$1 \text{ rad} = 10^{-2} \text{ J/kg}$$

The biological effect of radiation is expressed in *rems* (*r*adiation *e*quivalent for *m*an). The number of rems is found by multiplying the number of rads by a "damage" factor, *n*:

$$\text{no. of rems} = n(\text{no. of rads})$$

where *n* is 1 for gamma and beta emission, 5 for low-energy neutrons, and 10 to 20 for high-energy neutrons and alpha particles. Table A lists some of the effects to be expected when a person is exposed to a single dose of radiation at various levels.

The average exposure to radiation of people living in the United States is about 360 mrem (0.36 rem) per year. Notice (Figure A) that 82% of the radiation comes from natural sources; the greatest single source by far is radon (55%). The level of exposure to radon depends on location. In 1985 a man named Stanley Watras, who happened to work at a nuclear power plant, found that he was setting off the radiation monitors when he went to work in the

Table A Effect of Exposure to a Single Dose of Radiation

Dose (rems)	Probable Effect
0 to 25	No observable effect
25 to 50	Small decrease in white blood cell count
50 to 100	Lesions, marked decrease in white blood cells
100 to 200	Nausea, vomiting, loss of hair
200 to 500	Hemorrhaging, ulcers, possible death
500+	Fatal

morning. It turned out that the house he lived in had a radon level 2000 times the national average.

$^{222}_{86}$Rn, a radioactive isotope of radon, is a decay product of naturally occurring uranium-238. Because it is gaseous and chemically inert, radon seeps through cracks in concrete and masonry from the ground into houses. There its concentration builds up, particularly if the house is tightly insulated. Inhalation of radon-222 can cause health problems because its decay products, including Po-218 and Po-214, are intensely radioactive and readily absorbed in lung tissue. The Environmental Protection Agency (EPA) estimates that radon inhalation causes between 5000 and 20,000 of the 130,000 deaths from lung cancer annually in the United States. The EPA recommends that special ventilation devices be used to remove radon from basements if tests show (Figure B) that the radiation level exceeds 4×10^{-12} Ci/L.

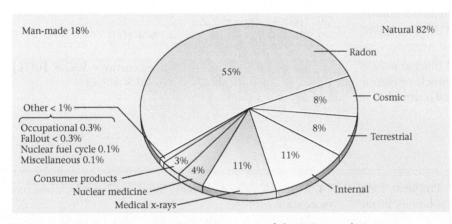

Figure A Sources of average radiation exposure of the U.S. population.

Figure B A commercially available home-test kit for radon.

Key Concepts

1. Determine the more stable isotope or atom.
   (**Example 18.1; Problems 1–4**)
2. Write balanced nuclear equations.
   (**Example 18.2; Problems 5–20**)
3. Relate activity to rate constant and number of atoms.
   (**Example 18.3; Problems 21–34**)
4. Relate activity to age of objects.
   (**Example 18.4; Problems 35–42**)

5. Relate Δm to ΔE in a nuclear reaction.
   (Examples 18.5; Problems 43, 44, 49–52)

6. Calculate binding energies.
   (Example 18.6; Problems 45–48)

Key Equations

Rate of decay $\qquad \ln \dfrac{X_o}{X} = kt \qquad k = 0.693/t_{1/2} \qquad A = kN$

Mass-energy $\qquad \Delta E = 9.00 \times 10^{10} \dfrac{\text{kJ}}{\text{g}} \times \Delta m$

Key Terms

activity	binding energy	fusion	nuclear equations
alpha particle	curies	gamma radiation	positron
beta particles	fission	K-electron capture	

Summary Problem

Consider the isotopes of gallium (Ga), a silvery metal that melts at 30°C. It is used primarily in different circuits: microwave, infrared, and high-speed switching.

(a) Write the nuclear symbol for Ga-67, which is used medically to scan for various tumors. How many protons are there in its nucleus? How many neutrons?

(b) Write the equation for the decomposition of Ga-67 by K-capture.

(c) Write nuclear equations to represent the bombardment of Ga-64 by a neutron, and the decay of the product nucleus by β-emission.

(d) Calculate ΔE in kJ when 1.00 g of Ga-71 (nuclear mass = 70.9476 amu) is bombarded by an α-particle producing As-72 (nuclear mass = 71.9493 amu) and neutrons.

(e) What are the mass defect and binding energy of Ga-71?

(f) The half-life of Ga-67 is 77.9 hours. What is the rate constant for Ga-67? What is the activity (in curies) of a 5.00×10^2-mg sample of Ga-67?

(g) How long will it take Ga-67 to decay and lose 95% of its radioactivity?

Answers

(a) $^{67}_{31}\text{Ga}$; 31 protons, 36 neutrons
(b) $^{67}_{31}\text{Ga} + {}^{0}_{-1}e \longrightarrow {}^{67}_{30}\text{Zn}$
(c) $^{64}_{31}\text{Ga} + {}^{1}_{0}n \longrightarrow {}^{65}_{31}\text{Ga} \longrightarrow {}^{0}_{-1}e + {}^{65}_{32}\text{Ge}$
(d) 3.32×10^7 kJ
(e) mass defect = 0.6249 g; binding energy = 5.62×10^{10} kJ
(f) $k = 8.90 \times 10^{-3}$ h^{-1}; $A = 3.00 \times 10^5$ Ci
(g) 337 h $\approx$ 14 days

Questions and Problems

Even-numbered questions and Challenge Problems have answers in Appendix 5 and fully worked solutions in the *Student Solutions Manual*.

18-1 Nuclear Stability

1. Which isotope in each of the following pairs should be stable?
   (a) $^{10}_{5}\text{B}$ or $^{9}_{5}\text{B}$ (b) $^{18}_{8}\text{O}$ or $^{16}_{8}\text{O}$ (c) $^{24}_{12}\text{Mg}$ or $^{23}_{12}\text{Mg}$
2. Which isotope in each of the following pairs should be more stable?
   (a) $^{28}_{14}\text{Si}$ or $^{29}_{14}\text{Si}$ (b) $^{6}_{3}\text{Li}$ or $^{8}_{3}\text{Li}$ (c) $^{23}_{11}\text{Na}$ or $^{20}_{11}\text{Na}$
3. For each pair of elements listed, predict which one has more stable isotopes:
   (a) I or Sn (b) Ga or Ni (c) Si or Cl

4. For each pair of elements listed, predict which one has more stable isotopes.
   (a) Ni or Cu (b) Se or Sb (c) Cd or Au

18-2 Radioactivity

Nuclear Equations

5. Smoke detectors contain a small amount of americium-241. Its decay product is neptunium-237. Identify the emission from americium-241.

6. Chromium-51, a positron emitter, is used in research to study red cell survival. It is delivered as a solution of sodium chromate. Write the nuclear equation for the decay of Cr-51.

7. Thorium-231 is the product of alpha emission. It is radioactive and emits beta radiation. Determine
   (a) the parent nucleus of Th-231.
   (b) the daughter product of Th-231 decay.

8. Uranium decays by alpha emission. Its daughter product emits β-radiation. Determine
   (a) the daughter product of U-235 decay.
   (b) the product obtained by β-emission of the daughter product in (a).

9. Write balanced nuclear reactions for the following:
   (a) Formation of Am-241 through β-emissions.
   (b) Formation of Kr-81 by K-electron capture.
   (c) Formation of Ra-223 by α-emission.

10. Write balanced nuclear equations for the following:
    (a) Formation of Mn-52 by positron emission.
    (b) Formation of Ac-228 by β-emission.
    (c) Formation of Np-232 by α-decay.

11. Write balanced nuclear equations for
    (a) the alpha emission resulting in the formation of Pa-233.
    (b) the loss of a positron by Y-85.
    (c) the fusion of two C-12 nuclei to give Na-23 and another particle.
    (d) the fission of Pu-239 to give Sn-130, another nucleus, and an excess of two neutrons.

12. Write balanced nuclear equations for
    (a) the loss of an alpha particle by Th-230.
    (b) the loss of a beta particle by Pb-210.
    (c) the fission of U-235 to give Ba-140, another nucleus, and an excess of two neutrons.
    (d) the K-capture of Ar-37.

13. When californium-249 is bombarded by a small isotope, rutherfordium, $^{257}_{104}$Rf, is formed together with 4 neutrons. The isotope formed quickly decays to Nobelium-253, $^{253}_{102}$No.
    (a) What small isotope is used in the bombardment?
    (b) What particle is emitted in the decay of Rf-257?

14. When Bi-209 is bombarded with Ni-64, one neutron and a new isotope, X, are formed. The isotope then goes through a series of alpha particle emissions.
    (a) Write the nuclear symbol for the isotope formed.
    (b) Write the nuclear symbol for the isotope formed after the third alpha particle emission.

15. Consider the "new" isotope $^{285}_{120}$Z. Compare the product nuclides (call them X and Q, respectively) after β and positron emission.

16. Follow the directions for Question 15 but compare the product nuclides after K-capture and positron emission.

17. Write balanced nuclear equations for the bombardment of
    (a) U-238 with a nucleus to produce Fm-249 and five neutrons.
    (b) Al-26 with an alpha particle to produce P-30.
    (c) Cu-63 with a nucleus producing Zn-63 and a neutron.
    (d) Al-27 with deuterium ($^2_1$H) to produce an alpha particle and another nucleus.

18. Write balanced nuclear equations for the bombardment of
    (a) Fe-54 with an alpha particle to produce another nucleus and two protons.
    (b) Mo-96 with deuterium ($^2_1$H) to produce a neutron and another nucleus.
    (c) Ar-40 with an unknown particle to produce K-43 and a proton.
    (d) a nucleus with a neutron to produce a proton and P-31.

19. Balance the following equations by filling in the blanks.
    (a) $^{235}_{92}$U $+ ^1_0n \longrightarrow ^{137}_{54}$_____ $+ 2^1_0n +$ _____
    (b) $^{232}_{90}$Th $+ ^{12}_6$ _____ $\longrightarrow$ _____$^1_0n + ^{240}_{96}$Cm
    (c) $^4_2$He $+ ^{96}_{42}$Mo $\longrightarrow ^{100}_{43}$_____ $+$ _____
    (d) _____ $+ ^2_1$H $\longrightarrow ^{210}_{84}$_____ $+ ^1_0n$

20. Balance the following nuclear equations by filling in the blanks:
    (a) Es-240 + neutron $\longrightarrow$ 2 neutrons +_____ + Gd-161
    (b) _____ $\longrightarrow \beta$-particle + Co-59
    (c) 4^1_1H $\longrightarrow$ _____ + 2 positrons
    (d) Mg-24 + neutron $\longrightarrow$ proton + _____

18-3 Rate of Radioactive Decay

21. How long (in hours) would it take to disintegrate 3.17×10^{17} atoms from a 12.0-mCi (millicurie) source?

22. A 15.0-mCi (millicurie) radioactive source runs for 2.63 hours. How many atoms have disintegrated?

23. A scintillation counter registers emitted radiation caused by the disintegration of nuclides. If each atom of a nuclide emits one count, what is the activity of a sample that registers 3.00×10^4 disintegrations in five minutes?

24. A Geiger counter counts 0.070% of all the particles emitted by a sample. What is the activity that registers 19.4×10^3 counts in one minute?

25. Yttrium-87 has a rate constant of 2.6×10^{-6} s^{-1}. What is the activity of a 5.00-mg sample?

26. Krypton-87 has a rate constant of 1.5×10^{-4} s^{-1}. What is the activity of a 2.00-mg sample?

27. Iodine-131 is used to treat thyroid cancer. It decays by beta emission and has a half-life of 8.1 days.
    (a) Write a balanced nuclear reaction for the decay of iodine-131.
    (b) What is the activity (in Ci) of a 2.50-mg sample of the isotope?

28. Copper-64 is used to study tumors of the brain. It has a half-life of 12.8 hours. What is the activity (in Ci) of 1.50 mg Cu-64?

29. Lead-210 has a half-life of 20.4 years. This isotope decays by beta particle emission. A counter registers 1.3×10^4 disintegrations in five minutes. How many grams of Pb-210 are there?

30. Bromine-82 has a half-life of 36 hours. A sample containing Br-82 was found to have an activity of 1.2×10^5 disintegrations/min. How many grams of Br-82 were present in the sample? Assume that there were no other radioactive nuclides in the sample.

31. Fluorine-18 has a decay constant of 6.31×10^{-3} min^{-1}. How many counts will one get on a Geiger counter in one minute from 1.00 mg of F-18? Assume the sensitivity of the counter is such that it intercepts 0.50% of the emitted radiation.

32. Chlorine-36 decays by β-emission. It has a decay constant of 2.3×10^{-6} y^{-1}. How many β-particles are emitted in one minute from a 1.000-mg sample of Cl-36? How many curies does this represent?

33. Co-60 is used extensively in radiotherapy as a source of γ-rays. Its half-life is 5.27 years.

(a) How many years will it take to decrease to 30.0% of its original activity?

(b) If a cobalt source is ten years old, what percent of its original activity remains?

34. Phosphorus-32 is used in biochemical studies to follow the pathway taken by metabolites. Its half-life is 14.3 days. A vial containing 10.0 μg (1 μg = 1 × 10^{-6} g) of P-32 is accidentally spilled into a sink.

(a) How long will it take P-32 in the drain to exhibit 68% of its original activity?

(b) How many atoms of P-32 remain in the sink after 755 days?

35. Carbon from a cypress beam obtained from the tomb of an ancient Egyptian king gave 9.2 disintegrations/minute of C-14 per gram of carbon. Carbon from living material gives 15.3 disintegrations/min of C-14 per gram of carbon. Carbon-14 has a half-life of 5730 years. How old is the cypress beam?

36. A "new" painting supposedly painted by Michelangelo (1475–1564) is authenticated by C-14 dating. The C-14 content ($t_{1/2}$ = 5730 y) of the canvas is 0.972 times that of a living plant. Could the painting have been done by Michelangelo, or at least while Michelangelo was alive?

37. Planks from an old boat were found in a cave near the shore. Analysis of the boat's wood gave 13.6 disintegrations/min/g of carbon. The half-life of C-14 is 5730 years. Analysis of a tree cut down when the cave was found showed 15.3 disintegrations/min/g of C-14. In what century was the boat built?

38. A sample of a wooden artifact gives 5.0 disintegrations/min/g carbon. The half-life of C-14 is 5730 years, and the activity of C-14 in wood just cut down from a tree is 15.3 disintegrations/min/g carbon. How old is the wooden artifact?

39. The radioactive isotope tritium, $_1^3$H, is produced in nature in much the same way as $_6^{14}$C. Its half-life is 12.3 years. Estimate the $_1^3$H ratio of the tritium of water in the area to the tritium in a bottle of wine claimed to be 25 years old.

40. Consider the information given about tritium in Problem 39. Estimate the age of a sample of Scotch whiskey that has a tritium content 62% that of the water in the area where the whiskey was produced.

41. A rock from an archaeological dig was found to contain 0.255 g of Pb-206 per gram of U-238. Assume that the rock did not contain any Pb-206 at the time of its formation and that U-238 decayed only to Pb-206. How old is the rock? (For U-238, $t_{1/2}$ = 4.5 × 10^9 y.)

42. A meteor has a Pb-206 : U-238 mass ratio of 0.813 : 1.00. Use the assumptions of Question 41 to estimate the age of the meteor.

18-4 Mass-Energy Relations

43. Plutonium-239 is used as the energy source for heart pacemakers and space probes. It decays by alpha emission.

(a) Calculate Δm in grams when one mole of Pu-239 decays.

(b) How much energy (in kilojoules) is given off by the decay of 2.00 mg of Pu-239?

44. Strontium-90 is a dangerous by-product of atomic testing because it mimics the action of calcium in the body. It decays in two β-emissions to give zirconium-90 (nuclear mass = 89.8824 g).

(a) Write a balanced nuclear equation for the overall decay of Sr-90.

(b) Calculate Δm in grams when one mole of Sr-90 decays to Zr-90.

(c) How much energy (in kJ) is given off by the decay of 6.50 mg of Sr-90?

45. For Be-10, calculate
(a) the mass defect.
(b) the binding energy.

46. For Al-28, calculate
(a) the mass defect.
(b) the binding energy in kJ/mol.

47. Which has the larger binding energy, F-19 or O-17?

48. Which has the larger binding energy, K-40 or Ca-40? Show by calculation.

18-5 Nuclear Fission

18-6 Nuclear Fusion

49. Consider the fusion of B-10 with an alpha particle. The products of the fusion are C-13 and a proton.

(a) Write a nuclear reaction for this process.

(b) How much energy is released when 1.00 g of B-10 is fused with an α-particle?

50. Show by calculation which process produces more energy per gram of material reacting.

fission of U-235: $_{92}^{235}$U + $_0^1 n \longrightarrow _{40}^{94}$Zr + $_{58}^{140}$Ce + 6 $_{-1}^0 e$ + 2 $_0^1 n$

fusion of deuterium: $_1^2$H + $_1^2$H $\longrightarrow _1^3$H + $_1^1$H

Nuclear masses for Ce-140 and Zr-94 are 139.8734 and 93.8841, respectively.

51. Consider the fission reaction in which U-235 is bombarded by neutrons. The products of the bombardment are Rb-89, Ce-144, beta particles, and more neutrons.

(a) Write a balanced nuclear equation for the bombardment.

(b) Calculate ΔE when one gram of U-235 undergoes fission.

(c) The detonation of TNT, an explosive, evolves 2.76 kJ/g. How many kilograms of TNT are required to produce the same amount of energy as one milligram of U-235?

52. Pentaerythritol tetranitrate ($C_5H_8N_4O_{12}$ [PETN]) was the explosive meant to be used by the "underwear bomber." Its combustion produces 6.47 kJ/g of energy. Use the reaction in Problem 51 to calculate the mass of PETN (in kg) required to produce the same amount of energy as that produced when one microgram (1 × 10^{-6} g) of U-235 undergoes fission.

Unclassified

53. How many disintegrations per second occur in a basement that is 40 × 40 × 10 feet if the radiation level from radon is the allowed 4 × 10^{-12} Ci/L?

54. Iodine-131 is used in the treatment of tumors in the thyroid gland. Its half-life is 8.1 days. Suppose that, due to a shipment delay, the I-131 in a hospital's pharmacy is 2.0 days old.

(a) What percentage of the I-131 has disintegrated?

(b) A patient is scheduled to receive 15.0 mg of I-131. What dosage (in milligrams) should the hospital pharmacist recommend for this patient if the 2-day-old bottle of I-131 is used?

55. An explosion of ammonium nitrate releases 37.0 kJ/mol of energy.

(a) How many tons of ammonium nitrate (1 ton = 2000 lbs) are required to release 1 million kJ of energy?

(b) How many tons of TNT ($\Delta E = -2.76$ kJ/g) are needed to release the same amount of energy?

(c) How many tons of U-235 are needed to obtain the same amount of energy?

56. Smoke detectors contain small amounts of americium-241. Am-241 decays by emitting α-particles and has a decay constant of 1.51×10^{-3} y^{-1}. If a smoke detector gives off ten disintegrations per second, how many grams of Am-241 are present in the detector?

57. The amount of oxygen dissolved in a sample of water can be determined by using thallium metal containing a small amount of the isotope Tl-204. When excess thallium is added to oxygen-containing water, the following reaction occurs.

$$2Tl(s) + \tfrac{1}{2}O_2(g) + H_2O \longrightarrow 2Tl^+(aq) + 2OH^-(aq)$$

After reaction, the activity of a 25.0-mL water sample is 745 counts per minute (cpm), caused by the presence of Tl$^+$-204 ions. The activity of Tl-204 is 5.53×10^5 cpm per gram of thallium metal. Assuming that O_2 is the limiting reactant in the above equation, calculate its concentration in moles per liter.

58. A 25-mL sample of 0.050 M AgNO$_3$ is mixed with 25 mL of 0.050 M NaI labeled with I-131. The NaI solution has an activity of 1.25×10^{10} counts/min·mL. The following reaction takes place:

$$Ag^+(aq) + I^-(aq) \longrightarrow AgI(s)$$

After the precipitate is removed by filtration, what is the activity of I$^-$ in the filtrate? The K_{sp} for AgI is 1×10^{-16}.

59. A 100.0-g sample of water containing tritium, ^3_1H, emits 2.89×10^3 beta particles per second. Tritium has a half-life of 12.3 years. What percentage of all the hydrogen atoms in the water sample is tritium?

60. Use the half-life of tritium given in Problem 59 to calculate the activity in curies of 1.00 mL of $^3_1H_2$ at STP.

61. One of the causes of the explosion at Chernobyl may have been the reaction between zirconium, which coated the fuel rods, and steam.

$$Zr(s) + 2H_2O(g) \longrightarrow ZrO_2(s) + 2H_2(g)$$

If half a metric ton of zirconium reacted, what pressure was exerted by the hydrogen gas produced at 55°C in the containment chamber, which had a volume of 2.0×10^4 L?

62. To measure the volume of the blood in an animal's circulatory system, the following experiment was performed. A 5.0-mL sample of an aqueous solution containing 1.7×10^5 counts per minute (cpm) of tritium was injected into the bloodstream. After an adequate period of time to allow for the complete circulation of the tritium, a 5.0-mL sample of blood was withdrawn and found to have 1.3×10^3 cpm on the scintillation counter. Assuming that only a negligible amount of tritium has decayed during the experiment, what is the volume of the animal's circulatory system?

63. Consider the fission reaction

$$^1_0n + ^{235}_{92}U \longrightarrow ^{89}_{37}Rb + ^{144}_{58}Ce + 3\,^0_{-1}e + 3\,^1_0n$$

How many liters of octane, C_8H_{18}, the primary component of gasoline, must be burned to $CO_2(g)$ and $H_2O(g)$ to produce as much energy as the fission of one gram of U-235 fuel? Octane has a density of 0.703 g/mL; its heat of formation is -249.9 kJ/mol.

64. A compound of Cr^{3+} and $C_2O_4^{2-}$ is made by a reaction that involves Na_2CrO_4 and oxalic acid, $H_2C_2O_4$, a reducing agent. The sodium chromate has an activity of 765 counts per minute per gram, from Cr-51. The oxalic acid has an activity of 512 counts per minute per gram. It is labeled with C-14. Because Cr-51 and C-14 emit different particles during decay, their activities can be counted independently. A sample of the compound was found to have a Cr-51 count of 314 cpm and a C-14 count of 235 cpm. How many oxalate ions are bound to one Cr^{3+} ion?

65. Radium-226 decays by alpha emission to radon-222. Suppose that 25.0% of the energy given off by one gram of radium is converted to electrical energy. What is the minimum mass of lithium that would be needed for the voltaic cell Li | Li$^+$ ‖ Cu^{2+} | Cu, at standard conditions, to produce the same amount of electrical work ($\Delta G°$)?

66. Polonium-210 decays to Pb-206 by alpha emission. Its half-life is 138 days. What volume of helium at 25°C and 1.20 atm would be obtained from a 25.00-g sample of Po-210 left to decay for 75 hours?

67. For how many years could all the energy needs of the world be supplied by the fission of U-235? Use the following assumptions:

- The world has about 1.0×10^7 metric tons of uranium ore, which are about 0.75% U-235.
- The energy consumption of the world is about 4.0×10^{15} kJ/y and does not change with time.
- The fission of U-235 releases about 8.0×10^7 kJ/g of U-235.

68. When a positron and an electron collide, they annihilate each other and produce two gamma photons, which carry the same amount of energy. What is the wavelength (in nanometers) of these photons?

Conceptual Questions

69. Classify the following statements as true or false. If false, correct the statement to make it true.

(a) The mass number increases in beta emission.

(b) A radioactive species with a large rate constant, k, decays very slowly.

(c) Fusion gives off less energy per gram of fuel than fission.

70. Explain how

(a) alpha and beta radiation are separated by an electric field.

(b) radioactive C-11 can be used as a tracer to study brain disorders.

(c) a self-sustaining chain reaction occurs in nuclear fission.

71. Suppose the $^{14}C/^{12}C$ ratio in plants a thousand years ago was 10% higher than it is today. What effect, if any, would this have on the calculated age of an artifact found by the C-14 method to be a thousand years old?

72. The following data are obtained for a radioactive isotope. Plot the data and determine the half-life of the isotope.

Time (h)	0.00	0.50	1.00	1.50	2.00	2.50
Activity (atoms/h)	14,472	13,095	11,731	10,615	9605	8504

73. Fill in the following table:

Event	Change in Atomic Number	Change in Mass Number
α-decay	_____	_____
β-emission	_____	_____
positron emission	_____	_____
K-electron capture	_____	_____
γ-decay	_____	_____

74. The principle behind the home smoke detector is described on page 472. Americium-241 is present in such detectors. It has a decay constant of 1.51×10^{-3} y^{-1}. You are urged to check the battery in the detector at least once a year. You are, however, never encouraged to check how much Am-241 remains undecayed. Explain why.

75. The cleavage of ATP (adenosine triphosphate) to ADP (adenosine diphosphate) and H_3PO_4 may be written as follows:

$$\text{adenosine}-O-\overset{\overset{\displaystyle O}{\|}}{\underset{\underset{\displaystyle OH}{|}}{P}}-O-\overset{\overset{\displaystyle O}{\|}}{\underset{\underset{\displaystyle OH}{|}}{P}}\overset{a}{\underset{b}{}}O-\overset{\overset{\displaystyle O}{\|}}{\underset{\underset{\displaystyle OH}{|}}{P}}-OH + HOH \longrightarrow$$

$$\text{adenosine}-O-\overset{\overset{\displaystyle O}{\|}}{\underset{\underset{\displaystyle OH}{|}}{P}}-O-\overset{\overset{\displaystyle O}{\|}}{\underset{\underset{\displaystyle OH}{|}}{P}}-OH + HO-\overset{\overset{\displaystyle O}{\|}}{\underset{\underset{\displaystyle OH}{|}}{P}}-OH$$

It is interesting to determine which bond (the P—O bond marked **a** or the O—P bond marked **b**) is cleaved by hydrolysis (reaction with water).

  (a) Outline an experiment (using radioactivity) that can be used to determine where the cleavage results.
  (b) Describe the results that would lead you to conclude that cleavage results at **a**.
  (c) Describe the results that would lead you to conclude that cleavage results at **b**.

Results show that the cleavage occurs at **b**.

Challenge Problems

76. An activity of 20 picocuries (20×10^{-12} Ci) of radon-222 per liter of air in a house constitutes a health hazard to anyone living there. The half-life of radon-222 is 3.82 days. Calculate the concentration of radon in air (moles per liter) that corresponds to a 20-picocurie activity level.

77. Plutonium-239 decays by the reaction

$$^{239}_{94}\text{Pu} \longrightarrow ^{235}_{92}\text{U} + ^{4}_{2}\text{He}$$

Its rate constant is 5.5×10^{-11}/min. In a one-gram sample of Pu-239,

  (a) how many grams decompose in 45 minutes?
  (b) how much energy in kilojoules is given off in 45 minutes?
  (c) what radiation dosage in rems (page 485) is received by a 75-kg man exposed to a gram of Pu-239 for 45 minutes?

78. It is possible to estimate the activation energy for fusion by calculating the energy required to bring two deuterons close enough to one another to form an alpha particle. This energy can be obtained by using Coulomb's law in the form $E = 8.99 \times 10^9 \, q_1q_2/r$, where q_1 and q_2 are the charges of the deuterons (1.60×10^{-19} C), r is the radius of the He nucleus, about 2×10^{-15} m, and E is the energy in joules.

  (a) Estimate E in joules per alpha particle.
  (b) Using the equation $E = mv^2/2$, estimate the velocity (meters per second) each deuteron must have if a collision between the two of them is to supply the activation energy for fusion (m is the mass of the deuteron in kilograms).

79. Consider the reaction

$$2\,^{2}_{1}\text{H} \longrightarrow ^{4}_{2}\text{He}$$

  (a) Calculate ΔE in kilojoules per gram of deuterium fused.
  (b) How much energy is potentially available from the fusion of all the deuterium in seawater? The percentage of deuterium in water is about 0.0017%. The total mass of water in the oceans is 1.3×10^{24} g.
  (c) What fraction of the deuterium in the oceans would have to be consumed to supply the annual energy requirements of the world (2.3×10^{17} kJ)?

80. Carbon-14 (C-14) with a half-life of 5730 years decays to nitrogen-14 (N-14). A sample of carbon dioxide containing carbon in the form of C-14 only is sealed in a vessel at 1.00-atmosphere pressure. Over time, the CO_2 becomes NO_2 through the radioactive decay process. The following equilibrium is established:

$$2NO_2(g) \rightleftharpoons N_2O_4(g)$$

If the equilibrium constant for the equation as written is 1×10^{-5}, what is the pressure of N_2O_4 after 20,000 years? Assume that the CO_2 does not participate in any chemical reactions.

81. Uranium in water decays to form Zn^{2+} and Sm by fission. Uranium has a half-life of 7×10^8 years. The zinc ions react with water and act as a weak acid according to the following equation:

$$Zn(H_2O)_4{}^{2+}(aq) \rightleftharpoons H^+(aq) + Zn(H_2O)_3(OH)^+(aq) \quad K_a = 3.3 \times 10^{-10}$$

Assuming that this is the only radioactive decay of any significance and that neither Sm nor U acts as a weak acid, how many years will it take a 1.00 M solution of U to achieve a pH of 6.0?

82. To determine the K_{sp} value of Hg_2I_2, a solid sample is used, in which some of the iodine is present as radioactive I-131. The count rate of the sample is 5.0×10^{11} counts per minute per mole of I. An excess amount of $Hg_2I_2(s)$ is placed in some water, and the solid is allowed to come to equilibrium with its respective ions. A 150.0-mL sample of the saturated solution is withdrawn and the radioactivity measured at 33 counts per minute. From this information, calculate the K_{sp} value for Hg_2I_2.

Complex Ions

When a tetrahedron is reflected or rotated, a structure indistinguishable from the original is obtained. Chemists call the tetrahedron a *symmetric* structure. Escher, in his painting *Symmetry No. 20*, shows a translational repeating symmetry. He flips or moves his objects, creating a pattern.

Copper's red 'cause d electrons
Preferentially get the blues
And absorption 'cross the band
 gaps
Gives the pigments gorgeous hues

—DR. JAMES D. LIVINGSTON
 Physics at Its Best

In previous chapters we have referred from time to time to compounds of the transition metals. Many of these have relatively simple formulas such as $CuSO_4$, $CrCl_3$, and $Fe(NO_3)_3$. These compounds are ionic. The transition metal is present as a simple cation (Cu^{2+}, Cr^{3+}, Fe^{3+}). In that sense, they resemble the ionic compounds formed by the main-group metals, such as $CaSO_4$ and $Al(NO_3)_3$.

It has been known for more than a century, however, that transition metals also form a variety of ionic compounds with more complex formulas such as

$$[Cu(NH_3)_4]SO_4 \qquad [Cr(NH_3)_6]Cl_3 \qquad K_3[Fe(CN)_6]$$

In these so-called *coordination compounds,* the transition metal is present as a complex ion, enclosed within the brackets. In the three compounds listed above, the following complex ions are present:

$$Cu(NH_3)_4{}^{2+} \qquad Cr(NH_3)_6{}^{3+} \qquad Fe(CN)_6{}^{3-}$$

The charges of these complex ions are balanced by those of simple anions or cations (e.g., $SO_4{}^{2-}$, $3Cl^-$, $3K^+$).

This chapter is devoted to complex ions and the important role they play in inorganic chemistry. We consider in turn

- the composition and names of complex ions and the coordination compounds that they form (Sections 19-1 and 19-2).
- the geometry of complex ions (Section 19-3).
- the electronic structure of the central metal ion (or atom) in a complex (Section 19-4).

19-1   Composition of Complex Ions

When ammonia is added to an aqueous solution of a copper(II) salt, a deep, almost opaque, blue color develops (Figure 19.1). This color is due to the formation of the $Cu(NH_3)_4^{2+}$ ion, in which four NH_3 molecules are bonded to a central Cu^{2+} ion. The formation of this species can be represented by the equation

The nitrogen atom of each NH_3 molecule contributes a pair of unshared electrons to form a covalent bond with the Cu^{2+} ion. This bond and others like it, where both electrons are contributed by the same atom, are referred to as *coordinate covalent bonds*.

The $Cu(NH_3)_4^{2+}$ ion is commonly referred to as a **complex ion**. We use the term complex ion to indicate a charged species in which a metal atom is bonded to neutral molecules and/or anions referred to collectively as **ligands**. The number of bonds formed by the **central atom** is called its **coordination number**. In the $Cu(NH_3)_4^{2+}$ complex ion

- the central atom is Cu^{2+}.
- the ligands are NH_3 molecules.
- the coordination number is 4.

Species such as $Al(H_2O)_6^{3+}$ and $Zn(H_2O)_3(OH)^+$, found in previous chapters, are further examples of complex ions. The metals that show the greatest tendency to form complex ions are those that form small cations with a charge of +2 or greater. Typically, these are the metals toward the right of the transition series (in the first transition series, $_{24}Cr$ through $_{30}Zn$). Nontransition metals, including Al, Sn, and Pb, form a more limited number of stable complex ions.

Cations of these metals invariably exist in aqueous solution as complex ions. Consider, for example, the zinc(II) cation. In a water solution of $Zn(NO_3)_2$, the $Zn(H_2O)_4^{2+}$ ion is present. Treatment with ammonia converts this to $Zn(NH_3)_4^{2+}$; addition of sodium hydroxide forms $Zn(OH)_4^{2-}$.

An ion such as $Cu(NH_3)_4^{2+}$ cannot exist by itself in the solid state. The +2 charge of this ion must be balanced by anions with a total charge of −2. A typical compound containing the $Cu(NH_3)_4^{2+}$ ion is

$[Cu(NH_3)_4^{2+}]Cl_2$:   1 $Cu(NH_3)_4^{2+}$ ion, 2 Cl^- ions

Compounds such as the one above, which contain a complex ion, are referred to as **coordination compounds**. The formula of the complex ion is set off by brackets, [ ], to make the structure of the compound clear.

The hydrated copper complex ion $[Cu(H_2O)_4]^{2+}$ is light blue.

When ammonia is added to 0.2 M Cu^{2+}, the $[Cu(NH_3)_4]^{2+}$ complex ion forms. The ammonia-containing ion is an intense deep blue, almost violet.

© Cengage Learning/Charles D. Winters

Figure 19.1 Colors of copper complexes.

Table 19.1 Complexes of Pt^{2+} with NH_3 and Cl^-

Complex	O.N. of Pt	Ligands	Total Charge of Ligands	Charge of Complex
$Pt(NH_3)_4^{2+}$	+2	$4NH_3$	0	+2
$Pt(NH_3)_3Cl^+$	+2	$3NH_3, 1Cl^-$	−1	+1
$Pt(NH_3)_2Cl_2$	+2	$2NH_3, 2Cl^-$	−2	0
$Pt(NH_3)Cl_3^-$	+2	$1NH_3, 3Cl^-$	−3	−1
$PtCl_4^{2-}$	+2	$4Cl^-$	−4	−2

Charges of Complexes

The charge of a complex is readily determined by applying a simple principle (Figure 19.2):

charge of complex = O.N. central metal + charges of ligands

("O.N." in the above is the abbreviation for "oxidation number.") The application of this principle is shown in Table 19.1, where we list the formulas of several complexes formed by platinum(II), which shows a coordination number of 4. Notice that one of the species, $Pt(NH_3)_2Cl_2$, is a neutral complex rather than a complex ion; the charges of the two Cl^- ions just cancel that of the central Pt^{2+} ion.

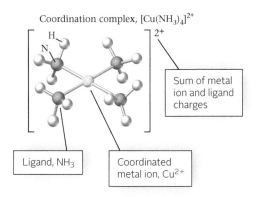

Coordination complex, $[Cu(NH_3)_4]^{2+}$

Sum of metal ion and ligand charges

Ligand, NH_3

Coordinated metal ion, Cu^{2+}

Figure 19.2 The $[Cu(NH_3)_4]^{2+}$ complex ion.

EXAMPLE 19.1

Consider the $Co(H_2O)_3Cl_3^-$ ion.

a What is the oxidation number of cobalt?

STRATEGY

Apply the relation

charge = oxidation number (O.N.) of the central atom + charge of ligands

SOLUTION

Charge of the complex ion	−1
Charge of the ligands	$H_2O = 0$; Cl: $3(-1) = -3$
Oxidation number of the metal	charge = O.N. + charge of ligands $-1 = $ O.N. $+ [0 + 3(-1)]$; O.N. = $+2$

b What is the formula of the coordination compound containing this anion and the Na^+ cation? The Ca^{2+} cation?

STRATEGY

1. You know the charge of the complex ion (−1) and the charge of the cations (+1 for Na; +2 for Ca). Apply the principle of electrical neutrality to write the formula of the coordination compound.

2. Use square brackets to enclose the formula of the complex ion.

SOLUTION

Compound with the Na^+ cation	$Na^+ \ Co(H_2O)_3Cl_3^{-1} \longrightarrow Na[Co(H_2O)_3Cl_3]$
Compound with the Ca^{2+} cation	$Ca^{2+} \ Co(H_2O)_3Cl_3^{-1} \longrightarrow Ca[Co(H_2O)_3Cl_3]_2$

Ligands; Chelating Agents

Unshared pair = lone pair.

In principle, any molecule or anion with an unshared pair of electrons ◂ⓘ can act as a Lewis base (see Chapter 13). In other words, it can donate a lone pair to a metal cation to form a coordinate covalent bond. In practice, a ligand usually contains an atom of one of the more electronegative elements (C, N, O, S, F, Cl, Br, I). Several hundred different ligands are known. Those most commonly encountered in general chemistry are NH_3 and H_2O molecules and CN^-, Cl^-, and OH^- ions.

Some ligands have more than one atom with an unshared pair of electrons and hence can form more than one bond with a central metal atom. Ligands of this type are referred to as *chelating agents;* the complexes formed are referred to as **chelates** (from the Greek *chela,* crab's claw). Two of the most common chelating agents are the oxalate anion (abbreviated *ox*) and the ethylenediamine molecule (abbreviated *en*), whose Lewis structures are

oxalate ion, $C_2O_4^{2-}$ ethylenediamine molecule $H_2N—(CH_2)_2—NH_2$

(The atoms that form bonds with the central metal are shown in color.)

Figure 19.3 shows the structure of the chelates formed by copper(II) with these ligands. Notice that in both of these complex ions, the coordination number of copper(II) is 4. The central cation is bonded to four atoms, two from each ligand.

For a ligand to act as a chelating agent, it must have at least two lone pairs of electrons. Moreover, these electron pairs must be far enough removed from one another to give a chelate ring with a stable geometry. In the chelates shown in Figure 19.3, the ring is five-membered, containing a Cu^{2+} ion and four nonmetal atoms. Most chelates contain five- or six-membered rings. Smaller or larger rings are less stable. For example, even though the hydrazine molecule

$$H—\ddot{N}—\ddot{N}—H$$
$$\quad\ \ |\quad |$$
$$\quad\ \ H\ \ H$$

has two lone pairs, it does not form chelates. To do so, it would have to form a three-membered ring with 60° bond angles.

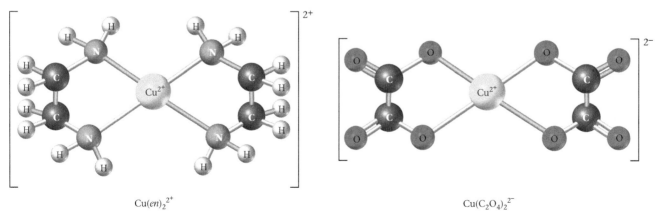

Cu(en)₂²⁺   Cu(C₂O₄)₂²⁻

Figure 19.3 **Chelates formed by Cu^{2+} with the ethylenediamine molecules (*en*) and the oxalate ions ($C_2O_4^{2-}$).**

Coordination Number

As shown in Table 19.2, the most common coordination number is 6. A coordination number of 4 is less common. A value of 2 is restricted largely to Cu^+, Ag^+, and Au^+. ◄ ⓘ

A few cations show only one coordination number in their complexes. Thus Co^{3+} always shows a coordination number of 6, as in

$$Co(NH_3)_6^{3+} \qquad Co(NH_3)_4Cl_2^+ \qquad Co(en)_3^{3+}$$

Other cations, such as Al^{3+} and Ni^{2+}, have variable coordination numbers, depending on the nature of the ligand.

ⓘ Odd coordination numbers are rare.

Table 19.2 Coordination Number and Geometry of Complex Ions*

Metal Ion	Coordination Number	Geometry	Example
Ag^+, Au^+, Cu^+	2	Linear	$Ag(NH_3)_2^+$
Cu^{2+}, Ni^{2+}, **Pd^{2+}**, **Pt^{2+}**	4	Square planar	$Pt(NH_3)_4^{2+}$
Al^{3+}, Au^+, **Cd^{2+}**, Co^{2+}, Cu^+, Ni^{2+}, **Zn^{2+}**	4	Tetrahedral	$Zn(NH_3)_4^{2+}$
Al^{3+}, Co^{2+}, **Co^{3+}**, **Cr^{3+}**, Cu^{2+}, **Fe^{2+}**, Fe^{3+}, Ni^{2+}, **Pt^{4+}**	6	Octahedral	$Co(NH_3)_6^{3+}$

*Cations that commonly show only one coordination number are in bold type.

EXAMPLE 19.2

Write formulas (including charges) of all complex ions formed by cobalt(III) with the hydroxide ion and/or ethylenediamine (*en*) molecules as ligands.

STRATEGY

1. See Table 19.2 for the coordination number(s) of Co^{3+}.

2. Recall that coordination number refers to the number of bonds made with the central (metal) atom.

3. Mix *en* molecules and OH^- ions as ligands to arrive at the maximum number of bonds specified by the coordination number.

4. Determine the charge of the complex ion by the relation

   charge = O.N. + charge of ligands

SOLUTION

1. Coordination number	6 = 6 bonds needed
2. Ligands	3 *en* (6 bonds) + 0 OH^- (0 bonds) 2 *en* (4 bonds) + 2 OH^- (2 bonds) 1 *en* (2 bonds) + 4 OH^- (4 bonds) 0 *en* (0 bonds) + 6 OH^- (6 bonds)
3–4. Complex ion with charge	$Co(en)_3$: charge = 3 + 3(0) = 3 $\longrightarrow$ $Co(en)_3^{3+}$ $Co(en)_2(OH)_2$: charge = 3 + 2(0) + 2(−1) = 1 $\longrightarrow$ $Co(en)_2(OH)_2^+$ $Co(en)(OH)_4$: charge = 3 + 1(0) + 4(−1) = −1 $\longrightarrow$ $Co(en)(OH)_4^-$ $Co(OH)_6$: charge = 3 + 6(−1) = −3 $\longrightarrow$ $Co(OH)_6^{3-}$

END POINT

Notice that moving from left to right, one *en* replaces two OH^- ions. This makes sense: an *en* molecule has two lone pairs that can bond to the metal; OH^- has only one.

19-2 Naming Complex Ions and Coordination Compounds

The nomenclature of compounds containing complex ions is more involved than that of the simple inorganic compounds considered in earlier chapters. We will see first how complex ions are named and then look at the nomenclature of coordination compounds.

Complex Cations and Neutral Complexes

To name a species like $Cu(NH_3)_4^{2+}$ or $Zn(H_2O)_2(OH)_2$, we need to specify

- the number and identity of each ligand (4 NH_3 molecules; 2 H_2O molecules, 2 OH^- ions).
- the oxidation number of the central metal atom: copper(II), zinc(II).

To accomplish this, we follow a set of rules.

1. *The names of anions that act as ligands are obtained by substituting* -o *for the normal ending.* Examples include

Cl^-	chloro	SO_4^{2-}	sulfato
OH^-	hydroxo	NO_3^-	nitrato
CN^-	cyano	CO_3^{2-}	carbonato

   Ordinarily, the names of molecules are not changed when they become ligands (e.g., ethylenediamine). There are three important exceptions:

H_2O	aqua	NH_3	ammine	CO	carbonyl

2. *The number of ligands of a particular type is ordinarily indicated by the Greek prefixes* di, tri, tetra, penta, hexa. Thus we have

$Cu(H_2O)_4^{2+}$	*tetra*aquacopper(II)
$Cr(NH_3)_6^{3+}$	*hexa*amminechromium(III)

   If the name of the ligand is itself complex (e.g., ethylenediamine), the number of such ligands is indicated by the prefixes *bis, tris, tetrakis, pentakis, hexakis*. The name of the ligand is enclosed in parentheses.

$Cr(en)_3^{3+}$	*tris*(ethylenediamine)chromium(III)

3. *When several types of ligand are present, they are named in alphabetical order (without regard to prefixes):*

$Cu(NH_3)_2(H_2O)_2^{2+}$	diamminediaquacopper(II)	ammine comes before aqua
$Cr(NH_3)_5Cl^{2+}$	pentaamminechlorochromium(III)	ammine comes before chloro

4. As you can deduce from the preceding examples, *the oxidation number of the central metal ion is indicated by a Roman numeral written at the end of the metal's name.*

Complex Anions

If the complex ion is an anion, such as $Zn(OH)_4^{2-}$, *this is indicated by inserting the suffix* ate *after the name of the metal.*

$Zn(OH)_4^{2-}$	tetrahydroxozinc*ate*(II)

In a few cases, the Latin name of the metal is used in anionic complexes.

Fe(CN)$_6$$^{3-}$ hexacyano*ferrate*(III)

Cu(Cl)$_4$$^{2-}$ tetrachloro*cuprate*(II)

Coordination Compounds

Coordination compounds, like simple ionic compounds, are named in a straightforward way. *The cation is named first, followed by the anion.*

[Cu(NH$_3$)$_4$]Cl$_2$ tetraamminecopper(II) chloride

[Fe(*en*)$_3$]PO$_4$ tris(ethylenediamine)iron(III) phosphate

K$_3$[Fe(CN)$_6$]   potassium hexacyanoferrate(III)

The common name of K$_3$[Fe(CN)$_6$] is potassium ferricyanide.

EXAMPLE 19.3

Name the following compounds:

[Cr(NH$_3$)$_4$Cl$_2$]Cl, K$_2$[PtCl$_4$], and [Co(*en*)$_3$](NO$_3$)$_3$

STRATEGY

1. Identify the complex ion (with the help of the brackets) and determine its charge.
2. Find the oxidation number of the metal atom.
3. Write the names of the ligands with their prefixes and assemble in alphabetical order.
4. Identify the anion and cation of the compound. Add the prefix *ate* if the complex ion is the anion.
5. Put all the parts together.

SOLUTION

[Cr(NH$_3$)$_4$Cl$_2$]Cl

1. Complex ion and charge	ion: Cr(NH$_3$)$_4$Cl$_2$ charge = +1
2. Oxidation number of Cr	+1 = O.N. + 4(0) + 2(−1); O.N. = +3
3. Ligands with prefix alphabetically	(NH$_3$)$_4$ = tetraammine; Cl$_2$ = dichloro tetraamminedichloro
4. Ions	cation: complex ion; anion: chloride
5. Name of compound	tetraamminedichlorochromium(III) chloride

K$_2$[PtCl$_4$]

1. Complex ion and charge	ion: PtCl$_4$ charge = −2 since there are two K$^+$ ions
2. Oxidation number of Pt	−2 = O.N. + 4(−1); O.N. = +2
3. Ligands with prefix	Cl$_4$ = tetrachloro
4. Ions	cation: potassium; anion: complex ion (end in *ate*)
5. Name of compound	potassium tetrachloroplatinate(II)

[Co(*en*)$_3$](NO$_3$)$_3$

1. Complex ion and charge	ion: Co(*en*)$_3$ charge = +3 since there are three NO$_3$$^-$ ions
2. Oxidation number of Co	−3 = O.N. + 3(0); O.N. = +3
3. Ligands with prefix	(*en*)$_3$ = tris(ethylenediamine)
4. Ions	cation: complex ion; anion: nitrate
5. Name of compound	tris(ethylenediamine)cobalt(III) nitrate

19-3 Geometry of Complex Ions

The physical and chemical properties of complex ions and of the coordination compounds they form depend on the spatial orientation of ligands around the central metal atom. Here we consider the geometries associated with the coordination numbers 2, 4, and 6. With that background, we then examine the phenomenon of **geometric isomerism,** in which two or more complex ions have the same chemical formula but different properties because of their different geometries.

Coordination Number = 2

Complex ions in which the central metal forms only two bonds to ligands are *linear;* that is, the two bonds are directed at a 180° angle. The structures of $CuCl_2^-$, $Ag(NH_3)_2^+$, and $Au(CN)_2^-$ may be represented as

$$(Cl-Cu-Cl)^- \qquad \begin{bmatrix} HH \\ H-N-Ag-N-H \\ HH \end{bmatrix}^+ \qquad (N\equiv C-Au-C\equiv N)^-$$

Coordination Number = 4

Four-coordinate metal complexes may have either of two different geometries (Figure 19.4). The four bonds from the central metal may be directed toward the corners of a regular **tetrahedron**. This is what we would expect from the VSEPR model (recall Chapter 7). Two common tetrahedral complexes are $Zn(NH_3)_4^{2+}$ and $CoCl_4^{2-}$.

Square planar complexes, in which the four bonds are directed toward the corners of a square, are more common. Certain complexes of copper(II) and nickel(II) show this geometry; it is characteristic of the complexes of Pd^{2+} and Pt^{2+}, including $Pt(NH_3)_4^{2+}$.

Coordination Number = 6

We saw in Chapter 7 that octahedral geometry is characteristic of many molecules (e.g., SF_6) in which a central atom is surrounded by six other atoms. (Remember, an *octa*hedron has eight sides, which is irrelevant here; it has *six* corners, which is important.) All complex ions in which the coordination number is 6 are

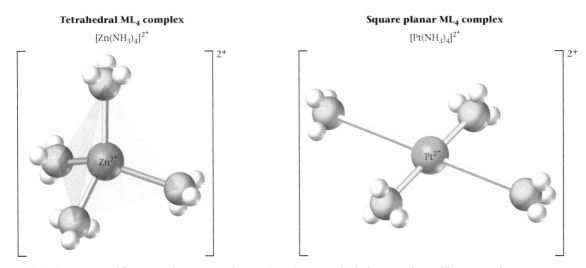

Tetrahedral ML₄ complex

$[Zn(NH_3)_4]^{2+}$

Square planar ML₄ complex

$[Pt(NH_3)_4]^{2+}$

Figure 19.4 Geometry of four-coordinate complexes. Complexes in which the central metal has a coordination number of 4 may be tetrahedral or square planar.

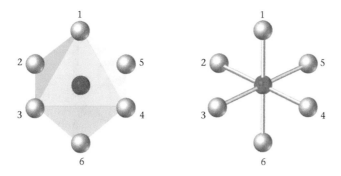

Figure 19.5 An octahedral, six-coordinate complex. The drawing at the left shows six ligands (represented by spheres) at the corners of an octahedron with a metal atom at the center. A simpler way to represent an octahedral complex is shown at the right.

octahedral. ◀ *i* The metal ion (or atom) is at the center of the **octahedron**; the six ligands are at the corners. This geometry is shown in Figure 19.5.

It is important to realize that

i The structure is often shown simply as ⟷.

- the six ligands can be considered equidistant from the central metal.
- an octahedral complex can be regarded as a derivative of a square planar complex. The two extra ligands are located above and below the square, on a line perpendicular to the square at its center.

Geometric Isomerism

Two or more species with different physical and chemical properties but the same formula are said to be **isomers** of one another. Complex ions can show many different kinds of isomerism, only one of which we will consider. *Geometric isomers* are ones that differ only in the spatial orientation of ligands around the central metal atom. Geometric isomerism is found in square planar and octahedral complexes. It cannot occur in tetrahedral complexes where all four positions are equivalent.

1. *Square planar.* There are two compounds with the formula $Pt(NH_3)_2Cl_2$, differing in water solubility, melting point, chemical behavior, and biological activity.* Their structures are

cis   *trans*

The complex in which the two chloride ions are at adjacent corners of the square, as close to one another as possible, is called the *cis* ◀ *i* isomer. It is made by reacting ammonia with the $PtCl_4^{2-}$ ion. In the *trans* ◀ *i* isomer, made by reacting $Pt(NH_3)_4^{2+}$ with HCl, the two chloride ions are at opposite corners of the square, as far apart as possible.

i *cis* = close together; *trans* = far apart.

Geometric isomerism can occur with any square complex of the type Mabcd, Ma_2bc, or Ma_2b_2, where M refers to the central metal and a, b, c, and d are different ligands. Conversely, geometric isomerism cannot occur with a square complex of the type Ma_4 or Ma_3b. Thus there are two different square complexes with the formula $Pt(NH_3)_2ClBr$ but only one with the formula $Pt(NH_3)_3Cl^+$.

cis   *trans*

*The *cis* isomer ("cisplatin") is an effective anticancer drug. This reflects the ability of the two Cl atoms to interact with the nitrogen atoms of DNA, a molecule responsible for cell reproduction. The *trans* isomer is ineffective in chemotherapy, presumably because the Cl atoms are too far apart to react with a DNA molecule.

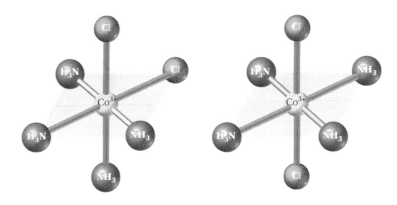

Figure 19.6 *Cis* and *trans* isomers of Co(NH$_3$)$_4$Cl$_2$$^+$.

2. Octahedral. To understand how geometric isomerism can arise in octahedral complexes, refer back to Figure 19.5. Notice that for any given position of a ligand, four other positions are at the same distance from that ligand, and a fifth is at a greater distance. ◂ ⓘ

> ⓘ It doesn't matter where you start; the octahedron is symmetrical.

Taking position 1 in Figure 19.5 as a point of reference, you can see that groups at 2, 3, 4, and 5 are equidistant from 1; 6 is farther away. In other words, positions 1 and 2, 1 and 3, 1 and 4, 1 and 5 are *cis* to one another; positions 1 and 6 are *trans*. Hence a complex ion like Co(NH$_3$)$_4$Cl$_2$$^+$ (Figure 19.6) can exist in two different isomeric forms. In the *cis* isomer, the two Cl$^-$ ions are at adjacent corners of the octahedron, as close together as possible. In the *trans* isomer they are at opposite corners, as far away from one another as possible.

EXAMPLE 19.4

How many geometric isomers are possible for the neutral complex [Co(NH$_3$)$_3$Cl$_3$]?

ISOMER I

STRATEGY

1. Create isomer I by putting two NH$_3$ molecules *trans* to each other.

2. Check to see how many different isomers you can obtain by placing the third NH$_3$ molecule at different corners of the octahedron.

3. Put the three chloride ions in the unoccupied corners.

SOLUTION

1. *trans* NH$_3$	NH$_3$ │ Co │ NH$_3$
2. Third NH$_3$	(a) or (b) or (c) or (d) It does not matter where the third NH$_3$ molecule (shaded in green) is. It will be *cis* to the NH$_3$ molecules already in place.
3. Cl ions	= Isomer I

continued

ISOMER II

STRATEGY AND SOLUTION

1. Create the second isomer by placing the two NH_3 molecules *cis* to each other.

$$\text{Co} \begin{array}{c} NH_3 \\ NH_3 \end{array}$$

2. Check to see how many different isomers you can obtain by placing the third NH_3 molecule at different corners of the octahedron. (Remember that the symmetry of a regular octagon requires that the distance across the diagonal of a square be the same as that from "top" to "bottom."

(a)	(b)	(c)	(d)

Structures (a) and (b) are identical to the first isomer created. There are two NH_3 molecules *cis* to each other and two NH_3 molecules *trans* to each other.

Structures (c) and (d) are identical to each other and different from isomer I. All three NH_3 molecules are *cis* to each other.

3. Choose either structure (c) or (d) and fill in the empty corners with Cl ions.

$$\begin{array}{c} Cl \\ Cl \\ Cl \end{array} \text{Co} \begin{array}{c} NH_3 \\ NH_3 \\ NH_3 \end{array} = \text{Isomer II}$$

Geometric isomerism can also occur in chelated octahedral complexes (Figure 19.7). Notice that an *ethylenediamine molecule,* here and indeed in all complexes, *can only bridge cis positions.* It is not long enough to connect to *trans* positions.

Figure 19.7 *Cis* and *trans* isomers of [Co(*en*)$_2$Cl$_2$]⁺. The ethylenediamine molecule is represented by —*en*—. It forms two bonds per molecule. The *cis* isomer is reddish-purple and the *trans* is dark green.

Figure 19.8 Brightly colored coordination compounds. Most coordination compounds are brilliantly colored, a property that can be explained by the crystal field model.

Marna G. Clarke

"Crystal field" isn't a very descriptive term, but it's part of the jargon. ⓘ

In transition metal cations, 3d is lower in energy than 4s. ⓘ

19-4   Electronic Structure of Complex Ions

Until about 20 years ago, the *valence bond model* discussed in Chapter 7 was widely used to explain electronic structure and bonding in complex ions. It assumed that lone pairs of electrons were contributed by ligands to form covalent bonds with metal atoms. This model had two major deficiencies. It could not easily explain the magnetic properties of complex ions. More important, it offered no explanation of the most striking characteristic of coordination compounds: their color (Figure 19.8).

We will discuss the **crystal field model** ◂ⓘ here. It assumes that the bonding between metal ions and ligands is essentially ionic. More specifically, it considers the effect of approaching ligands upon the energies of electronic levels in transition metal cations. We will apply this model to octahedral complexes. Before doing so, it may be helpful to review the electronic structure of uncomplexed transition metal cations, originally covered in Chapter 6.

Transition Metal Cations

Recall (pages 146–147) that in a simple transition metal cation

- there are no outer s electrons. Electrons beyond the preceding noble gas are located in an inner d sublevel (3d for the first transition series). ◂ⓘ
- electrons are distributed among the five d orbitals in accordance with Hund's rule, giving the maximum number of unpaired electrons.

To illustrate these rules, consider the Fe^{2+} ion. Because the atomic number of iron is 26, this +2 ion must contain $26 - 2 = 24e^-$. Of these electrons, the first 18 have the argon structure; the remaining six are located in the 3d sublevel. The abbreviated electron configuration is

$$Fe^{2+} \qquad [Ar]3d^6$$

CHEMISTRY THE **HUMAN** SIDE

The basic ideas concerning the structure and geometry of complex ions presented in this chapter were developed by one of the most gifted individuals in the history of inorganic chemistry, Alfred Werner. His theory of coordination chemistry was published in 1893 when Werner was 26 years old. In his paper Werner made the revolutionary suggestion that metal ions such as Co^{3+} could show two different kinds of valences. For the compound $Co(NH_3)_6Cl_3$, Werner postulated a central Co^{3+} ion joined by "primary valences" (ionic bonds) to three Cl^- ions and by "secondary valences" (coordinate covalent bonds) to six NH_3 molecules. Moreover, he made the inspired guess that the six secondary valences were directed toward the corners of a regular octahedron.

Werner spent the next 20 years obtaining experimental evidence to prove his theory. (At the University of Zürich, there remain several *thousand* samples of coordination compounds prepared by Werner and his students.) He was able to show, for example, that the electrical conductivities in water solution decreased in the order $[Co(NH_3)_6]Cl_3 > [Co(NH_3)_5Cl]Cl_2 > [Co(NH_3)_4Cl_2]Cl$ in much the same way as with simple salts, for example, $ScCl_3 > CaCl_2 > NaCl$. Another property

he studied was isomerism. In 1907, he was able to isolate a second geometric isomer of $[Co(NH_3)_4Cl_2]Cl$, in complete accord with his theory. Six years later, Werner won the Nobel Prize in Chemistry.

By all accounts, Werner was a superb lecturer. Sometimes as many as 300 students crowded into a hall with a capacity of 150 to hear him speak. So great was his reputation that students in theology and law came to hear him talk about chemistry. There was, however, a darker side to Werner that few students saw. A young woman badgered by Werner during an oral examination came to see him later to ask whether she had passed; he threw a chair at her. Werner died at age 52 of hardening of the arteries, perhaps caused in part by his addiction to alcohol and strong black cigars.

Alfred Werner
(1866–1919)

SPL/Science Source

These six electrons are spread over all five orbitals; the orbital diagram is

3d

$$Fe^{2+} \quad [Ar] \quad (\uparrow\downarrow)(\uparrow)(\uparrow)(\uparrow)(\uparrow)$$

The Fe^{2+} ion is *paramagnetic,* with four unpaired electrons.

EXAMPLE 19.5

Consider the Co^{3+} ion.

a Derive its abbreviated electron configuration.

STRATEGY AND SOLUTION

1. Find Z (number of protons = number of electrons for a neutral species) for Co in the periodic table.

   $Z = 27$

2. Write the abbreviated electron configuration.

   $[_{18}Ar]4s^23d^7$

3. Remove 3 electrons for the charged (+3) atom. Recall that electrons with the highest n are removed first.

   $[_{18}Ar]3d^6$

b How many unpaired electrons are there?

STRATEGY AND SOLUTION

1. Write the abbreviated orbital diagram following Hund's rule.

   $[_{18}Ar] (\uparrow\downarrow)(\uparrow)(\uparrow)(\uparrow)(\uparrow)$

2. Count the number of unpaired electrons.

   There are four unpaired electrons.

The shapes of the five d orbitals are shown in Figure 19.9. These orbitals are given the symbols

$$d_{z^2} \quad d_{x^2-y^2} \quad d_{xy} \quad d_{yz} \quad d_{xz}$$

In the uncomplexed transition metal cation, all of these orbitals have the same energy.

Octahedral Complexes

As six ligands approach a central metal ion to form an octahedral complex, they change the energies of electrons in the d orbitals. The effect (Figure 19.10) is to split the five d orbitals into two groups of different energy.

1. A higher energy pair, the $d_{x^2-y^2}$ and d_{z^2} orbitals
2. A lower energy trio, the d_{xy}, d_{yz}, and d_{xz} orbitals

Figure 19.9 The five d orbitals and their relation to ligands on the x-, y-, and z-axes.

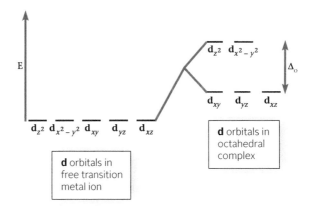

Figure 19.10 Change in the energy levels of d orbitals in complex formation. In a free transition metal ion, all five d orbitals have the same energy. When the ion forms an octahedral complex, two orbitals (d_{z^2}, $d_{x^2-y^2}$) have a higher energy than the other three (d_{xy}, d_{yz}, d_{xz}). The difference in energy between the two d orbital levels in the octahedral complex is known as the *crystal field splitting energy* and has the symbol Δ_o.

The difference in energy between the two groups is called the **crystal field splitting energy** and given the symbol Δ_o (the subscript "o" stands for "octahedral").

To see why this splitting occurs, consider what happens when six ligands (e.g., H_2O, CN^-, NH_3) approach a central metal cation along the *x*-, *y*-, and *z*-axes (Figure 19.9). The unshared electron pairs on these ligands repel the electrons in the d orbitals of the cation. The repulsion is greatest for the $d_{x^2-y^2}$ and d_{z^2} orbitals, which have their maximum electron density directly along the *x*-, *y*-, and *z*-axes, respectively. Electrons in the other three orbitals (d_{xy}, d_{yz}, d_{xz}) are less affected because their densities are concentrated between the axes rather than along them. ◀ⓘ

This gives the d_{xy}, d_{yz}, and d_{xz} orbitals a lower energy.

Depending on the magnitude of Δ_o, a cation may form either of two different kinds of complexes. This is illustrated for the Fe^{2+} ion in the two diagrams in Figure 19.11.

1. In the $Fe(CN)_6^{4-}$ ion shown at the left, Δ_o is so large that the six 3d electrons of the Fe^{2+} ion pair up in the lower energy orbitals. In this way, the Fe^{2+} ion achieves its most stable (lowest energy) electronic structure. The complex is diamagnetic, with no unpaired electrons.
2. In the $Fe(H_2O)_6^{2+}$ ion shown at the right, Δ_o is relatively small, and the six 3d electrons of the Fe^{2+} ion spread out over all the orbitals as expected by Hund's rule. In other words, the electron distribution in the complex is the same as in the Fe^{2+} ion itself (where Δ_o is 0!). The $Fe(H_2O)_6^{2+}$ ion is paramagnetic, with four unpaired electrons.

More generally, it is possible to distinguish between

■ **low-spin complexes,** containing the minimum number of unpaired electrons, formed when Δ_o is large; electrons are concentrated in the lower energy orbitals. ◀ⓘ

When Δ_o is small, the electron distribution is the same as in the simple cation; if Δ_o is large, Hund's rule is not strictly followed.

Figure 19.11 Electron distribution in low-spin and high-spin complexes of Fe^{2+}. Depending on the magnitude of Δ_o, either of two different complexes may be formed by Fe^{2+}.

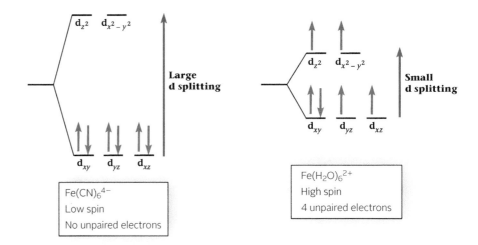

- **high-spin complexes,** containing the maximum number of unpaired electrons, formed when Δ_o is small; electrons are spread out among the d orbitals, exactly as in the uncomplexed transition metal cation.

For a given cation,

- the high-spin complex always contains more unpaired electrons than the low-spin complex and so is more strongly paramagnetic.
- the value of Δ_o is determined by the nature of the ligand. So-called "strong-field" ligands (e.g., CN^-), which interact strongly with d-orbital electrons, produce a large Δ_o. Conversely, "weak-field" ligands (e.g., H_2O) interact weakly with d-orbital electrons, producing a small Δ_o.

> When Δ_o is small, the electron distribution is the same as in the simple cation; if Δ_o is large, Hund's rule is not strictly followed.

> Remember:
> strong-field ligands → large Δ_o → low spin complex
> weak-field ligands → small Δ_o → high spin complex

EXAMPLE 19.6

Derive the electron distribution of the Fe^{3+} ion in the low-spin and high-spin octahedral complexes.

STRATEGY

1. Write the abbreviated electron configurations for Fe and then for Fe^{3+}.

2. For the high-spin complex, follow Hund's rule in distributing the electrons to *all* orbitals.

3. For the low-spin complex, follow Hund's rule in distributing the electrons first to the *lower* orbitals. If there are any electrons left, fill the upper orbitals, again following Hund's rule.

SOLUTION

1. Abbreviated electron configurations	Fe: $[_{18}Ar]4s^2 3d^6$; Fe^{3+}: $[_{18}Ar]3d^5$
2. High-spin complex	$(\uparrow\ )(\uparrow\ )$ $(\uparrow\ )(\uparrow\ )(\uparrow\ )$
3. Low-spin complex	$(\ \ )(\ \ )$ $(\uparrow\downarrow)(\uparrow\downarrow)(\uparrow\ )$

This model of the electronic structure of complex ions explains why high-spin and low-spin complexes occur only with ions that have four to seven electrons (d^4, d^5, d^6, d^7). With three or fewer electrons, only one distribution is possible; the same is true with eight or more electrons.

$$(\ \ )(\ \ ) \qquad (\uparrow)(\uparrow)$$
$$(\uparrow)(\uparrow)(\uparrow) \qquad (\uparrow\downarrow)(\uparrow\downarrow)(\uparrow\downarrow)$$
$$\text{3 electrons} \qquad\qquad \text{8 electrons}$$

> The color we see is what is left over after certain wavelengths are absorbed.

Color

Most coordination compounds are brightly colored except for those of cations whose d sublevels are completely empty (such as Sc^{3+}) or completely filled (such as Zn^{2+}). These colors are readily explained, at least qualitatively, by the model we have just described. The energy difference between two sets of d orbitals in a complex is ordinarily equal to that of a photon in the visible region. Hence by absorbing visible light, an electron may be able to move from the lower energy set of d orbitals to the higher one. This removes some of the component wavelengths of white light, so that the light reflected or transmitted by the complex is colored (Figure 19.12). The relation between absorbed and transmitted light can be deduced from the color wheel shown in Figure 19.13. For example, a complex that absorbs blue-green light appears orange, and vice versa.

© Cengage Learning/Charles D. Winters

Figure 19.12 Light absorption and color. The color of a solution is due to the color of the light *not* absorbed by the solution. Here a solution of Ni^{2+} ion in water absorbs red and blue light and so appears green.

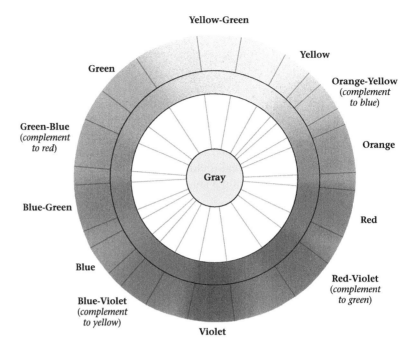

Figure 19.13 The color wheel. The color of a complex is complementary to (180° across from) the color of the light absorbed.

From the color (absorption spectrum) of a complex ion, it is sometimes possible to deduce the value of Δ_o, the crystal field splitting energy. The situation is particularly simple in $_{22}Ti^{3+}$, which contains only one 3d electron. Consider, for example, the $Ti(H_2O)_6{}^{3+}$ ion, which has an intense purple color. This ion absorbs at 510 nm, in the green region. The purple (red-violet) color of a solution of $Ti(H_2O)_6{}^{3+}$ is what is left over when the green component is subtracted from the visible spectrum. Using the equation relating energy to wavelength (Chapter 6), it is possible to calculate the energy of a photon with a wavelength of 510 nm:

$$E = \frac{hc}{\lambda} = \frac{(6.626 \times 10^{-34} \text{ J} \cdot \text{s})(2.998 \times 10^8 \text{ m/s})}{510 \times 10^{-9} \text{ m}} = 3.90 \times 10^{-19} \text{ J}$$

The energy in kilojoules per mole is

$$E = 3.90 \times 10^{-19} \text{ J} \times \frac{1 \text{ kJ}}{10^3 \text{ J}} \times \frac{6.022 \times 10^{23}}{1 \text{ mol}} = 2.35 \times 10^2 \text{ kJ/mol}$$

This is the energy absorbed in raising the 3d electron from a lower to a higher orbital. In other words, in the $Ti(H_2O)_6{}^{3+}$ ion, the two sets of d orbitals are separated by this amount of energy. The splitting energy, Δ_o, is 235 kJ/mol.

The smaller the value of Δ_o, the longer the wavelength of the light absorbed. This effect is shown in Table 19.3 and Figure 19.14. When weak-field ligands such

Table 19.3 Colors of Complex Ions of Co^{3+}

Complex	Color Observed	Color Absorbed	Approximate Wavelength (nm) Absorbed
$Co(NH_3)_6{}^{3+}$	Yellow	Violet	430
$Co(NH_3)_5NCS^{2+}$	Orange	Blue-green	470
$Co(NH_3)_5H_2O^{3+}$	Red	Green-blue	500
$Co(NH_3)_5Cl^{2+}$	Purple	Yellow-green	530
trans-$Co(NH_3)_4Cl_2{}^+$	Green	Red	680

Ray Boyington/S. Ruven Smith

Figure 19.14 A spectrochemical series. The five cations listed in Table 19.3 vary in color from yellow $[Co(NH_3)_6]Cl_3$ to green $[trans\text{-}Co(NH_3)_4Cl_2]Cl$.

This series is based on experiment, not theory.

as NCS^-, H_2O, or Cl^- are substituted for NH_3, the light absorbed shifts to longer wavelengths (lower energies). On the basis of observations like these, ligands can be arranged in order of decreasing tendency to split the d orbitals. A short version of such a *spectrochemical series* is

$$CN^- > NO_2^- > en > NH_3 > NCS^- > H_2O > F^- > Cl^-$$

| strong field | decreasing Δ_o | weak field |

CHEMISTRY **BEYOND** THE CLASSROOM

Chelates: Natural and Synthetic

Chelating agents, capable of bonding to metal atoms at more than one position, are abundant in nature. Certain species of soybeans synthesize and secrete organic chelating agents that extract iron from insoluble compounds in the soil, thereby making it available to the plant. Mosses and lichens growing on boulders use a similar process to obtain the metal ions they need for growth.

Many important natural products are chelates in which a central metal atom is bonded into a large organic molecule. In chlorophyll, the green coloring matter of plants, the central atom is magnesium; in the essential vitamin B_{12}, it is cobalt. The structure of heme, the pigment responsible for the red color of blood, is shown in Figure A. There is an

Figure A Structure of heme. In the hemoglobin molecule, the Fe^{2+} ion is at the center of an octahedron, surrounded by four nitrogen atoms, a globin molecule, and a water molecule.

Fe^{2+} ion at the center of the complex surrounded by four nitrogen atoms at the corners of a square. A fifth coordination position around the iron is occupied by the molecule globin, which in combination with heme, gives the protein referred to as hemoglobin.

The sixth coordination position of Fe^{2+} in heme is occupied by a water molecule that can be replaced reversibly by oxygen to give a derivative known as oxyhemoglobin, which has the bright red color characteristic of arterial blood.

$$\text{hemoglobin} + O_2 \rightleftharpoons \text{oxyhemoglobin} + H_2O$$

The position of this equilibrium is sensitive to the pressure of oxygen. In the lungs, where the blood is saturated with air ($P_{O_2} = 0.2$ atm), the hemoglobin is almost completely converted to oxyhemoglobin. In the tissues, the partial pressure of oxygen drops, and the oxyhemoglobin breaks down to release O_2 essential for metabolism. In this way, hemoglobin acts as an oxygen carrier, absorbing oxygen in the lungs and liberating it to the tissues.

Unfortunately, hemoglobin forms a complex with carbon monoxide that is considerably more stable than oxyhemoglobin. The equilibrium constant for the reaction

$$\text{hemoglobin} \cdot O_2(aq) + CO(g) \rightleftharpoons \text{hemoglobin} \cdot CO(aq) + O_2(g)$$

is about 200 at body temperature. Consequently, the carbon monoxide complex is formed preferentially in the lungs even at CO concentrations as low as one part per thousand. When this happens, the flow of oxygen to the tissues is cut off, resulting eventually in muscular paralysis and death.

Figure B Structure of the EDTA ion; the chelating atoms, six in all, have color screens.

Many chelating agents have been synthesized in the laboratory to take advantage of their ability to tie up *(sequester)* metal cations. Perhaps the best known synthetic chelating agent is the ethylenediaminetetraacetate (EDTA) ion shown in Figure B. EDTA can bond at as many as six positions, thereby occupying all the octahedral sites surrounding a central metal cation.

EDTA forms 1:1 complexes with a large number of cations, including those of some of the main-group metals. The complex formed by calcium with EDTA is used to treat lead poisoning. When a solution containing the Ca-EDTA complex is given by injection, the calcium is displaced by lead. The more stable Pb-EDTA complex is eliminated in the urine. EDTA has also been used to remove radioactive isotopes of metals, notably plutonium, from body tissues.

You may have noticed Ca-EDTA on the list of ingredients of many prepared foods, ranging from beer to mayonnaise. EDTA acts as a scavenger to pick up traces of metal ions that catalyze the chemical reactions responsible for flavor deterioration, loss of color, or rancidity. Typically, Ca-EDTA is added at a level of 30 to 800 ppm.

Still another use of EDTA is in the field of art restoration. Figure C shows "before and after" photographs of Rodin's bronze sculpture "The Thinker." This stood in front of the Rodin Museum in Philadelphia for about 60 years prior to its restoration in 1992. Over time the sculpture had picked up a thick coating of insoluble $CuSO_4 \cdot 3Cu(OH)_2(s)$. Application of a wet paste containing EDTA dissolved the coating by forming the soluble $Cu(EDTA)^{2-}$ complex.

$$CuSO_4 \cdot 3Cu(OH)_2(s) + 4EDTA^{4-}(aq) \longrightarrow$$
$$4Cu(EDTA)^{2-}(aq) + SO_4^{2-}(aq) + 6OH^-(aq)$$

Figure C *The Thinker* by Rodin, before (*left*) and after (*right*).

Key Concepts

1. Relate the composition of a complex ion to its charge, coordination number, and the oxidation number of the central metal.
   (**Examples 19.1 and 19.2; Problems 1–12**)

2. Write the name of complex ions and coordination compounds.
   (**Example 19.3; Problems 17–22**)

3. Sketch the geometry of a complex ion and identify geometric isomers.
(**Example 19.4; Problems 23–30**)

4. Give the electron configuration and/or orbital diagram of a transition metal cation.
(**Example 19.5; Problems 31–34**)

5. Derive orbital diagrams for high-spin and low-spin complexes.
(**Example 19.6; Problems 35–46**)

Key Terms

central atom	coordination number	isomers	tetrahedron
chelates	crystal field model	ligands	*trans* isomer
cis isomer	crystal field splitting energy	low-spin complexes	
complex ion	geometric isomerism	octahedron	
coordination compounds	high-spin complexes	square planar	

Summary Problem

Consider the complex $[Pt(NH_3)_2(CN)_4]$.

(a) Identify the ligands and their charges.

(b) What is the oxidation number of platinum?

(c) Suppose the ammonia molecules were replaced by OH^- ions. Write the formula of the resulting complex.

(d) What is the coordination number of platinum in the compound?

(e) Describe the geometry of the complex.

(f) How many possible geometric isomers are there?

(g) Write the abbreviated electron configuration for the platinum(IV) ion.

(h) Cyanide ligands are ligands with the strongest crystal field (large Δ_o). Show the electron distribution for plati-

num in this compound. Is this a high-spin or low-spin complex?

(i) Is the complex paramagnetic or diamagnetic?

Answers

(a) $NH_3 = 0$; $CN^- = -1$
(b) $+4$
(c) $[Pt(OH)_2(CN)_4{}^{2-}]$
(d) 6
(e) octahedral
(f) two
(g) $[Xe]\ 4f^{14}5d^6$
(h) **low spin**   ( )( )
    (↑↓)(↑↓) (↑↓)
(i) **diamagnetic**

Questions and Problems

Even-numbered questions and Challenge Problems have answers in Appendix 5 and fully worked solutions in the *Student Solutions Manual*.

19-1 Composition of Complex Ions

1. Consider the complex ion $[Fe(C_2O_4)_2(NH_3)Cl^{3-}]$.
   (a) Identify the ligands and their charges.
   (b) What is the oxidation number of nickel?
   (c) What is the formula for the potassium salt of this ion?

2. Consider the complex ion $[Cd(en)(SCN)_2(OH)_2{}^-]$.
   (a) Identify the ligands and their charges.
   (b) What is the oxidation number of cadmium?
   (c) What is the formula for the magnesium salt of this ion?

3. Nickel(III) forms many complexes, among them those with the following ligands. Give the formula and charge of each nickel complex described below.
   (a) four carbonyl molecules (CO) and two chloride ions
   (b) two ethylenediamine molecules

   (c) two thiocyanate molecules (SCN^-), two hydroxide ions, and two ammonia molecules

4. Platinum(II) forms many complexes, among them those with the following ligands. Give the formula and charge of each complex.
   (a) two ammonia molecules and one oxalate ion ($C_2O_4{}^{2-}$)
   (b) two ammonia molecules, one thiocyanate ion (SCN^-), and one bromide ion
   (c) one ethylenediamine molecule and two nitrite ions

5. What is the coordination number of the metal in the following complexes?
   (a) $[Co(en)_2(SCN)Cl^+]$ (b) $[Ni(CN)_4{}^{2-}]$
   (c) $[Zn(C_2O_4)_2{}^{2-}]$ (d) $[Ag(NH_3)Cl]$

6. What is the coordination number of the central metal atom in the following complexes?
   (a) $[Fe(H_2O)_6{}^{3+}]$ (b) $[Pt(NH_3)Br_3{}^-]$
   (c) $[V(en)Cl_4{}^{2-}]$ (d) $[Au(CN)_2{}^+]$

7. What is the oxidation number of the metal atom in the complex ions in Question 5?

8. What is the oxidation number of the metal atom in the complex ions in Question 6?

9. Write the formula for the carbonate salt (if the complex ion is a cation) or the barium salt (if the complex ion is an anion) for the ions in Question 5.

10. Follow the directions for Question 9 for the complex ions in Question 6.

11. Refer to Table 19.2 to predict the formula of the complex formed by
  (a) Cu^+ with NH_3
  (b) Pt^{2+} with the oxalate ion
  (c) Cadmium(II) with water molecules
  (d) Fe^{3+} with CN^-

12. Refer to Table 19.2 to predict the formula of the complex formed by
  (a) Pt^{4+} with NH_3 (b) Ag^+ with CN^-
  (c) Zn^{2+} with $C_2O_4^{2-}$ (d) Cd^{2+} with CN^-

13. What is the mass percent of sulfur in the $[Ag(SCN)_2^-]$ complex ion?

14. What is the mass percent of Cl in the sulfate salt of $[Ni(H_2O)_4Cl_2^+]$?

15. There are four iron atoms in each hemoglobin molecule. The mass percent of iron in a hemoglobin molecule is 0.35%. Estimate the molar mass of hemoglobin.

16. Vitamin B_{12} is a coordination compound with cobalt as its central atom. It contains 4.4% cobalt by mass and has a molar mass of 1.3×10^3 g/mol. How many cobalt atoms are in a molecule of vitamin B_{12}?

19-2 Naming Complex Ions and Coordination Compounds

17. Write formulas for the following ions or compounds:
  (a) zinchexachloroplatinate(IV)
  (b) dichlorobis(ethylenediamine)nickel(II)
  (c) diamminetriaquahydroxochromium(III) nitrate
  (d) ammonium pentachlorohydroxoferrate(III)

18. Write formulas for the following ions or compounds.
  (a) diamminetetraquachromium(II) chloride
  (b) tetrachlorosulfatocadmium(III)
  (c) sodium tetrahydroxonickelate(II)
  (d) dibromo*bis*(ethylenediamine)iron(III)

19. Name the following ions or compounds:
  (a) $[Ru(NH_3)_5Cl^{2+}]$ (b) $[Mn(NH_2CH_2CH_2NH_2)_3^{2+}]$
  (c) $K_2[PtCl_4]$ (d) $[Cr(NH_3)_5I]I_2$

20. Name the following ions or compounds:
  (a) $Na[Al(OH)_4]$ (b) $[Co(C_2O_4)_2(H_2O_2)^-]$
  (c) $[Ir(NH_3)_3Cl_3]$ (d) $[Cr(en)(NH_3)_2Br_2]_2SO_4$

21. Name the compounds or ions in Question 5.

22. Name the compounds or ions in Question 6.

19-3 Geometry of Complex Ions

23. Sketch the geometry of
  (a) $[Zn(NH_3)_2Cl_2]$ (tetrahedral)
  (b) *cis*-$[Co(H_2O)_4Cl_2]^+$
  (c) *trans*-$[Pt(NH_3)_2Br_2]^{2+}$
  (d) *trans*-$[Ni(ox)_2(OH)_2]^{3-}$
  (e) $[Au(CN)Br]^+$

24. Sketch the geometry of
  (a) *trans*-$[Cu(Br)_2(H_2O)_4]$
  (b) $[Zn(OH)_2I_2^{2-}]$ (tetrahedral)

  (c) $[Pt(NH_3)_4^{2+}]$ (square planar)
  (d) *cis*-$[Ni(H_2O)_2(C_2O_4)_2^{2-}]$
  (e) $[Au(CN)(NH_3)]$

25. The compound 1,2-diaminocyclohexane

(abbreviated "dech") is a ligand in the promising anticancer complex *cis*-$Pd(H_2O)_2(dech)^{2+}$. Sketch the geometry of this complex.

26. The acetylacetonate ion (acac$^-$)

forms complexes with many metal ions. Sketch the geometry of $Fe(acac)_3$.

27. Which of the following octahedral complexes show geometric isomerism? If geometric isomers are possible, draw their structures.
  (a) $[Co(en)Cl_4^-]$
  (b) $[Ni(C_2O_4)_2ClBr^{4-}]$
  (c) $[Cd(NH_3)_2Cl_4^{2-}]$

28. Follow the directions of Question 27 for the following:
  (a) $[Mn(H_2O)_2(NO_2)_4^{2-}]$
  (b) $[Pt(NH_3)_3(Cl_3)^+]$
  (c) $[Al(C_2O_4)(CO)_2(Br)_2^-]$

29. Draw all the structural formulas for the octahedral complexes of Co^{3+} with only *ox* and/or NH_3 as ligands.

30. Draw all the structural formulas for the octahedral complexes of $[Ni(H_2O)_3(OH)_2Cl]$.

19-4 Electronic Structure of Complex Ions

Electronic Structure of Metal Ions

31. Give the abbreviated electron configuration of the following metal ions:
  (a) Cr^{3+} (b) Pd^{2+} (c) V^{3+}
  (d) Ir^{3+} (e) Ru^{2+}

32. Give the electronic configuration for
  (a) Cd^{2+} (b) Fe^{2+} (c) Pt^{2+}
  (d) Mn^{2+} (e) Ni^{3+}

33. Write an abbreviated orbital diagram and determine the number of unpaired electrons in each species in Question 31.

34. Write an abbreviated orbital diagram and determine the number of unpaired electrons in each species in Question 32.

Electron Distributions and Crystal Field Energy

35. Give the electron distribution in low-spin and/or high-spin complexes of
  (a) Ru^{4+} (b) Pt^{2+}

36. Follow the directions of Question 35 for
  (a) Zn^{2+} (b) Ni^{3+}

37. For complexes of V^{3+}, only one distribution of electrons is possible. Explain.

38. Explain why Mn^{3+} forms high-spin and low-spin octahedral complexes but Mn^{4+} does not.

39. $[Cr(CN)_6^{4-}]$ is less paramagnetic than $[Cr(H_2O)_6^{2+}]$. Explain.

40. Why is $[Co(NH_3)_6^{3+}]$ diamagnetic while $[CoF_6^{3-}]$ is paramagnetic?

41. Give the number of unpaired electrons in octahedral complexes with strong-field ligands for

    (a) Rh^{3+} (b) Mn^{3+} (c) Ag^+

    (d) Pt^{4+} (e) Au^{3+}

42. For the species in Question 41, indicate the number of unpaired electrons with weak-field ligands.

43. $Ti(NH_3)_6^{3+}$ has a d-orbital electron transition at 399 nm. Find Δ_o at this wavelength.

44. MnF_6^{2-} has a crystal field splitting energy, Δ_o, of 2.60×10^2 kJ/mol. What is the wavelength responsible for this energy?

45. The wavelength of maximum absorption of $[Cu(NH_3)_4^{2+}]$ is 580 nm (orange-yellow). What color is a solution of $[Cu(NH_3)_4^{2+}]$?

46. A solution of $[Fe(CN)_6^{3-}]$ appears red. Using Figure 19.13, estimate the wavelength of maximum absorption.

Unclassified

47. A chemist synthesizes two coordination compounds. One compound decomposes at 280°C, the other at 240°C. Analysis of the compounds gives the same mass percent data: 52.6% Pt, 7.6% N, 1.63% H, and 38.2% Cl. Both compounds contain a +4 central ion.

    (a) What is the simplest formula of the compounds?

    (b) Draw structural formulas for the complexes present.

48. Analysis of a coordination compound gives the following results: 22.0% Co, 31.4% N, 6.78% H, and 39.8% Cl. One mole of the compound dissociates in water to form four moles of ions.

    (a) What is the simplest formula of the compound?

    (b) Write an equation for its dissociation in water.

Conceptual Problems

49. Explain why

    (a) oxalic acid removes rust stains.

    (b) there are no geometric isomers of tetrahedral complexes.

    (c) cations such as Co^{2+} act as Lewis acids.

    (d) $C_2O_4^{2-}$ is a chelating agent.

    (e) NH_3 can be a ligand but NH_4^+ is not.

50. Indicate whether each of the following is true or false. If the statement is false, correct it.

    (a) The coordination number of iron(III) in $[Fe(NH_3)_4(en)^{3+}]$ is 5.

    (b) $[Ni(CN)_6^{4-}]$ is expected to absorb at a longer wavelength than $[Ni(NH_3)_6^{2+}]$.

    (c) Complexes of Cr^{3+} are brightly colored, and those of Zn^{2+} are colorless.

    (d) Ions with seven or more d electrons cannot form both high- and low-spin octahedral complexes.

Challenge Problems

51. A child eats 10.0 g of paint containing 5.0% Pb. How many grams of the sodium salt of EDTA, $Na_4(EDTA)$, should he receive to bring the lead into solution as $Pb \cdot EDTA$?

52. A certain coordination compound has the simplest formula $PtN_2H_6Cl_2$. It has a molar mass of about 600 g/mol and contains both a complex cation and a complex anion. What is its structure?

53. Two coordination compounds decompose at different temperatures but have the same mass percent analysis data: 20.25% Cu, 15.29% C, 7.07% H, 26.86% N, 10.23% S, and 20.39% O. Each contains Cu^{2+}.

    (a) Determine the simplest formula of the compounds.

    (b) Draw the structural formulas of the complex ion in each case.

54. In the $[Ti(H_2O)_6^{3+}]$ ion, the splitting between the d levels, Δ_o, is 55 kcal/mol. What is the color of this ion? Assume that the color results from a transition between upper and lower d levels.

20 Chemistry of the Metals

The stabile *Equilibrium* is made up of metals. These were refined from ores using some of the processes discussed in this chapter.

The fields of Nature long
  prepared and fallow—
  the silent cyclic chemistry;
The slow and steady ages
  plodding—the
  unoccupied surface
  ripening—the rich ores
  forming beneath;

—WALT WHITMAN
Song of the Redwood Tree

Bruce Gray

An ore is a natural source from which a metal can be extracted profitably.

A s you can see from the periodic table on the inside cover or opening pages of this text, the overwhelming majority of elements, about 88%, are metals (shown in blue). In discussing the descriptive chemistry of the metals, we concentrate on

- the *main-group metals in Groups 1 and 2* at the far left of the periodic table. These are commonly referred to as the *alkali metals* (Group 1) and *alkaline earth metals* (Group 2). The group names reflect the strongly basic nature of the oxides (K_2O, CaO, . . .) and hydroxides (KOH, $Ca(OH)_2$, . . .) of these elements.
- the *transition metals,* located in the center of the periodic table. There are three series of transition metals, each consisting of ten elements, located in the fourth, fifth, and sixth periods. We focus on a few of the more important transition metals (Figure 20.1), particularly those toward the right of the first series.

Section 20-1 deals with the processes by which these metals are obtained from their principal ores. Section 20-2 describes the reactions of the alkali and alkaline earth metals, particularly those with hydrogen, oxygen, and water. Section 20-3 considers the redox chemistry of the transition metals, their cations (e.g., Fe^{2+}, Fe^{3+}), and their oxoanions (e.g., CrO_4^{2-}).

20-1 Metallurgy

The processes by which metals are extracted from their **ores** ◄ *i* fall within the science of **metallurgy.** As you might expect, the chemical reactions involved depend on the type of ore (Figure 20.2). We consider some typical processes used to obtain metals from chloride, oxide, sulfide, or "native" ores.

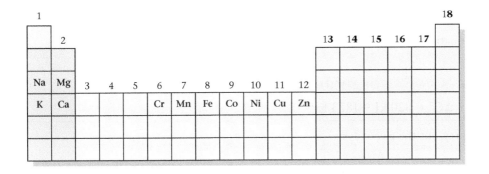

Figure 20.1 Metals and the periodic table. The periodic table groups discussed in this chapter are Groups 1 and 2, the alkali and alkaline earth metals (shaded in blue), and the transition metals (shaded in yellow). Symbols are shown for the more common metals.

Key: ☐ chlorides ▨ oxides ☐ silicates
☐ carbonates ☐ sulfides
▨ phosphates ▨ native metals

Figure 20.2 Principal ores of the Group 1, Group 2, and transition metals.

Chloride Ores: Na from NaCl

Sodium metal is obtained by the electrolysis of molten sodium chloride ◄ⓘ (Figure 20.3). The electrode reactions are quite simple:

$$\text{cathode:} \quad 2Na^+(l) + 2e^- \longrightarrow 2Na(l)$$

$$\text{anode:} \quad \underline{2Cl^-(l) \longrightarrow Cl_2(g) + 2e^-}$$

$$2NaCl(l) \longrightarrow 2Na(l) + Cl_2(g)$$

> ⓘ $Na^+(l)$ and $Cl^-(l)$ are the ions present in molten NaCl.

The cell is operated at about 600°C to keep the electrolyte molten; ◄ⓘ calcium chloride is added to lower the melting point. About 14 kJ of electrical energy is required to produce one gram of sodium, which is drawn off as a liquid (mp of Na = 98°C). The chlorine gas produced at the anode is a valuable byproduct.

> ⓘ It's cheaper to electrolyze NaCl(aq), but you don't get sodium metal that way.

EXAMPLE 20.1	

Consider the electrolysis of NaCl(l) at 600°C:

$$2NaCl(l) \longrightarrow 2Na(l) + Cl_2(g)$$

where $\Delta H° = 820$ kJ and $\Delta S° = 0.180$ kJ/K.

a	What is $\Delta G°$ at the electrolysis temperature?

ANALYSIS	
Information given:	equation for the reaction ($2NaCl(l) \longrightarrow 2Na(l) + Cl_2(g)$) $\Delta H°(820$ kJ); $\Delta S°(0.180$ kJ/K) $T(600°C = 873$ K)
Asked for:	$\Delta G°$ at 600°C

continued

STRATEGY	
Substitute into the Gibbs-Helmholtz equation (Chapter 16): $$\Delta G° = \Delta H° - T\Delta S°$$	
SOLUTION	
$\Delta G°$	$\Delta G° = 820 \text{ kJ} - [873 \text{ K}(0.180 \text{ kJ/K})] = 663 \text{ kJ}$

b	What is the voltage required to carry out the electrolysis?

ANALYSIS	
Information given:	from part (a) $\Delta G° = 663$ kJ
Information implied:	Faraday constant, $F(9.648 \times 10^4 \text{ J/mol} \cdot \text{V})$
Asked for:	voltage required ($E°$)

STRATEGY AND SOLUTION	
Substitute into the equation relating $E°$ and $\Delta G°$ (Chapter 17):	
$E°$	$$E° = \frac{-\Delta G°}{nF} = \frac{-663 \times 10^3 \text{ J}}{(2 \text{ mol})(9.648 \times 10^4 \text{ J/mol} \cdot \text{V})} = -3.44 \text{ V}$$

At least 3.44 V must be applied to carry out the electrolysis.

END POINT

Notice that the value of $\Delta G°$ calculated in part (a), 663 kJ, is for two moles of Na (2 mol × 22.99 g/mol = 45.98 g). The value of the free energy change per gram is

$$663 \text{ kJ}/45.98 \text{ g Na} = 14.4 \text{ kJ/g Na}$$

This is consistent with the statement in the text that "about 14 kJ of electrical energy is required to produce one gram of sodium. . . ."

Figure 20.3 Electrolysis of molten sodium chloride. Calcium chloride ($CaCl_2$) is added to lower the melting point. The iron screen prevents sodium and chlorine from coming into contact with each other.

Oxide Ores: Al from Al_2O_3, Fe from Fe_2O_3

Oxides of very reactive metals such as calcium or aluminum are reduced by electrolysis. In the case of aluminum, bauxite ore, Al_2O_3, is used.

$$2Al_2O_3(l) \longrightarrow 4Al(l) + 3O_2(g)$$

Cryolite, Na_3AlF_6, is added to Al_2O_3 to produce a mixture melting at about 1000°C. (A mixture of AlF_3, NaF, and CaF_2 may be substituted for cryolite.) The cell is heated electrically to keep the mixture molten so that ions can move through it, carrying the electric current. About 30 kJ of electrical energy is consumed per gram of aluminum formed. The high energy requirement explains in large part the value of recycling aluminum cans.

The process for obtaining aluminum from bauxite was worked out in 1886 by Charles Hall (1863–1914), just after he graduated from Oberlin College. ◀ ⓘ The problem that Hall faced was to find a way to electrolyze Al_2O_3 at a temperature below its melting point of 2000°C. His general approach was to look for ionic compounds in which Al_2O_3 would dissolve at a reasonable temperature. After several unsuccessful attempts, Hall found that cryolite was the ideal "solvent." Curiously enough, the same electrolytic process was worked out by Paul Héroult (1863–1914) in France, also in 1886.

With less active metals, a chemical reducing agent can be used to reduce a metal cation to the element. The most common reducing agent in metallurgical processes is carbon, in the form of coke or, more exactly, carbon monoxide formed from the coke.

The most important metallurgical process involving carbon is the reduction of hematite ore, which consists largely of iron(III) oxide, Fe_2O_3, mixed with silicon dioxide, SiO_2. Reduction occurs in a blast furnace (Figure 20.4a) typically 30 m high and 10 m in diameter. The furnace is lined with refractory brick, capable of withstanding temperatures that may go as high as 1800°C. The solid charge, admitted at the top of the furnace, consists of iron ore, coke, and limestone ($CaCO_3$). To get the process started, a blast of compressed air or pure O_2 at 500°C is blown into the furnace through nozzles near the bottom. Several different reactions occur, of which three are most important:

ⓘ Chemistry majors can be very productive.

a **b**

Figure 20.4 Iron and steel production. (a) A blast furnace for the production of pig iron. (b) A basic oxygen furnace, where the carbon content of pig iron is lowered by heating it with oxygen to form steel.

Courtesy of Bethlehem Steel Corporation

1. *Conversion of carbon to carbon monoxide.* In the lower part of the furnace, coke burns to form carbon dioxide, CO_2. As the CO_2 rises through the solid mixture, it reacts further with the coke to form carbon monoxide, CO. The overall reaction is

$$2C(s) + O_2(g) \longrightarrow 2CO(g) \qquad \Delta H = -221 \text{ kJ}$$

The heat given off by this reaction maintains a high temperature within the furnace.

2. *Reduction of Fe^{3+} ions to Fe.* ◄ⓘ The carbon monoxide reacts with the iron(III) oxide in the ore:

$$Fe_2O_3(s) + 3CO(g) \longrightarrow 2Fe(l) + 3CO_2(g)$$

Molten iron, formed at a temperature of 1600°C, collects at the bottom of the furnace. Four or five times a day, it is drawn off. The daily production of iron from a single blast furnace is about 1500 metric tons.

3. *Formation of slag.* The limestone added to the furnace decomposes at about 800°C:

$$CaCO_3(s) \longrightarrow CaO(s) + CO_2(g)$$

The calcium oxide formed reacts with impurities in the iron ore to form a glassy material called *slag*. The main reaction is with SiO_2 to form calcium silicate, $CaSiO_3$:

$$CaO(s) + SiO_2(s) \longrightarrow CaSiO_3(l)$$

The slag, which is less dense than molten iron, forms a layer on the surface of the metal. This makes it possible to draw off the slag through an opening in the furnace above that is used to remove the iron. The slag is used to make cement and as a base in road construction.

The product that comes out of the blast furnace, called "pig iron," is highly impure. On the average, it contains about 4% carbon along with lesser amounts of silicon, manganese, and phosphorus. To make steel from pig iron, the carbon content must be lowered below 2%. Most of the steel produced in the world today is made by the basic oxygen process (Figure 20.4b). The "converter" is filled with a mixture of about 70% molten iron from the blast furnace, 25% scrap iron or steel, and 5% limestone. Pure oxygen under a pressure of about 10 atm is blown through the molten metal. The reaction

$$C(s) + O_2(g) \longrightarrow CO_2(g)$$

occurs rapidly; at the same time, impurities such as silicon are converted to oxides, which react with limestone to form a slag. When the carbon content drops to the desired level, the supply of oxygen is cut off. At this stage, the steel is ready to be poured. The whole process takes from 30 min to 1 h and yields about 200 metric tons of steel in a single "blow."

This is the key reaction; it produces molten iron.

EXAMPLE 20.2

Write balanced equations for the reduction of each of the following oxide ores by carbon monoxide:

ZnO, MnO_2, and Fe_3O_4 (FeO · Fe_2O_3)

STRATEGY

The metal oxides are reduced to metal and CO is oxidized to CO_2.

SOLUTION

ZnO	$ZnO(s) + CO(g) \longrightarrow Zn(s) + CO_2(g)$
MnO_2	$MnO_2(s) + 2CO(g) \longrightarrow Mn(s) + 2CO_2(g)$
Fe_3O_4	$Fe_3O_4(s) + 4CO(g) \longrightarrow 3Fe(s) + 4CO_2(g)$

Sulfide Ores: Cu from Cu₂S

Sulfide ores, after preliminary treatment, most often undergo **roasting**, that is, heating with air or pure oxygen. With a relatively reactive transition metal such as zinc, the product is the oxide

$$2ZnS(s) + 3O_2(g) \longrightarrow 2ZnO(s) + 2SO_2(g)$$

which can then be reduced to the metal with carbon. With sulfides of less reactive metals such as copper or mercury, the free metal is formed directly on roasting. The reaction with cinnabar, the sulfide ore of mercury, is

$$HgS(s) + O_2(g) \longrightarrow Hg(g) + SO_2(g)$$

Among the several ores of copper, one of the most important is chalcocite, which contains copper(I) sulfide, Cu_2S, in highly impure form. Rocky material typically lowers the fraction of copper in the ore to 1% or less. The Cu_2S is concentrated by a process called *flotation* (Figure 20.5), which raises the fraction of copper to 20%–40%. The concentrated ore is then converted to the metal by blowing air through it at a high temperature, typically above 1000°C. (Pure O_2 is often used instead of air.) The overall reaction that occurs is a simple one:

$$Cu_2S(s) + O_2(g) \longrightarrow 2Cu(s) + SO_2(g)$$

The solid produced is called "blister copper." It has an irregular appearance due to air bubbles that enter the copper while it is still molten. Blister copper is impure, containing small amounts of several other metals.

Copper is purified by electrolysis. The anode, which may weigh as much as 300 kg, is made of blister copper. The electrolyte is 0.5 to 1.0 M $CuSO_4$, adjusted to a pH of about 0 with sulfuric acid. The cathode is a piece of pure copper, weighing perhaps 150 kg. The half-reactions are

oxidation: $\quad Cu(s,\ impure) \longrightarrow Cu^{2+}(aq) + 2e^-$

reduction: $\quad Cu^{2+}(aq) + 2e^- \longrightarrow Cu(s,\ pure)$

The overall reaction, obtained by adding these two half-reactions, is

$$Cu(s,\ impure) \longrightarrow Cu(s,\ pure)$$

Figure 20.5 Flotation process for concentrating a sulfide ore. Low-grade sulfide ores, including Cu_2S, are often concentrated by flotation. The finely divided sulfide particles are trapped in soap bubbles; the rocky waste sinks to the bottom and is discarded.

Because Cu can be produced by simple "roasting," it was one of the first metals known.

Thus the net effect of electrolysis is to transfer copper metal from the impure blister copper used as one electrode to the pure copper sheet used as the other electrode. Electrolytic copper is 99.95% pure.

EXAMPLE 20.3

A major ore of bismuth is bismuth(III) sulfide, Bi_2S_3. By roasting in air, it is converted to the corresponding oxide; sulfur dioxide is a byproduct. What volume of SO_2, at 25°C and 1.00 atm, is formed from one metric ton (10^6 g) of ore containing 1.25% Bi_2S_3?

ANALYSIS

Information given:	T(25°C); P(1.00 atm) mass of the ore (1 metric ton = 1×10^6 g) % Bi_2S_3 in ore (1.25%)
Information implied:	R value in L · atm/mol · K molar mass of Bi_2S_3
Asked for:	Volume of SO_2 formed

STRATEGY

1. Write a balanced equation for the reaction. Recall that roasting in air means reacting with oxygen gas.

2. Find the mass of Bi_2S_3 in the ore.

3. Calculate the moles of SO_2 formed.

   mass $Bi_2S_3 \longrightarrow$ mol $Bi_2S_3 \longrightarrow$ mol SO_2

4. Use the ideal gas law to find the volume of SO_2.

SOLUTION

1. Equation	$2Bi_2S_3(s) + 9O_2(g) \longrightarrow 2Bi_2O_3(s) + 6SO_2(g)$
2. Mass of Bi_2S_3	$(1 \times 10^6$ g ore$)(1.25$ g $Bi_2S_3/100$ g of ore$) = 1.25 \times 10^4$ g Bi_2S_3
3. mol SO_2	1.25×10^4 g $Bi_2S_3 \times \dfrac{1 \text{ mol } Bi_2S_3}{514.2 \text{ g}} \times \dfrac{6 \text{ mol } SO_2}{2 \text{ mol } Bi_2S_3} = 72.9$ mol SO_2
4. Volume SO_2	$V = \dfrac{nRT}{P} = \dfrac{(72.9 \text{ mol})(298 \text{ K})(0.0821 \text{ L} \cdot \text{atm/mol} \cdot \text{K})}{1.00 \text{ atm}} = 1.78 \times 10^3$ L

END POINT

That volume (1.78×10^3 L) is equal to 1.78 cubic meters. The roasting of sulfide ores is a major source of the gaseous pollutant SO_2, which contributes to acid rain (Chapter 14).

Figure 20.6 Modern gold jewelry.
The gold was probably obtained by the cyanide process.

Tiffany & Co.

Native Metals: Au

A few very unreactive metals, notably silver and gold, are found in nature in elemental form, mixed with large amounts of rocky material. For countless centuries, people have extracted gold by taking advantage of its high density (19.3 g/mL). In ancient times, gold-bearing sands were washed over sheepskins, which retained the gold; this is believed to be the source of the "Golden Fleece" of Greek mythology. The forty-niners in California obtained gold by swirling gold-bearing sands with water in a pan. Less dense impurities were washed away, leaving gold nuggets or flakes at the bottom of the pan.

Nowadays, the gold content of ores is much too low for these simple mechanical separation methods to be effective. Instead, the ore is treated with very dilute (0.01 M) sodium cyanide solution (Figure 20.6), through which air is blown. The following redox reaction takes place:

$$4Au(s) + 8CN^-(aq) + O_2(g) + 2H_2O \longrightarrow 4Au(CN)_2^-(aq) + 4OH^-(aq)$$

The oxidizing agent is O_2, which takes gold to the +1 state. The cyanide ion acts as a complexing ligand, forming the stable $Au(CN)_2^-$ ion. Metallic gold is recovered from solution by adding zinc; the gold in the complex ion is reduced to the metal.

$$Zn(s) + 2Au(CN)_2^-(aq) \longrightarrow Zn(CN)_4^{2-}(aq) + 2Au(s)$$

20-2   Reactions of the Alkali and Alkaline Earth Metals

The metals in Groups 1 and 2 are among the most reactive of all elements (Table 20.1). Their low ionization energies and high E_{ox}° values explain why they are so readily oxidized to cations, +1 cations for Group 1, +2 for Group 2. The alkali metals and the heavier alkaline earth metals (Ca, Sr, Ba) are commonly stored under dry mineral oil or kerosene to prevent them from reacting with oxygen or water vapor in the air. Magnesium is less reactive; it is commonly available in the form of ribbon or powder. Beryllium, as one would expect from its position in the periodic table, is the least metallic element in these two groups. It is also the least reactive toward water, oxygen, or other nonmetals.

Not under water!

EXAMPLE 20.4

Write balanced equations for the reaction of sodium with hydrogen and barium with oxygen.

STRATEGY

Use Table 20.1 to deduce the formulas of the products. Note that barium forms two products with oxygen.

SOLUTION

Na with H_2	$2Na(s) + H_2(g) \longrightarrow 2NaH(s)$
Ba with O_2	$2Ba(s) + O_2(g) \longrightarrow 2BaO(s)$
	$Ba(s) + O_2(g) \longrightarrow BaO_2(s)$

BaO_2 is barium peroxide.

Table 20.1   Reactions of the Alkali and Alkaline Earth Metals

Reactant	Product	Comments
	Alkali Metals (M)	
$H_2(g)$	$MH(s)$	On heating in hydrogen gas
$X_2(g)$	$MX(s)$	X = F, Cl, Br, I
$N_2(g)$	$M_3N(s)$	Only Li reacts; product contains N^{3-} ion
$S(s)$	$M_2S(s)$	On heating
$O_2(g)$	$M_2O(s)$	Li; product is an oxide (O^{2-} ion)
	$M_2O_2(s)$	Na; product is a peroxide (O_2^{2-} ion)
	$MO_2(s)$	K, Rb, Cs; product is a superoxide (O_2^- ion)
$H_2O(l)$	$H_2(g)$, M^+, OH^-	Violent reaction with Na, K
	Alkaline Earth Metals (M)	
$H_2(g)$	$MH_2(s)$	All except Be; heating required
$X_2(g)$	$MX_2(s)$	Any halogen
$N_2(g)$	$M_3N_2(s)$	All except Be; heating required
$S(s)$	$MS(s)$	On heating
$O_2(g)$	$MO(s)$	All; product contains O^{2-} ion
	$MO_2(s)$	Ba; product contains O_2^{2-} ion
$H_2O(l)$	$H_2(g)$, M^{2+}, OH^-	Ca, Sr, Ba

Reaction with Hydrogen

The compounds formed by the reaction of hydrogen with the alkali and alkaline earth metals contain H^- ions; for example, sodium hydride consists of Na^+ and H^- ions. These white crystalline solids are often referred to as saline hydrides because of their physical resemblance to NaCl. Chemically, they behave quite differently from sodium chloride; for example, they react with water to produce hydrogen gas. Typical reactions are

$$NaH(s) + H_2O(l) \longrightarrow H_2(g) + Na^+(aq) + OH^-(aq)$$
$$CaH_2(s) + 2H_2O(l) \longrightarrow 2H_2(g) + Ca^{2+}(aq) + 2OH^-(aq)$$

In this way, saline hydrides can serve as compact, portable sources of hydrogen gas for inflating life rafts and balloons.

Reaction with Water

The alkali metals react vigorously with water (Figure 20.7) to evolve hydrogen and form a water solution of the alkali hydroxide. The reaction of sodium is typical:

$$2Na(s) + 2H_2O(l) \longrightarrow 2Na^+(aq) + 2OH^-(aq) + H_2(g) \qquad \Delta H° = -368.6 \text{ kJ}$$

The heat evolved in the reaction frequently causes the hydrogen to ignite.

Among the Group 2 metals, Ca, Sr, and Ba react with water in much the same way as the alkali metals. The reaction with calcium is

$$Ca(s) + 2H_2O(l) \longrightarrow Ca^{2+}(aq) + 2OH^-(aq) + H_2(g)$$

Figure 20.8 shows the reaction of a calcium compound, CaH_2, with water. Beryllium does not react with water at all. Magnesium reacts very slowly with boiling water but reacts more readily with steam at high temperatures:

$$Mg(s) + H_2O(g) \longrightarrow MgO(s) + H_2(g)$$

This reaction, like that of sodium with water, produces enough heat to ignite the hydrogen. Firefighters who try to put out a magnesium fire by spraying water on it have discovered this reaction, often with tragic results. The best way to extinguish burning magnesium is to dump dry sand on it.

Reaction with Oxygen

Note from Table 20.1 that several different products are possible when an alkali or alkaline earth metal reacts with oxygen. The product may be a normal oxide (O^{2-} ion), a **peroxide** (O_2^{2-} ion), or **superoxide** (O_2^- ion). ◀ *ⓘ*

The superoxide ion has an unpaired electron; KO_2 is paramagnetic.

$$\left[:\ddot{\overset{..}{O}}: \right]^{2-} \qquad \left[:\ddot{\overset{..}{O}} - \ddot{\overset{..}{O}}: \right]^{2-} \qquad \left[:\ddot{\overset{..}{O}} - \ddot{O}\cdot \right]^-$$
oxide ion   peroxide ion   superoxide ion

Figure 20.7 Reaction of sodium with water. When a *very small* piece of sodium is added to water, it reacts violently with the water. The OH^- ions formed in the reaction turn the acid-base indicator phenolphthalein from colorless to pink.

Marna G. Clarke

© Cengage Learning/Charles D. Winters

Figure 20.8 Reaction of calcium hydride (CaH_2) with water.

Lithium is the only Group 1 metal that forms the normal oxide in good yield by direct reaction with oxygen. The other Group 1 oxides (Na_2O, K_2O, Rb_2O, Cs_2O) must be prepared by other means. In contrast, the Group 2 metals usually react with oxygen to give the normal oxide. Beryllium and magnesium must be heated strongly to give BeO and MgO (Figure 20.9). Calcium and strontium react more readily to give CaO and SrO. Barium, the most reactive of the Group 2 metals, catches fire when exposed to moist air. The product is a mixture of the normal oxide BaO (Ba^{2+}, O^{2-} ions) and the peroxide BaO_2 (Ba^{2+}, O_2^{2-} ions). The higher the temperature, the greater the fraction of BaO in the mixture.

The oxides of these metals react with water to form hydroxides:

$$Li_2O(s) + H_2O(l) \longrightarrow 2LiOH(s)$$
$$CaO(s) + H_2O(l) \longrightarrow Ca(OH)_2(s)$$

Figure 20.9 Burning magnesium.
A piece of magnesium ribbon bursts into flame when heated in the air; the product of the reaction is MgO(s).

The oxides of the Group 1 and Group 2 metals are sometimes referred to as "basic anhydrides" (bases without water) because of this reaction. The reaction with CaO is referred to as the "slaking" of lime; it gives off 65 kJ of heat per mole of $Ca(OH)_2$ formed. A similar reaction with MgO takes place slowly to form $Mg(OH)_2$, the antacid commonly referred to as "milk of magnesia."

When sodium burns in air, the principal product is yellowish sodium peroxide, Na_2O_2:

$$2Na(s) + O_2(g) \longrightarrow Na_2O_2(s)$$

Addition of sodium peroxide to water gives hydrogen peroxide, H_2O_2:

$$Na_2O_2(s) + 2H_2O(l) \longrightarrow 2Na^+(aq) + 2OH^-(aq) + H_2O_2(aq)$$

Through this reaction, sodium peroxide finds use as a bleaching agent in the pulp and paper industry.

The heavier alkali metals (K, Rb, Cs) form the superoxide when they burn in air. For example,

$$K(s) + O_2(g) \longrightarrow KO_2(s)$$

Potassium superoxide is used in self-contained breathing devices for firefighters and miners. It reacts with the moisture in exhaled air to generate oxygen:

$$4KO_2(s) + 2H_2O(g) \longrightarrow 3O_2(g) + 4KOH(s)$$

The carbon dioxide in the exhaled air is removed by reaction with the KOH formed:

$$KOH(s) + CO_2(g) \longrightarrow KHCO_3(s)$$

A person using a mask charged with KO_2 can rebreathe the same air for an extended period of time. This allows that person to enter an area where there are poisonous gases or oxygen-deficient air.

EXAMPLE 20.5

Consider the compounds strontium hydride, radium peroxide, and cesium superoxide.

| a | Give the formulas of these compounds. |

STRATEGY

The formulas can be deduced from the charges of the ions. Sr and Ra are in Group 2 and thus have a charge of +2. Cs is in Group 1 and has a charge of +1. Hydride is H^-, peroxide is O_2^{2-}, and superoxide is O_2^-.

continued

SOLUTION	
strontium hydride	$Sr^{2+} + H^- = SrH_2$
radium peroxide	$Ra^{2+} + O_2^{2-} = RaO_2$
cesium superoxide	$Cs^+ + O_2^- = CsO_2$

b | Write equations for the formation of these compounds from the elements.

STRATEGY

Write the formula of the products (from (a)) on the product side, put the elements that make them up on the reactant side, and write the states and balance.

SOLUTION	
strontium hydride	$Sr(s) + H_2(g) \longrightarrow SrH_2(s)$
radium peroxide	$Ra(s) + O_2(g) \longrightarrow RaO_2(s)$
cesium superoxide	$Cs(s) + O_2(g) \longrightarrow CsO_2(s)$

c | Write equations for the reactions of strontium hydride and radium peroxide with water.

STRATEGY

Recall from the preceding discussion that

$$hydride + H_2O \longrightarrow H_2 + metal\ cation + OH^-$$

$$peroxide + H_2O \longrightarrow H_2O_2 + metal\ cation + OH^-$$

SOLUTION	
Reaction with water:	$SrH_2(s) + 2H_2O(l) \longrightarrow 2H_2(g) + Sr^{2+}(aq) + 2OH^-(aq)$
	$RaO_2(s) + 2H_2O(l) \longrightarrow H_2O_2(l) + Ra^{2+}(aq) + 2OH^-(aq)$

20-3 | Redox Chemistry of the Transition Metals

The transition metals, unlike those in Groups 1 and 2, typically show several different oxidation numbers in their compounds. This tends to make their redox chemistry more complex (and more colorful). Only in the lower oxidation states ($+1$, $+2$, $+3$) are the transition metals present as cations (e.g., Ag^+, Zn^{2+}, Fe^{3+}). In higher oxidation states ($+4$ to $+7$) a transition metal is covalently bonded to a nonmetal atom, most often oxygen.

Reactions of the Transition Metals with Oxygen

Table 20.2 lists the formulas of the oxides formed when the more common transition metals react with oxygen. With one exception (Cu_2O), the transition metal is present as a $+2$ and/or a $+3$ ion. In Co_3O_4 and in other oxides of general formula M_3O_4, there are two different types of cations: $+2$ and $+3$. To be specific, there are twice as many $+3$ as $+2$ cations; we might show the composition of Co_3O_4 as

$$1Co^{2+} : 2Co^{3+} : 4O^{2-}$$

We should point out that many oxides of the transition metals beyond those listed in Table 20.2 can be prepared indirectly. For example, although silver does not react directly with oxygen, silver(I) oxide, Ag_2O, can be made by treating a solution of a silver salt with strong base.

$$2Ag^+(aq) + 2OH^-(aq) \longrightarrow Ag_2O(s) + H_2O$$

In another case, cobalt(II) oxide can be prepared by heating the carbonate in the absence of air.

$$CoCO_3(s) \longrightarrow CoO(s) + CO_2(g)$$

Table 20.2 Products of Reactions of the Transition Metals with Oxygen

Metal	Product
Cr	Cr_2O_3
Mn	Mn_3O_4
Fe	Fe_2O_3, Fe_3O_4
Co	Co_3O_4
Ni	NiO
Cu	Cu_2O, CuO
Zn	ZnO
Ag	—
Cd	CdO
Au	—
Hg	HgO

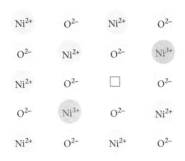

Figure 20.10 Nonstoichiometric nickel(II) oxide. For each missing Ni^{2+} ion, two others are converted to Ni^{3+} ions to maintain charge balance.

Figure 20.11 A superconductor. A pellet of superconducting material, previously cooled to 77 K with liquid nitrogen, floats above a magnet.

Many oxides of the transition metals can be *nonstoichiometric;* the atom ratios of the elements differ slightly from whole-number values. Stoichiometric nickel(II) oxide, in which the atom ratio is exactly 1:1 (i.e., $Ni_{1.00}O_{1.00}$), is a green, nonconducting solid. If this compound is heated to 1200°C in pure oxygen, it turns black and undergoes a change in composition to something close to $Ni_{0.97}O_{1.00}$. This nonstoichiometric solid was one of the early semiconductors; it has the structure shown in Figure 20.10. A few Ni^{2+} ions are missing from the crystal lattice; to maintain electrical neutrality, two Ni^{2+} ions are converted to Ni^{3+} ions for each missing Ni^{2+}.

Nonstoichiometry is relatively common among "mixed" metal oxides, in which more than one metal is present. In 1986 it was discovered that certain compounds of this type showed the phenomenon of *superconductivity;* on cooling to about 100 K, their electrical resistance drops to zero (Figure 20.11). A typical formula here is $YBa_2Cu_3O_x$, where x varies from 6.5 to 7.2, depending on the method of preparation of the solid.

Reaction of Transition Metals with Acids

Any metal with a positive standard oxidation voltage, E°_{ox}, can be oxidized by the H^+ ions present in a 1 M solution of a strong acid. All the transition metals in the left column of Table 20.3 react spontaneously with dilute solutions of such strong acids as HCl, HBr, and H_2SO_4. The products are hydrogen gas and a cation of the transition metal. A typical reaction is that of nickel:

$$Ni(s) + 2H^+(aq) \longrightarrow Ni^{2+}(aq) + H_2(g)$$

Table 20.3 Ease of Oxidation of Transition Metals

Metal		Cation	E°_{ox} (V)	Metal		Cation	E°_{ox} (V)
Mn	$\longrightarrow$	Mn^{2+}	+1.182	Cu	$\longrightarrow$	Cu^{2+}	−0.339
Cr	$\longrightarrow$	Cr^{2+}	+0.912	Ag	$\longrightarrow$	Ag^+	−0.799
Zn	$\longrightarrow$	Zn^{2+}	+0.762	Hg	$\longrightarrow$	Hg^{2+}	−0.852
Fe	$\longrightarrow$	Fe^{2+}	+0.409	Au	$\longrightarrow$	Au^{3+}	−1.498
Cd	$\longrightarrow$	Cd^{2+}	+0.402				
Co	$\longrightarrow$	Co^{2+}	+0.282				
Ni	$\longrightarrow$	Ni^{2+}	+0.236				

Courtesy of Edmund Scientific

Figure 20.12 Oxidation of nickel.
Nickel reacts slowly with hydrochloric acid to form $H_2(g)$ and Ni^{2+} ions in solution. Evaporation of the solution formed gives green crystals of $NiCl_2 \cdot 6H_2O$.

The end result of this reaction is shown in Figure 20.12. With metals that can form more than one cation, such as iron, the reaction product with H^+ in the absence of air is ordinarily the cation of lower charge, for example, Fe^{2+}:

$$Fe(s) + 2H^+(aq) \longrightarrow Fe^{2+}(aq) + H_2(g)$$

Metals with negative values of $E°_{ox}$, listed at the right of Table 20.3, are too inactive to react with hydrochloric acid. The H^+ ion is not a strong enough oxidizing agent to convert a metal such as copper ($E°_{ox} = -0.339$ V) to a cation. However, copper can be oxidized by nitric acid. The oxidizing agent is the nitrate ion, NO_3^-, which may be reduced to NO_2 or NO:

$$3Cu(s) + 8H^+(aq) + 2NO_3^-(aq) \longrightarrow 3Cu^{2+}(aq) + 2NO(g) + 4H_2O$$
$$E° = E°_{ox} \text{ Cu} + E°_{red} \text{ NO}_3^- = -0.339 \text{ V} + 0.964 \text{ V} = +0.625 \text{ V}$$

EXAMPLE   20.6

Write balanced equations for the reactions, if any, at standard conditions of chromium with hydrochloric acid and silver with nitric acid.

STRATEGY

1. The metals can only be oxidized. Use Table 20.3 to obtain $E°_{ox}$.

2. An aqueous solution of HCl contains only H^+ and Cl^- ions.
   Only H^+ can be reduced to H_2. Chloride ions cannot be reduced.
   Find $E°$ for the reaction between H^+ and the metal. The reaction will occur if $E° > 0$.

3. An aqueous solution of HNO_3 contains only H^+ and NO_3^- ions.
   Both H^+ and NO_3^- can be reduced, H^+ to H_2 and NO_3^- to NO.
   Find $E°$ for the reaction between H^+ and the metal and for NO_3^- and the metal. The reaction will occur if $E° > 0$.

4. Write balanced half-reactions to get the overall equation for the reaction.

SOLUTION

Cr and HCl

1. Metal $Cr \longrightarrow Cr^{2+}$: $E°_{ox} = 0.912$ V

2. HCl $H^+ \longrightarrow H_2$: $E°_{red} = 0.000$ V

3. $E°$ $E° = 0.912$ V $+ 0.000$ V $= 0.912$ V   The reaction will occur.

4. Equation
$$Cr(s) \longrightarrow Cr^{2+}(aq) + 2e^-$$
$$2H^+(aq) + 2e^- \longrightarrow H_2(g)$$
$$\overline{Cr(s) + 2H^+(aq) \longrightarrow Cr^{2+}(aq) + H_2(g)}$$

Ag and HNO_3

1. Metal $Ag \longrightarrow Ag^+$: $E°_{ox} = -0.799$ V

2. HNO_3 $H^+ \longrightarrow H_2$: $E°_{red} = 0.000$ V
   $NO_3^- \longrightarrow NO$: $E°_{red} = 0.964$ V

continued

3. $E°$	Ag + NO_3^- will occur. ($E° = 0.165$ V) Ag + H^+ will not occur. ($E° = -0.799$ V)
4. Equation	$3(Ag(s) \longrightarrow Ag^+(aq) + e^-)$ $\underline{NO_3^-(aq) + 4H^+(aq) + 3e^- \longrightarrow NO(g) + 2H_2O}$ $3Ag(s) + NO_3^-(aq) + 4H^+(aq) \longrightarrow 3Ag^+(aq) + NO(g) + 2H_2O$

Although gold is not oxidized by nitric acid, it can be brought into solution in *aqua regia*, ◄ⓘ a 3:1 mixture by volume of 12 *M* HCl and 16 *M* HNO_3:

$$Au(s) + 4H^+(aq) + 4Cl^-(aq) + NO_3^-(aq) \longrightarrow AuCl_4^-(aq) + NO(g) + 2H_2O$$

ⓘ "Aqua regia" = royal water.

The nitrate ion of the nitric acid acts as the oxidizing agent. The function of the hydrochloric acid is to furnish Cl^- ions to form the very stable complex ion $AuCl_4^-$.

Equilibria Between Different Cations of a Transition Metal

Several transition metals form more than one cation. For example, chromium forms Cr^{2+} and Cr^{3+}; copper forms Cu^+ and Cu^{2+}. Table 20.4 lists values of $E_{red}°$ for several such systems. Using the data in this table and in Table 17.1, it is possible to decide on the relative stabilities of different transition metal cations in water solution.

Cations for which $E_{red}°$ is a large, positive number are readily reduced and hence tend to be unstable in water solution. A case in point is the Mn^{3+} ion, which reacts spontaneously with water:

$$2Mn^{3+}(aq) + H_2O(l) \longrightarrow 2Mn^{2+}(aq) + \tfrac{1}{2}O_2(g) + 2H^+(aq)$$
$$E° = E_{red}° \ Mn^{3+} + E_{ox}° \ H_2O = +1.559 \text{ V} - 1.229 \text{ V} = +0.330 \text{ V}$$

As a result of this reaction, Mn^{3+} cations are never found in water solution. ◄ⓘ Manganese(III) occurs only in insoluble oxides and hydroxides such as Mn_2O_3 and $MnO(OH)$.

ⓘ Co^{3+} is also rare, except in complexes.

The cations in the center column of Table 20.4 (Cr^{2+}, Mn^{2+}, . . .) are in an intermediate oxidation state. They can be either oxidized to a cation of higher charge ($Cr^{2+} \longrightarrow Cr^{3+}$) or reduced to the metal ($Cr^{2+} \longrightarrow Cr$). With certain cations of this type, these two half-reactions occur simultaneously. This reaction type is called **disproportionation**. Consider, for example, the Cu^+ ion. In water solution, copper(I) salts *disproportionate*, undergoing reduction (to copper metal) and oxidation (to Cu^{2+}) at the same time:

$$2Cu^+(aq) \longrightarrow Cu(s) + Cu^{2+}(aq); \qquad E° = +0.518 \text{ V} - 0.161 \text{ V} = +0.357 \text{ V}$$

As a result, the only stable copper(I) species are insoluble compounds such as CuCN or complex ions such as $Cu(CN)_2^-$.

The Cu^+ ion is one of the few species in the center column of Table 20.4 that disproportionates in water, undergoing simultaneous reduction to the metal and oxidation to a cation of higher charge. However, cations of this type may be unstable for a quite different reason. Water ordinarily contains dissolved air; the O_2 in air may oxidize the cation. When a blue solution of chromium(II) salt is exposed to air, the color quickly changes to violet or green as the Cr^{3+} ion is formed by the reaction

$$
\begin{array}{ll}
2Cr^{2+}(aq) \longrightarrow 2Cr^{3+}(aq) + 2e^- & E_{ox}° = +0.408 \text{ V} \\
\underline{\tfrac{1}{2}O_2(g) + 2H^+(aq) + 2e^- \longrightarrow H_2O} & \underline{E_{red}° = +1.229 \text{ V}} \\
\\
2Cr^{2+}(aq) + \tfrac{1}{2}O_2(g) + 2H^+(aq) \longrightarrow 2Cr^{3+}(aq) + H_2O & E° = +1.637 \text{ V}
\end{array}
$$

Table 20.4 Ease of Reduction E°_{red} (V) of Transition Metal Cations

Chromium	Cr^{3+}	$\xrightarrow{-0.408\text{ V}}$	Cr^{2+}	$\xrightarrow{-0.912\text{ V}}$	Cr
Manganese	Mn^{3+}	$\xrightarrow{+1.559\text{ V}}$	Mn^{2+}	$\xrightarrow{-1.182\text{ V}}$	Mn
Iron	Fe^{3+}	$\xrightarrow{+0.769\text{ V}}$	Fe^{2+}	$\xrightarrow{-0.409\text{ V}}$	Fe
Cobalt	Co^{3+}	$\xrightarrow{+1.953\text{ V}}$	Co^{2+}	$\xrightarrow{-0.282\text{ V}}$	Co
Copper	Cu^{2+}	$\xrightarrow{+0.161\text{ V}}$	Cu^{+}	$\xrightarrow{+0.518\text{ V}}$	Cu
Gold	Au^{3+}	$\xrightarrow{+1.400\text{ V}}$	Au^{+}	$\xrightarrow{+1.695\text{ V}}$	Au
Mercury	Hg^{2+}	$\xrightarrow{+0.908\text{ V}}$	Hg_2^{2+} ◀ *i*	$\xrightarrow{+0.796\text{ V}}$	Hg

i The Hg_2^{2+} ion has the structure $(Hg-Hg)^{2+}$.

As a result of this reaction, chromium(II) salts are difficult to prepare and even more difficult to store.

The Fe^{2+} ion ($E^\circ_{ox} = -0.769$ V) is much more stable toward oxidation than Cr^{2+}. However, iron(II) salts in water solution are slowly converted to iron(III) by dissolved oxygen. ◀ *i* In acidic solution, the reaction is

i Solid Fe^{2+} salts are often contaminated with Fe^{3+}.

$$2Fe^{2+}(aq) + \tfrac{1}{2}O_2(g) + 2H^+(aq) \longrightarrow 2Fe^{3+}(aq) + H_2O$$
$$E^\circ = E^\circ_{ox}\ Fe^{2+} + E^\circ_{red}\ O_2 = -0.769\text{ V} + 1.229\text{ V} = +0.460\text{ V}$$

A similar reaction takes place in basic solution. Iron(II) hydroxide is pure white when first precipitated, but in the presence of air it turns first green and then brown as it is oxidized by O_2:

$$2Fe(OH)_2(s) + \tfrac{1}{2}O_2(g) + H_2O(l) \longrightarrow 2Fe(OH)_3(s)$$

EXAMPLE 20.7

Using Table 20.4, find

(a) three different cations, in addition to Mn^{3+}, that react with H_2O to form $O_2(g)$ ($E^\circ_{ox}\ H_2O = -1.229$ V).

(b) another cation, in addition to Cu^+, that disproportionates in water.

(c) two other cations, in addition to Cr^{2+} and Fe^{2+}, that are oxidized by $O_2(g)$ dissolved in water ($E^\circ_{red}\ O_2 = +1.229$ V).

STRATEGY AND SOLUTION

Use Table 20.4 and look for two half-reactions (one an oxidation, the other a reduction) that give $E^\circ > 0$ when E°_{red} and E°_{ox} are combined.

(a) E°_{red} must be larger than 1.229 V since $E^\circ_{ox} = -1.229$ V.
  The ions are: Co^{3+}: $E^\circ_{red} = 1.953$ V
  Au^{3+}: $E^\circ_{red} = 1.400$ V
  Au^{+}: $E^\circ_{red} = 1.695$ V

(b) Recall that disproportionation means that the same ion or element is reduced and oxidized. Look at the ions in the middle column and add E°_{ox} and E°_{red} on either side of the ion. Don't forget to change the sign for E°_{ox}.

$Cr^{2+} \longrightarrow Cr$: $E^\circ_{red} = -0.912$ V	$Cr^{2+} \longrightarrow Cr^{3+}$: $E^\circ_{ox} = 0.408$ V	$E^\circ < 0$
$Mn^{2+} \longrightarrow Mn$: $E^\circ_{red} = -1.182$ V	$Mn^{2+} \longrightarrow Mn^{3+}$: $E^\circ_{ox} = -1.559$ V	$E^\circ < 0$
$Fe^{2+} \longrightarrow Fe$: $E^\circ_{red} = -0.409$ V	$Fe^{2+} \longrightarrow Fe^{3+}$: $E^\circ_{ox} = -0.769$ V	$E^\circ < 0$
$Au^{+} \longrightarrow Au$: $E^\circ_{red} = 1.695$ V	$Au^{+} \longrightarrow Au^{3+}$: $E^\circ_{ox} = -1.400$ V	$E^\circ > 0$

Au^+ is the other cation.

continued

(c) E°_{ox} must be larger than -1.229 V.

Use the species in the center column as reactants to be oxidized.

You may not use the species on the right-hand column because they are not cations. (The question asks for cations.)

You may not use the cations on the left-hand column because they cannot be oxidized.

$$Mn^{2+} \longrightarrow Mn^{3+}: \quad E^\circ_{ox} = -1.559 \text{ V} < -1.229 \text{ V}$$

$$Co^{2+} \longrightarrow Co^{3+}: \quad E^\circ_{ox} = -1.953 \text{ V} < -1.229 \text{ V}$$

$$Cu^+ \longrightarrow Cu^{2+}: \quad E^\circ_{ox} = -0.161 \text{ V} > -1.229 \text{ V}$$

$$Au^+ \longrightarrow Au^{3+}: \quad E^\circ_{ox} = -1.400 \text{ V} < -1.229 \text{ V}$$

$$Hg_2^{2+} \longrightarrow Hg^{2+}: \quad E^\circ_{ox} = -0.908 \text{ V} > -1.229 \text{ V}$$

Cu^+ and Hg_2^{2+} are the other cations.

Oxoanions of the Transition Metals (CrO_4^{2-}, $Cr_2O_7^{2-}$, MnO_4^-)

Chromium in the +6 state forms two different oxoanions, the yellow chromate ion, CrO_4^{2-}, and the red dichromate ion, $Cr_2O_7^{2-}$ (Figure 20.13). The chromate ion is stable in basic or neutral solution; in acid, it is converted to the dichromate ion:

$$\underset{\text{yellow}}{2CrO_4^{2-}(aq)} + 2H^+(aq) \rightleftharpoons \underset{\text{red}}{Cr_2O_7^{2-}(aq)} + H_2O(l) \qquad K = 3 \times 10^{14}$$

The dichromate ion in acidic solution is a powerful oxidizing agent,

$$Cr_2O_7^{2-}(aq) + 14H^+(aq) + 6e^- \longrightarrow 2Cr^{3+}(aq) + 7H_2O(l) \qquad E^\circ_{red} = +1.33 \text{ V}$$

As you might expect from the half-equation for its reduction, the oxidizing strength of the dichromate ion decreases as the concentration of H^+ decreases (increasing pH).

The $Cr_2O_7^{2-}$ ion can act as an oxidizing agent in the solid state as well as in water solution. In particular, it can oxidize the NH_4^+ ion to molecular nitrogen. When a pile of ammonium dichromate is ignited, a spectacular reaction occurs (Figure 20.14).

$$\underset{\text{red}}{(NH_4)_2Cr_2O_7(s)} \longrightarrow N_2(g) + 4H_2O(g) + \underset{\text{green}}{Cr_2O_3(s)}$$

The ammonium dichromate resembles a tiny volcano as it burns, emitting hot gases, sparks, and a voluminous green dust of chromium(III) oxide. ◀ ⓘ

Chromates and dichromates are gradually disappearing from chemistry teaching laboratories because of concern about their toxicity. Long-term exposure of industrial workers to dust containing chromates has, in a few cases, been implicated in lung cancer. More commonly, repeated contact with chromate salts leads to skin disorders; a few people are extremely allergic to CrO_4^{2-} and $Cr_2O_7^{2-}$ ions, breaking into a rash on first exposure.

ⓘ This makes a great lecture demonstration.

Figure 20.13 Oxidation states of chromium. The first two bottles contain +3 chromium. Each bottle in the right-hand pair contains +6 chromium.

The compound ammonium dichromate, $(NH_4)_2Cr_2O_7$, has a reddish-orange color due to the presence of the $Cr_2O_7^{2-}$ ion.

When $(NH_4)_2Cr_2O_7$ is ignited...

...it decomposes to give finely divided Cr_2O_3 (which is green), nitrogen gas, and water vapor.

Figure 20.14 **Ammonium dichromate volcano.**

The permanganate ion, MnO_4^-, has an intense purple color easily visible even in very dilute solution.* Crystals of solid potassium permanganate, $KMnO_4$, have a deep purple, almost black color. This compound is used to treat such diverse ailments as "athlete's foot" and rattlesnake bites. These applications depend on the fact that the MnO_4^- ion is a very powerful oxidizing agent. This is especially true in acidic solution, where MnO_4^- is reduced to Mn^{2+}:

$$MnO_4^-(aq) + 8H^+(aq) + 5e^- \longrightarrow Mn^{2+}(aq) + 4H_2O \qquad E^\circ_{red} = +1.512 \text{ V}$$

In basic solution, MnO_4^- is reduced to MnO_2, with a considerably smaller value of E°_{red}:

$$MnO_4^-(aq) + 2H_2O + 3e^- \longrightarrow MnO_2(s) + 4OH^-(aq) \qquad E^\circ_{red} = +0.596 \text{ V}$$

However, even in basic solution, MnO_4^- can oxidize water:

$$4MnO_4^-(aq) + 2H_2O \longrightarrow 4MnO_2(s) + 3O_2(g) + 4OH^-(aq)$$
$$E^\circ = E^\circ_{red} \, MnO_4^- + E^\circ_{ox} \, H_2O = +0.596 \text{ V} - 0.401 \text{ V} = +0.195 \text{ V}$$

This reaction accounts for the fact that laboratory solutions of $KMnO_4$ slowly decompose, producing a brownish solid (MnO_2) and gas bubbles (O_2).

---

*The purple color of old bottles exposed to the sun for a long time is due to MnO_4^- ions. These are formed when ultraviolet light oxidizes manganese compounds in the glass.

CHEMISTRY **BEYOND** THE CLASSROOM

Essential Metals in Nutrition

Four of the main-group cations are essential in human nutrition (Table A). Of these, the most important is Ca^{2+}. About 90% of the calcium in the body is found in bones and teeth, largely in the form of hydroxyapatite, $Ca(OH)_2 \cdot 3Ca_3(PO_4)_2$. Calcium ions in bones and teeth exchange readily with those in the blood; about 0.6 g of Ca^{2+} enters and leaves your bones every day. In a normal adult this exchange is in balance, but in elderly people, particularly women, there is sometimes a net loss of bone calcium, leading to the disease known as osteoporosis.

Although sodium is an essential metal, it has been statistically associated with hypertension. About 50% of people with high blood pressure (10% of the general population)

Foods high in iron.

are sensitive to high concentrations of Na^+ ions. For that reason, it is recommended that the daily consumption of Na^+ be limited to 2.4 g. The simplest way to cut back on Na^+ is to read the label carefully before buying processed foods, which account for about 75% of our sodium intake. For example, a single serving of packaged scalloped potatoes can contain as much as 0.5 g of Na^+ ion as opposed to only 0.006 g for a large baked potato.

Of the ten trace elements known to be essential to human nutrition, seven are transition metals. For the most part, transition metals in biochemical compounds are present as complex ions, chelated by organic ligands. You will recall (Chapter 19) that hemoglobin has such a structure with Fe^{2+} as the central ion of the complex. The Co^{3+} ion occupies a similar position in vitamin B_{12}, octahedrally coordinated to molecules in somewhat similar fashion to those in hemoglobin.

As you can judge from Table A, transition metal cations are frequently found in enzymes. The Zn^{2+} ion alone is known to be a component of at least 70 different enzymes. One of these, referred to as "alcohol dehydrogenase," is concentrated in the liver, where it acts to break down alcohols. Another zinc-containing enzyme is involved in the normal functioning of oil glands in the skin, which accounts for the use of Zn^{2+} compounds in the treatment of acne.

Although Zn^{2+} is essential to human nutrition, compounds of the two elements below zinc in the periodic table, Cd and Hg, are extremely toxic. This reflects the fact that Cd^{2+} and Hg^{2+}, in contrast to Zn^{2+}, form very stable complexes with ligands containing sulfur atoms. As a result, these two cations react with and thereby deactivate enzymes containing —SH groups.

Table A Essential Metal Ions

	Major Species (Main-Group Cations)			
Ion	Amount in Body	Daily Requirement	Function in Body	Rich Sources*
Na^+	63 g	~2 g	Principal cation *outside* cell fluid	Table salt, many processed and preserved foods
K^+	150 g	~2 g	Principal cation *within* cell fluid	Fruits, nuts, fish, coffee, wheat bran
Mg^{2+}	21 g	0.35 g	Activates enzymes for body processes	Chocolate, nuts, coffee, wheat bran
Ca^{2+}	1160 g	0.8 g	Bone and tooth formation	Milk, cheese, broccoli, canned salmon (with bones)

	Trace Species (Transition Metal Cations)			
Ion	Amount in Body	Daily Requirement	Function in Body	Rich Sources*
Fe^{2+}, Fe^{3+}	5000 mg	15 mg	Component of hemoglobin, myoglobin	Liver, meat, clams, spinach
Zn^{2+}	3000 mg	15 mg	Component of many enzymes, hormones	Oysters, crab, meat, nuts
Cu^{2+}, Cu^+	100 mg	2 mg	Iron metabolism, component of enzymes	Lobster, cherries, chocolate
Mn^{2+}	15 mg	3 mg	Metabolism of carbohydrates, lipids	Beet greens, nuts, blueberries
Mo(IV,V,VI)	10 mg	0.3 mg	Fe, N metabolism; component of enzymes	Legumes, ice cream
Cr^{3+}	5 mg	0.2 mg	Glucose metabolism; affects action of insulin	Corn, clams, nuts
Co^{2+}, Co^{3+}	1 mg	0.005 mg	Component of vitamin B_{12}	Liver, shellfish

*Other, more mundane, sources can be found in any nutrition textbook.

Key Concepts

1. Calculate $\Delta G°$ from thermodynamic data.
   (**Example 20.1; Problems 5, 6, 35**)
2. Write balanced equations to represent
   ■ metallurgical processes.
   (**Examples 20.2 and 20.3; Problems 1–4, 7, 8**)
   ■ reactions of Group 1 and Group 2 metals.
   (**Examples 20.4 and 20.5; Problems 13–16**)
   ■ redox reactions of transition metals.
   (**Example 20.6; Problems 19–24**)
3. Determine $E°$ and reaction spontaneity from standard potentials.
   (**Example 20.7; Problems 25–28**)

Key Terms

disproportionation
metallurgy

ores
peroxide

superoxide

Summary Problem

Consider the alkaline earth metal strontium and the transition metal manganese.

(a) Write a balanced equation for the reaction of strontium with oxygen; with water.

(b) Give the formulas of strontium hydride, strontium nitride, and strontium sulfide.

(c) Electrolysis of strontium chloride gives strontium metal and chlorine gas. What mass of the chloride must be electrolyzed to form one kilogram of strontium? One liter of $Cl_2(g)$ at STP?

(d) Give the formulas of three manganese compounds containing manganese in different oxidation states.

(e) Write a balanced equation for the reaction of manganese metal with hydrochloric acid; Mn^{3+} with water; Mn with O_2.

(f) Write a balanced equation for the reduction of the principal ore of manganese, pyrolusite, MnO_2, with CO.

(g) For the reaction

$$MnO(s) + H_2(g) \longrightarrow Mn(s) + H_2O(g), \Delta H° = +143 \text{ kJ},$$
$$\Delta S° = +0.0297 \text{ kJ/K}$$

Would it be feasible to reduce MnO to the metal by heating with hydrogen?

(h) Given that $E°_{red}$ $Mn^{3+} \longrightarrow Mn^{2+} = +1.559$ V, $E°_{red}$ $Mn^{2+} \longrightarrow Mn = -1.182$ V, and $E°_{red}$ $O_2(g) \longrightarrow H_2O = +1.229$ V, show by calculation whether Mn^{2+}

in water solution will disproportionate; whether it will be oxidized to Mn^{3+} by dissolved oxygen; whether it will be reduced by hydrogen gas to the metal.

(i) Given

$$MnO_4^-(aq) + 2H_2O + 3e^- \longrightarrow MnO_2(s) + 4OH^-(aq)$$
$$E°_{red} = +0.59 \text{ V}$$

write a balanced equation for the reaction of MnO_4^- in basic solution with Fe^{2+} to form $Fe(OH)_3$. Calculate E_{red} for the MnO_4^- ion at pH 9.0, taking $[MnO_4^-] = 1$ M.

Answers

(a) $2Sr(s) + O_2(g) \longrightarrow 2SrO(s)$
$Sr(s) + 2H_2O \longrightarrow Sr^{2+}(aq) + 2OH^-(aq) + H_2(g)$

(b) SrH_2; Sr_3N_2; SrS

(c) 1.809 kg; 7.07 g

(d) $MnCl_2$, Mn_2O_3, $KMnO_4$

(e) $Mn(s) + 2H^+(aq) \longrightarrow Mn^{2+}(aq) + H_2(g)$
$2Mn^{3+}(aq) + H_2O \longrightarrow$
$$2Mn^{2+}(aq) + \tfrac{1}{2}O_2(g) + 2H^+(aq)$$
$3Mn(s) + 2O_2(g) \longrightarrow Mn_3O_4(s)$

(f) $MnO_2(s) + 2CO(g) \longrightarrow Mn(s) + 2CO_2(g)$

(g) no; $T_{calc} \approx 4800$ K

(h) no; in all cases, $E°$ is negative

(i) $MnO_4^-(aq) + 3Fe^{2+}(aq) + 5OH^-(aq) + 2H_2O \longrightarrow$
$$MnO_2(s) + 3Fe(OH)_3(s) \quad E_{red} = +0.99 \text{ V}$$

Questions and Problems

Even-numbered questions and Challenge Problems have answers in Appendix 5 and fully worked solutions in the *Student Solutions Manual*.

20-1 Metallurgy

1. Write a balanced equation to represent the electrolysis of molten sodium chloride. What volume of Cl_2 at STP is formed at the anode when 1.00 g of sodium is formed at the cathode?

2. Write a balanced equation to represent the electrolysis of aluminum oxide. If 2.00 L of O_2 at 25°C and 751 mm Hg is formed at the anode, what mass of Al is formed at the cathode?

3. Write a balanced equation to represent
(a) the roasting of nickel(II) sulfide to form nickel(II) oxide.
(b) the reduction of nickel(II) oxide by carbon monoxide.

4. Write a balanced equation to represent the roasting of copper(I) sulfide to form "blister copper."

5. Show by calculation whether the reaction in Problem 3(b) is spontaneous at 25°C and 1 atm.

6. Calculate $\Delta G°$ at 200°C for the reaction in Problem 4.

7. Write a balanced equation for the reaction that occurs when
(a) finely divided gold is treated with $CN^-(aq)$ in the presence of $O_2(g)$.
(b) finely divided zinc metal is added to the solution formed in (a).

8. Write a balanced equation for the reaction that occurs when
(a) iron(III) oxide is reduced with carbon monoxide.
(b) the excess carbon in pig iron is removed by the basic oxygen process.

9. How many cubic feet of air (assume 21% by volume of oxygen in air) at 25°C and 1.00 atm are required to react with coke to form the CO needed to convert one metric ton of hematite ore (92% Fe_2O_3) to iron?

10. Zinc is produced by electrolytic refining. The electrolytic process, which is similar to that for copper, can be represented by the two half-reactions

$$Zn(\text{impure}, s) \longrightarrow Zn^{2+} + 2e^-$$
$$Zn^{2+} + 2e^- \longrightarrow Zn(\text{pure}, s)$$

For this process, a voltage of 3.0 V is used. How many kilowatt hours are needed to produce one metric ton of pure zinc?

11. When 2.876 g of a certain metal sulfide is roasted in air, 2.368 g of the metal oxide is formed. If the metal has an oxidation number of +2, what is its molar mass?

12. Chalcopyrite, $CuFeS_2$, is an important source of copper. A typical chalcopyrite ore contains about 0.75% Cu. What volume of sulfur dioxide at 25°C and 1.00 atm pressure is produced when one boxcar load (4.00×10^3 ft^3) of chalcopyrite ore ($d = 2.6$ g/cm^3) is roasted? Assume all the sulfur in the ore is converted to SO_2 and no other source of sulfur is present.

20-2 Reactions of Alkali Metals and Alkaline Earth Metals

13. Give the formula and name of the compound formed by strontium with
   (a) nitrogen (b) bromine
   (c) water (d) oxygen

14. Give the formula and name of the compound formed by potassium with
   (a) nitrogen (b) iodine (c) water
   (d) hydrogen (e) sulfur

15. Write a balanced equation and give the names of the products for the reaction of
   (a) magnesium with chlorine.
   (b) barium peroxide with water.
   (c) lithium with sulfur.
   (d) sodium with water.

16. Write a balanced equation and give the names of the products for the reaction of
   (a) sodium peroxide and water.
   (b) calcium and oxygen.
   (c) rubidium and oxygen.
   (d) strontium hydride and water.

17. To inflate a life raft with hydrogen to a volume of 25.0 L at 25°C and 1.10 atm, what mass of calcium hydride must react with water?

18. What mass of KO_2 is required to remove 90.0% of the CO_2 from a sample of 1.00 L of exhaled air (37°C, 1.00 atm) containing 5.00 mole percent CO_2?

20-3 Redox Chemistry of Transition Metals

19. Write a balanced equation to show
   (a) the reaction of chromate ion with strong acid.
   (b) the oxidation of water to oxygen gas by permanganate ion in basic solution.
   (c) the reduction half-reaction of chromate ion to chromium(III) hydroxide in basic solution.

20. Write a balanced equation to show
   (a) the formation of gas bubbles when cobalt reacts with hydrochloric acid.
   (b) the reaction of copper with nitric acid.
   (c) the reduction half-reaction of dichromate ion to Cr^{3+} in acid solution.

21. Write a balanced redox equation for the reaction of mercury with *aqua regia*, assuming the products include $HgCl_4^{2-}$ and $NO_2(g)$.

22. Write a balanced redox equation for the reaction of cadmium with *aqua regia*, assuming the products include $CdCl_4^{2-}$ and $NO(g)$.

23. Balance the following redox equations.
   (a) $Cu(s) + NO_3^-(aq) \longrightarrow$
       $$Cu^{2+}(aq) + NO_2(g) \quad \text{(acidic)}$$
   (b) $Cr(OH)_3(s) + ClO^-(aq) \longrightarrow$
       $$CrO_4^{2-}(aq) + Cl^-(aq) \quad \text{(basic)}$$

24. Balance the following redox equations.
   (a) $Fe(s) + NO_3^-(aq) \longrightarrow Fe^{3+}(aq) + NO_2(g)$ (acidic)
   (b) $Cr(OH)_3(s) + O_2(g) \longrightarrow CrO_4^{2-}(aq)$ (basic)

25. Show by calculation which of the following metals will react with hydrochloric acid (standard concentrations).
   (a) Cd (b) Cr (c) Co
   (d) Ag (e) Au

26. Show by calculation which of the metals in Problem 25 will react with nitric acid to form NO (standard concentrations).

27. Of the cations listed in Table 20.4, show by calculation which one (besides Cu^+) will disproportionate at standard conditions.

28. Using Table 17.1 (Chapter 17) calculate $E°$ for
   (a) $2Co^{3+}(aq) + H_2O \longrightarrow$
       $$2Co^{2+}(aq) + \tfrac{1}{2}O_2(g) + 2H^+(aq)$$
   (b) $2Cr^{2+}(aq) + I_2(s) \longrightarrow 2Cr^{3+}(aq) + 2I^-(aq)$

29. Using Table 20.4, calculate, for the disproportionation of Fe^{2+},
   (a) the equilibrium constant, K.
   (b) the concentration of Fe^{3+} in equilibrium with 0.10 M Fe^{2+}.

30. Using Table 20.4, calculate, for the disproportionation of Au^+,
   (a) K.
   (b) the concentration of Au^+ in equilibrium with 0.10 M Au^{3+}.

Unclassified

31. A sample of sodium liberates 2.73 L of hydrogen at 752 mm Hg and 22°C when it is added to a large amount of water. How much sodium is used?

32. A self-contained breathing apparatus contains 248 g of potassium superoxide. A firefighter exhales 116 L of air at 37°C and 748 mm Hg. The volume percent of water in exhaled air is 6.2. What mass of potassium superoxide is left after the water in the exhaled air reacts with it?

33. Taking K_{sp} $PbCl_2 = 1.7 \times 10^{-5}$ and assuming $[Cl^-] = 0.20$ M, calculate the concentration of Pb^{2+} at equilibrium.

34. The equilibrium constant for the reaction

$$2CrO_4^{2-}(aq) + 2H^+(aq) \rightleftharpoons Cr_2O_7^{2-}(aq) + H_2O$$

is 3×10^{14}. What must the pH be so that the concentrations of chromate and dichromate ion are both 0.10 M?

35. Using data in Appendix 1, estimate the temperature at which Fe_2O_3 can be reduced to iron, using hydrogen gas as a reducing agent (assume $H_2O(g)$ is the other product).

36. A 0.500-g sample of zinc-copper alloy was treated with dilute hydrochloric acid. The hydrogen gas evolved was collected by water displacement at 27°C and a total pressure of 755 mm Hg. The volume of the water displaced by the gas is 105.7 mL. What is the percent composition, by mass, of the alloy? (Vapor pressure of H_2O at 27°C is 26.74 mm Hg.) Assume only the zinc reacts.

37. One type of stainless steel contains 22% nickel by mass. How much nickel sulfide ore, NiS, is required to produce one metric ton of stainless steel?

38. Silver is obtained in much the same manner as gold, using NaCN solution and O_2. Describe with appropriate equations the extraction of silver from argentite ore, Ag_2S. (The products are SO_2 and $Ag(CN)_2^-$, which are reduced with zinc.)

39. Iron(II) can be oxidized to iron(III) by permanganate ion in acidic solution. The permanganate ion is reduced to manganese(II) ion.

(a) Write the oxidation half-reaction, the reduction half-reaction, and the overall redox equation.

(b) Calculate $E°$ for the reaction.

(c) Calculate the percentage of Fe in an ore if a 0.3500-g sample is dissolved and the Fe^{2+} formed requires for titration 55.63 mL of a 0.0200 M solution of $KMnO_4$.

40. Of the cations listed in the center column of Table 20.4, which one is the

(a) strongest reducing agent?

(b) strongest oxidizing agent?

(c) weakest reducing agent?

(d) weakest oxidizing agent?

Challenge Problems

41. A sample of 20.00 g of barium reacts with oxygen to form 22.38 g of a mixture of barium oxide and barium peroxide. Determine the composition of the mixture.

42. Rust, which you can take to be $Fe(OH)_3$, can be dissolved by treating it with oxalic acid. An acid-base reaction occurs, and a complex ion is formed.

(a) Write a balanced equation for the reaction.

(b) What volume of 0.10 M $H_2C_2O_4$ would be required to remove a rust stain weighing 1.0 g?

43. A 0.500-g sample of steel is analyzed for manganese. The sample is dissolved in acid and the manganese is oxidized to permanganate ion. A measured excess of Fe^{2+} is added to reduce MnO_4^- to Mn^{2+}. The excess Fe^{2+} is determined by titration with $K_2Cr_2O_7$. If 75.00 mL of 0.125 M $FeSO_4$ is added and the excess requires 13.50 mL of 0.100 M $K_2Cr_2O_7$ to oxidize Fe^{2+}, calculate the percent by mass of Mn in the sample.

44. Calculate the temperature in °C at which the equilibrium constant (K) for the following reaction is 1.00.

$$MnO_2(s) \longrightarrow Mn(s) + O_2(g)$$

45. A solution of potassium dichromate is made basic with sodium hydroxide; the color changes from red to yellow. Addition of silver nitrate to the yellow solution gives a precipitate. This precipitate dissolves in concentrated ammonia but re-forms when nitric acid is added. Write balanced net ionic equations for all the reactions in this sequence.

Chemistry of the Nonmetals

National Gallery, London / Art Resource, NY

The electrical charge that creates this bolt of lightning also changes some of the oxygen in the air to ozone.

For what can so fire us,
Enrapture, inspire us,
As Oxygen? What so
  delicious to quaff?
It is so stimulating,
And so titillating,
E'en grey-beards turn
  freshy, dance, caper,
  and laugh.

—JOHN SHIELD
Oxygen Gas

Approximately 18 elements are classified as nonmetals; they lie above and to the right of the "stairway" that runs diagonally across the periodic table (Figure 21.1). As the word "nonmetal" implies, these elements do not show metallic properties; in the solid state they are brittle as opposed to ductile, insulators rather than conductors. Most of the nonmetals, particularly those in Groups 15 to 17 of the periodic table, are molecular in nature (e.g., N_2, O_2, F_2). The noble gases (Group 18) consist of individual atoms attracted to each other by weak dispersion forces. Carbon in Group 14 has a network covalent structure.

As indicated in Figure 21.1, this chapter concentrates on the more common and/or more reactive nonmetals, namely,

- nitrogen and phosphorus in Group 15.
- oxygen and sulfur in Group 16.
- the halogens (F, Cl, Br, I) in Group 17.

We will consider

- the properties of these elements and methods of preparing them (Section 21-1).
- their hydrogen compounds (Section 21-2).
- their oxides (Section 21-3).
- their oxoacids and oxoanions (Section 21-4).

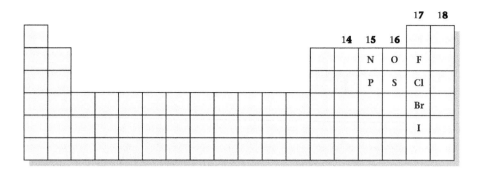

Figure 21.1 Nonmetals and the periodic table. The location of all nonmetals in the periodic table is shown in yellow. Symbols are given for the nonmetallic elements discussed in this chapter.

21-1 The Elements and Their Preparation

Table 21.1 lists some of the properties of the eight nonmetals considered in this chapter. Notice that all of these elements are molecular; those of low molar mass (N_2, O_2, F_2, Cl_2) are gases at room temperature and atmospheric pressure (Figure 21.2). Stronger dispersion forces cause the nonmetals of higher molar mass to be either liquids (Br_2) or solids (I_2, P_4, S_8).

Chemical Reactivity

Of the eight nonmetals listed in Table 21.1, *nitrogen* is by far the least reactive. Its inertness is due to the strength of the triple bond holding the N_2 molecule together (bond enthalpy (B.E.) for $N\equiv N$ = 941 kJ/mol). This same factor explains why virtually all chemical explosives are compounds of nitrogen (e.g., nitroglycerin, trinitrotoluene, ammonium nitrate, lead azide). These compounds detonate exothermically to form molecular nitrogen. The reactions with ammonium nitrate and lead azide are

$$2NH_4NO_3(s) \longrightarrow 4H_2O(l) + O_2(g) + 2N_2(g) \qquad \Delta H° = -412 \text{ kJ}$$

$$Pb(N_3)_2(s) \longrightarrow Pb(s) + 3N_2(g) \qquad \Delta H° = -476 \text{ kJ}$$

Fluorine is the most reactive of all elements, in part because of the weakness of the F—F bond (B.E. F—F = 153 kJ/mol), but mostly because it is such a

Table 21.1 Properties of Nonmetallic Elements

	Nitrogen	Phosphorus	Oxygen	Sulfur	Fluorine	Chlorine	Bromine	Iodine
Outer electron configuration	$2s^2 2p^3$	$3s^2 3p^3$	$2s^2 2p^4$	$3s^2 3p^4$	$2s^2 2p^5$	$3s^2 3p^5$	$4s^2 4p^5$	$5s^2 5p^5$
Molecular formula	N_2	P_4	O_2	S_8	F_2	Cl_2	Br_2	I_2
Molar mass (g/mol)	28	124	32	257	38	71	160	254
State (25°C, 1 atm)	gas	solid	gas	solid	gas	gas	liquid	solid
Melting point (°C)	−210	44	−218	119	−220	−101	−7	114
Boiling point (°C)	−196	280	−183	444	−188	−34	59	184
Bond enthalpy* (kJ/mol)	941	200	498	226	153	243	193	151
$E°_{red}$	—	—	—	—	+2.889 V	+1.360 V	+1.077 V	+0.534 V

*In the element (triple bond in N_2, double bond in O_2).

Figure 21.2 **Chlorine (Cl₂), bromine (Br₂), and iodine (I₂).** The bulbs contain *(left to right)* gaseous chlorine and the vapors in equilibrium with liquid bromine and solid iodine.

powerful oxidizing agent ($E°_{red} = +2.889$ V). ◄ ⓘ Fluorine combines with every element in the periodic table except He and Ne. With a few metals, it forms a surface film of metal fluoride, which adheres tightly enough to prevent further reaction. This is the case with nickel, where the product is NiF_2. Fluorine gas is ordinarily stored in containers made of a nickel alloy, such as stainless steel (Fe, Cr, Ni) or Monel (Ni, Cu). Fluorine also reacts with many compounds including water, which is oxidized to a mixture of O_2, O_3, H_2O_2, and OF_2.

ⓘ Fluorine is the most electronegative element.

Chlorine is somewhat less reactive than fluorine. Although it reacts with nearly all metals (Figure 21.3), heating is often required. This reflects the relatively strong bond in the Cl_2 molecule (B.E. Cl—Cl = 243 kJ/mol). Chlorine disproportionates ◄ ⓘ in water, forming Cl^- ions (O.N. Cl = −1) and HClO molecules (O.N. Cl = +1).

ⓘ A species disproportionates when it is oxidized and reduced at the same time.

$$Cl_2(g) + H_2O \rightleftharpoons Cl^-(aq) + H^+(aq) + HClO(aq)$$

The hypochlorous acid, HClO, formed by this reaction is a powerful oxidizing agent ($E°_{red} = +1.630$ V); it kills bacteria, apparently by destroying certain enzymes essential to their metabolism (Figure 21.4). The taste and odor that we

When copper is heated...

...and plunged into a cylinder containing chlorine gas, the copper reacts vigorously, giving off sparks.

a   **b**

Figure 21.3 **Reaction of chlorine with copper.** The equation for the reaction is

$$Cu(s) + Cl_2(g) \longrightarrow CuCl_2(s)$$

Figure 21.4 **Chlorinated water.** The oxidizing power of chlorine-containing chemicals keeps swimming pool water free of disease-causing microorganisms.

associate with "chlorinated water" are actually due to compounds such as CH_3NHCl, produced by the action of hypochlorous acid on bacteria.

EXAMPLE	21.1

Consider the reaction between chlorine gas and water.

$$Cl_2(g) + H_2O \rightleftharpoons Cl^-(aq) + H^+(aq) + HClO(aq)$$

a   Write the expression for the equilibrium constant K.

STRATEGY

Recall that in the K expression

- gases enter as partial pressures in atmospheres.
- aqueous species enter as concentrations in molarity.
- water is not included.
- products are written in the numerator raised to their coefficient in the balanced equation.
- reactants are written in the denominator raised to their coefficient in the balanced equation.

SOLUTION

K expression	$K = \dfrac{[HClO][Cl^-][H^+]}{P_{Cl_2}}$

b   At the temperature of the reaction, $K = 2.7 \times 10^{-5}$. What is the concentration of HClO in equilibrium with chlorine gas at 1.0 atm?

ANALYSIS

Information given:	P_{Cl_2}(1.0 atm), $K(2.7 \times 10^{-5})$
Information implied:	from (a) K expression
Asked for:	[HClO]

STRATEGY

Substitute into the K expression obtained in (a).

Let $x =$ [HClO]

Note that the stoichiometric ratios of HClO, H^+, and Cl^- are 1:1:1, thus [HClO] = [H^+] = [Cl^-] = x

SOLUTION

[HClO]	$2.7 \times 10^{-5} = \dfrac{(x)(x)(x)}{1.0} \longrightarrow x^3 = 2.7 \times 10^{-5} \longrightarrow x = 0.030\ M$

The oxidizing power of the halogens makes them hazardous to work with. Fluorine is the most dangerous, but it is very unlikely that you will ever come across it in a teaching laboratory. You are most likely to encounter chlorine as its saturated water solution, called "chlorine water." Remember that the pressure of chlorine gas over this solution (if it is freshly prepared) is 1 atm and that chlorine was used as a poison gas in World War I. Use small quantities of chlorine water and don't breathe the vapors. Bromine, although not as strong an oxidizing agent as chlorine, can cause severe burns if it comes in contact with your skin, particularly if it gets under your fingernails.

Of the four halogens, iodine is the weakest oxidizing agent. "Tincture of iodine," a 10% solution of I_2 in alcohol, is sometimes used as an antiseptic. Hospitals most often use a product called "povidone-iodine," a quite powerful iodine-containing antiseptic and disinfectant, which can be diluted with water to the desired strength. These applications of molecular iodine should not delude you into thinking that the solid is harmless. On the contrary, if $I_2(s)$ is allowed to remain in contact with your skin, it can cause painful burns that are slow to heal.

Occurrence and Preparation

Of the eight nonmetals considered here, three (nitrogen, oxygen, and sulfur) occur in nature in elemental form. *Nitrogen* and *oxygen* are obtained from air, where their mole fractions are 0.7808 and 0.2095, respectively. When liquid air at $-200°C$ (73 K) is allowed to warm, the first substance that boils off is nitrogen (bp N_2 = 77 K). After most of the nitrogen has been removed, further warming gives oxygen (bp O_2 = 90 K). About 2×10^{10} kg of O_2 and lesser amounts of N_2 are produced annually in the United States from liquid air.

At the close of the Civil War in 1865, oil prospectors in Louisiana discovered (to their disgust) elemental *sulfur* in the caprock of vast salt domes up to 20 km² in area. The sulfur lies 60 to 600 m below the surface of the earth. The process used to mine sulfur is named after its inventor, Herman Frasch (1851–1914), an American chemical engineer (born in Germany). A diagram of the **Frasch process** is shown in Figure 21.5. The sulfur is heated to its melting point (119°C) by pumping superheated water at 165°C down one of three concentric pipes. Compressed air is used to bring the sulfur to the surface. The air and sulfur form a frothy mixture that rises through the middle pipe. On cooling, the sulfur solidifies, filling huge vats that may be 0.5 km long. The sulfur obtained in this way has a purity approaching 99.9%.

Figure 21.5 Frasch process for mining sulfur. Superheated water at 165°C is sent down through the outer pipe to form a pool of molten sulfur (mp = 119°C) at the base. Compressed air, pumped down the inner pipe, brings the sulfur to the surface. Sulfur deposits are often 100 m or more beneath the earth's surface, covered with quicksand and rock.

The *halogens* are far too reactive to occur in nature as the free elements. Instead, they are found as halide anions:

- F$^-$ in the mineral calcium fluoride, CaF$_2$ (fluorite).
- Cl$^-$ in huge underground deposits of sodium chloride, NaCl (rock salt), underlying parts of Oklahoma, Texas, and Kansas.
- Br$^-$(aq) and I$^-$(aq) in brine wells in Arkansas (conc. Br$^-$ = 0.05 M) and Michigan (conc. I$^-$ = 0.001 M), respectively.

The fluoride and chloride ions are very difficult to oxidize (E°_{ox} F$^-$ = −2.889 V; E°_{ox} Cl$^-$ = −1.360 V). Hence the elements fluorine and chlorine are ordinarily prepared by electrolytic oxidation, using a high voltage. As pointed out in Chapter 17, chlorine is prepared by the electrolysis of aqueous sodium chloride:

$$2Cl^-(aq) + 2H_2O \longrightarrow Cl_2(g) + H_2(g) + 2OH^-(aq)$$

The process used to prepare fluorine was developed by Henri Moissan (1852–1907) in Paris in 1886. It won him one of the early (1906) Nobel Prizes in chemistry. The electrolyte is a mixture of HF and KF in a 2:1 mole ratio. At 100°C, fluorine is generated by the decomposition of hydrogen fluoride:

$$2HF(l) \longrightarrow H_2(g) + F_2(g)$$

Potassium fluoride furnishes the ions required to carry the electric current.

Bromide and iodide ions are easier to oxidize (E°_{ox} Br$^-$ = −1.077 V; E°_{ox} I$^-$ = −0.534 V), so bromine and iodine can be prepared by chemical oxidation. Commonly, the oxidizing agent is chlorine gas (E°_{red} = +1.360 V):

$$Cl_2(g) + 2Br^-(aq) \longrightarrow 2Cl^-(aq) + Br_2(l)$$
$$Cl_2(g) + 2I^-(aq) \longrightarrow 2Cl^-(aq) + I_2(s)$$

21-2 Hydrogen Compounds of Nonmetals

Table 21.2 lists some of the more important hydrogen compounds of the nonmetals. (Those of carbon are discussed in Chapter 22.) The physical states listed are those observed at 25°C and 1 atm. The remainder of this section discusses the chemical properties of the compounds shown in boldface in the table.

Ammonia, NH$_3$

Ammonia is one of the most important industrial chemicals; more than ten million tons of NH$_3$ are produced annually in the United States. You will recall (Chapter 12) that it is made by the Haber process

$$N_2(g) + 3H_2(g) \rightleftharpoons 2NH_3(g) \qquad 450°C, 200–600 \text{ atm, solid catalyst}$$

Table 21.2 Hydrogen Compounds of the Nonmetals

Group 15	Group 16	Group 17
Ammonia, NH$_3$(g)	Water, H$_2$O(l)	**Hydrogen fluoride, HF(g)**
Hydrazine, N$_2$H$_4$(l)	**Hydrogen peroxide, H$_2$O$_2$(l)**	
Hydrazoic acid, HN$_3$(l)		
Phosphine, PH$_3$(g)	**Hydrogen sulfide, H$_2$S(g)**	**Hydrogen chloride, HCl(g)**
Diphosphine, P$_2$H$_4$(l)		
		Hydrogen bromide, HBr(g)
		Hydrogen iodide, HI(g)

Ammonia is used to make fertilizers and a host of different nitrogen compounds, notably nitric acid, HNO_3.

The NH_3 molecule acts as a *Brønsted-Lowry base* in water, accepting a proton from a water molecule:

$$NH_3(aq) + H_2O \rightleftharpoons NH_4^+(aq) + OH^-(aq) \qquad K_b = 1.8 \times 10^{-5}$$

Ammonia can also act as a *Lewis base* ◄ⓘ when it reacts with a metal cation to form a complex ion

The NH$_3$ molecule donates an electron pair to Ag$^+$.

$$2NH_3(aq) + Ag^+(aq) \longrightarrow Ag(NH_3)_2^+(aq)$$

or with an electron-deficient compound such as boron trifluoride:

Lewis base   Lewis acid

Ammonia is often used to precipitate insoluble metal hydroxides such as $Al(OH)_3$. The OH^- ions formed when ammonia reacts with water precipitate the cation from solution as the hydroxide. The overall equation for the reaction is

$$Al^{3+}(aq) + 3NH_3(aq) + 3H_2O \longrightarrow Al(OH)_3(s) + 3NH_4^+(aq)$$

Nitrogen cannot have an oxidation number lower than -3, which means that when NH_3 takes part in a redox reaction, it always acts as a *reducing agent*. Ammonia may be oxidized to elementary nitrogen or to a compound of nitrogen. An important redox reaction of ammonia is that with hypochlorite ion:

$$2NH_3(aq) + ClO^-(aq) \longrightarrow N_2H_4(aq) + Cl^-(aq) + H_2O$$

Hydrazine, N_2H_4, is made commercially by this process. Certain byproducts of this reaction, notably NH_2Cl and $NHCl_2$, are both toxic and explosive, so solutions of household bleach and ammonia should never be mixed with one another.

Hydrogen Sulfide, H$_2$S

In water solution, hydrogen sulfide acts as a *Brønsted-Lowry acid*; it can donate a proton to a water molecule:

$$H_2S(aq) + H_2O \rightleftharpoons HS^-(aq) + H_3O^+(aq) \qquad K_a = 1.0 \times 10^{-7}$$

Like ammonia, hydrogen sulfide can act as a precipitating agent toward metal cations. The reaction with Cd^{2+} is typical:

$$Cd^{2+}(aq) + H_2S(aq) \longrightarrow CdS(s) + 2H^+(aq)$$

Like ammonia, hydrogen sulfide (O.N. S $= -2$) can act only as a reducing agent when it takes part in redox reactions. Most often the H_2S is oxidized to elementary sulfur, as in the reaction

$$2H_2S(aq) + O_2(g) \longrightarrow 2S(s) + 2H_2O$$

The sulfur formed is often very finely dispersed, which explains why aqueous solutions of hydrogen sulfide in contact with air have a milky appearance.

If you've worked with H_2S in the laboratory, you won't soon forget its rotten-egg odor. In a sense, it's fortunate that hydrogen sulfide has such a distinctive odor. This gas is highly toxic, as poisonous as HCN. At a concentration of 10 parts per million, H_2S can cause headaches and nausea; at 100 ppm it can be fatal.

EXAMPLE 21.2

When a solution containing Cu^{2+} is treated with hydrogen sulfide, a black precipitate forms. When another portion of the solution is treated with ammonia, a blue precipitate forms. This precipitate dissolves in excess ammonia to form a deep blue solution containing the $Cu(NH_3)_4{}^{2+}$ ion. Write balanced net ionic equations to explain these observations.

STRATEGY AND SOLUTION

1. The black precipitate is CuS.

$$Cu^{2+}(aq) + H_2S(aq) \longrightarrow CuS(s) + 2H^+(aq)$$

2. An aqueous solution of ammonia consists of NH_3 and H_2O. Since the blue precipitate must be $Cu(OH)_2$, water has to be a reactant.

$$Cu^{2+}(aq) + 2NH_3(aq) + 2H_2O \longrightarrow Cu(OH)_2(s) + 2NH_4{}^+(aq)$$

3. The precipitate, $Cu(OH)_2$, reacts with excess ammonia to form $Cu(NH_3)_4{}^{2+}$.

$$Cu(OH)_2(s) + 4NH_3(aq) \longrightarrow Cu(NH_3)_4{}^{2+}(aq) + 2OH^-(aq)$$

Hydrogen Peroxide

In hydrogen peroxide, oxygen has an oxidation number of -1, intermediate between the extremes for the element, 0 and -2. This means that H_2O_2 can act as either an oxidizing agent, in which case it is reduced to H_2O, or as a reducing agent, when it is oxidized to O_2. In practice, hydrogen peroxide is an extremely strong oxidizing agent:

$$H_2O_2(aq) + 2H^+(aq) + 2e^- \longrightarrow 2H_2O \qquad E°_{red} = +1.763 \text{ V}$$

but a very weak reducing agent:

$$H_2O_2(aq) \longrightarrow O_2(g) + 2H^+(aq) + 2e^- \qquad E°_{ox} = -0.695 \text{ V}$$

Hydrogen peroxide tends to decompose in water, which explains why its solutions soon lose their oxidizing power. The reaction involved is *disproportionation*, combining the two half-reactions referred to above:

$$H_2O_2(aq) + 2H^+(aq) + 2e^- \longrightarrow 2H_2O \qquad\qquad E°_{red} = +1.763 \text{ V}$$
$$H_2O_2(aq) \longrightarrow O_2(g) + 2H^+(aq) + 2e^- \qquad E°_{ox} = -0.695 \text{ V}$$
$$\overline{\rule{0pt}{0pt}\qquad\qquad 2H_2O_2(aq) \longrightarrow O_2(g) + 2H_2O \qquad\qquad E° = +1.068 \text{ V}}$$

Figure 21.6 Disproportionation of H_2O_2, catalyzed by MnO_2.

Half of the H_2O_2 molecules are reduced to H_2O; half are oxidized to O_2.

© Cengage Learning/Charles D. Winters

This reaction (Figure 21.6) is catalyzed by a wide variety of materials, including I^- ions, MnO_2, metal surfaces (Pt, Ag), and even by traces of OH^- ions dissolved from glass.

You are most likely to come across hydrogen peroxide as its water solution. Solutions available in a drugstore for medicinal use contain 3% to 6% by mass of H_2O_2. Industrially, concentrations as high as 86% H_2O_2 are available. All water solutions of hydrogen peroxide contain stabilizers to prevent disproportionation from taking place during storage.

EXAMPLE 21.3

Using Table 17.1, decide whether hydrogen peroxide will react with (oxidize or reduce) the following ions in acid solution at standard concentrations:

I^-, Sn^{2+}, and Co^{2+}

STRATEGY

1. Since H_2O_2 disproportionates, it can either be reduced (and oxidize the ions) or be oxidized (and reduce the ions).

2. For a redox reaction to occur where H_2O_2 is reduced,

$$E°_{red} \, H_2O_2(1.763 \text{ V}) + E°_{ox} \text{ ion} > 0$$

continued

3. For a redox reaction to occur where H_2O_2 is oxidized,

$$E^\circ_{ox}\ H_2O_2(-0.695\ V) + E^\circ_{red}\ ion > 0$$

SOLUTION

I⁻

oxidation	$(E^\circ_{ox}\ I^- = -0.534\ V) + (E^\circ_{red}\ H_2O_2 = 1.763\ V) > 0$ H_2O_2 can oxidize I^- to I_2.
reduction	Table 17.1 shows no reduction potential for I^-.

Sn²⁺

oxidation	$(E^\circ_{ox}\ Sn^{2+} = -0.154\ V) + (E^\circ_{red}\ H_2O_2 = 1.763\ V) > 0$ H_2O_2 can oxidize Sn^{2+} to Sn^{4+}.
reduction	$(E^\circ_{red}\ Sn^{2+} = -0.141\ V) + (E^\circ_{ox}\ H_2O_2 = -0.695\ V) < 0$ H_2O_2 cannot reduce Sn^{2+} to Sn.

Co²⁺

oxidation	$(E^\circ_{ox}\ Co^{2+} = -1.953\ V) + (E^\circ_{red}\ H_2O_2 = 1.763\ V) < 0$ H_2O_2 cannot oxidize Co^{2+} to Co^{3+}.
reduction	$(E^\circ_{red}\ Co^{2+} = -0.282\ V) + (E^\circ_{ox}\ H_2O_2 = -0.695\ V) < 0$ H_2O_2 cannot reduce Co^{2+} to Co.

Hydrogen Fluoride and Hydrogen Chloride

The most common hydrogen halides are HF (U.S. production $= 3 \times 10^8$ kg/yr) and HCl (3×10^9 kg/yr). They are most familiar as water solutions, referred to as hydrofluoric acid and hydrochloric acid, respectively. Recall (Chapter 13) that hydrofluoric acid is weak, incompletely dissociated in water, whereas HCl is a strong acid.

$$HF(aq) \rightleftharpoons H^+(aq) + F^-(aq) \qquad K_a = 6.9 \times 10^{-4}$$
$$HCl(aq) \longrightarrow H^+(aq) + Cl^-(aq) \qquad K_a \longrightarrow \infty$$

The difference in K_a values is due largely to the difference in bond strength. The H—F bond (bond enthalpy = 565 kJ/mol) is much stronger than the H—Cl bond (bond enthalpy = 431 kJ/mol).

Hydrofluoric and hydrochloric acids undergo very similar reactions with bases such as OH^- or CO_3^{2-} ions. The equations for these reactions look somewhat different because of the difference in acid strength. Thus for the reaction of hydrochloric acid with a solution of sodium hydroxide, the equation is simply

$$H^+(aq) + OH^-(aq) \longrightarrow H_2O$$

With hydrofluoric acid, the equation is

$$HF(aq) + OH^-(aq) \longrightarrow H_2O + F^-(aq) \blacktriangleleft \textit{\textbf{i}}$$

HF appears in this equation because aqueous hydrofluoric acid is weak, containing many more HF molecules than H^+ ions. Similarly, for the reaction of these two acids with a solution of sodium carbonate:

hydrochloric acid: $\quad 2H^+(aq) + CO_3^{2-}(aq) \longrightarrow CO_2(g) + H_2O$

hydrofluoric acid: $\quad 2HF(aq) + CO_3^{2-}(aq) \longrightarrow CO_2(g) + H_2O + 2F^-(aq)$

Concentrated hydrofluoric acid reacts with glass, which can be considered to be a mixture of SiO_2 and ionic silicates such as calcium silicate, $CaSiO_3$:

$$SiO_2(s) + 4HF(aq) \longrightarrow SiF_4(g) + 2H_2O$$
$$CaSiO_3(s) + 6HF(aq) \longrightarrow SiF_4(g) + CaF_2(s) + 3H_2O$$

As you might guess, HF solutions are never stored in glass bottles; plastic is used instead. ◀ ⓘ Hydrofluoric acid is sometimes used to etch glass. The glass object is

ⓘ Hydrochloric acid contains H^+ and Cl^- ions; hydrofluoric acid contains mostly HF molecules.

ⓘ How do you suppose HF(aq) was stored B.P. (before plastics)?

Figure 21.7 A test for limestone. Geologists use a few drops of hydrochloric acid to identify limestone, which is calcium carbonate.

first covered with a thin protective coating of wax or plastic. Then the coating is removed from the area to be etched and the glass is exposed to the HF solution. Thermometer stems and burets can be etched or light bulbs frosted in this way.

Hydrogen fluoride is a very unpleasant chemical to work with. If spilled on the skin, it removes Ca^{2+} ions from the tissues, forming insoluble CaF_2. A white patch forms that is agonizingly painful to the touch. To make matters worse, HF is a local anesthetic, so a person may be unaware of what is happening until it is too late.

Rest assured that you will never have occasion to use hydrofluoric acid in the general chemistry laboratory. Neither are you likely to come in contact with HBr or HI, both of which are relatively expensive. Hydrochloric acid, on the other hand, is a workhorse chemical of the teaching laboratory. It is commonly available as "dilute HCl" (6 M) or "concentrated HCl" (12 M). You will use it as a source of H^+ ions for such purposes as

- dissolving insoluble carbonates, like limestone (Figure 21.7), or hydroxides.

$$Ag_2CO_3(s) + 2H^+(aq) \longrightarrow 2Ag^+(aq) + CO_2(g) + H_2O$$
$$Zn(OH)_2(s) + 2H^+(aq) \longrightarrow Zn^{2+}(aq) + 2H_2O$$

- converting a weak base such as NH_3 to its conjugate acid.

$$NH_3(aq) + H^+(aq) \longrightarrow NH_4^+(aq)$$

- generating $H_2(g)$ by reaction with a metal.

$$Zn(s) + 2H^+(aq) \longrightarrow Zn^{2+}(aq) + H_2(g)$$

EXAMPLE 21.4

Write balanced net ionic equations to explain why

(a) aluminum hydroxide dissolves in hydrochloric acid.

(b) carbon dioxide gas is evolved when calcium carbonate is treated with hydrochloric acid.

(c) carbon dioxide gas is evolved when calcium carbonate is treated with hydrofluoric acid.

STRATEGY AND SOLUTION

(a) When a precipitate is said to dissolve, it means that the solid is broken up into its component ions.

$$Al(OH)_3(s) \longrightarrow Al^{3+}(aq) + 3OH^-(aq)$$

The OH^- ion then reacts with the appropriate ion contributed by HCl, in this case H^+

$$OH^-(aq) + H^+(aq) \longrightarrow H_2O$$

The reaction is

$$Al(OH)_3(s) + 3H^+(aq) \longrightarrow Al^{3+}(aq) + 3H_2O$$

(b) Recall that a carbonate (CO_3^{2-}) reacts with an acid by releasing CO_2 and water. In this case, the reacting species is H^+ because HCl is a strong acid (Chapter 4).

$$CaCO_3(s) + 2H^+(aq) \longrightarrow Ca^{2+}(aq) + CO_2(g) + H_2O$$

(c) As in (b), CO_2 is evolved, but the reacting species is HF because it is a weak acid.

$$CaCO_3(s) + 2HF(aq) \longrightarrow Ca^{2+}(aq) + CO_2(g) + 2F^-(aq) + H_2O$$

21-3 Oxygen Compounds of Nonmetals

Table 21.3 lists some of the more familiar nonmetal oxides. Curiously enough, only 5 of the 21 compounds shown are thermodynamically stable at 25°C and 1 atm (P_4O_{10}, P_4O_6, SO_3, SO_2, I_2O_5). The others, including all of the oxides of nitrogen and chlorine, have positive free energies of formation at 25°C and 1 atm. For example, ΔG_f° $NO_2(g) = +51.3$ kJ; ΔG_f° $ClO_2(g) = +120.6$ kJ. Kinetically, these compounds stay around long enough to have an extensive chemistry.

Table 21.3 Nonmetal Oxides*

Group 15	Group 16	Group 17
$N_2O_5(s)$, $N_2O_4(g)$, $NO_2(g)$		$OF_2(g)$, $O_2F_2(g)$
$N_2O_3(d)$, $NO(g)$, $N_2O(g)$		
$P_4O_{10}(s)$, $P_4O_6(s)$	$SO_3(l)$, $SO_2(g)$	$Cl_2O_7(l)$, $Cl_2O_6(l)$
		$ClO_2(g)$, $Cl_2O(g)$
		$BrO_2(d)$, $Br_2O(d)$
		$I_2O_5(s)$, $I_4O_9(s)$, $I_2O_4(s)$

*The states listed are those observed at 25°C and 1 atm. Compounds that decompose below 25°C are listed as (d). Oxides shown in boldface are discussed in the text.

Nitrogen dioxide, a reddish-brown gas, is a major factor in the formation of photochemical smog (Figure 21.8). Chlorine dioxide, a yellow gas, is widely used as an industrial bleach and water purifier, even though it tends to explode at partial pressures higher than 50 mm Hg.

Molecular Structures of Nonmetal Oxides

The Lewis structures of the oxides of nitrogen are shown in Figure 21.9. Two of these species, NO and NO_2, are paramagnetic, with one unpaired electron. When nitrogen dioxide is cooled, it dimerizes; the unpaired electrons combine to form a single bond between the two nitrogen atoms:

$$2NO_2(g) \rightleftharpoons N_2O_4(g)$$

A similar reaction occurs when an equimolar mixture of NO and NO_2 is cooled. Two odd electrons, one from each molecule, pair off to form an N—N bond:

$$NO_2(g) + NO(g) \rightleftharpoons N_2O_3(g)$$

At −20°C, dinitrogen trioxide separates from the mixture as a blue liquid.

Figure 21.8 **Smog over Los Angeles.** From a distance, you can clearly see the layer of smog containing reddish-brown nitrogen dioxide.

Dimerization occurs when two identical molecules combine.

Figure 21.9 **Lewis structures of the oxides of nitrogen.** Many other resonance forms are possible.

N_2O_5 (dinitrogen pentoxide)

N_2O_4 (dinitrogen tetraoxide)

N_2O_3 (dinitrogen trioxide)

NO_2 (nitrogen dioxide)

NO (nitrogen oxide or nitric oxide)

N_2O (dinitrogen oxide or nitrous oxide)

Nitric oxide (NO) is a minor but villainous component of the atmosphere. It is involved in the formation of both smog (Chapter 11) and acid rain (Chapter 14). You may be surprised to learn that small amounts of NO are also produced in the human body, where it has a generally beneficial effect. In particular, it has the ability to dilate blood vessels, lowering blood pressure and reducing the likelihood of strokes or heart attacks. Beyond that, NO is effective in treating what television commercials refer to as "erectile dysfunction"; it increases blood flow to the penis.

Perhaps the best known oxide of nitrogen is N_2O, commonly called nitrous oxide or "laughing gas." Nitrous oxide is frequently used as an anesthetic, particularly in dentistry. It is also the propellant gas used in whipped cream containers; N_2O is nontoxic, virtually tasteless, and quite soluble in vegetable oils. The N_2O molecule, like all those in Figure 21.9, can be represented as a resonance hybrid.

EXAMPLE 21.5

Consider the N_2O molecule shown in Figure 21.9. Draw another resonance form of N_2O and determine its bond angle and polarity.

STRATEGY

Recall the discussion in Chapter 7 (Sections 7-1, 7-2, and 7-3) where the principles of resonance, molecular geometry, and polarity are considered.

SOLUTION

resonance forms	$:\ddot{O}=N=\ddot{N}:$ or $:O\equiv N-\ddot{N}:$
bond angle	The molecule is linear. Its bond angle is 180°.
polarity	The molecule has two different atoms bonded to the central atom N. The molecule is polar.

The structures of SO_2 and SO_3 were referred to in Chapter 7. These molecules are often cited as examples of resonance; sulfur trioxide, for example, has three equivalent resonance structures:

As you might expect, the SO_3 molecule is nonpolar, with 120° bond angles.

Of the two oxides of phosphorus, P_4O_6 and P_4O_{10}, the latter is the more stable; it is formed when white phosphorus burns in air: ◀ *i*

$$P_4(s) + 5O_2(g) \longrightarrow P_4O_{10}(s)$$

Note from Figure 21.10 that in both oxides, as in P_4 itself, the four phosphorus atoms are at the corners of a tetrahedron. The P_4O_6 molecule can be visualized as derived from P_4 by inserting an oxygen atom between each pair of phosphorus atoms. In P_4O_{10}, an extra oxygen atom is bonded to each phosphorus.

i In a limited supply of air, some P_4O_6 is formed.

Reactions of Nonmetal Oxides with Water

Many nonmetal oxides are acidic in the sense that they react with water to form acids. Looking at the reaction

$$SO_3(g) + H_2O(l) \longrightarrow H_2SO_4(l)$$

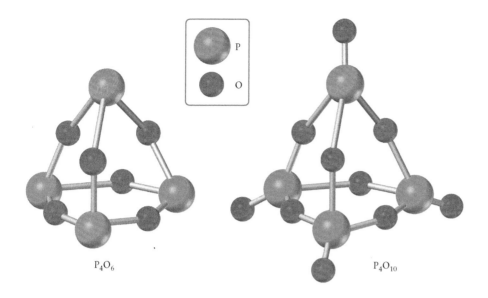

P₄O₆   P₄O₁₀

Figure 21.10 Ball-and-stick models of the oxides of phosphorus.

you can see that sulfur trioxide is an *acidic oxide*. Notice that in this reaction, the nonmetal does not change oxidation number; sulfur is in the +6 state in both SO_3 and H_2SO_4. Other acidic oxides include N_2O_5 and N_2O_3:

+5 nitrogen: $N_2O_5(s) + H_2O(l) \longrightarrow 2HNO_3(l)$

+3 nitrogen: $N_2O_3(g) + H_2O(l) \longrightarrow 2HNO_2(aq)$

The products are nitric acid, HNO_3, and an aqueous solution of nitrous acid, HNO_2.

One of the most important reactions of this type involves the +5 oxide of phosphorus, P_4O_{10}. Here the product is phosphoric acid, H_3PO_4:

$$P_4O_{10}(s) + 6H_2O(l) \longrightarrow 4H_3PO_4(s)$$

This reaction is used to prepare high-purity phosphoric acid and salts of that acid for use in food products (Figure 21.11). Phosphoric acid, H_3PO_4, is added in small amounts to soft drinks to give them a tart taste. It is present to the extent of about 0.05 mass percent in colas, 0.01 mass percent in root beers.

Figure 21.11 Household products containing phosphoric acid or its salts.

EXAMPLE 21.6

Give the formula of the oxide that reacts with water to form H_3PO_3, $HClO$, and H_2SO_3.

STRATEGY

Refer to Table 21.3. Find an oxide in which the central atom has the same oxidation number as the acid.

SOLUTION

H_3PO_3	oxidation number: +3; oxide with oxidation number +3: P_4O_6
$HClO$	oxidation number: +1; oxide with oxidation number +1: Cl_2O
H_2SO_3	oxidation number: +4; oxide with oxidation number +4: SO_2

21-4 Oxoacids and Oxoanions

Table 21.4 lists some of the more important oxoacids of the nonmetals. In all these compounds, the ionizable hydrogen atoms are bonded to oxygen, not to the central nonmetal atom. Dissociation of one or more protons from the oxoacid gives the corresponding oxoanion (Figure 21.12).

Table 21.4 Oxoacids of the Nonmetals

Group 15	Group 16	Group 17
HNO_3, HNO_2*		
H_3PO_4, H_3PO_3	H_2SO_4, H_2SO_3*	$HClO_4$, $HClO_3$*, $HClO_2$*, $HClO$*
		$HBrO_4$*, $HBrO_3$*, $HBrO$*
		HIO_4, H_5IO_6, HIO_3, HIO*

*These compounds cannot be isolated from water solution.

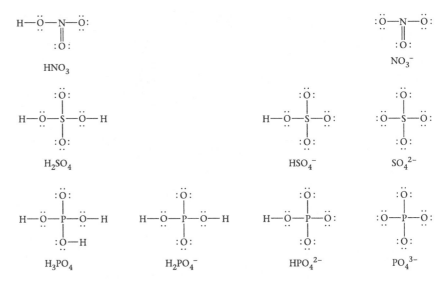

Figure 21.12 Lewis structures of the oxoacids HNO_3, H_2SO_4, H_3PO_4, and the oxoanions derived from them.

In this section we discuss the principles that allow you to predict the relative acid strengths of oxoacids such as those listed in Table 21.4. Then we consider their strengths as oxidizing and/or reducing agents. Finally, we take a closer look at the chemistry of three important oxoacids: HNO_3, H_2SO_4, and H_3PO_4.

Acid Strength

The acid equilibrium constants of the oxoacids of the halogens are listed in Table 21.5. Notice that the value of K_a increases with

- *increasing oxidation number of the central atom*
  ($HClO < HClO_2 < HClO_3 < HClO_4$)
- *increasing electronegativity of the central atom*
  ($HIO < HBrO < HClO$)

Table 21.5 Equilibrium Constants of Oxoacids of the Halogens

Oxid. State		K_a		K_a		K_a
+7	$HClO_4$	~10^7	$HBrO_4$	~10^6	HIO_4*	1.4×10^1
+5	$HClO_3$	~10^3	$HBrO_3$	3.0	HIO_3	1.6×10^{-1}
+3	$HClO_2$	1.0×10^{-2}	—	—	—	—
+1	$HClO$	2.8×10^{-8}	$HBrO$	2.6×10^{-9}	HIO	2.4×10^{-11}

*Estimated; in water solution the stable species is H_5IO_6, whose first K_a value is 5×10^{-4}.

These trends are general ones, observed with other oxoacids of the nonmetals. Recall, for example, that nitric acid, HNO_3 (O.N. N = +5), is a strong acid, completely ionized in water. In contrast, nitrous acid, HNO_2 (O.N. N = +3), is a weak acid ($K_a = 6.0 \times 10^{-4}$). The electronegativity effect shows up with the strengths of the oxoacids of sulfur and selenium:

$$K_{a1}\ H_2SO_3 = 1.7 \times 10^{-2} \quad K_{a1}\ H_2SeO_3 = 2.7 \times 10^{-3} \quad (\text{E.N. S} = 2.6, \text{Se} = 2.5)$$

Trends in acid strength can be explained in terms of molecular structure. In an oxoacid molecule, the hydrogen atom that dissociates is bonded to oxygen, which in turn is bonded to a nonmetal atom, X. The ionization in water of an oxoacid H—O—X can be represented as

$$H{-}O{-}X(aq) \rightleftharpoons H^+(aq) + XO^-(aq)$$

For a proton, with its +1 charge, to separate from the molecule, the electron density around the oxygen should be as low as possible. This will weaken the O—H bond and favor ionization. The electron density around the oxygen atom is decreased when:

- **X is a highly electronegative atom such as Cl.** This draws electrons away from the oxygen atom and makes hypochlorous acid stronger than hypoiodous acid.
- **Additional, strongly electronegative oxygen atoms are bonded to X.** These tend to draw electrons away from the oxygen atom bonded to H. Thus we would predict that the ease of dissociation of a proton, and hence K_a, should increase in the following order, from left to right:

$$X{-}O{-}H < \underset{\underset{O}{|}}{O{-}X}{-}O{-}H < \underset{\underset{O}{|}}{O{-}X}{-}O{-}H < \overset{\overset{O}{\|}}{\underset{\underset{O}{|}}{O{-}X}}{-}O{-}H$$

O.N. X: +1   +3   +5   +7

EXAMPLE 21.7

Consider sulfurous acid, H_2SO_3.

a   Show its Lewis structure and that of the HSO_3^- and SO_3^{2-} ions.

STRATEGY AND SOLUTION

Reread the discussion on writing Lewis structures in Chapter 7.

sulfurous acid   hydrogen sulfite ion   sulfite ion

b   How would its acid strength compare with that of H_2SO_4? H_2TeO_3?

STRATEGY AND SOLUTION

1. Predict acid strength on the basis of electronegativity when the central atoms of the acids being compared are different.

$$H_2SO_3 \qquad\qquad \text{vs} \qquad\qquad H_2TeO_3$$

$$\text{S} \quad\text{more electronegative than}\quad \text{Te}$$

H_2SO_3 is a stronger acid than H_2TeO_3.

continued

---

2. Predict acid strength on the basis of oxidation number when the central atoms of the acids being compared are identical.

$$H_2SO_3 \qquad\qquad vs \qquad\qquad H_2SO_4$$

oxidation number of S: $+4$ $\qquad\qquad$ oxidation number of S: $+6$

H_2SO_4 is a stronger acid than H_2SO_3.

Oxidizing and Reducing Strength

Many of the reactions of oxoacids and oxoanions involve oxidation and reduction. There are certain general principles that apply, regardless of the particular species involved.

Many oxoanions are very powerful oxidizing agents.

1. *A species in which a nonmetal is in its highest oxidation state can act only as an oxidizing agent, ◄ ⓘ never as a reducing agent.* Consider, for example, the ClO_4^- ion, in which chlorine is in its highest oxidation state, $+7$. In any redox reaction in which this ion takes part, chlorine must be reduced to a lower oxidation state. When that happens, the ClO_4^- ions act as an oxidizing agent, taking electrons away from something else. The same argument applies to

   - the SO_4^{2-} ion (highest O.N. S $= +6$).
   - the NO_3^- ion (highest O.N. N $= +5$).

   Note that, in general, *the highest oxidation number of a nonmetal is given by the second digit of its group number* (**17** for Cl, **16** for S, **15** for N).

2. *A species in which a nonmetal is in an intermediate oxidation state can act as either an oxidizing agent or a reducing agent.* Consider, for example, the ClO_3^- ion (O.N. Cl $= +5$). It can be oxidized to the perchlorate ion, in which case ClO_3^- acts as a reducing agent:

$$ClO_3^-(aq) + H_2O \longrightarrow ClO_4^-(aq) + 2H^+(aq) + 2e^- \qquad E_{ox}^\circ = -1.226 \text{ V}$$

   Alternatively, the ClO_3^- ion can be reduced, perhaps to a Cl^- ion. When that occurs, ClO_3^- acts as an oxidizing agent:

$$ClO_3^-(aq) + 6H^+(aq) + 6e^- \longrightarrow Cl^-(aq) + 3H_2O \qquad E_{red}^\circ = +1.442 \text{ V}$$

   Notice that the ClO_3^- ion is a much stronger oxidizing agent ($E_{red}^\circ = +1.442$ V) than reducing agent ($E_{ox}^\circ = -1.226$ V). This is generally true of oxoanions and oxoacids in an intermediate oxidation state, at least in acidic solution. Compare, for example,

   HClO $\quad E_{red}^\circ$ (to Cl_2) $= +1.630$ V; $\qquad E_{ox}^\circ$ (to $HClO_2$) $= -1.157$ V
   HNO₂ $\quad E_{red}^\circ$ (to NO) $= +1.036$ V; $\qquad E_{ox}^\circ$ (to NO_2) $= -1.056$ V

3. *Sometimes, with a species such as ClO_3^-, oxidation and reduction occur together, resulting in disproportionation:*

$$4ClO_3^-(aq) \longrightarrow 3ClO_4^-(aq) + Cl^-(aq) \qquad E^\circ = +0.216 \text{ V}$$

   In general, *a species in an intermediate oxidation state is expected to disproportionate if the sum $E_{ox}^\circ + E_{red}^\circ$ is a positive number.*

4. *The oxidizing strength of an oxoacid or oxoanion is greatest at high [H^+] (low pH). Conversely, its reducing strength is greatest at low [H^+] (high pH).* This principle has a simple explanation. Looking back at the half-equations

above, you can see that

- when ClO_3^- acts as an oxidizing agent, the H^+ ion is a reactant, so increasing its concentration makes the process more spontaneous.
- when ClO_3^- acts as a reducing agent, the H^+ ion is a product; to make the process more spontaneous, $[H^+]$ should be lowered.

EXAMPLE 21.8

Calculate E_{red} and E_{ox} at 25°C for the ClO_3^- ion in neutral solution, at pH 7.00, assuming all other species are at standard concentration ($E_{red}^\circ = +1.442$ V; $E_{ox}^\circ = -1.226$ V). Will the ClO_3^- ion disproportionate at pH 7.00?

ANALYSIS

Information given:	ClO_3^-: E_{red}° (1.442 V); E_{ox}° (−1.226 V) pH (7.00); T (25°C) All species besides ClO_3^- are at 1.00 M.
Asked for:	Will ClO_3^- disproportionate at the given pH?

STRATEGY

1. Write a half-equation for the reduction of ClO_3^-.

2. Calculate E_{red} by substituting into the Nernst equation for 25°C.

$$E_{red} = E_{red}^\circ - \frac{0.0257}{n} \ln Q$$

3. Write a half-equation for the oxidation of ClO_3^-.

4. Calculate E_{ox} by substituting into the Nernst equation for 25°C.

$$E_{ox} = E_{ox}^\circ - \frac{0.0257}{n} \ln Q$$

5. Find E.

$$E = E_{red} + E_{ox}$$

If $E > 0$, ClO_3^- will disproportionate.

SOLUTION

1. Reduction half-reaction	$ClO_3^-(aq) + 6H^+(aq) + 6e^- \longrightarrow Cl^-(aq) + 3H_2O$
2. E_{red}	$E_{red} = 1.442 - \dfrac{0.0257}{6} \ln \dfrac{[Cl^-]}{[ClO_3^-][H^+]^6} = 1.442 - \dfrac{0.0257}{6} \ln \dfrac{(1)}{(1)(1.00 \times 10^{-7})^6}$ $E_{red} = 1.028$ V
3. Oxidation half-reaction	$ClO_3^-(aq) + H_2O \longrightarrow ClO_4^-(aq) + 2H^+(aq) + 2e^-$
4. E_{ox}	$E_{ox} = -1.226 - \dfrac{0.0257}{2} \ln \dfrac{[ClO_4^-][H^+]^2}{[ClO_3^-]} = -1.226 - \dfrac{0.0257}{2} \ln \dfrac{(1)(1 \times 10^{-7})^2}{(1)}$ $E_{ox} = -0.812$ V
5. E	$E = E_{red} + E_{ox} = 1.028 + (-0.812) = 0.216$ V Disproportionation should occur.

END POINT

Notice that E is the same as E°, $+0.216$ V. You could have predicted that (and saved a lot of work!), because the overall equation for the disproportionation

$$4ClO_3^-(aq) \longrightarrow 3ClO_4^-(aq) + Cl^-(aq)$$

does not involve H^+ or OH^- ions.

Nitric Acid, HNO$_3$

Commercially, nitric acid is made by a three-step process developed by the German physical chemist Wilhelm Ostwald (1853–1932). The starting material is ammonia, which is burned in an excess of air at 900°C, using a platinum-rhodium catalyst:

$$4NH_3(g) + 5O_2(g) \longrightarrow 4NO(g) + 6H_2O(g)$$

The gaseous mixture formed is cooled and mixed with more air to convert NO to NO$_2$:

$$2NO(g) + O_2(g) \longrightarrow 2NO_2(g)$$

Finally, nitrogen dioxide is bubbled through water to produce nitric acid:

$$3NO_2(g) + H_2O(l) \longrightarrow NO(g) + 2HNO_3(aq)$$

Nitric acid is a strong acid, completely ionized to H$^+$ and NO$_3^-$ ions in dilute water solution:

$$HNO_3(aq) \longrightarrow H^+(aq) + NO_3^-(aq)$$

Many of the reactions of nitric acid are those associated with all strong acids. For example, dilute (6 M) nitric acid can be used to dissolve aluminum hydroxide:

$$Al(OH)_3(s) + 3H^+(aq) \longrightarrow Al^{3+}(aq) + 3H_2O$$

or to generate carbon dioxide gas from calcium carbonate:

$$CaCO_3(s) + 2H^+(aq) \longrightarrow Ca^{2+}(aq) + CO_2(g) + H_2O$$

Referring back to Example 21.4, you will find that these equations are identical with those written for the reactions of hydrochloric acid with Al(OH)$_3$ and CaCO$_3$. It is the H$^+$ ion that reacts in either case: Cl$^-$ and NO$_3^-$ ions take no part in the reactions and hence do not appear in the equation.

Concentrated (16 M) nitric acid is a strong oxidizing agent; the nitrate ion is reduced to nitrogen dioxide. This happens when 16 M HNO$_3$ reacts with copper metal (Figure 21.13):

$$Cu(s) + 4H^+(aq) + 2NO_3^-(aq) \longrightarrow Cu^{2+}(aq) + 2NO_2(g) + 2H_2O$$

Dilute nitric acid (6 M) gives a wide variety of reduction products, depending on the nature of the reducing agent. With inactive metals such as copper ($E_{ox}^\circ = -0.339$ V), the major product is usually NO (O.N. N = +2):

$$3Cu(s) + 2NO_3^-(aq) + 8H^+(aq) \longrightarrow 3Cu^{2+}(aq) + 2NO(g) + 4H_2O$$

With very dilute acid and a strong reducing agent such as zinc ($E_{ox}^\circ = +0.762$ V) reduction may go all the way to the NH$_4^+$ ion (O.N. N = -3):

$$4Zn(s) + NO_3^-(aq) + 10H^+(aq) \longrightarrow 4Zn^{2+}(aq) + NH_4^+(aq) + 3H_2O$$

As you would expect, the oxidizing strength of the NO$_3^-$ ion drops off sharply as pH increases; the reduction voltage (to NO) is +0.964 V at pH 0, +0.412 V at pH 7, and -0.140 V at pH 14. The nitrate ion is a very weak oxidizing agent in basic solution.

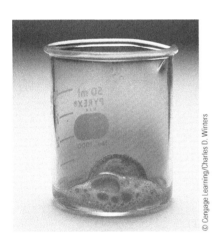

Figure 21.13 A copper penny dissolving in nitric acid. Copper metal is comparatively inactive, but it reacts with concentrated nitric acid. The brown fumes are NO$_2$(g), a reduction product of HNO$_3$. The copper is oxidized to Cu^{2+} ions, which impart their color to the solution. (The penny is an old one made of solid copper. Newer pennies have a coating of copper over a zinc core.)

© Cengage Learning/Charles D. Winters

EXAMPLE 21.9

Write a balanced net ionic equation for the reaction of nitric acid with insoluble copper(II) sulfide; the products include Cu^{2+}, S(s), and NO$_2$(g).

STRATEGY

1. Recall the general procedure for writing and balancing redox equations from Chapter 17.

2. Note that HNO$_3$ is a strong acid, so it should be represented as H$^+$ and NO$_3^-$ ions.

continued

SOLUTION	
Skeleton half-equations	oxidation: $CuS(s) \longrightarrow S(s)$ reduction: $NO_3^-(aq) \longrightarrow NO_2(g)$
Balanced half-reactions	oxidation: $CuS(s) \longrightarrow S(s) + Cu^{2+}(aq) + 2e^-$ reduction: $NO_3^-(aq) + 2H^+(aq) + e^-(aq) \longrightarrow NO_2(g) + H_2O$
Overall reaction	$CuS(s) \longrightarrow S(s) + Cu^{2+}(aq) + 2e^-$ $2(NO_3^-(aq) + 2H^+(aq) + e^-(aq) \longrightarrow NO_2(g) + H_2O)$
	$CuS(s) + 2NO_3^-(aq) + 4H^+(aq) \longrightarrow Cu^{2+}(aq) + S(s) + 2NO_2(g) + 2H_2O$

Concentrated nitric acid (16 M) is colorless when pure. In sunlight, it turns yellow (Figure 21.14) because it decomposes to $NO_2(g)$:

$$4HNO_3(aq) \longrightarrow 4NO_2(g) + 2H_2O + O_2(g)$$

The yellow color that appears on your skin if it comes in contact with nitric acid has quite a different explanation. Nitric acid reacts with proteins to give a yellow material called xanthoprotein.

You can always smell NO_2 over 16 M HNO_3.

Sulfuric Acid, H_2SO_4

Sulfuric acid is made commercially by a three-step *contact process*. First, elemental sulfur is burned in air to form sulfur dioxide:

$$S(s) + O_2(g) \longrightarrow SO_2(g)$$

The sulfur dioxide is converted to sulfur trioxide by bringing it into contact with oxygen on the surface of a solid catalyst:

$$SO_2(g) + \tfrac{1}{2}O_2(g) \longrightarrow SO_3(g)$$

The catalyst used is vanadium (V) oxide, V_2O_5, at a temperature of 450°C to 600°C. Sulfur trioxide is an acidic oxide that reacts with water to form sulfuric acid.

$$SO_3(g) + H_2O(l) \longrightarrow H_2SO_4(l)$$

An aqueous solution of nitric acid...

...slowly turns yellow because of the $NO_2(g)$ formed by decomposition.

Nitric acid also reacts with proteins (casein in milk [test tube] and albumen in eggs) to give a characteristic yellow color.

Figure 21.14 Nitric acid. Its decomposition and reaction with proteins.

© Cengage Learning/Charles D. Winters

Sulfuric acid is a strong acid, completely ionized to H^+ and HSO_4^- ions in dilute water solution. The HSO_4^- ion, in contrast, is only partially ionized.

$$H_2SO_4(aq) \longrightarrow H^+(aq) + HSO_4^-(aq) \qquad K_a \longrightarrow \infty$$
$$HSO_4^-(aq) \rightleftharpoons H^+(aq) + SO_4^{2-}(aq) \qquad K_a = 1.0 \times 10^{-2}$$

Most of the H^+ ions come from the first ionization; in 0.10 M H_2SO_4, $[H^+]$ = 0.11 M.

Sulfuric acid is a relatively weak oxidizing agent; most often it is reduced to sulfur dioxide:

$$SO_4^{2-}(aq) + 4H^+(aq) + 2e^- \longrightarrow SO_2(g) + 2H_2O \qquad E^\circ_{red} = +0.155 \text{ V}$$

This is the case when copper metal is oxidized by hot concentrated sulfuric acid:

$$Cu(s) + 4H^+(aq) + SO_4^{2-}(aq) \longrightarrow Cu^{2+}(aq) + 2H_2O + SO_2(g)$$

When dilute sulfuric acid (3 M) reacts with metals, it is ordinarily the H^+ ion rather than the SO_4^{2-} ion that is reduced. For example, zinc reacts with dilute sulfuric acid to form hydrogen gas:

$$Zn(s) + 2H^+(aq) \longrightarrow Zn^{2+}(aq) + H_2(g)$$

Concentrated sulfuric acid, in addition to being an acid and an oxidizing agent, is also a dehydrating agent. Small amounts of water can be removed from organic liquids such as gasoline by extraction with sulfuric acid. Sometimes it is even possible to remove the elements of water from a compound by treating it with 18 M H_2SO_4. This happens with table sugar, $C_{12}H_{22}O_{11}$; the product is a black char that is mostly carbon (Figure 21.15).

$$C_{12}H_{22}O_{11}(s) \longrightarrow 12C(s) + 11H_2O$$

When concentrated sulfuric acid dissolves in water, a great deal of heat is given off, nearly 100 kJ per mole of H_2SO_4. Sometimes enough heat is evolved to bring the solution to the boiling point. To prevent this and to avoid splattering, the acid should be added slowly to water, with constant stirring. If it comes in contact with the skin, concentrated sulfuric acid can cause painful chemical burns. Never add water to concentrated H_2SO_4!

When sulfuric acid is added to sugar (sucrose), an exothermic reaction occurs.

The elements of water are removed from the sugar, $C_{12}H_{22}O_{11}$,...

...leaving a mass of black carbon.

© Cengage Learning/Charles D. Winters

Figure 21.15 Sugar plus sulfuric acid.

Phosphoric Acid, H_3PO_4

In water solution, $+5$ phosphorus can be present as

- the H_3PO_4 molecule ($K_a = 7.1 \times 10^{-3}$), which is a (relatively strong) weak acid. A $0.10\ M$ solution of H_3PO_4 has a pH of 1.6.
- the $H_2PO_4^-$ ion ($K_a = 6.2 \times 10^{-8}$; $K_b = 1.4 \times 10^{-12}$) is a (very) weak acid. A $0.10\ M$ solution of NaH_2PO_4 has a pH of 4.
- the HPO_4^{2-} ion ($K_a = 4.5 \times 10^{-13}$; $K_b = 1.6 \times 10^{-7}$) is a (very) weak base. A $0.10\ M$ solution of Na_2HPO_4 has a pH of 10.
- the PO_4^{3-} ion ($K_b = 2.2 \times 10^{-2}$) is a (relatively strong) weak base. A $0.10\ M$ solution of Na_3PO_4 has a pH of 12.7.

Of the three compounds NaH_2PO_4, Na_2HPO_4, and Na_3PO_4, two are used as cleaning agents: NaH_2PO_4 in acid-type cleaners, Na_3PO_4 in strongly basic cleaners. The principal use of Na_2HPO_4 is in the manufacture of cheese. Around 1915, J. L. Kraft (1874–1953) discovered that this compound is an excellent emulsifying agent. To this day no one quite understands why this happens.

A great many "phosphates" are used in commercial fertilizers. Perhaps the most important of these is calcium dihydrogen phosphate, $Ca(H_2PO_4)_2$. In relatively pure form, this compound is known as "triple superphosphate of lime." A $1:2$ mol mixture of $Ca(H_2PO_4)_2$ and gypsum, $CaSO_4 \cdot 2H_2O$, is commonly referred to as "superphosphate of lime."

CHEMISTRY **BEYOND** THE CLASSROOM

Arsenic and Selenium

Arsenic and selenium, which fall directly below phosphorus and sulfur in the periodic table, are of interest for a variety of reasons. Arsenic is a true metalloid. A metallic form, called gray arsenic, has an electrical conductivity approaching that of lead. Another **allotrope**, yellow arsenic, is distinctly nonmetallic; it has the molecular formula As_4, analogous to white phosphorus, P_4. Selenium is properly classified as a nonmetal, although one of its allotropes has a somewhat metallic appearance and is a semiconductor. Another form of selenium has the molecular formula Se_8, analogous to sulfur.

The principal use of elemental arsenic is in its alloys with lead. Lead shot, formed by allowing drops of molten metal to fall through air, contains from 0.5% to 2.0% arsenic. Selenium has long been used as an additive in glassmaking. Particles of colloidally dispersed selenium give the ruby red color seen in traffic lights and stained glass windows (Figure A).

Physiological Properties

The "arsenic poison" referred to in true-crime dramas is actually the oxide of arsenic, As_2O_3, rather than the element itself. Less than 0.1 g of this white, slightly soluble powder can be fatal. The classic symptoms of acute arsenic poisoning involve various unpleasant gastrointestinal disturbances, severe abdominal pain, and burning of the mouth and throat.

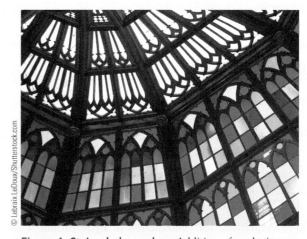

Figure A Stained glass colors. Addition of a selenium compound produces red glass, and an addition of a cobalt compound produces blue glass.

In the modern forensic chemistry laboratory (Figure B) arsenic is detected by analysis of hair samples, where the element tends to concentrate in chronic arsenic poisoning. A single strand of hair is sufficient to establish the presence or absence of the element. The technique most commonly used is neutron activation analysis, described in Chapter 18. If the concentration found is greater than about 0.0003%, poisoning is indicated; normal arsenic levels are much lower than this.

This technique was applied in the early 1960s to a lock of hair taken from Napoleon Bonaparte (1769–1821) on

Figure B A forensic toxicology laboratory. The young woman is working on a research project in the laboratory at Hartford Hospital.

St. Helena. Arsenic levels of up to 50 times normal suggested he may have been a victim of poisoning, perhaps on orders from the French royal family.

Arsenic is a known human carcinogen, found in drinking water in many parts of the world. The World Health Organization (WHO) has set the upper limit for arsenic in drinking water at ten parts per billion (10 ppb). The legal limit in the United States, set by the Environmental Protection Agency (EPA) is 50 ppb.

It has long been known that high concentrations of selenium are toxic. The cattle disease known picturesquely as "blind staggers" arises from grazing on grass growing in soil with a high selenium content. It is now known, however, that selenium is an essential element in human nutrition. A selenium-containing enzyme, glutathione peroxidase, in combination with vitamin E, destroys harmful free radicals in the body.

There is considerable evidence to suggest that selenium compounds are anticarcinogens. For one thing, tests with laboratory animals show that the incidence and size of malignant tumors are reduced when a solution containing Na_2SeO_3 is injected at the part per million level. Beyond that evidence, statistical studies show an inverse correlation between selenium levels in the soil and the incidence of certain types of cancer.

Key Concepts

1. Carry out equilibrium calculations for solution reactions. (**Example 21.1; Problems 43–48**)
2. Apply the Gibbs-Helmholtz equation. (**Example 21.2; Problems 49–54**)
3. Write balanced net ionic equations for solution reactions. (**Examples 21.3 and 21.5; Problems 15–26**)
4. Draw Lewis structures for compounds of the nonmetals. (**Examples 21.6 and 21.8; Problems 27–34**)
5. Relate oxoacids to the corresponding oxides and compare their acid strengths. (**Examples 21.7 and 21.8; Problems 7, 8, 63**)
6. Carry out electrochemical calculations involving $E°$, the Nernst equation, and/or electrolysis. (**Examples 21.4 and 21.9; Problems 55–62**)

Summary Problem

Consider bromine, whose chemistry is quite similar to that of chlorine.

(a) Write equations for the reaction of bromine with I^- ions; for the reaction of Br_2 with water (disproportionation).

(b) Write equations for the preparation of bromine by the electrolysis of aqueous NaBr; by the reaction of chlorine with bromide ions in aqueous solution.

(c) Write equations for the reaction of hydrobromic acid with OH^- ions; with CO_3^{2-} ions; with ammonia.

(d) Consider the species Br^-, Br_2, BrO^-, and BrO_4^-. In a redox reaction, which of these species can act only as an oxidizing agent? only as a reducing agent? Which can act as either oxidizing or reducing agents?

(e) When aqueous sodium bromide is heated with a concentrated solution of sulfuric acid, the products include liquid bromine and sulfur dioxide. Write a balanced equation for the redox reaction involved.

(f) Write Lewis structures for the following species: Br^-, BrO^-, BrO_4^-. What is the bond angle in the BrO_4^- ion? For which of these ions is the conjugate acid the weakest?

(g) Give the formula for the acidic oxide of HBrO. Knowing that K_a of HBrO is 2.6×10^{-9}, calculate the ratio $[HBrO]/[BrO^-]$ at pH 10.00.

(h) For the reduction of HBrO to Br^-, what is the change in voltage when the pH increases by one unit?

Answers

(a) $Br_2(l) + 2I^-(aq) \longrightarrow 2Br^-(aq) + I_2(s)$
$Br_2(l) + H_2O \longrightarrow HBrO(aq) + H^+(aq) + Br^-(aq)$

(b) $2Br^-(aq) + 2H_2O \longrightarrow Br_2(l) + H_2(g) + 2OH^-(aq)$
$2Br^-(aq) + Cl_2(aq) \longrightarrow Br_2(l) + 2Cl^-(aq)$

(c) $H^+(aq) + OH^-(aq) \longrightarrow H_2O$
$2H^+(aq) + CO_3^{2-}(aq) \longrightarrow CO_2(g) + H_2O$
$H^+(aq) + NH_3(aq) \longrightarrow NH_4^+(aq)$

(d) BrO_4^-; Br^-; Br_2 or BrO^-

(e) $2Br^-(aq) + SO_4^{2-}(aq) + 4H^+(aq) \longrightarrow$
$$Br_2(l) + SO_2(g) + 2H_2O$$

(f) $[:\!\ddot{B}\ddot{r}\!:]^-$ $[:\ddot{B}r\!-\!\ddot{O}:]^-$ $\left[:\ddot{O}\!-\!\overset{\displaystyle :\ddot{O}:}{\underset{\displaystyle :\ddot{O}:}{Br}}\!-\!\ddot{O}:\right]^-$ 109.5°; BrO^-

(g) Br_2O; 0.038

(h) -0.0296 V

Questions and Problems

Even-numbered questions and Challenge Problems have answers in Appendix 5 and fully worked solutions in the *Student Solutions Manual*.

21-1 The Elements and Their Preparation
21-2 Hydrogen Compounds of Nonmetals
21-3 Oxygen Compounds of Nonmetals
21-4 Oxoacids and Oxoanions

Formulas, Equations, and Reactions

1. Name the following species.
   (a) HIO_4 (b) BrO_2^- (c) HIO (d) $NaClO_3$
2. Name the following compounds.
   (a) $HBrO_3$ (b) KIO
   (c) $NaClO_2$ (d) $NaBrO_4$
3. Write the formula for each of the following compounds.
   (a) chloric acid   (b) periodic acid
   (c) hypobromous acid   (d) hydriodic acid
4. Write the formula for each of the following compounds.
   (a) potassium bromite   (b) calcium bromide
   (c) sodium periodate   (d) magnesium hypochlorite
5. Write the formula of a compound of each of the following elements that *cannot* act as an oxidizing agent.
   (a) N   (b) S   (c) Cl
6. Write the formula of an oxoanion of each of the following elements that *cannot* act as a reducing agent.
   (a) N   (b) S   (c) Cl
7. Give the formula for the acidic oxide of
   (a) HNO_3 (b) HNO_2 (c) H_2SO_4
8. Write the formula of the acid formed when each of these acidic oxides reacts with water.
   (a) SO_2 (b) Cl_2O (c) P_4O_6
9. Write the formulas of the following compounds.
   (a) ammonia   (b) laughing gas
   (c) hydrogen peroxide (d) sulfur trioxide
10. Write the formulas for the following compounds.
    (a) sodium azide   (b) sulfurous acid
    (c) hydrazine   (d) sodium dihydrogen phosphate
11. Write the formula of a compound of hydrogen with
    (a) nitrogen, which is a gas at 25°C and 1 atm.
    (b) phosphorus, which is a liquid at 25°C and 1 atm.
    (c) oxygen, which contains an O—O bond.
12. Write the formula of a compound of hydrogen with
    (a) sulfur.
    (b) nitrogen, which is a liquid at 25°C and 1 atm.
    (c) phosphorus, which is a poisonous gas at 25°C and 1 atm.
13. Give the formula of
    (a) an anion in which S has an oxidation number of −2.
    (b) two anions in which S has an oxidation number of +4.
    (c) two different acids of sulfur.
14. Give the formula of a compound of nitrogen that is
    (a) a weak base.   (b) a strong acid.
    (c) a weak acid.   (d) capable of oxidizing copper.

15. Write a balanced net ionic equation for
    (a) the electrolytic decomposition of hydrogen fluoride.
    (b) the oxidation of iodide ion to iodine by hydrogen peroxide in acidic solution. Hydrogen peroxide is reduced to water.
16. Write a balanced net ionic equation for
    (a) the oxidation of iodide to iodine by sulfate ion in acidic solution. Sulfur dioxide gas is also produced.
    (b) The preparation of iodine from an iodide salt and chlorine gas.
17. Write a balanced net ionic equation for the disproportionation reaction
    (a) of iodine to give iodate and iodide ions in basic solution.
    (b) of chlorine gas to chloride and perchlorate ions in basic solution.
18. Write a balanced net ionic equation for the disproportionation reaction of
    (a) hypochlorous acid to chlorine gas and chlorous acid in acidic solution.
    (b) chlorate ion to perchlorate and chlorite ions.
19. Complete and balance the following equations. If no reaction occurs, write NR.
    (a) $Cl_2(g) + I^-(aq) \longrightarrow$ (b) $F_2(g) + Br^-(aq) \longrightarrow$
    (c) $I_2(s) + Cl^-(aq) \longrightarrow$ (d) $Br_2(l) + I^-(aq) \longrightarrow$
20. Complete and balance the following equations. If no reaction occurs, write NR.
    (a) $Cl_2(g) + Br^-(aq) \longrightarrow$ (b) $I_2(s) + Cl^-(aq) \longrightarrow$
    (c) $I_2(s) + Br^-(aq) \longrightarrow$ (d) $Br_2(l) + Cl^-(aq) \longrightarrow$
21. Write a balanced equation for the preparation of
    (a) F_2 from HF.   (b) Br_2 from NaBr.
    (c) NH_4^+ from NH_3.
22. Write a balanced equation for the preparation of
    (a) N_2 from $Pb(N_3)_2$.   (b) O_2 from O_3.
    (c) S from H_2S.
23. Write a balanced equation for the reaction of ammonia with
    (a) Cu^{2+} (b) H^+ (c) Al^{3+}
24. Write a balanced equation for the reaction of hydrogen sulfide with
    (a) Cd^{2+} (b) OH^- (c) $O_2(g)$
25. Write a balanced net ionic equation for the reaction of nitric acid with
    (a) a solution of $Ca(OH)_2$.
    (b) $Ag(s)$; assume the nitrate ion is reduced to $NO_2(g)$.
    (c) $Cd(s)$; assume the nitrate ion is reduced to $N_2(g)$.
26. Write a balanced net ionic equation for the reaction of sulfuric acid with
    (a) $CaCO_3(s)$.
    (b) a solution of NaOH.
    (c) Cu; assume the SO_4^{2-} ion is reduced to SO_2.

Molecular Structure

27. Give the Lewis structure of
    (a) NO_2 (b) NO   (c) SO_2 (d) SO_3
28. Give the Lewis structure of
    (a) Cl_2O (b) N_2O (c) P_4 (d) N_2

29. Which of the molecules in Question 27 are polar?
30. Which of the molecules in Question 28 are polar?
31. Give the Lewis structure of
    (a) HNO_3 (b) H_2SO_4 (c) H_3PO_4
32. Give the Lewis structures of the conjugate bases of the species in Question 31.
33. Give the Lewis structure of
    (a) the strongest oxoacid of bromine.
    (b) a hydrogen compound of nitrogen in which there is an —N—N— bond.
    (c) an acid added to cola drinks.
34. Give the Lewis structure of
    (a) an oxide of nitrogen in the +5 state.
    (b) the strongest oxoacid of nitrogen.
    (c) a tetrahedral oxoanion of sulfur.

Stoichiometry

35. The average concentration of bromine (as bromide) in seawater is 65 ppm. Calculate
    (a) the volume of seawater ($d = 64.0$ lb/ft^3) in cubic feet required to produce one kilogram of liquid bromine.
    (b) the volume of chlorine gas in liters, measured at 20°C and 762 mm Hg, required to react with this volume of seawater.
36. A 425-gal tank is filled with water containing 175 g of sodium iodide. How many liters of chlorine gas at 758 mm Hg and 25°C will be required to oxidize all the iodide to iodine?
37. Iodine can be prepared by allowing an aqueous solution of hydrogen iodide to react with manganese dioxide, MnO_2. The reaction is

$$2I^-(aq) + 4H^+(aq) + MnO_2(s) \longrightarrow Mn^{2+}(aq) + 2H_2O + I_2(s)$$

If an excess of hydrogen iodide is added to 0.200 g of MnO_2, how many grams of iodine are obtained, assuming 100% yield?
38. When a solution of hydrogen bromide is prepared, 1.283 L of HBr gas at 25°C and 0.974 atm is bubbled into 250.0 mL of water. Assuming all the HBr dissolves with no volume change, what is the molarity of the hydrobromic acid solution produced?
39. When ammonium nitrate explodes, nitrogen, steam, and oxygen gas are produced. If the explosion is carried out by heating one kilogram of ammonium nitrate sealed in a rigid bomb with a volume of one liter, what is the total pressure produced by the gases before the bomb ruptures? Assume that the reaction goes to completion and that the final temperature is 500°C (3 significant figures).
40. Sulfur dioxide can be removed from the smokestack emissions of power plants by reacting it with hydrogen sulfide, producing sulfur and water. What volume of hydrogen sulfide at 27°C and 755 mm Hg is required to remove the sulfur dioxide produced by a power plant that burns one metric ton of coal containing 5.0% sulfur by mass? How many grams of sulfur are produced by the reaction of H_2S with SO_2?
41. A 1.500-g sample containing sodium nitrate was heated to form $NaNO_2$ and O_2. The oxygen evolved was collected over water at 23°C and 752 mm Hg; its volume was 125.0 mL. Calculate the percentage of $NaNO_3$ in the sample. The vapor pressure of water at 23°C is 21.07 mm Hg.
42. Chlorine can remove the foul smell of H_2S in water. The reaction is

$$H_2S(aq) + Cl_2(aq) \longrightarrow 2H^+(aq) + 2Cl^-(aq) + S(s)$$

If the contaminated water has 5.0 ppm hydrogen sulfide by mass, what volume of chlorine gas at STP is required to remove all the H_2S from 1.00×10^3 gallons of water ($d = 1.00$ g/mL)? What is the pH of the solution after treatment with chlorine?

Equilibria

43. The equilibrium constant at 25°C for the reaction

$$Br_2(l) + H_2O \rightleftharpoons H^+(aq) + Br^-(aq) + HBrO(aq)$$

is 1.2×10^{-9}. This is the system present in a bottle of "bromine water." Assuming that HBrO does not ionize appreciably, what is the pH of the bromine water?
44. Calculate the pH and the equilibrium concentration of HClO in a 0.10 M solution of hypochlorous acid. K_a HClO = 2.8×10^{-8}.
45. At equilibrium, a gas mixture has a partial pressure of 0.7324 atm for HBr and 2.80×10^{-3} atm for both hydrogen and bromine gases. What is K for the formation of two moles of HBr from H_2 and Br_2?
46. Given

$$HF(aq) \rightleftharpoons H^+(aq) + F^-(aq) \quad K_a = 6.9 \times 10^{-4}$$
$$HF(aq) + F^-(aq) \rightleftharpoons HF_2^-(aq) \quad\quad K = 2.7$$

calculate K for the reaction

$$2HF(aq) \rightleftharpoons H^+(aq) + HF_2^-(aq)$$

47. What is the concentration of fluoride ion in a water solution saturated with BaF_2, $K_{sp} = 1.8 \times 10^{-7}$?
48. Calculate the solubility in grams per 100 mL of BaF_2 in a 0.10 M $BaCl_2$ solution.

Thermodynamics

49. Determine whether the following redox reaction is spontaneous at 25°C and 1 atm:

$$2KIO_3(s) + Cl_2(g) \longrightarrow 2KClO_3(s) + I_2(s)$$

Use data in Appendix 1 and the following information: $\Delta H_f^°$ $KIO_3(s)$ = -501.4 kJ/mol, $S°$ $KIO_3(s)$ = 151.5 J/mol·K. What is the lowest temperature at which the reaction is spontaneous?
50. Follow the directions for Problem 49 for the reaction

$$2KBrO_3(s) + Cl_2(g) \longrightarrow 2KClO_3(s) + Br_2(l)$$

The following thermodynamic data may be useful:

$\Delta H_f^°$ $KBrO_3$ = -360.2 kJ/mol; $S°$ $KBrO_3$ = 149.2 J/mol·K
51. Consider the equilibrium system

$$HF(aq) \rightleftharpoons H^+(aq) + F^-(aq)$$

Given $\Delta H_f^°$ HF(aq) = -320.1 kJ/mol,

$\Delta H_f^°$ $F^-(aq)$ = -332.6 kJ/mol; $S°$ $F^-(aq)$ = -13.8 J/mol·K;
    K_a HF = 6.9×10^{-4} at 25°C

calculate $S°$ for HF(aq).
52. Applying the tables in Appendix 1 to

$$4HCl(g) + O_2(g) \longrightarrow 2Cl_2(g) + 2H_2O(l)$$

determine

(a) whether the reaction is spontaneous at 25°C and 1 atm.

(b) K for the reaction at 25°C.

53. Consider the reaction

$$4NH_3(g) + 5O_2(g) \longrightarrow 4NO(g) + 6H_2O(g)$$

(a) Calculate $\Delta H°$ for this reaction. Is it exothermic or endothermic?

(b) Would you expect $\Delta S°$ to be positive or negative? Calculate $\Delta S°$.

(c) Is the reaction spontaneous at 25°C and 1 atm?

(d) At what temperature, if any, is the reaction at equilibrium at 1 atm pressure?

54. Data are given in Appendix 1 for white phosphorus, $P_4(s)$. $P_4(g)$ has the following thermodynamic values: $\Delta H_f° = 58.9$ kJ/mol, $S° = 280.0$ J/K·mol. What is the temperature at which white phosphorus sublimes at 1 atm pressure?

Electrochemistry

55. In the electrolysis of a KI solution, using 5.00 V, how much electrical energy in kilojoules is consumed when one mole of I_2 is formed?

56. If an electrolytic cell producing fluorine uses a current of 7.00×10^3 A (at 10.0 V), how many grams of fluorine gas can be produced in two days (assuming that the cell operates continuously at 95% efficiency)?

57. Sodium hypochlorite is produced by the electrolysis of cold sodium chloride solution. How long must a cell operate to produce 1.500×10^3 L of 5.00% NaClO by mass if the cell current is 2.00×10^3 A? Assume that the density of the solution is 1.00 g/cm^3.

58. Sodium perchlorate is produced by the electrolysis of sodium chlorate. If a current of 1.50×10^3 A passes through an electrolytic cell, how many kilograms of sodium perchlorate are produced in an eight-hour run?

59. Taking $E°_{ox}$ $H_2O_2 = -0.695$ V, determine which of the following species will be reduced by hydrogen peroxide (use Table 17.1 to find $E°_{red}$ values).

(a) $Cr_2O_7^{2-}$ (b) Fe^{2+} (c) I_2 (d) Br_2

60. Taking $E°_{red}$ $H_2O_2 = +1.763$ V, determine which of the following species will be oxidized by hydrogen peroxide (use Table 17.1 to find $E°_{ox}$ values).

(a) Co^{2+} (b) Cl^- (c) Fe^{2+} (d) Sn^{2+}

61. Consider the reduction of nitrate ion in acidic solution to nitrogen oxide ($E°_{red} = 0.964$ V) by sulfur dioxide that is oxidized to sulfate ion ($E°_{red} = 0.155$ V). Calculate the voltage of a cell involving this reaction in which all the gases have pressures of 1.00 atm, all the ionic species (except H$^+$) are at 0.100 M, and the pH is 4.30.

62. For the reaction in Problem 61 if gas pressures are at 1.00 atm and ionic species are at 0.100 M (except H$^+$), at what pH will the voltage be 1.000 V?

Unclassified

63. Choose the strongest acid from each group.

(a) HClO, HBrO, HIO   (b) HIO, HIO$_3$, HIO$_4$

(c) HIO, HBrO$_2$, HBrO$_4$

64. What intermolecular forces are present in the following?

(a) Cl_2 (b) HBr   (c) HF

(d) $HClO_4$ (e) MgI_2

65. Write a balanced equation for the reaction of hydrofluoric acid with SiO$_2$. What volume of 2.0 M HF is required to react with one gram of silicon dioxide?

66. State the oxidation number of N in

(a) NO_2^- (b) NO_2 (c) HNO_3 (d) NH_4^+

67. The density of sulfur vapor at one atmosphere pressure and 973 K is 0.8012 g/L. What is the molecular formula of the vapor?

68. Give the formula of a substance discussed in this chapter that is used

(a) to disinfect water.   (b) in safety matches.

(c) to prepare hydrazine.   (d) to etch glass.

69. Why does concentrated nitric acid often have a yellow color even though pure HNO$_3$ is colorless?

70. Explain why

(a) acid strength increases as the oxidation number of the central nonmetal atom increases.

(b) nitrogen dioxide is paramagnetic.

(c) the oxidizing strength of an oxoanion is inversely related to pH.

(d) sugar turns black when treated with concentrated sulfuric acid.

Challenge Problems

71. Suppose you wish to calculate the mass of sulfuric acid that can be obtained from an underground deposit of sulfur 1.00 km^2 in area. What additional information do you need to make this calculation?

72. The reaction

$$4HF(aq) + SiO_2(aq) \longrightarrow SiF_4(aq) + 2H_2O$$

can be used to release gold that is distributed in certain quartz (SiO$_2$) veins of hydrothermal origin. If the quartz contains 1.0×10^{-3}% Au by weight and the gold has a market value of $425 per troy ounce, would the process be economically feasible if commercial HF (50% by weight, $d = 1.17$ g/cm^3) costs 75¢ a liter? (1 troy ounce = 31.1 g.)

73. The amount of sodium hypochlorite in a bleach solution can be determined by using a given volume of bleach to oxidize excess iodide ion to iodine; ClO$^-$ is reduced to Cl$^-$. The amount of iodine produced by the redox reaction is determined by titration with sodium thiosulfate, Na$_2$S$_2$O$_3$; I$_2$ is reduced to I$^-$. The sodium thiosulfate is oxidized to sodium tetrathionate, Na$_2$S$_4$O$_6$. In this analysis, potassium iodide was added in excess to 5.00 mL of bleach ($d = 1.00$ g/cm^3). If 25.00 mL of 0.0700 M Na$_2$S$_2$O$_3$ was required to reduce all the iodine produced by the bleach back to iodide, what is the mass percent of NaClO in the bleach?

74. What is the minimum amount of sodium azide, NaN$_3$, that can be added to an automobile airbag to give a volume of 20.0 L of N$_2(g)$ on inflation? Make any reasonable assumptions required to obtain an answer, but state what these assumptions are.

Organic Chemistry

The wine in the goblet can be produced by the fermentation of glucose (present in all the fruits shown in the painting) to ethyl alcohol.

The Philadelphia Museum of Art / Art Resource, NY

What is life? It is the flash of a firefly
in the night.
It is the breath of a buffalo in the
wintertime.
It is the little shadow which runs across
grass
And loses itself in the sunset.

—DYING WORDS OF CROWFOOT (1890)

Organic chemistry deals with the compounds of carbon, of which there are literally millions. More than 90% of all known compounds contain carbon atoms. There is a simple explanation for this remarkable fact. Carbon atoms bond to one another to a far greater extent than do atoms of any other element. Carbon atoms may link together to form chains or rings.

The bonds may be single (one electron pair), double (two electron pairs), or triple (three electron pairs).

There are a wide variety of different organic compounds that have quite different structures and properties. However, all these substances have certain features in common:

1. *Organic compounds are ordinarily molecular rather than ionic.* Most of the compounds we discuss consist of small, discrete molecules. Many of them are gases or liquids at room temperature.

2. *Each carbon atom forms a total of four covalent bonds.* ◄ ⓘ This is illustrated by the structures above. A particular carbon atom may form four single bonds, two single bonds and a double bond, two double bonds, or one single bond and a triple bond. One way or another, though, the bonds add up to four.

3. *Carbon atoms may be bonded to each other or to other nonmetal atoms, most often hydrogen, a halogen, oxygen, or nitrogen.* In most organic compounds:

   ■ a hydrogen or halogen atom (F, Cl, Br, I) forms one covalent bond, —H, —X
   ■ an oxygen atom forms two covalent bonds, —O— or =O
   ■ a nitrogen atom forms three covalent bonds, —N—, =N—, or ≡N

ⓘ Carbon always follows the octet rule in stable compounds.

In this chapter, we consider

- the simplest type of organic compound, called a **hydrocarbon,** which contains only two kinds of atoms, hydrogen and carbon. Hydrocarbons can be classified as alkanes (Section 22-1), alkenes and alkynes (Section 22-2), and aromatics (Section 22-3).
- organic compounds containing oxygen or nitrogen atoms in addition to carbon and hydrogen (Section 22-4).
- the phenomenon of isomerism, which is very common among organic compounds. This topic is introduced in Section 22-1 but discussed more generally in Section 22-5.
- different types of organic reactions (Section 22-6).

Throughout this chapter, we will represent molecules by *structural formulas,* which show all the bonds present. Thus we have

$$\begin{array}{ccc}
\quad\ \text{H}\ \ \text{H} & \quad\ \text{H}\ \ \text{H} & \quad\ \text{H}\quad\ \ \text{H} \\
\quad\ |\ \ \ | & \quad\ |\ \ \ | & \quad\ |\quad\ \ | \\
\text{H}-\text{C}-\text{C}-\text{H} & \text{H}-\text{C}-\text{C}-\text{O}-\text{H} & \text{H}-\text{C}-\text{O}-\text{C}-\text{H} \\
\quad\ |\ \ \ | & \quad\ |\ \ \ | & \quad\ |\quad\ \ | \\
\quad\ \text{H}\ \ \text{H} & \quad\ \text{H}\ \ \text{H} & \quad\ \text{H}\quad\ \ \text{H} \\
\text{ethane} & \text{ethanol} & \text{dimethyl ether}
\end{array}$$

To save space we often write condensed structural formulas such as

$$\begin{array}{ccc}
\text{CH}_3\text{CH}_3 & \text{CH}_3\text{CH}_2\text{OH} & \text{CH}_3-\text{O}-\text{CH}_3 \\
\text{or } \text{C}_2\text{H}_6 & \text{or } \text{C}_2\text{H}_5\text{OH} & \text{or } (\text{CH}_3)_2\text{O}
\end{array}$$

22-1 Saturated Hydrocarbons: Alkanes

One large and structurally simple class of hydrocarbons includes those substances in which all the carbon-carbon bonds are single bonds. These are called **saturated hydrocarbons,** or **alkanes.** In the alkanes the carbon atoms are bonded to each other in chains, which may be long or short, straight or branched.

The outer surfaces of these molecules contain mainly H atoms.

The ratio of hydrogen to carbon atoms is a maximum in alkanes, hence the term "saturated" hydrocarbon. The general formula of an alkane containing n carbon atoms is

$$\text{C}_n\text{H}_{2n+2}$$

The simplest alkanes are those for which $n = 1$ (CH$_4$), $n = 2$ (C$_2$H$_6$), or $n = 3$ (C$_3$H$_8$):

$$\begin{array}{ccc}
\quad\ \text{H} & \quad\ \text{H}\ \ \text{H} & \quad\ \text{H}\ \ \text{H}\ \ \text{H} \\
\quad\ | & \quad\ |\ \ \ | & \quad\ |\ \ \ |\ \ \ | \\
\text{H}-\text{C}-\text{H} & \text{H}-\text{C}-\text{C}-\text{H} & \text{H}-\text{C}-\text{C}-\text{C}-\text{H} \\
\quad\ | & \quad\ |\ \ \ | & \quad\ |\ \ \ |\ \ \ | \\
\quad\ \text{H} & \quad\ \text{H}\ \ \text{H} & \quad\ \text{H}\ \ \text{H}\ \ \text{H} \\
\text{methane} & \text{ethane} & \text{propane}
\end{array}$$

Around the carbon atoms in these molecules, and indeed in any saturated hydrocarbon, there are four single bonds involving sp³ hybrid orbitals. As would be expected from the VSEPR model, these bonds are directed toward the corners of a regular tetrahedron. The bond angles are approximately 109.5°, the tetrahedral angle. This means that in propane (C$_3$H$_8$) and in the higher alkanes, the carbon atoms are arranged in a "zigzag" pattern (Figure 22.1).

Figure 22.1 The three simplest alkanes. The bond angles in methane, ethane, and propane are all close to 109.5°, the tetrahedral angle.

$$\begin{array}{ccc}
\text{CH}_4 & \text{CH}_3\text{CH}_3 & \text{CH}_3\text{CH}_2\text{CH}_3 \\
\text{methane} & \text{ethane} & \text{propane}
\end{array}$$

CH$_3$CH$_2$CH$_2$CH$_3$
butane

CH$_3$CHCH$_3$
|
CH$_3$
2-methylpropane

Figure 22.2 Butane and 2-methylpropane, the isomers of C$_4$H$_{10}$.

Two different alkanes are known with the molecular formula C$_4$H$_{10}$. In one of these, called butane, the four carbon atoms are linked in a "straight (unbranched) chain." In the other, called 2-methylpropane, there is a "branched chain." The longest chain in the molecule contains three carbon atoms; there is a CH$_3$ branch from the central carbon atom. The geometries of these molecules are shown in Figure 22.2. The structures are

$$
\begin{array}{cccc}
H & H & H & H \\
| & | & | & | \\
H-C-C-C-C-H \\
| & | & | & | \\
H & H & H & H \\
\end{array}
$$
butane

$$
\begin{array}{ccc}
H & CH_3 & H \\
| & | & | \\
H-C-C\!-\!-\!-C-H \\
| & | & | \\
H & H & H \\
\end{array}
$$
2-methylpropane

Compounds having the same molecular formula but different molecular structures are called **structural isomers.** ◄ ⓘ Butane and 2-methylpropane are referred to as structural isomers of C$_4$H$_{10}$. They are two distinct compounds with their own characteristic physical and chemical properties.

ⓘ In structural isomers, the atoms are bonded in different patterns; the skeletons are different.

EXAMPLE 22.1

Draw structures for the isomers of C$_5$H$_{12}$.

STRATEGY

1. Start by writing all five-carbon atoms in a straight chain. Call this structure isomer (I).

2. Write a four-carbon chain structure bonding the fifth carbon atom to one of the C atoms in the straight chain.

   Note that bonding the fifth C atom to either end of the straight chain gives you isomer (I). You have to attach the C atom next to the terminal atom (on either end).

3. Write a three-carbon chain. Do not attach the remaining two carbon atoms as a chain to either end of the three-carbon chain. You will get Isomer (I). Attaching one carbon atom to each end of the three-carbon chain also gives you isomer (I).

SOLUTION

Isomer (I)	isomer (I) five-C chain:	$-\overset{\mid}{\underset{\mid}{C}}-\overset{\mid}{\underset{\mid}{C}}-\overset{\mid}{\underset{\mid}{C}}-\overset{\mid}{\underset{\mid}{C}}-\overset{\mid}{\underset{\mid}{C}}-$
Isomer (II)	isomer (II) four-C chain:	$-\overset{\mid}{\underset{\mid}{C}}-\overset{\mid}{\underset{\mid}{C}}-\overset{\mid}{\underset{\mid}{C}}-\overset{\mid}{\underset{\mid}{C}}-$ (with $-C-$ branch)
Isomer (III) ◄ ⓘ	isomer (III) three-C chain:	$-\overset{\mid}{\underset{\mid}{C}}-\overset{\mid}{\underset{\mid}{C}}-\overset{\mid}{\underset{\mid}{C}}-$ (with $-C-$ branches above and below)

ⓘ Remember, these molecules are three-dimensional.

If you want a really tedious job, try drawing the 366,319 structural isomers of $C_{20}H_{42}$.

Structural isomers have different properties. For example, the three isomers of C_5H_{12} in Example 22.1 boil at different temperatures.

Structure	Normal Boiling Point
I	36°C
II	28°C
III	10°C

In general, long "skinny" isomers like I tend to have higher boiling points than highly branched isomers like III. Dispersion forces become weaker as chain branching increases. This principle is used in the seasonal blending of gasoline. A greater percentage of more volatile, highly branched alkanes is included in winter, making combustion occur more readily at lower temperatures.

Nomenclature

As organic chemistry developed, it became apparent that some systematic way of naming compounds was needed. About 80 years ago, the International Union of Pure and Applied Chemistry (IUPAC) devised a system that is usable for all organic compounds. To illustrate this system, we will show how it works with alkanes.

For straight-chain alkanes such as

$$CH_3-CH_2-CH_3 \qquad CH_3-CH_2-CH_2-CH_3$$
propane \qquad\qquad\qquad butane

the IUPAC name consists of a single word. These names, for up to eight carbon atoms, are listed in Table 22.1.

With alkanes containing a branched chain, such as

$$CH_3-\overset{\displaystyle H}{\underset{\displaystyle CH_3}{C}}-CH_3$$

2-methylpropane

the name is more complex. A branched-chain alkane such as 2-methylpropane can be considered to be derived from a straight-chain alkane by replacing one or more hydrogen atoms by alkyl groups. The name consists of two parts:

- a *suffix* that identifies the parent straight-chain alkane. To find the suffix, count the number of carbon atoms in the longest continuous chain. For a three-carbon chain, the suffix is *propane;* for a four-carbon chain it is *butane,* and so on.
- a *prefix* that identifies the branching alkyl group (Table 22.1) and indicates by a number the carbon atom where branching occurs. In 2-methylpropane,

Table 22.1 Nomenclature of Alkanes

Straight-Chain Alkanes		Alkyl Groups	
Methane	CH_4	Methyl	CH_3-
Ethane	CH_3CH_3	Ethyl	CH_3-CH_2-
Propane	$CH_3CH_2CH_3$	Propyl	$CH_3-CH_2-CH_2-$
Butane	$CH_3(CH_2)_2CH_3$	Isopropyl	
Pentane	$CH_3(CH_2)_3CH_3$		$CH_3-\overset{\displaystyle H}{\underset{\displaystyle CH_3}{C}}-$
Hexane	$CH_3(CH_2)_4CH_3$		
Heptane	$CH_3(CH_2)_5CH_3$		
Octane	$CH_3(CH_2)_6CH_3$	Butyl	$CH_3-CH_2-CH_2-CH_2-$

referred to above, the methyl group is located at the second carbon from the end of the chain:

$$C_1 - \underset{|}{C_2} - C_3$$

Following this system, the IUPAC names of the isomers of pentane are

$$CH_3-CH_2-CH_2-CH_2-CH_3 \qquad CH_3-\underset{\underset{CH_3}{|}}{\overset{\overset{H}{|}}{C}}-CH_2-CH_3 \qquad CH_3-\underset{\underset{CH_3}{|}}{\overset{\overset{CH_3}{|}}{C}}-CH_3$$

pentane   2-methylbutane   2,2-dimethylpropane

Notice that

- if the same alkyl group is at two branches, the prefix *di-* is used (2,2-dimethylpropane). If there were three methyl branches, we would write trimethyl, and so on.
- the number in the name is made as small as possible. Thus, we write 2-methylbutane, numbering the chain from the left

$$C_1 - \underset{|}{C_2} - C_3 - C_4$$

rather than from the right.

EXAMPLE 22.2

Assign IUPAC names to

$$CH_3-\underset{\underset{CH_3}{|}}{\overset{\overset{CH_3}{|}}{C}}-CH_2-CH_3 \quad \text{and} \quad CH_3-CH_2-\underset{\underset{\underset{CH_3}{|}}{CH_2}}{\overset{\overset{H}{|}}{C}}-CH_2-CH_3$$

STRATEGY

1. Find the longest carbon chain. That is the parent chain, which is written at the end (see Table 22.1).
2. Count the alkyl groups attached to the straight chain and identify them by referring to the right column of Table 22.1. Write the names of these alkyl groups before the name of the straight chain.
3. Number the carbon atoms from left to right to indicate the carbon atoms to which the alkyl groups are bonded.

SOLUTION

$$CH_3-\underset{\underset{CH_3}{|}}{\overset{\overset{CH_3}{|}}{C}}-CH_2-CH_3$$

1. Longest chain	four carbon atoms = butane
2. Alkyl groups	two one-carbon groups = methyl = dimethyl
3. Number of C atom bonded	both groups at the C_2 position
4. Name	2,2-dimethylbutane

$$CH_3-CH_2-\underset{\underset{\underset{CH_3}{|}}{CH_2}}{\overset{\overset{H}{|}}{C}}-CH_2-CH_3$$

1. Longest chain	five carbon atoms = pentane

continued

2. Alkyl groups	one two-carbon group = ethyl
3. Number of C atom bonded	C_3
4. Name	3-ethylpentane

Sources and Uses of Alkanes

Natural gas, transmitted around the United States and Canada by pipeline, consists largely of methane (80%–90%), with smaller amounts of C_2H_6, C_3H_8, and C_4H_{10}. Cylinders of "bottled gas," used with campstoves, barbecue grills, and the like, contain liquid propane (C_3H_8) and butane (C_4H_{10}). Homes that do not have access to a gas line sometimes use large propane tanks for heating (Figure 22.3). The pressure remains constant as long as any liquid is present, then drops abruptly to zero, indicating that it's time for a refill.

The higher alkanes are most often obtained from petroleum, a dark brown, viscous liquid dispersed through porous rock deposits. Distillation of petroleum gives a series of fractions of different boiling points (Figure 22.4). The most important of these is gasoline; distillation of a liter of petroleum gives about 250 mL of "straight-run" (with no chemical additives) gasoline. It is possible to double the yield of gasoline by converting higher- or lower-boiling fractions to hydrocarbons in the gasoline range (C_5 to C_{12}).

Gasoline-air mixtures tend to ignite prematurely, or "knock," rather than burn smoothly. The *octane number* of a gasoline is a measure of its resistance to knock. It is determined by comparing the knocking characteristics of a gasoline

Figure 22.3 This cylinder contains propane gas and is used for home heating.

Figure 22.4 Petroleum distillation. (a) A distillation tower at a petroleum refinery. (b) A diagram showing the boiling points of the petroleum fractions separated by distillation.

C_5H_{10}
cyclopentane

C_6H_{12}
cyclohexane

Figure 22.5 Cycloalkanes. Cyclopentane and cyclohexane, the two most common cycloalkanes.

sample with those of isooctane (the common name of one of the isomers of octane) and heptane:

$$CH_3—\underset{\underset{CH_3}{|}}{\overset{\overset{CH_3}{|}}{C}}—CH_2—\underset{\underset{CH_3}{|}}{\overset{\overset{H}{|}}{C}}—CH_3 \qquad CH_3—CH_2—CH_2—CH_2—CH_2—CH_2—CH_3$$

isooctane   heptane

Isooctane, which is highly branched, burns smoothly with little knocking and is assigned an octane number of 100. Heptane, being unbranched, knocks badly. It is given an octane number of zero. Gasoline with the same knocking properties as a mixture of 90% isooctane and 10% heptane is rated as "90 octane."

To obtain premium gasoline of octane number above 80, it is necessary to use additives of one type or another. Until the mid-1970s, the principal antiknock agent was tetraethyllead, $(C_2H_5)_4Pb$. Its use was phased out because it poisons catalytic converters and contaminates the environment with lead compounds. To replace tetraethyllead, oxygen-containing organic compounds such as ethanol, C_2H_5OH, are added to gasoline to promote smooth, complete combustion.

Petroleum contains hydrocarbons other than the open-chain alkanes considered to this point. These include **cycloalkanes** in which 3 to 30 CH_2 groups are bonded into closed rings. The structures of the two most common hydrocarbons of this type are shown in Figure 22.5. Cyclopentane and cyclohexane, where the bond angles are close to the ideal tetrahedral angle of 109.5°, are stable liquids with boiling points of 49°C and 81°C, respectively.

Cycloalkanes, like alkanes, are saturated hydrocarbons containing only single bonds. Notice, though, that as a result of ring formation, each cycloalkane molecule contains two fewer hydrogen atoms than the corresponding alkane.

Number of C Atoms	3	4	5	6	n
Formula of alkane	C_3H_8	C_4H_{10}	C_5H_{12}	C_6H_{14}	C_nH_{2n+2}
Formula of cycloalkane	C_3H_6	C_4H_8	C_5H_{10}	C_6H_{12}	C_nH_{2n}

22-2 Unsaturated Hydrocarbons: Alkenes and Alkynes

In an *unsaturated hydrocarbon,* at least one of the carbon-carbon bonds in the molecule is a multiple bond. As a result, there are fewer hydrogen atoms in an unsaturated hydrocarbon than in a saturated one with the same number of carbons. We will consider two types of unsaturated hydrocarbons:

■ **alkenes,** in which there is one carbon-carbon double bond in the molecule:

$$\overset{\diagup}{\underset{\diagdown}{}}C=C\overset{\diagdown}{\underset{\diagup}{}}$$

Replacing a single bond with a double bond eliminates two hydrogen atoms. Hence the general formula of an alkene is

$$C_nH_{2n}$$

as compared with C_nH_{2n+2} for an alkane.

- **alkynes,** in which there is one carbon-carbon triple bond in the molecule:

$$-C\equiv C-$$

Again, two hydrogen atoms are "lost" when a double bond is converted to a triple bond. Hence the general formula of an alkyne is

$$C_nH_{2n-2}$$

EXAMPLE 22.3

What is the molecular formula of the alkane, alkene, and alkyne containing six carbon atoms, and of the alkane containing ten hydrogen atoms?

STRATEGY

Apply the formulas:

C_nH_{2n+2} for alkanes
C_nH_{2n} for alkenes
C_nH_{2n-2} for alkynes

SOLUTION

Six C atoms	alkane: $n = 6$; H atoms $= 2n + 2 = 2(6) + 2 = 14 \longrightarrow C_6H_{14}$
	alkene: $n = 6$; H atoms $= 2n = 2(6) = 12 \longrightarrow C_6H_{12}$
	alkyne: $n = 6$; H atoms $= 2n - 2 = 2(6) - 2 = 10 \longrightarrow C_6H_{10}$
Alkane with ten H atoms	alkane with 10 H atoms: $2n + 2 = 10$; $n = 4 \longrightarrow C_4H_{10}$

Alkenes

The simplest alkene is ethene, C_2H_4 (common name, ethylene). Its structural formula is

$$\overset{\displaystyle H}{\underset{\displaystyle H}{}}C=C\overset{\displaystyle H}{\underset{\displaystyle H}{}}$$

ethene

You may recall that we discussed the bonding in ethene in Chapter 7. The double bond in ethene and other alkenes consists of a sigma bond and a pi bond. The ethene molecule is planar. There is no rotation about the double bond, since that would require "breaking" the pi bond. The bond angle in ethene is 120°, corresponding to sp^2 hybridization about each carbon atom. The geometries of ethene and the next member of the alkene series, C_3H_6, are shown in Figure 22.6.

Ethene is produced in larger amounts than any other organic chemical, about 1.7×10^7 metric tons annually in the United States. It is made by heating ethane to about 700°C in the presence of a catalyst.

$$C_2H_6(g) \longrightarrow C_2H_4(g) + H_2(g)$$

Ethene is used to make a host of organic compounds; it is also the starting material for the preparation of polyethylene (Chapter 23). Since it is a plant hormone, ethene finds application in agriculture. It is used to ripen fruit that has been picked green to avoid spoilage in shipping. Exposure to ethene at very low concentrations produces the colors we associate with ripe bananas and oranges.

$H_2C = CH_2$
ethene

$CH_3CH = CH_2$
propene

Figure 22.6 The two simplest alkenes. Ethylene is planar, as is the $-CH=CH_2$ region of propene.

The names of alkenes are derived from those of the corresponding alkanes with the same number of carbon atoms per molecule. There are two modifications:

■ the ending -*ane* is replaced by -*ene*.

$$CH_3—CH_3 \qquad CH_2{=}CH_2$$
ethane $\qquad\qquad$ ethene

■ where necessary, a number is used to designate the double-bonded carbon; the number is made as small as possible.

$$CH_2{=}CH—CH_2—CH_3 \qquad CH_3—CH{=}CH—CH_3$$
1-butene $\qquad\qquad\qquad$ 2-butene

$$CH_2{=}C—CH_2—CH_3 \qquad CH_3—C{=}C—CH_3$$
$$\qquad | \qquad\qquad\qquad\qquad | \quad |$$
$$\quad CH_3 \qquad\qquad\qquad\qquad H_3C \quad H$$
2-methyl-1-butene $\qquad\qquad$ 2-methyl-2-butene

EXAMPLE 22.4

Give the structural formula for the alkene with the IUPAC name 3-ethyl-2-pentene.

STRATEGY AND SOLUTION

1. Start with the parent hydrocarbon. Pent means 5, so draw a five-carbon chain.

$$—C_1—C_2—C_3—C_4—C_5—$$

2. The suffix -*ene* means that there is a double bond, and the number 2 before the hydrocarbon means that the double bond is inserted at C_2 position.

$$—C_1—C_2{=}C_3—C_4—C_5—$$

3. The number 3 before the word *ethyl* means that an ethyl group is bonded to C_3.

$$—C_1—C_2{=}C_3—C_4—C_5—$$
$$\qquad\qquad | $$
$$\qquad\quad C_2H_5$$

4. Add the missing hydrogen atoms to get the final answer.

$$CH_3—CH{=}C—CH_2—CH_3 \qquad (C_7H_{14})$$
$$\qquad\qquad | $$
$$\qquad C_2H_5$$

Alkynes

The IUPAC names of alkynes are derived from those of the corresponding alkenes by replacing the suffix -*ene* with -*yne*. Thus we have

$$H—C{\equiv}C—H \qquad H—C{\equiv}C—CH_3$$
ethyne $\qquad\qquad\qquad$ propyne

$$H—C{\equiv}C—CH_2—CH_3 \qquad CH_3—C{\equiv}C—CH_3$$
1-butyne $\qquad\qquad\qquad\qquad$ 2-butyne

The most important alkyne by far is the first member of the series, commonly called acetylene. Recall from Chapter 7 that the C_2H_2 molecule is linear, with 180° bond angles. The triple bond consists of a sigma bond and two pi bonds; each carbon atom is sp-hybridized. The geometries of acetylene and the next member of the series, C_3H_4, are shown in Figure 22.7.

Figure 22.7 The two simplest alkynes. Both molecules contain four atoms in a straight line.

HC≡CH
acetylene

$CH_3C{\equiv}CH$
methylacetylene

Figure 22.8 Welding with an oxyacetylene torch.

Thermodynamically, acetylene is unstable with respect to decomposition to the elements:

$$C_2H_2(g) \longrightarrow 2C(s) + H_2(g) \qquad \Delta G° = -209.2 \text{ kJ at } 25°C$$

At high pressures, this reaction can occur explosively. For that reason, cylinders of acetylene do not contain the pure gas. Instead the cylinder is packed with an inert, porous material that holds a solution of acetylene gas in acetone.

You are probably most familiar with acetylene as a gaseous fuel used in welding and cutting metals (Figure 22.8). When mixed with pure oxygen in a torch, acetylene burns at temperatures above 2000°C. The heat comes from the reaction

$$C_2H_2(g) + \tfrac{5}{2}O_2(g) \longrightarrow 2CO_2(g) + H_2O(l) \qquad \Delta H = -1300 \text{ kJ}$$

The reaction gives off a brilliant white light, which served as a source of illumination in the headlights of early automobiles.

22-3 Aromatic Hydrocarbons and Their Derivatives

Aromatic hydrocarbons, sometimes referred to as *arenes*, can be considered as derivatives of benzene, C_6H_6. Benzene is a transparent, volatile liquid (bp = 80°C) that was discovered by Michael Faraday in 1825. Its formula, C_6H_6, suggests a high degree of unsaturation, yet its properties are quite different from those of alkenes or alkynes.

As pointed out in Chapter 7, the atomic orbital (valence bond) model regards benzene as a resonance hybrid of the two structures

In these "line-angle" formulas it is understood that there is a carbon atom at each vertex of the hexagon; hydrogen atoms are not shown. This model is consistent with many of the properties of benzene. The molecule is a planar hexagon with bond angles of 120°. The hybridization of each carbon is sp². However, this structure is misleading in one respect. Chemically, benzene does not behave as if double bonds were present.

A more satisfactory model of the electron distribution in benzene, based on molecular orbital theory (Appendix 4), assumes that

- each carbon atom forms three sigma bonds, one to a hydrogen atom and two to adjacent carbon atoms.
- the three electron pairs remaining are spread symmetrically over the entire molecule to form "delocalized" pi bonds. ◀ ❶

This model is often represented by the structure

where it is understood that

- a carbon atom is at each corner of the hexagon.
- an H atom is bonded to each carbon atom.
- the circle in the center of the molecule represents the six delocalized electrons.

At one time, benzene was widely used as a solvent, both commercially and in research and teaching laboratories. That use has largely disappeared because of its toxicity. Chronic exposure to benzene vapor leads to various blood disorders and, in extreme cases, aplastic anemia and leukemia. It appears that the culprits are oxidation products of the aromatic ring, formed in an attempt to solubilize benzene and thus eliminate it from the body.

There are three pi bonds in the benzene molecule.

Derivatives of Benzene

Among the more common monosubstituted benzenes are the following:

benzoic
acid   aniline   phenol   toluene

Phenol was the first commercial antiseptic; its introduction into hospitals in the 1870s led to a dramatic decrease in deaths from postoperative infections. Its use for this purpose has long since been abandoned because phenol burns exposed tissue, but many modern antiseptics are phenol derivatives. Toluene has largely replaced benzene as a solvent because it is much less toxic. Oxidation of toluene in the body gives benzoic acid, which is readily eliminated and has none of the toxic properties of the oxidation products of benzene. Indeed, benzoic acid or its sodium salt (Na^+, $C_6H_5COO^-$ ions) is widely used as a preservative in foods and beverages, including fruit juices and soft drinks.

When there are two groups attached to the benzene ring, three isomers are possible. These are designated by the prefixes *ortho-*, *meta-*, and *para-*, often abbreviated as *o-*, *m-*, and *p-*.

o-dichlorobenzene   *m*-dichlorobenzene   *p*-dichlorobenzene

Numbers can also be used; these three compounds may be referred to as 1,2-dichlorobenzene, 1,3-dichlorobenzene, and 1,4-dichlorobenzene, respectively. When three or more substituents are present, numbers become mandatory.

1,3,5-trinitrobenzene   2,4,6-trinitrotoluene (TNT)

In the structure at the right, it is understood that the CH_3 group of toluene is at position 1. Notice that the numbers used are as small as possible.

EXAMPLE 22.5

Name the following compounds as derivatives of phenol.

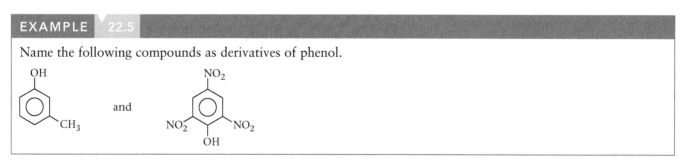

---

continued

SOLUTION

Parent compound is called phenol.

CH_3 (methyl) is in the *meta* position.
name: *meta*-methylphenol;   common name: *meta*-cresol

The parent compound is phenol.
Number the carbon atom to which the OH group is bonded C_1.
The NO_2 (nitro) groups are in the C_2, C_4, and C_6 positions whether you count in a clockwise or counterclockwise manner.
name: 2,4,6-trinitrophenol;   common name: picric acid

Condensed Ring Structures

In another type of aromatic hydrocarbon, two or more benzene rings are fused together. Naphthalene is the simplest compound of this type. Fusion of three benzene rings gives two different isomers, anthracene and phenanthrene.

| naphthalene | anthracene | phenanthrene | benzo[a]pyrene |
| $C_{10}H_8$ | $C_{14}H_{10}$ | $C_{14}H_{10}$ | $C_{20}H_{12}$ |

It is important to realize that in these structures no hydrogen atoms are bonded to the carbons at the juncture of two rings. The eight hydrogen atoms in naphthalene are located as shown below.

Certain compounds of this type are potent carcinogens. One of the most dangerous is benzo[a]pyrene, which has been detected in cigarette smoke. It is believed to be a cause of lung cancer, to which smokers are susceptible.

22-4 Functional Groups

Many organic molecules can be derived from hydrocarbons by substituting a **functional group** for a hydrogen atom. The functional group can be a nonmetal atom or a small group of atoms that is bonded to carbon. Table 22.2 lists the types of functional groups commonly found in organic compounds.

Alcohols and Ethers

Alcohols have the general structural formula

$$R—O—H$$

Table 22.2 Common Functional Groups

Group	Class	Example	Name*
—OH	Alcohols	C_2H_5OH	Ethanol (ethyl alcohol)
—O—	Ethers	CH_3—O—CH_3	Dimethyl ether
$\overset{\text{O}}{\overset{\|}{-\text{C}-\text{H}}}$	Aldehydes	$CH_3-\overset{\text{O}}{\overset{\|}{\text{C}}}-H$	Ethanal (acetaldehyde)
$\overset{\text{O}}{\overset{\|}{-\text{C}-}}$	Ketones	$CH_3-\overset{\text{O}}{\overset{\|}{\text{C}}}-CH_3$	Propanone (acetone)
$\overset{\text{O}}{\overset{\|}{-\text{C}-\text{OH}}}$	Carboxylic acids	$CH_3-\overset{\text{O}}{\overset{\|}{\text{C}}}-OH$	Ethanoic acid (acetic acid)
$\overset{\text{O}}{\overset{\|}{-\text{C}-\text{O}-}}$	Esters	$CH_3-\overset{\text{O}}{\overset{\|}{\text{C}}}-OCH_3$	Methyl acetate
$-\overset{\|}{\underset{}{\text{N}}}-$	Amines	CH_3NH_2	Aminomethane (methylamine)

*Common names are shown in red.

where R is an **alkyl group.** Alcohols are named by substituting the suffix *-ol* for the *-ane* suffix of the corresponding alkane. For the first four alcohols we have (common names in red):

$$CH_3OH \qquad CH_3—CH_2OH \qquad CH_3—CH_2—CH_2OH \qquad CH_3—\underset{\underset{OH}{|}}{CH}—CH_3$$

methanol (methyl alcohol)   ethanol (ethyl alcohol)   1-propanol (propyl alcohol)   2-propanol (isopropyl alcohol)

Ethers have the general structure

$$R—O—R'$$

where the two alkyl groups can be the same (e.g., R = R′ = CH_3) or different (R = CH_3; R′ = C_2H_5). Ethers are ordinarily named by citing the two alkyl groups:

$$CH_3—O—CH_3 \qquad CH_3—O—C_2H_5$$
dimethyl ether   ethyl methyl ether

Table 22.3 compares two physical properties, boiling point and water solubility, of alcohols, ethers, and alkanes of similar molar masses. Notice that

- alcohols have boiling points much higher than those of alkanes of comparable molar mass. This reflects the fact that alcohol molecules, like H_2O, are strongly hydrogen-bonded to one another. Ethers cannot do this, since they do not contain OH groups, so they have boiling points much lower than those of comparable alcohols.
- alcohols and ethers of low molar mass are generally soluble in water. Ethers, like alcohols, can form hydrogen bonds with water, making use of the OH groups in the water molecule.

About 3×10^9 kg of methanol are produced annually in the United States from *synthesis gas,* a mixture of carbon monoxide and hydrogen:

$$CO(g) + 2H_2(g) \xrightarrow[\text{250 atm, 350°C}]{\text{ZnO,Cr}_2\text{O}_3} CH_3OH(g)$$

Methanol is also formed as a byproduct when charcoal is made by heating wood in the absence of air. For this reason, it is sometimes called wood alcohol. Methanol is used in jet fuels and as a solvent, gasoline additive, and starting material for several industrial syntheses. It is a deadly poison; ingestion of as little as 25 mL can be fatal. The antidote in this case is a solution of sodium hydrogen carbonate, $NaHCO_3$.

Table 22.3 Physical Properties of Simple Alkanes, Alcohols, and Ethers

Name	Structure	Molar Mass	Boiling Point	Water Solubility
Propane	$CH_3CH_2CH_3$	44 g/mol	−42°C	Insoluble
Ethanol	CH_3CH_2OH	46 g/mol	78°C	Soluble
Dimethyl ether	CH_3OCH_3	46 g/mol	−23°C	Soluble
Butane	$CH_3CH_2CH_2CH_3$	58 g/mol	0°C	Insoluble
1-Propanol	$CH_3CH_2CH_2OH$	60 g/mol	97°C	Soluble
Ethyl methyl ether	$CH_3CH_2OCH_3$	60 g/mol	11°C	Soluble

Ethanol, the most common alcohol, can be made by the fermentation of sugar (Figure 22.9) or grain—hence its common name "grain alcohol." It is the active ingredient of alcoholic beverages, in which it is present at various concentrations (4%–8% in beer, 12%–15% in wine, and 40% or more in distilled spirits). The "proof" of an alcoholic beverage is twice the volume percentage of ethanol; an 86-proof bourbon whiskey contains 43% (by volume) ethanol. The flavor of alcoholic beverages is due to impurities; ethanol itself is tasteless and colorless.

Industrial ethanol is made by the reaction of ethylene with water, using sulfuric acid as a catalyst.

$$\underset{\substack{H \\ | \\ H}}{\overset{\substack{H \\ | \\ H}}{C}}=\underset{\substack{H \\ | \\ H}}{\overset{\substack{H \\ | \\ H}}{C} } \quad + \quad H-O-H \quad \xrightarrow{H^+} \quad H-\underset{\substack{| \\ H}}{\overset{\substack{H \\ |}}{C}}-\underset{\substack{| \\ H}}{\overset{\substack{H \\ |}}{C}}-OH$$

Frequently, ethanol that is not destined for human consumption is "denatured" by adding small quantities of methanol or benzene. This avoids the high federal tax on beverage alcohol; denatured alcohol is poisonous.

Figure 22.9 **Making wine.** Wine (*far right*) is produced from the glucose in grape juice (left) by fermentation. The reaction is

$$C_6H_{12}O_6(aq) \longrightarrow 2C_2H_5OH(aq) + 2CO_2(g)$$

The purpose of the bubble chamber in the fermentation jug (*center*) is to allow the carbon dioxide to escape but prevent oxygen from entering and oxidizing ethyl alcohol to acetic acid.

Certain alcohols contain two or more —OH groups per molecule. Perhaps the most familiar compounds of this type are ethylene glycol and glycerol:

$$H-\overset{\overset{\displaystyle H}{|}}{\underset{\underset{\displaystyle OH}{|}}{C}}-\overset{\overset{\displaystyle H}{|}}{\underset{\underset{\displaystyle OH}{|}}{C}}-H \qquad H-\overset{\overset{\displaystyle H}{|}}{\underset{\underset{\displaystyle OH}{|}}{C}}-\overset{\overset{\displaystyle H}{|}}{\underset{\underset{\displaystyle OH}{|}}{C}}-\overset{\overset{\displaystyle H}{|}}{\underset{\underset{\displaystyle OH}{|}}{C}}-H$$

ethylene glycol   glycerol

Ethylene glycol is widely used as an antifreeze. Glycerol is formed as a byproduct in making soaps. It is a viscous, sweet-tasting liquid used in making drugs, antibiotics, plastics, and explosives (nitroglycerin).

The only ether that you are likely to encounter in the laboratory is diethyl ether, often referred to simply as "ether." It was first used as an anesthetic in the 1840s. Today we use other compounds for that purpose, ◀ ⓘ for at least a couple of reasons. For one thing, ether vapor is highly flammable. For another, liquid ether, on standing in contact with air, tends to form compounds called hydroperoxides, which are potent explosives.

ⓘ "Laughing gas," N_2O, is sometimes used by dentists.

$$CH_3CH_2-O-CH_2CH_3(l) + O_2(g) \longrightarrow CH_3CH_2-O-\underset{\underset{\displaystyle O-OH}{|}}{C}HCH_3$$

diethyl ether   a hydroperoxide

Aldehydes and Ketones

All these compounds contain the **carbonyl group**

$$\overset{\displaystyle \diagdown\diagup}{\underset{\displaystyle \|}{\underset{\displaystyle O}{C}}}$$

In a **ketones**, two hydrocarbon groups (alkyl or aromatic) are bonded to the central carbon atom, giving the general structure

$$\overset{\displaystyle R\diagdown\diagup R'}{\underset{\displaystyle \|}{\underset{\displaystyle O}{C}}}$$

where (as with ethers) R and R′ can be the same or different. In the simplest ketone, commonly called *acetone,* R = R′ = CH_3. The structural formula of acetone is ordinarily written as

$$CH_3-\underset{\underset{\displaystyle O}{\|}}{C}-CH_3 \qquad \text{acetone (IUPAC name = propanone)}$$

with the understanding that the bond angle around the central carbon atom is approximately 120°.

In **aldehydes,** at least one of the two "groups" bonded to carbon must be a hydrogen atom:

$$\overset{\displaystyle R\diagdown\diagup H}{\underset{\displaystyle \|}{\underset{\displaystyle O}{C}}}$$

R can be any hydrocarbon group or, in the simplest case, *formaldehyde,* a second hydrogen atom:

$$\overset{\displaystyle H\diagdown\diagup H}{\underset{\displaystyle \|}{\underset{\displaystyle O}{C}}} \qquad \text{formaldehyde (IUPAC name = methanal)}$$

Figure 22.10 "Aromatic" aldehydes. Benzaldehyde, vanillin, and cinnamaldehyde are aldehydes with pleasant aromas that are found in almonds, vanilla beans, and cinnamon, respectively.

Acetone (bp = 56°C) is completely water-soluble and dissolves a wide variety of organic compounds as well. Accordingly, it is one of the most important industrial solvents. You may be most familiar with it as the solvent in nail polish remover.

Normally, only small amounts of acetone are produced in the human body. Diabetics produce larger amounts, sometimes enough to give their breath a sweet odor.

Formaldehyde (bp = −21°C) is ordinarily found in the laboratory in the form of a concentrated water solution (37% HCHO) known as formalin. At one time it was widely used as a preservative for biological specimens. That application has declined now that formaldehyde has been shown to be carcinogenic. Industrially, it is still used as a component of adhesives used in making plywood and fiberboard.

Formaldehyde, as a gas or in water solution, has a distinctly unpleasant odor. Higher "aromatic" aldehydes such as those shown in Figure 22.10 are a lot more pleasant to deal with; they are used in flavorings and even perfumes.

EXAMPLE 22.6

Classify each of the following as an alcohol, ether, aldehyde, or ketone.

STRATEGY

Identify the functional group and use Table 22.2. Note that more than one functional group may be involved.

SOLUTION

CH_3—O—CH_2—CH_2—CH_3	Functional group: —O— The molecule is an ether.
	Functional groups: C=O and —OH The molecule is an aldehyde and an alcohol.
	Functional groups: C=O and —OH The molecule is a ketone and an alcohol.

Carboxylic Acids and Esters

Carboxylic acids have the following general structure:

$$R—\underset{\underset{O}{\|}}{C}—OH$$

As usual, R is a hydrocarbon group or, in the simplest case, a hydrogen atom. The acidic hydrogen atom is the one bonded to oxygen. The IUPAC name of a carboxylic acid is obtained by substituting the suffix "-oic acid" for the final "e" in the name of the corresponding alkane. In practice, such names are seldom used. For example, the first two members of the series are commonly referred to as formic acid and acetic acid.

$$H—\underset{\underset{O}{\|}}{C}—OH \qquad CH_3—\underset{\underset{O}{\|}}{C}—OH$$

methanoic acid   ethanoic acid
formic acid   acetic acid

According to legend, formic acid used to be made by distilling ants (Latin, *formica*). Acetic acid is the active ingredient of vinegar, responsible for its sour taste. "White" vinegar is made by adding pure acetic acid to water, forming a 5% solution. "Brown" or "apple cider" vinegar is made from apple juice; the ethyl alcohol formed by fermentation is oxidized to acetic acid. ◂*i*

> *i* ─────────────
> Homemade wine often contains some vinegar.

The most important property of carboxylic acids is implied by their name: they act as weak acids in water solution.

$$RCOOH(aq) \rightleftharpoons H^+(aq) + RCOO^-(aq)$$

Carboxylic acids vary considerably in strength. Among the strongest is trichloroacetic acid ($K_a = 0.20$); a 0.10 M solution of $Cl_3C—COOH$ is about 73% ionized. Trichloroacetic acid is an ingredient of over-the-counter preparations used to treat canker sores and remove warts.

Treatment of a carboxylic acid with the strong base NaOH forms the sodium salt of the acid. With acetic acid, the acid-base reaction is

$$CH_3—\underset{\underset{O}{\|}}{C}—OH\ (aq) + OH^-(aq) \longrightarrow CH_3—\underset{\underset{O}{\|}}{C}—O^-(aq) + H_2O$$

Evaporation gives the salt sodium acetate, which contains Na^+ and CH_3COO^- ions. Many of the salts of carboxylic acids have important uses. Sodium and calcium propionate (Na^+ or Ca^{2+} ions, $CH_3CH_2COO^-$ ions) are added to bread, cake, and cheese to inhibit the growth of mold.

Soaps are sodium salts of long-chain carboxylic acids such as stearic acid:

$$CH_3(CH_2)_{16}—\underset{\underset{O}{\|}}{C}—OH \qquad Na^+,\ CH_3(CH_2)_{16}—\underset{\underset{O}{\|}}{C}—O^-$$

stearic acid   sodium stearate, a soap

The cleaning action of soap reflects the nature of the long-chain carboxylate anion. The long hydrocarbon group is a good solvent for greases and oils, and the ionic COO^- group gives water solubility.

Reactions between an alcohol and a carboxylic acid form **esters**, containing the functional group

$$—O—\underset{\underset{O}{\|}}{C}—$$

often abbreviated $—COO—$. The reaction between methyl alcohol and acetic acid is typical

$$CH_3—OH\ +\ HO—\underset{\underset{O}{\|}}{C}—CH_3 \xrightarrow{H^+} CH_3—O—\underset{\underset{O}{\|}}{C}—CH_3 + H_2O$$

methyl alcohol   acetic acid   methyl acetate

The common name of an ester consists of two words. The first word (methyl, ethyl, . . .) is derived from that of the alcohol; the second word (formate, acetate, . . .)

Table 22.4 Properties of Esters

Ester	Structure	Odor, Flavor
Ethyl formate	$CH_3CH_2-O-\overset{\displaystyle O}{\underset{\|}{C}}-H$	Rum
Isobutyl formate	$(CH_3)_2CHCH_2-O-\overset{\displaystyle O}{\underset{\|}{C}}-H$	Raspberry
Methyl butyrate	$CH_3-O-\overset{\displaystyle O}{\underset{\|}{C}}-(CH_2)_2CH_3$	Apple
Ethyl butyrate	$CH_3CH_2-O-\overset{\displaystyle O}{\underset{\|}{C}}-(CH_2)_2CH_3$	Pineapple
Isopentyl acetate	$(CH_3)_2CH(CH_2)_2-O-\overset{\displaystyle O}{\underset{\|}{C}}-CH_3$	Banana
Octyl acetate	$CH_3(CH_2)_7-O-\overset{\displaystyle O}{\underset{\|}{C}}-CH_3$	Orange
Pentyl propionate	$CH_3(CH_2)_4-O-\overset{\displaystyle O}{\underset{\|}{C}}-CH_2CH_3$	Apricot

is the name of the acid with the *-ic* suffix replaced by *-ate*. Thus ethyl formate (Table 22.4) is made from ethyl alcohol and formic acid:

$$CH_3CH_2OH \qquad HCOOH$$

ethyl alcohol   formic acid

Many esters have pleasant odors; they are commonly found in natural and synthetic fragrances.

EXAMPLE   22.7

Show the structure of the following:

a   the three-carbon alcohol with an —OH group at the end of the chain and the three-carbon carboxylic acid.

STRATEGY

1. Write the parent compound chain.
2. Put the functional group in the specified position.
3. Fill in the structural formula with H atoms.

SOLUTION

Alcohol
1. Parent compound chain   C—C—C
2. Functional group   C—C—C—OH
3. Hydrogen atoms $CH_3-CH_2-CH_2-OH$

Carboxylic acid
1. Parent compound chain   C—C—C
2. Functional group $C-C-\overset{\displaystyle O}{\underset{\|}{C}}-OH$

3. Hydrogen atoms $CH_3CH_2-\overset{\displaystyle O}{\underset{\|}{C}}-OH$

continued

b	the ester formed when the alcohol and the carboxylic acid react.

STRATEGY AND SOLUTION

Join the acid and the alcohol and remove OH from the functional group (COOH) of the acid and the H from the functional group (OH) of the alcohol.

$$CH_3CH_2CH_2-OH \qquad HO-\underset{\underset{O}{\|}}{C}-CH_2CH_3$$

$$\downarrow$$

$$CH_3CH_2CH_2-O-\underset{\underset{O}{\|}}{C}-CH_2CH_3$$

Amines

Amines were discussed earlier (Chapter 13). Here we will concentrate on *primary* amines, in which two hydrogens are bonded to a nitrogen atom, giving as a general formula

$$R-NH_2$$

where R may be an alkyl or aromatic group:

$$CH_3NH_2$$

methylamine   aniline

In Chapter 23, we'll look at the chemistry of amino acids, which contain both NH_2 and COOH groups.

Methylamine, the first member of the series,

- is a flammable gas at 25°C and 1 atm; its boiling point and critical temperature are −6°C and 157°C, respectively.
- has a somewhat unpleasant fishy smell. Other primary amines with more obnoxious odors include

$$H_2N-(CH_2)_4-NH_2 \qquad H_2N-(CH_2)_5-NH_2$$

putrescine   cadaverine

- is a weak base with a K_b (4×10^{-4}) about 20 times that of ammonia (1.8×10^{-5}) and a million times the value for aniline (4×10^{-10}).
- reacts essentially completely with strong acid:

$$CH_3NH_2(aq) + H^+(aq) \longrightarrow CH_3NH_3^+(aq)$$

but gives an equilibrium mixture with acetic acid:

$$CH_3NH_2(aq) + CH_3COOH(aq) \rightleftharpoons CH_3NH_3^+(aq) + CH_3COO^-(aq)$$

EXAMPLE	22.8	GRADED

Consider methylamine, CH_3NH_2.

a	Calculate the equilibrium constant K when it reacts with H^+ in the following reaction:

$$CH_3NH_2(aq) + H^+(aq) \longrightarrow CH_3NH_3^+(aq)$$

ANALYSIS

Information given:	reaction $(CH_3NH_2(aq) + H^+(aq) \longrightarrow CH_3NH_3^+(aq))$
Information implied:	K_a for $CH_3NH_3^+$
Asked for:	K

continued

1. Note that the given equation is the reverse of the equation for the dissociation of a weak acid.

2. Find K_a for $CH_3NH_3^+$ in Appendix 1 and use the reciprocal rule (Chapter 12).

SOLUTION

1. K_a expression	$CH_3NH_3^+(aq) \rightleftharpoons CH_3NH_2(aq) + H^+(aq) \qquad K_a = 2.4 \times 10^{-11}$
2. K	$CH_3NH_2(aq) + H^+(aq) \rightleftharpoons CH_3NH_3^+(aq) \qquad K = 1/K_a = 1/2.4 \times 10^{-11}$ $= 4.2 \times 10^{10}$

b Calculate the equilibrium constant K for its reaction with acetic acid:

$$CH_3NH_2(aq) + CH_3COOH(aq) \rightleftharpoons CH_3NH_3^+(aq) + CH_3COO^-(aq)$$

ANALYSIS

Information given:	reaction ($CH_3NH_2(aq) + CH_3COOH(aq) \longrightarrow CH_3NH_3^+(aq) + CH_3COO^-(aq)$) From (a): K (4.2×10^{10}) for the reaction $CH_3NH_2(aq) + H^+(aq) \rightleftharpoons CH_3NH_3^+(aq)$
Information implied:	K_a for CH_3COOH
Asked for:	K

STRATEGY

1. Break up the given equation into two reactions

$CH_3NH_2 \rightleftharpoons CH_3NH_3^+$ (1)

$CH_3COOH \rightleftharpoons CH_3COO^-$ (2)

2. Find K for each reaction.

3. Combine the two equations and their Ks. Apply the Rule of Multiple Equilibria (Chapter 12).

SOLUTION

K for equation (1)	From (a): $K_1 = 4.2 \times 10^{10}$
K for equation (2)	From Appendix 1: $K_2 = K_a = 1.8 \times 10^{-5}$
Combine both equations	$CH_3NH_2(aq) + H^+(aq) \rightleftharpoons CH_3NH_3^+(aq) \qquad K_1 = 4.2 \times 10^{10}$ $CH_3COOH(aq) \rightleftharpoons CH_3COO^-(aq) + H^+(aq) \qquad K_2 = 1.8 \times 10^{-5}$ $CH_3NH_2(aq) + CH_3COOH(aq) \rightleftharpoons CH_3NH_3^+(aq) + CH_3COO^-(aq)$
K	$K = K_1 \times K_2 = (4.2 \times 10^{10})(1.8 \times 10^{-5}) = 7.5 \times 10^5$

22-5 Isomerism in Organic Compounds

Isomers are distinctly different compounds, with different properties, that have the same molecular formula. In Section 22-1, we considered structural isomers of alkanes. You will recall that butane and 2-methylpropane have the same molecular formula, C_4H_{10}, but different structural formulas. In these, as in all structural isomers, the order in which the atoms are bonded to each other differs.

Structural isomerism is common among all types of organic compounds. For example, there are:

1. Three structural isomers of the alkene C_4H_8:

2. Three structural isomers with the molecular formula C_3H_8O. Two of these are alcohols; the third is an ether.

$$CH_3CH_2\!-\!\overset{\displaystyle H}{\underset{\displaystyle H}{\overset{\displaystyle |}{\underset{\displaystyle |}{C}}}}\!-\!OH \qquad CH_3\!-\!\overset{\displaystyle H}{\underset{\displaystyle OH}{\overset{\displaystyle |}{\underset{\displaystyle |}{C}}}}\!-\!CH_3 \qquad CH_3\!-\!O\!-\!CH_2CH_3$$

EXAMPLE 22.9

Consider the molecule $C_3H_6Cl_2$, which is derived from propane, C_3H_8, by substituting two Cl atoms for H atoms. Draw the structural isomers of $C_3H_6Cl_2$.

STRATEGY AND SOLUTION

1. All these compounds have three carbon atoms, hence the same skeleton.

$$C\!-\!C\!-\!C$$

   Note that the terminal C atoms are equivalent. The middle C atom is not.

2. Bond 2 Cl atoms to the second C atom for the first isomer, and 2 Cl atoms to the terminal C atom for the second isomer.

$$C\!-\!\overset{\displaystyle Cl}{\underset{\displaystyle Cl}{\overset{\displaystyle |}{\underset{\displaystyle |}{C}}}}\!-\!C \quad \text{and} \quad C\!-\!C\!-\!\overset{\displaystyle Cl}{\underset{\displaystyle Cl}{\overset{\displaystyle |}{\underset{\displaystyle |}{C}}}}$$

3. Bond a Cl atom to each terminal C atom for the third isomer.

$$\overset{\displaystyle Cl}{\overset{\displaystyle |}{C}}\!-\!C\!-\!\overset{\displaystyle Cl}{\overset{\displaystyle |}{C}}$$

4. Bond a Cl atom to a terminal C atom and the other Cl atom to the middle C atom for the fourth isomer.

$$\overset{\displaystyle Cl}{\overset{\displaystyle |}{C}}\,\overset{\displaystyle Cl}{\overset{\displaystyle |}{C}}\!-\!C$$

5. Add the appropriate number of H atoms to the C atoms so that there are four bonds around each C atom.

$$CH_3\!-\!CH_2\!-\!\overset{\displaystyle Cl}{\underset{\displaystyle Cl}{\overset{\displaystyle |}{\underset{\displaystyle |}{C}}}}\!-\!H \qquad CH_3\!-\!\overset{\displaystyle Cl}{\underset{\displaystyle Cl}{\overset{\displaystyle |}{\underset{\displaystyle |}{C}}}}\!-\!CH_3 \qquad H\!-\!\overset{\displaystyle H}{\underset{\displaystyle Cl}{\overset{\displaystyle |}{\underset{\displaystyle |}{C}}}}\!-\!CH_2\!-\!\overset{\displaystyle H}{\underset{\displaystyle Cl}{\overset{\displaystyle |}{\underset{\displaystyle |}{C}}}}\!-\!H \qquad CH_3\!-\!\overset{\displaystyle H}{\underset{\displaystyle Cl}{\overset{\displaystyle |}{\underset{\displaystyle |}{C}}}}\!-\!\overset{\displaystyle H}{\underset{\displaystyle Cl}{\overset{\displaystyle |}{\underset{\displaystyle |}{C}}}}\!-\!H$$

Geometric (cis-trans) Isomers

As we saw earlier, there are three structural isomers of the alkene C_4H_8. You may be surprised to learn that there are actually *four* different alkenes with this molecular formula. The "extra" compound arises because of a phenomenon called **geometric isomerism**. There are two different geometric isomers of the structure shown on the left, on the previous page, under (1).

cis isomer
bp = 4°C

trans isomer
bp = 1°C

In the *cis* **isomer,** the two CH_3 groups (or the two H atoms) are as close to one another as possible. In the *trans* **isomer,** the two identical groups are farther apart. The two forms exist because there is no free rotation about the carbon-to-carbon double bond. The situation is analogous to that with *cis-trans* isomers of square planar complexes (Chapter 19). In both cases, the difference in geometry is responsible for isomerism; the atoms are bonded to each other in the same way.

Geometric, or *cis-trans,* isomerism is common among alkenes. It occurs when both of the double-bonded carbon atoms are joined to two different atoms or groups. The other two structural isomers of C_4H_8 shown under (1) on page 566 do not show *cis-trans* isomerism. In both cases the carbon atom at the left is joined to two identical hydrogen atoms.

EXAMPLE 22.10

Draw all the isomers of the molecule $C_2H_2Cl_2$ in which two of the H atoms of ethylene are replaced by Cl atoms.

STRATEGY AND SOLUTION

1. Write the structure for ethylene.

2. Substitute two Cl atoms for the two H atoms on the right.

Substituting two Cl atoms for the two H atoms on the left gives the identical structure just written.

3. Substitute Cl atoms for the two H atoms at the top (1 H atom per C atom).

Substituting two Cl atoms for the two H atoms on the bottom gives the identical structure.

4. Substitute a Cl atom for the H atom on the top right and the other Cl atom for the H atom on the bottom left.

These are all the isomers that can be drawn.

END POINT

1. The first structure does not show any *cis-trans* isomerism because the two Cl atoms are bonded to the same C atom.

2. The second structure is the *cis*-isomer.

3. The third structure is the *trans*-isomer.

Cis and *trans* isomers differ from one another in their physical and, to a lesser extent, their chemical properties. They may also differ in their physiological behavior. For example, the compound *cis*-9-tricosene

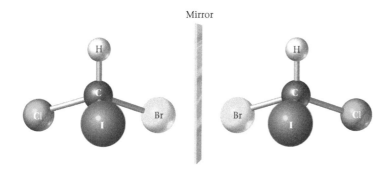

Figure 22.11 **Optical isomers.** These isomers of CHClBrI are mirror images of each other.

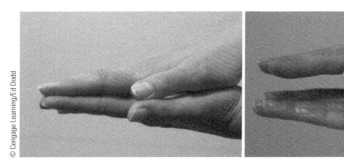

Figure 22.12 **Mirror images.** The right hand cannot be superimposed on the left hand. They are mirror images of each other.

is a sex attractant secreted by the female housefly. The *trans* isomer is totally ineffective in this capacity.

Optical Isomers

Optical isomers arise when at least one carbon atom in a molecule is bonded to four different atoms or groups. Consider, for example, the methane derivative CHClBrI. As you can see from Figure 22.11, there are two different forms of this molecule, which are mirror images of one another. The mirror images are not superimposable; that is, you cannot place one molecule over the other so that identical groups are touching. In this sense, the two isomers resemble right and left hands (Figure 22.12).

There is another way to convince yourself that the two structures shown in Figure 22.11 are different isomers. Imagine yourself directly above the hydrogen atom, looking down at the rest of the molecule. In the structure at the left, the atoms Cl, Br, I, in that order, are arranged in a clockwise pattern. With the structure at the right, in order to move from Cl to Br to I, you have to move counterclockwise.

A molecule such as CHClBrI, which exists in two different forms that are not superimposable mirror images, is said to be *chiral*. The two different forms are referred to as **enantiomers,** or optical isomers. Any molecule in which four different groups are bonded to carbon will be chiral; the carbon atom serves as a **chiral center.** Molecules may contain more than one chiral center, in which case there can be more than two enantiomers.

EXAMPLE	22.11

In the following structural formulas, locate each chiral carbon atom.

$$CH_3 - \overset{\overset{\displaystyle Cl}{|}}{\underset{\underset{\displaystyle Cl}{|}}{C}} - \overset{\overset{\displaystyle H}{|}}{\underset{\underset{\displaystyle OH}{|}}{C}} - CH_3, \quad HO - \overset{\overset{\displaystyle H}{|}}{\underset{\underset{\displaystyle O}{\|}}{C}} - \overset{\overset{\displaystyle H}{|}}{\underset{\underset{\displaystyle NH_2}{|}}{C}} - \overset{\overset{\displaystyle H}{|}}{\underset{\underset{\displaystyle H}{|}}{C}} - OH, \quad \text{and} \quad Cl - \overset{\overset{\displaystyle H}{|}}{\underset{\underset{\displaystyle OH}{|}}{C}} - \overset{\overset{\displaystyle CH_3}{|}}{\underset{\underset{\displaystyle OH}{|}}{C}} - Cl$$

continued

STRATEGY

Look for carbon atoms that have four different atoms or groups bonded to them. As soon as you see two (or more) identical atoms or groups, you can stop counting. The C atom is not chiral.

SOLUTION

C_a has three hydrogen atoms bonded to it. It is not chiral.

C_b has two chlorine atoms bonded to it. It is not chiral.

C_c has four different groups/atoms bonded to it:

$$1 \text{ H}, 1 \text{ OH}, 1 \text{ CH}_3, 1 \text{ CH}_3\text{—CCl}_2$$

It is chiral.

C_d has three hydrogen atoms bonded to it. It is not chiral.

C_a has only three groups/atoms bonded to it. It is not chiral.

C_b has four different groups/atoms bonded to it:

$$1 \text{ H}, 1 \text{ NH}_2, 1 \text{ COOH}, 1 \text{ CH}_2\text{OH}$$

It is chiral.

C_c has two hydrogen atoms bonded to it. It is not chiral.

C_a has four different groups/atoms bonded to it:

$$1 \text{ H}, 1 \text{ OH}, 1 \text{ Cl}, 1 \text{ C—Cl}$$

It is chiral.

C_b has four different groups/atoms bonded to it:

$$1 \text{ CH}_3, 1 \text{ OH}, 1 \text{ Cl}, 1 \text{ C—Cl}$$

It is chiral.

C_c has three hydrogen atoms bonded to it. It is not chiral.

The term "optical isomerism" comes from the effect that enantiomers have on plane-polarized light, such as that produced by a Polaroid lens (Figure 22.13). When this light is passed through a solution containing a single enantiomer, the plane is rotated from its original position. One isomer rotates it to the right (clockwise) and the other to the left (counterclockwise). If both isomers are present in equal amounts, we obtain what is known as a **racemic mixture.** In this case, the two rotations offset each other and there is no effect on plane-polarized light.

Figure 22.13 A polarimeter. The sample tube contains an optically active compound. The analyzing filter has been turned clockwise to restore the light field and measure the rotation caused by the sample.

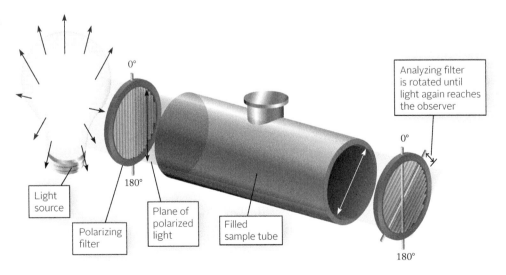

Except for their effect on plane-polarized light, two enantiomers of a chiral compound have identical physical properties. For example, the two isomers of lactic acid shown below have the same melting point, 52°C, and density, 1.25 g/mL.

$$\text{(I)} \quad \text{HO} - \overset{\displaystyle \text{COOH}}{\underset{\displaystyle \text{CH}_3}{\overset{|}{\underset{|}{\text{C}}}}} - \text{H} \qquad \text{(II)} \quad \text{H} - \overset{\displaystyle \text{COOH}}{\underset{\displaystyle \text{CH}_3}{\overset{|}{\underset{|}{\text{C}}}}} - \text{OH}$$

enantiomers of lactic acid

On the other hand, because enantiomers differ in their three-dimensional structure, they often interact with biochemical molecules in different ways. As a result, they may show quite different physiological properties. Consider, for example, amphetamine, often used illicitly as an "upper" or "pep pill." Amphetamine consists of two enantiomers:

enantiomers of amphetamine

Enantiomer I, called Dexedrine, is by far the stronger stimulant. It is from two to four times as active as Benzedrine, the racemic mixture of the two isomers.

In the 1990s, a large number of so-called chiral drugs containing a single enantiomer came on the market. One of the first of these was naproxen (trade name Aleve®).

The naproxen that you buy over the counter at your local pharmacy consists of a single enantiomer, which is an effective analgesic (pain killer). The other enantiomer is not a pain killer; even worse, it can cause liver damage.

Of the 100 best-selling drugs on the market today, about half consist of a single enantiomer. Chiral drugs are used to treat all kinds of illnesses, from arthritis to asthma, from heartburn to heart disease. In general, these drugs are prepared by either of two different methods. One possibility is to resolve a racemic mixture into its components. This was first done in the mid-nineteenth century by Louis Pasteur (1822–1895). He used a magnifying glass and a pair of tweezers to separate crystals of enantiomers. Today resolution can be accomplished much more quickly by using modern separation techniques such as liquid chromatography (Chapter 1).

A different approach to making chiral drugs is *asymmetric synthesis.* An optically inactive precursor is converted to the drug by a reaction that uses a special catalyst, usually an enzyme (Chapter 11). If all goes well, the product is a single enantiomer with the desired physiological effect. In 2001, William S. Knowles (1917–2012), Ryoji Noyori (1938–), and K. Barry Sharpless (1941–) won the Nobel Prize in Chemistry for work in this area.

22-6 Organic Reactions

The reactions that take place in organic chemistry differ in important ways from those discussed earlier in this text. In particular, the species taking part in organic reactions are most often molecules rather than ions. Again, most organic reactions take place between pure substances or in nonpolar solvents as opposed to water.

Generally speaking, organic reactions occur more slowly than inorganic ones. ◀ⓘ In this section we will look at four general types of organic reactions known as addition, elimination, condensation, and substitution.

Addition Reactions

In **addition reactions**, a small molecule (e.g., H_2, Cl_2, HCl, H_2O) adds across a double or triple bond. ◀ⓘ A simple example is the addition of hydrogen gas to ethene in the presence of a nickel catalyst.

$$H-\overset{\underset{\displaystyle H}{|}}{C}=\overset{\underset{\displaystyle H}{|}}{C}-H + H-H \xrightarrow{\text{Ni}} H-\overset{\overset{\displaystyle H}{|}}{\underset{\underset{\displaystyle H}{|}}{C}}-\overset{\overset{\displaystyle H}{|}}{\underset{\underset{\displaystyle H}{|}}{C}}-H$$

With ethyne (acetylene), one or two moles of H_2 may add, depending upon the catalyst and the conditions used.

Hydrogenation has found commercial application in the conversion of liquid to solid fats. Vegetable oils contain a relatively high proportion of double bonds. Treatment with hydrogen under pressure in the presence of a catalyst converts double bonds to single bonds and produces solids such as margarine.

As we saw earlier (Section 22-4), ethanol can be made from ethene by "adding" water in the presence of an acid catalyst. This process is referred to as **hydration**:

$$C_2H_4(g) + H_2O(l) \xrightarrow{H^+} C_2H_5OH(l)$$

Still another species that adds readily to ethene is elementary bromine (Figure 22.14):

$$H-\overset{\underset{\displaystyle H}{|}}{C}=\overset{\underset{\displaystyle H}{|}}{C}-H + Br-Br \longrightarrow H-\overset{\overset{\displaystyle H}{|}}{\underset{\underset{\displaystyle Br}{|}}{C}}-\overset{\overset{\displaystyle H}{|}}{\underset{\underset{\displaystyle Br}{|}}{C}}-H$$

Figure 22.14 Test for an alkene. At the left is a solution of bromine in carbon tetrachloride. Addition of a few drops of an alkene causes the red color to disappear as the alkene reacts with the bromine.

Elimination and Condensation Reactions

Elimination reactions are, in a sense, the reverse of addition reactions. They involve the elimination of two groups from adjacent carbon atoms, converting a saturated molecule into one that is unsaturated. An example is the dehydration of ethanol, which occurs when it is heated with sulfuric acid: ◀ⓘ

$$H-\overset{\overset{\displaystyle H}{|}}{\underset{\underset{\displaystyle H}{|}}{C}}-\overset{\overset{\displaystyle H}{|}}{\underset{\underset{\displaystyle OH}{|}}{C}}-H \longrightarrow H-\overset{\underset{\displaystyle H}{|}}{C}=\overset{\underset{\displaystyle H}{|}}{C}-H + H-OH$$

$$\text{ethanol} \qquad\qquad \text{ethene}$$

This is, of course, the reverse of the addition reaction by which ethanol is formed.

Somewhat similar reactions, referred to as **condensation reactions**, occur when two molecules combine by splitting out a small molecule such as H_2O. The molecules that combine may be the same or different:

$$CH_3-OH + HO-CH_3 \longrightarrow CH_3-O-CH_3 + H-OH$$
$$\text{methanol} \qquad \text{methanol} \qquad\qquad \text{dimethyl ether}$$

$$CH_3-OH + HO-\overset{\overset{\displaystyle}{}}{\underset{\underset{\displaystyle O}{\|}}{C}}-CH_3 \longrightarrow CH_3-O-\overset{\overset{\displaystyle}{}}{\underset{\underset{\displaystyle O}{\|}}{C}}-CH_3 + H-OH$$
$$\text{methanol} \qquad \text{acetic acid} \qquad\qquad \text{methyl acetate}$$

The first reaction offers a general way of preparing ethers from alcohol. The second reaction is that of esterification, referred to earlier.

Substitution Reactions

Substitution reactions are those in which an atom or group of atoms in a molecule is replaced by a different atom or group. Substitution reactions are very common in organic chemistry. Examples include the following:

- chlorination of alkanes. When a mixture of methane and chlorine gases is exposed to light, the following reaction occurs:

$$CH_4(g) + Cl_2(g) \longrightarrow CH_3Cl(g) + HCl(g)$$

- a very similar reaction with benzene:

$$C_6H_6(l) + Cl_2(g) \xrightarrow{FeCl_3} C_6H_5Cl(l) + HCl(g)$$

- the nitration of benzene:

$$C_6H_6(l) + HNO_3(l) \xrightarrow{H_2SO_4} C_6H_5NO_2(l) + H_2O(l)$$

In this case the nitro group, NO_2, substitutes for a hydrogen atom in the benzene ring.

- formation of an alcohol from an alkyl halide:

$$C_2H_5Br(l) + OH^-(aq) \longrightarrow C_2H_5OH(aq) + Br^-(aq)$$

Here a bromine atom is replaced by an $-OH$ group.

EXAMPLE 22.12

Classify each of the following as an addition, elimination, condensation, or substitution reaction.

(a) $C_2H_5OH(aq) + H^+(aq) + Br^-(aq) \longrightarrow C_2H_5Br(l) + H_2O$

(b) $C_6H_{12}(l) + Cl_2(g) \longrightarrow C_6H_{12}Cl_2(l)$

(c) $C_7H_8(l) + 2Cl_2(g) \longrightarrow C_7H_6Cl_2(l) + 2HCl(g)$

(d) $C_5H_{11}Cl(l) \longrightarrow C_5H_{10}(l) + HCl(g)$

SOLUTION

(a) substitution (Br for OH)   (b) addition (C_6H_{12} is an alkene)

(c) substitution (Cl for H)   (d) elimination (HCl lost)

CHEMISTRY **BEYOND** THE CLASSROOM

Cholesterol

An organic compound that we hear a great deal about nowadays is cholesterol, whose structure is shown in Figure A. Your body contains about 140 g of cholesterol; it is synthesized in the liver at the rate of 2 to 3 g/day. Cholesterol is essential to life for two reasons. It is a major component of all cell membranes, and it serves as the starting material for the synthesis of sex hormones, adrenal hormones, bile acids, and vitamin D.

Cholesterol, which is water-insoluble, is transported through the blood in the form of soluble protein complexes known as lipoproteins. There are two types of complexes: low-density (LDL), which contain mostly cholesterol, and high-density (HDL), which contain relatively little cholesterol. Commonly, these complexes are referred to as "LDL cholesterol" and "HDL cholesterol," respectively.

LDL, or "bad," cholesterol builds up as a plaque-like deposit on the interior walls of arteries. This process used to be called hardening of the arteries; today it is referred to as *atherosclerosis*. It can lead to cardiovascular diseases, including strokes and heart attacks. In contrast, HDL or "good" cholesterol retards or even reduces arterial deposits.

Figure A Cholesterol.

The relative amounts of LDL and HDL cholesterol in your bloodstream depend, at least in part, on your diet. In particular, they depend on the total amount and the type of fat that you eat. Fats (triglycerides) are esters of glycerol with long-chain carboxylic acids. The general structure of a fat can be represented as

$$
\begin{array}{c}
\text{H} \quad\quad \text{O} \\
| \quad\quad\; \| \\
\text{H}-\text{C}-\text{O}-\text{C}-\text{R} \\
| \quad\quad\; \text{O} \\
| \quad\quad\; \| \\
\text{H}-\text{C}-\text{O}-\text{C}-\text{R}' \\
| \quad\quad\; \text{O} \\
| \quad\quad\; \| \\
\text{H}-\text{C}-\text{O}-\text{C}-\text{R}'' \\
| \\
\text{H}
\end{array}
$$

where R, R′, and R″ are hydrocarbon residues typically containing 13 to 21 carbon atoms. Depending on the nature of these residues, we can distinguish among:

- *saturated fats,* in which R, R′, and R″ contain no multiple bonds. These are "bad" in the sense that they lead to high levels of LDL relative to HDL cholesterol.
- *monounsaturated fats,* in which each residue contains one double bond. These are much better nutritionally unless, God forbid, the hydrogen atoms on opposite sides of the double bonds are trans to one another. So-called *trans* fats are bad news.
- *polyunsaturated fats,* in which the hydrocarbon residues contain more than one double bond, are good guys, particularly if a double bond is located on the third or sixth carbon atom from the —CH$_3$ end of the residue. These so-called omega fats are the best of all.

Figure B shows the composition of various fats (solids) and oils (liquids). Now you know why sunflower seeds are better for you than coconuts.

POLYUNSATURATED FATS

Saturated fat   Linoleic acid   Monounsaturated fat

Alpha-linolenic acid (an Omega-3 Fatty Acid)

Canola oil
Safflower oil
Sunflower oil
Corn oil
Olive oil
Soybean oil
Peanut oil
Lard*
Beef tallow*
Palm oil
Butterfat*
Coconut oil

0   20   40   60   80   100
Percent of fatty acids

Figure B The composition of some common fats and oils. Fatty acids provide the R, R′, and R″ hydrocarbon residues in the general structure of a fat or oil shown in the text. Fats and oils all contain a mixture of saturated, monounsaturated, and polyunsaturated fatty acids.

*Cholesterol content (mg/Tbsp): Lard 12; Beef Tallow 14; Butterfat 33. No cholesterol in any vegetable-based oil.

Key Concepts

1. Draw
   - structural isomers
   (Examples 22.1 and 22.9; Problems 29–38)
   - geometric isomers
   (Example 22.10; Problems 39–42)
   - optical isomers, containing chiral carbon atoms
   (Example 22.11; Problems 43–46)
2. Name
   - alkanes
   (Example 22.2; Problems 5–10)

■ unsaturated hydrocarbons
(**Example 22.4; Problems 11, 12**)
■ aromatic compounds
(**Example 22.5; Problems 13, 14**)
3. Distinguish between alkanes, cycloalkanes, alkenes, and alkynes.
(**Example 22.3; Problems 1–4**)
4. Distinguish between alcohols, ethers, aldehydes, ketones, carboxylic acids, esters, and amines.
(**Example 22.6; Problems 15–18**)

5. Draw structural formulas for compounds containing the functional groups listed in Table 22.2.
(**Example 22.7; Problems 19–22**)
6. Determine equilibrium constants for the reaction of amines with strong or weak acids.
(**Example 22.8; Problems 27, 28**)
7. Classify an organic reaction as addition, elimination, condensation, or substitution.
(**Example 22.12; Problems 47–50**)

Key Terms

addition reactions	amines	elimination reactions	ketones
alcohols	aromatic hydrocarbons	enantiomers	optical isomers
aldehydes	carboxylic acids	esters	racemic mixture
alkanes	chiral center	ethers	structural isomers
alkenes	*cis* isomer	functional group	substitution reactions
alkyl group	condensation reactions	geometric isomerism	*trans* isomer
alkynes	cycloalkanes	hydrocarbon	

Summary Problem

Consider the alkyl halide that has the structure

$$H_3C-\underset{\underset{Cl}{|}}{\overset{\overset{H}{|}}{C}}-\underset{\underset{H}{|}}{\overset{\overset{CH_3}{|}}{C}}-CH_3$$

(a) Write the name of the alkyl halide and that of the alkane from which it is derived.
(b) Draw the structure of the compound formed when HCl is eliminated from the alkyl halide. Write its name and classify it as an alkane, alkene, or alkyne.
(c) Draw the structures and write the names of all the alkanes that are structurally isomeric to the alkane equivalent to (a).
(d) Write the names of all the alkenes that are structurally isomeric to (b).
(e) Does the alkyl halide exhibit optical isomerism? If so, how many chiral carbon atoms does it have?
(f) Take an isomer from (c) and add a functional group to make it
 1. an aldehyde
 2. a ketone
 3. an amine
 4. an ether
 5. a carboxylic acid
 Write the structural formula of each species obtained.
(g) Draw the structure of the species formed when this alkyl halide undergoes substitution with an OH⁻ ion.
(h) Draw the condensed structural formula and write the name of the ester formed when 1-pentanol reacts with formic acid, HCOOH.

Answers

(a) **2-chloro-3-methylbutane; 2-methylbutane**
(b)
$$H_3C-\overset{\overset{H}{|}}{C}=\overset{\overset{CH_3}{|}}{C}-CH_3 \quad \text{2-methyl-2-butene; alkene}$$
(c) $CH_3CH_2CH_2CH_2CH_3$ pentane

$$H_3C-\underset{\underset{CH_3}{|}}{\overset{\overset{CH_3}{|}}{C}}-CH_3 \quad \text{2,2-dimethylpropane}$$

(d) **1-pentene;   2-methyl-1-butene;   3-methyl-1-butene; *cis*-2-pentene; *trans*-2-pentene**
(e) **yes; 1**
(f) **(1)**
$$CH_3(CH_2)_3\overset{\overset{H}{|}}{C}=O$$
(2)
$$H_3C-\underset{\underset{H}{|}}{\overset{\overset{CH_3}{|}}{C}}-\overset{\|}{\underset{O}{C}}-CH_3$$
(3)
$$CH_3(CH_2)_3\underset{\underset{H}{|}}{\overset{\overset{H}{|}}{C}}-NH_2$$
(4)
$$H_3C-\underset{\underset{H}{|}}{\overset{\overset{H}{|}}{C}}-O-\underset{\underset{H}{|}}{\overset{\overset{CH_3}{|}}{C}}-CH_3$$
(5)
$$CH_3(CH_2)_3\overset{\overset{OH}{|}}{C}=O$$

(g)
$$H_3C-\underset{\underset{OH}{|}}{\overset{\overset{H}{|}}{C}}-\underset{\underset{H}{|}}{\overset{\overset{CH_3}{|}}{C}}-CH_3$$

(h) pentyl formate
$$CH_3(CH_2)_3CH_2-O-\underset{\underset{O}{\|}}{C}-H$$

Questions and Problems

Even-numbered questions and Challenge Problems have answers in Appendix 5 and fully worked solutions in the *Student Solutions Manual*.

22-1 Saturated Hydrocarbons: Alkanes

22-2 Unsaturated Hydrocarbons: Alkenes and Alkynes

22-3 Aromatic Hydrocarbons and Their Derivatives

1. Classify each of the following hydrocarbons as alkanes, alkenes, or alkynes.
    (a) C_6H_{14} (b) C_3H_4 (c) C_9H_{18}
2. Classify each of the following hydrocarbons as alkanes, alkenes, or alkynes.
    (a) $C_{12}H_{24}$ (b) C_7H_{12} (c) $C_{13}H_{28}$
3. Write the formula for
    (a) an alkene with two carbon atoms.
    (b) an alkane with 22 hydrogen atoms.
    (c) an alkyne with three carbon atoms.
4. Write the formula for
    (a) an alkyne with 16 hydrogen atoms.
    (b) an alkene with 44 hydrogen atoms.
    (c) an alkane with ten carbon atoms.
5. Name the following alkanes.

    (a) $CH_3-CH_2-CH-CH_3$
          |
         CH_3

    (b) $CH_3-CH_2-CH-CH_3$
          |
         CH_2
          |
         CH_3

    (c) $CH_3-CH-CH-CH_3$
        | |
       CH_3 CH_2
          |
         CH_3

    (d) $CH_3-CH-CH_2-CH-CH_3$
        | |
       CH_3 $H-C-CH_3$
             |
            CH_3

6. Name the following alkanes.

    (a) $CH_3-(CH_2)_5-CH-CH_3$
            |
           CH_3

    (b) $(CH_3)_4C$

    (c) $CH_3-CH-CH_2-C(CH_3)_3$
        |
       CH_3

         H
         |
    (d) $CH_3-C-(CH_2)_2-CH-CH_3$
        | |
       CH_2 CH_3
        |
       CH_3

7. Write structural formulas for the following alkanes.
    (a) 3-ethylpentane   (b) 2,2-dimethylbutane
    (c) 2-methyl-3-ethylheptane   (d) 2,3-dimethylpentane
8. Write structural formulas for the following alkanes.
    (a) 2,2,4-trimethylpentane   (b) 2,2-dimethylpropane
    (c) 4-isopropyloctane   (d) 2,3,4-trimethylheptane
9. The following names are incorrect; draw a reasonable structure for the alkane and give the proper IUPAC name.
    (a) 5-isopropyloctane   (b) 2-ethylpropane
    (c) 1,2-dimethylpropane
10. Follow the directions of Problem 9 for the following names.
    (a) 2,2-dimethylbutane   (b) 4-methylpentane
    (c) 2-ethylpropane
11. Name the following alkenes.
    (a) $CH_2=C-(CH_3)_2$
    (b) $(CH_3)_2-C=C-(CH_3)_2$
    (c) $CH_3-CH=CH-CH_2CH_3$
    (d) $CH_3-C=CH_2$
        |
       CH_2
        |
       CH_3
12. Write structural formulas for the following alkynes.
    (a) 2-pentyne   (b) 4-methyl-2-pentyne
    (c) 2-methyl-3-hexyne   (d) 3,3-dimethyl-1-butyne
13. Name the following compounds as derivatives of nitrobenzene.

14. Name the following compounds as derivatives of toluene.

22-4 Functional Groups

15. Classify each of the following as a carboxylic acid, ester, and/or alcohol.
    (a) $HO-CH_2-CH_2-CH_2-OH$

(b)

(c) $CH_3-\overset{\overset{\displaystyle H}{|}}{\underset{\underset{\displaystyle OH}{|}}{C}}-COOH$

16. Classify each of the following as a carboxylic acid, ester, and/or alcohol.

(a) $CH_3-(CH_2)_3-OH$

(b) $CH_3-CH_2-\overset{\overset{\displaystyle }{\underset{\underset{\displaystyle O}{||}}{C}}}-O-CH_2-CH_3$

(c) $CH_3-CH_2-O-\overset{\overset{\displaystyle }{\underset{\underset{\displaystyle O}{||}}{C}}}-(CH_2)_6-COOH$

17. In each of the following, name the functional group (i.e., ester, aldehyde, etc.).

(a) $CH_3-CH=CH-CH_2-OH$

(b) $CH_3-\langle\bigcirc\rangle-\overset{\overset{\displaystyle }{\underset{\underset{\displaystyle O}{||}}{C}}}-OH$

(c) $\langle\bigcirc\rangle-\overset{\overset{\displaystyle }{\underset{\underset{\displaystyle O}{||}}{C}}}-CH_3$

(d) $CH_3-CH_2-\overset{\overset{\displaystyle }{\underset{\underset{\displaystyle O}{||}}{C}}}-H$

(e) $\langle\bigcirc\rangle-CH_2-CH_2-NH_2$

18. In each of the following, name the functional group (i.e., ester, aldehyde, etc.).

(a) $CH_2=CH-O-CH_3$

(b) $CH_3-CH_2-\overset{\overset{\displaystyle }{\underset{\underset{\displaystyle O}{||}}{C}}}-O-CH_3$

(c) $CH_3-CH_2-CH_2-NH_2$

(d) $\langle\bigcirc\rangle-\overset{\overset{\displaystyle }{\underset{\underset{\displaystyle O}{||}}{C}}}-OH$

(e) $CH_3-\overset{\overset{\displaystyle }{\underset{\underset{\displaystyle O}{||}}{C}}}-\langle\bigcirc\rangle$

19. Give the structure of

(a) a four-carbon straight-chain alcohol with an $-OH$ group not at the end of the chain.

(b) a five-carbon straight-chain carboxylic acid.

(c) the ester formed when these two compounds react.

20. Give the structure of

(a) a three-carbon alcohol with the $-OH$ group on the center carbon.

(b) a four-carbon branched-chain carboxylic acid.

(c) the ester formed when these two compounds react.

21. Write the structural formula of the ester formed by methyl alcohol with

(a) formic acid. (b) acetic acid.

(c) $CH_3-(CH_2)_6-COOH$.

22. Write the structural formulas of all the esters that can be formed by ethylene glycol with formic and acetic acids. (One or both of the $-OH$ groups of ethylene glycol can react.)

23. Arrange these compounds in order of increasing boiling point. Values in °C are 0, 11, and 97.

(a) $CH_3CH_2CH_2OH$

(b) $CH_3CH_2-O-CH_3$

(c) $CH_3CH_2CH_2CH_3$

24. Arrange these compounds in order of increasing boiling point.

(a) 1-butanol, butane, diethylether

(b) hexane, 1-hexanol, dipropylether

25. The K_b for ethylamine ($CH_3CH_2NH_2$) is 4.3×10^{-4}. What is the pH of a 0.250 M solution of ethylamine? Compare with the pH of a 0.250 M solution of ammonia ($K_b = 1.86 \times 10^{-5}$).

26. When aniline, $C_6H_5NH_2$ ($K_b = 7.4 \times 10^{-10}$), reacts with a strong acid, its conjugate acid, $C_6H_5NH_3^+$, is formed. Calculate the pH of a 0.100 M solution of $C_6H_5NH_3^+$ and compare it with the pH of acetic acid ($K_a = 1.86 \times 10^{-5}$).

27. When ethylamine, a weak base ($K_b = 4.3 \times 10^{-4}$), reacts with formic acid, a weak acid ($K_a = 1.8 \times 10^{-4}$), the following reaction takes place:

$$CH_3CH_2NH_2(aq) + HCOOH(aq) \rightleftharpoons$$
$$CH_3CH_2NH_3^+(aq) + HCOO^-(aq)$$

Calculate K for this reaction.

28. When the conjugate acid of aniline, $C_6H_5NH_3^+$, reacts with the acetate ion, the following reaction takes place:

$$C_6H_5NH_3^+(aq) + CH_3COO^-(aq) \rightleftharpoons$$
$$C_6H_5NH_2(aq) + CH_3COOH(aq)$$

If K_a for $C_6H_5NH_3^+$ is 1.35×10^{-5} and K_a for CH_3COOH is 1.86×10^{-5}, what is K for the reaction?

22-5 Isomerism in Organic Compounds

29. Draw the structural isomers of the alkane C_6H_{14}.

30. Draw the structural isomers of the alkene C_4H_8.

31. Draw the structural isomers of C_4H_9Cl in which one hydrogen atom of a C_4H_{10} molecule has been replaced by chlorine.

32. Draw the structural isomers of $C_3H_6Cl_2$ in which two of the hydrogen atoms of C_3H_8 have been replaced by chlorine atoms.

33. There are three compounds with the formula C_6H_4ClBr in which two of the hydrogen atoms of the benzene molecule have been replaced by halogen atoms. Draw structures for these compounds.

34. There are three compounds with the formula $C_6H_3Cl_3$ in which three of the hydrogen atoms of the benzene molecule have been replaced by chlorine atoms. Draw structures for these compounds.

35. Write structures for all the structural isomers of double-bonded compounds with the molecular formula C_5H_{10}.

36. Write structures for all the structural isomers of compounds with the molecular formula C_4H_6ClBr in which Cl and Br are bonded to a double-bonded carbon.

37. Draw structures for all the alcohols with molecular formula $C_5H_{12}O$.

38. Draw structures for all the saturated carboxylic acids with four carbon atoms per molecule.

39. Of the compounds in Problem 35, which ones show geometric isomerism? Draw the *cis-* and *trans-* isomers.

40. Of the compounds in Problem 36, which ones show geometric isomerism? Draw the *cis-* and *trans-* isomers.

41. Maleic acid and fumaric acid are the *cis-* and *trans-* isomers, respectively, of $C_2H_2(COOH)_2$, a dicarboxylic acid. Draw and label their structures.

42. For which of the following is geometric isomerism possible?

(a) $(CH_3)_2C{=}CCl_2$ (b) $CH_3ClC{=}CCH_3Cl$

(c) $CH_3BrC{=}CCH_3Cl$

43. Which of the following can show optical isomerism?

(a) 2-bromo-2-chlorobutane

(b) 2-methylpropane

(c) 2,2-dimethyl-1-butanol

(d) 2,2,4-trimethylpentane

44. Which of the following compounds can show optical isomerism?

(a) dichloromethane

(b) 1,2-dichloroethane

(c) bromochlorofluoromethane

(d) 1-bromoethanol

45. Locate the chiral carbon(s), if any, in the following molecules.

(a) $HO{-}\overset{\displaystyle O}{\underset{}{C}}{-}\overset{\displaystyle H}{\underset{\displaystyle OH}{C}}{-}\overset{\displaystyle H}{\underset{\displaystyle OH}{C}}{-}\overset{\displaystyle O}{\underset{}{C}}{-}H$ (b) $CH_3{-}\overset{\displaystyle O}{\underset{}{C}}{-}\overset{\displaystyle O}{\underset{}{C}}{-}OH$

(c) $CH_3{-}CH_2{-}\overset{\displaystyle H}{\underset{\displaystyle NH_2}{C}}{-}COOH$

46. Locate the chiral carbon(s), if any, in the following molecules.

(a) $CH_3{-}\overset{\displaystyle H}{\underset{\displaystyle OH}{C}}{-}\overset{\displaystyle H}{\underset{\displaystyle OH}{C}}{-}H$ (b) $H{-}\overset{\displaystyle}{\underset{\displaystyle H}{C}}{=}\overset{\displaystyle}{\underset{\displaystyle H}{C}}{-}CH_2{-}OH$

(c) $CH_3{-}\overset{\displaystyle Cl}{\underset{\displaystyle Cl}{C}}{-}\overset{\displaystyle F}{\underset{\displaystyle H}{C}}{-}Cl$

22-6 Organic Reactions

47. Classify the following reactions as addition, substitution, elimination, or condensation.

(a) $C_2H_2(g) + HBr(g) \longrightarrow C_2H_3Br(l)$

(b) $C_6H_5OH(s) + HNO_3(l) \longrightarrow$
$$C_6H_4(OH)(NO_2)(s) + H_2O(l)$$

(c) $C_2H_5OH(aq) + HCOOH(aq) \longrightarrow$
$$C_2H_5{-}O{-}\overset{\displaystyle}{\underset{\displaystyle O}{C}}{-}H(l) + H_2O(l)$$

48. Classify the following reactions according to the categories listed in Question 47.

(a) $PCl_3(g) + Cl_2(g) \longrightarrow PCl_5(g)$

(b) $C_3H_7OH(l) \longrightarrow C_3H_6(g) + H_2O(l)$

(c) $CH_3OH(l) + C_2H_5OH(l) \longrightarrow$
$$CH_3{-}O{-}C_2H_5(l) + H_2O(l)$$

49. Name the products formed when the following reagents add to 1-butene.

(a) H_2 (b) Br_2

50. Name the products obtained when the following reagents add to 2-methyl-2-butene.

(a) Cl_2 (b) I_2

Unclassified

51. Which of the following are expected to show bond angles of 109.5°? 120°? 180°?

(a) $H_3C{-}CH_3$ (b) $H_3C{-}\overset{\displaystyle H}{\underset{}{C}}{=}CH_2$

(c) $H_3C{-}C{\equiv}C{-}CH_3$

52. What does the "circle" represent in the structural formula of benzene?

53. Calculate $[H^+]$ and the pH of a 0.10 M solution of chloroacetic acid ($K_a = 1.5 \times 10^{-3}$).

54. Explain what is meant by the following.

(a) a saturated fat

(b) a soap

(c) the "proof" of an alcoholic beverage

(d) denatured alcohol

55. The general formula of an alkane is C_nH_{2n+2}. What is the general formula of an

(a) alkene?   (b) alkyne?

(c) alcohol derived from an alkane?

56. Write a balanced net ionic equation for the reaction of

(a) $(CH_3)_2NH$ with hydrochloric acid.

(b) acetic acid with a solution of barium hydroxide.

(c) 2-chloropropane with a solution of sodium hydroxide.

Challenge Problems

57. The structure of cholesterol is given in the text (see Figure A, page 573). What is its molecular formula?

58. Draw structures for all the alcohols with molecular formula $C_6H_{14}O$.

59. What mass of propane must be burned to bring one quart of water in a saucepan at 25°C to the boiling point? Use data in Appendix 1 and elsewhere and make any reasonable assumptions.

60. Write an equation for the reaction of chloroacetic acid ($K_a = 1.5 \times 10^{-3}$) with trimethylamine ($K_b = 5.9 \times 10^{-5}$). Calculate the equilibrium constant for the reaction. If 0.10 M solutions of these two species are mixed, what will be their concentrations at equilibrium?

23

Organic Polymers, Natural and Synthetic

The glasslike sculpture is made of a polymer, which allows it to stand up to the outdoor weather. The repeating design, though random, recognizes the randomness and repetitiveness of the structure of polymers.

No single thing abides; but all things flow
Fragment to fragment clings—the
   things thus grow
Until we know and name them. By
   degrees
They melt, and are no more the things
   we know.

—TITUS LUCRETIUS CARUS (92–52 B.C.)
   No Single Thing Abides
   (Translated by W. H. Mallock)

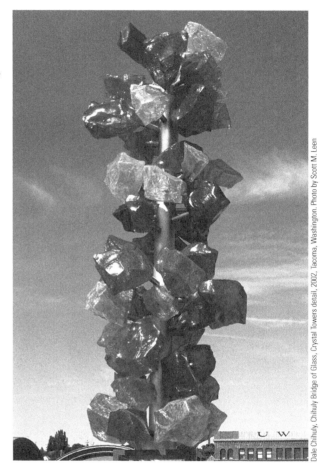

Dale Chihuly, Chihuly Bridge of Glass, Crystal Towers detail, 2002, Tacoma, Washington. Photo by Scott M. Leen

Industrial chemists work more with polymers than with any other class of materials.

In previous chapters we have discussed the chemical and physical properties of many kinds of substances. For the most part, these materials were made up of either small molecules or simple ions. In this chapter, we will be concerned with an important class of compounds containing large molecules. We call these compounds **polymers**.

A large number of small molecular units, called **monomers**, combined together chemically make up a polymer. A typical polymer molecule contains a chain of monomers several thousand units long. The monomer units that comprise a given polymer may be the same or different.

We start with synthetic organic polymers. Since about 1930, a variety of synthetic polymers have been made available by the chemical industry. The monomer units are joined together either by addition (Section 23-1) or by condensation (Section 23-2). They are used to make cups, plates, fabrics, automobile tires, and even artificial hearts.

The remainder of this chapter deals with natural polymers. These are large molecules, produced by plants and animals, that carry out the many life-sustaining processes in a living cell. The cell membranes of plants and the woody structure of trees are composed in large part of cellulose, a polymeric carbohydrate. We will look at the structures of a variety of different carbohydrates in Section 23-3. Another class of natural polymers are the proteins. Section 23-4 deals with these polymeric materials, which make up our tissues, bone, blood, and even hair.

Synthetic Addition Polymers

When monomer units add directly to one another, the result is an **addition polymer.** Table 23.1 lists some of the more familiar synthetic addition polymers. You will notice that each of these is derived from a monomer containing a carbon-carbon double bond. Upon polymerization, the double bond is converted to a single bond:

$$\underset{}{\overset{}{>}}C=C\overset{}{\underset{}{<}} \longrightarrow -\overset{|}{\underset{|}{C}}-\overset{|}{\underset{|}{C}}-$$

and successive monomer units add to one another.

Polyethylene

Perhaps the most familiar addition polymer is *polyethylene,* a solid derived from the monomer ethylene. ◄ ⓘ We might represent the polymerization process as

These polymers are usually called plastics.

$$n\left(\begin{array}{c}H \\ \diagdown \\ C=C \\ \diagup \qquad \diagdown \\ H \qquad\qquad H\end{array}\right) \longrightarrow \left(\begin{array}{cc} H & H \\ | & | \\ -C-C- \\ | & | \\ H & H \end{array}\right)_n$$

ethylene   polyethylene

where *n* is a very large number, of the order of 2000.

Table 23.1 Some Common Addition Polymers

Monomer	Name	Polymer	Uses
$\overset{H}{\underset{H}{>}}C=C\overset{H}{\underset{H}{<}}$	Ethylene	Polyethylene	Bags, coatings, toys
$\overset{H}{\underset{H}{>}}C=C\overset{H}{\underset{CH_3}{<}}$	Propylene	Polypropylene	Beakers, milk cartons
$\overset{H}{\underset{H}{>}}C=C\overset{H}{\underset{Cl}{<}}$	Vinyl chloride	Polyvinyl chloride (PVC)	Floor tile, raincoats, pipe, phonograph records
$\overset{H}{\underset{H}{>}}C=C\overset{H}{\underset{CN}{<}}$	Acrylonitrile	Polyacrylonitrile (PAN)	Rugs; Orlon and Acrilan are copolymers with other monomers
$\overset{H}{\underset{H}{>}}C=C\overset{H}{\underset{}{<}}$	Styrene	Polystyrene	Cast articles using a transparent plastic
$\overset{H}{\underset{H}{>}}C=C\overset{CH_3}{\underset{\overset{\displaystyle C-OCH_3}{\|\,O}}{<}}$	Methyl methacrylate	Plexiglas, Lucite, acrylic resins	High-quality transparent objects, latex paints
$\overset{F}{\underset{F}{>}}C=C\overset{F}{\underset{F}{<}}$	Tetrafluoroethylene	Teflon	Gaskets, insulation, bearings, pan coatings

Figure 23.1 Branched and linear polyethylene. The polyethylene bottle at the left is made of pliable, branched polyethylene. The one at the right is made of semirigid, linear polyethylene.

Marna G. Clarke

Depending upon the conditions of polymerization, the product may be—

■ *branched* polyethylene, which has a structure of the type

Here, neighboring chains are arranged in a somewhat random fashion, producing a soft, flexible solid (Figure 23.1). The plastic bags at the vegetable counters of supermarkets are made of this material.

■ *linear* polyethylene, which consists almost entirely of unbranched chains:

$$-\overset{\overset{\displaystyle H}{|}}{\underset{\underset{\displaystyle H}{|}}{C}}-\overset{\overset{\displaystyle H}{|}}{\underset{\underset{\displaystyle H}{|}}{C}}-\overset{\overset{\displaystyle H}{|}}{\underset{\underset{\displaystyle H}{|}}{C}}-\overset{\overset{\displaystyle H}{|}}{\underset{\underset{\displaystyle H}{|}}{C}}-\overset{\overset{\displaystyle H}{|}}{\underset{\underset{\displaystyle H}{|}}{C}}-\overset{\overset{\displaystyle H}{|}}{\underset{\underset{\displaystyle H}{|}}{C}}-\overset{\overset{\displaystyle H}{|}}{\underset{\underset{\displaystyle H}{|}}{C}}-\overset{\overset{\displaystyle H}{|}}{\underset{\underset{\displaystyle H}{|}}{C}}-$$

Neighboring chains in linear polyethylene line up nearly parallel to each other. This gives a polymer that approaches a crystalline material. It is used for bottles, toys, and other semirigid objects.

The two forms of polyethylene differ slightly in density. Linear polyethylene is referred to in the recycling business as *high-density polyethylene*, represented by the symbol "HDPE #2" on the bottom of a plastic bottle. The corresponding symbol for branched polyethylene is "LDPE #4," indicating low-density polyethylene. (The smaller the number, the easier it is to recycle.)

EXAMPLE 23.1

The addition polymer polyvinyl chloride (PVC) has the structure

$$-\overset{\overset{\displaystyle H}{|}}{\underset{\underset{\displaystyle H}{|}}{C}}-\overset{\overset{\displaystyle H}{|}}{\underset{\underset{\displaystyle Cl}{|}}{C}}-\overset{\overset{\displaystyle H}{|}}{\underset{\underset{\displaystyle H}{|}}{C}}-\overset{\overset{\displaystyle H}{|}}{\underset{\underset{\displaystyle Cl}{|}}{C}}-\overset{\overset{\displaystyle H}{|}}{\underset{\underset{\displaystyle H}{|}}{C}}-\overset{\overset{\displaystyle H}{|}}{\underset{\underset{\displaystyle Cl}{|}}{C}}-$$

a Draw the structure of the monomer from which PVC is made.

STRATEGY AND SOLUTION

Find the repeating unit from which the polymer chain is constructed.

repeating unit:
$$-\overset{\overset{\displaystyle H}{|}}{\underset{\underset{\displaystyle H}{|}}{C}}-\overset{\overset{\displaystyle H}{|}}{\underset{\underset{\displaystyle Cl}{|}}{C}}-$$

monomer:
$$H-\overset{\overset{\displaystyle H}{|}}{C}=\overset{\overset{\displaystyle H}{|}}{C}-Cl$$

b How many monomer units are in a PVC polymer that has a molar mass of 1.33×10^5 g/mol?

ANALYSIS

Information given:	From part (a): monomer ($H_2C = CHCl$) molar mass of polymer (1.33×10^5 g/mol)
Information implied:	molar mass of monomer
Asked for:	number of monomer units

continued

STRATEGY
1. Determine the molar mass of the monomer.
2. Divide the molar mass of polymer by the molar mass of the monomer.

SOLUTION	
1. Molar mass of the monomer	62.49 g/mol
2. Number of units	$\dfrac{1.33 \times 10^5 \text{ g/mol}}{62.49 \text{ g/mol}} = 2.13 \times 10^3$

Polyvinyl Chloride ◂ⓘ

Vinyl chloride, in contrast to ethylene, is an unsymmetrical molecule. We might refer to the CH_2 group in vinyl chloride as the "head" of the molecule and the CHCl group as the "tail":

vinyl chloride

head   tail

In principle at least, vinyl chloride molecules can add to one another in any of three ways to form:

1. A *head-to-tail polymer*, in which there is a Cl atom on every other C atom in the chain:

$$
\begin{array}{cccccccccc}
H & H & H & H & H & H & H & H & H & H \\
| & | & | & | & | & | & | & | & | & | \\
-C & -C & -C & -C & -C & -C & -C & -C & -C & -C- \\
| & | & | & | & | & | & | & | & | & | \\
H & Cl & H & Cl & H & Cl & H & Cl & H & Cl
\end{array}
$$

2. A *head-to-head, tail-to-tail polymer*, in which Cl atoms occur in pairs on adjacent carbon atoms in the chain:

$$
\begin{array}{cccccccccc}
H & H & H & H & H & H & H & H & H & H \\
| & | & | & | & | & | & | & | & | & | \\
-C & -C & -C & -C & -C & -C & -C & -C & -C & -C- \\
| & | & | & | & | & | & | & | & | & | \\
H & Cl & Cl & H & H & Cl & Cl & H & H & Cl
\end{array}
$$

3. A *random polymer*:

$$
\begin{array}{cccccccccc}
H & H & H & H & H & H & H & H & H & H \\
| & | & | & | & | & | & | & | & | & | \\
-C & -C & -C & -C & -C & -C & -C & -C & -C & -C- \\
| & | & | & | & | & | & | & | & | & | \\
H & Cl & H & Cl & Cl & H & H & Cl & H & Cl
\end{array}
$$

Similar arrangements are possible with polymers made from other unsymmetrical monomers (Example 23.2).

ⓘ PVC is a stiff, rugged, cheap polymer.

EXAMPLE 23.2

Consider propene:

H   H
 \\ /
  C=C
 /   \\
H   CH₃

Sketch the structure of a head-to-tail polymer derived from it and sketch another structure assuming a head-to-head, tail-to-tail polymer.

continued

SOLUTION

head-to-tail polymer

$$\begin{array}{cccccccccc} H & H & H & H & H & H & H & H & H & H \\ | & | & | & | & | & | & | & | & | & | \\ -C & -C & -C & -C & -C & -C & -C & -C & -C & -C- \\ | & | & | & | & | & | & | & | & | & | \\ H & CH_3 & H & CH_3 & H & CH_3 & H & CH_3 & H & CH_3 \end{array}$$

head-to-head, tail-to-tail polymer

$$\begin{array}{cccccccccc} H & H & H & H & H & H & H & H & H & H \\ | & | & | & | & | & | & | & | & | & | \\ -C & -C & -C & -C & -C & -C & -C & -C & -C & -C- \\ | & | & | & | & | & | & | & | & | & | \\ H & CH_3 & CH_3 & H & H & CH_3 & CH_3 & H & H & CH_3 \end{array}$$

Teflon

Teflon $(C_2F_2)_n$, where n is of the order of 10^4 to 10^5, is a versatile but expensive polymer (\$10–20/kg). It is virtually impervious to chemical attack, even at high temperatures. Apparently the fluorine atoms form a protective shield around the central carbon chain:

$$\begin{array}{cccccc} F & F & F & F & F & F \\ | & | & | & | & | & | \\ -C & -C & -C & -C & -C & -C- \\ | & | & | & | & | & | \\ F & F & F & F & F & F \end{array}$$

For much the same reason, Teflon is essentially invulnerable to attack by aqueous or organic solvents. Beyond that, Teflon is an excellent electrical insulator. Indeed, its largest industrial use is in that area.

You are probably most familiar with Teflon as a nonstick coating for kitchen utensils (Figure 23.2). The slippery surface of Teflon-coated fry pans and muffin tins results from Teflon's extremely low coefficient of friction.

© Cengage Learning/Charles D. Winters

Figure 23.2 Teflon. The Teflon coating helps prevent food from sticking to this cookware.

23-2   Synthetic Condensation Polymers

As we pointed out in Chapter 22, a condensation reaction is one in which two molecules combine by splitting out a small molecule such as water. The reaction of an alcohol, ROH, or an amine, RNH_2, with a carboxylic acid $R'COOH$ is a condensation:

$$\underset{\text{alcohol}}{R-OH} + \underset{\text{acid}}{HO-\underset{O}{\overset{\|}{C}}-R'} \longrightarrow \underset{\text{ester}}{R-O-\underset{O}{\overset{\|}{C}}-R'} + H_2O$$

$$\underset{\underset{H}{|}}{\underset{\text{amine}}{R-N-H}} + \underset{\text{acid}}{HO-\underset{O}{\overset{\|}{C}}-R'} \longrightarrow \underset{\underset{H}{|}\ \underset{O}{\|}}{\underset{\text{amide}}{R-N-C-R'}} + H_2O$$

A **condensation polymer** is formed in a similar way; the molecule split out is most often water. In order to produce a condensation polymer, *the monomers involved must have functional groups at both ends of the molecule.* Most often these groups are

$$-NH_2 \quad -OH \quad -COOH$$

The products formed are referred to as polyesters and polyamides.

Polyesters

Consider what happens when an alcohol with two —OH groups, HO—R—OH, reacts with a dicarboxylic acid, HOOC—R′—COOH. In this case the ester formed still has a reactive group at both ends of the molecule.

$$HO-R-OH \ + \ HO-\underset{\overset{\|}{O}}{C}-R'-\underset{\overset{\|}{O}}{C}-OH \longrightarrow$$

dihydroxy   dicarboxylic
alcohol   acid

$$HO-R-O-\underset{\overset{\|}{O}}{C}-R'-\underset{\overset{\|}{O}}{C}-OH \ + \ H_2O$$

ester with active
end groups

The COOH group at one end of the ester molecule can react with another alcohol molecule. The OH group at the other end can react with an acid molecule. This process can continue, leading eventually to a long-chain polymer containing 500 or more ester groups. The general structure of the **polyester** can be represented as

$$-\underset{\overset{\|}{O}}{C}-R'-\underset{\overset{\|}{O}}{C}-O-R-O-\underset{\overset{\|}{O}}{C}-R'-\underset{\overset{\|}{O}}{C}-O-R-O-$$

section of a polyester molecule

Perhaps the most important polyester is poly(ethylene terephthalate), commonly known as PET (or "PETE 1" on plastic beverage bottles). The annual production of PET in the United States is of the order of 10^8 kg (10^5 metric tons). Much of this is converted into fabric (trade name, Dacron) or magnetically coated film (Mylar).

The two monomers used to make PET are

$$HO-CH_2-CH_2-OH \qquad HO-\underset{\overset{\|}{O}}{C}-\bigcirc-\underset{\overset{\|}{O}}{C}-OH$$

ethylene glycol   terephthalic acid

EXAMPLE 23.3

Write the structural formula of the ester formed when one molecule of ethylene glycol reacts with one molecule of terephthalic acid, and draw a section of the "PET" polymer.

SOLUTION

ester

$$HO-CH_2-CH_2-O-\underset{\overset{\|}{O}}{C}-\bigcirc-\underset{\overset{\|}{O}}{C}-OH$$

"PET" polymer

$$-\underset{\overset{\|}{O}}{C}-\bigcirc-\underset{\overset{\|}{O}}{C}-O-CH_2-CH_2-O-\underset{\overset{\|}{O}}{C}-\bigcirc-\underset{\overset{\|}{O}}{C}-O-CH_2-CH_2-O-$$

Polyamides

When a diamine (molecule containing two NH_2 groups) reacts with a dicarboxylic acid (two COOH groups), a **polyamide** is formed. This condensation polymerization is entirely analogous to that used to make polyesters. In this case, the NH_2 group of the diamine reacts with the COOH group of the dicarboxylic acid:

$$NH_2-R-\underset{\underset{H}{|}}{N}-H + HO-\underset{\underset{O}{\|}}{C}-R'-\underset{\underset{O}{\|}}{C}-OH \longrightarrow$$

$$NH_2-R-\underset{\underset{H}{|}}{N}-\underset{\underset{O}{\|}}{C}-R'-\underset{\underset{O}{\|}}{C}-OH + H_2O$$

Condensation can continue to form a long-chain polymer.

EXAMPLE 23.4

Consider the following polyamide:

$$-\underset{\underset{O}{\|}}{C}-(CH_2)_4-\underset{\underset{O}{\|}}{C}-\underset{\underset{H}{|}}{N}-(CH_2)_6-\underset{\underset{H}{|}}{N}-\underset{\underset{O}{\|}}{C}-(CH_2)_4-\underset{\underset{O}{\|}}{C}-\underset{\underset{H}{|}}{N}-(CH_2)_6-\underset{\underset{H}{|}}{N}-$$

Give the structures of the two monomers, dicarboxylic acid and diamine, used to make this polymer.

STRATEGY

1. Break the polymer into two fragments. Each fragment should be one of the monomers.
2. Add an OH group to the —C=O terminal carbon atoms of the first monomer.
3. Add a H atom to the N terminal atoms of the second monomer.

SOLUTION

Monomer 1	$-\underset{\underset{O}{\|}}{C}-(CH_2)_4-\underset{\underset{O}{\|}}{C}-$
Add an OH group.	Dicarboxylic acid: $HO-\underset{\underset{O}{\|}}{C}-(CH_2)_4-\underset{\underset{O}{\|}}{C}-OH$
Monomer 2	$-\underset{\underset{H}{|}}{N}-(CH_2)_6-\underset{\underset{H}{|}}{N}-$
Add an H atom.	Diamine: $H-\underset{\underset{H}{|}}{N}-(CH_2)_6-\underset{\underset{H}{|}}{N}-H$

The polymer in Example 23.4 is "Nylon-66" (the numbers indicate the number of carbon atoms in the diamine and dicarboxylic acid molecules). This polymer was first made in 1935 by Wallace Carothers (1896–1937) working at DuPont (Figure 23.3). Since then, other nylons, all polyamides, have been synthesized.

The elasticity of nylon fibers is due in part to hydrogen bonds between adjacent polymer chains. These hydrogen bonds join carbonyl oxygen atoms on one chain to NH groups on adjacent chains (Figure 23.4).

Figure 23.3 **Nylon.** Nylon can be made by the reaction between hexamethylenediamine, H_2N—$(CH_2)_6$—NH_2, and adipic acid, $HOOC$—$(CH_2)_4$—$COOH$. It forms at the interface between the two reagents.

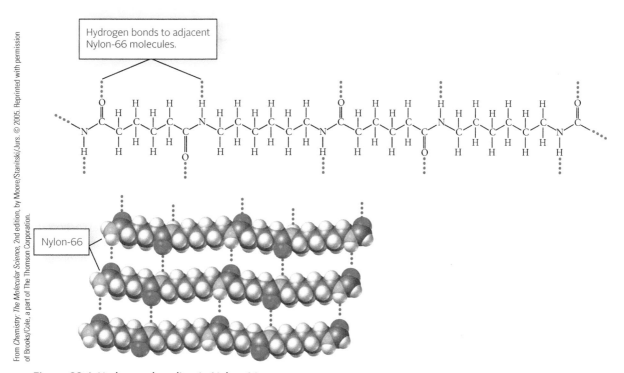

Hydrogen bonds to adjacent Nylon-66 molecules.

Nylon-66

Figure 23.4 **Hydrogen bonding in Nylon-66.**

23-3 Carbohydrates

Carbohydrates, which comprise one of the three basic classes of foodstuffs, contain carbon, hydrogen, and oxygen atoms. Their general formula, $C_n(H_2O)_m$, is the basis for their name. They can be classified as

1. **Monosaccharides,** which cannot be broken down chemically to simpler carbohydrates. The most familiar monosaccharides contain either six carbon atoms per molecule (glucose, fructose, galactose, . . .) or five (ribose, arabinose, . . .).

2. **Disaccharides,** which are dimers formed when two monosaccharide units combine with the elimination of H_2O. The monosaccharides may be the same (two glucose units in maltose) or different (a glucose and fructose unit in sucrose).

3. **Polysaccharides,** which are condensation polymers containing from several hundred to several thousand monosaccharide units. Cellulose and starch are the most common polysaccharides.

Glucose

By far the most important monosaccharide is glucose. As a pure solid, it has the "straight-chain" structure shown in the margin. In water solution, glucose exists primarily as six-membered ring molecules, shown below. The rings are not planar but have a three-dimensional "chair" form. The carbon atoms at each corner of the ring are not shown in the structure; they are represented by the numbers 1 to 5. The sixth atom in the ring is oxygen. There is a CH_2OH group bonded to carbon atom 5. There is an H atom bonded to each of the ring carbons; an OH group is bonded to carbon atoms 1 through 4. The heavy lines indicate the front of the ring, projected toward you.

The bonds to the H and OH groups from the ring carbons are oriented either in the plane of the ring (*equatorial,* shown in blue) or perpendicular to it *(axial).* In β-glucose, all four OH groups are equatorial. This keeps these rather bulky groups out of each other's way. In α-glucose, the OH group on carbon atom 1 is axial; all the others are equatorial. In water solution, glucose exists as an equilibrium mixture, about 37% in the alpha form and 63% in the beta form; the straight-chain structure accounts for less than 0.1% of the dissolved glucose.

D-glucose

α-glucose   β-glucose

As you can see from their structures, α- and β-glucose have several chiral carbon atoms. Both isomers are optically active; they are *not* enantiomers (mirror images of one another) because they differ in configuration only at carbon atom 1. As it happens, both α- and β-glucose rotate the plane of polarized light to the right (clockwise).

EXAMPLE   **23.5**

Consider the monosaccharide glucose.

a   What is the molecular formula of glucose?

STRATEGY AND SOLUTION

Count the number of carbon, hydrogen, and oxygen atoms.

$C_6H_{12}O_6$

b   How many chiral carbon atoms are there in a molecule of α-glucose? β-glucose? The straight-chain isomer?

SOLUTION

For both α- and β-glucose, only C_6 is not chiral. It has two hydrogen atoms bonded to it.

The straight chain structure has four chiral carbon atoms (C_2—C_5).

END POINT

The general formula for carbohydrates is $C_n(H_2O)_m$. For glucose, $n = m = 6$.

Glucose is also called "blood sugar." It is absorbed readily into the bloodstream and is normally found there at concentrations ranging from 0.004 to 0.008 mol/L. If the concentration of glucose drops below 0.003 M, a condition called hypoglycemia is created, with symptoms ranging from nervousness to loss of consciousness. If the glucose level rises above 0.01 M, as can easily happen with diabetics, it is excreted in the urine.

Although glucose can exist as a simple sugar, it is most often found in nature in combined form, as a disaccharide or polysaccharide. Several glucose-containing disaccharides are known. We will consider two of these, maltose and sucrose.

Maltose and Sucrose

Maltose, a decomposition product of starch, is a dimer of two glucose molecules. These are combined head-to-tail; carbon atom 1 of one molecule is joined through an oxygen atom to carbon atom 4 of the second molecule. To form maltose, the two OH groups on these carbon atoms react, condensing out H_2O and leaving the O atom bridge.

This reaction can be reversed either by using the enzyme maltase as a catalyst or by heating with acid. Two molecules of glucose are formed when this happens.

Sucrose, the compound we call "sugar," is the most common disaccharide. One of the monomer units in sucrose is α-glucose. The other is fructose, a monosaccharide found in honey and natural fruit juices.

In the body, this reaction is reversed by the enzyme sucrase. This occurs in digestion, which makes glucose and fructose available for absorption into the blood. Honey bees also carry an enzyme that can hydrolyze sucrose. Honey consists mostly of a 1 : 1 mol mixture of glucose and fructose with a small amount of unreacted sucrose.

EXAMPLE 23.6

Another common disaccharide is lactose, the sugar found in milk. Lactose is a dimer of glucose and galactose, a mono-saccharide that is identical to glucose except that the positions of the H and OH groups at carbon atom 4 are switched.

a   Draw the structure of β-galactose.

STRATEGY AND SOLUTION

The structure of β-galactose is identical to that of β-glucose except at C$_4$. In this carbon atom, the placement of the H atom and OH group is reversed. Equatorial groups are shown in color.

b   Draw the structure of lactose, which is broken down by the enzyme lactase to give equimolar amounts of β-galactose and β-glucose. Linkage occurs between carbon atom 1 of galactose and 4 of glucose.

SOLUTION

Table 23.2 Sweetness Relative to Sucrose

Lactose	0.16	Glucose	0.74	Aspartame (Equal)	180
Galactose	0.32	Sucrose	1.00	Saccharin (Sweet'N Low)	300
Maltose	0.33	Fructose	1.74	Sucralose (Splenda)	600

Table 23.2 gives the relative sweetness of the mono- and disaccharides considered here, along with sucrose substitutes.

Starch

Starch is a polysaccharide found in many plants, where it is stored in roots and seeds. ◄ⓘ It is particularly abundant in corn and potatoes, the major sources of commercial starch. Perhaps as much as 50% of our food energy comes from starch, mostly in the form of wheat products.

Starch is actually a mixture of two types of α-glucose polymers. One of these, called amylose, is insoluble in water and comprises about 20% of natural starch. It consists of long single chains of 1000 or more α-glucose units joined head-to-tail (Figure 23.5). The other glucose polymer found in starch is amylopectin, which is soluble in water. The linkage between α-glucose units is the same as in amylose. However, amylopectin has a highly branched structure. Short chains of 20 to 25 glucose units are linked through oxygen atom bridges. The oxygen atom joins a number 1 carbon atom at the end of one chain to a number 6 carbon atom on an adjacent chain.

ⓘ

Starch and cellulose are both condensation polymers of glucose.

Amylose

Figure 23.5 In starch, glucose molecules are joined head-to-tail through oxygen atoms. A thousand or more glucose molecules may be linked in this way, either in long single chains (amylose) or branched chains (amylopectin).

Figure 23.6 The bonding between glucose rings in cellulose is through oxygen bridges in the β position for each ring to the left of the bridge. This structure allows for ordered hydrogen bonding between chains and formation of long, strong fibers. Cotton and wood fiber have structures like this.

When you eat "starchy" foods, they are broken down into glucose by enzymes. The process starts in your mouth with the enzyme amylase found in saliva. This explains why, if you chew a piece of bread long enough, it starts to taste sweet. The breakdown of starch molecules continues in other parts of the digestive system. Within 1 to 4 hours after eating, all the starch in food is converted into glucose.

Cellulose

Cellulose contains long, unbranched chains of glucose units, about 10,000 per chain. It differs from starch in the way the glucose units are joined to each other. In starch, the oxygen bridge between the units is in the alpha position, as shown in Figure 23.5. In cellulose, it is in the beta position (Figure 23.6). Because humans lack the enzymes required to catalyze the hydrolysis of beta linkages, we cannot digest cellulose. The cellulose that we take in from fruits and vegetables remains undigested as "fiber." Ruminants, like cows and deer, and termites have bacteria in their intestines capable of breaking cellulose down into glucose.

The molecular structure of cellulose, unlike that of starch, allows for strong hydrogen bonding between polymer chains. This results in the formation of strong water-resistant fibers such as those found in cotton, which is 98% cellulose. Cotton actually has a tensile strength greater than that of steel. The major industrial source of cellulose is wood (~50% cellulose).

23-4 Proteins

The natural polymers known as **proteins** make up about 15% by mass of our bodies. They serve many functions. Fibrous proteins are the main components of hair, muscle, and skin. Other proteins found in body fluids transport oxygen, fats, and other substances needed for metabolism. Still others, such as insulin and vasopressin, are hormones. Enzymes, which catalyze reactions in the body, are chiefly protein.

Protein is an important component of most foods. Nearly everything we eat contains at least a small amount of protein. Lean meats and vegetables such as peas and beans are particularly rich in protein. In our digestive system, proteins are broken down into small molecules called *α-amino acids*. ◄ 𝒊 These molecules can then be reassembled in cells to form other proteins required by the body.

Proteins are condensation polymers of α-amino acids.

α-Amino Acids

The monomers from which proteins are derived have the general structure shown below. These compounds are called **α-amino acids.** They have an NH_2 group attached to the carbon atom (the α carbon) adjacent to a COOH group.

α-amino acid

The 20 α-amino acids shown in Table 23.3 are those most commonly found in natural proteins. They differ from one another in the nature of the R group attached to the alpha carbon. As you can see, R can be

- a H atom (glycine).
- a simple hydrocarbon group (alanine, valine, . . .).
- a more complex group containing one or more atoms of oxygen (serine, aspartic acid, etc.), nitrogen (lysine, etc.), or sulfur (methionine, etc.).

EXAMPLE 23.7

Consider the structures shown in Table 23.3.

(a) How many α-amino acids contain only one chiral carbon atom?

(b) How many contain only one —NH_2 group?

SOLUTION

(a) 17; the exceptions are glycine (0), isoleucine (2), and threonine (2).

(b) 17; the exceptions are proline (0), lysine (2), and glutamine (2).

In all the amino acids shown in Table 23.3 except glycine, the α-carbon is chiral. This means that these compounds are optically active. With alanine, for example, there should be two optical isomers:

Of the 20 common amino acids, ten are "essential" in the sense that they are required by humans for protein construction and cannot be synthesized in the body. These are designated by an asterisk in Table 23.3. Clearly the proteins we eat should contain these essential amino acids. Beyond that, a "high-quality" protein contains the various amino acids in the approximate ratio required by the body for protein synthesis. Generally speaking, animal proteins are of higher

Table 23.3 The Common α-Amino Acids (names and abbreviations below structures)

alanine Ala

glycine Gly

proline Pro

arginine* Arg

histidine* His

serine Ser

asparagine Asn

isoleucine* Ile

threonine* Thr

aspartic acid Asp

leucine* Leu

tryptophan* Trp

cysteine Cys

lysine* Lys

tyrosine Tyr

glutamic acid Glu

methionine* Met

valine* Val

glutamine Gln

phenylalanine* Phe

*These are essential amino acids that cannot be synthesized by the body and therefore must be obtained from the diet.

quality than plant proteins, with eggs at the head of the class. This is something to keep in mind if you're a vegetarian.

Acid-Base Properties of Amino Acids

You may recall from Example 22.8 that the reaction

$$CH_3COOH(aq) + CH_3NH_2(aq) \longrightarrow CH_3COO^-(aq) + CH_3NH_3^+(aq)$$

in which there is a proton transfer from a —COOH group to an —NH_2 group has a large equilibrium constant, 7×10^5. In other words, this reaction goes virtually to completion when solutions of acetic acid and methylamine are mixed.

A very similar reaction occurs with an amino acid; there is a proton transfer from the —COOH group to the —NH₂ group:

amino acid → zwitterion

Here, too, the equilibrium constant is large and the reaction goes essentially to completion. Putting it another way, an amino acid does not exist in the form shown in Table 23.3, either in water solution or in the solid state. Instead it occurs as a **zwitterion,** which behaves like a polar molecule. Within it, there is a +1 charge on the nitrogen atom and a −1 charge at one of the oxygen atoms of the COOH group. Overall, the zwitterion has no net charge.

In water, zwitterions are stable only over a certain pH range. At low pH (high H⁺ concentration), the COO⁻ group picks up a proton. The product, shown at the left below, has a +1 charge. At high pH (low H⁺ concentration), the NH₃⁺ group loses a proton. The species formed, shown at the right below, has a −1 charge.

cation   zwitterion   anion

In a very real sense, a zwitterion is analogous to a species such as the HCO_3^- ion, which accepts a proton at low pH and donates a proton at high pH:

$$H_2CO_3(aq) \xleftarrow{\ H^+\ } HCO_3^-(aq) \xrightarrow{\ -H^+\ } CO_3^{2-}(aq)$$

low pH   high pH

Both the zwitterion and the HCO_3^- ion are amphiprotic, capable of acting as either a Brønsted acid or a Brønsted base.

To treat acid-base equilibria involving zwitterions, it is convenient to consider the cation stable at low pH to be a diprotic acid (analogous to H_2CO_3), which ionizes in two steps. Using the symbols C^+, Z, and A^- to stand for the cation, zwitterion, and anion, respectively, we have

$$C^+(aq) \rightleftharpoons H^+(aq) + Z(aq); \qquad K_{a1} = \frac{[H^+][Z]}{[C^+]}$$

$$Z(aq) \rightleftharpoons H^+(aq) + A^-(aq); \qquad K_{a2} = \frac{[H^+][A^-]}{[Z]}$$

For glycine,

$$K_{a1} = 4.6 \times 10^{-3}; \qquad K_{a2} = 2.5 \times 10^{-10}$$

EXAMPLE 23.8

Using these acid dissociation constants for glycine, calculate the ratios [Z]/[C⁺] and [Z]/[A⁻] at pH 2.00, 6.00, and 10.00.

STRATEGY

Substitute into the general K_a expression where

$$C^+ = \underset{\underset{NH_3^+}{|}}{\overset{\overset{H}{|}}{H-C-COOH}}; \quad Z = \underset{\underset{NH_3^+}{|}}{\overset{\overset{H}{|}}{H-C-COO^-}}; \quad A^- = \underset{\underset{NH_2}{|}}{\overset{\overset{H}{|}}{H-C-COO^-}}$$

For the first dissociation:

$$C^+(aq) \rightleftharpoons H^+(aq) + Z(aq) \qquad K_{a1} = 4.6 \times 10^{-3}$$

For the second dissociation:

$$Z(aq) \rightleftharpoons H^+(aq) + A^-(aq) \qquad K_{a2} = 2.5 \times 10^{-10}$$

SOLUTION

pH = 2.00

$[H^+] = 1.0 \times 10^{-2}$

$\dfrac{[Z]}{[C^+]}$

$4.6 \times 10^{-3} = \dfrac{[H^+][Z]}{[C^+]} \longrightarrow \dfrac{[Z]}{[C^+]} = \dfrac{4.6 \times 10^{-3}}{1.0 \times 10^{-2}} = 0.46$

$\dfrac{[Z]}{[A^-]}$

$2.5 \times 10^{-10} = \dfrac{[H^+][A^-]}{[Z]} \longrightarrow \dfrac{[Z]}{[A^-]} = \dfrac{1.0 \times 10^{-2}}{2.5 \times 10^{-10}} = 4.0 \times 10^7$

pH = 6.00

$[H^+] = 1.0 \times 10^{-6}$

$\dfrac{[Z]}{[C^+]}$

$\dfrac{[Z]}{[C^+]} = \dfrac{4.6 \times 10^{-3}}{1.0 \times 10^{-6}} = 4.6 \times 10^3$

$\dfrac{[Z]}{[A^-]}$

$2.5 \times 10^{-10} = \dfrac{[H^+][A^-]}{[Z]} \longrightarrow \dfrac{[Z]}{[A^-]} = \dfrac{1.0 \times 10^{-6}}{2.5 \times 10^{-10}} = 4.0 \times 10^3$

pH = 10.00

$[H^+] = 1.0 \times 10^{-10}$

$\dfrac{[Z]}{[C^+]}$

$\dfrac{[Z]}{[C^+]} = \dfrac{4.6 \times 10^{-3}}{1.0 \times 10^{-10}} = 4.6 \times 10^7$

$\dfrac{[Z]}{[A^-]}$

$2.5 \times 10^{-10} = \dfrac{[H^+][A^-]}{[Z]} \longrightarrow \dfrac{[Z]}{[A^-]} = \dfrac{1.0 \times 10^{-10}}{2.5 \times 10^{-10}} = 0.40$

END POINTS

1. At low pH, [Z] < [C⁺], so there is more cation than zwitterion and essentially no anion.
2. At intermediate pH, the zwitterion is the principal species present.
3. At high pH, the anion [A⁻] is the principal species.

From Example 23.8 or Figure 23.7, it is clear that the zwitterion of glycine is the principal species over a wide pH range, certainly from pH 3 to pH 9. The maximum concentration of zwitterion occurs at about pH 6; this is referred to as the **isoelectric point** of glycine. At this pH, glycine does not migrate in an electric field, since the zwitterion is a neutral species. At low pH, the cation moves to the cathode; at high pH the anion migrates to the anode. In general, for an amino acid like glycine, where there is only one —COOH group and one —NH₂ group:

$$\text{pH at isoelectric point} = \frac{pK_{a1} + pK_{a2}}{2}$$

where K_{a1} and K_{a2} are the successive dissociation constants of the cation of the amino acid.

Figure 23.7 Relative concentrations of different glycine species as a function of pH. Between about pH 3 and 9, the zwitterion is the principal species; it has its maximum concentration at pH 6, the isoelectric point. Below pH 2, the cation dominates; above pH 10, the anion is the principal species.

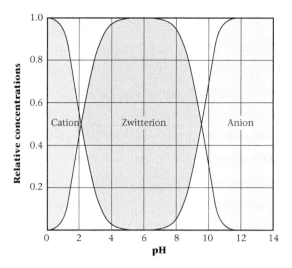

Polypeptides

As noted in Section 23-2, molecules containing NH_2 and COOH groups can undergo condensation polymerization. Amino acids contain both groups in the same molecule. Hence, two amino acid molecules can combine by the reaction of the COOH group in one molecule with the NH_2 group of the other molecule. If the acids involved are different, two different structural isomers are possible:

$$H_2N-CH_2-\overset{O}{\underset{\|}{C}}-OH + H-\overset{H}{\underset{\underset{CH_3}{|}}{\underset{|}{N}}}-\overset{H}{\underset{|}{C}}-COOH \longrightarrow H_2N-CH_2-\overset{O}{\underset{\|}{C}}-\overset{H}{\underset{|}{N}}-\overset{H}{\underset{\underset{CH_3}{|}}{C}}-COOH + H_2O$$

glycine   alanine   glycylalanine (Gly-Ala)

$$H_2N-\overset{H}{\underset{\underset{CH_3}{|}}{C}}-\overset{O}{\underset{\|}{C}}-OH + H-\overset{H}{\underset{|}{N}}-CH_2-COOH \longrightarrow H_2N-\overset{H}{\underset{\underset{CH_3}{|}}{C}}-\overset{O}{\underset{\|}{C}}-\overset{H}{\underset{|}{N}}-CH_2-COOH + H_2O$$

alanine   glycine   alanylglycine (Ala-Gly)

Each of these isomers is a dipeptide, formed from two amino acids by condensing out a water molecule and containing the group

$$-\overset{O}{\underset{\|}{C}}-\overset{H}{\underset{|}{N}}-$$

This linkage is similar to that in nylon.

commonly called a **peptide linkage.** ◄ 🛈 Dipeptides are named by writing first (far left) the amino acid that retains a free NH_2 group; the amino acid with a free COOH group is written last (far right).

EXAMPLE 23.9

Draw the structural formula of glycylserine (Gly-Ser).

STRATEGY AND SOLUTION

1. Write the structural formula of the two amino acids side by side. Put glycine on the left with the NH_2 group on the left and the COOH group on the right. Put the serine group on the right in a similar fashion.

$$H_2N-CH_2-\overset{O}{\underset{\|}{C}}-OH \quad \overset{H_2O \quad H \quad CH_2OH}{\underset{\underset{H}{|} \quad \underset{H}{|}}{N-C-COOH}}$$

glycine   serine

continued

2. The two amines react by condensing out a water molecule made up of the OH group from glycine and the H atom from serine.

$$CH_2OH$$
$$H_2N-CH_2-\underset{\underset{O}{\|}}{C}-\underset{\underset{H}{|}}{N}-\underset{\underset{H}{|}}{C}-COOH + H_2O$$

glycylserine (Gly-Ser)

The dipeptide formed in Example 23.9 has a reactive group at both ends of the molecule (NH_2 at one end, COOH at the other). Hence it can combine further to give tripeptides (three amino acid units), tetrapeptides, . . . , and eventually long-chain polymers called polypeptides. A protein containing a large number of amino acid units is a **polypeptide**:

$$-\underset{\underset{H}{|}}{N}-\underset{\underset{R_1}{|}}{\overset{\overset{H}{|}}{C}}-\overset{\overset{O}{\|}}{C}-\underset{\underset{H}{|}}{N}-\underset{\underset{R_2}{|}}{\overset{\overset{H}{|}}{C}}-\overset{\overset{O}{\|}}{C}-\underset{\underset{H}{|}}{N}-\underset{\underset{R_3}{|}}{\overset{\overset{H}{|}}{C}}-\overset{\overset{O}{\|}}{C}-\underset{\underset{H}{|}}{N}-\underset{\underset{R_4}{|}}{\overset{\overset{H}{|}}{C}}-\overset{\overset{O}{\|}}{C}-$$

(peptide linkages in color).

Proteins differ from the other polymers we have discussed in that they may contain up to 20 different monomer units. This means that there are a huge number of possible proteins. Using the amino acids listed in Table 23.3, we could make

$20 \times 20 = 400$ different dipeptides.
$20 \times 20 \times 20 = 8000$ different tripeptides.
20^n polypeptides containing n monomer units.

For a relatively simple protein containing only 50 monomer units, we have

$20^{50} = 1 \times 10^{65}$ possibilities.

Proteins; Primary Structure

To determine the **primary structure** of a protein that may contain a hundred or more amino acid units, it is necessary to

1. identify the amino acids present.
2. *determine the relative amounts of each amino acid.*
3. establish the *sequence* ◄ i of amino acid units in the protein.

The amino acid sequence in hemo-globin, with 574 units, is known.

Tasks (1) and (2) are relatively easy to accomplish. The sample is first heated with hydrochloric acid to break all of the "peptide linkages" (amide bonds) in the protein. The resulting solution is then passed through a chromatographic column (recall the discussion in Chapter 1). This separates the different amino acids and allows you to determine their identities and concentrations.

Task (3) is more difficult. It uses a series of reagents each of which is capable of breaking only certain amide bonds. One of these reagents is the enzyme trypsin, which breaks only those bonds formed by the carboxyl groups in arginine and lysine. It would break the polypeptide

Ala-Gly-Ser-Lys-Gly-Leu-Arg-Met

into three fragments:

Ala-Gly-Ser-Lys;   Gly-Leu-Arg;   Met

Remember that, by convention, the NH_2 group of each amino acid is written at the left; the COOH group is at the right. Other reagents split the chain at different points. By correlating the results of different hydrolysis experiments, it is possible to deduce the primary structure of the protein (or polypeptide). Example 23.10 shows how this is done in a particularly simple case.

EXAMPLE 23.10

A certain polypeptide is shown by acid hydrolysis to contain only three amino acids: serine (Ser), alanine (Ala), and methionine (Met) with mole fractions of $\frac{1}{2}$, $\frac{1}{4}$, and $\frac{1}{4}$, respectively. Enzymatic hydrolysis yields the following fragments: Ser-Ala; Ser-Met; Ala-Ser. Deduce the primary structure of the polypeptide.

STRATEGY AND SOLUTION

Align the fragments vertically so that they overlap one another

Ser—Ala
      Ala—Ser
            Ser—Met

There are four amino acid units: Ser – Ala – Ser – Met

Since there are two serine units, one alanine unit, and one methionine unit, their mole fractions are

$$\frac{2}{4} = \frac{1}{2} \text{ for serine} \qquad \frac{1}{4} \text{ for alanine} \qquad \frac{1}{4} \text{ for methionine}$$

In the body, proteins are built up by a series of reactions that in general produce a specific sequence of amino acids. Even tiny errors in this sequence may have serious effects. Among the genetic diseases known to be caused by improper sequencing are hemophilia, sickle cell anemia, and albinism. Sickle cell anemia is caused by the substitution of *one* valine unit for a glutamic acid unit in a chain containing 146 monomers. ◄ *i*

Fortunately nature doesn't make too many mistakes. *i*

Proteins; Secondary and Tertiary Structures

Protein chains (Figure 23.8) can align themselves so that certain patterns are repeated. These repeating patterns establish what we call the **secondary structure** of the protein. The nature of the pattern is determined in large part by hydrogen bonding. Oxygen atoms on C=O groups can interact with H atoms in nearby N—H groups to form these bonds. This can occur within a single protein chain or between neighboring chains.

A common secondary structure is an α-helix, which maximizes hydrogen bonding between peptide linkages. At the left of Figure 23.8 is the outline of the protein chain forming the helix. The more complete structure is given at the right. This shows where the hydrogen bonds form between amino acid units in the protein chain. Notice that the bulky R groups are all on the outside of the helix, where they have the most room. The actual structure of the helix is nearly independent of the nature of the R groups. The dimensions of the helix correspond

Figure 23.8 α-Helix structures of proteins. The main atom chain in the helix is shown schematically on the left. The sketch on the right more nearly represents the actual positions of atoms and shows where intrachain hydrogen bonding occurs. Wool and many other fibrous proteins have the α-helix structure.

closely to those observed in such fibrous proteins as wool, hair, skin, feathers, and fingernails.

The three-dimensional conformation of a protein is called its **tertiary structure.** An α-helix can be either twisted, folded, or folded and twisted into a definite geometric pattern. These structures are stabilized by dispersion forces, hydrogen bonding, ◀ⓘ and other intermolecular forces.

Collagen, the principal fibrous protein in mammalian tissue, has a tertiary structure made up of twisted α-helices. Three polypeptide chains, each of which is a left-handed helix, are twisted into a right-handed super helix to form an extremely strong tertiary structure. It has remarkable tensile strength, which makes it important in the structure of bones, tendons, teeth, and cartilage.

ⓘ This is a rather special kind of H bond.

CHEMISTRY **BEYOND** THE CLASSROOM

DNA Fingerprinting

Within every living cell, there is an organic natural polymer called deoxyribonucleic acid or, more commonly, DNA. The DNA molecule is enormous; its molar mass is estimated to be of the order of 2×10^{10} g. The molecule takes the form of a narrow, tightly coiled band, which, if straightened out, would have a length of about 1 m.

Figure A shows the primary structure of the DNA molecule. Notice that

- the backbone of the molecule consists of alternating units derived from phosphoric acid, H_3PO_4, and a 5-carbon sugar called deoxyribose.
- bonded to each sugar is a nitrogen base (amine). The base may be thymine (shown at the top in orange), adenine (red), guanine (blue), or cytosine (green). These are often abbreviated as the letters T, A, G, and C, respectively
- the structural unit of DNA, repeated over and over along the chain, is built from an H_3PO_4 molecule, a sugar molecule, and a nitrogen base, which may be any one of the four shown. Figure A shows four such structural units called *nucleotides;* an actual molecule of DNA would contain literally billions of such nucleotides.

The secondary structure of DNA is shown in Figure B. This "double helix" model was first proposed in 1953 by James Watson (1928–) and Francis Crick (1916–2004), who used the x-ray crystallographic data of Rosalind Franklin (1920–1958) and Maurice Wilkins (1916–2004). Beyond that, they were intrigued by the results of analyses that showed that in DNA the ratio of adenine to thymine molecules is almost exactly 1 : 1, as is the ratio of cytosine to guanine:

$$A/T = C/G = 1.00$$

This behavior is readily explained by the double helix; an A molecule in one strand is always hydrogen-bonded to a T molecule in the second strand. Similarly, a C molecule in one strand is situated properly to form a hydrogen bond with a G molecule in the other strand. In 1962, Watson, Crick, and Wilkins received the Nobel Prize in medicine.

From *Organic Chemistry,* 1st edition, by Brown. © 1995. Reprinted with permission of Brooks/Cole, a part of The Thomson Corporation

Thymine (T)
Adenine (A)
Guanine (G)
Cytosine (C)

Figure A A tetranucleotide section of a single-stranded DNA.

From *Chemistry: The Molecular Science*, 2nd edition, by Moore/Stanitski/Jurs. © 2005. Reprinted with permission of Brooks/Cole, a part of The Thomson Corporation

Figure B The double-helical strand of a DNA molecule. The diagram at the left shows the hydrogen bonding between base pairs adenine-thymine and cystosine-guanine that hold the strands together.

The double helix model provides a simple explanation for cell division and reproduction. In the reproduction process, the two DNA chains unwind from each other. As this happens, a new matching chain of DNA is synthesized on each of the original ones, creating two double helices. Since the base pairs in each new double helix must match in the same way as in the original, the two new double helices must be identical to the original. Exact replication of genetic data is thereby accomplished, however complex that data may be.

The many millions of DNA molecules in the cells of your body are identical to each other. That is, the base sequence in all of these molecules is the same. In contrast, the base sequence of the DNA molecules in every other person in the world differs at least slightly from yours. In that sense, your "DNA fingerprint" is unique.

Alec Jeffreys (1950–), an English geneticist, discovered in the 1980s how to apply this principle to forensics. To do this, it is necessary to locate that portion of the DNA molecule in which the base sequence differs significantly from one individual to another. That part of the molecule is cut out by a "restrictive enzyme" in much the same way that trypsin splits a protein molecule into fragments. The DNA sample obtained in this way from a suspect can be compared with that derived from blood, hair, semen, saliva, and so on, found at the scene of a violent crime.

At least in principle, comparison of DNA fingerprints can determine the guilt or innocence of a suspect beyond a reasonable doubt. If the two fingerprints differ, the suspect is innocent. Conversely, if the fingerprints match perfectly, the odds that the suspect is guilty are about 80 billion to one.

In practice, the situation isn't quite that simple. DNA samples taken from a victim are almost certain to be contaminated with DNA from fungi or bacteria. Certain dyes can combine with restriction enzymes, causing them to cut in the wrong places. Finally, DNA may decay in a warm or moist environment.

A DNA fingerprint can be used for many purposes other than solving violent crimes. In particular, it can serve to identify deceased individuals. In June of 1998 the "Vietnam Unknown" buried in the Tomb of the Unknown Soldier at Arlington National Cemetery was identified by DNA technology. He was shown to be First Lieutenant Michael Blassie, shot down over Vietnam in May of 1972. DNA samples taken from his mother matched those obtained from his body. A month later Blassie, a native of St. Louis, Missouri, was reburied in a national cemetery located in that city.

Key Concepts

1. Relate the structure of an addition polymer to that of the corresponding monomer.
   (Examples 23.1, 23.2; Problems 1–6, 37, 38)
2. Relate the structure of a condensation polymer to that of the corresponding monomer.
   (Examples 23.3, 23.4; Problems 9–14, 37, 38)
3. Relate the molar mass of a polymer to the number of monomer units.
   (Example 23.1; Problems 1, 2, 7, 17, 18)
4. Draw the structures of mono- and disaccharides.
   (Example 23.6; Problems 19, 20)

5. Identify the chiral carbon atoms in a carbohydrate or α-amino acid.
   (Examples 23.5, 23.7; Problems 21, 22)
6. Carry out equilibrium calculations relating pH to [Z], [C$^+$], and [A$^-$].
   (Example 23.8; Problems 29, 30)
7. Relate the structure of a polypeptide to that of the amino acids from which it is formed.
   (Example 23.9; Problems 23–26)
8. Deduce the structure of a polypeptide from hydrolysis data.
   (Example 23.10; Problems 31, 32)

Key Terms

addition polymer
α-amino acids
carbohydrates
condensation polymer
disaccharides

isoelectric point
monomers
monosaccharides
peptide linkage
polyamide

polyester
polymers
polypeptide
polysaccharides
primary structure

proteins
secondary structure
tertiary structure
zwitterion

Summary Problem

All of these questions pertain to six-carbon molecules.

(a) Sketch the head-to-tail and tail-to-tail addition polymers derived from

$$\underset{H}{\overset{H}{>}}C=C\underset{(CH_2)_3CH_3}{\overset{H}{<}}$$

(b) If the molar mass of the polymer is 1.2×10^4 g/mol, how many monomer units are present? What is the mass percent of carbon in the polymer?

(c) Write the structure of the condensation polymer formed when a straight-chain, saturated, six-carbon dialcohol reacts with a straight-chain, saturated, six-carbon dicarboxylic acid.

(d) Write the structure of the disaccharide formed when α-glucose and β-glucose, both of formula $C_6H_{12}O_6$, combine to form an isomer of maltose.

(e) Draw the structures of the two polypeptides derived from the six-carbon amino acids lysine and leucine.

Answers

(a)

$$-\overset{H}{\underset{H}{C}}-\overset{H}{\underset{R}{C}}-\overset{H}{\underset{H}{C}}-\overset{H}{\underset{R}{C}}- \text{ and } -\overset{H}{\underset{H}{C}}-\overset{H}{\underset{R}{C}}-\overset{H}{\underset{R}{C}}-\overset{H}{\underset{H}{C}}-;\qquad R = -(CH_2)_3CH_3$$

(b) 1.4×10^2; 85.6

(c) $-O-(CH_2)_6-O-\underset{O}{\overset{\Vert}{C}}-(CH_2)_4-\underset{O}{\overset{\Vert}{C}}-$

(d) See maltose structure (page 585); invert OH and H on carbon atom at lower right.

(e)

$$H-\overset{H_3C}{\underset{H_3C}{C}}-\overset{H}{\underset{H}{C}}-\overset{H}{\underset{NH_2}{C}}-\overset{O}{\overset{\Vert}{C}}-N-\overset{H}{\underset{(CH_2)_3-CH_2NH_2}{C}}-COOH \qquad \text{Leu-Lys}$$

$$NH_2-(CH_2)_4-\overset{H}{\underset{NH_2}{C}}-\overset{O}{\overset{\Vert}{C}}-N-\overset{H}{\underset{CH_2-CH-(CH_3)_2}{C}}-COOH \qquad \text{Lys-Leu}$$

Questions and Problems

Even-numbered questions and Challenge Problems have answers in Appendix 5 and fully worked solutions in the *Student Solutions Manual*.

23-1 Synthetic Addition Polymers

1. Consider a polymer made from tetrachloroethylene.
   (a) Draw a portion of the polymer chain.
   (b) What is the molar mass of the polymer if it contains 3.2×10^3 tetrachloroethylene molecules?
   (c) What are the mass percents of C and Cl in the polymer?

2. Consider Teflon, the polymer made from tetrafluoroethylene.
   (a) Draw a portion of the Teflon molecule.
   (b) Calculate the molar mass of a Teflon molecule that contains 5.0×10^4 CF_2 units.
   (c) What are the mass percents of C and F in Teflon?

3. Sketch a portion of the acrylonitrile polymer (Table 23.1), assuming it is a
   (a) head-to-tail polymer.
   (b) head-to-head, tail-to-tail polymer.

4. Styrene, $H_2C=\overset{H}{\underset{}{C}}-$⟨○⟩, forms a head-to-tail addition polymer.
Sketch a portion of a polystyrene molecule.

5. The polymer whose structure is shown below is made from two different monomers. Identify the monomers.

6. Show the structure of the monomer used to make the following addition polymers.

(a) and (b) structures shown

7. Consider the polymers referred to in Table 23.1. If each one contains the same number of monomer units, which one has the largest molar mass?

8. Of the polymers referred to in Table 23.1 which one contains the highest percentage by mass of carbon?

23-2 Synthetic Condensation Polymers

9. A rather simple polymer can be made from ethylene glycol, $HO—CH_2—CH_2—OH$, and oxalic acid, $HO—C—C—OH$ (with two $=O$ groups).

Sketch a portion of the polymer chain obtained from these monomers.

10. Lexan is a very rugged polyester in which the monomers can be taken to be carbonic acid, $HO—C—OH$ (with $=O$)

and

structure shown with CH_3 groups

Sketch a section of the Lexan chain.

11. *Para*-aminobenzoic acid is an "essential vitamin" for many bacteria:

$H_2N—$ (ring) $—C—OH$ (with $=O$)

Sketch a portion of a polyamide polymer made from this monomer.

12. Nylon-66 is made from a single monomer:

$H_2N—(CH_2)_5—C—OH$ (with $=O$)

Sketch a section of the polymer chain in Nylon-66.

13. The following condensation polymer is made from a single monomer. Identify the monomer.

$—N—CH_2—C—N—CH_2—C—$ (with H, O, H, O)

14. Identify the monomers from which the following condensation polymers are made.

(a) $—N—CH_2—CH_2—N—C—CH_2—C—$ (with H, H, O, O)

(b) $—C—$(ring)$—C—O—CH_2—C—O—$ (with O, O, CH_3)

23-3 Carbohydrates

15. Write a chemical equation, using molecular formulas, for the reaction of sucrose with water to form glucose and fructose.

16. Write a chemical equation, using molecular formulas, for the reaction of maltose with water to form glucose.

17. Cellulose consists of about 10,000 $C_6H_{10}O_5$ units linked together.

(a) What are the mass percents of C, H, and O in cellulose?
(b) What is the molar mass of cellulose?

18. Starch has the same empirical formula as cellulose and a molar mass of about 1.0×10^5 g/mol.

(a) What are the mass percents of C, H, and O in starch?
(b) How many $C_6H_{10}O_5$ units are linked together in a starch molecule?

19. Mannose has the same molecular formula as glucose and the same geometry except at carbon-2, where the H and OH groups are interchanged. Draw the structures of α- and β-mannose.

20. Draw the structure of the disaccharide formed by two moles of α-mannose (see Question 19).

21. How many chiral carbon atoms are there in α-glucose? Fructose?

22. How many chiral carbon atoms are there in sucrose? Maltose?

23-4 Proteins

23. Give the structural formula of two different dipeptides formed between arginine and serine.

24. Give the structural formulas of two different dipeptides formed between leucine and lysine.

25. (a) How many tripeptides can be made from glycine, alanine, and leucine, using each amino acid only once per tripeptide?
(b) Write the structural formulas of these tripeptides and name them in the shorthand abbreviation used for showing amino acid sequences.

26. A tripeptide contains valine, lysine, and phenylalanine residues.

(a) How many tripeptides are possible from these amino acids?
(b) Draw a structural formula for a possible form of the tripeptide and name it, using the shorthand form.

27. Consider the cysteine molecule shown in Table 23.3. Write structural formulas for

(a) the zwitterion of cysteine.
(b) the cation formed in acid.
(c) the anion formed in base.

28. Follow the directions of Question 27 for serine.

29. For alanine, $K_{a1} = 5.1 \times 10^{-3}$, $K_{a2} = 1.8 \times 10^{-10}$. Calculate the ratios $[Z]/[C^+]$ and $[Z]/[A^-]$ at pH

(a) 2.00.   (b) 6.00.   (c) 10.50.

What is the principal species at each pH?

30. Using the information given in Problem 29, calculate the pH

(a) when $[Z] = [C^+]$.
(b) when $[Z] = [A^-]$.
(c) at the isoelectric point.

31. On complete hydrolysis, a polypeptide gives two alanine, one leucine, one methionine, one phenylalanine, and one valine residue. Partial hydrolysis gives the following fragments: Ala-Phe, Leu-Met, Val-Ala, Phe-Leu. It is known that the first amino acid in the sequence is valine and the last one is methionine. What is the complete sequence of amino acids?

32. Suppose that, in the polypeptide referred to in Question 31, the first amino acid is alanine and the last one is also alanine. What is the complete sequence of amino acids?

Unclassified

33. Which of the following monomers could form an addition polymer? A condensation polymer?

(a) C_2H_6 (b) C_2H_4

(c) $HO-CH_2-CH_2-OH$ (d) $HO-CH_2-CH_3$

34. How would you explain to a young science student how to decide whether a given compound might be useful as a monomer for addition polymerization? Condensation polymerization?

35. Explain the difference between

(a) a synthetic and natural polymer.

(b) a polyester and polyamide.

(c) α- and β-glucose.

36. Explain the difference between

(a) linear and branched polyethylene.

(b) glucose and fructose.

(c) maltose and sucrose.

37. Draw the structures of the monomers that could be used to make the following polymers:

(a)
```
      Br  H   Br  H   Br  H
      |   |   |   |   |   |
   —C—C—C—C—C—C—
      |   |   |   |   |   |
      H   H   H   H   H   H
```

(b)
```
          H           H
          |           |
   —N—C—C—N—C—C—
      |   |   ‖   |   |   ‖
      H   H   O   H   H   O
```

(c)
```
      O                O   H
      ‖                ‖   |
   —C—(CH₂)₄—C—N—(CH₂)₆—N—
```

38. What monomers would be used to make the following polymers?

(a)
```
      O              O
      ‖              ‖
   —C—⟨◯⟩—C—O—(CH₂)₂—O—
```

(b)
```
      H   H   H   H   H   H
      |   |   |   |   |   |
   —C—C—C—C—C—C—
      |   |   |   |   |   |
      Cl  Cl  Cl  Cl  Cl  Cl
```

(c)
```
      O   H              O   H
      ‖   |              ‖   |
   —C—N—(CH₂)₅—C—N—(CH₂)₅—
```

39. Sketch the tetrapeptide obtained from four molecules of the α-amino acid glycine.

40. Sketch a portion of a head-to-head, tail-to-tail, and a head-to-tail polymer made from acrylonitrile, $H_2C=CHCN$.

41. Using bond energies, estimate ΔH for the hydrolysis of maltose to glucose.

42. Sketch the form in which leucine would exist in acid solution; in basic solution.

43. How many tripeptides could one make from glycine, valine, and lysine, using any number of each amino acid?

44. A 1.00-mg sample of a pure protein yielded on hydrolysis 0.0165 mg of leucine and 0.0248 mg of isoleucine. What is the minimum possible molar mass of the protein? (MM leucine = MM isoleucine = 131 g/mol)

45. Describe what is meant by

(a) the primary structure of a protein.

(b) the secondary structure of a protein.

(c) the tertiary structure of a protein.

46. Glycolysis is the process by which glucose is metabolized to lactic acid according to the equation

$$C_6H_{12}O_6(aq) \longrightarrow 2C_3H_6O_3(aq)$$

$$\Delta G° = -198 \text{ kJ at pH 7.0 and 25°C}$$

Glycolysis is the source of energy in human red blood cells. In these cells, the concentration of glucose is 5.0×10^{-3} M, while that of lactic acid is 2.9×10^{-3} M. Calculate ΔG for glycolysis in human blood cells under these conditions. Use the equation $\Delta G = \Delta G° + RT \ln Q$, where Q is the concentration quotient, analogous to K.

47. Plants synthesize carbohydrates from CO_2 and H_2O by the process of photosynthesis. For example,

$$6CO_2(g) + 6H_2O(l) \longrightarrow C_6H_{12}O_6(aq) + 6O_2(g)$$

$\Delta G° = 2.87 \times 10^3$ kJ at pH 7.0 and 25°C. What is K for the reaction at 25°C?

Challenge Problems

48. Glycerol, $C_3H_5(OH)_3$, and orthophthalic acid,

```
   ⟨◯⟩—COOH
        |
       COOH
```

form a cross-linked polymer in which adjacent polymer chains are linked together; this polymer is used in floor coverings and dentures.

(a) Write the structural formula for a portion of the polymer chain.

(b) Use your answer in (a) to show how cross-linking can occur between the polymer chains to form a water-insoluble, network covalent solid.

49. Determine the mass percents of the elements in Nylon whose structure is shown in Example 23.4.

50. Using bond energies, estimate ΔH for protein formation, per mole of amino acid added to the chain. Does this value seem reasonable?

51. One of the earliest, and still one of the most important, polymers is the material known as Bakelite, which is a condensation polymer of phenol, C_6H_5OH, and formaldehyde, $H_2C=O$. Formaldehyde will react with phenol to produce the following compounds when the ratio is 1:1:

```
   ⟨◯⟩—OH              OH—⟨◯⟩
        |        and          |
       CH₂OH                 CH₂OH
```

These species, which can be taken to be monomers, on being heated condense with each other and themselves, eliminating water (formed from the OH groups on CH_2OH and ring hydrogen atoms) and linking benzene rings by CH_2 groups. If the phenol:formaldehyde ratio is 1:1, a linear polymer forms. If the ratio is 1:2, two $H_2C=O$ molecules react with each ring, and the chains cross-link at every benzene ring, forming the infusible, insoluble, hard, brittle solid we know as Bakelite. Sketch the linear chain polymer and the cross-linked solid.

52. Aspartic acid acts as a triprotic acid with successive dissociation constants of 8.0×10^{-3}, 1.4×10^{-4}, and 1.5×10^{-10}. Depending upon pH, aspartic acid can exist in four different forms in water solution. Draw these forms and calculate the pH range over which each form is the principal species.

Units, Constants, and Reference Data

SI Units

As pointed out in Chapter 1, most quantities are commonly expressed in a variety of different units. For example, temperature can be reported in °F, °C, or K. Volume is sometimes quoted in liters or cubic centimeters, at other times in quarts, cubic feet, and so on. In 1960, the General Conference of Weights and Measures attempted to end this proliferation of units. It recommended adopting a self-consistent set of units based on the metric system. In the so-called International System of Units (SI) each quantity is represented by a single unit or decimal multiple thereof, using the prefixes listed in Table 1.2. For example, the fundamental SI unit of length is the meter; the length of an object can be expressed in meters, kilometers (1 km = 10^3 m), or nanometers (1 nm = 10^{-9} m), but *not* in inches, angstroms, or other non-SI units.

Table 1 lists the SI base and derived units for quantities commonly referred to in general chemistry. Perhaps the least familiar of these units to the beginning chemistry student are the ones used to represent force, pressure, and energy.

The *newton* is defined as the force required to impart an acceleration of one meter per second squared to a mass of one kilogram. Recall that Newton's second law can be stated as

$$\text{force} = \text{mass} \times \text{acceleration}$$

The *pascal* is defined as the pressure exerted by a force of one newton acting on an area of one square meter (recall that pressure = force/area).

The *joule* is defined as the work done when a force of one newton ($kg \cdot m/s^2$) acts through a distance of one meter (recall that work = force × distance).

Over the past 40 years, the International System of Units has met with a decidedly mixed reception, at least in the United States. On the one hand, scientists have adopted the joule as the base unit of energy, replacing the calorie. On the other hand, volumes are seldom if ever expressed in m^3. The cubic meter is simply too large a unit; the medium-sized test tube used in the general chemistry laboratory has a volume of about 0.00001 m^3. The density of water is commonly expressed as 1.00 g/cm^3, not as

$$\frac{1.00 \text{ g}}{cm^3} \times \frac{1 \text{ kg}}{10^3 \text{ g}} \times \frac{(10^2 \text{ cm})^3}{1 \text{ m}^3} = 1.00 \times 10^3 \text{ kg/m}^3$$

(A cubic meter of water weighs about one metric ton!)

Table 1 SI Units

Base Units			Derived Units		
Quantity	Unit	Symbol	Quantity	Unit	Symbol
length	meter	m	volume	cubic meter	m^3
mass	kilogram	kg	force	newton	N
time	second	s	pressure	pascal	Pa
temperature	kelvin	K	energy	joule	J
amount of substance	mole	mol	electric charge	coulomb	C
electric current	ampere	A	electric potential	volt	V

Fundamental Constants

Acceleration of gravity (standard)	9.8066 m/s^2
Atomic mass unit (amu)	1.6605×10^{-24} g
Avogadro constant	6.0221×10^{23}/mol
Electronic charge	1.6022×10^{-19} C
Electronic mass	9.1094×10^{-28} g
Faraday constant	9.6485×10^{4} C/mol
Gas constant	0.082058 L·atm/mol·K
	8.3145 J/mol·K
Planck constant	6.6261×10^{-34} J·s
Rydberg constant	2.1799×10^{-18} J
Velocity of light	2.9979×10^{8} m/s
π	3.1416
e	2.7183
$\ln x$	$2.3026 \log_{10}x$

Vapor Pressure of Water (mm Hg)

T (°C)	vp	T (°C)	vp
0	4.58	35	42.2
5	6.54	40	55.3
10	9.21	45	71.9
12	10.52	50	92.5
14	11.99	55	118.0
16	13.63	60	149.4
17	14.53	65	187.5
18	15.48	70	233.7
19	16.48	80	355.1
20	17.54	90	525.8
21	18.65	92	567.0
22	19.83	94	610.9
23	21.07	96	657.6
24	22.38	98	707.3
25	23.76	100	760.0
26	25.21	102	815.9
27	26.74	104	875.1
28	28.35	106	937.9
29	30.04	108	1004.4
30	31.82	110	1074.6

Thermodynamic Data

	$\Delta H_f°$ (kJ/mol)	$S°$ (kJ/K·mol)	$\Delta G_f°$ (kJ/mol) at 25°C		$\Delta H_f°$ (kJ/mol)	$S°$ (kJ/K·mol)	$\Delta G_f°$ (kJ/mol) at 25°C
Ag(s)	0.0	+0.0426	0.0	$ClO_3^-(aq)$	−104.0	+0.1623	−8.0
$Ag^+(aq)$	+105.6	+0.0727	+77.1	$ClO_4^-(aq)$	−129.3	+0.1820	−8.5
AgBr(s)	−100.4	+0.1071	−96.9	Cr(s)	0.0	+0.0238	0.0
AgCl(s)	−127.1	+0.0962	−109.8	$CrO_4^{2-}(aq)$	−881.2	+0.0502	−727.8
AgI(s)	−61.8	+0.1155	−66.2	$Cr_2O_3(s)$	−1139.7	+0.0812	−1058.1
$AgNO_3(s)$	−124.4	+0.1409	−33.4	$Cr_2O_7^{2-}(aq)$	−1490.3	+0.2619	−1301.1
$Ag_2O(s)$	−31.0	+0.1213	−11.2	Cu(s)	0.0	+0.0332	0.0
Al(s)	0.0	+0.0283	0.0	$Cu^+(aq)$	+71.7	+0.0406	+50.0
$Al^{3+}(aq)$	−531.0	−0.3217	−485.0	$Cu^{2+}(aq)$	+64.8	−0.0996	+65.5
$Al_2O_3(s)$	−1675.7	+0.0509	−1582.3	CuO(s)	−157.3	+0.0426	−129.7
Ba(s)	0.0	+0.0628	0.0	$Cu_2O(s)$	−168.6	+0.0931	−146.0
$Ba^{2+}(aq)$	−537.6	+0.0096	−560.8	CuS(s)	−53.1	+0.0665	−53.6
$BaCl_2(s)$	−858.6	+0.1237	−810.4	$Cu_2S(s)$	−79.5	+0.1209	−86.2
$BaCO_3(s)$	−1216.3	+0.1121	−1137.6	$CuSO_4(s)$	−771.4	+0.1076	−661.9
BaO(s)	−553.5	+0.0704	−525.1	$F_2(g)$	0.0	+0.2027	0.0
$BaSO_4(s)$	−1473.2	+0.1322	−1362.3	$F^-(aq)$	−332.6	−0.0138	−278.8
$Br_2(l)$	0.0	+0.1522	0.0	Fe(s)	0.0	+0.0273	0.0
$Br^-(aq)$	−121.6	+0.0824	−104.0	$Fe^{2+}(aq)$	−89.1	−0.1377	−78.9
C(s)	0.0	+0.0057	0.0	$Fe^{3+}(aq)$	−48.5	−0.3159	−4.7
$CCl_4(l)$	−135.4	+0.2164	−65.3	$Fe(OH)_3(s)$	−823.0	+0.1067	−696.6
$CHCl_3(l)$	−134.5	+0.2017	−73.7	$Fe_2O_3(s)$	−824.2	+0.0874	−742.2
$CH_4(g)$	−74.8	+0.1862	−50.7	$Fe_3O_4(s)$	−1118.4	+0.1464	−1015.5
$C_2H_2(g)$	+226.7	+0.2008	+209.2	$H_2(g)$	0.0	+0.1306	0.0
$C_2H_4(g)$	+52.3	+0.2195	+68.1	$H^+(aq)$	0.0	0.0000	0.0
$C_2H_6(g)$	−84.7	+0.2295	−32.9	HBr(g)	−36.4	+0.1986	−53.4
$C_3H_8(g)$	−103.8	+0.2699	−23.5	HCl(g)	−92.3	+0.1868	−95.3
$CH_3OH(l)$	−238.7	+0.1268	−166.3	$HCO_3^-(aq)$	−692.0	+0.0912	−586.8
$C_2H_5OH(l)$	−277.7	+0.1607	−174.9	HF(g)	−271.1	+0.1737	−273.2
CO(g)	−110.5	+0.1976	−137.2	HI(g)	+26.5	+0.2065	+1.7
$CO_2(g)$	−393.5	+0.2136	−394.4	$HNO_3(l)$	−174.1	+0.1556	−80.8
$CO_3^{2-}(aq)$	−677.1	−0.0569	−527.8	$H_2O(g)$	−241.8	+0.1887	−228.6
Ca(s)	0.0	+0.0414	0.0	$H_2O(l)$	−285.8	+0.0699	−237.2
$Ca^{2+}(aq)$	−542.8	−0.0531	−553.6	$H_2O_2(l)$	−187.8	+0.1096	−120.4
$CaCl_2(s)$	−795.8	+0.1046	−748.1	$H_2PO_4^-(aq)$	−1296.3	+0.0904	−1130.3
$CaCO_3(s)$	−1206.9	+0.0929	−1128.8	$HPO_4^{2-}(aq)$	−1292.1	−0.0335	−1089.2
CaO(s)	−635.1	+0.0398	−604.0	$H_2S(g)$	−20.6	+0.2057	−33.6
$Ca(OH)_2(s)$	−986.1	+0.0834	−898.5	$H_2SO_4(l)$	−814.0	+0.1569	−690.1
$CaSO_4(s)$	−1434.1	+0.1067	−1321.8	$HSO_4^-(aq)$	−887.3	+0.1318	−755.9
Cd(s)	0.0	+0.0518	0.0	Hg(l)	0.0	+0.0760	0.0
$Cd^{2+}(aq)$	−75.9	−0.0732	−77.6	$Hg^{2+}(aq)$	+171.1	−0.0322	+164.4
$CdCl_2(s)$	−391.5	+0.1153	−344.0	HgO(s)	−90.8	+0.0703	−58.6
CdO(s)	−258.2	+0.0548	−228.4	$I_2(s)$	0.0	+0.1161	0.0
$Cl_2(g)$	0.0	+0.2230	0.0	$I^-(aq)$	−55.2	+0.1113	−51.6
$Cl^-(aq)$	−167.2	+0.0565	−131.2	K(s)	0.0	+0.0642	0.0

continued

Thermodynamic Data *continued*

	$\Delta H_f°$ (kJ/mol)	$S°$ (kJ/K·mol)	$\Delta G_f°$ (kJ/mol) at 25°C		$\Delta H_f°$ (kJ/mol)	$S°$ (kJ/K·mol)	$\Delta G_f°$ (kJ/mol) at 25°C
$K^+(aq)$	−252.4	+0.1025	−283.3	$NaF(s)$	−573.6	+0.0515	−543.5
$KBr(s)$	−393.8	+0.0959	−380.7	$NaOH(s)$	−425.6	+0.0645	−379.5
$KCl(s)$	−436.7	+0.0826	−409.1	$Ni(s)$	0.0	+0.0299	0.0
$KClO_3(s)$	−397.7	+0.1431	−296.3	$Ni^{2+}(aq)$	−54.0	−0.1289	−45.6
$KClO_4(s)$	−432.8	+0.1510	−303.2	$NiO(s)$	−239.7	+0.0380	−211.7
$KNO_3(s)$	−494.6	+0.1330	−394.9	$O_2(g)$	0.0	+0.2050	0.0
$Mg(s)$	0.0	+0.0327	0.0	$OH^-(aq)$	−230.0	−0.0108	−157.2
$Mg^{2+}(aq)$	−466.8	−0.1381	−454.8	$P_4(s)$	0.0	+0.1644	0.0
$MgCl_2(s)$	−641.3	+0.0896	−591.8	$PCl_3(g)$	−287.0	+0.3117	−267.8
$MgCO_3(s)$	−1095.8	+0.0657	−1012.1	$PCl_5(g)$	−374.9	+0.3645	−305.0
$MgO(s)$	−601.7	+0.0269	−569.4	$PO_4^{3-}(aq)$	−1277.4	−0.222	−1018.7
$Mg(OH)_2(s)$	−924.5	+0.0632	−833.6	$Pb(s)$	0.0	+0.0648	0.0
$MgSO_4(s)$	−1284.9	+0.0916	−1170.7	$Pb^{2+}(aq)$	−1.7	+0.0105	−24.4
$Mn(s)$	0.0	+0.0320	0.0	$PbBr_2(s)$	−278.7	+0.1615	−261.9
$Mn^{2+}(aq)$	−220.8	−0.0736	−228.1	$PbCl_2(s)$	−359.4	+0.1360	−314.1
$MnO(s)$	−385.2	+0.0597	−362.9	$PbO(s)$	−219.0	+0.0665	−188.9
$MnO_2(s)$	−520.0	+0.0530	−465.2	$PbO_2(s)$	−277.4	+0.0686	−217.4
$MnO_4^-(aq)$	−541.4	+0.1912	−447.2	$S(s)$	0.0	+0.0318	0.0
$N_2(g)$	0.0	+0.1915	0.0	$S^{2-}(aq)$	+33.1	−0.0146	+85.8
$NH_3(g)$	−46.1	+0.1923	−16.5	$SO_2(g)$	−296.8	+0.2481	−300.2
$NH_4^+(aq)$	−132.5	+0.1134	−79.3	$SO_3(g)$	−395.7	+0.2567	−371.1
$NH_4Cl(s)$	−314.4	+0.0946	−203.0	$SO_4^{2-}(aq)$	−909.3	+0.0201	−744.5
$NH_4NO_3(s)$	−365.6	+0.1511	−184.0	$Si(s)$	0.0	+0.0188	0.0
$N_2H_4(l)$	+50.6	+0.1212	+149.2	$SiO_2(s)$	−910.9	+0.0418	−856.7
$NO(g)$	+90.2	+0.2107	+86.6	$Sn(s)$	0.0	+0.0516	0.0
$NO_2(g)$	+33.2	+0.2400	+51.3	$Sn^{2+}(aq)$	−8.8	−0.0174	−27.2
$NO_2^-(aq)$	−104.6	+0.1230	−32.2	$SnO_2(s)$	−580.7	+0.0523	−519.6
$NO_3^-(aq)$	−205.0	+0.1464	−108.7	$Zn(s)$	0.0	+0.0416	0.0
$N_2O_4(g)$	+9.2	+0.3042	+97.9	$Zn^{2+}(aq)$	−153.9	−0.1121	−147.1
$Na(s)$	0.0	+0.0512	0.0	$ZnI_2(s)$	−208.0	+0.1611	−209.0
$Na^+(aq)$	−240.1	+0.0590	−261.9	$ZnO(s)$	−348.3	+0.0436	−318.3
$NaCl(s)$	−411.2	+0.0721	−384.2	$ZnS(s)$	−206.0	+0.0577	−201.3

Equilibrium Constants of Weak Acids (K_a)

H_3AsO_4	5.7×10^{-3}	HNO_2	6.0×10^{-4}	$N_2H_5^+$	1.0×10^{-8}
$H_2AsO_4^-$	1.8×10^{-7}	H_3PO_4	7.1×10^{-3}	$Al(H_2O)_6^{3+}$	1.2×10^{-5}
$HAsO_4^{2-}$	2.5×10^{-12}	$H_2PO_4^-$	6.2×10^{-8}	$Ag(H_2O)_2^+$	1.2×10^{-12}
$HBrO$	2.6×10^{-9}	HPO_4^{2-}	4.5×10^{-13}	$Ca(H_2O)_6^{2+}$	2.2×10^{-13}
$HCHO_2$	1.9×10^{-4}	H_2S	1.0×10^{-7}	$Cd(H_2O)_4^{2+}$	4.0×10^{-10}
$HC_2H_3O_2$	1.8×10^{-5}	H_2SO_3	1.7×10^{-2}	$Fe(H_2O)_6^{3+}$	6.7×10^{-3}
HCN	5.8×10^{-10}	HSO_3^-	6.0×10^{-8}	$Fe(H_2O)_6^{2+}$	1.7×10^{-7}
H_2CO_3	4.4×10^{-7}	HSO_4^-	1.0×10^{-2}	$Mg(H_2O)_6^{2+}$	3.7×10^{-12}
HCO_3^-	4.7×10^{-11}	H_2Se	1.5×10^{-4}	$Mn(H_2O)_6^{2+}$	2.8×10^{-11}
$HClO_2$	1.0×10^{-2}	H_2SeO_3	2.7×10^{-3}	$Ni(H_2O)_6^{2+}$	2.2×10^{-10}
$HClO$	2.8×10^{-8}	$HSeO_3^-$	5.0×10^{-8}	$Pb(H_2O)_6^{2+}$	6.7×10^{-7}
HF	6.9×10^{-4}	$CH_3NH_3^+$	2.4×10^{-11}	$Sc(H_2O)_6^{3+}$	1.1×10^{-4}
HIO	2.4×10^{-11}	NH_4^+	5.6×10^{-10}	$Zn(H_2O)_4^{2+}$	3.3×10^{-10}
HN_3	2.4×10^{-5}				

Equilibrium Constants of Weak Bases (K_b)

AsO_4^{3-}	4.0×10^{-3}	N_3^-	4.2×10^{-10}	$HSeO_3^-$	3.7×10^{-12}
$HAsO_4^{2-}$	5.6×10^{-8}	NH_3	1.8×10^{-5}	$AlOH^{2+}$	8.3×10^{-10}
$H_2AsO_4^-$	1.8×10^{-12}	N_2H_4	1.0×10^{-6}	$AgOH$	8.3×10^{-3}
BrO^-	3.8×10^{-6}	NO_2^-	1.7×10^{-11}	$CaOH^+$	4.5×10^{-2}
CH_3NH_2	4.2×10^{-4}	PO_4^{3-}	2.2×10^{-2}	$CdOH^+$	2.5×10^{-5}
CHO_2^-	5.3×10^{-11}	HPO_4^{2-}	1.6×10^{-7}	$FeOH^{2+}$	1.5×10^{-12}
$C_2H_3O_2^-$	5.6×10^{-10}	$H_2PO_4^-$	1.4×10^{-12}	$FeOH^+$	5.9×10^{-8}
CN^-	1.7×10^{-5}	HS^-	1.0×10^{-7}	$MgOH^+$	2.7×10^{-3}
CO_3^{2-}	2.1×10^{-4}	SO_3^{2-}	1.7×10^{-7}	$MnOH^+$	3.6×10^{-4}
HCO_3^-	2.3×10^{-8}	HSO_3^-	5.9×10^{-13}	$NiOH^+$	4.5×10^{-5}
ClO_2^-	1.0×10^{-12}	SO_4^{2-}	1.0×10^{-12}	$PbOH^+$	1.5×10^{-8}
ClO^-	3.6×10^{-7}	HSe^-	6.7×10^{-11}	$ScOH^{2+}$	9.1×10^{-11}
F^-	1.4×10^{-11}	SeO_3^{2-}	2.0×10^{-7}	$ZnOH^+$	3.0×10^{-5}
IO^-	4.2×10^{-4}				

Formation Constants of Complex Ions (K_f)

$AgBr_2^-$	2×10^7	$Cu(CN)_2^-$	1×10^{24}	$PdBr_4^{2-}$	6×10^{13}
$AgCl_2^-$	1.8×10^5	$Cu(NH_3)_4^{2+}$	2×10^{12}	$PtBr_4^{2-}$	6×10^{17}
$Ag(CN)_2^-$	2×10^{20}	$FeSCN^{2+}$	9.2×10^2	$PtCl_4^{2-}$	1×10^{16}
AgI_2^-	5×10^{10}	$Fe(CN)_6^{3-}$	4×10^{52}	$Pt(CN)_4^{2-}$	1×10^{41}
$Ag(NH_3)_2^+$	1.7×10^7	$Fe(CN)_6^{4-}$	4×10^{45}	$c\text{-}Pt(NH_3)_2Cl_2$	3×10^{29}
$Al(OH)_4^-$	1×10^{33}	$HgBr_4^{2-}$	1×10^{21}	$t\text{-}Pt(NH_3)_2Cl_2$	3×10^{28}
$Cd(CN)_4^{2-}$	2×10^{18}	$HgCl_4^{2-}$	1×10^{15}	$c\text{-}Pt(NH_3)_2(H_2O)_2^{2+}$	4×10^{23}
CdI_4^{2-}	4.0×10^5	$Hg(CN)_4^{2-}$	2×10^{41}	$c\text{-}Pt(NH_3)_2I_2$	2×10^{33}
$Cd(NH_3)_4^{2+}$	2.8×10^7	HgI_4^{2-}	1×10^{30}	$t\text{-}Pt(NH_3)_2I_2$	5×10^{32}
$Cd(OH)_4^{2-}$	1.2×10^9	$Hg(NH_3)_4^{2+}$	2×10^{19}	$c\text{-}Pt(NH_3)_2(OH)_2$	1×10^{39}
$Co(NH_3)_6^{2+}$	1×10^5	$Ni(CN)_4^{2-}$	1×10^{30}	$Pt(NH_3)_4^{2+}$	2×10^{35}
$Co(NH_3)_6^{3+}$	1×10^{23}	$Ni(NH_3)_6^{2+}$	9×10^8	$Zn(CN)_4^{2-}$	6×10^{16}
$Co(NH_3)_5Cl^{2+}$	2×10^{28}	PbI_4^{2-}	1.7×10^4	$Zn(NH_3)_4^{2+}$	3.6×10^8
$Co(NH_3)_5NO_2^{2+}$	1×10^{24}	$Pb(OH)_3^-$	8×10^{13}	$Zn(OH)_4^{2-}$	3×10^{14}

Solubility Product Constants (K_{sp})

$AgBr$	5×10^{-13}	$Co(OH)_2$	2×10^{-16}	$NiCO_3$	1.4×10^{-7}
$AgC_2H_3O_2$	1.9×10^{-3}	$Co_3(PO_4)_2$	1×10^{-35}	$Ni_3(PO_4)_2$	1×10^{-32}
Ag_2CO_3	8×10^{-12}	$CuCl$	1.7×10^{-7}	NiS	1×10^{-21}
$AgCl$	1.8×10^{-10}	$CuBr$	6.3×10^{-9}	$PbBr_2$	6.6×10^{-6}
Ag_2CrO_4	1×10^{-12}	CuI	1.2×10^{-12}	$PbCO_3$	1×10^{-13}
AgI	1×10^{-16}	$Cu_3(PO_4)_2$	1×10^{-37}	$PbCl_2$	1.7×10^{-5}
Ag_3PO_4	1×10^{-16}	CuS	1×10^{-36}	$PbCrO_4$	2×10^{-14}
Ag_2S	1×10^{-49}	Cu_2S	1×10^{-48}	PbF_2	7.1×10^{-7}
$AgSCN$	1.0×10^{-12}	$FeCO_3$	3.1×10^{-11}	PbI_2	8.4×10^{-9}
AlF_3	1×10^{-18}	$Fe(OH)_2$	5×10^{-17}	$Pb(OH)_2$	1×10^{-20}
$Al(OH)_3$	2×10^{-31}	$Fe(OH)_3$	3×10^{-39}	PbS	1×10^{-28}
$AlPO_4$	1×10^{-20}	FeS	2×10^{-19}	$Pb(SCN)_2$	2.1×10^{-5}
$BaCO_3$	2.6×10^{-9}	GaF_3	2×10^{-16}	$PbSO_4$	1.8×10^{-8}
$BaCrO_4$	1.2×10^{-10}	$Ga(OH)_3$	1×10^{-35}	$Sc(OH)_3$	2×10^{-31}
BaF_2	1.8×10^{-7}	$GaPO_4$	1×10^{-21}	SnS	3×10^{-28}
$BaSO_4$	1.1×10^{-10}	Hg_2Br_2	6×10^{-23}	$SrCO_3$	5.6×10^{-10}
Bi_2S_3	1×10^{-99}	Hg_2Cl_2	1×10^{-18}	$SrCrO_4$	3.6×10^{-5}
$CaCO_3$	4.9×10^{-9}	Hg_2I_2	5×10^{-29}	SrF_2	4.3×10^{-9}
CaF_2	1.5×10^{-10}	HgS	1×10^{-52}	$SrSO_4$	3.4×10^{-7}
$Ca(OH)_2$	4.0×10^{-6}	Li_2CO_3	8.2×10^{-4}	$TlCl$	1.9×10^{-4}
$Ca_3(PO_4)_2$	2.5×10^{-33}	$MgCO_3$	6.8×10^{-6}	$TlBr$	3.7×10^{-6}
$CaSO_4$	7.1×10^{-5}	MgF_2	7×10^{-11}	TlI	5.6×10^{-8}
$CdCO_3$	6×10^{-12}	$Mg(OH)_2$	6×10^{-12}	$Tl(OH)_3$	2×10^{-44}
$Cd(OH)_2$	5×10^{-15}	$Mg_3(PO_4)_2$	1×10^{-24}	Tl_2S	1×10^{-20}
$Cd_3(PO_4)_2$	1×10^{-33}	$Mn(OH)_2$	2×10^{-13}	$ZnCO_3$	1.1×10^{-10}
CdS	1×10^{-29}	MnS	5×10^{-14}	$Zn(OH)_2$	4×10^{-17}

Properties of the Elements

Element	At. No.	mp (°C)	bp (°C)	E.N.	Ion. Ener. (kJ/mol)	At. Rad. (nm)	Ion. Rad. (nm)
H	1	−259	−253	2.2	1312	0.037	(−1)0.208
He	2	−272	−269		2372	0.05	
Li	3	186	1326	1.0	520	0.152	(+1)0.060
Be	4	1283	2970	1.6	900	0.111	(+2)0.031
B	5	2300	2550	2.0	801	0.088	
C	6	3570	subl.	2.5	1086	0.077	
N	7	−210	−196	3.0	1402	0.070	
O	8	−218	−183	3.5	1314	0.066	(−2)0.140
F	9	−220	−188	4.0	1681	0.064	(−1)0.136
Ne	10	−249	−246		2081	0.070	
Na	11	98	889	0.9	496	0.186	(+1)0.095
Mg	12	650	1120	1.3	738	0.160	(+2)0.065
Al	13	660	2327	1.6	578	0.143	(+3)0.050
Si	14	1414	2355	1.9	786	0.117	
P	15	44	280	2.2	1012	0.110	
S	16	119	444	2.6	1000	0.104	(−2)0.184
Cl	17	−101	−34	3.2	1251	0.099	(−1)0.181
Ar	18	−189	−186		1520	0.094	
K	19	64	774	0.8	419	0.231	(+1)0.133
Ca	20	845	1420	1.0	590	0.197	(+2)0.099
Sc	21	1541	2831	1.4	631	0.160	(+3)0.081
Ti	22	1660	3287	1.5	658	0.146	
V	23	1890	3380	1.6	650	0.131	
Cr	24	1857	2672	1.6	653	0.125	(+3)0.064
Mn	25	1244	1962	1.5	717	0.129	(+2)0.080
Fe	26	1535	2750	1.8	759	0.126	(+2)0.075
Co	27	1495	2870	1.9	758	0.125	(+2)0.072
Ni	28	1453	2732	1.9	737	0.124	(+2)0.069
Cu	29	1083	2567	1.9	746	0.128	(+1)0.096
Zn	30	420	907	1.6	906	0.133	(+2)0.074
Ga	31	30	2403	1.6	579	0.122	(+3)0.062
Ge	32	937	2830	2.0	762	0.122	
As	33	814	subl.	2.2	944	0.121	
Se	34	217	685	2.5	941	0.117	(−2)0.198
Br	35	−7	59	3.0	1140	0.114	(−1)0.195
Kr	36	−157	−152	3.3	1351	0.109	
Rb	37	39	688	0.8	403	0.244	(+1)0.148
Sr	38	770	1380	0.9	550	0.215	(+2)0.113
Y	39	1509	2930	1.2	616	0.180	(+3)0.093
Zr	40	1852	3580	1.4	660	0.157	
Nb	41	2468	5127	1.6	664	0.143	
Mo	42	2610	5560	1.8	685	0.136	
Tc	43	2200	4700	1.9	702	0.136	
Ru	44	2430	3700	2.2	711	0.133	
Rh	45	1966	3700	2.2	720	0.134	

continued

continued

Element	At. No.	mp (°C)	bp (°C)	E.N.	Ion. Ener. (kJ/mol)	At. Rad. (nm)	Ion. Rad. (nm)
Pd	46	1550	3170	2.2	805	0.138	
Ag	47	961	2210	1.9	731	0.144	(+1)0.126
Cd	48	321	767	1.7	868	0.149	(+2)0.097
In	49	157	2000	1.7	558	0.162	(+3)0.081
Sn	50	232	2270	1.9	709	0.140	
Sb	51	631	1380	2.0	832	0.141	
Te	52	450	990	2.1	869	0.137	(−2)0.221
I	53	114	184	2.7	1009	0.133	(−1)0.216
Xe	54	−112	−107	3.0	1170	0.130	
Cs	55	28	690	0.8	376	0.262	(+1)0.169
Ba	56	725	1640	0.9	503	0.217	(+2)0.135
La	57	920	3469	1.1	538	0.187	(+3)0.115
Ce	58	795	3468	1.1	528	0.182	(+3)0.101
Pr	59	935	3127	1.1	523	0.182	(+3)0.100
Nd	60	1024	3027	1.1	530	0.182	(+3)0.099
Pm	61	1027	2727	1.1	536	0.181	
Sm	62	1072	1900	1.1	543	0.180	
Eu	63	826	1439	1.1	547	0.204	(+2)0.097
Gd	64	1312	3000	1.1	592	0.179	(+3)0.096
Tb	65	1356	2800	1.1	564	0.177	(+3)0.095
Dy	66	1407	2600	1.1	572	0.177	(+3)0.094
Ho	67	1461	2600	1.1	581	0.176	(+3)0.093
Er	68	1497	2900	1.1	589	0.175	(+3)0.092
Tm	69	1356	2800	1.1	597	0.174	(+3)0.091
Yb	70	824	1427	1.1	603	0.193	(+3)0.089
Lu	71	1652	3327	1.1	524	0.174	(+3)0.089
Hf	72	2225	5200	1.3	654	0.157	
Ta	73	2980	5425	1.5	761	0.143	
W	74	3410	5930	1.7	770	0.137	
Re	75	3180	5885	1.9	760	0.137	
Os	76	2727	4100	2.2	840	0.134	
Ir	77	2448	4500	2.2	880	0.135	
Pt	78	1769	4530	2.2	870	0.138	
Au	79	1063	2966	2.4	890	0.144	(+1)0.137
Hg	80	−39	357	1.9	1007	0.155	(+2)0.110
Tl	81	304	1457	1.8	589	0.171	(+3)0.095
Pb	82	328	1750	1.9	716	0.175	
Bi	83	271	1560	1.9	703	0.146	
Po	84	254	962	2.0	812	0.165	
At	85	302	334	2.2			
Rn	86	−71	−62		1037	0.14	
Fr	87	27	677	0.7			
Ra	88	700	1140	0.9	509	0.220	
Ac	89	1050	3200	1.1	490	0.20	
Th	90	1750	4790	1.3	590	0.180	
Pa	91	1600	4200	1.4	570		
U	92	1132	3818	1.4	590	0.14	

Exponents and Logarithms

The mathematics you will use in general chemistry is relatively simple. You will, however, be expected to

- make calculations involving exponential numbers, such as 6.022×10^{23} or 1.6×10^{-10}.
- work with logarithms or inverse logarithms, particularly in problems involving pH:

$$pH = -\log_{10} (\text{conc. } H^+)$$

This appendix reviews each of these topics.

Exponential Notation

Chemists deal frequently with very large or very small numbers. In one gram of the element carbon there are

$$50,140,000,000,000,000,000,000 \text{ atoms of carbon}$$

At the opposite extreme, the mass of a single atom is

$$0.00000000000000000000001994 \text{ g}$$

Numbers such as these are very awkward to work with. For example, neither of the numbers just written could be entered directly on a calculator. Operations involving very large or very small numbers can be simplified by using *exponential (scientific)* notation. To express a number in exponential notation, write it in the form

$$C \times 10^n$$

where C is a number between 1 and 10 (for example, 1, 2.62, 5.8) and n is a positive or negative integer such as 1, -1, -3. To find n, count the number of places that the decimal point must be moved to give the coefficient, C. If the decimal point must be moved to the *left*, n is a *positive* integer; if it must be moved to the *right*, n is a *negative* integer. Thus

$$26.23 = 2.623 \times 10^1 \quad \text{(decimal point moved 1 place to left)}$$
$$5609 = 5.609 \times 10^3 \quad \text{(decimal point moved 3 places to left)}$$
$$0.0918 = 9.18 \times 10^{-2} \quad \text{(decimal point moved 2 places to right)}$$

Multiplication and Division

A major advantage of exponential notation is that it simplifies the processes of multiplication and division. To *multiply, add exponents:*

$$10^1 \times 10^2 = 10^{1+2} = 10^3 \qquad 10^6 \times 10^{-4} = 10^{6+(-4)} = 10^2$$

To *divide, subtract* exponents:

$$10^3/10^2 = 10^{3-2} = 10^1 \qquad 10^{-3}/10^6 = 10^{-3-6} = 10^{-9}$$

It often happens that multiplication or division yields an answer that is not in standard exponential notation. For example,

$$(5.0 \times 10^4) \times (6.0 \times 10^3) = (5.0 \times 6.0) \times 10^4 \times 10^3 = 30 \times 10^7$$

The product is not in standard exponential notation because the coefficient, 30, does not lie between 1 and 10. To correct this situation, rewrite the coefficient as 3.0×10^1 and then add exponents:

$$30 \times 10^7 = (3.0 \times 10^1) \times 10^7 = 3.0 \times 10^8$$

In another case,

$$0.526 \times 10^3 = (5.26 \times 10^{-1}) \times 10^3 = 5.26 \times 10^2$$

Exponential Notation on the Calculator

On all scientific calculators it is possible to enter numbers in exponential notation. The method used depends on the brand of calculator. Most often, it involves using a key labeled (EXP), (EE), or (EEX). Check your instruction manual for the procedure to be followed. To make sure you understand it, try entering the following numbers:

$$2.4 \times 10^6 \qquad 3.16 \times 10^{-8} \qquad 6.2 \times 10^{-16}$$

Multiplication, division, addition, and subtraction can be carried out directly on your calculator. Try the following exercises:

(a) $(6.0 \times 10^2) \times (4.2 \times 10^{-4}) = ?$

(b) $\dfrac{6.0 \times 10^2}{4.2 \times 10^{-4}} = ?$

(c) $3.6 \times 10^{-4} + 4 \times 10^{-5} = ?$

(d) $3.6 \times 10^{-4} - 4 \times 10^{-5} = ?$

The answers, expressed in exponential notation and following the rules of significant figures, are (a) 2.5×10^{-1} (b) 1.4×10^6 (c) 4.0×10^{-4} (d) 3.2×10^{-4}.

Exponential (or ordinary) numbers are readily raised to powers using a calculator. To square a number, use the x^2 key. For example,

$$(3 \times 10^4)^2 = 9 \times 10^8$$

For other powers, use the y^x key. To cube a number, y, enter that number, press the y^x key, enter 3, and read the answer. That way, you should find that

$$(2 \times 10^{-5})^3 = 8 \times 10^{-15}$$

Roots of exponentials are obtained in a similar way. To extract a square root, use the $\sqrt{x}$ key. For other roots, use the y^x key. To obtain a cube root, enter the number, press the y^x key, enter 1/3 (0.333333333) and read the answer:

$$(2.0 \times 10^{-5})^{1/3} = 2.7 \times 10^{-2}$$

(Many calculators have a $\sqrt[x]{y}$ key, in which case, enter $x = 3$ rather than 1/3.)

Try the following operations on your calculator.

$(3.12)^5 = ?$	answer: 2.96×10^2
$(3.12 \times 10^{-2})^5 = ?$	answer: 2.96×10^{-8}
$(3.12)^{1/5} = ?$	answer: 1.26
$(3.12 \times 10^{-2})^{1/5} = ?$	answer: 0.500

Logarithms and Inverse Logarithms

The logarithm of a number n to the base m is defined as the power to which m must be raised to give the number n. Thus

$$\text{if } m^x = n, \text{ then } \log_m n = x$$

In other words, a *logarithm* is an *exponent*, which may be a whole number (e.g., 1, −1) but more often is not (e.g., 1.500, −0.301).

In general chemistry, you will encounter two kinds of logarithms.

1. *Common logarithms*, where the base is 10, ordinarily denoted as $\log_{10}$. If $10^x = n$, then $\log_{10} n = x$. Examples include the following:

$$\log_{10} 100 = 2.00 \qquad (10^2 = 100)$$
$$\log_{10} 1 = 0.000 \qquad (10^0 = 1)$$
$$\log_{10} 0.001 = -3.000 \qquad (10^{-3} = 0.001)$$

2. *Natural logarithms,* where the base is the quantity $e = 2.718. \ldots$ Many of the equations used in general chemistry are expressed most simply in terms of natural logarithms, denoted as ln. If $e^x = n$, then $\ln n = x$.

$$\ln 100 = 4.605 \qquad (\text{i.e.,} \ 100 = e^{4.605})$$
$$\ln 1 = 0 \qquad (\text{i.e.,} \ 1 = e^{0})$$
$$\ln 0.001 = -6.908 \qquad (\text{i.e.,} \ 0.001 = e^{-6.908})$$

Notice that

(a) $\log_{10} 1 = \ln 1 = 0$. The logarithm of 1 to any base is zero, because any number raised to the zero power is 1. That is

$$10^0 = e^0 = 2^0 = \cdots = n^0 = 1$$

(b) Numbers larger than 1 have a positive logarithm; numbers smaller than 1 have a negative logarithm. For example,

$$\log_{10} 100 = 2 \qquad \ln 100 = 4.605$$
$$\log_{10} 10 = 1 \qquad \ln 10 = 2.303$$
$$\log_{10} 0.1 = -1 \qquad \ln 0.1 = -2.303$$
$$\log_{10} 0.01 = -2 \qquad \ln 0.01 = -4.605$$

(c) The common and natural logarithms of a number are related by the expression

$$\ln n = 2.303 \log_{10} n$$

That is, the natural logarithm of a number is 2.303 . . . times its base-10 logarithm.

(d) Finally, if $x \leq 0$, both $\log x$ and $\ln x$ are undefined.

An *inverse logarithm* (antilogarithm) is the number corresponding to a given logarithm. In general,

$$\text{if } m^x = n, \text{ then inverse } \log_m x = n$$

For example,

$$10^2 = 100 \qquad \text{inverse } \log_{10} 2 = 100$$
$$10^{-3} = 0.001 \qquad \text{inverse } \log_{10} (-3) = 0.001$$

In other words, the numbers whose base-10 logarithms are 2 and -3 are 100 and 0.001, respectively. The same reasoning applies to natural logarithms.

$$e^{4.606} = 100 \qquad \text{inverse } \ln 4.606 = 100$$
$$e^{-6.908} = 0.001 \qquad \text{inverse } \ln -6.908 = 0.001$$

The numbers whose natural logarithms are 4.606 and -6.908 are 100 and 0.001, respectively. Notice that if the inverse logarithm is positive, the corresponding number is larger than 1; if it is negative, the number is smaller than 1.

Finding Logarithms and Inverse Logarithms on a Calculator

To obtain a base-10 logarithm using a calculator, all you need to do is enter the number and press the ⌊LOG⌋ key. This way you should find that

$$\log_{10} 2.00 = 0.301 \ldots$$
$$\log_{10} 0.526 = -0.279 \ldots$$

Similarly, to find a natural logarithm, you enter the number and press the ⌊LN X⌋ key.

$$\ln 2.00 = 0.693 \ldots$$
$$\ln 0.526 = -0.642 \ldots$$

To find the logarithm of an exponential number, you enter the number in exponential form and take the logarithm in the usual way. This way you should find that

$$\log_{10} 2.00 \times 10^3 = 3.301 \ldots \qquad \ln 2.00 \times 10^3 = 7.601 \ldots$$
$$\log_{10} 5.3 \times 10^{-12} = -11.28 \ldots \qquad \ln 5.3 \times 10^{-12} = -25.96 \ldots$$

The base-10 logarithm of an exponential number can be found in a somewhat different way by applying the relation

$$\log_{10} (C \times 10^n) = n + \log_{10} C$$

Thus

$$\log_{10} 2.00 \times 10^3 = 3 + \log_{10} 2.00 = 3 + 0.301 = 3.301$$
$$\log_{10} 5.3 \times 10^{-12} = -12 + \log_{10} 5.3 = -12 + 0.72 = -11.28$$

The method used to find inverse logarithms depends on the type of calculator. On certain calculators, you enter the number and then press, in succession, the (INV) and either (LOG) or (LN X) keys. With other calculators, you press the (10^x) or (e^x) key. Either way, you should find that

$$\text{inverse } \log_{10} 1.632 = 42.8 \ldots \qquad \text{inverse } \ln 1.632 = 5.11 \ldots$$
$$\text{inverse } \log_{10} - 8.82 = 1.5 \times 10^{-9} \qquad \text{inverse } \ln -8.82 = 1.5 \times 10^{-4}$$

Significant Figures in Logarithms and Inverse Logarithms

For base-10 logarithms, the rules governing significant figures are as follows:

1. *In taking the logarithm of a number, retain after the decimal point in the log as many digits as there are significant figures in the number.* (This part of the logarithm is often referred to as the *mantissa*; digits that precede the decimal point comprise the *characteristic* of the logarithm.) To illustrate this rule, consider the following:

$$\log_{10} 2.00 = 0.301 \qquad \log_{10} (2.00 \times 10^3) = 3.301$$
$$\log_{10} 2.0 = 0.30 \qquad \log_{10} (2.0 \times 10^1) = 1.30$$
$$\log_{10} 2 = 0.3 \qquad \log_{10} (2 \times 10^{-3}) = 0.3 - 3 = -2.7$$

2. *In taking the inverse logarithm of a number, retain as many significant figures as there are after the decimal point in the number.* Thus

$$\text{inverse } \log_{10} 0.301 = 2.00 \qquad \text{inverse } \log_{10} 3.301 = 2.00 \times 10^3$$
$$\text{inverse } \log_{10} 0.30 = 2.0 \qquad \text{inverse } \log_{10} 1.30 = 2.0 \times 10^1$$
$$\text{inverse } \log_{10} 0.3 = 2 \qquad \text{inverse } \log_{10} (-2.7) = 2 \times 10^{-3}$$

These rules take into account the fact that, as mentioned earlier,

$$\log_{10} (C \times 10^n) = n + \log_{10} C$$

The digits that appear before (to the left of) the decimal point specify the value of n, that is, the power of 10 involved in the expression. In that sense, they are not experimentally significant. In contrast, the digits that appear after (to the right of) the decimal point specify the value of the logarithm of C; the number of such digits reflects the uncertainty in C. Thus

$$\log_{10} 209 = 2.320 \qquad \text{(3 sig. fig.)}$$
$$\log_{10} 209.0 = 2.3201 \qquad \text{(4 sig. fig.)}$$

The rules for significant figures involving natural logarithms and inverse logarithms are somewhat more complex than those for base-10 logs. However, for simplicity we will assume that the rules listed above apply here as well. Thus

$$\ln 209 = 5.342 \qquad \text{(3 sig. fig.)}$$
$$\ln 209.0 = 5.3423 \qquad \text{(4 sig. fig.)}$$

Operations Involving Logarithms

Because logarithms are exponents, the rules governing the use of exponents apply as well. The rules that follow are valid for all types of logarithms, regardless of the base. We illustrate the rules with natural logarithms; that is where you are most likely to use them in working with this text.

Multiplication: $\ln(xy) = \ln x + \ln y$

Example: $\ln(2.50 \times 1.25) = \ln 2.50 + \ln 1.25 = 0.916 + 0.223 = 1.139$

Division: $\ln(x/y) = \ln x - \ln y$

Example: $\ln(2.50/1.25) = 0.916 - 0.223 = 0.693$

Raising to a Power: $\ln(x^n) = n \ln x$

Example: $\ln(2.00)^4 = 4 \ln 2.00 = 4(0.693) = 2.772$

Extracting a Root: $\ln(x^{1/n}) = \dfrac{1}{n} \ln x$

Example: $\ln(2.00)^{1/3} = \dfrac{\ln 2.00}{3} = \dfrac{0.693}{3} = 0.231$

Taking a Reciprocal: $\ln(1/x) = -\ln x$

Example: $\ln(1/2.00) = -\ln 2.00 = -0.693$

Molecular Orbitals

In Chapter 7, we used valence bond theory to explain bonding in molecules. It accounts, at least qualitatively, for the stability of the covalent bond in terms of the overlap of atomic orbitals. By invoking hybridization, valence bond theory can account for the molecular geometries predicted by electron-pair repulsion. Where Lewis structures are inadequate, as in SO_2, the concept of resonance allows us to explain the observed properties.

A major weakness of valence bond theory has been its inability to predict the magnetic properties of molecules. We mentioned this problem in Chapter 7 with regard to the O_2 molecule, which is paramagnetic, even though it has an even number (12) of valence electrons. The octet rule, or valence bond theory, would predict that all the electrons in O_2 should be paired, which would make it diamagnetic.

This discrepancy between experiment and theory (and many others) can be explained in terms of an alternative model of covalent bonding, the molecular orbital (MO) approach. Molecular orbital theory treats bonds in terms of orbitals characteristic of the molecule as a whole. To apply this approach, we carry out three basic operations.

1. The atomic orbitals of atoms are combined to give a new set of molecular orbitals characteristic of the molecule as a whole. *The number of molecular orbitals formed is equal to the number of atomic orbitals combined.* When two H atoms combine to form H_2, two s orbitals, one from each atom, yield two molecular orbitals. In another case, six p orbitals, three from each atom, give a total of six molecular orbitals.

2. The molecular orbitals are arranged in order of increasing energy. The relative energies of these orbitals are ordinarily deduced from experiment. Spectra and magnetic properties of molecules are used.

3. The *valence electrons* in a molecule are distributed among the available molecular orbitals. The process followed is much like that used with electrons in atoms. In particular, we find the following:

(a) *Each molecular orbital can hold a maximum of two electrons.* When an orbital is filled the two electrons have opposed spins, in accordance with the Pauli principle (Chapter 7).

(b) *Electrons go into the lowest energy molecular orbital available.* A higher orbital starts to fill only when each orbital below it has its quota of two electrons.

(c) *Hund's rule is obeyed.* When two orbitals of equal energy are available to two electrons, one electron goes into each, giving two half-filled orbitals.

Diatomic Molecules of the Elements

To illustrate molecular orbital theory, we apply it to the diatomic molecules of the elements in the first two periods of the periodic table.

Hydrogen and Helium (Combination of 1s Orbitals)

Molecular orbital (MO) theory predicts that two 1s orbitals will combine to give two molecular orbitals. One of these has an energy lower than that of the atomic orbitals from which it is formed (Figure 1). Placing electrons in this orbital gives a species that is more stable than the isolated atoms. For that reason the lower molecular orbital in Figure 1 is called a *bonding orbital*. The other molecular orbital has a higher energy than the corresponding atomic orbitals. Electrons entering it are in an unstable, higher energy state. It is referred to as an *antibonding orbital*.

The electron density in these molecular orbitals is shown at the right of Figure 1. Notice that the bonding orbital has a higher density between the nuclei. This accounts for its stability. In the antibonding orbital, the chance of finding the electron between the nuclei is very small. The electron density is concentrated at the far ends of the "molecule." This means that the nuclei are less shielded from each other than they are in the isolated atoms.

The electron density in both molecular orbitals is symmetrical about the axis between the two nuclei. This means that both of these are sigma orbitals. In MO

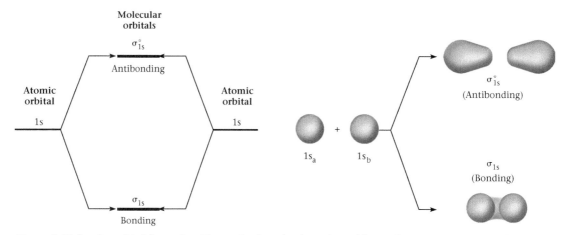

Figure 1 Molecular orbital formation. Two molecular orbitals are formed by combining two 1s atomic orbitals.

notation, the 1s bonding orbital is designated as σ_{1s}. The antibonding orbital is given the symbol σ_{1s}^*. An asterisk designates an antibonding orbital.

In the H_2 molecule, there are two 1s electrons. They fill the σ_{1s} orbital, giving a single bond. In the He_2 molecule, there would be four electrons, two from each atom. These would fill the bonding and antibonding orbitals. As a result, the number of bonds (the *bond order*) in He_2 is zero. The general relation is

$$\text{no. of bonds} = \text{bond order} = \frac{B - AB}{2}$$

where B is the number of electrons in bonding orbitals and AB is the number of electrons in antibonding orbitals. In H_2, $B = 2$ and $AB = 0$, so we have one bond. In He_2, $B = AB = 2$, so the number of bonds is zero. The He_2 molecule should not and does not exist.

Second Period Elements (Combination of 2s and 2p Orbitals)

Among the diatomic molecules of the second period elements are three familiar ones, N_2, O_2, and F_2. The molecules Li_2, B_2, and C_2 are less common but have been observed and studied in the gas phase. In contrast, the molecules Be_2 and Ne_2 are either highly unstable or nonexistent. Let us see what molecular orbital theory predicts about the structure and stability of these molecules. We start by considering how the atomic orbitals containing the valence electrons (2s and 2p) are used to form molecular orbitals.

Combining two 2s atomic orbitals, one from each atom, gives two molecular orbitals. These are very similar to the ones shown above. They are designated as σ_{2s} (sigma, bonding, 2s) and σ_{2s}^* (sigma, antibonding, 2s).

Consider now what happens to the 2p orbitals. In an isolated atom, there are three such orbitals, oriented at right angles to each other. We call these atomic orbitals, p_x, p_y, and p_z (top of Figure 2). When two p_x atomic orbitals, one from each atom, overlap head-to-head, they form two sigma orbitals. These are a bonding orbital, σ_{2p}, and an antibonding orbital, σ_{2p}^*. The situation is quite different when the p_z orbitals overlap. Because they are oriented parallel to one another, they overlap side-to-side (Figure 2c). The two molecular orbitals formed in this case are pi orbitals; one is a bonding orbital, π_{2p}, the other a nonbonding orbital, π_{2p}^*. In an entirely similar way, the p_y orbitals of the two atoms interact to form another pair of pi molecular orbitals, π_{2p} and π_{2p}^* (these orbitals are not shown in Figure 2).

The relative energies of the molecular orbitals available for occupancy by the valence electrons of the second period elements are shown in Figure 3. This order applies at least through N_2.*

---

*It appears that beyond N_2 the σ_{2p} orbital lies below the two π_{2p} orbitals. This does not affect the filling order, because in O_2 all of these orbitals are filled.

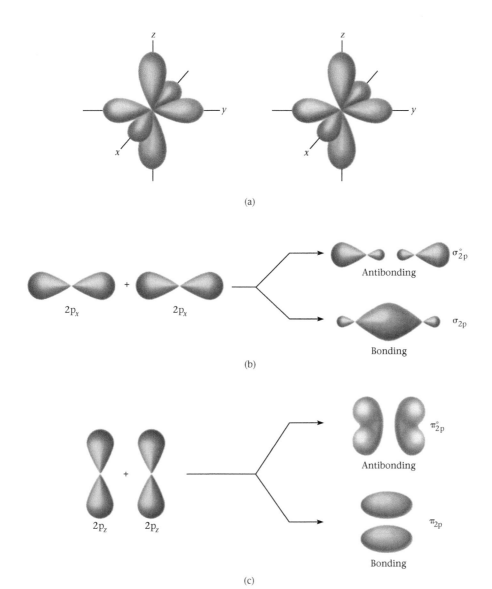

Figure 2 When p orbitals from two different atoms (a) overlap, there are two quite different possibilities. If they overlap head-to-head (b), two sigma molecular orbitals are produced. If, on the other hand, they overlap side-to-side (c), two pi molecular orbitals result.

To obtain the MO structure of the diatomic molecules of the elements in the second period, we fill the available molecular orbitals in order of increasing energy. The results are shown in Table 1. Note the agreement between MO theory and the properties of these molecules. In particular, the number of unpaired electrons predicted agrees with experiment. There is also a general correlation between the predicted bond order:

$$\text{bond order} = \frac{B - AB}{2}$$

and the bond energy. We would expect a double bond (C_2 or O_2) to be stronger than a single bond (Li_2, B_2, and F_2). A triple bond, as in N_2, should be still stronger.

A major triumph of MO theory is its ability to explain the properties of O_2. It explains how the molecule can have a double bond and, at the same time, have two unpaired electrons.

For example, using MO theory we can predict the electronic structure, bond order, and number of unpaired electrons in the peroxide ion.

Recalling that oxygen is in Group **16** of the periodic table, the number of valence electrons is 14 ($2 \times 6 + 2$), and the orbital diagram is

	σ_{2s}	σ^*_{2s}	π_{2p}	π_{2p}	σ_{2p}	π^*_{2p}	π^*_{2p}	σ^*_{2p}
O_2^{2-}	($\uparrow\downarrow$)	($\uparrow\downarrow$)	($\uparrow\downarrow$)	($\uparrow\downarrow$)	($\uparrow\downarrow$)	($\uparrow\downarrow$)	($\uparrow\downarrow$)	( )

Figure 3 Relative energies, so far as filling order is concerned, for the molecular orbitals formed by combining 2s and 2p atomic orbitals.

Table 1 **Predicted and Observed Properties of Diatomic Molecules of Second Period Elements**

	Occupancy of Orbitals							
	σ_{2s}	σ_{2s}^*	π_{2p}	π_{2p}	σ_{2p}	π_{2p}^*	π_{2p}^*	σ_{2p}^*
Li$_2$	(↑↓)	( )	( )	( )	( )	( )	( )	( )
Be$_2$	(↑↓)	(↑↓)	( )	( )	( )	( )	( )	( )
B$_2$	(↑↓)	(↑↓)	(↑)	(↑)	( )	( )	( )	( )
C$_2$	(↑↓)	(↑↓)	(↑↓)	(↑↓)	( )	( )	( )	( )
N$_2$	(↑↓)	(↑↓)	(↑↓)	(↑↓)	(↑↓)	( )	( )	( )
O$_2$	(↑↓)	(↑↓)	(↑↓)	(↑↓)	(↑↓)	(↑)	(↑)	( )
F$_2$	(↑↓)	(↑↓)	(↑↓)	(↑↓)	(↑↓)	(↑↓)	(↑↓)	( )
Ne$_2$	(↑↓)	(↑↓)	(↑↓)	(↑↓)	(↑↓)	(↑↓)	(↑↓)	(↑↓)

	Predicted Properties		Observed Properties	
	Number of Unpaired e⁻	Bond Order	Number of Unpaired e⁻	Bond Energy (kJ/mol)
Li$_2$	0	1	0	105
Be$_2$	0	0	0	Unstable
B$_2$	2	1	2	289
C$_2$	0	2	0	628
N$_2$	0	3	0	941
O$_2$	2	2	2	494
F$_2$	0	1	0	153
Ne$_2$	0	0	0	Nonexistent

Thus there are eight electrons in bonding orbitals, six in antibonding orbitals. The bond order is $\frac{8-6}{2} = 1$. There are no unpaired electrons. These conclusions are in agreement with the Lewis structure of the peroxide ion: $(:\ddot{O}—\ddot{O}:)^{2-}$.

Polyatomic Molecules; Delocalized π Electrons

The bonding in molecules containing more than two atoms can also be described in terms of molecular orbitals. We will not attempt to do this; the energy level structure is considerably more complex than the one we considered. However, one point is worth mentioning. In polyatomic species, *a pi molecular orbital can be spread over the entire molecule* rather than being concentrated between two atoms.

This principle can be applied to species such as the nitrate ion, whose Lewis structure is

$$:\ddot{O}—N—\ddot{O}:$$
$$\|$$
$$:\ddot{O}:$$

Valence bond theory (Chapter 7) explains the fact that the three N—O bonds are identical by invoking the idea of resonance, with three contributing structures. MO theory, on the other hand, considers that the skeleton of the nitrate ion is established by the three sigma bonds while the electron pair in the pi orbital is *delocalized*, shared by all of the atoms in the molecule. According to MO theory, a similar interpretation applies with all of the "resonance hybrids" described in Chapter 7, including SO_2, SO_3, and CO_3^{2-}.

Another species in which delocalized pi orbitals play an important role is benzene, C_6H_6. There are 30 valence electrons in the molecule, 24 of which are required to form the sigma bond framework:

The remaining six electrons are located in three π orbitals, which according to MO theory extend over the entire molecule. Figure 4 is one way of representing this structure; more commonly it is shown simply as

Figure 4 In benzene, three electron pairs are not localized on particular carbon atoms. Instead, they are spread out over two lobes of the shape shown, one above the plane of the benzene ring and the other below it.

Metals; Band Theory

In Chapter 9, we considered a simple picture of metallic bonding, the electron–sea model. The molecular orbital approach leads to a refinement of this model known as *band theory.* Here, a crystal of a metal is considered to be one huge molecule. Valence electrons of the metal are fed into delocalized molecular orbitals, formed in the usual way from atomic orbitals. A huge number of these MOs are grouped into an energy band; the energy separation between adjacent MOs is extremely small.

For purposes of illustration, consider a lithium crystal weighing one gram, which contains roughly 10^{23} atoms. Each Li atom has a half-filled 2s atomic orbital (elect. conf. Li = $1s^2 2s^1$). When these atomic orbitals combine, they form an equal number, 10^{23}, of molecular orbitals. These orbitals are spread over an energy band covering about 100 kJ/mol. It follows that the spacing between adjacent MOs is of the order of

$$\frac{100 \text{ kJ/mol}}{10^{23}} = 10^{-21} \text{ kJ/mol}$$

Because each lithium atom has one valence electron and each molecular orbital can hold two electrons, it follows that the lower half of the valence band (shown in color in Figure 5) is filled with electrons. The upper half of the band is empty. Electrons near the top of the filled MOs can readily jump to empty MOs only an infinitesimal distance above them. This is what happens when an electrical field is applied to the crystal; the movement of electrons through delocalized MOs accounts for the electrical conductivity of lithium metal.

The situation in beryllium metal is more complex. We might expect all of the 2s molecular orbitals to be filled because beryllium has the electron configuration $1s^2 2s^2$. However, in a crystal of beryllium, the 2p MO band overlaps the 2s (Figure 5). This means that, once again, there are vacant MOs that differ only infinitesimally in energy from filled MOs below them. This is indeed the basic requirement for electron conductivity; it is characteristic of all metals, including lithium and beryllium.

Materials in which there is a substantial difference in energy between occupied and vacant MOs are poor electron conductors. Diamond, where the gap between the filled valence band and the empty conduction band is 500 kJ/mol, is an insulator. Silicon and germanium, where the gaps are 100 kJ/mol and 60 kJ/mol, respectively, are semiconductors.

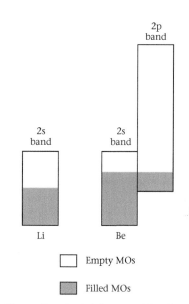

Figure 5 In both lithium and beryllium metal, there are vacant MOs only slightly higher in energy than filled MOs. This is the basic requirement for metallic conductivity.

Answers to Even-Numbered and Challenge Questions and Problems

Chapter 1

2. (a) element
   (b) mixture
   (c) compound
   (d) mixture
4. (a) heterogeneous mixture
   (b) heterogeneous mixture
   (c) solution
6. (a) chromatography
   (b) distillation or chromatography
8. (a) Cu   (b) C   (c) Br   (d) Al
10. (a) manganese   (b) gold   (c) lead   (d) silver
12. (a) yardstick   (b) buret/pipet   (c) thermometer
14. 177°C; 4.50×10^2 K
16. 35°C; 95°F
18. (a) 3   (b) 4   (c) 4   (d) 1   (e) 5
20. (a) 17.25 cm   (b) 169 lb
    (c) 5.00×10^2 °C   (d) 198 oz
22. (a) 4.0206×10^3 mL
    (b) 1.006 g
    (c) 1.001×10^2 °C
24. b and c
26. 5 is exact; 4000 is ambiguous; 17 has two and 18.5 has three
28. (a) 10.6   (b) 80.1   (c) 0.40   (d) 0.0923
    (e) 1.4×10^{-25}
30. 0.04 m^3
32. (a) less than   (b) greater than   (c) equal
34. (a) 215.6 mL   (b) 13.16 in^3 (c) 0.2278 qt
36. (a) 1.15078 mi   (b) 1852 m   (c) 25 mph
38. (a) USD 3.09   (b) USD 43   (c) 2.1×10^2 mi
40. 1.85×10^{-3} g/mL
42. 3.97×10^{14} L
44. 393.8 mg
46. 1.07 g/mL
48. 1.6 g/mL
50. 508 m
52. 3.88×10^4 g; 85.4 lb
54. supersaturated; 5 g
56. (a) homogeneous; 4.7 g more   (b) 9.1 g
58. 479 rems; 1.6×10^3 mammograms
60. 320.0 g
62. 6.5×10^{-4} in
64. 8.2 g/mL
66. (a) I   (b) I   (c) I   (d) E   (e) E
68. See pages 16–19
70. 3
72. (a) A   (b) $\approx$23°C   (c) no; (see answer to (a))

73. (a)

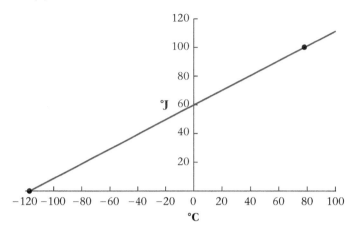

   (b) 0.51
   (c) 60
   (d) °J = 0.512(°C) + 60
74. 320°F = 160°C
75. 1.2 km^2
76. 21.9 cm
77. 8.1×10^{-3} g Pb
78. 6.87 g/cm^3

Chapter 2

2. See page 24
4. (a) none   (b) conservation of mass
   (c) none
6. Rutherford; see pages 25–26
8. (a) 86 p^+ (b) 134 n
10. (a) $^{54}_{26}$Fe; $^{56}_{26}$Fe
    (b) They differ in the number of neutrons. Fe-54 has 28, whereas Fe-56 has 30.
12. (a) 95   (b) 241   (c) $^{241}_{95}$Am
14. (a) A is arsenic: 33p^+, 42n, 33e^-
    (b) L is vanadium: 23p^+, 28n, 23e^-
    (c) Z is xenon: 54p^+, 77n, 54e^-
16. (a) isobars: Cr-54 and Fe-54; Fe-58 and Ni-58
    isotopes: Fe-54 and Fe-58
    (b) number of protons
    (c) Cr-54 and Ni-58
18. (a) < (c) < (b) < (d)
20. O-16
22. b
24. 20.2 amu
26. 64.94 amu; $^{65}_{29}$Cu
28. lightest isotope 78.6%; heaviest isotope 11.4%
30. (a) two: HCl-35 and HCl-37   (b) 36 and 38

(c)

32. 6×10^{13} atoms
34. (a) 7.00×10^{20} atoms (b) 1.631×10^{-22} g
36. (a) 8.400×10^{3} neutrons (b) 3.535×10^{11} neutrons
38. 8.96×10^{23} atoms
40. (a) sulfur (b) scandium (c) selenium
    (d) silicon (e) strontium
42. (a) nonmetal (b) transition metal (c) nonmetal
    (d) metalloid (e) metal
44. (a) 5 (b) 3 (c) 2
46. (a) 1 (b) 7 (c) 4
48. (a) CH_3COOH, $C_2H_4O_2$ (b) CH_3Cl
50. (a) $128p^+$, $128e^-$ (b) $48p^+$, $50e^-$
    (c) $18p^+$, $18e^-$ (d) $16p^+$, $18e^-$
52. Al-27; metal
    Te-128; 5;
    nonmetal; 18; 5; 80
    C-14; 2
54. (a) and (d)
56. (a) SiS_2 (b) PH_3 (c) P_2O_5
    (d) RnF_4 (e) NCl_3
58. (a) tellurium dichloride (b) tetraiodine nanoxide
    (c) xenon trioxide (d) tetrasulfur tetranitride
60. (a) BaI_2, Ba_3N_2 (b) FeO, Fe_2O_3
62. (a) K_2HPO_4 (b) Mg_3N_2 (c) $PbBr_4$
    (d) $ScCl_3$ (e) $Ba(C_2H_3O_2)_2$
64. (a) lead(II) acetate
    (b) aluminum hydroxide
    (c) strontium hydrogen carbonate
    (d) copper(II) phosphate
    (e) rubidium oxide
66. (a) $HNO_3(aq)$ (b) K_2SO_4 (c) $Fe(ClO_4)_3$
    (d) $Al(IO_3)_3$ (e) $H_2SO_3(aq)$
68. $HBrO_2(aq)$; cobalt(III) periodate; $PtCl_4$;
    nitrous acid; N_2O_3
70. (a) Ge (b) W (c) Sr (d) Bi
72. (a) C_2H_7N (b) $C_2H_5NH_2$
74. (a) always true (b) usually true (c) never true
76. 9.64×10^3 cm^3
78. a
80. only #4.
82.

84. $^{234}_{90}$Th
86. $^{126}_{52}$Te^{2-}
88. c
89. (a) Ratio of C in ethane to C in ethylene/g H is $3:2$.
    (b) CH_3, CH_2; C_2H_6, C_2H_4
90. 3.71 g/cm^3; lots of space between atoms
91. 2.3443×10^{-23} g
92. (a) 2.5×10^{24} molecules (b) 2.3×10^{-20}
    (c) $\approx 2.9 \times 10^2$ molecules
93. 8.2 L

Chapter 3

2. (a) 5.93×10^{17} g
   (b) 7.65×10^{-22} mol
4. (a) $18 \, e^-$ (b) 1.084×10^{25} (c) 2.411×10^{23}
6. (a) 1.66 mol (b) 7.79×10^{25}
8. (a) 190.2 (b) 84.01 (c) 396.63
10. (a) 0.3226
    (b) 0.001824
    (c) 0.01110
12. (a) 108 g (b) 537 g (c) 451 g
14.

	Number of Grams	Number of Moles	Number of Molecules	Number of O Atoms
(a)	0.1364 g	7.100×10^{-4}	4.276×10^{20}	2.993×10^{21}
(b)	239.8 g	1.248	7.515×10^{23}	5.261×10^{24}
(c)	13.8 g	7.17×10^{-2}	4.32×10^{22}	3.02×10^{23}
(d)	0.00253 g	1.32×10^{-5}	7.93×10^{18}	5.55×10^{19}

16. 0.70 M
18. (a) $[K^+] = 0.757 \, M$; $[O^{2-}] = 0.379 \, M$
    (b) $[Na^+] = [HCO_3^-] = 0.425 \, M$
    (c) $[Sc^{3+}] = 0.0684 \, M$; $[IO_2^-] = 0.205 \, M$
    (d) $[Mg^{2+}] = 0.407 \, M$; $[PO_4^{3-}] = 0.272 \, M$
20. (a) Dissolve 2.50×10^2 g of $Ni(NO_3)_2$ in 2.00 L of solution
    (b) Dissolve 184 g of $CuCl_2$ in 2.00 L of solution
    (c) Dissolve 241 g of $C_6H_8O_6$ in 2.00 L of solution
22. (a) 303.9 mL (b) 1.074×10^2 mL
24. 0.295 M
26. 67.49% C; 4.602% H; 9.841% N; 5.619% O; 12.45% Cl
28. 5.48 g
30. 54.0%
32. 38.40% C; 1.50% H; 52.3% Cl; 7.80% O
34. molar mass = 1.50×10^2; R = vanadium
36. Ni_2S_3; nickel(III) sulfide
38. (a) $C_8H_{20}Pb$ (b) $C_6H_8O_7$ (c) $N_2H_6Cl_2Pt$
40. $C_8H_8O_3$
42. $C_7H_5O_3SN$
44. Simplest formula: CH_4N Molecular formula: $C_2H_8N_2$
46. (a) 47.24% (b) 8.366 g
48. (a) $C_6H_{12}O_6(s) + 6O_2(g) \longrightarrow 6CO_2(g) + 6H_2O$
    (b) $XeF_4(g) + 2H_2O(l) \longrightarrow Xe(g) + O_2(g) + 4HF(g)$
    (c) $4NaCl(s) + 2H_2O(g) + 2SO_2(g) + O_2(g) \longrightarrow$
    $ 2Na_2SO_4(s) + 4HCl(g)$
50. (a) $2Sc(s) + 3S(s) \longrightarrow Sc_2S_3(s)$
    (b) $2Sc(s) + 3Cl_2(g) \longrightarrow 2ScCl_3(s)$

(c) $2Sc(s) + N_2(g) \longrightarrow 2ScN(s)$
(d) $4Sc(s) + 3O_2(g) \longrightarrow 2Sc_2O_3(s)$

52. (a) $2H_2S(g) + SO_2(g) \longrightarrow 3S(s) + 2H_2O(g)$
(b) $2CH_4(g) + 2NH_3(g) + 3O_2(g) \longrightarrow 2HCN(g) + 6H_2O(g)$
(c) $Fe_2O_3(s) + 3H_2(g) \longrightarrow 2\,Fe(s) + 3H_2O(g)$
(d) $UO_2(s) + 4HF(g) \longrightarrow UF_4(s) + 2H_2O(g)$
(e) $C_2H_5OH(l) + 3O_2(g) \longrightarrow 2CO_2(g) + 3H_2O(l)$

54. (a) 3.131 (b) 3.264
(c) 1.222 (d) 0.1816

56. (a) 882.2 g (b) 16.7 g
(c) 2.368 g (d) 38.60 g

58. (a) $SiO_2(s) + 2C(s) \longrightarrow Si(s) + 2CO(g)$
(b) 0.7120 mol
(c) 197.0

60. (a) 98.4 g (b) 76 L

62. (a) 294 mL (b) 5.8×10^2 g

64. (a) 0.3166 mol (b) 68.60 g/mol

66. (a) $Cl_2(g) + 3F_2(g) \longrightarrow 2ClF_3(g)$
(b) F_2 (c) 2.45 mol (d) 0.52 mol

68. 9.4×10^2 g

70. (a) $4NH_3(g) + 5O_2(g) \longrightarrow 4NO(g) + 6H_2O(g)$
(b) 7.502 g NO (c) 3.34 g (d) 82.9%

72. 2.76×10^3 g; 526 mL

74. (a)

(b) Mg (c) Mg
(d) acid (e) 4
(f) 122 mL; 11 mL

76. 54.3 mL

78. $2AB_3 + 3C \longrightarrow 3CB_2 + 2A$

80.

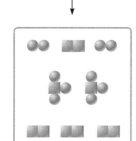

82. (a) Both boxes A and B have the same number of atoms as As and O_2.
(b) There are more discrete particles in Box A.
(c) Both boxes have the same mass.

84. 2.509×10^{26} atoms

86. (a) EQ (b) LT (c) LT
(d) $4PH_3(g) + 8O_2(g) \longrightarrow P_4O_{10}(s) + 6H_2O(g)$; GT
(e) GT

88. 893 g/mol

89. 6.01×10^{23} atoms

90. 3.657 g CaO; 2.972 g Ca_3N_2

91. 34.7%

92. (a) V_2O_3; V_2O_5 (b) 2.271 g

93. 28%

94. 13.2 g $NaClO_3$; 15.6 g $NaHCO_3$; 15.3 g Na_2CO_3; 55.9 g NaCl

95. 3.23×10^{24}

Chapter 4

2. (a) $BaCl_2$; soluble (b) $Mg(OH)_2$; insoluble
(c) $Cr_2(CO_3)_3$; insoluble (d) K_3PO_4; soluble

4. (a) NaOH (b) Na_2CO_3 (c) Na_3PO_4

6. (a) $Fe^{3+}(aq) + 3OH^-(aq) \longrightarrow Fe(OH)_3(s)$
(b) $Ba^{2+}(aq) + SO_4^{2-}(aq) \longrightarrow BaSO_4(s)$; $Mg^{2+}(aq) + 2OH^-(aq) \longrightarrow Mg(OH)_2(s)$

8. (a) $Ag^+(aq) + Cl^-(aq) \longrightarrow AgCl(s)$
(b) $Co^{2+}(aq) + 2OH^-(aq) \longrightarrow Co(OH)_2(s)$
(c) no reaction
(d) $Cu^{2+}(aq) + CO_3^{2-}(aq) \longrightarrow CuCO_3(s)$
(e) $Ba^{2+}(aq) + SO_4^{2-}(aq) \longrightarrow BaSO_4(s)$

10. (a) $3Ba^{2+}(aq) + 2PO_4^{3-}(aq) \longrightarrow Ba_3(PO_4)_2(s)$
(b) $Zn^{2+}(aq) + 2OH^-(aq) \longrightarrow Zn(OH)_2(s)$
(c) no reaction
(d) $Co^{3+}(aq) + PO_4^{3-}(aq) \longrightarrow CoPO_4(s)$

12. (a) 0.04976 M (b) 1.43 M (c) 0.04195 M

14. (a) $2Al^{3+}(aq) + 3CO_3^{2-}(aq) \longrightarrow Al_2(CO_3)_3(s)$
(b) 0.108 M (c) 0.379 g

16. (a) $Al^{3+}(aq) + 3OH^-(aq) \longrightarrow Al(OH)_3(s)$
(b) 0.553 g
(c) 0.0430 M

18. (a) weak acid (b) strong acid
(c) weak base (d) weak base

20. (a) HClO (b) $HCHO_2$ (c) $HC_2H_3O_2$
(d) H^+ (e) H_2SO_3

22. (a) C_7H_9N (b) OH^-
(c) C_8H_6NH (d) NH_3

24. (a) $HC_2H_3O_2(aq) + OH^-(aq) \longrightarrow H_2O + C_2H_3O_2^-(aq)$
(b) $(C_2H_5)_2NH(aq) + H^+(aq) \longrightarrow (C_2H_5)_2NH_2^+(aq)$
(c) $HF(aq) + OH^-(aq) \longrightarrow H_2O + F^-(aq)$

26. (a) $H^+(aq) + C_5H_5N(aq) \longrightarrow C_5H_5NH^+(aq)$
(b) $H^+(aq) + OH^-(aq) \longrightarrow H_2O$
(c) correct
(d) $NH_3(aq) + H^+(aq) \longrightarrow NH_4^+(aq)$
(e) correct

28. 15.5 mL

30. (a) 0.2395 M
(b) 9.320 M
(c) 16.40 M

32. 1.80×10^2 g/mol

34. 5.630%; Yes
36. 74.3%
38. one
40. (a) C = −4; H = +1
    (b) O = −2; C = +4
    (c) O = −2; I = +7
    (d) H = +1; N = −2
42. (a) H = +1, I = +5, O = −2
    (b) Na = +1, Mn = +7, O = −2
    (c) Sn = +4, O = −2
    (d) F = −1, O = −2, N = +3
    (e) Na = +1, O = $-\frac{1}{2}$
44. (a) oxidation   (b) reduction
    (c) reduction   (d) oxidation
46. (a) reduction   (b) reduction
    (c) oxidation   (d) oxidation
48. (a) As oxidized; Mn reduced
        MnO_4^-—oxidizing agent; As_2O_3—reducing agent
    (b) N oxidized; C reduced
        CO_3^{2-}—oxidizing agent; N_2H_4— reducing agent
50. 30.32%
52. 27.5 g
54. 0.0941 M
56. yes
58. 0.379%
60. (a) 4 HCl : 1 HNO_3
    (b) 7.9 mL HNO_3; 42 mL HCl
62. 1.03 g
64. (a) SA/WB   (b) WA/SB   (c) PPT
    (d) PPT   (e) NR
66. (a) (1)   (b) (3)   (c) (2)
68. (a) weak   (b) non
    (c) strong   (d) weak
70. (a) F   (b) T   (c) F   (d) F   (e) T
71. 0.794 g CaC_2O_4; yes
72. 0.29 L
73. 1.09 g Cu; 3.09 g Ag
74. 0.0980 M Fe^{2+}; 0.0364 M Fe^{3+}
75. oxalic acid
76. 5.27 g
77. 65.97 g $MgCl_2$; 165.0 g $NaNO_3$; 69.06 g $BaCl_2$
78. (a) 1.11 g
    (b) $[OH^-]$ = 0; $[Ba^{2+}]$ = 0.125 M; $[Al^{3+}]$ = 0.0415 M;
        $[NO_3^-]$ = 0.375 M

Chapter 5
2. V = 2.5 × 10³ L; n_{He} = 2.14 × 10⁴ mol;
   T = 298 K
4. (a) 3.20 × 10² mm Hg   0.422 atm   6.20 psi
   (b) 1.50 × 10³ mm Hg   1.98 atm   201 kPa
   (c) 599 mm Hg   11.6 psi   79.8 kPa
   (d) 1.600 atm   23.5 psi   162.1 kPa
6. 323°C
8. (a) 2.00 atm   (b) 3.21 atm
10. 31.5 psi (gauge pressure); 46.2 psi (actual pressure)
12. 1.02 × 10³ mm Hg
14. 0.76 mol
16. 93.3 g; 93.3 mL
18. 1217 g

20. (a) 0.0470 mol; 2.07 g
    (b) −33°C; 3.72 g
    (c) 36.9 L; 125.3 g
    (d) 11.2 atm; 1.072 mol
22. (a) 1.19 g/L
    (b) 2.10 g/L
    (c) 1.90 g/L
24. 0.0146 g/L in Mars, 1.80 g/L on earth
26. (a) 98.8 g/mol
    (b) $COCl_2$
28. (a) 28.6 g/mol
    (b) exhaled air = 1.12 g/L; ordinary air = 1.13 g/L
30. element = H; compound = PH_3 − phosphine
32. (a) $2NF_3(g) + 3H_2O(g) \longrightarrow NO(g) + NO_2(g) + 6HF(g)$
    (b) 1.74 L
34. (a) 3.72 L
    (b) 2.64 L
36. 3.2 × 10³ g
38. 2.83 L
40. (a) $2C_9H_{18}O_6(s) + 21O_2(g) \longrightarrow 18H_2O(g) + 18CO_2(g)$
    (b) 13.8 atm
42. 3.04 atm; HCl
44. CH_4: 1.06 atm
    C_2H_6: 0.11 atm
    C_3H_8: 0.030 atm
46. 134 mL
48. 2.19 atm
50. (a) $2NH_4NO_3(s) \longrightarrow 2N_2(g) + O_2(g) + 4H_2O(g)$
    (b) 2.38 atm
    (c) P_{N_2} = 0.681 atm; P_{O_2} = 0.341 atm; P_{H_2O} = 1.37 atm
52. (a) 729 mm Hg   (b) 4.52 × 10⁻⁴
    (c) 2.00 × 10⁻²   (d) 231 mm Hg
    (e) 976 mm Hg
54. (a) speed of effusion: Xe < CH_2F_2 < F_2 < CH_4
    (b) time: CH_4 < F_2 < CH_2F_2 < Xe
56. 3.74 times faster
58. (a) yes   (b) 100.1 g/mol
60. 15.8 s
62. (a) 291 m/s   (b) 145.3 m/s
64. (a) more
    (b) less
66. (a) 84 g/L
    (b) 65.5 g/L
    (c) ≈ 340 atm
68. (a) $2C_8H_{18}(l) + 25O_2(g) \longrightarrow 16CO_2(g) + 18H_2O(l)$
    (b) 1.5 × 10⁴ L
70. (a) 1.000
    (b) 0.5535
    (c) 1.000
    (d) 1.000
72. H_2SO_4
74. C = 32.0%;   H = 6.72%;   N = 18.7%;   O = 42.6%;
    $C_2H_5NO_2$
76. (a) 0.00387 mol   (b) 0.032 g
78. (a) Both gases have the same number of moles.
    (b) CO_2
    (c) CO_2
    (d) Both gases have the same average translational energy.
    (e) Both gases have the same partial pressure.

80. (a) same   (b) N_2 (A)   (c) same
    (d) NH_3 (B)   (e) P stays the same   (f) 1/2
82. (a) $tank_A = tank_B$
    (b) $tank_A = 2(tank_B)$
    (c) $tank_A = tank_B$
84. (a) A   (b) B   (c) B
86. (a) C   (b) 2.5 atm   (c) 4.0 atm
    (d) 4.0 atm (same)

(e) 4.0 atm

88. Ideal gas pressures: at 40°C = 25.0 mm Hg; at 70°C = 27.3 mm Hg; at 100°C = 29.7 mm Hg
    Vapor pressures: at 40°C = 55.3 mm Hg; at 70°C = 233.7 mm Hg; at 100°C = 760 mm Hg
    Vapor pressure implies the presence of some liquid together with the vapor. As T increases, more liquid becomes gas, increasing n and thus P.
89. 74.3% O_2; 17.2% N_2; 8.43% CO_2
90. 3.0 ft from NH_3 end
91. (a) 1.05 atm   (b) 1.27 atm   (c) 0.624 M
92. 78.7%
93. 6.62 m
94. 0.897 atm

Chapter 6

2. (a) 5.83×10^{14} s^{-1} (b) 3.86×10^{-19} J/photon
   (c) 233 kJ/mol
4. (a) 232 nm   (b) ultraviolet
   (c) 8.55×10^{-19} J/photon
6. (a) 127 nm   (b) ultraviolet
8. 48 kJ
10. (a) 434 nm   (b) visible   (c) yes
12.

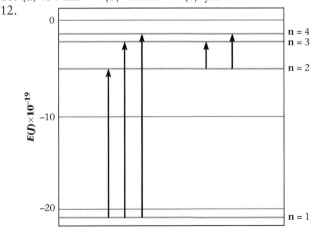

14. (a) 2.624×10^3 nm   (b) infrared
16. 7455 nm
18. (a) $m_\ell = -2, -1, 0, +1, +2$
    (b) $m_\ell = 0$
    (c) $\ell = 0; m_\ell = 0$ $\ell = 1, m_\ell = -1, 0, +1$
    $\ell = 2; m_\ell = -2, -1, 0, +1, +2$
    $\ell = 3, m_\ell = -3, -2, -1, 0, +1, +2, +3$
    $\ell = 4, m_\ell = -4, -3, -2, -1, 0, +1, +2, +3, +4$
20. (a) 3s   (b) 4d   (c) 4f   (d) 2s
22. (a) 3p   (b) 5s   (c) 6f
24. (a) 9   (b) 3   (c) 7   (d) 5
26. (a) 2   (b) 2   (c) 10
28. (b) No 1p sublevel   (d) m_ℓ cannot exceed 1
    (e) $m_\ell = 0$
30. (a) $1s^2 2s^2 2p^6 3s^2 3p^4$
    (b) $1s^2 2s^2 2p^6 3s^2 3p^6 4s^2 3d^1$
    (c) $1s^2 2s^2 2p^6 3s^2 3p^2$
    (d) $1s^2 2s^2 2p^6 3s^2 3p^6 4s^2 3d^{10} 4p^6 5s^2$
    (e) $1s^2 2s^2 2p^6 3s^2 3p^6 4s^2 3d^{10} 4p^6 5s^2 4d^{10} 5p^3$
32. (a) $[_{10}Ne]3s^2$ (b) $[_{54}Xe]6s^2 4f^{14} 5d^6$
    (c) $[_{18}Ar]4s^2 3d^{10} 4p^2$ (d) $[_{18}Ar]4s^2 3d^3$
    (e) $[_{54}Xe]6s^2 4f^{14} 5d^{10} 6p^5$
34. (a) Eu   (b) Zr   (c) P   (d) Ne
36. (a) 6/12   (b) 12/25   (c) 18/42
38. (a) impossible   (b) ground   (c) impossible
    (d) excited   (e) excited
40. 1s   2s   2p   3s   3p   4s   3d
    (a) (↑↓) (↑↓) (↑↓)(↑↓)(↑↓) (↑)
    (b) (↑↓) (↑↓) (↑↓)(↑)(↑)
    (c) (↑↓) (↑↓) (↑↓)(↑↓)(↑↓) (↑↓) (↑↓)(↑↓)(↑↓) (↑↓) (↑↓)(↑↓)(↑)(↑)(↑)
    (d) (↑↓) (↑↓) (↑↓)(↑↓)(↑↓) (↑↓) (↑↓)(↑↓)(↑)
42. (a) Li   (b) Cl   (c) Na
44. (a) Sn, Sb, Te   (b) K, Rb, Cs, Fr   (c) Ge, As, Sb, Te
    (d) none
46. (a) 1   (b) 0   (c) 3
48. (a) Ca   (b) K, Ga   (c) none   (d) none
50. (a) F: $1s^2 2s^2 2p^5$ F$^-$: $1s^2 2s^2 2p^6$
    (b) Sc: $1s^2 2s^2 2p^6 3s^2 3p^6 4s^2 3d^1$ Sc^{3+}: $1s^2 2s^2 2p^6 3s^2 3p^6$
    (c) Mn^{2+}: $1s^2 2s^2 2p^6 3s^2 3p^6 3d^5$
    Mn^{5+}: $1s^2 2s^2 2p^6 3s^2 3p^6 3d^2$
    (d) O$^-$: $1s^2 2s^2 2p^5$ O^{2-}: $1s^2 2s^2 2p^6$
52. (a) 0   (b) 0   (c) 0   (d) 0
54. (a) Cl < S < Mg   (b) Mg < S < Cl
    (c) Mg < S < Cl
56. (a) K   (b) Cl   (c) Cl
58. (a) P   (b) V^{4+} (c) K$^+$ (d) Co^{3+}
60. (a) Kr < K < Rb < Cs   (b) Ar < Si < Al < Cs
62. (a) 3.558×10^7 nm   (b) microwave
    (c) 3.363×10^{-3} kJ/mol
64. (a) Si   (b) Na   (c) Y   (d) At   (e) O
66. (a) (2) and (4)   (b) (1) and (3)   (c) (1)
    (d) (2)   (e) (1)
68. (a) spin   (b) less energy   (c) 5   (d) yes
70. (a) See pages 130–133.
    (b) See page 125, Figure 6.2.
    (c) See page 138, Figure 6.12.
72. (a) true   (b) false; inversely proportional to n^2
    (c) false; as soon as 4p is full

74. (a) True
   (b) True
   (c) False; absorbed
76. +2954 kJ
77. $\Delta E = 2.180 \times 10^{-18}\left(\dfrac{1}{4} - \dfrac{1}{n^2}\right)$

$\lambda = \dfrac{hc}{\Delta E} = \dfrac{(6.626 \times 10^{-34})(2.998 \times 10^{8})(1 \times 10^{9})}{2.180 \times 10^{-18}\left(\dfrac{1}{4} - \dfrac{1}{n^2}\right)}$

$= \dfrac{91.12}{\left(\dfrac{1}{4} - \dfrac{1}{n^2}\right)} = \dfrac{364.5\ n^2}{n^2 - 4}$

78. $1s^4\ 1p^4$
79. (a) s sublevel: $m_\ell = 0 = 3e^-$;
   p sublevel: $m_\ell = -1, 0, 1 = 9e^-$;
   d sublevel: $m_\ell = -2, -1, 0, 1, 2 = 15e^-$
   (b) $n = 3$, $\ell = 0, 1, 2$; total electrons = 27
   (c) $1s^3 2s^3 2p^2$ $1s^3 2s^3 2p^9 3s^2$
80. (a) 3.42×10^{-19} J
   (b) 581 nm

Chapter 7

2. (a) H—N̈—H
   (b) :F̈—K̇—F̈:
   (c) (:N≡O:)$^+$
   (d) (:Ö—B̈r—Ö:)$^-$

4. (a)
   (b)
   (c) [:S̈—C≡N:]$^-$
   (d)

6. (a) (:C≡C:)$^{2-}$
   (b) :F̈—N=Ö:
   (c)
   (d)

8. (H—O≡C:)$^+$

10. (a)
   (b)
   (c)

12.

14.

16. (a) Br_2 (b) CH_4 (c) CO   (d) SiF_4

18. (a)
   (b)
   (c)
   (d)

20. (a) :C̈l—Be—C̈l: (b) :Ö—S̈e—Ö:
   (c)
   (d)

22. (a)
   (b)
   (c) (Ö=C=N̈)$^-$ ⟷ (:Ö—C≡N:)$^-$ ⟷ (:O≡C—N̈:)$^-$

24. (a)
   (b)   and   and
   (c) No. Different skeleton.

26.

28. (a) 0   (b) 0   (c) 0
30. Structure I
32. (a) bent   (b) bent
   (c) trigonal pyramid   (d) linear
34. (a) linear   (b) bent
   (c) tetrahedron   (d) bent
36. (a) square pyramid   (b) square planar
   (c) octahedral   (d) trigonal bipyramid

38. (a) 120°   (b) 120°
    (c) 120° around N; 109.5° around O (left)
    (d) 109.5° around C and N
40. (a)

$$H-\overset{\overset{\displaystyle H}{|}}{\underset{\underset{\displaystyle H}{|}}{C}}-\overset{\overset{\displaystyle |}{|}}{\underset{\underset{\displaystyle :O:}{||}}{C}}-\ddot{O}-\ddot{O}-\overset{\overset{\displaystyle |}{||}}{\underset{\underset{\displaystyle :O:}{|}}{C}}-\overset{\overset{\displaystyle H}{|}}{\underset{\underset{\displaystyle H}{|}}{C}}-H$$

   (b) around CH_3—109.5°; around CO—120°C; around O—109.5°
42. 1 = 120°; 2 = 109.5°; 3 = 120°
44. a, b, c are dipoles
46. a, b, d
48. First structure is polar; dipoles do not cancel.
50. (a) sp^2 (b) sp^3 (c) sp^3 (d) sp
52. (a) sp (b) sp^2 (c) sp^3 (d) sp^2
54. (a) sp^3d^2 (b) sp^3d^2 (c) sp^3d^2 (d) sp^3d
56. (a) $6e^-$ pairs; sp^3d^2 (b) $6e^-$ pairs; sp^3d^2
    (c) $5e^-$ pairs; sp^3d
58. C_1 and $C_2 = sp^2$, C_3 and $N = sp$
60. (a) sp^3 (b) sp^2 (c) sp (d) sp^2
62. (a) sp^3 (b) sp^3 (c) sp^3
64. 6 σ, 3 π
66. (a) 4 σ (b) 3 σ, 1 π (c) 2 σ, 2 π (d) 4 σ, 1 π
68. (a)

$$\left[:\ddot{O}-\overset{\overset{\displaystyle :\ddot{O}:}{|}}{\underset{\underset{\displaystyle :\ddot{O}:}{|}}{S}}-\ddot{O}-\overset{\overset{\displaystyle :\ddot{O}:}{|}}{\underset{\underset{\displaystyle :\ddot{O}:}{|}}{S}}-\ddot{O}: \right]^{2-}$$

   (b) Each S atom has a formal charge of 2.
   (c)

$$\left[\ddot{O}=\overset{\overset{\displaystyle :O:}{||}}{\underset{\underset{\displaystyle :O:}{||}}{S}}-\ddot{O}-\overset{\overset{\displaystyle :O:}{||}}{\underset{\underset{\displaystyle :O:}{||}}{S}}=\ddot{O} \right]^{2-}$$

   (d) Each S atom has a formal charge of −1.
70. AX_2E_2 2   2   bent sp^3 polar
    AX_3 3   0   trigonal planar sp^2 nonpolar
    AX_4E_2 4   2   square planar sp^3d^2 nonpolar
    AX_5 5   0   trigonal bipyramid sp^3d nonpolar
72. (a) C   (b) N
74. PH_3 and H_2S—unshared pairs on central atom
76. $SnCl_2$ and SO_2; unshared pairs occupy more space
78. x = 3; T-shaped; polar; sp^3d; 90°, 180°; 3 σ bonds
79.

$$\underset{\underset{\displaystyle H}{}}{\overset{\overset{\displaystyle H}{}}{N}}-\underset{\underset{\displaystyle H}{}}{\overset{\overset{\displaystyle \ddot{}}{}}{N}}\underset{\displaystyle H}{} \quad \text{bent; 109.5°; polar}$$

80. 6; octahedron; sp^3d^2
81. (a)

$$\left[:\ddot{O}-\overset{\overset{\displaystyle :\ddot{O}:}{|}}{\underset{\underset{\displaystyle :\ddot{O}:}{|}}{S}}-\ddot{O}: \right]^{2-} \qquad \left[\ddot{O}=\overset{\overset{\displaystyle :\ddot{O}:}{}}{\underset{\underset{\displaystyle :\ddot{O}:}{|}}{S}}=\ddot{O} \right]^{2-}$$

   (b) tetrahedral for both   (c) sp^3; sp^3
   (d) 1st structure: O = −1; S = 2
       2nd structure: S = 0; O = −1; O = −1, O = 0

82. (a)

$$:\ddot{Cl}-\overset{\overset{\displaystyle :\ddot{O}:}{|}}{\underset{\underset{\displaystyle :\ddot{Cl}:}{|}}{P}}-\ddot{Cl}: \quad \text{Formal charges: P = +1; Cl = 0; O = −1}$$

(b)

$$:\ddot{Cl}-\overset{\overset{\displaystyle :O:}{||}}{\underset{\underset{\displaystyle :\ddot{Cl}:}{|}}{P}}-\ddot{Cl}: \quad \text{Formal charges: P = 0; Cl = 0; O = 0}$$

Chapter 8

2. 2.06°C
4. 0.140 J/g · °C
6. 3.7×10^2 J
8. (a) no   (b) −669 J   (c) 24.2°C
10. 1.3×10^2 mL
12. (a) 18.6 mL   (b) 34.0°C
14. (a) 81.3 kJ   (b) 86.2 kJ   (c) −167.5 kJ
    (d) -3.48×10^3 kJ/mol
16. (a) −218 kJ   (b) 77.6 kJ
    (c) 1.40×10^2 kJ (d) 5.67 kJ/°C
18. 27.0°C
20. 2.00 mL
22. (a) $KClO_3(s) \longrightarrow KCl(s) + \frac{3}{2}O_2(g)$ $\Delta H° = -44.7$ kJ
    (b) yes
    (c)

    (d) −1.09 kJ   (e) 41.1 g
24. (a) 13.4 kJ   (b) 0.965 kJ
26. (a) $2C_3H_5(NO_3)_3(l) \longrightarrow 3N_2(g) + 6CO_2(g) +$
    $5H_2O(g) + \frac{1}{2}O_2(g);$ $\Delta H = -2.84 \times 10^3$ kJ
    (b) −911 kJ
28. (a) $C_{57}H_{104}O_6(s) + 83O_2(g) \longrightarrow 57CO_2(g) +$
    $52H_2O(l)$ $\Delta H° = -3.022 \times 10^4$ kJ
    (b) 1.78×10^3 kJ
    (c) 0.804 lb
30. benzene
32. −48.9 kJ
34. (a) $2Mg(s) + SiO_2(s) + 2C(s) + 2Cl_2(g) \longrightarrow 2CO(g) +$
    $2MgCl_2(s) + Si(s)$
    (b) −592.7 kJ
    (c) yes
36. −35.6 kJ
38. (a) $K(s) + \frac{1}{2}Cl_2(g) + \frac{3}{2}O_2(g) \longrightarrow KClO_3(s)$
                                  $\Delta H = -397.7$ kJ
    (b) $C(s) + 2Cl_2(g) \longrightarrow CCl_4(l)$ $\Delta H = -135.4$ kJ
    (c) $\frac{1}{2}H_2(g) + \frac{1}{2}I_2(g) \longrightarrow HI(g)$ $\Delta H = 26.5$ kJ
    (d) $2Ag(s) + \frac{1}{2}O_2(g) \longrightarrow Ag_2O(s)$ $\Delta H = -31.0$ kJ
40. (a) −1134.7 kJ   (b) −101.9 kJ
42. (a) $N_2H_4(l) + O_2(g) \longrightarrow N_2(g) + 2H_2O(g)$
                                  $\Delta H = -534.2$ kJ
    (b) 14.0 kJ of heat are evolved
44. (a) −153.9 kJ   (b) −1036.0 kJ   (c) −714.8 kJ
46. (a) −50.3 kJ   (b) −84.5 kJ
48. (a) $CaCO_3(s) + 2NH_3(g) \longrightarrow CaCN_2(s) + 3H_2O(l)$
                                  $\Delta H° = 90.1$ kJ

    (b) −351.6 kJ/mol

50. −680 kJ/mol
52. −3.36 kJ
54. 1.20×10^2 L-atm
56. (a) −18 J   (b) +64 J
58. (a) yes   (b) −102 J
60. (a) 29.6 kJ   (b) 2.76 kJ   (c) 26.8 kJ
62. (a) $CH_3OH(l) + \frac{3}{2}O_2(g) \longrightarrow CO_2(g) + 2H_2O(g)$
$$\Delta H = -638.4 \text{ kJ}$$

(b) $\Delta E = -642.1$ kJ
64. 1.00 BTU/lb-°F
66. 29 min
68. -8.451×10^4 kJ
70. $\Delta E = -885$ kJ; $\Delta H = -890$ kJ
72. 0.793
74. 1.65×10^{-6} °C
76. (a) yes   (b) −   (c) no   (d) yes
78. Piston goes down.
80. $c_A > c_B$
82. c, d, e
84. 2.1×10^{28} photons
85. (a) 1.69×10^5 J   (b) 505 g
86. 22%
87. (a) −851.5 kJ   (b) 6.6×10^3 °C   (c) yes
88. 76.0%
89. 7.5 g
90. 3.50×10^2 kJ

Chapter 9

2. (a) 50°C: 297 mm Hg; 60°C: 307 mm Hg
   (b) 297 mm Hg > 269 mm Hg
       307 mm Hg < 389 mm Hg
   (c) 269 mm Hg at 50°C   307 mm Hg at 60°C
4. (a) 0.75 mg   (b) 0.14 mm Hg
6. (a) 2.4 mg   (b) 2.6 mg   (c) 0.40 mm Hg; yes
8. (a) 29.5 kJ/mol   (b) 39.8°C
10. 97°C
12. 17°C
14. 29 kJ/mol
16. (a) liquid   (b) vapor   (c) liquid
18. (a) vapor   (b) 20°C   (c) liquid
    (d) triple point   (e) normal boiling point
    (f) deposition   (g) no   (h) yes
20. (a) solid   (b) liquid   (c) liquid
22. (a)

(b) ≈30 atm   (c) yes

24. (a)

(b) 600°C   (c) sublimes
26. He < Ne < Ar < Xe
28. dipole forces: b, d; dispersion forces: all
30. b and c
32. (a) Na_2CO_3 is ionic.
    (b) N_2H_4 has hydrogen bonding.
    (c) Benzoic acid has larger dispersion forces.
    (d) CO has dipole forces.
34. a and b
36. (a) LiCl—ionic bonds
    (b) CH_3OH—hydrogen bonds
    (c) H_2O—hydrogen bonds
    (d) Cl_2—larger dispersion forces
38. (a) H-bonds   (b) dispersion forces
    (c) dispersion and dipole forces   (d) dispersion forces
40. (a) molecular   (b) ionic   (c) metallic
42. (a) metallic, network covalent
    (b) network covalent, metallic   (c) metallic
44. (a) metallic   (b) molecule   (c) network covalent
    (d) ionic   (e) molecule
46. (a) NO   (b) CaO   (c) SiO_2 (d) O_2
48. (a) C atoms   (b) Si and C atoms
    (c) ions   (d) C_2H_2 molecules
50. face-centered
52. $V = 0.05631$ nm^3; $r = 0.1355$ nm
54. (a) 0.698 nm   (b) 0.987 nm
56. (a) 0.700 nm   (b) 0.404 nm
58. 1 Cs^+, 1 Cl^-
60. (a) 1.43×10^{-8} cm   (b) 1.22×10^{-23} cm^3
    (c) 3.66 g/cm^3 (d) 2.71 g/cm^3
62. (a) 52%   (b) 74%   (c) 68%
64. 0.0436 atm
66. (a) F   (b) T   (c) F   (d) T
68. (a) GT   (b) MI   (c) LT   (d) EQ   (e) MI
70. (a) I   (b) G   (c) L/G
72. (a) Covalent bonds are between nonmetal atoms and they hold a molecule together. An ionic bond is a bond between ions.
    (b) boiling point—temperature and pressure for a liquid to become vapor;
        critical point—temperature and pressure where liquid cannot exist
    (c) deposition—gas to solid; sublimation—solid to gas

(d) freezing point—temperature and pressure for a liquid to become solid;

triple point—temperature and pressure where boiling, freezing, and subliming all occur

74. (a) A   (b) A   (c) ≈34°C
    (d) gas   (e) ≈200 mm Hg
75. 6.65×10^3 kJ
76. 6.05×10^{23} atoms/mol
77. (a) liquid and vapor   (b) 26.7 mm Hg
    (c) 3.4 atm
78. 41%
79. 80 atm; 0.60°C; see *Scientific American*, February 2000
80. $\dfrac{r_{\text{cation}}}{r_{\text{anion}}} = 0.414$
81. $P_{C_3H_8}$ is the vapor pressure of propane, which decreases exponentially with T. P_{N_2} is gas pressure, which decreases linearly with T.

Chapter 10

2. (a) 41.2%   (b) 58.8%   (c) 0.357
4. 0.15 M
6. 4.8×10^{-7} mol; 0.10 ppm
8. (a) 0.05413 M (b) 7.076 mL   (c) 139 g
10. (a) 3.18 m; 36.4%; 6.36×10^5
    (b) 4.411 m; 44.28%; 0.0736
    (c) 1.17 m; 8.26×10^5; 0.0206
    (d) 13.06%; 8.694×10^5; 0.01481
12. (a) Add 46.5 g of $Ba(OH)_2$ to enough water to make 750.0 mL of solution.
    (b) Add enough water to 55.1 mL of 0.493 M $Ba(OH)_2$ to make 750.0 mL of solution.
14. (a) 109.3 g
    (b) K_2S: 0.4953 M; K^+: 0.9906 M; S^{2-}: 0.4953 M
16. (a) 15.7 M (b) 1.89 M
18. (a) 0.678 m; 8.56%
    (b) 2.39 M; 2.88 m
    (c) 3.59 M; 41.4%
20. 1.23 m
22. (a) benzene — both nonpolar
    (b) octane; glycerol is more similar to water (OH groups)
    (c) $CHCl_3$ — similar dispersion forces
    (d) CBr_4 — both nonpolar
24. (a) ethanol   (b) NH_3
    (c) propionic acid   (d) sodium iodide
26. (a) $\Delta H = -13.0$ kJ   (b) no
28. (a) 1.3×10^{-6} M/mm Hg
    (b) 9.1×10^{-4} M
    (c) 1.1 g
30. (a) 62 mg   (b) 56 mg   (c) 1.0×10^1%
32. −11°C
34. (a) 26.1 mm Hg   (b) 27.0 mm Hg   (c) 23.0 mm Hg
36. Make a solution made up of 161.6 g water and 103.4 g of urea.
38. 2.46 mm Hg
40. (a) 5.7 g; 3.3 g   (b) 9.3 g; 5.4 g
42. (a) 15 m (b) 2.1×10^2 mL ethylene glycol
44. 9.08°C/m
46. 183.1 g/mol
48. 403 g/mol

50. $C_6H_4N_2O_4$
52. 1.50×10^5 g/mol
54. boiling point: $Al(ClO_3)_3$ > $(NH_4)_2Cr_2O_7$ > $MgSO_4$ > CH_3OH

    freezing point: CH_3OH > $MgSO_4$ > $(NH_4)_2Cr_2O_7$ > $Al(ClO_3)_3$
56. 9.74 atm
58. (a) 2.0   (b) (ii)
60. Li
62. (a) 2.002 M (b) 2.17 m
    (c) 196 atm   (d) −16.1°C
64. (a) 4.9°C/m (b) 3
66. (a) 5.5×10^{-7} mol/L   (b) 0.12
68. 3.0 atm
70. 4 M Ca^{2+}; 8 M Cl^-

72. (a) solution 2   (b) same   (c) solution 2
    (d) same   (e) solution 1
74. HF molecules
76. b
78. (a) less   (b) greater   (c) greater   (d) less
80. (a) Check electrical conductivity.
    (b) Solubility of gas decreases with increasing temperature.
    (c) Number of moles of water is large so X_{solute} is small.
    (d) Colligative property; the presence of a solute decreases vapor pressure.
82. 1.1 g/mL
83. Add 1.03×10^3 g of water.
84. 43.1%
85. 48%
86. 0.0018 g/cm³—intoxicated
87. (a) 2.08 M
    (b) 1.872 mol
    (c) 47.4 L
88. 0.055
89. $V_{\text{gas}} = \dfrac{n_{\text{gas}} \times RT}{P_{\text{gas}}}$; $n_{\text{gas}} = k \times P_{\text{gas}}$; $V_{\text{gas}} = kRT$

    V_{gas} depends only on temperature.

Chapter 11

2. (a) rate $= \dfrac{-\Delta[N_2O]}{2\Delta t}$ (b) rate $= \dfrac{\Delta[O_2]}{\Delta t}$
4. (a)
Time (min)	0	10	20	30	40	50
Moles AB_2	0.384	0.276	0.217	0.140	0.098	0.071

   (b) 7.7×10^{-3} mol/min
   (c) 1.73×10^{-3} mol/min
6. (a) 0.058 mol/L·s   (b) 0.029 mol/L·s
   (c) 0.048 mol/L·s
8. (a) $N_2(g) + 3H_2(g) \longrightarrow 2NH_3(g)$

   (b) rate $= \dfrac{\Delta[NH_3]}{2\Delta t}$ (c) 0.0186 mol/L·min

10. (a)

(b) 3.7×10^{-5} mol/L·s   (c) 4.8×10^{-5} mol/L·s

(d) instantaneous rate < average rate

12. (a) 0, 0, 0   (b) 2, 1, 3   (c) 0, 2, 2   (d) 1, 1, 2

14. (a) mol/L·min   (b) $L^2/mol^2 \cdot$ min

(c) L/mol·min   (d) L/mol·min

16. (a) 0   (b) rate $= k[Y]^0$ (c) $k = 0.045$ M/h

18. (a) 0.0756 mol/L·h   (b) 0.300 M

(c) 0.143 M (d) 0.00801 L/mol·h

20. (a) rate $= 2.5 \times 10^{-4}$ mol/L·min

(b) 2.5×10^{-4} mol/L·min

(c) any concentration

22. (a) 4.05 $L^2/mol^2 \cdot$ min   (b) 0.230 M

(c) 0.0915 M

24. (a) 1   (b) rate $= k[Y]$ (c) 1.44 min^{-1}

26. (a) second-order with respect to NO_2; zero-order with respect to CO; second-order overall

(b) rate $= k[NO_2]^2$

(c) $k = 0.297$ L/mol·s

(d) 0.0526 mol/L·s

28. (a) first-order with respect to O_2; second-order with respect to NO

(b) rate $= k[NO]^2[O_2]$

(c) 99.3 $L^2/mol^2 \cdot$ min

(d) 4.59×10^{-4} M/min

30. (a) first-order in CH_3COCH_3, first-order in H^+, zero-order in I_2

(b) rate $= k[CH_3COCH_3][H^+]$

(c) 2.6×10^{-5} L/mol·s

(d) 6.8×10^{-5} mol/L·s

32. (a) rate $= k[CH_3CO_2CH_3][OH^-]$ (b) 0.26 L/mol·s

(c) 0.51 mol/L·s

34. second-order

36. (a)

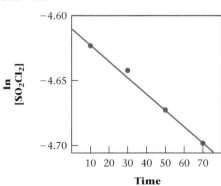

(b) 0.0015 min^{-1} (c) 462 min

(d) 567 min   (e) 2.30×10^{-5} M/min

38. (a) 4.05×10^{-3} s^{-1} (b) 171 s   (c) 0.152 M

40. (a) 0.758 g   (b) 3.22 months   (c) 4.9 months

42. (a) 0.0331 h^{-1} (b) 1.53×10^{-3} mol/L·h

(c) 2.4×10^1 h

44. 112 g

46. 16 days

48. 15 hours

50. (a) 2.37 mg/min   (b) 156 min   (c) 2.37 mg/min

(d) 2.37 mg/min

52. (a) 10.0 min   (b) 1.2 h

54. (a) 0.0153 L/mol·s   (b) 327 s   (c) 47.6 s

(d) 2.51×10^{-4} mol/L·s

56. 2.9 L/mol·min

58. 57%

60. $E_a = 2.2 \times 10^2$ kJ/mol

62.

64. Decrease by 31%

66. (a) At 25°C: 148 chirps/min; at 35°C: 220 chirps/min

(b) 3.0×10^1 kJ   (c) 49%

68. (a) 0.062 s^{-1} (b) 35°C

70. 86 kJ/mol

72.

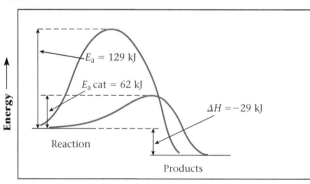

74. 357 kJ

76. (a) rate $= k[Cl]^2$

(b) rate $= k[N_2O_2][O_2]$

(c) rate $= k[I^-][HClO]$

78. yes

80. (1) rate $= k \times [NO_3] \times [NO]$;

$$k_2 \times [NO_3] = k_1 \times [NO] \times [O_2]$$

$$\text{rate} = \frac{kk_1}{k_2} \times [NO]^2 \times [O_2]$$

(2) rate $= k \times [N_2O_2] \times [O_2]$ $k_2 \times [N_2O_2] = k_1 \times [NO]^2$

$$\text{rate} = \frac{kk_1}{k_2} \times [NO]^2 \times [O_2]$$

82. 0.097 mol/L·s
84. ≈ 15 times faster
86. 7×10^{17} years; very unlikely

88. $\ln\left(\dfrac{k_2}{k_1}\right) = \dfrac{1}{RT}(E_{a_{uncat}} - E_{a_{cat}})$

90. (a) decreases   (b) increases   (c) stays the same
    (d) decreases   (e) increases
92. 1 represents the decrease of [X] with time
    2 represents the increase of [Z] with time
    3 represents the increase of [Y] with time
94. (a) A   (b) C   (c) A
96. (a) 8 kJ   (b) A   (c) -360 kJ   (d) 368 kJ
    (e) B   (f) activated complex
98. Expt. 1: rate $= 0.05$ mol/L·min
    Expt. 2: $k = 0.5$/min; rate $= 0.1$ mol/L·min; $E_a = 32$ kJ
    Expt. 3: rate $= 0.1$ mol/L·min; E_a cannot be
    determined
    Expt. 4: rate $= 0.06$ mol/L·min; $E_a = 32$ kJ
99. (a) 174 kJ   (b) 46 L/mol·s   (c) 1.8 mol/L·s
100. 12 min

101. $\dfrac{-d[A]}{a\,dt} = k[A]$; $\dfrac{-d[A]}{[A]} = ka\,dt$; $\ln\dfrac{[A_o]}{[A]} = akt$

102. rate $= k[A]^2[B][C]$

103. (a) $\dfrac{-d[A]}{dt} = k[A]^2$; $\dfrac{-d[A]}{[A]^2} = k\,dt$; $\dfrac{1}{[A]} - \dfrac{1}{[A_o]} = kt$

(b) $\dfrac{-d[A]}{dt} = k[A]^3$; $\dfrac{-d[A]}{[A]^3} = k\,dt$;

$\dfrac{1}{2[A]^2} - \dfrac{1}{2[A_o]^2} = kt$;

$\dfrac{1}{[A]^2} - \dfrac{1}{[A_o]^2} = 2kt$

104. 0.90 g; 54 mg no more than three times a day
105. A: 134 mm Hg; X: 244 mm Hg; Y: 122 mm Hg

Chapter 12

2. (a) ≈ 75 s   (b) greater than; equal to
4.

Time (min)	0	1	2	3	4	5	6
P_A (atm)	1.000	0.778	0.580	0.415	0.355	0.325	0.325
P_B (atm)	0.400	0.326	0.260	0.205	0.185	0.175	0.175
P_C (atm)	0.000	0.148	0.280	0.390	0.430	0.450	0.450

6. (a) $K = P_{CO_2}$

(b) $K = \dfrac{(P_{CO_2})^2(P_{H_2})^5}{P_{C_2H_6}}$

(c) $K = \dfrac{(P_{NH_3})^4(P_{O_2})^5}{(P_{NO})^4(P_{H_2O})^6}$

(d) $K = \dfrac{1}{P_{NH_3}}$

8. (a) $K = \dfrac{(P_{Cl_2})[I^-]^2}{[Cl^-]^2}$

(b) $K = \dfrac{[CH_3NH_3^+]}{[CH_3NH_2][H^+]}$

(c) $K = \dfrac{[Au(CN)_4^{2-}]}{[Au^{2+}][CN^-]^4}$

10. (a) $C_3H_6O(l) \rightleftharpoons C_3H_6O(g)$ $K = P_{C_3H_6O}$
    (b) $7H_2(g) + 2NO_2(g) \rightleftharpoons 2NH_3(g) + 4H_2O(g)$

$$K = \dfrac{(P_{NH_3})^2(P_{H_2O})^4}{(P_{H_2})^7(P_{NO_2})^2}$$

(c) $H_2S(g) + Pb^{2+}(aq) \rightleftharpoons PbS(s) + 2H^+(aq)$

$$K = \dfrac{[H^+]^2}{(P_{H_2S})[Pb^{2+}]}$$

12. (a) $2H_2O(g) + 2SO_2(g) \rightleftharpoons 2H_2S(g) + 3O_2(g)$
    (b) $IF(g) \rightleftharpoons \frac{1}{2}F_2(g) + \frac{1}{2}I_2(g)$
    (c) $Cl_2(g) + 2Br^-(aq) \rightleftharpoons 2Cl^-(aq) + Br_2(l)$
    (d) $2NO_3^-(aq) + 8H^+(aq) + 3Cu(s) \rightleftharpoons$
    $2NO(g) + 3Cu^{2+}(aq) + 4H_2O(g)$

14. (a) $K' = \dfrac{P_{NOCl}}{(P_{NO})(P_{Cl_2})^{1/2}}$

(b) $K'' = \dfrac{(P_{NO})(P_{Cl_2})}{(P_{NOCl})^2}$

(c) $K'' = \dfrac{1}{(K')^2}$

16. (a) 67   (b) 0.015
18. (a) 9×10^{15}
20. 0.48
22. 0.838
24. (a) $NOBr(g) \rightleftharpoons NO(g) + \frac{1}{2}Br_2(g)$ (b) $K = 0.090$
26. 3.39
28. (a) No; $Q \neq K$ (b) $\longrightarrow$
30. (a) $\longleftarrow$ (b) $\longleftarrow$ (c) $\longrightarrow$
32. $\longrightarrow$
34. 2.3 atm
36. $P_{I_2} = 0.32$ atm; $P_{Cl_2} = 0.96$ atm
38. $P_{NO} = 0.42$ atm; $P_{NO_2} = 0.32$ atm
40. $P_Q = P_R = 0.188$ atm; $P_Z = 0.308$ atm
42. (a) $P_{NO} = P_{SO_3} = 2.25$ atm; $P_{NO_2} = 0.25$ atm
    (b) $P = 5.00$ atm before and after; true only if $\Delta n_g = 0$
44. (a) 0.46 atm   (b) 14.0 g
46. $P_{H_2} = P_{I_2} = 0.021$ atm
48. (a) reactants   (b) products   (c) same
50. (a) (1) increase (2) increase (3) no effect (4) increase
    (5) increase
    (b) none; (5) will decrease it
52. (a) $\longrightarrow$ (b) $\longrightarrow$ (c) $\longrightarrow$

54. (a) 0.23   (b) $P_{X_2} = 4.3$ atm; $P_{X_2R} = 0.98$ atm
   (c) $P_{X_2} = 2.41$ atm; $P_{X_2R} = 0.56$ atm
56. (a) 5.9
   (b) $P_{SO_2} = 1.93$ atm; $P_{Cl_2} = 0.73$ atm; $P_{SO_2Cl_2} = 0.22$ atm
58. 0.132
60. -57 kJ
62. (a) $P_{S(CH_2CH_2Cl)_2} = 0.168$ atm; $P_{SCl_2} = 1.07$ atm;
   $P_{C_2H_4} = 2.51$ atm
   (b) 0.0249
64. 5.40 g
66. (a) 0.283   (b) 0.21
68. (a) $3A(g) \rightleftharpoons 2B(g)$
   (b) no; pressures still changing at 250 s
70. b
72. Will decrease and then level off sooner at a somewhat higher level
74. Endothermic; more product at a higher temperature
76. Balanced equation; temperature
77. 27

78. $K = \dfrac{(P_C)^c(P_D)^d}{(P_A)^a(P_B)^b} = \dfrac{([C]RT)^c([D]RT)^d}{([A]RT)^a([B]RT)^b}$

   $= \dfrac{[C]^c[D]^d}{[A]^a[B]^b}(RT)^{(c+d)-(a+b)}$

   $\Delta n_g = (c+d) - (a+b); \; K = K_c(RT)^{\Delta n_g}$

79. $P_{N_2} = 0.33$ atm; $P_{H_2} = 0.99$ atm; $P_{NH_3} = 0.34$ atm
80. 0.0442
81. 0.52 atm
82. 1.1
83. 25; 0.65 atm
84. (a) 0.22 atm   (b) 0.4 g
85. 9.57
86. 2.29

Chapter 13

2. (a) Brønsted-Lowry acid: H_2O, HCN
   Brønsted-Lowry base: CN^-, OH^-
   acid-base pairs: H_2O, OH^-; HCN, CN^-
   (b) Brønsted-Lowry acid: H_3O^+, H_2CO_3
   Brønsted-Lowry base: HCO_3^-, H_2O
   acid-base pairs: H_3O^+, H_2O; H_2CO_3, HCO_3^-
   (c) Brønsted-Lowry acid: $HC_2H_3O_2$, H_2S
   Brønsted-Lowry base: HS^-, $C_2H_3O_2^-$
   acid-base pairs: $HC_2H_3O_2$, $C_2H_3O_2^-$; H_2S, HS^-
4. as acid: $HAsO_4^{2-}$ as base: CH_3O^-, CO_3^{2-}, $HAsO_4^{2-}$
6. (a) $HAsO_4^{2-}$ (b) $Fe(H_2O)_4(OH)_2$ (c) ClO_3^-
   (d) NH_3 (e) $C_2H_3O_2^-$
8. (a) acid: $HCO_3^-(aq) \rightleftharpoons H^+(aq) + CO_3^{2-}(aq)$
   (b) base: $HCO_3^-(aq) + H_2O \rightleftharpoons OH^-(aq) + H_2CO_3(aq)$
10. (a) $Zn(H_2O)_3OH^+(aq) + H_2O \rightleftharpoons$
          $Zn(H_2O)_2(OH)_2(aq) + H_3O^+(aq)$
   (b) $HSO_4^-(aq) + H_2O \rightleftharpoons SO_4^{2-}(aq) + H_3O^+(aq)$
   (c) $HNO_2(aq) + H_2O \rightleftharpoons NO_2^-(aq) + H_3O^+(aq)$
   (d) $Fe(H_2O)_6^{2+}(aq) + H_2O \rightleftharpoons$
          $Fe(H_2O)_5OH^+(aq) + H_3O^+(aq)$
   (e) $HC_2H_3O_2(aq) + H_2O \rightleftharpoons C_2H_3O_2^-(aq) + H_3O^+(aq)$
   (f) $H_2PO_4^-(aq) + H_2O \rightleftharpoons HPO_4^{2-}(aq) + H_3O^+(aq)$
12. (a) $(CH_3)_3N(aq) + H_2O \rightleftharpoons (CH_3)_3NH^+(aq) + OH^-(aq)$
   (b) $HC_2O_4^-(aq) + H_2O \rightleftharpoons H_2C_2O_4(aq) + OH^-(aq)$

   (c) $CN^-(aq) + H_2O \rightleftharpoons HCN(aq) + OH^-(aq)$
   (d) $S^{2-}(aq) + H_2O \rightleftharpoons HS^-(aq) + OH^-(aq)$
   (e) $BrO_3^-(aq) + H_2O \rightleftharpoons HBrO_3(aq) + OH^-(aq)$
   (f) $H_2C_6O_5H_7^-(aq) + H_2O \rightleftharpoons H_3C_6O_5H_7(aq) + OH^-(aq)$
14. (a) 2.57 acidic
   (b) -0.18 acidic
   (c) 12.84 basic
   (d) 8.19 basic
16. (a) $[H^+] = 5.8$; $[OH^-] = 1.7 \times 10^{-15}$
   (b) $[H^+] = 7.8 \times 10^{-10}$; $[OH^-] = 1.3 \times 10^{-5}$
   (c) $[H^+] = 1.5 \times 10^{-4}$; $[OH^-] = 6.7 \times 10^{-11}$
   (d) $[H^+] = 8.3 \times 10^{-13}$; $[OH^-] = 0.012$
18. (a) $1.4 \times 10^{-6}\,M$ $7.2 \times 10^{-9}\,M$ 5.86   8.14   No
   (b) $5.2 \times 10^{-14}\,M$ $0.19\,M$ 13.28   0.72   Yes
   (c) $2.3 \times 10^{-12}\,M$ $4.3 \times 10^{-3}\,M$ 11.63   2.37   Yes
   (d) $2.75\,M$ $3.64 \times 10^{-15}\,M$ -0.44 14.44   No
20. Solution R is more basic; solution Q has a smaller pH.
22. (a) A: $4.8 \times 10^{-13}\,M$ B: $1.4 \times 10^{-12}\,M$
     C: $6.9 \times 10^{-7}\,M$
   (b) B: 11.85   C: 6.16
   (c) A is basic, B is basic, C is acidic
24. (a) cola: $7.9 \times 10^{-4}\,M$; green tea: $1.6 \times 10^{-6}\,M$
   (b) ratio: 5.0×10^2; cola is 500 times more acidic than green tea
26. (a) $[H^+] = 1.90\,M$; $pH = -0.28$; same pH in 10.0-mL sample
   (b) $[H^+] = 0.021\,M$; $pH = 1.67$ in 1.28 L of solution; $[H^+] = 0.21\,M$; $pH = 0.67$ in 0.128 L of solution
28. (a) $[OH^-] = 0.0293\,M$; $[H^+] = 3.4 \times 10^{-13}\,M$; $pH = 12.47$
   (b) $[OH^-] = 0.199\,M$; $[H^+] = 5.02 \times 10^{-14}\,M$; $pH = 13.30$
30. 0.792
32. 11.94
34. (a) $HSO_3^-(aq) \rightleftharpoons H^+(aq) + SO_3^{2-}(aq)$

          $K_a = \dfrac{[H^+][SO_3^{2-}]}{[HSO_3^-]}$

   (b) $HPO_4^{2-}(aq) \rightleftharpoons H^+(aq) + PO_4^{3-}(aq)$

          $K_a = \dfrac{[H^+][PO_4^{3-}]}{[HPO_4^{2-}]}$

   (c) $HNO_2(aq) \rightleftharpoons NO_2^-(aq) + H^+(aq)$

          $K_a = \dfrac{[H^+][NO_2^-]}{[HNO_2]}$

36. (a) 3.74   (b) 7.17   (c) 10.31
38. (a) $C < D < B < A$ (b) A
40. $HCl > HF > HClO > HCN$
42. $HCl < HF < HClO < HCN$
44. 8.7×10^{-12}
46. 1.91×10^{-10}
48. 1.0×10^{-4}
50. 3.77
52. (a) 2.0×10^{-2} (b) 2.8×10^{-2}
54. (a) $3.07 \times 10^{-3}\,M$ (b) $3.3 \times 10^{-12}\,M$
   (c) 2.51   (d) 1.3%
56. 2.29; 1.3%

58.

$H_3C_6H_5O_7(aq) \rightleftharpoons H^+(aq) + H_2C_6H_5O_7^-(aq)$ $K_{a1} = 7.5 \times 10^{-4}$

$H_2C_6H_5O_7^-(aq) \rightleftharpoons H^+(aq) + HC_6H_5O_7^{2-}(aq)$ $K_{a2} = 1.7 \times 10^{-5}$

$HC_6H_5O_7^{2-}(aq) \rightleftharpoons H^+(aq) + C_6H_5O_7^{3-}(aq)$ $K_{a3} = 4.0 \times 10^{-7}$

$H_3C_6H_5O_7(aq) \rightleftharpoons 3H^+(aq) + C_6H_5O_7^{3-}(aq)$ $K = 5.1 \times 10^{-15}$

60. pH = 2.00; $[HX^-] = 1.0 \times 10^{-2}$ M; $[X^{2-}] = 9.7 \times 10^{-8}$ M

62. pH = 1.05; $[HSeO_3^-] = 0.088$ M,

$[SeO_3^{2-}] = 5.0 \times 10^{-8}$ M

64. (a) $NH_3(aq) + H_2O \rightleftharpoons NH_4^+(aq) + OH^-(aq)$

$$K_b = \frac{[NH_4^+][OH^-]}{[NH_3]}$$

(b) $HS^-(aq) + H_2O \rightleftharpoons H_2S(aq) + OH^-(aq)$

$$K_b = \frac{[H_2S][OH^-]}{[HS^-]}$$

(c) $(CH_3)_3N(aq) + H_2O \rightleftharpoons (CH_3)_3NH^+(aq) + OH^-(aq)$

$$K_b = \frac{[(CH_3)_3NH^+][OH^-]}{[(CH_3)_3N]}$$

66. $NO_2^- < CHO_2^- < SO_3^{2-}$

68. $Ba(OH)_2 > KOH > NaCN > HCO_3^-$

70. (a) 6.7×10^{-11} (b) 6.7×10^{-10}

72. $[OH^-] = 1.3 \times 10^{-4}$ M; pH = 10.11

74. (a) $C_6H_5N(aq) + H_2O \rightleftharpoons C_6H_5NH^+(aq) + OH^-(aq)$

(b) 1.5×10^{-9} (c) 8.94

76. 2.1×10^2 g

78. (a) $(NH_4)_2SO_3$, NH_4CN (b) Na_2CO_3, K_2CO_3

(c) NaBr, KBr   (d) $FeCrO_4$, $(NH_4)_2CrO_4$

80. (a) acidic   (b) neutral   (c) acidic   (d) basic

(e) basic

82. (a) $Fe(H_2O)_6^{3+}(aq) + H_2O \rightleftharpoons$

$[Fe(H_2O)_5OH]^{2+}(aq) + H_3O^+(aq)$

(b) BaI_2 is neutral

(c) $NH_4^+(aq) + H_2O \rightleftharpoons NH_3(aq) + H_3O^+(aq)$

$K_a = 5.6 \times 10^{-10}$

$NO_2^-(aq) + H_2O \rightleftharpoons OH^-(aq) + HNO_2(aq)$

$K_b = 1.7 \times 10^{-11}$

(d) $HPO_4^{2-}(aq) + H_2O \rightleftharpoons PO_4^{3-}(aq) + H_3O^+(aq)$

$K_a = 4.5 \times 10^{-13}$

$HPO_4^{2-}(aq) + H_2O \rightleftharpoons H_2PO_4^-(aq) + OH^-(aq)$

$K_b = 1.6 \times 10^{-7}$

(e) $PO_4^{3-}(aq) + H_2O \rightleftharpoons HPO_4^{2-}(aq) + OH^-(aq)$

84. $ZnBr_2 < KClO_4 < LiF < NaOH < Sr(OH)_2$

86. $NH_4^+(aq) + NO_2^-(aq) \rightleftharpoons NH_3(aq) + HNO_2(aq)$

$K = 9.3 \times 10^{-7}$

Equilibrium is in the direction of the reactants; reaction unlikely.

88. 68 L

90. 5.5 g

92. 4.6 g

94. 9.8×10^{-9}

96. c

98. (a) LT   (b) LT   (c) EQ

    (d) MI   (e) LT   (f) GT

100. (1) weak acid; (2) strong acid

102. Dissolve 0.10 mol in water, measure pH. If pH > 7, solid is basic; pH < 13.0, it is a weak base. If pH < 7, solid is acidic; if pH > 1.0, it is a weak acid.

103. 1.52

104. (a) -5.58 kJ   (b) -6.83 kJ

105. 13.06

106. 1.7×10^2 mL

107. $-1.64°C$

108. 12.62

109. HA

110. 5.2×10^{-7}

Chapter 14

2. (a) $H^+(aq) + C_2H_3O_2^-(aq) \longrightarrow HC_2H_3O_2(aq)$

(b) $H^+(aq) + OH^-(aq) \longrightarrow H_2O$

(c) $HOCl(aq) + CN^-(aq) \longrightarrow HCN(aq) + OCl^-(aq)$

(d) $HNO_2(aq) + OH^-(aq) \longrightarrow NO_2^-(aq) + H_2O$

4. (a) $[Fe(H_2O)_6^{3+}](aq) + OH^-(aq) \longrightarrow$

$[Fe(OH)(H_2O)_5^{2+}](aq) + H_2O$

(b) $HCO_3^-(aq) + OH^-(aq) \longrightarrow CO_3^{2-}(aq) + H_2O$

(c) $NH_4^+(aq) + OH^-(aq) \longrightarrow NH_3(aq) + H_2O$

6. (a) 5.6×10^4 (b) 1.0×10^{14} (c) 48

(d) 6.0×10^{10}

8. (a) 6.7×10^{11} (b) 4.8×10^3 (c) 5.6×10^4

10. (a) $[OH^-] = 9.67 \times 10^{-9}$; pH = 5.99

(b) $[OH^-] = 9.60 \times 10^{-8}$; pH = 6.98

(c) $[OH^-] = 1.95 \times 10^{-7}$; pH = 7.29

(d) $[OH^-] = 3.1 \times 10^{-7}$; pH = 7.50

(e) $[OH^-] = 4.75 \times 10^{-7}$; pH = 7.68

12. 4.25

14. (a) 8.51   (b) pH remains the same.

16. 7.55

18. d

20. 5.08

22. 9.59

24. (a) $HC_2H_3O_2/C_2H_3O_2^-$ (b) HCN/CN^-

(c) HCO_3^-/CO_3^{2-}

26. (a) 0.18:1   (b) 0.10

(c) 9.2×10^2 mL   (d) 4.8×10^2 mL

28. (a) 0.429 M OH^-; 0.0249 M H^+

(b) 0.429 M OH^-; 0.247 M H^+

(c) 0.429 M OH^-; 0.504 M H^+

(d) 0.429 M OH^-; 0.811 M H^+

(e) 0.429 M OH^-; 1.22 M H^+

30. (a) 4.288

(b) 3.659

(c) 4.857

32. (a) 4.288   (b) 1.23   (c) 12.76

(d) In both problems (a) is the same. The pH in (b) and (c) differ dramatically.

(e) Dilution does not affect pH but drastically reduces the buffer capacity.

34. (a) 4.47   (b) 4.67   (c) 0.22

36. (a) 0.64   (b) 37%   (c) 65%

38. (a) methyl orange   (b) phenolphthalein

(c) all three will work   (d) methyl orange

40. (a) 3.2×10^{-10} (b) $8.5 - 10.5$

(c) blue

42. (a) 35.5 mL   (b) 7   (c) $[K^+] = [Cl^-] = 0.109\ M$
44. (a) $HSO_3^-(aq) + OH^-(aq) \longrightarrow SO_3^{2-}(aq) + H_2O$
    (b) 0.109 M
    (c) $[Na^+] = [K^+] = [SO_3^{2-}] = 0.0749\ M$;
    $[OH^-] = [HSO_3^-] = 1.1 \times 10^{-4}\ M$
    (d) 10.04
46. (a) $BrO_3^-(aq) + H^+(aq) \rightleftharpoons HBrO(aq)$
    (b) 64.6 mL
    (c) 4.68
    (d) $[H^+] = [BrO^-] = 2.07 \times 10^{-5}\ M$;
    $[K^+] = [NO_3^-] = [HBrO] = 0.171\ M$
48. (a) $H^+(aq) + OH^-(aq) \rightleftharpoons H_2O$
    (b) H_2O, K^+, Br^-
    (c) 32.5 mL
    (d) 1. 0.57; 2. 1.06; 3. 7
50. (a) $C_{17}H_{19}O_3N(aq) + H^+(aq) \rightleftharpoons C_{17}H_{19}O_3NH^+(aq)$
    (b) $C_{17}H_{19}O_3NH^+$, Cl^- (c) 71.8 mL   (d) 10.52
    (e) 7.87   (f) 4.54
52. (a) 2.86   (b) 4.18   (c) 8.30
54. 14.2 L
56. (a) 45.1 g/mol   (b) 5.9×10^{-4} (c) 1.7×10^{-11}
58. 62 mL
60. (a) 2.30   (b) 2.10
62. (a)

 buffer

(b)

 buffer

(c)

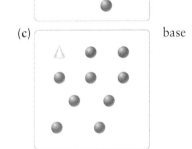 base

64. (a) EQ   (b) GT   (c) GT   (d) MI   (e) LT
66. a and d
68. Measure freezing point of solution:
    1 L HCl has 2.0×10^{-3} mol particles (H^+, Cl^- ions)
    1 L $HCHO_2$ has $\approx 6.0 \times 10^{-3}$ mol particles ($HCHO_2$ molecules)
    1 L $C_6H_5NH_3^+$ has $\approx 8.0 \times 10^{-2}$ mol particles ($C_6H_5NH_3^+$, Cl^- ions)
    Or, one could measure the pH at the equivalence point after titration with a strong base.
    $HCl < HCHO_2 < C_6H_5NH_3^+$
70. 2.87
71. 33 mL

72. 1.1×10^2 g/mol
73. $\dfrac{[NH_4^+]}{[NH_3]} = 5.7 \times 10^2$
74. (a) 1.61   (b) 3.22   (c) 8.44
    (d) 5.92   (e) 10.89
75. $x = 2$
76. (a) 0.8239   (b) 0.80
77. 4.6×10^{-6}
78. 6.0 g
79. 15.1 g
80. (a) Yes, excess OH^- in solution.
    (b) 1.172 M   (c) 2.26 mL

Chapter 15

2. (a) $3 \times 10^{-3}\ M$ (b) $5 \times 10^{-5}\ M$
4. 1.8×10^{-4}
6. (a) $AgCl(s) \rightleftharpoons Ag^+(aq) + Cl^-(aq)$
    $K_{sp} = [Ag^+] \times [Cl^-]$
    (b) $Al_2(CO_3)_3(s) \rightleftharpoons 2Al^{3+}(aq) + 3CO_3^{2-}(aq)$
    $K_{sp} = [Al^{3+}]^2 \times [CO_3^{2-}]^3$
    (c) $MnS_2(s) \rightleftharpoons Mn^{4+}(aq) + 2S^{2-}(aq)$
    $K_{sp} = [Mn^{4+}] \times [S^{2-}]^2$
    (d) $Mg(OH)_2(s) \rightleftharpoons Mg^{2+}(aq) + 2OH^-(aq)$
    $K_{sp} = [Mg^{2+}] \times [OH^-]^2$
8. (a) $PbO_2(s) \rightleftharpoons Pb^{4+}(aq) + 2O^{2-}(aq)$
    (b) $Hg_3(PO_4)_2(s) \rightleftharpoons 3Hg^{2+}(aq) + 2PO_4^{3-}(aq)$
    (c) $Ni(OH)_3(s) \rightleftharpoons Ni^{3+}(aq) + 3OH^-(aq)$
    (d) $Ag_2SO_4(s) \rightleftharpoons 2Ag^+(aq) + SO_4^{2-}(aq)$
10. (a) $[Li^+] = 0.016\ M$ (b) $[NO_2^-] = 0.024\ M$
    (c) $[Sn^{2+}] = 2.9 \times 10^{-19}\ M$
12. (a) 4.0×10^{-5} (b) 0.18 M   (c) 2×10^{-11}
14. (a) $[Ba^{2+}] = [C_2O_4^{2-}] = 1.3 \times 10^{-3}\ M$
    (b) $[OH^-] = 2.9 \times 10^{-8}\ M$
    (c) $[Pb^{2+}] = 2 \times 10^{-15}\ M$
16. (a) $3 \times 10^{-4}\ M$ (b) $1 \times 10^{-10}\ M$ (c) $6 \times 10^{-6}\ M$
18. (a) 8.5×10^{-5} (b) 1.7×10^{-13} (c) 8.2×10^{-33}
20. (a) 6×10^{-3} g/L   (b) 5×10^{-8} g/L   (c) 1×10^{-3} g/L
22. 5.49×10^{-16}
24. (a) yes   (b) no
26. 9.1×10^{-15}
28. (a) $[Cd^{2+}] = 1.4 \times 10^{-5}\ M$
    (b) 8.64
    (c) 1.0×10^1 %
30. yes; $1.0 \times 10^{-2}\ M$
32. (a) yes
    (b) $[Ag^+] = 1.6 \times 10^{-2}\ M$; $[CO_3^{2-}] = 3 \times 10^{-8}\ M$;
    $[Na^+] = 6.22 \times 10^{-4}\ M$; $[NO_3^-] = 1.7 \times 10^{-2}\ M$
34. (a) $Pb(OH)_2$ (b) 9.26
36. (a) $PbCrO_4$ (b) $3 \times 10^{-12}\ M$
38. Both $BaCO_3$ and $CaCO_3$ will precipitate.
40. (a) $CaF_2(s) + 2H^+(aq) \rightleftharpoons Ca^{2+}(aq) + 2HF(aq)$
    (b) $CuCO_3(s) + 2H^+(aq) \rightleftharpoons Cu^{2+}(aq) + H_2CO_3(aq)$
    (c) $Ti(OH)_3(s) + 3H^+(aq) \rightleftharpoons Ti^{3+}(aq) + 3H_2O$
    (d) $Sn(OH)_6^{2-}(aq) + H^+(aq) \rightleftharpoons Sn(OH)_5^-(aq) + H_2O$
    (e) $Cd(NH_3)_4^{2+}(aq) + H^+(aq) \rightleftharpoons Cd(NH_3)_3^{2+}(aq) + NH_4^+(aq)$
42. (a) $Cu(OH)_2(s) + 4NH_3(aq) \longrightarrow Cu(NH_3)_4^{2+}(aq) + 2OH^-(aq)$
    (b) $Cd^{2+}(aq) + 4NH_3(aq) \longrightarrow Cd(NH_3)_4^{2+}(aq)$
    (c) $Pb^{2+}(aq) + 2NH_3(aq) + 2H_2O \longrightarrow Pb(OH)_2(s) + 2NH_4^+(aq)$

44. (a) $Al^{3+}(aq) + 4OH^-(aq) \rightleftharpoons Al(OH)_4^-(aq)$
 (b) $Al^{3+}(aq) + PO_4^{3-}(aq) \rightleftharpoons AlPO_4(s)$
 (c) $Al(OH)_3(s) + 3H^+(aq) \rightleftharpoons Al^{3+}(aq) + 3H_2O$
46. $Co(OH)_2(s) + 2H^+(aq) \rightleftharpoons Co^{2+}(aq) + 2H_2O$;
 $K = 2 \times 10^{12}$
48. (a) 1.6×10^3 (b) yes
50. (a) 4×10^{-7} (b) 0.03 mol/L
52. 2×10^2
54. (a) 1.2×10^{-2} (b) $3\ M$
56. 6.8×10^{-4}
58. 8.2×10^{-4}
60. (a) $\longleftarrow$ (b) $\longleftarrow$ (c) $\longleftarrow$
62. (a) $3.8 \times 10^{-4}\ M$ (b) no (c) $0.022\ M$
64. ≈ 2 mg
66.

(1)

(2)

(3)

68. a
70. yes
72. (a) (3) (b) (5) (c) (2) (d) (1)
73. In HCN: $2 \times 10^{-3}\ M$; In H_2O: $2 \times 10^{-6}\ M$
74. 1.9×10^{-4} mol/L
75. $1.4 \times 10^{-4}\ M$
76. (a) $1 \times 10^{-5}\ M$ (b) $Al(OH)_3$, $Fe(OH)_3$
 (c) virtually all (d) 1.6 g
77. (a) 8×10^5 (b) 4×10^3
78. 5.7×10^{31}

Chapter 16

2. (a) spontaneous (b) nonspontaneous
 (c) nonspontaneous
4. (a) no (b) no (c) yes
6. a
8. (a) $O_2(g)$ with 758 mm Hg pressure
 (b) glucose(s)
 (c) $Hg(l)$
10. (a) $-$ (b) $+$ (c) $+$
12. (a) can't say (b) $+$ (c) $-$ (d) $+$
14. (a) $-$ (b) $+$ (c) $+$
16. (c) < (b) < (a) < (d)
18. (a) 180.8 J/K (b) 605.9 J/K
 (c) 2.9 J/K (d) -93.1 J/K
20. (a) -976.1 J/K (b) -816.0 J/K (c) -97.4 J/K
22. (a) -0.1319 kJ/K (b) -0.2712 kJ/K
 (c) -0.5330 kJ/K

24. (a) -294 kJ (b) 140.9 kJ (c) -498 kJ
26. (a) -980.6 kJ; spontaneous
 (b) -893.6; spontaneous
 (c) -394.7 kJ; spontaneous
 (d) -298; spontaneous
28. (a) -793.6 kJ (b) 224.2 kJ (c) -1226.6 kJ
30. (a) -184.1 kJ/mol (b) -166.4 kJ/mol
 (c) -53.5 kJ/mol
32. $\Delta G° = -6.0 \times 10^1$ kJ; plausible
34. at 27°C: -359.3 kJ at 39°C: -359.5 kJ
36. (a) 0.132 kJ/K; yes (b) 0.0956 kJ/mol·K
 (c) -510.9 kJ/mol
38. (a) -0.918 kJ/K (b) -0.159 kJ/mol·K
40. (a) spontaneous at high T; nonspontaneous at low T
 (b) spontaneous at low T; nonspontaneous at high T
 (c) spontaneous at any T; temperature has no effect
42. (a) spontaneous above 1157°C
 (b) spontaneous below -142°C
 (c) spontaneous at any T
44. 197°C
46. (a) $\Delta G = 492$ kJ $- T(0.327$ kJ/K$)$
 (b) 1504 K; spontaneous above 1504 K
48. Heated with C (coke)
50. 282 K
52. not at equilibrium at any temperature when $P = 1$ atm
54. 332 K = 59°C
56. (a) no; $\Delta G = 20.3$ kJ
 (b) yes; $\Delta G = -12.3$ kJ
58. (a) 368.0 kJ (b) 323 kJ
60. (a) 55.7 kJ (b) $1.1 \times 10^{-5}\ M$
 (c) yes; $K_{sp} = 1.1 \times 10^{-10}$
62. (a) 52.6 kJ (b) 6×10^{-10}
64. (a) 105 kJ (b) 104 kJ/mol
66. (a) -7.43 kJ (b) 0.15
68. 6.9×10^{-4}
70. 35.3 kJ/mol
72. 130.4 kJ
74. (a) $\Delta G° = 201.3$ kJ (nonspontaneous)
 (b) $\Delta G° = -98.9$ kJ
76. $C_6H_{12}O_6(s) + 6O_2(g) + 80ADP + 80HPO_4^{2-}(aq) +$
 $160H^+(aq) \longrightarrow 80ATP + 86H_2O + 6CO_2(g)$
78. 510 K = 238°C
80. ΔG for overall reaction is -2016 kJ
82. 4.4×10^{-3}
84. (a) False
 (b) False
 (c) False
 (d) False
86. (a) 0 K (b) 1 (c) larger
88. (a) Entropy decreases because $S°_{gas} > S°_{solid}$ or $S°_{liq}$
 (b) $\Delta S°$ is the difference between entropy of products and reactants.
 (c) A solid has fewer options for placement of particles, i.e., more orderly.
90. (a) Increase in T decreases $\Delta G°$ (b) yes (c) no
 (d) ≈ 350 K = 77°C (e) 7.6×10^{-3}
92. (a) LT (b) LT (c) MI
93. 0.2827 kJ/mol
94. $P_{H_2} = P_{I_2} = 0.07$ atm $P_{HI} = 0.46$ atm
95. (a) 6.00 kJ (b) 0 (c) 0.0220 kJ/K
 (d) 0.43 kJ (e) -0.45 kJ

96. (a) 4.2 kJ   (b) $\approx 5.2 \times 10^2$ g
97. 1430 K
98. (1) $\Delta G° = 146 - 0.1104\ T$; becomes spontaneous above 1050°C
   (2) $\Delta G° = 168.6 - 0.0758\ T$; becomes spontaneous above 1950°C
99. -0.10 kJ/mol·K
100. 26.7 kJ
101. 464 K
102. 0.2598 kJ/mol

Chapter 17

2. (a) $CH_3OH(aq) + H_2O \longrightarrow CO_2(g) + 6e^- + 6H^+(aq)$
  (b) $NO_3^-(aq) + 8e^- + 10H^+ \longrightarrow NH_4^+(aq) + 3H_2O$
  (c) $Fe^{3+}(aq) + 3e^- \longrightarrow Fe(s)$
  (d) $V^{2+}(aq) + 6OH^-(aq) \longrightarrow VO_3^- + 3e^- + 3H_2O$
4. (a) $2H^+(aq) + H_2O_2(aq) + 2Ni^{2+}(aq) \longrightarrow$
                        $2Ni^{3+}(aq) + 2H_2O$
  (b) $14H^+(aq) + Cr_2O_7^{2-}(aq) + 3Sn^{2+}(aq) \longrightarrow$
             $2Cr^{3+}(aq) + 3Sn^{4+}(aq) + 7H_2O$
6. (a) $P_4(s) + 12Cl^-(aq) + 12H^+(aq) \longrightarrow$
                        $4PH_3(g) + 6Cl_2(g)$
  (b) $2MnO_4^-(aq) + 6H^+(aq) + 5NO_2^-(aq) \longrightarrow$
             $2Mn^{2+}(aq) + 3H_2O + 5NO_3^-(aq)$
  (c) $3HBrO_3(aq) + 2Bi(s) \longrightarrow 3HBrO_2(aq) + Bi_2O_3(s)$
  (d) $2CrO_4^{2-}(aq) + 3SO_3^{2-}(aq) + 10H^+(aq) \longrightarrow$
             $2Cr^{3+}(aq) + 3SO_4^{2-}(aq) + 5H_2O$
8. (a) $3Ca(s) + 8H_2O + 2VO_4^{3-}(aq) \longrightarrow$
            $3Ca^{2+}(aq) + 2V^{2+}(aq) + 16OH^-(aq)$
  (b) $C_2H_4(g) + 10H_2O + 6BiO_3^-(aq) \longrightarrow$
            $2CO_2(g) + 6Bi^{3+}(aq) + 24OH^-(aq)$
  (c) $2H_2O + 2PbO_2(s) \longrightarrow$
            $2Pb^{2+}(aq) + O_2(g) + 4OH^-(aq)$
  (d) $16Cl^-(aq) + 9H_2O + 3IO_3^-(aq) \longrightarrow$
            $8Cl_2(g) + I_3^-(aq) + 18OH^-(aq)$
10. (a) $2NO(g) + 5H_2(g) \longrightarrow 2NH_3(g) + 2H_2O(g)$
  (b) $H_2O_2(aq) + 2ClO^-(aq) + 2H^+(aq) \longrightarrow$
                $O_2(g) + Cl_2(g) + 2H_2O$
  (c) $Zn(s) + 2VO^{2+}(aq) + 4H^+(aq) \longrightarrow$
             $Zn^{2+}(aq) + 2V^{3+}(aq) + 2H_2O$
12. (a) $3Cd(s) + 2Sb^{3+}(aq) \longrightarrow 3Cd^{2+}(aq) + 2Sb(s)$
  (b) $2Cu^+(aq) + Mg^{2+}(aq) \longrightarrow 2Cu^{2+}(aq) + Mg(s)$
  (c) $2Cr^{3+}(aq) + ClO_3^-(aq) + 4H_2O \longrightarrow$
            $Cr_2O_7^{2-}(aq) + Cl^-(aq) + 8H^+(aq)$
14. (a) Sn anode, Ag cathode; e^- move from Sn to Ag. Anions move to Sn and cations to Ag.
  (b) Pt both anode and cathode. Anode has H_2 gas and H^+; cathode has chloride ions, Hg and Hg_2Cl_2. e^- move from anode to cathode. Anions move to the anode; cations move to the cathode.
  (c) Pb anode and PbO_2 cathode. e^- move from Pb to PbO_2. Anions move to Pb; cations to PbO_2.
16. (a) anode to cathode
  (b) Pt
  (c) $MnO_4^-(aq) + 8H^+(aq) + 5e^- \longrightarrow$
                 $Mn^{2+}(aq) + 4H_2O$
18. anode: $2I^-(aq) \longrightarrow I_2(s) + 2e^-$
  cathode: $Br_2(l) + 2e^- \longrightarrow 2Br^-(aq)$
  overall: $Br_2(l) + 2I^-(aq) \longrightarrow 2Br^-(aq) + I_2(s)$
  cell notation: $Pt \mid I^- \mid I_2 \parallel Br_2, Br^- \mid Pt$

20. (a) Br^- (b) Ni   (c) Hg_2^{2+}
22. $K^+ < Mn^{2+} < S < Cl_2 < Co^{3+}$
24. oxidizing agents: $Fe^{2+} < Ni^{2+} < H^+ < Cu^+$
  reducing agents: $Fe^{2+} < Cu^+ < Zn$
  both oxidizing and reducing agents: Cu^+, Fe^{2+}
26. (a) Co^{2+} (b) Mn   (c) Co^{3+} (d) Mn^{2+}
  (e) yes   (f) no   (g) Co^{3+}, Fe^{3+} (h) Cd, Pb, Mn
28. (a) O_2 (b) Na, Ca   (c) Ca, Ba, K
30. (a) 2.305 V   (b) 1.83 V   (c) -0.846 V
32. (a) 0.562 V   (b) 0.534 V
34. (a) -0.645 V   (b) 2.02 V   (c) 0.743 V
36. (a) -0.300 V   (b) $+1.377$ V   (c) $+0.213$ V; same
38. a
40. (1) $2Co^{3+}(aq) + 2I^-(aq) \longrightarrow 2Co^{2+}(aq) + I_2(s)$
                            $E° = 1.419$ V
  (2) $6Co^{3+}(aq) + 2Cr^{3+}(aq) + 7H_2O \longrightarrow$
     $6Co^{2+}(aq) + 14H^+(aq) + Cr_2O_7^{2-}(aq)$ $E° = 0.623$ V
  (3) $Cr_2O_7^{2-}(aq) + 14H^+(aq) + 6I^-(aq) \longrightarrow$
      $3I_2(s) + 2Cr^{3+}(aq) + 7H_2O$ $E° = 0.796$ V
42. (a) Nitrate will oxidize Cu to Cu^{2+} but not Au to Au^{3+}.
  (b) no product
  (c) Both acidic and basic mediums will work.
44. (a) no reaction
  (b) $MnO_2(s) + 4H^+(aq) + 2Hg(l) \longrightarrow$
    $Mn^{2+}(aq) + 2H_2O + Hg_2^{2+}(aq)$ $E° = +0.433$ V
  (c) no reaction
46. Na and Pb
48. (a) no reaction
  (b) no reaction
  (c) $2Cr^{2+}(aq) + S(s) + 2H^+(aq) \longrightarrow$
                    $2Cr^{3+}(aq) + H_2S(aq)$
50. (a) $\Delta G° = 15.9$ kJ; $E° = -0.041$ V
  (b) $\Delta G° = -45.2$ kJ; $K = 8.3 \times 10^7$
  (c) $E° = 0.015$ V; $K = 10$
52. (a) -104 kJ   (b) -208 kJ   (c) -312 kJ
54. $E° = 0.054$ V; $\Delta G° = -52$ kJ; $K = 1.3 \times 10^9$
56. (a) -444.8 kJ   (b) -1059 kJ   (c) 325.3 kJ
58. (a) 1×10^{19} (b) 1×10^{18}
60. 1.57×10^{25}
62. 0.29 V

64. (a) 0.152 V   (b) $E = 0.152 - \dfrac{0.0257}{2} \ln \dfrac{[Mn^{2+}]}{[Br^-]^2[H^+]^4}$
  (c) -0.221 V
66. (a) 0.305 V

  (b) $E = 0.305 - \dfrac{0.0257}{6} \ln \dfrac{(P_{NO})^2[OH^-]^8}{[S^{2-}]^3[NO_3^-]^2}$
  (c) 0.299 V
68. (a) 0.319 V   (b) -0.369 V
70. 1.2
72. $[H^+] = 8 \times 10^{-7}$ M
74. (a) yes; $E° = 0.899$ V   (b) yes; $E = 0.534$ V
  (c) no; $E = -0.058$ V   (d) 7.51
76. (a) 0.127 V   (b) 1.6×10^{-3} M   (c) 1.6×10^{-5}
78. (a) $1.37 \times 10^{23}\ e^-$ (b) 2.20×10^4 C
  (c) 2.30 g H_2; 8.08 g Cl_2
80. (a) 7.3×10^3 C   (b) Cu
82. (a) 5.92 g   (b) 6.91 min
84. (a) 8.4×10^7 J   (b) \$2.1
86. 0.550 g Cd; 0.810 g Ni_2O_3
88. (a) -1.596 V   (b) 1.0 g

90. 96 g
92. 0.896 V
94. no ($E = -0.400$ V)
96. c
98. b
100. (a) (1) $2Tl(s) + Tl^{3+}(aq) \longrightarrow 3Tl^+(aq)$
    (2) $2Tl(s) + Tl^{3+}(aq) \longrightarrow 3Tl^+(aq)$
    (3) $2Tl(s) + Tl^{3+}(aq) \longrightarrow 3Tl^+(aq)$
    (b) (1) 1.62 V   (2) 0.54 V   (3) 1.08 V
    (c) (1) -313 kJ   (2) -313 kJ   (3) -313 kJ
    (d) $\Delta G°$ is a state property; $E°$ is not.
102. Y > X > Z
103. $\approx 31.8\%$ Sn and 68.2% Cu
104. 2.007 V
105. (a) 0.621 V   (b) increases; decreases
    (c) 1×10^{-21}
    (d) $[Zn^{2+}] = 2.0$ M; $[Sn^{2+}] = 2 \times 10^{-21}$ M
106. (a) 1.12×10^5 J; 0.38×10^5 J; 1.50×10^5 J
    (b) -0.389 V
107. 0.414 V
108. 12 days
109. 79% for Zn; 21% for Sn

Chapter 18

2. (a) $^{28}_{14}Si$ (b) $^{6}_{3}Li$ (c) $^{23}_{11}Na$
4. (a) Ni   (b) Se   (c) Cd
6. $^{51}_{24}Cr \longrightarrow ^{0}_{1}e + ^{51}_{23}V$
8. (a) Th-231   (b) Pa-231
10. (a) $^{52}_{26}Fe \longrightarrow ^{52}_{25}Mn + ^{0}_{1}e$
    (b) $^{228}_{88}Ra \longrightarrow ^{228}_{89}Ac + ^{0}_{-1}e$
    (c) $^{236}_{95}Am \longrightarrow ^{232}_{93}Np + ^{4}_{2}He$
12. (a) $^{230}_{90}Th \longrightarrow ^{4}_{2}He + ^{226}_{88}Ra$
    (b) $^{210}_{82}Pb \longrightarrow ^{0}_{-1}e + ^{210}_{83}Bi$
    (c) $^{235}_{92}U + ^{1}_{0}n \longrightarrow ^{140}_{56}Ba + 3^{1}_{0}n + ^{93}_{36}Kr$
    (d) $^{37}_{18}Ar + ^{0}_{-1}e \longrightarrow ^{37}_{17}Cl$
14. (a) $^{272}_{111}Rg$
    (b) $^{260}_{105}Db$
16. The product ($^{282}_{114}Z$) is the same.
18. (a) $^{54}_{26}Fe + ^{4}_{2}He \longrightarrow 2^{1}_{1}H + ^{56}_{26}Fe$
    (b) $^{96}_{42}Mo + ^{2}_{1}H \longrightarrow ^{1}_{0}n + ^{97}_{43}Tc$
    (c) $^{40}_{18}Ar + ^{4}_{2}He \longrightarrow ^{43}_{19}K + ^{1}_{1}H$
    (d) $^{31}_{16}S + ^{1}_{0}n \longrightarrow ^{1}_{1}H + ^{31}_{15}P$
20. (a) $^{78}_{35}Br$ (b) $^{59}_{26}Fe$ (c) $^{4}_{2}He$ (d) $^{24}_{11}Na$
22. 5.25×10^{12} atoms
24. 1.25×10^{-5} Ci
26. 5.6×10^4 Ci
28. 5.74×10^3 Ci
30. 5.1×10^{-14} g
32. 7.3×10^7 β-particles; 3.3×10^{-5} Ci
34. (a) 8.0 days   (b) 22 atoms
36. No. The canvas is about 235 years old, which means it was used around 1775.
38. 9.2×10^3 yr
40. 8.5 years
42. 4.3×10^9 yr

44. (a) $^{90}_{38}Sr \longrightarrow 2^{0}_{-1}e + ^{90}_{40}Zr$
    (b) -0.0034 g/mol
    (c) 2.21×10^4 kJ
46. (a) 0.24992 g/mol   (b) 2.25×10^{10} kJ/mol
48. Ca-40
50. fusion
52. 1.19×10^{-2}
54. (a) 16%   (b) 18 mg
56. 8.4×10^{-11} g
58. 2.5×10^3 counts/min/mL
60. 2.60 Ci
62. 6.5×10^2 mL
64. $2C_2O_4^{2-}/Cr^{3+}$
66. 38 mL
68. 2.42×10^{-3} nm
70. (a) Alpha particles (He nuclei) attracted to negative pole; beta particles (electrons) attracted to positive pole.
    (b) Can follow path of C-11 in organic compounds in brain, using radiation counter.
    (c) Neutrons produced can continue fission reaction.
72. $t_{1/2} = 3.3$ hours
74. 99.8% remains after one year. Decay is slow.
76. 5.8×10^{-19} mol/L
77. (a) 2.5×10^{-9} g   (b) 5.5×10^{-3} kJ
    (c) 73 rems
78. (a) 1×10^{-13} J   (b) 6×10^6 m/s
79. (a) -5.72×10^8 kJ/g   (b) -1.3×10^{28} kJ
    (c) 1.8×10^{-11}
80. 8.3×10^{-6} atm
81. 3×10^6 yr
82. 4.3×10^{-29}

Chapter 19

2. (a) $en = 0$; SCN $= -1$, OH $= -1$
    (b) $+3$
    (c) $Mg[Cd(en)(SCN)_2(OH)_2]$
4. (a) $Pt(NH_3)_2(C_2O_4)$ (b) $Pt(NH_3)_2(SCN)Br$
    (c) $Pt(en)(NO_2)_2$
6. (a) 6   (b) 4   (c) 6   (d) 2
8. (a) 3   (b) 2   (c) 2   (d) 3
10. (a) $[Fe(H_2O)_6]_2(CO_3)_3$
    (b) $Ba[Pt(OH)_3Br_3]_2$
    (c) $Ba[V(en)Cl_4]$
    (d) $[Au(CN)_2]_2CO_3$
12. (a) $Pt(NH_3)_6^{4+}$ (b) $Ag(CN)_2^-$
    (c) $Zn(C_2O_4)_2^{2-}$ (d) $Cd(CN)_4^{2-}$
14. 28.4%
16. one
18. (a) $[Cr(NH_3)_2(H_2O)_4]Cl_2$ (b) $[Cd(Cl)_4(SO_4)^{3-}]$
    (c) $Na_2[Ni(OH)_4]$ (d) $[Fe(Br)_2(en)_2^+]$
20. (a) sodium tetrahydroxoaluminate(III)
    (b) diaquadioxalatocobaltate(III)
    (c) triamminetrichloroiridium(III)
    (d) diamminedibromoethylenediaminechromium(III) sulfate
22. (a) hexaaquairon(III)
    (b) amminetribromoplatinate(II)

(c) tetrachloroethylenediaminevanadate(II)
(d) dicyanogold(III)

24.

(a)   (b)

(c)

$$\begin{array}{ccc} & NH_3 & NH_3 \\ & \diagdown & \diagup \\ & Pt & \\ & \diagup & \diagdown \\ H_3N & & NH_3 \end{array}$$

(d)

(e) $H_3N \!-\! Au \!-\! CN$

26.

28.

(a)

(b)

(c)

30.

32. (a) $[_{36}Kr]4d^{10}$
(b) $[_{18}Ar]3d^6$
(c) $[_{54}Xe]4f^{14}5d^8$
(d) $[_{18}Ar]3d^5$
(e) $[_{18}Ar]3d^7$

34. (a) $[_{36}Kr]$ (↑↓)(↑↓)(↑↓)(↑↓)(↑↓) no unpaired electrons
(b) $[_{18}Ar]$ (↑↓)(↑)(↑)(↑)(↑) 4 unpaired electrons
(c) $[_{54}Xe]4f^{14}$ (↑↓)(↑↓)(↑↓)(↑)(↑) 2 unpaired electrons
(d) $[_{18}Ar]$ (↑)(↑)(↑)(↑)(↑) 5 unpaired electrons
(e) $[_{18}Ar]$ (↑↓)(↑↓)(↑)(↑)(↑) 3 unpaired electrons

36. (a) for both high and low spin: (↑↓)(↑↓)
(↑↓)(↑↓)(↑↓)
(b) for low spin: (↑)( ) for high spin: (↑)(↑)
(↑↓)(↑↓)(↑↓) (↑↓)(↑↓)(↑)

38. Mn^{3+} contains four 3d electrons, so it can form
( )( ) (↑)( )
(↑↓)(↑)(↑) (↑)(↑)(↑)
low spin high spin
Mn^{4+}, with only three 3d electrons, cannot do this.

40. NH_3 has a large Δ_o; low spin; no unpaired electrons.
F^- has a small Δ_o; high spin; unpaired electrons present.

42. (a) 4 (b) 4 (c) 0 (d) 4 (e) 2
44. 4.60×10^2 nm
46. ≈ 500 nm; green

48. (a) $CoN_6H_{18}Cl_3$
(b) $Co(NH_3)_6Cl_3(s) \rightleftharpoons Co(NH_3)_6^{3+}(aq) + 3Cl^-(aq)$
50. (a) 6, not 5 (b) shorter, not longer
(c) true (d) eight, not seven
51. 0.92 g
52. $[Pt(NH_3)_4] [PtCl_4]$ or $[Pt(NH_3)_3Cl] [Pt(NH_3)Cl_3]$
53. (a) $CuC_4H_{22}N_6SO_4$

(b)

54. red-violet

Chapter 20

2. $2Al_2O_3(l) \longrightarrow 4Al(l) + 3O_2(g)$; 2.91 g
4. $Cu_2S(s) + O_2(g) \longrightarrow 2Cu(s) + SO_2(g)$
6. -211.9 kJ
8. (a) $Fe_2O_3(s) + 3CO(g) \longrightarrow 2Fe(l) + 3CO_2(g)$
(b) $C(s) + O_2(g) \longrightarrow CO_2(g)$
10. 2.5×10^3 kWh
12. 1.7×10^6 L
14. (a) potassium nitride, K_3N
(b) potassium iodide, KI
(c) potassium hydroxide, KOH
(d) potassium hydride, KH
(e) potassium sulfide, K_2S
16. (a) $Na_2O_2(s) + 2H_2O \longrightarrow 2Na^+(aq) + 2OH^-(aq) + H_2O_2(aq)$
sodium and hydroxide ions, hydrogen peroxide
(b) $2Ca(s) + O_2(g) \longrightarrow 2CaO(s)$; calcium oxide
(c) $Rb(s) + O_2(g) \longrightarrow RbO_2(s)$; rubidium superoxide
(d) $SrH_2(s) + 2H_2O \longrightarrow Sr^{2+}(aq) + 2OH^-(aq) + 2H_2(g)$
strontium and hydroxide ions, hydrogen gas
18. 0.126 g
20. (a) $Co(s) + 2H^+(aq) \longrightarrow Co^{2+}(aq) + H_2(g)$
(b) $3Cu(s) + 2NO_3^-(aq) + 8H^+(aq) \longrightarrow$
$3Cu^{2+}(aq) + 2NO(g) + 4H_2O$
(c) $Cr_2O_7^{2-}(aq) + 6e^- + 14H^+(aq) \longrightarrow$
$2Cr^{3+}(aq) + 7H_2O$
22. $3Cd(s) + 12Cl^-(aq) + 2NO_3^-(aq) + 8H^+(aq) \longrightarrow$
$3CdCl_4^{2-}(aq) + 2NO(g) + 4H_2O$
24. (a) $Fe(s) + 3NO_3^-(aq) + 6H^+(aq) \longrightarrow$
$Fe^{3+}(aq) + 3NO_2(g) + 3H_2O$
(b) $4Cr(OH)_3(s) + 3O_2(g) + 8OH^-(aq) \longrightarrow$
$4CrO_4^{2-}(aq) + 10H_2O$
26. (a) Cd $(E° = 1.366$ V) (b) Cr $(E° = 1.708$ V)
(c) Co $(E° = 1.246$ V) (d) Ag $(E° = 0.165$ V)
(e) no reaction
28. (a) 0.724 V (b) 0.942 V
30. (a) 9×10^9 (b) 2×10^{-4} M
32. 208 g
34. 6.7
36. 53.8% Zn, 46.2% Cu
38. $2Ag_2S(s) + 8CN^-(aq) + 3O_2(g) + 2H_2O \longrightarrow$
$4Ag(CN)_2^-(aq) + 2SO_2(g) + 4OH^-(aq)$
$2Ag(CN)_2^-(aq) + Zn(s) \longrightarrow Zn(CN)_4^{2-}(aq) + 2Ag(s)$
40. (a) Cr^{2+} (b) Au^+ (c) Co^{2+} (d) Mn^{2+}
41. 2% BaO_2

42. (a) $Fe(OH)_3(s) + 3H_2C_2O_4(aq) \longrightarrow$
$Fe(C_2O_4)_3{}^{3-}(aq) + 3H_2O + 3H^+(aq)$
(b) 0.28 L
43. 2.80%
44. 2.83×10^3 K
45. $Cr_2O_7{}^{2-}(aq) + 2OH^-(aq) \longrightarrow 2CrO_4{}^{2-}(aq) + H_2O$
$2Ag^+(aq) + CrO_4{}^{2-}(aq) \longrightarrow Ag_2CrO_4(s)$
$Ag_2CrO_4(s) + 4NH_3(aq) \longrightarrow$
$2Ag(NH_3)_2{}^+(aq) + CrO_4{}^{2-}(aq)$
$2Ag(NH_3)_2{}^+(aq) + 4H^+(aq) + CrO_4{}^{2-}(aq) \longrightarrow$
$Ag_2CrO_4(s) + 4NH_4{}^+(aq)$

Chapter 21

2. (a) bromic acid   (b) potassium hypoiodite
(c) sodium chlorite   (d) sodium perbromate
4. (a) $KBrO_2$ (b) $CaBr_2$
(c) $NaIO_4$ (d) $Mg(ClO)_2$
6. (a) $NO_3{}^-$ (b) $SO_4{}^{2-}$ (c) $ClO_4{}^-$
8. (a) H_2SO_3 (b) $HClO$ (c) H_3PO_3
10. (a) NaN_3 (b) H_2SO_3
(c) N_2H_4 (d) NaH_2PO_4
12. (a) H_2S (b) N_2H_4 (c) PH_3
14. (a) NH_3, N_2H_4 (b) HNO_3
(c) HNO_2 (d) HNO_3
16. (a) $2I^-(aq) + SO_4{}^{2-}(aq) + 4H^+(aq) \longrightarrow$
$I_2(s) + SO_2(g) + 2H_2O$
(b) $2I^-(aq) + Cl_2(g) \longrightarrow I_2(s) + 2Cl^-(aq)$
18. (a) $3HClO(aq) \longrightarrow Cl_2(g) + HClO_2(g) + H_2O$
(b) $2ClO_3{}^-(aq) \longrightarrow ClO_4{}^-(aq) + ClO_2{}^-(aq)$
20. (a) $Cl_2(g) + 2Br^-(aq) \longrightarrow 2Cl^-(aq) + Br_2(l)$
(b) NR   (c) NR   (d) NR
22. (a) $Pb(N_3)_2(s) \longrightarrow 3N_2(g) + Pb(s)$
(b) $2O_3(g) \longrightarrow 3O_2(g)$
(c) $2H_2S(g) + O_2(g) \longrightarrow 2S(s) + 2H_2O$
24. (a) $Cd^{2+}(aq) + H_2S(aq) \longrightarrow CdS(s) + 2H^+(aq)$
(b) $H_2S(aq) + OH^-(aq) \longrightarrow H_2O + HS^-(aq)$
(c) $2H_2S(aq) + O_2(g) \longrightarrow 2H_2O + 2S(s)$
26. (a) $2H^+(aq) + CaCO_3(s) \longrightarrow CO_2(g) + H_2O + Ca^{2+}(aq)$
(b) $H^+(aq) + OH^-(aq) \longrightarrow H_2O$
(c) $Cu(s) + 4H^+(aq) + SO_4{}^{2-}(aq) \longrightarrow$
$Cu^{2+}(aq) + 2H_2O + SO_2(g)$
28. (a) $:\ddot{Cl}—\ddot{O}—\ddot{Cl}:$ (b) $:\ddot{O}—N\equiv N:$

(c) [P diamond structure with central P]   (d) $:N\equiv N:$

30. a and b

32. (a) $\left[:\ddot{O}—N=\ddot{O}: \atop \quad| \atop \quad:\ddot{O}: \right]^-$ (b) $\left[H—\ddot{O}—\overset{\textstyle :\ddot{O}:}{\underset{\textstyle :\ddot{O}:}{S}}—\ddot{O}: \right]^-$

(c) $\left[H—\ddot{O}—\overset{\textstyle :\ddot{O}:}{\underset{\textstyle :\ddot{O}:}{P}}—\ddot{O}—H \right]^-$

34. (a) $:\ddot{O}—N—\ddot{O}—N—\ddot{O}: \atop \quad\; \| \qquad\quad \| \atop \quad\; :O: \qquad\; :O:$ (b) $H—\ddot{O}—N=\ddot{O}: \atop \qquad\qquad\; | \atop \qquad\qquad :\ddot{O}:$

(c) $\left[:\ddot{O}—\overset{\textstyle :\ddot{O}:}{\underset{\textstyle :\ddot{O}:}{S}}—\ddot{O}: \right]^{2-}$

36. 14.3 L
38. 0.204 M
40. 7.7×10^4 L; 1.5×10^5 g
42. 12 L; 3.53
44. 4.28, 0.10 M
46. 1.9×10^{-3}
48. 0.012 g/100 mL
50. yes; 0 K
52. (a) yes   (b) 2×10^{16}
54. 510 K
56. 2.26×10^5 g
58. 27.4 kg
60. b, c, d
62. 4.6
64. (a) dispersion   (b) dispersion, dipole
(c) dispersion, H—bonds   (d) dispersion, H—bonds
(e) no intermolecular forces; not a molecule
66. (a) +3   (b) +4   (c) +5   (d) −3
68. (a) HClO   (b) S, $KClO_3$
(c) NH_3, NaClO   (d) HF
70. (a) See text pages 540–541.   (b) has an unpaired electron
(c) H^+ is a reactant.   (d) C is a product.
71. density of sulfur; depth of deposit, purity of S
72. no
73. 1.30%
74. Assume reaction is $NaN_3(s) \longrightarrow Na(s) + \frac{3}{2}N_2(g)$
Assume 25°C, 1 atm pressure; mass of NaN_3 is 35 g.

Chapter 22

2. (a) alkene   (b) alkyne   (c) alkane
4. (a) C_9H_{16} (b) $C_{22}H_{44}$ (c) $C_{10}H_{22}$
6. (a) 2-methyloctane   (b) 2,2-dimethylpropane
(c) 2,2,4-trimethylpentane
(d) 2,5-dimethylheptane
8.

(a) $CH_3—\overset{\textstyle CH_3}{\underset{\textstyle CH_3}{C}}—CH_2—\overset{\textstyle H}{\underset{\textstyle CH_3}{C}}—CH_3$

(b) $CH_3—\overset{\textstyle CH_3}{\underset{\textstyle CH_3}{C}}—CH_3$

(c) $CH_3—(CH_2)_2—\overset{\textstyle }{\underset{\textstyle CH_3—C—CH_3 \atop H}{C}}—(CH_2)_3—CH_3$

(d) $CH_3—\overset{\textstyle CH_3}{\underset{\textstyle H}{C}}—\overset{\textstyle H}{\underset{\textstyle CH_3}{C}}—\overset{\textstyle CH_3}{\underset{\textstyle H}{C}}—(CH_2)_2—CH_3$

10.

(a)
$$CH_3-\underset{\underset{CH_3}{|}}{\overset{\overset{CH_3}{|}}{C}}-CH_2-CH_3$$

2,2-dimethylbutane

(b)
$$CH_3-(CH_2)_2-\underset{\underset{CH_3}{|}}{CH}-CH_3$$

2-methylpentane

(c)
$$CH_3-\underset{\underset{CH_2}{|}}{CH}-CH_3$$
        $$|$$
        $$CH_3$$

2-methylbutane

12. (a) $CH_3-C\equiv C-CH_2-CH_3$

(b)
$$CH_3-C\equiv C-\underset{\underset{CH_3}{|}}{CH}-CH_3$$

(c)
$$CH_3-\underset{\underset{}{\overset{\overset{CH_3}{|}}{CH}}}-C\equiv C-CH_2-CH_3$$

(d)
$$H-C\equiv C-\underset{\underset{CH_3}{|}}{\overset{\overset{CH_3}{|}}{C}}-CH_3$$

14. (a) o-chlorotoluene   (b) m-bromotoluene
    (c) 1,2,5-tribromotoluene
16. (a) alcohol   (b) ester   (c) ester, acid
18. (a) ether   (b) ester   (c) amine
    (d) carboxylic acid   (e) ketone

20. (a)
$$CH_3-\underset{\underset{OH}{|}}{\overset{\overset{H}{|}}{C}}-CH_3$$
    (b)
$$CH_3-\underset{\underset{CH_3}{|}}{\overset{\overset{H}{|}}{C}}-\overset{\overset{O}{\|}}{C}-OH$$

(c)
$$CH_3-\underset{\underset{H}{|}}{\overset{\overset{CH_3}{|}}{C}}-\overset{\overset{}{\|}}{\underset{O}{C}}-O-\underset{\underset{CH_3}{|}}{\overset{\overset{H}{|}}{C}}-CH_3$$

22.

$$H-\underset{\underset{H}{|}}{\overset{\overset{H}{|}}{C}}-O-\overset{\overset{O}{\|}}{C}-H$$
$$H-\overset{\overset{H}{|}}{C}-OH$$
$$\underset{H}{|}$$

$$H-\underset{\underset{H}{|}}{\overset{\overset{H}{|}}{C}}-O-\overset{\overset{O}{\|}}{C}-H$$
$$H-\underset{\underset{H}{|}}{\overset{\overset{}{|}}{C}}-O-\overset{\overset{O}{\|}}{C}-H$$

$$H-\underset{\underset{H}{|}}{\overset{\overset{H}{|}}{C}}-O-\overset{\overset{O}{\|}}{C}-H$$
$$H-\underset{\underset{H}{|}}{\overset{\overset{}{|}}{C}}-O-\overset{\overset{O}{\|}}{C}-CH_3$$

$$H-\underset{\underset{H}{|}}{\overset{\overset{H}{|}}{C}}-OH$$
$$H-\underset{\underset{H}{|}}{\overset{\overset{}{|}}{C}}-O-\overset{\overset{O}{\|}}{C}-CH_3$$

$$H-\underset{\underset{H}{|}}{\overset{\overset{H}{|}}{C}}-O-\overset{\overset{O}{\|}}{C}-CH_3$$
$$H-\underset{\underset{H}{|}}{\overset{\overset{}{|}}{C}}-O-\overset{\overset{O}{\|}}{C}-CH_3$$

24. (a) butane < diethylether < 1-butanol
    (b) hexane < dipropylether < 1-hexanol
26. pH of $C_6H_5NH_3^+$ = 2.93; pH of CH_3COOH = 2.86; pHs are comparable.

28. 0.73

30.
$$\overset{}{\underset{}{}}C=C-\overset{|}{C}-\overset{|}{C}-\quad -\overset{|}{C}-C=C-\overset{|}{C}-\quad -\overset{|}{C}-\overset{|}{\underset{\underset{C}{|}}{C}}-\overset{|}{C}-$$

32.
$$-\overset{|}{\underset{|}{C}}-\overset{|}{\underset{|}{C}}-\overset{|}{\underset{Cl}{C}}-Cl \quad -\overset{|}{\underset{|}{C}}-\overset{\overset{Cl}{|}}{\underset{Cl}{C}}-\overset{|}{\underset{|}{C}}-$$

$$-\overset{|}{\underset{Cl}{C}}-\overset{|}{\underset{Cl}{C}}-\overset{|}{C}- \quad -\overset{|}{C}-\overset{|}{\underset{Cl}{C}}-\overset{|}{\underset{Cl}{C}}-$$

34.
(structures of trichlorobenzene isomers)

36.
$$Br-\overset{\overset{Cl}{|}}{C}=\overset{|}{C}-\overset{|}{C}-\overset{|}{C}-, \quad Br-\overset{\overset{Cl}{|}}{C}=\overset{|}{C}-\overset{|}{C}-\overset{|}{C}-,$$

$$Cl-\overset{|}{C}=\overset{\overset{Br}{|}}{C}-\overset{|}{C}-\overset{|}{C}-, \quad -\overset{|}{C}-\overset{\overset{Br}{|}}{C}=\overset{\overset{Cl}{|}}{C}-\overset{|}{C}-,$$

$$-\overset{|}{\underset{Cl}{C}}-\overset{\overset{\overset{|}{C}}{|}}{C}=C-Br$$

38.
$$-\overset{|}{C}-\overset{|}{C}-\overset{|}{C}-\overset{\overset{}{|}}{\underset{O}{C}}-OH, \quad -\overset{\overset{HO\diagdown\diagup O}{C}}{}\overset{|}{C}-\overset{|}{C}-\overset{|}{C}-$$

40. All of the compounds in Problem 36 show cis-trans isomerism except

$$-\overset{\overset{\overset{|}{C}}{|}}{\underset{Cl}{C}}-\overset{|}{C}=\overset{|}{C}-Br$$

where there are two —CH_3 groups attached to the same carbon.
42. b, c
44. c and d
46. (a) —center carbon   (b) none
    (c) —carbon atom at right
48. (a) addition   (b) elimination   (c) condensation
50. (a) 2,3-dichloro-2-methylbutane
    (b) 2,3-diodo-2-methylbutane
52. Three pairs of electrons spread around the ring
54. (a) no multiple bonds
    (b) sodium salt of a long-chain carboxylic acid
    (c) twice the volume of alcohol present
    (d) ethanol with additives (mainly methanol) that make it unpalatable, sometimes even poisonous
56. (a) $(CH_3)_2NH(aq) + H^+(aq) \longrightarrow (CH_3)_2NH_2^+(aq)$
    (b) $CH_3COOH(aq) + OH^-(aq) \longrightarrow$
        $$CH_3COO^-(aq) + H_2O$$

(c) $CH_3-\underset{\underset{Cl}{|}}{\overset{\overset{H}{|}}{C}}-CH_3(l) + OH^-(aq) \longrightarrow$

$CH_3-\underset{\underset{OH}{|}}{\overset{\overset{H}{|}}{C}}-CH_3(aq) + Cl^-(aq)$

57. $C_{27}H_{46}O$

58.
C—C—C—C—C—C with OH on 5th C

C—C—C—C—C—C with OH on 4th C

C—C—C—C—C—C with OH on 3rd C

C—C—C—C—C—OH with C branch on 4th C

C—C—C—C—C with OH on 3rd C, C branch on 3rd C

C—C—C—C—C with OH on 3rd, C branch

OH
C—C—C—C—C with C branch

HO—C—C—C—C—C with C branch

OH
C—C—C—C—C with C branch

C—C—C—C—C with C—OH branch

C—C—C—C—C with C and OH branches

C—C—C—C—C—OH with C branch

OH
C—C—C—C with C C branches

C—C—C—C—OH with C C branches

C—C—C—C with C—OH branch, C branch

C—C—C—C with C OH branches

C
C—C—C—C with C—OH branch

C
C—C—C—C with C OH branches

C
C—C—C—C—OH with C branch

59. Approximately 6 g (Assume $H_2O(l)$ is a product; neglect heat capacity of pan.)

60. $ClCH_2COOH(aq) + (CH_3)_3N(aq) \longrightarrow$
$ClCH_2COO^-(aq) + (CH_3)_3NH^+(aq)$
$K = 8.8 \times 10^6$; $3.4 \times 10^{-5}\ M$

Chapter 23

2. (a)
$-\underset{\underset{F}{|}}{\overset{\overset{F}{|}}{C}}-\underset{\underset{F}{|}}{\overset{\overset{F}{|}}{C}}-\underset{\underset{F}{|}}{\overset{\overset{F}{|}}{C}}-\underset{\underset{F}{|}}{\overset{\overset{F}{|}}{C}}-$

(b) 2.5×10^6 g/mol
(c) 24.02% C, 75.98% F

4.
$-\underset{\underset{H}{|}}{\overset{\overset{H}{|}}{C}}-\underset{\overset{H}{|}}{\overset{\overset{H}{|}}{C}}-\underset{\underset{H}{|}}{\overset{\overset{H}{|}}{C}}-\underset{\overset{H}{|}}{\overset{\overset{H}{|}}{C}}-$
(with phenyl groups on 2nd and 4th carbons)

6. (a) $H_2C=CHF$
(b) $CH_3-\underset{\underset{H}{|}}{C}=\underset{\underset{H}{|}}{C}-CH_3$

8. Polystyrene

10. $-O-\underset{\underset{O}{\|}}{C}-O-\bigcirc-\underset{\underset{CH_3}{|}}{\overset{\overset{CH_3}{|}}{C}}-\bigcirc-O-$

12. $-\underset{\underset{H}{|}}{N}-(CH_2)_5-\underset{\underset{O}{\|}}{C}-\underset{\underset{H}{|}}{N}-(CH_2)_5-\underset{\underset{O}{\|}}{C}-$

14. (a) $H_2N-CH_2-CH_2-NH_2$ and
$HOOC-CH_2-COOH$

(b) $HOOC-\bigcirc-COOH$ and $HO-CH_2-\underset{\underset{CH_3}{|}}{\overset{\overset{H}{|}}{C}}-OH$

16. $C_{12}H_{22}O_{11}(aq) + H_2O \longrightarrow 2C_6H_{12}O_6(aq)$

18. (a) 44.44% C, 6.22% H, 49.34% O   (b) 6.2×10^2

20.

22. 9; 10

24. Leu-Lys:
$(CH_3)_2-CH-CH_2-\underset{\underset{NH_2}{|}}{C}-\underset{\underset{}{\overset{\overset{O}{\|}}{C}}}-\underset{\underset{H}{|}}{N}-\underset{\underset{(CH_2)_4-NH_2}{|}}{\overset{\overset{H}{|}}{C}}-COOH$

Lys-Leu:
$NH_2-(CH_2)_4-\underset{\underset{NH_2}{|}}{\overset{\overset{H}{|}}{C}}-\underset{\underset{O}{\|}}{C}-\underset{\underset{H}{|}}{N}-\underset{\underset{CH_2-CH(CH_3)_2}{|}}{\overset{\overset{H}{|}}{C}}-COOH$

26. (a) 6
(b)
$(CH_3)_2CH-\underset{\underset{NH_2}{|}}{\overset{\overset{H}{|}}{C}}-\underset{\underset{O}{\|}}{C}-\underset{\underset{H}{|}}{N}-\underset{\underset{(CH_2)_4-NH_2}{|}}{\overset{\overset{H}{|}}{C}}-\underset{\underset{}{\overset{\overset{O}{\|}}{C}}}-\underset{\underset{H}{|}}{N}-\underset{\underset{CH_2-\bigcirc}{|}}{\overset{\overset{H}{|}}{C}}-COOH$
Val-Lys-Phe

28.
(a) $HO-CH_2-\underset{\underset{NH_3^+}{|}}{\overset{\overset{H}{|}}{C}}-COO^-$

(b) $HO-CH_2-\underset{\underset{NH_3^+}{|}}{\overset{\overset{H}{|}}{C}}-COOH$

(c) $HO-CH_2-\underset{\underset{NH_2}{|}}{\overset{\overset{H}{|}}{C}}-COO^-$

30. (a) 2.29   (b) 9.74   (c) 6.0
32. Ala-Phe-Leu-Met-Val-Ala
34. Double bond for addition; two functional groups for condensation
36. (a) Linear has straight chain with no branches.
    (b) See page 585
    (c) See page 585

38. (a) $HOOC-\bigobig-COOH$ and $HO-(CH_2)_2-OH$

    (b) $ClHC=CHCl$ (c) $H_2N-(CH_2)_5-COOH$

40.
head to head: $-\underset{\underset{H}{|}}{\overset{\overset{H}{|}}{C}}-\underset{\underset{CN}{|}}{\overset{\overset{H}{|}}{C}}-\underset{\underset{CN}{|}}{\overset{\overset{H}{|}}{C}}-\underset{\underset{H}{|}}{\overset{\overset{H}{|}}{C}}-$

head to tail: $-\underset{\underset{CN}{|}}{\overset{\overset{H}{|}}{C}}-\underset{\underset{H}{|}}{\overset{\overset{H}{|}}{C}}-\underset{\underset{CN}{|}}{\overset{\overset{H}{|}}{C}}-\underset{\underset{H}{|}}{\overset{\overset{H}{|}}{C}}-$

42.
$(CH_3)_2CH-CH_2-\underset{\underset{NH_3^+}{|}}{\overset{\overset{H}{|}}{C}}-COOH;$

$(CH_3)_2CH-CH_2-\underset{\underset{NH_2}{|}}{\overset{\overset{H}{|}}{C}}-COO^-$

44. 1.59×10^4 g/mol
46. -214 kJ

48. (a) $-\underset{\underset{O}{\|}}{C}\bigobig\underset{\underset{O}{\|}}{C}-O-CH_2-\underset{\underset{OH}{|}}{\overset{\overset{H}{|}}{C}}-CH_2-O-$

    (b) Orthophthalic acid condenses with OH groups in two adjacent chains.
49. 63.68% C, 9.80% H, 14.14% O, 12.38% N
50. -17 kJ

51.
linear:

crossed-linked:

52.
$HOOC-CH_2-\underset{\underset{NH_3^+}{|}}{\overset{\overset{H}{|}}{C}}-COOH,$
I

$HOOC-CH_2-\underset{\underset{NH_3^+}{|}}{\overset{\overset{H}{|}}{C}}-COO^-,$
II

$HOOC-CH_2-\underset{\underset{NH_2}{|}}{\overset{\overset{H}{|}}{C}}-COO^-,$
III

$^-OOC-CH_2-\underset{\underset{NH_2}{|}}{\overset{\overset{H}{|}}{C}}-COO^-$
IV

I, pH < 2.10; II, pH 2.10–3.85; III, pH 3.85–9.82; IV, pH > 9.82

Conversion Factors

Electrical Units

1 C	=	$1 \text{ A} \cdot \text{s}$
1 V	=	1 J/C
1 W	=	1 J/s
1 mol e^-	=	$9.648 \times 10^4 \text{ C}$

Energy

1 cal	=	4.184 J
$1 \text{ L} \cdot \text{atm}$	=	0.103 kJ
1 kWh	=	$3.600 \times 10^6 \text{ J}$
1 BTU	=	$1.055 \times 10^3 \text{ J}$

Length

1 nm	=	$10^{-9} \text{ m} = 10 \text{ Å}$
1 in	=	2.54 cm
1 m	=	39.37 in
1 mile	=	1.609 km

Mass

1 metric ton	=	10^3 kg
1 lb	=	453.6 g

Pressure

1 atm	=	$760 \text{ mm Hg} = 14.70 \text{ lb/in}^2$
1 atm	=	$1.01325 \times 10^5 \text{ Pa} = 1.01325 \text{ bar}$
1 torr	=	1 mm Hg

Volume

1 m^3	=	10^3 L
1 cm^3	=	1 mL
1 ft^3	=	28.32 L
1 L	=	1.057 qt

Table of Atomic Masses

A bracketed value denotes the mass of the longest-lived isotope.

	Symbol	Z	Atomic Mass		Symbol	Z	Atomic Mass
Actinium	Ac	89	[227.0277]	Mendelevium	Md	101	[258]
Aluminum	Al	13	26.981538	Mercury	Hg	80	200.59
Americium	Am	95	[243]	Molybdenum	Mo	42	95.94
Antimony	Sb	51	121.760	Neodymium	Nd	60	144.24
Argon	Ar	18	39.948	Neon	Ne	10	20.1797
Arsenic	As	33	74.92160	Neptunium	Np	93	[237.0482]
Astatine	At	85	[209.9871]	Nickel	Ni	28	58.6934
Barium	Ba	56	137.327	Niobium	Nb	41	92.90638
Berkelium	Bk	97	[247]	Nitrogen	N	7	14.00674
Beryllium	Be	4	9.012182	Nobelium	No	102	[259]
Bismuth	Bi	83	208.98038	Osmium	Os	76	190.23
Bohrium	Bh	107	[270]	Oxygen	O	8	15.9994
Boron	B	5	10.811	Palladium	Pd	46	106.42
Bromine	Br	35	79.904	Phosphorus	P	15	30.973761
Cadmium	Cd	48	112.411	Platinum	Pt	78	195.078
Calcium	Ca	20	40.078	Plutonium	Pu	94	[244]
Californium	Cf	98	[251]	Polonium	Po	84	[208.9824]
Carbon	C	6	12.0107	Potassium	K	19	39.0983
Cerium	Ce	58	140.116	Praseodymium	Pr	59	140.90765
Cesium	Cs	55	132.90545	Promethium	Pm	61	[145]
Chlorine	Cl	17	35.4527	Protactinium	Pa	91	231.03588
Chromium	Cr	24	51.9961	Radium	Ra	88	[226.0254]
Cobalt	Co	27	58.933200	Radon	Rn	86	[222.0176]
Copernicum	Cn	112	[285]	Rhenium	Re	75	186.207
Copper	Cu	29	63.546	Rhodium	Rh	45	102.90550
Curium	Cm	96	[247]	Roentgenium	Rg	111	[280]
Darmstadtium	Ds	110	[281]	Rubidium	Rb	37	85.4678
Dubnium	Db	105	[268]	Ruthenium	Ru	44	101.07
Dysprosium	Dy	66	162.50	Rutherfordium	Rf	104	[265]
Einsteinium	Es	99	[252]	Samarium	Sm	62	150.36
Erbium	Er	68	167.26	Scandium	Sc	21	44.955910
Europium	Eu	63	151.964	Seaborgium	Sg	106	[271]
Fermium	Fm	100	[257]	Selenium	Se	34	78.96
Flerovium	Fl	114	[289]	Silicon	Si	14	28.0855
Fluorine	F	9	18.9984032	Silver	Ag	47	107.8682
Francium	Fr	87	[223.0197]	Sodium	Na	11	22.989770
Gadolinium	Gd	64	157.25	Strontium	Sr	38	87.62
Gallium	Ga	31	69.723	Sulfur	S	16	32.066
Germanium	Ge	32	72.61	Tantalum	Ta	73	180.9479
Gold	Au	79	196.96655	Technetium	Tc	43	[98]
Hafnium	Hf	72	178.49	Tellurium	Te	52	127.60
Hassium	Hs	108	[277]	Terbium	Tb	65	158.92534
Helium	He	2	4.002602	Thallium	Tl	81	204.3833
Holmium	Ho	67	164.93032	Thorium	Th	90	232.0381
Hydrogen	H	1	1.00794	Thulium	Tm	69	168.93421
Indium	In	49	114.818	Tin	Sn	50	118.710
Iodine	I	53	126.90447	Titanium	Ti	22	47.867
Iridium	Ir	77	192.217	Tungsten	W	74	183.84
Iron	Fe	26	55.845	Ununoctium*	Uuo	118	[294]
Krypton	Kr	36	83.80	Ununpentium*	Uup	115	[288]
Lanthanum	La	57	138.9055	Ununseptium*	Uus	117	[294]
Lawrencium	Lr	103	[262]	Ununtrium*	Uut	113	[284]
Lead	Pb	82	207.2	Uranium	U	92	238.0289
Lithium	Li	3	6.941	Vanadium	V	23	50.9415
Livermorium	Lv	116	[293]	Xenon	Xe	54	131.29
Lutetium	Lu	71	174.967	Ytterbium	Yb	70	173.04
Magnesium	Mg	12	24.3050	Yttrium	Y	39	88.90585
Manganese	Mn	25	54.938049	Zinc	Zn	30	65.39
Meitnerium	Mt	109	[276]	Zirconium	Zr	40	91.224

*Not authenticated by IUPAC